NUMERICAL METHODS IN GEOMECHANICS / INNSBRUCK

PROCEEDINGS OF THE SIXTH INTERNATIONAL CONFERENCE ON NUMERICAL METHODS IN GEOMECHANICS / INNSBRUCK / 11-15 APRIL 1988

Numerical Methods in Geomechanics Innsbruck 1988

Edited by
G.SWOBODA
Institute of Structural Engineering, University of Innsbruck

VOLUME ONE:
Main lectures
1 *Numerical methods and programming*
2 *Constitutive laws of geotechnical materials*
3 *Flow and consolidation*

Published on behalf of the International Committee for Numerical Methods in Geomechanics by
A.A.BALKEMA / ROTTERDAM / BROOKFIELD / 1988

ORGANIZING COMMITTEES

International Conference Committee

Prof. G.Swoboda (Chairman) University of Innsbruck, Austria
Prof. C.S.Desai (Co-Chairman) University of Arizona, USA
Prof. H.Duddeck (Co-Chairman) Technical University of Braunschweig, FR Germany
Prof. W.Wittke (Co-Chairman), Technical University of Aachen, FR Germany
Prof. T.Adachi, Kyoto University, Japan
Prof. D.Aubry, Ecole Centrale des Arts, France
Dr. G.Beer, CSIRO, Division Geomechanics, Australia
Prof. Y.K.Cheung, University of Hong Kong, Hong Kong
Prof. G.Clough, Virginia State University, USA
Dr. J.T.Christian, Stone & Webster Engineer Corp., USA
Prof. W.D.L.Finn, University of British Columbia, Canada
Prof. G.Gioda, Politecnico di Milano, Italy
Prof. G.Gudehus, Technical University of Karlsruhe, FR Germany
Prof. Y.Ichikawa, Nagoya University, Japan
Prof. K.Kovári, ETH-Hönggerberg, Switzerland
Prof. Y.M.Lin, Northeast University, People's Republic of China
Prof. S.Ukhov, Moscow Civil Engineering Institute, USSR
Prof. F.Medina, Universidad de Chile, Chile
Prof. Z.Mróz, Polish Academy of Science, Poland
Prof. J.Prevost, Princeton University, USA
Prof. J.Smith, University of Manchester, UK
Prof. C.Tanimoto, Kyoto University, Japan
Prof. S.Valliappan, University of New South Wales, Australia
Prof. A.Varadarajan, ITT, Delhi, India
Prof. S.J.Wang, Academia Sinica, People's Republic of China
Prof. N.E.Wiberg, Chalmers University of Technology, Sweden
Prof. O.C.Zienkiewicz, University of Wales, UK

International Committee for Numerical Methods in Geomechanics

Prof. T.Adachi, Japan
Prof. D.Aubry, France
Prof. A.S.Balasubramaniam, Thailand
Prof. J.R.Booker, Australia
Dr. C.A.Brebbia, UK
Prof. Y.K.Cheung, Hong Kong
Dr. J.T.Christian, USA
Dr. A.Cividini, Italy
Prof. C.S.Desai, USA (Chairman)
Prof. J.M.Duncan, USA
Prof. Z.Eisenstein, Canada
Prof. A.J.Ferrante, Brazil
Prof. W.D.L.Finn, Canada
Dr. J.Geertsma, Netherlands
Prof. J.Ghaboussi, USA
Prof. K.Höeg, Norway
Prof. K.Ishihara, Japan
Prof. T.Kawamoto, Japan
Prof. K.Kovári, Switzerland
Prof. S.Prakash, India
Prof. J.M.Roesset, USA
Prof. I.M.Smith, UK
Prof. V.I.Solomin, USSR
Prof. G.Swoboda, Austria
Prof. S.Valliappan, Australia
Prof. C.Viggiani, Italy
Prof. S.Wang, People's Republic of China
Prof. N.E.Wiberg, Sweden
Prof. W.Wittke, FR Germany
Prof. O.C.Zienkiewicz, UK

The texts of the various papers in this volume were set individually by typists under the supervision of each of the authors concerned.

Published by

A.A.Balkema, P.O.Box 1675, 3000 BR Rotterdam, Netherlands

A.A.Balkema Publishers, Old Post Road, Brookfield, VT 05036, USA

For the complete set of three volumes ISBN 90 6191 809 X
For volume 1: ISBN 90 6191 810 3
For volume 2: ISBN 90 6191 811 1
For volume 3: ISBN 90 6191 812 X

Printed in the Netherlands

Numerical Methods in Geomechanics (Innsbruck 1988), Swoboda (ed.)
© 1988 Balkema, Rotterdam. ISBN 90 6191 809 X

Contents

2 *Constitutive laws of geotechnical materials*

Numerical Methods in Geomechanics (Innsbruck 1988), Swoboda (ed.)
© 1988 Balkema, Rotterdam. ISBN 90 6191 809 X

Preface

These proceedings comprise the papers presented at the 6th International Conference on Numerical Methods in Geomechanics in Innsbruck from 11-15 April 1988. The 480 papers submitted were of such a high scientific standing that it was difficult indeed to not accept them all for publication.

This conference has set itself the goal of narrowing the gap that exists between the theory and practice of numerical methods in geomechanics. Numerical methods enjoy an ever increasing role in the design of large structures in geotechnical engineering. Despite the enormous expansion in the use of nonlinear theories for describing the complex properties of soil and rock, design engineers are still faced with many unsolved problems. This conference thus aims to familiarize the design engineer with the latest advances in his field.

The conferences held since 1972 in Vicksburg, USA are a good indication of how rapidly technological developments take place; interest grew steadily from Blacksburg, USA (1976) to Aachen, FR Germany (1979), Edmonton, Canada (1982) and, most recently, Nagoya, Japan (1985). In continuation of this 16-year tradition, the 1988 conference focuses its attention on new topics such as ice mechanics, modeling of joints and of infinite domains, in order to bring the latest trends in research to a broader audience. The paper and posters received from 38 different countries show how intensely this field is researched throughout the world.

We are indebted to the International Committee for Numerical Methods for their guidance and experience in setting up this conference. The scientific content of the conference was the responsibility of the Conference Committee, to whose members, in particular Prof. Desai (University of Arizona, Tucson), Prof. Duddeck (University of Braunschweig), Prof. Wittke (University of Aachen), Prof. Gioda (University of Milan), Prof. Kawamoto (Nagoya University), Prof. Ichikawa (Nagoya University), we express our grateful appreciation for their support, valuable advice and especially for their painstaking review of papers. This enormous undertaking also owes its success to the assistance of additional reviewers: Prof. W.Ambach (University of Innsbruck), Prof. J.R.Booker (University of Sydney), Prof. J.Carter (University of Sydney), Dr. A.Cividini (University of Milan), Dr. R.Cowling (Mount Isa Mines, Australia), Prof. L.Gaul (University of Hamburg), Dr. M.John (ILF Consulting Engineers, Innsbruck), Prof. T.Kawamoto (Nagoya University), Prof. H.Kulhawy (Cornell University, New York), Dr. J.Meek (University of Queensland, Australia), Prof. W.Schober (University of Innsbruck), Prof. B.Schrefler (University of Padua), Prof. G.I.Schuëller (University of Innsbruck), Dr. H.F.Schweiger (University of Graz), Prof. G.Seeber (University of Innsbruck), Prof. M.Zaman (University of Oklahoma).

The day-to-day planning for the conference was performed by the Local Organizing Committee: Dr. H.Wagner (Mayreder Consult, Linz), Dr. M.John (ILF Consulting Engineers, Innsbruck), Dipl.Ing. F.Laabmayr (Consulting Engineer, Salzburg), Prof. K.Moser (University of Innsbruck), Dr. H.Passer (Consulting Engineer, Innsbruck), Dr. J.Pelz (City of Vienna), Dr. W.Pircher (TIWAG,

Innsbruck), Prof. W.Schober (University of Innsbruck), Prof. G.Seeber (University of Innsbruck) who are thanked for having taken on such a multitude of tasks, and here particularly Professor Kurt Moser, whose long-standing support has decisively influenced developments in this technical field. The major portion of the conference preparations was, however, in the capable hands of my co-workers, Ingo Mader, Wolfgang Mertz and Gerald Zenz, whose assistance was made possible through a grant from the Austrian 'Fonds zur Förderung der Wissenschaft' (projects S30-2C and P 5723), which we gratefully acknowledge.

The conference is generously sponsored by the following bodies: International Committee for Numerical Methods in Geomechanics, University of Innsbruck, Austrian Society for Geomechanics (Österreichische Gesellschaft für Geomechanik), Austrian Federal Chamber of Engineers (Bundesingenieurkammer), Austrian Society of Industrial Construction Firms (Vereinigung Industrieller Bauunternehmen Österreichs).

A final word of thanks is owed to the conference participants for their interest in the conference and their scientific contribution.

G.Swoboda
January 1988

Main lectures

Numerical Methods in Geomechanics (Innsbruck 1988), Swoboda (ed.)
© 1988 Balkema, Rotterdam. ISBN 90 6191 809 X

Behavior and simulation of soil tunnel with thin cover

Toshihisa Adachi
Department of Transportation Engineering, Kyoto University, Japan
Tadashi Kikuchi
Kanto Regional Office, Japan Railway Construction Public Corporation, Tokyo
Hiroshi Kimura
Nishi-Funabashi Construction Office, Japan Railway Construction Public Corporation, Tokyo

ABSTRACT: The use of conventional tunneling method has been increasingly implemented even in sandy ground with thin cover and groundwater head. Firstly in this paper, an analytical method was derived in due consideration of mechanical interaction between groundwater and surrounding ground and the efficiency of groundwater lowering method in tunneling was qualitatively investigated as analytical example. Secondary, analyses were carried out to simulate the behavior of a sandy ground tunnel which was excavated on trial by using two different divisions of heading in order to find out a suitable excavation method to minimize the ground movement.

1 INTRODUCTION

As a result of expansion of the major population regions in Japan, it has become necessary to construct new railway systems. It is becoming more common that the underground railways are planned and constructed because of the environment-preservation especially in commercial and residential sections.

In the underground railway construction, it is better to take the earth covering as thinner as possible from the point of view of the construction cost and the maintenance cost for drainage, ventilation and traffic safety as well as the convenience for passengers. Since the major cities of Japan are located in wide alluvial plains where soft ground with a high groundwater table prevails, tunnels should be able to be built safely without damaging adjacent or overlying buildings, streets or utilities. In the past, these urban tunnels have been usually constructed by cut and cover method or shield method. In recent years, however, the use of the conventional method has been increasingly implemented in sandy ground tunneling in lieu of shield method and this has resulted in the reduction of the tunneling costs to a great extent.

When tunneling in earth ground with thin cover, it is required to investigate the following problems carefully;

1)the stability to the tunnel face and the surrounding ground,

2)problems related to the underground water conditions, and

3)the effect of tunneling on adjacent or overlying structures or utilities.

To work out the problems, it is most important to understand the actual behavior of the surrounding ground during tunnel excavation and to establish an analytical method which can simulate the behavior as accurately as possible.

The tunnel construction causes discontinuity especially in cohesionless sandy ground. In the previous work(Adachi et al.(1985)) a way to simulate the discontinuous movement in sandy ground was proposed by using joint elements(Goodman and St.John(1977)) in analysis and showed its efficiency.

In this paper, firstly the analytical method is extended to be in due consideration of the mechanical interaction between groundwater and surrounding ground. As an example of application, the effect of the groundwater lowering method on the ground stabilization is investigated.

To minimize the effects of the construction on overlying and adjacent properties is one of the important factors. Namely, in order to construct urbane tunnels successfully, it is more important to minimize the deformation of surrounding ground(e.g., surface settlement etc.) than to minimize the earth pressure on tunnel supports or tunnel lining, in often discussed by using the so-called characteristic line method (NATM).

On that subject, secondly to show the effect of the division of heading on the ground deformation, analytical results of a tunnel which passes through a sandy layer at relatively shallow depth(10m) are compared with the observed results.

2 EFFECT OF GROUNDWATER LOWERING METHOD

When tunnels pass through soil layers with a high groundwater head, various procedures have been used to lower the groundwater level. In this section, the effect of groundwater lowering method on tunneling will be discussed by using a developed analytical method.

In addition, the effects of the ground permeability and the depth of earth covering on the ground stability are investigated.

2.1 Analytical method

Fig.1 shows the initial and boundary conditions for the problems discussed in this section. A tunnel with 8m of diameter is excavated 15.5m of earth covering under four different initial groundwater conditions indicated by Case 1 to Case 4, that is, Case 1; the initial groundwater table is at the ground surface, Case 2; it is at 7.5m above the tunnel crown, Case 3; it changes parabolically from the tunnel springline into the surrounding ground as corresponding to a case of partial dewatering, Case 4; it is at the base rock. The soil layer through which a tunnel passes is assumed to be sands($k=10^{-2}$ cm/sec) or silts($k=10^{-4}$ cm/sec) in this study.

The framework of the analytical method adopted here is the same as that developed by Akai and Tamura(1978) and used in a tunnel problem(Adachi et al.(1979)) except introducing joint element in the analysis (Adachi et al.(1985), and Adachi(1985)). This method is thoroughly derived on the basis of the Biot's equations for consolidation(Biot(1943)) as follows:

equilibrium equations

$$\dot{\sigma}'_{ij,j} + \dot{p}_{w,i} + \dot{f}_i = 0 \qquad (1)$$

constitutive equations

$$\dot{\sigma}'_{ij} = D_{ijkl}\varepsilon_{kl} \qquad (2)$$

strain-displacement relation

$$\varepsilon_{ij} = (u_{i,j} + u_{j,i})/2 \qquad (3)$$

Darcy's law for the movement of groundwater

$$v_i = -k_{ij}h_{,j} \qquad (4)$$

equation of continuity

$$\dot{\varepsilon}_v = v_{i,i} \qquad (5)$$

in which σ'_{ij} is the component of effective stress tensor, p_w the pore water pressure, γ the density of soil skeleton, f_i the component of body force, D_{ijkl} the component of tensor for the stress-strain relation, ε_{ij} the component of strain tensor, u_i the component of displacement vector, v_i the component of velocity vector of ground-water, k_{ij} the component of permeability tensor, h the total water head and the volumetric strain.

The elasto-plastic or elasto-viscoplastic constitutive law can be taken in this analytical method, however, the elastic constitutive law is adopted for the finite elements in this study.

The finite element mesh used in the problems is as shown in Fig.2. To simulate the discontinuity developed in the surrou-

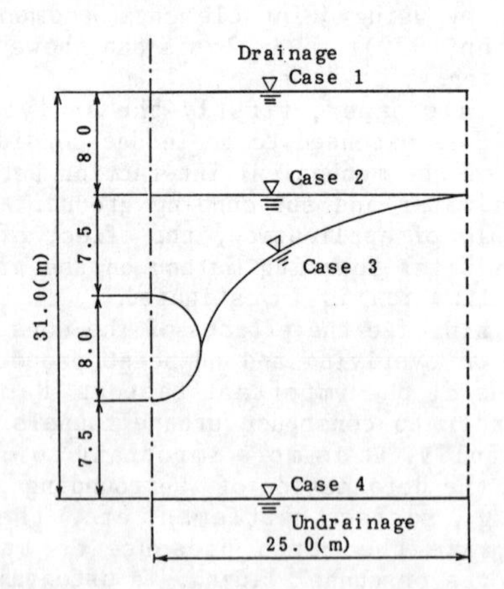

Fig.1 Initial and boundary conditions of ground water

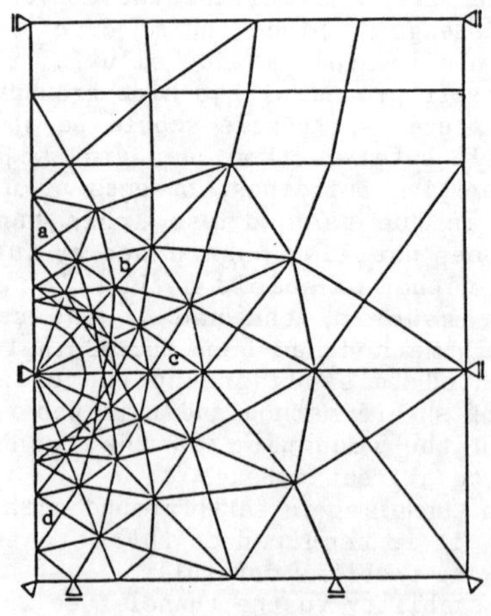

Fig.2 Finite element mesh

nding ground during tunnel excavation, Goodman's joint elements were arranged on some borders between triangular or rectangular elements. The direction of the joint elements at the tunnel wall was so arranged to be made at an angle of $(45^\circ - \phi/2)$ to that of the maximum principal stress and then joint elements were stretched parabolically into the surrounding ground. The thick solid lines in the mesh correspond to the joint elements.

Mohr-Coulomb's criterion was used as the plastic yield condition for joint element as follows;

$$\begin{aligned} \tau_y &= c' + \sigma'_n \tan\phi & \sigma'_n &\geq 0 \\ \tau_y &= 0 & \sigma'_n &< 0 \end{aligned} \qquad (6)$$

where τ_y is the yield stress, σ'_n the effective normal stress, c' the effective cohesive strength of joint element and ϕ' the effective internal friction angle of joint element.

The material parameters used in the analysis are summarized in Table 1, in which k_n and k_s are the joint stiffness parameters for the normal and tangential directions, respectively.

Table 1 Material parameters used in analyses

Unit Weight	$\gamma(tf/m^3)$	2.0
Young's Modulus	$E(tf/m^2)$	2,000
Poisson's Ratio	ν	1/3
Cohesive Strength	$c(tf/m^2)$	0
Internal Friction Angle	$\phi(°)$	30
Coefficient of Earth Pressure at Rest	K_o	0.5
Permeability Coefficient	k(cm/sec)	10^{-2}, 10^{-4}
Joint Stiffness	$k_n(tf/m^2)$ $k_s(tf/m^2)$	150,000 150,000

The tunnel excavation in the analysis was simulated by releasing the initial nodal stresses from the nodal points on the tunnel periphery by 4 stress release steps. The complete stress release(100%) was assumed to be taken place for one day.

In order to investigate qualitatively the effect of dewatering, the problems were analyzed for the case of excavation without support.

2.2 Analytical results and discussions

Effect of groundwater table

The effect of the underground water table were investigated for tunneling in a sandy layer with the permeability coefficient of 10^{-2} cm/sec. Fig.3 shows the distributions of the initial mean effective stress and the initial pore water pressure p_w in the adjacent surrounding ground along the tunnel periphery. Naturally, in the Case 1 with high groundwater table, the mean effective stress value is small and the pore water pressure is relatively high, while those values are reversed in Case 3 and Case 4 with lower groundwater table. Considering the strength dependency on the effective confining stress, it can be easily estimated that the stable ground condition will be preserved under the lower groundwater table condition.

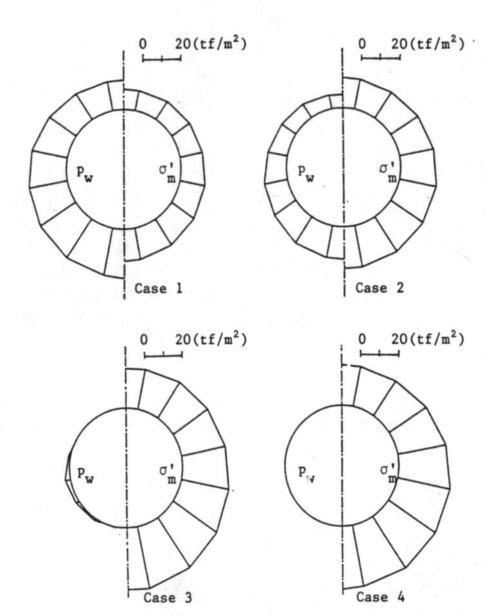

Fig.3 Initial mean effective stress and pore water pressure around tunnel periphery

In Fig.4, the equi-displacement lines and the displacement vectors calculated at 75% of the stress release are given for four cases. Although the computed results at 100% of the stress release is not given here, in Case 1 the soils especially above the tunnel flake off and fall down to the tunnel opening due to the liquefaction. From Fig.4, the effect of underground water head on the ground movements is more remarkable above than beneath the tunnel. It is obvious also that the larger displacements can be seen under high groundwater table condition.

The plastic yielded joint elements developed in the ground at 75% of the stress

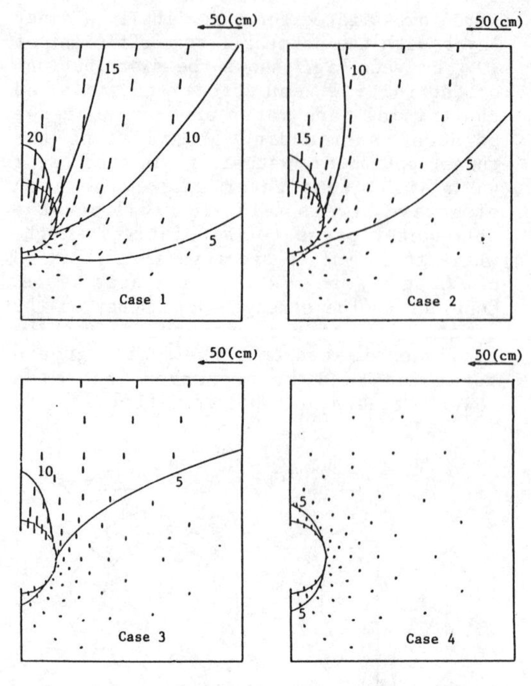

Fig.4 Displacement vectors and equi-displacement lines in surrounding ground at 75% of stress released state

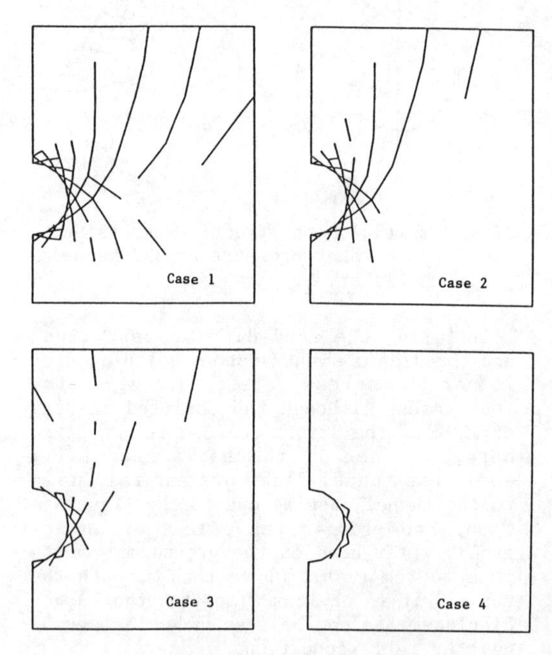

Fig.5 Plastic yielded joint elements at 75% of stress released state

release are shown in Fig.5. In the cases with high groundwater table, i.e., case 1 and 2, the yielded joint elements become to be interconnected from the tunnel wall to the ground surface, while in Case 4 the yielded joint elements appear only around the tunnel springline.

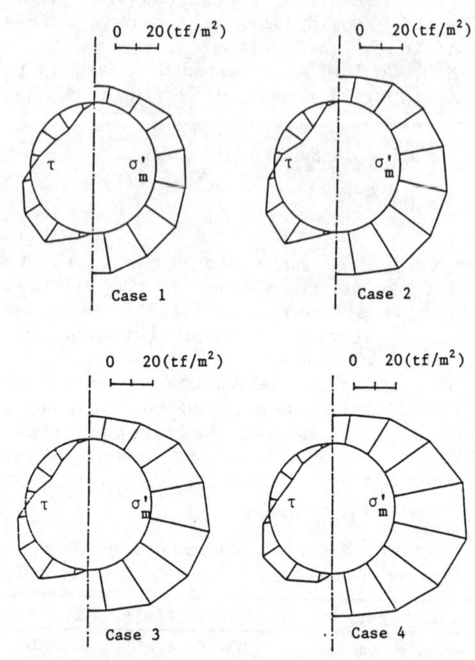

Fig.6 Mean effective stress and shear stress around tunnel periphery

Fig.6 shows the mean effective stress and the shear stress distributions in the adjacent ground around the tunnel periphery. The groundwater head is more effective on the mean effective stress than on the shear stress. Namely, the mean effective stress is sensitive to the change of the groundwater head but the shear stress is not altered very much. The distributions of stress ratio are obtained as shown in Fig.7. The larger values are given and naturally ground becomes unstable in the case of high groundwater head.

Fig.8 shows the change of the mean effective stress distribution in the ground between the ground surface and the tunnel crown with the progress of the stress release. In the cases with a high groundwater head, the magnitude of the mean effective stress is smaller even before the excavation started(0% of the stress release), and it is worth noting that in Case 1 the mean effective stress becomes nearly zero value at the complete stress

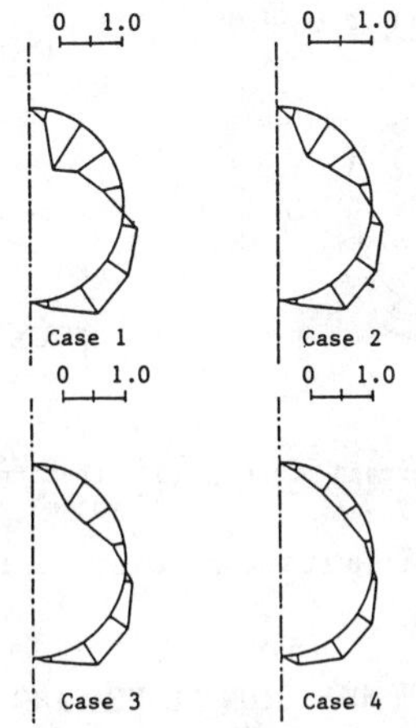

Fig.7 Stress ratio(τ/σ'_m) around tunnel periphery

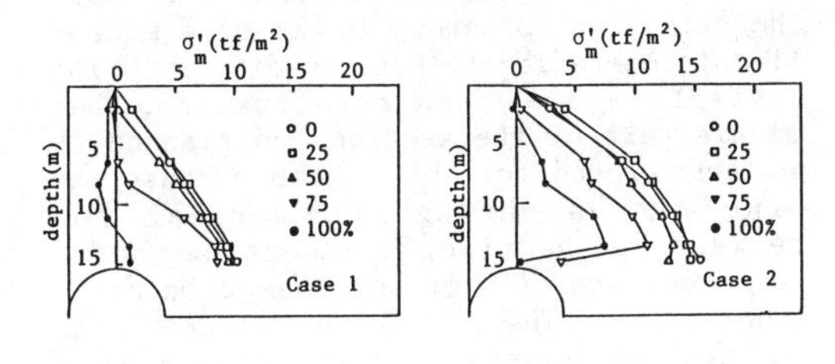

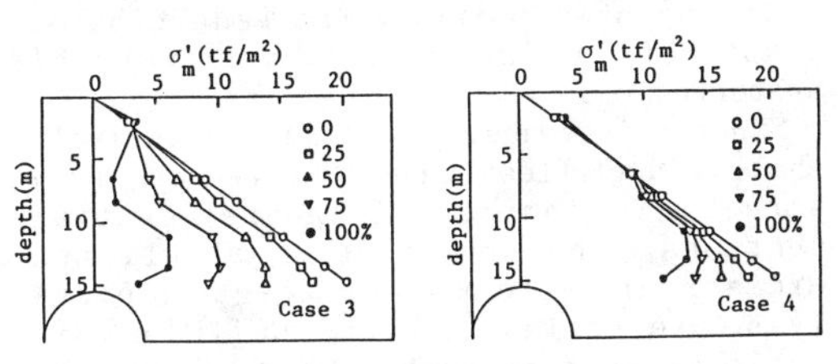

Fig.8 Change of mean effective stress distribution in the ground above tunnel crown during progress of stress release

release state(100%). This satisfied the condition for the liquefaction of the ground above the tunnel.

Effect of ground permeability

In order to investigate the effect of ground permeability on the tunneling stability, a problem of Case 2 condition was analyzed by using the permeability coefficient of 10^{-4} cm/sec and the results were compared with the previously obtained results with $k=10^{-2}$ cm/sec. The results at the stress release of 75% are shown in Fig.9, that is, the displacement vectors and the equi-displacement lines in

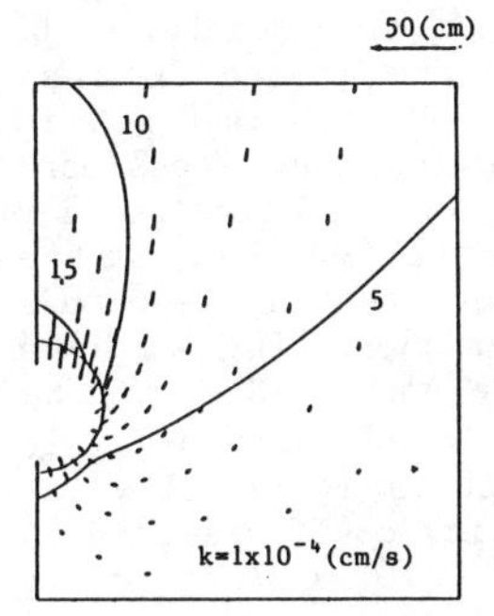

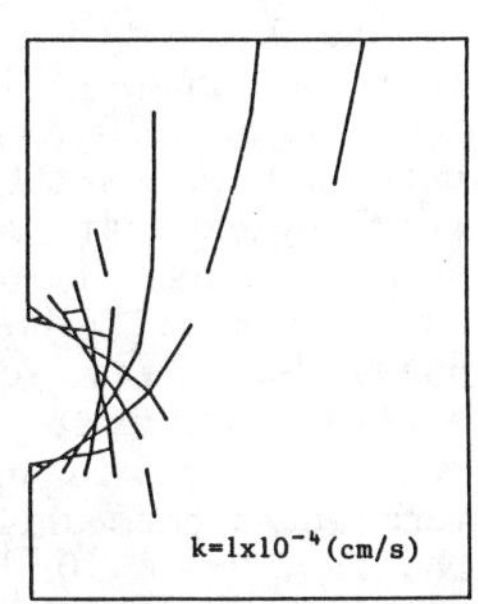

(a) displacement vectors and equi-displacement lines (b) plastic yielded joint elements

Fig.9 Results for $k=10^{-2}$ cm/sec

Fig.9(a), while Fig.9(b) gives the plastic yielded joint elements.

What is evident on comparing Fig.4(b) with Fig.9(a) and Fig.5(b) with Fig.9(b) is that there is not so much effect of the ground permeability on the stability. If examined in detail, however, a higher tunneling stability can be preserved when the permeability coefficient takes a smaller value.

Fig.10 shows the pore water pressure

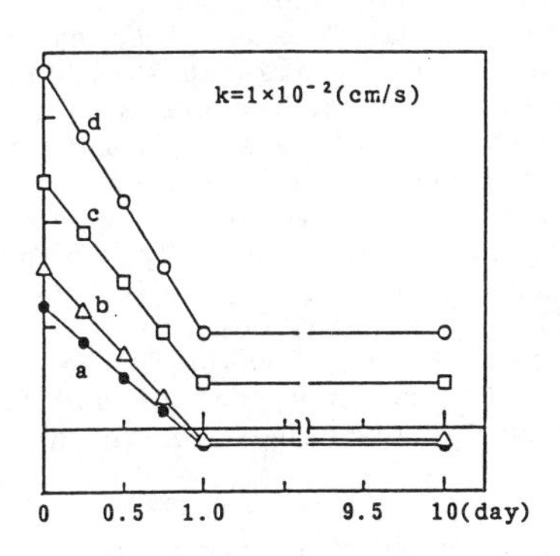

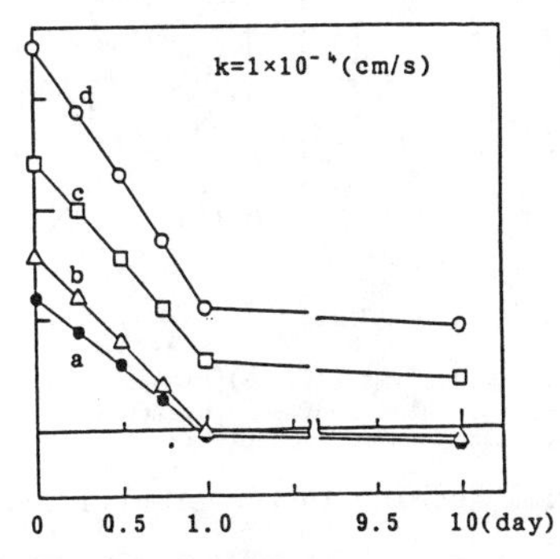

Fig.10 Change of pore water pressure with time at 4 points in surrounding ground

change with time at the four points, a, b, c, and d located in the ground 2m away from the tunnel wall as shown in Fig.2. in the case of $k=10^{-4}$ cm/sec, the excess pore water pressure still remains after one day from the excavation started, i.e., at the stress release of 100%, but the magnitude is very small. When tunneling with an usual daily advance rate(100% of the stress release completed for one day in the analysis) through sandy or silty layers whose permeability coefficient is in the region of 10^{-2} - 10^{-4}cm/sec it is concluded that the effect on the ground stability is very slight.

Effect of earth covering

To investigate the effect of earth covering on the tunneling stability by comparing with the results of Case 2, we analyzed a problem under the following conditions, that is, the depth of cover is about 1D(7.5m) with a groundwater table at the ground surface.

The computed results are illustrated in Fig.11. Namely Fig.11(a) shows the displacement vectors and the equi-displacement lines, while the plastic yielded joint elements are given in Fig.11(b), it is obvious that when the cover become thinner, the displacement becomes larger, and the number of plastic yielded joint elements increase. The stresses and the stress ratio(τ/σ'_m) distributions in the adjacent ground around the tunnel periphery at the stress release of 75% are given in Fig.12(a) and (b). By comparing Fig.12(a) with Fig.6(b) and Fig.12(b) with Fig.8(b), it can be seen that the stress ratio takes a larger value in the case of thinner cover. Thus, the ground become more unstable when the cover is thinner.

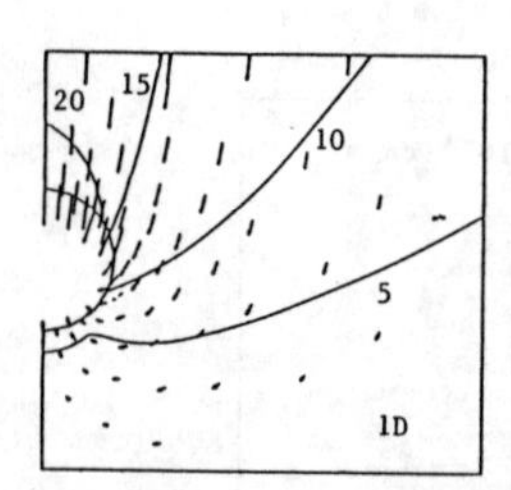

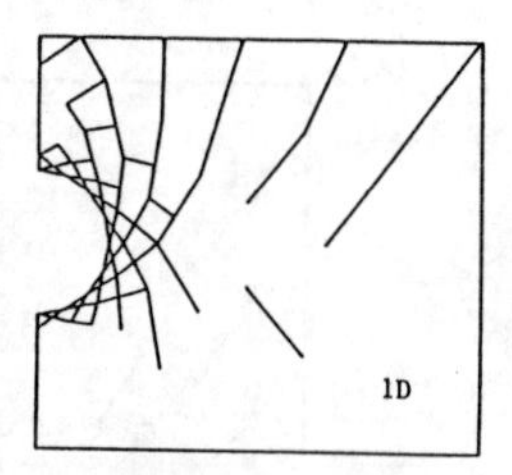

(a) displacement vectors and equi-displacement lines (b) plastic yielded joint elements

Fig.11 Results for cover of 1D(7.5m)

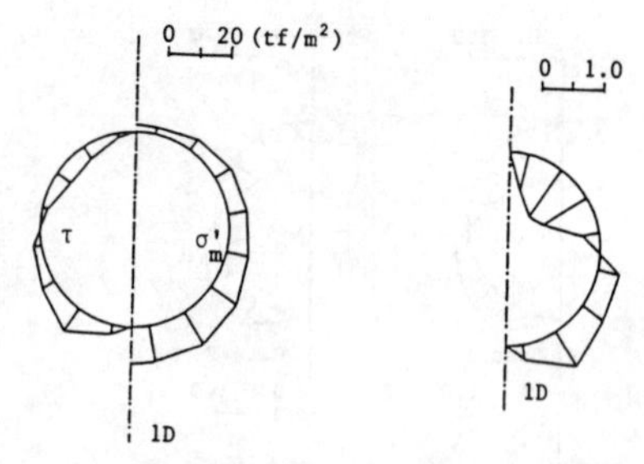

(a) stress around tunnel periphery (b) stress ratio around tunnel periphery

Fig.12 Results for cover of 1D(7.5m)

3 EFFECTS OF DIVISION OF HEADING ON TUNNEL STABILITY

In the Tokyo metropolitan area, a new railway line(Toyo-Kosoku) which will form the extension for about 16.4km of a subway line in operation to Katsuta-city in Chiba Prefecture is under construction. The middle part of the section was planned to go underground for about 2.8km because the line passes through commercial and residential section. This section therefore consists of a tunnel, named Narashinodai Tunnel. The tunnel was decided to be constructed by using conventional method from the standpoint of the reduction of construction cost and the environmental preservation.

Since this tunnel passes at a relatively shallow depth(less than 10m) beneath busy streets in commercial and residential sections, the construction should not excessively affect on adjacent or overlying properties. In order to find a suitable excavation method to minimize the ground movement, a section was excavated on trial by using two different division of heading.

The analytical results will be compared with the measured results and the observations obtained from the trial tunneling and the analyses will be discussed in this section.

Geological conditions

The soil conditions and topography along the tunnel alignment are shown in Fig.13. The cover over the tunnel ranges zero to 10m. The soils vary from tills, loams, clays and sands with a groundwater table near the tunnel springline. The tunnel passes through an upper sandy layer, D_{s1} with N-value less than 30.

From soil explorations and soil tests, the following material parameters of D_{s1}

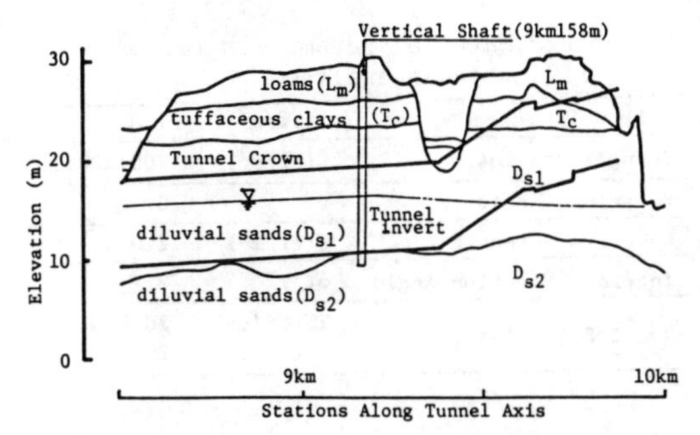

Fig.13 Soil conditions and topography along the tunnel alignment

were obtained; unit weight $\gamma=1.8tf/m^3$, Young's modulus E=2300 - $5000tf/m^2$, the cohesive strength $c=2.0tf/m^2$ and the internal friction angle $\phi=30°$.

Tunneling procedures

The tunnel in the standard section is horseshoe shaped, 10m wide and 8.5m high. The trial tunneling was carried out from a vertical shaft (Kita-Narashinodai Shaft locates at 9.185 km indicated in Fig.13) for about the first 40m by a division of heading, so-named CD-method(with center diaphragm) and for next 25m by CDS-method (with center diaphragm and struts) as illustrated in Fig.14.

In CD-method, tunneling was carried out by taking out section ① and providing primary support which consists of H-shape steel support(150mm) and shotcrete of 20cm thick as well as center diaphragm made of H-shape steel support(125mm) and shotcrete of 12cm thick as shown in Fig.14. The similar procedures were continued in tunneling the sections ②, ③, ④, ⑤, ⑥. The excavation of the right half section, that is, the sections ④ to ⑥, were performed 20m behind the working face of the left half section.

On the other hand, what distinguishes CDS-method from CD-method is in the heading sequence and the installation of struts between the primary support and the center diaphragm at each excavation stage. It should be noted also that the cycle length to reach full tunnel section was 15m(taking about 20 days) in case of CDS-method, while it was 30m(about 30days) in CD-method. It can be said that tunneling by CDS-method was performed smoothly.

At the present time, since the center diaphragm is not taken away, how the removal of diaphragm affects on the ground movement as well as on the stress change

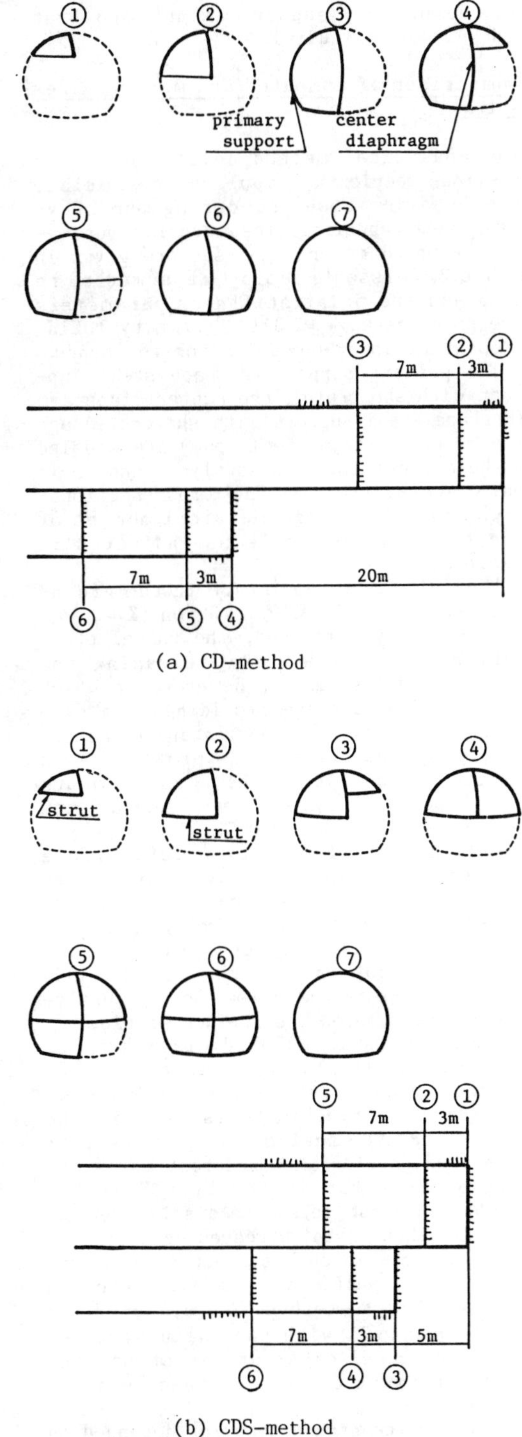

Fig.14 Division of heading of excavation sequence

in the primary support is an important subject to be studied.

Comparison of computed and measured results

The analytical method developed in the previous section was applied to simulate the behavior observed during the above mentioned tunneling. The material parameters adopted in the analyses are given in Table 2. Poisson's ratio is assumed to be 0.35 and the joint stiffness parameters are given as $k_n=k_s=2.0\times10^4$ tf/m^2 by taking account of Young's modulus for the sands.

The primary support (H-shape steel support with shotcrete), the center diaphragm (H-shape steel support with shotcrete) and the struts(H-shape steel beam) are modeled as beam elements in the analyses and their parameters, that is EA(A:cross-sectional area) and EI(I:the geometrical moment of inertia) are set to be as indicated in Table 3.

The initial stress state in the ground was assumed to be K_o-condition (K_o=0.54). As previously mentioned, the tunnel excavation was simulated by releasing the initial nodal stresses, however, at which stress released stage providing beam elements as by way of primary support, center diaphragm and struts is a problem. As a result of consideration, beam elements were assumed to be introduced at the stress released rate of 67%.

The lateral distributions of surface settlement are shown in Fig.15 to compare the measured and calculated results at the excavation stage⑥. Relatively close agreement between measured and calculated values was obtained in the case by CDS-method, while in the CD-method the analysis underestimated the surface settlement. The measure and calculated distributions of vertical displacement in the ground between ground surface and tunnel crown are given in Fig.16. It is seen in the figure that the analysis for the case by CD-method considerably underestimates the displacement but it for by CDS-method slightly overestimates the displacements.

Fig.17 shows axial stresses developed at some portions in each support element and the measured values are also indicated in the case by CDS-method. The measured values agree well with the calculated results in tunnel crown arch section, but good agreement can not be seen in other portions.

Although the ground behavior depended on the difference in division of heading, a large difference can not be predicted in so far as the analysis. A reason why the

Table 2 Material parameters for sand used in analyses

Unit Weight	γ(tf/m^3)	1.8
Young's Modulus	E(tf/m^2)	200+200σ_{mo}
Poisson's Ratio	ν	0.35
Cohesive Strength	c(tf/m^2)	2.0
Internal Friction Angle	ϕ(°)	32
Joint Stiffness	k_n(tf/m^2) k_s(tf/m^2)	20,000 20,000

Table 3 Parameters for primary support, center diaphragm and struts used in analyses

	Primary support	Center diaphragm	Struts
EA(tf)	235,200	188,370	63,630
EI(tf·m^2)	620	338	178

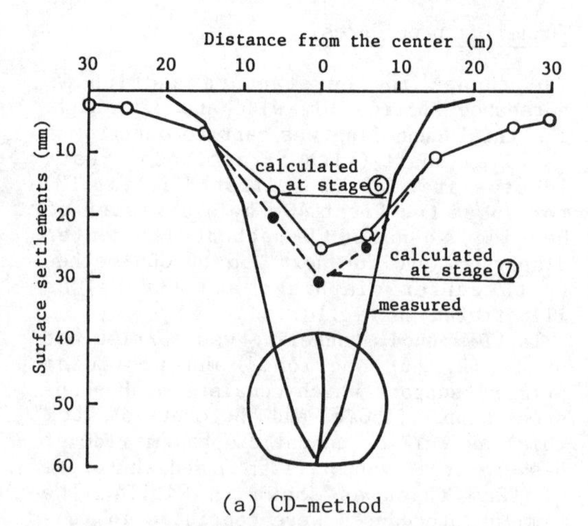

(a) CD-method

Distance from the center (m)

30 20 10 0 10 20 30

Surface settlement (mm)

0 10 20 30 40

measured

calculated at stage ⑦

calculated at stage ⑥

(b) CDS-method

Fig.15 Surface settlements

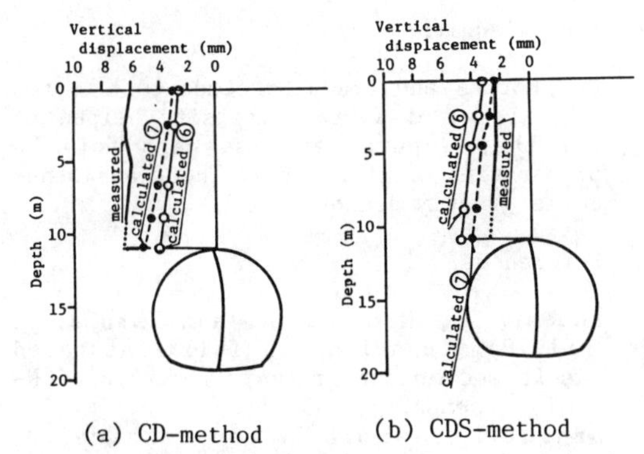

(a) CD-method (b) CDS-method

Fig.16 Vertical displacement in the ground above tunnel crown

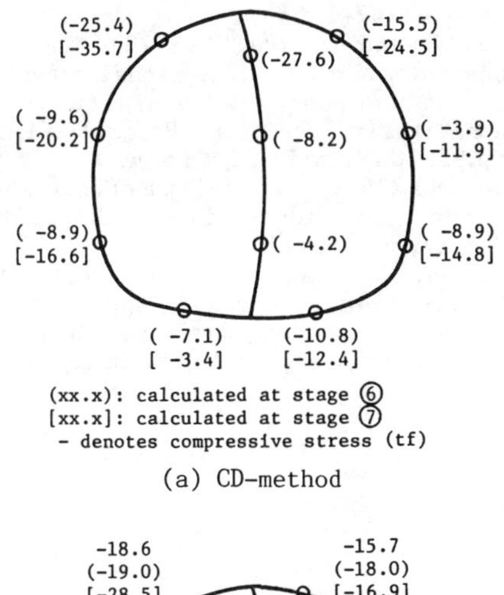

(a) CD-method

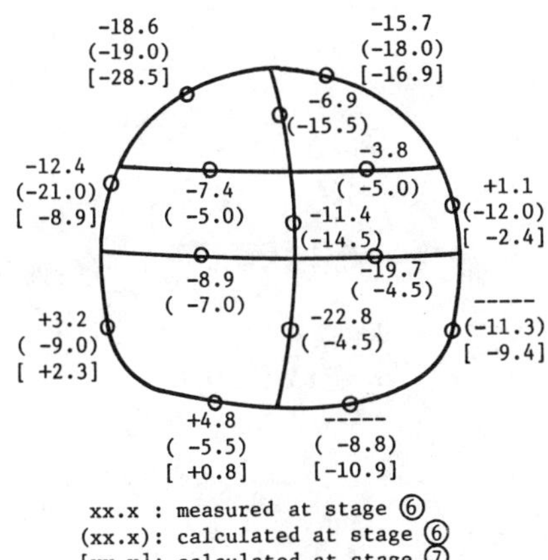

(b) CDS-method

Fig.17 Axial stress developed in support elements

larger movement took place in the case by CD-method can be considered due to its slower tunneling advance rate. In order to take account of this effect an analysis was carried out by taking the time to provide the supports at 100% of stress released condition. The calculated results together with the measured values are shown in Fig.18. The results give a better estimation than that by the previous analysis. Namely, this tells us that it is very important to excavate and form the inner structural shell around the tunnel periphery as quick as possible. Also it is

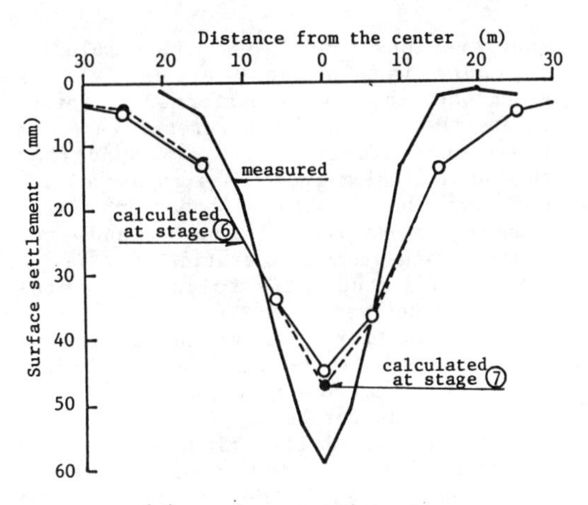

(a) surface settlements

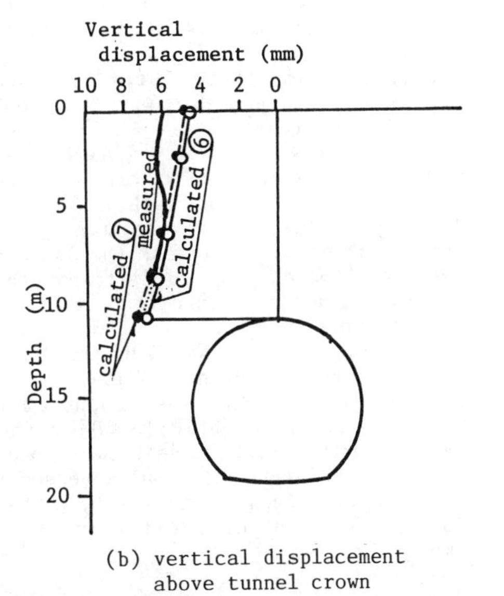

(b) vertical displacement above tunnel crown

Fig.18 Re-evaluation in the case of CD-method

necessary in analyses to take account of degrees of skill in tunneling.

Next, the influences caused by removal of the center diaphragm should be estimated. The calculated results corresponding this tunneling stage ⑦ have been indicated in the figures already given. it is expected from the figures that the ground above tunnel crown will raise about 5mm in the case by CDS-method, but further surface settlement will occur in CD-method, and the axial stresses in the primary support will increase when the center diaphragm removed.

4 CONCLUSIONS

Under various conditions, the simulation of tunneling in sandy ground with a groundwater head were performed to investigate the effect of underground water on the ground stability. It is found that this analytical method can simulate qualitatively the mechanical behavior of surrounding ground with various groundwater heads during tunnel excavation.

From this study, the folloing conslusions are derived.

(1)The stability of surrounding ground decreases (while the stress ratio (τ / σ'_m)increases) with increase of the groundwater head.

(2)Under a usual daily advance rate (the stress release is completed for one day in the analysis), the difference in the ground permeability coefficient in the region between 10^{-2} (sands) and 10^{-4} cm/sec (silts) affects on the stability very slightly.

(3)Under the same groundwater head, the ground becomes unstable when the depth of cover decreases.

From the study on a sandy ground tunnel which was excavated on trial by using two different divisions ofheading, the following conclusions are obtained.

(4)In so far as the analyses, the influences resulted due to two different tunneling method are found to be a little. The difference observed in the actual tunneling might be caused by the difference of tunnel advance rate.

(5)It is expected that the ground above tunnel raises a little in CDS-method while further surface settlement occur in CD-method, and the axial stresses in the primary support increase by certain amount when the center diaphragm removed.

Acknowledgment

Dr.M.Horita and graduate students K.Kojima and Y.Yuasa of Kyoto University helped us for the computer analyses and for the prepatation of this paper. Their assistance is greatly acknowledged.

References

Adachi, T., Mochida, Y. and Tamura, T. (1979). Tunneling in fully-saturated soft sedimentary rocks, Proc. 3rd ICONMIG, Aachen, pp.599-610.

Adachi,T., Tamura T. and Yashima, A. (1985). Behavior and simulation of sandy ground tunnel, Proc. 11th ICSMFE, San Francisco, pp.709-712.

Adachi,T.(1985). Some supporting methods for tunneling in Japan and their analytical studies, Proc. 5th ICONMIG, Nagoya, pp.1747-1754.

Akai, K. and Tamura, T.(1978). Numerical analysis of multi-dimensional consolidation accompanied with elasto-plastic constitutive equation, Proc. JSCEA, No., pp.95-104-104(in Japanese).

Biot, M.A.(1941). General theory of three-dimensional consolidation, J.Appl.Phys., Vol.12, pp.155-164.

Goodman, R.E. and St. John, C. (1977). Finite element analysis for discontinuous rocks, Proc. Numerical Methods in Geotechnical engg., McGraw-Hill, pp.148-175.

Numerical Methods in Geomechanics (Innsbruck 1988), Swoboda (ed.)
© 1988 Balkema, Rotterdam. ISBN 90 6191 809 X

Progress in BEM applications in geomechanics via examples

P.K.Banerjee & Gary F.Dargush
State University of New York, Buffalo, USA

ABSTRACT: The boundary element method (BEM) has become very popular in the solution of geomechanics problems principally because it is an efficient numerical method for bulky solids having boundaries extending to infinity. In this paper some of the major recent developments for inelastic, consolidation and dynamic analyses are outlined. In particular, a series of realistic examples is presented to demonstrate that BEM has been developed for the analysis of practical problems.

1. INTRODUCTION

Boundary integral formulations for most classical differential equations of engineering have existed for nearly 50 years. With the emergence of digital computers in the early 1960's the method began to gain popularity as the 'boundary integral equation method', 'panel method', 'integral equation method', and 'surface source method'. The name was changed to 'Boundary Element Method' (BEM) by Banerjee and Butterfield (1975), so as to make it more appealing to the engineering analysis community. Since then a very large number of text books, monographs and conference proceedings have appeared.

In 1982, NASA initiated an ambitious program of research and development on BEM. At that time, NASA asked Pratt and Whitney Aircraft and State University of New York at Buffalo to combine resources toward the development of a large scale BEM system for the linear and nonlinear static, dynamic, and thermal analyses of solids, comparable to the NASTRAN finite element code. Since then the scope of the work has been widened considerably to include, for example, transient heat transfer, transient thermoviscoplasticity, steady state and transient thermoviscous flow. This large engineering analysis system (Banerjee, Wilson and Miller, 1985, 1988), BEST-Boundary Element Solution Technique, the development of which is still continuing, is currently the result of about 42 man-years effort, and includes many of the latest boundary element integration, solution, and substructuring technologies. In this paper some examples of the application of this system in the area of geomechanics are described.

2. BEM FORMULATIONS

2.1 New BEM formulations for body forces

In most problems of engineering the governing differential equation can be written in an operator notation as

$$L(u) + b = 0 \qquad (1)$$

The solution of (1) can be accomplished as the sum of the complimentary solution u^c and a particular integral u^p, where u^c and u^p satisfy

$$L(u^c) = 0 \qquad (2a)$$

$$L(u^p) + b = 0 \qquad (2b)$$

The basic boundary only formulation for the stress analysis via equation (2a) can be written as (Banerjee and Butterfield, 1981):

$$\alpha_{ij}\, u_i^c(\xi) = \int_S [G_{ij}(x,\xi)t_i^c(x) - F_{ij}(x,\xi)u_i^c(x)]dS \qquad (3)$$

which when discretized takes the form

$$\mathbf{G}\ \mathbf{t}^c - \mathbf{F}\ \mathbf{u}^c = 0\ . \tag{4}$$

Recalling that the total solution is

$$\mathbf{u} = \mathbf{u}^c + \mathbf{u}^p$$

$$\mathbf{t} = \mathbf{t}^c + \mathbf{t}^p,$$

one can express (4) as

$$\mathbf{G}\ \mathbf{t} - \mathbf{F}\ \mathbf{u} = \mathbf{G}\mathbf{t}^p - \mathbf{F}\ \mathbf{u}^p\ . \tag{5}$$

By incorporating the boundary conditions the above equation can be rewritten as

$$\mathbf{A}\mathbf{x} = \mathbf{b} + \mathbf{G}\mathbf{t}^p - \mathbf{F}\mathbf{u}^p \tag{6}$$

Equation (6) can now be solved if the particular solution t^p and u^p are known from the solution of equation (2b).

The particular solution can be arbitrary as long as it satisfies the required differential equation (2b). These solutions are often algebraic polynomials that are obtained either by trial and error or by the method of undetermined coefficients. Although this method of particular integrals was discussed by Watson (1979) and Banerjee and Butterfield (1981) its explicit use in dealing with body forces due to self weight, centrifugal loading and nonconservative body forces in thermoelasticity and elastoplasticity did not take place until recently (Pape and Banerjee, 1987; Henry et al, 1987; Henry, 1987; Banerjee et al, 1988).

2.2 BEM for elastoplastic analysis

Although the BEM formulation for elastoplastic analysis using particular integrals is available (Henry, 1987), for the results presented in this paper the analysis is carried out using the conventional BEM formulation (Banerjee and Butterfield, 1981; Banerjee and Raveendra, 1986, 1987; Henry and Banerjee, 1987):

$$\alpha_{ij}\dot{u}_i(\xi) = \int_S [G_{ij}(x,\xi)\dot{t}_i(x) - F_{ij}(x,\xi)\dot{u}_i(x)]dS + \int_V B_{ikj}(x,\xi)\dot{\sigma}^o_{ik}(x)dV \tag{7}$$

$$\dot{\sigma}_{jk}(\xi) = \int_S [G^{\sigma}_{ijk}(x,\xi)\dot{t}_i(x) - F^{\sigma}_{ijk}(x,\xi)\dot{u}_i(x)]dS + \int_V B^{\sigma}_{ipjk}(x,\xi)\dot{\sigma}^o_{ip}(x)dV + J_{ipjk}\dot{\sigma}^o_{ip}(\xi) \tag{8}$$

Equations (7) and (8) together with the constitutive equation can be used to develop either (a) an iterative initial stress or initial strain algorithm (Banerjee and Butterfield, 1981; Banerjee and Raveendra, 1986) or (b) an incrementally direct variable stiffness type algorithm (Banerjee and Raveendra, 1987; Henry and Banerjee, 1988). Two examples of elastoplastic analysis are presented below.

Flexible strip footing (Raveendra, 1984)

The inelastic deformation of a smooth flexible strip footing of width (2b) under plane strain conditions is examined using the discretization shown in Figure 1. The non-dimensionalized load-displacement response is shown in

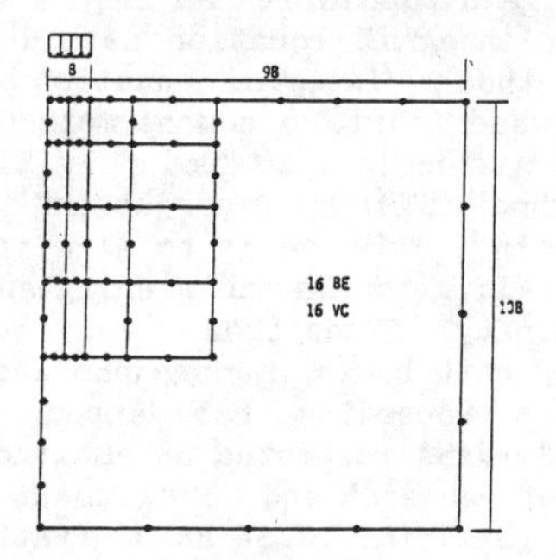

Figure 1. Discretization of flexible strip footing

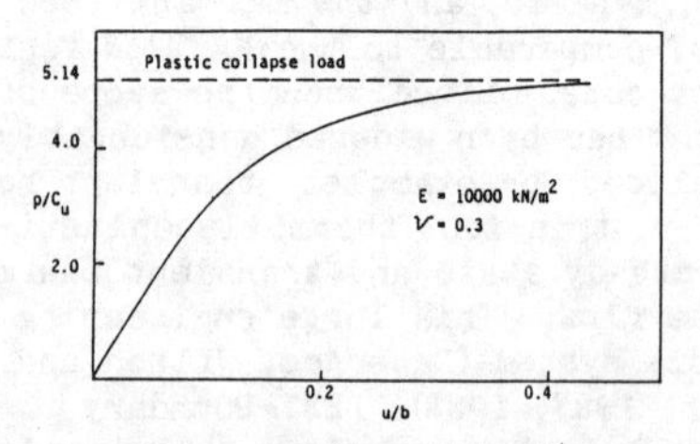

Figure 2. Load-displacement behavior of strip footing

Figure 2. The exact solution for the collapse load (both upper and lower bounds) in terms of (undrained) shear strength (C_u) is well known to be $(\pi+2)C_u$.

The collapse load obtained from the present analysis using a large number of cell patterns ranged between 5.10 C_u and 5.12 C_u, indicating the very superior accuracy of the analysis. Unlike the finite element method where the collapse load predictions often varies with the mesh, the present analysis predicts very stable results.

Flexible circular footing (Henry, 1987)

This problem concerns a flexible circular footing on an elastic-ideally plastic half space which has a modulus of elasticity of $E = 10{,}000$ kN/m^2, Poisson's ratio of $\nu = 0.25$ and a yield stress of $\sigma_o = 173.2$ kN/m^2 ($Cu = \sigma_o/\sqrt{3}$). A Von Mises criterion and associated flow rule is assumed.

For this axisymmetric analysis, two modeling regions are used. One region encloses the anticipated plastic zone and the other defines the remainder of the half space which is assumed to remain elastic. In Figure 3a,b, elements modeling the infinite half space are shown starting at the axis of symmetry and continuing along the surface up to a finite distance where it is assumed that additional elements (modeling the infinite boundary) do not affect the functions in the area of interest.

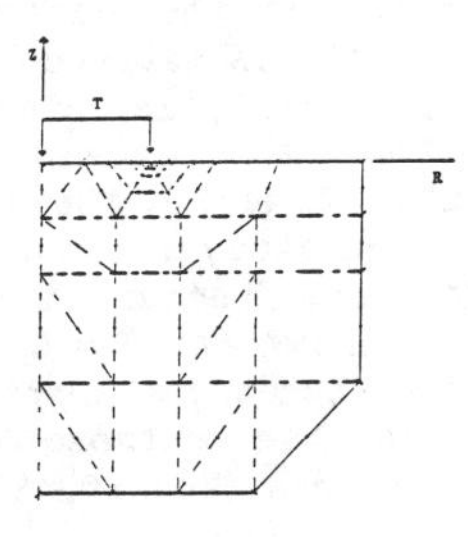

Figure 3a. Mesh 2: Plastic region for circular footing

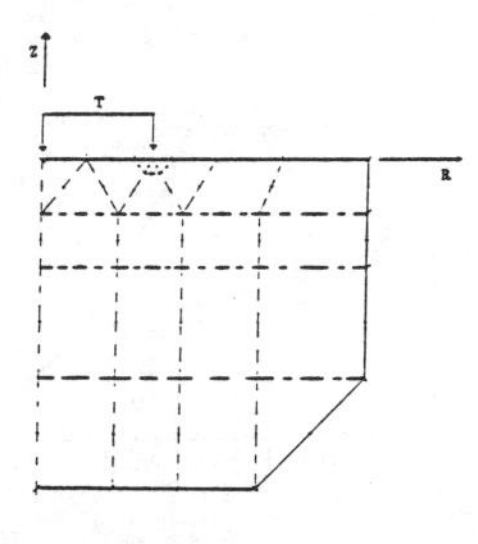

Figure 3b. Mesh 1: Discretization for circular footing

In order to find the convergence of the correct solution, a number of cell patterns were constructed for the plastic region. Two of these meshes that were fine enough to produce satisfactory results are shown in Figure 3a and 3b. The load-displacement behavior at the center of the footing is shown in Figure 4 along side the results obtained by Cathie and Banerjee (1980) where the exact analytical collapse load for this problem is $6C_u$. The results of Cathie and Banerjee (1980) are stiffer than the results of the present paper. The reason for this is the original analysis utilized cells with constant variations of initial stress rates in contrast to the quadratic shape functions used in the present analysis. More importantly, the first investigators calculated stress rates via a linear shape function representation for displacement rates (as in the finite element method), whereas the present analysis uses an accurate integral representation. Once again, it can be seen that cell pattern has no effect.

2.3 BEM for the consolidation analysis

For saturated or unsaturated porous medium, BEM formulations have been recently developed and implemented in a large engineering analysis system (Dargush, 1987) for two and three-dimensional, as well as, axisymmetric multi-region problems. For linear transient analysis, the BEM formulation is given by

$$\alpha_{ij}u_i'(\xi,t) = \int_S [G_{ij}'(x,\xi,t,\tau)*t_i'(x,\tau) - F_{ij}'(x,\xi,t,\tau)*u_i'(x,\tau)]dS \qquad (9)$$

where u_i' and t_i' represent the generalized vectors containing (displacements and pore water pressure) and (tractions and flux), respectively. In the above equation G_{ij}' and F_{ij}' are derived from the fundamental solutions for generalized displacements and generalized tractions due to a point force and a point source in an infinite poroelastic solid (Rudnicki, 1987; Dargush, 1987). The use of * in the above equation implies Riemann convolution, i.e.,

$$\phi * \psi = \int_0^t \phi(x,\tau)\psi(x,\tau)d\tau \ . \qquad (10)$$

Consolidation under a strip load

A number of analytical, as well as, finite element solutions are available for the consolidation of a single poroelastic stratum beneath a strip load. The geometry for this problem is completely defined in the x-y plane by the depth of the layer H, and the total breadth of the strip load 2a. The lower boundary remains impervious, while free drainage is permitted along the entire upper surface. For convenience, non-dimensional forms of the material parameters are utilized. In particular, let E (Young's modulus) = 1.0, ν (Poisson's ratio) = 0.0, κ (permeability) = 1.0, ν_u (undrained Poisson's ratio) = 0.5, and B (saturation) = 1.0. Notice, for this set of properties, that the diffusivity is unity. The presentation of results is simplified further by also assuming unit values for both the half-width a and the applied traction.

The boundary element mesh for the particular case of H/a = 5 is depicted in Figure 5. It contains twelve elements, 25 boundary nodes, and nine interior points. Symmetry is imposed along x = 0. As in the finite element analysis cited here, the region is truncated at a horizontal dimension of L/a = 10. While theoretically the vertical elements along x = 10 are unnecessary, closure of the region permits accurate determination of the strongly singular diagonal blocks of the F matrix via rigid body considerations.

First, results are presented for H/a $\to \infty$. Figure 6 contains the pore pressure time history for a point, on the centerline, a unit depth below the surface. The well-known Mandel-Cryer effect, of increasing pore pressure during the early stages of the process, is evident in the graph. Actually, only the final portion of the rising pore pressure response was captured in the boundary element analysis, since a very large time step was selected. Even with this large step, however, the analytical solution is accurately reproduced. In Figure 7, the pore pressure distribution with depth is shown for T = 0.1. Obviously, the boundary element solutions are comparable to those obtained via finite elements (Yousif, 1984).

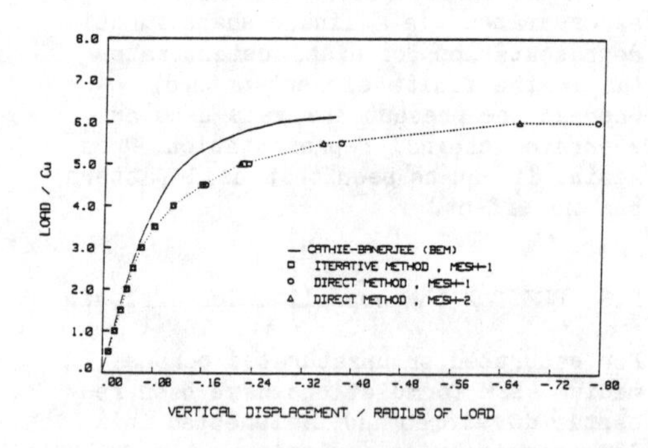

Figure 4. Load displacement behavior under a circular footing

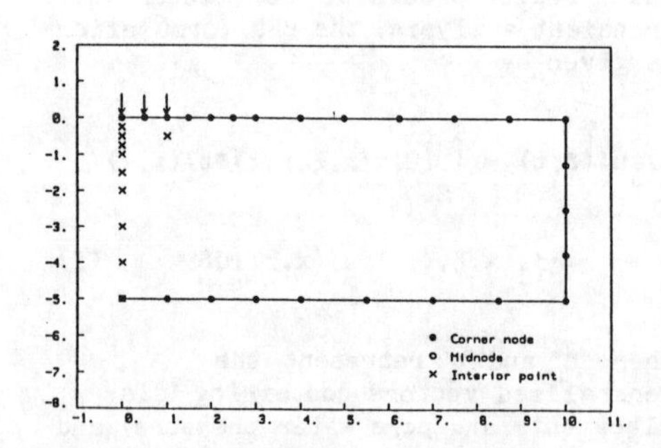

Figure 5. Consolidation under a strip load - mesh

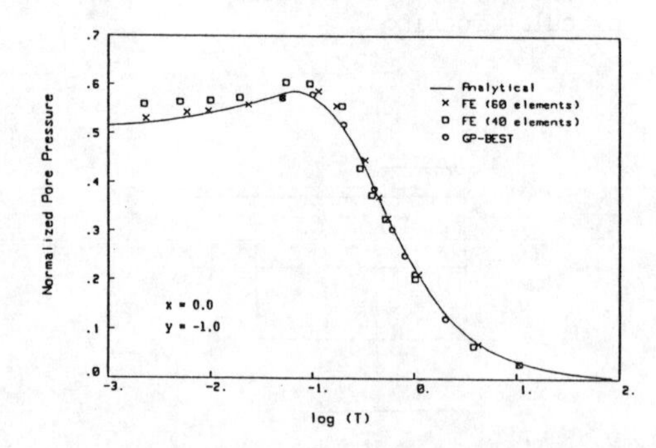

Figure 6. Consolidation under a strip load - pore pressure dissipation

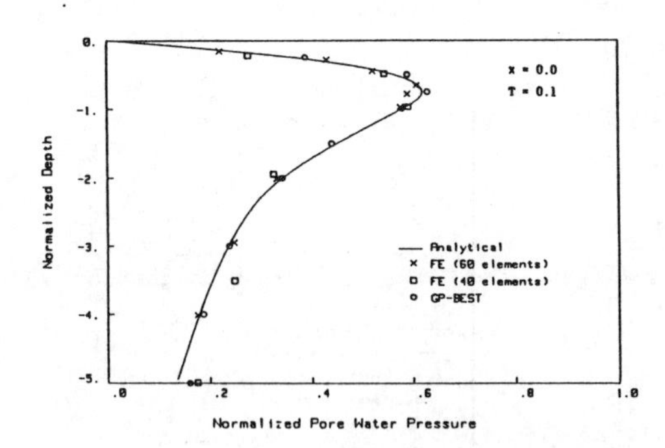

Figure 7. Consolidation under a strip load - pore pressure distribution

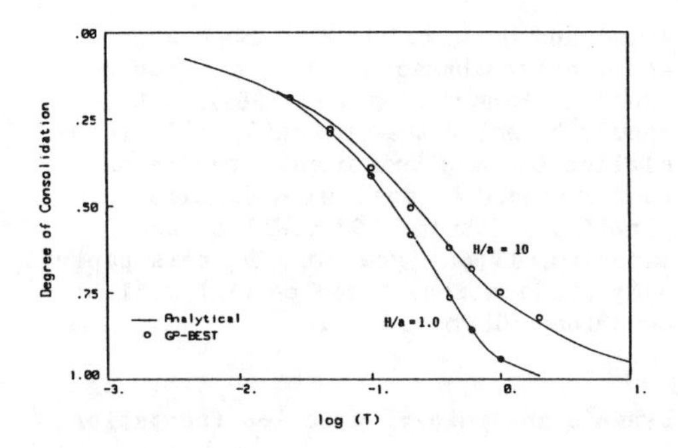

Figure 9. Consolidation under a circular load - effects of larger thickness

Consolidation under a circular load

A very similar problem involves the consolidation of a soil layer under a circular load. The parameter definitions are identical to those detailed in the previous section and will not be repeated here. The only exception is that a is now the radius of the circular load. Two H/a ratios, 10 and 1.0, are examined in the analysis.

The boundary element idealization for H/a = 10 is topologically equivalent to that shown in Figure 5, although now the coordinates are r and z. On the other hand, for H/a = 1.0, a two region model is adopted in order to effectively reduce the aspect ratio of the domain. Figure 8 depicts this model. Region I contains ten elements and 21 nodes, while Region II consists of eight elements connecting sixteen nodes. BEM results for the degree of consolidation are compared with the analytical solution in Figure 9. The correlation is good for H/a = 10, but, for H/a = 1.0, the agreement is almost exact. The

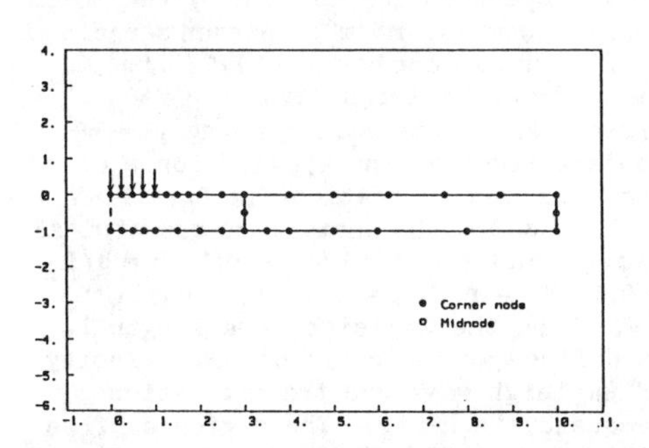

Figure 8. Consolidation under a circular load - 2-region mesh

additional refinement in the vicinity of the load, included in the two region model of Figure 8, is the probable cause for the precise results obtained in the latter case.

2.4 BEM for dynamic analysis

Two types of BEM formulations are available for dynamic analyses. The periodic dynamic formulation for displacements $u_i(x,\omega)$ can be expressed as (Banerjee and Butterfield, 1981):

$$\alpha_{ij}u_i(x,\omega) = \int_S [G_{ij}(x,\xi,\omega)t_i(x,\omega) - F_{ij}(x,\xi,\omega)u_i(x,\omega)]dS \qquad (11)$$

where ω is the circular frequency in radians.

For transient dynamic analysis, the equivalent boundary element representation for $u_i(x,t)$ is given by

$$\alpha_{ij}u_i(x,t) = \int_S [G_{ij}(x,\xi,t,\tau)*t_i(x,t) - F_{ij}(x,\xi,t,\tau)*u_i(x,\tau)]dS \qquad (12)$$

where once again * represents Riemann convolution.

The kernel functions in the above equations are, respectively, those of Kelvin's solution for a periodic point force and Stokes' solution for a unit impulse within an infinite solid. The complete algorithm for the numerical solutions of (11) and (12) can be found in Banerjee and Ahmad (1985); Banerjee,

Ahmad and Manolis (1986). Some major applications based on (12) can also be found in Banerjee et al (1986). It should be noted that recently this formulation for the transient dynamics has been extended to deal with dynamic plasticity (Ahmad, 1986; Wilson and Banerjee, 1986). However, in this paper only applications based on (11) will be described (Chen, 1987).

Dynamic analysis of a buried foundation

This problem was analyzed recently by Rizzo and Shippy (1985) using their implementation of BEM which was subsequently criticized by others who wrongfully concluded that BEM is unsatisfactory for these types of problems. Rizzo and Shippy themselves blamed fictitious eigenfrequency for the inaccuracies of their results.

A number of meshes used are shown in Figure 10 and the results are compared with the exact analytical solution and previous BEM results in Figure 11. It can be seen from Figure 11 that the results from the present analysis agree very well with the analytical and finite element results. At this point, it is important to mention that in the FEM analysis (Day and Frazier, 1979) 1000 axisymmetry elements were used to obtain the correct results, whereas in the present BEM analysis the problem is modeled in three-dimensions with only 16 surface elements.

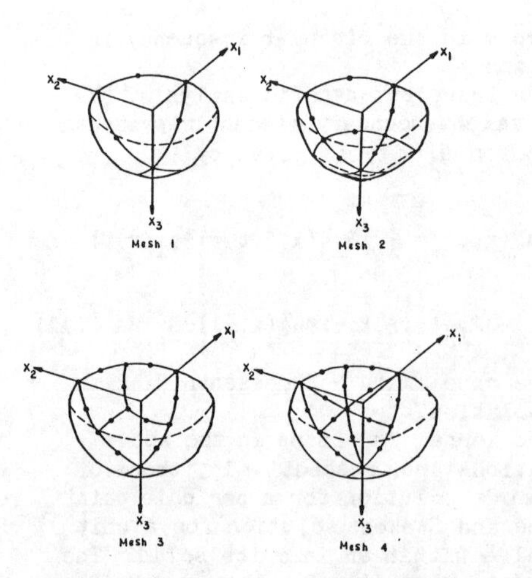

Figure 10. Discretizations of hemispherical foundation-soil interfaces

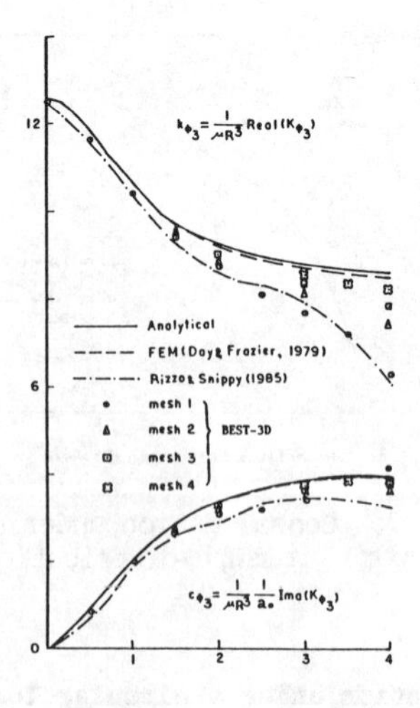

Figure 11. Comparison of torsional stiffness coefficients for a hemispherical formulation

Application to the vibration isolation problem (Chen, 1987)

As a final comprehensive example, results using BEST for a number of vibration isolation cases are compared with the actual field observation of Woods (1968) and results obtained numerically by Dasgupta et al (1986).

Figures 12a-b show the front and top views of the schematic diagram for a rigid square foundation surrounded by a rectangular open trench and excited by vertical oscillating force. The energy due to excitation is screened at the source of vibration. Soil properties and geometrical dimensions of the trench and the foundation are the same as those used by Dasgupta et al (1986), who were the first to analyze this problem. The soil properties are defined by the shear modulus $G = 132$ MN/m^2, Poisson's ratio $\nu = 0.25$, mass density $\rho = 17.5$ KN/m^3, velocity of Rayleigh wave $C_R = 250$ m/sec., and hysteretic damping $\beta = 6\%$. The magnitude of the applied force on the massless foundations is $P_0 = 1$ KN. Additionally, the normalized geometrical dimensions are $T = t/L_R = 0.5$, $B = b/L_R = 0.6$, $R_1 = r_1/L_R = 0.4$, $R_2 = r_2/L_R = 0.4$, where the Rayleigh wave length L_R is defined as the ratio of the velocity of Rayleigh wave and the excitation frequency in Hertz. The region of free field in the BEM discretization is taken as far as fifty-six (56) times the width of the trench, which in turn is about

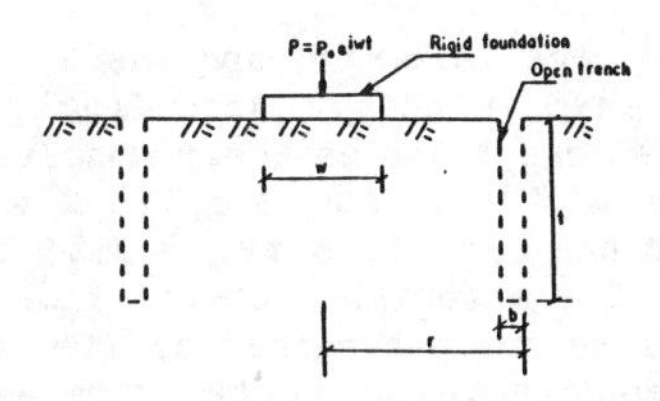

Figure 12a. Front view of the schematic diagram for a foundation with open trench

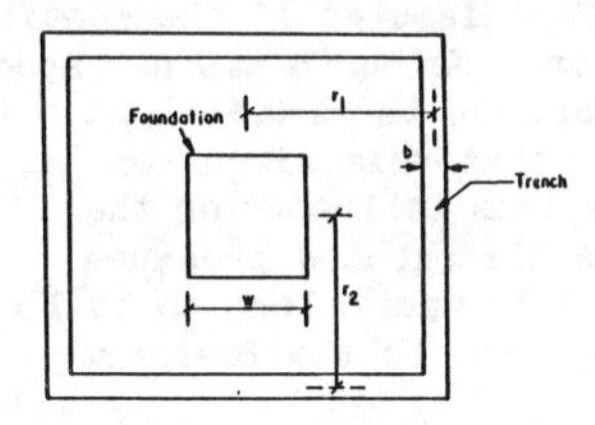

Figure 12b. Top view of the schematic diagram for a foundation with open trench

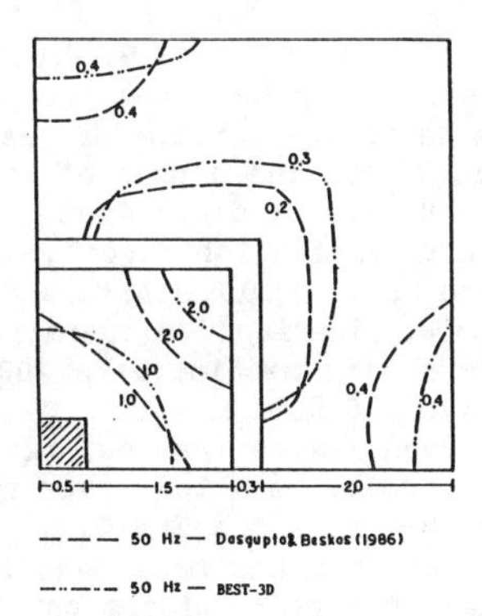

Figure 13. Comparison of amplitude reduction factors

8.5 times the separation from center of the foundation to centerline of the trench. Figure 13 shows a comparison of the contour diagrams of amplitude reduction factor for excitation frequency of 50 Hertz with published data from Dasgupta et al (1986). The amplitude reduction factor is defined as the ratio of vertical displacement amplitude after trench installation to vertical displacement amplitude before trench installation. Comparison shows that the present implementation yields somewhat different results. However, Dasgupta, et al used constant elements whereas the present analysis uses quadratic variation.

Woods (1968) carried out a series of field tests in his attempt to define the screened zone and degree of amplitude reduction within the screened zone for trenches of a few specific shapes and sizes. He classified the foundation isolation problem into two categories, namely active isolation and passive isolation. Active isolation is the employment of barriers close to or surrounding the source of vibrations to reduce the amount of wave energy radiated away from the source. Passive isolation is the employment of barriers at points remote from the source of vibrations but near a site where the amplitude of vibration must be reduced.

These tests were performed at a selected site of a two-layer system as shown on Figure 14. A layer of uniform silty fine sand (SM) with dry density ρ = 104 lb/ft^3, water content w = 7%, void ratio e = 0.61, pressure wave velocity v_p = 940 ft/sec, and the Rayleigh wave velocity v_R = 413 ft/sec rests on a layer of sandy silt (ML) with dry density ρ = 91 lb/ft^3, water content w = 23%, void ratio e = 0.68, and pressure wave velocity v_p = 1750 ft/sec (Woods, 1968). The field tests for the active isolation were carried out for four lengths of the trench forming an arc of 90, 180, 270, and 360 degrees. The depth of the trench was varied between 0.5, 1.0, and 2.0 feet, while the width was 0.25 feet for all trenches. The distance from the center of footing to the centerline of trench is 1.0 foot. The field tests for the passive isolation were carried out for six lengths of

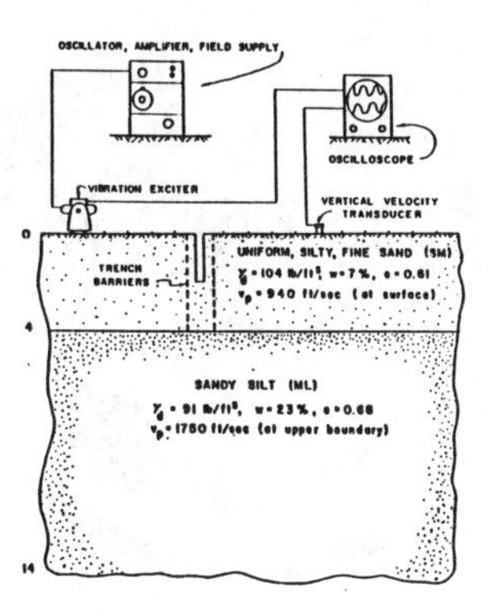

Figure 14. Field site soil properties and schematic of instrumentation (after Woods, 1968)

trench, namely 1.0, 2.0, 3.0, 4.0, 6.0, and 8.0 feet. Distances from the center of footing to the centerline of trench were 5 and 10 feet and depths of trench were 1.0, 2.0, 3.0, and 4.0 feet. A constant input excitation force vector of 18 lb was used in all active and passive tests. The applied operational frequencies of the machine were 200, 250, 300, and 350 Hertz.

In the present investigation, the soil profile is modeled as a two-layer system defined by the shear modulus G, Poisson's ratio ν , and mass density ρ. The material properties of the top layer are basically determined by the relations among the pressure wave velocity, the shear wave velocity, and the Rayleigh wave velocity. Accordingly, the values are taken as shear modulus G = 647200 lb/ft^2, Poisson's ratio ν = 0.35, and the mass density ρ = 3.25 lb-sec/ft^3. The material properties of the bottom layer are determined by assuming the Poisson's ratio to be the same as that of the top layer. Accordingly, the values are taken as shear modulus G = 1991150 lb/ft^2, Poisson's ratio ν = 0.35, and the mass density ρ = 2.84 lb-sec/ft^3. The diameter of the footing of vibration excitor which was not specified, is assumed to be 0.50 feet. It was thought that this dimension has relatively less influence on the results. A distributed pressure of 81.5 lb/ft^2 which is equivalent to 18 lb force is applied to the footing.

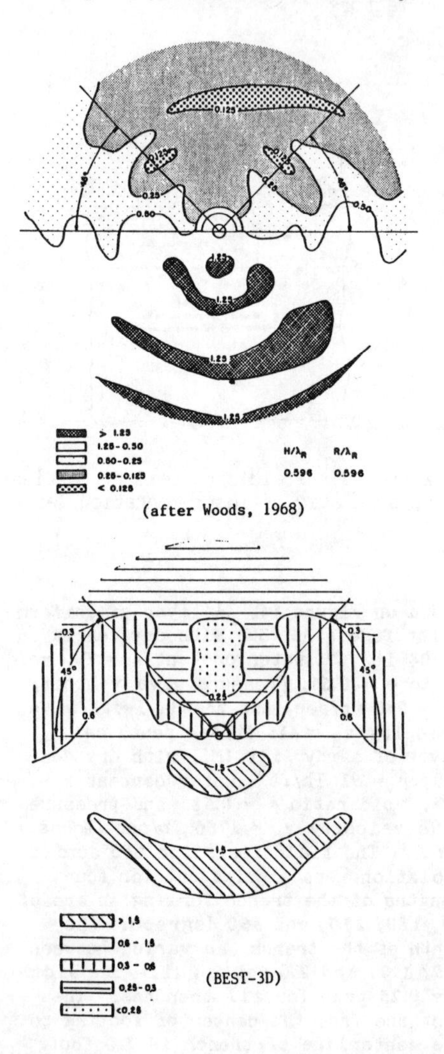

Figure 15. Amplitude reduction factor contour diagram (180° trench)

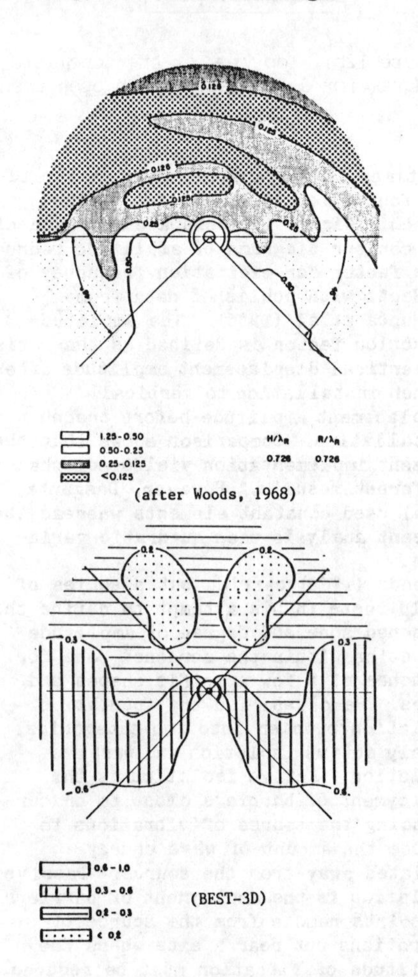

Figure 16. Amplitude reduction factor contour diagram (270° trench)

For the active isolation problem, a typical comparison of the BEM solutions and the experimental results are shown in Figures 15 and 16. Figure 15 shows the overview of a contour diagram for a half circle trench with depth of 1.0 foot under an operating frequency of 250 Hertz. Both approaches predict similar screened zones which are area symmetrical about a radius from the source of excitation through the center of the trench and bounded laterally by two radial lines extending from the center of the source of excitation through points 45 degrees from each end of the trench. The expanding of screened zones may be noted when the screened zones in Figure 16, which is obtained by extending the length of trench from half circle to 270 degrees, are compared with those of Figure 15.

For the passive isolation problem, comparisons of the BEM solutions and the experimental results are shown in Figure 17 which shows the contour diagram of amplitude reduction factor for a rectangular trench of six feet in length and 2 feet in depth under an operating frequency of 250 Hertz. This trench is located 5 feet from the center of source. The comparison shows that again the experimental results are predicted qualitatively by the BEM solution.

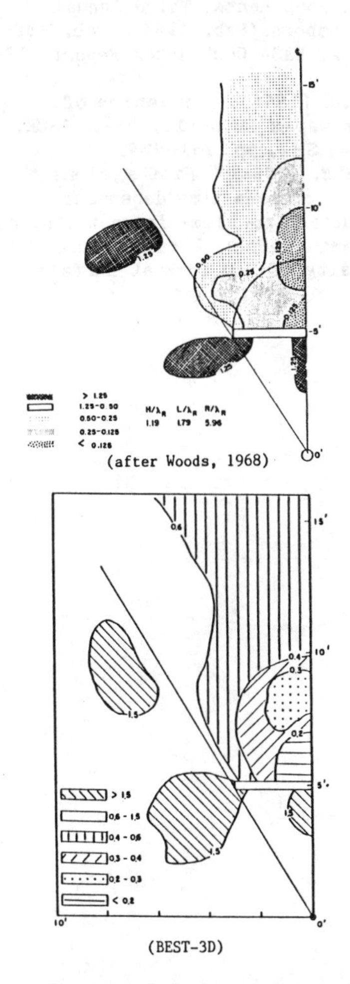

Figure 17. Amplitude reduction factor contour diagram (passive isolation)

3. CONCLUSIONS

A series of BEM applications in nonlinear stress analysis, consolidation analysis and dynamic analysis of problems of geomechanics is described. The developed analysis has been incorporated in a large general purpose BEM system.

4. REFERENCES

Ahmad, S. (1986). Linear and nonlinear dynamic analysis by boundary element method, Ph.D. Dissertation, State University of New York at Buffalo.

Banerjee, P.K. and Ahmad, S. (1985). Advanced three-dimensional dynamic analysis by boundary element methods, Advanced Topics in Boundary Element Analysis, AMD Vol. 72, ASME, New York, pp. 65-81.

Banerjee, P.K., Ahmad, S. and Manolis, G.D., (1986). Transient elastodynamic analysis of three-dimensional problems by boundary element methods, Earthquake Engineering and Structural Dynamics, Vol. 14, pp. 933-949.

Banerjee, P.K. and Butterfield, R. (1975). Boundary element methods in geomechanics, Chapter 16 in Finite Element in Geomechanics, ed. G. Gudehus, John Wiley and Sons, 1977. (Proc. of NMSMR, University of Karlsruhe, West Germany, September 1975).

Banerjee, P.K. and Butterfield, R. (1981). Boundary element methods in engineering science, McGraw Hill, London.

Banerjee, P.K. and Raveendra, S.T. (1986). Advanced boundary element analysis of two and three-dimensional problems of elastoplasticity, Int. Jour. Numerical Methods in Eng., Vol. 23, pp. 985-1002.

Banerjee, P.K. and Raveendra, S.T. (1987). A new boundary element formulation for two-dimensional elastoplastic analysis, Journal of Engineering Mechanics, ASCE, Vol. 113, No. 2, pp. 252-265.

Banerjee, P.K., Wilson, R. and Miller, N. (1985). The development of a large BEM system for the three-dimensional inelastic analysis, Proc. ASME Conf. on Advanced Topics in Boundary Element Analysis, AMD, Vol. 72, ASME, New York, November 1985.

Banerjee, P.K., Wilson, R.B. and Miller, N. (1988). Elastic and inelastic analysis of gas turbine engine structures by BEM, Int. Jour. Num. Methods in Engrg. (in print).

Cathie, D.N. and Banerjee, P.K. (1980). Boundary element method in axisymmetric plasticity, Innovative Numerical Analysis for the Applied Engineering Sciences, ed., R.P. Shaw et al, University of Virginia Press.

Chen, K.H. (1987). Dynamic analysis of embedded 3-D foundations by BEM, Ph.D. Dissertation, State University of New York at Buffalo.

Dargush, G.F. (1987). Boundary element methods for the analogous problems of thermomechanics and soil consolidation, Ph.D. Dissertation, State University of New York at Buffalo.

Dasgupta, G., Beskos, D.E. and Vardoulakis, I.G. (1986). 3-D vibration isolation using open trenches, 4th Conf. on Computational Mech., Atlanta, pp. 385-392.

Day, S.M. and Frazier, G.A. (1979). Seismic response of hemispherical foundation, Proc. ASCE, Vol. 105, EM1, pp. 29-41.

Henry, D.P. (1987). Advanced development of the boundary element method for the elastic and inelastic thermal stress analysis, Ph.D. Dissertation, State University of New York at Buffalo.

Henry, D.P. and Banerjee, P.K. (1988). A variable stiffness type boundary element formulation for axisymmetric elastoplasticity, Int. Jour. Numerical Methods in Engineering (in print).

Henry, D.P., Pape, D and Banerjee, P.K. (1987). New axisymmetric BEM formulation for body forces using particular integrals, Jour. Engineering Mechanics Division, ASCE, May, pp. 671-688.

Pape, D. and Banerjee, P.K. (1987). Treatment of body forces in 2D elastostatic BEM using particular integrals, Jour. of Applied Mechanics, ASME (in print).

Raveendra, S.T. (1984). Advanced development of BEM for two and three-dimensional and nonlinear analysis, Ph.D. Dissertation, State University of New York at Buffalo.

Rizzo, F.J. and Shippy, D.J. (1985). Boundary integral equation analysis for a class of earth-structure interaction problems, Final Report to NSF, Grant No. CEE80-13461, Dept. of Engrg. Mech., University of Kentucky, Lexington.

Watson, J.O. (1979). Two and three-dimensional problems of elastoplasticity, Chapter 2 in Development in BEM - Vol. 1, eds., P.K. Banerjee and R. Butterfield, Applied Science Publishers, London.

Wilson,, R.B. and Banerjee, P.K. (1986). 3-D inelastic analysis methods for hot section components, Third Annual Status Report (Feb. 1985 - Feb. 1986) - Vol. 2, NASA Contractor Report 179 517.

Woods, R.D. (1968). Screening of surface waves in soils, Proc. ASCE, Vol. 94, SM4, pp. 951-979.

Yousif, N.B. (1984). Finite element analysis of some time dependent construction problems in geotechnical engineering, Ph.D. Dissertation, State University of New York at Buffalo.

Numerical Methods in Geomechanics (Innsbruck 1988), Swoboda (ed.)
© 1988 Balkema, Rotterdam. ISBN 90 6191 809 X

Application of analytic and semi-analytic techniques to geotechnical analysis

J.R.Booker
University of Sydney, School of Civil Mining Engineering, Australia

ABSTRACT: This paper examines the application of analytic and semi analytic methods to geotechnical analysis. It is shown that such techniques can be applied to a wide variety of problems and thus because they can be implemented on microcomputer and because of their relative low cost can be used to provide valuable parametric and sensitivity studies.

1. INTRODUCTION

In recent times there have been considerable advances in the application of numerical techniques to Geotechnical Analysis. Perhaps the greatest advances have been made in finite element analysis which is now capable of, at least in principle, of following a highly realistic construction path, and of taking account of complex configurations and sophisticated constitutive behaviour. Sometimes it is not necessary or desirable to use the full power of the finite element method, for example in the analysis of underground openings in rock it would be prohibitively costly to perform a full three dimensional finite element analysis yet boundary element analyses of such openings are performed routinely and yield much valuable information. It follows that analytic and semi-analytic methods have an important role to play in geotechnical analysis. Of course these methods do not have the fine resolution to investigate complex detail that characterises finite element methods nevertheless their very lack of complexity often allows an uncluttered look at the processes in operation and allows an assessment of the relative importance of various contributing factors. Moreover such special purpose methods are usually relatively low cost and can be implemented on microcomputers. This means they are accessible to all engineers and can be used routinely to perform parametric and sensitivity studies.

It is not possible in this paper to make a comprehensive review of all such approaches, instead the lesser task of using some of the research that the author has been personally associated with to illustrate the scope of applicability of analytic and semi-analytic methods, has been undertaken.

2. ANALYTIC APPROACHES

2.1 Settlement of a Non-Homogeneous Elastic Half Space

The behaviour of a foundation on a deep clay deposit can often be modelled by assuming that the layer can be aproximated by an elastic half space that has a modulus $E = M_E z^\alpha$ which varies with the depth below the surface, z. Such deposits have been investigated for linear variation by Gibson (1967), Gibson, Brown and Andrews (1971), Brown and Gibson (1972), Awajobi and Gibson (1973), Gibson and Sills (1975), Calladine and Greenwood (1978) and more generally by Holl (1940) and Booker, Balaam and Davis (1985a, b).

It is possible to obtain some important analytic solutions, for example the observation that, if a vertical load P is applied at the origin, then the stress state within the half space will be inversely proportional to the square of the distance from the origin, leads to the following expression for surface deflection w:

$$w = \frac{BP}{M_E r^{1+\alpha}} \tag{1}$$

where r is the distance of the point on the surface from the point of application and value of B is given in Figure 1 for a range of Poisson's ratio ν and the exponent α.

Once the fundamental solution given by equation (1) is known it is possible to find a host of solutions using the principle of

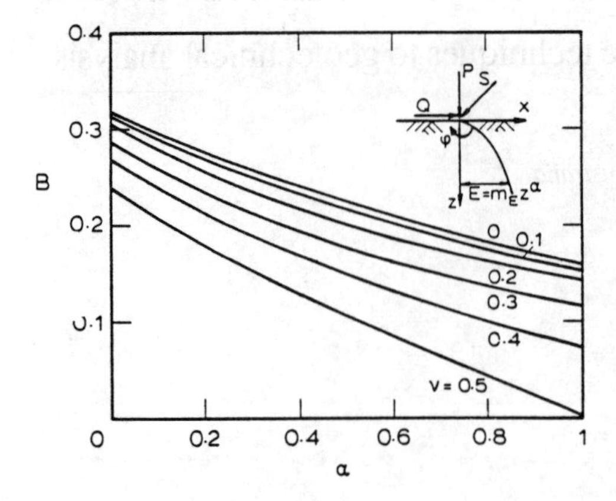

Figure 1. Variation of coefficient B with α and ν.

superposition. For example the surface settlement w of perfectly flexible circular footing subjected to a uniformly distributed load of intensity q is given by equation (2), the deflection profile is shown in Figure (2).

$$w = \frac{2}{1-\alpha}\,\frac{\pi a^2 q}{M_E\,a^{1+\alpha}} BF\left[\frac{\alpha+1}{2},\ \frac{\alpha-1}{2},\ 1,\frac{r^2}{a^2}\right] \quad (2)$$

where F (a, b, c, z) is Gauss's hypergeometric function.

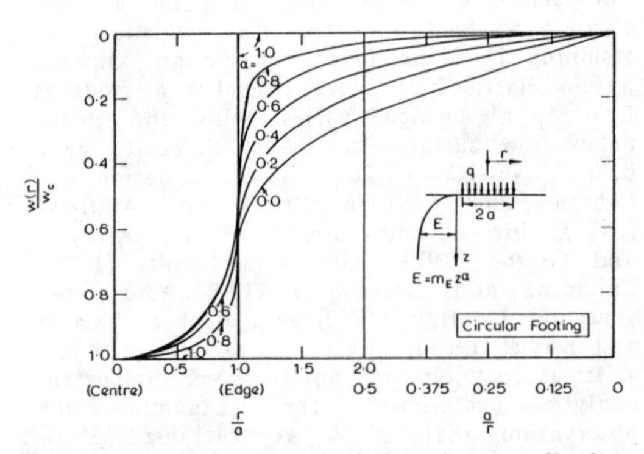

Figure 2. Settlement profile for a flexible circular footing.

If on the other hand a circular rigid footing of radius a is acted on by a central load of intensity P, then it is found that the vertical stress distribution beneath the footing is given by:

$$\sigma_{zz} = \frac{P(1+\alpha)}{2\pi a^2}\left[1-\frac{r^2}{a^2}\right]^{\frac{\alpha-1}{2}} \quad (3a)$$

while the deflection of the footing is given by

$$w = \frac{PB}{M_E\,a^{\alpha+1}}\,\frac{\frac{1}{2}\pi(\alpha+1)}{\sin\frac{1}{2}\pi(\alpha+1)} \quad (3b)$$

The internal and external deflection is shown in Figure 3. It is interesting to note that as $\alpha\to 1$ we obtain the Winkler behaviour discovered by Gibson (1967).

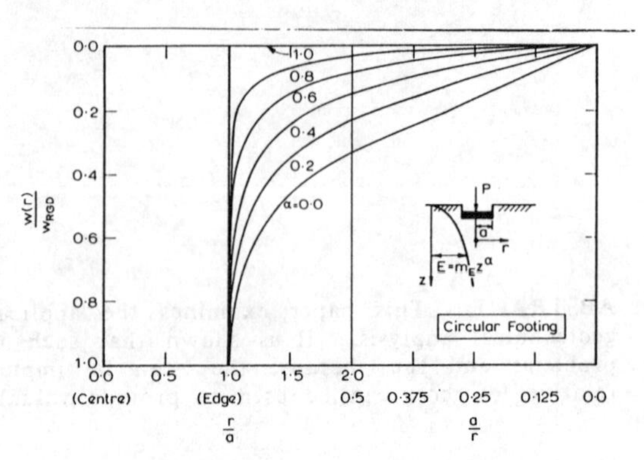

Figure 3. Settlement profile for a rigid circular footing.

2.2 Analysis of One Dimensional Primary and Secondary Consolidation

Consider the one dimensional consolidation of a specimen of depth H which is subject to a constant stress q at time t = 0 under conditions of one way drainage. If the material creeps under constant effective stress then the strain ϵ is given by

$$\epsilon = \sigma'(0)\,J(t) + \int_0^t J(t-\tau)\,\frac{d\sigma'(\tau)}{d\tau} \quad (4)$$

The equations of consolidation can be solved with the aid of a Laplace transform and it is found that the surface settlement S is given by

$$\bar{S} = \bar{m}_v q\,H\,\frac{\tan\beta H}{\beta H} \quad (5)$$

where the superior bar denotes a Laplace transform $\bar{m}_v = s\bar{J}$, $\bar{c} = k/\gamma_w \bar{m}_v$, $\beta^2 = s/\bar{c}$ and k is the permeability and γ_w is the unit weight of water.

Equation (5) gives the solution in the Laplace transform domain, the solution in the time domain may be obtained analytically Gibson and Lo (1961) or by numerical inversion Talbot (1979).

Some clay materials have a creep function which varies linearly with the logarithm of time for large time, De Ambrosis and Poulos (1976) proposed the following model

$$J(t) = A + B\,\ell n(1+\alpha t) \quad (6)$$

The solution for this case has been given by Booker (1982). Figure (4) shows the variation of dimensionless settlement S/qAH against dimensionless time $T = c_0t/H^2$, where $c_0 = k/\gamma_w A$, and shows that the deeper the layer the more important the effect of secondary consolidation.

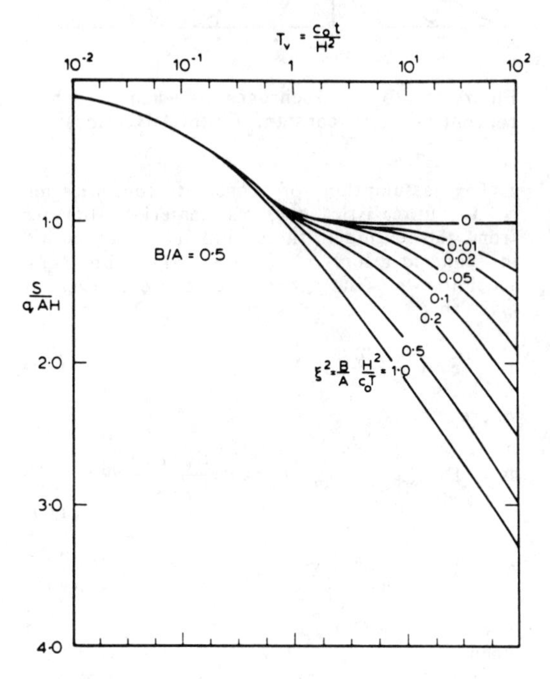

Figure 4. Time settlement behaviour B/A = 0.5.

Booker (1982) shows how it is possible to determine the primary and secondary consolidation parameters from a single test, a comparison of observed and predicted results based on data given by Wissa and Heiburgh (1969) is given in Figure (5).

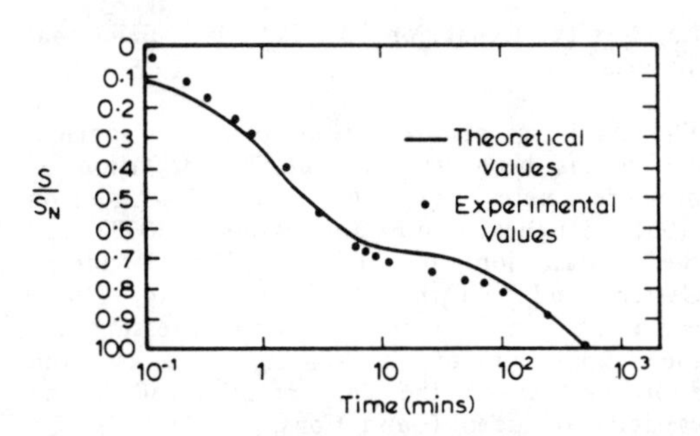

Figure 5. Comparison of observed and predicted behaviour.

2.3 Consolidation around a Spherical Heat Source

When a heat source such as a cannister of radioactive waste is placed in a saturated soil the temperature of the soil will increase. Since the coefficient of expansion of the pore water usually exceeds that of the skeletal material there will be a build up of pore water pressure this in turn will lead to a reduction of effective stress and the possibility of cracking. If the soil is sufficiently permeable the pore pressures generated as a consequence of the increase of temperature will dissipate due to the process of consolidation.

Booker and Savvidou (1984) have considered the case of a spherical heat source of radius R_0 generating a quantity of heat Q in unit time buried deep in a two phase thermoelastic soil having a coefficient of heat conduction K a diffusivity κ and a coefficient of consolidation c and shown that a distance R from the centre of the sphere the temperature θ and excess pore water pressure P are given by

$$\theta = \frac{Q}{4\pi KR} f(\kappa, R, t) \tag{7a}$$

$$p = \frac{X}{1 - c/\kappa} \frac{Q}{4\pi KR}[f(\kappa, R, t) - f(c, R, t)] \tag{7b}$$

where X is the increase in pore pressure per unit increase in temperature under undrained conditions

$$f = \mathrm{erfc}\beta - \mathrm{erfc}(\alpha+\beta)\exp(2\alpha\beta+\alpha^2)$$

$$f = \mathrm{erf}\,\beta - e^{2\alpha\beta+\beta^2}\,\mathrm{erfc}(\alpha+\beta) \tag{8}$$

$$\alpha^2 = \kappa t/R_0^2$$

$$\beta^2 = (R - R_0)^2/4\kappa t$$

Equation (7) can be used to develop the solutions for a point source Booker and Savvidou (1985) by allowing $R_0 \to 0$. This fundamental solution can then be used to generate other solutions. The solution for pore pressure P divided by the maximum pore pressure p_N that would arise if no consolidation occurred, generated by a cylinder of radius a and length 2h for the case in which the ratio of the coefficient of consolidation to the thermal diffusivity is 2 h/a = 10 and $T = \kappa t/a^2$, is shown in Figure 6. It will be observed that because of consolidation the pore pressure only reaches a fraction of the value it would reach under undrained conditions.

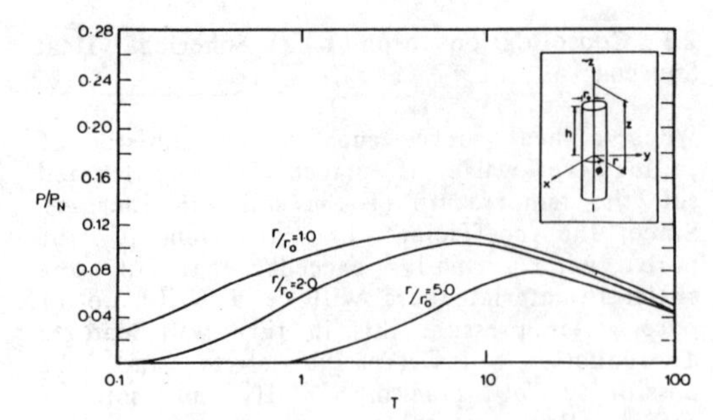

Figure 6. Variation of Pore Pressure with time for a cylindrical source.

2.4 Diffusion from a Cylindrical Waste Depository

One way of disposing toxic waste is to bury it deep underground. If the containing holding the waste breaks down it will diffuse into the environment or be transported by the flow of ground water. In the event of such occurrence it is important to be able to assess the impact on the environment.

Suppose waste is deposited in a cylindrical hole of radius a and that ground water is flowing with an advective velocity v in the x direction. One idealisation would be to assume that the concentration in the bore hole remains constant at its initial value c_0. The solution can be found by taking a Laplace transform.

The distribution of concentration for the case of zero and non zero advective velocity are shown in Figures (7 a, b) respectively. Comparison of these figures shows that the presence of an advective velocity has the effect of sweeping contaminent downstream and inducing an assymmetry in the contaminent distribution.

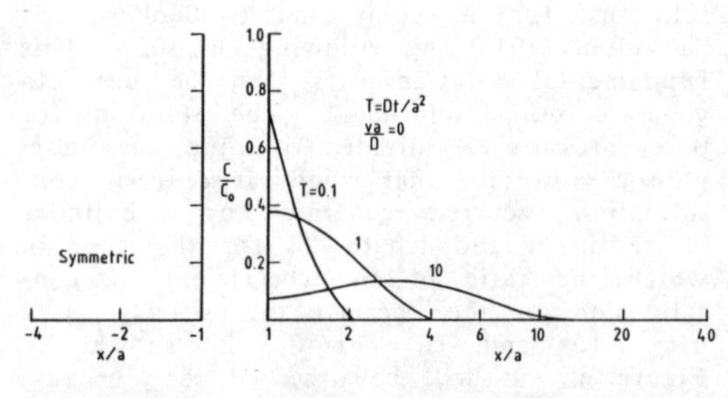

Figure 7a. Isochrones when source concentration is constant (No Advection).

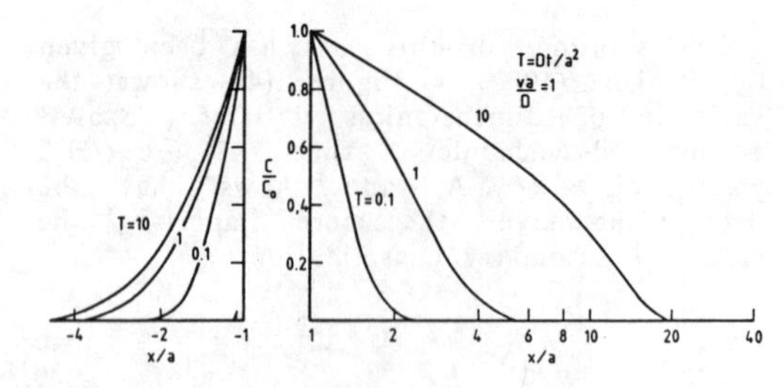

Figure 7b. Isochrones when source concentration is constant (With Advection).

The assumption of constant concentration is an unrealistic one, as material diffuses from the cylinder there will be a reduction in concentration. If we take this into account we find that the concentration is given by:

$$\bar{c} = \frac{c_0}{s} \frac{\Phi}{\Delta} e^{\lambda x} \tag{9}$$

where

$$\Phi = \sum_{\ell=0}^{\infty} (-1)^{\ell} \epsilon_{\ell} I_{\ell}(\lambda a) \frac{K_{\ell}(\mu r)}{K_{\ell}(\mu a)} \tag{10}$$

$$\Delta = 1 - \frac{2nD\mu}{as} \sum_{\ell=0}^{\infty} (-1)^{\ell} I_{\ell}^{2}(\lambda a) \frac{K'_{\ell}(\mu a)}{K_{\ell}(\mu a)}$$

where $\lambda = v/2D$, $\mu^2 = \lambda^2 + s/D$, $\epsilon_{\ell} = 1$ when $\ell = 0$, $\epsilon_{\ell} = 2$ when $\ell \neq 0$, D is the diffisuion coefficient and n is the porosity.

This solution is evaluated for the case of no advection in Figure (8a) and an advective velocity of $av/D = 1$ in Figure (8b). Clearly the more realistic assumption shows that concentration reduces far more quickly than was predicted previously, taking account of advection shows that an advective velocity will lead to contaminent being swept down stream and increase concentrations there as would be expected.

2.5 Cavity Expansion in Cohesive Frictional Material

The analysis of expansion of a cylindrical cavity has been used in the interpretation of pressure meter tests Gibson and Anderson (1961), Hughes Wroth and Windle (1977) and the installation of driven piles Randolph Carter and Wroth (1979). Vesic (1972) presented solutions for the limit pressure for the expansion of a spherical cavity and applied these to the determination of baring capacity of deep foundations.

Recently Carter, Booker and Yeung (1986) showed that it is possible to determine

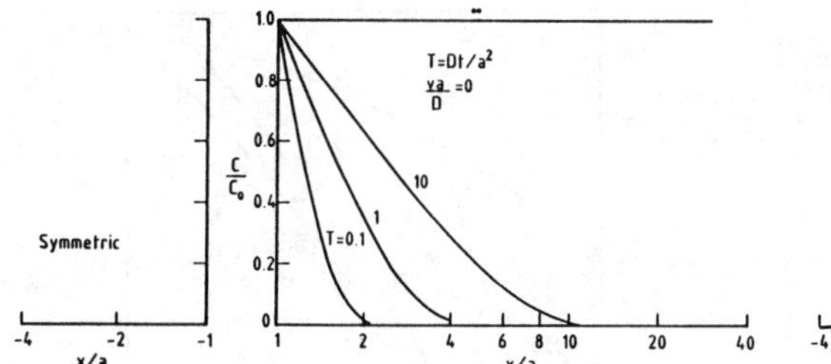

Figure 8a. Isochrones when source concentration varies (No Advection).

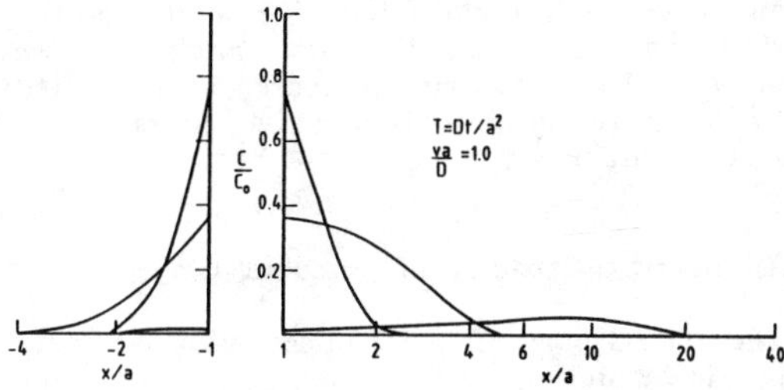

Figure 8b. Isochrones when source concentration varies (With Advection).

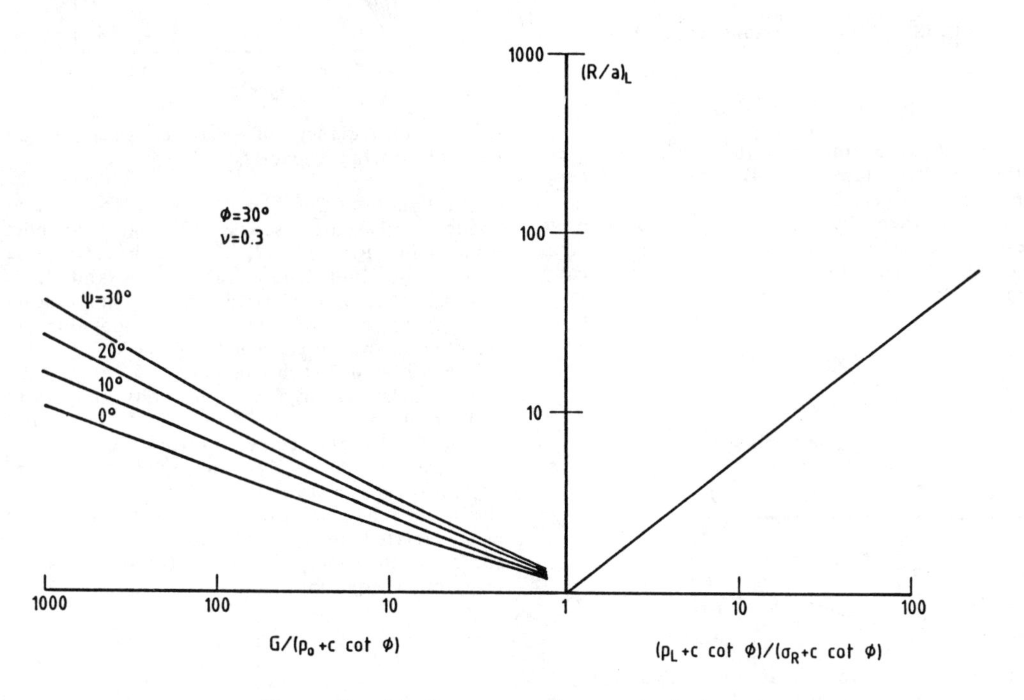

Figure 9. Limit pressure for a spherical cavity.

analytically the limit pressure P_L necessary for the expansion of both cylindrical and spherical cavity their solution depends upon the observation that after a large amount of expansion the ratio of the internal radius a of the cavity to the distance of the elastic plastic interface reaches a constant value. Thus it is found for a purely frictional material with angle of internal friction φ and dilatency angle ψ

$$\frac{2G}{P_o} = \left[\frac{N-1}{N+k}\right]\left[T\left[\frac{P_L}{\sigma_R}\right]^{\gamma} - Z\left[\frac{P_L}{\sigma_R}\right]\right] \quad (11)$$

where γ, Z, T are simple functions of material properties

$$\sigma_R = \left[\frac{1+k}{N+k}\right] N\ P_o$$

$$M = (1+\sin\psi)/(1-\sin\psi)$$

$$N = (1+\sin\varphi)/(1-\sin\varphi)$$

and G denotes the shear modulus, ν Poisson's ratio, P_o the insitu stress and k takes the value 1 for a cylinder and 2 for a sphere.

The variation of limit pressure for the expansion of spherical cavity a cohesive frictional material is shown in Figure (9) for an angle of internal friction $\varphi = 30°$ and a range of possible dilatency angles. We observe that the limit pressure is dominated by the ratio of initial effective stress P_0 to shear modulus.

2.6 Bearing Capacity of Fissured Clays

The behaviour of an anisotropic perfectly plastic material under conditions of plane strain has been discussed by Booker and Davis (1972). If we consider a purely cohesive material with the failure condition shown in Figure (10) Booker and Davis (1972) showed that

$$-dP + dS = 0 \qquad \text{along an } \alpha \text{ line}$$
$$+dP + dS = 0 \qquad \text{along a } \beta \text{ line} \tag{12}$$

where the α, β lines are the stress characteristics $dy/dx = \tan(\theta\text{-m-}\pi/4)$, $dy/dx = \tan(\theta\text{-m}+\pi/4)$ respectively.

It can then be shown that the average bearing capacity p_u of a surface strip footing is equal to the length of the heavy line marked on Figure (10).

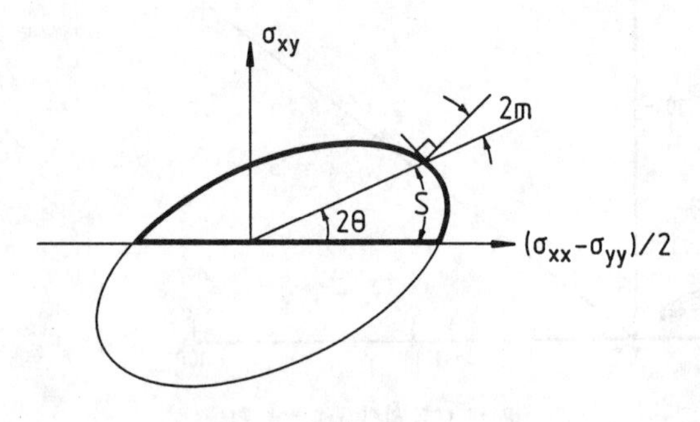

Figure 10. Failure surface for a anisotropic purely cohesive material.

If we apply this to the bearing capacity of the footing shown in the inset of Figure (11) resting on a purely cohesive material with intact strength c_s and weakened by a single set of fissures having a strength c_f and inclined at an angle ω to the vertical we find after Davis (1980) the bearing capacity factors shown in Figure (11).

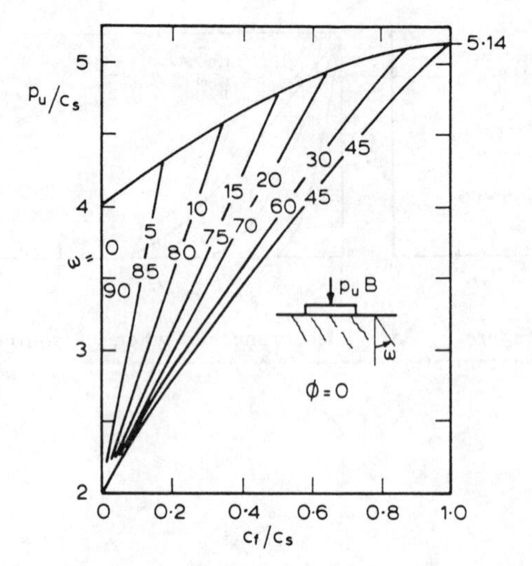

Figure 11. Effect of fissure strength on vertical bearing capacity.

3. SEMI-ANALYTIC TECHNIQUES

3.1 Response of Multiple Underream Anchors

The concept of substructuring provides a powerful method of solving soil structure interaction. Consider, after Rowe and Booker (1980), a series of rigid anchors shown in Figure (12), if each of these anchors is divided into sub region and it is assumed that a uniformly distributed traction acts over each of these, it is possible to establish an influence matrix for the deformation of each anchor. Clearly each anchor must undergo a rigid body movement and must be in equilibrium under the rod tractions and the forces exerted upon it by the soil. Similarly the deflections of adjacent anchors can be used to calculate the strain in the tie rod connecting them and hence related to the rod tension through the stiffness equation. This leads to a set of interaction equations

$$\begin{bmatrix} \mathbf{J} & -\mathbf{B} & \mathbf{O} \\ -\mathbf{B}^T & \mathbf{O} & \mathbf{C} \\ \mathbf{O} & \mathbf{C}^T & \mathbf{\Phi} \end{bmatrix} \begin{bmatrix} \mathbf{F} \\ \mathbf{R} \\ \mathbf{T} \end{bmatrix} = \begin{bmatrix} \mathbf{O} \\ \mathbf{K} \\ \mathbf{O} \end{bmatrix} \tag{13}$$

where **F** is the force acting on the anchor elements, **R** is the vector of rigid body motions, **T** is the vector of rod tractions in the rods, **K** is the applied force, **J** is influence matrix for soil deflection, **B** is matrix giving rigid body deflections, **C** is matrix giving net force exerted by tie rods on each anchor and **Φ** is the flexibility matrix for the rods

This analysis allows the investigation of a wide variety of anchor configurations and spacings, Figure (13) shows the effect of interaction of underreams for a range of spacings when the anchors are placed deep in a homogeneous elastic soil with a Young's modulus E and Poisson's ratio ν.

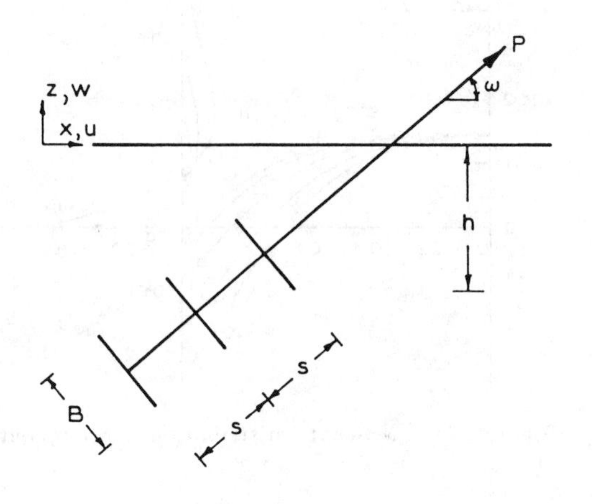

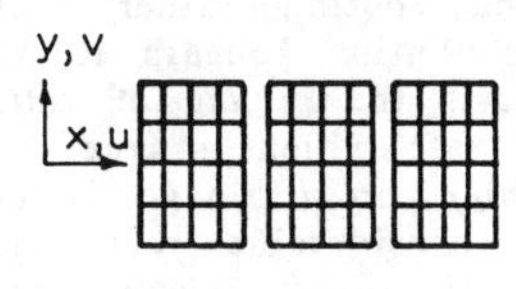

Figure 12. Anchor configuration.

3.2 Analysis of a Liquid Storage Tank on an Elastic Foundation

As mentioned previously the substructuring concept leads to an efficient way of solving a wide variety of soil structure interaction problems. For example if we consider the interaction of a cylindrical liquid storage tank of radius a, Booker and Small (1983), and a deep elastic foundation we can conveniently divide this system into tank walls, base plate and soil foundation. These substructures can be found by developing flexibilities for the individual members and equating the rim deflection and rotation of the base plate and wall the deflections of soil and base plate. The flexibility matrix of the wall can be generated exactly, while the development of the flexibility matrix for base plate and wall is aided by approximating the vertical stress σ_{zz} exerted by the tank on the soil in the form:

$$\sigma_{zz} = \sum_{n=0}^{N} F_n \varphi_n(r) \tag{14a}$$

where $\varphi_0 = (1 - r^2/a^2)^{-\frac{1}{2}}$,

$\varphi_n = (1 - r^2/a^2)^n$ otherwise

The coefficients F_n can be thought of as generalised forces and can be associated with generalised deflections δ_n defined

$$\delta_n = 2\pi \int_0^a r\, w(r)\, \varphi_n(r)\, dr \tag{14b}$$

It is then possible by using the Greens function for a circular plate and using

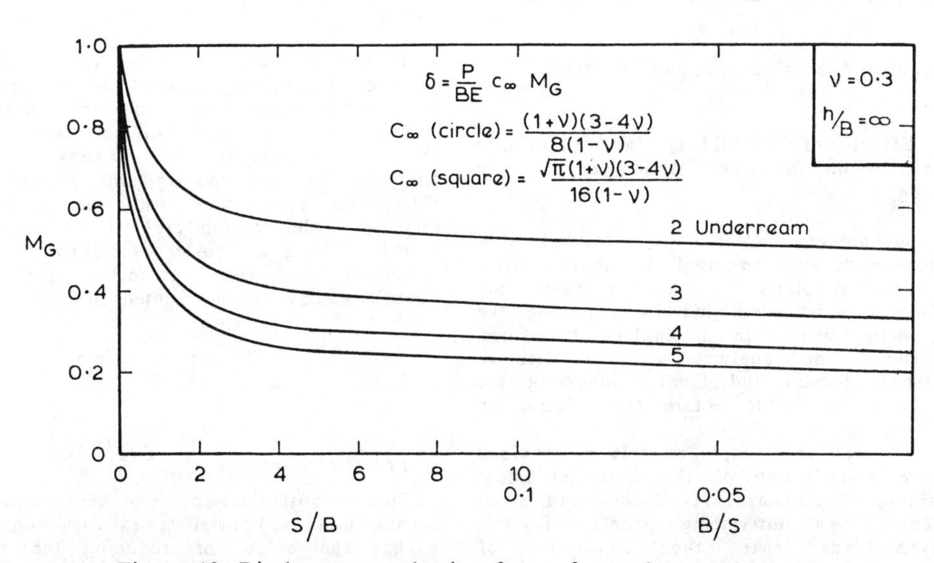

Figure 13. Displacement reduction factor for a deep anchor system.

Boussinesqs solution for a vertical point load acting on a half space to obtain explicit expressions for the elements of the individual flexibility matrix. The tank configuration and notation is shown in Figure (14), in discussing results it is useful to introduce a stiffness factor $K = E_p(1 - \nu^2)(t/a)^3/E_s$. The moment distribution in the wall is shown in Figure 15 and shows the intensifying that arises from increasing the wall thickness.

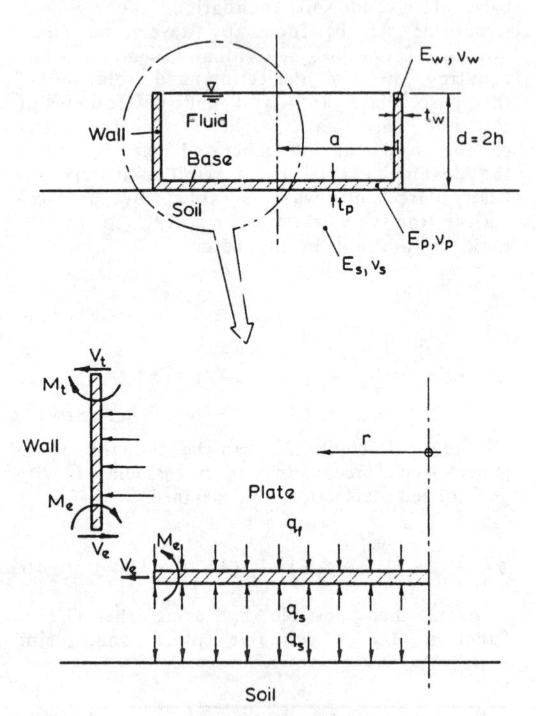

Figure 14. Storage tank configuration.

3.3 Effect of Flexibility and Drainage Conditions on the Consolidation of Circular Footings

The techniques described in previous sub-sections can also be used to analyse time dependent problems. In such cases the analysis can be simplified by removing the time dependence with a Laplace transform, performing the analysis in the Laplace transform domain and finally inverting the Laplace transform to obtain the solution in the time domain.

An interesting example of this is provided by the investigation of the time settlement behaviour of circular rafts Booker and Small (1986a). An interesting aspect of this analysis was that the condition of impermeability beneath the footing was simulated by adjusting the values of the flows outside the footing rather than the pore pressure distribution beneath it. Figure (16) shows contrasts the degree of settlement of rigid circular raft which is free to drain with one in which no drainage is possible and shows that considerable retardation of consolidation occurs in the latter case, in fact an impermeable raft will take approximately three times as long to reach 50% cnsolidation.

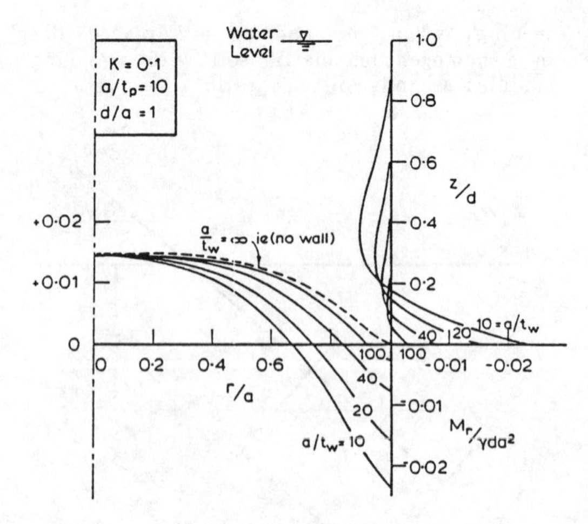

Figure 15. Moment distribution throughout tank.

3.4 The Analysis of Deformations caused by Loadings Applied to the Walls of a Circular Tunnel

A method of determining the elastic modulus of the surrounding ground is to apply a jack to the walls of a tunnel or bore hole.

For a circular tunnel or bore hole, Carter and Booker (1984) this process may be analysed by decomposing the applied load acting on the tunnel walls in a Fourier series in the circumferential (θ) direction and by taking a Fourier transform in the z direction, so that typically the field quantities may be represented in the form

$$(u_r, u_\theta, iu_z)^T = \int_{-\infty}^{\infty} \Delta\, e^{i\alpha z} d\alpha \tag{15}$$

$$(\sigma_{rr}, \sigma_{\theta\theta}, i\sigma_{rz}) = \int_{-\infty}^{\infty} T\, e^{i\alpha z} d\alpha$$

This transformation together with the expansion as a Fourier series with respect to θ has the effect of reducing the partial differential equations of elasticity to

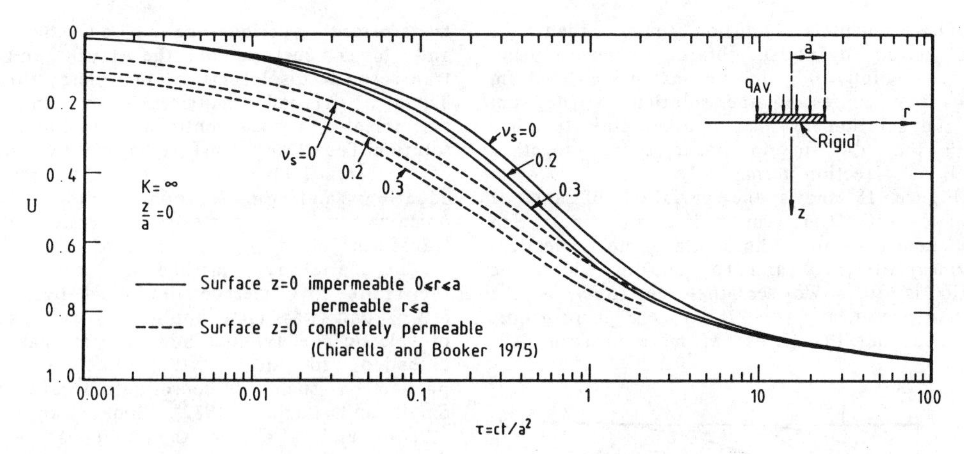

Figure 16. Degree of settlement for a rigid circular footing.

ordinary differential equations, these may be solved and use to generate the flexibility relationship.

$$\alpha\Delta_n = \frac{1}{G}\,\Phi\,F_n \tag{16}$$

where Φ is a known function of the parameter $r\alpha$, G is the shear modulus and the subscript n denotes the 'nth' Fourier coefficient.

If the distribution of load is known then its Fourier components and transformations can be found and so equation (16) can be used to obtain the solution in transform space. The actual solution can then be found by numerical integration of equation (15a) and by summation of the Fourier components.

The calculated deflection is shown in Figure 17. If these values are compared with those obtained by assuming that the load acts over a half space, it can be seen that this usual form of interpretation can lead to a significant overestimate of the ground stiffness.

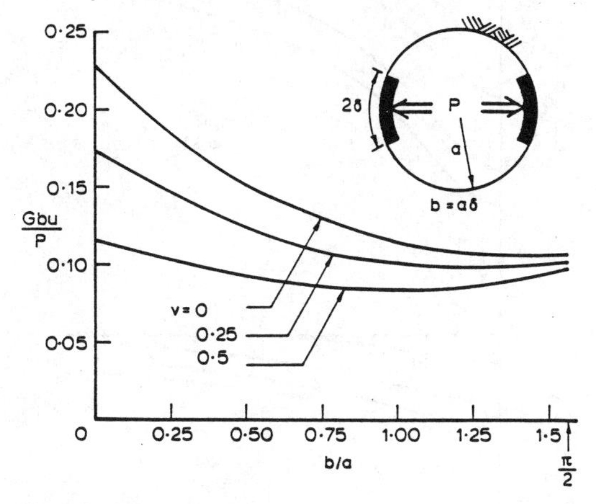

Figure 17. Radial displacement of a pair of curvelinear square rigid plates.

3.5 Consolidation caused by a Point Sink embedded in a Deep Layer

In geotechnical, hydraulic and petroleum engineering it is sometimes necessary to pump water or some other fluid from the ground. In general this leads to an increase in effective stress which causes consolidation of the ground and may lead to large scale subsidence.

Probably the best known examples of this phenomenon occur in Bangkok, Venice and Mexico City where widespread subsidence has been caused by withdrawal of water from aquifers for industrial and domestic purposes. Recorded settlments in Mexico City have reached rates of 5 to 6 cm per year (Scott, 1978).

This process can often been analysed by taking a Laplace transform to remove time dependancy and by introducing repeated Fourier transforms. Suppose for definiteness that a point sink is buried at depth h below the surface of a porous elastic half space, Booker and Carter (1987). Introduction of a Fourier transform

$$(U,P) = \frac{1}{4\pi^2}\int_{-\infty}^{\infty}\int_{-\infty}^{\infty} e^{i(\alpha x+\beta y)}(u,p)\,dx\,dy \tag{17}$$

reduces the set partial differential equations of consolidation to a set or ordinary linear differential equations with constant coefficients.

The complete solution may then be synthesized by first obtaining the solution for a point sink in an extensive medium and then expressing the solution as the sum of the solution to such a point sink together with that due to an image source together with a correction term.

Figure 18 shows the variation of surface profile with time for a material in which the ratio of horizontal to vertical permeability is 2 and for one in which the ratio is 10. We see that not only is the subsidence less for the more anisotropic material but that it is far more uniform.

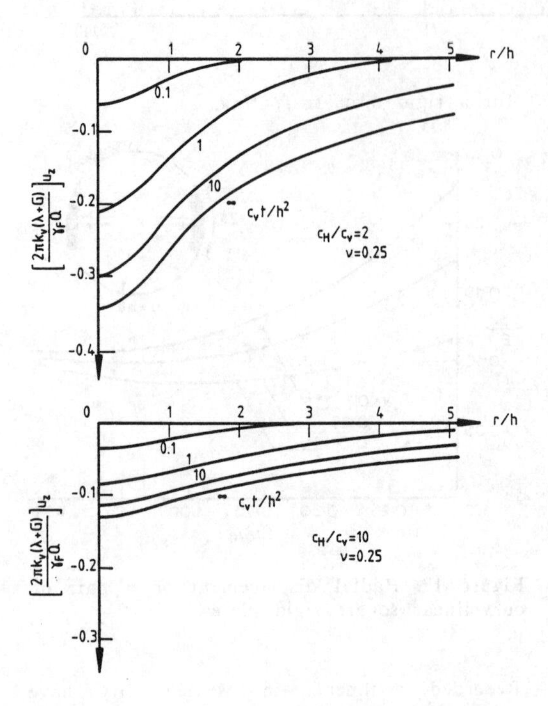

Figure 18. Settlement profiles induced by pumping.

3.6 Finite Layer Techniques

It is possible to formalise the methods of integral transforms described in the previous section so that it becomes a convenient and powerful method for analysing the behaviour of horizontally layered soils.

The essence of the method is to use multiple integral transforms to remove variation in the horizontal (x - y) plane. The governing partial differential equations then become ordinary differential equations with constant coefficients. These equations can be solved in straightforward fashion and use to establish a relation between the transformed tractions acting on the upper and lower surface of the layer and the transformed displacements of these surfaces. These 'stiffness' matrices can then be assembled in conventional fashion, the solution found in transform space by solution of a relatively small set of stiffness equations and the solution in the physical domain found by numerical inversion of the transforms.

The finite layer method has its origin in the finite strip method developed by Cheung (1976) and was first applied to the analysis of elastic behaviour, however it has been extended to deal with time dependent phenomena such as consolidation and creep Small and Booker (1982), Booker and Small (1986b) and also for the analysis of such diverse phenomena as heat flow and pollution migration Rowe and Booker (1985).

To illustrate the approach we will first consider the strains induced by a circular load acting on a layered system consisting of an incompressible viscoelastic material overlaying an elastic material. Figure (19) shows both vertical and horizontal strains, it will be observed the strains in the underlying elastic layer do not vary significantly with time but tht there is a significant time dependence of strain in the upper layer which is undergoing undrained creep.

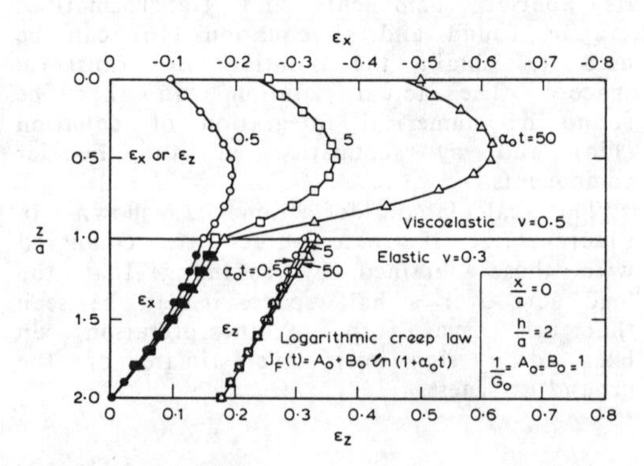

Figure 19. Strains in a viscoelastic layered system.

A second illustration is provided by the pollution migration Rowe and Booker (1985) from a hypothetical 200 m landfill which is underlain by a permeable acquifer in which water flows with a velocity v_6 Figure (20) contrasts a two dimensional and one dimensional analysis. It shows that not only does the presence of an advective velocity in the acquifer significantly reduce the base concentration but that two dimensional effects become significant after lower base velocities.

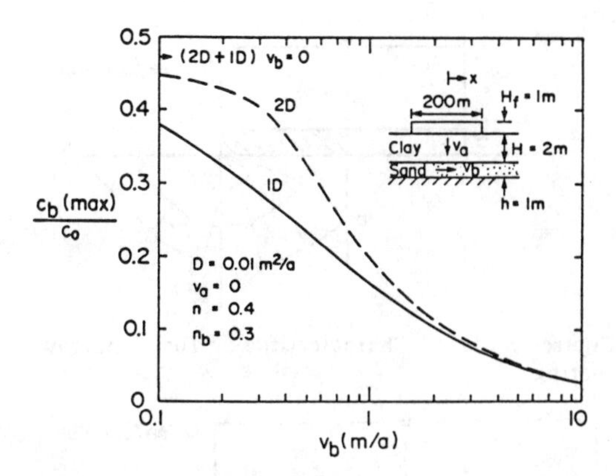

Figure 20. Variation of maximum base concentration with base velocity.

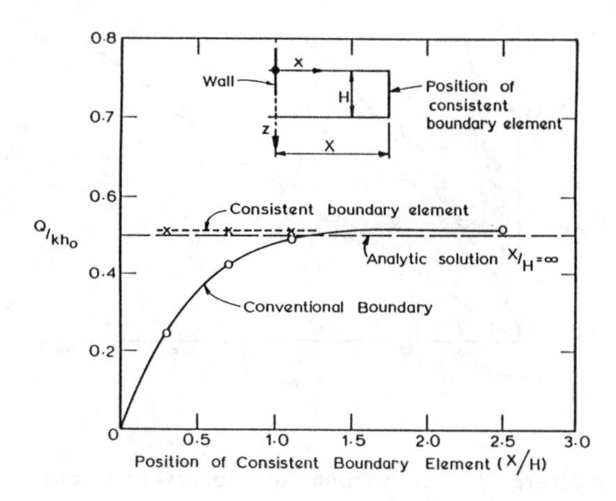

Figure 21. Effect of positioning of boundary element.

3.7 Infinite Elements for Finite Element Analysis

One difficulty in finite element analyses is the placement of boundaries when modelling extensive media. Analytic and semi-analytic methods can be used to augment finite element analyses by developing stiffness matrices for zones of infinite extent.

One simple approach is that developed by Booker and Small (1981) essentially they take a single column of elements they then add to this the stiffness of an identical column and eliminate the degrees of freedom at any internal nodes, this leads to a stiffness matrix of a column having twice the width of the original column, the procedure is then repeated so tht the width of the column doubles with each repetition. it is found this procedure converges very rapidly to give the stiffness matrix for an infinite number of columns.

The method may be applied to static elasticity, dynamic elasticity (consistent boundary element) or to flow. Figure 21 shows the flow under a retaining wall calculated by placement of an infinite element some distance from the boundary or by the conventional placement of impermeable boundaries. It can be seen that although the calculated flow is independent of the placement of the unfinite elemnt it depends significantly in the positioning of a conventional boundary.

3.8 Solutions of Nonlinear Problems

The previous examples in this section have been restricted in considering the solution of linear problems in which it is possible to make significant use of the principle of superposition.

It is however possible to deal with non-linear problems, for example Balaam and Booker (1985) have examined the response of an extensive rigid raft overlaying a site reinforced with stone columns. They assumed that the stress state in the stone columns was triaxial and yield was possible and that the deformation in the surrounding material was, because of the restriction of possible movement, elastic. This is illustrated in Figure (22). A comparison of this approximate solution and a full finite element solution is shown in Figure (23) and the differences are barely discernible.

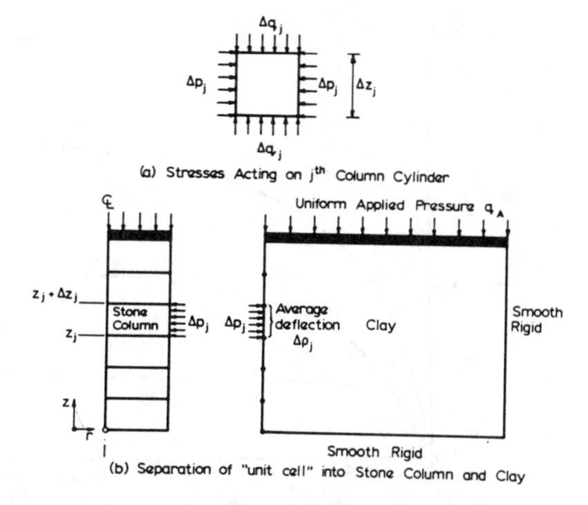

Figure 22. Stone column-raft idealisation.

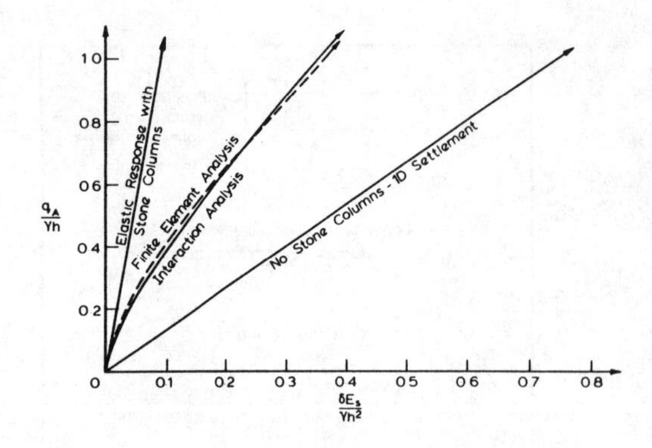

Figure 23. Comparison of approximate and finite element analysis.

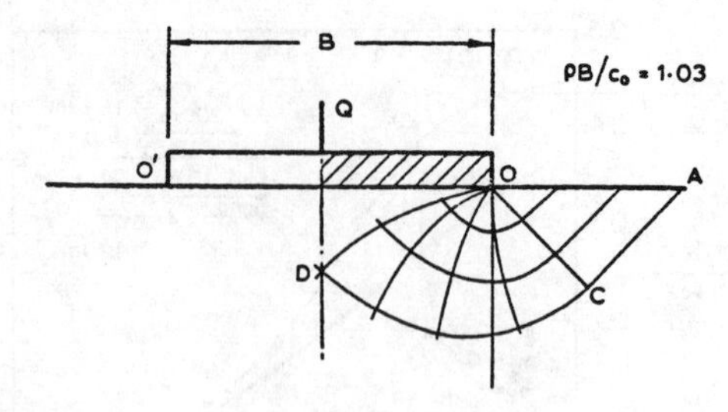

Figure 24a. Characteristics for narrow footing.

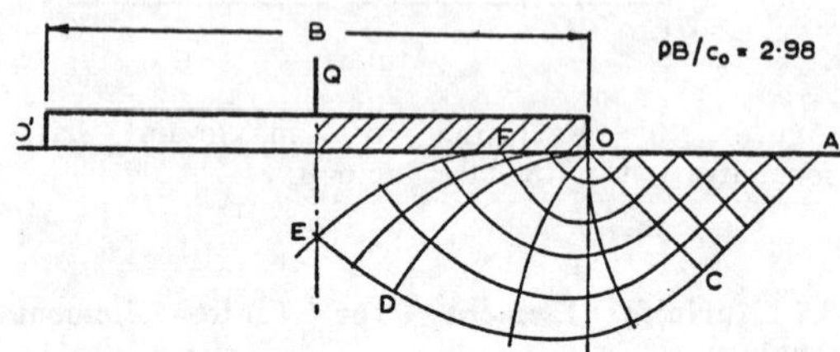

Figure 24b. Characteristics for wide footing.

The classical theory of plasticity can be used to obtain collapse loads by the determination and integration along stress characteristics. Figure (24) shows the stress characteristic beneath a strip footing resting on a purely cohesive soil that has an undrained strength $c = c_0 + \rho z$ which varies with depth z below the surface, Davis and Booker (1973). The bearing capacity for such footings is shown in the form of a correction factor in Figure (25). It is perhaps worth observing that these solutions can be obtained readily by the method of characteristics but the limiting case as $c_0/\rho B \rightarrow 0$ is most difficult to obtain by finite element analysis.

4. CONCLUSIONS

The first step in the effective analysis of a Geotechnical problem is in a proper idealisation, that is one which recognises the dominant and important features of the problem. The next step is that of choosing the appropriate method of analysis. This paper due has sought to show that for a wide range of problems this is possible by the use of an analytic or semi-analytic approach.

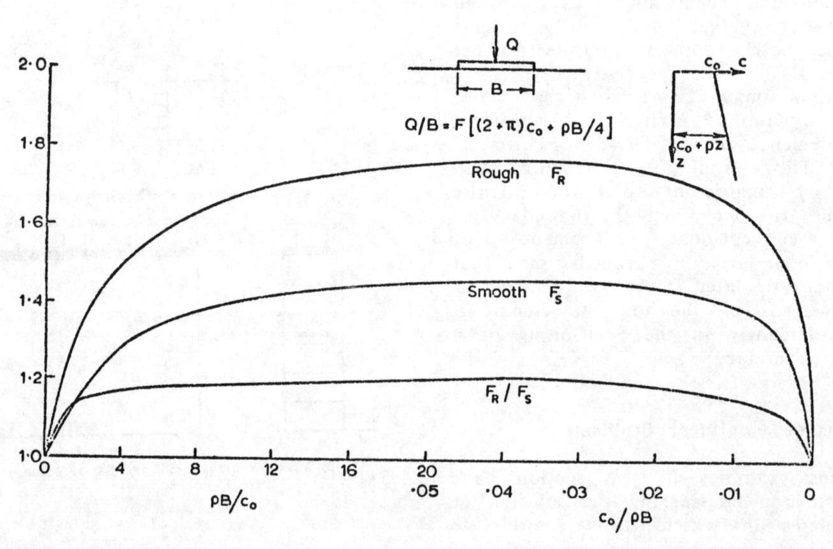

Figure 25. Bearing capacity factor.

5. ACKNOWLEDGEMENTS

The work presented in this paper depends directly or indirectly on collaboration with a large number of valued colleagues. In particular the author we like to express his indebtedness and his thanks to Dr N P Balaam, Dr P T Brown, Dr J P Carter, Professor H G Poulos, Dr M S Rahman, Dr M F Randolph, Dr R K Runesson, Dr C Savvidou, Dr J C Small and in particular to the late Professor E H Davis.

6. REFERENCE

Awojobi, A.O. and Gibson, R.E. (1973). Plane strain and axially symmetric problems of a linearly non-homogeneous elastic half-space. Q. Jnl. Mech. Appl. Math. 26, pp. 285-302.

Balaam, N.P. and Booker, J.R. (1985). The effect of stone column yield on settlement of rigid foundations in stabilized clay. International Journal for Numerical and Analytical Methods in Geomechanics, Vol. 9, pp. 331-334.

Booker, J.R. and Davis, E.H. (1972). A general treatment of plastic anisotropy under conditions of plane strain. Journal of the Mechanics and Physics of Solids, Vol. 20, pp. 239-250.

Booker, J.R. and Small, J.C. (1981). Finite element analysis of problems with infinitely distant boundaries. International Journal for Numerical and Analytical Methods in Geomechanics, Vol. 5, pp. 345-368.

Booker, J.R. (1982). One dimensional primary and secondary consolidation of a soil which creeps indefinitely. University of Sydney, School of Civil and Mining Engineering, Research Report No. R420.

Booker, J.R. and Small, J.C. (1983). The analysis of liquid storage tanks on deep elastic foundations. International Journal for Numerical and Analytical Methods in Geomechanics, Vol. 7, pp. 187-207.

Booker, J.R. and Savvidou, C. (1984). Consolidation around a spherical heat source. International Journal of Solids and Structures, Vol. 20, No. 11/12, pp. 1079-1090.

Booker, J.R., Balaam, N.P. and Davis, E.H. (1985). The behaviour of an elastic non-homogeneous half space part 1 - Line and point loads. International Journal for Numerical and Analytical Methods in Geomechanics, pp. 353-367.

Booker, J.R., Balaam, N.P. and Davis, E.H. (1985). The behaviour of an elastic non-homogeneous half space part 2 - Circular and strip footings. International Journal for Numerical and Analytical Methods in Geomechanics, Vol. 9, pp. 369-381.

Booker, J.R. and Savvidou, C. (1985). Consolidation around a point heat source. International Journal for Numerical and Analytical Methods in Geomechanics, Vol. 9, pp. 173-184.

Booker, J.R. and Small, J.C. (1986a). The behaviour of an impermeable flexible raft on a deep layer of consolidating soil. International Journal for Numerical and Analytical Methods in Geomechanics, Vol. 10, pp. 311-327.

Booker, J.R. and Small, J.C. (1986). Finite layer analysis of viscoelastic layered material. International Journal for Numerical and Analytical Methods in Geomechanics, Vol. 10, pp. 415-430.

Booker, J.R. and Carter, J.P. (1987). Elastic consolidation around a point sink embedded in a half space with anisotropic permeability. International Journal for Numerical and Analytical Methods in Geomechanics, Vol. 11, No. 1, pp. 61-77.

Brown, P.T.and Gibson, R.E. (1972). Surface settlement of a deep elastic stratum whose modulus increases linearly with depth. Canadian Geotechnical Jnl. 9, pp. 467-476.

Calladine, C.R. and Greenwood, J.A. (1978). Line and point loads on a non-homogeneous incompressible elastic half-space. Q. Jnl. Mech. Appl. Math. 28, pp. 507-529.

Carter, J.P. and Booker, J.R. (1984). Determination of the deformation modulus of rock from tunnel and borehole loading tests. Proceedings of the fourth Australia-New Zealand Geomechanics Conference, Perth, pp. 509-513.

Carter, J.P. and Booker, J.R. (1984). Determination of the deformation modulus of rock from tunnel and borehole loading tests. Proceedings of the fourth Australia-New Zealand Geomechanics Conference, Perth, pp. 509-513.

Carter, J.P., Booker, J.R. and Yeung, S.K. (1986). Cavity expansion in cohesive frictionless soils. Geotechnique 36, No. 3, pp. 349-358.

Cheung, Y.K. (1976). Finite strip method in structural analysis, Pergamon.

Davis, E.H. and Booker, J.R. (1973). The effect of increasing strength with depth on the bearing capacity of clays, Geotechnique 23, No. 4, pp. 551-563.

Davis, E.H. (1980). Some plasticity solutions relevant to the bearing capacity of rock and fissured clay. University of Sydney, School of Civil and Mining Engineering, Research Report, R370.

De Ambrosis, L.P. and Poulos, H.G. (1976). Use of a phenomenological model to analyse soil creep. Geotechnical Engineering, Vol. 6, pp. 95-118.

Gibson, R.E. and Anderson, W.F. (1961). In situ measurement of soil properties with the pressuremeter. Civil Engineering and Public Works Review 56, pp. 615-618.
Gibson, R.E. and Lo, K.Y. (1961). A theory of consolidation of soils exhibiting secondary consolidation. Norwegian Geotech. Inst., Pub. No. 41, pp. 1-16.
Gibson, R.E., (1967). Some results concerning displacements and stresses in a non-homogeneous elastic half-space. Géotechnique 17, pp. 58-67.
Gibson, R.E., Brown, P.T. and Andrews, K.R.F. (1971). Some results concerning displacements in a non-homogeneous elastic layer. Zeit. ang. Math. Phys. 22, pp. 855-864.
Gibson, R.E. and Sills, G.C. (1975). Settlement of a strip load on a non-homogeneous orthotropic incompressible elastic half-space. Q. Jnl. Mech. Appl. Math. 28, pp. 233-243.
Holl, D.L. (1940). Stress transmission in earths. Proc. High. Res. Board, Vol. 20, pp. 709-721.
Hughes, J.M.O., Wroth, C.P. and Windle, D. (1977). Pressuremeter tests in sands. Geotechnique, 27, pp. 455-477.
Randolph, M.F., Carter, J.P. and Wroth, C.P. (1979). Driven piles in clay - The effects of installation and subsequent consolidation. Geotechnique, 29, pp. 361-393.
Rowe, R.K. and Booker, J.R. (1980). The analysis of multiple underream anchors. Proceedings of Third Australia-New Zealand Geomechanics Conference, Wellington, pp. 2-247-2-252.
Rowe, R.K. and Booker, J.R. (1985). 2-D pollutant migration in soils of finite depth. The University of Sydney, School of Civil and Mining Engineering, Research Report No. R484.
Scott, R.J. (1978). Subsidence - A Review. Evaluation and prediction of subsidence, Ed. S.K. Saxena, ASCE, New York, pp. 1-25.
Small, J.C. and Booker, J.R. (1982). Finite layer analysis of primary and secondary consolidation. Proceedings of the Fourth International Conference on Numerical Methods in Geomechanics, Edmonton, pp. 365-371.
Talbot, A. (1979). The accurate numerical inversion of Laplace transforms. J. Inst. Maths. Applications, Vol. 23, pp. 97-129.
Vesic, A.S. (1972). Expansion of cavities in infinite soil mass. J. Soil Mech. Found. Divn., ASCE, Vol. 98, SM3, pp. 265-290.
Wissa, A.E.Z. and Heiberg, S. (1969). A new one-dimensional consolidation test. School of Civil Engineering, Massachusetts Institute of Technology, Research Report 69-9, Soils Publication No. 229.

Numerical Methods in Geomechanics (Innsbruck 1988), Swoboda (ed.)
© 1988 Balkema, Rotterdam. ISBN 90 6191 809 X

Incrementally multi-linear and non-linear constitutive relations: A comparative study for practical use

F.Darve, B.Chau & H.Dendani
*Institut de Mécanique de Grenoble, France**

ABSTRACT : Many constitutive relations are used today to describe the behaviour of geomaterials. Some International Workshops (Montreal, 1980 ; Grenoble, 1982 ; Cleveland, 1987) try to exhibit indications concerning their predictive capacity but it is generally difficult to distinguish the different influences of the fundamental structure of the relations and of some contingent characteristics as for example the values of constitutive parameters. The purpose of this paper is to compare constitutive relations with different structures (incrementally linear, octo-linear and non-linear), but with all their other characteristics equal besides. It will be showed on examples that for some paths the structure is not really influential and, on the contrary, for other paths it has a drastic influence.

1 INTRODUCTION

For large engineering structures (dams, offshore platforms, nuclear power plants, tunnels,...) it is now quite common to calculate stress-strain fields with Finite Elements codes. In these codes the main physical assumptions concern the used constitutive relation. As many constitutive equations have been now developed, the complexity of the actual situation makes comparisons, discussions, analyses,... not easy at all, because contingent points or details are exhibited without showing essential assumptions as the one related to the structure of the constitutive relations.

But what means the "structure of a constitutive relation" ? To simplify the presentation of this brief paper, we will only consider non-viscous materials without thermomechanical effect. To describe the behaviour of such media is equivalent to consider the relation between the strain rate $\underset{\sim}{D}$ and an objective derivative of stress $\overset{\circ}{\underset{\sim}{\sigma}}$, or between the incremental strain and stress tensors with :

$$d\underset{\sim}{\varepsilon} = \underset{\sim}{D} \cdot dt$$

$$d\underset{\sim}{\sigma} = \overset{\circ}{\underset{\sim}{\sigma}} \cdot dt$$

where dt is the time increment.
Let us call this function $\underset{\sim}{F}$:

$$d\underset{\sim}{\varepsilon} = \underset{\sim}{F}(d\underset{\sim}{\sigma}) \quad (1)$$

To study the structure of constitutive relations for non-viscous materials means to study the kind of non-linearity of the tensorial function $\underset{\sim}{F}$.

What kind of structures are possible and which are the corresponding physical behaviours ?

$\underset{\sim}{F}$ can be :

a) a linear function : $d\underset{\sim}{\varepsilon} = \underset{\sim}{M}\, d\underset{\sim}{\sigma}$
This is the case in elasticity or strict hypo-elasticity.

b) a bi-linear function :

$$d\underset{\sim}{\varepsilon} = \underset{\sim}{M}^{(e)} d\underset{\sim}{\sigma} \quad \text{for} \quad \underset{\sim}{A} \cdot d\underset{\sim}{\sigma} < 0$$

$$d\underset{\sim}{\varepsilon} = \underset{\sim}{M}^{(ep)} d\underset{\sim}{\sigma} \quad \text{for} \quad \underset{\sim}{A} \cdot d\underset{\sim}{\sigma} > 0$$

as in conventional elasto-plasticity with one plastic potential. The equation : $\underset{\sim}{A} \cdot d\underset{\sim}{\sigma} = 0$ represents an hyper-plane in the $d\underset{\sim}{\sigma}$-space. If $d\underset{\sim}{\sigma}$ belongs to the semi-infinite space of "unloading" ($\underset{\sim}{A} \cdot d\underset{\sim}{\sigma} < 0$), $\underset{\sim}{M}$ will take an "elastic" value and if $d\underset{\sim}{\sigma}$ belongs to the "loading" domain ($\underset{\sim}{A} \cdot d\underset{\sim}{\sigma} > 0$), $\underset{\sim}{M}$ will take an "elasto-plastic" value.

c) a multi-linear function : the space is divided into "tensorial zones"

(*) GRECO "Rhéologie des Géomatériaux".

(see Darve and Labanieh (1982)), in which $\underset{\sim}{M}$ keeps the same value independently of the direction of $d\underset{\sim}{\sigma}$ in the tensorial zone.

d) a non-linear function : there exists no more zone in which the relation between $d\underset{\sim}{\varepsilon}$ and $d\underset{\sim}{\sigma}$ is linear.

In this paper we will compare an incrementally linear constitutive relation (class 1), an octo-linear one (class 3) and a non-linear one (class 4).

2 THREE CONSTITUTIVE RELATIONS WITH THREE DIFFERENT STRUCTURES

After eq. (1) and for non-viscous materials, the tensorial function $\underset{\sim}{F}$ is anisotropic, non-linear, homogeneous of degree one (due to rate-independency condition) and depends on the previous stress-strain history by some memory parameters.

The homogeneity condition implies, by applying Euler's Identity :

$$d\varepsilon_{ij} \equiv \frac{\partial F_{ij}}{\partial (d\sigma_{kl})} d\sigma_{kl} = M_{ijkl}(d\hat{\underset{\sim}{\sigma}})\; d\sigma_{kl} \qquad (2)$$

with : $d\hat{\underset{\sim}{\sigma}} = d\underset{\sim}{\sigma} \,/\, \| d\underset{\sim}{\sigma} \|$

and $\| d\underset{\sim}{\sigma} \| = \sqrt{d\sigma_{ij} \cdot d\sigma_{ij}}$

Therefore, the rate-independency condition implies the existence of the constitutive tensor $\underset{\sim}{M}$ and the fact that $\underset{\sim}{M}$, with respect to $d\underset{\sim}{\sigma}$, depends only on the "direction" of $d\underset{\sim}{\sigma}$, characterized by the unit tensor $d\hat{\underset{\sim}{\sigma}}$.

2.1 The incrementally non-linear constitutive relation

With vectorial notations (α, β, γ = 1,..., 6) the general expression of incrementally non-linear constitutive relations of second order is :

$$d\varepsilon_{\alpha} = M^{1}_{\alpha\beta} \cdot d\sigma_{\beta} + \frac{1}{\| d\underset{\sim}{\sigma} \|} M^{2}_{\alpha\beta\gamma} \cdot d\sigma_{\beta} \cdot d\sigma_{\gamma} \qquad (3)$$

(see Darve (1984) and Darve, Flavigny, Rojas (1985)).

In this paper we will consider only paths in fixed stress-strain principal axes. Eq. (3), with some assumptions of physical nature (see the previous references) and written in fixed axes, becomes :

$$\begin{bmatrix} d\varepsilon_1 \\ d\varepsilon_2 \\ d\varepsilon_3 \end{bmatrix} = \frac{\underset{\sim}{N}^{+} + \underset{\sim}{N}^{-}}{2} \begin{bmatrix} d\sigma_1 \\ d\sigma_2 \\ d\sigma_3 \end{bmatrix} + \frac{\underset{\sim}{N}^{+} - \underset{\sim}{N}^{-}}{2\, \| d\underset{\sim}{\sigma} \|} \begin{bmatrix} (d\sigma_1)^2 \\ (d\sigma_2)^2 \\ (d\sigma_3)^2 \end{bmatrix} \qquad (4)$$

where $\underset{\sim}{N}^{+}$ and $\underset{\sim}{N}^{-}$ are determined only from the knowledge of soil behaviour for "generalized triaxial paths" (Darve, 1974). These paths are such that stress-strain principal directions are fixed and confounded, two principal stresses σ_j and σ_l are constant and the third one σ_k increases ("compression" denoted by index "+") or decreases ("extension" denoted by "-").

For these paths, let f^{+}, g^{+}, h^{+} be the following functions (f^{-}, g^{-}, h^{-} defined in the same manner) :

$$\begin{aligned} \sigma_k &= f^{+}(\varepsilon_k, \sigma_j, \sigma_l) \\ \varepsilon_j &= g^{+}(\varepsilon_k, \sigma_j, \sigma_l) \\ \varepsilon_l &= h^{+}(\varepsilon_k, \sigma_j, \sigma_l) \end{aligned} \qquad (5)$$

and U_k^{+}, V_k^{j+} and V_k^{l+} be the following partial derivatives :

$$\begin{aligned} U_k^{+} &= \left(\frac{\partial f^{+}}{\partial \varepsilon_k}\right)_{\sigma_j, \sigma_l} \\ V_k^{j+} &= -\left(\frac{\partial g^{+}}{\partial \varepsilon_k}\right)_{\sigma_j, \sigma_l} \\ V_k^{l+} &= -\left(\frac{\partial h^{+}}{\partial \varepsilon_k}\right)_{\sigma_j, \sigma_l} \end{aligned} \qquad (6)$$

The expressions of $\underset{\sim}{N}^{+}$ and $\underset{\sim}{N}^{-}$ are then given by :

$$\underset{\sim}{N}^{+} = \begin{bmatrix} \frac{1}{U_1^{+}} & \frac{-V_2^{1+}}{U_2^{+}} & \frac{-V_3^{1+}}{U_3^{+}} \\ \frac{-V_1^{2+}}{U_1^{+}} & \frac{1}{U_2^{+}} & \frac{-V_3^{2+}}{U_3^{+}} \\ \frac{-V_1^{3+}}{U_1^{+}} & \frac{-V_2^{3+}}{U_2^{+}} & \frac{1}{U_3^{+}} \end{bmatrix} \qquad (7)$$

For an elastic material, $\underset{\sim}{N}^{+}$ and $\underset{\sim}{N}^{-}$ are identical; therefore the non-linear term in eq. (4) is null and eq. (4) and (7) represent the usual elastic constitutive relation. But generally $\underset{\sim}{N}^{+}$ and $\underset{\sim}{N}^{-}$ take very different values.

2.2 The incrementally octo-linear constitutive relation

Its expression is given by eq. (8) :

$$\begin{bmatrix} d\varepsilon_1 \\ d\varepsilon_2 \\ d\varepsilon_3 \end{bmatrix} = \frac{\underset{\sim}{N}^{+} + \underset{\sim}{N}^{-}}{2} \begin{bmatrix} d\sigma_1 \\ d\sigma_2 \\ d\sigma_3 \end{bmatrix} + \frac{\underset{\sim}{N}^{+} - \underset{\sim}{N}^{-}}{2} \begin{bmatrix} |d\sigma_1| \\ |d\sigma_2| \\ |d\sigma_3| \end{bmatrix}$$

where $|d\sigma_k|$ means the absolute value of $d\sigma_k$.

This relation can be called "octo-linear", because after the signs of $d\sigma_k$, eight different matrices are relating ($d\varepsilon_1$, $d\varepsilon_2$, $d\varepsilon_3$) to ($d\sigma_1$, $d\sigma_2$, $d\sigma_3$). This structure is the same as for an elasto-plastic law with three plastic potentials (Darve and Labanieh, 1982).

2.3 The incrementally linear constitutive relation

Finally we will exhibit a linear relation by assuming that the same constitutive matrix (between the eight ones given by eq. (8)) is valid for a whole path : the one related to the first incremental stress.

That leads to a relation with a structure identical to an hypoelastic formulation.

3 CALIBRATION OF THESE CONSTITUTIVE RELATIONS

The calibration of our models needs six drained triaxial tests in compression and in extension for three different values of the radial pressure.

The functions f, g, h (eq. (5)) are given by analytical expressions in which some constitutive constants (their number is 16) appear and are calculated by some procedures (see B. Chau, 1988). These expressions have been recently improved by H. Dendani (1988).

Firstly we consider these formulations in the case of axisymmetrical compression-extension tests (conventional triaxial path), and secondly we develop them for generalized triaxial paths ($\sigma_2 \neq \sigma_3$).

The behaviour of geomaterials for conventional triaxial path are described by the following relationships :

a)

$$\sigma_1 = f(\varepsilon_1, \sigma_3, \sigma_3) = \sigma_1^R + \sigma_3 B_s \frac{\exp(|A_1(\varepsilon_1 - \varepsilon_1^R)|) - 1}{\exp(|A_1(\varepsilon_1 - \varepsilon_1^R)|) - A_s} \quad (9)$$

for a monotonic Loading ε_1^R is the isotropic strain under isotropic stress ($\sigma_1^R = \sigma_1 = \sigma_3$), for cyclic loading ε_1^R and σ_1^R are the values of ε_{11} and σ_{11} at the last load reversal in the orthotropy axis 1.

$$B_s = \frac{(\sigma_1 - \sigma_3)_f}{\sigma_3}$$

Bs is a function of the friction angle Φ and the cohesion C.

As depends on the initial slope of σ_1 versus ε_1 curve, for cyclic Loading As and A_1 depend on the stress strain σ_1^R, ε_1^R, $\varepsilon_1^{R'}$ (figure -1)

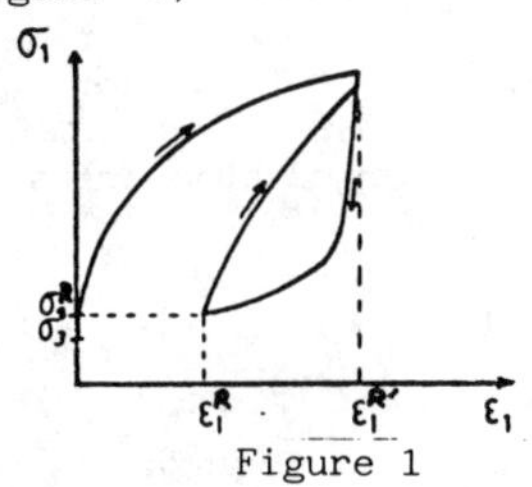

Figure 1

b) for axisymmetrical compression tests, the void ratio e is given analytically by :

$$e = e_o^R - e_m \left[1 - \exp(-A_e[\varepsilon_1 - \varepsilon_1^R])\right] + B_e(\varepsilon_1 - \varepsilon_1^R) - C_e(\varepsilon_1 - \varepsilon_1^R)\exp(-D_e[\varepsilon_1 - \varepsilon_1^R]) \quad (10)$$

e_o^R is the initial void ratio for a monotonic Loading or its value at the beginning of each load reversal. An extensive discussion of relationships (9) and (10) is presented by Darve, F. and Labanieh, S. 1982 and Darve, F. 1984.

c) for axisymmetrical extension tests e is given by :

$$e = e_o^R + [1 - \exp(A_2 \varepsilon_1)] G_e - F_e$$

$$F_e = \frac{\varepsilon_1}{80000 . \varepsilon_1^2 - 300\, \varepsilon_1 + A_3},$$

Fe is a function which depends on the initial Poisson's ratio and describes the initial dilatancy in triaxial extension tests.

$$G_e = A_4 \varepsilon_1 + A_5,$$

Ge is the asymptote of the $e - \varepsilon_1$ curve, which characterizes the dilatancy (figure 2), A_4 and A_5 are some of the constitutive constants.

A_2 is a parameter which influences the form of $e - \varepsilon_1$ curve.

In cyclic loading A_3 depends on strain history represented by discrete variables changing at load reversals.

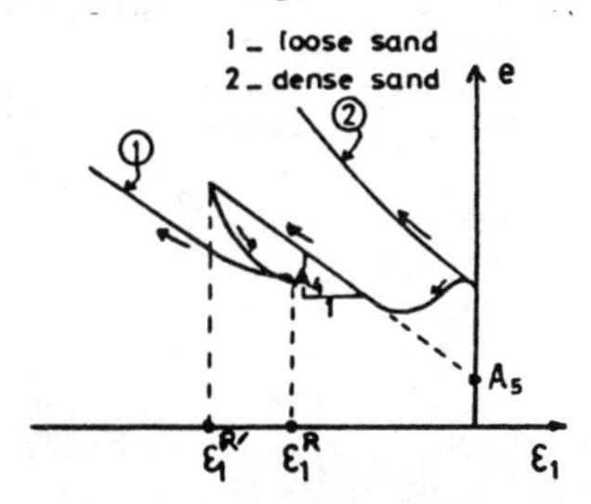

Figure 2

For axisymmetrical tests the lateral strain is given by :

$$\varepsilon_3 = g(\varepsilon_1, \sigma_3, \sigma_3) = \frac{1}{2}\left[\mathrm{Ln}\left(\frac{1+e}{1+e_0}\right) - \varepsilon_1\right]$$

The generalization of f(ε_1 , σ_3 , σ_3) into f(ε_1 , σ_2 , σ_3) is obtained by introducing the intermediate stress σ_2 in the limit condition :

$$\tan\Phi = A_4\left[1.09 - 0.09\exp\left(0.7\left[\frac{\sigma_2}{\sigma_3} - 1\right]\right)\right]$$
$$(\sigma_3 \leqslant \sigma_2 \leqslant \sigma_1)$$

and the generalization of g(ε_1 , σ_3 , σ_3) into g(ε_1 , σ_2 , σ_3) [and h(ε_1 , σ_2 , σ_3)] by distinguishing the two lateral strains ε_2 and ε_3 :

$$\frac{\varepsilon_2}{\varepsilon_3} = 8\left[1 - \frac{7}{8}\left(\frac{\sigma_2}{\sigma_3}\right)^2\right] \text{ for compressions,}$$

$$\frac{\varepsilon_3}{\varepsilon_2} = 8\left[1 - \frac{7}{8}\left(\frac{\sigma_2}{\sigma_3}\right)^2\right] \text{ for extensions.}$$

Fig. 3 presents an example of calibration with a sand : the incrementally non-linear constitutive relation and the incrementally octo-linear one are wholly confounded.

4 COMPARISONS FOR PROPORTIONAL PATHS

As we see on Fig. 4, for monotonous proportional paths (as the undrained one) all the three constitutive relations give good and close results, but for unloading the hypoelastic relation (i.e. the linear one) exhibits aberrant predictions because the pore pressure decreases after the load reversal.

5 COMPARISONS FOR NON-PROPORTIONAL PATHS

On the contrary of the previous case, Fig. 5 presents comparisons for highly non-proportional paths, which are circles in a stress deviatoric plane. The circle is followed twice, once in a sense and after in the other sense. Two facts are remarkable on this figure :

a) the linear constitutive relation gives aberrant results,
b) the octo-linear and non-linear relations produce quite different predictions.

At the point A on fig. 5, we exhibited the graphical representation proposed by G. Gudehus (1979). Fig. 6 presents this result, which also indicates (and illustrates) differences between the octo-linear and non-linear relations.

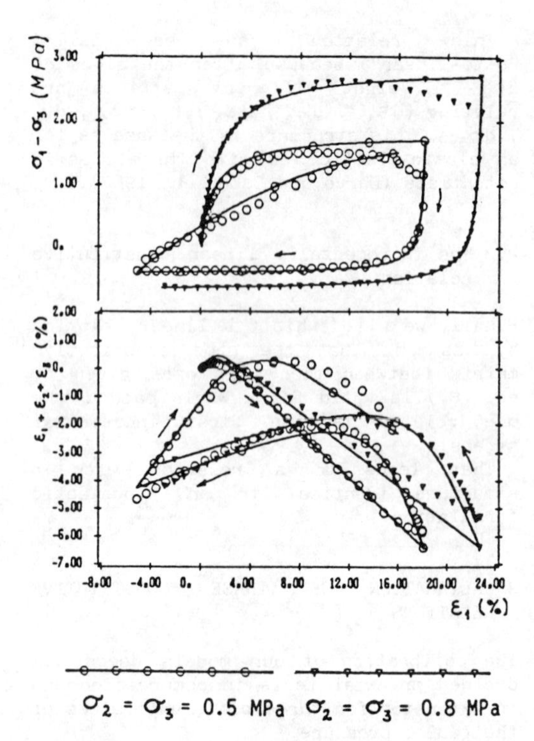

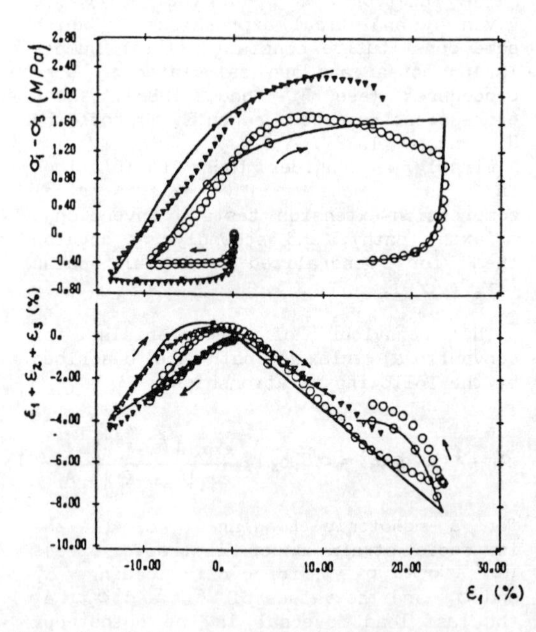

Fig. 3 : Calibration with drained triaxial data in compression and in extension

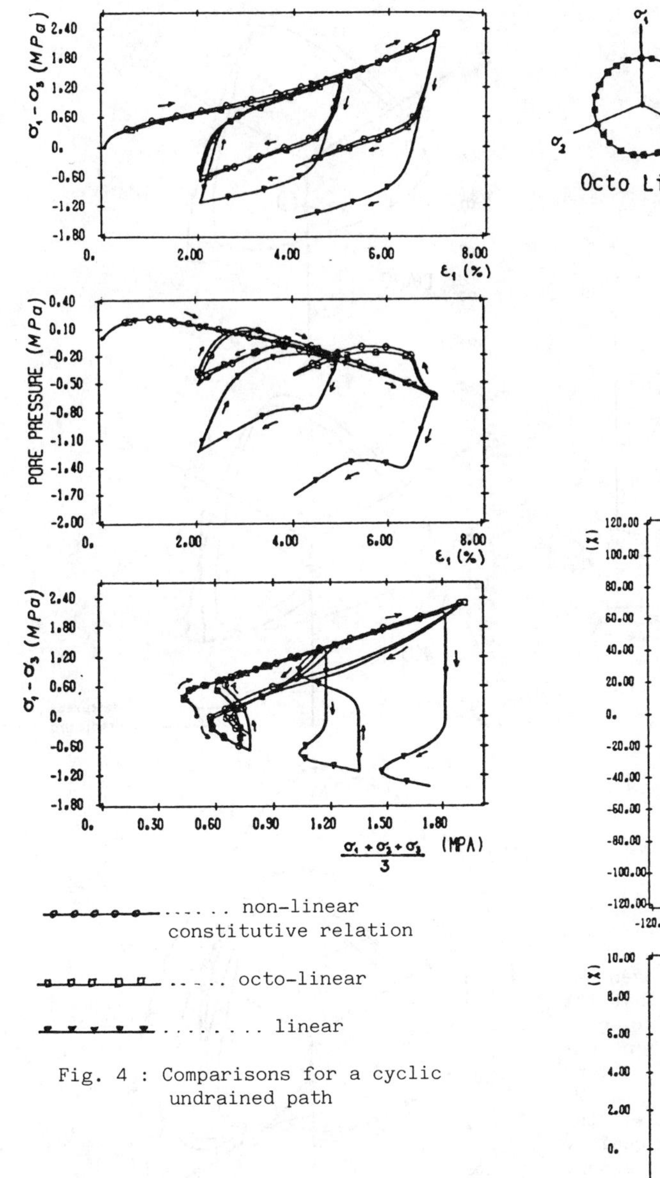

Fig. 4 : Comparisons for a cyclic undrained path

Finally Fig. 7 presents predictions given by Liang (1987), Ko and Sture (1987), and Darve and Dendani (1987) at the Cleveland's Workshop. As on Fig. 5, the differences between the four models are very large, showing once more that for non-proportional paths the influence of the structure of constitutive relations is determining on the quality of the predictions.

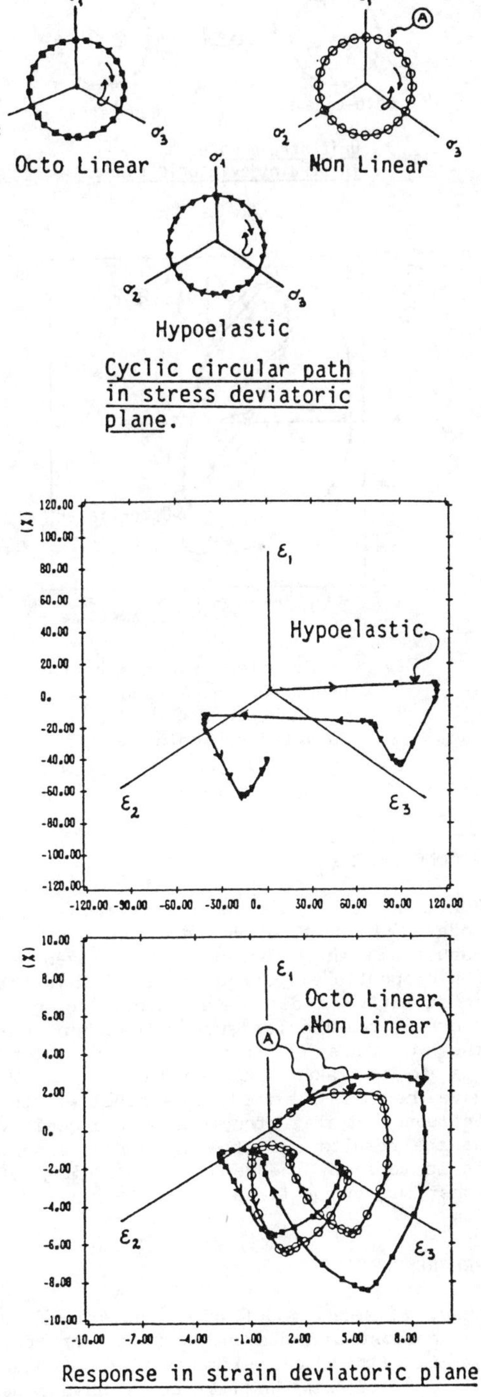

Fig. 5 : Comparisons for a cyclic circular stress path.

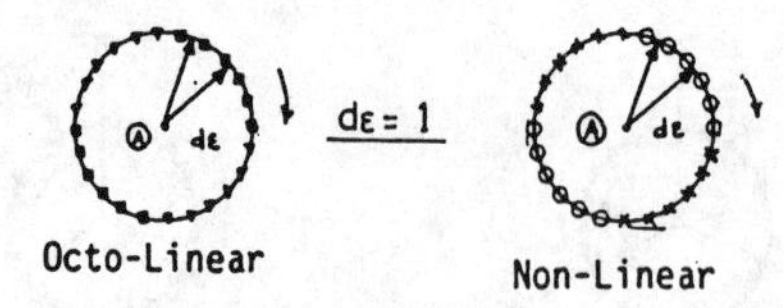

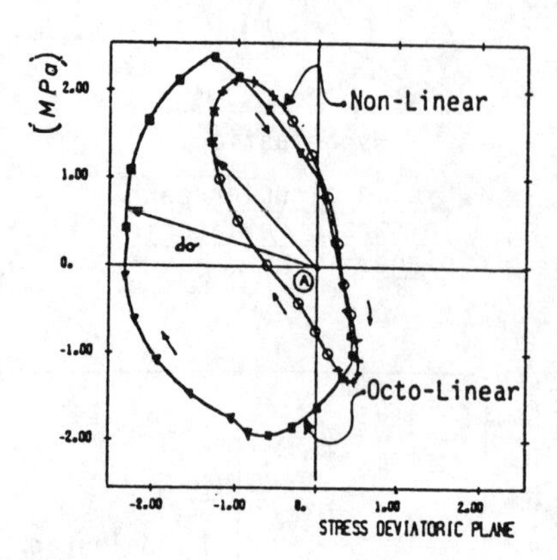

Fig. 6 : Graphical representation (see, for the method, Gudehus, 1979) platted for point A on Fig. 5.

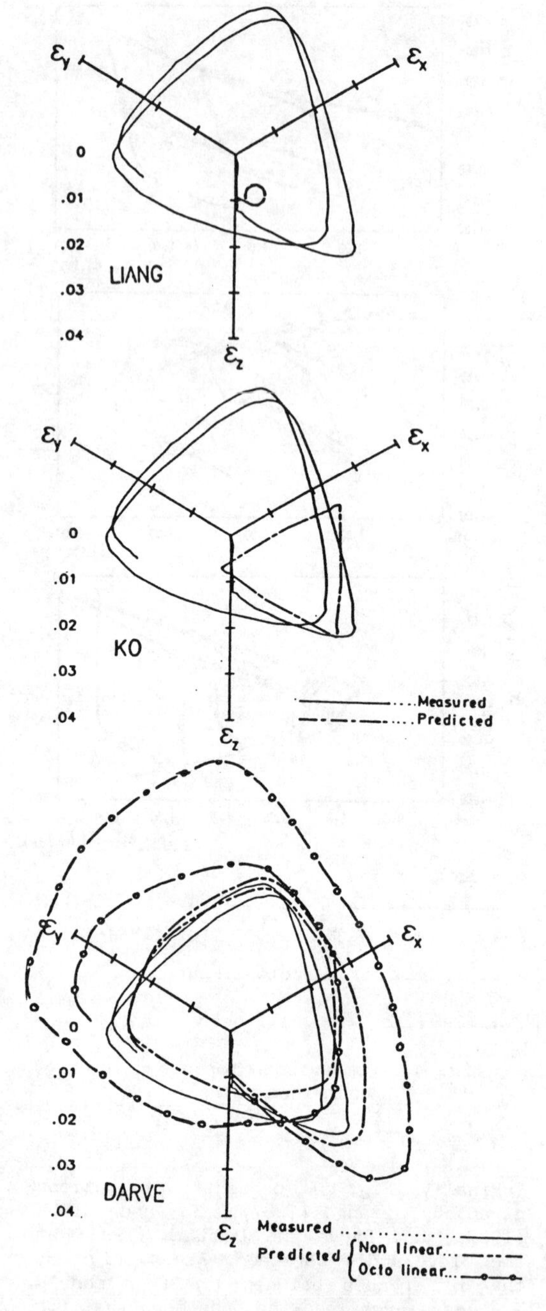

Fig. 7 : Predictions given by four different constitutive relations for the circular deviatoric stress path required at Cleveland's Workshop (1987)

6 CONCLUSIONS

For a practical use in a Finite Elements code, the constitutive relations must be robust with respect to eventual non-proportional paths. It means that the constitutive models must avoid to give aberrant results for such paths. For that the structure of constitutive relations has to be taken into consideration. We gave here few examples, exhibiting the influence of this structure on the quality of the results and showing how necessary it is to study the domain of validity of constitutive equations.

REFERENCES

Chau, B. 1988. Modélisation numérique du comportement des ouvrages en terre par la méthode des éléments finis. Thèse de doctorat. Institut de Mécanique de Grenoble. France.

Darve, F. 1974. Contribution à la détermination de la loi rhéologique incrémentale des sols. Thèse de doctorat. Institut de Mécanique de Grenoble. France.

Darve, F. 1984. An incrementally non-linear constitutive law : assumptions and predictions. Constitutive Relations for Soils. Gudehus, Darve and Vardoulakis eds., A.A. Balkema publ. : 385-404.

Darve, F. and Labanieh, S. 1982. Incremental constitutive law for sands and clays : simulations of monotonic and cyclic tests. Int. J. Num. Anal. Meth. in Geomech. 6 : 243-275.

Darve, F., Flavigny, E., Rojas, E. 1985. A class of incrementally non-linear constitutive relations and applications to clays. Computers and Geotechnics 2 : 43-66.

Darve, F., Dendani, H. 1987. Preliminary Proceedings of Cleveland's Workshop.

Dendani, H. 1988. Comportement de matériaux de barrages en terre : étude expérimentale et modélisations. Thèse de doctorat. Institut de Mécanique de Grenoble. France.

Gudehus, G. 1979. A comparison of some constitutive laws for soils under radially symmetric loading and unloading. Third Int. Conf. on Num. Meth. in Geomech., Wittke ed., A.A. Balkema publ. : 1309-1323.

Ko, H.Y., Sture, S., Klisinski. 1987. Preliminary Proceedings of Cleveland's Workshop on Constitutive Equations for Granular Non-cohesive Soils.

Liang, R. 1987. Preliminary Proceedings of Cleveland's Workshop.

Numerical Methods in Geomechanics (Innsbruck 1988), Swoboda (ed.)
© 1988 Balkema, Rotterdam. ISBN 90 6191 809 X

Unified approach for constitutive modelling for geologic materials and discontinuities

C.S.Desai
Department of Civil Engineering and Egineering Mechanics, University of Arizona, Tucson, USA

ABSTRACT: A unified hierarchical concept for modelling stress-strain behavior of solids and discontinuities is described together with verification with respect to laboratory tests from multiaxial testing of soils and rocks, and shear testing of joints and interfaces.

1 INTRODUCTION

Results from any solution technique, including the modern numerical procedures such as the finite element and boundary element methods, are dependent on the use of appropriate constitutive models for (geologic) materials and discontinuities, interfaces and (joints). It is important to develop constitutive models based on laboratory and/or field tests that are relevant to the practical boundary value problems. Development of rational and viable models should account for various physical and loading conditions and should undergo various phases: (1) mathematical and mechanics basis, (2) identification of significant parameters, (3) determination of the parameters from tests, and (4) verification with respect to laboratory tests and behavior of realistic boundary value problems.

The objectives of this chapter are to describe (1) a unified constitutive modelling approach for soils and rocks treated as solid continua, and for discontinuities (interfaces and joints) under quasistatic and cyclic loadings in the laboratory, (2) the required constants from laboratory tests, and (3) typical verification of the models.

2 UNIFIED MODELLING APPROACH FOR SOLIDS AND DISCONTINUITIES

This approach, proposed by the author and co-workers (1-3)*, allows progressive development of models starting with a basic (δ_o) model for isotropic plastic hardening with associative flow rule. The progression allows for adding of contributions of factors such as nonassociative response (δ_1) model (1-4), anisotropic hardening (δ_2) model (3, 5), damage and softening (δ_{o+r}) model (3, 6), pore water pressure (δ_{o+p}) model (7) and viscoplastic response (δ_{o+v}) model (8, 9). A schematic of these models is shown in Figure 1. Here, a brief description of four selected versions - δ_o, δ_1, δ_2 and δ_{2+p} - that have been applied to characterize behavior of various soils rocks and simulated joints in concrete are presented.

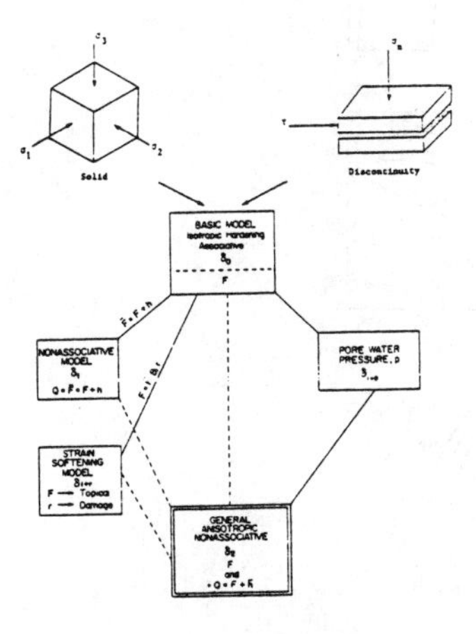

Fig. 1 Hierarchical approach for solids and discontinuities

*Indicates references at the end.

2.1 Solids

The δ_o model is based on a general yield function, F, given by

$$F = J_{2D} - (- \frac{\alpha}{\alpha_o^{n-2}} J_1^n + \gamma J_1^2)(1 - \beta S_r)^m \quad (1a)$$

$$= J_{2D} - F_b F_s = 0 \quad (1b)$$

where J_1 is the first invariant of the stress tensor, σ_{ij}, J_{2D} is the second invariant of the deviatoric stress tensor, S_{ij}, α, n, γ, β and m are the response functions and S_r = stress ratio such as $J_{3D}^{1/3}/J_{2D}^{1/2}$ and Lode angle, and J_{3D} is the third invariant of S_{ij}, α_o = 1 unit of stress. Typical plots of F in various stress spaces are shown in Figure 2. A simple form of the hardening function is given by

$$\alpha \; \frac{a_1}{\xi^{\eta_1}} \quad (2)$$

where $\xi = \int (d\varepsilon_{ij}^p \, d\varepsilon_{ij}^p)^{1/2}$ is the trajectory of plastic strains, and a_1 and η_1 are material hardening constants. Further details of constants and procedures for finding them are given elsewhere (3-5).

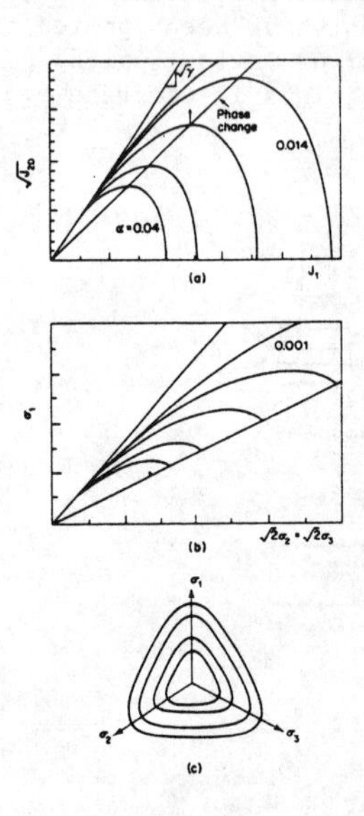

Fig. 2 Plot of yield function in various stress spaces: (a) in $J_1 - \sqrt{(J_{2D})}$ space; (b) in triaxial space; (c) in deviatoric space

The δ_1 model is obtained by introducing a correction function h to control the volumetric response, and by defining the plastic potential, Q, as (1,2,6)

$$\overline{F} = Q = F + h\,(J_i, \xi) \quad (3)$$

The correction function h is incorporated by correcting the hardening or growth function α to α_Q as

$$\alpha_Q = \alpha + k\,(\alpha_I - \alpha)(1 - r_v) \quad (4)$$

where k is a material constant, α_I is the value of α at the initiation of nonassociativeness and r_v is the ratio of the trajectory of volumetric plastic strains to the trajectory of total plastic strains, ξ_v/ξ.

For anisotropic hardening, the nonassociative concept is extended, this is referred to as the δ_2 model. Here a modified expression for Q is developed, and it is allowed to translate in a fixed field of yield surfaces F, Figure 3, during virgin (loading) and nonvirgin (unloading and reloading) phases of the deformation process. The expression for Q is given by

$$Q = \overline{J}_{2D} - (\alpha_o \overline{J}_1^n + \gamma \overline{J}_1^2)(1 - \beta \overline{S}_r)^m \quad (5)$$

Here the overbar denotes quantities associated with a modified stress tensor $\overline{\sigma}_{ij}$ given by

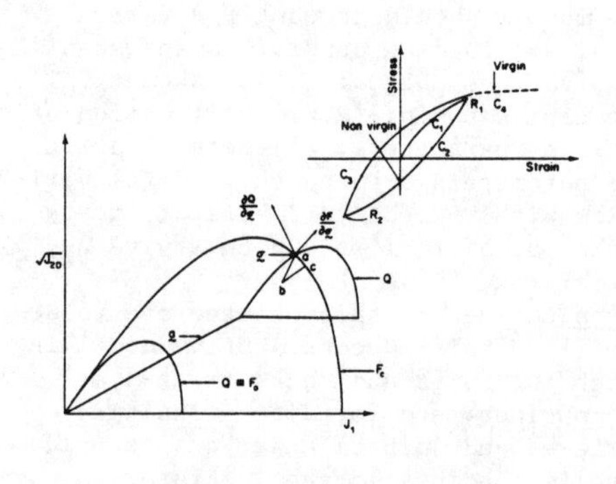

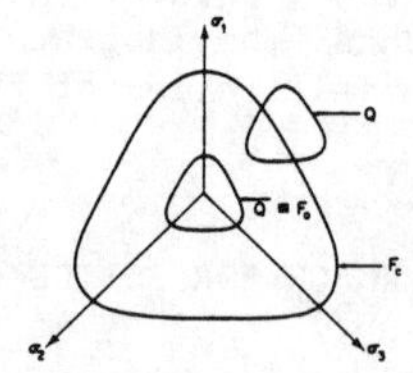

Fig. 3 Anisotropic hardening model, δ_2

$$\bar{\sigma}_{ij} = \sigma_{ij} - a_{ij} \quad (6)$$

For convenience, in this study, Q is assumed to remain unchanged in shape, size and orientation, with its size the same as that of the initial yield surface F_o, with constant value of hardening parameter α_o, and its location in the stress space is defined by a_{ij}.

A new measure of induced anisotropy, a_n, is proposed in addition to a_{ij} which is a function of the total trajectory of plastic strain

$$a_n = \frac{I_1^e}{\sqrt{I_2^e}} - \frac{I_1^p}{\sqrt{I_2^p}} = R_e - R_p \quad (7)$$

Here I_1 and I_2 = first and second invariants of the strain tensor ε_{ij}, the superscripts e and p refer to elastic and plastic strains, respectively, and the ratio $I_1/\sqrt{I_2}$ denotes a measure of orientation of the state of strain in the strain space.

It has been shown that variation a_n during loading, unloading and reloading is similar to that of the deviation from normality of the plastic strain increments, and the physical anisotropy measured in the laboratory (5). Interpolation and translation and rules during unloading and reloading are defined through four parameters h_1, h_2, h_3 for the former, and h_4 for the latter (3, 5).

For the δ_{2+p} model, the pore water pressure is included by defining effective stress tensor as (3, 7)

$$\sigma_{ij}' = \sigma_{ij} - p\,\delta_{ij} \quad (8)$$

where p = fluid pressure, σ_{ij} = total stress, the prime denotes effective stress and δ_{ij} = the Kronecker delta.

2.2 Discontinuities - interfaces, joints

In order to describe the behavior of discontinuities properly, it is appropriate to first describe the behavior of the individual joints. The model for joints is a special case of the above general hierarchical modelling procedure developed for solids. The specialized yielding function for joints is derived from the generalized yielding function, equation (1), as (10,11)

$$F = \tau^2 + \alpha\sigma_n^n - \gamma\sigma_n^2 = 0 \quad (9)$$

where τ is the shear stress, σ_n is the normal stress and the hardening function is given by

$$\alpha = \frac{a}{\xi^b} \quad (10)$$

where ξ is the trajectory of plastic (relative) displacements and a and b are constants.

For the nonassociative model, the potential function Q is defined as

$$Q = \tau^2 + \alpha_Q\,\sigma_n^n - \gamma\sigma_n^2 = 0 \quad (11)$$

and

$$\alpha_Q = \alpha + k\,(\alpha_I - \alpha)\,(1 - r_v) \quad (12)$$

For including both the shear and normal deformations, an alternative function α can be developed:

$$\alpha = \alpha_\tau\,\alpha_n \quad (13a)$$

with

$$\alpha_n = \gamma\,e^{(n-2)\,(C_s V_r + d_s)} \quad (13b)$$

$$\alpha_\tau = 1 - \frac{a_s}{\gamma}\,U_r^{b_s} \quad (13c)$$

where V_r, U_r are the vertical and horizontal relative displacement, respectively, γ, a_s, b_s, C_s, d_s and n are the material constants. This is called the "split" joint model since the hardening function is a product of two terms which represent the normal and tangential hardening behavior, respectively (12).

3 INTEGRATION OF STRESS-STRAIN EQUATIONS

By using the consistency and normality conditions of plasticity, the following incremental stress-strain relations are derived:

$$\{d\sigma\} = [C^{ep}]\,\{d\varepsilon\} \quad (14)$$

where $\{d\sigma\}$ and $\{d\varepsilon\}$ are the vectors of incremental stresses and strains, respectively, and the elastoplastic constitutive matrix $[C^{evp}]$ for various models is given by

$[C^{ep}]$: symmetric for δ_o model
$[C^{ep}]$: nonsymmetric for δ_1 model
$[C^{ep}]$: nonsymmetric for δ_2 model

To verify various models, equations (12) are integrated along a given stress path stating from an initial hydrostatic (or normal) stress state. Laboratory data from tests with cylindrical and cubical specimens for various rocks, and from quasistatic and cyclic tests for joint using a new cyclic multi degree-of-freedom (CYMDOF) shear device (13) is used. Typical verifications are presented below.

4 VERIFICATIONS

The models have been applied and verified for about eighteen different soils, rocks and concrete, and joints and interfaces. Here only typical results are presented, details for finding constants and numerical implementations in finite element procedures are given elsewhere (3,5,7,14). Tables 1 and 2 give values of the constants for the soils and joints considered in this paper.

4.1 δ_o Model: Soapstone

Soapstone is a soft metamorphic rock composed mainly of talc minerals. The tested soapstone is available in the southwest region of Oregon, U.S.A. This soft rock is found to be relatively homogeneous and isotropic (15).

Figure 4 shows predictions from the δ_o model for a typical stress path, conventional triaxial compression, CTC, ($\sigma_1 > \sigma_2 = \sigma_3$) in comparison with the observed stress-strain and volumetric responses.

4.2 δ_1 Model: Ottawa Sand

Comparisons between predictions and observations for a triaxial extension (TE) path in multiaxial tests with cubical (10 x 10 x 10 cm) specimens with δ_o = 10 psi (89 kPa) for the Ottawa sand, initial density = 1.73 gm/cm, are shown in Figure 5.

4.3 δ_1 Model: Rock Salt

Rock salt is a polycrystalline aggregate composed essentially of halite crystals (9, 17). The tested rock salt is found in New Mexico and belongs to the Salado formation. Figure 6 shows comparisons between predicted and observed behavior for two CTC paths. It can be seen that the nonassociative model provides improved predictions for the stress-strain and especially the volumetric behavior as compared with those from the associative model.

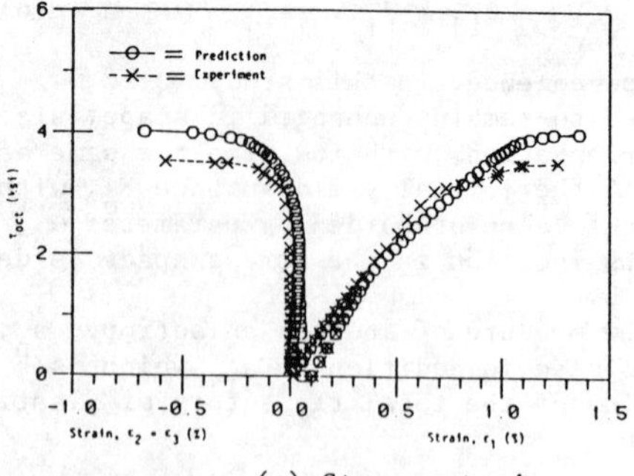

(a) Stress-strain

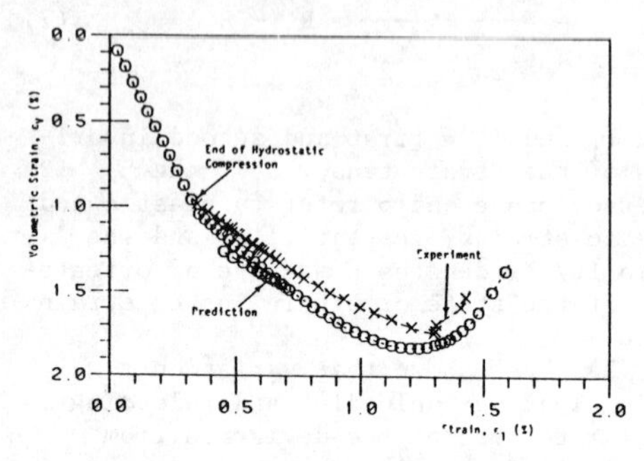

(b) Volumetric

Fig. 4 Comparison of responses of conventional triaxial compression (CTC) test for soapstone (σ_o = 3.0 ksi), (1.0 psi = 6.89 kPa).

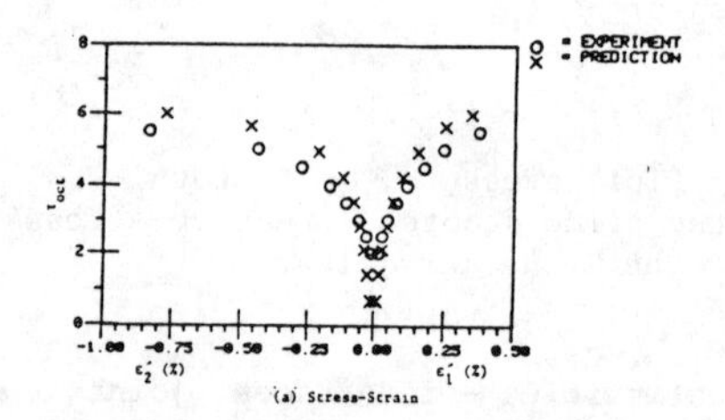

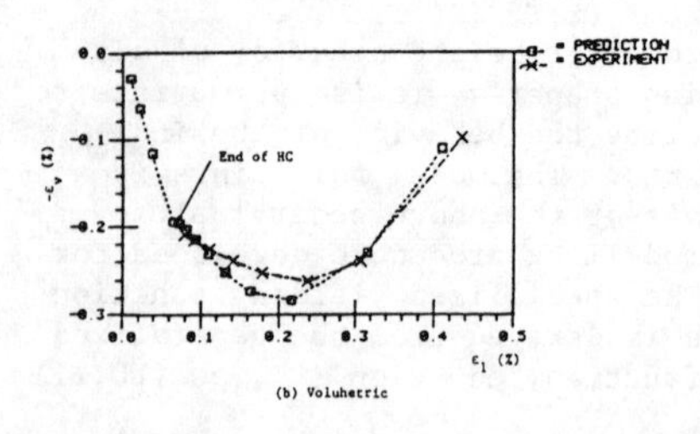

Fig. 5 Comparisons for TE test (σ_o = 10 psi) with nonassociative model: Ottawa sand (γ_o = 1.73 gm/cc)

Table 1. Material constants for various soils and rocks.

	Material	Soap-stone δ_o	Ottawa Sand δ_1	Rock Salt δ_1	Leighton Buzzard Sand δ_2	Munich Sand δ_2	Sand C δ_{2+p}
Elasticity	E (MPa)	9152	269	20685	1034	585	1410
	ν	0.079	0.37	0.27	0.29	0.25	0.15
Plasticity	Tensile Strength R^* (MPa)	1.067	0.0	1.79	0.0	0.0	0.0
	m	-0.50	-0.5	-0.5	-0.5	-0.50	-0.5
	Ultimate γ	0.047	0.12	0.095	0.096	0.105	0.636
	β_o	-0.75	β^+ = 0.49	β_o = -0.99	β^+ = 0.612	β^+ = 0.747	β^+ = 0.60
	β_1	0.047		β_1 = 0.0048			
	Phase Change n	7.0	3.0	3.0	2.5	3.2	3.0
	Hardening η_1	0.75	0.37	0.232	0.495	0.761	0.495
	α_1, α_o = 1 kPa	0.0018	2.51×10^{-3}	1.8×10^{-6}	5.2×10^{-4}	8.5×10^{-6}	0.168×10^{-4}
			Nonassociative k = 0.265	k = 0.275	Interpolation h_1 = 0.63, h_2 = .95, h_3 = 2.0; Translation h_4 = 3.8 kPa	h_1 = 0.23, h_2 = 3.4, h_3 = 0.70, h_4 = 2.76	

*Denotes tensile intercept
+β constant

Table 2. Constants for quasistatic and cyclic behavior of joints.

(a) Static and quasistatic

Constant	Forward				
	Surface			9°	
	Flat	5°	7°	Forward	Reverse
γ	0.36	0.42	0.78	0.31	0.45
n	2.50	2.50	2.50	2.50	2.50
a	0.02312	0.01063	0.03054	0.04736	0.024
b	-0.116	-0.293	-0.223	-0.162	-0.149
k	0.70	0.70	0.70	0.55	1.50

(b) Cyclic

9° Surface, σ_n = 20 psi, u_r^m = 0.1 in

	Phase 1 Cycle 1	Phase 1 (N+1) and 3	Phase 2 + 4
γ^*	N/(1.67+0.71N)	N/(1.67+.71N)	0.228
n^*	2.5	2.5	2.5
a	0.00455	*N/(200+50N)	0.0096
b	-0.364	-0.90	-0.206
κ	0.60	0.60	1.20

*γ and a are functions of number of cycles N.

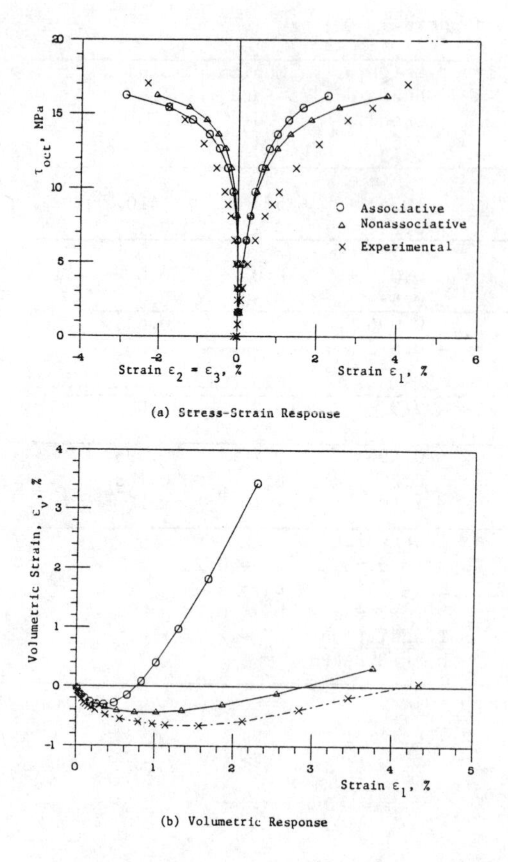

Fig. 6 Comparisons between predictions and observations for CTC test, σ_o = 3.45 MPa: rock salt

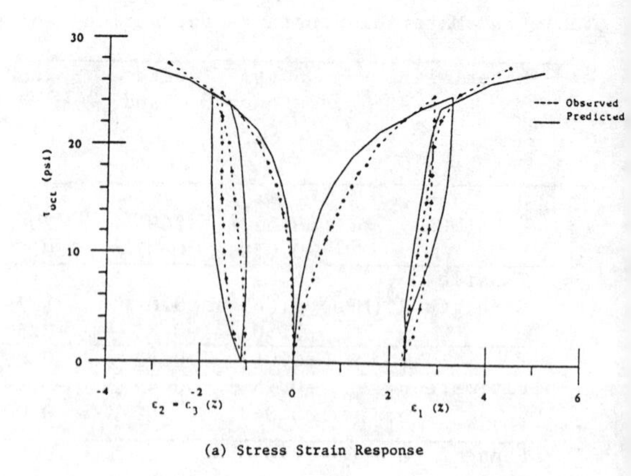

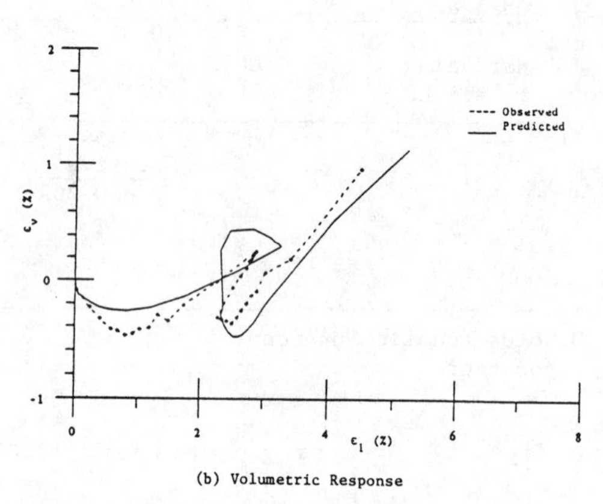

Fig. 7 Comparisons between predictions and observations for Munich sand under TE test (σ_o = 13 psi)

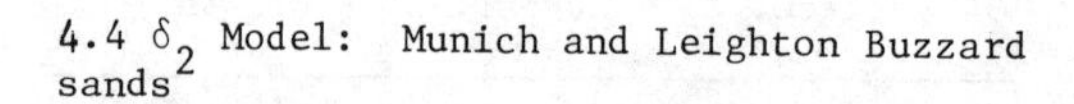

4.4 δ_2 Model: Munich and Leighton Buzzard sands

Figures 7 and 8 show comparisons between predictions and observations for two stress paths for Munich and Leighton Buzzard sands (5,7,14). Note that the δ_2 model, among other factors, captures well the nonlinear plastic response during unloading and reloading.

4.5 δ_{2+p} model: Sand C

Figures 9 and 10 show comparisons between back predictions and observations of the triaxial (cylindrical specimen) drained and undrained response of sand C tested by Castro (18). This sand is derived from natural beaches near Huachipato, Chile and is uniform, fine to medium with average specific gravity of 2.87. It is seen that the model provides good predictions of stress-strain and pore water responses.

4.6 Discontinuities - joints: δ_o, δ_1 and δ_2 models

As discussed earlier, the hierarchical model is specialized for interfaces and joints. A brief description of its application for characterizing quasistatic and cyclic response of joints in concrete is given below.

A comprehensive series of laboratory tests have been performed (10, 11) by using the newly developed cyclic multi degree-of-freedom shear (CYMDOF) device (13). Here tests are conducted with smooth joints and joints with different asperity angles under different normal loads and cyclic (sinusoidal) loadings (frequency ω = 1 Hz) are performed (10,11). Figure 11 shows the

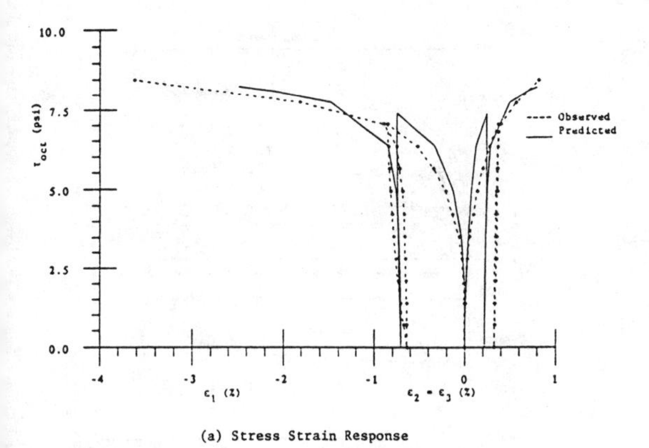

(a) Stress Strain Response

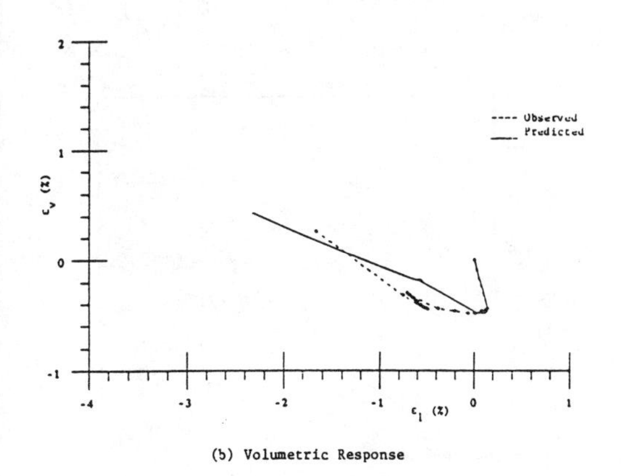

(b) Volumetric Response

Fig. 8 Comparisons between predictions and observations for Leighton Buzzard sand under TE (σ_o - 13 psi) test

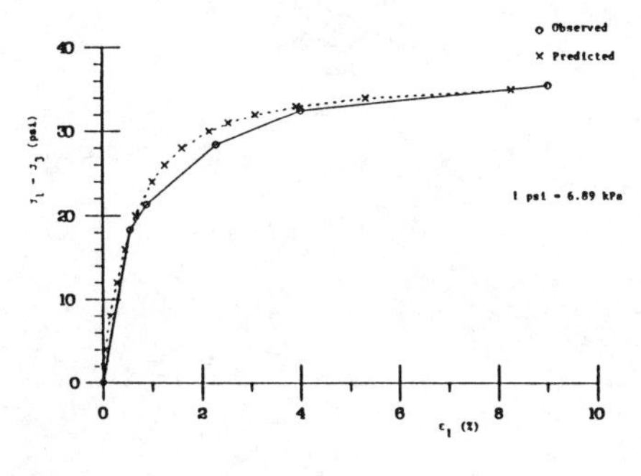

(a) Drained stress-strain response

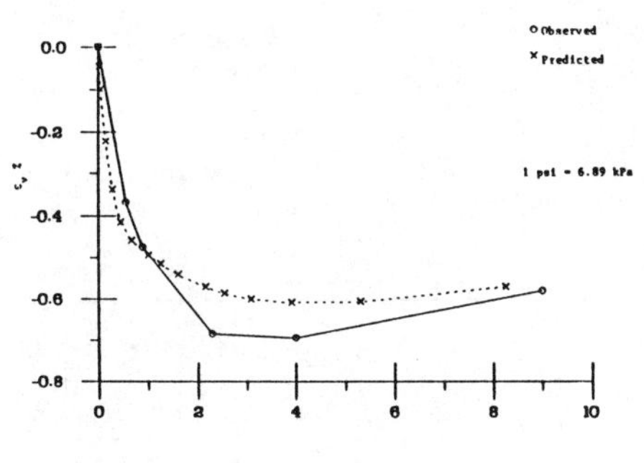

(b) Drained volumetric response

Fig. 9 Comparisons for sand C (σ_o = 98 kPa)

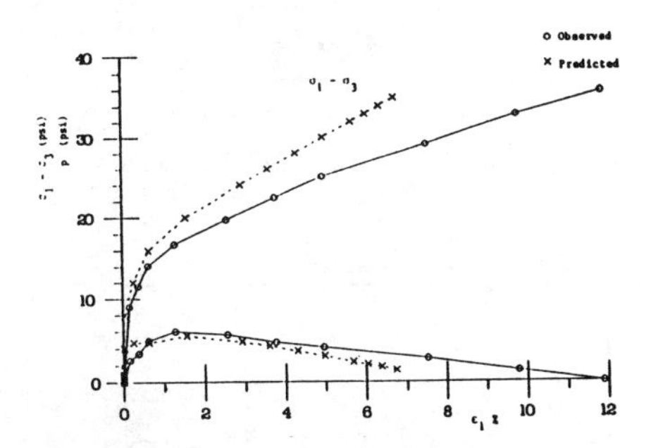

Fig. 10 Comparison of stress-strain-pore pressure response for undrained CTC test on sand C [σ_o = 14.21 psi (98 kPa)]

simulated forward and reverse motions up and down the asperities including various phases. Table 2 shows typical material constants for the δ_o, δ_1 model for tests involving loading, unloading and reloading and model for cyclic behavior of the joints. The anisotropy due to different responses up and down the asperities is allowed for by defining different constants for the two motions.

Figure 12 shows the comparisons for i = 5 degrees for shear stress (τ) vs. relative horizontal displacement U_r and normal displacement V_r vs. U_r for monotonic static loading. In Figure 13 are shown similar comparisons for static loading, unloading and reloading behavior for i = 9 degrees.

Figure 14 shows comparisons for cyclic loading (σ_n = 2 psi, i = 9 deg., ω = 1 Hz) for typical loading cycles N = 10 for τ vs. U_r and V_r vs. U_r responses.

It can be seen from Figures 11 to 13 that the model provides good back predictions for the normal (dilative) response. The latter for the cyclic loading is not very satisfactory and would require additional considerations.

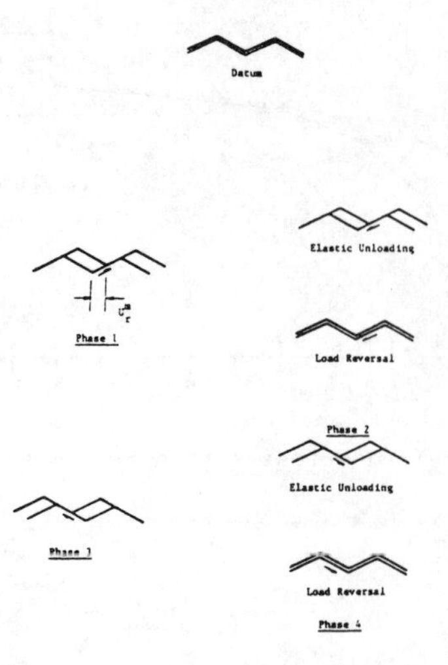

Fig. 11 Phases of cyclic loading, unloading and reloading

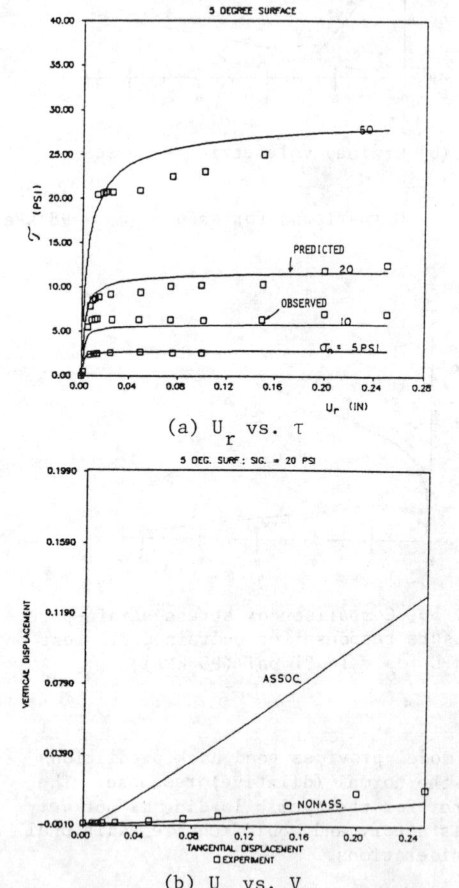

(a) U_r vs. τ

(b) U_r vs. V_r

Fig. 12 Comparisons: static loading $i = 5$ deg.

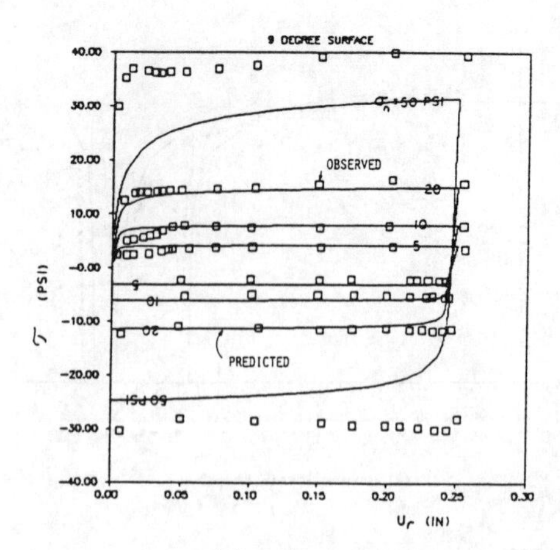

(a) τ vs. U_r

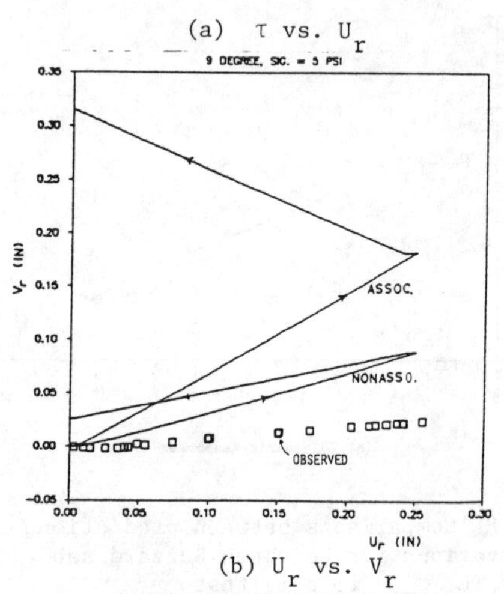

(b) U_r vs. V_r

Fig. 13 Comparisons: quasistatic loading, unloading, reloading. $i = 9$ deg.

5 CONCLUSIONS

A new and general hierarchical modelling approach is developed and verified for solid geologic materials and discontinuities. It can provide unified framework for general yet simplified models, that can be easily applied in practice.

6 ACKNOWLEDGMENTS

The research results presented herein were supported from Grant Nos. 8215344 and CEE 8320256 from the National Science Foundation and No. 830256 from the Air Force Office of Scientific Research, Bolling AFB, D.C. Drs. S. Somasundaram, M.R. Salami, H.M. Galagoda, K.L. Fishman and Q.S.E. Hashmi participated in the research investigations.

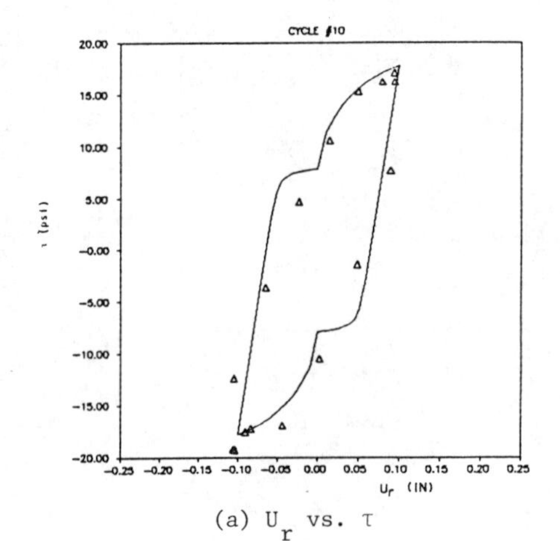

(a) U_r vs. τ

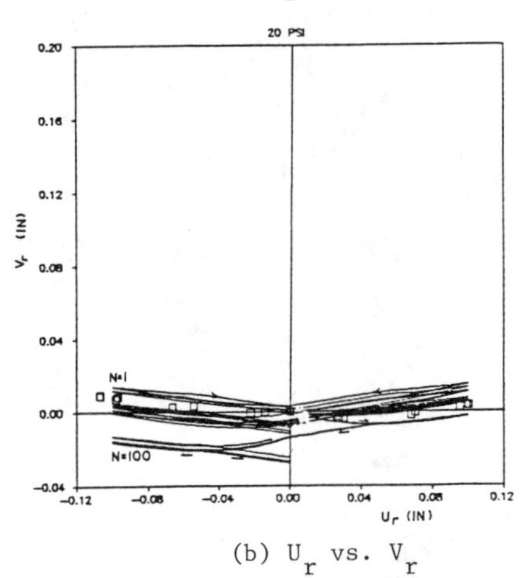

(b) U_r vs. V_r

Fig. 14 Comparisons: cyclic loading σ_n = 20 psi, i = 9 deg. N = 10

REFERENCES

1. Desai, C. S. 1980. A general basis for yield, failure and potential functions in plasticity. International Journal for Numerical and Analytical Methods in Geomechanics 4: 361-375.
2. Desai, C. S. and Faruque, M.O. 1984. Constitutive model for geologic materials. Journal Engineering Mechanics Division, ASCE 110, No. 9: 1391-1408.
3. Desai, C. S., Somasundaram, S. and Frantziskonis, G. N. 1986. A hierarchical approach for constitutive modelling of geologic materials. International Journal for Numerical and Analytical Methods in Geomechanics 10: 225-257.
4. Frantziskonis, G., Desai, C. S. and Somasundaram, S. 1986. Constitutive model for nonassociative behavior. Journal of Engineering Mechanics, ASCE 9: 932-946.
5. Somasundaram, S. and Desai, C. S. Modelling and testing for anisotropic behavior of soils. Journal of Engineering Mechanics Division, ASCE. Accepted (tentative).
6. Frantziskonis, G. and Desai, C. S. 1987. Constitutive model with strain softening. International Journal Solids and Structures 23, No. 6.
7. Galagoda, H. M. 1986. Nonlinear analysis of porous media and applications. PhD thesis,University of Arizona, Tucson, Arizona.
8. Desai, C. S. and Zhang, D. 1987. Viscoplastic model with generalized yield function. International Journal for Numerical and Analytical Methods in Geomechanics 11, No. 6.
9. Desai, C. S. and Varadarajan, A. 1986. A constitutive model for short-term behavior of rock salt. Journal of Geophysical Research. In press.
10. Fishman, K.L. and Desai, C.S. 1987. A constitutive model for hardening behavior of rock joints. Second International Conference on Constitutive Laws for Engineering Materials, Tucson, Arizona.
11. Fishman, K.L. 1987. Testing and modelling static and cyclic behavior of joints. PhD thesis University of Arizona, Tucson, Arizona.
12. Ma, Y. and Desai, C.S. Generalized modelling for normal and shear response of joints. Under preparation.
13. Desai, C.S. 1980. A dynamic multi-degree-of-freedom shear device. Report No. 8-36, Department of Civil Engineering, Virginia Polytechnic Institute and State University, Blacksburg, Virginia.
14. Hashmi, Q.S.E. 1986. Nonassociative plasticity model for cohesionless materials and its implementation in soil-structure interaction. PhD thesis, University of Arizona, Tucson, Arizona.
15. Desai, C. S. and Salami, M.R. 1987. Constitutive model including testing for soft rock. International Journal Rock Mechanics and Mining Engineering. In press.
16. Desai, C.S. and Salami, M.R. 1987. A constitutive model for rocks. Journal of Geotechnical Engineering Division, ASCE 13: 407-423.

17. Wawersik, W.R. and Hannum, D.W. 1980. Mechanical behavior of New Mexico rock salt in triaxial compression up to 200°C. Journal of Geophysical Research 85, No. B2: 891-900.
18. Castro, G. 1969. Liquefaction of sands. PhD thesis, Harvard University, Cambridge, Massachusetts, USA.

Numerical Methods in Geomechanics (Innsbruck 1988), Swoboda (ed.)
© 1988 Balkema, Rotterdam. ISBN 90 6191 809 X

Dynamic nonlinear hysteretic effective stress analysis in geotechnical engineering

W.D.Liam Finn & M.Yogendrakumar
University of British Columbia, Vancouver, Canada
N.Yoshida
Sato Kogyo Ltd, Tokyo, Japan

ABSTRACT: A method for dynamic nonlinear, hysteretic, effective stress analysis is presented which is applicable to embankment dams and soil-structure systems. The method is validated by data from a simulated earthquake test on a centrifuged model of a structure embedded in a saturated sand foundation. The utility of the analysis in engineering practice is demonstrated by the dynamic response analysis of a tailings dam on a nonhomogeneous foundation.

1. INTRODUCTION

The basic elements in the dynamic analysis of a soil-structure system are input motion, appropriate models of site and structure, constitutive relations for all materials present, and a stable, efficient, accurate, computational procedure.

Linear elastic analysis is appropriate for low levels of shaking in relatively firm ground. As the shaking becomes more intense, soil response becomes nonlinear. A great variety of constitutive relations are available for nonlinear response analysis ranging from equivalent linear elastic models to elastic-plastic models with both isotropic and kinematic hardening.

The most widely used methods for dynamic analysis are based on the equivalent linear model. Computer programs representative of this approach are SHAKE (Schnabel et al., 1972) for one-dimensional analysis (1-D) and FLUSH (Lysmer et al., 1975) for 2-D analysis. These programs perform total stress analyses only. Equivalent linear models can exhibit pseudo-resonance, an amplification of computed response that is a function of the nature of the model only. This phenomenon can lead to increased design requirements (Finn et al., 1978).

The dynamic response characteristics and stability of an earth structure during earthquakes are controlled by the effective stress regime in the structure. In saturated regions of the structure, porewater pressures are induced by seismic excitation. These pressures continuously modify the effective stresses during the earthquake and hence have a major impact on dynamic response and stability; in extreme cases, they can trigger flow slides.

It is clearly a very important step in the design process to make reliable estimates of seismically induced porewater pressures. A semi-empirical method of estimation was developed by Seed (1979a), which is widely used in practice. Since 1976, there has been growing interest in the development and application of effective stress methods of dynamic response analysis (Finn et al., 1976, 1986; Dikmen and Ghabbousi, 1984; Ishihara and Towhata, 1982; Prevost et al., 1981; Siddharthan and Finn, 1982; and Zienkiewicz et al., 1978). These methods model the important phenomenological aspects of dynamic response of saturated soils. However, because of a lack of data from suitably instrumented structures in the field it has not been possible to validate the quantitative predictive capabilities of the methods except in a few cases of level ground conditions (Finn et al., 1982; Iai et al, 1985).

A limited validation of these methods has been possible using data from element tests such as cyclic triaxial or simple shear tests (Finn and Bhatia, 1980). Although this type of validation is an important first step, it is inadequate because in these tests either the stress or strain field is prescribed and both are considered homogeneous. Therefore, the tests do not provide the rigorous trial of either the constitutive relations or the robustness of the computational procedure that data from an instrumented structure in the field with inhomogeneous stress and strain fields would make possible.

There are two procedures for modelling the complex response of field structures, a model test conducted on a shake table or in a centrifuge. In a centrifuged model, stresses at the same levels that exist in a full scale structure at corresponding points can be produced by creating an artificial gravity field of intensity Ng, where g is the acceleration due to the gravity of the earth and 1/N is the linear scale of the model. This ability to create prototype stresses in the model is important since soil properties are dependent on effective stresses. For this reason, seismic tests on a centrifuged model are considered superior to those conducted on a shaking table in a 1g environment. Since the static stress levels in both model and prototype are similar at corresponding points, each soil element in the centrifuged model may be expected to undergo the same response history as corresponding elements in the prototype for a given excitation (Barton, 1982).

The United States Nuclear Regulatory Commission (USNRC) through the U.S. Army Corps of Engineers sponsored a series of centrifuged model tests to provide data for the verification of the dynamic nonlinear effective stress method of analysis incorporated in the program TARA-3 (Finn et al., 1986). The tests were conducted on the large geotechnical centrifuge at Cambridge University in the United Kingdom. Details of the Cambridge centrifuge and associated procedures for simulated earthquake testing have been described by Schofield (1981). Some of the USNRC tests will be described and analyzed to demonstrate current capability in dynamic effective stress analysis and seismic porewater pressure estimation. Analyses of other tests may be found in Finn (1985) and Finn et al. (1984, 1985a, 1985b).

2. METHOD OF ANALYSIS BY TARA-3

An incrementally elastic approach has been adopted to model nonlinear behaviour using tangent shear and bulk moduli, G_t and B_t respectively. The incremental dynamic equilibrium forces $\{\Delta P\}$ are given by

$$[M]\{\Delta\ddot{x}\} + [C]\{\Delta\dot{x}\} + [K]\{\Delta x\} = \{\Delta P\} \quad (1)$$

where $[M]$, $[C]$ and $[K]$ are the mass, damping and stiffness matrices respectively, and $\{\Delta x\}$, $\{\Delta\dot{x}\}$, $\{\Delta\ddot{x}\}$ are the matrices of incremental relative displacements, velocities and accelerations. The viscous damping is of the Rayleigh type and the stiffness matrix is a function of the current tangent moduli. The use of shear and bulk moduli allows the elasticity matrix $[D]$ to be expressed as

$$[D] = B_t[\underline{Q}_1] + G_t[\underline{Q}_2] \quad (2)$$

where $[\underline{Q}_1]$ and $[\underline{Q}_2]$ are constant matrices for the plane strain conditions usually considered in analyses. This formulation reduces the computation time for formulating $[D]$ whenever G_t and B_t change in magnitude because of straining or porewater pressure changes.

2.1 Stress-Strain Behaviour

The behaviour of soil in shear is assumed to be nonlinear and hysteretic, exhibiting Masing behaviour (1926) during unloading and reloading. Therefore damping is primarily hysteretic.

The response of the soil to uniform all round pressure is assumed to be nonlinearly elastic and dependent on the mean normal stress. In this deformation mode, hysteresis is neglected.

The relationship between shear stress τ and shear strain γ for the initial loading phase under either drained or undrained loading conditions is assumed to be hyperbolic and given by

$$\tau = f(\gamma) = \frac{G_{max}\,\gamma}{(1 + (G_{max}/\tau_{max})\,|\gamma|)} \quad (3)$$

in which G_{max} = maximum shear modulus and τ_{max} = appropriate shear strength. This initial loading or skeleton curve is shown in Fig. 1(a). The unloading-reloading has been modelled using the Masing criterion. This implies that the equation for the unloading curve from a point (γ_r, τ_r) at which the loading reverses direction is given by

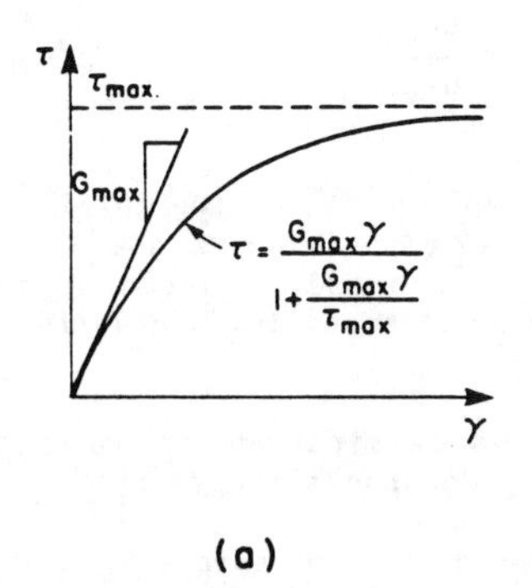

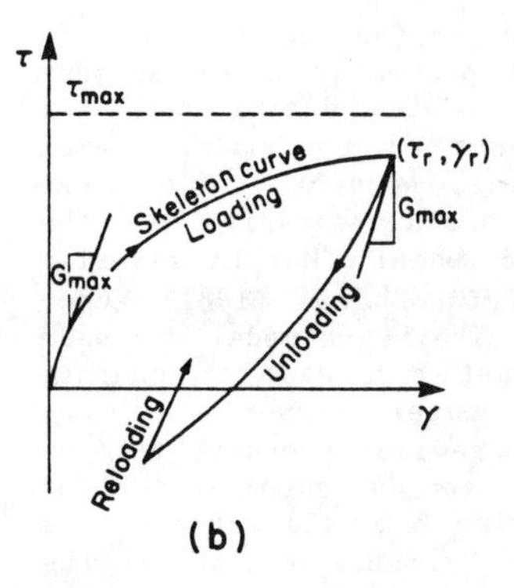

Figure 1. (a) Initial loading curve; (b) Masing stress strain curves for unloading and reloading.

$$\frac{\tau-\tau_r}{2} = \frac{G_{max}(\gamma-\gamma_r)/2}{1 + (G_{max}/2\tau_{max})\,|\gamma-\gamma_r|} \quad (4)$$

or

$$\frac{\tau-\tau_r}{2} = f\left(\frac{\gamma-\gamma_r}{2}\right) \quad (5)$$

The shape of the unloading-reloading curve is shown in Fig. 1(b).

Finn et al. (1976) proposed rules for extending the Masing concept to irregular loading. They suggested that unloading and reloading curves follow the skeleton loading curve when the magnitude of the previous maximum shear strain is exceeded.

The stiffness matrix [K] in Eqn. 1 is determined using the appropriate tangent shear modulus, G_t, derived from Eqn. 4 and the bulk modulus, B_t from

$$B_t = K_b P_a \left(\frac{\sigma_m}{P_a}\right)^n \quad (6)$$

in which K_b is the bulk modulus constant, P_a is atmospheric pressure, σ_m is the current mean normal effective stress and n is a constant for a given soil type. K_b and n are determined by triaxial tests (Duncan and Chang, 1970).

Both G_t and B_t depend on the current mean-normal effective stress $\sigma_m'=\sigma_m-u$, in which σ_m is the total mean normal stress and u the current seismically induced porewater pressure. Therefore, as the porewater pressure increases and reduces the mean effective stresses, these parameters must be adjusted accordingly. For example, it is commonly assumed that $G_{max}\alpha(\sigma_m')^{1/2}$, therefore

$$\frac{G}{G_{max}} = \left(\frac{\sigma_m'}{\sigma_{mo}'}\right)^{1/2} \quad (7)$$

where G = maximum shear modulus for the current cycle of loading (Finn et al., 1976).

If significant volumetric compaction occurs during seismic loading, the moduli should also be modified to reflect this strain hardening, following procedures outlined by Finn et al. (1976). The program continuously modifies the soil properties for the effects of porewater pressures and dynamic strains.

2.2 Residual Porewater Pressure Model

During seismic shaking two kinds of porewater pressures are generated in saturated sands; transient and residual. The transient pressures are due to changes in the applied mean normal stresses during seismic excitation. For saturated sands, the transient changes in porewater pressures are equal to changes in the mean normal stresses. Since they balance each other, the effective stress regime in the sand remains largely unchanged and so the stability and deformability of the sand is not seriously affected.

The residual porewater pressures are due to plastic deformations in the sand skeleton. These persist until dissipated by drainage or diffusion and therefore they exert a major influence on the strength and stiffness of the sand skeleton. Since the shear and bulk moduli are dependent on the effective stresses in the soil, excess porewater pressures must be continually updated during analysis, and their effects on the moduli taken progressively into account. Two porewater pressure models are available; the Martin-Finn-Seed model (Martin et al., 1975) and the Finn-Bhatia (1981) endochronic model. The M-F-S model was used in the subsequent analyses to generate the residual porewater pressures. Therefore computed porewater pressure records will show the steady accumulation of pressure with time but will not show the fluctuations in pressure caused by the transient changes in mean normal stresses.

In the Martin-Finn-Seed model the increments in porewater pressure ΔU that develop in a saturated sand under seismic shear strains are related to the volumetric strain increments $\Delta\varepsilon_{vd}$ that occur in the same sand under drained conditions with the same shear strain history. The original model applies only to level ground, so that there are no static shear stresses acting on horizontal planes prior to the earthquake. The M-F-S model was subsequently modified to include the effects of the initial static shear stresses present in 2-D analyses as described later.

The porewater pressure model is described by

$$\Delta U = \bar{E}_r \cdot \Delta\varepsilon_{vd} \qquad (8)$$

in which $\bar{E}_r$ = one-dimensional rebound modulus of sand at an effective stress σ'_v.

Under drained simple shear conditions, the volumetric strain increment $\Delta\varepsilon_{vd}$ is a function of the total accumulated volumetric strain ε_{vd} and the amplitude of the current shear strain γ, and is given by

$$\Delta\varepsilon_{vd} = C_1(\gamma - C_2\varepsilon_{vd}) + \frac{C_3\varepsilon_{vd}^2}{\gamma + C_4\varepsilon_{vd}} \qquad (9)$$

in which C_1, C_2, C_3 and C_4 are volume change constants that depend on the sand type and relative density and may be determined experimentally by means of drained cyclic simple shear tests on dry or saturated samples.

An analytical expression for the rebound modulus $\bar{E}_r$, at any effective stress level σ'_v, is given by Martin et al. (1975) as

$$\bar{E}_r = \frac{d\sigma'_v}{d\varepsilon_{vr}} = (\sigma'_v)^{1-m}/[m\ K_2(\sigma'_{vo})^{n-m}] \qquad (10)$$

in which σ'_{vo} is the initial value of the effective stress and K_2, m and n are experimental constants derived from rebound tests in a consolidation ring.

2.3 Determination of Porewater Pressure Constants in Practice

The direct measurement of the constants in the porewater pressure model requires cyclic simple shear equipment which is not yet in common use. Therefore, to facilitate the use of TARA-3 in practice, techniques have been developed to derive the constants from the liquefaction resistance curve of the soil. The liquefaction curve may be determined from cyclic triaxial tests and then corrected to simple shear conditions as described by Seed (1979b) or derived directly from Standard Penetration Test data (Seed et al., 1983). In the latter case the constants are derived by a regression process to ensure that the predicted liquefaction curve compares satisfactorily with the field liquefaction curve using the program SIMCYC2 (Yogendrakumar and Finn, 1986a). If the liquefaction curve has been derived by laboratory tests, the rate of porewater pressure increase is known. Then a program for regression analysis, C-PRO, (Yogendrakumar and Finn, 1986b) is used to select constants that match both the rate of porewater pressure generation and the liquefaction curve.

This process has been adapted to the TARA-3 model to include the effects of static shear. The volumetric strain constants are derived from porewater pressure data from cyclic loading tests with various levels of static shear stress or from appropriate liquefaction resistance curves reflecting the influence of initial static shear.

2.4 Slip Elements

For analysis involving soil-structure interaction it may be important to model slippage between the structure and soil. Slip may occur during very strong shaking or even under moderate shaking if high porewater pressures are developed under the structure. TARA-3 contains slip elements of the type developed by Goodman et al., (1968), to allow for relative movement between soil and structure in both sliding and rocking modes during earthquake excitation.

3. RESPONSE OF SATURATED EMBANKMENT WITH EMBEDDED STRUCTURE

A schematic view of a saturated embankment with an embedded structure is shown in Fig. 2. This configuration with a strong soil-structure interaction provides a very severe test of the capabilities of TARA-3 to model dynamic response. The structure is made from a solid piece of aluminum alloy and has dimensions 150mm wide by 108mm high in the plane of shaking. The length perpendicular to the plane of shaking is 470mm and spans the width of the model container. The structure is embedded a depth of 25mm in the sand foundation. Sand was glued to the base of the structure to prevent slip between structure and sand.

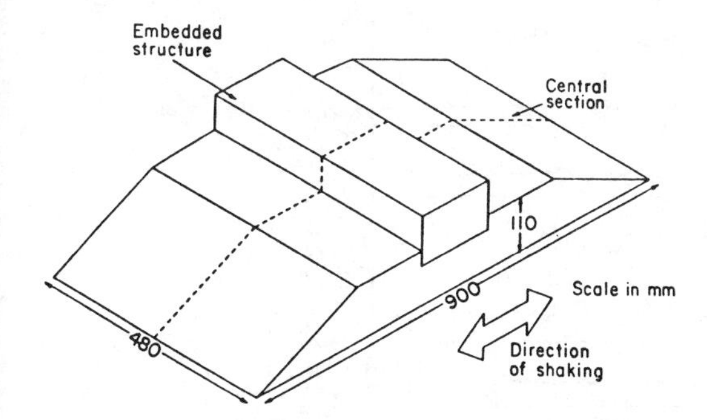

Figure 2. Centrifugal model of embedded structure.

The foundation was constructed of Leighton Buzzard Sand passing BSS No. 52 and retained on BSS No. 100. The mean grain size is therefore 0.225mm. The sand was placed as uniformly as possible to a nominal relative density $D_r = 52\%$.

During the test the model experienced a nominal centrifugal acceleration of 80 g. The model therefore simulated a structure approximately 8.6m high by 12m wide embedded 2m in the foundation sand.

De-aired silicon oil with a viscosity of 80 centistokes was used as a pore fluid. In the gravitational field of 80g, the structure underwent consolidation settlement which led to a significant increase in density under the structure compared to that in the free field. This change in density was taken into account in the analysis.

The locations of the accelerometers (ACC) and pressure transducers (PPT) are shown in Fig. 3. Analyses of previous centrifuge tests indicated that TARA-3 was capable of modelling acceleration response satisfactorily. Therefore, in the present test, more instrumentation was devoted to obtaining a good data base for checking the ability of TARA-3 to predict residual porewater pressures.

As may be seen in Fig. 3, the porewater pressure transducers are duplicated at corresponding locations on both sides of the centre line of the model except for PPT 2255 and PPT 1111. The purpose of this duplication was to remove any uncertainty as to whether a difference between computed and measured porewater pressures might be due simply to local inhomogeneity in density.

The porewater pressure data from all transducers are shown in Fig. 4. These records show the sum of the transient and residual porewater pressures. The peak residual pressure may be observed when the excitation has ceased at about 95 milliseconds. The pressures recorded at corresponding points on opposite sides of the centre line such as PPT 2631 and PPT 2338 are generally quite similar although there are obviously minor differences in the levels of both total and residual porewater pressures. Therefore it can be assumed that the sand foundation is remarkably symmetrical in its properties about the centre line of the model.

3.1 Computed and Measured Acceleration Responses

The soil-structure interaction model was converted to prototype scale before analysis using TARA-3 and all data are quoted at prototype scale. Soil properties were consistent with relative density.

The computed and measured horizontal accelerations at the top of the structure at the location of ACC 1938 are shown in Fig. 5. They are very similar in frequency content, each corresponding to the frequency of the input motion given by ACC 3441 (Fig. 4). The peak accelerations agree fairly closely.

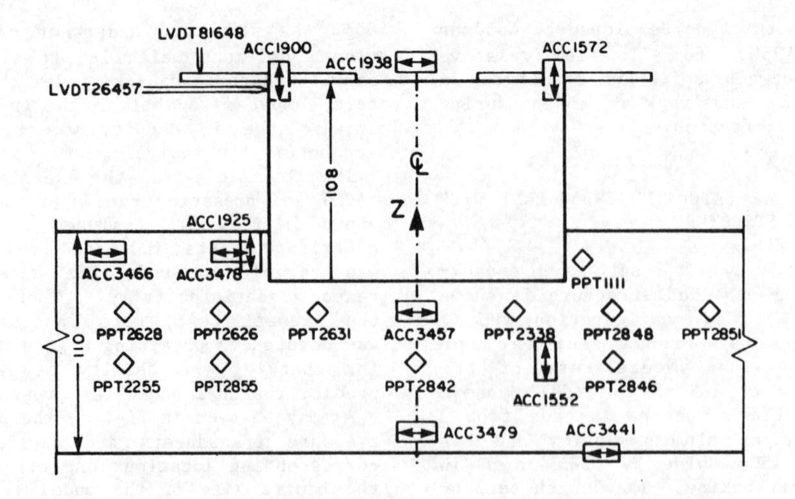

Figure 3. Instrumentation of centrifuged model.

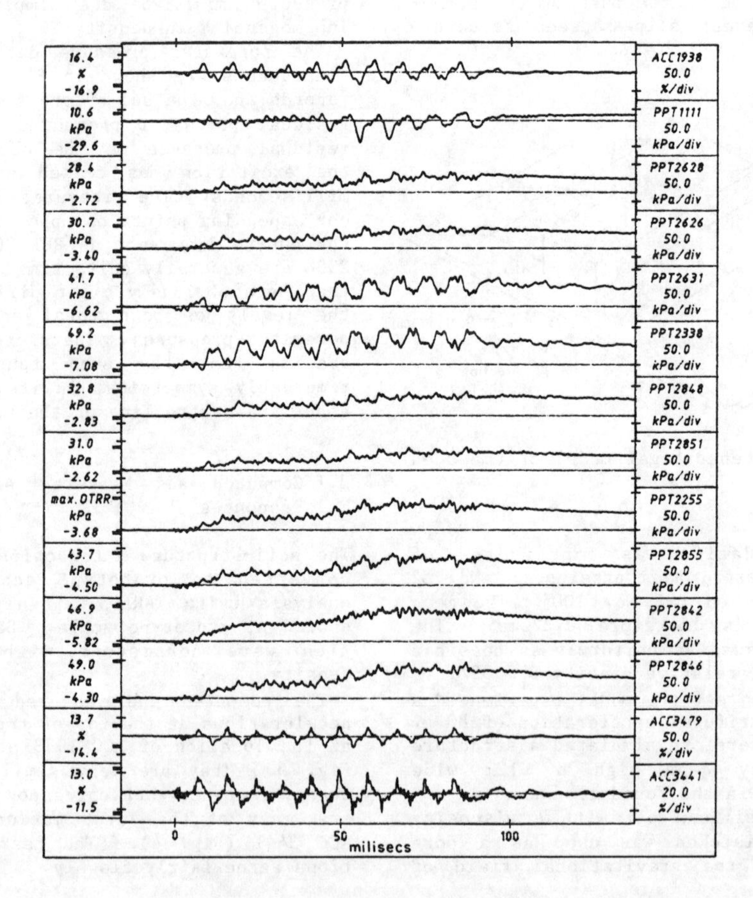

Figure 4. Complete porewater pressure data from centrifuge test.

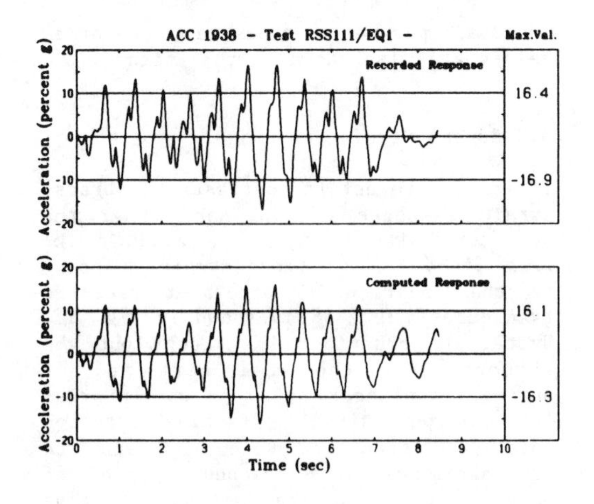

Figure 5. Recorded and computed horizontal accelerations at ACC 1938.

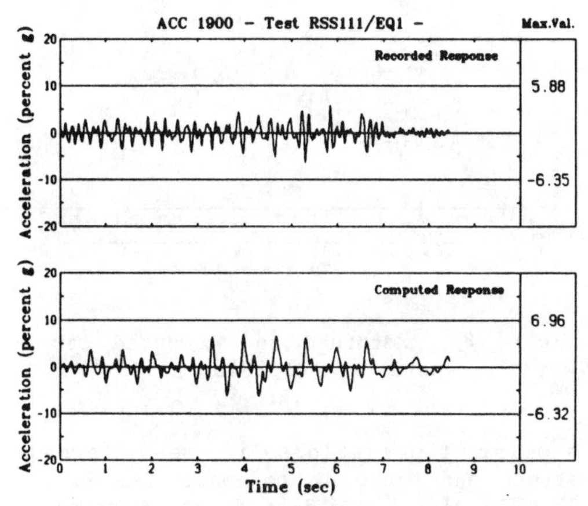

Figure 6. Recorded and computed vertical accelerations at ACC 1900.

The vertical accelerations due to rocking as recorded by ACC 1900 and those computed by TARA-3 are shown in Fig. 6. Again, the computed accelerations closely match the recorded accelerations in both peak values and frequency content. Note that the frequency content of the vertical accelerations is much higher than that of either the horizontal acceleration at the same level in the structure or that of the input motion. This occurs because the foundation soils are much stiffer under the normal compressive stresses due to rocking than under the shear stresses induced by the horizontal accelerations.

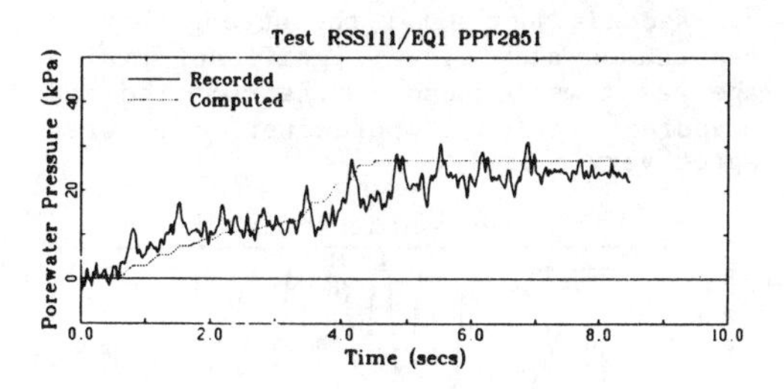

Figure 7. Recorded and computed porewater pressures at PPT 2851.

3.2 Computed and Measured Porewater Pressures

The porewater pressures in the free field recorded by PPT 2851 are shown in Fig. 7. In this case the changes in the mean normal stresses are not large and the fluctuations of the total porewater pressure about the residual value are relatively small. The peak residual porewater pressure, in the absence of drainage, is given directly by the pressure recorded after the earthquake excitation has ceased. In the present test, significant shaking ceased after 7 seconds. A fairly reliable estimate of the peak residual pressure is given by the record between 7 and 7.5 seconds. The recorded value is slightly less than the value computed by TARA-3 but the overall agreement between measured and computed pressures is quite good.

As the structure is approached, the recorded porewater pressures show the increasing influence of soil-structure interaction. The pressures recorded by PPT 2846 adjacent to the structure (Fig. 8) show somewhat larger oscillations than those recorded in the free field. This location is close enough to the structure to be affected by the cyclic normal stresses caused by rocking. The recorded peak value of the residual porewater pressure is given by the relatively flat portion of the record between 7 and 7.5 seconds. The computed and recorded values agree very closely.

Transducer PPT 2338 is located directly under the structure near the edge and was subjected to large cycles of normal stress due to rocking of the structure. These fluctuations in stress resulted in

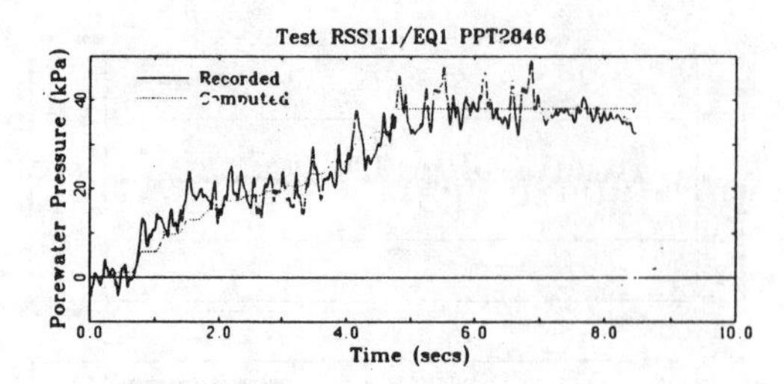

Figure 8. Recorded and computed porewater pressures at PPT 2846.

similar fluctuations in mean normal stress and hence in porewater pressure. This is clearly evident in the porewater pressure record shown in Fig. 9. The higher frequency peaks superimposed on the larger oscillations are due to dilations caused by shear strains. The peak residual porewater pressure which controls stability is observed between 7 and 7.5 seconds just after the strong shaking has ceased and before significant drainage has time to occur. The computed and measured residual porewater pressures agree very closely.

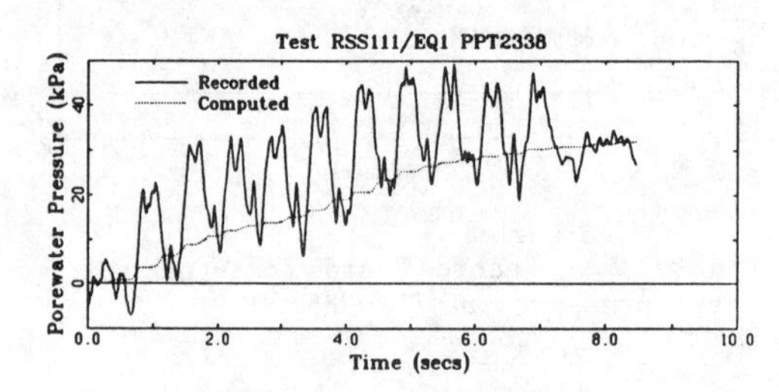

Figure 9. Recorded and computed porewater pressures at PPT 2338.

Contours of computed porewater pressures are shown in Fig. 10. They indicate very symmetrical distribution of residual porewater pressure. Recorded values are also shown in this figure.

3.3 Stress-Strain Response

It is of interest to contrast the stress-strain response of the sand under the structure with that of the sand in the free field. The stress-strain response at the location of porewater pressure transducer PPT 2338 is shown in Fig. 11. Hysteretic behaviour is evident but the response for the most part is not strongly nonlinear. This is not surprising as the initial effective stresses under the structure were high and the porewater pressures reached a level of only about 20% of the initial effective vertical stress. The response in the free field at the location of PPT 2851 (Fig. 12) is strongly nonlinear with large hysteresis loops indicating considerable softening due to high porewater pressures and shear strain. At this location the porewater pressures reached about 80% of the initial effective vertical pressure.

4. ANALYSIS OF DAMS

Since the development of TARA-3 in 1986, it has been used to estimate the seismic response of a number of dams. In particular, it has been used to determine the peak dynamic displacements and the post-earthquake permanent deformations. Typical results for the proposed Lukwi tailings dam in Papua New Guinea will be presented to show the kind of data that is provided by a true nonlinear effective stress method of analysis (Finn et al., 1987). First, however, the framework of a TARA-3 analysis as applied to dams will be presented.

TARA-3 conducts both static and dynamic analysis. A static analysis is first

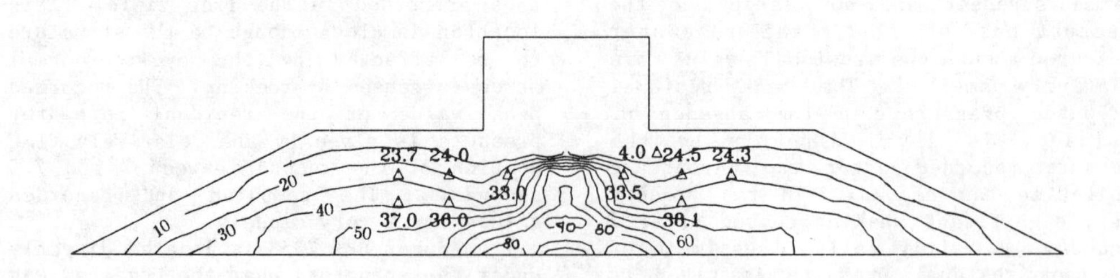

Figure 10. Contours of computed porewater pressures.

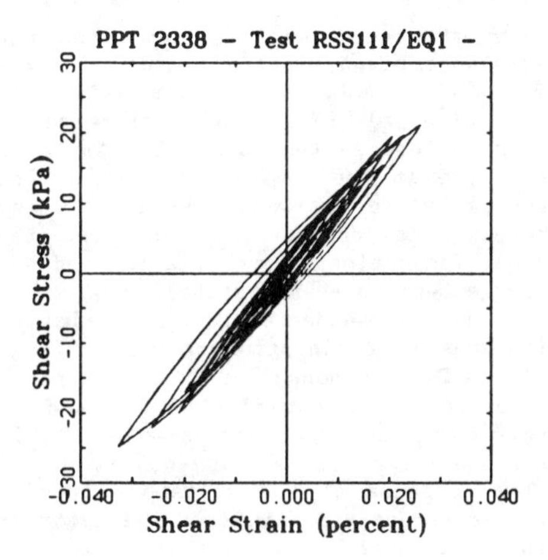

Figure 11. Stress strain response under the structure.

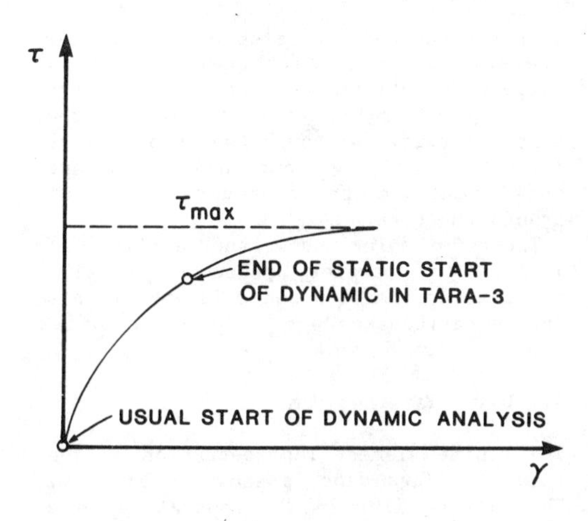

Figure 13. Different ways of initiating dynamic analysis.

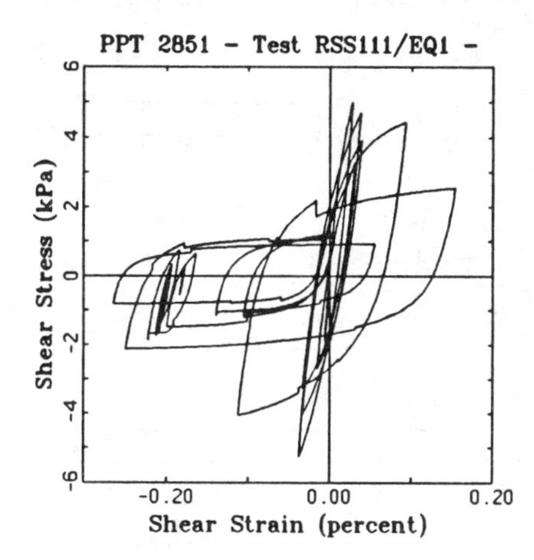

Figure 12. Stress strain response in the free field.

carried out to determine the stress and strain fields throughout the cross-section of the dam at the end of construction. The program can simulate the gradual construction of the dam.

Dynamic analysis in each element of the dam starts from the static stress-strain condition as shown in Fig. 13. This leads to accumulating permanent deformations in the direction of the smallest residual resistance to deformation. Methods of dynamic analysis commonly used in practice ignore the static strains in the dam and start from the origin of the stress-strain curve in all elements even in those which carry high shear stresses. TARA-3 also allows the analysis to start from the zero stress-strain condition, if it is desired to follow current practice.

As shaking proceeds, two phenomena occur; porewater pressures develop in saturated portions of the embankment and, in the unsaturated regions, volumetric strains and associated settlements develop. The program takes into account the effects of the porewater pressures on moduli and strength during dynamic analysis and estimates the additional deformations due to gravity acting on the softening soil. At the end of the earthquake, additional settlements occur due to consolidation as the seismically induced residual porewater pressures dissipate. The final deformed shape of the dam results from the sum of permanent deformations due to the hysteretic dynamic stress-strain response, constant volume deformations in saturated portions of the embankment, volumetric strains in unsaturated portions and deformations due to consolidation as the seismic porewater pressures dissipate. The final post-earthquake deformed shape of a saturated embankment computed by TARA-3 is shown in Fig. 14. This shows the classical spreading due to high porewater pressures.

The post-earthquake deformed shape of an embankment with a central core is shown in Fig. 15. The water table is about 1.7 m below the crest. Only the upstream segment to the left of the core is saturated and generates high porewater pressure during earthquake shaking. Large deformations occur upstream and the core is strongly deformed towards the upstream side. Although the deformations in this case are contained, they are sufficient to cause severe cracking around the core.

These examples show the ability of TARA-3 to predict phenomenologically observed deformation modes in embankments during earthquakes.

4.1 Lukwi Tailings Dam

The finite element representation of the Lukwi tailings dams is shown in Fig. 16. The sloping line in the foundation is a plane between two foundation materials. Upstream to the left is a limestone with shear modulus G = 6.4 × 10^6 kPa and a shear strength defined by c' = 700 kPa and ϕ' =45°. The material to the right is a siltstone with a low shearing resistance given by c' = 0 and ϕ' = 12°. The shear modulus is approximately G = 2.7 × 10^6 kPa. The difference in strength between the foundation soils is reflected in the dam construction. The upstream slope on the limestone is steep whereas the downstream slope on the weaker foundation is much flatter and has a large berm to ensure stability.

The dam was subjected to strong shaking with a peak acceleration of 0.33 g (Fig. 17). The response of the limestone foundation is almost elastic as shown in Fig. 18 by the shear stress-shear strain response for a typical element.

The response of the siltstone foundation is strongly nonlinear. The deformations increase progressively in the direction of the initial static shear stresses as shown in Fig. 19. Since the analysis starts from the initial post-construction stress-strain condition subsequent large dynamic stress impulses

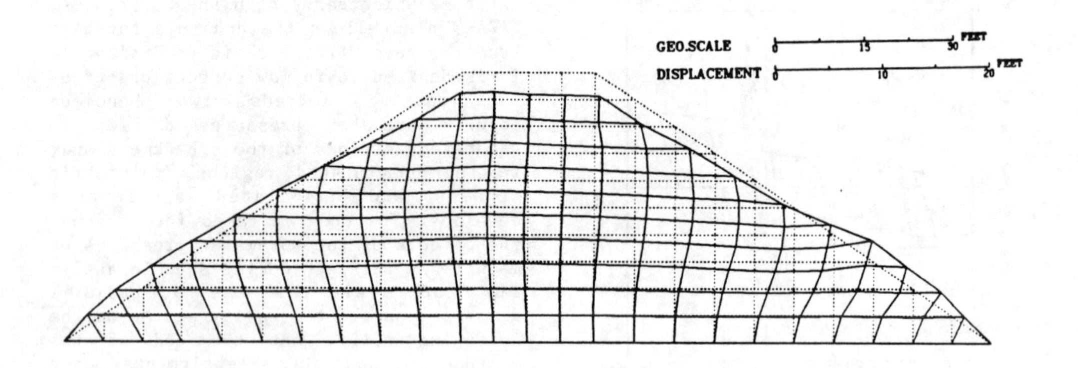

Figure 14. Deformed shape of uniform embankment after earthquake.

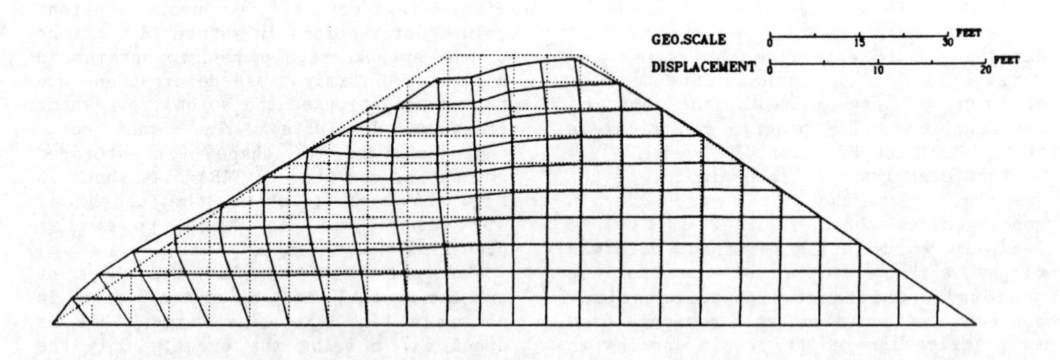

Figure 15. Deformed shape of central core embankment after earthquake.

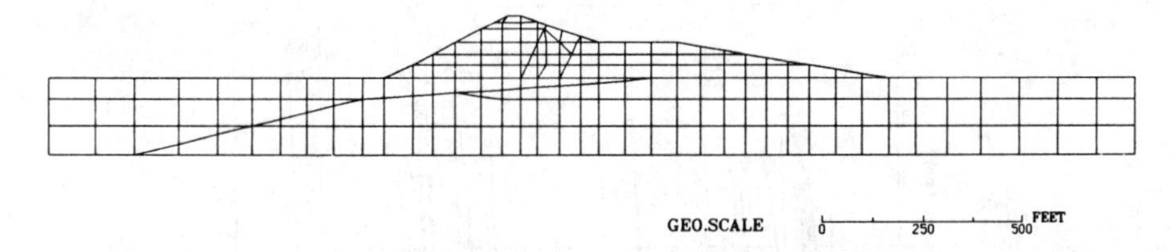

Figure 16. Finite element idealization of Lukwi tailings dam.

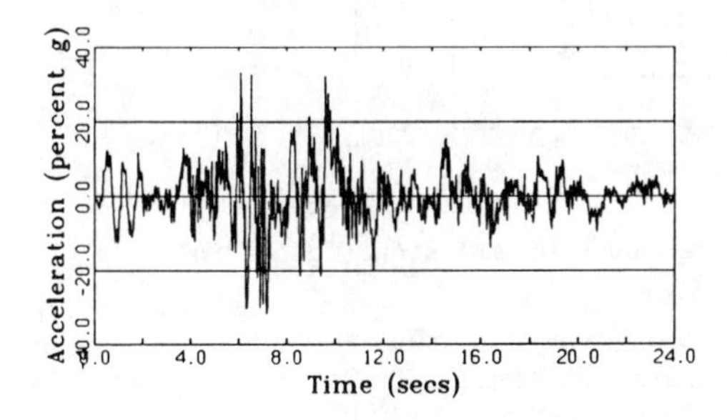

Figure 17. Input motion of analysis of Lukwi tailings dam.

move the response close to the highly nonlinear part of the stress-strain curve. It may be noted that the hysteretic stress-strain loops all reach the very flat part of the stress-strain curve, thereby ensuring successively large plastic deformations.

An element in the berm also shows strong nonlinear response with considerable hysteretic damping (Fig. 20).

The acceleration time history of a point near the crest in the steeper upstream slope is shown in Fig. 21. The displacement time history of the point is shown in Fig. 22. Note that the permanent deformation is of the order of 25 cm. Most of this was generated by a large permanent slip which occurred about 8 secs after the start of shaking.

The deformed shape of the central portion of the dam is shown to a larger scale in Fig. 23.

5. CONCLUSIONS

Phenomenological aspects of soil-structure interaction are clearly demonstrated in centrifuge tests such as high frequency rocking response, the effects of rocking on porewater pressure patterns and the distortion of free-field motions and porewater pressures by the presence of a structure.

The comparison between measured and computed responses for the centrifuge model of a structure embedded in a satur-

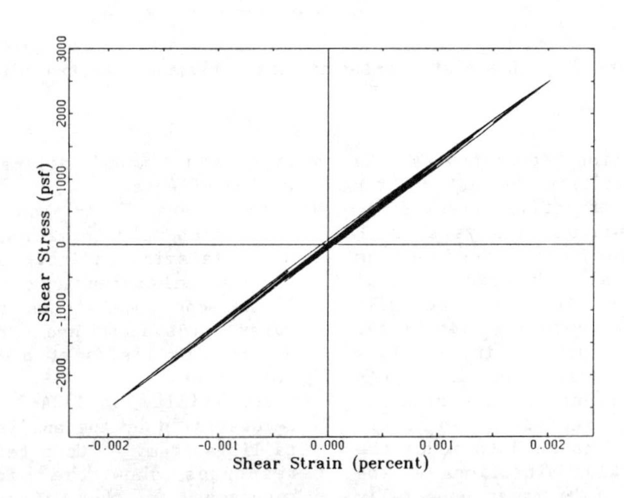

Figure 18. Shear stress-shear strain response of limestone foundation.

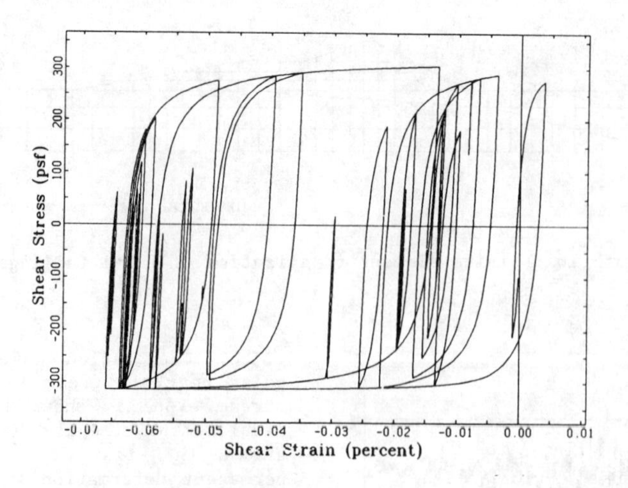

Figure 19. Shear stress-shear strain response of siltstone foundation.

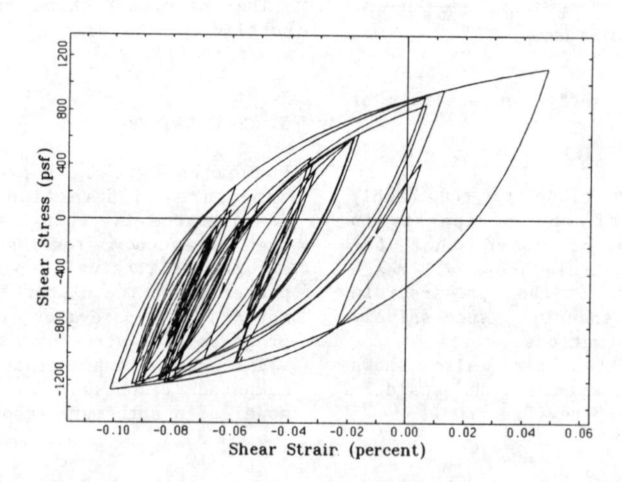

Figure 20. Shear stress-shear strain response in the berm.

ated sand foundation demonstrates the wide ranging capability of TARA-3 for performing complex effective stress soil-structure interaction analysis with acceptable accuracy for engineering purposes. Seismically induced residual porewater pressures are satisfactorily predicted even when there are significant effects of soil-structure interaction. Computed accelerations agree in magnitude, frequency content and distribution of peaks with those recorded. In particular, the program was able to model the high frequency rocking vibrations of the model structures. This is an especially difficult test of the ability of the program to model soil-structure interaction effects.

The program TARA-3 can compute directly the permanent deformations of earth dams under seismic loading. It can reproduce the lateral spreading characteristic of loose sand embankments with high porewater pressures and the asymmetrical deformation fields of dams with impermeable cores.

The utility of TARA-3 in practice was demonstrated by the analysis of the Lukwi tailings dam. Computed stress-strain responses show the widely different responses of the different foundation materials to the design earthquake. The

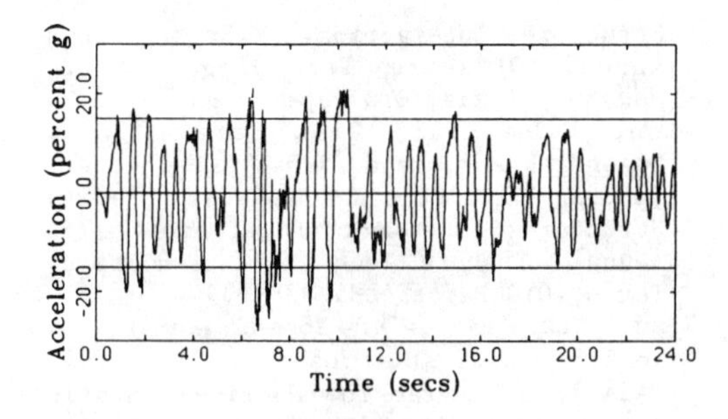

Figure 21. Computed accelerations of a point near the crest.

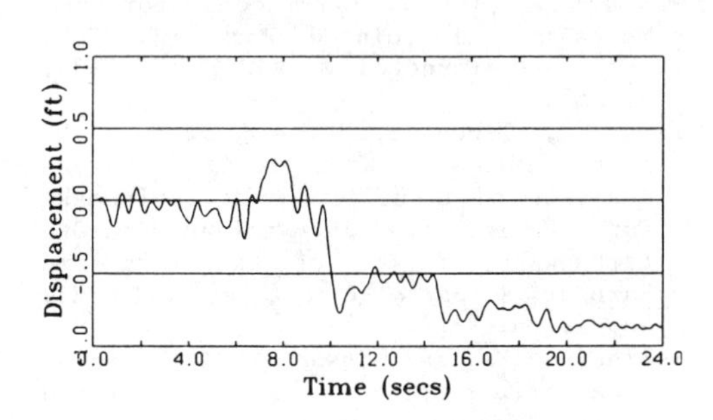

Figure 22. Displacement history of a point near the crest.

computed final deformed shape of the dam itself reflected clearly the influence of dam geometry and the different foundation materials.

The nonlinear effective stress analysis provides a very clear overall picture of the response of the dam to the design earthquake as well as providing the designer with all the details necessary in zones of potential concern.

6. ACKNOWLEDGEMENTS

This research was supported financially by the U.S. Government through the European Research Office of the U.S. Army and the National Science and Engineering Research Council of Canada under Grant No. 1498. The project was managed by W. Grabau and J.C. Comati of the European Research Office, U.S. Army London; R.H. Ledbetter of USAE Waterways Experiment Station, Vicksburg, Miss.; and L.L. Beratan, Office of Research, U.S. Nuclear Regulatory Commission. The centrifuge tests were conducted by R. Dean, F.H. Lee and R.S. Steedman of Cambridge University, U.K, under a separate contract. Technical assistance was provided by Chris Collinson who also operated the centrifuge. The tests were under the general direction of A.N. Schofield, Cambridge University and were monitored by the first author on behalf of USAE. Description of model test and the related figures are used by permission of Cork Geotechnics Ltd., Ireland. Data from the Lukwi tailings dam analysis are used with permission of Klohn Leonoff Consultants Ltd., Richmond, B.C., Canada.

REFERENCES

Barton, Y.O. (1982). Laterally Loaded Model Piles in Sand; Centrifuge Tests and Finite Element Analyses, Ph.D. Thesis, Cambridge University, Engineering Department.

Dickmen, S.U. and Ghaboussi, J. (1984). Effective Stress Analysis of Seismic Response and Liquefaction: Theory, Journal of the Geotech. Eng. Div., ASCE, Vol. 110, No. 5, Proc. Paper 18790, pp. 628-644.

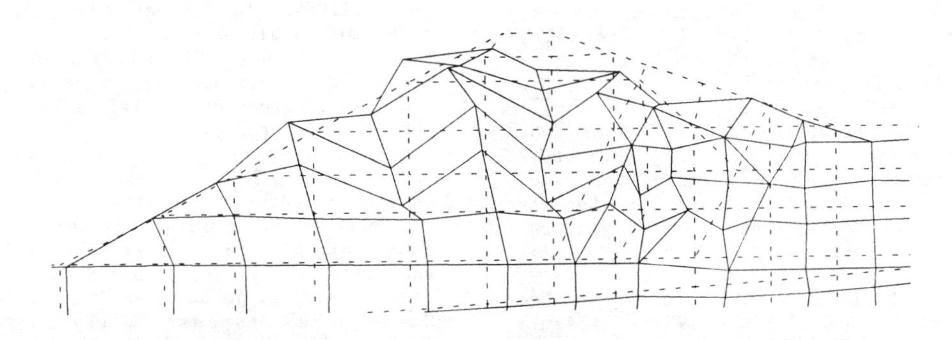

Figure 23. Deformed shape of the dam after the earthquake to enlarged scale.

Duncan, J.M. and C.-Y. Chang. (1970). Nonlinear Analysis of Stress and Strain in Soils, J. Soil Mech. Found. Div. ASCE, 96 (SM5), 1629-1651.

Finn, W.D. Liam. (1985). Dynamic Effective Stress Response of Soil Structures; Theory and Centrifugal Model Studies, Proc. 5th Int. Conf. on Num. Methods in Geomechanics, Nagoya, Japan, Vol. 1, 35-36.

Finn, W.D. Liam and Bhatia, S.K. (1980). Verification of Nonlinear Effective Stress Model in Simple Shear, Application of Plasticity and Generalized Stress-Strain in Geotechnical Engineering, ASCE, Editors, R.N. Yong and E.T. Selig, pp. 241-252.

Finn, W.D. Liam and S.K. Bhatia. (1981). Prediction of Seismic Porewater Pressures, Proceedings in the Tenth Int. Conf. on Soil Mechanics and Foundation Engineering, Vol. 3, A.A. Balkema Publishers, Rotterdam, Netherlands, pp. 201-206.

Finn, W.D. Liam, Iai, S. and Ishihara, K. (1982). Performance of Artificial Offshore Islands Under Wave and Earthquake Loading: Field Data Analyses, Proceedings of the 14th Annual Offshore Technology Conference, Houston, Texas, May 3-6, Vol. 1, OTC Paper 4220, pp. 661-672.

Finn, W.D. Liam, W.K. Lee and G.R. Martin. (1976). An Effective Stress Model for Liquefaction, Journal of the Geotechnical Engineering Division, ASCE, Vol. 103, No. GT6, Proc. Paper 13008, 517-533.

Finn, W.D. Liam, Lo, R.C. and Yogendrakumar, M. (1987). Dynamic Nonlinear Analysis of Lukwi Tailings Dam, Soil Dynamics Group, Dept. of Civil Engineering, University of British Columbia, Vancouver, B.C., Canada.

Finn, W.D. Liam, G.R. Martin and M.K.W. Lee. (1978). Comparison of Dynamic Analyses for Saturated Sands. Proceedings of the ASCE Specialty Conference on Earthquake Engineering and Soil Dynamics. Vol. I, ASCE, New York, N.Y., pp. 472-491.

Finn, W.D. Liam, R. Siddharthan, and R.H. Ledbetter. (1985a). Soil-Structure Interaction During Earthquakes, Proc. 11th Int. Conf. of the Int. Society of Soil Mech. and Found. Engineers, San Francisco, California, August 11-14.

Finn, W.D. Liam, R. Siddharthan, F. Lee and A.N. Schofield. (1984). Seismic Response of Offshore Drilling Islands in a Centrifuge Including Soil-Structure Interaction. Proc., 16th Annual Offshore Technology Conf., Houston, Texas, OTC Paper 4693.

Finn, W.D. Liam, R.S. Steedman, M. Yogendrakumar, and R.H. Ledbetter. (1985b). Seismic Response of Gravity Structures in a Centrifuge, Proc. 17th Annual Offshore Tech. Conf., Houston, Texas, OTC Paper 4885, 389-394.

Finn, W.D. Liam, M. Yogendrakumar, N. Yoshida, and H. Yoshida. (1986). TARA-3: A Program for Nonlinear Static and Dynamic Effective Stress Analysis, Soil Dynamics Group, University of British Columbia, Vancouver, B.C.

Goodman, R.E., R.L. Taylor and T.L. Brekke. (1968). A Model for the Mechanics of Jointed Rock, J. Soil Mech. and Found. Div. ASCE, 94 (SM3), 637-659.

Iai, S., Tsuchida, H. and Finn, W.D. Liam (1985). An Effective Stress Analysis of Liquefaction at Ishinomaki Port During the 1978 Miyagi-Ken-Oki Earthquake, Report of the Port and Harbour Research Institute, Vol. 24, No. 2, pp. 1-84.

Ishihara, K. and Towhata, I. (1982). Dynamic Response of Level Ground Based on the Effective Stress Method. Ch. 7, Soil Mechanics - Transient and Cyclic Loads, Edited by G.N. Pande and O.C. Zienkiewicz, John Wiley and Sons Ltd., New York, N.Y., pp. 133-172.

Lysmer, J., Udaka, T., Tsai, C.F. and Seed, H.B. 1975. FLUSH: A Computer Program for Approximate 3-D Analysis of Soil-Structure Interaction Problems. Report No. EERC 75-30, Earthquake Engineering Research Center, University of California, Berkeley, California.

Martin, G.R., W.D. Liam Finn, and H.B. Seed. (1975). Fundamentals of Liquefaction Under Cyclic Loading, Soil Mech. Series Rpt. No. 23, Dept. of Civil Engineering, University of British Columbia, Vancouver; also Proc. Paper 11284, J. Geotech. Eng. Div. ASCE, 101 (GT5): 324-438.

Masing, G. (1926). Eigenspannungen und Verfestigung beim Messing, Proceedings, 2nd Int. Congress of Applied Mechanics, Zurich, Switzerland.

Prevost, J.H., Cuny, B., Hughes, T.J.R. and Scott, R.F. (1981). Offshore Gravity Structures: Analysis, Journal of Geotechnical Engineering Division, ASCE, Vol. 107, No. GT2, pp. 143-165.

Schnabel, P.B., Lysmer, J. and Seed, H.B. (1972). SHAKE: A Computer Program for Earthquake Response Analysis of Horizontally Layered Sites, Report No. EERC 72-12, Earthquake Engineering

Research Center, University of California, Berkeley, California.

Schofield, A.N. (1981). Dynamic and Earthquake Geotechnical Centrifuge Modelling, Proceedings Int. Conf. on Recent Advances in Geot. Engineering and Soil Dynamics, St. Louis, Missouri, Vol. III: 1081-1100.

Seed, H.B. (1979a). Considerations in the Earthquake-Resistant Design of Earth and Rockfill Dams, 19th Rankine Lecture, Geotechnique 29, No. 3, pp. 215-263.

Seed, H.B. (1979b). Soil Liquefaction and Cyclic Mobility Evaluation for Level Ground During Earthquakes, Journal of Geotechnical Engineering Division, ASCE, Vol. 105, No. GT2, pp. 201-255.

Seed, H.B., I.M. Idriss and I. Arango. (1983). Evaluation of Liquefaction Potential Using Field Performance Data, Journal of Geotechnical Engineering Division, Vol. 109, No. 3, pp. 458-482.

Siddharthan, R. and W.D. Liam Finn (1982). TARA-2, Two dimensional Nonlinear Static and Dynamic Response Analysis, Soil Dynamics Group, University of British Columbia Vancouver, Canada.

Yogendrakumar, M. and W.D. Liam Finn. (1986a).. SIMCYC2: A Program for Simulating Cyclic Simple Shear Tests on Dry or Saturated Sands, Report, Soil Dynamics Group, Dept. of Civil Engineering, University of British Columbia, Vancouver, Canada.

Yogendrakumar, M. and W.D. Liam Finn. (1986b). C-PRO: A Program for Simulating Cyclic Simple Shear Tests on Dry and Saturated Sands, Report, Soil Dynamics Group, Dept. of Civil Engineering, University of British Columbia, Vancouver, Canada.

Zienkiewicz, O.C., Chang, C.T. and Hinton, E. (1978). Nonlinear Seismic Response and Liquefaction, Int. J. Num. and Analytical Meth. Geomechanics, 2(4), 381-404.

Numerical Methods in Geomechanics (Innsbruck 1988), Swoboda (ed.)
© 1988 Balkema, Rotterdam. ISBN 90 6191 809 X

Some numerical techniques for free-surface seepage analysis

Giancarlo Gioda
Technical University of Milan and University of Udine, Italy
Augusto Desideri
University of Rome, Italy

ABSTRACT: Some numerical techniques are discussed applicable to the solution of unconfined seepage problems, considering both steady state and transient conditions. In particular, the procedures based on the finite element method are considered, which can be subdivided into two main groups: the so called "constant" and "variable" mesh approaches. An alternative technique is also discussed that requires a bijective mapping to be established between the physical domain (the shape of which is a priori unknown, and varies with time for transient problems) and a fixed domain having a simple (e.g. square), time independent shape. Finally, some comments are presented on the stability problems that might show up when the point of intersection between the free surface and a pervious boundary exposed to the atmosphere has to be determined.

1. INTRODUCTION

In geotechnical engineering it is often necessary to solve unconfined seepage problems, like for instance when the stability of an earth dam has to be assessed or when a drainage system has to be designed. These are seepage analyses in which the fluid flow takes place through a domain part of whose boundary (the so called free surface) is a priori unknown and has to be determined, together with the distribution of hydraulic head, as part of the solution.

Depending on the problem to be solved, steady or transient analyses could be required. The first one is characterized by a free surface the shape of which is constant with time, whilst in the second case the free surface moves during time, eventually reaching a steady state configuration.

Due to the difficulties introduced by their geometrical non linearity, free surface problems can be solved analytically only in particularly simple cases. For handling more complex situations it is necessary to adopt some numerical, computer oriented procedure, based on the discretization of the flow domain. In fact, these techniques are becoming quite popular in engineering practice, and in many cases replace more traditional methods like the graphical ones, that require the hand drawing of flow nets [1], or those based on the physical Hele-Shaw viscous flow models.

Among the numerical procedures, the finite element method (see e.g. [2,3,4]) and, more recently, the boundary integral equation method (see e.g. [5-8]) are perhaps the most frequently used.

Here the discussion will be focused on the solution approaches based on the finite element method, which can be subdivided into two main groups. In both cases the non linear problem is solved through iterative procedures.

The approaches of the first group modify the mesh geometry during the solution process so that at the end of iterations part of the mesh boundary represents the "correct" shape of the free surface. They are in general characterized by high accuracy of the final results, but are also affected by some drawbacks. In particular, stability problems might show up in the presence of complex geometries (e.g. layered deposits with nearly horizontal interface between layers) or when the intersection between the free surface and a pervious boundary exposed to the atmosphere has to be determined.

The approaches of the second group operate on meshes of constant geometry and allow the free surface to pass through the elements by means of procedures that in many cases are conceptually similar to those adopted for non linear, or elasto-plastic, stress analyses.

These methods lead to results perhaps slightly less accurate that those obtained with the approach of the first category. On the other hand, they require in general less programming effort for their implementation and can be easily applied to problems involving non homogeneous or layered media.

In the following, after recalling some techniques belonging to the two mentioned groups, an approach will be illustrated which can be viewed as a compromise between them. In fact, this method is based on a bijective mapping between the physical flow region (the shape of which is unknown and varies with time for transient problems) and a fixed domain having a simple, time independent shape. The numerical integration of the governing equations is carried out on the fixed domain, and the solution consists of both the distribution of the hydraulic head and the functions describing the (time dependent) geometry of the physical domains.

Finally some comments are presented on the stability problems that may show up when the intersection between the free surface and a pervious boundary has to be determined, and that in some cases lead to apparent non uniqueness of solution.

2. GOVERNING EQUATION AND BOUNDARY CONDITIONS

Consider a two-dimensional seepage flow through a saturated porous medium, assuming that both pore fluid and soil grains are incompressible and that the deformability of the soil skeleton can be neglected. Under these hypotheses the flow continuity equation is written in the following form, where v_x and v_y are the components of the fluid discharge velocity, and q is the assigned flux per unit volume.

$$\frac{\partial v_x}{\partial x} + \frac{\partial v_y}{\partial y} = q \tag{1}$$

The majority of seepage phenomena in geomechanics involve laminar flow conditions. Consequently, Darcy's law can be adopted as a linear "constitutive" relationship between the components of the discharge velocity and those of the gradient of the hydraulic head h. If x and y coincide with the principal directions of permeability, and if y is the upward vertical coordinate, Darcy's law is expressed as

$$v_x = - k_x \, i_x = - k_x \frac{\partial h}{\partial x} \tag{2a}$$

$$v_y = - k_y \, i_y = - k_y \frac{\partial h}{\partial y} \tag{2b}$$

where

$$h = y + \frac{p}{\gamma} \tag{3}$$

In the above equations k_x and k_y are the coefficients of permeability; p is the pressure of the pore fluid and γ is its unit weight.

Substitution of eqs.(2) into eq.(1) leads to the following final relationship governing the problem at hand.

$$\frac{\partial}{\partial x} (k_x \frac{\partial h}{\partial x}) + \frac{\partial}{\partial y} (k_y \frac{\partial h}{\partial y}) = q \tag{4}$$

Note that eq.(1) is derived without introducing any assumption on the time dependency of flow. As a consequence eq.(4) holds for both steady and transient flows, neglecting for the later case the inertial effects.

For the typical problem shown in fig.1, in which the water levels in the upper and lower reservoirs are known functions of time t, the boundary conditions to be associated to eq.(4) are:

- pervious (constrained) boundaries:

$$h = h_1(t) \quad \text{on 1-2} \tag{5a}$$

$$h = h_2(t) \quad \text{on 4-5} \tag{5b}$$

- pervious (wet) boundary, or seepage face (3-4):

$$h = y(x) \tag{6}$$

- impervious boundary normal to the y axis (5-1):

$$\frac{\partial h}{\partial y} = 0 \quad (\text{or } v_y = 0) \tag{7}$$

- free surface (2-3):

$$h = y = f(x,t) \tag{8a}$$

$$v^{FS} = v_n / m \tag{8b}$$

where

$$v_n = [v_x \, (\partial F/\partial x) + v_y \, (\partial F/\partial y)]/M \tag{8c}$$

$$F = y - f(x,t) \tag{8d}$$

$$M = [(\partial F/\partial x)^2 + (\partial F/\partial y)^2]^{1/2} \tag{8e}$$

Eqs.(8a,b) represent the two conditions to be imposed on the free surface. They require that the hydraulic head be equal to the elevation of the free surface (eq.8a) and that the velocity of the free surface, v^{FS}, be equal to the velocity of the fluid normal to it, v_n, divided by the effective porosity m (eq.8b).

Note that the function y(x) in eq.(6) is known, since it describes the shape of the downstream slope of the dam. On the contrary, the function f(x,t) in eq.(8a), re-

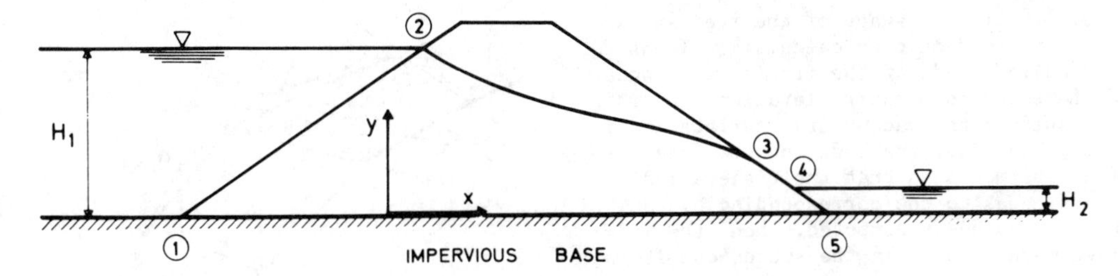

Fig. 1 Unconfined seepage flow through an earth dam.

presenting the shape of the free surface, is unknown. It can be also observed that in steady state regime the functions of time in eqs.(5) become constants and that the boundary condition (8b) reduces to

$$v_n = 0 \qquad (9)$$

When the finite element method is adopted in the solution of seepage problems, the flow domain is subdivided into elements within which the hydraulic head distribution h(x,y) depends on the nodal hydraulic heads $\underline{h}$ through a vector $\underline{b}$ of interpolation functions.

$$h(x,y) = \underline{b}(x,y)^T \, \underline{h} \qquad (10)$$

By writing the governing equation (4) in weak form, and taking into account eq. (10), a linear relationship is arrived at between nodal hydraulic heads $\underline{h}$ and assigned nodal fluxes $\underline{q}$,

$$\underline{K} \, \underline{h} = \underline{q} \qquad (11)$$

where the so called "flow" matrix $\underline{K}$ is expressed by the following integral over the volume V of the element.

$$\underline{K} = \int_V \begin{bmatrix} \frac{\partial \underline{b}}{\partial x} & \frac{\partial \underline{b}}{\partial y} \end{bmatrix} \begin{bmatrix} k_x & 0 \\ 0 & k_y \end{bmatrix} \begin{bmatrix} (\partial \underline{b}/\partial x)^T \\ (\partial \underline{b}/\partial y)^T \end{bmatrix} dv \qquad (12)$$

The system of linear equations governing the solution of the discrete problem is obtained by suitably assembling eq.(11) written for all the elements of the mesh.

The finite element solution of confined seepage problem (i.e. those problems in which all boundaries of the flow domain have known geometry) is straightforward. In fact, for most cases of interest in geomechanics the entries of vector $\underline{q}$ are equal to zero and only two types of boundary conditions are present that correspond, respectively, to the cases of pervious (constrained) and impervious boundaries.

The first one (the so called "essential" boundary condition) is introduced in eq.(11) by setting the hydraulic heads of the relevant boundary nodes equal to values computed according to eq.(3). This is done through the same procedure used in a stress analysis for imposing known values of the displacements of nodes belonging to a constrained surface. After introducing these conditions the assembled flow matrix becomes non singular and the system of linear equations can be solved.

On the contrary, no modification of eq. (11) is necessary in order to enforce the condition on impervious boundaries (the so called "natural" boundary condition). In fact, this condition is implicitly taken into account in the formulation of the discrete problem, exactly as the presence of unconstrained surfaces without applied tractions is in a stress analysis.

Unconfined analyses turn out to be more complex than the confined ones. In fact, in addition to the distribution of hydraulic head, also the location of the free surface (and hence the shape of the domain within which the seepage flow takes place) is unknown.

Two main groups of iterative procedures have been proposed for the solution of unconfined problems in the context of the finite element method. The bases of these techniques, which are referred to as "variable and "constant" mesh approaches, are outlined in the following sections.

3. "VARIABLE" MESH APPROACHES

The techniques belonging to this category (see e.g. [9,10,11]) consist of iterative processes that modify the geometry of the finite element mesh, part of whose boundary coincide with the free surface, until an adequate approximation of the correct shape of the flow domain is reached.

A relatively simple procedure of this kind was proposed for steady state problems by Taylor and Brown [11]. At the beginning of each iteration a confined analysis is performed, on the basis of the current geometry, in which the free surface is treated as an impervious boundary (eq.9). If the shape of the free surface is not correct this calculation leads to hydraulic heads at the free surface nodes different from their elevation. In order to fulfill the second free surface condition, eq.(8a), the nodes on the free surface are moved so that their elevation becomes equal to the corresponding hydraulic head previously computed. Then, the modified mesh is used in the subsequent iteration.

This solution method presents an ambiguity in defining the movement of the node at the intersection between the free surface and the wet boundary, or seepage face (i.e. point 3 in fig.1). In fact, this node (that can be referred to as "exit" node of the free surface) has in general a non vanishing flux, even though it belongs to the free surface. Consequently, during the confined analysis a "pervious" boundary condition (eq.8a) has to be imposed on it, and the procedure adopted for displacing the other free surface nodes cannot be applied. The authors observe, however, that this difficulty is minimized by reducing the mesh size adjacent to this point.

An improved solution method was proposed by Neuman and Witherspoon [12], which requires two confined analyses for each iteration. This method was subsequently extended to the analysis of transient problems [13].

In the first analysis the free surface is considered as a pervious boundary and the hydraulic head of its nodes is set equal to their elevation (eq.8a). From these results the fluxes at the nodes located on the seepage face (line 3-4 in fig.1) are evaluated. Then a second confined analysis is performed in which the free surface is impervious and the nodal fluxes previously computed are imposed on the seepage face. Finally, the mesh is modified by making the elevation of the free surface nodes equal to the corresponding hydraulic heads obtained from the second analysis. The scheme suggested for determining the new positions of the free surface nodes is shown in fig.2.

As observed by the authors, also this procedure presents some difficulties in treating the exit node of the free surface. In fact, during the first confined analysis, the flux evaluated for this node correspond to the contribution of both the free surface (which is a pervious boundary in this calculation) and the seepage face. In the second analysis, however, only the part of this flux through the seepage face (which is not straightforward to evaluate) should be imposed to the exit node. The authors observe, however, that if sufficiently small elements are used, this problem can be eliminated by imposing to the exit node a flux equal to one half of the flux obtained for the adjacent node on the wet boundary.

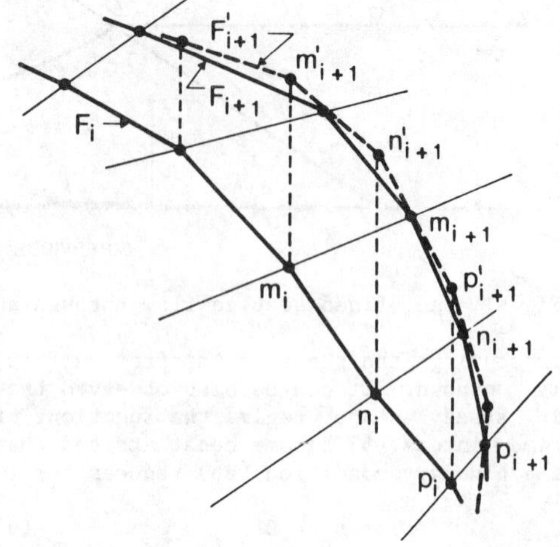

Fig. 2 Scheme for shifting the free surface nodes (after [12]).

Another provision apt to reduce possible spurious oscillations of the free surface [14] consists in describing its shape by means of a reduced number of parameters, smaller than the number of free surface nodes. This implies, however, that the free surface boundary condition can be satisfied only in an average sense.

A transient problem solved with a variable mesh technique is shown in fig.3. The results refer to the variation with time of the free surface within an earth dam caused by the rapid lowering of the reservoir.

A comparison between some variable mesh approaches was presented in [15], where five different methods were adopted in the analysis of the seepage flow through a homogeneous and isotropic square block dam. On the basis of the numerical results the author concluded that there are no significant differences in the free surface shapes, pore pressure distributions and total discharge flows obtained by the various methods considered. Consequently he did not suggest for practical application the methods based on relatively complex procedures, since they do not present appreciable advantages with respect to simpler approaches.

It has to be observed that the variable mesh techniques may present some drawbacks in certain applications. For instance, stability problems might show up when the free surface crosses a nearly horizontal boundary between different materials. In

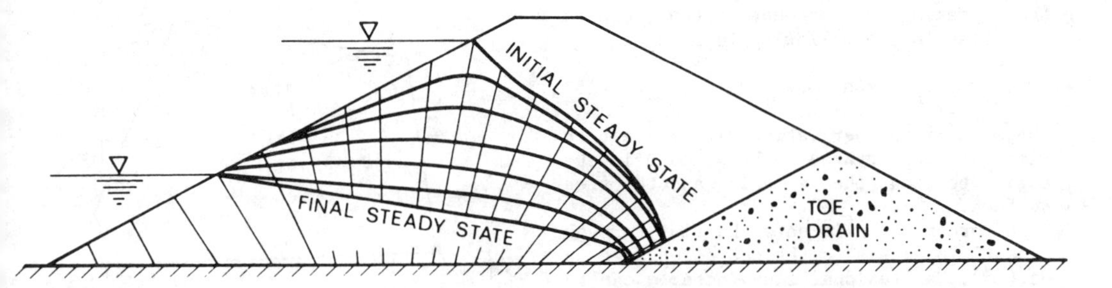

Fig. 3 Transient flow in an earth dam caused by the rapid drawdown of the reservoir (after [13]).

other instances the procedure for shifting the free surface nodes may fail, like e.g. when part of this surface becomes almost vertical and its nodes are subjected to nearly horizontal movements. Some comments on these problems were presented in [12].

4. "CONSTANT" MESH APPROACHES

As previously observed the variable mesh techniques might present some drawbacks in their application to complex cases. In order to overcome these disadvantages some approaches have been proposed (see e.g. [16-23]) in which the geometry of the finite element mesh is kept constant during the iterative solution process. Among them, two will be recalled here which involve solution procedures conceptually similar to those customarily adopted for non linear stress analyses.

4.1 Residual flow procedure.

This solution procedure allows the free surface to pass through the elements by applying to the relevant nodes of the mesh "residual" fluxes determined in such a way that the two boundary conditions on the free surface are simultaneously satisfied, at least in an average sense. To be used this method requires a mesh containing a part of the porous medium larger than that in which the seepage flow takes place.

In the case of steady state problems [16] the iterative solution process is initiated by performing a confined analysis in which boundary conditions of "pervious" and "impervious" types only are applied to the various portions of the mesh contour.

The nodal hydraulic heads obtained by this calculation are used to find the elements crossed by the line h=y(x) (denoted by FS in fig.4) that represents a first approximation of the free surface. The position of this line (on which only eq.8a is fulfilled) within the elements is determined by means of eq.(10).

In order to enforce also the second boundary condition (eq.9), it is necessary to evaluate the flow velocity v_n normal to the free surface segments. Taking into account eqs.(2) and (10) the following relationship is arrived at, where ϑ represents the angle between the free surface segment and the x direction.

$$v_n = -\begin{Bmatrix} \sin\vartheta \\ \cos\vartheta \end{Bmatrix}^T \begin{bmatrix} k_x & 0 \\ 0 & k_y \end{bmatrix} \begin{bmatrix} (\partial \underline{b}/\partial x)^T \\ (\partial \underline{b}/\partial y)^T \end{bmatrix}_{FS} \underline{h} \qquad (13)$$

The flux (associated to v_n) crossing the free surface can be seen, using a terminology suited for stress analysis problems, as a "distributed load" on segment FS. This load has to be reduced to zero in order to fulfill the boundary condition (9). This can be done by applying to the nodes of the corresponding element a set of "residual" nodal fluxes $\underline{q}$ that produce through segment FS a flux equal to the one associated to v_n, but with opposite sign.

$$\underline{q} = -\int_{FS} \underline{b}\, v_n\, T\, ds \qquad (14)$$

In eq.(14) T represents the "thickness" of the finite element discretization.

By applying the residual fluxes $\underline{q}$, evaluated for all the elements crossed by the free surface, to the relevant nodes of the mesh and by solving again eq.(11), a new vector of nodal hydraulic heads is determined. This will lead to a different, more refined, approximation of the free surface geometry and to a new residual flow vector. The iterative process is continued until the changes in geometry of the free surface becomes negligible.

When dealing with transient seepage problems [18], the flow velocity normal to the free surface is related to the velocity of the free surface itself by eq.(8b) that, taking into account eq.(8a), can be rewritten in the following form,

$$v_n - m\,\dot{h}\cos\vartheta = 0 \qquad (15)$$

where $\dot{h}$ is the derivative with respect to time of the hydraulic head and ϑ is the angle between the x axis and the free surface.

In order to eliminate the flux through line FS in excess to the one associated to eq.(15), a residual flux vector $\underline{q}$ can be applied to the nodes of the elements crossed by the free surface.

$$\underline{q} = -\int_{FS} \underline{b}\,(v_n - m\,\dot{h}\cos\vartheta)\,T\,ds \qquad (16)$$

Taking into account eqs.(13) and (10), the above equation can be expressed in the following equivalent form.

$$\underline{q} = \{\int_{FS} \underline{b} \begin{Bmatrix}\sin\vartheta\\ \cos\vartheta\end{Bmatrix}^T \begin{bmatrix}k_x & 0\\ 0 & k_y\end{bmatrix} \cdot$$

$$\cdot \begin{bmatrix}(\partial\underline{b}/\partial x)^T\\ (\partial\underline{b}/\partial y)^T\end{bmatrix}_{FS} T\,ds\}\,\underline{h} +$$

$$+ \{\int_{FS} m\,\underline{b}\,\underline{b}^T \cos\vartheta\,T\,ds\}\,\dot{\underline{h}} \qquad (17)$$

Euler backward scheme is convenient for the time integration of the transient problem. The solution for each time step is reached through an iterative procedure analogous to that adopted for steady state analyses.

4.2 Variable permeability procedure

When applied in steady state conditions these procedures replace the intrinsic geometrical non linearity of free surface problems with a non linear relationship between pore pressure and coefficient of permeability [17,24]. A small (close to zero) value of the permeability is assumed when the pore pressure becomes equal to, or decreases below, the atmospheric pressure (conditions that take place on the free surface or above it). The actual geometry of this surface is then evaluated through a procedure conceptually similar to those used to determine the boundary between elastic and plastic zones in non linear stress analyses.

Fig.5 shows a possible idealization of the relationship between coefficient of permeability and pore pressure that can be adopted in the calculations. Note that the permeability for negative pore pressure is slightly greater than zero in order to avoid possible numerical instabilities during solution.

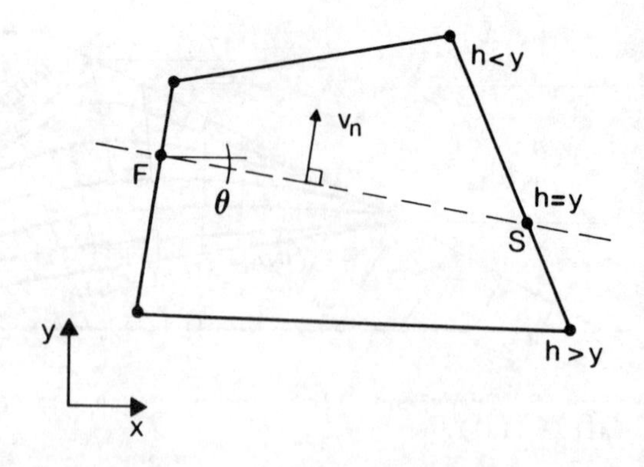

Fig. 4 Approximation of the free surface (dashed line FS) through a finite element.

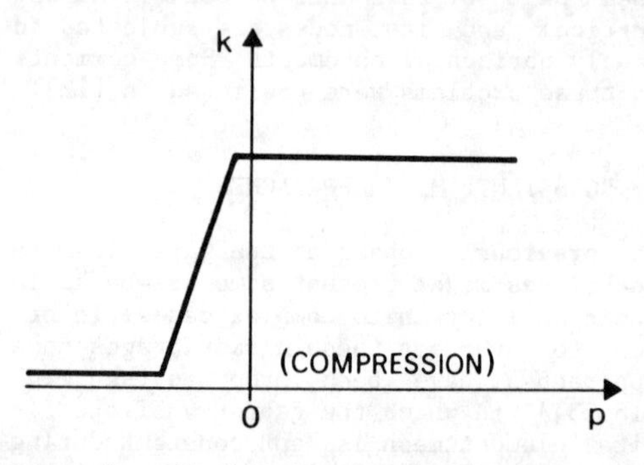

Fig. 5 Idealized relationship between pore pressure p and coefficient of permeability k.

The solution of the non linear material problem can be based on the well known modified Newton scheme. This leads to an iterative procedure in which the assembled flow matrix has to be evaluated and triangularized only once, with a consequent reduction of computational cost.

After determining an initial distribution of nodal hydraulic heads by means of a preliminary confined analysis, a first approximation of the free surface is obtained with the same procedure described in Sec. 4.1 (cf.fig.4). Then, the hydraulic head is evaluated (cf.eq.10) at the points within the elements used in the numerical integration of the flow matrix $\underline{K}$ (cf.eq.12)

The permeability associated to each integration point is evaluated according to the material law schematically shown in fig.5, and the corresponding assembled flow matrix $\underline{K}'$ is worked out.

Then, the variation of the nodal hydraulic heads for the i-th iteration of the solution process is determined by solving again eq.(11),

$$\underline{K}\ \Delta\underline{h}_i = \underline{q} - \underline{q}_i \tag{18}$$

where the nodal fluxes $\underline{q}_i$ are evaluated on the basis of matrix $\underline{K}'$ and of the hydraulic head computed at the end of the previous (i-1) iteration.

$$\underline{q}_i = \underline{K}'_{i-1}\ \underline{h}_{i-1} \tag{19}$$

Then, the hydraulic head vector is updated and another iteration can be performed.

$$\underline{h}_i = \underline{h}_{i-1} + \Delta\underline{h}_i \tag{20}$$

An extension of the above solution technique to the analysis of transient free surface problems has been discussed by Desai and Li [24].

The described constant mesh procedures present an ambiguity, similar to the one affecting the variable mesh methods, in determining the intersection between the free surface and a pervious (wet) boundary. Note, in fact, that since on the surface of seepage the hydraulic head is equal to the elevation y, the scheme shown in fig.2 cannot be adopted to estimate the position of the exit point of the free surface.

This drawback can be eliminated [25] by applying a flow-deflecting zone, represented by a thin layer of elements having a permeability markedly higher than that of the porous material, on the boundary where the seepage face occurs.

5. AN ALTERNATIVE APPROACH FOR TRANSIENT ANALYSES

In addition to the finite element constant mesh approaches, other numerical procedures have been proposed in order to overcome the drawbacks affecting the variable mesh methods. Among them, one based on a mapping technique and on a finite difference integration scheme [26,27,28] will be described here. The compressibility of the soil skeleton, that is taken into account in the mentioned papers, thus leading to the so called "uncoupled" approach for soil consolidation, is not introduced here for simplicity.

This procedure requires a bijective mapping to be introduced between the physical domain Ω (the shape of which is a priori unknown, and varies with time for transient problems) and a fixed integration domain Ω_o (having a conveniently simple shape, e.g. square). In other words, a relationship has to be established such that every point $P_o(\xi,\eta)$ in Ω_o corresponds to a unique point $P(x,y)$ in Ω, and vice versa (cf. fig.6). Due to its simple shape, a finite difference integration scheme can be adopted on the fixed domain. The solution is mapped on the physical domain obtaining the distribution of hydraulic head and the shape of the flow region.

The functions governing the mapping

$$x = x(\xi,\eta,t) \quad ; \quad y = y(\xi,\eta,t) \tag{21a,b}$$

can be arbitrarily chosen, provided that they respect the mentioned bijection. It will be shown subsequently that these functions are numerically determined as part of the solution.

5.1 Transformation of the governing relationship.

The differential equation governing both steady and transient flows is rewritten for convenience in the following form,

$$k_x\, h,_{xx} + k_y\, h,_{yy} = 0 \tag{22}$$

where a comma means partial derivation. The hydraulic head h is a function of the Cartesian coordinates x,y in the physical domain Ω, and of time t in the case of transient phenomena. Note that the time dependency of flow for transient problems is introduced in the final solution by the variation with time of the boundary conditions, and in particular of the position of the free surface.

In order to set up the solution procedure it is necessary to express the x and y partial derivatives in eq.(22) in terms of the coordinates ξ and η of the fixed domain Ω_o. The following relationship is easily established for the first derivatives,

$$\begin{Bmatrix} h,_\xi \\ h,_\eta \end{Bmatrix} = \underline{J} \begin{Bmatrix} h,_x \\ h,_y \end{Bmatrix} = \begin{bmatrix} x,_\xi & y,_\xi \\ x,_\eta & y,_\eta \end{bmatrix} \begin{Bmatrix} h,_x \\ h,_y \end{Bmatrix} \tag{23}$$

that can be inverted leading to

$$\begin{Bmatrix} h,_x \\ h,_y \end{Bmatrix} = \underline{J}^{-1} \begin{Bmatrix} h,_\xi \\ h,_\eta \end{Bmatrix} = \frac{1}{|\underline{J}|} \begin{bmatrix} y,_\eta & -y,_\xi \\ -x,_\eta & x,_\xi \end{bmatrix} \begin{Bmatrix} h,_\xi \\ h,_\eta \end{Bmatrix} \tag{24}$$

In the above equations, $\underline{J}$ is the Jacobian matrix and $|\underline{J}|$ is its determinant.

$$\det \underline{J} = |\underline{J}| = x,_\xi\, y,_\eta - x,_\eta\, y,_\xi \tag{25}$$

As to the second derivatives, the rela-

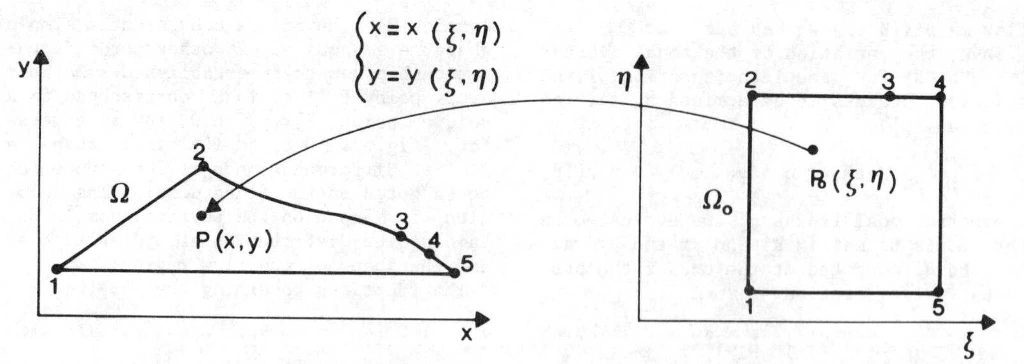

Fig. 6 Bijective mapping between physical domain Ω (cf.fig.1) and fixed domain Ω_o.

tionship between the derivatives in the ξ-η and x-y reference systems can be expressed in the following compact form,

$$\underline{h}''_o = \underline{A}\ \underline{h}'' + \underline{C}\ \underline{h}' \tag{26}$$

where

$\underline{h}''_o = \{h,_{\xi\xi}\ h,_{\xi\eta}\ h,_{\eta\eta}\}^T$

$\underline{h}'' = \{h,_{xx}\ h,_{xy}\ h,_{yy}\}^T$; $\underline{h}' = \{h,_x\ h,_y\}^T$

$\underline{A}$ is a 3x3 matrix,

$A_{11} = (x,_\xi)^2$; $A_{12} = 2x,_\xi\ y,_\xi$

$A_{13} = (y,_\xi)^2$

$A_{21} = x,_\xi\ x,_\eta$; $A_{22} = x,_\xi\ y,_\eta + x,_\eta\ y,_\xi$

$A_{23} = y,_\xi\ y,_\eta$

$A_{31} = (x,_\eta)^2$; $A_{32} = 2x,_\eta\ y,_\eta$

$A_{33} = (y,_\eta)^2$

and $\underline{C}$ is a 3x2 matrix.

$C_{11} = x,_{\xi\xi}$; $C_{12} = y,_{\xi\xi}$; $C_{21} = x,_{\xi\eta}$

$C_{22} = y,_{\xi\eta}$; $C_{31} = x,_{\eta\eta}$; $C_{32} = y,_{\eta\eta}$

By inverting matrix $\underline{A}$ the following equation is easily obtained.

$$\underline{h}'' = \underline{A}^{-1}\ \underline{h}''_o - \underline{A}^{-1}\ \underline{C}\ \underline{h}' \tag{27}$$

The determinant of matrix $\underline{A}$ is related to that of the Jacobian matrix $\underline{J}$,

$$|\underline{A}| = |\underline{J}|^3 = (x,_\xi\ y,_\eta - x,_\eta\ y,_\xi)^3 \tag{28}$$

and the entries A^*_{ij} of the inverted matrix $\underline{A}$ are:

$A^*_{11} = (y,_\eta)^2/|\underline{J}|^2$; $A^*_{12} = -2y,_\xi\ y,_\eta/|\underline{J}|^2$

$A^*_{13} = (y,_\xi)^2/|\underline{J}|^2$; $A^*_{21} = -x,_\eta\ x,_\eta/|\underline{J}|^2$

$A^*_{22} = (x,_\xi\ y,_\eta + x,_\eta\ y,_\xi)/|\underline{J}|^2$

$A^*_{23} = -x,_\xi\ x,_\xi/|\underline{J}|^2$; $A^*_{31} = (x,_\eta)^2/|\underline{J}|^2$

$A^*_{32} = -2x,_\xi\ x,_\eta/|\underline{J}|^2$; $A^*_{33} = (x,_\xi)^2/|\underline{J}|^2$

On the basis of the expressions of the second derivatives of h with respect to x and y obtained from eq.(27), eq.(22) can be re-written as

$$g_{11}\ h,_{\xi\xi} + 2\ g_{12}\ h,_{\xi\eta} + g_{22}\ h,_{\eta\eta} +$$
$$-(g_{11}\ x,_{\xi\xi} + 2\ g_{12}\ x,_{\xi\eta} + g_{22}\ x,_{\eta\eta})\ h,_x +$$
$$-(g_{11}\ y,_{\xi\xi} + 2\ g_{12}\ y,_{\xi\eta} + g_{22}\ y,_{\eta\eta})\ h,_y = 0 \tag{29}$$

where

$g_{11} = [k_x\ (y,_\eta)^2 + k_y\ (x,_\eta)^2]/|\underline{J}|^2$

$g_{12} = -(k_x\ y,_\xi\ y,_\eta + k_y\ x,_\xi\ x,_\eta)/|\underline{J}|^2$

$g_{22} = [k_x\ (y,_\xi)^2 + k_y\ (x,_\xi)^2]/|\underline{J}|^2$

In order to reduce eq.(29) in a form suitable for the solution of the seepage problem it is convenient to choose the mapping functions (21) in such a way that the coefficients of the first derivatives of h with respect to x and y in eq.(29) vanish. By imposing this condition, the following final system of differential equations is arrived at.

$$g_{11}\ h,_{\xi\xi} + 2\ g_{12}\ h,_{\xi\eta} + g_{22}\ h,_{\eta\eta} = 0 \tag{30a}$$

$$g_{11}\ x,_{\xi\xi} + 2\ g_{12}\ x,_{\xi\eta} + g_{22}\ x,_{\eta\eta} = 0 \tag{30b}$$

$$g_{11}\ y,_{\xi\xi} + 2\ g_{12}\ y,_{\xi\eta} + g_{22}\ y,_{\eta\eta} = 0 \tag{30c}$$

It is interesting to observe that eq. (30a), governing the hydraulic head distribution, has the same form of eqs. (30b,c) that govern the bijective mapping between the domains Ω_o and Ω. This leads to a non negligible reduction of the programming effort for the implementation of the solution algorithm. Note also that eqs.(30b,c) are non linear, since the coefficients g_{ij} depend on the derivatives of the unknown functions $x(\xi,\eta)$ and $y(\xi,\eta)$, consequently, the solution process will require some iterative procedure.

5.2 Boundary conditions and integration procedure.

Two sets of boundary conditions have to be taken into account in the solution of system (30). The first set consists of conditions on the mapping functions, thus concerning eqs.(30b,c), and define a one to one correspondence between the points on the boundary of the physical domain Ω and those on the boundary of the fixed domain Ω_o (see fig.6). These are "essential" boundary conditions expressed in the form.

$$\bar{x} = x(\bar{\xi},\bar{\eta}) \quad ; \quad \bar{y} = y(\bar{\xi},\bar{\eta}) \qquad (31a,b)$$

Here the known values of the coordinates are denoted by a superposed bar.

The second set of boundary conditions, the expressions of which in the x-y reference system are given by eqs.(5) to (9), concerns the hydraulic head h and eq. (30a). While the introduction of the "essential" conditions (eqs.5, 6 and 8a) does not present particular difficulties, some care is required by the "natural" boundary conditions (eqs.7, 8b and 9). In fact, they involve the derivatives of the hydraulic head in the x and y directions which have to be expressed as functions of the coordinates ξ and η of the fixed domain.

Consider eq.(7), concerning an impervious boundary coinciding with the x axis. This boundary is represented in the fixed domain by the η axis. On the basis of eq. (24) the boundary condition $v_y=0$ becomes

$$-h,_\xi \, x,_\eta + h,_\eta \, x,_\xi = 0 \qquad (32)$$

In the more general case of a boundary represented by a function $F(x,y)=0$ (cf.eq. 8d), two possible expressions of the flow velocity normal to it can be obtained:

a) if the boundary corresponds to a line η-const.=0 in the fixed domain, the normal velocity can be expressed as

$$v_n = \frac{h,_\eta \, a - h,_\xi \, b}{|\underline{J}| \, c} \qquad (33)$$

where

$$a = k_x \, (y,_\xi)^2 + k_y \, (x,_\xi)^2 \qquad (34a)$$

$$b = k_x \, y,_\xi \, y,_\eta + k_y \, x,_\xi \, x,_\eta \qquad (34b)$$

$$c = -[(x,_\xi)^2 + (y,_\xi)^2]^{1/2} \qquad (34c)$$

b) if the boundary in the fixed domain is represented by a line ξ-const.=0, the normal velocity becomes

$$v_n = \frac{h,_\xi \, a - h,_\eta \, b}{|\underline{J}| \, c} \qquad (35)$$

where

$$a = k_x \, (y,_\eta)^2 + k_y \, (x,_\eta)^2 \qquad (35a)$$

$$c = -[(x,_\eta)^2 + (y,_\eta)^2]^{1/2} \qquad (35b)$$

and the expression (34b) still holds for the constant b.

Since the space integration of the differential eq.(30) has to be performed on the fixed domain, that usually has square shape, a square or rectangular mesh and a simple finite difference scheme can be adopted to this purpose. As previously observed, this scheme has to be cast into an iterative procedure, due to the non linear nature of the problem at hand.

For transient problems the velocity of the free surface is determined on the basis of eq.(8b), in which the components of the discharge velocity are expressed by means of eqs.(2) and (24).

The time integration is based on Crank-Nicholson (implicit) scheme, that requires an iterative process for each time increment Δt. For the first iteration the geometry of the free surface at time $t+\Delta t$ is estimated assuming a constant fluid velocity during the time increment, equal to that at time t. Then a new map is determined by solving eqs.(30b,c). The hydraulic head distribution is evaluated through eq.(30a) considering the free surface as a pervious boundary. On this basis, a new approximation of the fluid velocity field and a new position of the free surface at time $t+\Delta t$ are determined. The iterations for the current time increment end when further changes of the free surface geometry becomes negligible.

For steady state problems, after assuming a first trial position of the free surface, eqs.(30b,c) are solved and the hydraulic head distribution is determined, through eq.(30a), by considering the free surface as an impervious boundary. Then the free surface nodes are moved so that their elevation becomes equal to the corresponding hydraulic head previously computed, and a new iteration is carried out. The technique adopted for moving the free surface nodes is equivalent to that depicted in fig.2. Details on the implementation of the solution procedure can be found in [26].

An illustrative example [28] solved with this technique is shown in fig.7, concerning the analysis of the lowering of the water table caused by tubular drains situated at a given depth below the ground surface. In fig.7 Ω is the domain within which the seeage flow is analysed, Γ_1 is an equipotential boundary coinciding with the drain contour, Γ_2 are impervious boundaries and Γ_3 represents the free surface. The shapes of the free surface, and the corresponding meshes, obtained for some values of the non dimensional time T are shown in fig.8.

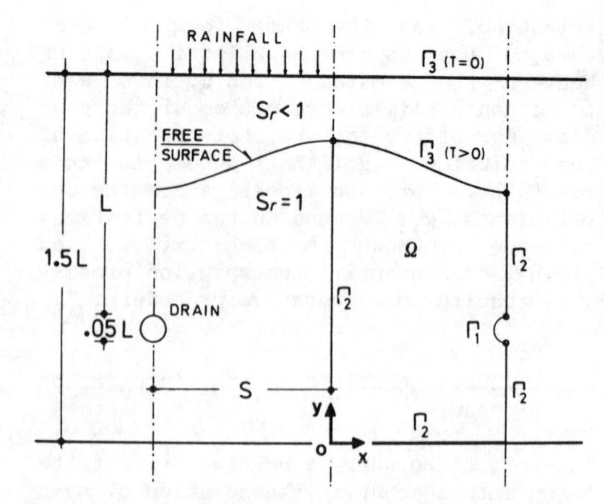

Fig. 7 Scheme for the analysis of seepage toward tubular drains (after [28]).

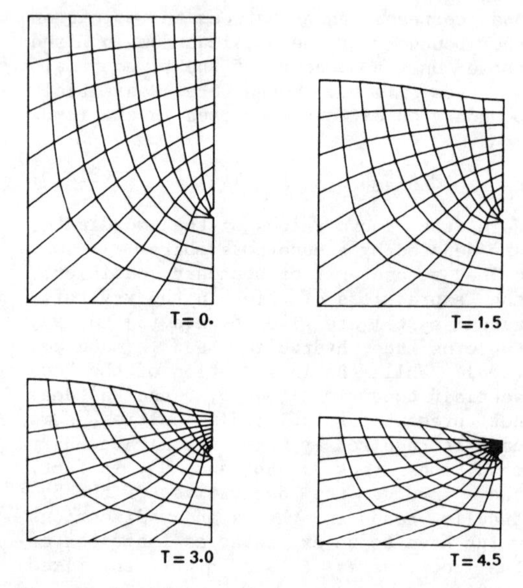

Fig. 8 Variation of the free surface with the non dimensional time T.

6. POSITION OF THE "EXIT" POINT OF THE FREE SURFACE

It has been previously observed that most of the techniques for unconfined analysis require some specific provision when the intersection between the free surface and the seepage face (i.e the "exit" point of the free surface) has to be determined. In some cases stability problems show up during the iterative solution process, that might lead to an apparent non uniqueness of solution.

In order to reach a better understanding, from the engineering view point, of the causes of instability a study was carried out [29] in which a solution technique based on a non-linear programming algorithm for function minimization was adopted.

The solution of the steady state problem is reached through a sequence of confined analyses where the free surface is treated as a pervious boundary (eq.8a). The total flux leaving or entering the free surface at its nodes is evaluated for each analysis. The minimization algorithm permits modifying the geometry of the free surface until this flux is reduced to zero, and hence also eq.(9) is fulfilled.

Note that the first condition (eq.8a) involves all the nodes belonging to the free surface. On the contrary, the second condition (eq.9) cannot be applied to the free surface nodes that belong also to pervious boundaries, since the flux at these nodes will not be in general zero.

In order to limit the number of problem variables, and to achieve a better control on the geometrical changes of the mesh, it is convenient to chose a priori the direction of shifting for each node and to relate the movement of the generic, say j-th, node to a scalar parameter α_j.

Following the above procedure, the solution of the unconfined problem should be reached through the unconstrained minimization of the following function Q

$$\min_{\underline{\alpha}} \left[Q(\underline{\alpha}) = \{\sum_{1}^{n} q_i^2(\underline{x})\}^{1/2}\right] \qquad (37)$$

where: $\underline{x} = \underline{x}(\underline{\alpha})$ and $\underline{\alpha} = \{\alpha_1\ \alpha_2\ \dots\ \alpha_m\}^T$. In eq.(37) $\underline{x}$ is the vector of coordinates of the nodes belonging to the free surface and q_i is the fluid flux at the i-th free surface node. This flux is computed on the basis of the hydraulic head vector obtained by the confined solution for the current geometry.

It can be observed that the number m of the free variables α_i in the above equation (i.e. the number of nodes the position of which is unknown) is greater than the number n of nodes in which the total flux is minimized. In fact, the position of exit node E of the free surface is a priori unknown, hence the corresponding parameter α_E is one of the problem unknowns, but the same node has also a non vanishing total flux, and consequently its flux q_E cannot be considered in the summation in eq.(37).

This discrepancy between m and n should not represent a problem. In fact, even though q_E is not among the terms of the summation, the objective function $Q(\underline{\alpha})$ does depend on the free parameter α_E. This can be seen considering that the position of the exit node has a marked influence on the flux of the neighbour free surface nodes, which do appear in the summation.

Various mathematical programming algorithms are suitable for the minimization in eq.(37), ranging from relatively simple direct search procedures to rather sophisticated quasi-Newton methods. Details on thes algorithms can be found, for instance, in [30].

The above solution procedure was adopted for some numerical tests concerning the square block dam shown in fig.9. The results of these tests clearly show that the function $Q(\underline{\alpha})$ does not present an unique minimum. In fact, infinite different vectors $\underline{\alpha}$ exist (each one corresponding to a different shape of the free surface) that bring to zero this function. Each solution is represented by a sort of wavy distortion, with arbitrary amplitude, of the correct free surface. The same tests show also that a unique minimum exists when the position of one of the free surface nodes is arbitrarily chosen. In fig.9 some of the solutions are presented which were obtained by a priori defining the position of the exit node of the free surface.

On these bases it seems reasonable to conclude that, in the context of the variable mesh methods, the minimization of the nodal flux entering or leaving the discrete model through the nodes of the free surface (except those belonging also to pervious boundaries) cannot uniquely define the correct shape of this surface, unless the position of one of its nodes (in addition to the "entrance" node) is already known. Consequently, some modification of the solution technique has to be introduced.

To this purpose, consider the fluid flux

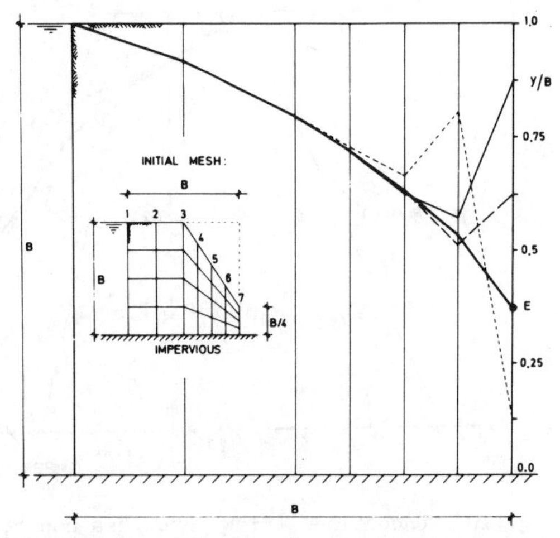

Fig. 9 Unconfined flow through a square block dam. Variation of the free surface shape with the assumed position of its "exit" node E. A heavy line represents the "optimal solution" (after [29]).

"tangent" to the free surface or, in other words, the distribution of nodal fluxes leaving and entering the elements facing the free surface at their nodes located on the free surface itself. The numerical results indicate that a non correct geometry of the free surface is associated to an irregular variation of the numerically evaluated flux tangent to the free surface. It has to be considered, however, that in the exact solution of unconfined seepage problems the stream lines are almost parallel to the free surface in its vicinity. This should lead in the discretized model to a regular or "smooth" variation of the nodal fluxes entering and leaving the elements facing the free surface.

The above observation suggest modifying the solution technique by introducing a "regularity" condition for the distribution of the free surface fluxes. This can be done by adding to the objective function in eq.(37) a term that tends to zero if a constant flux takes place through the free surface nodes.

$$\min_{\underline{\alpha}} \left(Q(\underline{\alpha}) = a_1 \{\sum_1^n {}_i\, q_i^2(\underline{x})\}^{1/2} + \right.$$

$$\left. + a_2 \{\sum_1^k {}_i\, [q_{i1}(\underline{x}) - q_{i2}(\underline{x})]^2\}^{1/2} \right) \quad (38)$$

In the above equation, k is the number of elements facing the free surface and q_{i1} and q_{i2} represent the fluxes computed at the two nodes of the i-th element belonging to the free surface.

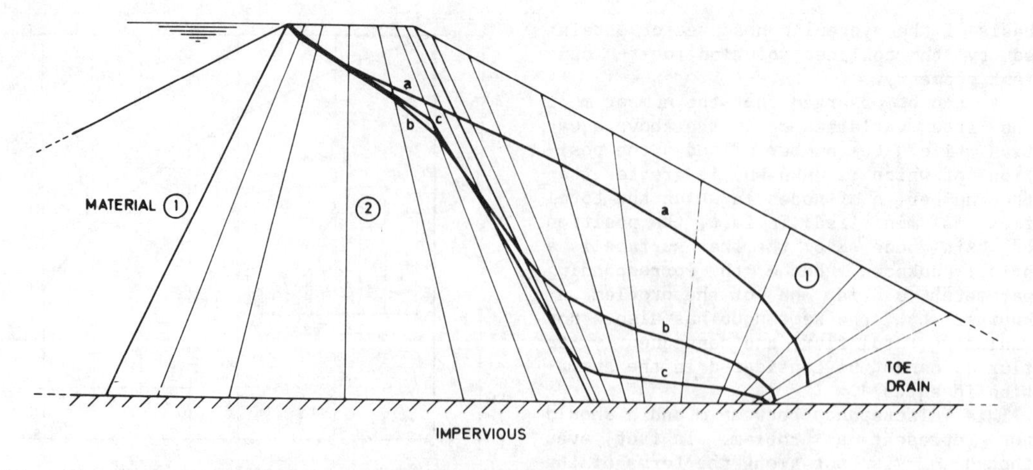

Fig. 10 Unconfined flow through a non homogeneous dam. Influence of the ratio between the coefficients of permeability of zones 1 and 2 on the shape of the free surface. $k_2/k_1 = 1/1$ (a); 1/10 (b); 1/100 (c) (after [29]).

The coefficients a_1 and a_2 were introduced in order to improve the rate of convergence. In fact, approaching the minimum, the first term of the objective function (i.e. the one already present in eq. 37) tends to zero, while the second term tends to a positive (small) value. This difference could reduce the rate of convergence of the solution algorithm during the final stages of the minimization process (i.e. in the vicinity of the optimal point), specially when coarse meshes are adopted. The performed numerical tests show that this negative effect can be eliminated by using values of a_1 slightly greater than those of a_2 (e.g. $a_1/a_2 \simeq 2$).

It has to be observed that a conceptual difference exists between the two parts of the objective function in eq.(38). The first term depends directly on one of the boundary condition of the continuous problem. The "regularity" condition expressed by the second term, however, depends only on the discrete model for numerical solution, and no direct relationship can be found between it and the conditions for the continuous problem. In fact, this additional term has the mere purpose of eliminating the approximation or indetermination (which is not present in the continuous problem) caused by the reduction to discrete terms of the boundary condition concerning the fluid flux through the free surface.

The modified objective function was used for solving the already mentioned square dam problem obtaining the optimal shape of the free surface represented by heavy solid lines in fig.9. Another example is shown in fig.10, concerning a non homogeneous trapezoidal dam with a toe drain resting on an impervious base.

These results indicate that a possible way to overcome the numerical instabilities, often met when using variable mesh approaches, consists in introducing a condition of global nature, i.e. a sort of regularity condition for the entire free surface. This condition is not similar to the local ones, sometimes adopted in order to stabilize otherwise unstable approaches, that in most cases concern the exit node only or the shape of the free surface in the vicinity of this node.

7. CONCLUDING REMARKS

Some finite element approaches for unconfined seepage analysis have been recalled, considering both constant mesh (C.M.) and variable mesh (V.M.) procedures. An alternative approach has been also discussed [27], which can be seen as a compromise between the techniques of the above two classes. In this case, in fact, the governing equations are integrated on a fixed domain (having a known, simple geometry) and the shape of the physical domain in which the seepage flow takes place is determined, by means of a mapping technique, as a part of the solution.

Based on the various studies mentioned in the previous sections, some comments on the finite element analysis of unconfined seepage problems can be made.

From the programming view point the C.M. methods present some advantages with respect to the V.M. procedures. In fact, they are basically similar to the methods commonly used for solving elasto-plastic problems and, consequently, their implementation can be based on already available finite element codes for non linear stress analysis (see e.g. [17]). This leads to a non negligible reduction of the programming effort with respect to V.M. procedures that usually require the preparation of hoc codes. In addition, some programming difficulties could be caused in the V.M. methods by the procedure for shifting the free surface nodes and by the provisions to be introduced when dealing with layered or non homogeneous media.

As to the accuracy of results, it appears that V.M. methods are able to define the shape of the free surface with a precision higher than that of the C.M. methods, when both are applied to the same finite element grid. On the other hand, it has to be considered that V.M. approaches are in general affected by stability and convergence problems that are less pronounced in the C.M. procedures.

For steady state analyses these stability problems can be particularly evident if the initial trial shape of the free surface is markedly different from the final solution. A similar drawback shows up also in transient analyses if the adopted value of the time increments is too large, i.e. when a marked difference exists between the shapes of the free surface at two subsequent solution steps.

Another case in which V.M. methods could present convergence problems, that in some instances lead to an apparent non uniqueness of solution, is when the point of intersection between the free surface and a pervious boundary has to be determined. A possible way to overcome this difficulty consists in introducing a sort of regularity condition for the overall shape of the free surface [29]. This is apparently more effective than to enforce local conditions directly on the intersection point or on the elements in its vicinity.

It has to be observed, however, that this provision has been applied only in the framework of a particular V.M. technique based on a mathematical programming algorithm for function minimization. Consequently its introduction in standard methods for unconfined analysis is not straightforward. Perhaps some further study on this point could lead to provisions having more general validity, and applicable to solution methods of different nature.

From the computational view point the V.M. approaches are apparently heavier than C.M. methods. This is due to the fact that V.M. methods modify the mesh during the solution process and, consequently, re-assemble at least a part of the flow matrix and triangularize it at each iteration. This is not required by C.M. methods when solution procedures of the modified Newton type are used.

In choosing between the various methods of analysis one should also consider that the V.M. methods, for their very nature, are restricted only to the solution of seepage flow problems. On the contrary, C.M. methods can be also used in coupled analyses [22,31], like for instance when the influence of the seepage flow on the deformation of the soil skeleton cannot be neglected and a consolidation analysis has to be performed.

From the above considerations its seems possible to conclude that C.M. methods are in general more "robust", and perhaps more "flexible", than V.M. techniques. They are particularly suitable for unconfined seepage analyses when a reasonable approximation of the free surface, and of the corresponding hydraulic head distribution, is required with a limited computational effort. If more refined results are needed, then a variable mesh analysis could be necessary. Some of the mentioned stability problems can be avoided in this case if the input data of the variable mesh analysis are defined on the basis of the geometry of the flow region obtained from the previous constant mesh calculations.

ACKNOWLEDGEMENTS

The financial support of the National Research Council (CNR) and of the Ministry of Education (MPI) of the Italian Government is gratefully acknowledged.

REFERENCES

[1] H.R. Cedergren, "Seepage, drainage and flow nets", J.Wiley & Sons, New York, 1967

[2] O.C. Zienkiewicz, P. Mayer, Y.K. Cheung, "Solution of anisotropic seepage by finite elements", J.of Engineering Mechanics Div., ASCE, Vol.92, No.EM1, pp.111-120 (1966)

[3] P.W. France, "Finite element analysis of three-dimensional groundwater flow problems", J.of Hydrology, Vol.21, pp.381-398 (1974)

[4] C.S. Desai, "Flow through porous media", in Numerical Methods in Geotechnical Engineering, C.S. Desai and J.T. Christian (edts.), Mcgraw-Hill, New York, 1977

[5] J.A. Liggett, "Location of free surface in porous media", J.of Hydraulic Div., ASCE, 13, HY4, 353-365 (1977)

[6] P.K. Banerjee,R. Butterfield, G.R. Tomlin, "Boundary element methods for two-dimensional problems of transient ground water flow", Int.J.Numer.Anal. Methods Geomechanics, Vol.5, pp.15-31 (1981)
[7] J.A. Liggett, P.L.F. Liu, "The boundary integral equation method for porous media flow", George Allen & Unwin Publishers, London, 1983
[8] Z.K. Lu, C.A. Brebbia, R.A. Adey, "Calculation of free surface seepage through zoned anisotropic dams", Proc.7th Boundary Element Conference, Springer-Verlag, Berlin, 1985
[9] C.S. Desai, "Seepage analysis of earth banks under drawdown", J.Soil Mechanics and Foundation Div., ASCE, Vol.98, No. SM11, pp.1143-1162 (1972)
[10] R.T.S. Chen, C.Y. Li, "On the solution of transient free-surface flow problems in porous media by the finite element method", J.of Hydrology, Vol.20, pp.49-63 (1973)
[11] R.L. Taylor, C.B. Brown, "Darcy flow solution with a free surface", J.of Hydraulic Div., ASCE, Vol.93, No.HY2, pp.25-33 (1967)
[12] S.P. Neuman, P.A.Witherspoon, "Finite element method of analyzing steady seepage with a free surface", Water Resources Research, Vol.6, No.3, pp.889-897 (1970)
[13] S.P. Neuman, P.A.Witherspoon, "Analysis of nonsteady flow with a free surface using the finite element method", Water Resources Research Vol.7, No.3, pp.611-623 (1971)
[14] C.J. Taylor, P.W. France, O.C. Zienkiewicz, "Some free surface transient flow problems of seepage and irrotational flow", in The Mathematics of Finite Elements and Applications, J.R. Whiteman (edt.), Academic Press, London (1973)
[15] L.T. Isaacs, "Location of seepage free surface in finite element analyses", Civil Engineering Transaction, Australia, CE-22, No.1, pp.9-16, 1980
[16] C.S. Desai, "Finite element residual schemes for unconfined flow", Int.J. Numer.Methods Eng., Vol.10, pp.1415-1418 (1976)
[17] K.J. Bathe, M.R.Khoshgoftaar, "Finite element free surface seepage analysis without mesh iteration", Int.J.Numer. Anal.Methods in Geomechanics, Vol.3, No.13, pp.13-22 (1979)
[18] K.J. Bathe, V. Sonnad, P. Domigan, "Some experiences using finite element methods for fluid flow problems" Proc.of 4th Int.Conf.on Finite Element Methods in Water Resources, Hannover, 1982
[19] C.S. Desai, G.C. Li, "Transient free surface flow through porous media using a residual procedure", Proc.4th Int.Conf.on Finite Elements in Fluids Tokyo, 1982
[20] A. Cividini, G. Gioda, "An approximate F.E. analysis of seepage with a free surface", Int.J.Numer.Anal. Methods Geomechanics, Vol.8, pp.549-566 (1984)
[21] D.R. Westbrook, "Analysis of inequality and residual flow procedures and an iterative scheme for free surface seepage", Int.J.Numer.Meth.in Eng., 21, No.10, pp.1791-1802 (1985)
[22] S.J. Lacy, "Numerical procedures for nonlinear transient analysis of two-phase soil systems", PhD. Dissertation, Department of Civil Engineering, Princeton University, 1986
[23] J.H. Prevost, S.J.Lacy, "Flow through porous media: locating the free surface", Proc.1st World Conf.on Computational Mechanics, Austin, 1986
[24] C.S. Desai, G.C. Li, "A residual flow procedure and application for free surface flow in porous media", Int.J. Advances in Water Resources, Vol.6, pp.27-35 (1983)
[25] E.N. Bromhead E.N., Discussion of "Finite element scheme for unconfined flow", by C.S. Desai, Int.J.Numer. Methods Engng., Vol.11,No.5, pp.908-910 (1977)
[26] A. Desideri, "A mathematical model for the analysis of free surface seepage flow" (in italian), Technical Report, Department of Structural and Geotechnical Engineering, University of Rome, 1983
[27] A. Burghignoli, A. Desideri, "Free surface flow in soils", Proc.of 5th Int.Conf.on Numerical Methods in Geomechanics, Nagoya, 1985
[28] A. Burghignoli, A. Desideri, "On the effectiveness of tubular drains", Proc.9th European Conf. on Soil.Mech. and Found.Engineering, Dublin, 1987.
[29] G. Gioda, C. Gentile, "A nonlinear programming analysis of unconfined steady state seepage", Int.J.Numer. Anal.Methods Geomechanics, Vol.11, pp.283-305 (1987)
[30] D.M. Himmelblau, "Applied non linear programming", Mc-Graw-Hill, New York, 1972
[31] G.C. Li, "Free surface flow and stress analysis of earth dams", PhD. Dissertation, Department of Civil Engineering, Virginia Tech., 1981

Numerical Methods in Geomechanics (Innsbruck 1988), Swoboda (ed.)
© 1988 Balkema, Rotterdam. ISBN 90 6191 809 X

Numerical methods versus statistical safety in geomechanics

G.Gudehus
University of Karlsruhe, FR Germany

ABSTRACT: The probabilistic concept yields a principally correct measure of safety. Some consequences for numerical methods are outlined by a series of examples related with shallow foundations and retaining structures. Errors due to systematic faults of theory and data are elucidated versus the influence of scattering of soil properties. It is shown that in some cases conventional models are sufficient in the light of scattering. More often there is a need of sophisticated models as a basis of simplified ones. Principal limitations of the present mechanical and statistical models are outlined. Observational methods appear to be a promising issue.

1 INTRODUCTION

Safety and economy are the general goals of engineering. Whereas a principally trivial measure of economy is at hand (dollars within a certain time, say), measures of safety are rather doubtful. It has been realized that conventional factors of safety do not represent measures; more precisely, the requirements of measure theory - unit, additivity, equivalence - are not satisfied. The probability of failure, p_f, is a correct measure, and likewise the risk $R = Cp_f$ wherein C denotes the costs resulting from failure. In spite of the principal objections against p_f (lack of data) and C (damage is not fully quantifiable) the probabilistic concept is a powerful aid towards rational and balanced safety assessments. Simplifications are at hand and frequently acceptable: replacement of p_f by the safety index β, or use of a set of partial safety factors.

It has been shown that this concept is useful in geotechnical engineering. It appears that safety regulations can thus be unified and even simplified, also in the international plane (such as Eurocode and ISO). Starting with load assumptions, structural engineering has gone rather far on this way. Geotechnical engineers hesitate to follow for two main reasons: mechanical soil models will never be fully correct, and the statistical data basis for soil will never be complete.

The present paper is an attempt to elucidate the role of numerical methods in this play. Examples related with shallow foundations and retaining structures serve to briefly introduce the relative importance of systematic errors, data scattering and finite element methods (FEM): It is shown that FEM can, if certain requirements are satisfied, serve as powerful tool for establishing simplified - and practically more adequate - safety assessments. The statistical concept helps to improve the balance of experimental and numerical efforts.

Some inevitable shortcomings have to be outlined. Systematic errors of mechanical models and soil experiments cannot strictly be measured. On the other hand, the statistical data basis has always to be extended by some kind of judgment, hopefully improved in the sense of Bayes' theorem. The cry for simple methods is justified not only by financial arguments, but also by the additional risk due to less transparent procedures. Observational methods could help to overcome the dilemma.

2 SHALLOW FOUNDATIONS

2.1 Sliding upon horizontal base

A block placed on a hard stratum (Fig.2.1a) may serve to briefly introduce the statistical safety concept (Gudehus 1987). Sliding failure will occur if the limit state inequality

$$f := N\mu - H \leq 0 \tag{2.1}$$

holds, wherein μ denotes the base friction coefficient. The scattering quantities μ and H may be normally distributed with mean values m_μ and m_H, and variation coefficients V_μ and V_H. The probability of failure

$$p_f := P\,[f \leq 0] \tag{2.2}$$

is determined from the safety index

$$\beta := \frac{m_f}{\sigma_f} = \frac{Nm_\mu - m_H}{\sigma_\mu^2 + \sigma_H^2} \tag{2.3}$$

by

$$p_f = \Phi^{-1}(-\beta) \; , \tag{2.4}$$

wherein σ denotes variance, and Φ Gauss' error function. A prescribed β is approximately obtained by satisfying the limit condition f = 0 with the design values

$$\mu_d = m_\mu \gamma_\mu \; ; \; \mu_H = m_H \gamma_H \tag{2.5}$$

with the partial safety factors

$$\gamma_\mu = 1 - \beta V_\mu / \sqrt{2} \; ; \; \gamma_H = 1 + \beta V_H / \sqrt{2} \; . \tag{2.6}$$

In case of log-normal distributions, Equ. 2.6 is replaced by

$$\gamma_\mu = \exp(-\beta V_\mu / \sqrt{2}) ; \; \gamma_H = \exp(+\beta V_H / \sqrt{2}) . \tag{2.7}$$

As the mechanical model, Equ. 2.1, is well-established, our comment is focussed on statistical aspects. The choice of the

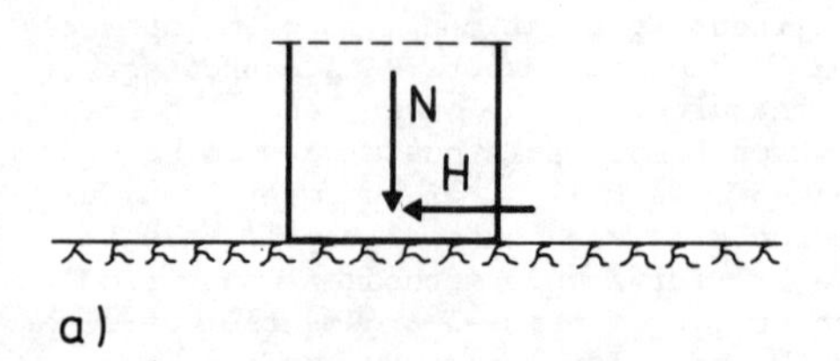

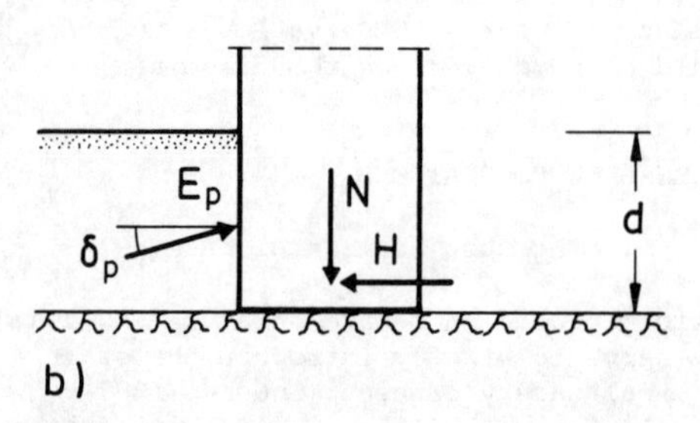

Figure 2.1 Sliding blocks

probability distribution cannot be justified by experimental data only. The Central Limit Theorem would favour normal distributions. Negative values of μ would be avoided by the log-normal approach. An extreme-value distribution could be more adequate for H (Ang & Tang 1975).

With the usual size of β and variation coefficients the difference of distribution functions is less important than the uncertainty of mean values and variances. Apart from systematic measurement errors, a kind of educated guess is inevitable. The statistical concept is not disproved, however: it provides a balanced, principally objective safety assessment.

With an embedded foundation (Fig. 2.1b), the limit state condition has to imply passive earth pressure:

$$f := N_\mu + \frac{1}{2}\gamma d^2 K_p \cos\delta_p - H \leq 0. \tag{2.8}$$

Taking K_p from limit equilibrium theory, this model is uncertain because of the unknown inclination δ_p. A stochastic uncertainty is added as the angle of soil friction, ϕ, will scatter. $f \leq 0$ is avoided with a prescribed p_f, or β via Equ. 2.4, if $f \geq 0$ is satisfied with the partial safety factors

$$\gamma_i \simeq \exp(\pm\, \beta \alpha_i V_i) \tag{2.9}$$

with the weight factors

$$\alpha_i = \frac{|\sigma_i \partial f / \partial \chi_i|}{\sqrt{\Sigma(\sigma_i \partial f / \partial \chi_i)^2}} \tag{2.10}$$

In Equ. 2.9 + holds for $\partial f/\partial \chi_i < 0$ (action) and - for $\partial f/\partial \chi_i > 0$ (resistance).

In the present case, the stochastic variables are $\chi_1 = \mu$, $\chi_2 = \phi$, $\chi_3 = H$. Assume that α_ϕ exceeds about 0.1 (found from Equs. 2.8 and 2.10 with estimated variances); otherwise the scattering of ϕ could be neglected here. Only then δ_p deserves further inspection. It should be possible to estimate δ_p, and also K_p, with the aid of FEM such as proposed by Nakai (1985). FEM would not serve then for immediate practical application but for improving, or better supporting, a simplified model.

2.2 Compression of a soft layer

Consider a shallow foundation safely embedded into sand or gravel with a relatively soft thin inclusion (Fig. 2.2a). Neglecting sand deformation, the requirement of limited settlement can be written

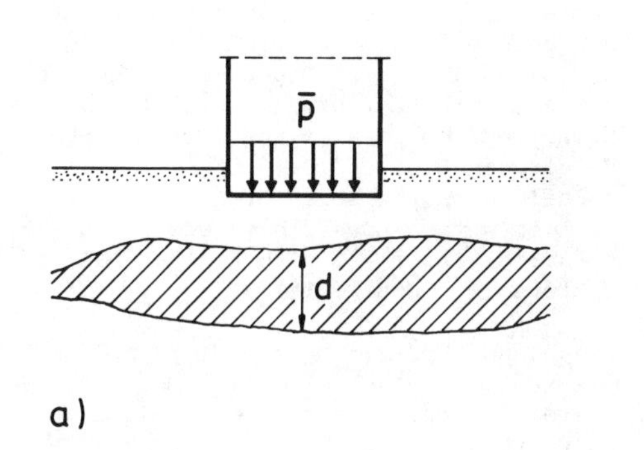

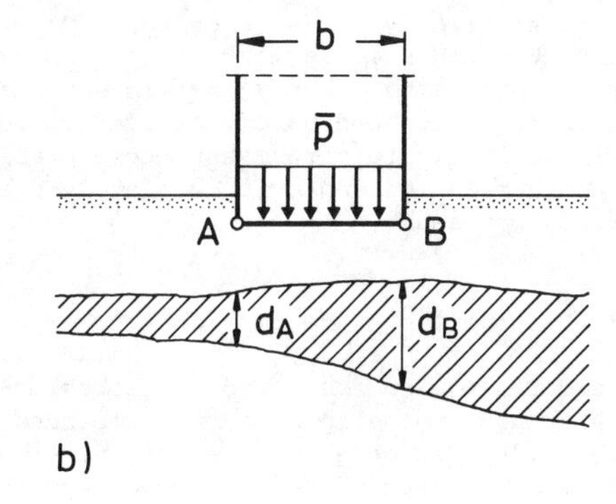

Figure 2.2 Foundations above compressible layer

as

$$f := s_a - \bar{p} i_\sigma d/E_s > 0 \qquad (2.11)$$

wherein s_a = allowable settlement, $\bar{p}$ = mean base pressure, i_σ = stress influence factor, d = layer thickness, E_s = compression modulus. Equ. 2.11 is a condition of serviceability. Taking d and E_s as log-normally distributed with mean m and variance σ, the probability of $f \leq 0$ is limited formally as in Sec. 2.1, by satisfying Equ. 2.11 with the design values

$$d_d = m_d \exp(\beta V_d/\sqrt{2}); \; E_{sd} = m_E \exp(-\beta V_E/\sqrt{2}). \qquad (2.12)$$

The mechanical model of Equ. 2.11 implies systematic errors: the influence of structure and soil stiffness on the stress in the soft layer is neglected, and the layer is not really deformed as in an oedometer. The relative model error referred to the exact settlement in case of non-scattering soil data may reach between 0.1 and 0.2 under favourable conditions. Taking a conventional $\beta \simeq 3$ and realistic variation coefficients $V_d \simeq 0.2$ and $V_E \simeq 0.3$, Equ. 2.12 yields a factor

$$\exp[\beta(V_d + V_E)/\sqrt{2}] \simeq \exp[3(0.2+0.3)/\sqrt{2}] \simeq 2.9;$$

this has to be multiplied with the calculated mean settlement, $\bar{p} i_\sigma d/E_s$, for satisfying Equ. 2.11 safely enough.

As the influence of scattering by far exceeds the model error, the latter need not be reduced in this case. The role of FEM could be, however, to delimit the allowable range of the simplified Equ. 2.11. Note that layer thickness and stiffness can also contain systematic errors (sampling, testing etc.) possibly exceeding the model error; this point cannot be further discussed here.

If the tilt of a building on similar ground is limited to a certain amount, ψ_a say, the serviceability limit condition reads

$$f := \psi_a + [i_{\sigma A} d_A/E_{sA} - i_{\sigma B} d_B/E_{sB}]\bar{p}/b > 0 \qquad (2.13)$$

with the respective values under points A and B. It is reasonable to assume the same variation coefficients V_d and V_E under A and B. Application of Equs. 2.9 and 2.10 leads to the partial safety factors

$$\gamma_{dA} = \exp\left[\beta V_d/\sqrt{2(1+\kappa^2)}\right];$$
$$\gamma_{dB} = \exp\left[-\beta V_d/\sqrt{2(1+1/\kappa^2)}\right], \qquad (2.14)$$

$$\gamma_{EA} = \exp\left[-\beta V_E/\sqrt{2(1+\kappa^2)}\right];$$
$$\gamma_{EB} = \exp\left[-\beta V_E/\sqrt{2(1+1/\kappa^2)}\right] \qquad (2.15)$$

with

$$\kappa := \frac{i_B m_{dB} V_d m_{EA}}{i_A m_{dA} V_A m_{EB}}.$$

Reminding the mechanical and statistical uncertainty one can simplify into

$$\gamma_d = \exp(\beta V_d/2); \; \gamma_E = \exp(-\beta V_E/2). \qquad (2.16)$$

Equ. 2.16 supports and renders more precise the conventional procedure. Again, the mechanical model is sufficient and could be delimited by FEM-calculations.

The situation is more delicate if bending moments of the structure are involved. The conventional model (Kany 1974) implies second order differences of settlements and flexural stiffness of the structure; these quantities are mechanically and statistically rather uncertain. Comparative FEM and stochastic calculations are not yet promising as they contain systematic errors which can hardly be overlooked. It is advisible to circumvent these problems by another design and/or a completely different approach.

2.3 Foundations on sand or gravel

Consider a square foundation on cohesionless ground, for simplicity not embedded (Fig. 2.3a). In case of a centric vertical load limit states of bearing capacity are described by

$$f := \frac{1}{2}\gamma N_\gamma s_\gamma b^3 - P \leq 0 \qquad (2.17)$$

Taking the shape factor s_γ as an empirical constant and neglecting variations of specific wight γ, one can consider only N_γ and P as stochastic variables. P may be normal with given m_P and V_P.

The bearing capactiy factor N_γ , depending on friction angle ϕ, is uncertain in a deterministic and a stochastic sense. Assuming ϕ as constant, N_γ is strongly influenced by the kinematic freedom of the foundation (Fig. 2.3b). The deterministic influence of further ground properties - such as dilatancy and pressure dependance of ϕ - should and could be clarified by FEM. The mechanical uncertainty seems to be between about 20 % and 50 %.

There is a spatial scattering of ϕ, possibly with preferred wave lengths (Fig. 2.3c). N_γ depends on a weighted average of ϕ, $\bar{\phi}$ say. An ensemble of statistically equivalent ground situations could have a log-normal distribution with mean $m_{\bar{\phi}}$ and variation $V_{\bar{\phi}}$. (Both parameters are partly subjective due to restricted sampling.) $V_{\bar{\phi}}$ is possibly size-dependent: wider foundation can better bridge spatial fluctuations, but could also reach wider waves. $V_{\bar{\phi}}$ between 0.05 and 0.15 is reported.

Failure in the sense of Equ. 2.17 is avoided with a prescribed safety index β by satisfying f = 0 with the partial safety factors (cf. Equ. 2.12)

$$\gamma_P \simeq \exp(\beta V_P/\sqrt{2}); \ \gamma_N \simeq \exp(-\beta V_N/\sqrt{2}),$$

wherein γ_N and V_N refer to N_γ. V_N depends on $V_{\bar{\phi}}$ via

$$V_N \simeq \frac{\partial N_\gamma}{\partial \phi}\frac{m_{\bar{\phi}}}{m_N} V_\phi . \qquad (2.18)$$

Typical figures are:

$m_{\bar{\phi}} = 35^\circ$, $m_N = 50$, $\partial N_\gamma/\partial\phi = 6/1^\circ$,
i.e. with $V_\phi = 0.1$ $\quad V_N = \frac{6}{1^\circ}\frac{35^\circ}{50}\,0.1 \simeq 0.4$,
and with $\beta = 3$ (say)

$$\gamma_N = \exp(-3 \cdot 0.4/\ 2) \simeq 0.4.$$

One can conclude that it is recommendable to

- care for a rather precise mechanical model,
- avoid systematic errors when determining m_ϕ,
- enable objective and subjective reduction of $V_{\bar{\phi}}$.

Note that the present standards, e.g. $\gamma_P/\gamma_N = 2$ in the German DIN 4017, do not differentiate as much so that the design can be too far on the safe or unsafe side.

The foundation width is more often determined by an allowable settlement, s_a say.

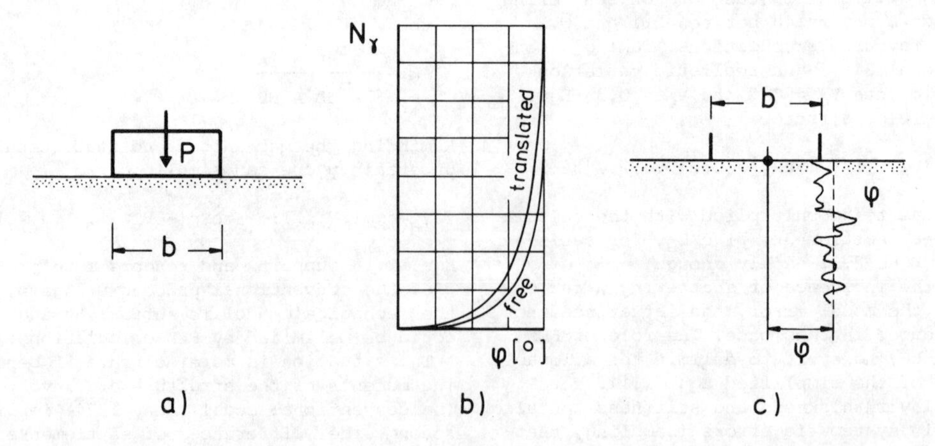

Figure 2.3 Square-foundation on sand (a), bearing capactiy factor (b), spatial scattering of friction angle (c)

Using Hettler's (1985) formula for the load-settlement relationship (cf. Fig. 2.4a), the condition of serviceability can be written as

$$f:= s_a - bC_e\kappa\left(\frac{P}{\gamma b^3}\right)^{\mu} > 0 . \qquad (2.19)$$

Herein, the factor κ and the exponent μ are empirical constants; typical values are $\kappa = 0.05$ and $\mu = 1.6$. The factor C_e depends on void ratio (Fig. 2.4b).

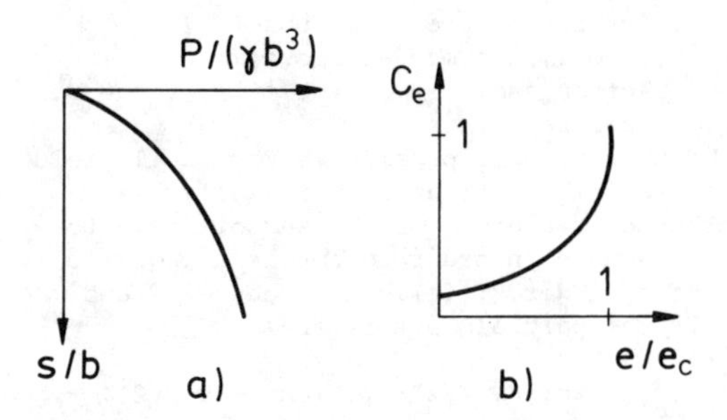

Figure 2.4 Load-settlement relationship(a), void ratio correction factor (b) with e_c = critical void ratio

We first consider systematic model errors. The influence of pressure level, not implied by Equ. 2.19, can be covered by a factor (Hettler & Gudehus 1985). A linear approach could be sufficient (Holzlöhner 1984). FEM-calculations can help to select an adequate simplified model if the following requirements are satisfied,

- constitutive relation corroborated by numerical and physical element tests,
- numerical procedure sufficiently free from discretization and arithmetic errors.

As the constitutive relation will scarely be realistic and feasible at the same time, a further support by model tests - with uniform density - will be indispensablw.

Coming to the stochastic side, with the aid of Equ. 2.19, we have to face variations of the void ratio factor C_e. As with N_γ above, C_e refers to the average within a certain soil volume. Hettler (1985) reports typical variations up to $V_c \simeq 0.3$. Following Equ. 2.7 this would lead to a safety factor

$$\gamma_c \simeq \exp(\beta V_c/\sqrt{2}) \simeq \exp(3\cdot 0.3/\sqrt{2}) \simeq 1.9$$

for C_e. This correction has about the same order of magnitude as the required model correction.

If tilting angles and/or bending of the structure are relevant the situation is much more difficult. The mechanical and statistical models mentioned above are not reliable enough for first or second order settlement differences. More promosing is a nonlinear subgrade reaction approach (Schlegel 1985). The range of applicability could be delimited by FEM plus physical model tests.

For many practical cases the subgrade reaction can be linearized (Gudehus 1984). A safety assessment on statistical basis is adequate in case of low scattering of subgrade moduli. This is obtained by homogeneizing fill and/or compaction. Field control procedures are principally at hand, but they should and could be enriched by statistics. Scattering of subgrade reaction will show typical wave lengths (cf. Fig. 2.3c, likewise caused by density fluctuations).

The analysis is illustrated for a bedded beam with two strut loads (Fig. 2.5a).

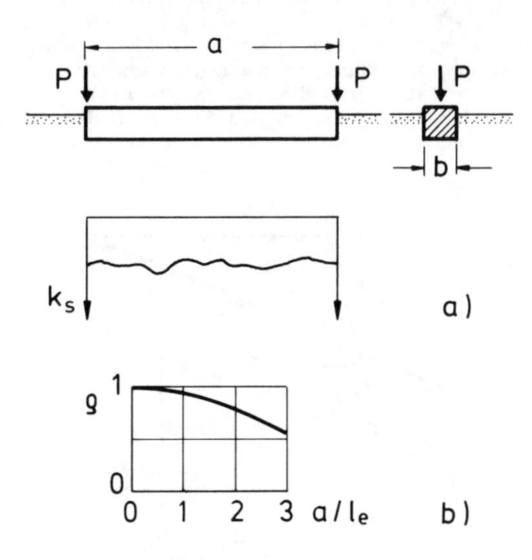

Figure 2.5 Elastic beam on scattering subgrade (a), correction factor for bending moment (b)

With a constant subgrade modulus k_s, bending failure occurs for

$$f:= M_f - \frac{1}{4} P a \rho , \qquad (2.20)$$

with the failure moment M_f (not further considered here), and a factor ρ according to Fig. 2.5b, wherein

$$l_e := \sqrt[4]{\frac{4\,EI}{bk_s}} \qquad (2.21)$$

denotes the elastic length. The spatial mean $\bar{k}_s$ of the subgrade modulus may have the ensemble parameters m_k and V_k (possibly depending on size a). We leave aside the influence of spatial variation of k_s on failure probability; it could be clarified by comparative calculations if scatter wave lengths have been observed. Taking M_f, P, and k_s as stochastic variables, the partial safety factor for the subgrade modulus can be estimated via Equs. 2.9, 2.10, 2.20, 2.21 and Fig. 2.5b.

It turns out that the stochastic correction of k_s is negligible versus systematic errors. This shows that the design can be insensitive with respect to inevitable fluctuations, provided their wave lengths are shown to be small enough for being bridged and that the model is mechanically justified.

3 RETAINING STRUCTURES

3.1 Gravity wall on yielding ground

We consider a gravity wall (e.g. concrete, gabions or filled concrete grid) on inclined cohesive ground supporting a fill (Fig. 3.1a). The combined failure mode indicated in the drawing can be described

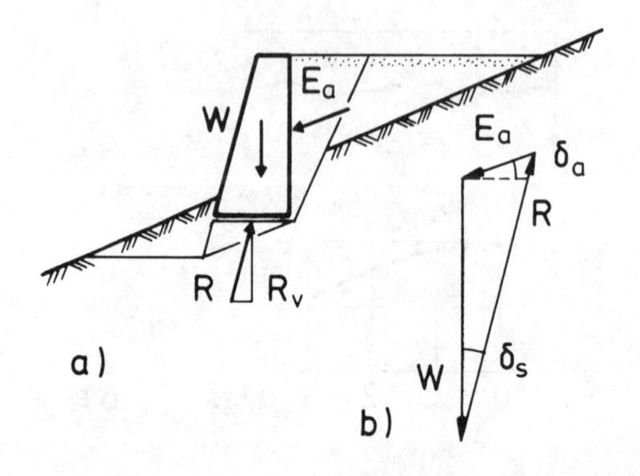

Figure 3.1 Gravity retaining wall (a) in limit equilibrium (b)

approximately by the equilibrium of weight W, active earth pressure E_a and base resistance R (Fig. 3.1b). The vertical component of R may be approximated by

$$R_v = bcN_c(\phi)i(\delta_s)\kappa(\delta_s) , \qquad (3.1)$$

thus neglecting the influence of soil weight under the wall and of eccentricity, but allowing for slope and load inclination, β and δ_s. The factors N_c, i, and κ may be given on the base of plastic limit analysis. The condition of limit equilibrium then reads

$$f := W + \frac{1}{2}\gamma h^2 K_a \cos\delta_a - bcN_c i\kappa = 0 \qquad (3.2)$$

wherein i depends on

$$\delta_s = \tan^{-1}[E_{ah}/(W+E_{av})]. \qquad (3.3)$$

The stochastic variables are
- cohesion c, with rather high scattering $V_c > 0.3$,
- friction angle of the subsoil ϕ, $V_\phi \simeq 0.1$, assumed as above,
- friction angle ϕ_f of the fill, $V_\phi = 0.1$ as before.

With usual soil parameters Equ. 2.11 yields a negligible weight factor $\alpha_{\phi f}$ for ϕ_f. Thus only c and ϕ of the subsoil have to be reduced in order to obtain a required safety index β; following Secs. 2.2 and 2.3 the partial factors are

$$\gamma_c \simeq \exp(-\beta V_c/\sqrt{2}); \; \gamma_\phi \simeq \exp(-\beta V/\sqrt{2}). \qquad (3.4)$$

Various systematic errors have to be considered. The mechanical model, Equ. 3.2, can be checked by static and kinematic theories of plasticity. The deviations will scarcely exceed about 20 % here. More important is the influence of inadequate sampling and testing, mainly due to weathering and fissures. Conventional methods need empirical corrections based on large scale tests and on back analysis of failure events. It is difficult to separate the influence of natural scattering, implying wave lengths of imperfection patterns.

Serviceability of the fill can get lost by uneven yielding of the retaining wall. There is no mechanical model at hand as simple as the ones of Equs. 2.11 and 2.19. Purely empirical statements, such as 'good retaining walls will not give by more than 1 % of their height', are almost irrelevant for safety. FEM-calculations are scarcely of use because of the errors mentioned above. Instead of working with arbitrary correction factors one should pass to an observational method (introduced in Sec. 3.2).

3.2 Strutted vertical wall

In our next example a vertical wall (steel or concrete) is placed into cohesionless ground and supported by top struts and the earth in front of the lower part after excavation (Fig. 3.2a). The simplest failure mechanism implies a Coulomb wedge, passive

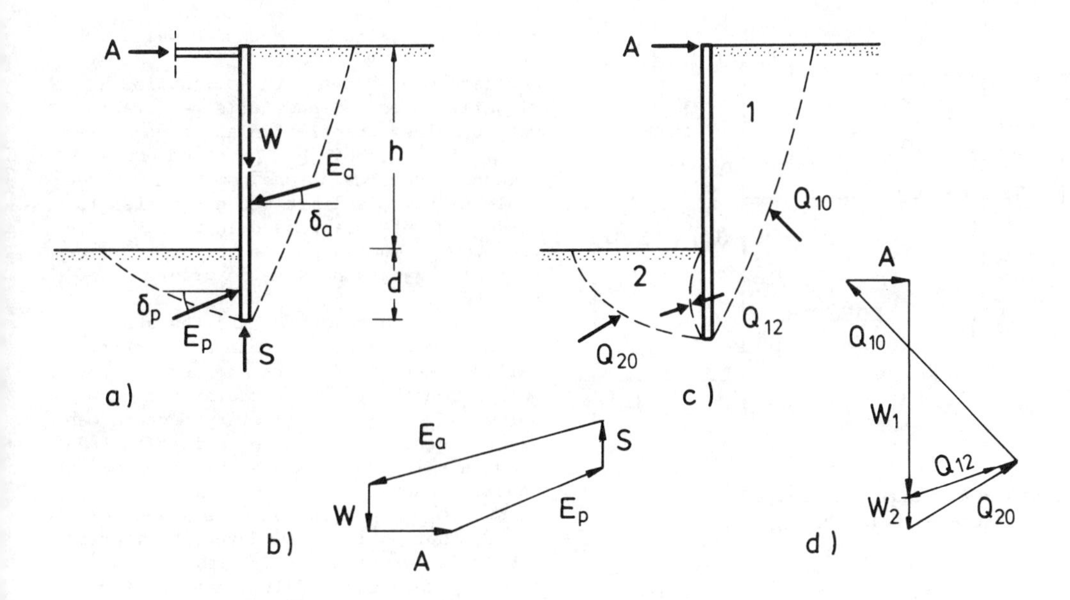

Figure 3.2 Strutted wall (a), simplified equilibrium (b), combined failure mechanism (c), equilibrium (d)

yield at the toe, and simultaneous yielding of the top struts. Equilibrium of the horizontal force components (Fig. 3.2b) yields

$$f:=\frac{1}{2}\gamma d^2 K_p \cos\delta_p - A - \frac{1}{2}\gamma(h+d)^2 K_a \cos\delta_a = 0, \tag{3.5}$$

wherein K_a and K_p depend on the inclinations δ_a and δ_p respectively. Equilibrium of the vertical components involves the unknown toe force S (not necessarily vertical). Even if S and wall weight W could be neglected there is only one equation for the two unknowns δ_a and δ_p. (Note that δ_a and δ_p are only bounded by angles of wall friction, but not equal to them).

An issue from this dilemma is possible through the kinematical method of plasticity theory (Fig. 3.2c). Instead of E_a, E_p, S and their unknown inclinations, slip surface resultants Q_{10}, Q_{12}, Q_{20} determined by the friction circle assumption are considered. The force polygon (Fig. 3.2d) leads to a single limit state equation. Slip circles have to be varied in order to find the maximum (i.e. most unfavourable) upper support force A.

A substitute set of inclinations δ_a, δ_p should be determined so that Equ. 3.5 can be used for practical design. As in previous sections, soil friction angle and strength of the struts have to be reduced by partial factors. The inevitable uncertainty of statistical parameters yields a bound for the required mechanical accuracy of Equ. 3.5.

We only mention other failure mechanisms, such as rotation around the top or bending rupture of the wall. They have to be analyzed likewise and avoided with the same safety index in order to obtain an optimal design.

Especially because of embedded neighbouring structures, serviceability can get lost by uneven displacements. The frequently used purely empirical estimation ('a wall of good quality in dense gravel does not yield more than 5 ^{o}/oo of its height', e.g.) is not apt for the statistical safety format. Linear subgrade reaction starting from the at rest state and bounded by the Rankine limit values is more promising (Fig. 3.3). As far as bending failure is concerned, the earth pressure parameters K_o, K_a, K_p are more important than the subgrade reaction (cf. Sec. 2.3). The prediction of lateral displacement is much more subject to errors. FEM calculations could help under certain rather ideal conditions (e.g. Kastner et al. 1985).

In addition to a good constitutive relation and secured numerical procedures, the initial stress state after placing the wall is required.

It is more adequate to measure wall displacements and strut forces and to adapt

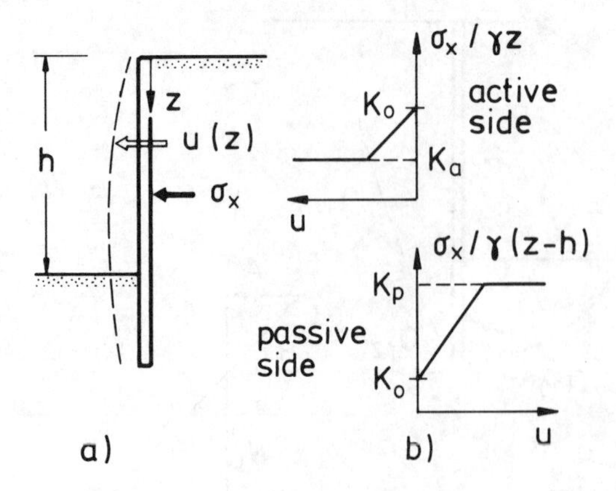

Figure 3.3 Idealized subgrade reaction (b) of a displaced wall (a)

the struts. This observational method can be improved by a successive back analysis (cf. Tominaga et al. 1985): Taking conventional earth pressure coefficients K_o, K_a, K_p, subgrade reaction factors can be calculated from the observed data. Updated predictions can be made with partial safety factors in order to keep the displacement requirements safely enough.

The procedures outlined above can easily be extended to walls with several layers of struts or inclined struts. Essential differences can arise with prestressing and ground anchors, however (Sec. 3.3).

3.3 Soil nailing and backtied wall

A steep covered cut with soil nails in slightly cohesive sand fails preferably with combined translation as shown in Fig. 3.4a. Gäßler (1987) has produced ample evidence by calculations, small and large scale tests; his limit state equation is condensed into diagrams such as in Fig.3.4b, wherein T_m denotes pull-out resistence per unit of length, and a·b the grid size of nail heads.

Gäßler has also shown that the statistical safety concept is applicable to soil nailing. Partial factors are derived for soil strength, pull-out resistance and surcharge. The partial safety factors can be estimated by simpler equations as the ones in this paper because of the remaining indeterminacy.

Numerous practical examples have revealed that a kind of updated observational method is adequate for determining the design parameters. Preliminary coefficients of variation stem from experience and geological considerations; they are updated through the observed variation of boring and sounding resistance, which still is only an educated guess. The mean of friction angle is found from improved triaxial tests. The mean of cohesion is better determined by back-analysis of a cut on the verge of failure. The mean of pull-out resistance stems from in situ tests; their results can be mixed with previous comparable results by means of Bayes' formula

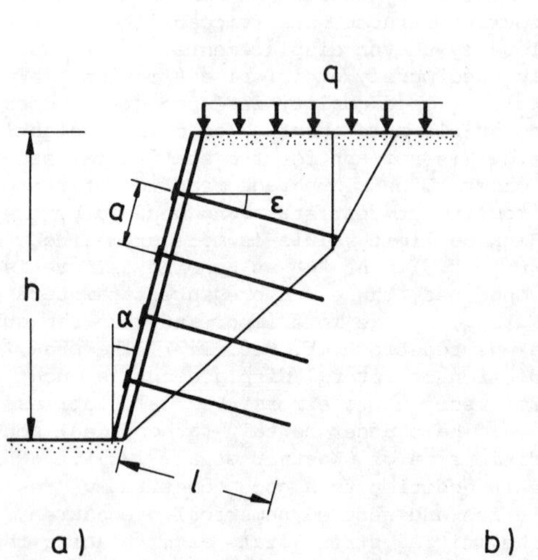

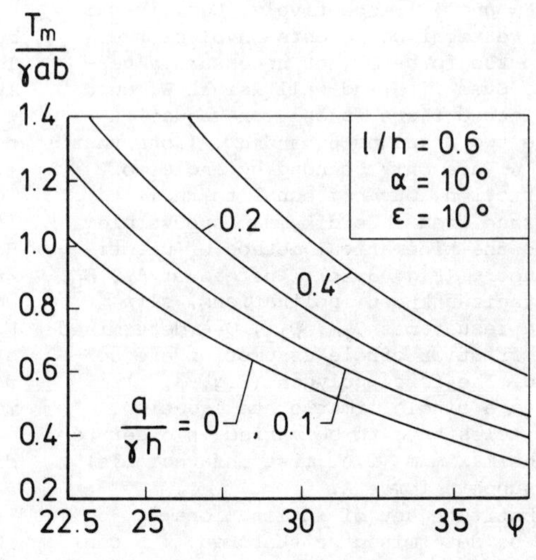

Figure 3.4 Soil nailing: failure mechanism (a) and limit condition (b)

(Ang & Tang 1975). Thus the design can be modified during the execution.

The serviceability is as yet judged in a purely empirical manner. Subgrade reaction models are scarcely helpful (except for designing the skin). FEM-calculations may be an issue if they are supported by large-scale tests. Such efforts should lead to an updated observational method.

Our last example is a multiply back-tied wall in cohesionless soil with sensitive embedded structures (Fig. 3.5a). Following Figs. 3.2 and 3.4 the limit equilibrium (Fig. 3.5b) can be approximated by

$$f := -(W+P)\sin(\theta_a-\phi)-E_{a1}\cos(\theta_a-\phi-\delta_a)$$

$$+E_p\cos(\theta_a-\phi-\delta_p)+\Sigma A_i\cos(\theta_a-\phi+\varepsilon) = 0, \tag{3.6}$$

wherein ΣA_i denotes the pull-out resistance of the anchor sections outside the sliding body. The relevant angles θ_a, δ_a, δ_p can be estimated by analyzing combined failure mechanisms (cf. Sec. 3.2). Design parameters can then be determined as in the previous examples.

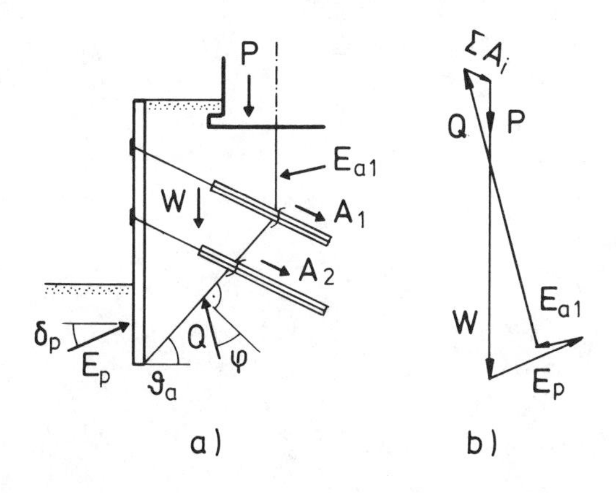

Figure 3.5 Back-tied wall: mechanism (a) and limit equilibrium (b)

More serious problems arise with the assessment of serviceability. The tie-backs are prestressed in order to reduce displacements (Equ. 3.6 is not influenced thus). Together with stepwise excavation the stress and strain paths of the soil become non-monotonic (likewise with prestressed struts). Because of the path dependance of soil, critical displacements cannot be covered by finite equations (such ad Equ. 2.11 and 2.19). The so-called Level II statistical concept breaks down as state functions cannot exist.

FEM-calculations can help to clarify such situations if the implied constitutive relations allow for path-dependance. They should also lead to simplified models for predictions updated with the aid of site measurements. A statistical analysis would be very complicated and not feasible; a stochastic process with a nonlinear differential mechanical model is beyond the present capacity.

4 CONCLUDING REMARKS

The series of examples could be extended to other geometries (e.g. dams and slopes) and loads (e.g. from free or ground water) without changing the message:

- If failure is described by a scalar equation, its probability due to scattering of data can rather easily be allowd for in design;
- the failure model can be supported numerically by plastic limit analysis or finite element calculations;
- systematic mechanical or statistical errors have to be excluded by experience;
- in situ observations during constructions can, with the aid of mechanical and statistical models, improve the safety assessment.

The whole procedure can break down because of path dependence (as exemplified with prestressed tie-backs). This is even more so with dynamic loading, especially if excess pore pressures can build up. Also not included are mechanisms with bifurcation and/or localisation possibly leading to unannounced collapse and involving scale effects.

A word on simple models may be allowed here. The engineer cannot normally use the most precise model. The price for simplicity is accuracy. Unfortunately it is principally impossible to measure the reliability of a prediction model; K.Popper has clearly written about this dilemma in his 'Logic of Scientific Discovery'. It may happen tomorrow that someone upsets - or better newly delimits - the usual variational principles of plasticity, including Coulomb's principle.

A. Einstein said 'Everything should be made as simple an possible,but not simpler'. The engineer cannot always wait for the simplest allowable model but he still has to take responsibility. Intricate models can also increase the risk as errors cannot easily be discovered. In this sense sophisticated models can be 'judgment killers'

(Mortensen 1983). Just as finite element calculations the statistical safety concept can improve or suppress judgment.

REFERENCES

Ang, A.H.S. & W.H. Tang 1975. Probability concepts in engineering planning and design. New York: Wiley.

Gäßler, G. 1987. Vernagelte Geländesprünge. Tragverhalten und Standsicherheit. Veröff. Inst. Bodenmech. u. Felsmech. Univ. Karlsruhe, Heft 106.

Gudehus, G. 1984. Vereinfachte Ermittlung der Dicke von Flachfundamenten aus Stahlbeton. Bauingenieur 59: 337-345.

Hettler, A. 1985. Setzungen von Einzelfundamenten auf Sand. Bautechnik 6: 189-197.

Hettler, A. & G. Gudehus 1985. A pressure-dependent correction for displacement results from 1g model tests with sand. Géotechnique 35, 4: 497-510.

Holzlöhner, U. 1984. Settlements of shallow foundations on sand. Soils and Found., Jap. Soc. Soil Mech. Found. Eng., 24, 4: 58-70.

Kany, M. 1974. Berechnung von Flächengründungen. Berlin: Ernst u. Sohn.

Kastner, R. et al. 1985. Soutènements flexibles, essais sur modèle, calculs. Proc. 11th Int. Conf. Soil Mech. Found. Eng., S.Francisco, Vol. 4: 2103-2106.

Mortensen, K. 1984. Is Limit State Design a Judgment Killer? Norweg. Geot. Inst., Publ. No. 148.

Nakai, T. 1985. Finite element computations for active and passive earth pressure problems of retaining wall. Soils and Found. 23: 98-112, Jap.Soc.Soil and Found. Eng.

Schlegel, T. 1985. Anwendung einer neuen Bettungsmodultheorie zur Berechnung biegsamer Gründungen auf Sand. Veröff. Inst. Bodenmech. u. Felsmech. Univ. Karlsruhe, Heft 98.

Tominaga, M. et al. 1985. A computerized construction control and its application to an excavation. Numerical Meth. in Geomech., Nagoya (Ed. Kawamoto & Ichikawa), 2: 927-934.

Numerical Methods in Geomechanics (Innsbruck 1988), Swoboda (ed.)
© 1988 Balkema, Rotterdam. ISBN 90 6191 809 X

Simulation of hydraulic fracture in poroelastic rock

A.R.Ingraffea & T.J.Boone
Cornell University, Ithaca, N.Y., USA

ABSTRACT: This paper has two themes: first, it discusses poroelastic effects on fracture propagation in rock and second, it demonstrates methods of using work station based computer graphics to display results from numeric simulations. The paper initially presents analytical solutions and notes their significance. A method for simulating fracture propagation using a finite element approximation to the coupled equations is outlined. A generalized Dugdale-Barrenblatt is used to model the fracture process zone using nonlinear fracture mechanics concepts. It is shown that rate dependent poroelastic effects can occur during fracture propagation. It is also shown that poroelastic effects can significantly alter the direction of fracture propagation.

1 INTRODUCTION

Research at the Program of Computer Graphics at Cornell University has dual objectives. The first is to conduct research into computational mechanics with an emphasis on fracture mechanics, and the second is to research methods for the application of computer graphics to engineering problems. In this paper and its presentation, poroelastic effects during fracture propagation are discussed and investigated with the aid of computer graphics.

It is most difficult within the confines of a conference paper to present computer graphic capabilities. One is limited to a few, black-and-white photographs and drawings. For most engineers these limitations reflect their readily accessible graphic capabilities. However, the engineering workstation is in the process of drastically altering this frame of reference. A generation of workstations is about to become available that will allow the engineers to view their work using complex, animated, 3-D, color images on a desk top device. It is hoped that, in the process of presenting our work at the conference, we can demonstrate some of these evolving capabilities. Clearly, these new engineering tools present a tremendous opportunity for the profession. For example, numerical simulation of fracture propagation in poroelastic materials is an inherently transient problem. One can now view, review, and analyze the results of such problems in a dynamic or transient manner at one's desk.

2 THEORY OF POROELASTICITY APPLIED TO FRACTURE PROPAGATION

The basic linear theory of poroelasticity was first formulated by Biot in 1941 [1]. Since then, Biot and others such as Rice and Cleary [2] have made significant contributions to its development. Numerical methods have also developed, primarily for application to soil mechanics problems [3,4,5,6]. Others have discussed poroelastic influences on fracture propagation [7,8].

Ruina [9] and Huang and Russell [10] have analytically solved fracture problems in poroelastic media. Ruina's solution is for a semi-infinite fracture moving at a constant rate in a saturated poroelastic material. The fracture is loaded as shown in Figure 1(a) by an applied stress, P along the crack faces. A zero flow condition is also imposed along the crack faces. Figure 2 shows a 3-D plot of Ruina's solution in the close vicinity of the crack-tip where it is assumed that the fracture is moving at a speed that is fast relative to pore pressure diffusion rate in the structure. In Figure 2 the crack tip is at the origin of the axes, and the crack extends along the negative X-axis. The 2-D structure lies in the X-Y plane. The pore pressure field is plotted in relief along the Z-axis above the X-Y plane. The contours also plot the pore pressure field.

Several important features are illustrated by this solution. First, along the crack faces the pore pressure is equal to the ambient pore pressure, in this case zero. This is indicative of the fact that there is no volumetric strain along the crack faces. Second, the slope of the

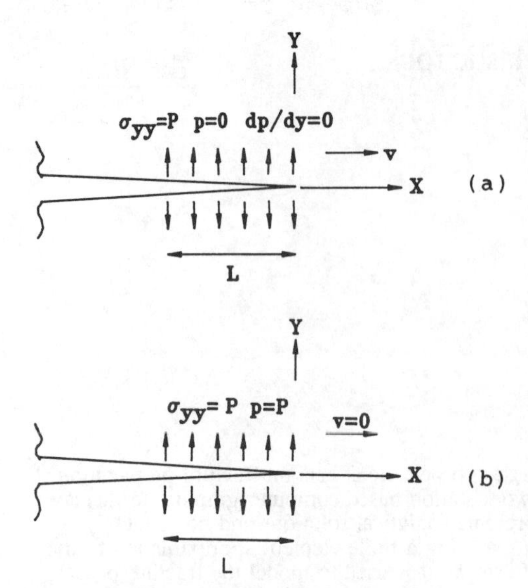

Figure 1. (a) Boundary conditions for Ruina's analytical solution [9]. (b) Boundary conditions for Huang and Russell's analytical solution [10].

pore pressure field normal to the crack faces is zero everywhere along the crack faces. This complies with the no-flow boundary condition. Ahead of the crack tip is a negative pore pressure which is indicative of a reduced effective (tensile) stress in that region relative to the fully drained solution. However, at the crack tip itself, the pore pressure equals the ambient value and the material is effectively drained. This is due to the fact that the volumetric strain gradient approaches infinity near the crack tip. Thus, if the material has a finite permeability then the crack tip must be drained.

Ruina gives the stress intensity for this case, based on an effective stress criterion, as:

$$K_{nom}/K_{Ic} = \{1-2B(1+\nu_u)/3\}^{-1} \quad (1)$$

where K_{nom} is the nominal stress intensity factor required to propagate a fracture and K_{Ic} is the plane strain fracture toughness that is applicable if the material response is totally drained, B is Skempton's pore pressure coefficient and ν_u is the undrained Poisson's ratio. The ratio of K_{nom}/K_{Ic} is typically in the range 2 to 5 for rock.

Figure 3 is a plot of Huang and Russell's solution which is similar in form to Figure 2. In this case the fracture is static and the boundary conditions are a pore pressure, P, applied from the origin to a distance, L, along the negative X-axis. Along the rest of the crack faces the pore pressure is prescribed as zero. Since this is a steady state solution, the rock mass is in a drained state. The stress singularity at the crack tip is of the standard form with a stress intensity given by:

$$K_I = P(8L/\pi)^{0.5}\{[0.5+\nu/2(1-\nu)]\alpha\} \quad (2)$$

$$\alpha = 3(\nu_u-\nu)/[B(1-2\nu)(1+\nu_u)] = 1-K/K_s \quad (3)$$

where P is the pressure applied to the crack faces, ν is the drained Poisson's ratio, ν_u is the undrained Poisson's ratio, α is a poroelastic constant, K is the bulk modulus of the skeletal structure, and K_s is the bulk

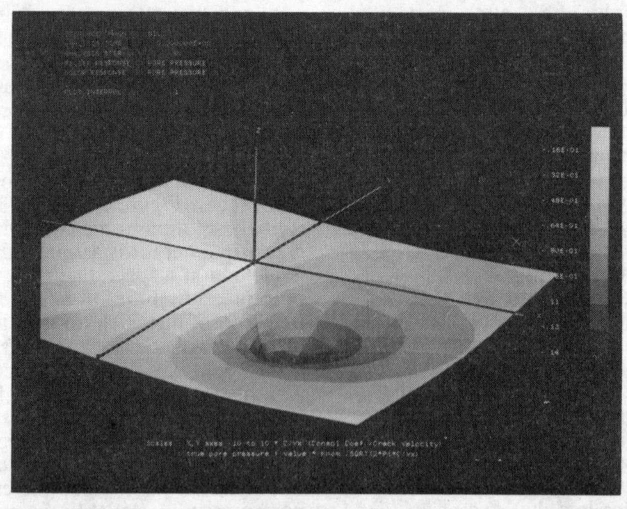

Figure 2. A 3-D plot of Ruina's analytical solution for a fracture moving at a rapid rate in a poroelastic medium. Pore pressure is plotted in relief. Colour contours also represent pore pressure.

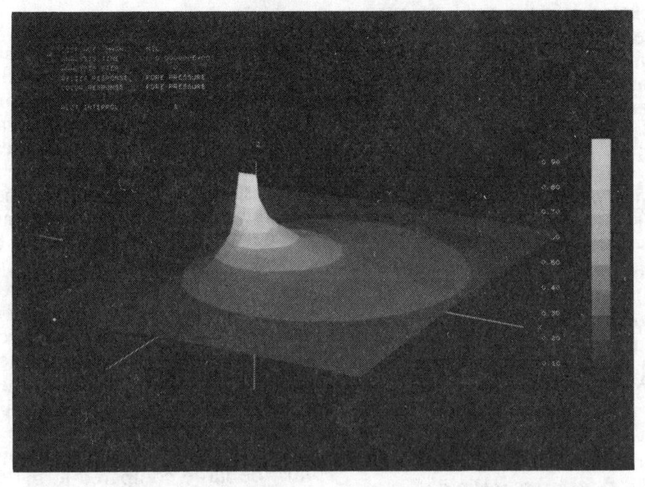

Figure 3. A 3-D plot of Huang and Russell's analytical solution for a static fracture a poroelastic medium. Pore pressure is plotted in relief. Colour contours also represent pore pressure.

modulus of the solid constituent. K_I is shown to be positive for all material properties with a bracketed term {} having a typical value of about 0.6. The result is actually independent of any material properties associated with the free flowing pore fluid. If the equivalent pressure load was applied as a stress along the crack face the stress intensity factor would be:

$$K_I = P(8L/\pi)^{0.5} \quad (4)$$

which is the upper limit of the Equation (2).

It should be noted that the boundary conditions along the crack face for this problem are applied pore pressures. In Ruina's solution the boundary condition is applied stress. However, at high crack speeds the applied stress boundary condition is equivalent to an applied pore pressure boundary condition. Therefore, Huang and Russell's and Ruina's solutions represent the extremes of very slow and very fast propagation of hydraulically driven fractures, respectively. In practical situations, at finite crack speeds, the pore pressure field will lie between these extremes.

These two linear elastic solutions are presented as indicative of poroelastic effects. Two important results are worth noting. The first is that under steady state conditions the stress intensity factor (given by Equation (2)) due to loading of an internally pressurized fracture in a permeable medium is typically about 60% of that for the same loading applied to an impermeable medium. The second is that negative pore pressures can develop ahead of the crack tip so that the effective stress is reduced in this region.

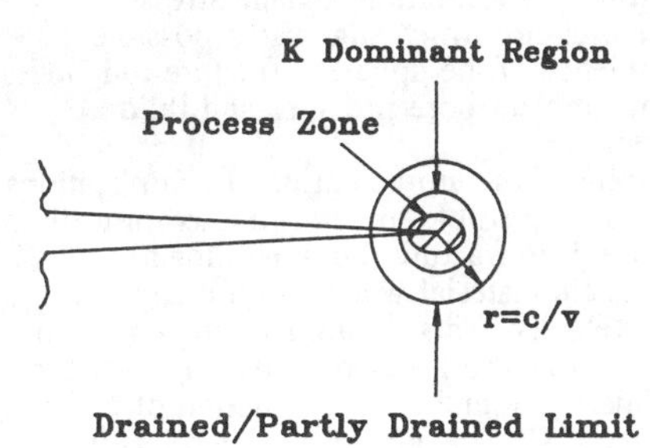

Figure 4. Idealized plot of three regions around a crack tip: nonlinear (process) zone, K dominant zone and the drained zone.

3 LIMITATIONS OF LINEAR ELASTIC FRACTURE MECHANICS (LEFM)

LEFM is applicable in limited circumstances. The primary limitation is that all dimensions in the structure must be large with respect to the size of the process zone (zone of inelastic behavior) at the crack tip. This implies that the process zone is contained within the region near the crack tip where the $r^{-0.5}$ stress singularity is dominant as shown schematically in Figure 4 [11]. Under these conditions the material parameter K_{Ic}, known as the plane strain fracture toughness, can be used to characterize resistance to fracture propagation.

In poroelastic materials a third ring must be added to Figure 4 which separates the drained from partially drained materials. The characteristic length associated with this ring is c/v [9]. Where c is the diffusivity coefficient of the rock and v is the crack tip velocity. For the case of a static or slowly moving crack this dimension is large relative to the process zone and K_{Ic} can be used to characterize the material resistance to fracture propagation provided other geometric constraints are met. For cracks moving at high velocities the drained region decreases to a vanishingly small region around the crack tip.

Ideally, one might be able to characterize fracture resistance for the case of a rapidly moving crack assuming that the characteristic length c/v is small relative to the size of the process zone. Equation (1) suggests a relationship between K_{nom} for a fracture moving at a high speed in a poroelastic media and K_{Ic} for a static fracture. This relationship is certainly idealized. It does not account for significant differences in the radial variation of stress near the crack tip for the two cases. This arises because the fluid pressure in a poroelastic media is only coupled with the volumetric strain rate and not the deviatoric strain rate. Another complicating factor is that fluid has a limiting 'tensile stress' or vapor pressure which suggests a possible dependence of the apparent fracture toughness on the ambient pore pressure and fluid temperature.

In spite of the aforementioned complications, it could be possible to determine empirically a characteristic fracture toughness for a poroelastic material which is a function of crack velocity. This situation is analogous to dynamic fracture problems where the fracture toughness is found to be a function of crack velocity [11]. The adjective dynamic is used here to describe problems where the kinetic energy is an important consideration.

It is worth expanding on the analogy between poroelastic fracture and dynamic fracture since the study of dynamic fracture is well advanced. There are some important similarities. First, the elastic solutions for the near tip stress fields preserve the $r^{-0.5}$ stress singularity found in the static solution. Second, the angular variations of the stress fields differ from the static distribution. And finally, in both cases there is an additional energy sink that does not exist in the static case: kinetic energy and pore fluid strain energy. The implication of this analogy is that in poroelasticity the relationship between G, the global potential energy release rate, and K, the stress intensity factor may be dependent on the crack velocity. An equation of the form below is found for the dynamic case [11]:

$$G = A(v)K^2/E \qquad (5)$$

Where A(v) is a function dependent on the elastic properties and the crack velocity. For the static, plane strain case A(v) is simply $(1-\nu^2)^{-1}$. Another implication is that contour integrals such as the J integral will not necessarily be path independent.

The practical importance of this discussion is that the apparent fracture toughness of rocks such as very permeable sandstones, at fracture propagation rates typical of hydraulic fracturing (0.01 m/s), can be characterized using static K_{Ic} values since the process zone can be assumed to be fully drained. However, the permeability of rock varies by several orders of magnitude so this is not generally true. If an empirical function relating the fracture toughness to crack velocity is available for a material the poroelastic process zone effects could be accommodated within linear elastic codes in the same manner in which dynamic fracture has been modeled [12]. One is still limited by the LEFM constraint that all dimensions in a structure must be large compared to the fracture process zone. Many practical problems such as breakdown at a borehole, cases where the fracturing fluid approaches the vicinity of the process zone and offshore dredging of rock cannot be analyzed under LEFM constraints. A more general approach is therefore desirable.

4 NONLINEAR FRACTURE MECHANICS

Over the past twenty years nonlinear fracture mechanics concepts have been developing in their application to rock mechanics. Instead of characterizing the resistance to fracture using a fracture toughness parameter such as K_{Ic}, it can be characterized using a stress versus

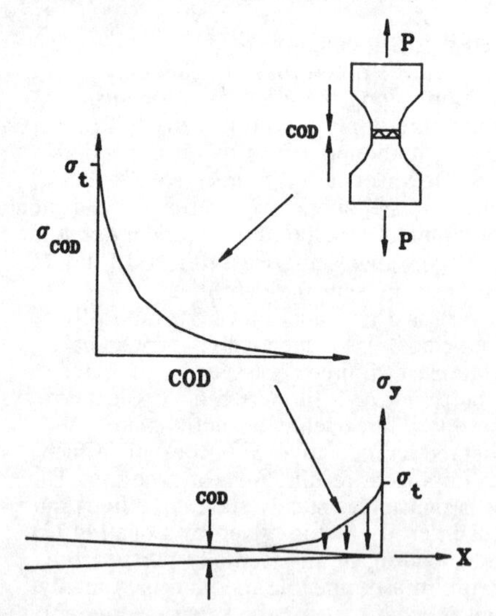

Figure 5. A schematic figure showing the relationship between a simple tension test, a Stress vs COD curve and the stress carried across the the fracture plane in the fracture process zone.

crack opening displacement curve as shown in Figure 5. This is in essence a Dugdale-Barrenblatt approach to fracture mechanics. Ideally, the stress versus COD curve can be determined from the results of a simple tension test on a rock specimen. The principles have previously been discussed in detail by the authors [13] and are illustrated schematically in Figure 5. The area under the stress vs. COD curve is equivalent to the fracture energy, G_c and also the J integral as noted by Rice [2]. Under plane strain conditions G_c and K_{Ic} are related through the Equation (5) where A(v) is $(1-\nu^2)^{-1}$ as noted previously.

Tension tests on numerous rock types, concrete, and mortar have produced stress vs displacement curves of similar form [14,15,16]. The authors have therefore suggested a stress vs. COD curve of the following form [17]:

$$\sigma_{COD} = \sigma_t \, e^{-k(COD)}, k=G_c/\sigma_t \qquad (6)$$

where σ_{COD} is the stress carried across the crack, σ_t is the tensile strength of the rock and k is a decay constant. The form of the relationship is chosen for its simplicity. The model incorporates the rock tensile strength and preserves the plane strain fracture energy G_c. These features allow a simulation to transition properly from fracture initiation to conditions where LEFM is applicable.

In extending these concepts to fractures propagating in poroelastic media the global energy release rate G may no longer represent a material parameter so it seems more appropriate to use a J integral criterion. As noted by Kanninen it has the 2 important features, (1) for linear elastic behavior it is identical to G and, (2) for elastic plastic behavior it characterizes the crack tip region [11]. Rice has shown that for Dugdale-Barrenblatt models as described above [18].

$$J = \int_0^{COD} \sigma_{COD} \, d(COD) \qquad (7)$$

The J criterion is appropriate for poroelastic analysis because it is a measure of the local energy dissipation at the crack tip. The simplest extension of rock mechanics principles, where effective stress is used as a strength criterion, to poroelastic fracture is to define J_c as follows:

$$J_c = \int_0^{COD} \sigma_{COD}' \, d(COD) \qquad (8)$$

where σ_{COD}' is the effective stress measured across the crack opening. Equation (6) now takes the form:

$$\sigma_{COD}' = \sigma_t \, e^{-k(COD)}, \; k=G_c/\sigma_t \qquad (9)$$

Since Equation (9) is equivalent to Equation (6) under drained conditions, and J_c as defined in Equation (8) is equivalent to G_c under the same conditions, J_c as defined in equation (8), can be determined directly from tests on dry rock samples.

The authors are not aware of any specific experimental work which has investigated how poroelasticity affects fracture toughness. Atkinson has compiled results of fracture toughness tests under confining pressure where specimens are jacketed, unjacketed, partially and fully saturated [19]. Specimens which are jacketed or only partially saturated show a proportional increase in the apparent fracture toughness with confining pressure. However, specimens where the pore pressure in the rock mass is equal to the confining pressure show no increase in apparent fracture toughness. This result is consistent with the model suggested above.

5 TRANSIENT SOLUTION OF THE FINITE ELEMENT APPROXIMATION

Zienkiewicz has described in detail the formulation of a method using finite elements for the solution of poroelastic problems [4,20]. If all inertial terms are neglected the matrix equations reduce to those given below :

$$[L]\{p\} + [K]\{u\} = \{f\} \qquad (10)$$

$$[L]^T\{\dot{u}\} + [H]\{p\} = \{q\} \qquad (11)$$

Here u are the nodal displacements, p the nodal pressures, f the nodal forces, q the nodal flows, [K] the stiffness matrix, [L] the coupling matrix, and [H] the flow matrix. In the first set of analyses presented herein, triangular and quadrilateral isoparametric elements are used with quadratic displacement fields and linear pore pressure fields. The fracture path is predetermined. Along this path six-noded zero thickness interface (Goodman) elements are inserted into the mesh [21]. These elements have zero opening until the maximum tensile strength is exceeded. At that point the elements assume the stress versus COD relationship given in Equation (9).

Previous experience solving static problems of this nature has shown Newton-Raphson and dynamic relaxation techniques to be effective solution methods [20]. Dynamic relaxation has proven to be the more reliable technique.

In solving Equations (10) and (11) standard finite difference techniques are typically used to advance the equations. The method employed herein is novel in that artificial mass is introduced in equation (10) as is done in dynamic relaxation so that it becomes:

$$[M']\{\ddot{u}\} + [C']\{\dot{u}\} + [K]\{u\} + [L]\{p\} = \{f\} \quad (12)$$

Where [M'] and [C'] are artificial mass and damping matrices, respectively. These matrices can be optimized as is done in standard dynamic relaxation techniques [24]. To maintain accuracy, it is important to ensure that the fundamental structural period is not affecting the accurate representation of the coupled decay modes. In situations where small time steps are required the 'true' dynamic form of Equation (12) can be employed.

The finite difference approximation of equation (12), using central difference operators is given below:

$$[M']\{u_{t+1} - 2u_t + u_{t-1}\}/\Delta t^2 + [K]\{u_t\} + [L]\{p_t\} = \{f_t\} \quad (13)$$

This equation is explicitly solved for u_{t+1}, where the subscripts represent the time step. The explicit formulation allows one to incorporate easily the nonlinear behavior of the interface elements. Equation (8) is also approximated using central difference operators and a scheme employed by Comini for the heat equation [23]. The approximation is:

$$[S]\{p_{t+1} - p_{t-1}\}/2\Delta t + [L]^T\{u_{t+1} - u_{t-1}\}/2\Delta t + [H]\{p_{t-1} + p_t + p_{t+1}\}/3 = \{q_t\} \quad (14)$$

This type of approximation has been shown to be unconditionally stable [25]. It has the additional advantage of easily accommodating nonlinear boundary conditions. However, it must be solved implicitly.

This explicit-implicit form of time stepping has been shown to be reasonably efficient relative to other techniques [26]. The time step is only limited by accuracy considerations since Equation (14) is unconditionally stable and the artificial mass matrix in Equation (13) can be adjusted accordingly. However, one must concede that large numbers of time steps are required for accurate solutions. Convergence is assured in the finite element sense as both the element size and time step approach zero. However, an additional requirement for convergence is that the ratio of the time step to element size also must approach zero.

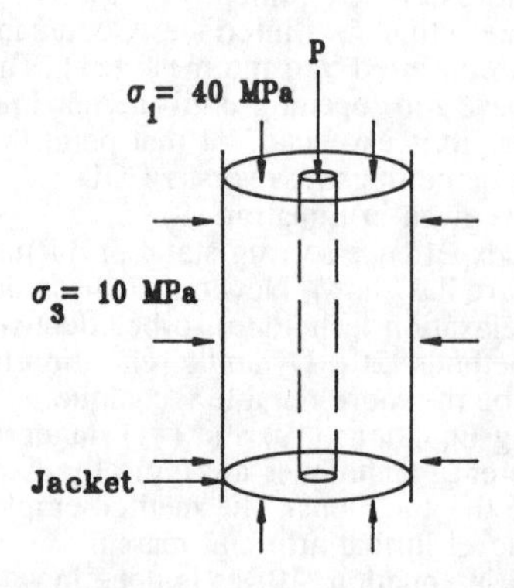

Figure 6. Cylindrical test specimens used by Zoback [27].

6 NUMERICAL SIMULATIONS

6.1 Pressurized Cylinders

A series of experiments has been conducted by Zoback [27] on specimens of Weber sandstone with the dimensions shown in Figure 6. The specimens were initially dry and jacketed. At time zero, water or oil was injected into the center hole of the specimen. The fluid pressure was then increased at a constant rate and the fluid allowed to permeate into the specimen. The specimens were also subjected to a confining pressure of 10 MPa. The pressure was observed to peak as a fracture abruptly extended through the specimen. The peak pressure was observed to be rate dependent. It has been suggested that this rate dependent effect is at least partially a result of poroelastic effects [2,7].

The finite element mesh used in the simulations is shown in Figure 7 along with the simplified boundary conditions used in the simulation and the material properties. Zero-flow conditions were enforced along the crack face. In the simulation the specimens

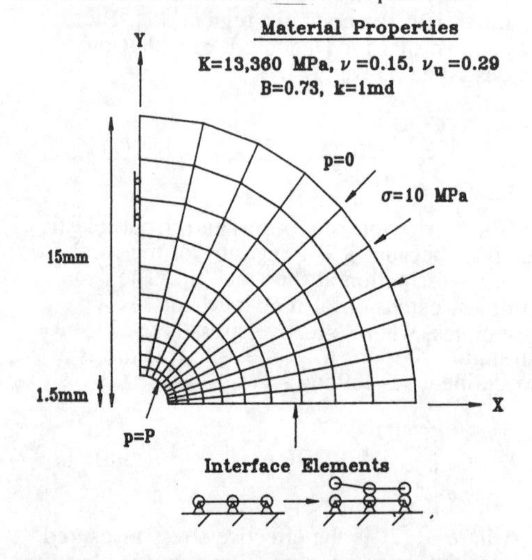

Figure 7. Finite element mesh and simplified boundary conditions used in simulating cylindrical test specimens. (k=permeability).

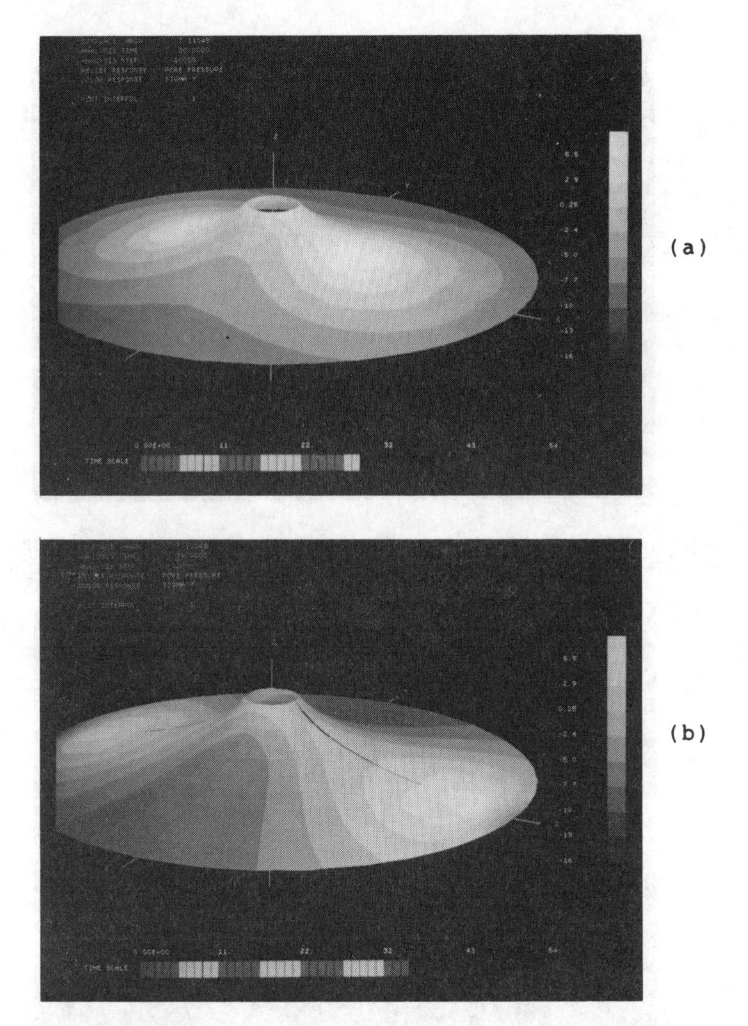

Figure 8. A 3-D plot of a cylindrical test specimen simulation at a pressurization rate of 1 MPa/s. Pore pressure is shown in relief. σ_{yy}' is contoured. (a) Borehole pressure, P=30 MPa, (b) P=36 MPa.

were assumed to be initially saturated with a zero pore pressure at the outer edge of the specimen. The fracture path was predetermined to run along the x-axis where the interface elements are located. Typically 25,000 time steps were used for the analyses.

The results of an analysis for a pressurization rate of 1 MPa/s are shown in Figures 8(a) and (b). The 2-D mesh lies in the X-Y plane. The pore pressure is plotted in the Z dimension. Superimposed on the pore pressure field are gray shade contours of the effective stress in the Y direction (σ_{yy}'). The pore pressure field is plotted over the displaced mesh so that the development of the fracture can be seen as well. On an engineering workstation these images can be displayed in rapid succession so that a dynamic effect is achieved. Color contouring also greatly enhances the quality of the image. While these plots are complex and may require the reader to take a few minutes to become initially oriented, they contain a considerable amount of information. They allow the user to monitor simultaneously progression of the stress, pore pressure and displacement fields.

Figures 8(a) and (b) show a pore pressure distribution that is essentially equivalent to the steady state field. The pressurization rate is obviously slow compared to the diffusion rate through the radius of the structure. The stress contours show the development of a concentration in the σ_{yy}' stress at the edges of the hole along the x-axis and subsequent

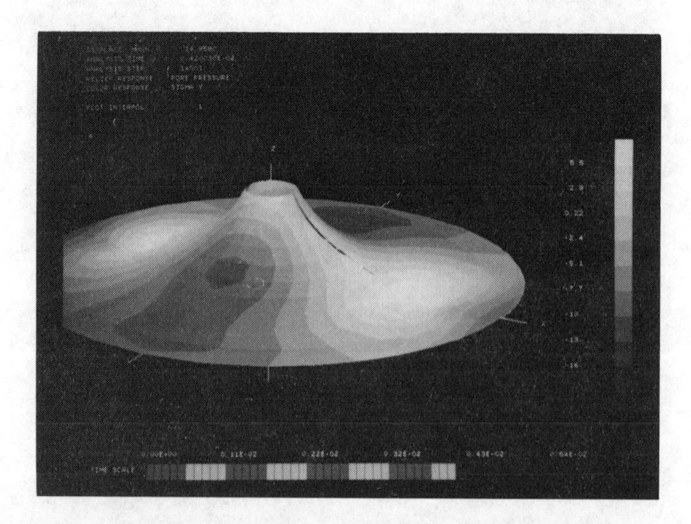

Figure 9. A 3-D plot of a cylindrical test specimen simulation at a pressurization rate of 10,000 MPa/s. Pore pressure is shown in relief. σ_{yy}' is contoured. Borehole pressure, P=42 MPa.

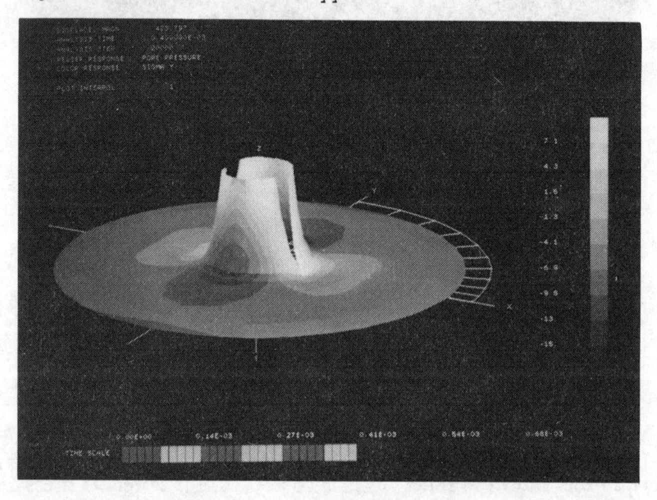

Figure 10. A 3-D plot of a cylindrical test specimen simulation at a pressurization rate of 100,000 MPa/s. Pore pressure is shown in relief. σ_{yy}' is contoured. Borehole pressure, P=40 MPa.

progression of the fracture process zone at the tip of the fracture. The pore pressure appears unaffected at the crack tip due to the relatively slow rate of fracture propagation.

Figures 9 and 10 show similar photographs of pore pressure and stress plots but for simulations with pressurization rates of 10,000 MPa/s and 100,000 MPa/s, respectively. It is apparent in Figure 10 that the pressurization rate is rapid relative to the diffusion rate through the specimen. A decrease in the pore pressure near the crack tip can be observed in both Figures 9 and 10. In Figure 10 it is observed that fracture progression is retarded with respect to fracture development at the same applied pressure for slower rates of pressurization. The rate of fracture propagation is limited by the rate of pore pressure diffusion through the rock mass.

The results of these simulations show that poroelastic effects can produce the rate dependent breakdown pressures at boreholes. However, a much larger pressurization rate was required to produce this effect numerically than was required in the experiments by Zoback. This can be attributed in part to the fact that Zoback's specimens were initially dry which would significantly alter the diffusion rate through the specimen.

6.2 Mixed-Mode Fracture Propagation

In the simulations that follow it is shown that stress fields induced by poroelastic effects can significantly alter the direction of fracture propagation. The program FRANC has been used to simulate quasistatic fracture propagation in a rock mass [28]. The program allows the user to conduct a series of fracture analyses and to propagate interactively fractures through a finite element mesh on an engineering workstation. LEFM principles, and the maximum circumferential tensile stress criterion have been used as a basis for propagating the fractures. This approach has been shown to predict accurately the direction of fracture propagation in rock [29].

A practical method of poroelastically altering the stress state in a rock mass is to inject or produce fluid from a well bore. Under steady state conditions the pore pressure variation around the borehole varies logarithmically:

$$\Delta p = \Delta p_h \frac{\ln b - \ln r}{\ln b - \ln a} \qquad (15)$$

where b is the radius of influence of the well, a is the well radius, r is the radial distance to a point in the rock mass, Δp is the change in pore pressure at a point, and Δp_h is the applied change in pore pressure at the well bore. This form of loading is discussed in detail elsewhere [8,2]. It suffices to note here that a positive increase in the borehole pressure ($\Delta p_h > 0$) produces tensile circumferential effective stress ($\sigma_{\Theta\Theta}'$) and compressive radial effective stress (σ_{rr}'). It should also be noted that the well radius 'a' could be an effective well radius if there are short fractures emanating from the well which are very permeable relative to the rock mass. These fractures could be mechanically induced.

In the examples that follow the analytical pore pressure field given by Equation (15) has been applied to the finite element meshes as a body force integrated over the domain. The fractures have been driven by a constant internal pressure over the length of the fracture. It is assumed that the fracture is propagating at a fast enough rate that there is negligible fluid flow into or within the rock mass as a result of the propagating fracture. The borehole has been pressurized over a relatively long period of time. The internal crack pressure has been adjusted to bring the fracture to the point of instability at each step of fracture propagation. The maximum circumferential tensile stress theory has been used as crack propagation criterion.

In this example two boreholes are located in close vicinity as shown in Figure 11. One borehole has been pressurized for an extended period of time. The borehole radius of 1m is an effective radius. If the pressure increase at the borehole is sufficient, the direction of

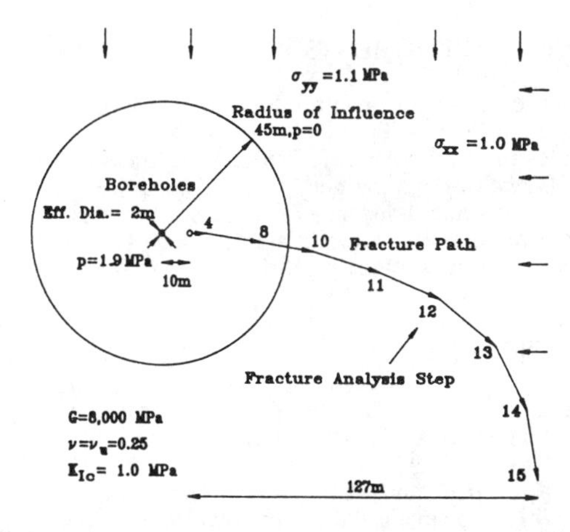

Figure 11. Boundary and Load conditions for a mixed-mode fracture propagation analysis. The final crack path is also shown.

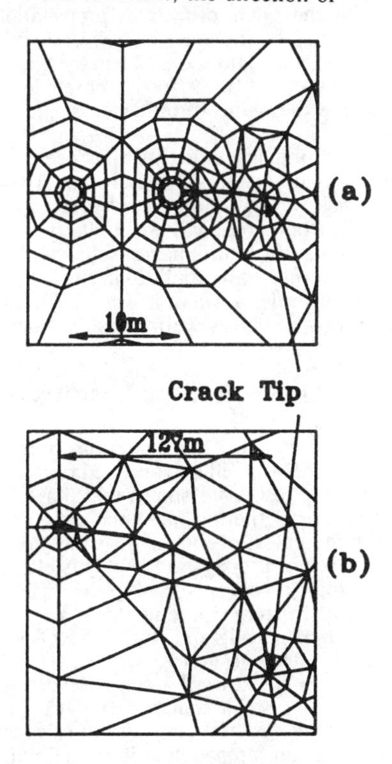

Figure 12. (a) A plot from a small-scale finite element mesh used in the first 8 fracture propagation steps. (b) A plot from a large-scale finite element mesh used for the final 7 propagation steps.

the maximum and minimum stresses in the vicinity of the second borehole will be reversed. The hydraulic fractures that would initiate from the second borehole would be perpendicular to the far field maximum compressive stress. One fracture would propagate towards the first borehole and the second in the opposite direction. Figure 12 shows a multi-step analysis of the latter fracture using a refined mesh and a large scale mesh. The first eight steps of the analysis were conducted using a refined smaller scale mesh. In this analysis the fracture was initiated at an angle of 5° from the symmetry line to induce asymmetry in the problem. It can be seen that the fracture eventually turns so that it is perpendicular to the minimum principal stress, but not until it has traveled a significant distance. The practical significance of this example is that in many cases it is preferable to drive fractures in a direction perpendicular to the maximum in situ compressive stress. These fractures may then intersect the primary set of natural fractures which often lie parallel to the maximum in situ compressive stress. A detailed mixed-mode fracture analysis is required in order to determine the extent of fracture propagation before it curves. There are also several scale effects that enter into the problem such as fracture toughness, the effective borehole radius, and the radius of influence of the well.

A second practical effect that has been numerically observed is that fractures driven by internal pressurization are diverted towards pressurized boreholes and are similarly repelled away from producing wells. This observation may have applicability to geothermal projects where it is desirable to connect two boreholes with hydraulic fractures.

7 DISCUSSION AND CONCLUSIONS

The primary intent of this paper has been to explore poroelastic effects on fracture propagation. Ruina has shown that, based on an effective stress criterion, the apparent fracture toughness of rock is dependent on the rate of fracture propagation [9]. This effect is analogous to concepts used in the analysis of dynamic fracture where inertia is a consideration. Huang and Russell have shown that for the steady state case the stress intensity at a crack tip of a pressurized crack is less in a permeable material than it is in an impervious material [10]. For most practical cases of fracture propagation in poroelastic materials it is therefore necessary to employ numerical simulations to quantify these effects.

There are very few experimental results that allow for comparison with or calibration of numerical models. It is therefore necessary to make numerous assumptions in developing a numerical fracture propagation model. Clearly, it is desirable to develop methods that are logical extensions of current practice. A numerical model, based on Biot's linear theory of poroelasticity, has been presented herein. It employs finite element methods developed initially for application to saturated soil problems [20]. An 'effective stress' fracture criterion has been implemented which is an extension of Dugdale-Barrenblatt models that have previously been applied to fracture of rock [13].

Numerical simulations of pressurized cylinder tests indicated that poroelastic effects may induce some rate dependent fracture propagation effects as predicted analytically [9]. To effectively simulate these experiments, it will be necessary to model the fluid flow in the fracture which is a current area of development. The second example illustrated how stress changes in a rock mass, induced poroelastically by pressurizing a borehole can alter the direction of a propagating fracture. The simulation illustrated that a fracture initiated from a borehole adjacent to a pressurized borehole can be driven distances perpendicular to the far-field, maximum principal compressive stress.

This paper largely presents work in progress. It is therefore little more than an introduction to the field of fracture propagation in poroelastic materials. It is an area of research that could assist in improving efficiency of rock excavation in saturated environments and provide novel techniques for hydraulic fracturing. It is also an area of research that can be greatly enhanced through the use of computer graphics tools.

ACKNOWLEDGMENTS

This research has been funded in part by the National Science Foundation of the United States under grant no. PYI 8351914, and in part by Dowell Schlumberger Ltd., Tulsa, OK. The comments and assistance of Mr. Paul Wawyznek, Dr. Emmanuel Detournay and Dr. J.C. Roegiers have been very much appreciated.

REFERENCES

[1] Biot, M.A.: "General Theory of Three-Dimensional Consolodation," **J. Appl. Phys., 12**, 155-164. (1941)

[2] Rice, J.R. and M.P. Cleary: "Some Basic Stress Diffusion Solutions for Fluid-Saturated Elastic Porous Media with Compressible Constituents," **Rev. Geophys. and Space Phys., 14**, 227-241. (1976)

[3] Sandhu, R.S.: "Finite Element Analysis of Coupled Deformation and Fluid Flow in Porous Media," **Numerical Methods in Geomechanics**, J. Martinus (ed.), Holland, 203-228. (1982)
[4] Zienkiewicz, O.C., V.A. Norris, L.A. Winnicki, D.J. Naylor, and R.W. Lewis: "A Unified Approach to the Soil Mechanics Problems of Offshore Engineering," **Numerical Methods in Offshore Engineering**, Zienkiewicz, Lewis and Stagg (ed.), Chichester, 361-412. (1978)
[5] Cheng, H.-D. A. and J.A. Ligget: "Boundary Integral Equation Method for Linear Porous Elasticity with Applications to Soil Consolidation," **Int. J. Num. Meth. Eng.**, **20**, 255-278. (1984)
[6] Cheng, H.-D. A. and J.A. Ligget: "Boundary Integral Equation Method for Linear Porous Elasticity with Applications to Fracture Propagation," **Int. J. Num. Meth. Eng.**, **20**, 279-296. (1984)
[7] Cleary, M.P.: "Rate and Structure Sensitivity in Hydraulic Fracturing of Fluid-saturated Porous Formations," **Proc. 20th U.S. Rock Mechanics Symposium**, Austin, TX, K. Gray (Ed.), (1979)
[8] Detouray E., J.D. McLennan and J.-C. Roegiers,: "Poroelastic Concepts Explain Some of the Hydraulic Fracture Concepts," **SPE 15262**, (1985).
[9] Ruina, A.: "Influence of Coupled Deformation-Diffusion Effects on the Retardation of Hydraulic Fracture," **Proc. 19th US Rock Mechanics Symposium**, 274-282. (1978)
[10] Huan85 Huang, N.C., and S.G. Russell: "Hydraulic Fracturing of a Saturated Porous Medium-II: Special Cases," **Theoret. Appl. Fracture Mech.**, **4**, 215-222. (1985)
[11] Kanninen, M.P. and C.H. Popelar: **Advanced Fracture Mechanics**, Oxford University Press, New York. (1985)
[12] Swenson D.V.: **Modelling Mixed-mode Dynami Crack Propagation**, PhD thesis, Cornell University. (1986)
[13] Boone T.J., P.A. Wawryznek and A.R. Ingraffea: "Simulation of the Fracture Processes in Rock with Application to Hydrofracture," **Int. J. Rock Mech. Min. Sci. & Geomech. Abstr.**, 255-265. (1986)
[14] Krech W.W.: "The Energy Balance Theory and Rock Fracture Energy Measurements for Uniaxial Tension," **Proc. 3rd Int. Cong. Int. Soc. Rock Mech.**, Nat. Acad. Sci., Washington, D.C., 167-173. (1974)
[15] Labuz J.F., S.P. Shah and C.H. Dowding: "Experimental Analysis of Crack Propagation in Granite. **Int. J. Rock Mech. Min. Sci. & Geomech. Abstr. 22**, 85-98. (1985)
[16] Gopalaratnam V.S. and S.P. Shah: "Softening Response of Plain Concrete in Direct Tension," **ACI J.**, **82**, 3, 310-323. (1985)
[17] Boone T.J., P.A. Wawryznek and A.R. Ingraffea: Discussion 82-27 (of [16]), **ACI J.**, **83**, 2, 316-318. (1986)
[18] Rice, J.R.: "A Path Independent Integral and the Approximate Analysis of Strain Concentration by Notches and Cracks," **J Appl. Mech.**, **35**, 379-386.(1968)
[19] Atkinson, B.K. and P.G. Meridith: "Experimental Fracture Mechanics Data for Rocks and Minerals," **Fracture Mechanics of Rock**, B.K. Atkinson (ed.), London, 477-525. (1987)
[20] Zienkiewicz, O.C.: "Basic Formulation of Static and Dynamic Behaviour of Soil and Other Porous Media," **Numerical Methods in Geomechanics**, J. Martinus (ed.), Holland, 39-56. (1982)
[21] Goodman R.E., R.L. Taylor and T.L. Brekke: "A Model for the Mechanics of Jointed Rock," **J. Soil Mech. Fdns. Div., Am. Soc. Civ. Engrs.**, **110**, 4, 871-890. (1984)
[22] Wawrzynek, P.A., T.J. Boone and A.R. Ingraffea: "Efficient Techniques for Modelling the Fracture Process Zone in Rock and Concrete," **Proc. 4th Int. Conf. Numerical Methods in Fracture Mechanics**, A.R. Luxmoore et al. (ed.), Swansea, 473-479. (1987)
[23] Comini G., S. Del Guidice, R.H. Lewis and O.C. Zienkiewicz: "Finite Element Solution of the Nonlinear Heat Conduction Problems with Special Reference to Phase Change," **Int. J. Num. Meth. Eng.**, **8**, 613-624. (1974)
[24] Underwood, P.: "Dynamic Relaxation - A Review," **Computational Methods for Transient Response Analysis**, T. Belytschko and T.J.R. Hughes, North-Holland, (1982).
[25] M. Lees: "A Linear Three-Level Difference Scheme for Quasilinear Parabolic Equations," **Maths. Comp.**,**20**, 516-522. (1966)
[26] Simon B.R., J.S.-S. Wu, O.C. Zienkiewicz, and D.K. Paul: "Evaluation of u-w and u-π Finite-Element Methods for the Dynamic Response of Saturated Porous Media using One-Dimensional Models," **Int. J. Num. Anal. Meth. Geomech.**, **10**, 5, 461-482. (1986)
[27] Zoback, M.D., F. Rummel, R. Jung and C.B. Raleigh: "Laboratory Hydraulic Fracturing Experiments in Intact and Pre-fractured Rock," **Int. J. Rock Mech. Min. Sci. & Geomech. Abstr.**, **14**, 49-58. (1977)
[28] Wawrzynek, P.: "Interactive Finite Element Analysis of Fracture Processes: An Integrated Approach," **MS Thesis**, Cornell University, Ithaca, NY, USA. (1987)
[29] Ingraffea, A.R.: "Theory of Crack Initiation and Propagation in Rock," **Fracture Mechanics of Rock**, B.K. Atkinson (ed.), London, 71-110. (1987)

Numerical Methods in Geomechanics (Innsbruck 1988), Swoboda (ed.)
© 1988 Balkema, Rotterdam. ISBN 90 6191 809 X

A modelling of jointed rock mass

T.Kawamoto
Nagoya University, Japan

ABSTRACT: The mechanical behaviour of rock mass is strongly affected by discontinuities such as faults and joints. In this paper, a damage mechanics theory is described to deal with some sets of discontinuities distributed in the rock mass, for example, joint sets. In this theory, the distributed discontinuities are characterised by a second-order symmetric tensor, called the damage tensor. By introducing the damage concept, it is possible to treat the deformation and fracturing of the rock mass within the framework of continuum mechanics. A numerical procedure is developed to implement the damage mechanics model by employing the finite element method. It is then applied to several laboratory tests on strength, deformability and seismic properties of rock mass models with distributed discontinuities and results of the analyses are compared with those of experiments to check the validity of the present damage mechanics approach.

1. INTRODUCTION

The evaluation of mechanical behaviour of rock mass and its modelling are essential requirements for the design of underground excavations as well as in many other rock mechanics problems. The discontinuities present in rock mass can cause a variety of problems in both surface and underground engineering operations. Therefore, it should be understood that rock mass is a structure, not a material, as it is composed of intact rock blocks and discontinuities of geologic origin.

As shown in Figure 1, the mechanical properties of rock mass with discontinuities were often evaluated through either emprical rock classifications based upon tests on rock material and geological investigation or in-situ test results ([2,6,9]). Although these are effective as an index in classifying the rock mass, it is not always possible to determine the mechanical properties for the design of rockmass structures.

In addition to the above methods, there are some approaches applying the mathematical models which consider the rock mass consisting of intact rock and discontinuities of geologic origin. In other words, these models evaluates the rock mass as an equivalent continuous medium considering the discontinuity sets as distributed discontinuities through the damage tensor ([22,23,28,29,30,31]) or fabric tensor([34]). A mechanical model for rock mass is generally developed for the analysis of rock structures using on eof the above approaches. However, in many cases, needs for the modification of the rock mass model (strength, deformability, etc.) often arise in relation with dimensions of discontinuities relative to those of the structures. When discontinuities are sparsely distributed, the rock mass about the structures can be modelled as a continuum medium. On the other hand, when the mass is highly jointed, and discontinuities of relatively large scale such as joint sets, faults, seams, fracture zones are present, the medium becomes highly discontinuous. A thorough treatment of these discontinuities as well as rock element is, therefore, required in the modelling of the rock mass. However, a gap between theory and practice, even at present, does exist, and there are still some rooms for emprical elements in the characterization of the rock mass. Nonetheless, attempts to introduce the rock mechanics approach to make the characterization of the mass rationalized have been recently increasing and very promising.

Discontinuities present in rock mass are characterised by orientation, spacing, persistence, filling material, seepage, number of discontinuity sets, and block size, etc. ([7]). These characteristics may be estimated by appropriate laboratory tests or field measurements. The current research activities mainly concentrated on the geometrical quantification of discontinuities and the mechanical properties of a single discontinuity plane ([1,16,39,40]). Nevertheless, researches on the mechanical behaviour of rock mass with several discontinuity sets are still rare and insufficient.

Several numerical models have been proposed on the basis of an explicit description of each discontinuity. Various joint elements are proposed to deal with a separable and contact body ([10,14,15,41]). Cundall's model is used for the time-dependent movement of several distinct rock blocks([8]). Pietruszczak and Mroz treat a shear band formation and softening of a rock specimen ([37]). Kawamoto et al extend this model to generate the shear band in a specific finite element and to assess a progressive failure of a rock slope([21]). These are probably the currently available models for discontinua.

When rock mass has a number of cracks which are relatively small in dimension compared with a structure, the distinct effect of each crack may be negligible. This implies that the rock mass can be idealised as a continuum body whether it is isotropic or anisotropic.

The problem becomes more important for the behaviour of rock mass when intermediate sized discontinuities are present. No appropriate theory or workable numerical method has yet been developed for this circumstance. Since these discontinuities involve complications, and their mechanical effects can not be estimated in a simple manner, a damage mechanics theory, presented herein, can handle this problem. That is, if the rock mass contains a number of cracks which are sufficiently small compared to the structure, the discontinuities can be regarded as damage of the mass. Then, the behaviour of the mass with such cracks can be conveniently be treated by describing the geometry of such cracks, and the properties of the intact

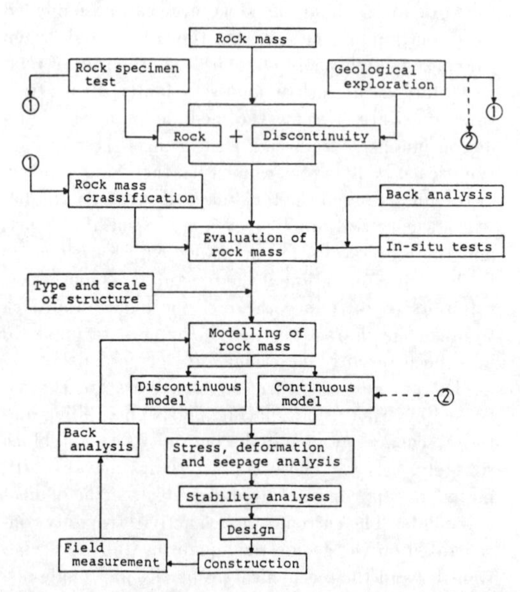

Figure 1. Flow diagram of evaluation of rock mass and analysis procedure.

rock, separately. Herein, we first present how to use the damage mechanics theory as an approach to model the discontinuous rock mass and its finite element formulation. Then, we show its applications to estimate the strength and deformability and seismic properties of the rock mass with distributed discontinuities and compare the computed results with experimental ones to check the validity of our approach.

2. THEORY OF DAMAGE MECHANICS AND ITS APPLICATION TO ROCK MECHANICS

2.1 Fundamental concept of damage mechanics

The damage mechanics was originated from the concept of damage in the form of numerous cracks and cavities due to the start and propagation of failure that result in the weakening of the material and is based upon the idea of representing their mechanical effect as a continuous variable field (called damage field) and to treat the mechanical behaviour of materials with numerous cracks and cavities within the framework of continuum mechanics.

The damage mechanics concept was first introduced by Kachanov to predict the time of creep rupture([20]). A scalar variable representing a measure of voids in a cross-section is employed as the damage variable, and the theory is only applicable to one-dimensional problems. Later, this theory was extended to the multi-dimensional case by many authors([18,33]). The damage mechanics approach is effective not only for creep problems of metals but also for softening behaviour of rocks and concrete. By using the damage concept, Dragon and Mroz proposed an elasto-plastic constitutive law of strain softening type, in which plastic dilation is directly related to the damage and employed as a softening parameter([11]). However, the softening behaviour of rocks and concrete is associated with structural changes. Frantziskonis and Desai proposed a damage model to describe such a softening behaviour([13]). They decompose such behaviour in two parts, a topical (continuum) part V_t and a damage part V_o, and define the average stress in the damage body by using the partial stress concept of the theory of mixtures. Then, each partial stress is related to a kinematically admissible strain field by a constitutive law, and are superposed. The propagation of damage is introduced through a power law of deviatoric plastic strain. Lemaitre proposed a similar type of stress-strain relations based on the strain equivalence hypothesis, which states that the strain behaviour of a damaged material is represented by the constitutive equation of the material with no damage, by simply replacing the stress by effective stress (net stress)([32]). However, distinguished from these approaches, the damage body treated herein is a discontinuous rock mass, and cracks regarded as damage are macroscopic ones such as joints. In general, almost all cracks in the rock mass are planar and distributed regularly. Such cracks can be characterized by a second-order symmetric tensor, called the damage tensor, which was previously proposed by Murakami and Ohno in their creep damage theory for metals([33]). The damage tensor represents a ratio of the effective resisting part of surface area in a damaged body, and it is set into a new framework of a continuum mechanics theory. Figure 2 shows the position of damage mechanics theory in the mechanics. It is

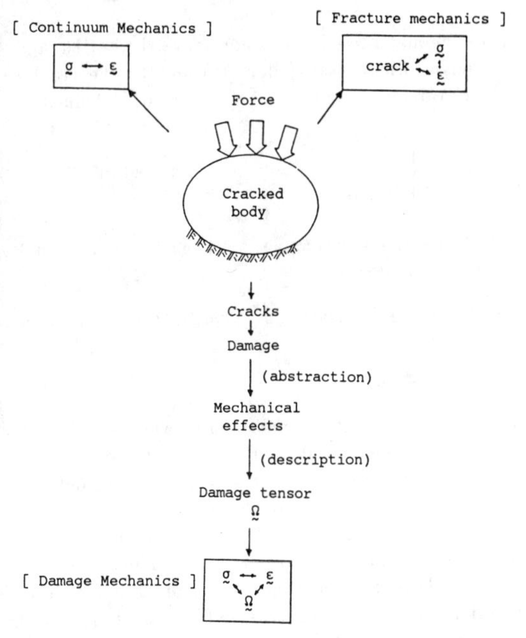

Figure 2. Damage mechanics, continuum mechanics and fracture mechanics.

possible to treat the mechanical behaviour of discontinuous rock mass within the framework of continuum mechanics using three field variables; stress field, displacement field and damage field variables. In classical continuum mechanics the stress field is related to the displacement field through the constitutive equation. On the other hand, this is done following the statically admissible stress field is transformed to the net stress field by the new field variable. Since the variable fields are three, another field variable called the propagation equation is necessary to relate the field variables to each other besides the constitutive equation (Fig.3).

2.2 Damage tensor for rock mass

Let us suppose that a set of planar defects are distributed in a material body with the areal density Ω

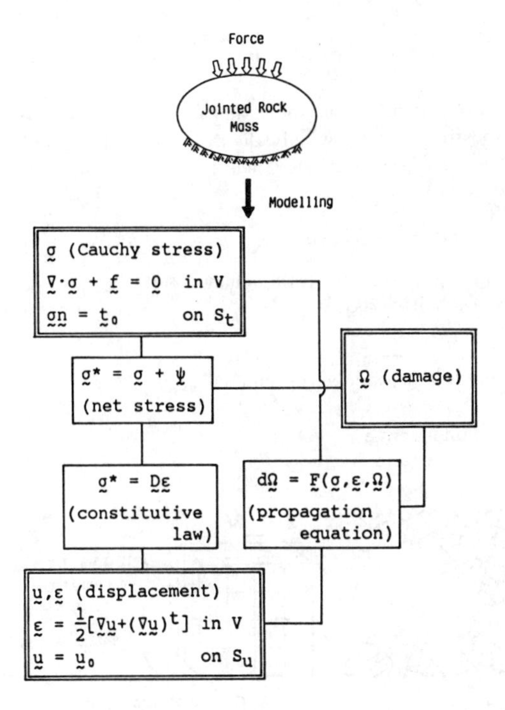

Figure 3. Diagram of Damage mechanics theory for discontinuous rock mass.

and the unit normal $\boldsymbol{n}$, then the damage tensor is simply defined as

$$\boldsymbol{\Omega} = \Omega(\boldsymbol{n} \otimes \boldsymbol{n}) \qquad (1)$$

where $\otimes$ denotes the tensor product.

For the rock mass, the jointing is idealised as follows:

1. The individual joints are of planar shape.
2. The rock mass is broken into the same size of blocks. The size is intrinsic to the rock formation (Fig.4). We call the block as the fundamental element. Discontinuities or cracks exist on the interfaces of these elements, and will propagate along the interfaces if the applied load is changed. The size of the fundamental element is determined by observing the spacing of the joint sets.

The fundamental element is idealised into a cube with a volume V and a side length l (Fig.4). Suppose that in the rock mass with its volume V, there exists N-cracks, and let the area of the k-th crack be a^k with a unit normal $\boldsymbol{n}^k$. Then, the damage tensor can be expressed as

$$\boldsymbol{\Omega} = \frac{l}{V}\sum_{k=1}^{N} a^k(\boldsymbol{n}^k \otimes \boldsymbol{n}^k) \tag{2}$$

However, informations on a^k and $\boldsymbol{n}^k$ for all cracks are impossible. We know only the average $\bar{a}^i$, N^i, l^i and $\boldsymbol{n}^i$ for the i–th joint set. Then, the damage tensor of this set is

$$\boldsymbol{\Omega}^i = \frac{l^i}{V^i} N^i \bar{a}^i(\boldsymbol{n}^i \otimes \boldsymbol{n}^i) \tag{3}$$

Summing up this, the total damage tensor of the rock mass is written as

$$\boldsymbol{\Omega} = \sum \boldsymbol{\Omega}^i \tag{4}$$

where n is the number of joint sets.

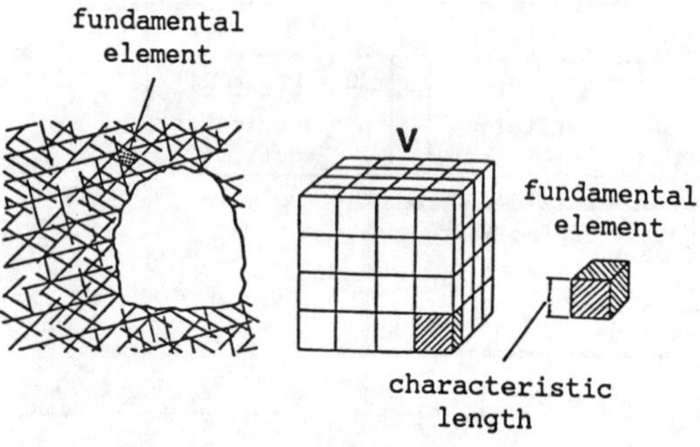

Figure 4. Jointed rock mass and fundamental element: (a) jointed rock mass; (b) idealized fundamental element.

2.3 Net stres for rock mass

Due to the plane cracks involved in the material body, the effective area on which the stress vector is acting is reduced, hence if the cracks does not transmit the force completely, the Cauchy stress $\boldsymbol{\sigma}$ can be changed into the net stress $\boldsymbol{\sigma}^*$ (Murakami and Ohno ([33])):

$$\boldsymbol{\sigma}^* = \boldsymbol{\sigma}\,(\,\boldsymbol{I} - \boldsymbol{\Omega}\,)^{-1} \tag{5}$$

where $\boldsymbol{I}$ is the unit tensor.

In the rock masses, however, a part of the shear stress transmitted throughout cracks as well as the compressive stress normal to the cracks. Then the net stress $\boldsymbol{\sigma}^*$ is modified as

$$\begin{aligned}\boldsymbol{\sigma}^* = \boldsymbol{T}^t\,[\,\boldsymbol{\sigma}'_t\,(\,\boldsymbol{I} - C_t\,\boldsymbol{\Omega}'\,)^{-1} \\ + \;\boldsymbol{\sigma}'_n\,\{\,\boldsymbol{H}\,(\,\boldsymbol{\sigma}'_n\,)(\,\boldsymbol{I} - \boldsymbol{\Omega}'\,)^{-1} \\ + \boldsymbol{H}\,(-\,\boldsymbol{\sigma}'_n\,)(\,\boldsymbol{I} - C_n\,\boldsymbol{\Omega}'\,)^{-1}\}]\,\boldsymbol{T}\end{aligned} \tag{6}$$

where $\boldsymbol{T}$ is the coordinate transformation tensor composed by eigenvectors of $\boldsymbol{\Omega}$, so that the diagonal tensor $\boldsymbol{\Omega}'$ is

$$\boldsymbol{\Omega}' = \boldsymbol{T}\boldsymbol{\Omega}\boldsymbol{T}^t \tag{7}$$

The stress $\boldsymbol{\sigma}$ is similarly transformed into the eigendirection of the damage tensor:

$$\boldsymbol{\sigma}' = \boldsymbol{T}\boldsymbol{\sigma}\boldsymbol{T}^t \tag{8}$$

This $\boldsymbol{\sigma}'$ is decomposed into the diagonal term $\boldsymbol{\sigma}'_n$ and the deviatoric term $\boldsymbol{\sigma}_t$ as

$$\boldsymbol{\sigma}' = \boldsymbol{\sigma}'_n + \boldsymbol{\sigma}'_t \tag{9}$$

where

$$\boldsymbol{\sigma}'_n = \begin{bmatrix} \sigma'_{11} & 0 & 0 \\ 0 & \sigma'_{22} & 0 \\ 0 & 0 & \sigma'_{33} \end{bmatrix}, \quad \boldsymbol{\sigma}'_t = \begin{bmatrix} 0 & \sigma'_{12} & \sigma'_{13} \\ \sigma'_{21} & 0 & \sigma'_{23} \\ \sigma'_{31} & \sigma'_{32} & 0 \end{bmatrix}$$

The diagonal stress $\boldsymbol{\sigma}'_n$ acts normal to the cracks, and the deviatoric stress $\boldsymbol{\sigma}'_t$ does tangent to the cracks. The characteristics function $\boldsymbol{H}$ in Eqn(6) is defined as

$$H_{ij}(x_{ij}) = \begin{cases} 0 & \text{if} \quad x_{ij} \le 0 \\ 1 & \text{if} \quad x_{ij} > 0 \end{cases} \qquad (i, j \text{ : not summed }) \tag{10}$$

2.4 Costitutive law, damage propagation and virtual work equation

The stress-strain relation is given between the strain $\boldsymbol{\epsilon}$ and the net stress $\boldsymbol{\sigma}^*$:

$$\boldsymbol{\epsilon} = \boldsymbol{\Phi}(\,\boldsymbol{\sigma}^*\,) \tag{11}$$

However, in laboratory experiments, we use an intact rock as specimens which are comparatively smaller than the fundamental element, and have no defects in the sense of intactness of the fundamental element. This implies that $\boldsymbol{\Omega} = \mathbf{o}$ for these specimens, then as observed in Eqn(6), we have $\boldsymbol{\sigma} = \boldsymbol{\sigma}^*$. Thus, the constitutive law obtained from laboratory uniaxial and triaxial tests (Kyoya, Ichikawa and Kawamoto ([23,28]) is directly used in the damage theory, and it is not needed to obtain the rock mass properties. This is an intrinsic characteristic of the damage theory in the rock mechanics.

If the intact rock is linear and elastic, Eqn(11) is written as follows:

$$\boldsymbol{\sigma}^* = \boldsymbol{D}\boldsymbol{\epsilon}\,, \quad \boldsymbol{\epsilon} = \boldsymbol{C}\boldsymbol{\sigma}^* \tag{12}$$

In the case of dynamic problems, the discontinuities has an effect on not only the net stress but also the scattering and damping of wave propagation. Therefore, the stress-strain relation is given as

$$\boldsymbol{\sigma}^* = \boldsymbol{D}\boldsymbol{\epsilon} + \boldsymbol{C}\dot{\boldsymbol{\epsilon}} \tag{13}$$

where $\boldsymbol{C}$ denotes the damping coefficients $\boldsymbol{C} = \boldsymbol{C}(\boldsymbol{\Omega})$.

The discontinuities or cracks in the rock mass will be further developed. this propagation equation is usually non-linear, and is written in the incremental form as

$$d\Omega = F(\sigma, \epsilon, \Omega) \tag{14}$$

The comprehensive form of the propagation equation (14) is however still unrecognised.

The virtual work equation is given in terms of the Cauchy stress σ as

$$\int_V \rho\ddot{u}\cdot\delta u + dV \int_V \sigma\cdot\delta\epsilon dV = \int_{S_t} t^o\cdot\delta u dS_t + \int_V f\cdot\delta u dV \tag{15}$$

where t^o is the external force is given on S_t, and f the body force vector. Now, the net stress given by Eqn(6) is written in the form

$$\sigma^* = \sigma + \psi \tag{16}$$

where

$$\psi = T^t\,[\,\sigma'_t\,(\,\phi_t - I\,) + \sigma'_n\,\{\,H\,(\,\sigma'_n\,)\,\phi + H\,(-\,\sigma'_n\,)\,\phi_n\,] - I\,\}]\,T$$

$$\phi = (\,I - \Omega'\,)^{-1},\quad \phi_n = (\,I - C_n\,\Omega'\,)^{-1},\quad \phi_t = (\,I - C_t\,\Omega'\,)^{-1}$$

Substituting the relation (12) into Eqn(15) we have the virtual work equation for the damage field:

$$\int_V \rho\ddot{u}\cdot\delta u + \int_V \sigma^*\cdot\delta\epsilon dV = \int_{S_t} t^o\cdot\delta u dS_t + \int_V f\cdot\delta u dV + \int_V \psi\cdot\delta\epsilon dV \tag{17}$$

3. FINITE ELEMENT DISCRETIZATION PROCEDURES AND METHOD OF ANALYSIS

The standard finite element discretization is introduced as follows:

$$\{u\} = [N]\{U\},\quad \{\epsilon\} = [B]\{U\} \tag{18}$$

where $\{U\}$ is the nodal displacement vector, $[N]$ the matrix of shape fuction, $[B]$ the strain-nodal displacement matrix, and $\{u\}$ and $\{\epsilon\}$ are the vector forms of the displacement the strain, respectively.

Substituting these into equation(17), the following simultaneous equations are obtained for the dynamic damage analysis:

$$[M]\{\ddot{U}\} + [C]\{\dot{U}\} + [K]\{U\} = \{F\} + \{F^*\} \tag{19}$$

in which,

$$[M] = \int_V \rho[N]^t[N]dV$$

$$[C] = \int_V [B]^t[C][B]dV$$

$$[K] = \int_V [B]^t[D][B]dV$$

$$\{F\} = \int_{S_t} [N]^t\{t^0\}dS_t + \int_V [N]^t\{f\}dV$$

$$\{F^*\} = \int_V [B]^t\{\psi\}dV$$

It is observed that the stiffness matrix $[K]$ depends only on the material properties of the intact rock and the mechanical effects of damage are represented by force term $\{F^*\}$.

For the case of static problems, $\{\ddot{U}\} = \{\dot{U}\} = \{0\}$, so the following equation is obtained:

$$[K]\{U\} = \{F\} + \{F^*\} \tag{20}$$

In the analysis of dynamic problems, the lumped mass matrix is used for the mass matrix $[M]$, and the damping matrix $[C]$ is assumed of the following form:

$$[C] = \alpha[M] \tag{21}$$

using the Rayleigh damping. The central differences method is used in discretisation of Eqn(19) in time domain, and the following expression is obtained:

$$\left(\tfrac{1}{\Delta t^2}[M] + \tfrac{1}{2\Delta t}[C]\right)\{U_{n+1}\} = \{F_n\} + \{F^*\} - \left([K] - \tfrac{2}{\Delta t}[M]\right)\{U_n\} - \left(\tfrac{1}{\Delta t^2}[M] - \tfrac{1}{2\Delta t}[C]\right)\{U_{n-1}\} \tag{22}$$

$\{U_0\}$ is the initial condition, and $\{U_{-1}\}$ is given by the following relation:

$$\{U_{-1}\} = \{U_0\} - \Delta t\{\dot{U}_0\} + \frac{\Delta t^2}{2}\{\ddot{U}_0\} \tag{23}$$

In the static problems, Eqn(20) can be resolved into the followings:

$$[K]\{U_F\} = \{F\} \tag{24}$$

$$[K]\{U_F^*\} = \{F^*\} \tag{25}$$

$$\{U\} = \{U_F\} + \{U_F^*\} \tag{26}$$

That is, by solving equation (24), the nodal displacement $\{U_F\}$ of the body with non-damage is first obtained. Then the Cauchy stress σ is calculated from this $\{U_F\}$ as

$$\{\sigma\} = [D][B]\{U_F\} \tag{27}$$

By using the Cauchy stress term σ and the damage tensor Ω, the net stress σ^* is computed. Then, the additional nodal force vector $\{F^*\}$ is calculated by Eqns(16) and (19), and the corresponding displacement $\{U_F^*\}$ is obtained by solving Eqn(25). Finally, adding up $\{U_F^*\}$ to $\{U_F\}$, the global displacement $\{U\}$ for the damaged body is computed.

4. DATA PROCESSING OF JOINTING

The damage tensor of a rock mass which involves a couple of joint sets can be determined by Eqns(3) and (4), if the number of joints per unit volume (density), spacing, trace length and direction of jointing is known for each joint set. It is however difficult to identify the individual joint set on outcrops. We here present two methods to estimate the outcrop data with the help of a microcomputer.

Direct Observation Method

Suppose that there exists one set of joints in a rock mass, and the jointing can be observed on the mutually orthogonal surfaces X_1, X_2, and X_3 (Fig.5). We measure the angle and length of each joint as shown in Fig.5, and count the number of joints on each surface. Let the number per unit area be ordered as $N_p \geq N_q \geq N_r$, and the average joint length on each surface $L_p \geq L_q \geq L_r$, respectively. Then, the average area $\bar{a}^i$ in Eqn(3) is supposed to be as

$$\bar{a}^i = L_p L_q$$

and the number N^i involved in this mass is

$$N^i = V^{\frac{1}{3}} \left\{ \frac{N_p N_q}{L_p L_q \sqrt{(1-n_p^2)(1-n_q^2)}} \right\}^{\frac{1}{2}}$$

where n_p and n_q are the p-th and q-th components of the unit tensor $\boldsymbol{n}$ which is normal to the jointing plane. Substituting these into Eqn(3), the damage tensor of this set is determined as

$$\boldsymbol{\Omega} = \frac{l}{V^{\frac{2}{3}}} \left\{ \frac{N_p N_q L_p L_q}{\sqrt{(1-n_p^2)(1-n_q^2)}} \right\}^{\frac{1}{2}} (\boldsymbol{n} \otimes \boldsymbol{n}) \quad (28)$$

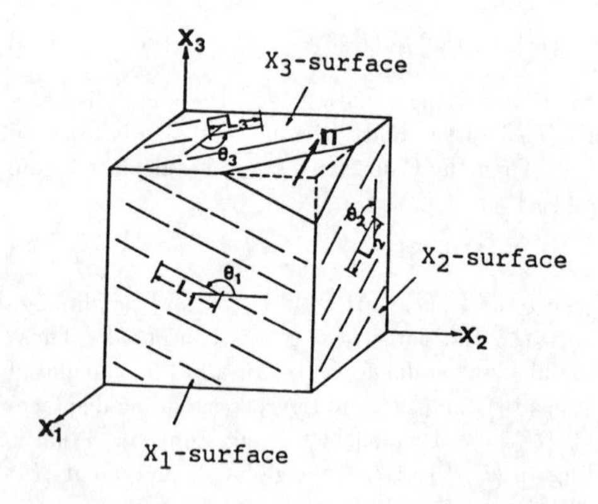

Figure 5. Single set of joints.

Figure 6. Stereophotograph of outcrop.

It should be noted that for the angles θ_1, θ_2 and θ_3, we have the relation

$$\tan\theta_1 \tan\theta_2 \tan\theta_3 = -1 \quad (29)$$

and the unit normal $\boldsymbol{n}$ is written as

$$\boldsymbol{n} = (\alpha\cos\theta_p\cos\theta_q,\ \alpha\sin\theta_p\sin\theta_q,\ -\alpha\cos\theta_p\sin\theta_q)$$
$$\alpha = (\cos^2\theta_p + \sin^2\theta_p\sin^2\theta_q)^{-\frac{1}{2}} \quad (30)$$

By using the relation (29), we may separate out the plural and complex jointing into individual sets.

Stereophotograph Method

If a set of stereophotographs of an outcrop is prepared, the direction of each joint can be specified by applying the Bundle Method with Self-Calibration (Kondoh and Shinji 1986), which is originally developed for the purpose of photographic surveying. Inputting at least three points of a joint plane from the set of pictures by the digitizing pad of a microcomputer yields in the determination of the unit normal of the plane. Fig.6 shows a set of stereophotographs taken in an exploration adit of a hydroelectric power station. In order to calibrate the spacing and trace length of joints, the standard square scale is also set in the pictures.

It is however difficult to determine the size of joints by a set of stereophotographs, since the pictures can provide only the data of one face. If the data for several planes are combined, it becomes possible to specify the size.

5. VERIFICATION OF NUMERICAL ANALYSIS

In order to verify the developed damage mechanics theory and the computation procedure, the results of numerical analyses are compared with the series of model tests.

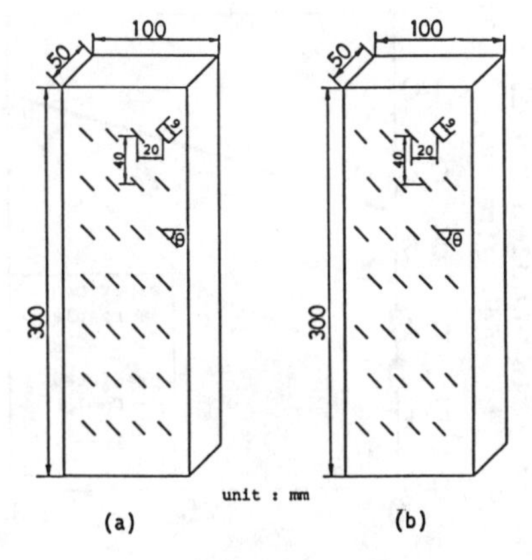

Figure 7. Specimens for uniaxial compression test: (a) columnar arranged cracks; (b) zigzag arranged cracks.

5.1 Uniaxial compression tests

Uniaxial compression tests were carried out using rectangular prismatic specimens with several patterns of planar cracks, as in Figure 7. Parameters to be adopted are (a) intact materials, (b) types of crack pattern and (c) angles of set of cracks.

Materials used are (a) a gypsum plaster mortar which consists of a mixture of gypsum plaster and water with proportions of 1:1, and (b) a cement mortar which is a mixture of cement, fine sand and water with proportions of 1:1:0.46 by weight respectively. Material properties of these two mortars are given in Table 1 and Table 2 respectively. These properties are obtained from a series of triaxial tests. In the tables, C and ϕ denote the cohesion and the internal friction angle of the Mohr-Coulomb's criterion at peak, respectively, while α and K_0 are constants for the Drucker-Prager's criterion at initial yielding:

$$\alpha\bar{\sigma} + \bar{s} = K_0 \tag{31}$$

where $\bar{\sigma} = \frac{1}{\sqrt{3}} tr(\boldsymbol{\sigma})$ is the mean stress and $s = (\boldsymbol{s} \cdot \boldsymbol{s})^{\frac{1}{2}}$ is the norm of the deviatoric stress tensor $\boldsymbol{s} = \boldsymbol{\sigma} - tr(\boldsymbol{\sigma})\boldsymbol{I}/3$.

Cracks were made by inserting thin steel strips (thickness 0.2mm); thus they were open and planar. For plaster mortar specimens, the number of cracks is 52, and their inclination θ was varied from 0°, 30°, 45°, 60° to 90°. The cracks were regularly arranged in an arrays as illustrated in Figure 7. On the other hand, cement mortar specimens contained 28, 36 and 52 cracks, and inclinations were 0°, 45° and 90°. Their crack arrays were regularly arranged in either a columnar pattern (Figure 7(a)) or a zigzag pattern (Figure 7(b)).

Table 1. Material properties of plaster mortar

Elastic constants		Peak strength (Mohr-Coulomb)	
Young's modulus E(MPa)	Poisson's ratio ν	Friction angle ϕ	Cohesion C(MPa)
1110	0.17	8.0°	1.32

Table 2. Material properties of cement mortar

Elastic constants		Peak strength (Mohr-Coulomb)		Initial strength (Drucker-Prager)	
Young's modulus E(MPa)	Poisson's ratio ν	Friction angle ϕ	Cohesion C(MPa)	α	K_0
14020	0.125	47.0°	11	0.77	10.72

In analyses, materials were assumed to be elastic under a plane stress state, and the propagation of damage was not considered ($d\boldsymbol{\Omega} = \mathbf{o}$). Coefficients C_t for the net stress (16) was set as 1.0, and C_n was varied as 0.0, 0.5, and 1.0. The damage tensor of each specimen are directly determined from Eqn (3).

Let Young's moduli E_θ of these specimens be estimated by $E_\theta = P/(A\delta)$, in which P is the applied load, A the area of cross-section of the specimen and δ the vertical displacement at the top of the specimen. They are normalized by E_θ/E_0, where E_0 is the Young's modulus of the intact specimen. Results are presented in Figure 8 for the plaster mortar, and in Figure 9 for the cement mortar. These agree with the experimental data if $C_n = 1.0$. This is because the cracks are open and the condition $C_n = 1.0$ corresponds to open cracks.

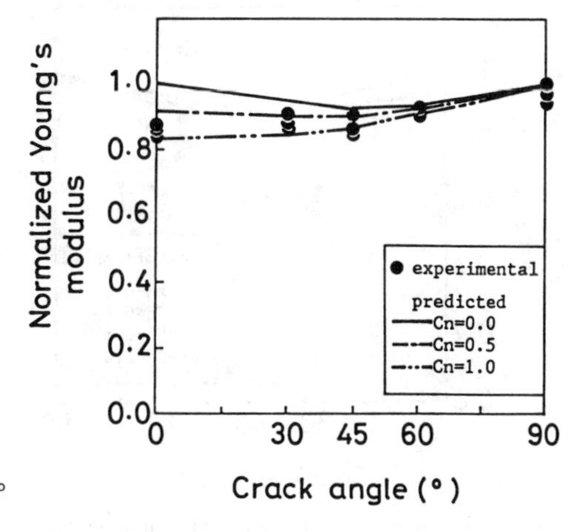

Figure 8. Normalized Young's moduli of plaster mortar specimens.

Next, let the apparent peak strength of the specimen with cracks be calculated as follows: The net stress σ_x^*, σ_y^* and σ_{xy}^* which satisfy the Mohr-Coulomb's failure condition for the intact material are calculated

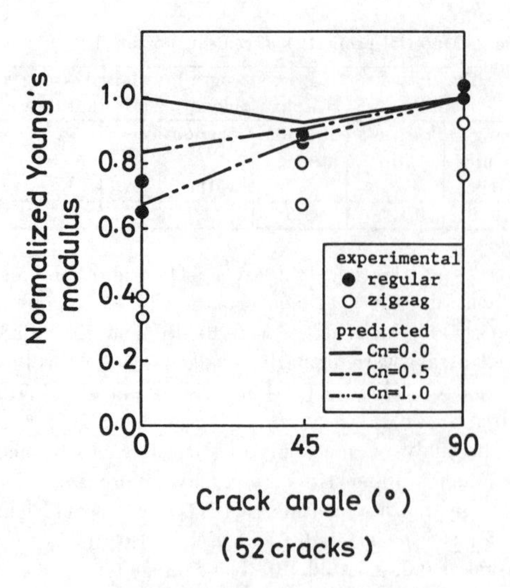

Figure 9. Normalized Young's moduli of cement mortar specimens with 52 cracks.

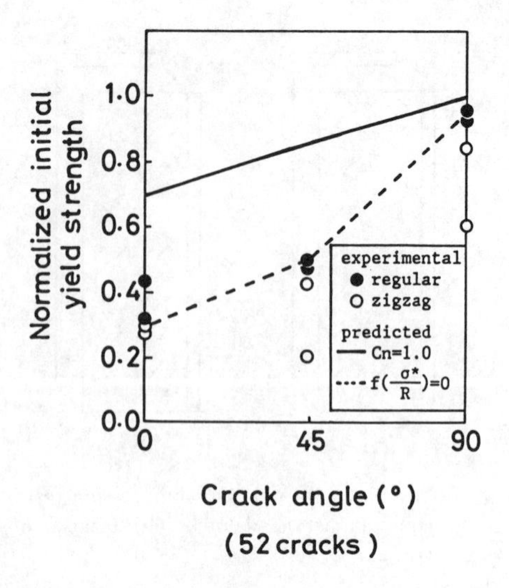

Figure 11. Normalized initial yield strengths of cement mortar specimens with 52 cracks.

from an applied stress $\sigma_\theta = P/A$ by using Eqn(16). Then, the apparent peak strength is defined by the applied stress σ_θ, and is normalized by the uniaxial strength σ_0 of the intact specimen. In Figure 10, the values predicted by this method for the plaster mortar are compared with experimental results which are obtained directly by dividing the strength of cracked specimen by the intact one.

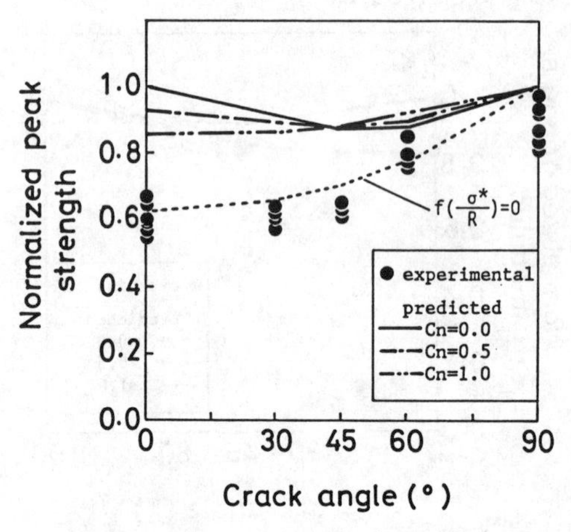

Figure 10. Normalized peak strengths of plaster mortar specimens.

Similarly, the apparent initial strength is calculated by using Drucker-Prager initial yield condition, and for the cement mortar. The predicted values are compared with the experimental ones in Figure 11. It is observed that these apparent strengths are relatively lower than the predicted values. We suppose that the failure or yielding phenomenon of rock-like materials is induced by propagation of cracks, which is resulted mainly from the local stress concentration in the vicinity of pre-existing cracks and interactions of cracks, while in the above predictions the local effects are not considered. Thus, this simple method is insufficient for estimating the strength, so that the propagation equation should be introduced. However, on the micro-scale, every rock involves cracks, even if it looks intact, and the micro-cracks are growing into lager cracks after yielding. This interaction phenomenon is highly nonlinear, and it is difficult to formulate the initiation and growth condition of the damage. In order to avoid these difficulties, we introduce a reduction coefficient R into the yield or failure criterion such as

$$f(\frac{\sigma^*}{R}) = 0, \quad 0 < R < 1 \tag{32}$$

For the Mohr-Coulomb criterion it is given by

$$\sigma_1^* - \sigma_3^* = 2RC\cos\phi - (\sigma_1^* + \sigma_3^*)\sin\phi \tag{33}$$

and for the Drucker-Prager condition it is given by

$$\alpha\bar{\sigma}^* + s^* = RK_0 \tag{34}$$

Thus, if the yield function is homogeneous of order one with respect to the net stress, the surface is proportionally reduced by the coefficient R.

The experimental results shown in Figures 10 and 11 suggest that the form of R should be

$$R(\boldsymbol{\sigma}^*, \boldsymbol{\Omega}) = a - b\,[\frac{L}{l+L}(1-\frac{1}{3}tr\boldsymbol{\Omega})^{-\frac{1}{2}}]\sin\theta^* \tag{35}$$

where a and b are constants, l is the average spacing of the cracks, L the average trace length of the cracks, and θ the angle between the unit normal vector of the plane damage and the direction of the maximum of the principal net stress. The term $(1 - tr\boldsymbol{\Omega}/3)$ implies a ratio of the mean effective resisting part. By the least squares method, the constants a and b in Eqn(35) are determined as $a = 0.95, b = 0.08$ for the peak strength of plaster mortar specimens: this results shown in Figure 10 by dashed line. Similarly, for the initial yielding of cement mortar specimens, the constants are $a = 0.95, b = 1.7$, and results using Eqn(34) is also shown in Figure 11.

5.2 Direct shear tests

We next apply the damage mechanics theory to direct shear tests, and try to predict anisotropic strength induced by distributed cracks. The experiments are schematically illustrated in Figure 12. The specimens are made of a plaster mortar, and its material properties are given in Table 3. Cracks are also formed by steel strips, so that they are open.

Table 3. Material properties of plaster mortar

Elastic constants		Peak strength (Mohr-Coulomb)	
Young's modulus E(MPa)	Poisson's ratio ν	Friction angle ϕ	Cohesion C(MPa)
3730	0.16	56.3°	3.33

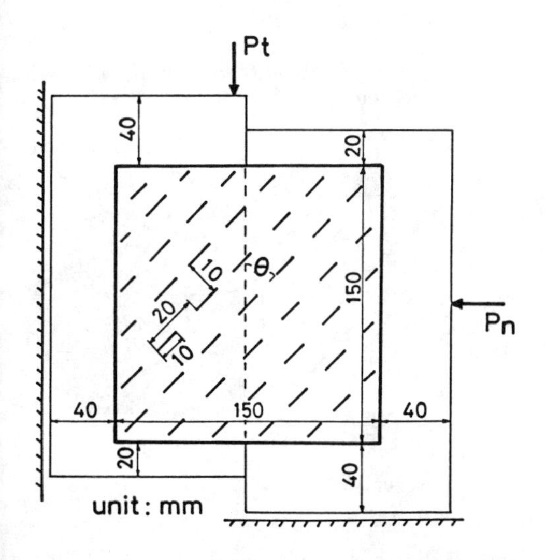

Figure 12. Direct shear test and crack pattern.

In the computations we assume that the material is elastic, and in order to understand a safety margin for failure, the failure criterion (33) is also introduced where the contants are $a = 0.95$ and $b = 0.08$. The constant C_t in the net stress equation (16) is chosen to be 1.0, and C_n is changed as 0.0, 0.5 and 1.0. The load at failure is obtained if the net stress at the centre of the specimen satisfies the failure criterion (33).

Experimental and predicted results are presented in Figure 13. Except for $P = 25N$ there is a mere disagreement; it is observed that the case $C_n = 0.5$ agrees reasonably well with experiments. This may result from some cracks becoming closed during the shearing process. Thus, if the initiation and propagation of a damage are considered by such a simple method, it is possible to predict the anisotropic strength characteristic of such medium.

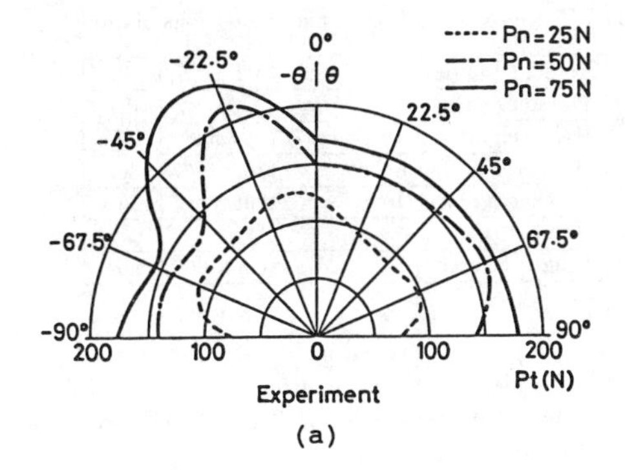

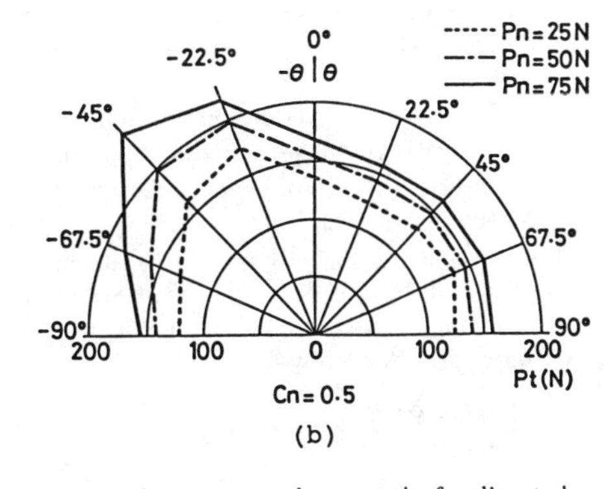

Figure 13. Apparent peak strengths for direct shear test: (a) experimantal; (b) predicted for the case $C_n = 0.5$.

5.3 Seismic tests

Next, we check the velocity of the damage model for dynamic problems. Seismic tests were carried out using disc-like specimens with twenty-four cracks as illustrated in Fig.14. The specimens had sixteen corners. The materials were made of gypsum plaster mortar ,the properties of which are given in Table 4.

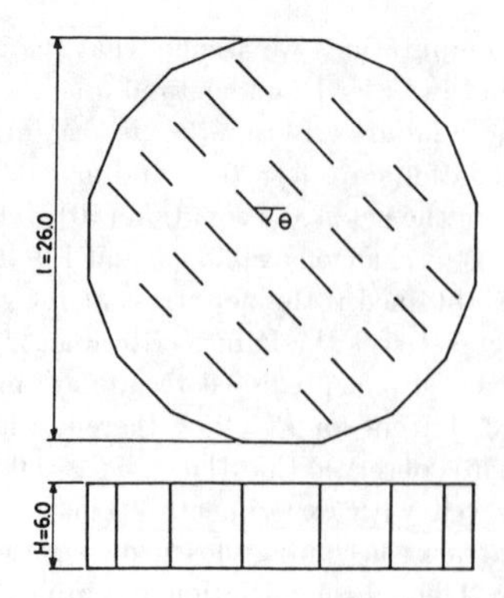

Figure 14. Specimen for seismic test.

Table 4. Material properties and dimensions of model

Young's modulus	E	(MPa)	2400
Poisson's ratio	ν		0.1
Unit weight	γ	(gf/cm^3)	1.17
Diameter of Specimen	D	(cm)	26
Thickness of Specimen	H	(cm)	6
Number of cracks			24
Length of cracks	l	(cm)	3

By using a P-wave transducer, characteristics of wave transmission in various directions were observed. Fig.15 show the wave form and its Fourier spectrum of acceleration generated by the P-wave transducer. The length of the wave transmitting in a specimen is estimated as about 30mm using the material constants given in Table 4.

A series of analyses were performed to simulate the experiments using the damage mechanics approach. Boundary conditions and finite element meshes used in the analyses are shown in Fig.16.

The inclination of cracks with respect to the direction of wave transmission is denoted by θ as shown in Fig.16(a). The damage tensors for each inclination of cracks are determined as follows:

$$\theta = 0^\circ \quad \boldsymbol{\Omega} = \begin{bmatrix} 0.000 & 0.000 \\ 0.000 & 0.402 \end{bmatrix} \quad \theta = 22.5^\circ \ \boldsymbol{\Omega} = \begin{bmatrix} 0.059 & 0.142 \\ 0.142 & 0.343 \end{bmatrix}$$

$$\theta = 45^\circ \ \boldsymbol{\Omega} = \begin{bmatrix} 0.201 & 0.201 \\ 0.201 & 0.201 \end{bmatrix} \quad \theta = 67.5^\circ \ \boldsymbol{\Omega} = \begin{bmatrix} 0.343 & 0.142 \\ 0.142 & 0.059 \end{bmatrix}$$

$$\theta = 90^\circ \ \boldsymbol{\Omega} = \begin{bmatrix} 0.402 & 0.000 \\ 0.000 & 0.000 \end{bmatrix}$$

In the analyses, coefficients C_t and C_n are set as 1.0, and the constant α in Eqn(21) was chosen to be 0 and 8000. Time interval Δt in Eqn(22) was chosen to be 0.5 micro second. The acceleration wave shown in Fig.16(a) was input from one side of the model and the response of acceleration at the opposite side was observed.

Acceleration waves transmitted through the cracked specimens and their Fourier spectrums are shown in Fig.17 for experiments, and in Fig.18 for the damage analyses for $\alpha = 8000$. It is observed that the spectrums computed agree reasonably well with those of experiments.

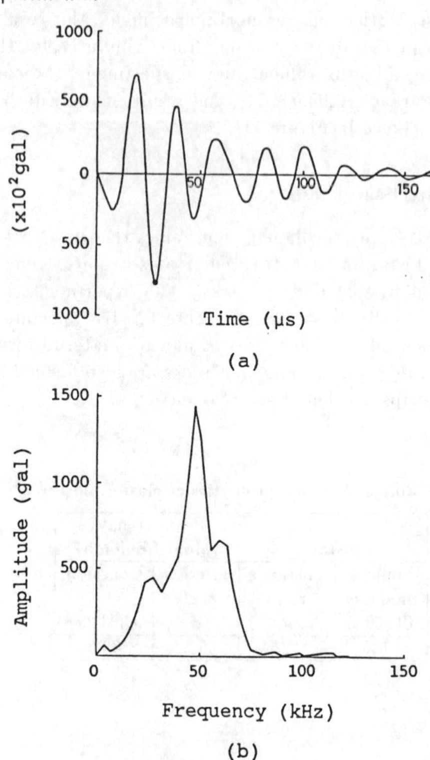

Figure 15. Input acceleration: (a) wave form; (b) Fourier spectrum.

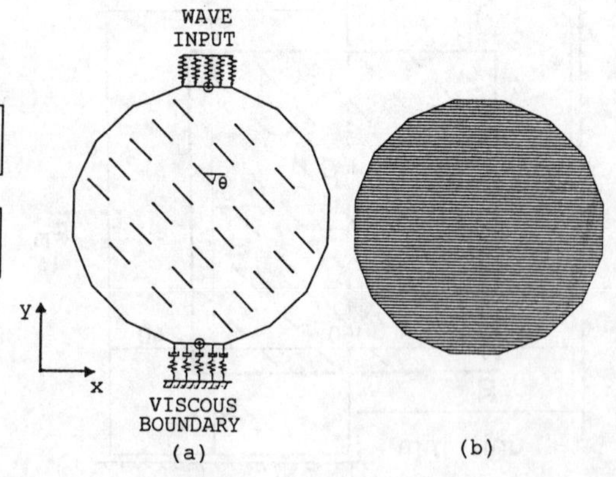

Figure 16. Finite element model: (a) boundary conditions; (b) mesh divisioin.

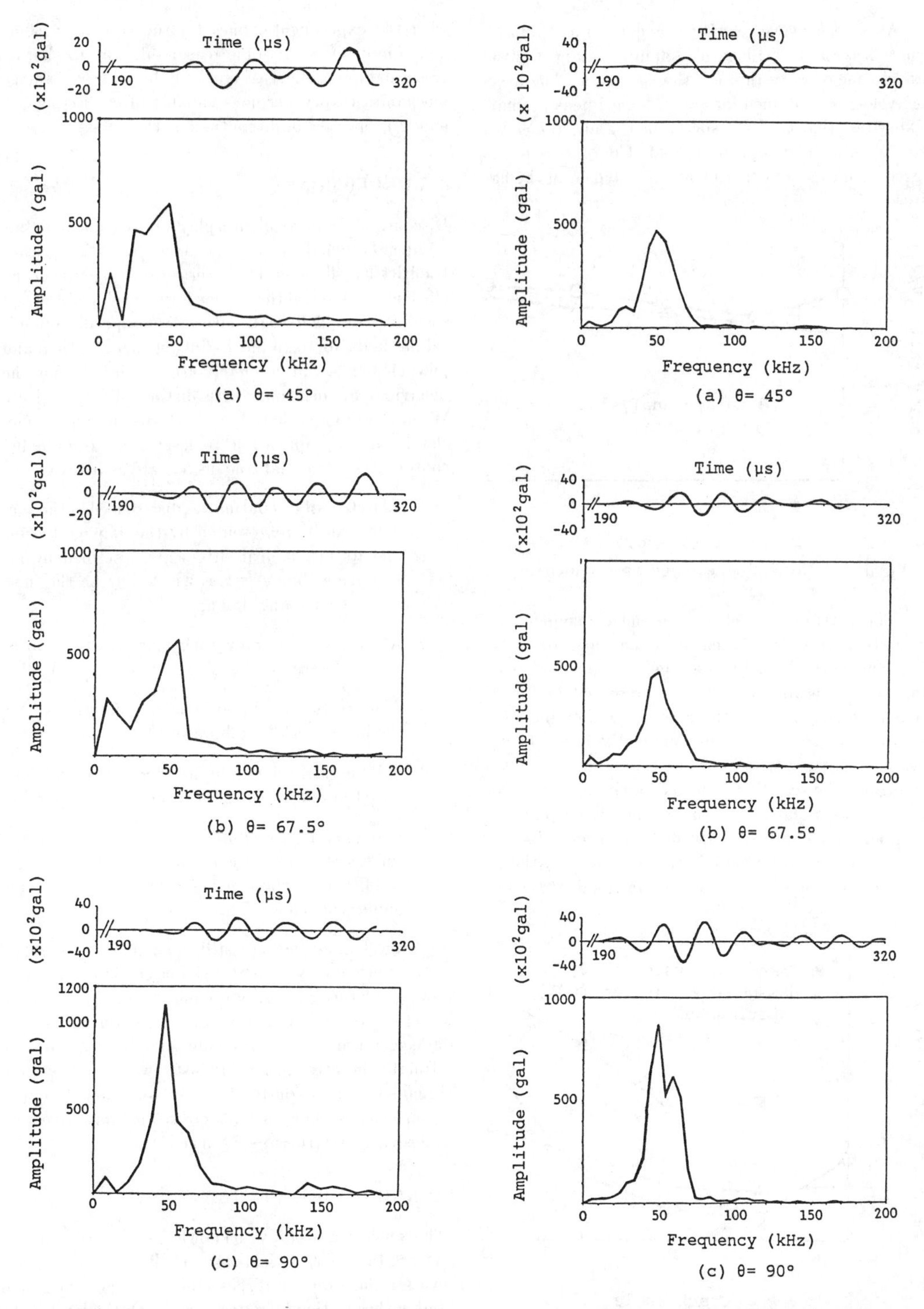

Figure 17. Acceleration responses in experiments: (a) $\theta = 45°$; (b) $\theta = 67.5°$; (c) $\theta = 90°$.

Figure 18. Acceleration responses predicted by damage analysis: (a) $\theta = 45°$; (b) $\theta = 67.5°$; (c) $\theta = 90°$.

An average wave velocity V is defined as $V = h/t$, where h is the travel length (260 mm) in and t is the travel time of wave through the specimen. The average velocities obtained for cracked specimens are normalised by that of intact specimen V_0, and are shown in Fig.19 in the case of $\alpha = 8000$. Computed results agree with those of experiments as noted from the figure.

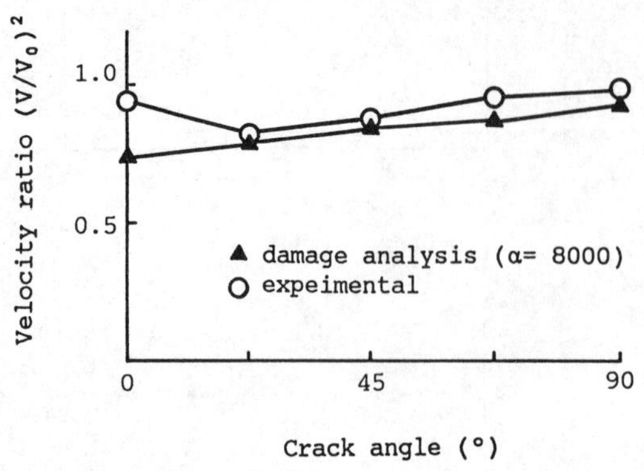

Figure 19. Normalized average wave velocity.

The maximum amplitude of the transmitted acceleration wave was picked up as an index of damping, and normalised by the one obtained for an intact specimen A_0. Results are presented in Fig.20. It is observed that, for the case of no-damping ($\alpha = 0$), the damage analyses predict relatively higher values as compared with those of experiments. This is thought to be resulting from inexplicit representation of of cracks in damage mechanics model while, in experiment, waves are attenuated through reflection and scattering caused by actual cracks. However, the results of the analyses are almost in good agreement with the experimental ones for the case $\alpha = 8000$. This implies that, if some reasonable way to characterize damping effect of cracks is found, the damage mechanics theory becomes capable of evaluating dynamic properties of discontinuous rock mass.

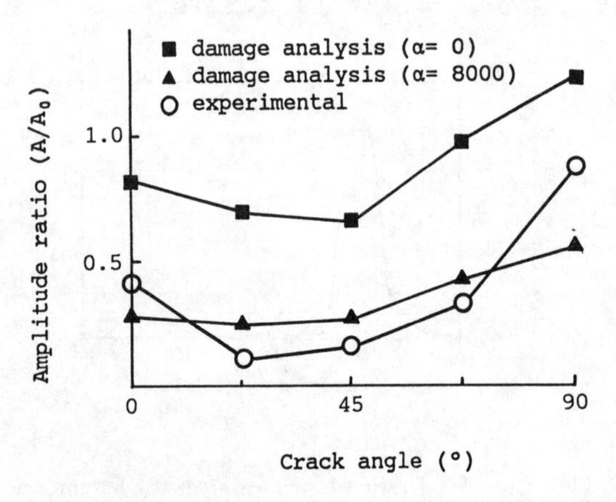

Figure 20. Attenuation of maximum amplitude by cracks.

6. CONCLUSIONS

Discontinuities in rock mass play a major role in determining the behaviour of rock structures. The discontinuities usually exist in patterns or sets; however, no efficient mechanical theory has been developed to treat such distributed discontinuities. Some specific numerical methods, for example distinct element method and joint elements, can be merely used for illustrating the behaviour of rock mass with distinct discontinuities. We have presented here a basic theory of damage mechanics and its applications to describe distributed discontinuities. Essential conclusions are as follows:

1. The state of discontinuties distributed in the rock mass can be represented by the damage tensor, and its mechanical effects are described by introducing the net stress which acts on the 'net' surface excluding damage.

2. A method to survey rock mass discontinuities and to determine the damage tensor is proposed.

3. The constitutive equation is only determined for the intact rock by laboratory tests.

4. A numerical scheme by finite elements is introduced in analysing the rock mass behaviour.

5. Using artificial rock-like materials which involve some sets of patterned cracks, we illustrate the validity of the damage mechanics theory through numerical calculations.

Several questions are still open in the theory of damage mechanics to treat distributed discontinuities: how to fix the propagation equation of the damage; how to develop a simple method in determining the damage tensor by using computer graphics; how to estimate the attenuation of elastic wave propagation through sets of discontinuities. We hope that the damage mechanics theory will introduce an alternative way to design rock structures in near future.

ACKNOWLEDGEMENTS

The author gratefully acknowledges his co-workers, Assoc. Prof. Dr. Y. Ichikawa and Res. Assoc. T. Kyoya for the valuable discussions, comments and help and wishes to thank to Res. Assoc. Ö. Aydan for the help during the preparation of this paper and Mr. S. Han for the help with computations.

REFERENCES

[1] Bandis, S., Lumsden, A.C. and Barton, N.R. (1981):"Experimental studies of scale effect on shear behaviour of rock joints", Int. J. Rock Mech. Min. Sci. & Geomech. Abstr., Vol.18, pp.1-21.

[2] Barton, N., Lien, R. and Lunde, J. (1974):"Engineering classification of rock mass for the design of tunnel tupport", Rock Mechanics, Vol.6, pp.189-236.

[3] Barton, N., et al. (1978): "Suggested methods for the quantitative description of discontinuities in rock masses ", Int. J. Rock Mech. Min. Sci. and Geomech. Abstr., Vol.15, No.6, pp.319-368.

[4] Bažant, Z.P. and Belytschko, T. (1987):"Strain-softening continuum damage", Procs. 2nd Int. Conf. on Constitutive Laws for Engineering Materials, Vol.1, pp.11-33.

[5] Bieniawski, Z.T. (1967): "Mechanics of brittle fracture of rock", Part I to Part III, Int. J. Rock Mechs. and Min. Sci., Vol.4, pp.396-430.

[6] Bieniawski, Z.T. (1974):"Geomechanics classification of rock and its application in tunneling", Procs. Int. Cong. Rock Mech., Vol.2, A, pp.27-32.

[7] Brown, E.T. (1981): "Rock characterization tesing and monitoring:ISRM Suggested Methods", Pergamon Press, New York.

[8] Cundall, P.A. (1971): "A computer model for simulating progressive large-scale movements in blocky rock systems", Procs. Int. Symp. Rock Mechs., ISRM, Nancy, pp.11-18.

[9] Deer, D.U., Hendron, A.J.Jr., Patton, F.D. and Cording, E.J. (1967):"Design of surface and near-surface construction in rock", in Failure and Breakage of Rock, Proc. 8th U.S. Symp. on Rock Mech., pp.237-307.

[10] Desai, C.S., Zaman, M.M., Lightner, J.G. and Siriwardane (1984): "Thin-layer element for interfaces and joints", Int. J. Numer. Anal. Meths. in Geomechanics, Vol.8, No.1, pp.19-43.

[11] Dragon, A. and Mroz, Z. (1979): "A continuum model for plastic brittle behaviour of rock and concrete", Int. J. Eng. Sci., Vol.17, pp.121-137.

[12] Franklin, J.A. (1987): "Photographic measurements of jointing and fragmentation", Proc. 2nd Int. Symp. Field Meas. Geomech. Kobe, Vol.1, pp.1-11.

[13] Frantziskonis, G. and Desai, C.S. (1987): "Constitutive model with strain-softening", J. Solids Struct., Vol.23, No.6.

[14] Ghaboussi, J., Wilson E.L., and Isenberg, J. (1973): "Finite element for rock joints and interfaces", J. Soil Mech. Found. Div., ASCE, (SM10), pp.833-848.

[15] Goodman, R.E., Taylor, R.L. and Brekke, T.L. (1968): "A model for the jointed rock", J. Soil Mech. Found. Div., ASCE, (SM3), pp.637-659.

[16] Goodman, R.E. (1974): "The mechanical properties of joints", Procs. of 3rd Int Congr. Rock Mechs, ISRM, Denver, 1A, pp.127-140.

[17] Hallbauer, D.K., Wagner, H. and Cook, N.G.W. (1973): "Some observations concerning the microscopic and mechanical behaviour of quartzite specimens in stiff, triaxial compression test", Int. J. Rock Mechs. and Min. Sci., Vol.10, pp.713-776.

[18] Hayhurst, D.R. (1972): "Creep rupture under multi-axial state of stress", J. Mech. Phys. Solids, Vol.20, No.6, pp.381-390

[19] Ichikawa, Y., Nakamura, Y., Kyoya, T. and Kawamoto, T. (1987): "Identification of damage field of rock mass", Proc. 2nd Int. Conf. Education, Practice and Promotion of Comp. Meth. in Eng. Using Small Computers, (to be appeared)

[20] Kachanov, L.M. 1958: "The theory of creep", English translation (A.J. Kennedy), Ch. IX,X, Nat. Lending Lib., Boston.

[21] Kawamoto, T., Obara, Y., Yamabe, T., Ichikawa, Y., and Shimizu, Y. (1982): "Elasto-plastic analysis by cracked triangle element", Proc. Int. Conf. FEM, Shanghai, pp.756-760.

[22] Kawamoto, T., Ichikawa, Y. and Kyoya, T. (1986): "Rock mass discontinuities and damage mechanics", Computational Mechanics 86, Proc. Int. Conf. on Computational Mechanics, Tokyo, Vol.2, Springer-Verlag, pp.IX-27.

[23] Kawamoto, T., Ichikawa, Y., and Kyoya, T. (1987): "Deformation and fracturing behaviour of discontinuous rock mass and damage mechanics theory", Int. J. Num. Anal. Meth. Geomech., to be appeared.

[24] Kazi, A. and Şen, Z. (1985):"Volumetric R.Q.D: An index of rock quality", Proc. Int. Symp. on Fundamentals of Rock Joints, pp.95-102.

[25] Kikuchi, K., Mimuro, T., Kobayashi, T., Izumi, Y., and Mito, Y. (1987): "A joint survey and determinations of joint distribution", Proc. 2nd Int. Symp. Field Meas. Geomech. Kobe 1987, Vol.1, pp.231-240.

[26] Kondoh, T., and Shinji, M. (1986): "Back analysis of assessing for alope stability based on displacement measurements", Proc. Int. Symp. Eng. in Complex Rock Formations, Beijing 1986, pp.809-815.

[27] Krajcinovic, D. and Fonseka, G.U. (1981):"The continuous damage theory of brittle materials", J. Applied Mechanics, Vol.48, pp.809-824.

[28] Kyoya, T., Ichikawa, Y., and Kawamoto, T. (1985): "A damage mechanics theory for discontinuous rock mass", Proc. 5th Int. Conf. Num. Meth. Geomech. Nagoya 1985, Vol.1, Balkema, pp.469-480.

[29] Kyoya, T., Ichikawa, Y., and Kawamoto, T. (1985): "An application of the damage tensor for estimating mechanical properties of rock mass", Proc.JSCE, No.358/III-3, pp.27-35 (in Japanese).

[30] Kyoya, T., Ichikawa, Y. Kusabuka, M. and Kawamoto, T. (1986): "A damage mechanics analysis for underground excavation in jointed rock mass", Proc. of Int. Symp. Eng. in Complex Rock Formations, Beijing, pp.506-513. [A

[31] Kyoya, T., Ichikawa, Y. and Kawamoto, T. (1987): "Deformation and fracturing process of discontinuous rock masses and damage mechanics", Procs. 6th Int. Cong. on Rock Mech., Vol.2, pp.1039-1042.

[32] Lemaitre, J. (1985): "A continuous damage mechanics model for ductile fracture", J. Eng. mater. Technol., Vol.107, pp.83-89.

[33] Murakami, S., and Ohno, N. (1981): "A continuum theory of creep and creep damage", in Creep and Structures ed. by A.R.S. Ponter and D.R. Hayhurst, Proc. 3rd Symp. Leichester, Springer, pp.422-433.

[34] Oda, M. (1982): "Fabric tensor for discontinuous geological materials", Soils and Found., Vol.22, No.4, pp.96-108.

[35] Olsson, O., Falk, L., Forslund, O., Lundmark, L. and Sandberg, E. (1985):"Radar investigations of fracture zones in crystalline rock", Proc. Int. Symp. on Fundamentals of Rock Joints, pp.515-523.

[36] Panek, L.A. (1985):"Estimating fracture trace length from censored measurements on multiple scanlines", Proc. Int. Symp. on Fundamentals of Rock Joints, pp.13-24.

[37] Pietruszczak, S.T. and Mroz, Z. (1981): "Finite element analysis of deformation of strain-softening materials", Int. J. Numer/Meths. in Engng., 17, pp.327-334.

[38] Simo, J.C., Ju, J.W., Taylor, R.L. and Pister, K.S. (1987): "On strain-based continuum damage models", Procs. 2nd Int. Conf. on Constitutive Laws for Engineering Materials, Vol.1, pp.233-245.

[39] Swan, G. and Zongqi, S. (1985):"Prediction of shear behaviour of joints using profiles", Rock Mechanics and Rock Engineering, Vol.18, pp.183-212.

[40] Yoshinaka, R. and Yamabe, T. (1986):"Joint stiffness and the deformation behaviour of discontinuous rock", Int. J. Rock Mech. Min. Sci. & Geomech. Abstr., Vol.23, pp.19-28.

[41] Zienkiewicz, O.C., Best, B., Dullage, C. and Stagg, K.G. (1970): "Analysis of non-linear problem in rock mechanics with particular reference to jointed rock systems", Proc. 2nd Int. Congress of ISRM, Beograd, Vol.3, 8-14, pp.501-509.

Numerical Methods in Geomechanics (Innsbruck 1988), Swoboda (ed.)
© 1988 Balkema, Rotterdam. ISBN 90 6191 809 X

Some recent developments in interactive computer graphics for 3-D nonlinear geotechnical FEM analysis

Fred H.Kulhawy, Kirk L.Gunsallus, Anthony R.Ingraffea & Patrick C.-W.Wong
Cornell University, Ithaca, N.Y., USA

ABSTRACT: The geotechnical engineering analysis and design process should be creative. Computer graphics can significantly enhance this process by simplifying the overwhelming input and output concerns. This paper provides a status report on the developing computer graphics system for FEM analysis at Cornell University. The framework and hardware of the system are discussed first, followed by the preprocessing and postprocessing operations controlling the analysis process. Appropriate screen images from this software are included for illustration. The paper further describes the material and interface modeling capabilities that the system supports to allow practical geotechnical problems to be solved.

1 INTRODUCTION

The engineering analysis and design process for complex problems often relies on the results of finite element analyses. However, the preparation of input data and presentation of output results can be very frustrating for the creative engineer. To avoid being overwhelmed by input and output concerns, the engineer often delegates the analysis to computer or systems people and simply responds to their presentations. Another alternative is to utilize simplified closed form solutions or estimations based on similar problems examined in the literature. In either case, the design engineer is not dealing directly with the analysis of the actual problem.

Computer technologies have evolved which change the above scenario into a "user-friendly" environment, in which the design engineer can analyze problems in a truly interactive mode, make changes and explore variations easily, and do all of these in a limited time frame, measured in hours or less. Previous papers have described the requirements of an interactive analysis/design system and the status of one such system being developed at Cornell University for geotechnical problems (Ingraffea, et al. 1981; Kulhawy, et al. 1982; Kulhawy, et al. 1985). The following will review these points briefly.

An effective computer-aided analysis system should be simple to use, handle complex situations, and provide monitoring facilities. The problem at hand should be the focus of attention, not the computer, and the engineer should be able to concentrate fully on the creative process of analysis and design. Preprocessing, or initial input, must stress simplicity, flexibility, and optimal graphics usage. Postprocessing, or output presentation, also should be simple and emphasize graphics, but it must provide the versatility to allow the engineer to select the desired results and display them in a variety of formats. Lastly, the system should possess a monitoring facility to allow all I/O data to be checked and interpreted easily.

The analysis system developed at Cornell University has been fulfilling these requirements by stressing engineer-computer interaction. The major components of the system include an alphanumeric terminal, a digitizing tablet and pen, a black and white vector-refresh display, a color high resolution raster display, and the main processor. The relationships among the components are illustrated in Figure 1. The terminal controls access to and initiation of different phases of the system and plays a minor role in the overall procedure. The digitizing tablet is the main input link between the engineer and the computer. There is a one to one correspondence between movement of the pen across the surface of the tablet and the movement of the cursor observed on the display screen. Selecting numbers, input

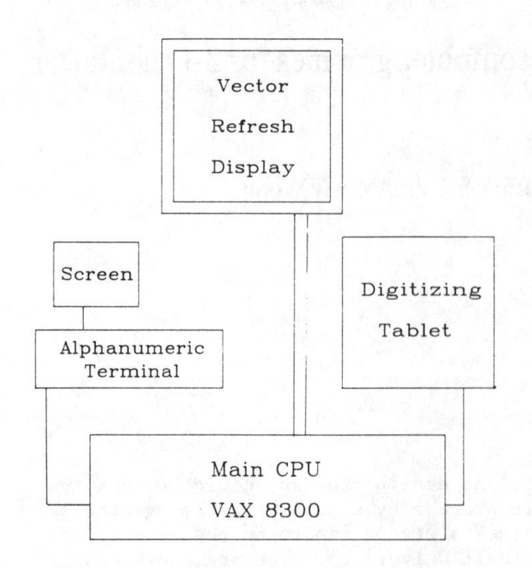

Figure 1. Components of FEM system.

items, and menu options then becomes a simple exercise, while flexibility and accuracy are maintained. The vector-refresh device is primarily used in preprocessing, and it is well-suited to the wire-frame meshing of geometries necessary for finite element analysis. The screen is updated several times a second, allowing real-time motion such as image zooming, translation, rotation, and visual cues like blinking. These effects are particularly important for three-dimensional problems. The raster display is the primary output device. The use of color to display analysis results aids the engineer's understanding of the problem and facilitates identification of critical design areas or regions with (input) errors. Lastly, the processor monitors the communication between the devices and does the calculations required to achieve results. The analysis system described in previous papers used a VAX 11-780 as the main processor. However, research into truly three-dimensional behavior for complex geometries required increased speed. Accordingly, the analysis system was rebuilt so that the majority of the calculations would be executed on a FPS-164, which is a floating point array processor. The current configuration includes 3 MAXX boards giving further increases in speed. Input and output is handled with a VAX 11-750. Table 1 shows typical increases in solution time.

Based on experience gathered with previous efforts and initiated because of the acquisition of faster hardware, several changes to the software that supports the analysis system have been implemented. This paper illustrates these changes by stepping through the procedure for an analysis of a drilled shaft foundation and reports on developments and enhancements made to the system, with particular emphasis on new material models and interface modeling.

Table 1. Representative Comparisons between FPS and VAX run times.

Number of Equations	FPS time (sec)	VAX time (sec)	VAX/FPS
2131	122	6204	50.8
5336	239	22032	92.2
9261	634	113872	179.6

(Wawrzynek, 1985)

2 PREPROCESSING FOR A DRILLED SHAFT FOUNDATION SIMULATION

The steps involved in preprocessing a FEM analysis problem include geometry definition with mesh generation, attribute specification, and analysis preparation. The subtasks within each of these categories can be seen in Figure 2.

The geometry definition phase of preprocessing is the same as described in previous papers. Initially, a grid is established to encompass the geometry of the model to be created. Allowing for variable grid spacing permits the engineer to utilize different units and sizes for different problems and maintain desired levels of accuracy. Subsequently, the outline of the 2-D reference sections are defined. The outlines can have different numbers and sizes of segments. For the drilled shaft example described herein, symmetry is used and only one quarter of the problem needs to be meshed. The outline for this example is shown in Figure 3. Once the outline is described, the mesh is generated. Elements can be linear or quadratic, and the variation of their sizes can be specified.

A typical 2-D cross-section for the drilled shaft example is given in Figure 4. This cross-section comprises both 6 and 8 noded elements (the final 3-D mesh will contain 15 and 20 noded elements). The shaft is made up of the first four rows of elements in the lower left corner. The 2-D sections then are located in space. Several options are available for specifying the location, including parallel to coordinate planes, mapped to a different size, and arbitrary orientations. The 3-D mesh is lofted between adjacent 2-D sections. As

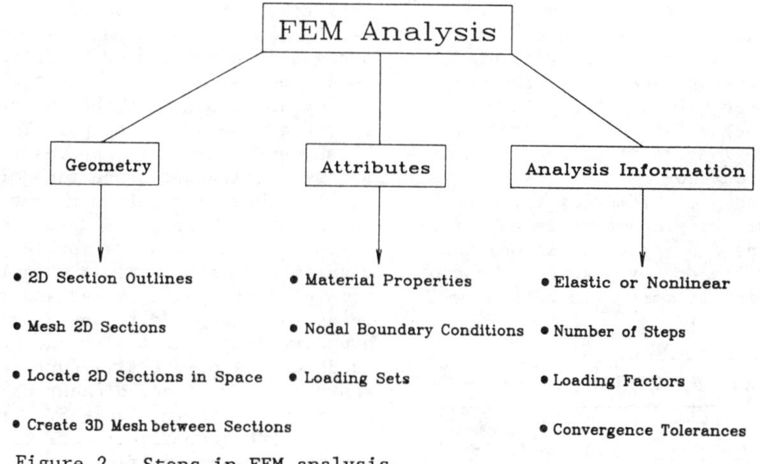

Figure 2. Steps in FEM analysis.

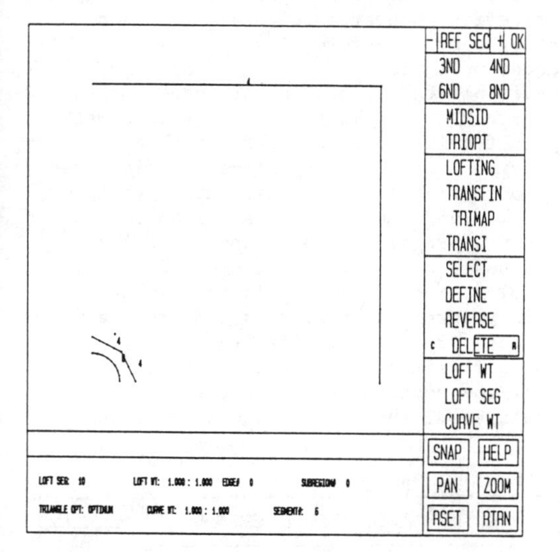

Figure 3. 2-D section outline for drilled shaft.

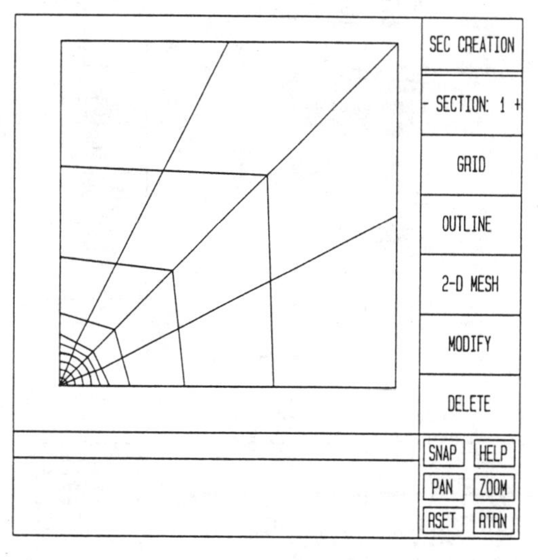

Figure 4. 2-D section for drilled shaft.

in creation of the 2-D section, the variation of element size with depth (lofting weight) is user-controlled. Figure 5 shows the drilled shaft problem with part of the 3-D mesh created and illustrates several 2-D sections located in space.

Following geometry creation, the attributes of the problem are specified. The attributes include the material properties of the elements, nodal restraints, and loads. For material properties, the engineer first selects that option from the menu and inputs appropriate values. Subsequently, the values can be assigned to regions, layers, or elements under user control. Once an element has been selected, its number flashes at the center of the element whenever that material property set is active. This feature allows the engineer to verify that the chosen elements have the correct material properties.

The property specification page is shown later in this paper. New material models have been added to the system to enhance the analysis capabilities. However, the material property section of the previous preprocessing program was not developed to accommodate the new models. Therefore, a new program was created, which currently handles the input for hyperbolic and bounding surface parameters and allows for others to be added in the future. As shown in Figure 6, the new program displays the input parameters and allows the user to specify their values. Assigning material property sets for the new models to elements or regions is still handled in the original preprocessor.

The next step is to identify the nodes to be restrained. Displacements in any of the axis directions (in any combination) for individual, colinear, or planar regions of nodes can be set to zero. In keeping with user-friendliness, ease of use, and graphical feedback, the boundary specification process is accomplished by selecting the nodes, edges, or sides with the digitizing tablet and pen, while the image is displayed on the vector scope. The nodes blink once they have been restrained. Figure 7 illustrates one face of the drilled shaft simulation when the nodes are restrained in the x direction.

Lastly, the loads are applied. Again, flexibility is maintained by allowing acceleration, point, line, and surface loads to be applied to any part of the boundary using the tablet and graphical feedback. A compressive surface load is shown applied to one element at the top of a drilled shaft in Figure 8.

After the geometry is defined and the attributes are specified, the problem must be described in analysis form. Within this task, the type of analysis, load step specification, convergence tolerances, and other control information are generated under full user control. The options available to the analyst are shown in Figure 9, which is the analysis information page from the preprocessor. Each of the previous preprocessing steps are discussed more completely in earlier papers (Kulhawy, et al. 1982; Kulhawy, et al. 1985).

A new program has been developed for the postprocessing of three-dimensional problems once the analysis is complete. This program is similar to its predecessor in many ways. Initially, the results of the analysis are translated into a form that the postprocessor can use. Once translated, the responses can be viewed.

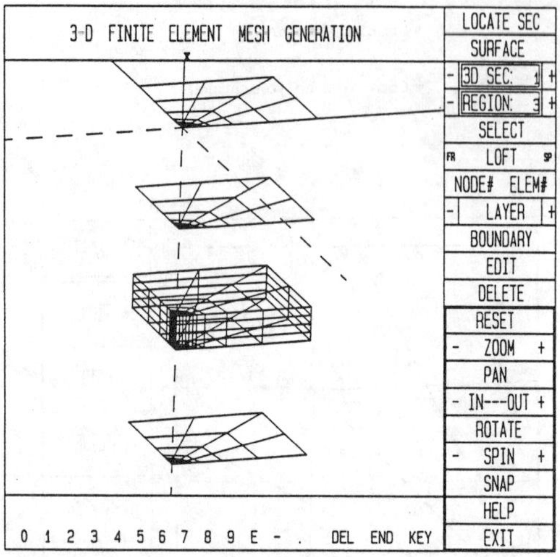

Figure 5. 3-D mesh during creation.

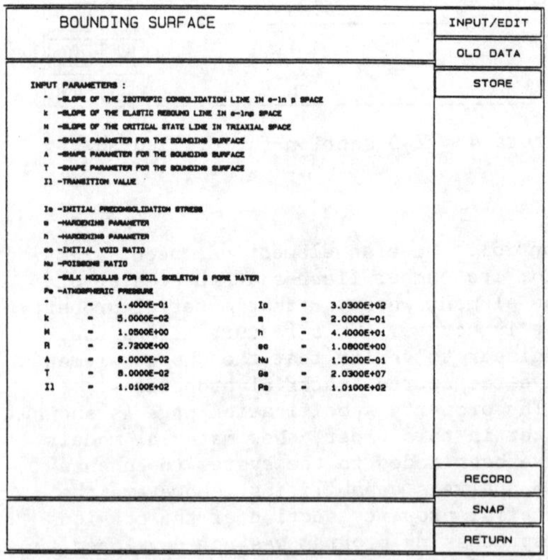

Figure 6. Auxiliary materials input page.

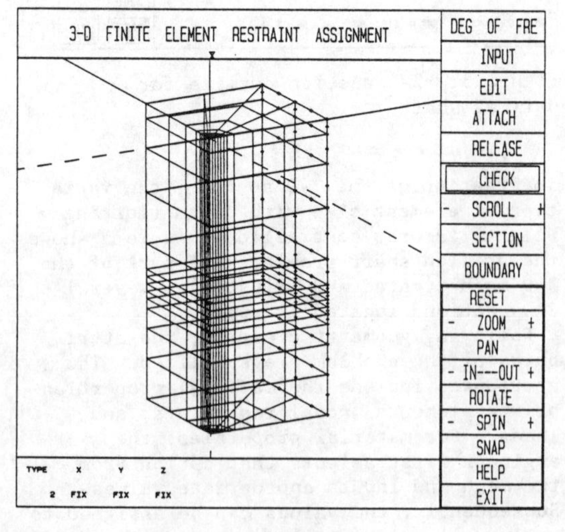

Figure 7. Nodal restraints.

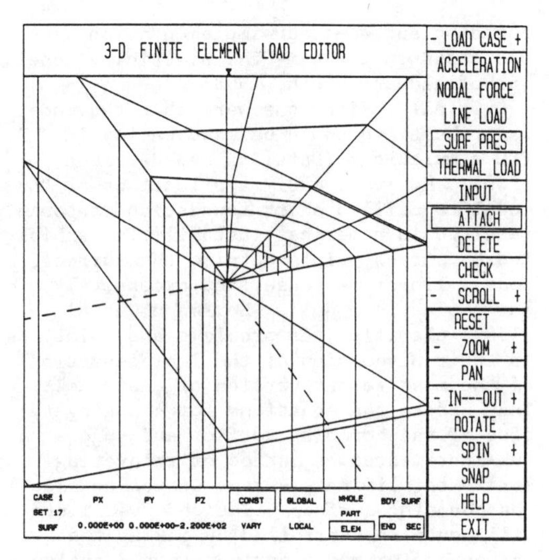

Figure 8. Compressive surface load.

3-D F.E. INTERFACE problem : shaft

ANALYSIS PARAMETERS

Problem Title	None
Process Type	Batch
Analysis Type	
Linear Elastic	
Buckling	Inactive
Buckling Type	
Buckling Mode	
Geometric Nonlinear	Inactive
Material Nonlinear	Active
Marching Control	Load
Maximum Number of Steps	4
Iteration Tolerance	2.0000E-02
Iteration Limit per Step	16
Follower Pressure	Inactive
Substructure analysis	Inactive
Output Selection	
Geometric Data	On
Nodal Data	On
Element Data	On
Load Data	On
Displacement	On
Stress	On
Nodes	Selective
Interface to Analysis	
Filename	shaft
Analysis Procedure	FACS
Create Files to Analysis	

RESET
- ZOOM +
PAN
- IN---OUT +
ROTATE
- SPIN +
SNAP
HELP
EXIT

0 1 2 3 4 5 6 7 8 9 E - . DEL END KEY

Figure 9. Analysis information page.

The first step of this process is to read in the data and then adjust the viewing characteristics of the problem, which could include object rotation and translation, location of light source, and selection of color scales. Plate 1 shows the drilled shaft problem oriented for response viewing with a frontal light source. (All photograph plates are grouped together at the end of the paper.) Next, the desired result is selected, such as stresses, displacements, and error estimates. These results can be displayed in a variety of formats. For example, Plate 2 shows the options available for stresses. In general, the responses are displayed on a raster device with color contours delimiting the variations of the response. In addition, explicit numerical data can be obtained for any given location by choosing the appropriate node and requesting the value. Also, the displaced shape of the mesh can be viewed. Other capabilities of the program include subobject extraction, line plotting, and viewing flexibility. These capabilities are discussed more completely later in the paper. The vertical and shear stress contours for the drilled shaft example are shown in Plates 3 and 4. Examples of other features of the postprocessor are included in later sections of this paper.

3 MATERIAL MODELS

Several nonlinear material models have been incorporated into the FEM analysis system to investigate the degree of sophistication warranted in different classes of problems. In addition to the elastic and hyperbolic models discussed in previous papers, the system now supports normal and shear stress dependent cutoff formulations and three types of plasticity models: von Mises, Drucker-Prager, and bounding surface. The types of differences observed are illustrated in Figure 10 and Plates 5 and 6. Figure 10 shows load-displacement curves for a typical drilled shaft simula-

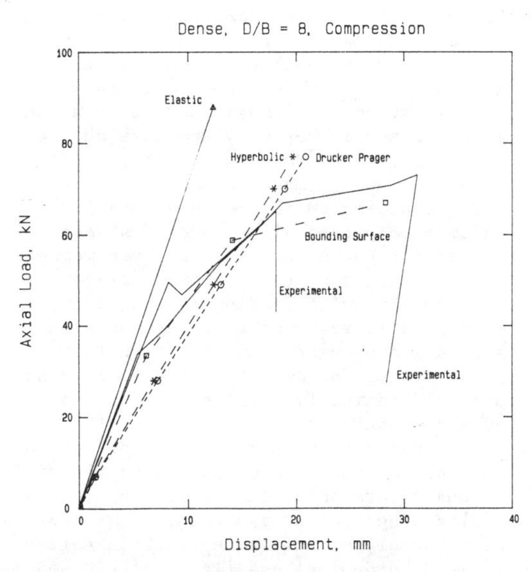

Figure 10. Typical load-displacement curves.

tion using different soil models. The simulation uses 520 elements and 2637 nodes to represent a shaft inside a dense soil loaded in compression. The shaft has a 75 mm diameter and a depth/width ratio of 8. The results of a laboratory experiment replicating the conditions of the simulation are included for comparison. Note the difference in nonlinearity and displacement for the different models. A comparison of the displaced shapes for these analyses using elastic and hyperbolic materials is shown in Plates 5 and 6. The displacements for the elastic analysis extend radially much further into the soil mass than do those for the hyperbolic analysis. This type of response could be expected because soil failure would cause more displacement in the immediate area of the shaft where the loads are the highest. Future papers will discuss comparisons of these methods in depth.

Most of these models have been discussed extensively in the literature in the past. However, the bounding surface model is still relatively new and practical applications are still scarce. The remainder of this section will review the basis and general characteristics of the bounding surface model briefly, discuss its implementation, and present an example of its use for analyzing drilled shaft foundations.

The bounding surface model was developed by Dafalias and is discussed extensively in Dafalias and Herrmann (1980). It is an outgrowth of original work done independently by Dafalias and Popov (1975) and by Krieg (1975). Since implementation of this model, Dafalias and his co-workers have extended its capabilities and robustness. The current status of the model, as well as its mathematical basis and formulation, is discussed in depth by Anandarajah and Dafalias (1986), Dafalias (1986), and Dafalias and Herrmann (1986).

The bounding surface refers to a limit which encompasses all allowable stress states for the material at a given moment in its loading history. This surface changes size with loading and unloading. More plastic response is experienced for stress states nearer the limit surface, while points far away from the limit behave almost elastically. There has been no purely elastic zone, although Dafalias and Herrmann (1986) have recently implemented an elastic zone, and unloading is elastic. Strong points about the model include cyclic response (strain accumulates because of no purely elastic zone), strain softening and hardening, and flexibility. Any legitimate limit can be utilized for the bounding surface; our implementation uses one based on the CAM-CLAY plasticity model (Schofield and Wroth, 1968).

Original indications were that the model was robust and could be used easily in 3-D FEM simulations (Dafalias and Herrmann, 1980). However, early applications of the model to drilled shaft foundation response resulted in numerical instabilities exhibited by stress points outside the surface, nearly linear response, and excessively large plastic moduli. Herrmann, et al. (1987) clarified the problems and solutions in their discussion of the implementation of the most recent version of the model. They update the equations for changing the size of the bounding surface and suggest a sub-incrementation method of recovering stress to eliminate stress states outside the bounding surface. The observed instabilities were controlled by making the corresponding modifications to the analysis code. For the other nonlinear material models, a 4th order Runge-Kutta procedure is used to recover stress increments from strain increments. This nonlinear recovery procedure gives good estimates of stress for large, nonlinear strain increments. This fact, plus the observation that the first implementations of the bounding surface model worked well for many cases other than the drilled shaft foundation simulation, led to further attempts to utilize the Runge-Kutta recovery scheme (instead of the sub-incrementation method). Several other parameters that can typically cause numerical problems in FEM analysis were modified in the investigation, including element distortion, curved edges, initial stress levels, steep loading gradients, and disparate adjacent material properties. Adjustments in these factors could not control the instabilities in the model, leading to the conclusion that the sub-incrementation method is required for good results in general 3-D FEM analyses with the bounding surface model. Limited investigations of the sub-incrementation method for large load steps indicate that the accuracy of the result is not affected by using large load steps. Figure 11 shows the load-displacement curve for a drilled shaft analyzed using 1, 2, 3, 5, and 10 increments. All points lie on the same curve.

4 INTERFACE MODELING

Wherever there are two dissimilar materials coming into contact with each other, an interface is created. The behavior of the interface under static or dynamic loading

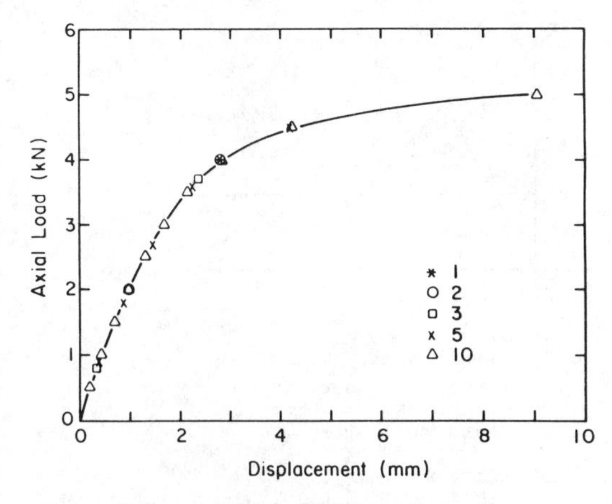

Figure 11. Comparison of number of analysis steps using bounding surface model.

conditions may be the same as, or different from, that of the parent materials creating the interface. Therefore, the physical properties of the interface usually are related to those of the parent materials. One example is the soil-structure interface.

When a load is imposed on a structure, relative motions can be induced at the soil-structure interface, including translation, rotation, and rocking. Because of these motions, the interface could experience four different modes of deformation: stick (no slip), slip, debonding, and rebonding (Desai, 1981). The stick mode occurs for perfect bonding between the soil and the structure and no relative slip. The slip mode evolves with relative slip while full soil-structure contact is maintained. Debonding occurs when there is a physical separation between the soil and the structure. Rebonding occurs when the originally debonded soil re-establishes contact with the structure, such as with a reversal of load. In some classes of geotechnical problems, the soil-structure interface behavior dictates the global system response and, in the extreme, failure of the system can be attributed to the failure of the interfaces. The interface between a drilled shaft foundation and its surrounding soil, for example, experiences all four modes of deformation when the shaft is subjected to a repeated lateral load, as shown in Figure 12. Therefore, it is important that the interface behavior be modeled as realistically as possible during a numerical simulation.

There are basically two different groups of isoparametric interface/joint elements commonly used in finite element analysis. Goodman, et al. (1968) proposed a zero thickness, isoparametric joint element to model rock joints. The relative displacements on one side of the interface are taken as the nodal degrees-of-freedom of the element, and the thickness of the element across the interface is assumed to be zero. Desai (1981) suggested a thin layer element, which is essentially an isoparametric continuum element with limits on the element thickness. From a recent study (Desai, et al. 1984), the recommended limits on average width to thickness ratio are from 10 to 100.

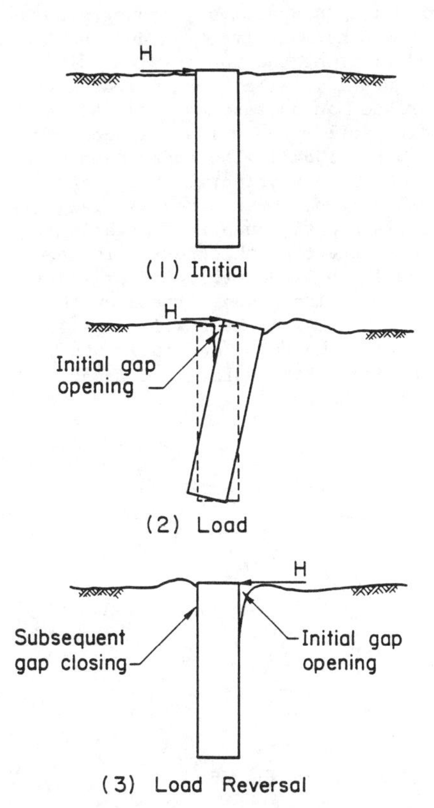

Figure 12. Interface behavior of drilled shaft under lateral load.

There are several advantages of the thin layer element over the zero thickness joint element. First, it is more realistic, because there is always a thin layer of soil participating in the interface shearing process. Second, the thin layer element can be incorporated easily into an existing finite element computer code without further modification, since its element formulation is the same as that for a continuum element. Third, the stresses recovered in the thin layer element at the end of the analysis are meaningful, which is not always the case for the zero thickness joint element. Also, simple to complex constitutive models suitable for an interface can be implemented readily without any difficulty. For these reasons, the thin layer element has been adopted for our code.

To implement the three-dimensional thin layer elements, a local element coordinate system has been adopted with two orthogonal in-plane axes and an out-of-plane normal axis. This element is characterized by three principal properties: Young's modulus along the normal direction and the two shear moduli along the two in-plane directions (Figure 13). The Young's modulus of the interface can be taken as that of the surrounding soil without loss of accuracy (Desai, et al. 1984). The shear moduli, on the other hand, are primarily functions of the normal stress, shear stresses, relative shear displacements, number of shearing cycles, and interface thickness. These shear moduli can be determined from static and/or dynamic direct shear tests on the interface. A local element stiffness formulation using the three moduli is used, and uncoupling of the normal and shear behavior is assumed.

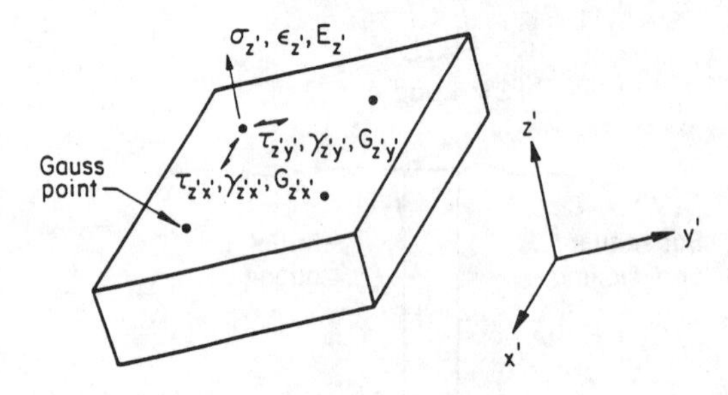

Figure 13. Thin layer element with local gauss point parameters.

Two trilinear constitutive models have been implemented to describe the out-of-plane normal and in-plane shear behavior of the interface. The normal strain-softening model consists of a linear curve extending up to a tensile strength level, a downward linear curve representing strain-softening, and a fully-softened plateau (Figure 14). By varying the input parame-

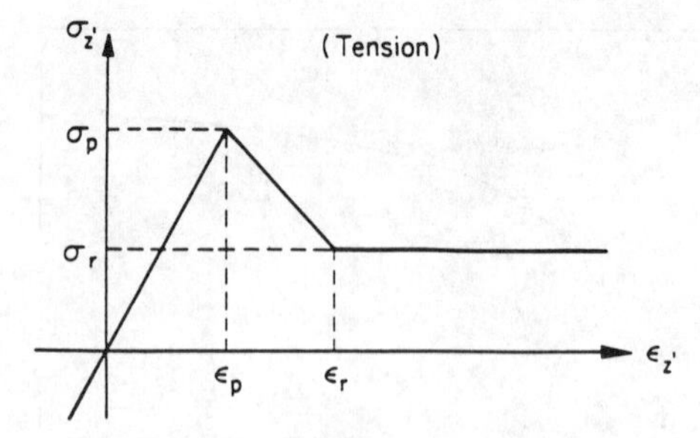

Figure 14. Model for interface normal behavior.

ters, the user can construct curves of different shapes. For most practical geotechnical problems, the peak tensile strength and the fully-softened strength of the interface can be taken as zero. The geometry of the shear strain-softening model is similar to that of the normal strain-softening model. However, there is no distinction in behavior between shearing along any one particular direction (Figure 15). In addition, the peak and fully-softened shear strength can be made normal stress-dependent using Mohr-Coulomb criteria.

To simulate the four modes of deformation, a stress redistribution scheme (Desai, et al. 1984) is used in the analysis program. For example, if the normal stress across the interface exceeds the tensile capacity given by the constitutive model, debonding is initiated. The excess normal stress above the tensile capacity is converted into nodal forces, and another analysis is performed with these unbalanced nodal forces and a reduced normal stiffness for the interface. This iterative procedure continues until the normal stress across the interface is approximately equal to the value determined by the constitutive model. The normal strain accumulated during these iterations with a reduced normal stiffness emulates the debonding effect. A similar procedure is used for the in-plane shear behavior.

The preprocessor mentioned previously facilitates the input of the control parameters for the interface constitutive models. At the material property menu page (Figure 16), the user can input interactively the appropriate values into the material property table. An on-line interactive help module, which contains definitions of the symbols used in the table, can be accessed by the user through the computer terminal. All of these features enable efficient input, visual confirmation, and easy modifications of the model parameters.

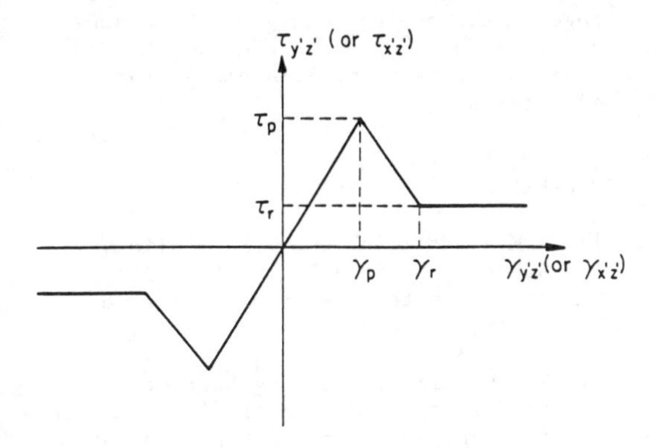

$\tau_{y'z'}$, $\tau_{x'z'}$ - in-plane stresses
$\gamma_{y'z'}$, $\gamma_{x'z'}$ - corresponding shear strains

Input parameters: - τ_p
- gamma factor (γ_r/γ_p)
- phi factor (ϕ_r/ϕ_p)

Figure 15. Model for interface shear behavior.

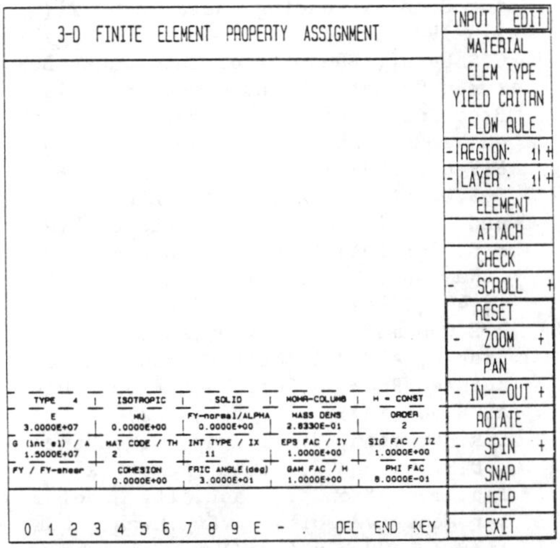

Figure 16. Material property page.

5 EXAMPLE APPLICATION OF THIN LAYER ELEMENT

An application of this thin layer element is for the analysis of a drilled shaft foundation under uplift. Several 3-D nonlinear FEM analyses were performed using a mesh which represents a quarter of the system. The soil elements and concrete shaft elements were assumed to be linear elastic while the thin layer elements along the vertical side and tip (bottom) of the shaft were given a shear strain-softening model and a tension cutoff model. With the postprocessor mentioned previously, the effects of incorporating thin layer elements can be seen easily. Plate 7 gives a pictorial representation of the different materials in the system through the use of different colors. Material type 1 represents the soil. Material types 2 and 3 are the core and base, respectively, of the concrete drilled shaft. Material types 4 and 6 represent the thin layer elements along the side of the shaft and at the shaft tip, respectively, and they are connected by transition elements belonging to material type 5.

Plate 8 shows the vertical displacement contours of the above system. In this figure, there is a concentration of contours in the thin layer elements along the side of the shaft, indicating a large variation of vertical displacement across the interface. This situation is a result of the relative slip between the shaft and the soil.

A useful feature of the postprocessor is subobject extraction, in which elements belonging to the same material type can be extracted from the original object to form a subobject, which can be viewed separately. Plate 9 shows a subobject of thin layer elements which are located originally at the shaft tip (Plate 7). The vertical displacement contours are displayed also on top of the subobject. Large vertical displacement occurs at the upper portion of the thin layer elements because debonding has occurred around the mid-section of these elements and their upper and lower portions are "numerically" separated from each other. Thus, the upper portion moves upward as a rigid body together with the shaft without any resistance.

The deformed mesh of the system displayed in Plate 10 shows that there is a large normal strain accumulated in the thin layer elements at the bottom of the shaft and this is, in fact, a realistic simulation of the debonding situation. This figure also shows that there is large relative shear displacement in the thin layer elements along the side of the shaft, which represents a good approximation of relative slip between the shaft and the soil. Furthermore, the combined effect of relative slip and debonding in the thin layer elements near the top of the shaft results in a situation where adjacent soil elements are peeled away from the shaft. This phenomenon is in agreement with field observations (Kulhawy, et al. 1983).

Another useful feature of the postprocessor is the plot-line function, with which the user can place an arbitrary straight plot-line on the external surfaces of the object being viewed and then ask for a plot of variation of stress or strain along the chosen line. Plate 11 shows the variation of shear stress in the thin layer elements along the side of the shaft at an intermediate step of the analysis. The x-axis represents decreasing depth and the y-axis represents shear stress. It can be seen that the thin layer elements near the tip of the shaft are in post-peak behavior, with shear stresses equal to the fully softened interface shear strength. The excess shear stresses are redistributed and picked up by the thin layer elements immediately above them.

From limited analyses to date, it appears that the thin layer element with the two proposed interface constitutive models realistically can model the displacement response and stress distribution in geotechnical problems in which soil-structure interfaces play a major role.

6 CONCLUDING COMMENTS

The interactive computer graphics system described in this paper is still in the development stage. The 3-D analysis capabilities are well-advanced, and sophisticated forms of nonlinearity are being addressed.

A limitation of our current system is in the areas of documentation, transportability, and software updating. Anyone working in computer-aided analysis or design has encountered similar problems. One can either strive for a system that is robust and well-proven or for a system that is at the edge of the state-of-the-art. The first case leads to hardware and software that is muscle-bound, slow, and soon outdated. The second case leads to continuous software development and hardware upgrading, resulting in a unique system. We have chosen the second course of action so that we can address state-of-the-art problems in geotechnical engineering.

7 ACKNOWLEDGMENTS

The research described in this paper has been the result of many researchers and sponsors working in the Program of Computer Graphics at Cornell. Their collective efforts have led to the general system and graphics capabilities we have used. The specific work on material models and soil-structure interaction has been sponsored by the Electric Power Research Institute under contract RP1493-4; V.J. Longo is the project manager. This support is acknowledged gratefully. L. Mayes prepared the text and A. Avcisoy drafted the figures.

REFERENCES

[1] A.M.Anandarajah and Y.F.Dafalias, Bounding Surface Plasticity.III: Application to Anisotropic Cohesive Soils, J.Eng.Mech.-ASCE 112(12), 1292-1318 (1986)

[2] Y.F.Dafalias, Bounding Surface Plasticity.I: Mathematical Foundation and Hypoplasticity, J.Eng.Mech.-ASCE 112(9), 966-987, (1986)

[3] Y.F.Dafalias and L.R.Herrmann, Development and Numerical Implementation of a Bounding Surface Constitutive Model for Cohesive Soils, Rpt. 80-2, U.Calif., Davis (1980)

[4] Y.F.Dafalias and L.R.Herrmann, Bounding Surface Plasticity.II: Application to Isotropic Cohesive Soils, J.Eng.Mech.-ASCE 112(12), 1276-1291 (1986)

[5] Y.F.Dafalias and E.P.Popov, A Model for Nonlinearly Hardening Materials for Complex Loading, Acta Mech. 21(3), 173-192 (1975)

[6] C.S.Desai, Behavior of Interfaces Between Structural and Geologic Media, Proc.Intl.Conf.Recent Adv.Geot. Earthquake Eng. and Soil Dynamics, Rolla, 619-638 (1981)

[7] C.S.Desai, M.M.Zaman, J.G.Lightner, and H.J.Siriwardane, Thin-Layer Element for Interfaces and Joints, Intl. J.Num.Anal.Methods in Geomech. 8(1), 19-43 (1984)

[8] R.E.Goodman, R.L.Taylor, and T.L. Brekke, A Model for the Mechanics of Jointed Rock, J.Soil Mech.Fndns. Div.-ASCE 94 (SM3), 637-659 (1968)

[9] L.R.Herrmann, V.Kaliakin, C.K.Shen, K.D.Mish, and Z.-Y.Zhu, Numerical Implementation of Plasticity Model for Cohesive Soils, J.Eng.Mech.-ASCE 113(4), 500-519 (1987)

[10] A.R.Ingraffea, F.H.Kulhawy, and J.F. Abel, Interactive Computer Graphics for Analysis of Geotechnical Structures, Proc.1st Intl.Conf.Computing in Civil Eng., New York, 864-875 (1981)
[11] R.D.Krieg, A Practical Two-Surface Plasticity Theory, J.App.Mech. 42(3), 641-646 (1975)
[12] F.H.Kulhawy, A.R.Ingraffea, Y.-P. Huang, T.-Y.Han, and M.A.Schulman, Interactive Computer Graphics in Geomechanics, Proc.4th ICONMIG, Edmonton, 1181-1192 (1982)
[13] F.H.Kulhawy, A.R.Ingraffea, T.-Y. Han, and Y.-P.Huang, Interactive Computer Graphics in 3-D Nonlinear Geotechnical FEM Analysis, Proc.5th ICONMIG, Nagoya, 1673-1681 (1985)
[14] F.H.Kulhawy, C.H.Trautmann, J.F.Beech, T.D.O'Rourke, W.McGuire, W.A.Wood, and C.Capano, Transmission Line Structure Foundations for Uplift-Compression Loading, Rpt. EL-2780, Elec.Power Res. Inst., Palo Alto (1983)
[15] A.N.Schofield and C.P.Wroth, Critical State Soil Mechanics, McGraw-Hill, London, 1968
[16] P.A.Wawrzynek, 3-D FE Analysis on the FPS-264 Attached Processor, Internal Rpt., Cornell, Ithaca (1985)

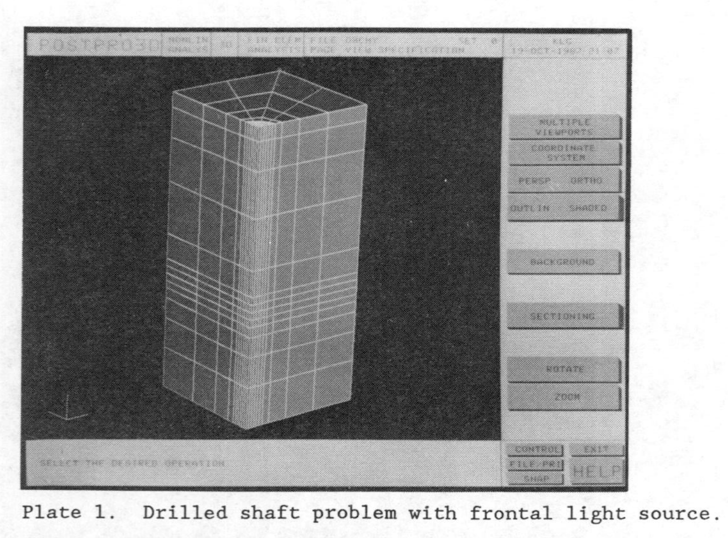

Plate 1. Drilled shaft problem with frontal light source.

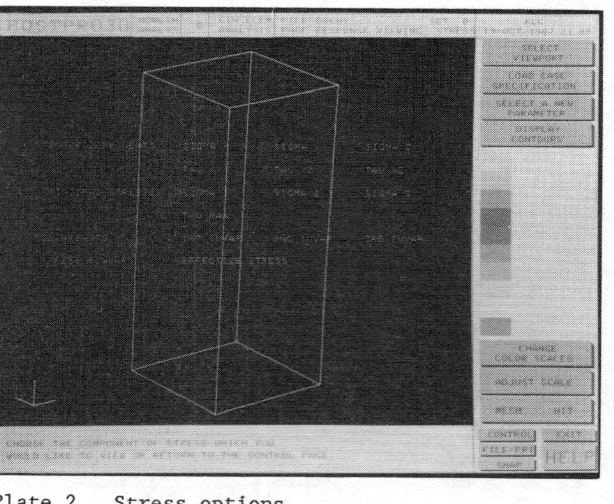

Plate 2. Stress options.

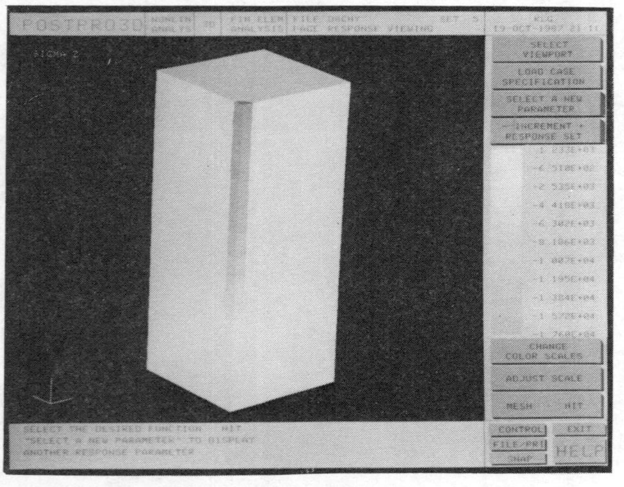

Plate 3. Vertical stres profile.

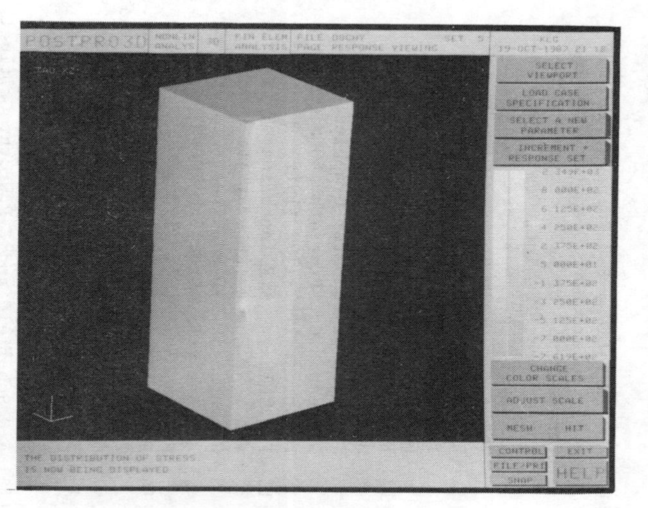

Plate 4. Shear stress profile.

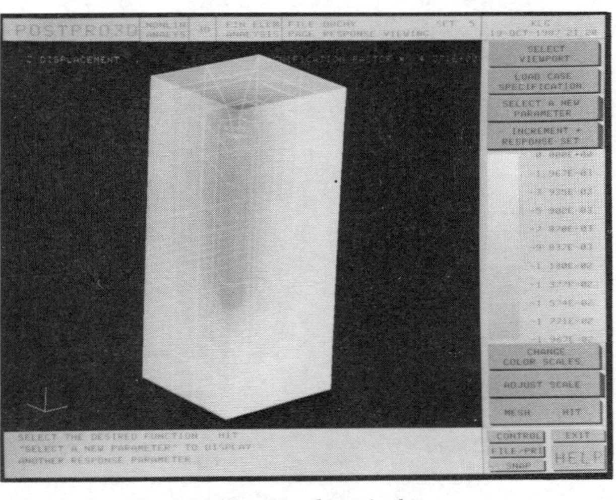

Plate 5. Elastic displaced shape.

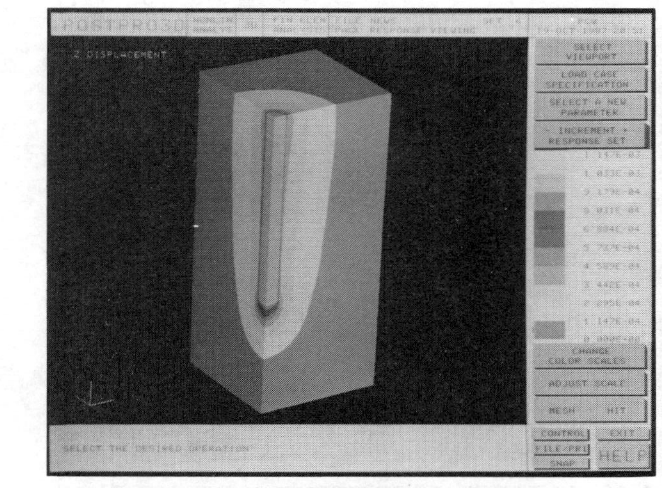

Plate 6. Hyperbolic displaced shape.

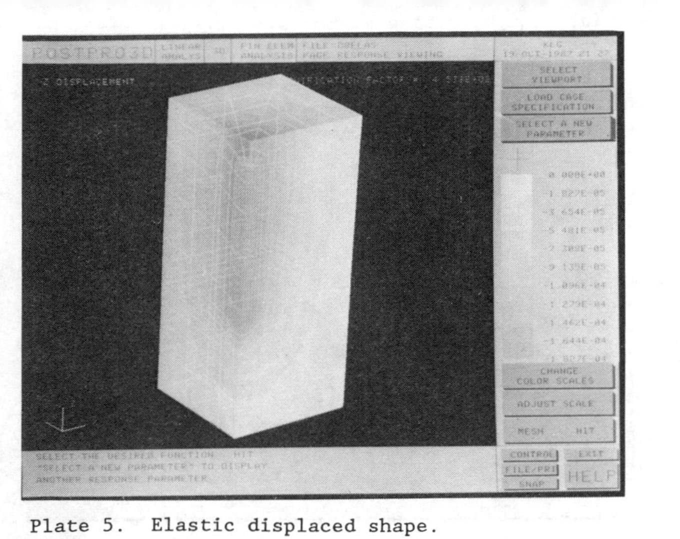

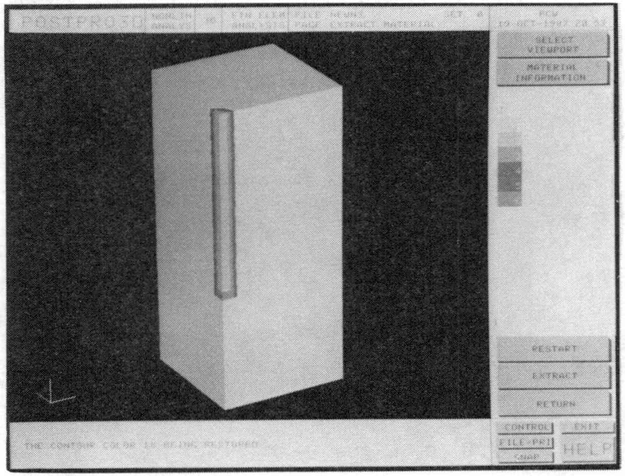

Plate 7. Material type representation.

Plate 8. Vertical displacement contours.

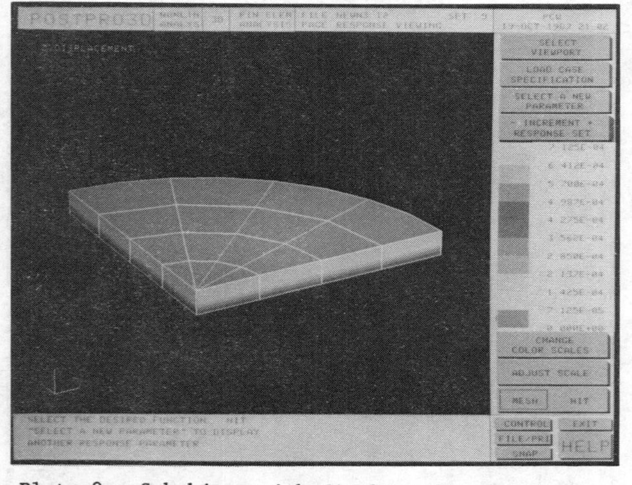

Plate 9. Subobject with displacement contours.

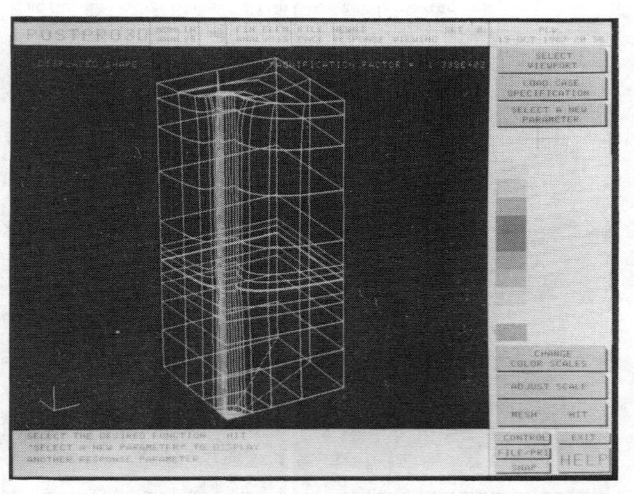

Plate 10. Deformed mesh.

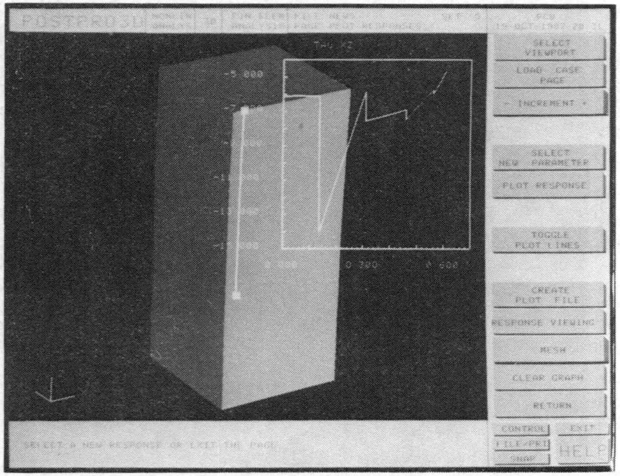

Plate 11. Plot-line of shear stress in thin layer elements.

Numerical Methods in Geomechanics (Innsbruck 1988), Swoboda (ed.)
© 1988 Balkema, Rotterdam. ISBN 90 6191 809 X

Static and dynamic approaches to rock burst phenomena

Z.Mróz
Institute of Fundamental Technological Research, Warsaw, Poland

ABSTRACT

Rock burst phenomenon occurs when static stability conditions of rock mass are violated and the dynamic failure process occurs. The onset of dynamic process is usually preceded by extensive cracking near the excavation face, delamination of roof layers and by resulting considerable redistribution of stresses. General stability conditions of mining excavation were discussed in several papers, cf. Salamon (1970), Petukchov and Linkov (1979), Zubelewicz and Mróz (1983). The material model assumed in such analysis should account for both stable and unstable ranges of response, that is exhibiting two deformation ranges for which the incremental work is positive or negative definite, that is

$$\text{stability:}\quad d\tilde{\sigma}\cdot d\tilde{\varepsilon} > 0\,,$$

$$\text{instability:}\quad d\underline{\sigma}\cdot d\underline{\varepsilon} < 0 \qquad (1)$$

The effect of damage and development of cracking is simulated in the constitutive model by assuming degradation of elastic moduli and of plastic hardening. In the case of joints, a separate relation is formulated between contact stresses dt_n, dt_t and joint displacements u_n and u_t. The stability condition would then generally require

$$\int d\tilde{\sigma}\cdot d\tilde{\varepsilon}dV + \int d\tilde{t}\cdot d\tilde{u}dS_j > 0 \qquad (2)$$

where the first term represents the incremental work within the domain of rock affected by excavation and the second term represents the effect of joint slip. In the case of propagation of major cracks, their stability condition should be included in the analysis.

To study the onset of instability, a fairly simple material model may be assumed, namely the Coulomb material with the initial cohesion c_o decreasing in the course of deformation until its residual value c_r is reached and with the constant angle of internal friction ϕ. Even such a simple model provides a rational tool to examine specific cases and determine the onset of instability. In order to study a post-critical dynamic failure process the effect of material viscosity should be accounted for.

Two approaches to analysis of rock burst phenomena are presented. The first approach is aimed at determining the limit equilibrium point beyond which the uncontrollable deformation process begins. It should be noted that the system behavior is controlled by the configuration parameters such as opening span or height and the instability point is reached when critical sizes of these parameters are reached. The incremental elasto-plastic analysis can therefore be carried out with increasing step-by-step excavation size. Using the finite element discretization for an incremental problem, we have

$$\tilde{K}d\tilde{\delta} = d\tilde{F} \qquad (3)$$

with $\tilde{K}$ denoting the global tangent stiffness matrix, and $\tilde{\delta}$, $\delta\tilde{F}$ being the nodal displacement and force increments. The instability point is then specified by the condition

$$\det[\tilde{K}] = 0 \quad \text{or} \quad \int d\tilde{\sigma}\cdot d\tilde{\varepsilon}dV = 0 \qquad (4)$$

The determinant of the global stiffness matrix can be determined for each incremental solution and plotted against the size parameter, thus enabling to specify the critical parameter value for which $\det[\tilde{K}] = 0$. However, this method was found out to be tedious in practical application as very small size increments are required near the instability point. As an alternative method, a regular progression condition was assumed. As a control variable, the size x_r of a damaged zone corresponding to post-critical regime was assumed whereas the opening size a is regarded as a response variable. The stability and instability conditions are respectively

$$\text{stability:} \quad \frac{da}{dx_r} > 0 ,$$

$$\text{instability:} \quad \frac{da}{dx_r} < 0 \tag{5}$$

The static solution may now be continued beyond the instability zone by imposing progression of the damaged zone and reducing the size parameter.

The pillar and seam stability is studied by using the condition (5) and applying the simplified model allowing for analytical treatment of the problem cf. Mróz and Nawrocki (1987). The roof strata are simulated as shear beam and the pillar is treated as an elasto-plastic, softening layer. The finite element solution to pillar problem is also presented and the instability point is determined.

The second dynamic approach is considered as more natural and economic providing information not only on the onset of instability but also on the whole dynamic failure process. Instead of static equilibrium equations (3), the equations of motion are considered

$$\tilde{K}\tilde{\delta} + \tilde{C}_d\dot{\tilde{\delta}} + \tilde{M}\ddot{\tilde{\delta}} = \tilde{F}(t) \tag{6}$$

where $\tilde{K}$ denotes the stiffness matrix, $\tilde{C}_d$ is the damping matrix and $\tilde{M}$ is the mass matrix. The static solution is first obtained for a specified configuration and within the elasto-plastic range of deformation this solution is constructed incrementally for increasing size of excavation. At some stage, a small dynamic disturbance is introduced into the system by specifying either initial velocity field $\dot{\tilde{v}}_o(x)$ of prescribed total kinetic energy K_o or by initiating a dynamic process by sudden removal of portions of rock. The subsequent dynamic solution will indicate whether growth of the kinetic energy occurs within the system and if so, the associated failure mode can be investigated. If, however, the initial disturbance is not amplified, a further increase in excavation size may be imposed and the corresponding static solution may be constructed.

In carrying out dynamic analysis of rock, any finite element discretization can be applied. However, it was found that the analysis is considerably simplified when special rigid elements are used and both elastic and plastic properties are lumped into some interaction modes lying on element interfaces. The local failure process at interacting nodes occurs when the specified stress condition is satisfied. Subsequently, the elements may separate under tension or contact each other under compression. The contact element technique was applied successfully to study dynamic failure process in rocks by Zubelewicz and Mróz (1983).

Examples of static and dynamic approaches to rock instability problems are presented for the cases of a seam and a set of pillars for which post-critical softening is assumed to simulate material degradation.

REFERENCES

Salamon, M.D.G. 1970. Stability, instability and design of pillar workings. Int. J. Rock Mech. Min. Sci. 7: 613-631.

Petukchov, I. M. and Linkov, A. M. 1979. The theory of post-failure deformations and the problem of stability in rock mechanics. Int. J. Rock Mech. Min. Sci. 16: 57-76.

Zubelewicz, A. and Mróz. Z. 1983. Numerical simulation of rock burst processes treated as problems of dynamic instability. Rock Mech. and Rock Eng. 16: 253-274.

Mróz, Z. and Nawrocki, P. 1987. Deformation and stability of an elastic-plastic rock pillar. Submitted for publication.

Numerical Methods in Geomechanics (Innsbruck 1988), Swoboda (ed.)
© 1988 Balkema, Rotterdam. ISBN 90 6191 809 X

Two 'Class A' predictions of offshore foundation performance

I.M.Smith
University of Manchester, UK

ABSTRACT: Constitutive models for soil and the computational means for their implementation in analyses have been available for twenty years or more. Recently the opportunity has arisen for genuine predictions to be made of the performances of two very different types of offshore foundation. The paper discusses the 'predictive' process, its uncertainties, and the likelihood of 'good' agreement with eventual observations.

1.0 THE PREDICTIVE PROCESS

The word 'prediction' is one of the most mis-used terms in our science. In the present paper it is used exclusively in the sense of a forecast of a future event, the outcome of which is unknown to anyone at the time the prediction is made. Lambe (1973) called this a 'class A' prediction and, in classifying 5 different types of 'prediction' (A,B,B1,C,C1), illustrated the complexity of the predictive process. For example, his classes C and C1 are 'predictions' made after the event, and are inevitably surrounded by scepticism, see Table 1 (taken from Lambe, 1973).

Table 1 : Classification of prediction

Prediction type	When prediction made	Results at time prediction made
A	Before event	Not known
B	During event	Not known
B1	During event	Known
C	After event	Not known
C1	After event	Known

In the context of our subject, it is quite common for results from a constitutive model to be termed 'predictions' of the results of laboratory tests on soil elements, whereas what is in fact going on is a curve-fitting, or calibration, process. The author has been as guilty of this loose usage as many others.

Even amongst forecasts one can distinguish attempts which are more daring or genuinely 'predictive' than others. For example, weather forecasting is, in a sense a genuinely predictive process, but is based on data and evoluntionary analyses of the earth's atmosphere which make comparatively minor excursions into the unknown, and involve daily updating of the database. In many senses, therefore, weather forecasting could be described as essentially a 'class B' predictive exercise. (Nonetheless, these predictions can be dramatically wrong in extreme circumstances, such as the recent hurricane in Southern England.)

Similar minor excursions from the unknown take place in geomechanics when constitutive models for soil are calibrated against certain test data, and then used to predict the results of somewhat similar tests. Several such exercises have been conducted in the past, most recently at Case Western Reserve University, U.S.A., in July 1987.

2.0 SOME 'CLASS B' PREDICTIONS IN GEOMECHANICS

The Case Western Reserve prediction exercise (proceedings to be published 1988) involved a database of some six drained 'triaxial' tests on each of two quartzitic sands (named 'Hostun' and 'Reid Bedford'), and conducted using two different apparatus (cubical and hollow cylindrical specimens respectively). The database therefore totalled 24 drained tests.

Based on these, the results of twelve somewhat different drained tests, six performed in each apparatus, were to be predicted. For example, the calibration tests were done using $b(=\sigma'_2-\sigma'_3/\sigma'_1\sigma'_3)$ of 0 and 1 ('triaxial' compression and 'triaxial' extension respectively), whereas results of tests at different intermediate b values were to be predicted. These are clearly relatively minor excursions into the unknown. On the other hand, two tests (one on each material) involved a rather esoteric journey in stress space describing a circle in the π-plane, while two more involved cyclic loading. These last four constitute much bolder predictive leaps and could possibly be classed as 'A' predictions.

Concentrating for the moment on the 'B' category, predictors were nevertheless faced with various dilemmas. Figure 1 shows experimental data obtained from cubical triaxial compression tests at three different confining pressures on Hostun sand. Also shown are curve fits obtained using the constitutive model due to Molenkamp (Molenkamp, 1981, Smith et al, 1988b). It can immediately be seen that:

a) The test data are inconsistent, the test at a confining pressure of 350 kPa being more compressive than the test at 500 kPa. No consistent constitutive relation could conform to this behaviour.
b) The fits to soil stiffness, using the crude two-parameter curves employed, led to errors of the order of 100% (for example at a confining stress of 200 kPa and a deviator stress of 400 kPa) in some circumstances.
c) Errors in fitted volumetric strains could also be of the order of 100% at worst.

To compound all this, the data from two different test apparatus were markedly different. The deposited sands showed significantly different anisotropies and a predictor using an isotropic constitutive model could have averaged the data in many different ways. The author (Smith et al, 1988b) chose to calibrate the Molenkamp model based on data from the cubical test apparatus only (for one thing it exhibited far less anisotropy) and using only compression test results.

Two of the author's poorest resulting class 'B' predictions are shown in Figures 2 and 3. The former (relating to a hollow cylinder test) shows stiffness discrepancies at $\tau_{\theta z}$ of 200 kPa which are at least 100%, while the latter (relating to a cube test) shows large discrepancies in predicted volumetric strains. No other predictor was markedly more successful and many were less successful. Nevertheless, in a general sense, these class 'B' predictions were encouraging. The scatter of experimental data is exemplified by Figure 4, which shows peak stress ratios (mobilised friction angles) in the π-plane.

The Molenkamp model adheres to the curve marked 'Lade' (Lade, 1977), whereas the scatter of data could fit Mohr-Coulomb, or indeed a surface located somewhat outside Lade's. No smooth surface has yet been suggested which would be in any sense a 'best fit' to the measured data from the two apparatus testing the same sand.

Although it is beyond the scope of the present discussion, it may be remarked that the predictions (class 'A'?) of the more complex stress paths were radically worse than those described above.

3.0 FEATURES NOT YET ADDRESSED IN PREDICTIONS OF ELEMENT TEST RESULTS

In the predictive exercise described in Section 2, the element tests were all conducted on drained sands, and although the materials were dense, the tests were stopped short of strains which would have led to the spread of inhomogeneities through the specimens. Figure 5 shows the strain distribution computed in a specimen with imperfect end conditions, due to strain-softening effects, and Figure 6 shows typical test data (Desrues, 1984). Clearly, the problems of calculation of localisation of deformation in dense materials will have to be addressed in future.

It is also crucial that 'undrained' and 'partially drained' situations are analysed. These constitute the most interesting cases in practice, where predictions of excess porewater pressures must be made (Hicks and Smith, 1986).

Ultimately, undrained and partially drained problems which also involve localisation must be treated.

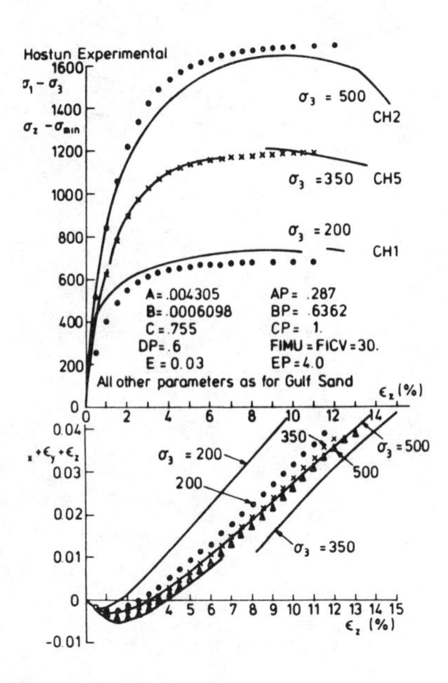

Fig.1 : Measured (solid line) and computed (dotted) curves for triaxial compression data on sand.

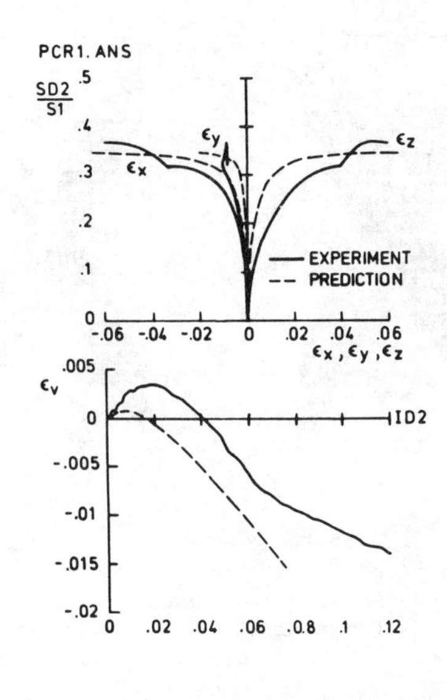

Fig.3 : Measured and predicted results for cube test on sand.

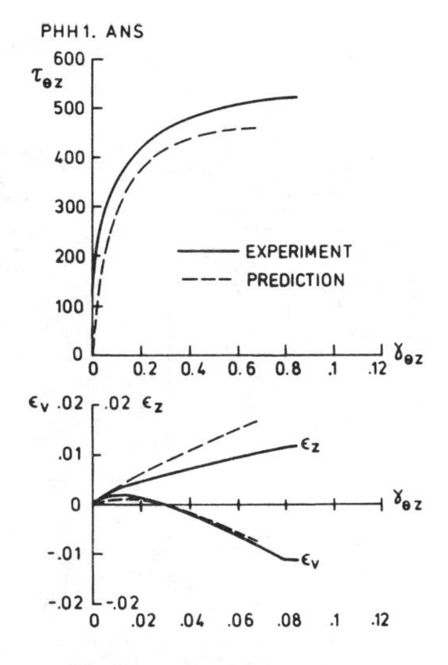

Fig.2 : Measured and predicted results for hollow cylinder test on sand.

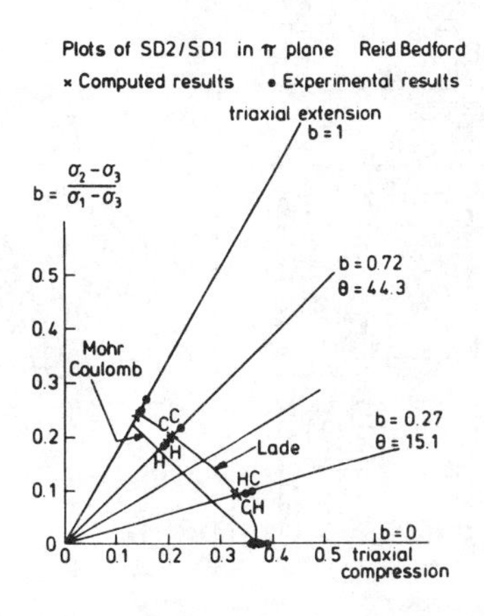

Fig.4 : Measured and computed failure points in the π-plane for sand.

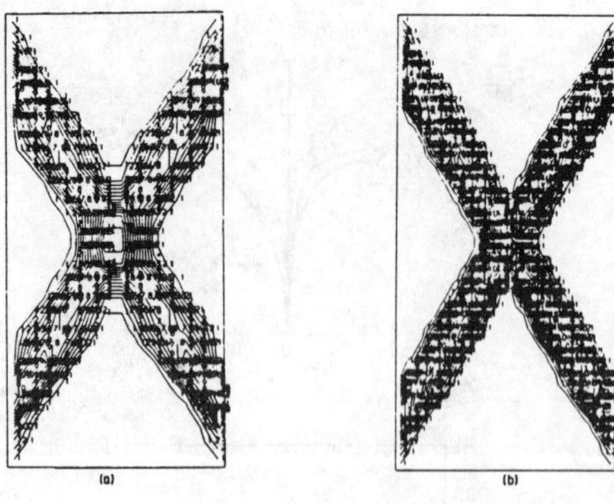

Fig.5 : Computed shear strain contours for rough-ended plane strain specimen of sand (a) coarser mesh (b) finer mesh

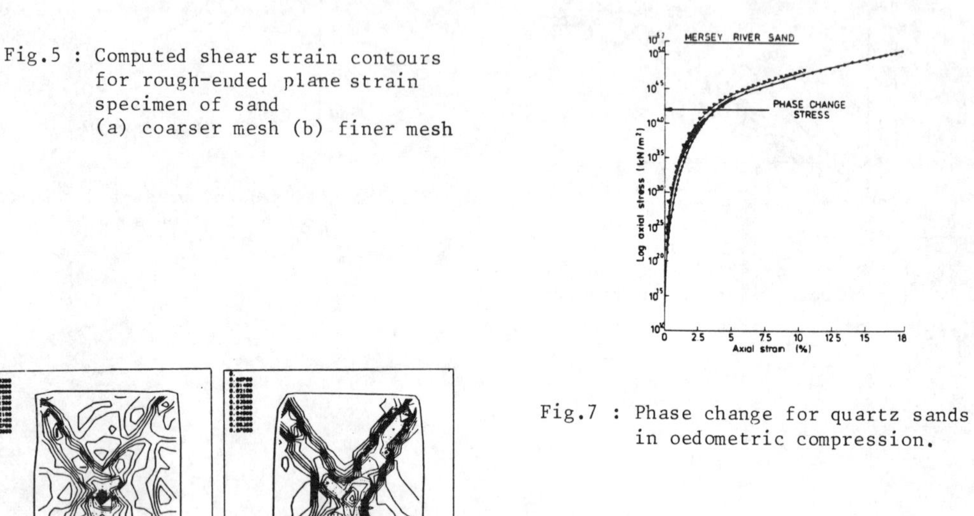

Fig.7 : Phase change for quartz sands in oedometric compression.

Fig.6 : Measured strain contours for plane strain specimen of sand.

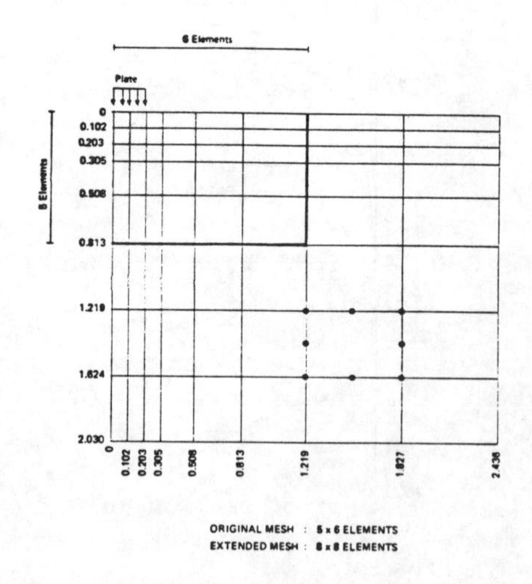

Fig.8 : Axisymmetric finite element mesh for plate load test analysis.

3.1 Phase change

Element tests previously described have involved specimens of dense sand in a "pluviated" or normally consolidated state. At 'moderate' pressures and temperatures, these materials deform without change of state (i.e. crushing of the particles, fusion and so on). As pressures increase, quartz sands such as 'Hostun' and 'Reid-Bedford', previously referred to, will suffer a change of state, as illustrated in Figure 7, for two quartzitic materials (Mersey River Sand and Leighton Buzzard Sand). Both exhibit a "crushing stress" of the order of 15 MPa in oedometric compression. 'Granular' materials, with softer skeletons, for example some calcareous deposits, will crush at pressures an order of magnitude less than this (say 1 to 5 MPa) while soils with yet softer skeletons, traditionally called "clays" can exist naturally in a highly compressible state ("normally consolidated") or exhibit compressibility behaviour analogous to phase change due to "preconsolidation" at pressures traditionally believed to be dependent upon the preloading history (0.3 MPa for some strata of 'Boston clay', up to 2 MPa for 'London clay' and so on).

In some constitutive models for soil, for example the author's implementation of the Molenkamp model, such phase changes are catered for mainly by the concept of initial yield surfaces. In other "continuous plasticity" models the concept of yield surfaces is absent. Nevertheless, phase change simulation will be important in complete descriptions of soil behaviour.

4.0 PREDICTIONS OF FIELD EVENTS BASED ON CALIBRATIONS FROM ELEMENT TESTS

In the previous sections it was shown that predictive exercises can be carried out using tests on 'perfect' soil elements in the drained state. So far, the undrained state for sands has not been examined in this way, nor have the problems of localisation. Until such time as these tests can be reliably done in the laboratory, field events provide the best means of checking predictive capacity for practical purposes. It was shown in Section 2 that stiffnesses (and therefore deflections) might, at most, be in error by a factor of 2 in any specific stress path test. In the field, an infinity of stress paths are being followed, and one would guess that, on average, deflections would be predictable to better than a factor of 2.

Recently, two field situations have been analysed, and indeed deflections have been predicted with an accuracy far better than the factor 2. In neither of these situations were dilation and strain-softening dominant features of the problems and so localisation, in the sense of shear band propagation, (Shuttle and Smith, 1987) was not calculated or observed.

4.1 Drained displacements of a rigid circular foundation under axial load

In this example, (Smith et al, 1988a), plate load tests were conducted on a layered foundation. At the time of the plate tests, there was a limited data-base available, consisting of results of drained oedometer and triaxial compression tests. Using these, the Molenkamp soil model was calibrated and used in finite element analyses emplying meshes typically illustrated in Figure 8. The predicted displacements differed from those measured in the test by at most 20% over a range of plate pressures up to 10 MPa. Subsequently, the model was used to compute likely undrained and partially drained responses of typical foundation configurations, as will be illustrated at the Conference.

4.2 Undrained/partially drained displacements of a laterally loaded caisson in ice

In this example, rather detailed laboratory test data were available before the event. Again the Molenkamp model was calibrated, in this case involving curve fits to data from drained isotropic compression tests, drained isotropically consolidated triaxial compression tests and undrained isotropically consolidated triaxial compression tests. Typical examples of these three calibration exercises are shown in Figures 9, 10 and 11. After calibration, the soil properties were incorporated in a finite element model of the laterally loaded caisson (Hicks and Smith, 1987), as shown in Figure 12. Several analyses were conducted for drained, undrained and partially drained conditions and for various 'best' and 'worst' assumptions about the soil. Figure 13 shows the predicted load-displacement behaviour of the caisson, resulting in an estimated 70mm displacement under the given load. When this load was subsequently encountered in the field, the resulting displacement was of the order of 80mm, an error of less than 15%.

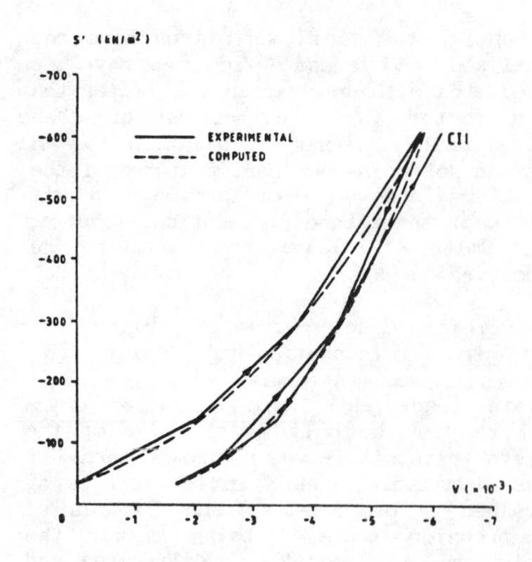

Fig.9 : Calibration of model for quartz sand - drained isotropic compression.

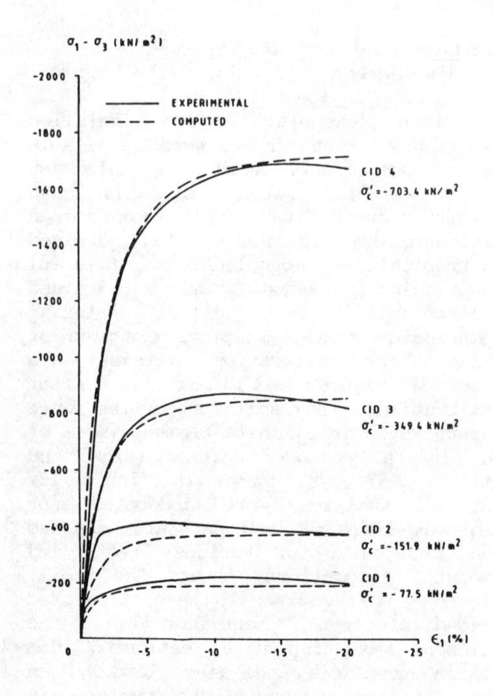

Fig.10 : Calibration of model for quartz sand - drained triaxial compression.

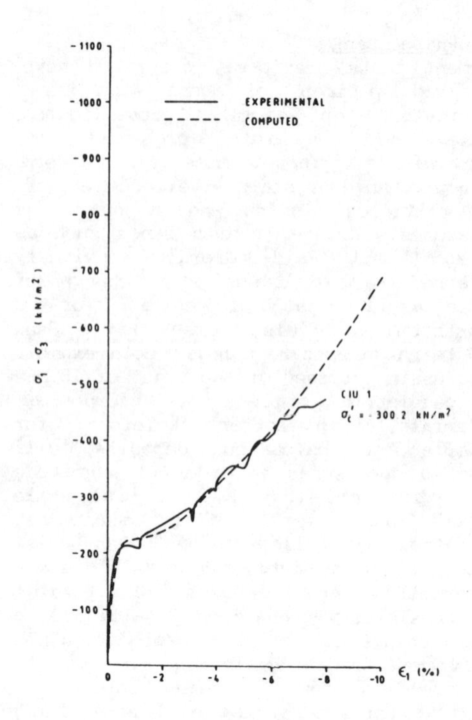

Fig.11 : Calibration of model for quartz sand - undrained triaxial compression

5.0 CONCLUSIONS

Some aspects of the 'predictive' process have been outlined. Many of the exercises which are called 'predictions' in the geomechanics literature are not so in any etymologically correct sense. Lambe (1973) attempted to resolve this problem by classifying 'predictions' in 5 classes, of which only 'A' is genuine. This paper has described two such class 'A' predictions of field deformations in an offshore context. In both cases the agreement obtained with actual measurements was to within about 20%. This may seem paradoxical since the paper also shows the 'good' agreement in calibrating constitutive models for soil against laboratory tests can involve errors of up to 100%. However, an averaging process seems to operate over the many stress paths present in the field so that the maximum possible error is never achieved (in much the same way as computers operate to far greater than their theoretical numerical accuracy). Of course, if a field problem involves very specific stress paths, appropriate tests should be used in the calibration of the constitutive model.

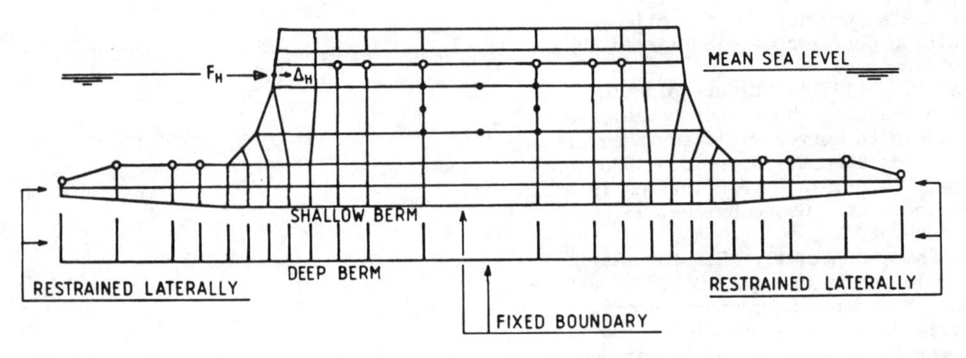

Fig.12 : Finite element idealisation of caisson.

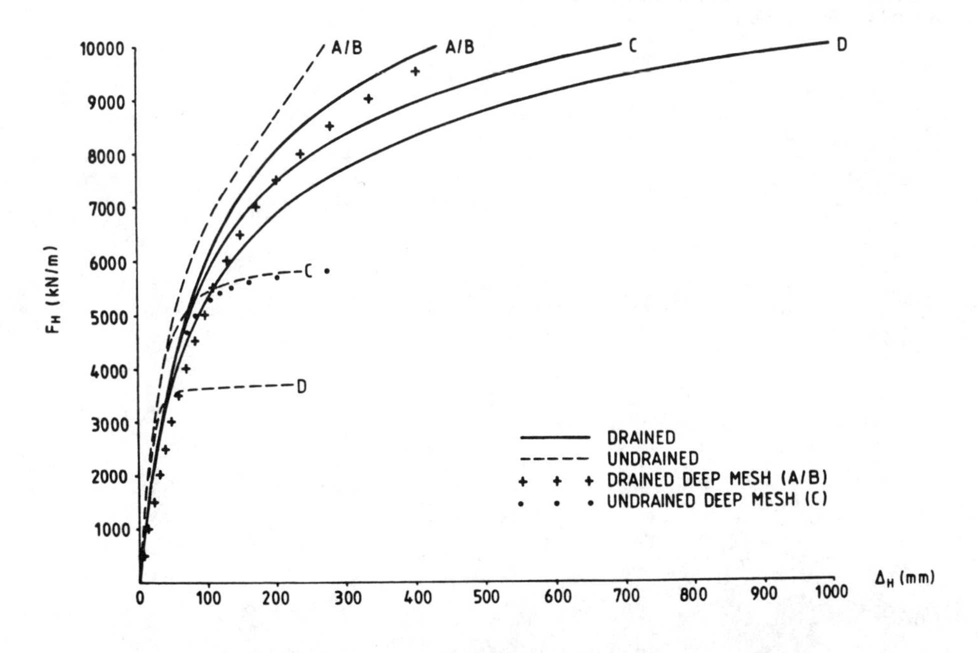

Fig.13 : Computed caisson displacements.

6.0 REFERENCES

Desrues, J. (1984). Localisation de la deformation plastique dans les materieux granulaires, These d'etat, Universite de Grenoble.

Hicks, M.A. and Smith, I.M. (1987), 'Class A' prediction of Arctic caisson performance. Submitted for publication.

Lade, P.V. (1977). Elasto-plastic stress-strain theory for cohesionless soils with curved yield surfaces. Int. J.Solids Struct. 13, pp1019-1035.

Lambe, T.W. (1973). Predictions in Soil Engineering. Geotechnique, 23, 2, pp149-202.

Molenkamp, F. (1981). Elasto-plastic double hardening model MONOT. Delft Soil Mechanics Laboratory Report.

Shuttle, D.A. and Smith, I.M. (1987). Numerical simulation of 'Shear Band' formulation in soils. Submitted for publication.

Smith, I.M., Hicks, M.A., Kay, S. and Cuckson, J. (1988a). Undrained and partially drained behaviour of end bearing piles and bells founded in untreated calcarenite. Proc.Conf. Calcareous Soils, Perth, Western Australia.

Smith, I.M., Shuttle, D.A., Hicks, M.A. and Molenkamp, F. (1988b). Prediction No.32, International Workshop on Constitutive Equations for Granular, Non-Cohesive Soils, Case Western Reserve University, Ohio, U.S.A.

Numerical Methods in Geomechanics (Innsbruck 1988), Swoboda (ed.)
© 1988 Balkema, Rotterdam. ISBN 90 6191 809 X

Numerical modelling of tunnel blasting

S.Valliappan
University of New South Wales, Australia
K.K.Ang
National University of Singapore

ABSTRACT:This paper describes the details of the finite element study of the problem of vibrations induced due to blasting associated with the excavation of underground tunnels.The discussion presented is based on the extensive parametric studies of the vibration effects on existing tunnel due to the excavation of a new tunnel using explosives. The various effects such as the rise time of the blast pulse, explosive detonation velocity and the cover between the tunnels have been taken into account for the purpose of this discussion.The primary interest in this investigation is the prediction of the magnitudes of the peak particle velocities experienced at the existing tunnel.

1 INTRODUCTION

The problem of wave propagation resulting from the detonation of explosives is an important aspect of rock blasting that is of considerable interest.Blasting vibrations,although temporary in nature,can cause serious permanent damage to nearby existing structures. Considerable research has been carried out in the past to predict the vibration response of rock mass subjected to explosive detonation.Various approaches including empirical methods based on full-scale blasting tests, mathematical modelling as well as numerical modelling have been adopted for these investigations. The work done in the area of empirical methods for underground tunnel blasting includes amongst others the references [1-3].However,only a few explicit mathematical solutions have been obtained using elastic wave theory,and these solutions are only for simple cases of blasting under idealized situations[4,5].Numerical techniques such as the finite difference and the finite element methods have been used for modelling the blasting, with varying degrees of success.Numerical modelling of blasting based on two dimensional approximation using the finite difference method has been reported in references[6-8].The finite element method has been used to study the wave propagation problems arising from rock blasting in ref.[9-14].However, it should be pointed out that almost all the finite element solutions have been obtained for two dimensional models due to the limited availability of computer resources.

In this paper,the finite element results for tunnel blasting are presented based on two dimensional 'pseudo-plane' strain concept which models the 3-D situation with reasonable accuracy.

2 TUNNEL BLASTING

When excavation for underground tunnels using explosives has to be carried out close to the ground surface or in the vicinity of existing substructures such as another tunnel, the prediction and control of vibrations is important in order to prevent any damage.The complex nature of tunnel blasting has often restricted the prediction of vibrations to empirical approaches.Field investigations for understanding the dynamic behaviour of underground tunnels due to blasting have been made [15]. The availability of present-day powerful computers and numerical techniques can be exploited to predict the vibrations produced by tunnel blasting. However, the development of a comprehensive and precise model of tunnel blasting is formidable due to the complexity of the problem augmented by the great number of variables influencing the problem. Inevitably, the approach adopted by many investigators is one of using a simplified numerical model with several assumptions, which may still be regarded as a more positive approach [16]. The advantage of numerical modelling is that by merely varying some of the parameters of the problem and carrying out more simulations, a great deal of information on a wide variety of blasting situations can be easily and quickly obtained as compared to the tedious job of gathering data from field measurements. From these numerical simulations, the critical parameters can be identified and then if necessary, useful field tests can be made.

In order to develop a reasonably accurate finite element model of tunnel blasting, various aspects of significance have to be considered. One is the description of the blast pulse acting on the walls of the borehole, such as the maximum pressure, the rise time, the duration and the pulse history. This is critical for designing the finite element mesh as well as for estimating an adequate time step size for numerical integration.

A precise description of pressure-time history for blast loading is difficult to obtain. Therefore, a simpler approach is to consider the pressure pulse applied at the walls of the borehole to be an appropriate known function of time, such as a double exponential time decay function [17,18]. In view of the difficulties involved in getting suitable description of the actual pressure - time function and the lack of generlity in experimentally determined results, the adoption of simplified approximate functions is attractive.

Several equations from the family of double exponential type functions which can simulate a variety of explosive induced loads have been proposed in [18]. These functions can be expressed as

$$P(t)=P_o f_o (e^{-\alpha t} - e^{-\beta t})$$

$$f_o = (e^{-\alpha t_R} - e^{-\beta t_R})^{-1}$$

where $P(t)$ is the pressure acting at any point on the borehole wall at a given instant of time, α and β are arbitrary constants defining the particular form of the function and P_o is the peak pressure. f_o is dimensionless and assumes a value such that $P(t)/P_o$ attains a maximum value of one at $t=t_R$, where t_R is rise time,

$$t_R = \frac{\ln(\alpha/\beta)}{\alpha-\beta}$$

The pulse rise time has an important effect as far as numerical modelling is concerned. It can be expected that blast pulses of smaller rise times will tend to generate higher dominant frequency waves and consequently smaller size finite elements must be used in order to propagate these high frequency wave components, for example, if the rise time is 5μs, then elements of the order of 10mm may be required for accurate analysis. Thus, the overall dimensions of the finite element mesh for the practical cases of tunnel blasting will be extremely large. Therefore, it is inevitable not to adopt a higher rise time than the actual one. In a linear elastic analysis, the results corresponding to smaller rise times may be qualitatively extrapolated from the results for the higher rise times.

For a further description of the behaviour of the explosive in the blasthole, a method for simulating the travelling detonation front is required. If the velocity of detonation of the explosive is infinitely large, then all the points on the blasthole walls in contact with the entire length of charge will be loaded with the same magnitude of pressure at any instant of time. The simulation of the travelling detonation front in the explosive may be done with 'programmed burn technique' [19]. In fact, the velocity of detonation determines the detonation pressure which tends to be larger for higher velocities.

The frequencies of dominant waves propagating in a medium play an important part in the dynamic finite element analysis. For choosing an optimal mesh and time step,it is necessary to know apriori,the cut-off frequency which is the highest frequency of the dominant wave that is propagating.The size of elements is chosen in order to allow the transmission of only those wave components of frequencies lesser than the cut-off. The exclusion of the effects of waves of frequencies higher than the cut-off does not lead to any appreciable error.Since the cut-off frequency depends on the frequency content of the input loading function,a knowledge of the important frequencies in the loading pulse is useful.The identification of these frequencies may be achieved by performing a discrete fourier transform (DFT) of the function describing the blast pulse loading.It should be noted that the frequency content and hence the maximum important frequency of the pulse depends substantially on the rise time of the pulse.A pulse with a small rise time can be expected to contain high dominant frequencies and conversely,a pulse possessing a gentler rise to its peak can be expected to contain lower dominant frequencies.Ref.[20] gives the details of the frequecy domain analysis carried out for the results discussed in this paper.

Another aspect of blasting that needs to be considered is the proper modelling of tunnel round.The modelling of tunnel round is difficult since the distribution pattern of blastholes over the tunnel face is generally complicated.The blasting of tunnel round may be idealized to the case of a single equivalent blast,in so far as predicting the peak blast vibration.The equivalent charge to be used in the analysis could be the charge with zero-delay interval or the largest charge as recommended in ref.[21].Alternatively,the maximum cooperating charge proposed in ref.[22] could be used.If the concern is only in the prediction of the maximum blast vibration induced in the ground at relatively distant locations from the tunnel face,then it is justified to idealize the pattern in a tunnel round by a single hole,but with an equivalent charge. The idealized single hole placed at the centre of the tunnel,in essence, models the detonation of zero-delay charge.

3 FINITE ELEMENT MODEL

The problem of a tunnel which is being excavated,by means of blasting, at right angles to the axis of an existing tunnel is truly three dimensional in nature.However,a complete 3-D dynamic analysis, either linear or nonlinear will place a heavy demand on computer resources both in terms of core memory and the execution time.The use of an equivalent plane strain analysis is generally accepted as a suitable alternative to full 3-D.However,it must be pointed out that great caution must be exercised in using the plane strain model for the wave propagation problems.In 2-D modelling,the waves get trapped within the unit slice of the model and attenuation is mainly due to material damping if considered.In actual situation wave energy dissipates much faster due to the propagation of waves in the third direction. Thus the conventional plane strain analysis is not suitable for wave propagation problems.Therefore, the authors proposed a 'pseudo-plane strain' analysis [23,24] which leads to an approximate but more realistic simulation of the rate of attenuation of stress and velocity with distance from borehole that would be expected normally.

The tunnel blasting is essentially a dynamic problem involving wave propagation.Thus,in the basic equation of motion,both mass and damping matrices play an important part in addition to the stiffness matrix.A critical study of the effectiveness of the type of mass formulation in modelling the wave propagation problems has been presented in [25] and it has been recommended that the consistent mass matrix be used for this purpose.The efficiency of various boundary damping techniques has also been discussed in [25] and the viscous damping has been found to be suitable. Regarding the constraints for spatial discretization, due to low-pass filtering action of the finite elements, the element sizes must be limited to certain critical values depending on the nature of the applied loads,the type of mass formulation and the order of finite elements.If mesh grading is adopted, certain restrictions have to be imposed to avoid errors being introduced from spurious reflections of waves propagating in a graded mesh. A technique for grading the mesh through the use of modified shape functions for isoparametric elements is presented in [26]. An efficient and reliable method of integrating the finite element equations of equilibrium with respect to time is essential since it is important to keep the errors arising from the time discretization, a minimum. For this purpose, a new one step implicit algorithm which includes a term defining an impulse load vector has been developed in [27]. This impulse load vector permits the use of time increments which can be solely controlled by accuracy requirements. In addition, variabe time steps can be used effectively [20].

4 NUMERICAL RESULTS

The problem considered in this paper is the case of a new tunnel blast excavated below an existing tunnel. The new tunnel is considered to cross perpendicularly to the axis of the existing tunnel and the cross-section of the tunnels is assumed to be an inverted U. The width and the height of the tunnels have been assumed to be 5m and 6m respectively. The finite element grid used for the dynamic phase of the analysis, which satisfies the criteria mentioned in the previous paragraph is shown in Fig.1. To model the infinite extent of the region beyond the boundaries of the numerical model, energy absorbing boundary dampers of the unified viscous type [28] have been placed along the boundaries. It is to be noted that the use of mesh grading has been extensively applied in order to reduce the size of the model.

The type of explosive used for blasting is an important parameter. For this purpose, the detonation velocity and the pulse rise time have been varied in different cases. Since the results have been normalized with respect to maximum blast pressure, the weight of the explosives has not been varied. Unless otherwise specified, the pull length of the tunnel round has been taken to be 2m and the direction of firing is assumed to be in the same direction as the tunnel advance.

Since peak particle velocity has generally been accepted as the most

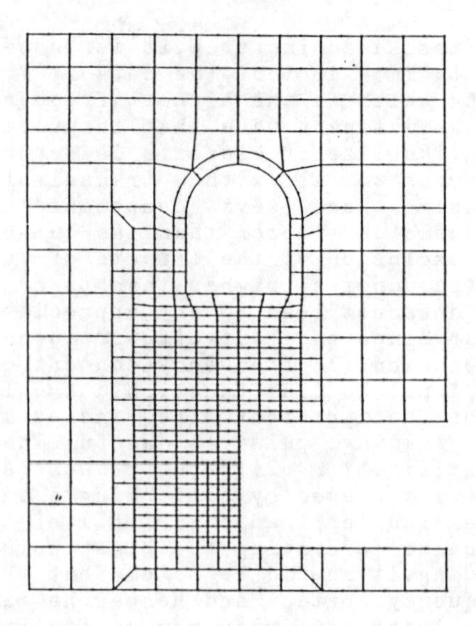

Figure 1. Finite element mesh.

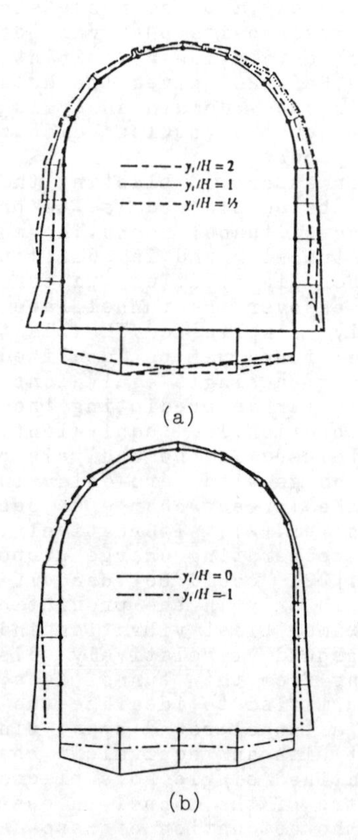

Figure 2. Distribution of peak particle velocity at existing tunnel

suitable criteion for assessing the damage due to blasting, the results will be presented in the form of peak particle velocities.

The depth of overburden from the ground surface to the crown of the existing tunnel as well as the cover between the two tunnels is assumed to be equal to one tunnel height. The problem has been analysed with the advancing face of the new tunnel located at five positions with respect to the existing tunnel, thereby simulating the progress of the tunnel construction. The five positions correspond to y_t /H equal to 2, 1, $\frac{1}{3}$, 0 and -1, where y_t is the horizontal distance between the advancing face and the centre of the existing tunnel and H is the tunnel height.

The three types of explosives considered are with detonation velocity of 4000 m/s and rise times of 100, 200 and 1000 μs.

The distribution of peak particle velocity around the perimeter of the existing tunnel corresponding to the rise time of 100 μs is plotted in Fig.2. It can be observed that for the range of advancing positions considered, the floor of the existing tunnel tends to be affected more than the crown, as would be expected.

The results indicating the effect of rise time are presented in Fig.3, in the form of non-dimensional plots of peak particle velocity response at the existing tunnel section, for the five positions of the new tunnel. The peak particle velocity has been plotted non-dimensionally by multiplying the velocitry with the term $\rho c_p / P_e$ where the numerator is the characteristic impedance of the rock and the denominator is the maximum blast pressure. The five parts of the graph correspond to the five positions of advancing face. It can be seen from the curves presented that the largest magnitude of ppv are obtained for the smallest rise time, 100 μs and there is very little difference between 100 μs and 200 μs , whereas the pulse with 1000 μs is seen to produce considerably smaller vibrations at all parts of the tunnel and for all positions of the advancing face. The effect of rise time is clearly evident from Fig.4 which shows the largest peak particle velocity produced is closely related to the rise time. Also, it can be noticed that the critical ppv is registered when the face of the new tunnel is at a short distance before the crossing point.

For the purpose of studying the influence of the detonation velocity, three values were considered - 2000, 4000 m/s and one that is infinitely large. The results showing the peak particle velocity response at the existing tunnel section for the case where the advancing face is located directly beneath the centre of the existing tunnel are presented in Fig.5 for pulse rise times of 100, 200 and 1000μs. It can be seen from the curves that the vibrations produced are strongly affected by the detonation velocity, being larger for higher values of detonation velocity. This could be due to a greater cooperation of waves arriving at the existing tunnel. However, this effect is less for large rise time. It can be said that the detonation velocity has a greater effect on the ppv than the rise time provided the rise times are small enough.

In order to study the effect of cover between the two tunnels, it was reduced to half the height of the tunnel and the results for the two cases of cover thickness are shown in Fig.6. It can be seen that larger vibrations are produced for small cover thickness as expected. It can also be noted that the largest ppv for the case of small cover is nearly six times that for the case with twice the thickness.

The effect of blasting in a nonhomogeneous medium was considered by analysing two cases. In the first case, the medium was considered to compose of two different layers with the existing tunnel in one layer and the new tunnel in another layer. The results showing the peak particle velocity at the existing tunnel for three ratios of elastic modulus are shown in Fig.7. It can be seen that the effect of vibration is severe for the case when the existing tunnel is located in a medium with higher impedance. In the second case considered both tunnels were assumed to be located in layers with the same properties but an inclusion was introduced in between the two tunnels. The results for various modular ratios are presented in Fig.8. It can be noted that for a very soft inclusion, the peak particle velocities are reduced significantly.

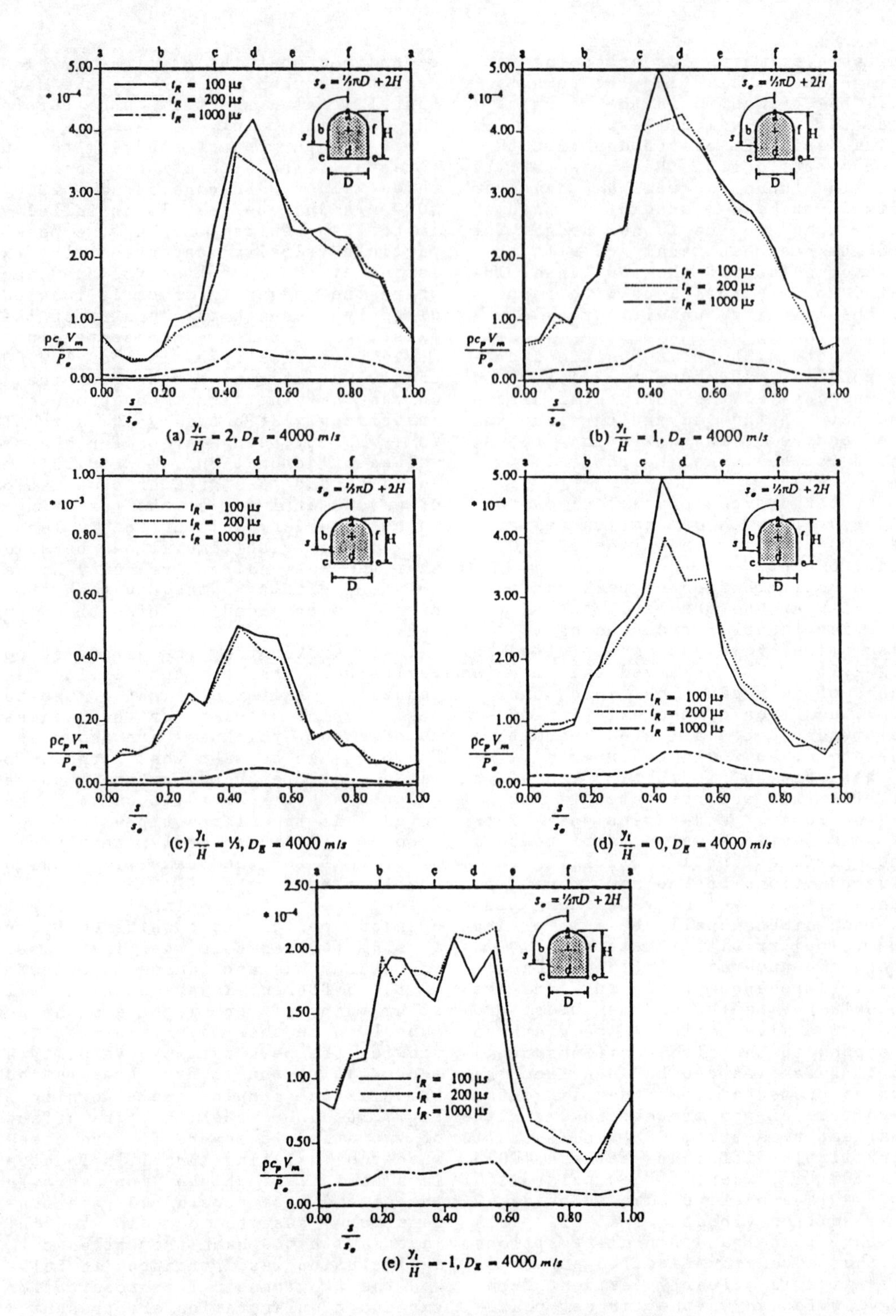

Figure 3. Influence of pulse rise time on peak particle velocity response.

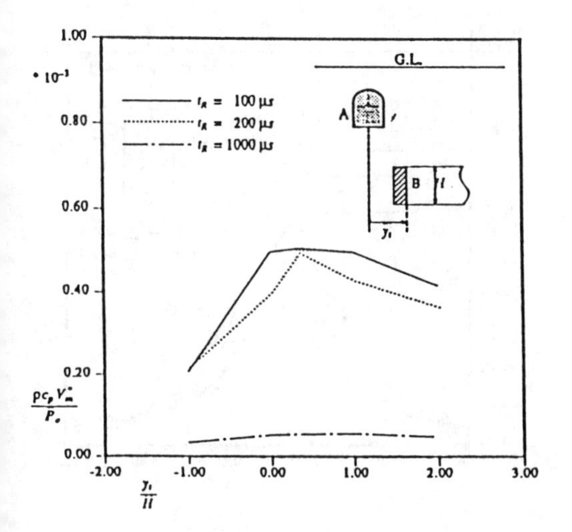

Figure 4. Influence of rise time on largest ppv.

In order to investigate the effect of blasting on lined tunnels, the case of a tunnel lined with concrete and located above a new tunnel which is being blast excavated, was considered. Two different thicknesses for the lining were assumed and the ppv response is plotted in Fig.9. It can be seen from the plot, that the effect of vibration is more on the unlined tunnel than the lined and further there is very little difference in the ppv response for the two thicknesses considered. This indicates that there may be an optimum thickness for the lining beyond which the vibration effect cannot be reduced. Further, it can be noted that higher peak particle velocities are registered at the outer rather than the inner surface of the lining.

5 CONCLUSIONS

The problem of ground vibrations induced as a result of blasting for the excavation of an underground tunnel has been discussed. The development of a suitable finite element model for this problem is indicated. The inherently complex nature of tunnel blasting compounded with the great number of variables influencing the problem makes it difficult to develop a precise and comprehensive model. For example, a precise description of pressure-time function for the blast

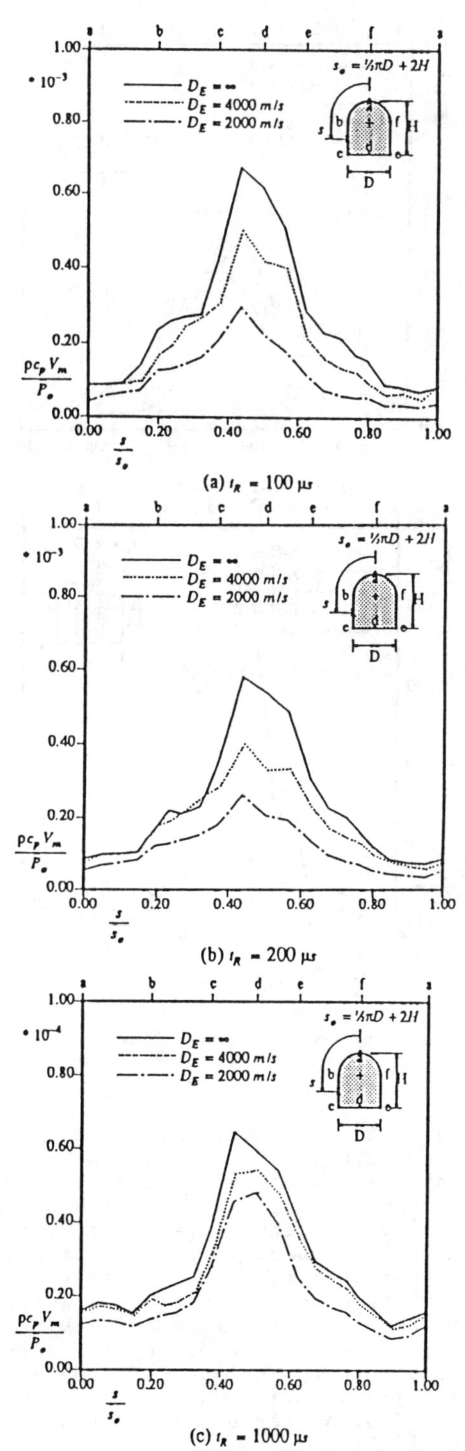

Figure 5. Effect of detonation velocity on peak particle velocity response.

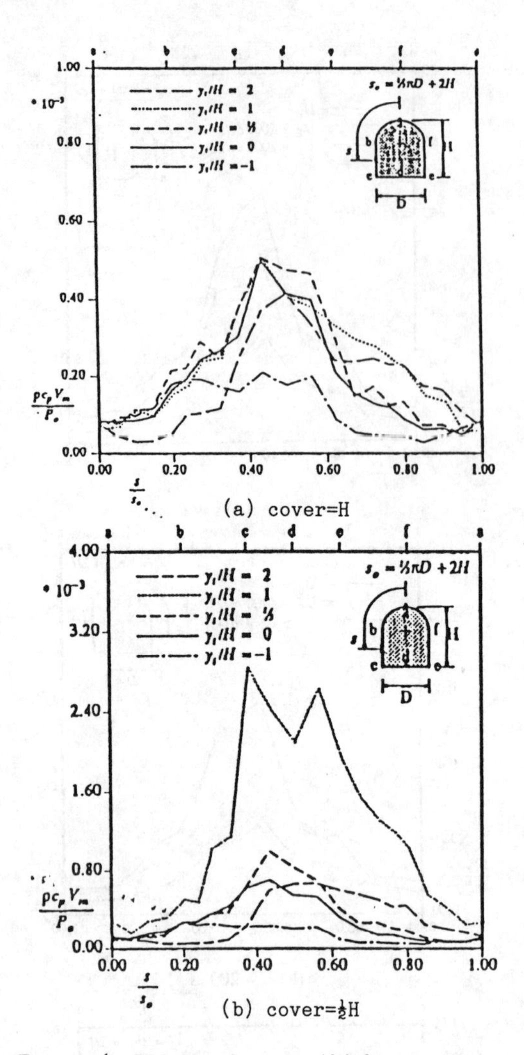

Figure 6. Effect of cover thickness on ppv.

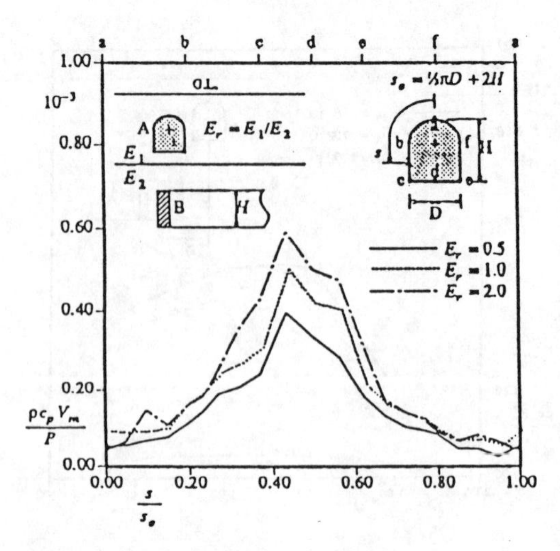

Figure 7. Peak particle velocity response for two layered medium.

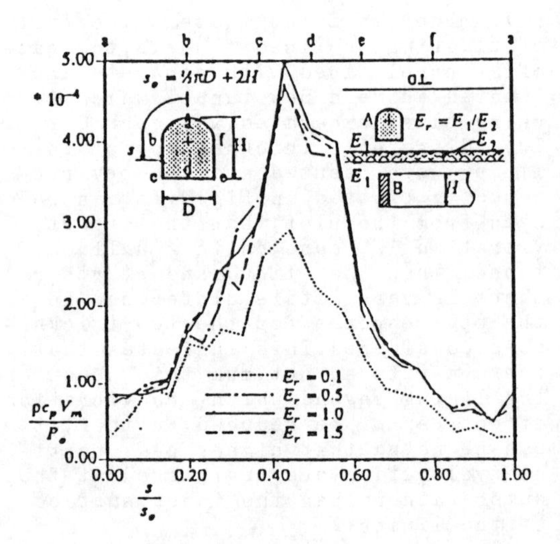

Figure 8. Peak particle velocity response for the case of a medium with inclusion.

loading is difficult.Therefore,in the present study,a double exponential decay time function has been assumed to define the blast pressure history.

In order to investigate the effect of blasting on existing structures, the case of a tunnel excavated below an existing tunnel has been studied. The results in the form of peak particle velocity response have been presented for various cases.Some of the major conclusions that have been drawn from a series of parametric studies are:

The effect of rise time of the blast pulse on the vibration response is appreciable.However,it is not highly significant between small rise times.

The explosive detonation velocity has a more pronounced effect on peak particle velocity than the rise time. However,the degree of influence from detonation velocity is less for large rise times.

Larger vibrations are produced at the existing tunnel for smaller cover of rock between the existing and the new tunnels.

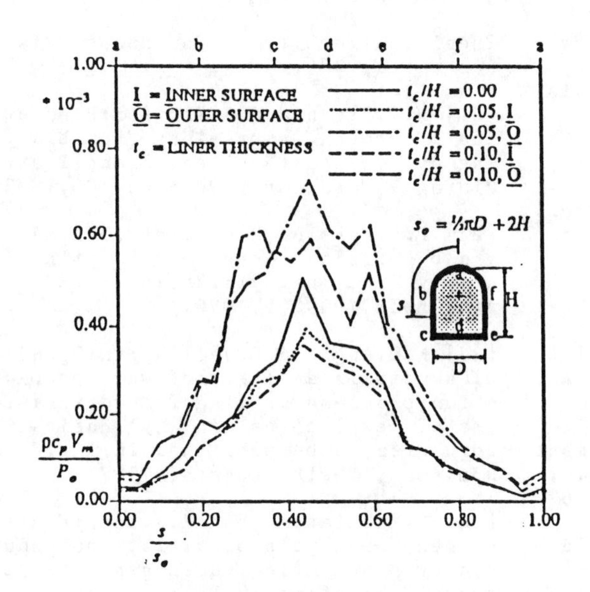

Material Properties for Figure 9:

Rock: E = 25 GPa.

ν = 0.25

ρ = 2000kg/m^3

Concrete: E = 42 GPa.

ν = 0.20

ρ = 2320kg/m^3

Explosive:

Detonation velocity=4000m/s

Rise time = 100 μs

Figure 9. Peak particle velocity response for the case of tunnel with concrete lining.

Larger vibrations are registered in an unlined tunnel compared to the one which has concrete lining.Further,it appears that there may be an optimum thickness of the liner beyond which the effect of vibration cannot be reduced.

It may finally be concluded that the dynamic finite element method can be used to solve the tunnel blasting problem,provided careful considerations are given to various aspects discussed in this paper with respect to proper modelling.

6 ACKNOWLEDGEMENTS

The authors wish to acknowledge the financial support provided by the Australian Research Grants Committee and the sponsorship of the second author by the Australian Development Assistance Bureau.

7 REFERENCES

[1] N.R.Ambraseys and A.J.Hendron: Dynamic behaviour of rock masses,in Rock Mechanics in Engineering Practice,Stagg and Zienkiewicz(eds),John Wiley & Sons,N.203-236,1968.

[2] D.E.Siskind,M.S.Stagg,J.W.Kopp and C.H.Dowding:Structure response and damage produced by ground vibration from surface mine blasting,U.S. Bureau of Mines,RI 8507,1980.

[3] J.J.Snodgrass and D.E.Siskind: Vibrations from underground blasting, U.S.Bureau of Mines,RI 7939,1974.

[4] R.F.Favreau:Generation of strain waves in rock by an explosion in a spherical cavity,J.Geophysical Research,74,4267-4280,1969.

[5] G.D.Free: Dynamic Stress Waves Produced in Rock by Explosive Blasts, Ph.D.Thesis,University of Queensland, Austalia,1979.

[6] L.G.Margolin and T.F.Adams:Numerical simulation of fracture,Proc.1st Int.Sym.on Rock Fragmentation by Blasting,Sweden,347-360,1983.

[7] D.E.Grady and M.E.Kipp:Continuum modelling of explosive fracture in oil shale,Int.J.Rock Mech.Mining Sci. 17,147-157,1980.

[8] J.O.Hallquist: DYNA-3D,Nonlinear dynamic analysis of solids in three dimensions,Pub.No.UCIB-19592,Lawrence Livermore Lab.,Calif.,1982.

[9] S.A.Shipley,H.G.Leistner and R.E. Jones:Elastic wave propagation - a comparison between finite element

predictions and exact solutions,Proc. Int.Sym.on Wave Propagation and Dynamic Properties of Earth Materials, New Mexico,509-519,1967.

[10] S.Valliappan, I.K.Lee,V.Murti, K.K.Ang and A.H.Ross:Numerical modelling of rock fragmentation,Proc. 1st Int.Sym.on Rock Fragmentation by Blasting,Sweden,375-391,1983.

[11] S.Valliappan,I.K.Lee and Y.V.A. Rao: Numerical modelling of rock fracture by explosives,Proc.4th Int. Conf.on Applied Num.Modeling,Tainan, 201-207,1984.

[12] C.J.Costantino: Finite element approach to stress wave problems,J. Engg.Mech.Div.,ASCE,93,154-175,1967.

[13] P.J.Digby,L.Nilsson and M.Oldenburg:Finite element simulation of time dependent fracture and fragmentation processes in rock blasting,Int. J.Num.Analyt.Meth.Geomech.,9,317-329, 1985.

[14] D.V.Swenson and L.M.Taylor: A finite element model for the analysis of tailored pulse simulation of boreholes,Int.J.Num.Analyt.Meth.Geomech., 7,469-484,1983.

[15] S.Sakurai and Y.Kitamura: Vibration of tunnel due to adjacent blasting operation,Proc.Int.Symp.of Field Measurements in Rock Mechanics,Zurich, 1977.

[16] R.F.Favreau: Rock displacement velocity during a bench blast,Proc. 1st Int.Symp.on Rock Fragmentation by Blasting,Sweden,753-776,1983.

[17] W.I.Duvall:Strain-wave shapes in rock near explosions,Geophysics, 18,310-323,1953.

[18] T.P.Bligh:Gaseous detonations at very high pressures and their applications to a rock breaking device,Ph.D.Thesis,University of Witwatersrand,1972.

[19] T.F.Adams,R.B.Demuth,L.G.Margolin and B.D.Nichols: Simulation of rock blasting with the shale code, Proc.1st.Int.Symp.on Rock Fragmentation by Blasting,Sweden,361-373,1983.

[20] K.K.Ang: Finite Element Analysis of Wave Propagation Problems,Ph.D. Thesis,University of New South Wales, Australia,1986.

[21] W.J.Birch and R.Chaffer:Prediction of ground vibrations from blasting on opencast sites, Trans.Inst. Mining & Metallurgy,92,A103-106,1983.

[22] P.A.Persson,R.Holmberg,G.Lande and B.Larrson:Underground blasting in a city,Proc.Int.Symp.Rockstore 80, stockholm,199-206,1980.

[23] K.K.Ang and S.Valliappan:Pseudo-plane strain analysis of wave propagation problems arising from detonations of explosives in cylindrical boreholes,to be published in Int.J. Num.Analyt.Meth.Geomech.,1987.

[24] S.Valliappan and K.K.Ang: Finite element analysis of vibrations induced by propagating waves generated by tunnel blasting,to be published in Rock Mech.& Rock Engg.,1987.

[25] S.Valliappan and K.K.Ang:Dynamic analysis applied to rock mechanics problems, Proc. 5th Int.Conf.on Num. Meth.in Geomech.,Nagoya,119-132,1985.

[26] K.K.Ang and S.Valliappan: Mesh grading technique using modified isoparametric shape functions and its application to wave propagation problems,Int.J.Num.Meth.in Engg.,23,331-348,1986.

[27] S.Valliappan and K.K.Ang: η Method of numerical integration,to be published in Int.J.Comp.Mechanics,1987.

[28] S.Valliappan,W.White and I.K.Lee: Energy absorbing boundary for anisotropic material,Proc.2nd Int.Conf.on Num. Meth.in Geomech.,Blacksburg,1013-1024, 1976.

Numerical Methods in Geomechanics (Innsbruck 1988), Swoboda (ed.)
© 1988 Balkema, Rotterdam. ISBN 90 6191 809 X

Stability and bifurcation in geomechanics

I.G.Vardoulakis
University of Minnesota, Minneapolis, USA

ABSTRACT: The concepts of stability and bifurcation in geomechanics are illustrated by discussing the two most typical failure mechanisms occurring in soil bodies, namely shear banding and liquefaction. For the sake of simplicity the analysis is done on the basis of a two-dimensional flow theory of plasticity for frictional and dilatant granular materials.

INTRODUCTION

Roscoe et al. (1963) in their paper on "Evaluation of Test Data for Selecting a Yield Criterion for Soils" conclude: "...It has been shown that the distribution of deformation within triaxial specimens during drained compression tests is quite different from that occurring during drained extension tests. It is suggested that the results of these two types of tests cannot be compared for investigating the suitability of any failure criterion for soils. Before this can be achieved, more reliable methods must be developed for submitting soil specimens to uniform strain even as they approach failure...". Following these ideas several investigators have developed since a great variety of testing apparatuses, for example the 'Independent Stress Control Cell' (Green 1971), the 'Biaxial Apparatus' (Hambly 1972) or the 'True-Triaxial Apparatus' (Goldscheider and Gudehus 1973). As implied in the above quotation from Roscoe et al. (1963), it was expected that by using more refined equipment and by carefully preparing and lubricating specimens with a small height/diameter ratio, the overall inhomogeneities would be suppressed; cf. Bishop and Green (1973).

Mandel (1964) has indicated that localization of the deformation into shear-bands could be suppressed if the sample is subjected to pure kinematic control. However, this situation is very unrealistic in nature; it is even impossible to attain in the laboratory if the material is fine grained. Hambly (1972) reported that in strain-controlled, plane-strain tests with normally consolidated kaolin, shear-bands were always observed in the sample when the stress path reached the Mohr-Coulomb limit surface; see Figure 1 (a) after Kuntsche (1982). Deformation patterning is also observed in the triaxial test with lubricated ends as shown in Figure 1 (b) in a normally consolidated undrained kaolin specimen (Kuntsche 1982). The mean distance between shear-bands in a pattern like the one shown in Figures 1 (a) and (b), is usually seen as an autocorrelation length between threshold imperfections (e.g. local fluctuations of the water content and/or fluctuations in the silty fraction) which serve as sites for strain nucleation. The finer the mesh of this shear-band pattern is, the more uniform the specimen is initially (Lade 1982). Such observations, while furnishing a statistical (descriptive) evaluation of the situation, they provide neither insight into the mechanism of pattern formation nor they relate the occurrence and wavelength of the microscopic pattern with that of the macroscopic one.

These observations mean that localized inhomogeneities like shear-bands can be suppressed neither by refinements on the boundaries and on the samples' preparation, nor by pure kinematic control. The latter has been shown theoretically by Rice (1976). Hettler, Gudehus, and Vardoulakis (1982) (see

Hettler and Vardoulakis 1984) have developed a new triaxial apparatus by which the effect of bulging was completely suppressed by testing very short specimens of large overall dimensions. Localization of deformations could not be avoided, however, with a system of shear bands developing in a confined fashion.

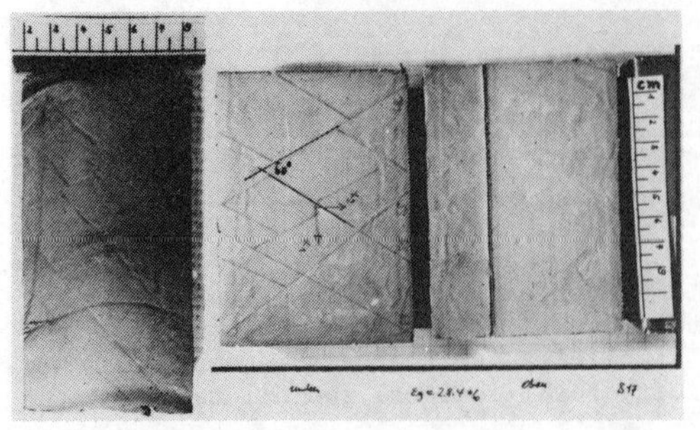

Figure 1. Deformation patterning in normally consolidated kaolin specimens: (a) plane strain-controlled test; (b) triaxial compression test (Kuntsche 1982).

Experiments with 'perfect' boundary conditions and 'perfectly' homogeneous material do not generally secure homogeneous deformation, as various bifurcation modes of the deformation are possible and do actually develop. Bifurcation of the deformation process means that at some critical state the deformation process does not follow its 'straight-ahead' continuation but turns to an entirely different mode. Typical examples of bifurcation phenomena are buckling, barrelling, necking, shear banding and liquefaction phenomena that are observed in soil specimens with lubricated ends. Mathematically bifurcation means that the equations describing the equilibrium continuation do not provide a unique solution. Equilibrium bifurcation analyses in deformable solids are mostly based on the fundamental work of Hill (1958).

One distinguishes between continuous (diffuse) and discontinuous (localized) bifurcation modes. Axisymmetric diffuse bifurcation phenomena in sand specimens were experimentally studied by Roscoe et al. (1963), Kirkpatrick and Belshaw (1968), Deman (1975), Bishop and Green (1973), Reads and Green (1976), and Hettler and Vardoulakis (1984). Axisymmetric, diffuse bifurcation modes in the triaxial compression on dry sand samples have been studied theoretically first by Vardoulakis (1979, 1981a). In a following paper Vardoulakis (1983), found that for rigid-plastic, frictional dilatant material diffuse bulging in the triaxial compression test and diffuse necking in the extension test are always possible at small positive hardening rates, in the vicinity of the plastic limit state. These results are in accordance with the experimental observations. Diffuse bifurcations of sand specimens under plane-strain conditions of were also studied by Vardoulakis (1981b), who found that the only significant mode in the biaxial test with mixed boundary conditions is antisymmetric buckling that occurs before shear-banding.

Experimental work on localization phenomena in granular media has been reported by a number of investigators; cf. Vardoulakis (1978, 1980), Scarpelli and Wood (1982), Lade (1982), Desrues (1984), Hettler and Vardoulakis (1984) and Vardoulakis and Graf (1985). Under plane strain conditions, shear-banding in dense dry sands is observed in the hardening regime of the stress-ratio strain curve (Vardoulakis 1978). Under axisymmetric conditions localization in the compression test is taking place in the softening regime (Hettler and Vardoulakis 1984) whereas and in the extension test localized necking is occurring in the hardening regime (Reads and Green 1976). These results have been recovered theoretically by Vardoulakis (1980, 1983) within the frame of deformation theory of plasticity for rigid-plastic, dilatant frictional materials.

This Paper summarizes some results from the author's recent work on stability and bifurcation. In the next section the constitutive equations a two-dimensional flow theory of plasticity for frictional-dilatant materials are outlined and briefly discussed. The following two sections focus on the two major manifestations of failure in granular soils, namely shear-banding and liquefaction. The paper concludes with some thoughts on the concept of failure in soil mechanics.

FLOW THEORY OF PLASTICITY

The current trend in constitutive modeling in soil mechanics is to mathematically describe a plethora of non-standard effects, such as non-coaxiality, anisotropic hardening, cyclic loading through phenomenological generalizations of constitutive equations appropriate to locally homogeneous deformations (see

Gudehus et al. 1984 and Desai et al. 1987). As opposed to this, we will maintain here the classical structure of soil plasticity (e.g. Mohr-Coulomb, non-associate elastoplastic model). Here the point of view is advanced that a most important class of soil behavior such as shear banding and liquefaction phenomena can only be modelled by capturing the heterogeneity of deformation rather than by further refining constitutive equations for homogeneous deformations. The increasing complexity of constitutive models has rendered impossible to assess with reasonable accuracy their shortcomings. In this connection, we want to emphasize that the experimental data used for calibration and validation of these models are assumed to correspond to locally or globally homogeneous deformations whereas, in fact, bifurcation may have already occurred thus rendering the corresponding analyses questionable.

In this section we will discuss a simple two-dimensional elastoplastic constitutive model for frictional, dilatant granular materials. This model is a special case of the elastoplastic, isotropic hardening, pressure-sensitive model with non-associate flow-rule (Mròz 1963) that was first discussed in the context of material stability and shear-band formation by Mandel (1963, 1964).

Constitutive Equations

Let σ'_{ij} ($\Delta\sigma'_{ij}$) and $\Delta\varepsilon_{ij}$ be the effective Cauchy stress tensor (its increment) and the strain increment tensor, respectively. These tensors are decomposed into a deviatoric and a spherical part:

$$\sigma'_{ij} = s_{ij} + \sigma'_{kk}\,\delta_{ij}/2$$
$$\Delta\varepsilon_{ij} = \Delta e_{ij} + \Delta\varepsilon_{kk}\,\delta_{ij}/2\ . \quad (1)$$

Furthermore the strain increments are decomposed in elastic and plastic part:

$$\Delta\varepsilon_{ij} = \Delta\varepsilon^{e}_{ij} + \Delta\varepsilon^{p}_{ij}\ . \quad (2)$$

In the simplest version of a flow theory for strain hardening frictional, dilatant granular materials, elastic strain increments are given by Hook's law and plastic strain increments are defined by means of a Coulomb yield surface **F** and a plastic potential surface **G** as follows (Figure 2):

$$\mathbf{F} = \tau/p' - \mu\ ;\ \mathbf{G} = \tau/p' - \beta\ , \quad (3)$$

where p' and τ be the mean effective pressure and the shearing stress intensity, respectively:

$$p' = -\sigma'_{kk}/2\ ;\ \tau = (s_{ij}s_{ji}/2)^{1/2}\ . \quad (4)$$

$\mu(n_o,\gamma^p)$, $\beta(n_o,\gamma^p)$ are the mobilized-friction and dilatancy functions, which are assumed to be functions of the initial porosity n_o of the granular material and of the accumulated plastic shear strain $\gamma^p=\int d\gamma^p$, where

$$d\gamma^p = (2de^p_{ij}de^p_{ji})^{1/2}\ . \quad (5)$$

The friction and dilatancy coefficients are also expressed in terms of the so-called mobilized Mohr-Coulomb friction angle ϕ_m ($\mu=\sin\phi_m$) and Hansen-Lundgren dilatancy angle ψ_m ($\beta=\sin\psi_m$) .

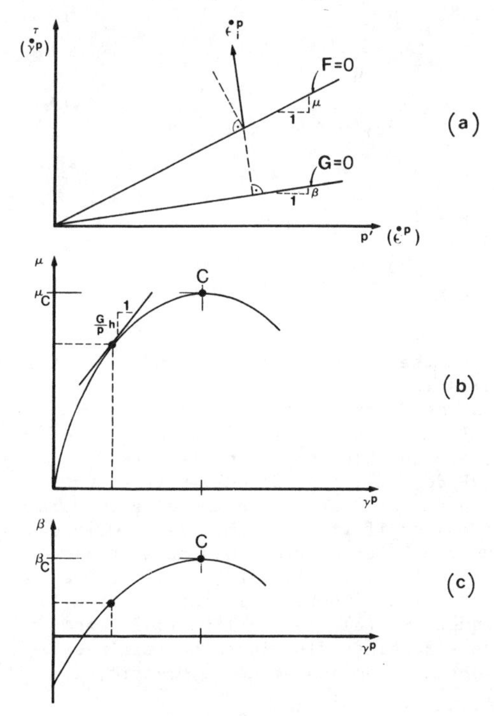

Figure 2. (a) Yield surface and plastic potential surface; (b) mobilized friction function; (c) mobilized dilatancy function

Plastic strain increments are defined by a set of constitutive assumptions that constitute the flow-rule, which in classical flow theory of plasticity reads as follows: (A1) Plastic strain increments are coaxial to the Cauchy stress. (A2) Plastic strain increments are normal to the plastic potential surface **G**. If **G** coincides with **F** (μ=β) then the material obeys an associated flow-rule, otherwise it obeys a non-associated flow-rule.

The constitutive equations for an elastoplastic frictional material can be easily derived from the above assumptions, resulting to the following expressions between the effective stress increment and the stain increment:

$$\Delta\sigma'_{ij} = L_{ijkl}\,\Delta\varepsilon_{kl}$$
$$L_{ijkl} = G(L^{e}_{ijkl} - L^{p}_{ijkl}) \qquad (6)$$

where

$$L^{e}_{ijkl} = (\kappa - 1)\,\delta_{ij}\delta_{kl} + 2\,\delta_{ik}\delta_{jl}$$
$$\kappa = K/G = 1/(1 - 2\nu)$$
$$L^{p}_{ijkl} = \frac{<1>}{H}\,A^{G}_{ij}\,A^{F}_{kl}$$
$$A^{G}_{ij} = s_{ij}/\tau + \kappa\,\beta\,\delta_{ij} \qquad (7)$$
$$A^{F}_{ij} = s_{ij}/\tau + \kappa\,\mu\,\delta_{ij}$$
$$H = 1 + h - h_{T} > 0$$
$$h = -\frac{p'}{G}\frac{d\mu}{d\gamma^{p}} \qquad h_{T} = -\kappa\,\mu\,\beta\,.$$

In the above equations G and K are the elastic shear and compression moduli and ν is the Poisson ratio for plane-strain deformations. $<1>=1$ if loading of the yield surface is taking place [**F**=0, $(\partial F/\partial\sigma_{ij})L^{e}_{ijkl}\dot{\varepsilon}_{kl}>0$]; $<1>=0$ if either the stress state is in the elastic domain (F<0) or if it is on the yield surface and unloading or neutral loading is taking place [**F**=0, $(\partial F/\partial\sigma_{ij})L^{e}_{ijkl}\dot{\varepsilon}_{kl}\leq 0$]. Due to the switch function <1> the constitutive equations (6) are non-linear. In case of non-associate flow-rule the stiffness tensor becomes also non-symmetric; i.e. $L_{ijkl} \neq L_{klij}$.

Within the framework of hardening plasticity, the plastic work is given by the following expression:

$$\Delta W^{p} = \sigma_{kl}\Delta\varepsilon^{p}_{lk} = <1>p(\mu - \beta)\Delta\gamma^{p}. \qquad (8)$$

Let μ_C=max{$\mu(\gamma^p)$} and β_C (C for Coulomb) be the values of the friction and of the dilatancy coefficient in the plastic limit state C_C (h=h_C=0). The experimental evidence suggests that for granular soils the dilatancy function assumes its maximum also at C_C; i.e. β_C=max{$\beta(\gamma^p)$}. On the other hand ΔW^p must be strictly positive at the plastic limit state where the production of plastic strains is zero (Mròz 1963), which, according to equation (8) means that $\mu_C>\beta_C$, and consequently for monotonous functions:

$$\mu > \beta\,. \qquad (9)$$

Above inequality means that granular materials are modeled by a flow theory of plasticity with non-associated flow-rule.

Linear Comparison Solids

In dealing with the linear bifurcation problem, Raniecki and Bruhns (1981) generalized Hill's (1958) theory of uniqueness and stability for elastoplastic solids obeying the normality flow-rule so as to consider non-associative behavior as well. Within this theory one can define two linear comparison solids such that solutions to the linear bifurcation problem based on the constitutive equations for these comparison solids yield upper or lower bounds for the true bifurcation load. The hypoelastic comparison solid that can be defined by equations (6) with a stiffness tensor always corresponding to loading conditions has been proven to yield an upper bound to the true bifurcation load; in this case the comparison solid is defined by the following stiffness tensor

$$L_{ijkl} = G(L^{e}_{ijkl} - \frac{1}{H}A^{G}_{ij}\,A^{F}_{kl}). \qquad (10)$$

Lower bounds can be obtained by the constitutive equations of a linear comparison solid that is defined through

$$L_{ijkl} = G(L^{e}_{ijkl} - \frac{1}{4rH}(A^{G}_{ij} + rA^{F}_{ij})(A^{G}_{kl} + rA^{F}_{kl})), \qquad (11)$$

where r>0, is a free parameter that can be varied so as to maximize the lower bound estimate for the bifurcation load. One should notice that the stiffness matrix for this linear comparison solid is symmetric. If one assumes an associate flow-rule, then the two linear comparison solids coincide for r=1, and the bifurcation load obtained from a linear bifurcation analysis coincides with the true bifurcation load (Hill 1958).

SHEAR-BAND FORMATION

Formulation

Theoretical analyzes of shear-band formation are usually based on the Thomas-Hill-Mandel shear-band model (Thomas 1961, Hill 1962, Mandel 1963, 1964). According to this model a shear-band must be perceived as a thin material layer that is bounded by two material discontinuity surfaces of the incremental displacement gradient. The usual approach to this problem involves the consideration of:

1. Geometrical compatibility conditions, stating that across the shear-band boundaries the incremental displacement field is continuous and only its gradient jumps:

$$[\Delta u_i] = 0 \; ; \; [\Delta u_{i,j}] = \zeta_i n_j \; . \qquad (12)$$

2. Statical compatibility conditions, assuring continuity of the initial stress and equilibrium of the effective stress-increment across a shear-band boundary (fully drained boundary conditions):

$$[\sigma'_{ij}] = 0 \; ; \; [\Delta\sigma'_{ik}] \, n_k = 0 \; . \qquad (13)$$

In the above compatibility conditions $[\cdot]$ denotes the jump of a quantity across the shear-band boundary with unit normal vector n_i; ζ_i is the jump of the normal derivative of the incremental displacement vector.

From these compatibility equations and the observation that the stiffness tensor for a linear comparison solid is continuous across the shear-band boundary follows that weak stationary discontinuities for the displacement gradient exist only if

$$\det(L_{ijkl} n_j n_l) = 0 \; , \qquad (14)$$

Equation (14) is the characteristic equation in terms of the direction cosines n_i of a statically and kinematically admissible discontinuity surface. If the characteristic equation provides real solutions for n_i, discontinuity surfaces for the displacement gradient exist and may also develop in due course of the deformation. The state C_B (B for band) of shear-band bifurcation is assumed to be that state at which equation (14) is firstly met during a monotonous deformation process.

Upper-Bound Comparison Solid

Let $\mu_B = \sin\phi_B$ and $\beta_B = \sin\psi_B$ be the values of the friction and dilatancy functions at C_B. By using the constitutive equations (10) for the upper-bound linear comparison solid, the critical hardening rate h_B at C_B can be computed, resulting to (Mandel 1964):

$$h_B = \frac{(\mu_B - \beta_B)^2}{8(1-\nu)} \; . \qquad (15)$$

In granular materials inequality (9) applies, which according to equation (15) suggests that shear banding in plane-strain deformations always takes place in the hardening regime ($h_B>0$). The corresponding shear-band inclination angle Θ with respect to the absolutely minimum principal stress direction is approximately given by the following formula

$$\Theta_{1,2} = \pm\Theta_B \; ; \; \Theta_B \approx 45° + (\phi_B + \psi_B)/4 \; , \qquad (16)$$

that was first proposed by Arthur et al. (1977) on the basis of experimental observations, and was subsequently proven theoretically by Vardoulakis (1980).

At the limit state C_C the rate of plastic hardening vanishes, $h=h_C=0$, and friction, and dilatancy angles assume their maxima: $\phi_C = \max\{\phi_m(\gamma^p)\}$; $\psi_C = \max\{\psi_m(\gamma^p)\}$. At C_C two symmetric solutions for the shear-band orientation exist (Vardoulakis 1976):

$$\begin{aligned} \Theta_{1,2} &= \pm\Theta_C \; ; \; \Theta_C = 45° + \phi_C/2 \\ \Theta_{3,4} &= \pm\Theta_R \; ; \; \Theta_R = 45° + \psi_C/2 \; . \end{aligned} \qquad (17)$$

The first solution is the classical Coulomb solution whereas the second is called the Roscoe solution. It is important to realize that the theoretical shear-band orientation, equation (16), is approximately the mean of the Coulomb and Roscoe classical solutions. An inevitable experimental error present in any test may well obscure the results, and give support to either solution depending on the experimenters' bias. In case of associate plasticity C_B coincides with C_C; i.e. if $\mu=\beta$ then

$$h_B = h_C = 0 \; ; \; \Theta_B = \Theta_C = \Theta_R = \Theta_A . \qquad (18)$$

This means that for associated flow-rule Coulomb's failure criterion can be derived directly from a bifurcation analysis.

Non-Associate Rigid-Plastic Laws

Starting with Hill (1950), non-associated flow-rules in frictional materials has been widely discussed in the context of rigid, perfectly-plastic, or elastic, perfectly-plastic material behavior (cf. Shield 1953). The idea was to approximate true material behavior with the relatively simple constitutive law of perfect plasticity. All these models lead to two sets of distinct characteristics, the so-called statical and kinematical characteristics. The orientation of these two sets of characteristics with respect to the absolutely minimum principal stress is given by equations (17). Accordingly, the domains of solution for stresses and velocities do not coincide. The existence of two sets of characteristics triggered extensive experimental investigations aiming at determination as whether or not the zero-extension lines in soils coincide with the static characteristics of the perfectly plastic solid; cf. Bransby and Milligan (1975).

A simple shear-band analysis for a continuously hardening non-associative material, proves that shear banding in plane strain occurs at positive hardening rates. Perfect plasticity presumes, however, that plastic deformation and formation of slip lines occurs at zero hardening rate. Thus, the perfectly plastic non-associative model cannot be a good approximation of the hardening non-associative plastic model. This point has been overlooked in the literature, where perfect plasticity was adopted regardless of the flow-rule. Apparently, the adoption of perfect plasticity for soils has been borrowed from metal plasticity, where the associative flow rule satisfactorily describes plastic deformation; it results directly from incompressibility and pressure insensitivity of metals. In all other materials, perfect plasticity is justified only if the flow-rule is associative.

Experimental Results

Figure 3 shows the 'raw' experimental data from a biaxial test on a dry specimen of a fine-grained Dutch dune sand (Vardoulakis et al. 1985), namely the axial force P and of the lateral displacement, $u_1=u_{1L}+u_{1R}$, vs the axial displacement u_2. The initial dimensions of the tabular shaped specimen were: $l_{10}=41.03$ mm, $l_{20}=140.93$ mm, $l_{30}=79.10$ mm. The overall initial porosity was $n_o=0.383$ ($n_{min}=0.362$; $n_{max}=0.482$; $G_s=2.66$).

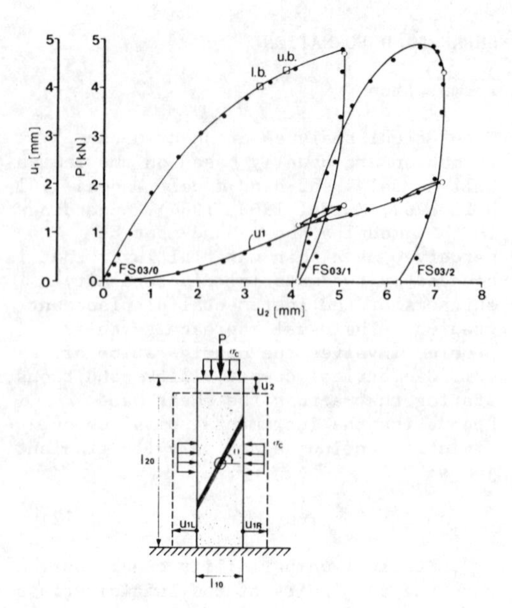

Figure 3. Biaxial test on a fine grained sand (Vardoulakis et al. 1985).

During this test the confining pressure was kept constant, $\sigma_c=294.3$ kPa. The test consists of one major loading path with two unloading loops. Before first loading and after each unloading X-ray radiographs have been taken in order to investigate the homogeneity of the specimen. The specimen contained a small density imperfection of loose sand. At the state of maximum axial load a shear band was seen to emerge from the density disturbance. The evaluation of the X-ray plates prompted the suggestion that homogeneous deformations are disrupted at shearing strain intensities:

$$\gamma_E^{(\ell)} \leq \gamma \leq \gamma_E^{(u)}, \tag{19}$$

where $\gamma_E^{(\ell)}\approx 0.06$, and $\gamma_E^{(u)}\approx 0.10$, are measured lower and upper bounds for the bifurcation strain, respectively, in a series of experiments with the same sand; see Vardoulakis and Graf (1985) and Vardoulakis et al. (1985). The lower bound of the bifurcation strain, $\gamma_E^{(\ell)}$, corresponds to that value of the overall strain for which a first indication for shear-band formation could be seen in the X-ray plate. The measured inclination angle of the shear-band was $\Theta_E=62.5°$; see Table 1. Eventually the growing density inhomogeneity reached the specimens faces, leading to separation by a fully formed shear-band. The upper bound for the overall strain that corresponded to separation is denoted by $\gamma_E^{(u)}$.

Numerical Results

Calibration of the considered elastoplastic model by using the above presented data base from loading, unloading and reloading stress-paths resulted to G=32.6 MPa, ν=0.1 and the following curve-fits for the mobilized friction and dilatancy functions (Vardoulakis 1987):

$$\mu = \gamma^p/(c_1 + c_2\gamma^p) \ ; \ \beta \approx c_3 \ , \qquad (20)$$

where c_1=0.00104, c_2=1.4273, c_3=0.3696.

An upper bound for the bifurcation strain and of the corresponding shear-band inclination angle, $\gamma_B^{(u)}$, $\Theta_B^{(u)}$, can be computed directly from equations (15) and (16). Lower bounds, $\gamma_B^{(\ell)}$ and $\Theta_B^{(\ell)}$, on the other hand are determined numerically by evaluating characteristic equation (14) at C_B and by using the constitutive equations (11) of the corresponding linear comparison solid. The results of these computations are summarized in Table 1 and are also depicted in Figure 3. It should be noticed that in Table 1 the highest lower-bound solution in terms of the bifurcation strain is shown, that corresponds to an optimum value of the parameter r=0.861.

Table 1: Experimental and theoretical results on shear band formation in the biaxial test

		γ	h	ϕ	ψ	Θ
Ex.	(l.b.)	0.06	(0.010)	44.6°	21.7°	62.5°
Th.	(u.b.)	0.05	0.014	43.2°	21.7°	61.2°
	(l.b.)	0.04	0.023	42.9°	21.7°	60.5°

Table 1 demonstrates a puzzling result that deserves further attention, namely that an upper-bound estimate for the shear-band bifurcation strain is less then a lower-bound for the shear strain at which shear-band formation is observed experimentally ($\gamma_B^{(u)}<\gamma_E^{(\ell)}$). As it will be explained in the next section, the solution to this problem lies into the recognition that shear-band formation is a post-bifurcation phenomenon within the frame of a classical material description, and that this description breaks dawn at the bifurcation point.

Localization and Size Effect

There is ample experimental evidence that shear-bands in granular materials engage a significant number of grains. According to a proposition made by Roscoe (1970) that was based on direct experimental observations, the width of shear-bands is about 10 times the average grain diameter; see also Scarpelli and Wood (1982). In Figure 4 X-ray photographs of shear-bands are shown that are formed in the biaxial tests reported by Vardoulakis and Graf (1985) and Vardoulakis et al. (1985). Figure 4 (a) corresponds to a medium-grained sand from Karlsruhe, whereas Figure 4 (b) corresponds to a fine-grained sand from Holland. Table 2 summarizes the evaluation of these plates. In this table $d_{50\%}$ denotes the mean grain size of the tested sand and $2d_E$ the measured shear-band thickness. Accordingly, these experiments suggest a shear-band thickness that is about 16 times the mean grain diameter.

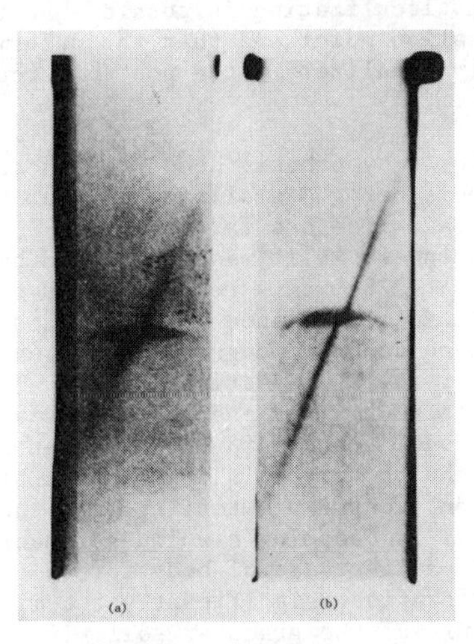

Figure 4. Shear-band emerging out of a density inhomogeneity: (a) medium grained Karlsruhe sand (Vardoulakis and Graf 1985); (b) fine grained Holland sand (Vardoulakis et al. 1985).

Table 2: Measured shear-band thickness

	$d_{50\%}$ [mm]	$2d_E$ [mm]	$2d_E/d_{50\%}$ [−]
Fine sand (FS)	0.20	3.7	18.5
Medium sand (D)	0.33	4.3	13.0

As outlined above, the classical approach to the shear-band problem involves the consideration of a particular constitutive model and the examination of the existence of discontinuity planes for the incremental displacement gradient which in turn identified with the shear-band boundaries. If the formulation of the problem does not contain a material property with the dimension of length then it is not possible to produce a statement about the shear-band thickness. Following Roscoe's hypothesis, Mühlhaus and Vardoulakis (1987) related this internal length to the mean grain size of the material by means of a constitutive model for granular materials with Cosserat structure (E.& F. Cosserat 1909). This study showed that the predicted shear-band thickness at the bifurcation point of the classical description is infinite and that it approaches the measured value $2d_E$ for overall strains corresponding to the experimentally observed upper-bound $\gamma_E^{(u)}$. This finding should be understood as that no localization is possible at the bifurcation point and that the deformation rapidly localizes in the post-bifurcation regime.

Roscoe's hypothesis about the shear-band thickness being a small multiple of the mean grain size has far reaching consequences. As it was emphasized by Gudehus (1978), this result should be of significant importance in understanding the phenomenon of progressive failure in granular soils. Investigation of progressive failure phenomena in small scale model tests with the same material as in the prototype is then impossible. In addition, centrifuge testing with the same material is becoming also questionable, because an increase of body forces is not going to affect significantly the micro-structure of the granular soil. If in contrary one tries to scale down the grain size according to the model scale than one may switch from a sand to a powder that has significantly different rheologic behavior.

LIQUEFACTION

Experimental Observations

One of the most successful concepts applied in the evaluation of tests on water-saturated granular soils is Terzaghi's (1936) effective stress principle. As opposed to micromechanical considerations (Biot 1956, Bowen 1980), Terzaghi's definition of effective stress, σ'_{ij}, is heuristic. The total equilibrium stress is decomposed into an effective stress assigned to the soil skeleton and into a pore-water pressure p:

$$\sigma_{ij} = \sigma'_{ij} - p\,\delta_{ij} \quad (p > 0) , \qquad (21)$$

where compression is taken as negative. Both total stress σ_{ij} and pore-water pressure p are measurable quantities, and thus equation (21) can be understood as an operational definition of effective stress. The validity of Terzaghi's effective-stress principle is widely corroborated through drained and undrained tests performed on a variety of water-saturated granular materials and for stress levels corresponding to the ones encountered in soil mechanics problems (see Bishop and Skinner 1977).

Fully drained tests evaluated according to Terzaghi's effective stress principle reveal the frictional character of the shear strength of soils like sands and normally consolidated clays. The application of the effective stress principle to the evaluation of undrained shear tests is sometimes confusing, since the results seemingly contradict those from drained tests. Drained tests support a failure criterion of the Mohr-Coulomb type, which states that failure occurs at a state C_C where the principal effective stress ratio is at its maximum. Undrained tests, on the other hand, do not exclude the possibility of a Tresca type of failure criterion, which states that failure occurs at a state C_T (T for Tresca) where the principal stress difference is at its maximum. This state of affairs was described as early as 1960 in the papers by Whitman, and Bjerrum and Simons, that deal with the shear strength of normally consolidated clays. For example, a normally, anisotropically consolidated highly plastic clay reaches in undrained tests the Tresca state, C_T, under definite pore-water pressure generation; see Figure **5** (**a**) (Graham and Li 1985). Figure **6** (**a**) indicates that for a material like this the pore-water pressure generation becomes unstable C_T. This instability is typical for normally, anisotropically consolidated, highly plastic, structured clays for which arbitrarily high values for Skempton's pore-pressure parameter at failure (A_f) are obtained; Figure **7**. Finally, it is worth noticing that overconsolidated specimens fail in a completely different fashion than normally, anisotropically consolidated specimens. Overconsolidated specimens reach first the state C_C of maximum effective stress obliquity with the pore-pressure following a descending branch. (Figures **5 b** and **6 b**). In this case the background drained behavior is dilatant and failure is manifested by shear banding.

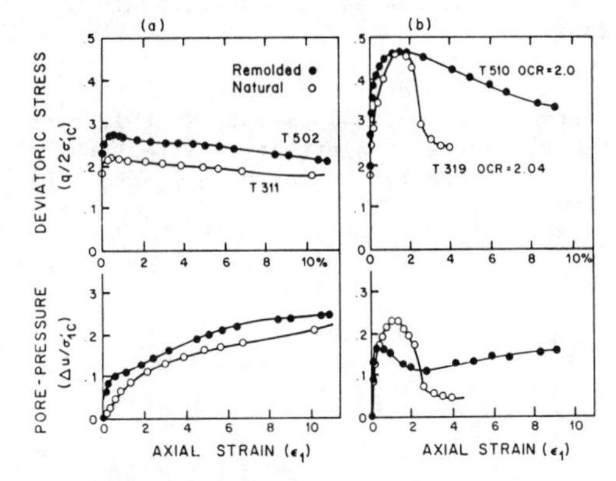

Figure 5. Stress-strain and pore-water pressure curves from undrained triaxial compression tests on anisotropically consolidated specimens:(a) normally consolidated;(b) overconsolidated (Graham and Li 1985).

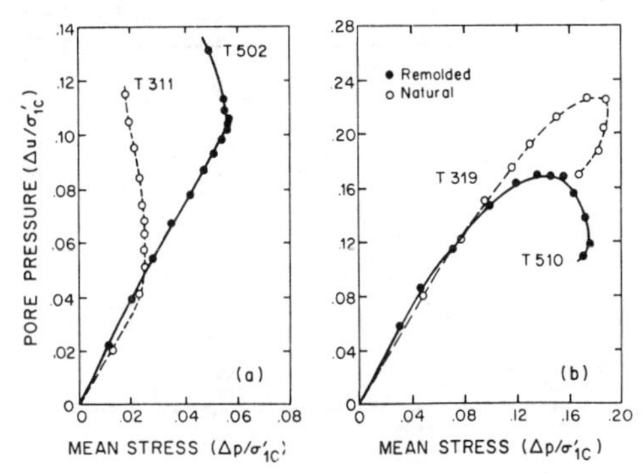

Figure 6. Pore-water pressure changes vs mean stress changes:(a) normally consolidated;(b) overconsolidated (Graham and Li 1985).

Similar, but more dramatic, results are frequently reported for loose sands, which during undrained shear tests liquefy at a state of maximum principal stress difference. As shown in Figure **8** taken from Castro (1969) (see also Vaid and Chern 1983) loose sand samples actually do liquefy at the state of maximum deviatoric stress (Peters 1984) whereas medium dense and dense sand samples undergo a stable dilatant hardening process, for the soil presumably never reaches the limit condition.

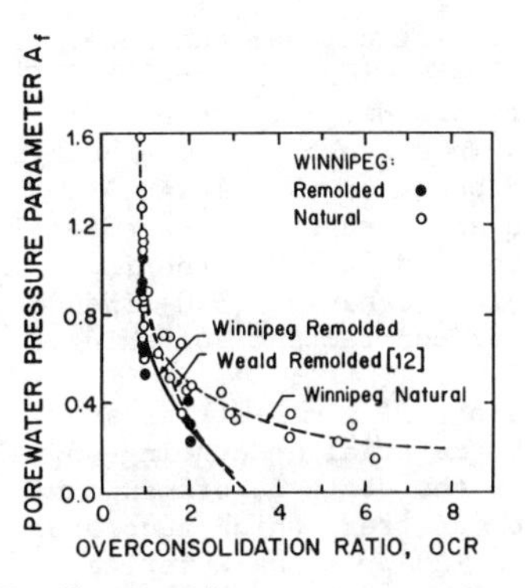

Figure 7. Skempton's pore-water pressure coefficient A_f vs overconsolidation ratio (Graham and Li 1985).

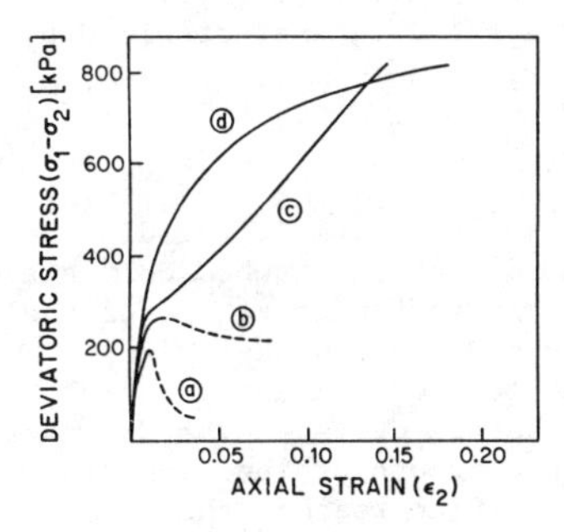

Figure **8**. Triaxial compression tests on water-saturated sand:(a) undrained test on loose sand,D_r=30%;(b) undrained test on critically dense sand,D_r=44%;(c) undrained test on dense sand,D_r=47%;(d) drained test on loose sand,D_r=33% (Castro 1969).

Definition of Limit-Stress States

In undrained homogeneous deformations of water-saturated granular materials consisting of incompressible grains and fluid, the total volumetric strain is everywhere zero, and consequently:

$$\Delta\varepsilon^{p}_{kk} = -\Delta\varepsilon^{e}_{kk} \quad . \tag{22}$$

This means that plastic dilatation or contraction of the soil skeleton due to slippage among the grains must be compensated for by elastic volume changes of the skeleton. From equation (22) and for an elastoplastic solid skeleton we get the following expressions for the increments of the mean effective stress and of the deviatoric stress:

$$\Delta p' = K\beta\Delta\gamma^{p} \ ; \ \Delta\tau = G(h - h_T)\Delta\gamma^{p} \ . \tag{23}$$

Due to the frictional character of the material behavior, changes in effective pressure are directly responsible for changes in strength. If the background drained behavior is dilatant ($\beta>0$), shear induces an increase of the effective pressure p'(dilatant hardening), whereas, contractant behavior ($\beta<0$) induces the opposite (contractant softening).

An important limit-stress state of a granular material undergoing undrained shear is the state C_T of maximum deviatoric stress which according to equation $(23)_2$ is characterized by the condition

$$h = h_T = - \kappa \mu \beta . \quad (24)$$

On the other hand, the limit-stress state C_C of maximum effective stress obliquity τ/p', is characterized by the condition:

$$h = h_C = 0 . \quad (25)$$

As shown in Figure 9, for dilatant material C_T can be reached only after C_C, i.e. in the softening regime of the background drained behavior ($h_T<0$), whereas, for contractant material C_T is reached before C_C in the hardening regime of the background drained behavior ($h_T>0$). If the material reaches the limit state C_C with diminishing contractancy ($\beta \to 0^-$) then C_T and C_C coincide:

$$h = h_T = h_C = 0 , \quad (26)$$

and the corresponding limit state is called a critical state (C_{cs}).

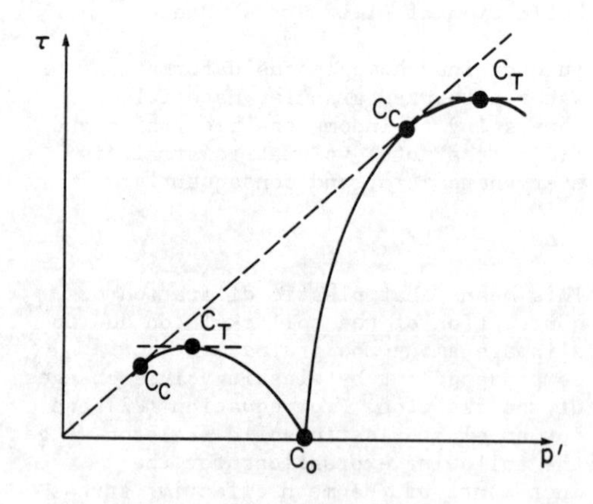

Figure 9. Schematic undrained stress-paths

Considering that the slope of the effective stress-path in the (τ,p')-plane is given by the expression:

$$\eta = d\tau/dp' = - \mu (h - h_T)/h_T , \quad (27)$$

we distinguish among the following two limit-stress states for contractant material:

a) There is a definite state C_T with $h_T>0$ where condition (24) is met. In this case the tangent to the effective stress path at C_T is parallel to the p'-axis ($\eta_T = 0$); Figure 10.

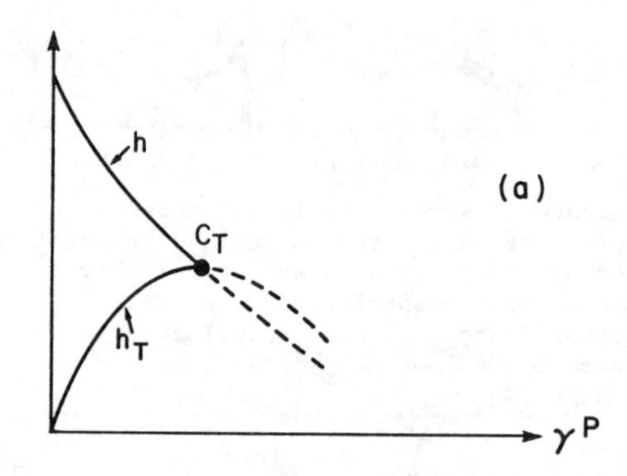

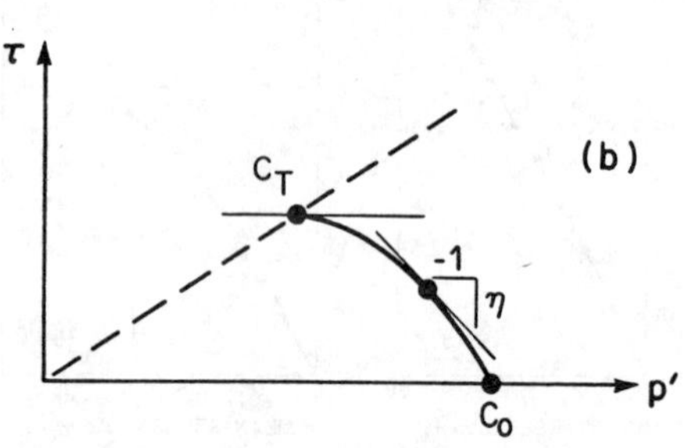

Figure 10. Definition of the Tresca limit state ($\beta_T<0$): (a) schematic evolution of h and h_T; (b) undrained stress-path.

b) C_T is reached asymptotically and thus coincides with the critical state C_{cs}. In this case η_T becomes indeterminate and the tangent to the effective stress path at C_T is not necessarily parallel to the p'-axis, Figure 11. [This interpretation of a horizontal tangent to the undrained stress-path at C_T is different from the one usually presented in critical state soil mechanics, where C_T is always perceived as being a critical state].

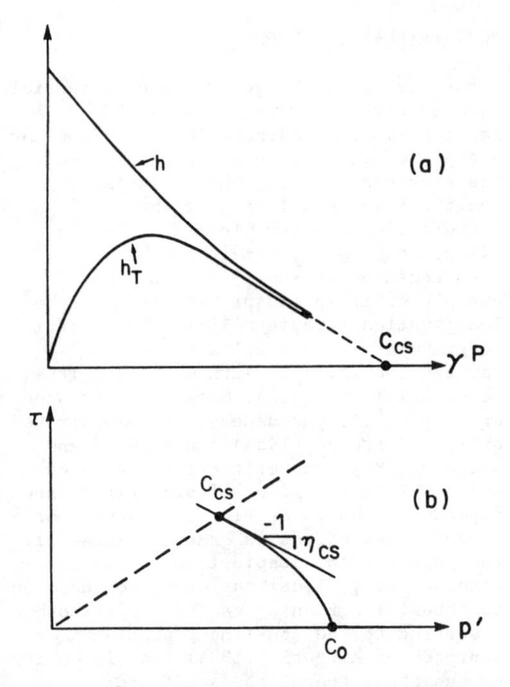

Figure 11. Definition of the critical state ($\beta_T \rightarrow 0^-$): (a) schematic evolution of h and h_T; (b) undrained stress-path.

Stability Analysis

Stability and bifurcation in undrained deformations of water-saturated granular media were discussed by the author in two recent papers (Vardoulakis 1985, 1986). The stability problem is defined as follows: at any given state of an undrained deformation, small perturbations Δu_i and Δp of the displacements of the soil skeleton and of the pore-water pressure are considered and their evolution in time under constant boundary stresses is studied. The considered stability problem is governed by:

1. Equations of motion for both soil skeleton and fluid:

$$\Delta\sigma'_{ij,j} = -\Delta b_i + (1-n)\rho_s \Delta u_{i,tt}$$
$$-\Delta p_{,i} = \Delta b_i + n\rho_f \Delta u_{i,tt} + \rho_f \Delta q_{i,t} \qquad (28)$$

where ρ_s and ρ_f are the bulk densities of the solid and the fluid and n the porosity of the skeleton, $-\Delta b_i$ is the seepage volume force that is exerted by the flowing fluid on the soil skeleton and Δq_i the relative specific discharge vector.

2. Continuity of flow, holding for incompressible fluid and grains:

$$-\Delta q_{i,i} = \Delta\varepsilon_{kk,t} \; . \qquad (29)$$

3. Constitutive equations for the soil skeleton and the fluid flow. In particular the behavior of the soil skeleton can be described by the constitutive equations (10) of the upper-bound linear comparison solid and fluid flow in a porous medium is governed by Darcy's law for the seepage force:

$$\Delta b_i = b\,\Delta q_{,i} \qquad (30)$$

where $b=(\rho_f\, v_k)/k$, v_k and k are the kinematic viscosity of the fluid and the intrinsic permeability of the soil skeleton, respectively.

Stability of undrained shear can be easily studied in the case of a long strip of water-saturated granular material under dead loading conditions and impervious boundaries (Rice 1975, Vardoulakis 1986; Figure 12). If inertia terms are neglected from the equations (28), the stability problem results into a consolidation-type equation for Δp:

$$\Delta p_{,yy} = c\,\Delta p_{,t} \; . \qquad (31)$$

For contractant material and in the vicinity of C_T the 'diffusivity', $c=b(h-h_T)/h$. Harmonic solutions for Δp evolve as exp(imy+ft), with the growth coefficient of the instability being inversely proportional to c; i.e.

$$f = -c_1(h-h_T)^{-1} \quad (c_1>0). \qquad (32)$$

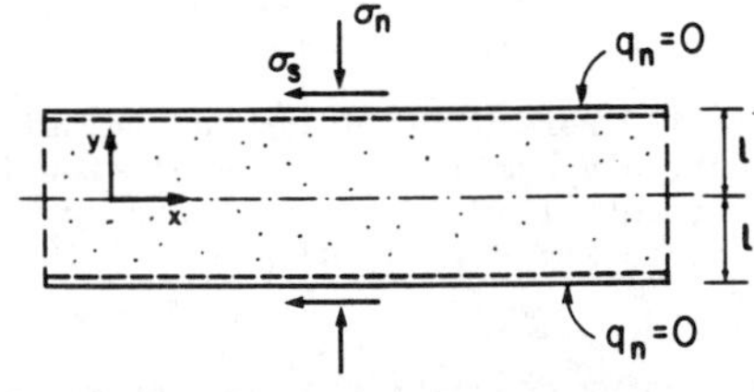

Figure 12. Undrained shear of an infinite strip of water saturated granular soil.

Equation (31) demonstrates the fact that the considered instabilities are related to inhomogeneous deformations with internal fluid flow. According to equation (32) the evolution of growth coefficient f is anomalous, since at C_T f is switching from $-\infty$ to $+\infty$ and for states beyond C_T, f is decreasing. This anomaly points to an error which is introduced by applying Rice's inertia-free formulation in the case of contractant material. In studying slow diffusion processes, usually the assumption is made that the effects of inertia are negligible as compared to the effects of dissipation, and thus acceleration terms are neglected. In the vicinity of C_T, however, this assumption is wrong, because for sufficiently small values of c the effects of inertia become important. This criticism does not apply to Rice's original work, since he studied the stability of untrained shear of dilatant materials. In this case $c \rightarrow \pm \infty$ and $f \rightarrow \pm 0$ at C_C, and inertia terms are insignificant. Vardoulakis (1986) found that in the case of contractant material the dynamic formulation yields the following asymptotic solution for the growth coefficient of the instability at states beyond the limit state C_T:

$$f = c_2 |h - h_T|^{1/2} . \quad (c_2 > 0) \qquad (33)$$

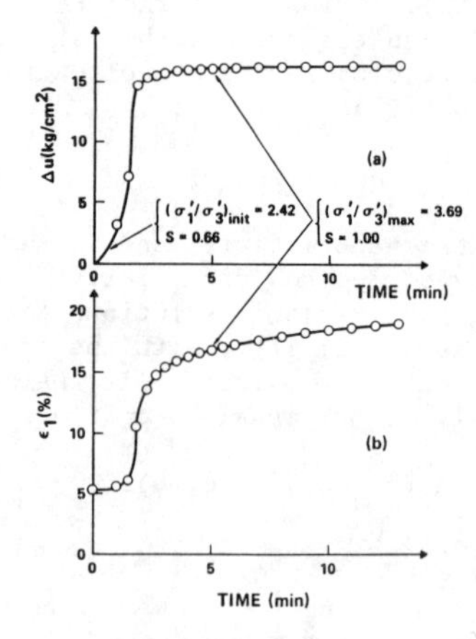

Figure 13. Pore-water pressure explosion and unstable nominal strain evolution in stress-controlled undrained tests on loose Sacramento River sand (Lade et. al 1987).

This finding demonstrates the catastrophic nature of liquefaction phenomena, since the instability evolves at C_T as e^{ft} with f growing with infinite pace. Figure 12 from Lade et al. (1987) demonstrates the pore-water pressure explosion and the unstable nominal strain evolution after the instability point C_T has been surpassed. The considered instability has its origin in the sensitivity of the soil-water system towards changes in the flow-rule of the skeleton. Contractive behavior means a strong violation of the normality flow-rule, and under undrained conditions this results in instabilities in the generation of pore-water pressure.

POST-FAILURE BEHAVIOR

Post-failure, the governing differential equations change type and/or yield improperly posed mathematical problems. An example of ill-posedness was presented in the previous section, where Rice's inertia-free formulation of the liquefaction problem yielded a classic ill-posed problem, namely the backward heat equation (Payne 1975). An other example of ill-posed problem is localization (Krawietz 1981). Localization in granular soils results to extreme rarefaction which in turn manifests itself to material softening. Material softening aggravates ill-posedeness, as demonstrated by Wu and Freund (1984) and Sandler and Wright (1984), who critically discussed the stabilizing effect of viscosity. Rate dependence should be viewed, however, as a second order effect in granular materials that are not in a rapid flow regime. Alternatively, localization in conjunction to material softening can be addressed by resorting to the non-linear gradient approach of Aifantis (1984) that is built on the phase transition and fluid interfaces treatment of Aifantis and Serrin (1983). Triantafyllidis and Aifantis (1986) and Zbib and Aifantis (1987) in particular, used this approach to obtain shear-band widths in hyperelastic and rigidly plastic materials, respectively. The gradient approach allows for solutions for localization with finite shear-band thickness inside the softening regime without a change of the type of the governing equations. Similar approaches to the shear-band width problem and the material behavior in the softening regime have also been proposed recently by Coleman and Hodgdon (1985) and Schreyer and Chen (1986). Specifically, Coleman and Hodgdon did not use a gradient-dependent yield condition, while Schreyer and Chen used a first (instead of a second) gradient modification of the flow stress. It appears, however, that both these formulations possess certain undesirable features.

In all these cases severe mathematical questions arise that are presently not at all satisfactorily answered, as this is demonstrated by the recent work of Schaeffer (1987) and Pitman and Schaeffer (1987) who raise some very intriguing questions on instability and ill-posedness in the context of plasticity theory.

ACKNOWLEDGEMENT

The author wants to thank the National Science Foundation for supporting this research (Grant No. MSM-8511843).

REFERENCES

E.C.Aifantis, On the micro-structural origin of certain inelastic models, J. Eng. Mat. & Tech. 106, 326-330 (1984)
E.C.Aifantis and J.B.Serrin, Equilibrium solutions in the mechanical theory of fluid microstructures, J. of Coll. & Interf. Sc. 96, 530-547 (1983)
J.R.F.Arthur, T.Dunstan, Q.A.J.Al-Ani and A.Assadi, Plastic deformation and failure of granular media, Géotechnique, 27, 53-74 (1977)
M.A.Biot, Theory of propagation of elastic waves in fluid-saturated porous solid, J. Acoust. Soc. Am. 28, 168-178 (1956)
A.W.Bishop and G.E.Green, The influence of end restraint on the compression strength of cohesionless soil. Géotechnique, 23, 243-266 (1973)
A.W.Bishop and A.E.Skinner, The influence of high pore-pressure on the strength of cohesionless soils, Phil. Trans. Roy. Soc. London, 284, 91-130 (1977)
L.Bjerrum and N.E.Simons, Comparison of shear strength characteristics of normally consolidated clays, ASCE Reg. Conf. on Shear Strength of Cohesive Soils, Univ. of Colorado, 711-726 (1960)
R.M.Bowen, Incompressible porous media models by use of the theory of mixtures, Int. J. Engng. Sci. 18, 1128-1148 (1980)
P.L.Bransby and G.W.E.Milligan, Soil deformations near cantilever sheet pile walls, Géotechnique, 25, 175-195 (1975)
G.Castro: Liquefaction of sands, Ph.D. Thesis, Harvard University, 1969
B.D.Coleman and M.L.Hodgdon, On shear bands in ductile materials, Arch. Rat. Mech. Anal. 90, 219-247 (1985)
E.&F.Cosserat, Théorie des Corps Déformable, Hermann, Paris, 1909
F.Deman, Achsensymmetrische Spannungs-und Verformumngsfelder von trockenem Sand, Dissertation, Univ. Karlsruhe, 1975
C.S.Desai, E.Krempl, P.D.Kiousis and T.Kundu, Constitutive Laws for Engineering Materials, Elsevier, New York, 1987
J.Desrues, J., Localization de la deformation plastique dans les materieux granulaires, These d'etat, Univ. de Grenoble, 1984
M.Goldscheider and G.Gudehus 1973, Rectilinear extension of dry sand: testing apparatus and experimental results. Proc. 8th ICSMFE, Moscow, 1/21, 143-149 (1973)
J.Graham and E.C.C.Li, Comparison of natural and remolded plastic clay, ASCE J. of Geotech. Engrng. 111,865-881 (1985).
G.E.Green, Strength and deformation of sand measured in an independent stress control cell, Proc. Roscoe Memorial Symp. Cambridge, 285-323 (1971)
G.Gudehus, Engineering approximations for some stability problems in geomechanics, in: Adv. Anal. Geotech. Instabilities, Univ. of Waterloo Press, 13, 1-24, 1978
G.Gudehus, F.Darve and I.Vardoulakis, Constitutive Relations for Soils, Balkema, Rotterdam, 1984
E.C.Hambly, Plane strain behavior of remolded normally consolidated kaolin, Géotechnique, 22, 301-317 (1972)
A.Hettler, G.Gudehus and I Vardoulakis, Stress-strain behavior of sand in triaxial tests,in: Constitutive Relations for Soils, Balkema, Rotterdam, 55-66, 1984
A.Hettler and I.Vardoulakis, Behavior of dry sand tested in a large triaxial apparatus, Géotechnique, 34,183-198 (1984)
R.Hill, Mathematical Theory of Plasticity, Clarendon Press, Oxford, 1950
R.Hill, A general theory of uniqueness and stability in elastic-plastic solids, J. Mech. Phys. Solids, 6, 236-249 (1958).
R.Hill, Acceleration waves in solids, J. Mech. Phys. Solids 10: 1-16 (1962)
A.Krawietz, Positive Elliptizität und Stabilität der ebenen Verformung, Acta Mechanica, 41, 223-253 (1981)
W.M.Kirkpatrick and D.J.Belshaw 1968, On the interpretation of the triaxial test, Géotechnique, 18, 336-350 (1968)
K.Kuntsche, Materialverhalten von wassergesättigtem Ton bei ebenen und zylindrischen Verformungen, Dissertation Univ. Karlsruhe, 1982.
P.V.Lade, Localization effects in triaxial tests on sand, in: Deformation and Failure of Granular Materials, Balkema, Rotterdam, 461-471, 1982
P.V.Lade, R.B.Nelson and Y.M.Ito, Instability of granular materials with nonassociated flow, ASCE J. Engng. Mech. submitted (1987)
J.Mandel, Propagation des surfaces de discontinuite dans un milieu elasto-plastique, in: Stress Waves in Anelastic Solids, Springer, Berlin, 331-341, 1964
J.Mandel, J. 1964, Conditions de stabilite et postulat de Drucker, in: Rheology and Soil Mechanics, Springer, Berlin, 58-68, 1965.
Z.Mróz, Nonassociated flow laws in plasticity, Journal de Mechanique, 1, 21-42 (1963)
H.B.Mühlhaus and I.Vardoulakis, The thickness of shear bands in granular materials, Gèotechnique, 37, 271-283 (1987).
L.E.Payne, Improperly posed problems in partial differential equations, SIAM Publ.

Reg. Conf. Ser. in Appl. Math. 1-76, 1975
J.F.Peters, Instability of loose saturated sands under monotonic loading, 5th Eng. Mech. Div. Spec. Conf. 2, 945-948 (1984)
E.B.Pitman and D.G.Schaeffer, Stability of time dependent compressible granular flow in two dimensions, Comm. on Pure and Appl. Math. 15, 421-447 (1987)
B.Raniecki and O.T.Bruhns, Bounds to bifurcation stresses in solids with non-associated plastic flow law of finite strain, J. Mech. Phys. Solids, 29, 153-172 (1981)
D.W.Reads and G.E.Green 1976. Independent stress control and triaxial extension tests on sand, Géotechnique, 26, 551-576 (1976)
J.R.Rice, On the stability of dilatant hardening for saturated rock masses. J. of Geophys. Res. 80, 1531-1536 (1976)
J.R. Rice, The localization of plastic deformation, Proc. 14th IUTAM Cong. Delft 207-220 (1976)
K.H.Roscoe, The influence of strains in Soil Mechanics, Géotechnique, 20, 129-170 (1970)
K.H.Roscoe, A.N.Schofield and A. Thurairajah 1963, An evaluation of test data for selecting a yield criterion for soils, ASTM Spec. Publ. No. 361, 111-128, 1963
I.S.Sandler and J.P Wright, Strain-Softening, in: Mechanics of Elastic and Inelastic Solids 6, Nijhoff, Dordrecht, 285-296, 1984
G.Scarpelli and D.M.Wood, Experimental observations of shear band patterns in direct shear tests, in: Deformation and Failure of Granular Materials, Balkema, Rotterdam, 473-484, 1982
D.G.Schaeffer, Instability in the evolution equations describing incompressible granular flow, Journal of Differential Equations 66: 19-50 (1987)
R.T.Shield, Mixed boundary value problems in soil mechanics, Q. Appl. Math. 11, 61-75 (1953)
L.H.Schreyer and Z.Chen, One-dimensional softening with localization, J. Appl. Mech. 53, 791-797 (1986)
K.v.Terzaghi 1936, The shearing resistance of saturated soils, Proc. 1st ICSMFE, Cambridge, 1, 54-56, (1936)
T.Y.Thomas, Plastic Flow and Fracture in Soilds, Academic Press, 1961
N.Triantafyllidis and E.C.Aifantis, A gradient approach to localization of deformation - I. Hyperelastic materials. J. of Elasticity,16, 225-238 (1986)
Y.P.Vaid and J.C.Chern, Effect of static shear on resistance to liquefaction, Soils and Foundations,23, 47-60 (1983)
I.Vardoulakis, Equilibrium theory of shear bands in plastic bodies, Mech. Res. Comm. 3, 209-214 (1976)
I.Vardoulakis, Equilibrium bifurcation of granular earth bodies, in: Adv. Anal. Geotech. Instabilities, Univ. of Waterloo Press, 13, 65-120, 1978
I.Vardoulakis, Bifurcation analysis of the triaxial test on sand samples, Acta Mechanica, 32, 35-54 (1979)
I.Vardoulakis, Shear band inclination and shear modulus of sand in biaxial tests, Int. J. Num. Anal. Meth. in Geomech. 4, 3-119 (1980)
I.Vardoulakis, Constitutive properties of dry sand observable in the triaxial test, Acta Mechanica,38, 219-239 (1981a)
I.Vardoulakis, Bifurcation analysis of the plane rectilinear deformation on dry sand samples, Int. J. Solids Struct. 17, 1085-1101 (1981b)
I.Vardoulakis, Rigid granular plasticity model and bifurcation in the triaxial test, Acta Mechanica, 49, 57-79 (1983)
I.Vardoulakis, Stability and bifurcation of undrained, plane rectilinear deformations on water-saturated granular soils, Int. J. Num. Anal. Meth. in Geomech. 9, 399-414 (1985)
I.Vardoulakis, Dynamic stability of undrained simple-shear on water-saturated granular soils, Int. J. Num. Anal. Meth. in Geomech. 10, 177-190 (1986)
I.Vardoulakis, Theoretical and experimental bounds for shear-band bifurcation strain in biaxial tests on dry sand, Res Mechanica, in print (1987)
I.Vardoulakis and B.Graf 1982. Imperfection and sensitivity of the biaxial test on dry sand, in: Deformation and Failure of Granular Materials, Balkema, Rotterdam, 485-491, 1982
I.Vardoulakis and B.Graf,Calibration of constitutive models for granular materials using data from biaxial experiments, Géotechnique, 35, 299-317 (1985)
I.Vardoulakis, B.Graf and A.Hettler, Shear-band formation in a fine-grained sand, 5th Int. Conf. on Num. Meth. in Geomech. 1, 517-521, Balkema, Rotterdam, 517-521, 1985.
R.V. Whitman, Some considerations and data regarding the shear strength of cohesive soils, ASCE Reg. Conf. on Shear Strength of Cohesive Soils, Univ. of Colorado, 581-614 (1960)
F.J.Wu and L.B Freund, Deformation trapping due to thermoplastic instability in one dimensional wave propagation, J. Mech. Phys. Solids, 32, 119-132 (1984)
H.Zbib and E.C.Aifantis, On the localization and post-localization phenomena of plastic deformation, Res Mechanica, in print (1987).

Numerical Methods in Geomechanics (Innsbruck 1988), Swoboda (ed.)
© 1988 Balkema, Rotterdam. ISBN 90 6191 809 X

Simple models for soil behaviour and applications to problems of soil liquefaction

O.C.Zienkiewicz & A.H.C.Chan
Institute for Numerical Methods in Engineering, University College of Swansea, UK
M.Pastor
Madrid, Spain
(Visiting Lecturer at the Institute for Numerical Methods in Engineering)

1. INTRODUCTION

Quantitative prediction of the effects of earthquakes on important structures and their foundations is of obvious importance to the engineer concerned with their safety.

If non-linear behaviour and cumulative damage effects can occur, a full modelling by computational techniques is generally necessary. This is particularly true if soil interaction with possible liquefaction occurs.

Saturated soils must be considered thus as two-phase materials with a constant interaction between the skeleton and the interstitial migratory water.

Soil skeleton can contract when subjected to cyclic shear stresses. This will result in an increase of the pore water pressure, and hence the fluid will flow. In turn, pore pressure increase reduces the contact stresses between particles and hence weakens the soil.

Liquefaction is a limiting case of this interaction, in which effective mean confining pressure is close to zero.

To study these effects in a quantitative manner, it is necessary to:

a. Describe a mathematical model for fluid-skeleton interaction;
b. Define qualitatively the behaviour of the soil under varying stress and strain (constitutive modelling);
c. Develop efficient computational means of numerical approximation by finite element methods.

It is obvious that meaningful answers can only be obtained if all three aspects are simultaneously tackled.

Although the basis of a mathematical model for fluid-skeleton interaction was established by Biot [1,2], much work had to be carried out to devise economical and efficient solutions for non-linear situations [3-5].

A correct constitutive equation is of paramount importance to achieve a satisfactory solution of the problem. We will present here two simple models for sands and clays capable of reproducing most of the fundamental features of soil behaviour under monotonic and cyclic loading.

Last, but not least, accurate numerical schemes should be used to model interaction and constitutive equations, together with appropriate boundary conditions. We will approximate the full set of interaction equations written in terms of solid and fluid displacements by a simplified system in which skeleton displacements and fluid pressure are the only variables. This approach is valid for problems of moderate and low frequency.

To assess the predictive capability of the model, we will apply it to reproduce tests carried out on the Cambridge Centrifuge, for which full data on input of earthquake, displacements and pore pressure history are available.

2. CONSTITUTIVE EQUATIONS FOR SOILS - A MULTIMECHANISM APPROACH

2.1 Generalized Plasticity Theory

The classical form of Plasticity Theory allows accurate modelling of soil behaviour under monotonic loading [6-8], but fails to predict soil response to stress paths within the yield surface, such as those under repeated loading and unloading.

A key factor in reproducing liquefaction and cyclic mobility phenomena is to introduce plastic deformation for such paths. For this reason, we have developed models in the framework of Generalized Plasticity, in which the concept of yield surface is not explicitly used, and which permits a direct definition of parameters [9].

Deformation of the material can be considered as the result of deformations produced by M separate mechanisms, all of them subjected to the same state of stress.

Increment of strain is written as

$$d\underline{\varepsilon} = \sum_{m=1}^{M} d\underline{\varepsilon}^{(m)} \qquad (1)$$

and can be related to the increment of stress (co-rotational) by

$$d\underline{\varepsilon} = \sum_{m=1}^{M} \underline{C}^{(m)} : d\underline{\sigma} \qquad (2)$$

where $\underline{C}^{(m)}$ is a fourth order constitutive tensor dependent on the direction of loading $d\underline{\sigma}$ on a set of internal variables $\underline{\alpha}$ and on the current state of stress $\underline{\sigma}$ and strain $\underline{\varepsilon}$.

Dependence on the direction of loading can be introduced by defining a direction $\underline{n}^{(m)}$ for every mechanism, such that

$$d\underline{\varepsilon} = \underline{C}_L^{(m)} : d\underline{\sigma} \qquad (3)$$

if $d\underline{\sigma} : \underline{n} > 0$ (loading) (4)

and $d\underline{\varepsilon} = \underline{C}_u^{(m)} : d\underline{\sigma}$ (5)

for $d\underline{\sigma} : \underline{n} < 0$ (unloading) (6)

Continuity between loading and unloading requires that constitutive tensors $C_L^{(m)}$ and $C_u^{(m)}$ are of the form [9]

$$C_L^{(m)} = \underline{C}^{e(m)} + \frac{1}{H_L} \underline{n}_{gL}^{(m)} \otimes \underline{n}^{(m)}$$

$$C_u^{(m)} = \underline{C}^{e(m)} + \frac{1}{H_u} \underline{n}_{gu}^{(m)} \otimes \underline{n}^{(m)} \qquad (7)$$

where $\underline{n}_{gL/u}$ are arbitrary unit tensors.

Material behaviour under neutral loading is reversible.

$$\underline{C}_L^{(m)} : d\underline{\sigma} = - \underline{C}_u^{(m)} : (- d\underline{\sigma}) \qquad (8)$$

and can therefore be regarded as elastic.

$$d\underline{\varepsilon}^{(m)} = d\underline{\varepsilon}^{e(m)} + d\underline{\varepsilon}^{p(m)} \qquad (9)$$

We note that irreversible-plastic-formations have been introduced without need of specifying any yield or plastic potential surfaces.

Total increment of strain is given by the summation of all mechanisms, i.e.,

$$d\underline{\varepsilon} = \sum_{m=1}^{M} \underline{C}^{e(m)} : d\underline{\sigma} + \sum_{m=1}^{M} \frac{1}{H_{L/u}^{(m)}} \underline{n}_{gL/u}^{(m)} \otimes \underline{n}^{(m)}) = d\underline{\sigma} \qquad (10)$$

or

$$d\underline{\varepsilon} = \underline{C}^{ep} : d\sigma \qquad (11)$$

Inversion of $\underline{C}^{ep}$ will give the equivalent $\underline{D}^{ep}$ but this can only be done if all plastic moduli are non-zero.

However, it can be shown that $\underline{D}^{ep}$ can be directly determined as [10]

$$\underline{D}^{ep} = \underline{D}^{e} - \underline{D}^{e}\, \underline{n}_{gL/u}\, \underline{B}^{-1}\, \underline{n}^{T}\, \underline{D}^{e} \qquad (12)$$

$$\begin{aligned} \underline{n} &= \{\underline{n}^{(m)}\} \\ \underline{n}_{gL/u} &= \{\underline{n}_{gL/u}^{(m)}\} \\ \underline{B} &= \underline{A} + \underline{n}^{T}\, \underline{D}^{e}\, \underline{n}_{gL/u} \\ \underline{A} &= \mathrm{diag}\,(H_{L/u}^{(m)}) \end{aligned} \qquad (13)$$

Classical Plasticity, Bounding Surface or Kinematic Hardening Models can be shown to be particular cases of outlined theory [11-13].

To characterize material response, we need to provide suitable laws for directions $\underline{n}_{gL/u}$ and $\underline{n}$ and scalar $\underline{H}$ for every mechanism

2.2 Isotropic Models for Soils under Transient Loading

In the authors' opinion, single-mechanism based models can provide accurate modelling of soil behaviour. However, multi-mechanism alternatives should be considered if rotation of principal stress axes is expected to occur.

Such isotropic models can be casted in terms of stress invariants, I_1', J_2', J_3' or, alternatively, in terms of p', q and Θ defined as

$$p' = - I_1'$$

$$q = \sqrt{3J_2'} \qquad (14)$$

$$-\frac{\pi}{6} < \Theta = \frac{1}{3} \sin^{-1}\left(\frac{3\sqrt{3}}{2} \frac{J_3'}{J_2'^{3/2}}\right) < \frac{\pi}{6}$$

where superscript "'" refers to effective stress parameters.

Tensorial notation has so far been followed. An equivalent vectorial form is widely used in numerical computation. All expressions given can be easily transformed into such a vectorial notation.

To derive a simple constitutive equation, we begin by assuming that elastic strain increments are much smaller than plastic, and define soil dilatancy, d_g by

$$d_g = \frac{d\varepsilon^p_v}{d\varepsilon^p_s} = \frac{d\varepsilon_v}{d\varepsilon_s} \quad (15)$$

where

$$d\varepsilon_v = tr(d\underline{\varepsilon})$$

$$d\varepsilon_s = \frac{2}{3}\,\frac{1}{2}(d\underline{e} : d\underline{e})^{\frac{1}{2}} \quad (16)$$

$$d\underline{e} = dev(d\underline{\varepsilon})$$

Experiments performed by Balasubramanian [14] on clays and Frossard [15] on sands suggest that dilatancy can be assumed to be a linear function of the stress ratio,

$$d_g = (1+\alpha)(M_g-\eta) \quad (17)$$

where M_g is the stress ratio at which shear of the soil takes place at constant volume. It can be assumed to depend on Lode's angle as [10]

$$M_g = \frac{18\ M_{gc}}{18 + 3M_{gc}(1-\sin 3\theta)} \quad (18)$$

where M_{gc} is the value obtained in triaxial tests (compression.

Direction $\underline{n}_{gL}$ is then obtained as

$$(n_{gv},\ n_{gs})^T$$

$$n_{gv} = \frac{d_g}{(1 + d_g^2)^{\frac{1}{2}}} \quad (19)$$

$$n_{gs} = \frac{1}{(1 + d_g^2)^{\frac{1}{2}}}$$

Direction $\underline{n}$ can be assumed of the same form,

$$\underline{n} = (n_v, n_s)^T$$

$$n_v = \frac{d}{(1 + d^2)^{\frac{1}{2}}} \quad (20)$$

$$n_s = 1/(1 + d^2)^{\frac{1}{2}}$$

$$d_f = (1 + \alpha)(M - \eta)$$

It can be seen that "associated forms" of plastic flow can be obtained by choosing $M_f = M_g$, and indeed experiments carried out by Atkinson [16] show this is the case for clays.

Plastic modulus can now be introduced as

$$H_L = H_o p'\ f_1(\eta) + f_2(\xi)\ f_{DM}\left(\frac{\zeta}{\zeta_{max}}\right) \quad (21)$$

where $f_1(\eta)$ is given for clays as

$$f_1(\eta) = \left|1 - \frac{\eta}{M_g}\right|^{\mu} \frac{(1+d_o^2)}{(1+d^2)}\ \text{sign}\left(1 - \frac{\eta}{M_g}\right) \quad (22)$$

$$d_o = (1-\alpha)\ M_g$$

and

$$f_1(\eta) = \left(1 - \frac{\eta}{\eta_f}\right)^4 \left(1 - \frac{\eta}{M_g}\right) \quad (23)$$

$$\eta_f = \left(1 + \frac{1}{\alpha}\right) M$$

for sands $(M \neq M_g)$

The function $f_2(\xi)$ is given by

$$f_2(\xi) = \beta_o\beta_1\ \exp\ (-\ \beta_o\xi) \quad (24)$$

where

$$d\xi = |d\varepsilon_s|$$

Discrete memory effects are taken into account by $f_{DM}\ (\zeta/\zeta_{max})$ where ζ is a mobilized stress function of which maximum value reached is ζ_{max}.

$$\zeta = p'\left(1 - \left(\frac{1+\alpha}{\alpha}\right)\frac{\eta}{M}\right)^{-\ 1/\alpha} \quad (25)$$

$$f_{DM} = (\zeta_{max}/\zeta)^{\gamma} \quad (26)$$

where γ is a new parameter.

Finally, we have to provide direction $\underline{n}_{gu}$ and H_u for unloading. It has been shown elsewhere [12] that convenient expressions for sands are

$$\underline{n}_{gu} = (n_{guv}, n_{gus})^T$$

$$n_{guv} = n_{gLv} \quad (27)$$

$$n_{gus} = -\ n_{gLs}$$

$$H_u = H_{uo}\left(\frac{M_g}{\eta_u}\right)^{\gamma_u} \quad \text{for} \quad \left|\frac{M_g}{\eta_u}\right| > 1$$

$$H_u = H_{uo} \quad \left|\frac{M_g}{\eta_u}\right| \leq 1 \quad (28)$$

where η_u is the stress ratio from which unloading takes place.

The proposed model can predict very loose sand liquefaction under undrained shearing, exhibiting a peak in deviatoric stress after which the strength reduces to zero while the pore pressure increases continuously (fig.1) [17]. The stress path approaches the origin and the soil reaches a liquefied state in which it loses its resistance to shear.

It has to be remarked that the stress ratio is continuously increasing, and no softening takes place.

Liquefaction of very loose sands under cyclic loading is caused by an accumulation of pore pressure followed by liquefaction during the loading part of a cycle as described before, (fig.2) [17].

Finally, cyclic mobility of sands is a progressive failure process in which the stress path approaches the Critical State line. Deformations during unloading make the stress path turn towards the origin, therefore causing an increase in strain amplitude as can be seen in fig.3 [18].

2.3 Multimechanism Models

Any model casted in terms of stress or strain invariants will predict no response to a pure rotation of principal stress axes. However, it can produce accurate results if rotation is accompanied by an increase of the deviatoric strain [19] as occurs in the hollow cylinder device.

Multimechanism models [20-22] can be used to account for pure rotation of principal stress axes as the stress state varies in some mechanisms.

Considering three mechanisms defined on planes XY, YZ and XZ, we can define plane invariants as follows

$$q^{(k)} = \left| \left(\frac{\sigma_i - \sigma_j}{2} \right)^2 + \tau_{ij}^2 \right|^{\frac{1}{2}} \tag{29}$$

$$p^{(k)} = \frac{\sigma_i + \sigma_j}{2} \tag{30}$$

with plastic conjugate strain parameters ε_v, ε_s.

It can be easily seen that pure rotation of principal stresses direction results in a variation of $q^{(k)}$ and $p^{(k)}$, therefore producing irreversible strains.

Now we can introduce suitable laws for $\underline{n}^{(k)}$, $\underline{n}_{gL/u}^{(k)}$ and $H^{(k)}$ as done in (17)-(18), and, indeed, we can introduce similar expressions for these items.

3. THE U-P SOLUTION TO THE BIOT EQUATION

Biot [1,2] introduced the governing equations for soil saturated pre-fluid interaction. In this section, the Biot equations with some simplifications which leads to the U-P formulation is presented. For the solid phase, the equation governing is the equilibrium equation.

$$\sigma_{ji,j} + \rho g_i = \rho \ddot{u}_i \tag{31}$$

with

$$d\sigma''_{ij} = d\sigma_{ij} + \alpha\, \delta_{ij}\, dp \tag{32}$$

$$d\sigma''_{ij} = D_{ijk\ell}(d\varepsilon_{k\ell} - d\varepsilon^{\circ}_{k\ell}) \tag{33}$$

where

$$\alpha = 1 - K_T/K_S \tag{34}$$

For small strain and deformation

$$d\varepsilon_{ij} = (dU_{i,j} + dU_{j,i})/2 \tag{35}$$

The equilibrium equation has to be solved in conjunction with appropriate boundary conditions for displacement U_j, total stress σ_{ij}, and fluid pressure, p. The equation (32) is the modified effective stress equation. Coefficient α is introduced to account for the compressibility of the soil grains. K_T and K_S are the bulk moduli of the soil skeleton and soil grain respectively. $\varepsilon^{\circ}_{k\ell}$ is the "initial strain" and $D_{ijk\ell}$ is the tangential constitutive tensor and for the numerical results, the Pastor-Zienkiewicz Mark III model [13] is used.

Under dynamic or static transient conditons, the mass acceleration relative to the soil skeleton can be neglected. The relative velocity of the fluid can be eliminated [23]. Hence, we can write an extended (D'Arcy) flow equation in terms of pressure p and soil displacement $\underline{u}$ only:

$$-(k\, p,i),i + \alpha\varepsilon_{ii} - (k\, \rho_f\, g_i),i + (k\, \rho_f\, \ddot{u}_i),i + \dot{p}/Q=0 \tag{36}$$

In the above equation, k is the permeability coefficient, g_i gravity acceleration. Once again, the appropriate boundary condition is needed. Q is a parameter defining the compressibility of the fluid, solid, etc., and this is given as [3]

$$1/Q = n/K_f + (\alpha - n)/K_s \tag{37}$$

where n is the porosity and K_f is the Bulk modulus of the fluid. Frequently, it is assumed that 1/Q is zero, but generally, this compressibility is non-negligible. The acceleration term in eq. (36) has negligible influence in medium and slow transient condition, so it is left out in further derivation.

With usual spatial Finite Element discretization assumed for u and p as

$$\underline{u} = \underline{N}_u \hat{\underline{u}} \quad \text{and} \quad p = \underline{N}_p \hat{\underline{p}} \tag{38}$$

the application of Galerkin procedures [14] results in the following discrete equation with which the appropriate natural boundary conditions are incorporated in the forcing terms.

$$\underline{M}\,\ddot{\hat{\underline{u}}} + \int_\Omega \underline{B}^T \underline{\sigma}''\, d\Omega - \underline{Q}\,\hat{\underline{P}} = \underline{f}_1 \tag{39a}$$

$$\underline{H}\,\hat{\underline{p}} + \underline{Q}^T \dot{\hat{\underline{u}}} + \underline{S}\,\dot{\hat{\underline{p}}} = \underline{f}_2 \tag{39b}$$

where $\underline{M}$ is the mass matrix and $\underline{B}$ the strain matrix, $\underline{H}$ the permeability matrix, $\underline{Q}$ connecting matrix and $\underline{S}$ the compressibility matrix. Using the SS_{pj} time stepping algorithm [25-27], the equation system can be written in a form

$$\underline{\psi}\,(\underline{\gamma}) = \underline{\psi}\,(\underline{\alpha},\underline{\beta}) = 0 \tag{40}$$

where $\hat{\underline{u}}_{n+1} = \hat{\underline{u}}_n + \dot{\hat{\underline{u}}}_n \Delta t + \underline{\alpha}_n \Delta t^2/2$

$$\hat{\underline{p}}_{n+1} = \hat{\underline{p}}_n + \beta_n \Delta t$$

and the gradient matrix for nonlinear iteration is defined as

$$\underline{A}^* = \partial\underline{\psi}/\partial\underline{\gamma} \tag{41}$$

$$\underline{A}^* = \begin{vmatrix} \underline{M} + \bar{\underline{K}}\theta_2 \Delta t^2/2 & -\underline{Q}\theta_1 \Delta t \\ -\underline{Q}^T \bar{\theta}_1 \Delta t & -\underline{H}\bar{\theta}_1 \Delta t - \underline{S} \end{vmatrix} \tag{42}$$

where $\bar{\underline{K}}$ is the tangential stiffness matrix and the equation is symmetric if $\theta_n = \bar{\theta}_\ell$. Details of the derivation are given in [28].

4. NUMERICAL RESULTS

Both the U-P formulation and the Pastor-Zienkiewicz Mark III model have been implemented on a research computer program DIANA -SWANDYNE. Its prediction is checked against centrifuge experiments done in Cambridge University. The parameters for the soil model are found from the triaxial test done with the same used in the experiment. The parameters are defined in Table I. These test data are incomplete and some parameters had to be assumed from previous tests, of which full model comparison has been presented.

In this example, a full scale run has been performed. The experimental set-up of the centrifuge is shown in fig.4, and one corresponding Finite Element mesh in fig.5. The bilinear shape function is used for both solid and fluid phase. The other material parameters, as extracted from the centrifuge data, are shown in Table 2, together with the line-stepping data of Finite Element Analysis. The vertical deflection of the crest of the dyke is shown in fig.6. The result compares excellently with the experiment. The shape, the final values and the initial major increase matches the experiment perfectly. The readers are reminded that the result is obtained on the first run without any parameter importance. The extent of total settlement is of most concern in earthquake analysis if total failure did not occur. Also shown in fig.6 is the input motion of the earthquake as defined as the base section of the centrifuge experimental compartment.

Comparisons with ten pure pressure transducers have been done. Due to the lack of space, only two are shown in fig.7 (a) and (b). They correspond to points A and D in fig.5. They show a typical behaviour of the ten comparisons. The rising time and the magnitude compare well with the experiment. However in some cases, the oscillations of the numerical results are excessive. The magnitude has been improved by a slight adjustment in the soil parameters and the oscillation is reduced if further damping is added.

The result of the accelerometer at point L is shown in fig.7 (c). The initial cycles show good agreement, however, the experiment shows less shear wave transmission to the top layer than the numerical prediction. But in general, very good results have been obtained. Further comparison, using other elements and other centrifuge tests, have been done and further trials are in progress.

CONCLUSIONS

From the results achieved, it can be concluded that the mathematical model for fluid-skeleton interaction used, the constitutive relations for soil behaviour under general transient loading conditions and the numerical schemes proposed are an efficient tool to analyse liquefaction and other dynamic problems in soils.

ACKNOWLEDGEMENTS

The authors express their gratitude to Andrew Schofield and Associates from Cambridge University for kindly providing the Centrifuge Model experimental data. Support from the British Council and the Ministerio de Education y Ciencia of Spain is gratefully acknowledged. The financial support received by the third author from Japanese Industries and the Croucher Foundation is also gratefully acknowledged.

REFERENCES

1. M. A. BIOT. Theory of three-dimensional consolidation, J.Appl.Phys., 12, 155-164, 1941.
2. M. A. BIOT - Mechanics of deformation and acoustic propagation in porous media, J.Appl.Phys., 33, 1483-1498, 1960.
3. O. C. ZIENKIEWICZ and T. SHIOMI - Dynamic behaviour of saturated porous media; the generalized Biot formulation and its numerical solution, Int.J.Num. Anal.Meth.Geomech., Vol.8, 71-96, 1985.
4. O. C. ZIENKIEWICZ - The coupled problems of soil-pore fluid-external fluid interaction: Basis for a general geomechanics code, Fifth Int.Conf.Num.Meth. Geomech., 1731-1740, Nagoya, Japan, 1985.
5. O. C. ZIENKIEWICZ, C. T. CHANG and P. BETTESS - Drained, undrained and consolidating and dynamic behaviour assumptions in soils, Geotechnique, Vol.30, No.4, 385-395, 1980.
6. D. C. DRUCKER and M. PRAGER - Soil Mechanics and Plastic Analysis or limit design, Quart.Appl.Math., 10, 157-165, 1952.
7. K. H. ROSCOE, A. N. SCHOFIELD and C. P. WROTH - On the yielding of soils, Geotechnique, 8, 22-53, 1958.
8. R. NOVA and D. M. WOOD - A constitutive model for sand, Int.J.Num.Anal.Meth. Geomech. 3, 255-278, 1979.
9. O. C. ZIENKIEWICZ and Z. MROZ - Generalized plasticity formulation and applications to geomechanics, Mechanics of Engineering Materials, Ch.33, 655-679 eds. C. S. Desai and R. H. Gallagher, Wiley, 1984.
10. M. PASTOR, O. C. ZIENKIEWICZ and A. H. C. CHAN - Generalized plasticity and the modelling of soil behaviour, submitted to Int.J.Num.Anal.Meth. Geomech. 1987.
11. O. C. ZIENKIEWICZ, K. H. LEUNG and M. PASTOR - A simple model for transient soil loading in earthquake analysis: I - Basic model, Int.J.Num.Anal.Meth. Geomech. 9, 453-476, 1985.
12. M. PASTOR, O. C. ZIENKIEWICZ and K. H. LEUNG - II - Non-associative model for sands, Int.J.Num.Anal.Meth. Geomech. 9, 477-498, 1985.
13. M. PASTOR and O. C. ZIENKIEWICZ - A generalized plasticity, hierarchical model for sand under monotonic and cyclic loading, Proc.Int.Symp.Num.Mod. Geomech., Ghent, Belgium, Numerical Models in Geomechanics, eds. G. N. Pande and W. F. Van Impe, 131-150, 1986.
14. A. S. BALASUBRAMANIAN and A. R. CHAUDRY - Deformation and strength characteristics of soft Bangkok clay, J.Geotech. Eng.Div.A.S.C.E., Vol.104, No.GT9, 1153-1167, 1978.
15. E. FROSSARD - Une équation d'écoulement simple pour les materiaux granulaires. Geotechnique, 33, 1, 21-29, 1983.
16. J. H. ATKINSON and D. RICHARDSON - Elasticity and normality in soil. Experimental examinations, Geotechnique, 35, No.4, 443-449, 1985.
17. G. CASTRO - Liquefaction of sands, Ph.D. Thesis, Harvard University, 1969.
18. K. ISHIHARA, F. TATSUOKA and Y. YASUDA - Undrained deformation and liquefaction of sand under cyclic stress, Soils and Foundations, Vol.15, No.1, 29-44.
19. M. PASTOR, O. C. ZIENKIEWICZ and A. H. C. CHAN - Generalized plasticity model for three-dimensional sand behaviour, Int.Symp.Constitutive Equations for granular soils, ed. A. Saada, Cleveland, USA, 1978.
20. O. C. ZIENKIEWICZ and G. N. PANDE - Time-dependent multilaminate model for rocks. A numerical study of deformation and failure of rock masses, Int.J.Num. An.Meth.Geomech. Vol.1, 219-247, 1977.
21. G. N. PANDE and K. G. SHARMA - Multilaminate models for clays - A numerical evaluation of the influence of rotation of principal stress axes, Int.J.Num.An. Meth.Geomech. Vol.7, 397-418.
22. D. AUBRY, J. C. HUJEUX, F. LASSODIÈRE and Y. MEIMON - A double memory model with multiple mechanisms for cyclic soil behaviour, Int.Symp.Num.Models in Geomechanics, Zurich, 1982.
23. T. SHIOMI - Nonlinear behaviour of soils in earthquake, Ph.D. Thesis, C/Ph/73/83, University College of Swansea, 1983.
24. O. C. ZIENKIEWICZ - The Finite Element Method, McGraw-Hill, 1977.
25. O. C. ZIENKIEWICZ, W. L. WOOD, N. W. HINE and R. L. TAYLOR - A unified set of single step algorithms, part 1: General formulation and applications, Int.J.Num.Meth.Eng. Vol.20, 1529-1552, 1984.
26. W. L. WOOD - A unified set of single step algorithms, part 2: Theory, Int.J.Num.Meth.Eng. Vol.20, 2303-2309, 1984.
27. W. L. WOOD - Addendum to 'A unified set of single step algorithms, part 2: Theory, Int.J.Num.Meth.Eng. Vol.21, 1165, 1985.
28. A. CHAN - A unified Finite Element Solution to static and dynamic problems of geomechanics, Ph.D. Thesis, Univ. College of Swansea, 1987, to be submitted.

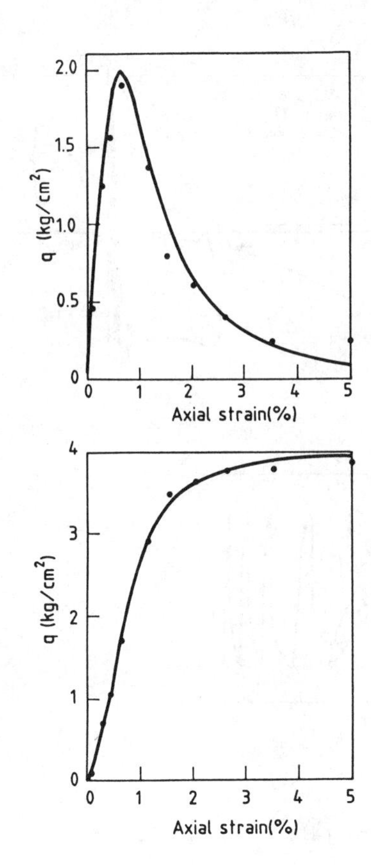

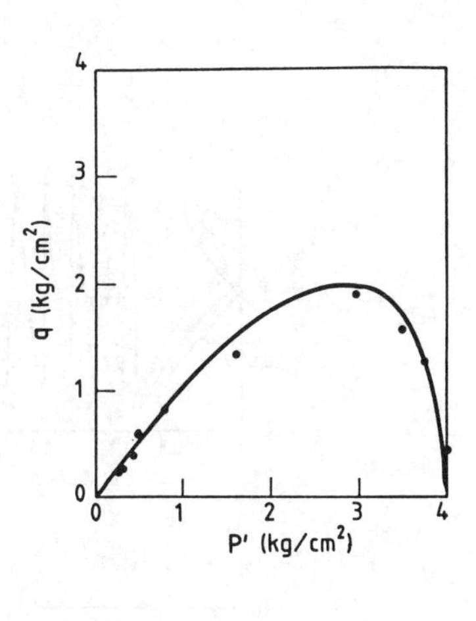

Experimental •
Predicted ——
D_R = 27%
P'_0 = 4kg/cm²

FIGURE 1 LIQUEFACTION OF A VERY LOOSE SAND UNDER MONOTONIC, UNDRAINED LOADING (EXPERIMENTAL DATA FROM CASTRO)

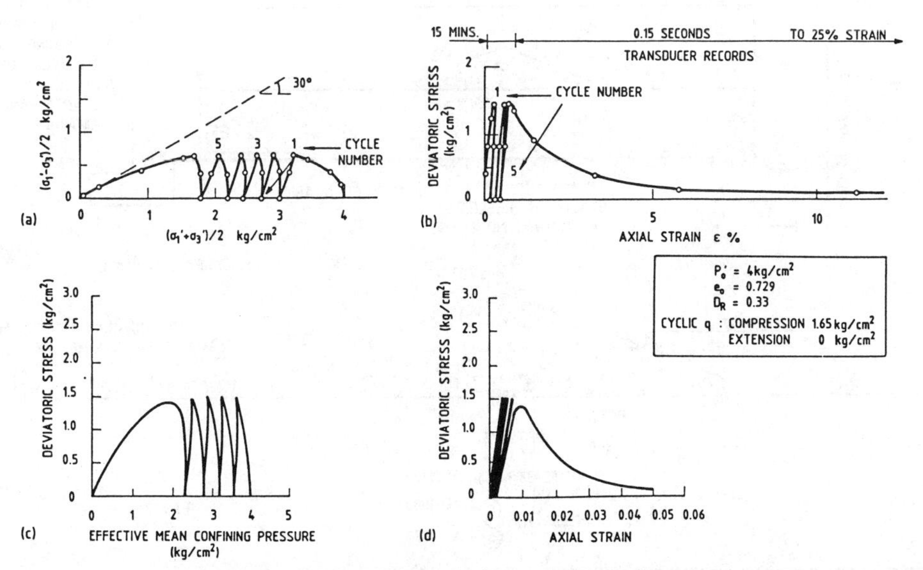

FIGURE 2 LIQUEFACTION OF BANDING SAND UNDER CYCLIC LOADING
(a) AND (b) DATA FROM CASTRO (1969)
(c) AND (d) PREDICTED

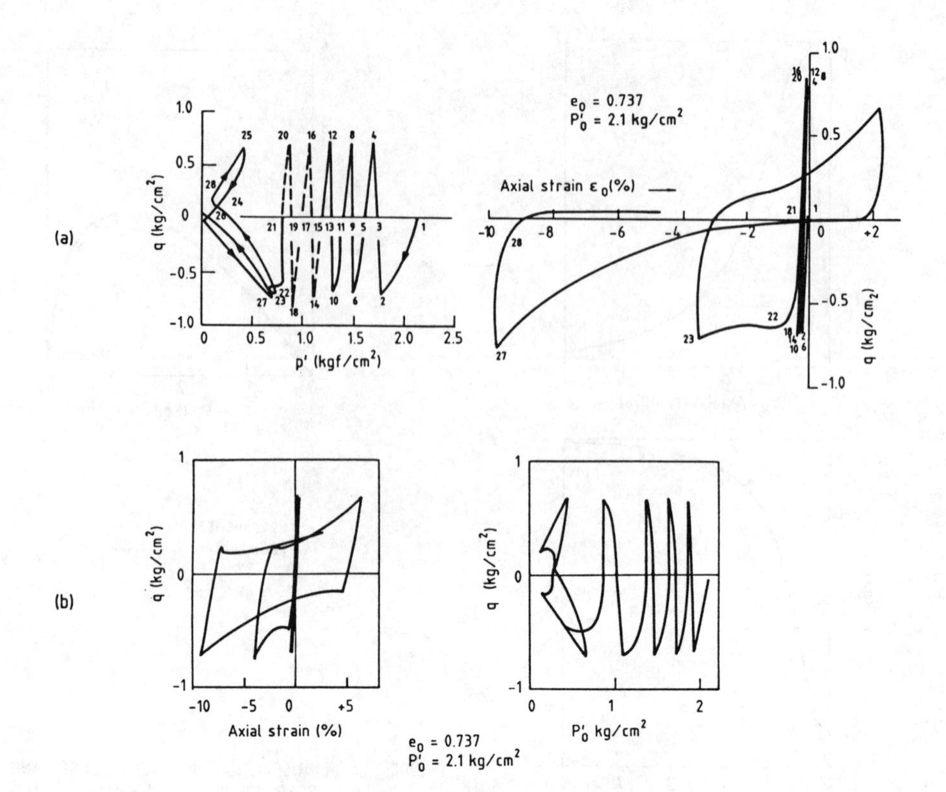

Figure 3 Cyclic Mobility (a) experimental; (b) predicted

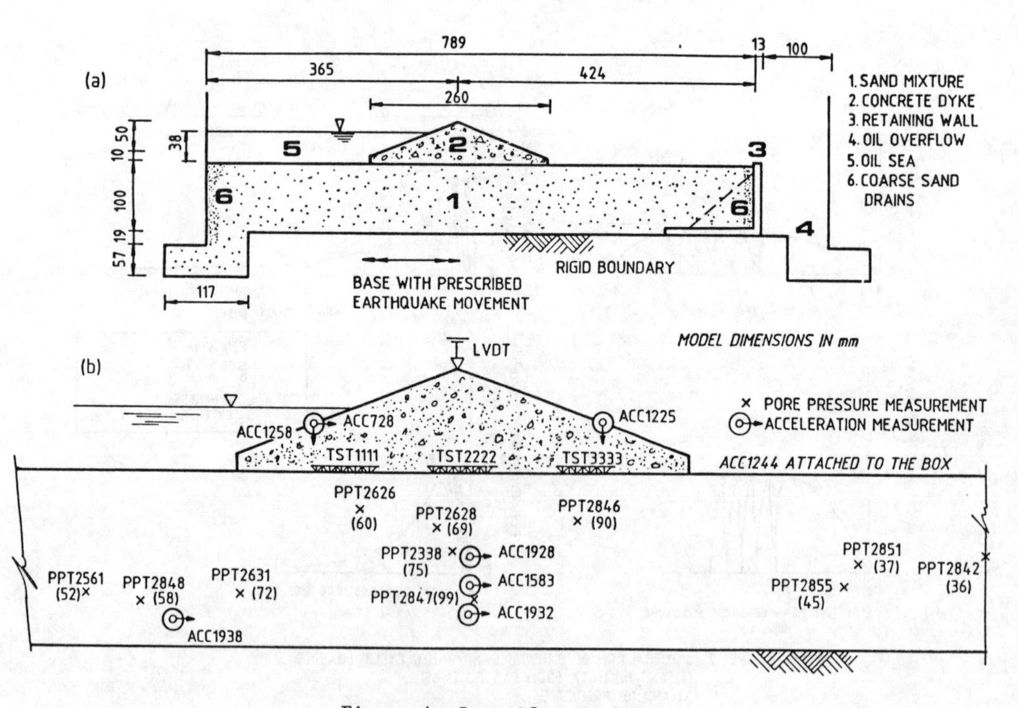

Figure 4 Centrifuge Model

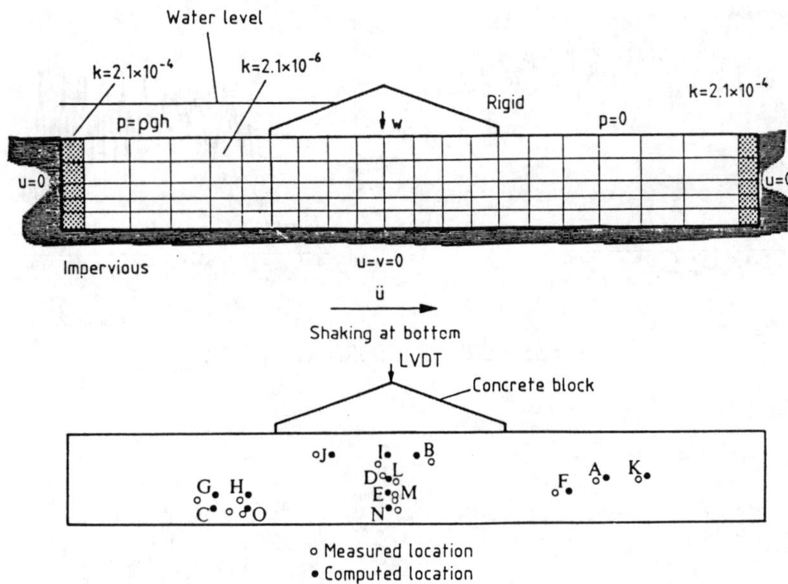

Figure 5 Finite Element Mesh

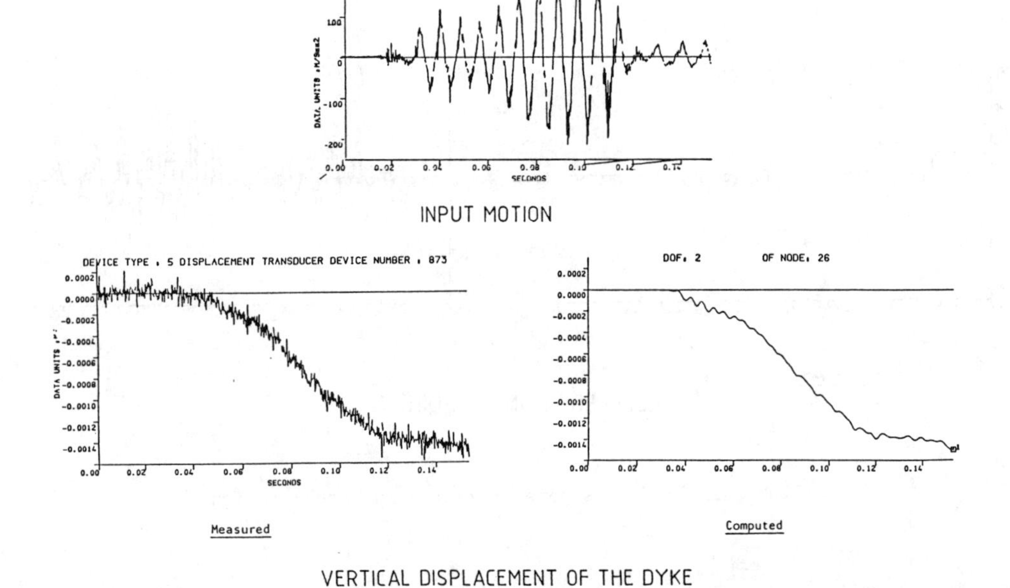

Figure 6 Vertical Displacement of the Dyke

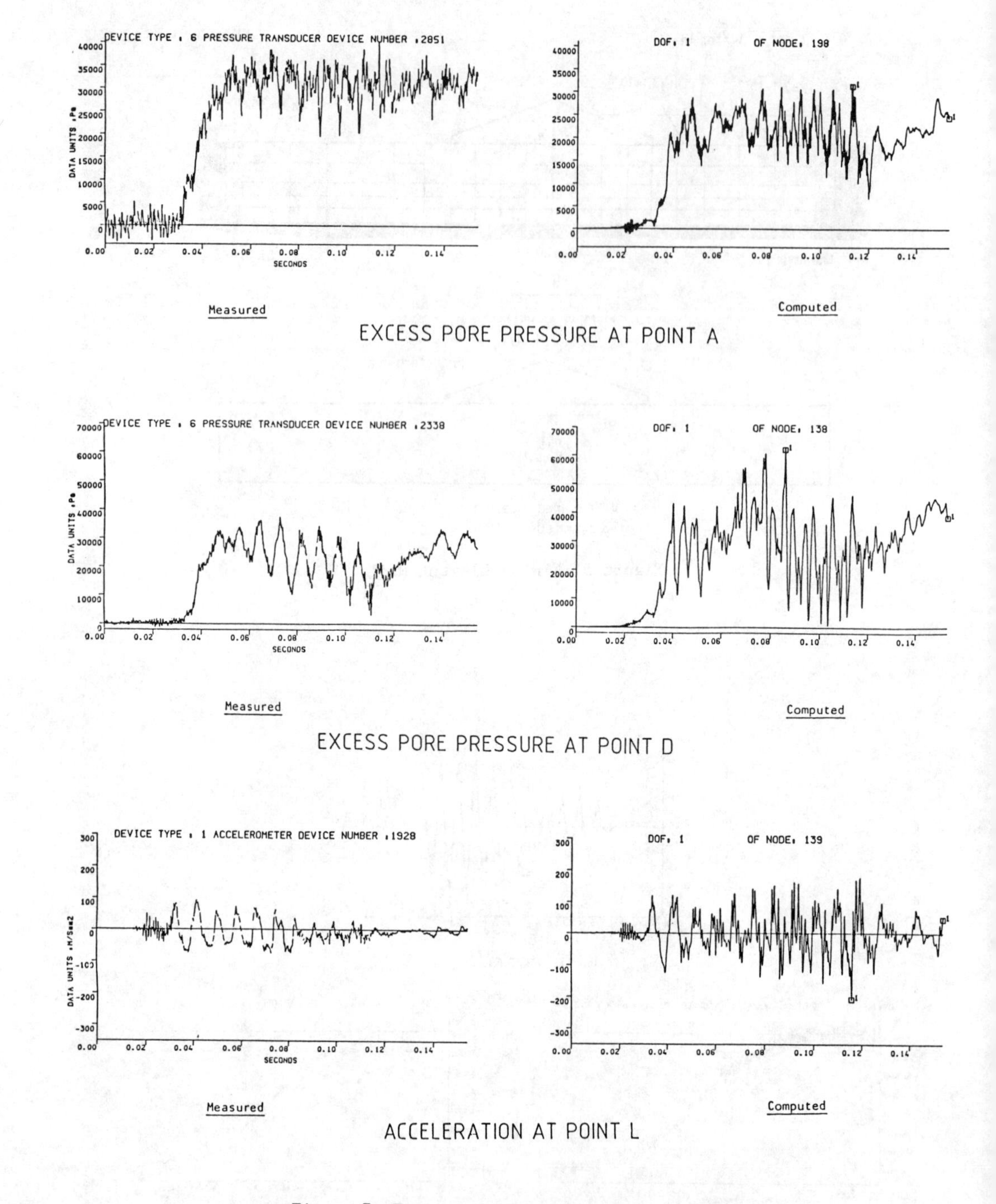

Figure 7 Excess pore pressures and acceleration

TABLE FOR MATERIAL DATA (Soil Model)

	TEST	13	14	PREDICTED
	ADJUSTED Po(kPa)	55.0	64.5	55.0
	Kevo	20,000	23,000	20,000
(1)	Kevo/Po	360	360	360
	Keso	30,000	35,000	30,000
(2)	Keso/Po	540	540	540
(3)	Mg	1.15	1.26	1.26
(4)	Mf	0.86 ($Mg.D_R$=0.46)	0.55 ($Mg.D_R$=0.51)	0.84 ($Mg.D_R$=0.756)
(5)	α_f, α_g*	0.45	0.45	0.45
(6)	β_0*	4.2	4.2	4.2
(7)	β_1*	0.2	0.2	0.2
(8)	H_0	500	600	700
(9)	H_{U0} (kPa)	—	—	60,000
(10)	γ_u*	—	—	2
(11)	γ_{DM}*	—	—	TEST I 2 / 0 TEST II
	e	0.875	0.873	0.80
	D_R**	40%	40.5%	60%

* As suggested in original paper

** e_{min}=0.65 (D_R=100%) e_{max}=1.025 (D_R=0%)

$$D_R = \frac{e_{max}-e}{e_{max}-e_{min}} \times 100\%$$

Table 1 Material Data

Material Data for Finite Element Analysis

Bulk Density (average soil-pore fluid) = 1943 kgm^{-3}

Density of the pore fluid = 1000 kgm^{-3}

Bulk Modulus of the pore fluid = 2.25×10^9 Pa

Biot-alpha = 1.0000

Porosity (n) = 0.42792

Initial void ratio (e) = 0.748

Permeability = 1.0×10^{-3} ms^{-1}

g-acceleration = 10.0 ms^{-2}

Gauss point for all cases (Gauss-Legendre) = 2x2

No incremental strain subdivision is performed

Initial Stress Method is used for nonlinear iterations

Convergence Criteria for the nonlinear iterations:

Residual force norm versus current external force norm (for each phase) $\leq$ 0.1%

10 seconds of the Elcentro N-S earthwuake is applied

(Maximum acceleration = 0.2g)

Time step size = 500 of 0.02 seconds (application of earthquake)
200 of 0.1 seconds
100 of 1.0 seconds
100 of 10.0 seconds

θ's for the SSpj scheme = 0.6,0.605

Table 2 Data for Finite Element Analysis

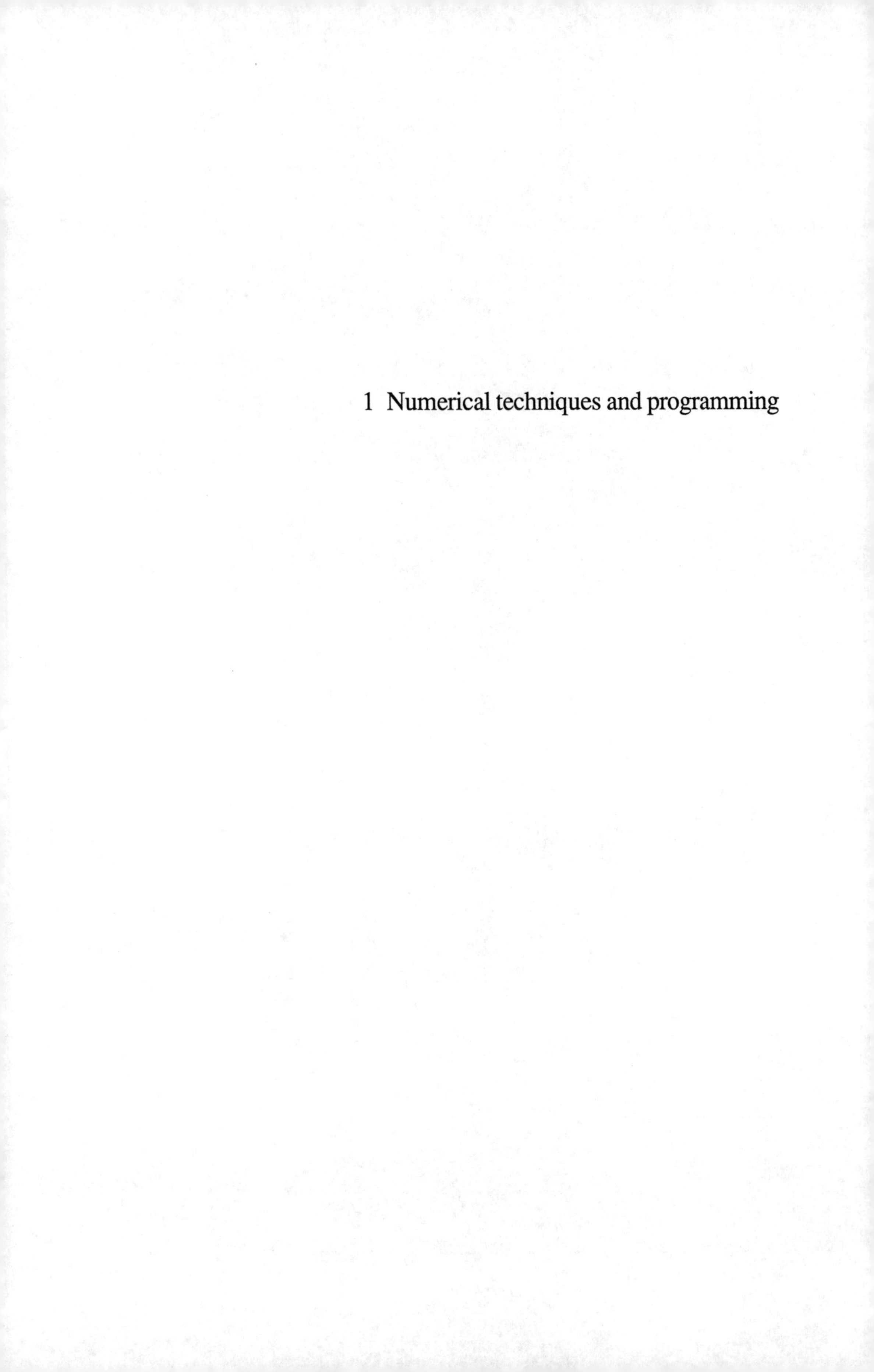

1 Numerical techniques and programming

Numerical Methods in Geomechanics (Innsbruck 1988), Swoboda (ed.)
© 1988 Balkema, Rotterdam. ISBN 90 6191 809 X

Experiences with the analysis of geotechnical problems solved by the FEM using different equation solvers on several computers

Hermann Schad & Frank Breinlinger
Institute for Geotechnics, University of Stuttgart, FR Germany

Abstract: A block equation solver for non–symmetric matrices according to the *Gauss method* has been developed and tested in order to prove its efficiency. A series of tests was performed to quantify the influence due to programming and compilation features. It could be shown that on vector computers program techniques are drastically influencing efficiency. If this fact is taken into account for the program development it will be possible to economically analyse sophisticated problems with physical and geometrical nonlinearities.

1 Introduction

The rapid development concerning vector computers opens a wide range of new possibilities for analysing nonlinear foundation engineering problems by FEM. The large central storages of vector computers enable the user to work almost without any memory restrictions and therefore programming has become more comfortable and efficient. However one cannot fully take advantage of this situation without changing programming habits and the handling of system dependent software. Therefore several tests have been performed to evaluate the influences on efficiency due to programming and compilation features. These influences are illustrated by a geotechnical example with complicated boundary conditions.

2 The Gauss algorithm for non-symmetric block matrices

From the experiences with practical foundation problems like the raft foundation with piles in figure 1 (SOMMER 1987) the decision was made to extend our program system to non–symmetric matrices by developing an appropriate hyper matrix code. Thus a better convergence behaviour can be expected when using nonlinear material models with non–associated flow rules or similar constitutive models.

This allows us to still take advantage of the freedom family concept which can be very useful for contact problems (SCHAD 1985 and SMOLTCZYK/BREINLINGER 1987) and for large problems where substructure techniques become necessary. Due to the hyper matrix code the program does not loose its adaptability to different computer systems. The factorization algorithm for a hyper matrix $[\mathbf{A}]$ with $m \times m$ submatrices $\mathbf{A}_{ij}$ is outlined in the following diagram.

```
Loop over the diagonal submatrices: k ← 1,m
Inversion of the submatrix A_kk
A_kk ← A_kk^-1
    Loop over the columns: j ← k+1,m
    C=A_kj
    A_kj =A_kk C
        Loop over the rows: i ← k+1,m
        D=A_ik C
        A_ij ←(A_ij−D)
```

The diagram shows the principle of the code and does not include checks for the existence of a single submatrix. This of course is very important for solving practical problems with mesh refinement resulting in an irregular pattern of the stiffness matrix (figure 1) in order to avoid unnecessary operations.

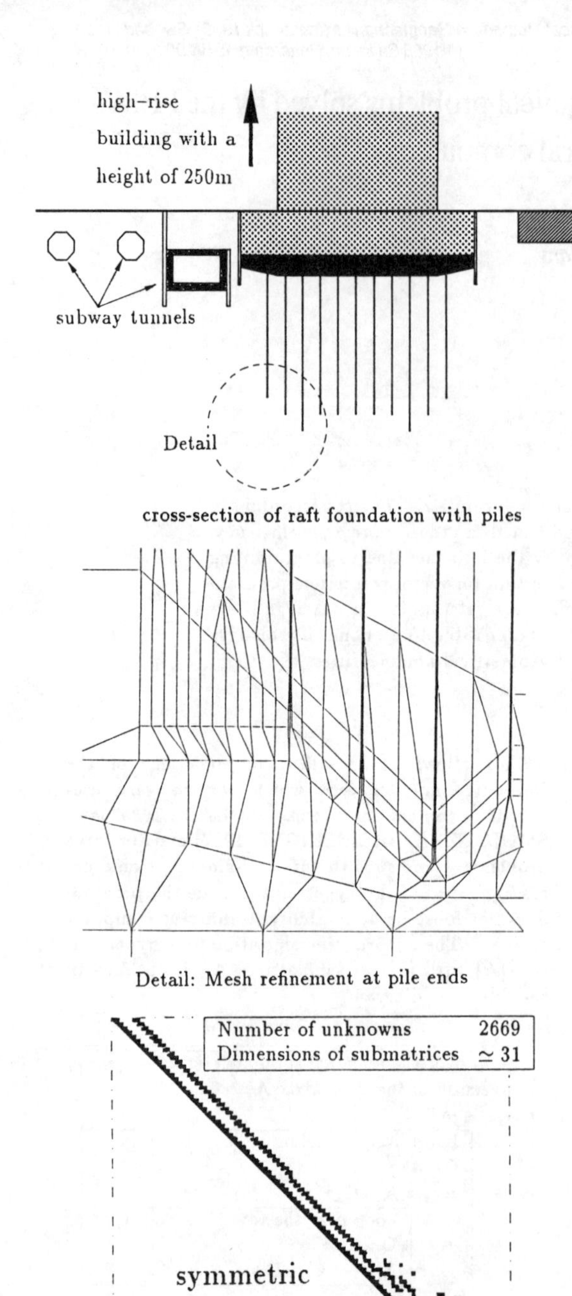

Figure 1: Raft foundation with piles in Frankfurt/M.

3 Accuracy of various equation solvers

The basis for the tests were especially generated fully occupied symmetric matrices which were much worse conditioned than stiffness matrices usually occurring in FEM problems. As a measure for the accuracy of the different equation solvers the Euklidian norm $||\Delta \mathbf{x}||_2$ with vector $\Delta \mathbf{x} = \mathbf{x}^* - \mathbf{x}$ was used. The exact solution $\mathbf{x}^*$ was calculated so that the right hand side could be computed by the equation $\mathbf{r} = \mathbf{A}\,\mathbf{x}^*$. The condition numbers (K) of the matrices which indicate the reciprocal conditions were computed with the LINPACK routine SGECO (DONGERRA 1979). The elements of the solution vector can usually be expected to have $log(K)$ fewer significant digits of accuracy than the elements of $\mathbf{A}$. The condition numbers of the tested matrices of the dimension $n \times n$ were:

n	7	17	33	63	100
$log(K)$	3.42	5.61	7.63	8.55	9.72
n	200	300	1000	2000	
$log(K)$	10.2	11.8	15.1	17.9	

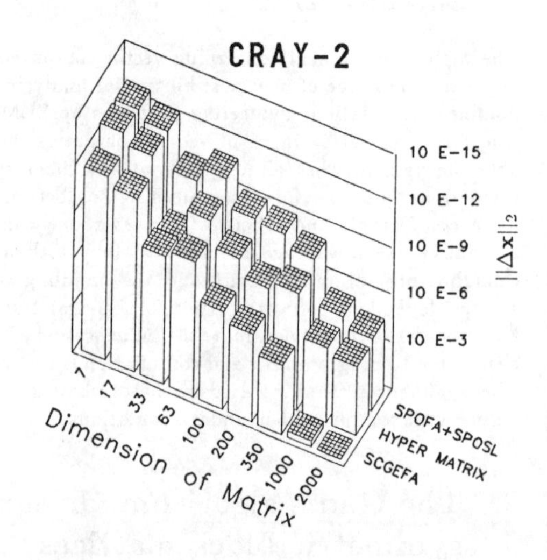

Figure 2: Accuracy of various equation solvers

Figure 2 shows a comparison of the three equation solvers used for accuracy tests:

- SPOFA is a LINPACK routine to compute the *Cholesky* decomposition of a positive definite matrix. The according routine for forward- and backward-substition is SPOSL.

- SCGEFA is a Fortran5 programmed routine for equation solving according to *Gauss* without pivoting according to MONRO 1982, page 227.
- HYPER MATRIX(HYGEFA) stands for a block equation solver according to *Gauss* for non-symmetric matrices (see section 2).

In case of using for HYGEFA the LINPACK routines SGEFA for factorization and SGEDI for inversion which have partial pivoting the hyper matrix code has almost the accuracy of the Cholesky method and is better than the Gauss method without pivoting.

4 Efficiency

Tests have been performed on a CYBER 170/835, a CONVEX C1 and on a CRAY-2 computer. As a measure for the efficiency MEGAFLOPS (Millions of floating point operation per second) were used. The considered floating point operations were multiplication, division, addition and subtraction.

Matrix multiplications **C**=**A B** are usually done by dot products

$$c_{ij} = \mathbf{a}_i^T \mathbf{b}_j \tag{1}$$

where $\mathbf{a}_i^T$ are the rows of **A** and $\mathbf{b}_j$ are the columns of **B**. This means that a scalar is the result of an operation with two vectors. According to the BLAS routines this matrix multiplication shall be called SDOT.

Alternatively multiplications (GENTZSCH 1987) can be carried out by

$$\mathbf{c}_j = \sum_{k=1}^{m} b_{kj}\mathbf{a}_k \tag{2}$$

where $\mathbf{a}_k$ are the columns of **A** and $\mathbf{c}_j$ are the columns of **C**. A vector now is the result of an operation with a scalar and a vector. Because the routine for this operation in BLAS is called SAXPY this method is named SAXPY in figure 3.

To quantify the influence due to programming and compilation, matrix multiplications have been performed using square matrices of different dimensions on the vector computer CONVEX C1. It showed (figure 3) that efficiency can only be obtained with optimizing and vectorizing compilers and programming according to equation 2. Even on a small ATARI 1040ST the code using SAXPY was about 10% faster than the code using SDOT.

Another important effect occuring on vector computers is the so called "memory conflict". If e.g. even increments are used for loops these conflicts usually result. During the matrix multiplication tests on the CONVEX C1, a decrease in speed due to "memory

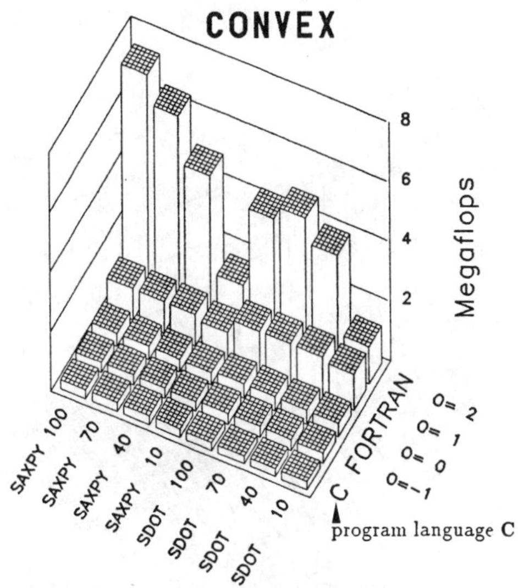

$O = -1$	no optimization and no vectorization
$O = \ 0$	local scalar optimization
$O = \ 1$	local and global scalar optimization
$O = \ 2$	local and global scalar optimization plus vectorization

Figure 3: Efficiency on CONVEX C1 considering matrix multiplications

conflicts" became only significant when the multiplication was done on the basis of SDOT and the dimensions of the matrices were 64×64 or 128×128. Using these dimensions, efficiency reduces to 0.5 as shown in the following table:

Dimension n	62	63	64	65	66
Megaflops	5.1	4.3	2.2	5.1	4.5
Dimension n	126	127	128	129	129
Megaflops	4.4	5.3	2.0	4.5	3.6

The tremendous influence due to programming on the CRAY-2 becomes obvious when comparing SPOFA+SPOSL with SGEFA+SGESL in figure 4. Although SGEFA is a highly optimized FORTRAN routine for the CRAY-2 the hyper matrix code which uses system routines for matrix multiplications has nearly the same efficiency. CALAHAN 1980 states his experiences which seem to be valid for the CRAY-2 as well as follows:

For example, the solution of a full matrix through Fortran will execute at an asymptotic rate one-third the maximum processing rate of the CRAY-1, due to the difficulty of managing the vector register cache memory through Fortran. .....As a result, the discipline-oriented researcher is initially inclined to

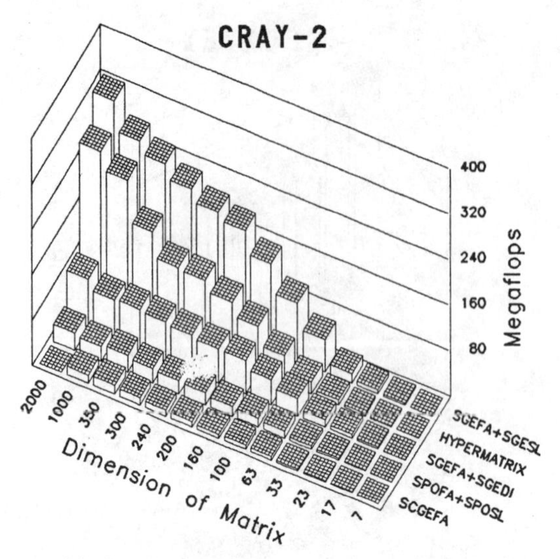

SGEFA+SGESL: Equation solving with LINPACK routines for LU-factorization and forward- backward substitution.

SGEFA+SGEDI: Equation solving with inversion and matrix multiplication by using the LINPACK routines SGEFA and SGEDI.

Figure 4: Various equation solvers on the CRAY-2

accept the above-mentioned 3:1 degradation obtainable from a vectorized Fortran code.

To obtain a comparison the same tests were performed on a CYBER 170/835 which is a classical main-frame computer. Figure 5 shows that on such a "scalar" computer the influence on efficiency due to coding is relatively small.

Comparing the efficiency of the equation solvers SCGEFA, SGEFA and HYGEFA (figures 4 and 5) it can be seen that on the main-frame (CYBER) the maximum is reached at matrices of 20×20 whereas on the CRAY-2 the efficiency is becoming better and better up to dimensions of about 1000.

On the CRAY-2 the block equation solver is slightly slower than the highly optimized routines SGEFA/SGESL. However at large dimensions ($n > 1000$) the difference becomes negligible. This effect results from the solution technique of block equation solvers where most of the operations are matrix multiplications with fixed block size which can be handled by the system routine MXM. Multiplying matrices of dimension 64×64, this routine achieves 280 Megaflops and therefore HYGEFA can

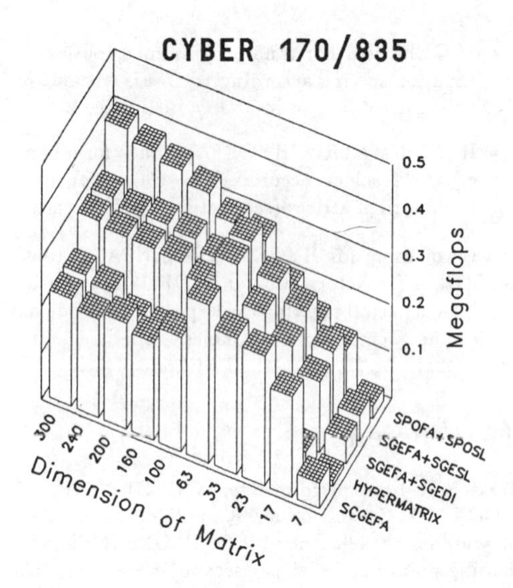

Figure 5: Various equation solvers on CYBER 170/835

never be as efficient using submatices of about these dimensions as SGEFA with more than 300 Megaflops ($n = 2000$). Nevertheless these efficiencies are quite acceptable since the maximum theoretical value is 448 Megaflops. For instance a common Fortran programmed matrix multiplication according to equation 1 yields 24 Megaflops and according to equation 2 it yields 56 Megaflops.

The CPU-time (seconds [s]) used during factorization of the coefficient matrix for the raft foundation with piles (figure 1) is outlined in the following table:

block equation solver	**dimensions of submatrices**		
	$\simeq 31$	$\simeq 63$	$\simeq 127$
Cholesky (symmetric)	3.54s	4.30s	5.55s
Gauss (non symm.)	8.56s	4.28s	5.37s

Looking at these figures one still has to keep in mind that the efficiency which can be achieved during factorization can hardly be obtained in other parts of the FEM analysis. This means that the portion of computing time due to factorization which in the past (on main–frames) was about 50 - 80% of the total solution time for linear calculations decreases to 10 – 20% on large vector computers.

5 Conclusions

The developed and tested block equation solver for non–symmetric matrices proves to be very efficient. The solver needed hardly more time than the corresponding solver for symmetrical systems due to the possible efficient use of system routines as well as due to the general architecture of vector computers.

It could be shown that the mathematical formulation as well as the program technique has a tremendous influence on the efficiency of vector computers. If these effects are taken into account for the program development it will be possible to economically analyse sophisticated problems with physical and geometrical nonlinearities.

References

Braun, K. A. (1976): Some Hypermatrix Algorithm in Linear Algebra. Lecture Notes in Economics and Mathematical Systems, Volume 134, Computing Methods in Applied Sciences and Engineering. Proceedings 1975. VIII, Springer Verlag Berlin Heidelberg New York.

Calahan, D.A. (1980): A Block-Oriented Equation Solver for the CRAY-1. SEL Report 136, System Engineering Laboratory, University of Michigan, Ann Arbor.

Dennis, J.E./ Schnabel, R.B. (1983): Quasi-Newton Methods for Nonlinear Problems. Prentice Hall Englewood Cliffs, N.J.

Dongarra, J.J. et al. (1979): LINPACK User's Guide, SIAM Philadelphia.

Monroe, D. M. (1982): Fortran 77, Edward Arnold Ltd, London. Support. ELSEVIER, Amsterdam Oxford New York Tokyo.

Gentzsch, W. (1987): Vektorisieren auf der CRAY-2. RUS Benutzer Informationen 5/87, 1 – 4.

Schad, H. (1985): Computing Costs for FEM Analysis of Foundation Engineering problems and Possible Ways of Increasing Efficiency. Int. Journal for Numerical and Analytical Methods in Geomechanics, Vol. 9, 261 – 275.

Smoltczyk, U./ Breinlinger, F. (1987): The FEM as Design Aid for Tunnels Built in Open Cuts. Young Geotechnical Engineers Conference, YGEC-1987, Copenhagen.

Sommer, H. (1986): Kombinierte Pfahl-Plattengründungen von Hochhäusern im Ton. Vorträge der Baugrundtagung 1986 in Nürnberg, 391 – 405.

Numerical Methods in Geomechanics (Innsbruck 1988), Swoboda (ed.)
© 1988 Balkema, Rotterdam. ISBN 90 6191 809 X

Some computational experiences of a geomechanical benchmark in rock salt

D.Piper & N.C.Knowles
W.S.Atkins Engineering Sciences, Epsom, Surrey, UK

ABSTRACT: Research into the geological storage of radioactive waste in salt formations has been active for nearly two decades. Geomechanical aspects have been studied extensively using finite element methods and in consequence a number of programs have been developed. The paper is based on a benchmark exercise to compare the predictive abilities of a number of these programs which was conducted under the auspices of the CEC's research program into radioactive waste management and storage.

The exercise has been under way for some 3 years and has progressed from program verification in the initial stage, to the latest benchmark which involves comparison of computational predictions with measured in-situ behaviour. Interest centres around the various constitutive models for rock salt. The paper summarises the project to date and presents some results from the latest benchmark.

1 INTRODUCTION

Research studies dealing with the geomechanical behaviour of repositories for radioactive waste in rock salt have now been pursued for nearly two decades. Such studies rely increasingly on fairly sophisticated computer calculations to predict the long term stress and deformation history within the host strata and a number of computer programs have now been developed [1]. Project COSA (COmparison of computer codes for SAlt) was organised to compare the capabilities of these codes.

It was set up by the Commission of the European Communities as part of its research programme into "Management and Storage of Radioactive Waste" following a meeting of CEC research contractors in December 1983 [2]. The first phase (COSA I) was completed in July 1986, the second phase (COSA II) commenced in November 1986 and will run for two years. Ten organisations have participated to date, with the British engineering consultants WS Atkins Engineering Sciences (formerly Atkins R&D) of Epsom acting as the non-participating (and hopefully impartial) coordinator and scientific secretary. The participants in COSA I were:

In Belgium: The company FORAKY, in association with the Centre d'Etude de l'Energie Nucleaire (CEN/SCK) and the Laboratoire de Genie Civil of the University of Louvain-la-Neuve (LGC).

In Germany: The Rheinisch-Westfalische Technische Hochschule (RWTH), Aachen; The Gesellschaft fur Strahlen-und Umweltforschung (GSF), Braunschweig; The Kernforschungszentrum Karlsruhe (KfK), Karlsruhe.

In Denmark: The Engineering Department of the RISØ Laboratory (RISØ), Roskilde.

In France: The Laboratoire de Mecanique des Solides of the Ecole Polytechnique (LMS), Palaiseau; The Centre de Mecanique des Roches of the Ecole des Mines (EMP), Fontainebleau; The Laboratoire d'Analyse Mecanique des Structures, Departement des Etudes Mecaniques et Thermiques, of the CEA-Saclay (CEA-DEMT).

In Italy: The Istituto Sperimentale Modelli e Strutture (ISMES), Bergamo.

In the Netherlands: The Energieonderzoek Centrum Nederlands (ECN), Petten.

With the exception of RISØ, the participants in COSA II are the same. In addition, the Technical University of Delft, Holland, has been engaged to supply expert advice as necessary on salt rheology.

In all, 13 "mechanical" computer codes have been used together with complimentary or associated codes for thermal calculations. All but one (CYSIPHE) are based on the finite element method. They range from relatively small special purpose programs specifically developed for this generic problem (e.g. GOLIA, CREEP) to large scale general purpose codes which are commercially available (e.g. ADINA, ANSYS). All participants have considerable experience of geomechanical finite element analysis and are regarded as "expert users" of their particular codes. The participants and the codes used are listed in Table 1.

Table 1. Participants and codes.

Participant	Country	Thermal Code	Mechanical Code
ECN	Holland	ANSYS	GOLIA MARC ANSYS
RISØ	Denmark		ADINA
RWTH	W Germany	FAST-RZ	MAUS
KfK	W Germany	ADINAT	ADINA
FORAKY, CEN/SCK & LGC	Belgium	HEAT	CREEP FLORA
LMS	France	ASTHER	ASTREA
CEA	France	DELFINE	INCA
EMP	France	CHEF	VIPLEF CYSIPHE
ISMES	Italy	GAMBLE*	GAMBLE*
GSF	W Germany	ANTEMP	ANSALT

* N.B. GAMBLE is a pseudonym for a general purpose commercial code.

2 ORGANISATION OF THE EXERCISE

"Benchmarking" is an increasingly popular means of improving confidence in the reliability of complex computer predictions of situations where an accepted solution is not available (e.g. from experimental or direct measurement). The term has a range of connotations; the key element being that a common problem is "solved" by several computer codes and the results are compared. Project COSA was intended to progress from a verification exercise in its first stage to more complex exercises in later stages where codes are asked to predict real life behaviour. Accordingly, three benchmark problems have been planned, of which two have been completed.

Benchmark 1 was deliberately chosen to be a simple hypothetical creep problem, complex enough to demonstrate the strengths and weaknesses of the particular codes, yet sufficiently well-posed for there to be semi-analytical solutions available for comparison. It was intended principally to assess numerical and mathematical performance, with particular emphasis on the temporal integration capabilities of the codes, and is documented elsewhere [3].

Benchmark 2 was based on a laboratory experiment, the intention being that the codes would attempt to replicate the observed behaviour. Interest nevertheless still centred on code performance and capability. For this reason the modelling of the problem was fully discussed a priori by the participants and a mutually agreeable "common" specification of the problem to be solved was drawn up by the coordinators. This specification defined the geometry, boundary conditions, loading history and material law. It was left to the individual participants to decide on the most appropriate spatial and temporal discretisation (i.e. number, type and arrangement of elements, time step and solution method) and other aspects peculiar to their codes. In short, all participants were to solve the same mathematical model in a manner most appropriate to the codes available to them.

A common mathematical model means, of course, that only those capabilities that were available in all codes could be used. The "quality" of the solutions is thus limited by the approximations necessary to allow it to be modelled by the "weakest" code. For this reason, attention focussed more on comparing the individual computer predictions with each other, rather than with the experimental behaviour.

Benchmark 3, the current exercise, is based on a series of large scale, relatively long term in-situ tests in salt. The aim here is to allow participants greater freedom to model the problem as they see fit, and to permit interpretation and characterisation of the material constitutive behaviour by individual participants, in a form suitable for their own computer codes.

3 BENCHMARK 2

3.1 Experimental details

The small-scale test which provided the basis of Benchmark 2 was conducted in the Mining Department of the Technical University of Delft, Holland. It consisted of a salt cube, with side 300mm, subjected to triaxial compression and heated internally by a heater placed in a cylindrical axial hole. A schematic diagram of the experiment is given in Figure 1. The salt was sampled from the "Na2" seam at the Asse mine in West Germany which has been the subject of considerable rheological study for several years and as such was familiar to most participants.

The test duration was 13.5 hours. Temperatures were monitored by 12 thermocouples placed in a plane of symmetry of the cube, and along the borehole wall. Convergence of the borehole was measured using an optical system sighted along the borehole axis, and external deformations of the cube were also recorded. The maximum platen force applied was 2500kN, and the maximum heater power was 1000 Watts. Full details of the experiment are given in [3].

Participants modelled the experiment in two parts. First, the transient temperature distributions were calculated up to the full time of 810 minutes. Predictions were compared with experimental measurements and then, satisfied that the thermal behaviour was represented adequately, creep calculations were conducted up to 600 minutes.

An axisymmetric representation was chosen. This was necessary because several codes could not model generalised plane strain (i.e. the condition that the strain normal to the plane is uniform), which was the other possible 2-D representation (3-D models were not actively considered on grounds of cost).

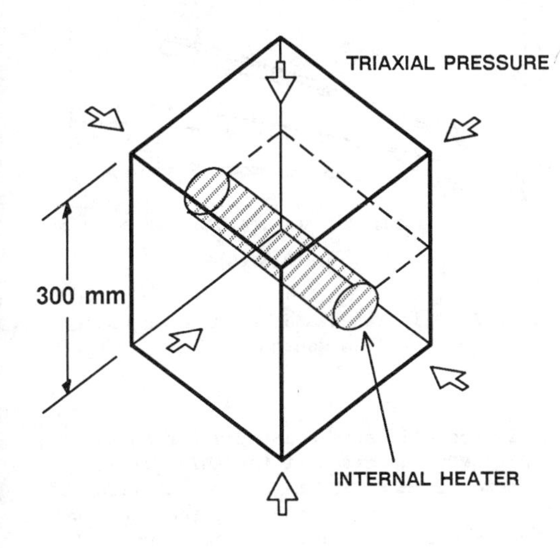

Figure 1. Schematic diagram of Benchmark 2 laboratory test.

3.2 Creep analysis

The basic rheological properties of salt were obtained from previous research at Asse in the form of primary and secondary creep equations. Primary (transient) creep dominates over the short duration of the test and this was described by

$$\dot{\epsilon}_{cr} = mB \exp(-mt)\, \sigma^n \exp(-Q/RT)$$

with $m = 0.35$ /day, $B = 0.21$,
$n = 5$, $Q = 44.8 \times 10^3$ J

However, some codes could handle only secondary type creep laws and the primary creep equation was therefore approximated to a time-independent form by setting $m = 0$.

3.3 Results

The prediction of the temperature field proved to be reasonably straightforward and there was generally close agreement between codes and also with the experimental measurements, particularly at the central (symmetry) plane. Solution of the creep part of the benchmark was more difficult and generally required considerably more computing resources than had been anticipated.

Displacement results for ten solutions are illustrated in Figure 2, showing the borehole convergence at position M near the central plane. Similarly wide variations were evident in the predictions of displacement at the other positions and also in stresses.

Detailed examination showed that "user errors", and in one case a "code discrepancy", account for many of the differences. Subsequently, a number of corrected analyses were performed and far better agreement between the individual solutions was obtained as shown in Figure 3.

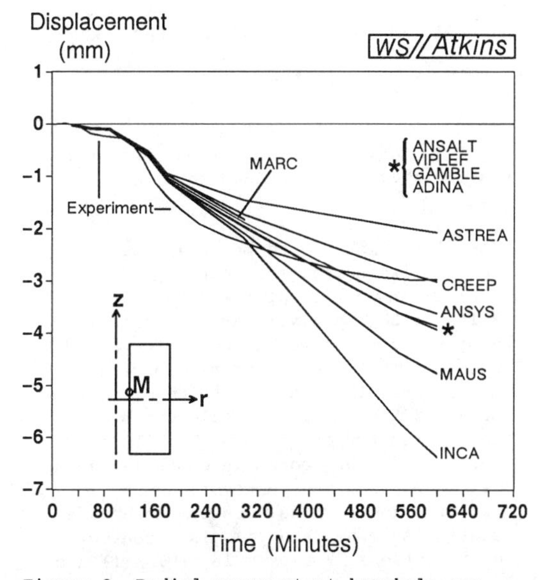

Figure 2. Radial movement at borehole surface (position M) - initial results.

4 BENCHMARK 3

The third benchmark is based on a series of experiments performed in the 300m dry-drilled borehole in the Asse mine, West Germany [4]. Specifically, three experiments are being considered, namely the isothermal free convergence of the borehole (IFC), the first heated pressure probe experiment (HPP I) and the first heated free convergence probe (HFCP I).

The borehole was drilled from the 750m level in the Asse mine with a nominal diameter of 315mm. The location of the three experiments in the 300m hole is given in Figure 4, along with the dates of each experiment.

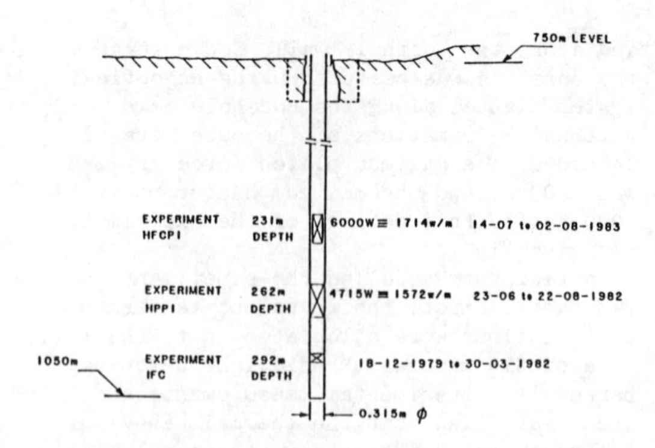

Figure 4. Location of the experiments in the 300m borehole.

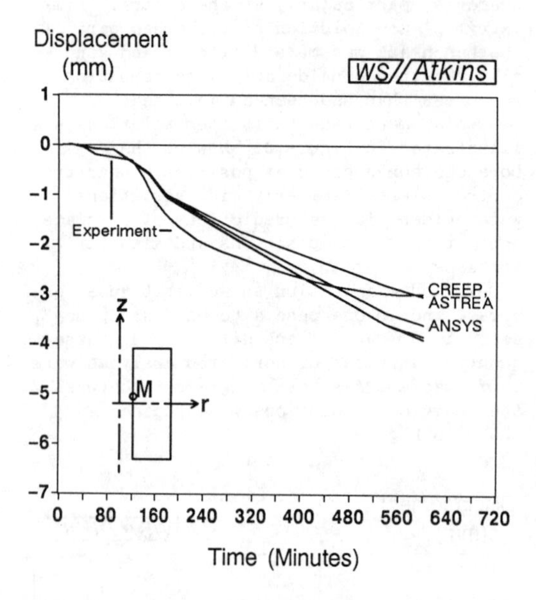

Figure 3. Radial movement at borehole surface (position M) - revised results.

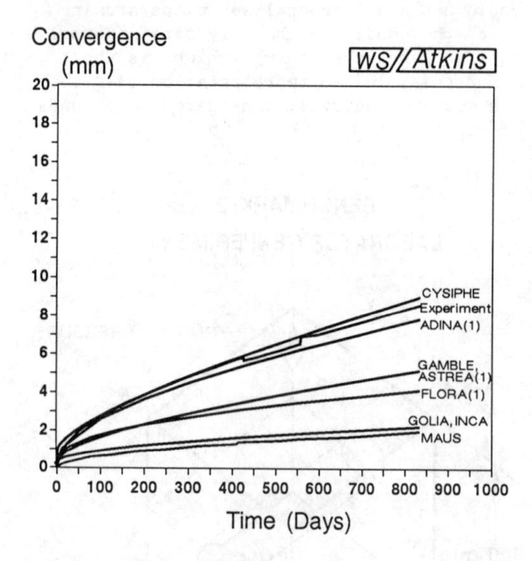

Figure 5. Calculated and measured borehole convergence at 292m depth.

The first task of the benchmark was to model the isothermal free convergence (IFC) of the lower portion of the borehole. The results of this analysis are compared with the measured convergence in Figures 5 and 6. The spread in the calculatons can be attributed to variations in the choice of constitutive law and the lithostatic state of stress. From the overburden, the lithostatic stress at the 292m borehole depth is calculated to be 22.4 MPa. However, there is evidence of a lower value, due to the excavations in the mine, and this is borne out by the predictions. For example, ADINA(1) and (2) use the same secondary creep law, but the lithostatic stress assumed for ADINA(1) is 20.4 MPa whereas that for ADINA(2) is 22.4 MPa. GAMBLE, ASTREA(1), GOLIA, INCA and MAUS also use this secondary creep law but with the creep coefficient reduced by half. GAMBLE and ASTREA(1) assume a stress of 22 MPa whilst the others use 17 MPa. CYSIPHE on the other hand employs a strain hardening law (due to Lemaitre), in an attempt to replicate the primary creep behaviour, with a lithostatic stress of 22 MPa.

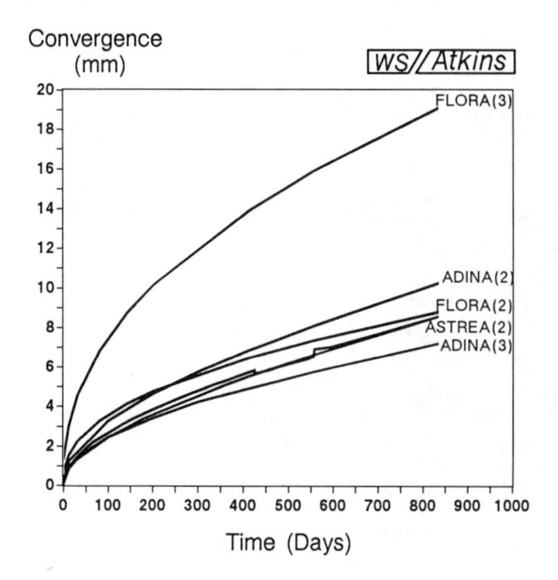

Figure 6. Calculated and measured borehole convergence at 292m depth.

5 CONCLUDING DISCUSSION

Estimates of the behaviour of repositories for radioactive waste necessarily rely on computer predictions, given the timescales involved. Confidence in long term predictions can only be based on comparisons with short term experiments and the degree of consistency between alternative computer calculations. The reliability of such calculations is therefore vital in the context of nuclear waste disposal.

Benchmark 1 showed that all codes were capable of reproducing satisfactory results for simple hypothetical problems. The same cannot be said about prediction of more complex realistic situations.

It was clear from the outset that the simplifying assumptions required to model Benchmark 2 would preclude the successful prediction of the experimental results. The departure of the calculations from the measured data shown in Figures 2 and 3 is therefore to be expected. Of greater significance is the wide variation in the computational results, as initially reported to the coordinators. Several participants produced broadly similar results and there is an implication that this "common" solution is the correct one for the specified model. However, all solutions are plausible in isolation and it appears that independent assessment by several teams will be required of complex non-linear phenomena.

In Benchmark 3, participants have been given increased latitude to model the problem according to the dictates of their experience and capabilities of their codes. For the isothermal free convergence, a range of solutions was obtained (Figures 5 and 6). While "analyst errors" similar to those in Benchmark 2 cannot be excluded, it appears that the variation is partly due to a lack of knowledge of the lithostatic stress. This illustrates the difficulty of making "blind" predictions of in-situ behaviour and emphasises the need for reliable stress (and other) measurements for input to the models. It will be interesting to observe how the participants respond to the lessons of the IFC in the remaining stages of Benchmark 3.

ACKNOWLEDGEMENTS

It is a pleasure to acknowledge the enthusiastic support and continuing assistance of Mr. B. Come of the CEC. In addition, the authors are indebted to all participants for their co-operation and generosity in submitting their work for critical comparison.

This paper is published with the permission of the Directors of WS Atkins Engineering Sciences.

REFERENCES

[1] T.W. Broyd, R.B. Dean, M.McD. Grant, G.D. Hobbs, N.C. Knowles, J.M. Putney and J. Wrigley, A directory of computer programs for assessment of radioactive waste disposal in geological formations, CEC, EUR Report 8669 EN (1985).

[2] B. Come, ed., Computer modelling of stresses in rock, CEC, EUR Report 9355 EN (1985).

[3] M.J.S. Lowe and N.C. Knowles, The Community Project COSA: Comparison of geomechanical computer codes for salt, CEC, EUR Report 10760 EN (1986).

[4] J. Prij, D. Jansen, W. Klerks, G.B. Luyten, A. de Ruiter and L.H. Vons, Measurements in the 300m deep dry-drilled borehole and feasibility study on the dry-drilling of a 600m deep borehole in the Asse II salt mine, CEC, EUR Report 10737 EN (1986).

Numerical Methods in Geomechanics (Innsbruck 1988), Swoboda (ed.)
© 1988 Balkema, Rotterdam. ISBN 90 6191 809 X

Improved algorithm for non-linear analysis by the finite element method

P.Humbert
Numerical Models Section, Laboratoire Central des Ponts et Chaussées, Paris, France
P.Mestat
Foundations Section, Laboratoire Central des Ponts et Chaussées, Paris, France

ABSTRACT: This paper reviews and describes some basic resolution methods for non linear finite element analysis using the displacement formulation. These methods are applied to elastoplastic analysis in small deformations. A number of applications are presented to show that significant improvements can be made using the D-F-P method combined with the initial stress method : these applications show that convergence is obtained three or four times faster than with the initial stress method.

The D-F-P scheme used appears as the most accurate, stable and time saving solution. Moreover, it is very simple to implement it in an existing finite element code.

INTRODUCTION

The development of finite element methods allows engineers to design many structures in any environment where their behaviour is tridimensional and highly non linear.

Many resolution procedures have been developed in structural non linear analysis. Most of these procedures require a large amount of computational time : hence, it is of considerable importance to obtain an efficient and economic resolution method in non linear analysis.

After a general review of resolution methods, we present a very simple procedure which is based on three approaches :

(i) the initial stress method [1]
(ii) the arc-length method [2]
(iii) the updated D-F-P method [3], [4].

A combination of these techniques gives a very efficient method for solving non linear problems.

Several applications will be presented that allow comparison with classical acceleration methods.

BASIC FORMULATION

Incremental methods in conjunction with some efficient iterative procedures are among the most popular resolution technique used.

The incremental finite element equations that govern the response of the finite element system in static analysis are given by K.J. Bathe [5]

$$\psi\left(u(t),\ \lambda(t)\right) = R\left(u(t)\right) - \lambda(t).\ P = 0 \quad (1)$$

where

$u(t)$ = vector of nodal point incremental displacements at time t;

$R(u)$ = vector of nodal point forces corresponding to the internal element stresses at time t;

$\lambda(t)$= scalar loading parameter at time t;

P = time-independant vector of external applied loads ;

ψ = vector of residual forces or out-of-balance forces.

The resolution of the equilibrium system (1) is based on linearization. Let (u_o,λ_o) a couple of nodal displacements u_o and scalar load factor λ_o not necessarily satisfying the equilibrium :

$$\|\phi(u_o, \lambda_o)\| \neq 0 \qquad (2)$$

Let further $\delta u = u - u_o$ and $\delta\lambda = \lambda - \lambda_o$ denote the displacement and loading parameter corrections. A linearization of $\psi(u_o + \delta u, \lambda_o + \delta\lambda)$ leads to the linear approximation :

$$\psi(u,\lambda) = B.(u-u_o) - (\lambda-\lambda_o).P + \psi(u_o,\lambda_o) \qquad (3)$$

where B is an n x n matrix if n is the number of equations in the equilibrium system (1).

The solution of $\psi(u,\lambda) = 0$ is used as a new approximation (u,λ) of the true solution, if it exists.

We can formulate the iterative procedure; let (u_i, λ_i) be a set at iteration i, if all matrices B_k for $o \leqslant k \leqslant i$ are inversible, the displacement u_{i+1} at iteration i+1 is given by

$$u_{i+1} = u_i - B_i^{-1}\left(\psi(u_i,\lambda_i)\right) - (\lambda_{i+1}-\lambda_i).P) \qquad (4)$$

Use of equations (4) involves computation of n components of $\psi(u,\lambda)$, of n^2 components of matrix B_i and the inversion of a linear equation system.

The system (4) has n+1 unknowns for n equations ; consequently, for construction of the set (u_{i+1}, λ_{i+1}), we need another scalar equation between u_{i+1} and λ_{i+1}. One of the following three relations can be used :

(i) <u>the simplest relation</u> :

$$\delta\lambda = 0 \qquad (5)$$

In this case, the incremental load is imposed and fixed for all the iterations. It is the classical iterative procedure for monotonic loading. (see fig. 1 for a one-dimensional representation).

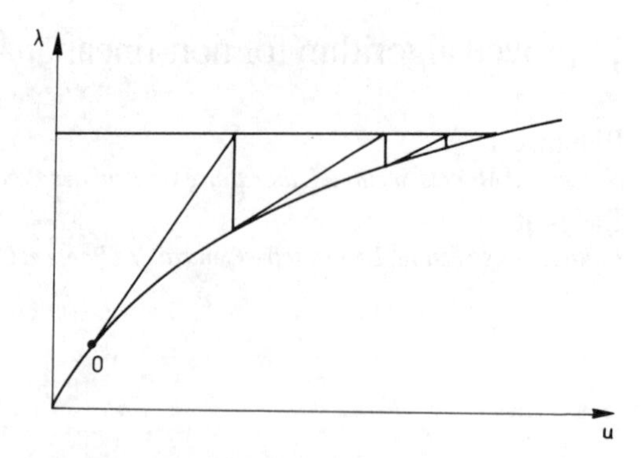

Fig. 1 : $\delta\lambda = 0$

(ii) <u>the linear relation</u> :

$$\delta\lambda = a^T . \delta u \qquad (6)$$

The incremental load varies linearly at each iteration : the linear relation was developped by Riks [6] and Wempner [7].

(See figure 2 - for a one-dimensional representation). a is a constant vector for each increment or iteration.

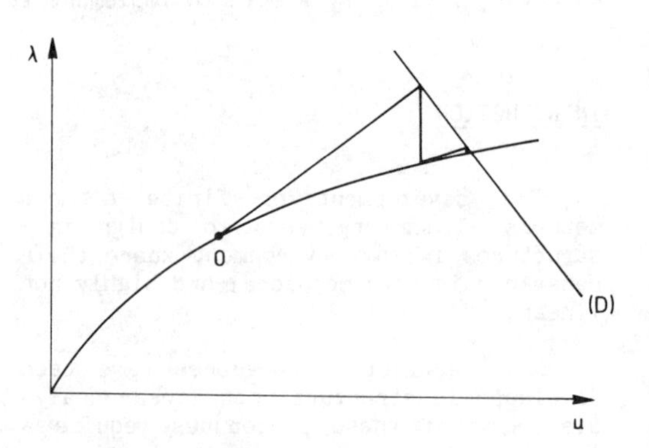

Fig. 2 : $\delta\lambda = a^T . \delta u$

(iii) <u>the elliptic relation</u>

$$(u-u_o)^T (u-u_o) + (\lambda-\lambda_o)^2 p^T.p = L^2 \qquad (7)$$

where (u_o, λ_o) is a converged solution for a former increment and L a prescribed length, called arc-length (see figure 3 - for a one-dimensional representation). This method was developed by M.A. Crisfield [2].

The last relation combined with a resolution method gives a very powerful method for solving non linear problems, in particular those with a peak in the stress-strain law.

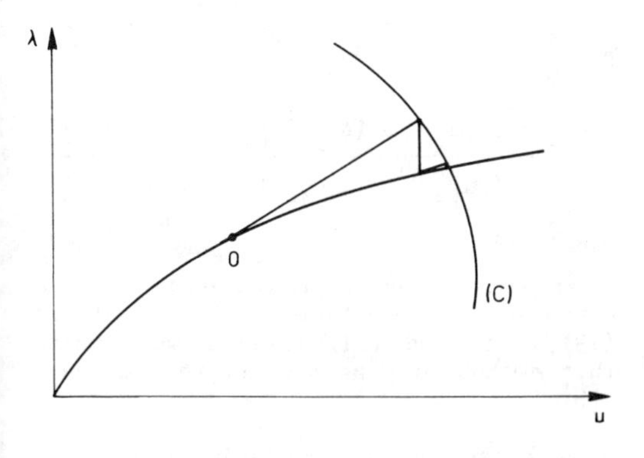

Fig. 3

For simplicity in further developments, we assume $\delta\lambda = 0$. Thus for each iteration, we compute u_{i+1} as

$$u_{i+1} = u_i - B_i^{-1} \cdot \psi(u_i) \qquad (8)$$

Construction of matrix B_i or B_i^{-1} .

A particular form of the linearization method is obtained by setting

$$B_i = \frac{\partial\psi}{\partial u} (u_k) \qquad (9)$$

where k is an integer less than or equal to i.

k = i gives the Newton-Raphson method of tangent stiffness method,

k < i gives the modified Newton-Raphson method,

k = o gives the initial stress method (a particular modified Newton-Raphson method).

The principal disadvantages in using the tangent stiffness method are : first, by the difficulty for computation of n^2 partial derivatives in (9) and secondly, the large amount of computational time required to assemble a new stiffness matrix and solve again the system of equations for each iteration i.

Hence, it is of considerable importance to obtain an effective and economic method for solving non linear problems. Among the proposed methods, we have considered :

i) the initial stress method combined with an acceleration procedure [1], [8]

ii) the updated methods (Broyden, B-F-G-S, D-F-P ...) [3] , [4] .

The initial stress method

The basic idea is to conserve in each iteration i, the initial tangent stiffness matrix B_o ; for such an approach, B_o is decomposed as :

$$B_i (u_i) = B_o - Kp (u_i) \qquad (10)$$

Kp is the non linear part of matrix B_i.

Substituting (10) into (8), we obtain with $\delta u_i = u_{i+1} - u_i$:

$$B_o \cdot \delta u_i = \psi (u_i) + Kp (u_i). \delta u_i \qquad (11)$$

The initial stress method developed by Nayak and Zienkiewicz [1] consists of a simple approximation for the last term in (11)

$$Kp(u_i). \delta u_i \simeq Kp (u_i).\delta u_{i-1} \qquad (12)$$

Thus, the system of equations for the initial stress method becomes :

$$B_o.\delta u_i = \psi(u_i)+Kp(u_i). \delta u_{i-1} = \phi(u_i) \qquad (13)$$

where δu_{i-1} is the vector of displacements during iteration i and u_i the vector of total displacements at the end of iteration i.

The initial stress approach is the most flexible method for non linear analysis but it often involves slow convergence. Thus, many authors have proposed acceleration procedures for improving the convergence of the iteration process.

Procedure with convergence acceleration

Among the proposed techniques, we have selected three convergence accelerations :

. line search method [2]

$$u_{i+1} = u_i + \alpha. \delta u_i \qquad (15)$$

. secant method [8]

$$u_{i+1} = u_i + \beta \, \delta u_{i-1} + \delta u_i \qquad (16)$$

. line search + secant method [8]

$$u_{i+1} = u_i + \alpha\,(\beta\,\delta u_{i-1} + \delta u_i) \qquad (17)$$

α and β are two scalar parameters ($\alpha = 1$ and $\beta = 0$ corresponds to the initial stress method without acceleration).

The following relations allow their computation :

(i) <u>expression of parameter α</u>

α can be determined by considering the ratio of the external work to the internal work of the system [8] :

$$\alpha = \frac{\text{External work}}{\text{Internal work}} = \frac{\phi(u_i^T).\delta u_i}{[\phi(u_i)-\phi(u_i+\delta u_i)].\delta u^T} \qquad (18)$$

(ii) <u>expression of parameter β</u>

(17) can be written as :

$$u_{i+1} = u_i + \delta u_i^* \qquad (19)$$

with $\delta u_i^* = \alpha(\beta\,\delta u_{i-1} + \delta u_i)$ and α is given by (18)

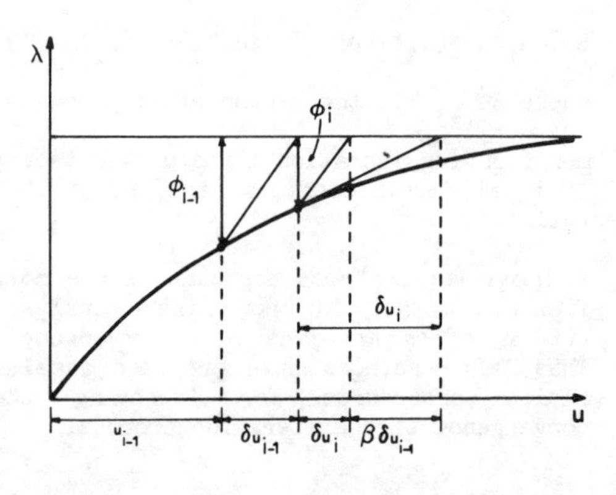

Fig. 4 - Basic secant procedure

Using the notations in figure 4 (one dimensional representation), the secant condition is :

$$(\delta u_i)^{*T}.[\phi(u_{i-1})-\phi(u_i)] = (\delta u_{i-1})^T.\phi(u_i) \qquad (20)$$

(18) and (20) give the value of β knowing α :

$$\beta = \frac{1}{\alpha}\,.\,\frac{(\delta u_{i-1})^T.\phi(u_i) - \alpha\,(\delta u_i)^T.[\phi(u_{i-1})-\phi(u_i)]}{(\delta u_{i-1})^T.\,[\phi(u_{i-1}) - \phi(u_i)]} \qquad (21)$$

Finally, for each iteration i, n+2 unknowns must be determined from equations (13), (18) and (21) (n+3 if we combine this method with an arc length procedure [8]).

An other idea for replacing the costly evaluation of the tangent stiffness matrix B_i is to update the matrix.

<u>The update methods</u>

The basic idea is to evaluate B_i or B_i^{-1} by an economic approximation.

Three updated methods are used : i) Broyden's method (1965) ; ii) the Davidon-Fletcher-Powell method (D-F-P) ; iii) the Broyden-Fletcher-Goldfarb-Shanno method (B-F-G-S) [3], [4] .

All the updated methods are based on the following sequence. Let w_i and v_i defined by,

$$w_i = u_{i+1} - u_i \qquad (22)$$

$$v_i = \phi(u_{i+1}) - \phi(u_i)$$

The classic iterative procedure leads to :

$$u_{i+1} = u_i - B_i^{-1}.\,\phi(u_i) \qquad (23)$$

The matrix B_i or B_i^{-1} is updated as

$B_i = B_{i-1} + \Delta B_i$; ΔB_i is computed from B_{i-1} to satisfy the quasi-Newton equation :

$$B_i\,.\,(u_{i+1} - u_i) = \phi(u_{i+1}) - \phi\,(u_i) \qquad (24)$$

B_i can be considered as an approximation of the tangent stiffness matrix resulting of the linearization in (1). Of course, it is more efficient to update B_i^{-1} directly ; in general, the construction of B_i from (24) is a particular case of following problem;

Knowing a nxn definite positive matrix $K \in \mathbb{R}^{nxn}$ and two vectors $w, v \in \mathbb{R}^n$

compute the matrix A or A^{-1} as

$$v = A.w = (K + \Delta A)\, w \qquad (25)$$

The choice is large and many formulas have been derived. But if we impose rank $(\Delta A) = 1$, the solution becomes unique and we obtain the Broyden's formula [3] : A^{-1} is given by,

$$\text{if } w^T.\, K\, v > 0 \qquad A^{-1} = K + \frac{(w - K\, v)\, w^T K}{w^T.K\, v} \qquad (26)$$

For the particular case of a symmetric tangent stiffness matrix, two updated methods of rank 2 for ΔA have been developed :

(i) B-F-G-S method , if $u^T . v > 0$

$$A^{-1} = K + \frac{ww^T}{w^T.v} - \frac{wv^T K}{w^T.\, v} - \frac{Kvw^T}{w^T.\, v} + \frac{v^T.\, Kv\, ww^T}{w^T.\, v} \qquad (27)$$

(ii) the D-F-P method, if $(w - Kv)^T.v > 0$

$$A^{-1} = K + \frac{(w - Kv)\,(w - Kv)^T}{(w-Kv)^T.v} \qquad (28)$$

It is important to note that in (26), (27), (28) K can be any definite positive matrix.

Programmation of the updated methods

The implementation of the updated methods in a finite element code requires for each iteration i the storage of one matrix B_i (Broyden, D-F-P) or the storage for each iteration $k < i$ of 2 n-vectors (for B-F-G-S in factorisation form [4], [5]).

But, the great disadvantage of these methods (Broyden - D-F-P) is the destruction of the banded nature of the matrix B_i : so, the storage becomes quickly considerable. On the other hand, for B-F-G-S, a continued application leads to the storage of more and more vectors.

In order to avoid the storage, Crisfield has proposed to set $K = B_o^{-1}$ for all iterations in (26), (27) or (28). B_o is the initial stiffness matrix.

The implementation in finite element codes can be achieved, taking the D-F-P method as can for example, in the following way :

. store $\delta u_{i-1} = B_o^{-1} . \psi(u_{i-1})$ (29)

. compute $\delta u_i = B_o^{-1} . \psi(u_i)$ (30)

. compute $v_i = \psi(u_{i-1}) - \psi(u_i)$ (31)

$$u_{i+1} = u_i + B_o^{-1}\, \psi(u_i) + \frac{(\delta u_i - B_o^{-1}\, v_i)\,(\delta u_i - B_o^{-1}\, v_i)^T\, \psi(u_i)}{(\delta u_i - B_o^{-1}\, v_i)^T\, v_i} \qquad (32)$$

The computational effort is reduced to one initial stress method iteration (δu_i), one storage (δu_{i-1}) and two scalar products.

In our finite element code CESAR-LCPC, the implementation of the D-F-P method needs 10 Fortran instructions lines for computation of u_{i+1} by (32).

Elastoplastic resolution

The numerical Nayak-Zienkiewicz scheme [1] is employed in the elasto-plastic analysis : the integration of the constitutive relations governing the material behaviour is achieved by an explicite method with a sub-incrementation technique.

Numerical examples

In this section, we present 4 examples that illustrate the efficiency of the modified D-F-P method by comparing it with others approaches.

The convergence of the discrete problem is measured with three criteria relevant respectively to :

(i) the norms of displacements

$$\varepsilon^D = \frac{\|u_i - u_{i-1}\|}{\|u_i\|} \qquad (33)$$

(ii) the norms of residual forces

$$e^R = \frac{\| \phi (u_i) \|}{\| \phi (u_o) \|} \tag{34}$$

(iii) the energy

$$\varepsilon^T = \frac{\left| \phi (u_i)^T . (u_{i+1}) - (u_i) \right|}{\left| \phi (u_o)^T . (u_i - u_o) \right|} \tag{35}$$

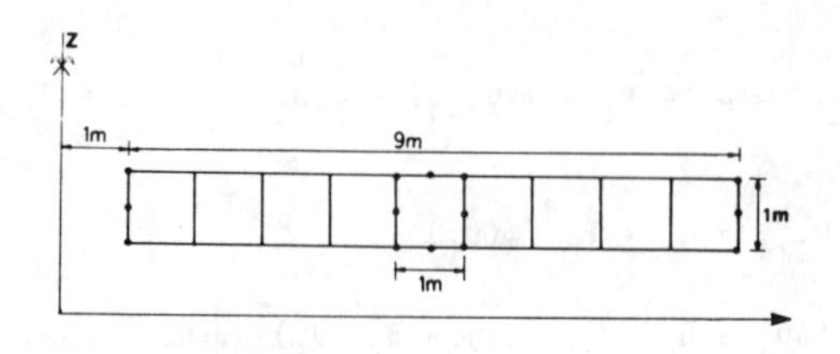

Fig 5 - Finite element mesh

For all these criteria, we impose $\varepsilon \leq .001$. All the calculations have been computed on a Norsk Data MD 570.

<u>Example 1</u>

<u>Thick cylinder under internal pressure in axisymmetric strain</u>

The first example is the thick cylinder in axisymmetric studied by Owen and Marques [10].

The inner and outer radii are 1 m and 10 m respectively.

The D-F-P modified method is the best method : fast and economic.

The geometry and mesh data are defined in figure 5 : the thick cylinder is modelled using nine 8-node isoparametric elements, and reduced integration (2x2) is used.

Table I - <u>Comparative study of solution methods</u>

r	Analytical Solution	1	2	3	4	5	6	7	8	9
1.00	.19189	.19049	.19140	.19132	.19105	.19134	.19133	.19142	.19143	.19128
1.50	.12275	.12184	.12244	.12239	.12221	.12240	.12239	.12245	.12246	.12236
2.00	.08820	.08753	.08798	.08794	.08780	.08795	.08794	.08798	.08799	.08792
2.50	.06782	.06729	.06765	.06761	.06751	.06762	.06762	.06765	.06766	.06760
Number of iterations	-	152	5	25	53	16	34	20	10	14
CPU-Time (sec)	-	26.68	4.80	5.98	10.46	4.80	7.32	5.90	3.44	4.10

N.B.
1 - initial stress method without acceleration
2 - tangent stiffness method
3 - initial stress method with acceleration parameter β
4 - " " " with acceleration parameter α
5 - " " " with acceleration parameters (α,β)
6 - B-F-G-S modified method
7 - " " " + acceleration parameter α
8 - D-F-P modified method
9 - " " " + acceleration parameter α

The elastic parameters of the material are $E = 10^7$ Pa and $\nu = 0.33$; we assume a perfectly plastic Tresca behaviour with C = 600 Pa.

The Tresca yield condition can be written as :

$$F(\sigma) = |\sigma_1 - \sigma_3| - 2C$$

where σ_1 and σ_3 are the extreme principal stresses.

The loading applied is an internal pressure of 2250 Pa.(corresponding to a plastic radius of 4.357 m). The displacement u(r) after convergence under this pressure is given in table I for several iterative methods.

The efficiency of the methods can be appreciated by comparing the results with the analytical solution.

Example 2

Thick cylinder in strain hardening material

Now, we consider the same infinitely long thick cylinder, as in example 1, but we assume that the material behaviour is governed by strain hardening Tresca criterion.

The yield condition is formulated as :

$$F(\sigma, \varepsilon^P) = |\sigma_1 - \sigma_3| - f(\bar{\varepsilon}^P) \quad (36)$$

with

$$\bar{\varepsilon}^P = \int_{To}^{T} J_2(\dot{\varepsilon}^P)\, dt = \int_{To}^{T} \dot{\lambda}\, dt = \lambda \quad (37)$$

$\dot{\lambda}$ is the plastic scalar multiplier, in the normality rule

$$\dot{\varepsilon}^P = \dot{\lambda} \cdot \frac{\partial F}{\partial \sigma} \quad (38)$$

These assumptions allow to construct a simple hardening model with 2 parameters a, b :

$$F(\sigma, \varepsilon^P) = |\sigma_1 - \sigma_3| - (a \cdot \lambda + b) \quad (39)$$

An analytical solution can be found for the expansion problem of a thick cylinder under internal pressure.

Table II - Comparative study of solution methods

	Analytical Solution	Initial stress Method		D-F-P modified Method	
Applied Pressure	$u(R_1)$	Nb iterations	$u(R_1)$	Nb iterations	$u(R_1)$
1000.	$.16735.10^{-3}$	16	$.17095.10^{-3}$	4	$.17109.10^{-3}$
1500.	$.43429.10^{-3}$	34	$.42656.10^{-3}$	7	$.42769.10^{-3}$
2000.	$.11428.10^{-2}$	112	$.11431.10^{-2}$	8	$.11448.10^{-2}$
2500.	$.33526.10^{-2}$	348	$.33373.10^{-2}$	8	$.33437.10^{-2}$
3000.	$.95896.10^{-1}$	18408	$.95237.10^{-1}$	12	$.96153.10^{-1}$
3500.	.29798	20824	.29732	9	.29904
4000.	.50007	20824	.49941	11	.49967
4500.	.70216	20824	.70150	9	.70197
5000.	.90425	20824	.90359	10	.90378
CPU.Time			1H 15mn		1mn 20 sec.

An application with $f(\lambda) = 5000.\lambda + 1200.$ and an internal pressure $p = 5000$ applied in nine increments has been achieved.

The comparison between the results in table II shows that for the last loading increments the D-F-P modified method converges 2000 times faster than the standard initial stress method.

In this particular case, the B-F-G-S modified method quickly diverges.

Example 3

Geotechnical study of a tunnel in plane strain

As a third example, we consider the numerical simulation of a tunnel excavation in a homogeneous non-cohesive soil. Figure 6 represents the geometry of the finite element mesh.

The soil is assumed to be an elastoplastic Mohr-Coulomb material :

$E = 200$ MPa, $\nu = .4$, $\gamma = .02$ MN/m , $c = .5$MPa, $\varphi = \psi = 30°$ (associated flow rule).

In the first phase, only the upper half section is burraved (see figure 7).

The effect of the excavation is simulated by application of nodal forces along the boundary ABC.

After convergence, the stresses and displacements for some nodes are given in table III for the initial stress method and for the D-F-P modified method.

The numbers of iterations are :

106 for the initial stress method
12 for the modified D-F-P method.

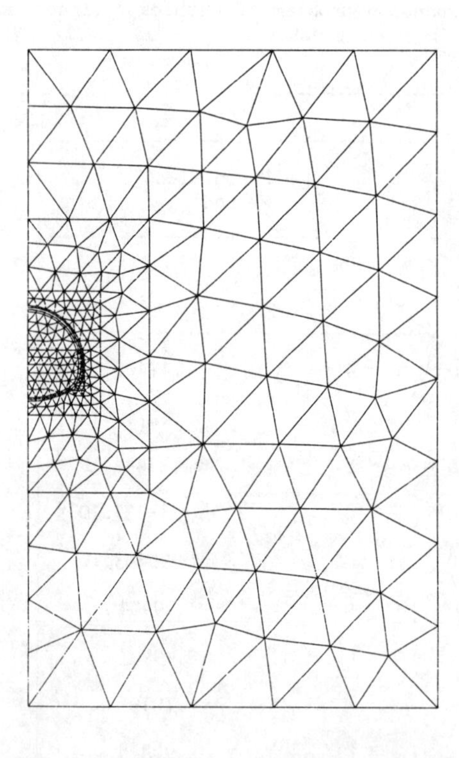

Figure 6 - Finite element mesh for the tunnel excavation simulation 1111 Nodes and 508 elements.

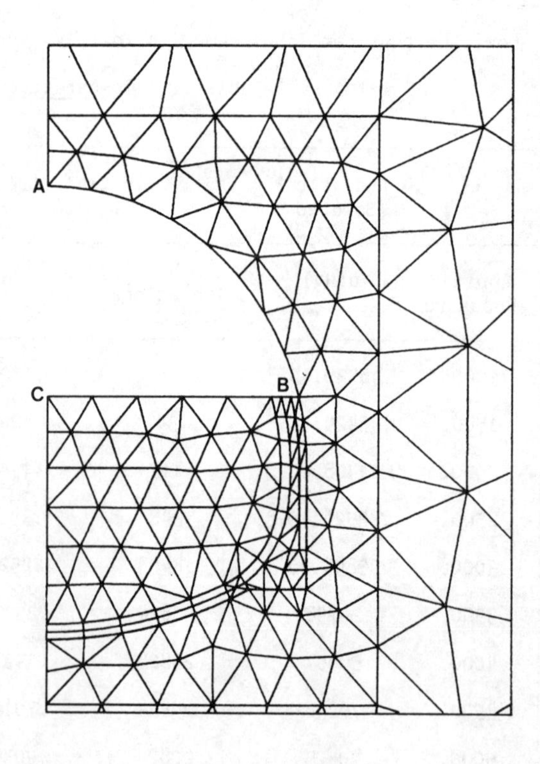

Figure 7 - Geometry for the first phase.

Table III - Comparative study of solution methods

	Elastic node		Plastic node	
Nodes	A		B	
Method	Initial Stress	D.F.P	Initial Stress	D.F.P
u	0	0	$-.2733.10^{-3}$	$-.2903.10^{-3}$
v	$-.3510.10^{-1}$	$-.3511.10^{-1}$	$.8922.10^{-3}$	$.8318.10^{-3}$
σ_1	$.5820.10^{-3}$	$.5862.10^{-3}$	$-.1096.10^{+1}$	$-.1087.10^{+1}$
σ_2	$-.6588.10^{-1}$	$-.6575.10^{-1}$	$-.3309.10^{+1}$	$-.3303.10^{+1}$
σ_{xx}	$-.6545.10^{-1}$	$-.6532.10^{-1}$	$-.1119.10^{+1}$	$-.1109.10^{+1}$
σ_{yy}	$-.1467.10^{-3}$	$.1517.10^{-3}$	$-.3286.10^{+1}$	$-.3281.10^{+1}$
σ_{zz}	$.9919.10^{-1}$	$.9925.10^{-1}$	$-.1831.10^{+1}$	$-.1825.10^{+1}$
σ_{xy}	$-.5361.10^{-2}$	$-.5351.10^{-2}$	-.2231	-.2199

All the examples presented demonstrate the efficiency of the D-F-P modified method in solving associated elastoplastic problems.

In geotechnics, though, many materials have a non associated flow rule : consequently the stiffness matrix B_i becomes non symetric, and the application of updated methods D-F-P or B-F-G-S is not adapted.

However, even the updated Broyden method and the secant method are not reliable, for non associated plasticity, as can be shown in the following example.

Example 4

Thick cylinder in non associated plasticity

We consider again the finite element mesh in figure 5. The material is assumed perfectly plastic : the Mohr-Coulomb Yield condition :

$$F(\sigma) = |\sigma_1 - \sigma_3| + (\sigma_1 + \sigma_3) \sin\varphi - 2 C \cos\varphi$$

and the plastic potential

$$G(\sigma) = |\sigma_1 - \sigma_3| + (\sigma_1 + \sigma_3) \sin\psi - 2C \cos\psi$$

We use the same material constants ; the computations are achieved with a fixed friction angle φ = 20° and with a varying dilatancy angle ψ .

The limit pressure P, independant of ψ , is applied as in the example 1 on the inner face - the displacement $u(R_1)$ is given in table IV for some resolution methods in fonction of the ψ - value.

The best results are obtained when the material is standard ($\varphi = \psi = 20°$) . But some results with D-F-P method seem to be good, when the non-associativity is weak ; other computations which are not in this paper confirm this fact.

Conclusions

All the examples presented show the great efficiency of the D-F-P modified resolution method in solving associated plasticity problems. Compared with the standard initial stress approach or with all the other approaches, this scheme is always the most economical in CPU time.

Table IV - Some results for non associated problems.
$\varphi = 20°$ P = 3684.

	ψ	0.	10.	20.	30.	40.
(1)	$u(R_1) \times 100$	1.083	.438	.315	.250	.215
	Number of iterations	16	13	11	12	13
(2)	$u(R_1) \times 100$	divergence	.512	.314	.254	.226
	Number of iterations		13	60	16	16
(3)	$u(R_1) \times 100$	.689	.425	.308	.246	.209
	Number of iterations	857	474	355	298	265
(4)	$y(R_1) \times 100$	.732	.450	.323	.256	.217

N.B. : (1) D-F-P method ; (3) Initial stress method
(2) Secant method ; (4) Analytical solution

Netherless, two major problems remain for founding a very general acceleration method ;

1) the loading-unloading problems,
2) the non associated problems.

To our knowledge, no efficient acceleration method exists, at present, for non-associated plasticity problems. Research efforts in this direction must be pursued.

Acknowledgments: A part of this research was carried out in connection with CNRS (GRECO 90, "Rheology of Geomaterials").

BIBLIOGRAPHY

[1] G.C. NAYAK and O.C. ZIENKIEWICZ, "Elasto-plastic stress analysis. A generalization for various constitutive relations including strain softening", Int. J. Num. Meth. Eng., 5, 113-135 (1972).

[2] M.A. CRISFIELD, "An arc-length method including line searches and accelerations", Int. J. Num. Meth. Eng., 19, 1269-1289 (1983).

[3] J.E. DENNIS and J.J. MORE, "Quasi Newton methods, motivation and theory". SIAM Review - Vol. 19, n°1, 46-89, January 1977.

[4] Y.S.RYU and J.S. ARORA, "Review of non linear finite element methods with substructures", Journal of Engineering Mechanics.Vol. 111, 11, 1361-1379, november 1985.

[5] K.J. BATHE, "Finite element procedures in engineering analysis", Prentice Hall Inc. (1981).

[6] E. RICKS, "The application of Newton's method to the problem of elastic stability", J. Appl. Mech., 39, 1060-1066 (1972).

[7] G.A. WEMPNER, "Discrete approximations related to non-linear theories of solids", Int. J. Solids. Struct., 7, 1581-1599 (1971).

[8] P. HUMBERT, S.H. LO and P. MESTAT, "A new acceleration procedure for finite element elastoplastic analysis", European Conference on Numerical Methods in Geomechanics. Stuttgart, September 1986.

[9] E. HINTON, D.R.J. OWEN and C. TAYLOR, "Recent advances in non-linear computational mechanics", Pineridge Press Swansea, U.K.1982.

[10] J.M.M.C. MARQUES and O.R.J. OWEN, "Some reflections on elasto-plastic stress calculation in finite element analysis", Computers and Structures, Vol. 18, 6, 1135-1139 (1984).

Numerical Methods in Geomechanics (Innsbruck 1988), Swoboda (ed.)
© 1988 Balkema, Rotterdam. ISBN 90 6191 809 X

A unified approach to the analysis of saturated-unsaturated elastoplastic porous media

B.A.Schrefler & L.Simoni
Istituto di Scienza e Tecnica delle Costruzioni, University of Padua, Italy

ABSTRACT: The paper presents a unified approach to the analysis of saturated-unsaturated porous media. The approach is based on a generalized formulation of Biot's theory. The governing equations are nonlinear due to the dependence of the material parameters on the degree of saturation. Furthermore, the soil skeleton is assumed to have an elastoplastic behaviour. The analysis involves also the tracking of the free surface. Attention is paid to the numerical problems arising when this moving discontinuity surface is within a finite element. The paper presents the governing equations and their numerical implementation. A numerical application is presented, which illustrates the capability of the computer code developed.

1 INTRODUCTION

The behaviour of unsaturated soils is a very important aspect in a large number of engineering problems. It has however received less attention than saturated soil behaviour.

Mechanics of saturated-unsaturated soils involves usually the determination of the free surface. The solution of this problem, even in uncoupled cases, leads to quasilinear differential equations which are very difficult to solve using analytical methods. Finite elements techniques were firstly applied in this area by Neuman (1975). The problem is limited to the flow field. More recently, Narasimhan and Witherspoon (1977) proposed a numerical model for simulating groundwater motion in variably saturated deformable heterogeneous porous media. This model considers a three dimensional field of fluid flow in conjunction with a one-dimensional vertical deformation of the soil which is handled according to Terzaghi's consolidation theory. An improvement of this model may be found in the work by Safai and Pinder (1980) who take into account a three dimensional deformation field coupled with a three dimensional hydrologic field. These authors neglected however some terms in the governing equations of both fields. Another numerical approach was presented by Lloret and Alonso (1980), where state surfaces were used to define the solid behaviour in the flow equation. This approach is however not coupled in the sense of Biot.

In the present paper the equations governing the saturated-unsaturated soil behaviour are obtained as a particular case of a general multiphase formulation, under the conditions that the gaseous phase is continuous in the unsaturated zone and remains at atmospheric pressure. This implies that the continuity equation for air is neglected. The used theory is a generalization of the Biot's self-consistent one, which considers fully coupled displacement and pressure fields. The resulting numerical model is also capable of handling nonlinear elastic or elastoplastic solid behaviour. In the presented formulation the problem of determining the free surface, is reduced to finding the internal isobar p_w=0 which separates the saturated from the unsaturated regions. This is very easy to solve in the finite element context. The free surface separates zones with different material properties, and its position is changing with time until steady state conditions are achieved. For these reasons the problem is non linear with moving boundaries. In such a case mesh readjustment seem to be an ideal technique. In the present paper, however, to avoid cumbersome work due to the nonlinear behaviour of the soil skeleton, mesh readjustment is substituted by a technique which makes it possible to handle finite elements with discontinuous properties.

Eventually, a numerical application is presented whose results are compared with those of a laboratory experiment. This represents a validation of the numerical model and its implementation in a computer code.

2 GOVERNING EQUATIONS

The soil skeleton is considered as a three-phase medium characterized by

(i) total stress vector σ;
(ii) displacement vector of the solid phase $\mathbf{u}$;
(iii) strain rate of the solid phase

$$d\varepsilon = \mathbf{L}^T \mathbf{du} \quad (1)$$

(iv) the liquid phase pore pressure p_w;
(v) the gaseous phase pore pressure p_g.

Even though a certain amount of purely volumetric strain is caused by pressure changes, the main deformation of the soil skeleton depends on the effective stress σ'.

$$\sigma = \sigma' - \mathbf{m}\,\bar{p} \quad (2)$$

where σ is the total stress, m is equal to unity for normal stress components and zero for the shear stress components and $\bar{p}$ is the pressure of the mixture surrounding the grains.

2.1 The fluid pressure in a multiphase flow regime

The evaluation of the fluid pressure in a multiphase flow was carried out by Lewis and Schrefler (1987) for the case of immiscible two-phase flow, using the averaging technique developed by Whitaker (1973). In the case of two phase flow, this yields

$$\bar{p} = S_w\ p_w + S_g\, p_g \quad (3)$$

where S_r indicates the saturation of either phase.

If the air is assumed to be at atmospheric pressure, which represents the reference pressure (p_{atm}=p_g=0), equation (3) reduces to

$$\bar{p} = S_w\ p_w \quad (4)$$

and hence

$$\sigma' = \sigma - \mathbf{m}\, S_w\ p_w \quad (5)$$

This statement may be assumed as an equivalent expression of the relation proposed by Bishop, McMurdie and Day (1960)

$$\sigma' = \sigma - \mathbf{m}\,\chi\, p \quad (6)$$

where Bishop's coefficient χ, usually determined by experiment, is assumed equal to the degree of saturation S_w. This assumption is valid for a series of soils as can be seen in Bishop and Blight (1963).

In this way the moisture suction is only partly converted to mechanical stress for soils of moderate to high saturation.

2.2 Deformation of the soil skeleton

The equilibrium equation of the soil is now derived taking into account an elastoplastic behaviour of the medium. The costitutive equation relating the effective stress σ' to the strain of the skeleton can be written in tangential form for a general nonlinear material, thus allowing plasticity to be incorporated.

$$d\sigma' = \mathbf{D}_T\,(d\varepsilon - d\varepsilon_c - d\varepsilon_p - d\varepsilon_o) \quad (7)$$

where the tangent matrix $\mathbf{D}_T$ is dependent on the level of the effective stress and also, if strain effects are considered, on the total strain of the skeleton; ε_c is the creep strain of the skeleton; ε_p is the volumetric strain caused by uniform compression of the particles and ε_o the initial strain.

The numerical model, as it is implemented in the computer code, can handle various costitutive models reported in Lewis and Schrefler (1987).

The equilibrium equation, in incremental form, relating the total stress σ to the body forces $\mathbf{b}$ and to the boundary tractions $\hat{\mathbf{t}}$ specified at the boundary Γ of the domain Ω is formulated in terms of the unknown displacement vector $\mathbf{u}$. Using the principle of virtual work and incorporating the effective stress relationship (2) and the constitutive relationship given in eq. (7) the general equilibrium statement can be written as

$$\int_\Omega \delta\varepsilon^T \mathbf{D}_T \frac{\partial \varepsilon}{\partial t} d\Omega + \int_\Omega \delta\varepsilon^T \mathbf{D}_T \mathbf{m} \frac{\partial \bar{p}}{\partial t} \frac{1}{3K_s} d\Omega - \int_\Omega \delta\,\varepsilon^T \mathbf{m} \frac{\partial \bar{p}}{\partial t} d\Omega$$

$$- \int_\Omega \delta\varepsilon^T \mathbf{D}_T \mathbf{c}\; d\Omega - \int_\Omega \delta\varepsilon^T \mathbf{D}_T \frac{\partial \varepsilon_o}{\partial t} d\Omega - \frac{\partial \hat{f}}{\partial t} = 0$$

where

$$\frac{\partial \hat{f}}{\partial t} = \int_\Omega \delta\mathbf{u}^T \frac{\partial \mathbf{b}}{\partial t} d\Omega + \int_\Gamma \delta\mathbf{u}^T \frac{\partial \hat{\mathbf{t}}}{\partial t} d\Gamma \quad (9)$$

represents the variation in time of the external force work due to boundary and body force loadings.

Using the Bishop's equation for the definition of the effective stress, the nonlinear term $S_w(p_w)$ has to be substituted by the term $\chi(S_w)$, which is also nonlinear.

2.3 Fluid phases behaviour in a deforming porous medium

For the transport of water and of gas the validity of Darcy's law is assumed.

$$q = -\frac{1}{\mu} \mathbf{k} \nabla \left(p + \rho g h\right) \quad (10)$$

If we consider a porous medium with two fluid phases flowing simultaneously, a relative permeability function $k_{rw}(S_w)$ needs to be introduced to modify the permeability matrix $\mathbf{k}$ in eq. (10). This function is usually obtained by experiment. In the same manner, the saturation of either phase is obtained from the capillary pressure versus saturation.

From a general point of view, it is possible to take into account miscible fluids using the approach proposed by Collins (1961).The phenomenon is handled by defining two parameters, which are dependent on the pressure of each phase:
- the formation volume factor $B_\pi(p_\pi)$, which relates the volume of the π-phase at the pressure p_π to its volume at standard conditions;
- the solution ration $R_{s\pi}(p_\pi)$, which depends on the solubility of gas in water.

Incorporating these parameters, we can write the continuity equation for each of the components of the mixture. The following contributions to the rate of fluid accumulation have to be taken into account:

a) rate of change of the total strain

$$\frac{\partial \varepsilon_v}{\partial t} = \mathbf{m}^T \frac{\partial \varepsilon}{\partial t} \tag{11}$$

b) rate of change of the grain volume due to pressure changes

$$\frac{(1-\phi)}{K_s} \frac{\partial \bar{p}}{\partial t} \tag{12}$$

where $\bar{p}$ is the average effective pressure previously defined;

c) rate of change of saturation

$$\phi \frac{\rho_\pi}{B_\pi} \frac{\partial S_\pi}{\partial t} \tag{13}$$

d) rate of change of fluid density

$$\phi S_\pi \frac{\partial}{\partial t}\left(\frac{\rho_\pi}{B_\pi}\right) \tag{14}$$

e) change of grain size due to effective stress changes

$$-\frac{1}{3K_s} \mathbf{m}^T \frac{\partial \sigma'}{\partial t} \, . \tag{15}$$

The continuity equation for water, with no source terms, therefore becomes

$$-\nabla^T \left\{ \mathbf{k} \frac{k_{rw}\rho_w}{\mu_w B_w} \nabla (p_w + \rho_w g h) \right\} + \phi \frac{\rho_w}{B_w} \frac{\partial S_w}{\partial t} + \phi S \frac{\partial}{\partial t}\left(\frac{\rho_w}{B_w}\right)$$

$$+ \rho_w \frac{S_w}{B_w} \left\{ \left(\mathbf{m}^T - \frac{\mathbf{m}^T \mathbf{D}_T}{3K_s}\right) \frac{\partial \varepsilon}{\partial t} + \frac{\mathbf{m}^T \mathbf{D}_T \mathbf{c}}{3K_s} \right. \tag{16}$$

$$\left. + \left[\frac{(1-\phi)}{K_s} - \frac{1}{(3K_s)^2} \mathbf{m}^T \mathbf{D}_T \mathbf{m}\right] \frac{\partial \bar{p}}{\partial t} \right\} = 0$$

For the gaseous phase, taking into account the possible source of gas through the solution ratio $R_{s\pi}$, the continuity equation is the following

$$-\nabla^T \left\{ \mathbf{k} \rho_g \left[\frac{k_{rg}}{\mu_g B_g} + \frac{R_{sw} k_{rw}}{\mu_w B_w} \right] \nabla (p_g + \rho_g g h) \right.$$

$$+ \phi \frac{\partial}{\partial t} \left\{ \rho_g \left[\frac{S_g}{B_g} + \frac{R_{sw} S_w}{B_w} \right] \right\}$$

$$+ \rho_g \left[\frac{S_g}{B_g} + \frac{R_{sw} S_w}{B_w} \right] \left\{ \left(\mathbf{m}^T - \frac{\mathbf{m}^T \mathbf{D}_T}{3K_s}\right) \frac{\partial \varepsilon}{\partial t} + \frac{\mathbf{m}^T \mathbf{D}_T \mathbf{c}}{3K_s} \right.$$

$$\left. + \left[\frac{1-\phi}{K_s} - \frac{1}{(3K_s)^2} \mathbf{m}^T \mathbf{D}_T \mathbf{m} \right] \frac{\partial \bar{p}}{\partial t} \right\} = 0 \tag{17}$$

In the last two equations, the coefficient $k_{r\pi}$ is a function of the saturation of the π-phase, and B_π is a function of p_π, resulting in a nonlinear problem.

The above continuity equations are subjected to the condition that

$$S_w + S_g = 1 \tag{18}$$

and have to be solved simultaneously with the equilibrium equation (8), due to the presence of the coupling terms. This represents a complete model for the solution of saturated-unsaturated porous media.

A simplified solution of the problem of saturated - unsaturated flow may be found, if the following assumptions are made:
- the air is continuous in the unsaturated zone (bubbles are not present).
- the air is everywhere at atmospheric pressure, which is assumed equal to zero;
- the effect of capillary pressure is neglected.

These assumptions are common to the aforementioned works of Neuman, Narasimhan and Witherspoon, Safai and Pinder.

In this case, the pressure head is taken to be positive in the saturated zone and negative in the unsaturated zone. The formation volume factor for water B_w is taken equal to unity. With this in mind, only the continuity equation for water is necessary. This is now written in a more suitable form.

Since the relationships $p_w = p_w(S_w)$ and $k_{rw} = k_{rw}(S_w)$ are generally known from laboratory experiments, we can obtain

$$\phi \frac{\partial S_w}{\partial t} = \phi \frac{\partial S_w}{\partial p_w} \frac{\partial p_w}{\partial t} = C_s \frac{\partial p_w}{\partial t} \tag{19}$$

where C_s is the specific moisture content.

Moreover equation (14), divided by ρ_w may be rearranged as

$$\frac{\phi}{\rho_w} S_w \frac{\partial \rho_w}{\partial t} = \phi \frac{S_w}{K_w} \frac{\partial p_w}{\partial t} \tag{20}$$

where K_w is the bulk modulus of water.

On substituting equation (19) and equation (20) into (16) yields

$$-\nabla^T \left\{ \mathbf{k} \frac{k_{rw}}{\mu_w} \nabla(p_w+\rho_w gh) \right. + S_w\left(\mathbf{m}^T - \frac{\mathbf{m}^T \mathbf{D}_T}{3K_s}\right)\frac{\partial \varepsilon}{\partial t}$$

$$+ S_w \frac{\mathbf{m}^T \mathbf{D}_T \mathbf{c}}{3K_s} + \left[C_s + \phi \frac{S_w}{K_w} + S_w\left(\frac{1-\phi}{K_S} - \frac{1}{(3K_s)^2}\mathbf{m}^T \mathbf{D}_T \mathbf{m}\right)\right.$$

$$\left.\left(S_w + \frac{C_s}{\phi} p_w\right)\right] \frac{\partial p_w}{\partial t} = 0 \qquad (21)$$

where compressibilities of the soil grains and of water are accounted for. These last contributions are usually negligible in soil mechanics but have importance in rock mechanics context and in concrete.

As shown by Lewis and Schrefler (1987) equation (21) is more general than the equation used by Neuman (1975). In equation (21) the coefficients k_{rw}, S_w, C_s appear, which are dependent on water pressure, resulting hence in a nonlinearity. It should be noted that equation (21) obviously coincides with the continuity equation in the saturated zone with the appropriate values assumed by coefficients k_{rw}, S_w, C_s.

This implies that a consolidation program for one phase flow can be adapted, with some changes, to also solve unsaturated or partly saturated situations.

2.4 Boundary conditions

The above equations are associated with the forced and "natural" boundary conditions which apply to the two interacting fields:

– *displacement field*:

a) prescribed traction on Γ_t

$$\mathbf{n}^T \sigma = \hat{\mathbf{t}} \qquad (22)$$

b) prescribed displacements on Γ_u

$$\mathbf{u} = \mathbf{u}^* \qquad (23)$$

– *pressure field:*

a) prescribed pressure on Γ_p

$$p = p^* \qquad (24)$$

b) prescribed flux on Γ_f

$$-k\frac{\partial p}{\partial n} = Q_e \qquad (25)$$

as usually done in saturated problems.

3 SEMI-DISCRETIZED SOIL-PORE FLUID EQUATIONS

Equations (8) and (21) are discretized in the space domain using the standard finite element procedure. The approximations used for the basic variables, i.e. the displacements and the pore pressure, are the following

$$\mathbf{u} = \mathbf{N_u}\,\bar{\mathbf{u}} \qquad \text{and} \qquad p = \mathbf{N_p}\,\bar{\mathbf{p}} \qquad (26)$$

where $\mathbf{N_u}$ are the interpolation functions for the displacement field;

$\mathbf{N_p}$ are the interpolation functions for the pressure field;

$\bar{\mathbf{u}}, \bar{\mathbf{p}}$ are the vectors containing the nodal values of the basic variables.

$\mathbf{N_u}$ and $\mathbf{N_p}$ may be different. Both must however be of C_o continuity. Obviously many choices of elements are possible. However, it is known that the use of an interpolation for **u** one order higher that that for p, is usually more accurate and also more economical, see Zienkiewicz et al. (1986). For instance, using a quadratic interpolation for **u** and a linear one for p, the coupled solution is not much more costly than that of the single displacement field, because unknowns are increased only by 25% in 2-D situations.

When using the same interpolation for **u** and p, it is convenient to employ quadratic shape functions in conjunction with a smoothing technique, as will be seen in the next paragraph.

The application of a Galerkin procedure to equations (8) and (21) results in the following discrete system of differential equations in time

$$\begin{aligned} &\mathbf{K}\frac{d\bar{\mathbf{u}}}{dt} + \bar{\mathbf{L}}\frac{d\bar{\mathbf{p}}_w}{dt} - \mathbf{C} - \frac{d\mathbf{f}}{dt} = 0 \\ &\bar{\mathbf{H}}\bar{\mathbf{p}}_w + \bar{\mathbf{S}}\frac{d\bar{\mathbf{p}}_w}{dt} + \mathbf{T}^T\frac{d\bar{\mathbf{u}}}{dt} - \bar{\bar{\mathbf{f}}} = 0 \end{aligned} \qquad (27)$$

The matrices relative to the various terms are listed in appendix I. The system (27) is coupled and non-symmetric.

4 NUMERICAL STRATEGIES

The solution of the system of differential equation (27), with first order time derivatives, presents some numerical difficulties, which are now shortly explained.

4.1 Oscillation in the pressure field

When using a quadratic approximation for the pressure field, as can be seen for instance in Reed (1984), spatial oscillation occurs. This fact could indicate e.g. unsaturate zones where this is not true. This oscillation may be eliminated employing the smoothing technique proposed by Majorana et al. (1985).

Even though linear approximation in p (element 84) results in a more economical solution, the quadratic approximation with smoothing (element 88), behaves often better in proximity of pressure discontinuities.

4.2 Finite element cut by the saturation line

As previously indicated, the free surface (saturation line) is an isobar which moves through the domain until steady state conditions are reached. Moreover this isobar is a discontinuity line for the parameters k_{rw}, Sw, and C_s.

Alternative to the use of a moving mesh, in which the discontinuity always coincides with a side of a finite element, are the following two techniques:
- Use of quadratic isoparametric elements with discontinuity. This technique was proposed by Steven (1980) and was used in our numerical application due to its straightforward implementation. As a first step the pore pressure is smoothed, then the position of the saturation line, which is assumed straight, is calculated; eventually the matrices obtained for each element by assuming the parameters corresponding to satured conditions, are corrected to account for the changes of the material properties. The last step is performed using the Gauss-Legendre quadrature formulae for triangles. It has to be noted that in this case the polynomial basis is defined over the entire element.
- The second approach was proposed by Mackinnon and Carey (1987). If the discontinuity intersects the element, this is subdivided by the saturation line into subelements. Temporary subelement nodal data are defined together with the introduction of interface constraints on the saturation line. Special constrained basis functions and subelement matrices are calculated and accumulated to form a constrained element matrix, with the condensation of the temporary nodes. This technique is simple to use when a linear approximation is employed for the pressure field.

4.3 Discretization in the time domain and treatment of the nonlinearities

The nonlinear system of differential equations (27) is integrated in the time domain by means of two- or three-level schemes.

The first runs were performed using the Lees' three-level scheme which presents the advantage of not requiring iterations in time. The matrices are calculated at the central point, where the state variables are known. However some troubles were experienced in the use of this method: the time marching procedure starts with a two-level scheme, oscillations in time appear, the time step is usually small and changes in the time steps result again in oscillation of the solution in time.

For these reasons, a two-level scheme was adopted (Crank - Nicholson), in which iterations are necessary to solve the problem in the time domain. Iterations are necessary anyway because of elastoplasticity. To reduce the number of iterations, an automatic adjustement of the time step is implemented in our program based on a convergence check and on the number of iterations performed: when this number is less than n_1 the time step for the next iteration is increased, when it is larger than n_2 the time step is reduced and the calculation is reset at an earlier time. n_1 and n_2 with the initial time step are given by the user.

In the first iteration of each time step, the pressure of the last solution is used as a predictor for the calculation of the matrices at $t+\theta$.

5 NUMERICAL EXAMPLE

It was very difficult to choose an appropriate test to validate the proposed numerical model and its implementation in the computer code. This is so because there exist no analytical solutions for this type of coupled problem, where deformations of the solid skeleton are studied together with the suturated-unsaturated flow of the fluid. For this reason our attention was focused on laboratory experiments. This was a difficult task, because it seems that the major part of the laboratory experiments on this type of phenomena are performed to check the hydraulic parameters only.

The choosen problem for a numerical test is the experiment conducted by Liakopoulos (1965) on the drainage of water from a vertical column of sand. This test was also used by Narasimhan and Witherspoon (1978) to check their numerical model.

A column of perspex, 1 meter high, was packed by Del Monte sand and instrumented to measure continuously the moisture tension at various points within the column. Prior to the start of the experiment ($t<0$) water was continuously added from the top of the column and was allowed to drain freely at the bottom through a filter. The flow was carefully regulated until the tensiometers read zero pore pressure, which corresponds to a unit vertical gradient of the potential (fig. 1). At this point (initial conditions), the inflow of water from the top was stopped and the upper boundary made impermeable to water. From then on ($t>0$) the tensiometer readings were recorded and the flow rates at the draining filter were measured.

The physical properties of Del Monte sand were measured by Liakopoulos (1965) by an independent set of experiments: the porosity was 29.75% and the parameters k_{rw} and S_w are given in fig. 2 as function of the pressure of the water. It has to be noted that S_w and its derivative C_S remain unchanged in the range $-0.2 < p_w < 0.0$ meters of water. This means that before the pores can be desaturated, the moisture suction must exceed the "air entry" value of 0.2 m. Hence the soil remains saturated until a suction of –0.2 m is reached. For numerical purposes the column was simulated by 5 to 20 finite elements of equal size, in order to assess the usefulness of the techniques of point 4.2, always obtaining pratically the same results (fig. 3).

As a first step of the numerical runs, the solid skeleton was assumed to be nondeformable, as usually done when dealing with unsaturated flow. The resulting solution and the experimental results agree at high time values, but at initial time values the computed suctions are considerably larger than the experimental ones. Narasimhan and Witherspoon (1978) obtained similar results.

In a second time the soil deformability was accounted for. Unfortunately in Liakopoulos' experiments the parameters defining the behaviour of the soil skeleton are not given. For this reason the influence of the elastic modulus on the numerical solution was investigated and it was found that this parameter has an important influence on the calculated flow field. The value $E=1300$ KN/m^2 gave acceptable results as shown in fig. 4. This value is within the range of the possible values of this material.

From fig. 4 can be noted that the numerical results at the onset of the experiment, are better than for intermediate time. This means that other nonlinearities become important with increasing desaturation, as observed by various authors. These nonlinearities may concern:

– the Bishop's parameter, which in our model was assumed equal to the degree of saturation.

– the porosity, as proposed by Matyas and Radhakrishna (1968);

– the Young's modulus of the soil skeleton, as noticed in Lloret an Alonso (1980).

The numerical model proposed in this paper may easily be expanded, both from the theoretical and the computational point of view, to account for these improvements.

6 CONCLUSIONS

The numerical model presented in this paper is capable of dealing both with fully saturated and saturated-unsaturated porous media. It solves the problem in coupled form.

The model is general, even though some simplifications are introduced, and can easily be improved by introducing new nonlinear relations between the parameters defining the problem. This task requires however laboratory experiments, where more emphasis is given to the coupling aspects which play an important role in the deformation of desaturating porous media.

NOTATION

$\mathbf{b}$ body force vector
C_S specific moisture content
$\mathbf{c}$ stain independent vector defining the creep strain rate
$\mathbf{D}_T$ tangential stiffness matrix
g gravity acceleration
h elevation above some datum
$\mathbf{k}$ absolute permeability matrix
$k_{r\pi}$ relative permeability of the π–phase
K_S bulk modulus of the solid phase
K_w bulk modulus of the water
$\mathbf{L}$ differential operator which relates displacementsto strains
$\mathbf{m}^T$ (1 1 1 0 0 0)
p_π pore pressure of the π-phase
Q_e outflow of the fluid per unit volume of the solid
S_π degree of saturation of the π-phase
t time variable
$\hat{\mathbf{t}}$ boundary traction vector
$\mathbf{u}$ displacement vector
ε total strain vector of the soil skeleton
ε_o represents all other strains not directly associated with stress change
μ dynamic viscosity of water
ρ_S density of the solid
ρ_w density of the fluid
σ' effective stress in the soil skeleton
ϕ porosity
χ Bishop's parameter

ACKNOWLEDGEMENTS

This research was partially supported by the MPI and CNR under the contract number c. t. 84.021050.7.

REFERENCES

Bishop, A.W. 1960. The principle of effective stress. Publ. Norwegian Geotech. Inst. Oslo. 32:1-5.

Bishop, A.W. and Blight, G.E. 1963. Some aspects of effective stress in saturated and partly saturated soils. Geotechnique 13:177-197.

Collins, R.E. 1961. Flow of fluids through porous material. New York, Reinhold.

Lewis, R.W. and Schrefler, B.A. 1987. The finite element method in the deformation and consolidation of porous media. London, Wiley.

Liakopoulos, A.C. 1965. Transient flow through unsaturated porous media. D. Eng. dissertation, Univ. of California, Berkeley.

Lloret, A. and Alonso, E.E. 1980. Consolidation of unsaturated soils including swelling and collapse behaviour. Geotechnique 30:449-477.

Mackinnon, R.J. and Carey, G.F. 1987. Treatment of material discontinuities in Finite element computations. Int. J. Num. Methods Eng. 24:393-417.

Majorana, C.E., Odorizzi, S.A. and Vitaliani, R. 1985. Direct determination of finite element local smoothing matrices. Comm. Appl. Num. Meth., 1:39-43.

Matyas, E.L. and Radhakrishna, H.S. 1968. Volume change characteristics of partially saturated soils. Geotechnique, 18:432-448.

McMurdie, J.L. and Day, P.R. 1960. Slow tests under soil moisture suction. Soil Sci. Soc. Amer. Proc., 24:441-444.

Narasimhan, T.N. and Witherspoon, P.A. 1977. Numerical model for saturated-unsaturated flow in deformable porous media. 1 Theory. Water resour. res., 13:657-664.

Narasimhan, T.N. and Witherspoon, P.A. 1978. Numerical model for saturated-unsaturated flow in deformable porous media. 3 Applications. Water resour. res., 14:1017-1034.

Neuman, S.P. 1975. Galerkin approach to saturated-unsaturated flow in porous media. In Gallagher, R.H., Oden, J.T., Taylor, C. and Zienkiewicz, O.C. (eds). Finite Elements in Fluids Vol. 1. London, Wiley.

Reed, M.B. 1984. An investigation of numerical errors in the analysis of consolidation by finite elements. Int. J. Num. Anal. Meth. Geomech., 8:243-257.

Safai, N.M. and Pinder, G.F. 1979. Vertical and horizontal land deformation in a desaturating porous medium. Adv. Water Resour., 2:19-25.

Steven, G.P.1982. Internally discontinuous finite elements for moving interface problems. Int. J. Num. Methods Eng., 18:569-582.

Whitaker, S. 1973. The transport equations for multiphase systems. Chem. Eng. Sci., 28:139-147.

Zienkiewicz, O.C., Qu, S., Taylor, R.L. and Nakazawa, S. 1986. The patch test for mixed formulation. Int. J. Num. Meth. Eng., 23:1873-1883.

APPENDIX I

List of matrices of equation (27) using the usual symbols.

$$\mathbf{K} = -\int_\Omega \mathbf{B}^T \mathbf{D}_T \mathbf{B}\, d\Omega$$

$$\bar{\mathbf{L}} = \int_\Omega \mathbf{B}^T S_w \left(\mathbf{m} - \frac{\mathbf{m}\mathbf{D}_T}{3K_s}\right) \bar{\mathbf{N}}\, d\Omega$$

$$+ \int_\Omega \mathbf{B}^T \frac{C_s}{\phi} \left(\mathbf{m} - \frac{\mathbf{m}\mathbf{D}_T}{3K_s}\right) \bar{\mathbf{N}} \bar{\mathbf{p}}_w \bar{\mathbf{N}}\, d\Omega$$

$$\mathbf{C} = -\int_\Omega \mathbf{B}^T \mathbf{D}_T \mathbf{c}\, d\Omega$$

$$\mathbf{df} = -\int_\Omega \mathbf{N}^T \mathbf{db}\, d\Omega - \int_\Gamma \mathbf{N}^T \mathbf{d}\hat{\mathbf{t}}\, d\Gamma - \int_\Omega \mathbf{B}^T \mathbf{D}_T \mathbf{d}\, \varepsilon_o\, d\Omega$$

$$\bar{\mathbf{H}} = \int_\Omega (\nabla \bar{\mathbf{N}})^T \frac{k_{rw}}{\mu_w} \mathbf{k}\, \nabla \bar{\mathbf{N}}\, \delta\Omega$$

$$\bar{\mathbf{S}} = \int_\Omega \bar{\mathbf{N}}^T \bar{s} \bar{\mathbf{N}}\, d\Omega + \int_\Omega \bar{\mathbf{N}}^T \bar{\bar{s}} \bar{\mathbf{N}} \bar{\mathbf{p}}_w \bar{\mathbf{N}}\, d\Omega$$

where

$$\bar{s} = C_s + \phi \frac{S_W}{K_w} + (S_w)^2 \left(\frac{1-\phi}{K_s} - \frac{1}{(3K_s)^2} \mathbf{m}^T \mathbf{D}_T \mathbf{m} \right)$$

$$\bar{\bar{s}} = \frac{S_w C_s}{\phi} \left(\frac{1-\phi}{k_s} - \frac{1}{(3K_s)^2} \mathbf{m}^T \mathbf{D}_T \mathbf{m} \right)$$

$$\mathbf{T} = \int_\Omega \bar{\mathbf{N}}^T S_w \left(\mathbf{m}^T - \frac{\mathbf{m}^T \mathbf{D}_T}{3K_s}\right) \mathbf{B}\, d\Omega$$

$$\bar{\bar{f}} = -\int_\Gamma \mathbf{N}^T q\, d\Gamma - \int_\Omega \frac{S_w}{3K_s} \mathbf{m}^T \mathbf{D}_T \mathbf{c}\, d\Omega$$

$$- \int_\Omega (\nabla \bar{\mathbf{N}})^T \frac{k_{rw} \mathbf{k}}{\mu_w} \nabla \rho_w g h\, d\Omega$$

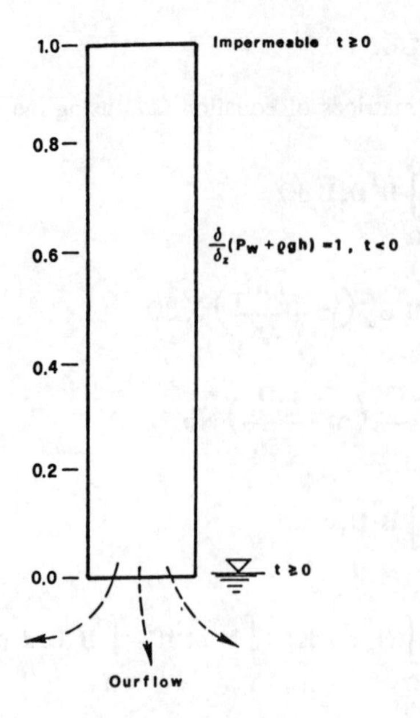

Figure 1. Test problem.

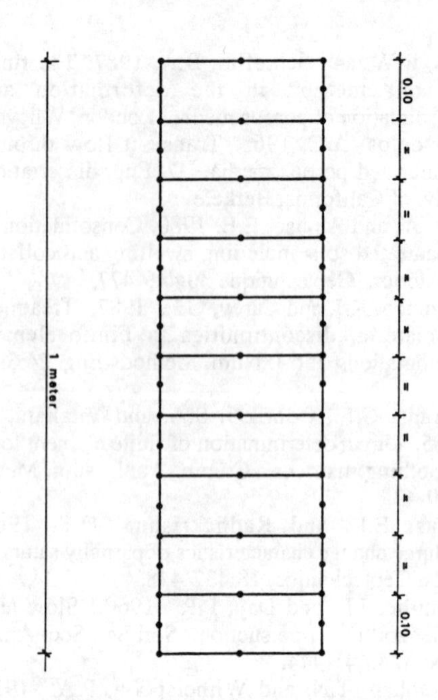

Figure 3. Finite element discretization.

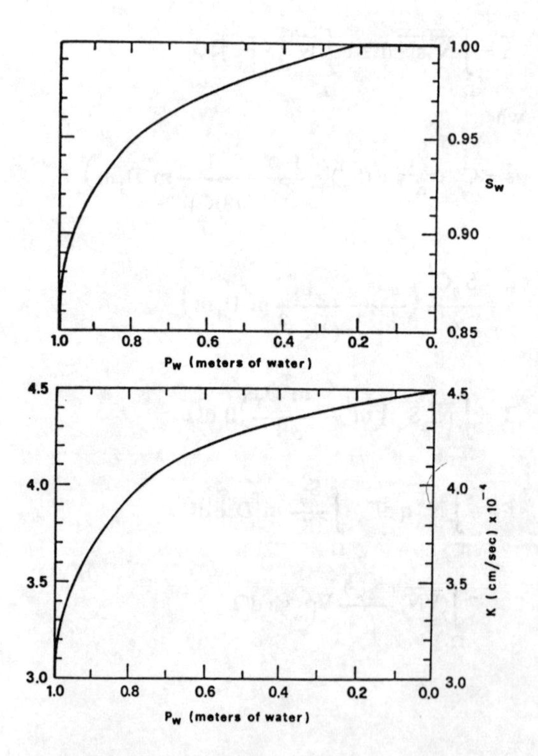

Figure 2. Hydraulic material properties.

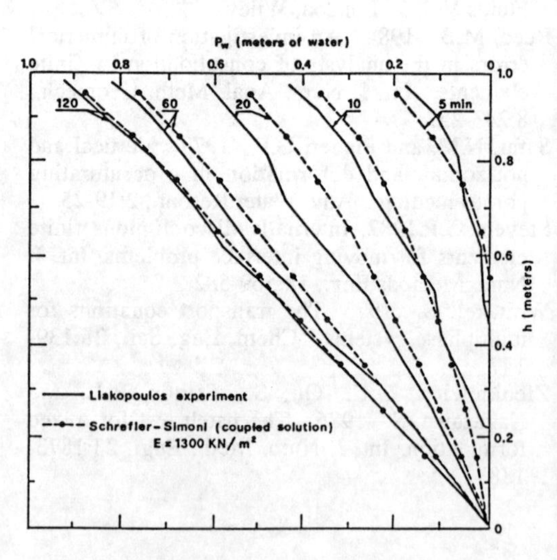

Figure 4. Comparison of numerical and experimental results for pore pressure.

Numerical Methods in Geomechanics (Innsbruck 1988), Swoboda (ed.)
© 1988 Balkema, Rotterdam. ISBN 90 6191 809 X

Uni-axial wave propagation through fluid-saturated elastic soil layer

L.W.Morland
School of Mathematics, University of East Anglia, Norwich, UK
R.S.Sandhu & W.E.Wolfe
Department of Civil Engineering, Ohio State University, Columbus, USA

ABSTRACT: The motions of fluid and solid in a saturated linear elastic soil are coupled through the dependence of fluid and matrix stresses on the deformation of both constituents, and through the interaction drag (momentum transfer). The coupled momentum equations for the fluid and matrix velocities in uni-axial motion have two distinct signal (characteristic) speeds which govern the propagation of discontinuities, and which correspond to the wave speeds of the motion for zero drag. The major coupling is through the diffusive effect of the interaction drag. Laplace transform analysis is used to exhibit the wave fronts initiated by step-velocity boundary loading and their successive reflexions from both surfaces of a soil layer. A shift theorem and aymptotic expansions determine the discontinuities of the velocities and their first and second derivatives at each wave front, defining a singular part of the solution. A decomposition into singular and smooth fields with derivatives up to the second continuous, yield coupled second order differential equations with continuous forcing terms for the smooth fields, which can be solved accurately by routine numerical methods. Comparisons of solutions determined by the above decomposition, by a direct finite element method, and by numerical inversion of the Laplace transform solutions, are presented to assess the merits of different numerical approximations.

1 INTRODUCTION

The response of fluid-saturated soils to dynamic loads is of considerable practical concern, from long seismic waves of 100+ m length to short explosive induced pulses of 1 m length. On the one hand, the interaction drag causes diffusive dissipation and pulse decay, but also the resistance to drainage induces increase of the pore pressure with consequent intergranular buoyancy and the possibility of liquefaction. Realistic models must account for complex internal response and multi-dimensional wave propagation, and solutions to boundary/initial value problems can be sought only by direct large-scale numerical computation. Confidence in such numerical results cannot be realised until the numerical codes have been tested exhaustively against exact solutions to problems containing in turn each of the main model ingredients. Furthermore, such solutions determine the likely effects of different features and their sensitivity to variations in natural parameters, helping to construct a physical picture of fluid-soil interactions.

Here we adopt the most simple constitutive model for a fluid-saturated porous solid; a linear elastic solid-ideal fluid mixture undergoing small compression and small matrix shear, of more direct application to porous rocks than to granular materials. We focus on one-dimensional compression waves induced by the surface loading and consider successive reflections when the base is rigid and impervious. For short pulses, the dimensionless drag coefficient can be small, order unity, or large over the wide ranges of permeability arising in practice, so weak, moderate, and strong coupling are all of interest. This linear elastic solid-ideal fluid problem has been treated by Garg, Nayfeh, and Good [1] for one-dimensional wave propagation along a semi-infinite body, so reflections from the rigid, impervious base, and in turn from the surface, are new features. Garg et al. [1] determine a Laplace transform solution which is inverted numerically for the case of equal step-function solid and fluid velocities imposed at the surface, and compare to finite-difference numerical solutions using an artificial viscosity to smear the step fronts.

We now show that the Laplace transform solution determines the fast and slow wave fronts and their successive reflections, and for each wave front, the decaying discontinuity amplitude (for step-function boundary conditions) and the discontinuities in successive derivatives. Requiring that boundary-induced discontinuity fronts do not amplify as they propagate, and that the velocity fields do not become unbounded as time increases, for a physically sensible solution, determines restrictions on the five elastic constants describing the elastic solid-fluid mixture.

By constructing singular fields which incorporate all discontinuities of the velocity fields and their first and second derivatives, an additive decomposition leaves twice differentiable fields which satisfy coupled, hyperbolic, second order differential equations with continuous forcing terms. This system can be solved accurately by routine finite-difference or finite element methods, and a simple explicit scheme has been used for several finite drag examples. Additionally, the same examples have been solved by a direct finite-element method and by numerical inversion of the complete Laplace transform, for comparison of accuracy and computation times.

2 MODEL EQUATIONS

Let superscripts s and f denote quantities associated with the solid matrix and fluid respectively. In the framework of mixture theory [2] the solid and fluid have overlapping (mean) velocity fields v^s, v^f such that $\rho^s v^s, \rho^f v^f$ define the respective mass flux vectors, where ρ^s, ρ^f are the partial densities. We assume infinitesimal (mean) displacement vectors u^s, u^f, so that

$$v^s = \partial u^s/\partial t \ , \quad v^f = \partial u^f/\partial t \ , \qquad (2.1)$$

where t denotes time, an infinitesimal partial matrix strain with components

$$\varepsilon^s_{ij} = \tfrac{1}{2}(\partial u^s_i/\partial x_j + \partial u^s_j/\partial x_i) \qquad (2.2)$$

in rectangular spatial co-ordinates x_i $(i = 1,2,3)$, and infinitesimal partial dilations $\varepsilon^s, \varepsilon^f$ given by

$$\varepsilon^s = \partial u^s_j/\partial x_j , \quad \partial \varepsilon^f/\partial t = \partial v^f_j/\partial x_j \ . \qquad (2.3)$$

Partial tractions are defined as forces on each constituent per unit mixture cross-section, with associated partial stress tensors σ^s, σ^f. Here $\sigma^f = -p^f 1$ where 1 is the unit tensor and p^f is the partial fluid pressure, since viscous stresses associated with global strain-rates are neglected in comparison with the interaction drag. If φ is the porosity (fluid volume fraction), then φ and $1 - \varphi$ are the respective ratios of partial to intrinsic density, and we make the conventional assumption that the ratios of partial to intrinsic traction (stress) are also φ and $1 - \varphi$. Constitutive relations between the partial variables may be formulated from constituent properties relating intrinsic variables and a partial strain-intrinsic strain relation, together with a constitutive relation for φ describing the sharing of mixture space when stresses are applied [3], [4].

An explicit isotropic linear elastic-ideal fluid theory [4] gives

$$-p^s = c_{11}\varepsilon^s + c_{12}\varepsilon^f , \quad -p^f = c_{21}\varepsilon^s + c_{22}\varepsilon^f , \qquad (2.4)$$

together with a matrix shear response

$$\sigma^s_{ij} + p^s 1_{ij} = 2G(\varepsilon^s_{ij} - \tfrac{1}{3}\varepsilon^s 1_{ij}) \ , \qquad (2.5)$$

where G is the mixture shear modulus. The four elastic moduli c_{11}, c_{12}, c_{21}, and c_{22} are determined by the intrinsic solid and fluid compressibilities $\varkappa_s, \varkappa_f$, and two other mixture compressibilities, such as the free-draining compressibility $\varkappa$, the undrained compressibility $\bar{\delta}$, and unjacketed compressibility δ [5]. It is common to postulate the coupled linear pressure-dilation relations (2.4) with a symmetry assumption $c_{21} = c_{12}$; we retain the general form (2.4). In the linear theory ρ^s, ρ^f, undergo infinitesimal changes, and are therefore replaced by their constant initial values in the momentum equations

$$\frac{\partial \sigma^s_{ij}}{\partial x_j} + \rho^s g_i + \frac{\mu\varphi^2}{k}(v^f_i - v^s_i) = \rho^s \frac{\partial v^s_i}{\partial t} \ ,$$

$$(2.6)$$

$$-\frac{\partial p^f}{\partial x_i} + \rho^f g_i - \frac{\mu\varphi^2}{k}(v^f_i - v^s_i) = \rho^f \frac{\partial v^s_i}{\partial t} \ ,$$

where g_i are the external body force components per unit mass, here gravity, μ is the fluid viscosity, and k is the matrix permeability; $\mu\varphi^2/k$ is the hydraulic conductivity. The interaction drag is assumed proportional to the velocity difference, and measures the transfer of momentum between the constituents. Note that the balances (2.6) assume that the mass flux velocities v^s, v^f also determine the momentum fluxes.

Confining attention to vertical motion with velocities w^s, w^f in a downward direction Oz, then

$$\partial\varepsilon^s/\partial t = \partial w^s/\partial z, \quad \partial\varepsilon^f/\partial t = \partial w^f/\partial z , \tag{2.7}$$

$$\sigma^s = \hat{c}_{11}\varepsilon^s + c_{12}\varepsilon^f, \quad \hat{c}_{11} = c_{11} + 4/3G, \tag{2.8}$$

where σ^s is the axial matrix stress, p^f is given by $(2.4)_2$, and the axial momentum equations become

$$\frac{\partial\sigma^s}{\partial z} + \rho^s g + \frac{\mu\varphi^2}{k}(w^f - w^s) = \rho^s \frac{\partial w^s}{\partial t} ,$$
$$- \frac{\partial p^f}{\partial z} + \rho^f g - \frac{\mu\varphi^2}{k}(w^f - w^s) = \rho^f \frac{\partial w^f}{\partial t} . \tag{2.9}$$

This system is equivalent to that of Garg et al. [1]if gravity is ignored and the symmetry $c_{21} = c_{12}$ is assumed. Following [1] we will treat the case of applied surface velocities

$$z = 0: \; w^s = w_s(t), \; w^f = w_f(t) , \tag{2.10}$$

and additionally incorporate a rigid imperveous base

$$z = \ell: \quad w^s = w^f = 0 . \tag{2.11}$$

If w_o is a velocity magnitude governed by the surface conditions (2.10), then dimensionless, order unity, velocities W^s, W^f are defined by

$$w^s = w_o W^s, \; w^f = w_o W^f . \tag{2.12}$$

A wave magnitude c_o is given by $\rho c_o^2 = \hat{c}_{11} + c_{12} + c_{21} + c_{22}$ where $\rho = \rho^s + \rho^f$ is the mixture density, so normalised moduli $C_{\ell m}$ are defined by

$$c_{\ell m} = \rho c_o^2 C_{\ell m} \quad (\ell, m = 1,2),$$
$$\hat{c}_{11} = \rho c_o^2 \hat{C}_{11} , \tag{2.13}$$

and normalised dynamic stresses p^s, p^f are given by

$$\sigma^s = \sigma^s_e - \rho w_o c_o P^s, \; p^f = p^f_e + \rho w_o c_o P^f \tag{2.14}$$

where σ^s_e, p^f_e are equilibrium stresses under gravity and zero surface tractions.

Choose a unit of length ℓ_o and set

$$z = \ell_o Z, \; \ell = \ell_o L, \quad t = \ell_o T/c_o \tag{2.15}$$

so that Z and T derivatives have equal status in the wave motion. If ℓ is a wave length ℓ_w defined by imposed surface conditions, then the Z and T derivatives of W^s and W^f are order unity, but since there is a distinct length scale ℓ_D associated with the momentum transfer, the balances must be analysed carefully. With step-function surface conditions there is only the length scale ℓ_D. Note that the layer thickness ℓ is purely geometric and does not influence derivative magnitudes, but may be convenient for numerical purposes.

The system $(2.4)_2$, (2.8), with (2.7), and (2.9) becomes

$$- \frac{\partial P^s}{\partial T} = \hat{C}_{11}\frac{\partial W^s}{\partial Z} + C_{12}\frac{\partial W^f}{\partial Z} , \; - \frac{\partial P^f}{\partial T} = C_{21}\frac{\partial W^s}{\partial Z} + C_{22}\frac{\partial W^f}{\partial Z} , \tag{2.16}$$

$$-\frac{\partial P^s}{\partial Z} + D(W^f - W^s) = r_s \frac{\partial W^s}{\partial T} , \tag{2.17}$$

$$-\frac{\partial P^f}{\partial Z} - D(W^f - W^s) = r_f \frac{\partial W^f}{\partial T} , \tag{2.18}$$

where

$$r_s = \rho^s/\rho, \; r_f = \rho^f/\rho, \; r_s + r_f = 1 . \tag{2.19}$$

D is a dimensionless drag parameter defined by

$$D = \ell_o \mu\varphi^2/\rho c_o k, \quad = \ell_o/\ell_D \tag{2.20}$$

where

$$\ell_D = \rho c_o k/\mu\varphi^2 \tag{2.21}$$

defines a length scale over which the momentum transfer takes place during the wave passage, equivalent to the transfer time $\rho k/\mu\varphi^2$ introduced by Garg et al. [1]. The surface and base conditions (2.10), (2.11) become

$$Z = 0: W^s = W_s(T), \; W^f = W_f(T), \tag{2.22}$$

$$Z = L: W^s = W^f = 0 . \tag{2.23}$$

Eliminating P^s, P^f between (2.16) and (2.18) gives the coupled hyperbolic velocity equations

$$M_s[W^s,W^f] = \hat{C}_{11}\frac{\partial^2 W^s}{\partial Z^2} + C_{12}\frac{\partial^2 W^f}{\partial Z^2} - r_s\frac{\partial^2 W^s}{\partial T^2} + D\left(\frac{\partial W^f}{\partial T} - \frac{\partial W^s}{\partial T}\right) = 0 , \tag{2.24}$$

$$M_f[W^s,W^f] + C_{21}\frac{\partial^2 W^s}{\partial Z^2} + C_{22}\frac{\partial^2 W^f}{\partial Z^2} - r_f\frac{\partial^2 W^f}{\partial T^2}$$

$$- D\left(\frac{\partial W^f}{\partial T} - \frac{\partial W^s}{\partial T}\right) = 0 \quad . \qquad (2.25)$$

Given a wave length ℓ_w and $\ell_o = \ell_w$ it is natural to define weak, moderate, and strong coupling (or small, moderate, large drag) by

$$D << 1, \quad D = O(1), \quad D >> 1 , \quad \text{respectively.} \qquad (2.26)$$
$$\ell_D >> \ell_w, \quad \ell_D = O(\ell_w), \quad \ell_D << \ell_w ,$$

Choosing $\ell_o = \ell_D$ sets D = 1, but the length and time units are not necessarily appropriate to an applied wave form, and ℓ_D will be very small compared to the propagation distance of interest in the large drag situation. In fact, regardless of the chosen scale, large drag gives rise to a surface boundary layer and a boundary layer at a wave front - that is, large derivatives and hence numerical difficulty.

3 LAPLACE TRANSFORM SOLUTION

Denoting the Laplace transform of a piecewise-continuous bounded function f(T) defined on T > 0 by $\bar{f}(s)$, analytic in Re(s) > 0, transforming the velocity equations (2.24), (2.25) gives the coupled ordinary differential equations

$$\hat{C}_{11}\frac{\partial^2\bar{W}^s}{\partial Z^2} + C_{12}\frac{\partial^2\bar{W}^f}{\partial Z^2} - r_s s^2\bar{W}^s + Ds(\bar{W}^f - \bar{W}^s) = 0, \qquad (3.1)$$
$$C_{21}\frac{\partial^2\bar{W}^s}{\partial Z^2} + C_{22}\frac{\partial^2\bar{W}^f}{\partial Z^2} - r_f s^2\bar{W}^f - Ds(\bar{W}^f - \bar{W}^s) = 0,$$

assuming a rest state at T = 0+ for 0 < Z < L. The solution satisfying the zero velocity condition (2.23) on Z = L is

$$\bar{W}^s = \sum_{r=1}^{2} A_r\{e^{-\theta_r Z} - e^{-\theta_r(2L-Z)}\} ,$$
$$\bar{W}^f = \sum_{r=1}^{2} B_r A_r\{e^{-\theta_r Z} - e^{-\theta_r(2L-Z)}\} , \qquad (3.2)$$

where

$$B_r = [Ds + r_s s^2 - \hat{C}_{11}\theta_r^2]/[Ds + C_{12}\theta_r^2], \qquad (3.3)$$

and θ_r are the roots of

$$\Delta\theta^4 - (as^2 + bDs)\theta^2 + r_s r_f s^4 + Ds^3 = 0 \qquad (3.4)$$

with positive real part as $|s| \to \infty$ in Re(s) > 0, and

$$\Delta = \hat{C}_{11}C_{22} - C_{12}C_{21}, \quad a = r_f\hat{C}_{11} + r_s C_{22},$$
$$b = \hat{C}_{11} + C_{22} + C_{12} + C_{21} \quad . \qquad (3.5)$$

Note that b = 1 by the normalisation (2.13). The surface velocity conditions (2.24) gives

$$A_1 = \frac{B_2\bar{W}_s - \bar{W}_f}{X_1(B_2 - B_1)} \ , \quad A_2 = \frac{\bar{W}_f - B_1\bar{W}_s}{X_2(B_2 - B_1)} \ , \qquad (3.6)$$

where

$$X_r(s) = 1 - e^{-2L\theta_r(s)} \quad (r = 1,2) \ . \qquad (3.7)$$

A zero drainage solution $W^f \equiv W^s$ corresponds to zero permeability or the infinite drag limit $D \to \infty$, when by (3.3) B = 1, and the roots θ_1^2, θ_1^2 of (3.4) are replaced by an infinite root and a finite root

$$D \to \infty : \theta_\infty^2 = s^2/q_\infty^2, \quad q_\infty^2 = b = 1 \ . \qquad (3.8)$$

Retaining the finite root, only one surface condition, consistent with $W^f = W^s$, can be applied. The exponential terms e^{-sZ/q_∞}, $e^{-s(2L-Z)/q_\infty}$, correspond to waves travelling with speed q_∞ in positive and negative z-directions. A free drainage solution corresponds to infinite permeability or zero drag D = 0, but the velocity equations (2.24) are coupled still through the constitutive coupling given by non-zero C_{12}, C_{21}, and there are two limit roots

$$D = 0: \theta_1^2 \to s^2 q_1^2, \quad \theta_2^2 \to s^2/q_2^2 \ , \qquad (3.9)$$

where

$$2r_s r_f(q_1^2, q_2^2) = a \pm \sqrt{a^2 - 4r_s r_f\Delta} \ , \qquad (3.10)$$

which corresponds to waves travelling with speeds q_1 and q_2. We postulate that the medium with zero drag must propagate real waves for all $0 \le r_s \le 1$, $r_f = 1 - r_s$; that is, q_1^2 and q_2^2 are real and postive. This requires

$$C_{11} \ge 0, \ C_{22} \ge 0, \ C_{12}C_{21} \ge 0, \ q_1 \ge q_2. \qquad (3.11)$$

The transforms (3.2) have no analytic inversion. Garg et al. [1] compare finite difference solutions of the original differential equations on a half-space subjected to unit step function surface velocities

$$W_s(T) = W_f(T) = H(T) \ , \qquad (3.12)$$

with numerical inversions of the transforms. Large discrepancies arise due to the smearing of the two discontinuity fronts by the finite difference scheme. We now show that the transforms (3.2) yield analytic expressions for the discontinuities in the velocities and derivatives at the fronts, so that finite difference calculations are needed only for very smooth remainder velocities.

Since $\mathrm{Re}(\theta_r)$ is positive in $\mathrm{Re}(s) > 0$ as $|s| \to \infty$ and θ_r has no branch point in $\mathrm{Re}(s) > 0$ (for a bounded solution as $T \to \infty$), then θ_r^2 has no zero and hence $|\exp(-2L\theta_r)| < 1$ in $\mathrm{Re}(s) > 0$. Thus

$$X_r^{-1} = \sum_{n=0}^{\infty} e^{-2Ln\theta_r} \tag{3.13}$$

converges in $\mathrm{Re}(s) > 0$, and hence on the inversion contour, and hence

$$\bar{W}_s = \sum_{r=1}^{2} \sum_{n=0}^{\infty} \hat{A}_r (e^{-\alpha_n \theta_r} - e^{-\beta_n \theta_r}) ,$$

$$\bar{W}_f = \sum_{r=1}^{2} \sum_{n=0}^{\infty} B_r \hat{A}_r (e^{-\alpha_n \theta_r} - e^{-\beta_n \theta_r}) , \tag{3.14}$$

where

$$\alpha_n = Z + 2Ln, \; \beta_n = 2L(n+1) - Z , \quad n = 0,1,\ldots, \tag{3.15}$$

and $\hat{A}_r = X_r A_r$ is rational in θ^2. Further, since θ_r is analytic in $\mathrm{Re}(s) > 0$, inverse power series solutions of the form

$$\theta_r = \frac{s}{q_r}(1 + \frac{\ell_r}{s} + \ldots) \text{ for } |s| > R > 0 , \quad (r = 1,2) . \tag{3.16}$$

can be constructed from the quartic equation (3.4), and in turn for the coefficients B_r, A_r given the surface velocity transform expansions

$$\bar{W}_s = \frac{1}{s}(g_o + \frac{g_1}{s} + \ldots), \; \bar{W}_f = \frac{1}{s}(f_o + \frac{f_1}{s} \ldots) \quad \text{for } |s| > R > 0 , \tag{3.17}$$

where g_o, $f_o \neq 0$ in the case of step functions. Finally

$$\bar{W}^s = \sum_{r=1}^{2} \sum_{n=0}^{\infty} \left\{ e^{-\frac{\alpha_n s}{q_r}} e^{-\frac{\alpha_n \ell_n}{q_r}} G_r(\alpha_n, s) - e^{-\frac{\beta_n s}{q_r}} e^{-\frac{\beta_n \ell_r}{q_r}} G_r(\beta_n, s) \right\} , \tag{3.18}$$

$$\bar{W}^f = \sum_{r=1}^{2} \sum_{n=0}^{\infty} \left\{ e^{-\frac{\alpha_n s}{q_r}} e^{-\frac{\alpha_n \ell_r}{q_r}} F_r(\alpha_n, s) - e^{-\frac{\beta_n s}{q_r}} e^{-\frac{\beta_n \ell_r}{q_r}} F_r(\beta_n, s) \right\} ,$$

where $G_r(k,s)$ and $F_r(k,s)$ are inverse power series with lead terms G_{ro}/s, F_{ro}/s.

Invoking the inversion shift theorem,

$$e^{-xs}\bar{f}(s) :- f(T-x)H(T-x), \; x > 0 , \tag{3.19}$$

shows that wave fronts propagate along the forward paths $Z = q_r T - 2Ln$ and backward paths $Z = 2L(n+1) - q_r T$, with respective amplitude decay factors $\exp(-\alpha_n \ell_r/q_r)$, $\exp(-\beta_n \ell_r/q_r)$. We postulate $\ell_r^n > 0$ to determine restrictions on the elastic moduli $C_{\ell m}$. Figure 1 illustrates the fast (q_1) and slow (q_2) wave fronts initiated at $Z = 0$, $T = 0$ and successively reflected from $Z = L$ and $Z = 0$. By constructing G_r and F_r explicitly to terms of order $(1/s^3)$ - details are presented in Morland, Sandhu, Wolfe and Hiremath [6] - the remainder term is order $(1/s^4)$ giving rise to a term of order $(T-x)^3 H(t-x)$ at a front $T = x$, as $T \to x$. That is, the remainder terms are twice continuously differentiable across each front. Thus, in (2.24), (2.25), setting

$$W^s = W_*^w + W_c^s , \; W^f = W_*^f + W_c^f , \tag{3.20}$$

where W_*^s, W_*^f are the explicit inversions, yields coupled second order hyperbolic equations for the twice continuously differentiable fields W_c^s, W_c^f:

$$M_s[W_c^s, W_c^f] = - M_s[W_*^s, W_*^f] = Q_s(Z,T) ,$$

$$M_f[W_c^s, W_c^f] = - M_f[W_*^s, W_*^f] = Q_f(Z,T) , \tag{3.21}$$

where Q_s and Q_f are continuous.

4 ILLUSTRATION

We have applied the decomposition method, solving the smooth equation (3.21) by a simple explicit finite difference scheme, for the surface conditions (3.12) and alternative surface conditions

$$W_s = H(T), \; W_f = [1 - (1-W_o)e^{-\alpha T}]H(T) \tag{4.1}$$

which describe an imposed velocity difference $1 - W_o$ at $T = 0$ which decays to zero over a time scale $1/\alpha$ due to the drag. The same problems have been solved by the numerical inversion formula [7]

$$f(T) = \frac{1}{\tau} e^{rT} \sum_{k=0}^{M}{}' \mathrm{Re}[\bar{f}(r + \frac{k\pi i}{2\tau})]\cos(\frac{k\pi T}{2\tau}),$$

$$0 \leq T \leq \tau \ , \qquad (4.2)$$

where Σ' signifies that one-half of the $k = 0$ term is counted, and the sum from $k = M$ to infinity is supposed negligible. The error estimate is of order $\exp(-2r\tau)$. Additionally, the problems have been solved by a finite element procedure described by Sandhu, Wolfe and Shaw [8]. We have compared the velocity profiles and histories for surface conditions (3.12) and (4.1) with the physical properties used by Garg et al. [1], for which

$$\varphi = 0.8, \ r_s = 0.9238, \ r_f = 0.0762, \quad c_0 = 3548.5 \ \mathrm{ms}^{-1} \ ,$$

$$\hat{C}_{11} = 0.926175, \ C_{12} = C_{21} = 0.031207, \quad C_{22} = 0.011412 \ , \qquad (4.3)$$

$$q_\infty = 1, \ q_1 = 1.00921, \ q_2 = 0.36584.$$

Numerous illustrations for various drag parameters are shown in Morland et al. [6], but here we show one example for moderate drag

$$D = 0.13068, \ \ell_o = 0.5 \ \mathrm{m}, \ \ell_D = 3.8259 \ \mathrm{m}, \qquad (4.4)$$

with surface conditions (4.1) with $W_o = 0.8$, $\alpha = 1$. For all examples the finite element method distorts the wave fronts, the numerical inversion gives a mean value at the discontinuity, and the decomposition method gives an accurate wave front (and elsewhere) solution. Good agreements between all methods are obtained away from the fronts. The largest discrepancies arise in the fluid velocity W^1, but as seen in a typical fluid velocity history shown in Fig. 2 for $Z = 0.4$, $0 \leq T \leq 7$ (time unit 3.93×10^{-3} s), during which six reflexions of the fast wave and two reflexions of the slow wave at $Z = 1$ and $Z = 0$ occur, the finite element calculations are very good. This adds confidence to their use with non-linear material models for which a linear analysis is not possible.

REFERENCES

[1] S. K. Garg, A. H. Nayfeh and A. H. Good, Compressible waves in fluid-saturated elastic-porous media, J. Appl. Phys. 45, 1968-1974 (1974).

[2] C. Truesdell, On the Foundations of Mechanics and Energetics, 293-304, Gordon and Breach, New York, 1965.

[3] L. W. Morland, A simple constitutive theory for a fluid-saturated porous solid, J. Geophys. Res. 7, 890-900 (1972).

[4] L. W. Morland, A theory of slow fluid flow through a thermoelastic matrix, Geophys. J. Roy. Astr. Soc. 55, 393-410 (1978).

[5] M. A. Biot and D. G. Willis, The elastic coefficients of the theory of consolidation, J. Appl. Mech. 79, 594-601 (1957).

[6] L. W. Morland, R. S. Sandhu, W. E. Wolfe and M. S. Hiremath, Wave propagation in a fluid-saturated elastic layer, Ohio State University Rep. 717885-87-5 (1987).

[7] H. Dubner and J. Abate, Numerical inversion of Laplace transforms by relating them to the finite Fourier cosine transform, J. Assoc. Comput. Mach. 15, 1- (1968).

[8] R. S. Sandhu, W. E. Wolfe and H. L. Shaw, Dynamic response of saturated soils using three-fluid formulation, Ohio State University Rep. 716894-86-3, (1986).

ACKNOWLEDGEMENT

This work was done in 1986 while Dr. L. W. Morland was on study leave in the Civil Engineering Department at Ohio State University, supported by the Air Force Office of Scientific Research under Grant AFOSR-83-0055. The numerical calculations were performed by research assistants Mr. M. S. Hiremath and Mr. R. Singh.

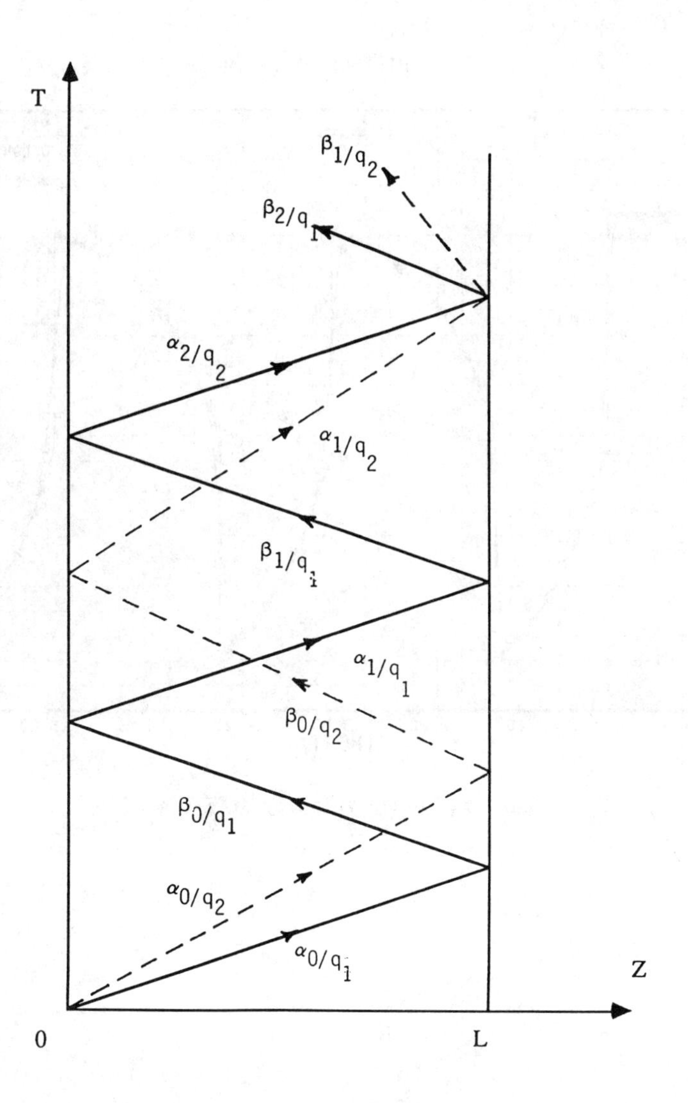

Figure 1: Fast and Slow Wave Fronts

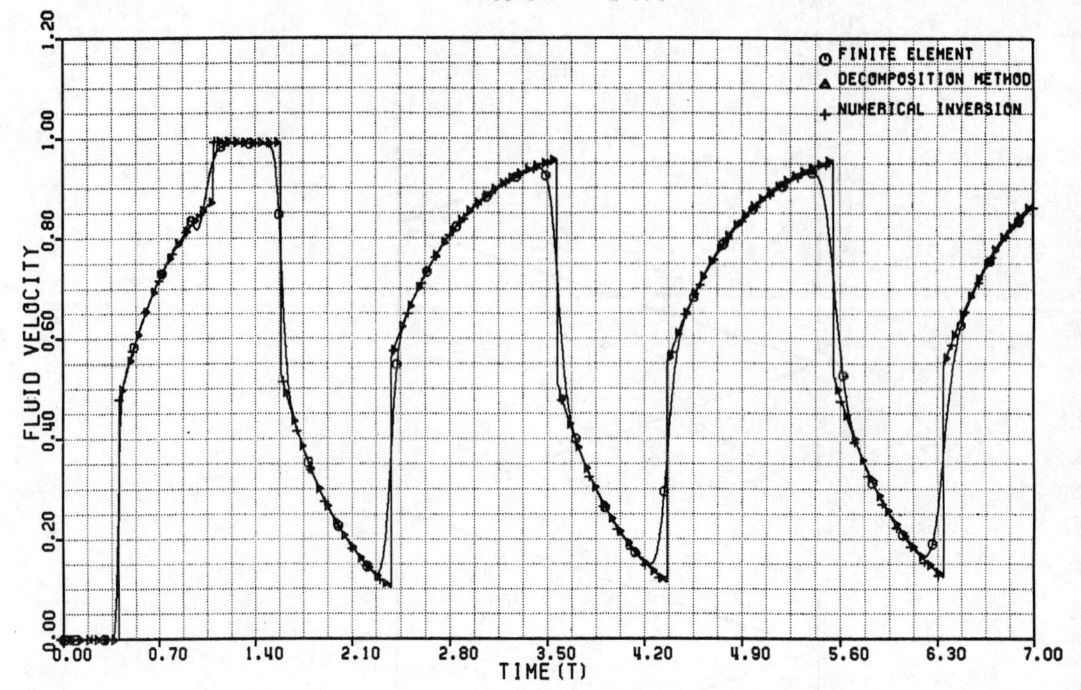

Figure 2: Fluid Velocity at Z=0.4

Numerical Methods in Geomechanics (Innsbruck 1988), Swoboda (ed.)
© 1988 Balkema, Rotterdam. ISBN 90 6191 809 X

Formulations and a methodology for computing the response of unsaturated soils to changes in temperature

H.H.Vaziri
Golder Associates, Seattle, Wash., USA

ABSTRACT: Theoretical expressions, in terms of elastic and thermal properties of multi-phase soil systems, have been derived for the change in pore fluid pressure under undrained and the change in soil stress under drained conditions. A methodology is proposed for linking these equations to coupled flow and deformation finite-element formulations that are applicable to partially saturated soils. The developed finite-element program satisfies the complex temperature, fluid flow and stress interactions under any specified displacement and fluid pressure boundary conditions. These facets are demonstrated by comparing the computational results with closed-form solutions and analyzing a practical problem that is typical of an enhanced recovery process often employed in extracting highly viscous bitumen from oil-sand deposits.

1. INTRODUCTION

Temperature variations can significantly influence the behavior of soils. Such effects are of practical concern in the geotechnical aspects of numerous engineering projects, such as power plants, underground storage of nuclear waste, underground power cables, liquefied natural gas storage and pipelines, geothermal energy development, and recovery of bitumen from heavy-oil reservoirs by the process of steam injection.

In general, variations in temperature affect pore fluid pressure, induce changes in volume, and modify some of the engineering properties of soil. The changes in volume include the expansion of solid particles, of the soil skeleton, and of the pore fluid. Invariably, the volume increase of the pore fluid is greater than that of the voids in the soil skeleton, producing an increase in pore fluid pressure and a consequent reduction in effective stress. If the thermal diffusivity of the soil is relatively more dominant than the hydraulic diffusivity, excess pore pressure will develop as a result of any temperature increase. The increase in pore pressure can potentially cause soil liquefaction in relatively impermeable soils subjected to a rapid increase in temperature. During subsequent consolidation, excess pore pressure dissipates, effective stresses increase, and displacements reduce to eventually become equal to those corresponding to the expansion of the solid grains alone.

A theoretical analysis of volume change in saturated soil subjected to temperature variations, and pore pressure changes due to undrained heating has been proposed by Campanella and Mitchell [1]. The formulations are valid for pressures above the gas/liquid saturation pressure (i.e., the pore pressure sufficient to prevent gas evolution), with the assumption that the solid particles are incompressible. Although in their analysis, Campanella and Mitchell made provision for solid particle compressibility, it is not treated in a consistent manner and their formulation is in fact strictly true only if the compressibility of the solids is zero. The approach proposed herein takes into account the compressibility of solid grains, as well as the compressibility of occluded gases in the fluid phase.

In this paper, expressions are derived for the change in pore pressure under undrained conditions, and the change in soil stress under fully restrained conditions, due to changes in temperature. The expressions provide the basis for the development of a finite-element code linking temperature changes to fully coupled stress/deformation and fluid-flow phenomena. The model is verified with an example for which a closed-form analytical solution is available. The finite-element code is then used to simulate the recovery process of bitumen from tar-sands by means of heat application.

2. THEORETICAL RATIONALE

The soil matrix can be visualized as an arrangement of solid grains, with liquid and gas phases occupying the voids between the grains. All soil components have finite compressibility characteristics. Consider an element of soil, as illustrated in Figure 1.

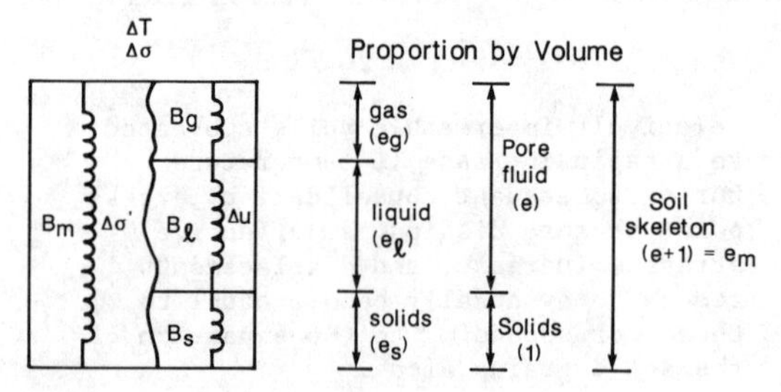

Figure 1. Conceptual model of soil constituents

By subjecting this element of soil to a change in temperature (ΔT) and a change in total stress (Δσ), a change in pore pressure (Δu) will be developed under undrained conditions. Under these conditions, the overall change in volume is a function of thermal expansion as well as the change in pore fluid pressure (Δu), and the change in effective stress (Δσ-Δu). To account first for pressure- and stress-related volume changes, the problem can be decomposed into two stages for analysis as shown in Figure 2.

In stage (a), a pore-pressure change, Δu, and an equal total stress change, Δσ, are applied. The volume change of each component can be defined in terms of the bulk modulus, B, and the volume of the component expressed as proportions of the volume of solids, e_s (see definitions in Figure 1):

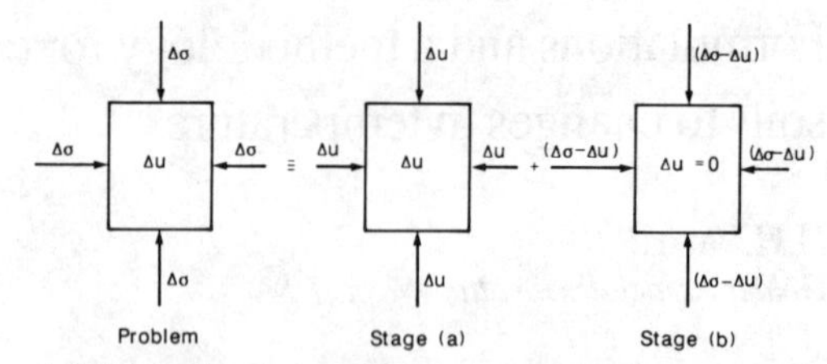

Figure 2. Diagramatic representation of the stress transfer in the soil system

$$\Delta e_g = e_g \Delta u / B_g \tag{1}$$

$$\Delta e_\ell = e_\ell \Delta u / B_\ell \tag{2}$$

$$\Delta e_s = \Delta u / B_s \quad \text{(Note } e_s = 1) \tag{3}$$

$$\Delta e_m = e_m \Delta u / B_s \tag{4}$$

In stage (b), a total stress change equal to (Δσ-Δu) is applied, with no change in pore pressure. The volume changes are then given by:

$$\Delta e_s = e_m(\Delta\sigma - \Delta u)/B_s \tag{5}$$

$$\Delta e_m = e_m(\Delta\sigma - \Delta u)/B_m \tag{6}$$

where B_m is the bulk modulus of the soil matrix under drained conditions.

Next, the changes in volume caused by thermal expansion must be considered. Under stress-free boundary conditions these are:

$$\Delta e_g = - e_g \alpha_g \Delta T \tag{7}$$

$$\Delta e_\ell = - e_\ell \alpha_\ell \Delta T \tag{8}$$

$$\Delta e_s = - \alpha_s \Delta T \tag{9}$$

$$\Delta e_m = - e_m \alpha_s \Delta T \tag{10}$$

where α is the thermal expansion coefficient.

In addition, as has been physically observed (Campanella and Mitchell [1]; Agar et al. [2]), a change in temperature may induce a change in interparticle forces or frictional resistance that necessitates some reorientation or movement of soil grains to permit the soil structure to carry the same stress. Denoting this component of volume change by, Δe_t, we can write:

$$\Delta e_t = e_m \alpha_t \Delta T \tag{11}$$

where α_t is the thermal expansion coefficient for the soil structure and can be positive or negative depending on the boundary conditions, state of temperature, thermal history, porosity and the soil type.

Using these formulations for volume change due to each component of the soil system we can proceed by considering two cases:

(a) Free case, where the soil element is subjected to a change in temperature only (i.e., no change in the external loads).

(b) Restrained case, where the soil element is subjected to a change in temperature and a change in boundary pressure, as may arise from any impedence of the thermally induced movements.

(a) Free case ($\Delta\sigma = 0$)

For this situation, as there is no change in the applied load, the change in effective stress is:

$$\Delta\sigma' = -\Delta u \tag{12}$$

Under undrained conditions, volumetric compatibility can be invoked between the soil matrix and its constituents:

$$\Delta e_m = \Delta e_s + \Delta e_\ell + \Delta e_g + \Delta e_t \tag{13}$$

After making the relevant substitutions and rearrangement of the terms, the following expression for the excess pore pressure can be obtained:

$$\Delta u = \frac{\Delta T(\alpha_g e_g + \alpha_\ell e_\ell + \alpha_s e_s - \alpha_s e_m - \alpha_t e_m)}{(e_g/B_g + e_\ell/B_\ell + e_s/B_s + e_m/B_m - 2e_m/B_s)} \tag{14}$$

Equation (14) provides the expression for predicting pore pressures under undrained conditions with no change in boundary stresses. The formulation is valid for unsaturated as well as saturated soils, and takes account of both solid and fluid compressibilities.

For saturated soils (i.e., e_g=0), equation (14) simplifies to:

$$\Delta u = \frac{n\Delta T(\alpha_\ell - \alpha_s) - \alpha_s \Delta T}{1/B_m + n/B_\ell - (1 + n)/B_s} \tag{15}$$

where n is the porosity; $n = e_\ell/e_m$.

Equation (15) differs from the one proposed by Campanella and Mitchell [1] since their analysis neglects the change in matrix volume due to the change in pore fluid pressure (Δu), as defined by equation (4).

(b) Restrained case ($\Delta w = 0$)

In this case, the soil element under consideration is restrained from deforming. Such a constraint can arise if fixity is enforced by prescribed boundary conditions. The main reason, however, for studying this particular situation is that it forms the basis of the thermo-elastic approach adopted in the finite-element method described in the next section. In this case, the change in temperature results in some change in pore pressure, as well as in effective stress caused by the boundary restraint.

The assumption of no volume change implies that:

$$\Delta e_s + \Delta e_\ell + \Delta e_g + \Delta e_t = 0 \tag{16}$$

and

$$\Delta e_m = 0 \tag{17}$$

After making the relevant substitutions:

$$\Delta u(e_s/B_s + e_\ell/B_\ell + e_g/B_g) - \Delta T(e_s\alpha_s + e_\ell\alpha_\ell + e_g\alpha_g - e_m\alpha_t) + e_m\Delta\sigma'/B_s = 0 \tag{18}$$

$$e_m\Delta u/B_s + e_m\Delta\sigma'/B_m - e_m\alpha_s\Delta T = 0 \tag{19}$$

Equations (18) and (19) can be solved for the unknowns Δu and $\Delta\sigma'$ to give:

$$\Delta u = \frac{\Delta T(e_s\alpha_s + e_\ell\alpha_\ell + e_g\alpha_g - e_m\alpha_t - e_m\alpha_s B_m/B_s)}{(e_s/B_s + e_\ell/B_\ell + e_g/B_g - e_m B_m/B_s{}^2)} \tag{20}$$

$$\Delta\sigma' = B_m\alpha_s\Delta T - B_m\Delta u/B_s \tag{21}$$

Equations (20) and (21) provide theoretical expressions for the change in pore pressure and effective stress when the soil is restrained from any overall volume change. Moreover, under drained conditions, when there is no excess pore pressure, effective stresses can be explicitly determined from equation (21) as a function of change in temperature.

3. FINITE ELEMENT FORMULATIONS

In this section, the basic equations governing the consolidation of a porous medium saturated with a compressible fluid are derived and a methodology for computing changes in pore pressure and effective stress caused by thermal variations is described. As will be shown, changes in temperature can be applied at any stage during the consolidation process under any desired pore-pressure boundary condition.

Using the conventional Cartesian tensor notation, Biot's [3] equations may be derived assuming: (a) the stresses are in equilibrium; (b) the effective stresses are related to the strains by some suitable relationship; (c) the flow of fluid through the soil is governed by Darcy's Law; and (d) continuity is satisfied for a compressible pore fluid. This leads to the following equations:

$$\sigma_{ij},_j - F_i = 0 \qquad (22)$$

$$\sigma'_{ij} = \sigma_{ij} - u\delta_{ij} = D_{ijk\ell}\epsilon_{k\ell} \qquad (23)$$

$$v_i = \frac{k_{ij}u,_j}{\mu\gamma_f} \qquad (24)$$

$$- v_{i},_i + \dot{w}_{i},_i - \dot{u}/B_e = 0 \qquad (25)$$

where σ_{ij} is the total stress tensor, F_i is the body force, $D_{ijk\ell}$ is the tensor relating stress and strain, w_i is the displacement vector, v_i is the superficial velocity vector, k_{ij} is the permeability tensor, γ_f is the fluid unit weight, μ is the dynamic viscosity, ϵ_{ij} is the strain vector ($\epsilon_{ij} = 1/2(w_i,_j+w_j,_i)$), B_e is the equivalent bulk modulus of the compressible phases that constitute the soil matrix ($1/B_e = n/B_f+(1-n)/B_s$ in which B_f is bulk modulus of the fluid phase).

To define the problem, both displacement and flow boundary conditions must be specified. With reference to Figure 3, these can be written as follows:

$$\sigma_{ij}n_j = \bar{T}_i \text{ on } S_t \text{ for } t\geq 0 \qquad (26)$$

$$w_i = \bar{w}_i \text{ on } S_w \text{ for } t\geq 0 \qquad (27)$$

$$u = \bar{u} \text{ on } S_u \text{ for } t>0 \qquad (28)$$

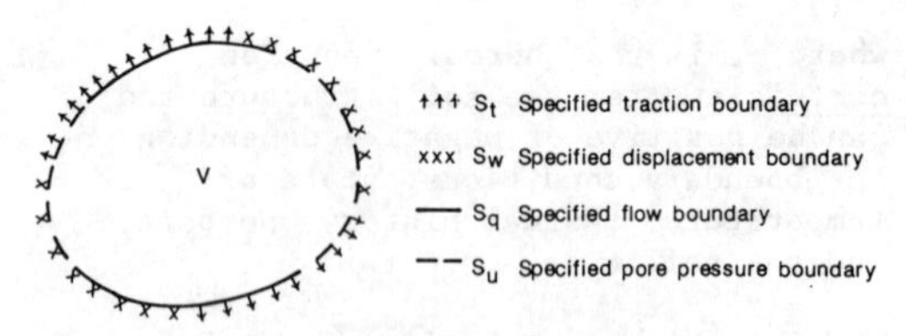

Figure 3. Elastic and hydraulic boundary conditions for a body of volume V

$$v_i n_i = 0 \text{ on } S_q \text{ for } t\geq 0 \qquad (29)$$

where S describes that part of the boundary surface subjected to the boundary conditions, n_i is the normal vector to the boundary surface, the bar symbol (-) indicates a prescribed quantity, and T_i are the applied tractions.

Making the appropriate substitutions into the equilibrium and continuity equations yields the following coupled governing equations:

$$(1/2(w_i,_j + w_j,_i)D_{ijk\ell}),_j + u,_j = F_i \qquad (30)$$

$$\frac{(-k_{ij}u,_j),_i}{\mu\gamma_f} + \dot{w}_{i,i} - \frac{\dot{u}}{B_e} = 0 \qquad (31)$$

To develop the finite-element formulation of these equations, it is assumed that the independent variables w_i and u can be approximated in space by means of assumed shape functions as follows:

$$\{w_i^*\} = [N_w]\{w_d\} \qquad (32)$$

$$\{u^*\} = [N_u]\{u_p\} \qquad (33)$$

where N_w and N_u are the assumed shape functions for the displacements and pore pressures respectively, * denotes the approximate values of the variables within the element, subscripts $_d$ and $_p$ indicate the nodal number used for displacement and pore pressure nodes in an element.

Employing Galerkin's procedure, it can be shown that the incremental finite-element solution of equations (30) and (31) subject to the boundary conditions shown by equations (26) to (28) is (Vaziri [4]):

$$\begin{bmatrix} [K] & [L] \\ [L]^T & -[E]\Delta t-[G] \end{bmatrix} \begin{Bmatrix} \{\Delta w\} \\ \{\Delta u\} \end{Bmatrix} = \begin{Bmatrix} \{\Delta A\} \\ [E]\Delta t\{u_t\} \end{Bmatrix} \qquad (34)$$

where

$$[K] = \sum_{n=1}^{N} \int_{v} [B_w]^T [D] [B_w] dv$$

$$[L] = \sum_{n=1}^{N} \int_{v} [B_w]^T \{m\} [N_u] dv$$

$$[E] = \sum_{n=1}^{N} \int_{v} \frac{1}{\gamma_f \mu} [B_u]^T \{k\} [B_u] dv$$

$$[G] = \sum_{n=1}^{N} \int_{v} \frac{1}{B_e} [N_u]^T [N_u] dv$$

$$\{A\} = \int_{s} [N_w]^T \{T\} ds + \sum_{n=1}^{N} \int_{v} [N_w]^T \{F\} dv$$

$[B_w]$ and $[B_u]$ are the derivatives of shape-function matrices $[N_w]$ and $[N_u]$, N refers to the total number of elements, $\{m\}$ is a vector made up of unity and zero to relate the appropriate soil matrix components to the pore fluid, and t designates time.

Having derived the coupled fluid-flow and stress formulations, we can now consider implementation of the thermally induced pore-pressure and stress solutions developed in the previous section. The method adopted for computing the soil response due to a change in temperature is schematically represented in Figure 4.

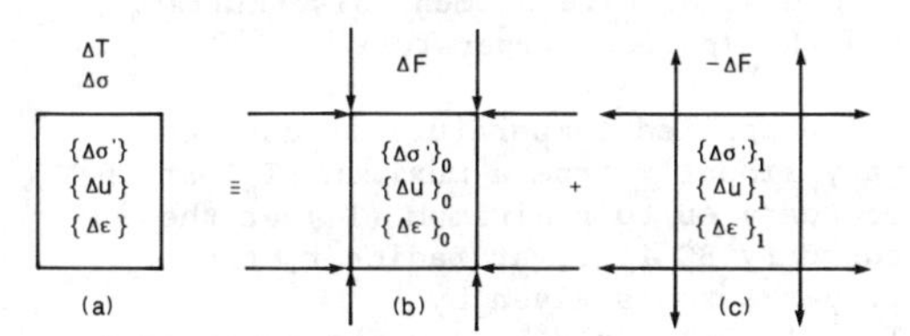

Figure 4. Equivalent representation of stresses and strains due to thermal effects

As discussed in the previous section, unless the element of soil is completely free from any constraints, it will be subjected to changes in boundary stress as a result of thermal expansion (Figure 4a). The finite-element modeling of this problem is most conveniently performed by first calculating the isotropic force system required to prevent any volume change in the soil element due to the temperature change (Figure 4b). The domain is then analyzed elastically, subject to an equal and opposite force system, allowing the thermo-elastic displacements and strains to be computed (Figure 4c).

For the situation depicted in Figure 4c, the changes in pore pressure and effective stress can be computed using equations (20) and (21), respectively. Thus, by knowing $\{\Delta u\}_o$ and $\{\Delta\sigma'\}_o$, the equivalent nodal forces $\{\Delta F\}$ to restrain the element can be computed. One approach is to invoke the principle of virtual work:

$$\{\bar{w}\}\{\Delta F\} = \int_{v} \{\bar{\epsilon}\}^T \{\Delta\sigma + \Delta\sigma'\}_o \tag{35}$$

where $\{\bar{w}\}$ and $\{\bar{\epsilon}\}$ are the virtual displacement and strain vectors.

Using the strain-displacement relationship, equation (35) can be simplified to:

$$\{\Delta F\} = \int_{v} [B_w]^T \{\Delta u + \Delta\sigma'\}_o \tag{36}$$

In analyzing problems where the nodal pore fluid pressures are prescribed, such as along drainage boundaries where the incremental change in pore pressures is forced to be zero, the appropriate adjustments to the thermally induced pore pressures must be made prior to computing the nodal forces. Under these circumstances equation (36) must be preceded by the following steps:

$$\{\Delta u_p\}_o = [N_u]^{-1} \{\Delta u\}_o \tag{37}$$

An averaging technique must be employed to correctly account for computing pore pressure at nodes that are shared by more than one element. Subsequently $\{\Delta u_p\}$ values at the nodes subjected to prescribed pore pressure boundary conditions will be adjusted to the specified values. Then the revised elemental pore pressures (designated by subscript r) can be calculated:

$$\{\Delta u_r\}_o = [N_u] \{\Delta u_p\}_o \tag{38}$$

Hence, the nodal forces representative of the thermally induced fluid pressure and stress changes, subject to prescribed fluid-pressure boundary conditions, can be calculated by using the revised pore pressures in equation (36):

$$\{\Delta F_r\} = \int_V [B_w]^T \{\Delta u_r + \Delta\sigma'\}_o \tag{39}$$

The next step is to apply forces equal and opposite to those given by equation (39) and compute the resulting displacements, strains $\{\Delta\epsilon\}_1$ stresses $\{\Delta\sigma'\}_1$ and elemental pore pressure $\{\Delta u\}_1$ and nodal pore pressure changes $\{\Delta u_p\}_1$. Finally the required stress, strain and pore pressure changes are calculated as follows:

$$\{\Delta\epsilon\} = \{\Delta\epsilon\}_1 \tag{40}$$

$$\{\Delta\sigma'\} = \{\Delta\sigma'\}_o + \{\Delta\sigma'\}_1 \tag{41}$$

$$\{\Delta u\} = \{\Delta u_r\}_o + \{\Delta u\}_1 \tag{42}$$

$$\{\Delta u_p\} = \{\Delta u_p\}_o + \{\Delta u_p\}_1 \tag{43}$$

It should be noted that the above procedure enables thermal analysis to be performed without the need for making significant modifications to existing finite-element codes. This is because the equivalent loads due to thermal effects are explicitly computed and then treated in the same manner as any specified applied load. A finite-element code called ENHANS, has been developed incorporating the above formulations. ENHANS also caters for nonlinear material behavior, nonlinear fluid-flow characteristics, and the evolution of dissolved gases; it is described in Vaziri [5] and is now commercially available.

4. VERIFICATION

Closed-form solutions provided by Timoshenko and Goodier [6] for a non-porous circular cylinder subjected to a temperature change have been chosen for verification purposes. The general solutions for plane strain (taking compressive stresses as positive) are:

$$w = \frac{(1+\nu)}{(1-\nu)}\alpha \frac{1}{r}\int_a^r Trdr + C_1 r + \frac{C_2}{r} \tag{44}$$

$$\sigma_r = \frac{\alpha E}{(1-\nu)}\frac{1}{r^2}\int_a^r Trdr - \frac{E}{(1+\nu)}\left[\frac{C_1}{(1-2\nu)} - \frac{C_2}{r^2}\right] \tag{45}$$

$$\sigma_\theta = \frac{\alpha E}{(1-\nu)}\frac{1}{r^2}\int_a^r Trdr + \frac{\alpha ET}{(1-\nu)} - \frac{E}{(1+\nu)}\left[\frac{C_1}{(1-2\nu)} + \frac{C_2}{r^2}\right] \tag{46}$$

$$\sigma_z = \frac{ET}{(1-\nu)} - \frac{2\nu EC_1}{(1+\nu)(1-2\nu)} \tag{47}$$

where E is the modulus of elasticity, ν is Poisson's ratio, α is the coefficient of linear thermal expansion, a is the internal radius, r is the radial distance, and C_1 and C_2 are constants which must be evaluated for the particular boundary conditions relevant to the problem.

The general problem under consideration is illustrated in Figure 5.

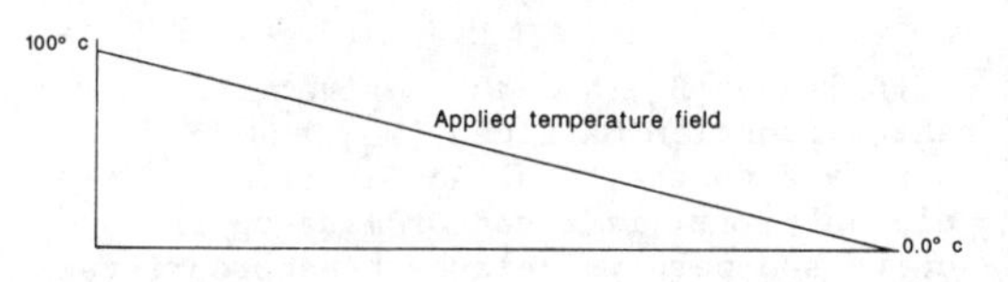

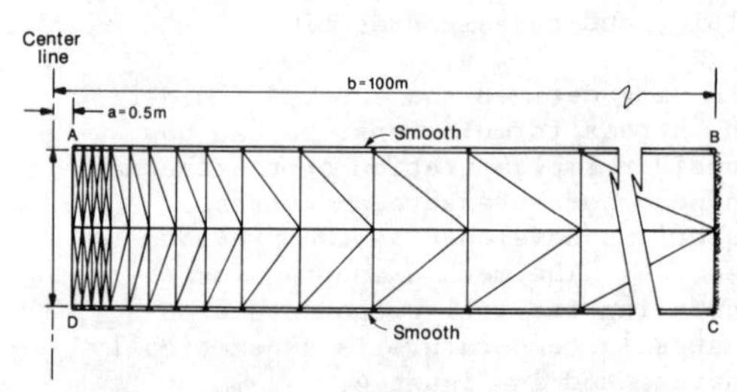

Figure 5. Finite-element discretization and the applied temperature

The applied temperature is assumed to vary linearly from a maximum (T_a) at the boundary AD to a minimum (T_b) at the boundary BC i.e., at radius r, the temperature is given by $T = (T_a - T_b)(b-r)/(b-a)$. The parameters chosen to describe the problem are; $E = 10^5$ kPa, $\nu = 0.25$, $\alpha = 10^{-4}\ K^{-1}$, $a = 0.5$ m, $b = 100$ m, $T_a = 100^oC$, $T_b = 0^oC$. Four cases have been studied as indicated in Table 1.

Figures 6, and 7 illustrate the variation of stress components with radial distance for Cases 2 and 4, respectively. Figure 8 shows the variation of displacements with the radial distance for the two cases.

Similar results were obtained for the other two cases. The close agreement between analytical and finite-element results verifies satisfactory implementation of the theory for the thermal expansion of a non-porous cylinder.

Table 1. Cases analyzed for verification purposes

Case	Boundary conditions		Constants	
	AD (r=a)	BC (r=b)	C1	C2
1. AD and BC Free	$\sigma_r = 0$	$\sigma_r = 0$	1.4E-4	6.9E-5
2. AD Fixed and BC Free	$w = 0$	$\sigma_r = 0$	1.4E-4	-3E-5
3. AD and BC Fixed	$w = 0$	$w = 0$	-2.7E-4	6.9E-5
4. Solid Cylinder with BC Fixed	$a = 0$	$w = 0$	-2.7E-4	0.0

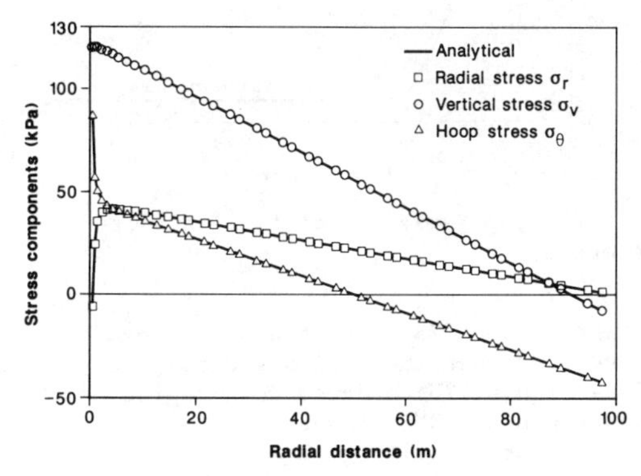

Figure 6. Radial distribution of thermally-induced stresses for a long cylinder (Case 2)

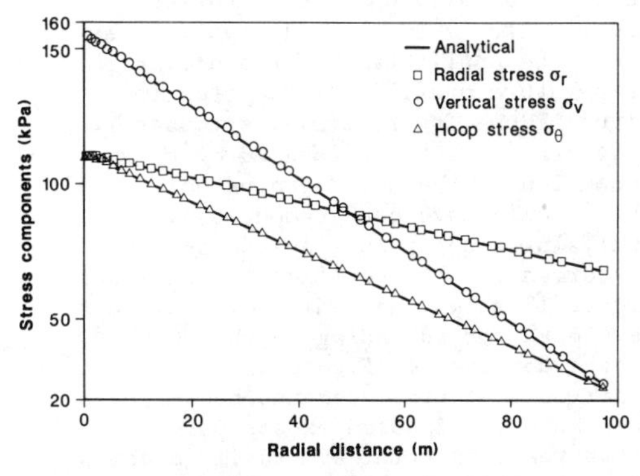

Figure 7. Radial distribution of stress components for a solid cylinder (Case 4)

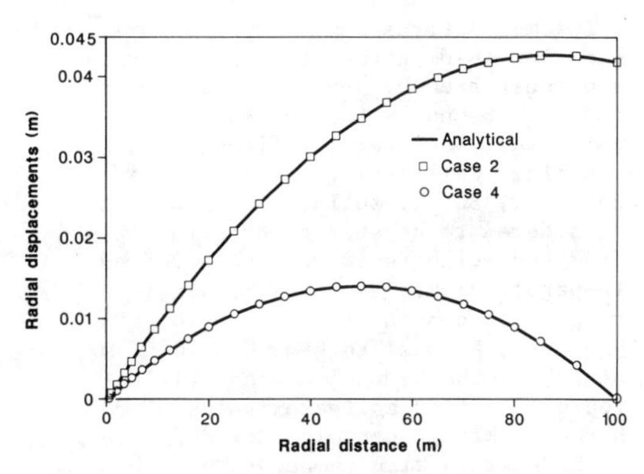

Figure 8. Radial variation of thermally-induced displacement for a long cylinder (Cases 2 & 4)

5. APPLICATION

Simulation of the steam-injection recovery process often employed in extracting viscous bitumen from oil-sand deposits constitutes a useful application for ENHANS. Some of the unusual engineering properties and characteristics of oil-sands are described in Vaziri [7]. In order to demonstrate some of the salient aspects of the analysis, without unduly complicating the understanding of the computational results, a simple geometrical configuration is considered, subject to realistic boundary conditions. Since the boundary conditions are relatively complex, elastic material behavior is assumed in this example to eliminate soil failure effects from the response of the oil-sand layer when subjected to time-dependent flow and temperature application. The problem to be analyzed is shown in Figure 9.

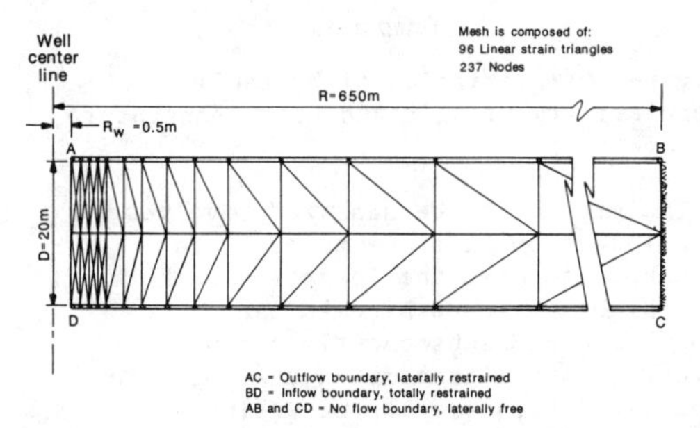

Figure 9. Finite element mesh of the wellbore through a layer of oil-sand

The mesh represents a layer of oil-sand confined between two rigid strata and penetrated by a wellbore 0.5 m in radius. Boundary AD, representing the well face, is laterally fixed to model the rigid well casing. The outside boundary, BC, is fully fixed since it is considered to be sufficiently far away from the wellbore to be unaffected by temperature changes. The material properties chosen are: $E = 10^6$ kPa, $\nu = 0.25$, $B_f = 2.7E6$ kPa, $K = 10^{-12}$ m/s at 10^oC. The hydraulic conductivity represents flow characteristics of bitumen through densely packed fine oil-sand under in situ temperature conditions. With the heat application, bitumen's viscosity is lowered, thus increasing its mobility and in turn increasing hydraulic conductivity. The variation shown in Figure 10 has been employed in the present analysis. In situ stress and pore pressure conditions, which are typical for strata at approximately 200 m depth, are: $\sigma_v' = 3500$ kPa, $\sigma_r' = 2000$ kPa, $\sigma_\theta' = 2000$ kPa, and $u = 500$ kPa.

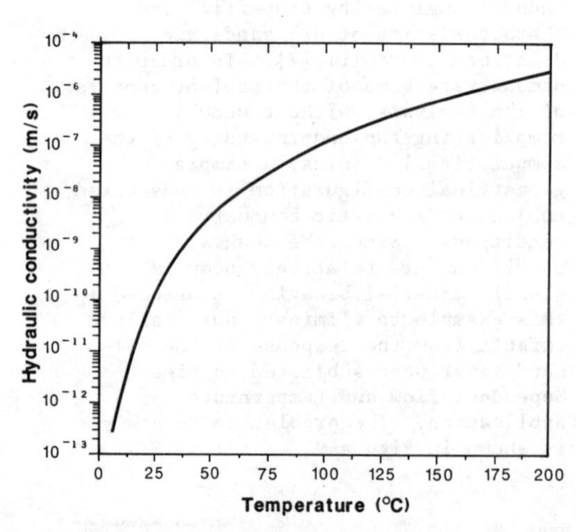

Figure 10. Variation of hydraulic conductivity of oil-sand with temperature

The problem to be analyzed involves the injection of steam through the borehole to heat the formation such that the highly viscous bitumen can be mobilized and subsequently produced. In order to avoid fracturing the formation, the fluid pressure at the wellbore is controlled initially at 1100 kPa (i.e., 600 kPa above in situ pressure) while the temperature is raised. Figure 11 shows the temperature profile that is developed incrementally over approximately 3 days. A fully transient consolidation analysis is simulated; Figure 12a shows the effective stress components and the pore fluid pressure soon (approximately 1 minute) after the application of the final temperature increment.

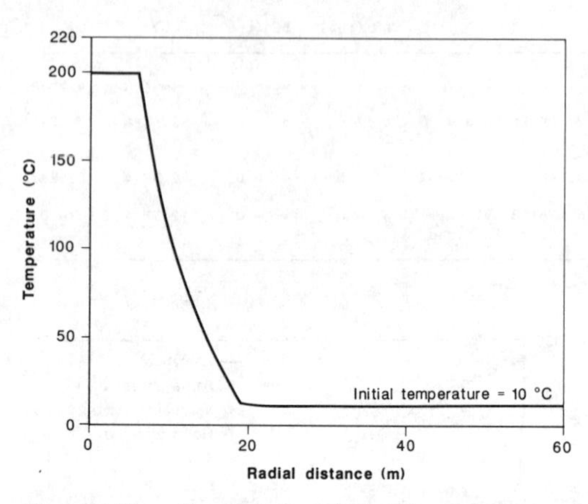

Figure 11. Variation of final applied temperature

After about 23 days, the pore pressure developed near the well has dissipated (see Figure 12b) because that region has sustained a 190^oC increase in temperature resulting in an enhancement of the hydraulic conductivity by approximately six orders of magnitude. At this point in the simulation, the wellface boundary condition is changed by setting the pore pressure equal to zero. The fully drained situation occurs after a further 37 days (60 days after the application of the ultimate temperature profile) as depicted in Figure 12c. The remaining stresses at this stage are those induced by thermal expansion of the solid grains only, which could have been computed by analyzing a non-porous system as performed for the verification analyses. Figure 13 shows the radial displacements developed corresponding to the three stages depicted in Figure 12. It can be observed that displacements decrease as the thermally induced excess pore pressures dissipate, eventually becoming equal to those caused by expansion of the solid soil grains only.

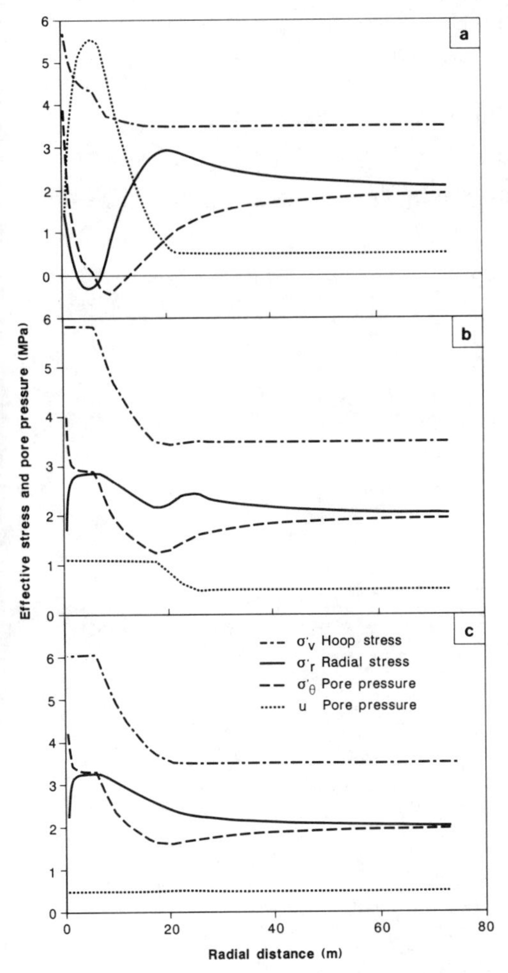

Figure 12. Radial distribution of effective stress and pore pressure after (a) one minute, (b) 23 days, (c) 60 days

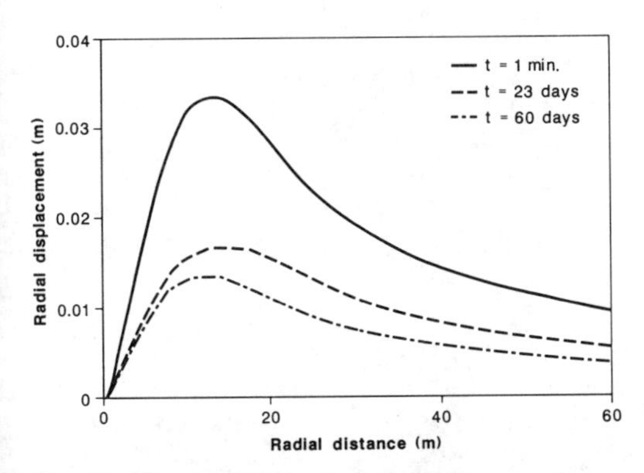

Figure 13. Variation of radial displacement at different times after the final temperature application

6. SUMMARY AND CONCLUSIONS

Heating of soils results in volume increases and pore fluid pressure increases if no drainage is allowed. Theoretical expressions, in terms of the elastic and thermal properties of soil, have been derived by maintaining volumetric compatibility between the soil matrix and its multi-phase constituents.

The proposed formulations are applicable to both saturated and unsaturated soils under any condition of drainage. The formulations coded into ENHANS include the soil matrix bulk modulus (B_m) whose value may vary in accordance to any incremental law employed for representing the stress-strain behavior. Situations where the solid and liquid compressibilities are considered to behave in a nonlinear fashion can also be accounted for. The proposed formulations allow for the effects of change in volume due to reorientation of soil particles when subjected to thermal effects. The magnitude of this component, however, requires some form of experimental measurement as its behavior cannot be theoretically linked to other fundamental soil properties. The formulations described herein have been incorporated into a coupled fluid-flow and deformation finite-element model (ENHANS) that employs nonlinear soil behavior and accounts for the gassy nature of the soil, as well as nonlinear fluid flow characteristics. This code has been applied to simulate the steam injection process used to recover oil from deep seated oil-sand deposits, to determine the time-dependent pore pressure and stress response. The example given demonstrates the changes in effective stress, pore pressure, and displacements as temperature is applied and the process of consolidation is allowed to continue. The analyses show that maximum displacements occur under short-term conditions in response to expansion of the pore fluid, and reduce to a minimum at the completion of consolidation, when the displacements become solely those due to expansion of solid grains.

ACKNOWLEDGEMENTS

Assistance provided by Golder Associates Inc. in presenting this work is gratefully acknowledged. I am indebted to Michael Kenrick who improved the clarity and general style of this paper. Perceptive review of the paper by Ian Miller resulted in some important and useful suggestions for enhancing the value of the paper.

BIBLIOGRAPHY

[1] R.G. Campanella and K.J. Mitchell, Influence of Temperature Variations on Soil Behavior, J. of SMFD, ASCE, SM3. 94, 709-734 (1968).

[2] J.G. Agar, N.R. Morgenstern and J.D. Scott, Thermal Expansion and Pore Pressure Generation in Oil Sands, Canadian Geotechnical Journal, 23, 327-333 (1986).

[3] M.A. Biot, General Theory of Three Dimensional Consolidation, J. of Applied Physics, 12, 155-164 (194).

[4] H.H. Vaziri, Mechanics of Fluid and Sand Production from Oil Sand Reservoirs, Petroleum Society of CIM, 37, 519-535 (1986).

[5] H.H. Vaziri, Nonlinear Temperature and Consolidation Analysis of Gassy Soils, Ph.D. Thesis, University of British Columbia, Vancouver, Canada, 548 p (1986).

[6] S. Timoshenko and J.N. Goodier, Theory of Elasticity, MCGraw Hill, New York, 1951.

[7] H. H. Vaziri, Finite Element Analysis of Oil Sands Subjected to Thermal Effects, Petroleum Society of IM, 37, 497-518 (1986).

Numerical Methods in Geomechanics (Innsbruck 1988), Swoboda (ed.)
© 1988 Balkema, Rotterdam. ISBN 90 6191 809 X

Several aspects of formation of structures of computer programs, selection and algorithmization of medium models in geomechanics

A.A.Yufin & V.I.Titkov
Moscow Academy of Civil Engineering, USSR
A.S.Morozov
'Hydroproject' Institute, Moscow, USSR
T.L.Berdzenishvili
Georgian Computer and Information Centre, Tbilisi, USSR

ABSTRACT: The paper discusses the principles of development of the computer code "STATAS", shows the computer analysis results of the outline displacement of the underground machine hall of the Khoudoni hydroelectric station based on the data of geomechanical triaxial compression tests of enclosing rocks as well as on the results of geophysical investigations.

Development of any computer code is started with defining a problem is to be designed. Potentialities of the code with a view of its future application in the period of operation are considered concurrently, i.e. its universal applicability, placement in architecture and certain algorithmic solutions for predicting the properties tending to upgrade the technical parameters of a program. The designers of DINAS computer code (Yufin, 1979; Yufin et al., 1985) kept in their mind these up-to-date requirements imposed on program products using the most advanced programming techniques.

The users of the program, depending on their qualification, are interested in obtaining the results for every of their problems as cheap and quick as possible. The DINAS computer code provides for various possibilites to meet the users' demands depending on the level of their qualification (first of all it depends on the extent of automation of the process of the input data preparation as well as the procedure of computation of objects tending to change with time which constitute the main class of problems of geomechanics connected with stage-by-stage construction of structures and open and underground excavations).

Everybody's contradictory requirements are taken into account by the designers intuitively. The following are the main factors determining the accuracy of obtained results (displacements, stresses, etc.).

1. Mathematical model of mechanical properties of the medium. Generally, this is a polynominal whose exponents and coefficients are determined by a user proceeding from his idea of the adequacy of the in-situ and laboratory studies and their interrelation. In principle, any accuracy of reproduction of experimental diagrams can be attained but actually it is limited either by the program or by experimental data adequacy and completeness. Quantitative evaluation of the accuracy of solution is of no practical importance in this case.

2. Solution method with an available mathematical model of the medium properties (in our case this is the FEM). The user establishes the parameters of the FEM mesh himself proceeding from his own requirements of the accuracy, scale factor, program potentialities and the computer he uses. The quality of the results depends not only on the quantitative factors (number of units and elements in the diagram) but also on the quality of the FEM mesh. Particular attention was paid to this question in the DINAS computer code. The quality of the results in case the program or computer resource is exhausted can be upgraded due to:

a) application of higher-order isoparametric elements;

b) dividing of the computational zone into the zones containing elements of different types and density;

c) introduction of iteration in dividing computational zone in case of high stress gradients;

d) upgrading the automation of mesh construction at the initial stage of defining the initial data;

3. Accuracy of input data. It was partly considered under item 1. Here description of the geometry of computational zone and boundary conditions is meant.

4. The quality of programming and algorithms. Round off error accumulation in the computer arithmetic operations decreases due to decrease in the number of arithmetic operations in the computational algorithm. First of all it requires the bandwidth of non-zero terms in the elasticity matrix of the system to be made smaller which, in its turn, requires development of an effective algorithm of renumeration of uniting points. Sometimes one is forced to use an expensive precision increasing procedure - the double precision arithmetics.

Even the highest precision computer codes are unlikely to ensure the required quality of the results, if the jobs under items 1 and 3 are made inadequately. Here it should be noted that joint application of elements of different types and orders in one computational scheme can (as the authors' experience showed) increase the computer capability in obtaining quality results but contradicts to the idea of effective automation of the computational scheme construction procedure. Moremover, such an approach requires highly skilled program users. The computer resource (computation time and memory) can also be increased due to decrease in the number of unknown quantities in the mathematical model, e.g., by introducing infinite or boundary elements at the boundaries of the zone. However, the determining parameters of infinite elements are problem-dependent and require special attention when making computations, while joint use of finite and boundary elements leads to the loss of generality of computational algorithms and does not result in any savings. That is why, the finite element method of computations is more preferable in the majority of cases than the combined scheme.

The economic efficiency of a program product depends not only and not so much on the computer resource utilization, but on the quantity and quality of manual operations for defining initial information and computing of obtained results which is of great importance.

The initial data for every problem are prepared and stored as program-compilated file sets in the computer peripheral memory (rather than cards or video-displayed lines). Since every input data array is compilated line by line, provision is made for every file to be corrected for data check-out.

Application of automation for defining initial data will decrease the time for manual (non-computer-related) designing and the number of errors in the initial information. The structure and content of program diagnostics are aimed at reducing manual operations and computer time for adjustment of the initial data.

The initial data in the problems to be solved are of such a character that it is impossible to eliminate mistakes using only debugging aids. Debugging as a rule is transferred to the first production runs. From the viewpoint of economy the greatest attention is paid to the completeness and quality of diagnostics.

High efficiency of the program operation with computer runs can be attained:

1. methodologically by selecting an optimum iteration process;

2. algorithmically by reducing the number of arithmetic operations, minimising calls to outer media, etc.;

3. programmably by using low level language - an assembler - in separate subroutines.

Since the output information in the system is highly diverse, the structure and language have been created to describe the required subsets of output data (which is defined by the user). This reduces the time for obtaining and computing the results. The results obtained by the authors using the above mathematical software have been widely published. The DINAS computer code is developed for wide application by relevant research and designing organizations.

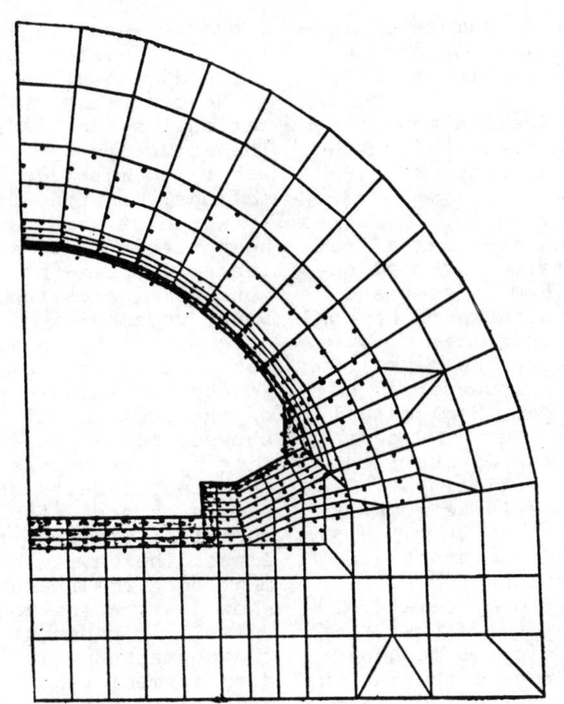

Fig. 1. Fragment of the finite-element mesh.

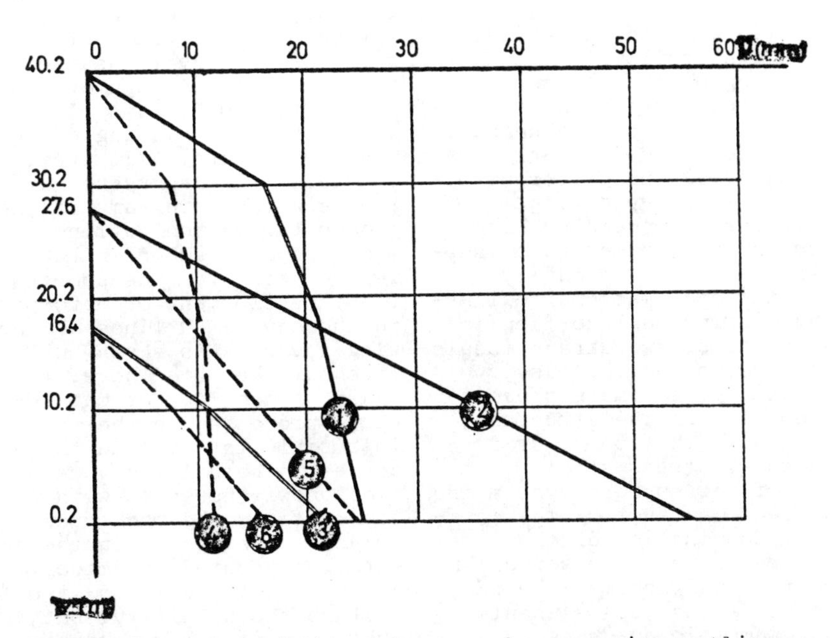

Fig. 2. Diagrams of the underground excavation outlines displacement: (═══) - without anchoring; (– – –) - with anchoring.

Let us consider an example of computer code application for designing an underground machine hall. It should be noted here that the software users' qualification mentioned above is of no importance since the computer code was used by its designers.

The analysis is made for the stress-strain state of the machine hall at the Khoudoni hydroelectric station (the Caucasus) 23.6 m in span and 47.8 m in height which is to be constructed in tuff-sandstones of Jurassic system composing the right bank of the Ingouri river downstream of the Khoudoni arch dam. Horizontally the machine hall is set at a depth of 120-150 m from the day surface in the rock mass. Vertically the thickness of the overlying rock is 80-120 m.

According to the results of geophysical investigations the rock mass is characterized by relatively high horizontal natural stresses of 10-15 MPa. The studies of mechanical properties of the rock composing the massif under such a stressed state were made using YOHD-20 triaxial apparatus (Morozov and others, 1985), resulting in stress-strain curves which constituted the basis for the numerical model.

The objectives of the numerical analysis were as follows:

- Forecasting of the stress-strain state of the rock massif when excavating the cavity with machine hall roof of predetermined "Ingouri" design (Yufin, 1979) and different alternatives of wall supporting;
- Development of recommendations on the design and parameters of supporting the walls of the cavity.

The computational mathematical model of the machine hall portion includes a set of stress-strain models of the rock composing the massif and computational scheme representing the geometry of the structure, the structure of the rock massif and its natural stress-strain state, the design of the support system and the sequence of excavating the cavity section. Combination of these requirements resulted in construction of a computational scheme (the FEM mesh) incorporating 1984 elements (1508 isoparametric quadrangular and 276 triangular elements) and 1894 uniting points. Therefore, with the problem to be solved nonlinearly each stage required solution of a system consisting of 3788 algebraic equations.

The central part of the FEM computational scheme is shown in Fig. 1. The actual topography of the area was taken into account in the computations. With the depth of the cavity which is comparable with its height this is of considerable importance. The size of the computational zone eliminates the effect of the boundary conditions on the results of computations near the structure boundaries. The computations take account of anisotropy and strength of the rock massif. The computations showed that in general the distribution of stresses is favourable and the zone of destressing in the massif is extended for distances of no more than 9 m from the cavity outlines. The cavity outlines displacement diagrams presented in Fig. 2 show that in the zone where the cavity wall is undercut by a fracture on the downstream side at El. 527.6 m displacement reaches 54.1 mm due to opening of the fracture and separate portions in the outlines do not satisfy the specified strength criterion.

The second series of computations was made with due regard to installation of anchor supports in the walls of the cavity as the excavation advances. A regular anchorage spaced at 2.0x2.0 and 15.0 m long was adopted consisting of reinforcing steel bars, AIII type, 40 mm in diameter, prestressed at 30.0 tf. The resultant diagrams of the underground excavation outlines displacement are also shown in Fig. 2 parallel to those for unsupported excavation. Development of deformations in the cavity outlines as the driving advances is shown in Fig. 3. It should be noted here, that because of the lack of data on rheological processes in the rock it was impossible to similate these processes in actual time scale. As practice shows, however, with increasing deformations the rock creep will decrease the stresses in the massif. With the adopted scheme of flexible anchorage of the walls it will not disturb the stability, but, on the contrary, it will decrease to an extent the stresses in the crown due to their redistribution.

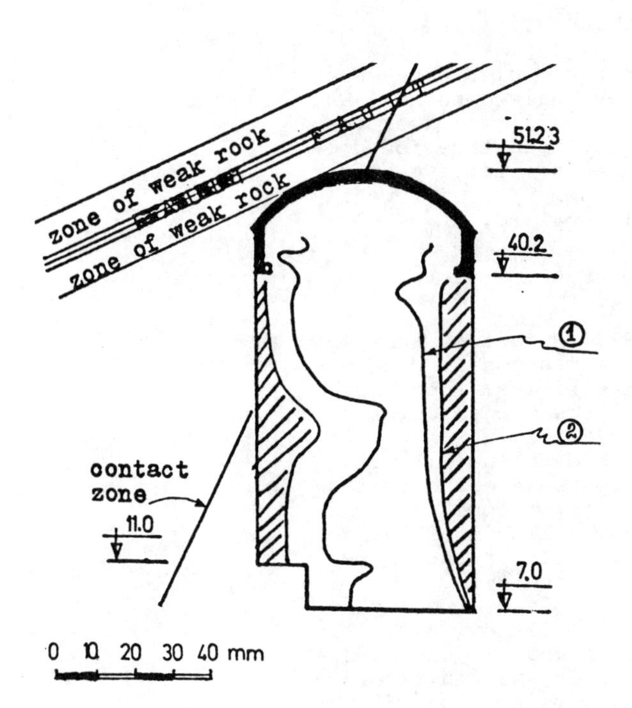

Fig. 3. Deformations in the cavity outlines as the driving advances: 1 - without anchoring; 2 - with anchoring.

The analysis of stress development in the cavity crown shown that when driving the cavity section displacement in the rock massif entail excessive compressive stresses in the crown, especially in the zone of its bending and in the vertical wall where the crane beam is secured. Crane loads do not have any essential effect on the stresses in the crown and should be taken into account only for the crane beam analysis.

The studies showed that 75-80% of bending-compressive stresses in the crown developing due to cutting the cavity section appear in the course of excavation of the first bench after the vertical wall-beams are concreted. The effect of displacements in the rock on the crown can be eliminated by presplitting at the walls around the cavity outlines for a height of the bench (8-10 m). In this case the stressed state of the crown depends on its dead weight and ground-water pressure (adopted equal to 0.1 MPa) and in general is considered to be favourable. The cavity crown thickness under consideration (80 cm) is characterized by sufficient strength even if vertical rock pressure takes place due to rock outburst which can be eliminated by providing temporary anchorage.

For the crown portion of the cavity anchors were adopted 4.5 m long made of reinforcing bars, AIII type, 28 mm in diameter.

The design seismicity of the region is of eight points which necessitated dynamic impact analysis. The main vibrations were supposed to be caused by longitudinal wave propagation. The wave front direction was changed in different series of computations. The computations were made in wave problem formulation. This direction effected inessentially on the resultant stresses because of prevailing effect of the topography of the day surface located nearby.

In conclusion it should be noted that the computed displacements of the machine hall of the Khoudoni hydroelectric station well correlated with the results obtained in in-situ investigations available nowadays.

At present and in their future work the authors have outlooks for modifying the mathematical methods of processing the data of in-situ and laboratory studies of rock properties (in particular using special-purpose personal computers) and for considerable unification of the mathematical model construction procedures including simplifying and decreasing the number of different types of models applied.

REFERENCES

Yufin, S.A. 1979. Numerical analysis of rock structures considering material nonlinearities. Proc. 20th Symposium on Rock Mechanics, Austin, Texas: 265-272.

Yufin, S.A. & Berdzenishvili, T.L. 1980. A multipurpose computer code for the solution of some problems of geophysics. Bulletin of the Academy of Sciences of the Georgian SSR, 97, 3: 605-608.

Yurfin, S.A., Postolskaya, O.K., Shvachko, I.R. & Titkov, V.I. 1985. Some aspects of underground structure mechanics in the finite element method analysis. Proc. Fifth International Conference on Numerical Methods in Geomechanics, Nagoya. A.A. Balkema/Rotterdam: 1093-1100.

Morozov, A.S., Postolskaya, O.K., Yufin, S.A. 1985. Structure sensitivity to different geological conditions as the base to optimization of rock mechanics research setup. Proc. International Symposium on the Role of Rock Mechanics in Excavations for Mining and Civil Works. Zacatecas, Mexico: 2IB. 12.

Yufin, S.A., Shvachko, I.R. & Morozov, A.S. 1987. Implementation of finite element model of heterogeneous anisotropic rock mass for the Tkibuli-Shaor coal deposit conditions pp. 1345-1349. Proc. 6 International Congress on Rock Mechanics, Montreal.

Numerical Methods in Geomechanics (Innsbruck 1988), Swoboda (ed.)
© 1988 Balkema, Rotterdam. ISBN 90 6191 809 X

A boundary integral code based on a variational formulation

K.Ben Naceur
Dowell Schlumberger, Tulsa, Okla., USA

Boundary integral formulations are powerful tools to solve 2 and 3 Dimensional elasticity and thermo-elasticity problems. Double density representations of the equations such as displacement discontinuities have also been successful at solving problems with discontinuity surfaces such as cracks. This paper presents a code based on a variational formulation of the double density representation. The technique (which uses finite elements instead of displacement discontinuity segments) was introduced by Nedelec [15], and Bamberger [1] for elastodynamic problems, and applied to static and propagating cracks (Touboul and BenNaceur [18]). The code uses a mesh generator for finite elements (Modulef [13]) to represent fracture and discontinuity surfaces, and has a library of linear and higher order elements for elasticity and thermo-elasticity (2D and 3D). A comparison between the performances of the code, classical finite elements, and boundary integral techniques is then given.

Introduction

The use of boundary integral equations (BIE) has become increasingly popular in geomechanics for linear diffusion and elasticity problems over the last 15 years. Codes based on boundary elements have been developed for 2- and 3-dimensional problems. They allow a minimum level of discretization of domains (at the boundary), while allowing a more elaborate analysis of the behavior inside the domain if necessary. Several versions of the BIE technique have been developed: they include the single-layer representations, based on a direct use of the Green's function associated with the operator (in elasticity, displacements and stresses are the primary unknowns), and the double-layer (DL) methods, which use instead the discontinuity of the unknown function across a series of hyperplanes (the boundary of the domain) as the natural unknown. The latter method is particularly useful in the case of problems involving discontinuity surfaces (such as fractures, or obstacles which may be diffraction surfaces for waves). Two types of BIE based on those double layer representations have been used in geomechanics: The (older) dislocation density method is based on the concept of Somigliana dislocation, and infinitesimal Burgers vectors (Eshelby [9]). The application of the technique to fracture mechanics has been extensively discussed by Bilby and Eshelby [5]. the displacement discontinuity method (Starfield and Fairhurst [17], Crouch [8], Crouch and Starfield [7]) uses a discrete representation of the jump of the displacement field across elementary surfaces, computes the inter-element influence leading to an algebraic system. It can be shown that the application of a simple *integration by parts* allows to derive the displacement discontinuity method from the dislocation density representation. Both techniques consist in discretizing the boundary of the domains (including discontinuity surfaces and fractures), and solving the corresponding boundary integral problem. Displacements and stresses can be easily computed from the primary unknown. Higher order discrete double layer have also been discussed by Crouch and Starfield [7], leading to the equivalent of higher order of finite elements. Problems involving fractures are particularly adapted, since the unknown stress intensity factors are directly expressed as a function of the displacement discontinuities. Some drawbacks associated with the use of classical double layers reside in the hyper singularity of the kernels used, as well as the need, when implementing the technique, to reconstruct the basic data structures, since both finite elements and direct boundary integral codes are not adapted to the formulation.

The use of *integro-variational* schemes is discussed here: the technique introduced in the late 70-s allows a derivation of a variational technique associated with the double layer representation of the problem. It differs essentially from the use of Galerkin's techniques for DL equations, since they involve regular kernels, and hence do not require a special treatment for singularities. The boundary of the domain is discretized into *finite elements*, and a variational form is minimized over a suitable set of functions based on those finite elements. The formulation allows the use of classical finite element modules to build a code, which is particularly adapted to fracture, and diffraction surfaces. The present code is based on a non-commercial finite element library (Modulef [13]), and the implementation of the technique is described here along with examples of applications.

General Integro-Variational Schemes

The integro-variational technique has been introduced by Nedelec [14] for the Laplace operator, and subsequently used in 2D elasticity (Bonnemay [6]), for Helmoltz's equation (Hamdi [11]), for 3D elasticity and elasto-dynamics (Bamberger [1]), for 3D fracture problems (Touboul and Ben-Naceur [18]), and for 2D and 3D thermo- and poro-elasticity (BenNaceur [4]). If a problem is defined as finding an unknown function u solution of a boundary problem (with the linear operators L and B):

$$Lu = 0 \qquad in\ \Omega \qquad (1)$$

$$B(u, \frac{\partial u}{\partial n}) = g \qquad in\ \Gamma \qquad (2)$$

where Ω is a n-dimensional domain (n=2 or 3), and Γ is the (sufficiently smooth) boundary of Ω, the principle of the integro-variational schemes resides in deriving first a double layer representation for the given problem. We shall assume in the following that the boundary relation is simply expressed as a function of the conjugate of the field (tractions in the case of displacements for elastic problems: $\tau = g$). The double layer relation can generally be expressed as:

$$u(M) = \int_\Gamma K_1(M,N).[u](N)dN - \int_\Gamma K_2(M,N)\tau(u)(N)dN \qquad (3)$$

where K_1 and K_2 are kernels corresponding to the given operator, τ is the B-conjugate of the function u, and [ ] refers to the discontinuity of a (vector) function over the surface Γ. Problems with using the previous relation reside in the numerical approximation of the boundary, due to the kernels singularity. The problem is transformed then into a variational form, with regular kernels, that can be handled using classical finite elements structures.

The case most frequently used will be considered here, i.e. diffusion and elasto-statics.

Diffusion Equation with a Neumann condition

The double layer representation is given by (see for example Nedelec [14]):

$$u(y) = -\int_\Gamma [u]\frac{\partial}{\partial n}(\frac{1}{|x-y|})d\gamma \qquad (4)$$

where [u] (noted ϕ in the following) represents the displacement discontinuity over Γ. Equation 4 has a singular kernel $\frac{\partial}{\partial n}(\frac{1}{|x-y|})$. It can be shown (Nedelec [14]) that a (non-singular) variational form can be associated to that representation, and the problem is to minimize:

$$J(\phi) = \frac{1}{2}\int_\Gamma\int_\Gamma \frac{1}{|x-y|}\mathbf{rot}\phi(\mathbf{x}).\mathbf{rot}\phi(\mathbf{x}')\mathbf{dxdx}'\ldots - \int_\Gamma g\phi(x)dx = \frac{1}{2}a(\phi,\phi) - b(\phi) \qquad (5)$$

Hence the variational form a becomes regular, and finite elements formulations can be used (with trial functions being defined only on the boundary of the domain).

Elasto-Static Problems

If L represents Navier's operator, and assuming that the stresses are continuous across a surface Γ, the displacement field can also have a double-layer representation (Kuprazde [12]):

$$u(x) = \int_\Gamma \tau^n(K(x-x')^t).[u](x')dx' \qquad (6)$$

where, again, [u] represents the displacement discontinuity across the surface Γ. K is the classical Kelvin's tensor, which represents the fundamental solution for 3D elasto-statics, and τ^n is the traction operator.

A variational formulation corresponding to equation 6 has been derived by Bamberger [1], to obtain a non-singular representation:

$$Min J(\phi) = \frac{1}{2} a_3(\phi, \phi) - b(\phi) \quad (7)$$

where the displacement discontinuity [u] = ϕ = (ϕ_1, ϕ_2, ϕ_3) is the uknown of the formulation. For a fracture parallel to the x1-x2 plane, a_3 is given by the double integral:

$$a_3(\phi, \psi) = \beta \sum_{j=1}^{3} \int_{\Gamma \times \Gamma} \frac{1}{4\pi|x - x'|} \vec{\nabla}\phi_j(x)\vec{\nabla}\psi_j(x') dx dx'$$
$$-\beta' \int_{\Gamma \times \Gamma} (\frac{\partial \phi_2}{\partial x_1} - \frac{\partial \phi_1}{\partial x_2})(x)(\frac{\partial \psi_2}{\partial x_1} - \frac{\partial \psi_1}{\partial x_2})(x') dx dx' \quad (8)$$

where β and β' are combination of elastic constants (for planar fracture problems, the second term of the RHS of the previous relation cancels, and β is simply proportional to the plane strain modulus $E/(1-\nu^2)$. The linear operator b is given (in the absence of volumic loads) by:

$$b(\phi) = \int_{\Gamma} g\phi dx \quad (9)$$

a_3 is a bilinear symmetric positive definite form, for which classical finite elements approximations can be written.

In two dimensions, the same type of technique applies, and one obtains (Bonnemay [6]) the following results, for a fracture (or a discontinuity surface) parallel to the x-axis:

$$Min \frac{1}{2} a_2(\phi, \phi) - b(\phi) \quad (10)$$

where a_2 is defined as a bilinear form:

$$a_2(\phi, \psi) = \sum_{i,j=1}^{2} a_2^{ij}(\phi_i, \psi_j) \quad (11)$$

$$a_2^{ij} = \int_{\Gamma \times \Gamma} G^{ij} \frac{d\phi_i}{dx} \frac{d\psi_j}{dx'} dx dx' \quad (12)$$

and G^{ij} are kernels with an integrable singularity. Again b is a linear form defined on the boundary Γ. Simple transformations allow to derive a representation for arbitrarily oriented fractures.

The more general case of elasto-dynamics formulations has been formulated by Bamberger [1], and estimates of convergence of the technique as a function of the mesh refinement have been discussed by Bonnemay [6] for the 2D elastostatics case. Comparisons between the different double layer representations (interpolation, accuracy, speed) can be found in Ben-Naceur [3].

For both 2- and 3-dimensional elasto-statics solutions, analytical formulae can be used to calculate the influence function (hence the stiffness matrix) if constant or linear elements are used (Ben-Naceur [3]). However in more general situations, the degree of interpolation may be larger than 1, and numerical integration schemes are to be used, which are already built-in finite element libraries.

IMPLEMENTATION WITHIN MODULEF

Modulef is a modular finite element library, including an extensive set of matrix solvers (direct and iterative). Its use is based on the notion of data structures, which include mesh description, elementary and assembled matrixes, second members ..., and modules which operate on the data structures (mesh generation, definition of the interpolation, solvers, pre- and post- processing ...). Each element in the library is described by a variational formulation, the type of interpolation, transformation from the current element to the reference one, and numerical integration formulae. Guidelines for the integration of new finite elements are given by George *et al.* [10] .

For the integro-variational schemes, the first step consists in generating the mesh, which consists in a series of hypersurfaces (lines in 2D, planes in 3D). The interpolation is defined for the different types of elements used here (1D linear, 1D quadratic, P1 triangle, P2 isoparametric triangle,

Q1 and Q2 (isoparametric) rectangles) while the calculation of the elementary matrixes is based on classical Gauss numerical integration schemes. In the same manner, the second member is assembled using the corresponding Modulef module.

Finally the post-processing of data defined at the boundary domain is performed using the Modulef library of routines (contours, maps, interpolation), while external modules have to be developed to calculate internal unknowns (displacements and stresses inside the domain), using the relation between the influence functions.

EXAMPLE 1: Fracture in 2 and 3 Dimensions

An example of the technique is given here: The classical problem of a rectangular fracture in an infinite medium under a uniform load is considered. Only normal displacements are to be computed for this example. The limiting case of an infinitely large fracture leads to Sneddon's two-dimensional solution for a pressurized segment [16], which can be solved using the two-dimensional formulation (ie with 1D finite elements) for the integro-variational formulation given by Equation 10). The 3D solution depends on the aspect ratio of the fracture height/length (see Figure 1): The 2D asymptote is obtained for $H/L \to 0$. Figure 2 shows a mesh used for different aspect ratios (respectively 0.1,1, 10). The displacements are computed along the center line, and Figure 3 shows the different results for the

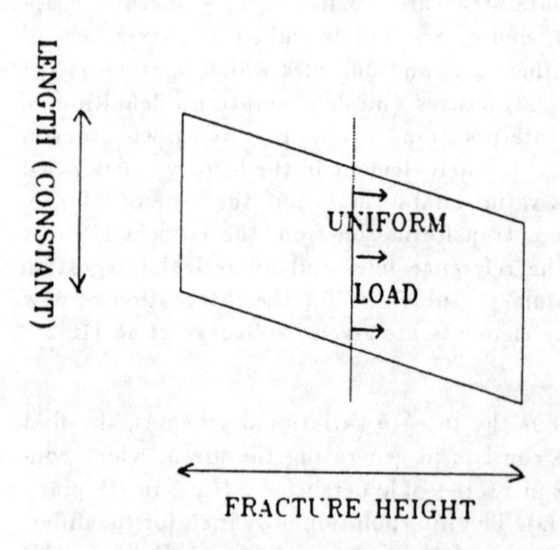

Figure 1: Planar Rectangular Fracture

3D problem compared with the 2D solution, and the analytical one.

The stress intensity factors can also be compared to the analytical results. Figure 4 show the stress intensity factor for the three geometries, a good agreement is obtained with the third case as expected. It should be noted that for both 2D and 3D integro-variational schemes, only linear elements were used, and the use of special crack-tip elements (Barsoum [2]) was unnecessary.

EXAMPLE 2: Parallel Fractures

The elastic influence between two uniformly pressurized parallel fractures is investigated 5 The displacements field at the center of each of the fracture (which is the same for obvious reasons) depend on the distances between the two ellipses and their dimensions, and they are compared to the value for a single fracture. The non-dimensional load on the fractures was assumed to be uniform, and the dimensionless ratio of the fracturing pressure to the plane strain modulus has a value of 0.0001 .

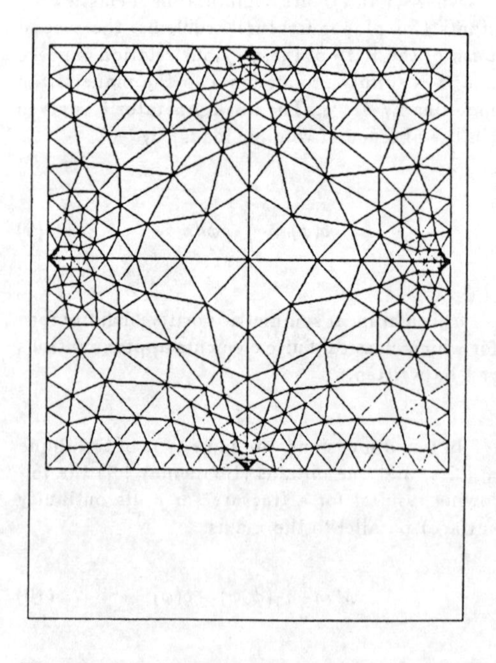

Figure 2: Mesh Used For Fracture Geometry

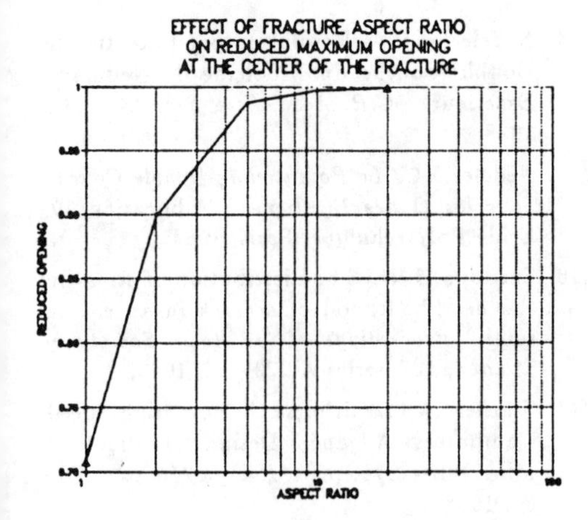

Figure 3: Effect of Aspect Ratio on Normalized Width

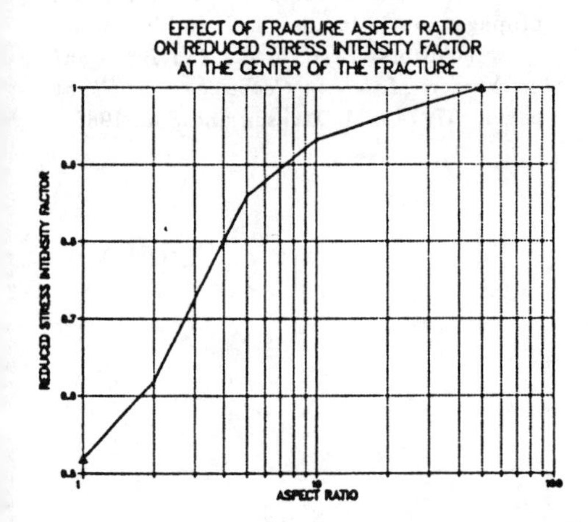

Figure 4: Stress Intensity Factors Along Centerline

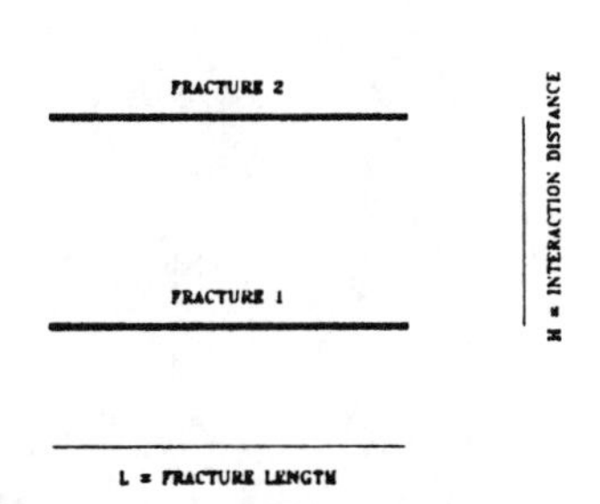

Figure 5: Interaction Between Parallel Fractures

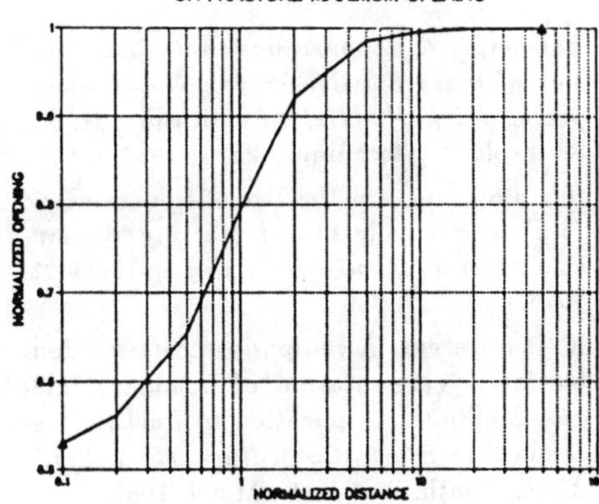

Figure 6: Normalized Width as a Function of Distance

The two-dimensional integro-variational technique is used for the set of two fractures, and the global system for the two sets of elements on both fractures is solved. Figure 6 shows for two fractures of unit length the normalized displacement at the center of the fracture (ratio to the opening for a single fracture) as a function of the distance between the two cracks.

Conclusions

The principle of a variational formulation for different double layer representations of boundary integral equations has been presented. The technique is particularly adapted for problems involving discontinuity surfaces (fractures, slip zones, diffraction planes), and leads to a symmetric positive set of equations. The performance of the technique is significantly increased over classical boundary integral methods, while its formulation allows an easy integration into an already existing finite element code (here **Modulef**). Elastic problems including fractures can be handled without the use of crack-tip elements, and different examples of applications were presented. Future extensions of the technique will include plasticity problems, heterogeneous media, and anisotropic constitutive equations.

ACKNOWLEDGMENTS

The author wish to thank Dr E. Touboul for his precursor work, and the management of Dowell Schlumberger for permission to publish this paper.

References

[1] Bamberger,A. *Approximation de la Diffraction d'Ondes Elastiques: une Nouvelle Approche (I), (II), (III).* Publication 91, 96, 98, Ecole Polytechnique, Paris, 1983.

[2] R.S. Barsoum. On the Use of Isoparametric Finite Elements in Linear Fracture Mechanics . *Int. J. Num. Meth. Eng.* , 10:25–38, 1976.

[3] K. Ben Naceur. A comparison between double layer representation of boundary integral equations: application to fracture mechanics. In *Fourth Conf. Num. Meth. Fract. Mech.*, South. Res. Inst., March 1987.

[4] Ben Naceur,K. Integro-variational formulation in thermo- and poro-elasticity. In *Int. Conf. Comp. Eng. Sci.*, Georgia Tech., April 1988.

[5] Bilby, B.A. and Eshelby, J.D. *Dislocations and the Theory of Fracture* , pages 99–182. Volume 1 of *Fracture*, Academic Press, Liebowitz Ed., 1968.

[6] Bonnemay,P. *Equations Integrales pour l'Elasticite Plane.* PhD thesis, Univ. de Paris VI, Paris, 1979.

[7] Crouch, S. L. and Starfield, A. M. *Boundary Element Methods in Solid Mechanics.* George Allen and Unwin Ltd., London, U.K., 1983.

[8] Crouch,S.L. . Solution of Plane Elasticity Problems by the Displacement Discontinuity Method . *Int. J. Num. Meth. Eng.* , 10:301–343, 1976.

[9] Eshelby, J.D. . . *Phil. Mag.* , 40:903–917, 1949.

[10] Perronnet,A. George, P.L and Vidrascu,M. *Integration d'un NOuvel Element Fini, Creation d'une Bibliotheque d'Elements Finis.* Publication, INRIA, Paris, 1987.

[11] Hamdi,M.A. *Equations Integrales pour les Equations de l'Acoustique.* PhD thesis, Université Technologique de Compiègne, France, 1982.

[12] V.D. Kuprazde. *Three-Dimensional Problems in the Mathematical Theory of Elasticity and Thermo-Elasticity.* North Holland, Amsterdam, 1979.

[13] Modulef. *Presentation du Club Modulef.* Publication, INRIA, Paris, 1987.

[14] Nedelec, J.C. . Resolution par Potentiel de Double Couche du Probleme de Neumann Exterieur . *C.R. Acad. Sci., Serie A* , 286, 1978.

[15] Nedelec,J.C. *Le Potentiel de Double Couche pour les Ondes Elastiques.* Publication 99, Ecole Polytechnique, Paris, 1983.

[16] Sneddon, I. N. The distribution of stress in the neighbourhood of a crack in an elastic solid. *Proceedings of the Royal Society of London*, 187–series A:229–260, 1946.

[17] Starfield,A.M.Fairhurst,C. How High-Speed Computers Advance Design of Practical Mine Piller Systems . *Energy Min. J.* , 78–84, 1968.

[18] Touboul, E. and Ben-Naceur, K. Variational solutions to boundary integral equations in elasticity and their application to three-dimensional computation of fracture propagation. In Pande, G. N. and Van Impe, W. F., editors, *Proc. of the 2nd Int. Conf. on Numer. Models in Geomechanics, Ghent*, pages 737–746, M. Jackson and Son, 1986.

Numerical Methods in Geomechanics (Innsbruck 1988), Swoboda (ed.)
© 1988 Balkema, Rotterdam. ISBN 90 6191 809 X

The use of Lees' algorithm in the analysis of some ground heat and mass transfer problems

H.R.Thomas & S.W.Rees
University College, Cardiff, UK

ABSTRACT: A numerical model incorporating Lees' algorithm, capable of predicting coupled simultaneous transient non-linear heat and moisture movement in unsaturated soils, is used to achieve solutions for two problems. Results obtained for the first, one of a soil column subject to fixed boundary conditions, are in agreement with previously held views derived from non-linear heat conduction analyses, that Lees' algorithm produces unconditionally stable and convergent results. The importance of maintaining adequate timestep size control to reduce oscillatory behaviour is however illustrated, together with the use of an averaging technique to achieve the same end. For the second problem, one of seasonal changes in temperature and moisture content in the ground, the results show that Lees' algorithm compares favourably with a Crank-Nicolson scheme and is capable of coping with time varying Dirichlet boundary conditions. In the present circumstances, the results give confidence to the use of Lees algorithm.

1. INTRODUCTION

Ground heat and mass transfer analysis is a growing area of application of numerical methods in geomechanics. Industrial demand for solutions to problems such as nuclear waste disposal, interseasonal heat storage, energy losses from buildings and soil shrinkage provides the impetus for such growth. Given the ever increasing complexity of the modelling required, both in terms of the scale of the problems to be considered and the various physical phenomena to be included, the efficiency of the solution algorithms employed remains of importance.

Within this overall context, the specific problem under consideration in this paper is the solution of coupled simultaneous transient non-linear heat and moisture transfer in unsaturated soil. The particular formulation employed and the development of the numerical algorithm have been described in detail elsewhere (Thomas 1987). Only the salient features will therefore be given here. Attention will be focussed however in this paper on the performance of the timestepping algorithm implemented, since in any transient analysis, this feature is of considerable importance.

The algorithm chosen for investigation is a three level, second order accurate scheme (Lees, 1960). Its advantage is that direct evaluation of varying material parameters may take place at the intermediate time level, thus eliminating the need for iterations within each time step. Bonacina and Comini (1973) investigated the local accuracy, stability and convergence of the scheme, within the context of the two-dimensional finite difference solution of non-linear heat conduction problems. Using von Neumann's approach, on the basis that the varying material parameters could be "frozen" to fixed values in various regions of the domain, they showed that the scheme was unconditionally stable and convergent. Their analysis was carried out for the case of boundary conditions of the first kind.

The first application of the algorithm in finite elements appears to be due to Comini, Del Guidice, Lewis and Zienkiewicz (1974), again for the case of non-linear heat conduction problems. The authors used extensive numerical experiments to examine the behaviour of the algorithm and concluded that the scheme was also unconditionally stable and convergent in the context of their new application.

Given the difficulties of generating definitive rulings on the stability and convergence of finite difference schemes in non-linear finite element transient analyses, the numerical experimentation approach adopted by the above authors seems to provide a readily amenable method of investigating the performance of a particular scheme. Such an approach is adopted here for the case of the application of Lees algorithm in coupled ground heat and moisture transfer.

Two problems are considered. In the first a one-dimensional column of soil subjected to surface evaporative moisture content losses is considered. Boundary conditions of the first kind are applied and the response of the algorithm to various timestep size changes is explored. In the second an attempt is made to model the field behaviour of a stratum of soil over a two year test period. In order to achieve this aim, the boundary conditions must be varied with time. An analysis is therefore performed using time varying Dirichlet boundary conditions and the effect of the introduction of such boundary conditions on the performance of the algorithm is viewed.

2. THEORY AND FINITE ELEMENT FORMULATION

The theoretical formulation upon which the work presented here is based, leads to the following coupled set of differential equations (Thomas, 1987).

$$C\frac{\partial T}{\partial t} = \nabla.((\lambda+L\varepsilon\rho_{\ell}D_T)\nabla T)+\nabla.((L\varepsilon\rho_{\ell}D_{\Theta})\nabla\Theta_{\ell}) + L\rho_{\ell}\frac{\partial(\varepsilon K)}{\partial z} \quad (1)$$

$$\frac{\partial\Theta_{\ell}}{\partial t} = \nabla.(D_T\nabla T) + \nabla.(D_{\Theta}\nabla\Theta_{\ell}) + \frac{\partial K}{\partial z}$$

In the above C is the volumetric heat capacity, T is the temperature, t is the time, λ is the thermal conductivity, L is the latent heat of vaporisation of water, ε is the phase conversion factor, ρ_{ℓ} is the density of water, D_T is the thermal moisture diffusivity, D_{Θ} is the isothermal moisture diffusivity, Θ_{ℓ} is the volumetric liquid content, K is the unsaturated hydraulic conductivity and z is the elevation.

The above equations are solved simultaneously, subject to Dirichlet or Neumann type boundary conditions, in order to obtain variations in time and space of Θ_{ℓ} and T.

A finite element solution of the problem, following a Galerkin weighted residual approach, led to the following matrix equation set.

$$\underline{D}_q\underline{T} + \underline{D}_{\varepsilon}\underline{\Theta}_{\ell} + \underline{C}_q\dot{\underline{T}} + \underline{J}_q = 0$$
$$\underline{D}_T\underline{T} + \underline{D}_{\Theta}\underline{\Theta}_{\ell} + \underline{C}_{\Theta}\dot{\underline{\Theta}}_{\ell} + \underline{J}_{\Theta} = 0 \quad (2)$$

These can be written more concisely as

$$\underline{D}\underline{\phi} + \underline{C}\dot{\underline{\phi}} + \underline{J} = 0 \quad (3)$$

In order to evaluate the time varying behaviour of $\underline{\phi}$, Lees algorithm is applied to equation $(\overline{3})$. This leads to the recurrence relationship

$$\underline{\phi}^{n+1} = -\left|\frac{\underline{D}}{3} + \frac{\underline{C}^n}{2\Delta t}\right|^{-1} x\left|\underline{D}^n\frac{\underline{\phi}^n}{3} + \underline{D}^n\frac{\underline{\phi}^{n-1}}{3} - \underline{C}^n\frac{\underline{\phi}^{n-1}}{2\Delta t} + \underline{J}^n\right| \quad (4)$$

where the superscript n refers to the time level and Δt is the timestep size. It can be seen that the non-linear matrices $\underline{D}$ and $\underline{C}$ are evaluated directly at the mid time level n, where conditions are known. A similar approach is adopted here for the evaluation of the vector $\underline{J}$, where again the need for an iterative calculation of $\underline{J}^{n+1}$ is avoided by setting the average of the fluxes over three timesteps equal to the flux at time n. This slight modification of the original algorithm is only of relevance here insofar as it affects the gravitational flow term, since both the problems considered are subjected to fixed boundary conditions, not flux.

The scheme requires two starting values of $\underline{\phi}$ for initiation. These may be assumed to be constant, for example, or evolved from an iterative approach (Thomas and Chan, 1986). Alternatively a two level scheme may be used to start a three level analysis.

3. APPLICATIONS

Problem 1

The first problem analysed is the same as previously considered by Thomas (1987). The one-dimensional heat and mass transfer behaviour of a column of soil, 50cm in height, is investigated using a mesh of 5 elements, each 10cm in height. Initial conditions of a uniform temperature of 10^oC were assumed, subject to a Dirichlet boundary condition of 25^oC. The initial volumetric moisture content varied with depth from 0.4 at the surface to 0.45 at full depth. Dirichlet boundary conditions of $\Theta_{\ell} = 0.2$ were applied in this case.

The values of the material parameters used in the analysis were chosen after an assessment of available data in the literature. Values representative of a fine soil with components of sand and silt were selected, according to the methods outlined by Thomas (1987). The numerical values used, for the range of parameters required, are given therein and for the sake of conciseness will not be repeated here.

In the absence of a theoretical solution to the problem an exercise was carried out to establish a "benchmark" against which the performance of Lees algorithm could be judged. It was decided that the unconditionally stable, non-oscillatory, fully implicit scheme, used in conjunction with a predictor corrector approach, would, in the limit, as the mesh was refined and the timestep size reduced, provide suitable benchmark results. Such an approach was therefore adopted and the results obtained are presented in Figure 1, for both temperature and moisture content variations.

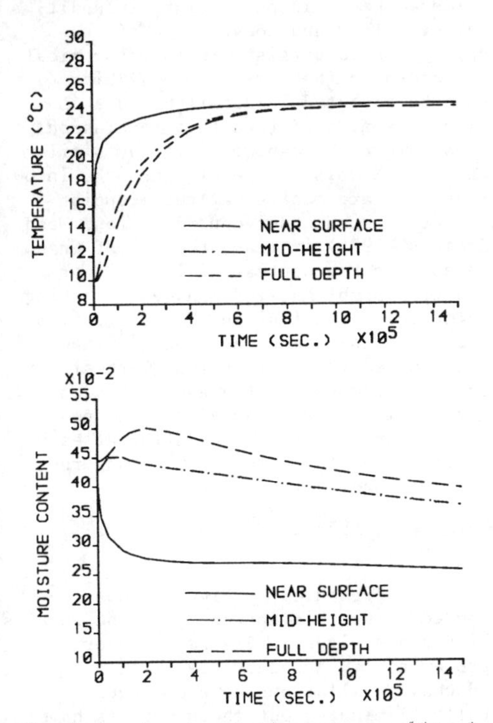

Fig. 1 Benchmark solutions for problem 1

The efficient use of any timestepping scheme, in the analysis of problems such as the one under consideration here, requires the inclusion of variable timestep sizes. In particular, while it may be necessary to limit timestep sizes to small values initially, in order to correctly reproduce the variations that are taking place, it is equally necessary to employ larger timestep sizes, as steady state is approached, in order to reduce total computational time. The timestep size control procedure adopted here can be summarised as follows: If the percentage change in the value of a variable over a time interval exceeds an upper tolerance (TOLU), at any node, the timestep size is halved. If the percentage change is less than a lower tolerance (TOLL), at all nodes, the timestep size is doubled. The use of this method, with an initial timestep size of 300 secs, a TOLU value of 5% and a TOLL value of 3% yielded Lees algorithm results that were virtually identical to the benchmark solution shown in Figure 1. The validity of this approach is hence confirmed.

An assessment of the performance of Lees algorithm was then carried out (i) by relaxing the tolerances on the timestep size control procedure so that more rapid increases in timestep sizes could be achieved and (ii) by increasing the initial timestep size, keeping the tolerance values the same at 3/5%.

Considering the effect of the relaxation of the tolerances first, the results obtained for two of the cases considered are shown in Figure 2. These are for upper and lower tolerances of 18/15% and 25/20% respectively. For clarity of presentation only the results at one node near the surface, are now shown.

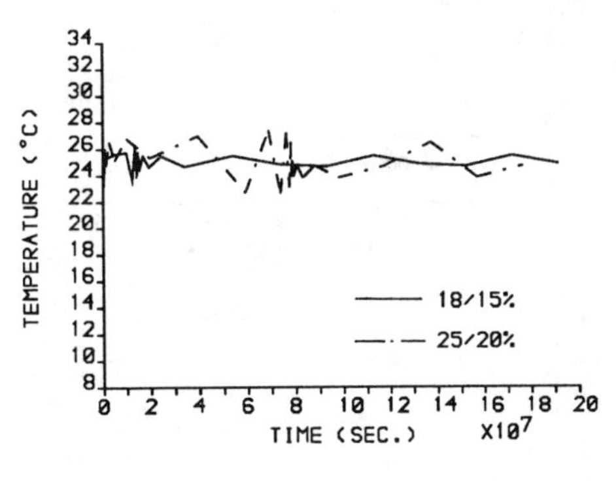

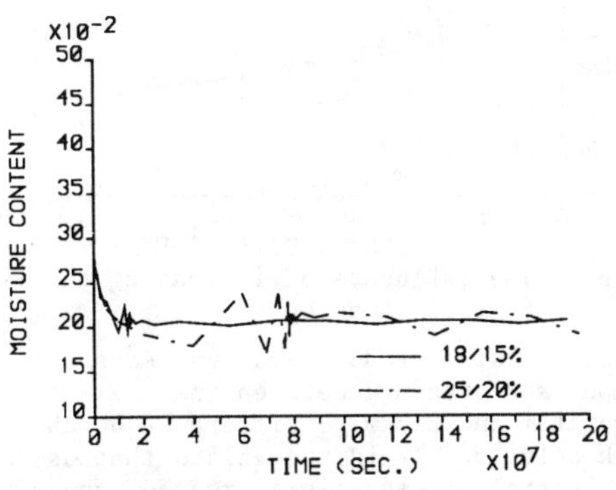

Fig. 2 The influence of relaxing timestep size control tolerances (TOLU/TOLL)

The features shown however also apply at other nodes in the mesh. It is apparent that as the tolerances are relaxed and the timestep sizes increased, oscillations about the benchmark solution are induced. The greater the relaxation in the tolerances, the greater the magnitude of the oscillations. Clearly therefore the accuracy of the solution deteriorates as this process continues. However it is instructive to note that for the two cases considered, no matter how great the induced oscillations, the results ultimately converge to the correct steady state solution albeit after a long period of time.

Considering next the influence of various initial timestep sizes, the results obtained for the various cases considered are shown in Figure 3. Of the options listed previously regarding the choice of two starting values for $\underline{\phi}$, the method adopted here is simply to assume that both values are the same. The validity of this approach has already been assessed, for an initial

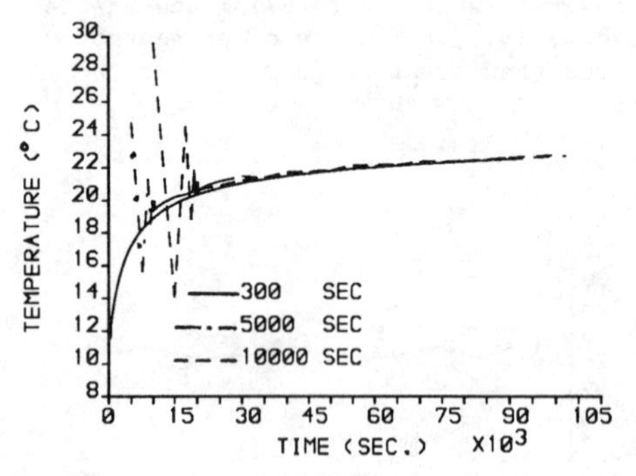

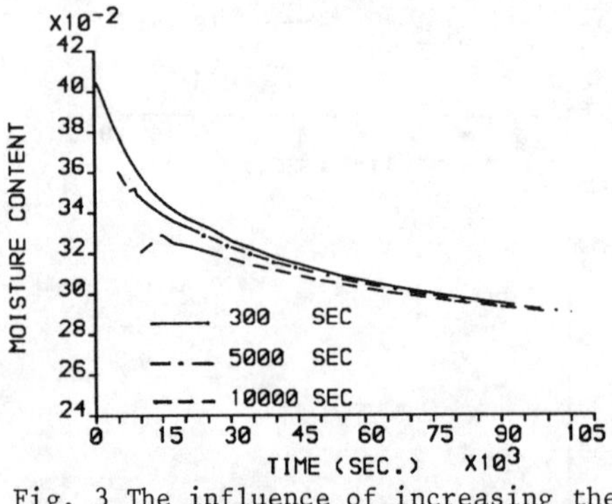

Fig. 3 The influence of increasing the initial timestep size

timestep size of 300 secs, by means of the comparison obtained between the Lees algorithm and Backward Difference benchmark solution. It is recognised that as the initial timestep is increased, the accuracy of this assumption diminishes and errors may well be introduced into the solution. In particular oscillations may be induced because of this effect. Despite this added tendency to de-stabilise the solution the results obtained conform with those observed previously when the timestep size control tolerances were relaxed. In particular, for all the cases considered, no matter how great the induced oscillations, the results converge to the correct steady state solution.

On the basis of the above numerical experiments, it appears that the conclusions previously reached, regarding the use of the algorithm in the finite element analysis of non-linear heat conduction problems, also apply for the example considered here. In particular, unconditional stability and convergence of the scheme seems to persist. Further numerical experiments on the same problem, which cannot be reported in detail because of lack of space, confirmed this conclusion.

Lees' algorithm can be viewed as similar to a Crank-Nicolson type of approach, insofar as both are centred at the midpoint of the time span under consideration. Wood and Lewis (1975) examined the use of Crank-Nicolson, in the context of linear heat conduction problems, and managed to achieve substantial reductions in the oscillations produced by the use of an averaging technique. Indeed the oscillations were almost completely removed by the approach adopted. A similar method was therefore examined here. In particular when updating at each new time step, the value of $\underline{\phi}^{n-1}$ was redefined according to

$$\underline{\phi}^{n-1} = (\underline{\phi}^{n} + \phi^{n-1} + \underline{\phi}^{n-2})/3 \qquad (5)$$

The work presented in Figure 2 was recomputed using this approach and the results achieved are given in Figure 4. Major improvements in accuracy are clearly produced. Oscillations have not been totally eliminated but their effects have been substantially reduced.

Problem 2

The second problem analysed is one of seasonal temperature and moisture content changes in the ground. Field monitoring experiments were conducted by British Gas at a site in Swindon, and results were available of the variation of temperatures and volumetric moisture content over the two years 1983/84. Thomas and Owen (1987) present the results of the comparison which was achieved, when calculations from a linear version of the present model were

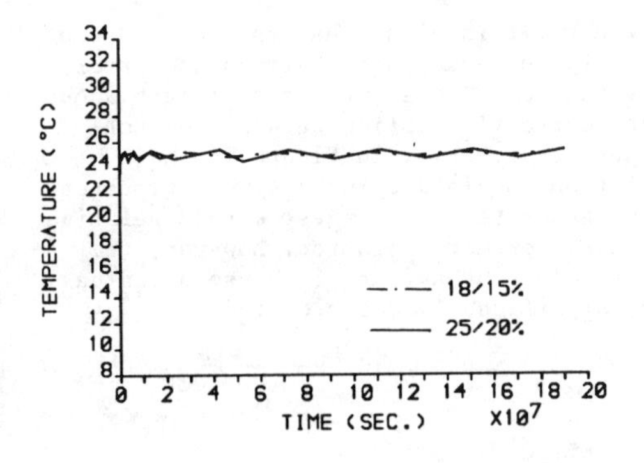

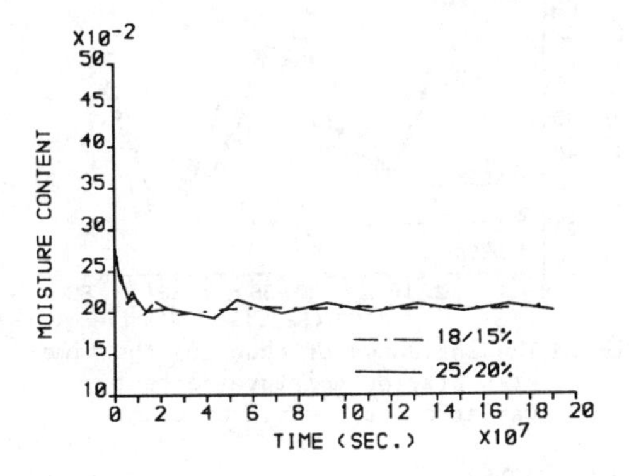

Fig. 4 The effect of using an averaging technique

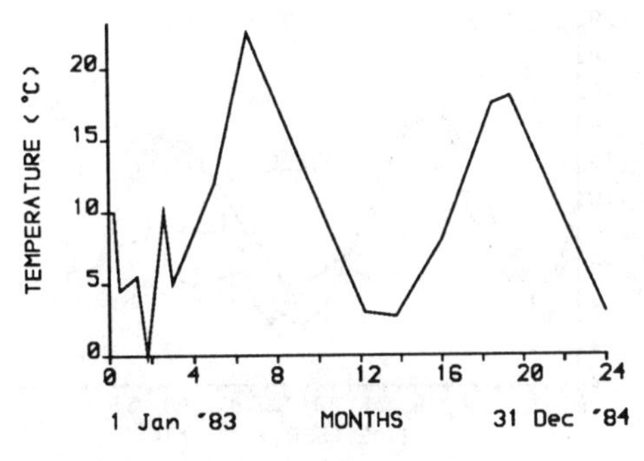

Fig. 5 Time varying temperature boundary conditions

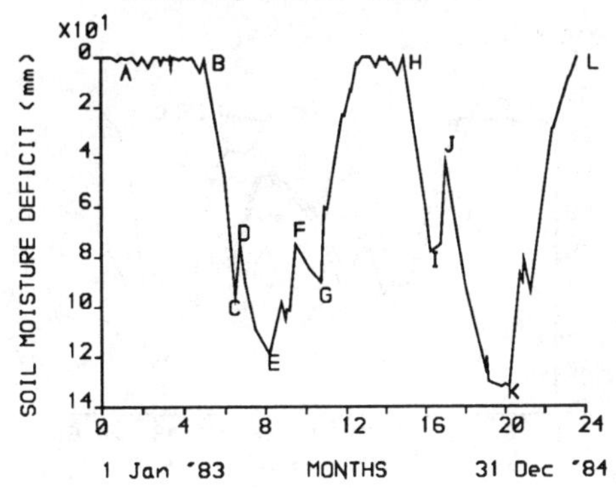

Fig. 6 Soil moisture deficit variation with time

compared with field values. Calculations using the non-linear model are currently being developed and as a first step in this process, comparisons were made between the previous linear numerical work and the current non-linear analysis, running in linear mode. This exercise permits therefore the characteristics of Lees algorithm, when subjected to time varying Dirichlet boundary conditions, to be viewed.

Full details of the field experiment and the modelling approach adopted are given in Thomas and Owen (1987). Only a brief summary will therefore be repeated below :

(i) Crank-Nicolson's timestepping algorithm was used.

(ii) Temperature boundary conditions were determined from Meteorological Office records of maximum and minimum air temperatures. The form of the variation applied is shown in Figure 5.

(iii) Volumetric moisture content boundary conditions were determined from soil moisture deficit data, also supplied by the Meteorological Office. The information supplied is shown diagrammatically in Figure 6. This was interpreted in the analysis as a series of fixed boundary conditions, alternating from saturated to dry, starting with AB as saturated. Each change from saturated to dry, or vice versa was applied over one day.

(iv) Initial conditions were available from the field data. Two values, at n and n-1, were therefore readily obtained.

(v) The field experiment was conducted in Kimmeridge clay. Representative material parameters were assumed.

(vi) A one-dimensional analysis of the upper 2.3m of the clay was performed. This region was sub-divided into a mesh of 7 elements, of variable thickness, ranging from 50mm at the surface, to 800mm at the base.

(vii) A constant timestep size of one week was tried and found to be acceptable.

Results obtained are shown in Figures 7 and 8 which give temperature variation at 0.9m depth and volumetric moisture content variation at 0.3m depth respectively. It can be seen that the results obtained compare favourably with those previously generated using the two level Crank-Nicolson algorithm. These results are reassuring insofar as two independent timestepping algorithms have produced the

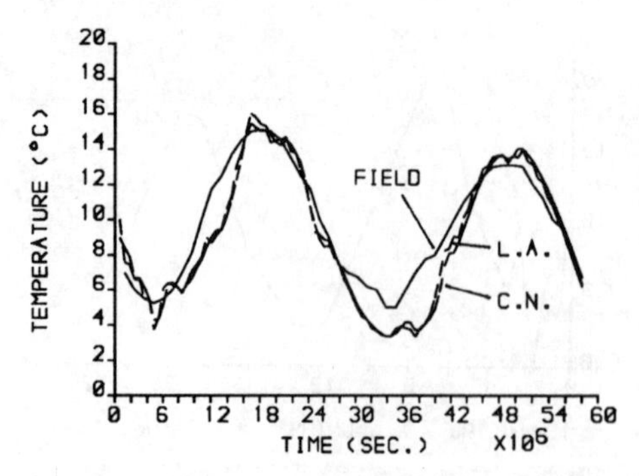

Fig. 7 Seasonal temperature variation at a depth of 0.9m

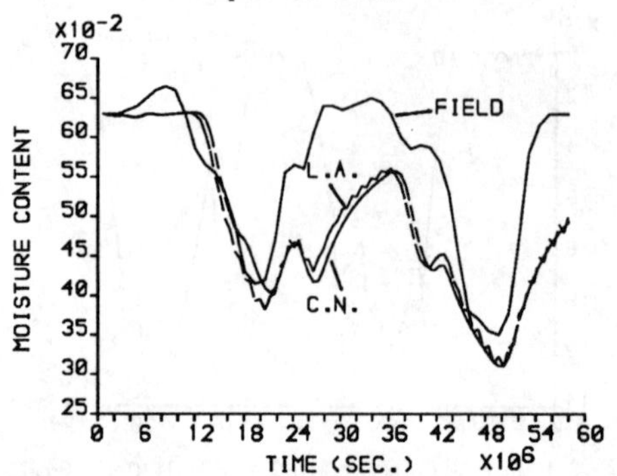

Fig. 8 Seasonal moisture content variation at a depth of 0.3m

same results. Lees' algorithm's ability to cope with this type of problem is therefore illustrated here.

The influence of increasing the timestep size was briefly examined. Figures 9 and 10 present temperature variation at 0.9m depth and volumetric content variation at 0.3m depth. The solutions illustrated were obtained using timestep sizes of 1, 2 and

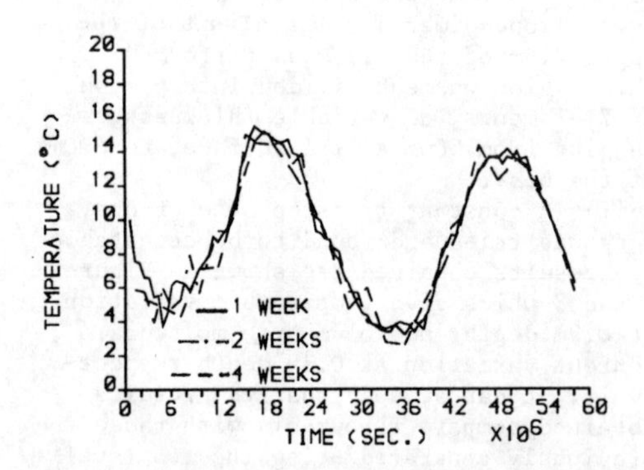

Fig. 9 The influence of changing the timestep size on temperature variation at a depth of 0.9m

4 weeks. It is clear that as the timestep size is increased oscillations and errors are induced. The solutions oscillate about the 'correct' solution path but do not appear to diverge. In Figure 10 the solution obtained using a 4 week timestep size generates the largest errors relative to the 'correct' solution. However, the oscillatory behaviour of the solution has not significantly deteriorated.

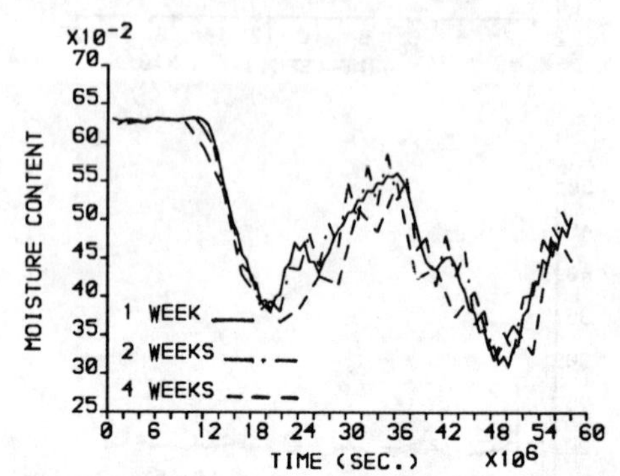

Fig.10 The influence of changing the timestep size on moisture content variation at a depth of 0.3m

CONCLUSIONS

For the first problem, benchmark solutions have been achieved using a Backward Difference approach. Solutions have also been achieved using Lees algorithm which show insignificant difference from the benchmark solutions. Numerical experiments have been carried out to show the influence of relaxing timestep size control tolerances and increasing initial timestep sizes. Results show that slack tolerances or large initial timestep sizes cause the onset of oscillatory behaviour from Lees' algorithm. The ability of an averaging technique to reduce oscillations has been illustrated. Lees'algorithm's previously observed characteristic of unconditional stability and convergence, was seen to also apply for the problem considered here.

Solutions for the second problem have been achieved using the non-linear model running in linear mode. The ability of Lees algorithm to cope with time varying Dirichlet boundary conditions has been illustrated. A comparison of solutions obtained using Lees algorithm and the Crank-Nicolson approach has been presented. The results show that Lees' Algorithm compares favourably with Crank-Nicolson for the problem analysed.

The infleunce of increasing the timestep size, using Lees' algorithm for the second problem has also been illustrated. The results show progressive oscillatory behaviour as the timestep size is increased. However, for the timestep sizes considered, the results show oscillations appear to pivot about the 'correct' solution and do not appear divergent.

ACKNOWLEDGEMENTS

British Gas' collaboration and SERC's financial support are gratefully acknowledged.

REFERENCES

Bonacina, C. and Comini, G. 1973. 'On the Solution of the Nonlinear Heat Conduction Equations by Numerical Methods'. Int.J.Heat Mass Transfer, 16, 581-589.

Comini, G., Del Guidice, S., Lewis, R.W. and Zienkiewicz, O.C. 1974. 'Finite Element Solution of Non-linear Heat Conduction Problems with special reference to Phase Change'. Int.J.Num. Meth.Engng., 8, 613-624.

Lees, M. 1966. 'A linear three-level difference scheme for quasilinear parabolic equations'. Maths.Comp., 20, 516-522.

Thomas, H.R. 1987. 'Non-linear analysis of heat and moisture transfer in unsaturated soils'. J.Eng.Mech., A.S.C.E., 113(8), 1163-1180.

Thomas, H.R. and Chan, P.S. 1986. 'An Algorithm for the prediction of Non-linear One-dimensional Consolidation'. Microsoftware for Engineers, 2(4), 225-232.

Thomas, H.R. and Owen, R.C. 1987. 'A Comparison of Field Monitored and Numerically Predicted Heat and Moisture Movement Results'. Submitted for publication.

Wood, W.L. and Lewis, R.W. 1975. 'A Comparison of Time Marching Schemes for the Transient Heat Conduction Equation', Int.J.Num.Meth.Engng., 9, 679-689.

Numerical Methods in Geomechanics (Innsbruck 1988), Swoboda (ed.)
© 1988 Balkema, Rotterdam. ISBN 90 6191 809 X

Hybrid finite element model utilized for fracture of concrete beams on elastic foundations

S.M.Sargand, G.A.Hazen & H.Zheng
Department of Civil Engineering, Ohio University, Athens, USA

ABSTRACT: Use of the crack band theory in a finite element model requires a high degree of accuracy in the computation of stresses and displacements, that presents use of a mesh with a low directional bias. A hybrid stress model is applied to three problems in the fracture of a concrete; beam on an elastic foundation, a simply supported beam and a panel. The results of this model are compared with those obtained from the displacement method and exact solutions.

1 INTRODUCTION

In recent years, the displacement model has been often utilized in the analysis of the fracture of concrete based on crack band theory. However, the accuracy of the computed stresses may not be sufficient for fracture problems. Also, when employing the displacement model the predicted displacements and stresses are influenced by the shape of elements.

The crack band theory is based on the assumption that the cracks are continuously distributed within a finite element and the fracture process is simulated by gradual reduction of the elastic modulus in the direction normal to the crack (Bazant and Oh (1983), Bazant (1985)). In this theory the material properties are modeled by a compliance matrix. To use the displacement finite element method the compliance matrix must be inverted to obtain the stiffness matrix. In the case where fracture direction is unknown and the finite element mesh has an inclination relative to fracture direction, it is difficult to determine accurately which element will crack next (Bazant (1985)).

The objective of this paper is to present the application of the stress hybrid finite element procedure in the study of crack propagation in conjunction with the analysis of a concrete beam on an elastic foundation. The procedure followed can provide more accurate stresses as well as displacements and can overcome the difficulty of directional bias encountered when the displacement finite element method is used. The application of the assumed stress hybrid finite element model in the crack band theory leads to a better understanding of the material response of an elastic foundation and an interface element to crack propagation. Various applications, in addition to the concrete beam on an elastic foundation, are also discussed.

2 CONSTITUTIVE EQUATIONS

In the crack band model, the softening behavior is simulated by means of reduction of stiffness in the direction normal to the crack in the band. The compliance matrix for this model is given as (Bazant and Oh (1983))

$$\begin{Bmatrix} \varepsilon_x \\ \varepsilon_y \\ \varepsilon_z \end{Bmatrix} = 1/E \begin{bmatrix} 1 & -\nu & -\nu \\ -\nu & 1 & -\nu \\ -\nu & -\nu & \mu^{-1} \end{bmatrix} \begin{Bmatrix} \sigma_x \\ \sigma_y \\ \sigma_z \end{Bmatrix} \qquad (1)$$

$$\mu^{-1} = \frac{E}{-E_t} \quad \frac{\varepsilon_z}{\varepsilon_o - \varepsilon_z} \qquad (2)$$

μ is the cracking parameter varying from 0 to 1, $\mu = 1$ implies a crack-free condition and $\mu = 0$ corresponds to a fully cracked state. The value of μ is computed from the tensile stress-strain diagram (Fig. 1).

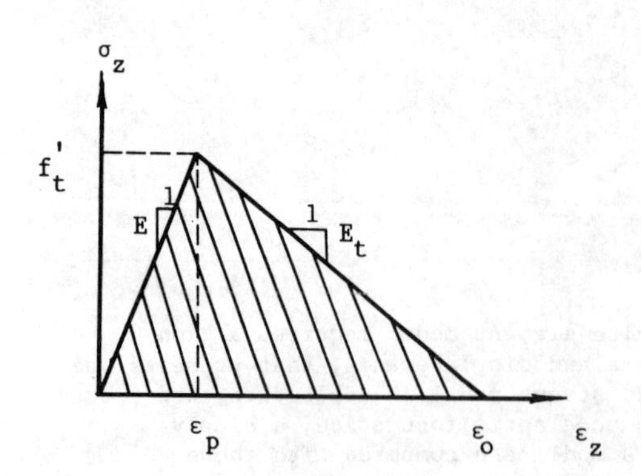

Fig.1 Stress-strain diagrams for crack band

In the stress hybrid method the compliance matrix can be used directly. During this study the load is applied incrementally by controlling the displacement increments at the loading location.

3 STRESS HYBRID METHOD

In the commonly used stress hybrid method, an equilibrium stress field is assumed within the element, and an independent displacement field is assumed on the element boundaries. Pian et al (1982, 1984) presented a new procedure to formulate the stress hybrid model, where they introduced the stress equilibrium conditions with the aid of Lagrange multipliers. Thus obviating the need for an exact equilibrium state. As a consequence the stress field can be expressed in terms of natural coordinates.

The stress hybrid method can be obtained from the following functional (Pian and Chen (1982)).

$$\Pi_R = \int_v [-\frac{1}{2}\{\sigma\}^T[S]\{\sigma\} + \{\sigma\}^T([D]\{u_q\}) - ([D]^T\{\sigma\})^T\{u_\lambda\}]dv \tag{3}$$

where [S] = compliance matrix; $\{\sigma\}$ = stresses, $\{u_q\}$ = compatible displacements in terms of nodal displacements; $\{u_\lambda\}$ = internal displacements; V = volume. The stresses, $\{\sigma\}$, neglecting the body forces, are expressed as

$$\{\sigma\} = [P]\{\beta\} \tag{4}$$

and the compatible displacements, $\{u_q\}$, the internal displacements, $\{u_\lambda\}$, are assumed as

$$[u_q] = [N]\{q\} \tag{5}$$

$$[u_\lambda] = [L]\{\lambda\} \tag{6}$$

Substitution of Eqs. 4-6 into Eq. 3 leads to an expression for Π_R as

$$\Pi_R = -\frac{1}{2}\{\beta\}^T[H]\{\beta\} + \{\beta\}^T[G]\{q\} - \{\beta\}^T[R_1]\{\lambda\} \tag{7}$$

variation of Π_R with respect to $\{\beta\}$ and $\{\lambda\}$ and substitution of results back into Eq. 7 leads to the expression for stiffness matrices as

$$[K] = [G]^T[H]^{-1}[G] - [G]^T[H]^{-1}[R_1]([R_1]^T[H]^{-1}[R_1])^{-1}[R_1]^T[H]^{-1}[G] \tag{8}$$

in which

$$[H] = \int_v [P]^T[S][P]dv; \ [G] = \int_v [P]^T[B]dv, \quad [R_1] = \int_v [E]^T[L]dv, \quad [B] = [D][N], \ [E] = [D]^T[P] \tag{9}$$

The calculation of the load vector is very similar to the displacement methods.

In this investigation, a two-dimensional isoparametric, four node element is used. The detail formulation of this element appears in Ref. (Pian and Sumihara (1984)).

4 APPLICATIONS

Beam under pure bending--Consider a deep beam loaded under pure bending (Fig. 2), the displacement u at point A is calculated by both hybrid finite element procedure and displacement finite element model for different aspect ratios.

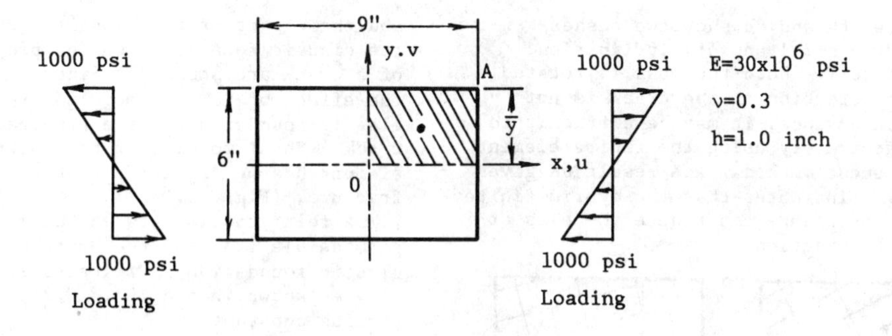

Fig.2 Pure bending beam and loading

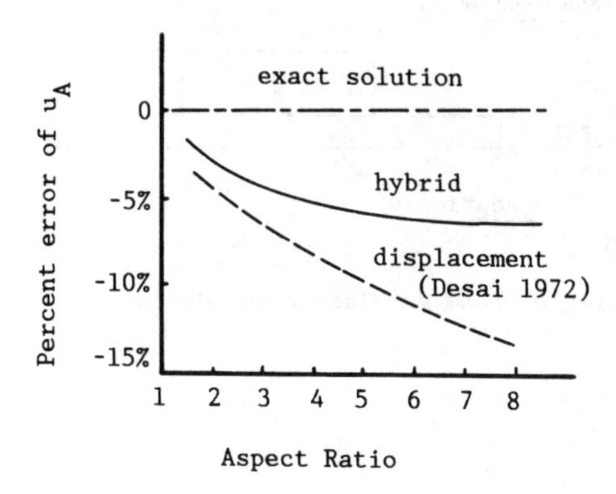

Fig.3 Inaccuracy vs aspect ratio

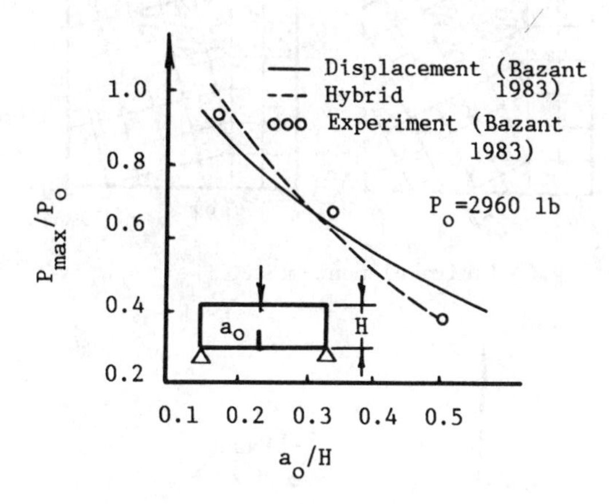

Fig.4 Comparison of the maximum load results

For very large structures, if the fracture direction is known, we can use larger elements outside the crack band and use elements with the width ω_c within the crack band to reduce the total number of elements. For this case, high aspect ratio elements can be used. The hybrid finite element procedure is able to provide more accurate high aspect ratio elements than the displacement finite element procedure (Fig. 3).

Notched Beam--A notched beam, 6 in. wide, 6 in. deep, and 20 in. long is investigated. The fracture energy G, is assumed to be 0.101 lb/in., the tensile strength is 300 psi, Young's modulus is 4190 ksi.

Figure 4 shows that the hybrid finite element procedure provides results closer to the experimental data than the displacement finite element method.

Rectangular center-cracked panel--Fig. 5 shows a rectangular, center-cracked panel. The top and the bottom of the panel are loaded at their centers by force P. For the panel E=3000 ksi, =0.18, G =0.12 lb/in, f ′=330 psi; and dimensions b=10 in., H=10 in., and B=1 in.

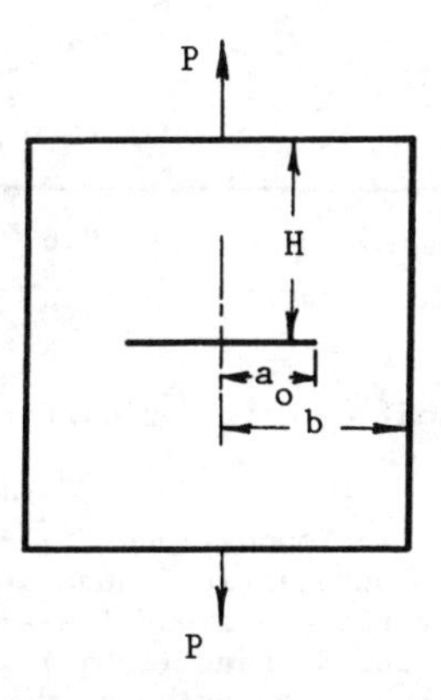

Fig.5 Center-cracked panel

Figures 6a and 6b show two meshes rotated to a 2:1 and 4:1 inclination, respectively. When the mesh is rotated and the direction of the crack is not known in advance, it may be difficult to solve accurately using the finite element displacement method. The result as given in Fig. 7 indicates that the hybrid finite element procedure can reduce the bias of the mesh direction.

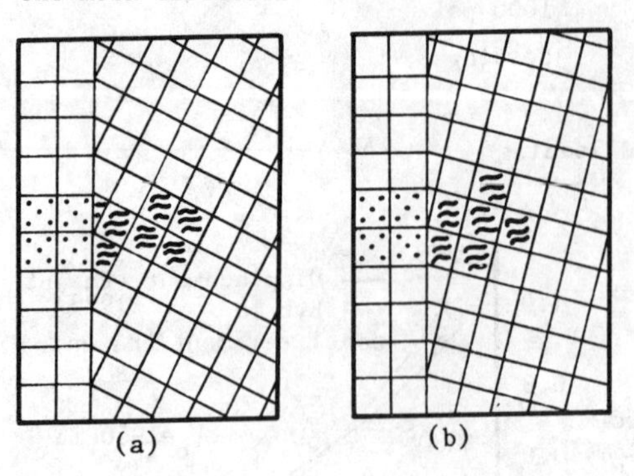

Fig.6 Finite element mesh

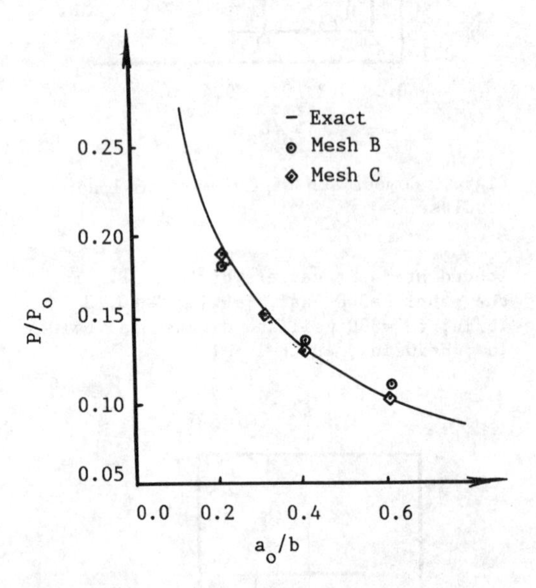

Fig.7 Numerical results for different mesh inclination

Beam on elastic foundation--Fig. 8 shows a notched beam on elastic foundation. A thin-layer interface element (Desai et al (1984), Desai and Sargand (1984)) is used to represent the interaction between the beam and elastic foundation. The effect of shear modulus of the interface element and the effect of the Young's modulus of the elastic foundation on the propagation of a crack are studied. Figure 9 shows the effect of Young's modulus of the elastic foundation to the propagation of a crack. Shear modulus of the interface element has no significant effect on the fracture, (Fig. 10.)

The relationship between the maximum permissible load and the depth of the elastic foundation is illustrated in Fig. 11. As shown when d_f/L is 0.25, the load remains constant.

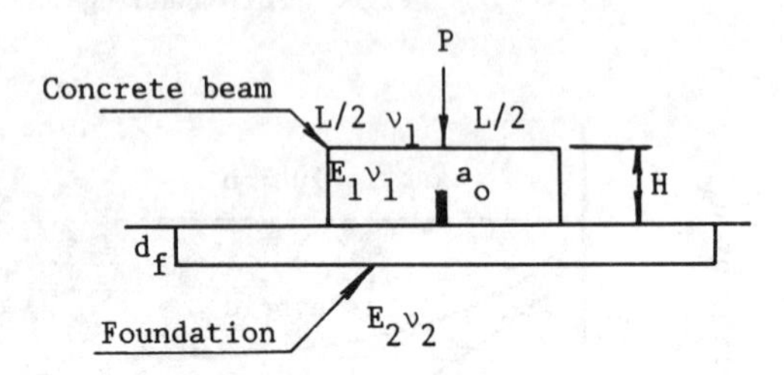

Fig.8 Beam on elastic foundation

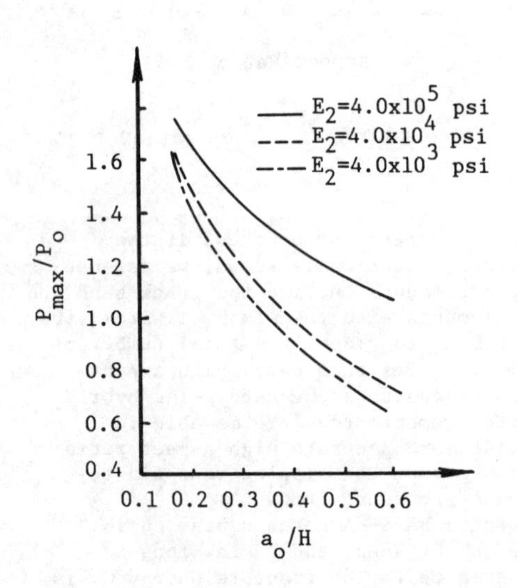

Fig.9 The effect of E_2 to the fracture propagation

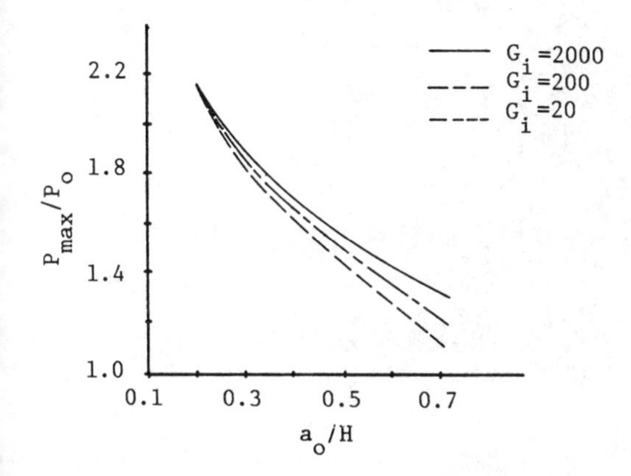

Fig.10 The effect of G_i to the fracture propagation

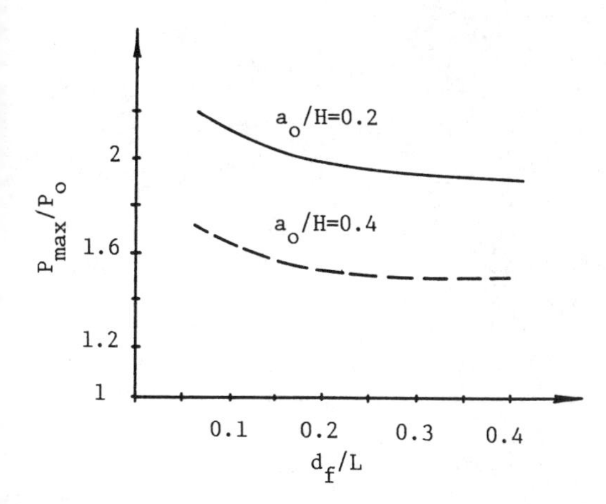

Fig.11 The maximum permissible load vs. the depth of the elastic foundation

5 CONCLUSIONS

The crack band theory in conjunction with the stress hybrid model was used to predict fracture in concrete. It is obvious from the solutions presented for the notched beam and the rectangular center-cracked panel, that the stress hybrid model can provide as good or an improvement over the results of the displacement method. The stress hybrid model is less orientation biased and can give reasonable predictions for a problem where the direction of crack is not known in advance. From the analysis of a beam on an elastic foundation results, it appears that the crack band theory can be used in design of pavement.

REFERENCES

Bazant, Z.P. and Cedolin, L. 1979. Blant crack band propagation in finite element analysis. Journal of the Engineering Mechanics Division, ASCE, 105, EM2, 297-315.

Bazant, Z.P. and Oh, B.H. 1983. Crack band theory for fracture of concrete. Materials and Structures (RILEM, Paris), 16, 155-177.

Bazant, Z.P. 1985. Fracture in concrete and reinforced concrete. Mechanics of Geomaterials, pp. 259-303.

Pian, T.H.H. and Tong, Pin, 1969. Basis of finite element methods for solid continua. Int. J. Numer. Methods Eng., 1, 3-28.

Pian, T.H.H. and Chen, D.P. 1982. Alternative ways for formulation of hybrid stress elements. Int. J. Numer. Methods Eng., 18, 1679-1684.

Pian, T.H.H. and Sumihara, K., 1984. Rational approach for assumed stress finite elements. Int. J. Numer. Methods Eng., 20, 1685-1695.

Desai, C.S. and Abel, J.F., 1972. Introduction to the finite element method. New York, Van Nostrand Reinhold Company.

Desai, C.S., Zaman, M.M., Lightner, J.G. and Siriwardane, H.J., 1984. Thin-layer element for interfaces and joints. Int. J. Numer. Anal. Methods Geomech., 8, 19-43.

Desai, C.S. and Sargand, S., 1984. Hybrid FE procedure for soil-structure interaction. Journal of the Geotechnical Engineering Division, ASCE, Vol. 110, No. 4.

Numerical Methods in Geomechanics (Innsbruck 1988), Swoboda (ed.)
© 1988 Balkema, Rotterdam. ISBN 90 6191 809 X

Limit analysis of plane problems in soil mechanics

S.W.Sloan
Department of Civil Engineering and Surveying, University of Newcastle, NSW, Australia

ABSTRACT: This paper describes techniques for computing rigorous upper and lower bound limit loads for soil mechanics problems under conditions of plane strain. The methods assume a perfectly plastic soil model, which is either purely cohesive or cohesive frictional, and employ finite elements in conjunction with the bound theorems of classical plasticity. Both formulations lead to linear programming problems which may be solved efficiently using an active set algorithm.

1. INTRODUCTION

The bound theorems of classical plasticity theory are powerful tools for analysing the stability of problems in soil mechanics. The lower bound theorem states that any statically admissible stress field will furnish a lower bound (or "safe") estimate of the true limit load; whereas the upper bound theorem states that the power dissipated by any kinematically admissible velocity field can be equated to the power dissipated by the external loads to obtain an upper bound (or "unsafe") estimate of the true limit load. The two theorems may thus be used to bracket the true limit load from above and below, but it is frequently difficult to derive tight bounds via traditional analytical techniques.

When using the lower bound theorem, it is often difficult to manually construct statically admissible stress fields for practical problems involving complicated loading and complex geometry. A statically admissible stress field is one which satisfies the stress boundary conditions, equilibrium and nowhere violates the yield criterion (the stresses must lie inside or on the yield surface in stress space). An alternative method for computing lower bounds, which uses finite elements and linear programming, has been presented by Lysmer (1). This procedure uses 3-noded triangular elements with the nodal variables being the unknown stresses. Statically admissible stress discontinuities are permitted to occur at the interfaces between adjacent triangles. The finite element formulation leads to a linear programming problem where the objective function, which is to be maximised, corresponds to the collapse load and is expressed in terms of the unknown stresses. The unknowns are subject to a set of linear constraints arising from the stress boundary conditions, equilibrium and yield criterion. The computation of lower bound limit loads by finite elements and linear programming has been studied by several authors including Anderheggen and Knopfel (2) and Bottero et al (3).

When using the upper bound theorem, the power dissipated by a kinematically admissible velocity field is equated to the power dissipated by the external loads to deduce a strict upper bound on the true limit load. A kinematically admissible velocity field is one which satisfies compatibility, the flow rule and the velocity boundary conditions. Numerical procedures for computing upper bound limit loads, which employ finite elements in conjunction with the upper bound theorem, have been given by Anderheggen and Knopfel (2) and Bottero et al (3). The latter authors, whose formulation is adopted in this paper, employ 3-noded triangular elements with the nodal variables being the unknown velocities. An additional set of unknowns, the plastic multipliers, are associated with each element. Kinematically admissible velocity discontinuities are permitted along specified planes within the grid. The finite element formulation of the upper bound theorem leads to a classical linear programming problem where the objective function, which is to be minimised, corresponds to the dissipated power and is expressed in terms of the unknown velocities and plastic multipliers. The unknowns are subject to a set of linear constraints arising from the imposition of the flow rule and velocity boundary conditions.

Advantages of the numerical formulations of the bound theorems include the ability to deal with complex geometries, complicated loading and a variety of boundary conditions. It is shown that the linear programming problems that arise from the finite element formulations may be solved efficiently using an active set algorithm (Best and Ritter (4), Sloan (5)). Detailed timing statistics are given to support this claim. One example is analysed to illustrate that the numerical procedures may be used to bracket the exact limit load with an accuracy which is sufficient for design purposes.

2. LOWER BOUND FORMULATION

The lower bound formulation used in this paper follows that of Bottero et al (3). Although the formulation proposed by Lysmer (1) requires a smaller number of variables for an equivalent mesh, and hence is potentially more efficient, it often leads to a constraint matrix with terms of widely varying magnitude. As the conditioning of the constraint matrix has a critical effect on the performance of linear programming algorithms, we prefer the formulation of Bottero et al (3) which generally results in a constraint matrix whose terms vary by only a few orders of magnitude.

The triangular element used to model the stress field under conditions of plane strain is shown in Figure 1. The variation of the stresses throughout each element is linear and each node is associated with 3 unknown stresses σ_x, σ_y and τ_{xy}. The stresses are given by:

$$\sigma_x = \sum_{i=1}^{3} N_i \, \sigma_{xi} \quad ; \quad \sigma_y = \sum_{i=1}^{3} N_i \, \sigma_{yi}$$

$$\tau_{xy} = \sum_{i=1}^{3} N_i \, \tau_{xyi} \tag{1}$$

where σ_{xi}, σ_{yi} and τ_{xyi} are the nodal stresses and N_i are linear shape functions. A mesh of linear stress triangles is shown in Figure 2.

Unlike the usual form of the finite element method, each node is unique to a particular element and more than one node may share the same co-ordinates. Statically admissible stress discontinuities are permitted at edges shared by adjacent triangles. If E denotes the number of triangles in the mesh, then there are 3E nodes and 9E unknown stresses.

In order to satisfy equilibrium, the stresses throughout each triangle must satisfy the equations

$$\frac{\partial \sigma_x}{\partial x} + \frac{\partial \tau_{xy}}{\partial y} = 0 \quad ; \quad \frac{\partial \sigma_y}{\partial y} + \frac{\partial \tau_{xy}}{\partial x} = \gamma \tag{2}$$

where tensile stresses are taken as positive, γ is the soil unit weight and a right-handed cartesian co-ordinate system is adopted. Differentiating equation (1) and substituting in (2) gives the equilibrium constraints on the nodal stresses as

$$\underset{2\times 9}{\{a_1^e\}} \; \underset{9\times 1}{\{\sigma^e\}} = \underset{2\times 1}{\{b_1^e\}} \tag{3}$$

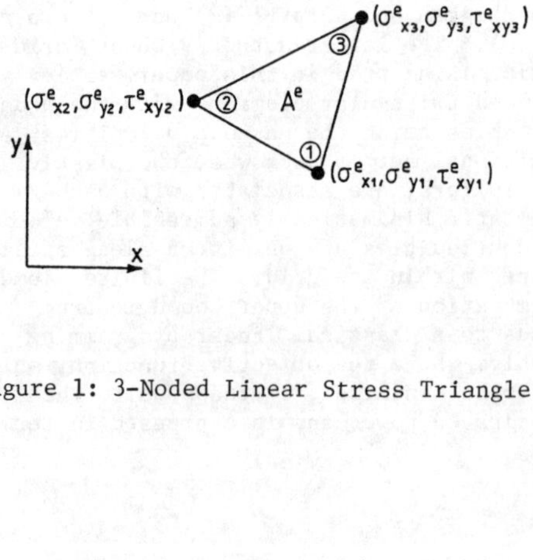

Figure 1: 3-Noded Linear Stress Triangle

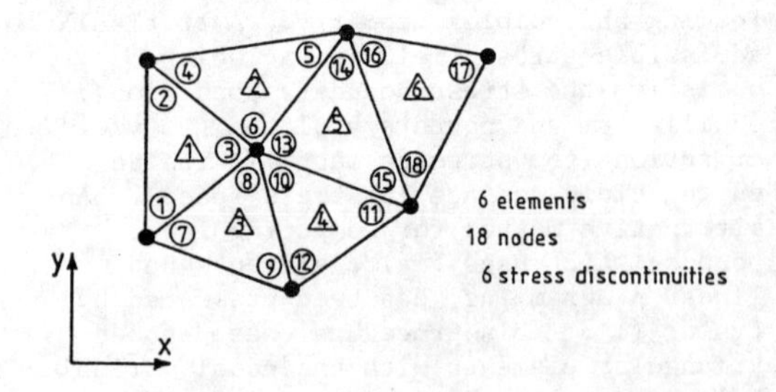

Figure 2: Lower Bound Mesh

where $[a_1^e]$ is a matrix of constants, $\{\sigma^e\}$ is a vector of the triangle nodal stresses and $\{b_1^e\}^T = \{0 \ \gamma\}$.

A statically admissible stress discontinuity permits the tangential stress to be discontinuous, but requires that continuity of the corresponding shear and normal components is preserved. These discontinuities, which are allowed at the edges shared by adjacent triangles, give rise to additional constraints on the nodal stresses. The normal and shear stresses acting on a plane inclined at an angle θ to the x-axis are given by

$$\sigma_n = \sin^2\theta \ \sigma_x + \cos^2\theta \ \sigma_y - \sin 2\theta \ \tau_{xy}$$
$$\tau = -\tfrac{1}{2}\sin 2\theta \ \sigma_x + \tfrac{1}{2}\sin 2\theta \ \sigma_y + \cos 2\theta \ \tau_{xy} \qquad (4)$$

Since the stresses are assumed to vary linearly along the edge of each triangle, discontinuity equilibrium is ensured by enforcing equation (4) at the two nodal pairs defining each edge. This gives rise to the equilibrium constraints for each discontinuity as

$$\underset{4\times12}{[a_2^d]} \ \underset{12\times1}{\{\sigma^d\}} = \underset{4\times1}{\{b_2^d\}} \qquad (5)$$

where $[a_2^d]$ is a matrix of constants, $\{\sigma^d\}$ is a vector of stresses for the discontinuity nodes and $\{b_2^d\}^T = \{0 \ 0 \ 0 \ 0\}$.

In order to satisfy the stress boundary conditions, the normal stresses and shear stresses must equal prescribed values along the boundaries of the mesh. Since the stresses are permitted to vary linearly along an edge of a triangle, the stress boundary conditions are satisfied exactly by enforcing equation (4) at the two nodes which define each edge. If both the normal stresses and shear stresses are prescribed for a boundary segment, this gives rise to equality constraints of the form

$$\underset{4\times6}{[a_3^\ell]} \ \underset{6\times1}{\{\sigma^\ell\}} = \underset{4\times1}{\{b_3^\ell\}} \qquad (6)$$

where $[a_3^\ell]$ is a matrix of constants, $\{\sigma^\ell\}$ is a vector of stresses associated with the two nodes defining the segment and $\{b_3^\ell\} = \{q_1 \ t_1 \ q_2 \ t_2\}$. $(q_1 \ t_1)$ and $(q_2 \ t_2)$ are the normal stresses and shear stresses which are fixed at the two nodes defining the segment.

Assuming tensile stresses are taken as positive and plane strain conditions, the Mohr-Coulomb yield criterion may be expressed as

$$F = (\sigma_x-\sigma_y)^2 + (2\tau_{xy})^2 - (2c.\cos\phi - - (\sigma_x+\sigma_y)\sin\phi)^2 = 0 \qquad (7)$$

Since we wish to formulate the lower bound theorem as a linear programming problem, it is necessary to approximate (7) by a yield criterion which is a linear function of the stresses. Letting $X = \sigma_x-\sigma_y$, $Y = 2\tau_{xy}$ and $R = 2c.\cos\phi - (\sigma_x+\sigma_y)\sin\phi$, the Mohr-Coulomb condition plots us a circle as shown in Figure 3. The linearised Mohr-Coulomb surface is an interior polygon with p sides and p vertices and each stress point must satisfy

$$A_k\sigma_x + B_k\sigma_y + C_k\tau_{xy} \le D \ ; \ k = 1,2..,p \qquad (8)$$

where

$$A_k = \cos\alpha_k + \sin\phi \cos\beta$$

$$B_k = \sin\phi \cos\beta - \cos\alpha_k$$

$$\beta = \pi/p$$

$$C_k = 2\sin\alpha_k$$

$$D = 2c.\cos\phi \cos\beta$$

$$\alpha_k = 2k\beta$$

A detailed derivation of the above may be found in Sloan (6). We note in passing that the inequality in (8) arises because each stress point must lie inside or on

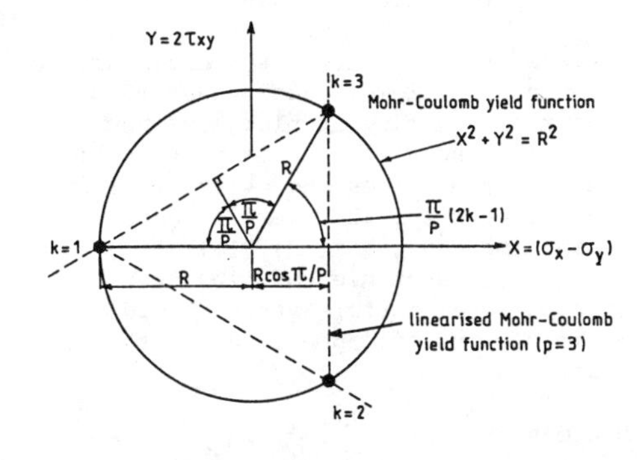

Figure 3: Linearised Mohr-Coulomb Yield Function

the linearised yield surface. To ensure that the solution obtained is a strict lower bound on the exact collapse load, the linearised yield surface is defined to lie inside the Mohr-Coulomb yield surface in stress space. The linearised yield condition is satisfied throughout a triangle by enforcing (8) at each of its nodes. This gives rise to the inequality constraints

$$\underset{p\times 3}{[a_4^i]}\ \underset{3\times 1}{\{\sigma^i\}} = \underset{p\times 1}{\{b_4^i\}} \qquad (9)$$

where $[a_4^i]$ is comprised of the coefficients A_k, B_k, C_k, $\{\sigma^i\}$ is a vector of nodal stresses and $\{b_4^i\}^T = \{2c.\cos\phi\ \cos\beta\ .....,\ 2c.\cos\phi\ \cos\beta\}$.

For most plane strain geotechnical problems, we wish to find a statically admissible stress field which maximises an integral of the form

$$Q = h \int_S \sigma_n \, dS$$

where Q is the collapse load, h is the out-of-plane thickness and σ_n is the normal stress acting over S. Since the stresses vary linearly throughout each element, $Q = (Lh/2)(\sigma_{n1} + \sigma_{n2})$ where L is the length of the edge and $(\sigma_{n1}, \sigma_{n2})$ are the normal stresses at the two nodes defining it. Using (4), assuming h = 1 and letting θ denote the inclination of the edge to the x-axis we obtain

$$Q = \{c^s\}^T \{\sigma^s\} \qquad (10)$$

where $\{c^s\}$ is a vector of constants (with units of area) and $\{\sigma^s\}$ is a vector of stresses associated with the two nodes defining the edge.

All of the steps that are necessary to formulate the lower bound theorem as a linear programming problem have now been covered. The next step is to assemble the constraint matrices and objective function coefficients for the overall mesh. Assembling equations (3), (5), (6), (9) and (10), the problem of finding a statically admissible stress field which maximises the collapse load may thus be stated as

$$\text{Minimise} \quad -\{c\}^T \{\sigma\}$$

$$\text{Subject to} \quad [A_1]\{\sigma\} = \{b_1\}, \quad [A_2]\{\sigma\} \le \{b_2\} \qquad (11)$$

where

$$[A_1] = \sum_e^E [a_1^e] + \sum_d^D [a_2^d] + \sum_\ell^L [a_3^\ell]$$

$$[A_2] = \sum_i^N [a_4^i]$$

$$\{b_1\} = \sum_e^E \{b_1^e\} + \sum_d^D \{b_2^d\} + \sum_\ell^L \{b_3^\ell\}$$

$$\{b_2\} = \sum_i^N \{b_4^i\}$$

$$\{c\} = \sum_s^S \{c^s\}$$

and E, N, D, L and S are, respectively, the number of triangles, nodes, discontinuities, boundary edges with prescribed tractions and loaded boundary edges.

3. UPPER BOUND FORMULATION

Following Bottero et al (3), the 3-noded triangle used for the upper bound limit analysis is shown in Figure 4. Each element is associated with 6 nodal velocities and p plastic multipliers (p is the number of sides in the linearised yield polygon). The velocities vary linearly throughout each triangle according to

$$u = \sum_{i=1}^{3} N_i u_i \; ; \quad v = \sum_{i=1}^{3} N_i v_i \qquad (12)$$

where (u_i, v_i) are the nodal velocities in the x- and y-directions and N_i are linear shape functions. In order to formulate the upper bound theorem as a linear programming problem, it is again necessary to approximate the Mohr-Coulomb yield function by a

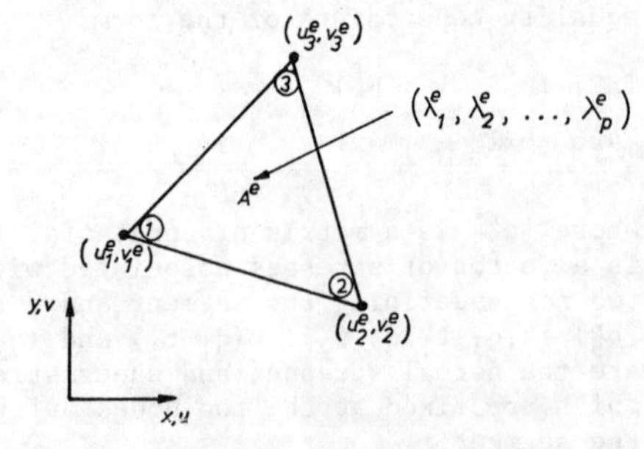

Figure 4: 3-noded Triangle for Upper Bound Limit Analysis

function which is linear in the stresses. The linearised yield criterion circumscribes the Mohr-Coulomb yield criterion in stress space, so that the bounding property of the solution is preserved, and is defined by

$$F_k = A_k\sigma_x + B_k\sigma_y + C_k\tau_{xy} - 2c.\cos\phi = 0$$
$$k = 1, 2, \ldots, p \qquad (13)$$

where

$$A_k = \cos\alpha_k + \sin\phi$$

$$B_k = \sin\phi - \cos\alpha_k$$

$$C_k = 2\sin\alpha_k$$

$$\alpha_k = 2\pi k/p$$

For the linearised yield criterion, an associated flow rule gives the plastic strain rates throughout each triangle as

$$\dot{\varepsilon}_x = \frac{\partial u}{\partial x} = \sum_{k=1}^{p} \lambda_k \frac{\partial F_k}{\partial \sigma_x} = \sum_{k=1}^{p} \lambda_k A_k$$

$$\dot{\varepsilon}_y = \frac{\partial v}{\partial y} = \sum_{k=1}^{p} \lambda_k \frac{\partial F_k}{\partial \sigma_y} = \sum_{k=1}^{p} \lambda_k B_k$$

$$\lambda_k \geq 0 \qquad (14)$$

$$\dot{\gamma}_{xy} = \frac{\partial v}{\partial x} + \frac{\partial u}{\partial y} = \sum_{k=1}^{p} \lambda_k \frac{\partial F_k}{\partial \tau_{xy}} = $$

$$= \sum_{k=1}^{p} \lambda_k C_k$$

where λ_k is the plastic multiplier associated with the kth side of the yield surface. Equations (12) and (14) lead to a set of equality constraints of the form

$$[a_{11}^e] \{u^e\} + [a_{12}^e] \{\lambda^e\} = \{0\} \qquad (15)$$

where

$[a_{11}^e]$, $[a_{12}^e]$ are matrices of constants

$\{u^e\}^T = \{u_1\ v_1\ u_2\ v_2\ u_3\ v_3\}$ and

$\{\lambda^e\}^T = \{\lambda_1^e\ \lambda_2^e\ \ldots\ldots\ \lambda_p^e\}$ with

$\lambda_k^e \geq 0$ for $k = 1, 2, \ldots, p$

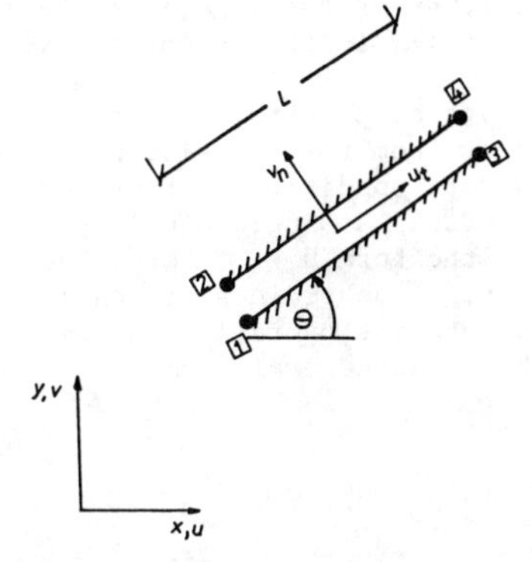

Figure 5: Velocity Discontinuity

In order to permit kinematically admissible velocity discontinuities, additional constraints need to be enforced on the nodal velocities. If u_t and v_n denote the jumps in tangential and normal velocity across a discontinuity, then we have for an associated flow rule $v_n = \tan\phi\, |u_t|$.

A typical segment of a velocity discontinuity, defined by the nodal pairs (1, 2) and (3, 4) is shown in Figure 5. For this arrangement we have at the nodal pair (1, 2)

$$u_t = \cos\theta\ (u_2 - u_1) + \sin\theta\ (v_2 - v_1)$$

$$v_n = \sin\theta\ (u_1 - u_2) + \cos\theta\ (v_2 - v_1)$$

Similar expressions exist for the nodal pair (3, 4).

In order to preserve a linear constraint matrix in the numerical formulation, it is necessary to specify a sign for each discontinuity, S, such that $|u_t| = S.u_t$. For a velocity discontinuity which is defined by several nodal pairs, the constraint $S.u_t \geq 0$ must be enforced at each nodal pair. This gives rise to an inequality constraint of the form

$$\underset{1\times4}{[a_2^q]}\ \underset{4\times1}{\{u^q\}} \leq \underset{1\times1}{\{0\}} \qquad (16)$$

where $[a_2^q]$ depends on the sign and orientation of the discontinuity and $\{u^q\}^T = \{u_1\ v_1\ u_2\ v_2\}$. Similarly the flow rule leads to one equality constraint of the form

$$\underset{1\times4}{[a_3^q]}\ \underset{4\times1}{\{u^q\}} = \underset{1\times1}{\{0\}} \qquad (17)$$

where $[a_3^q]$ depends on the sign, orientation and friction angle of the discontinuity.

The final type of constraint on the unknowns arises from the imposition of the velocity boundary conditions. Each boundary condition generates a single equality constraint of the form $u_i = \delta_1$ or $v_i = \delta_2$ at node i. These constraints are easily incorporated into the overall constraint matrix and have the general form

$$\underset{1\mathrm{x}1}{[a_4^\ell]} \; \underset{1\mathrm{x}1}{\{u^\ell\}} = \underset{1\mathrm{x}1}{\{\delta^\ell\}} \qquad (18)$$

where $\{u^\ell\}$ is the u- or v-velocity that is prescribed, $\{\delta^\ell\}$ is its value and $[a_4^\ell]$ is a unit matrix.

When using the upper bound theorem, we wish to minimise the dissipated power. In the linear programming formulation, the dissipated power thus becomes the objective function. For the velocity discontinuity segment defined by Figure 5, with length L, sign S, cohesion c, and inclination θ to the axis it may be shown (Sloan, (7)) that the dissipated power is given by

$$P^d = S.c.\frac{L}{2}\,[\cos\theta(u_2+u_4-u_1-u_3) + \\ + \sin\theta(v_2+v_4-v_1-v_3)]$$

This may be written as

$$P^d = \{c_1^d\}^T \; \{u^d\} \qquad (19)$$

where $\{c_1^d\}$ is a function of S, c, L and θ and $\{u^d\}^T = \{u_1\ v_1\ u_2\ v_2\ u_3\ v_3\ u_4\ v_4\}$. In addition to the power dissipated along velocity discontinuities, we permit power to be dissipated by volumetric distortion of the triangles. Using the definition of plastic power and equations (13) and (14), it may be shown (Sloan (7)) that the power dissipated in a triangle is given by

$$P^e = 2c.\cos\phi.A^e.\sum_{k=1}^{p}\lambda_k \quad ; \quad \lambda_k \geq 0$$

which may be written as

$$P^e = \{c_2^e\}^T\{\lambda^e\} \qquad (20)$$

where $\{c_2^e\}$ is a function of c, ϕ and the element area (A^e) and $\{\lambda^e\}^T = \{\lambda_1^e\ \lambda_2^e\ ..,\ \lambda_p^e\}$.

All of the steps necessary to formulate the upper bound theorem as a linear programming problem have now been covered. Assembling equations (15), (16), (17), (18), (19) and (20), the problem of finding a kinematically admissible velocity field which minimises the dissipated power may thus be stated

$$\min\ \{c_1\}^T\{u\} + \{c_2\}^T\{\lambda\} + \{c_3\}^T\{s\} \qquad (21)$$

subject to

$$[A_{11}]\{u\} + [A_{12}]\{\lambda\} = \{b_1\}$$

$$[A_2]\{u\} + [I]\{s\} = \{b_2\}$$

$$[A_3]\{u\} = \{b_3\}$$

$$[A_4]\{u\} = \{b_4\}$$

$$\{\lambda\},\ \{s\} \geq \{0\}$$

where

$$[A_{11}] = \sum_e^E [a_{11}^e] \quad ; \quad [A_{12}] = \sum_e^E [a_{12}^e]$$

$$[A_2] = \sum_q^Q [a_2^q] \quad ; \quad [A_3] = \sum_q^Q [a_3^q]$$

$$[A_4] = \sum_\ell^L [a_4^\ell]$$

$$\{c_1\} = \sum_d^D \{c_1^d\} \quad ; \quad \{c_2\} = \sum_e^E \{c_2^e\}$$

$$\{b_1\} = \{b_2\} = \{b_3\} = \{c_3\} = \{0\}$$

$$\{u\}^T = \{u_1\ v_1\ \ldots\ldots,\ u_N\ v_N\}$$

$$\{\lambda\}^T = \{\lambda_1^1\ \ldots\ \lambda_p^1,\ \lambda_1^2\ \ldots\ \lambda_p^2,\ \ldots,\ \lambda_1^E\ \ldots\ \lambda_p^E\}$$

$$\{s\}^T = \{s_1\ s_2\ \ldots\ldots,\ s_Q\}$$

and E, N, D, L and Q are, respectively, the number of triangles, nodes, discontinuity segments, prescribed velocities and nodal pairs defining velocity discontinuities. Note that the inequalities introduced by the sign conditions on the velocity discontinuities have been converted to equalities by the addition of non-negative slack variables and that [I] denotes the identity matrix.

4. SOLUTION PROCEDURE

The algorithm used to solve the upper and lower bound linear programming problems is a steepest edge active set algorithm (Sloan (5)). This is a modification of the active set procedure of Best and Ritter (4) which fully exploits the sparsity of the constraint matrices and uses the efficient factorisation routines developed by Reid (8). The advantages of the active set algorithm, when applied to the lower bound linear programming problem, have been discussed by Sloan (6). We note in passing that the upper bound linear programming problem may be solved efficiently by applying the active set method to the dual linear programming problem (Sloan, (7)).

5. UNDRAINED LOADING OF A TRAPDOOR

One of the classical problems in soil mechanics, for which the exact collapse load is unknown, is that of a purely cohesive soil layer underlain by a trapdoor. The problem, defined in Figure 6, had been studied by a number of authors including Davis (9) and Gunn (10). The soil above the trapdoor may fail in either an active or a passive mode. In the former case, the pressure supporting the trapdoor resists failure and the overall stability is summarised by the quantity $N = (q + \gamma H - q_f)/c_u$, where c_u is the undrained shear strength and γ is the unit weight. This stability number is a function of H/B. For H/B = 5, the best analytical bounds are due to Gunn (10) who obtained an upper bound of N = 6.53 and a lower bound of N = 4.61. The mesh used for the upper bound calculation is shown in Figure 7. The mesh is arranged such that four triangles are coalesced to form a single quadrilateral with the central node lying at the centroid. This permits the 3-noded triangle to satisfy the incompressibility condition under plane strain loading (Nagtegaal (11), Sloan and Randolph (12)). The meshes employed for the lower bound computation are illustrated in Figure 8. Since we are modelling a semi infinite domain, the extended mesh is used to check that the stress field can be extended to the right without affecting the collapse load.

The results for the upper and lower bound analyses are shown in Table 1. For a 24-sided approximation of the Tresca yield criterion, the numerical upper and lower bounds on the stability number are N = 6.38 and N = 5.71. Thus the exact collapse load is bracketted to within 12 percent, a substantial improvement on the analytical bounds of Gunn (10). The lower bound collapse load is almost identical for the extended mesh, indicating that the lower bound is valid for a semi infinite soil layer. Generally speaking, the lower bound calculation is much more efficient than the upper bound calculation. As a further check on the bounds produced by the linear programming formulations, the limit load for the trapdoor problem has also been estimated using a finite element analysis with 56 cubic strain triangles (see Sloan (7) for details). These elements are known to yield accurate estimates of collapse (Sloan and Randolph (12)), and furnished a stability number of N = 6.05. The CPU time for the finite element computations, which required 100 load increments and involved 493 nodes, was around 1200 seconds.

6. CONCLUSION

Two methods for computing rigorous upper and lower bounds on limit loads have been described. When applied to the undrained loading of a trapdoor, the techniques

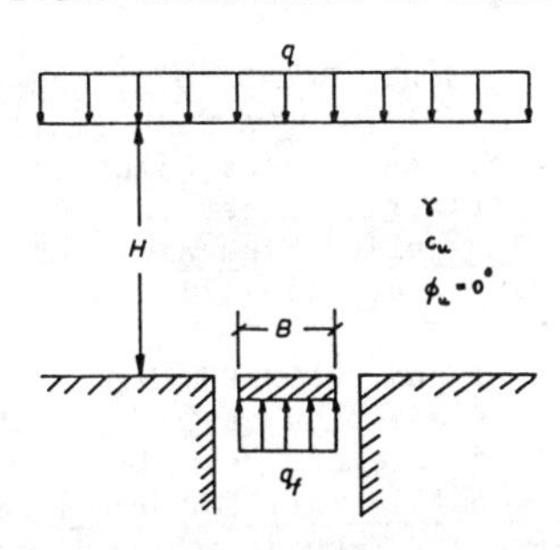

Figure 6: The Trapdoor Problem

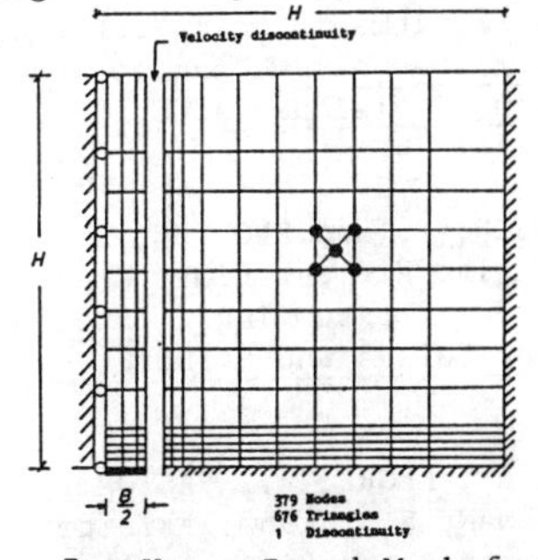

Figure 7: Upper Bound Mesh for Trapdoor Problem

Table 1: Results for Trapdoor Problem

analysis	p	n	m	i	t	N
lower bound	6	1422	4096	550	291	5.24
upper bound	6	2122	4828	2749	975	6.84
lower bound	12	1422	6940	683	366	5.63
upper bound	12	2122	8884	4209	1671	6.47
lower bound	24	1422	12628	765	465	5.71
lower bound*	24	1422	12628	780	480	5.70
upper bound	24	2122	16996	5943	3070	6.38

Notes: p = number of sides in linearised yield polygon
n = number of columns in constraint matrix
m = number of rows in constraint matrix
i = number of iterations for active set algorithm
t = CPU-SEC for VAX 8550 with FORTRAN 77 compiler
* = results for extended mesh

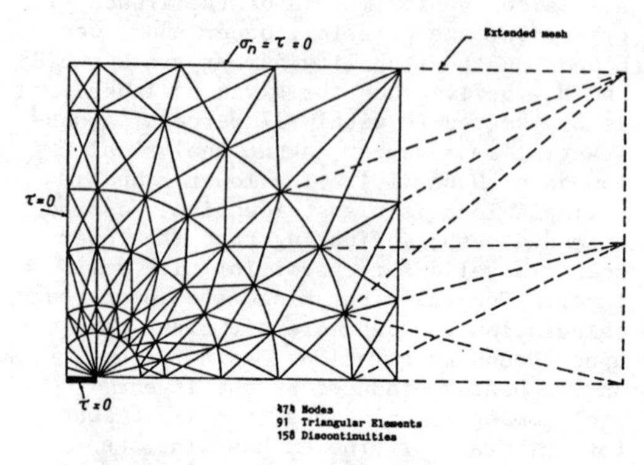

Figure 8: Lower Bound Mesh for Trapdoor Problem

improved the existing bounds on the exact collapse load substantially. These bounds have been verified by finite element analysis. The active set algorithm permits the resulting linear programming problems to be solved efficiently, and both methods are competitive with the finite element methods in terms of computation effort.

REFERENCES

1. LYSMER, J., 1970, Limit Analysis of Plane Problems in Soil Mechanics, Journal of the Soil Mechanics and Foundations Division, ASCE, 96, SM4, 1311-1334.

2. ANDERHEGGEN, E. and KNOPFEL, H., 1972, Finite Element Limit Analysis Using Linear Programming, International Journal of Solids and Structures, 8, 1413-1431.

3. BOTTERO, A., NEGRE, R., PASTOR, J. and TURGEMAN, S., 1980, Finite Element Method and Limit Analysis Theory for Soil Mechanics Problems, Computer Methods in Applied Mechanics and Engineering, 22, 131-149.

4. BEST, M.J. and RITTER, K., (1985), Linear Programming: Active Set Analysis and Computer Programs, Prentice-Hall, New Jersey.

5. SLOAN, S.W., 1987, A Steepest Edge Active Set Algorithm for Solving Sparse Linear Programming Problems, University of Newcastle, Civil Engineering Report No. 022.05.1987.

6. SLOAN, S.W., 1987, Lower Bound Limit Analysis Using Finite Elements and Linear Programming, University of Newcastle, Civil Engineering Report No. 020.021.1987, (to appear in International Journal for Numerical and Analytical Methods in Geomechanics).

7. SLOAN, S.W., 1987, Upper Bound Limit Analysis Using Finite Elements and Linear Programming, University of Newcastle, Civil Engineering Report No. 025.09.1987, (submitted to International Journal for Numerical and Analytical Methods in Geomechanics).

8. REID, J.K., 1976, FORTRAN Subroutines for Handling Sparse Linear Programming Bases, Harwell Report No. R8269, England.

9. DAVIS, E.H., 1968, Theories of Plasticity and the Failure of Soil Masses, Chapter 6 of Soil Mechanics - Selected Topics, Ed. I.K. Lee, Butterworths.

10. GUNN, M.J., 1980, Limit Analysis of Undrained Stability Problems Using a Very Small Computer, Proc. Symp. on Computer Applications to Geotechnical Problems in Highway Engineering, Cambridge University, Engineering Department, 5-30.

11. NAGTEGAAL, J.C., PARKS, D.H. and RICE, J.R., 1974, On Numerically Accurate Finite Element Solutions in the Fully Plastic Range, Computer Methods in Applied Mechanics and Engineering, 4, 153-177.

12. SLOAN, S.W. and RANDOLPH, M.F., 1982, Numerical Prediction of Collapse Loads Using Finite Element Methods, International Journal for Numerical and Analytical Methods in Geomechanics, 6, 47-76.

Numerical Methods in Geomechanics (Innsbruck 1988), Swoboda (ed.)
© 1988 Balkema, Rotterdam. ISBN 90 6191 809 X

Displacement discontinuities and interactive graphics for three-dimensional, hydraulic fracturing simulators

L.Vandamme
Dowell Schlumberger, Tulsa, Okla., USA
P.A.Wawrzynek
Cornell University, Ithaca, N.Y., USA

ABSTRACT: Hydraulic fracturing is a widely used well stimulation method for secondary recovery of oil and gas from underground reservoirs.

The first part of this paper describes a three-dimensional, hydraulic fracturing simulator based on the displacement discontinuity method. The model includes a coupling between the structural analysis of the rock in which the fracture propagates and the fluid-flow analysis inside the fracture. The features of this model are briefly discussed, especially its ability to propagate a fracture out of its plane in multilayered formations.

The second part of the paper describes the interaction capabilities between the model and the analyst. Special emphasis is placed on the use of interactive graphics techniques, and on the specific data structures best suited to accommodate these techniques.

1 INTRODUCTION

The purpose of hydraulic fracturing is to create a highly conductive channel in a reservoir rock to significantly increase the drainage rate toward the well. The design of such treatments is usually based on numerical models simulating hydraulically driven fractures.

The first numerical models for hydraulic fractures were two-dimensional [1,2]. In these models, only the fracture widths and lengths were computed; vertical fracture heights were not taken into account. The first attempts to relax this constraint, and hence be able to investigate containment of the fracture to the pay zone, were provided in so-called pseudo-three-dimensional models [3,4]. Recently, fully three-dimensional models have been designed to allow spatial fracture growth to be completely arbitrary.

2 THREE-DIMENSIONAL SIMULATOR

2.1 Structural analysis

The displacement discontinuity (DD) [5] technique is one of the most attractive methods for modeling fracture problems in geomechanics, because of the following features:

- boundaries at infinity are built in the solution;
- the discretization mesh need not be reorganized when propagating a fracture beyond its original geometry; and
- the basic unknowns of the problem correspond to the physical aperture and ride (normal and tangential relative displacements) of the fracture.

Moreover, using this technique, hydraulic fractures, preexisting fractures and stratigraphic interfaces can be conveniently modeled as single surfaces of discontinuity in an infinite medium.

The DD method requires the knowledge of fundamental solutions, which give the formulation for the state of stress and displacement in an infinite medium resulting from a unit-strength DD at a point. The fundamental solutions for the stresses, σ_{ij}, and displacements, u_i, at point x, due to a concentrated DD at point ξ, are expressed as follows:

$$u_i(x) = G^*_{ikn}(x,\xi)\delta_{kn}(\xi) \tag{1}$$

and

$$\sigma_{ij}(x) = T^*_{ijkn}(x,\xi)\delta_{kn}(\xi) \tag{2}$$

where δ_{kn} is a unit-strength DD defined as the following Dirac delta function:

$$\lim_{y_n \to 0^-} u_k - \lim_{y_n \to 0^+} u_k = \delta(\xi); \tag{3}$$

G^*_{ikn} and T^*_{ijkn} are influence functions, also called Green functions or kernels [6,7].

Thus, a fracture can be considered as a series of displacement discontinuities, having different unknown strengths, $\phi_{kn}(\xi)$, distributed on the surface of the fracture. The principle of superposition, valid within the framework of linear elasticity, allows one to add the influences of all of these individual DD's to obtain the altered stress field at any point of the infinite rock medium. This is done by integrating the fundamental solutions in Equations (1) and (2) along the surface of the fracture, Γ.

Discretizing the fracture surface into elements has the effect of describing the behavior of $\phi_{kn}(\xi)$ inside the elements in terms of nodal values, $\phi_{kn}(\xi_m)$, through shape functions, N_m:

$$\phi_{kn}(\xi) = \sum_{e=1}^{E}\sum_{m=1}^{M_e} \phi_{kn}(\xi_m)N_m(\xi) \tag{4}$$

where the summation signs scan all nodes (from 1 to M_e) of all elements (from 1 to E).

In this model, six-noded triangular elements are implemented, and the shape functions, N_m, follow a quadratic law everywhere (Figure 1), except near the fracture front. In that region, N_m has a square-root behavior to conform to the

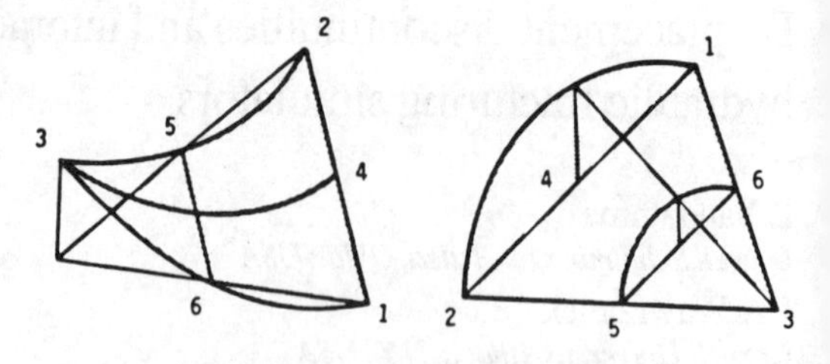

Figure 1: Quadratic shape functions at nodes 3 and 4 (from [7]).

theory of linear elastic fracture mechanics [8] (Figure 2).

The equations to be solved state that the computed stresses should equal the known stresses applied to the sides of the fracture (boundary conditions):

$$\sigma_{ij}(x) = \sum_{e=1}^{E}\sum_{m=1}^{M_e} [\phi_{kn}(\xi_m) \iint_\Gamma T^*_{ijkn}(x,\xi)N_m(\xi)\, d\Gamma]. \tag{5}$$

This system of equations is solved for the discontinuity strengths, ϕ_{kn}, which directly translate into the fracture aperture ($k = n$) and ride ($k \neq n$) distributions.

2.2 Fluid-flow analysis

In hydraulic fracturing applications, the driving force that causes fracture propagation also must be modeled. Hence, the discretization of the solid medium enclosing the fracture must be accompanied by a coupled analysis of the fluid flow inside the fracture.

Assuming that the fracture aperture, ϕ_{kk}, is known, the fluid-flow analysis will provide the pressure distribution, σ_{ii}, inside the fracture to be used as boundary condition for the structural analysis.

The fluid injected into the borehole is assumed to be Newtonian, and the injection rate is small enough that the flow in the fracture is laminar. As a first approximation, it is reasonable to treat flow in hydraulic fractures as steady-state flow between parallel plates:

$$q_i = \frac{h\phi^3_{kk}}{12\mu}\chi_{,i} \tag{6}$$

where q_i is the fluid flow per unit area of fracture cross section; ϕ_{kk} and h are the

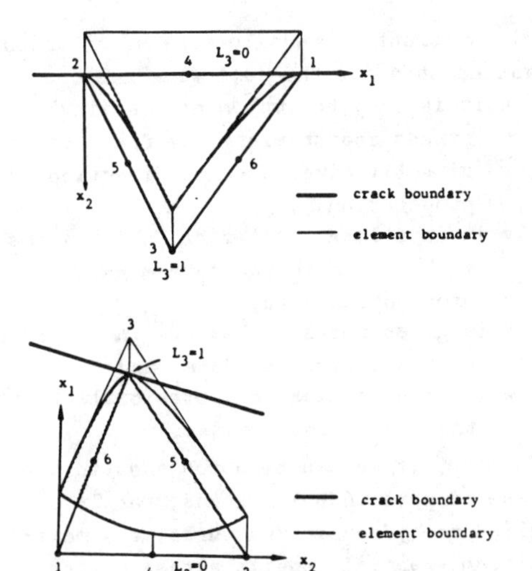

Figure 2: Square-root shape functions at nodes 3 and 4 (from [7]).

aperture and height of the fracture, respectively; μ is the viscosity of the injected fluid; and $\chi_{,i}$ is the pressure gradient in the direction of flow. The values of χ in the fluid-flow analysis must correspond to the values of σ_{ii} in the structural analysis.

The fluid-mass conservation equation states that all the fluid pumped into the borehole is spent either in an increase of the fracture volume or in fluid loss into the formation:

$$q_{i,i} = \frac{\partial V}{\partial t} + Q(\Delta t) \tag{7}$$

where V is the volume of the fracture and Q is a time-dependent, fluid-loss function.

A standard, finite-element scheme is used to solve the fluid flow inside the fracture. The boundary conditions are either the pressure or the flow rate at the borehole, and a no-flow condition at the fracture front. The time required to obtain the given fracture geometry is solved for by stating that the fracture is about to propagate. This condition is enforced by adjusting the time step, Δt, so that the stress intensity factor, a function of V that characterizes the singular stress field near the fracture front, is equal to the fracture toughness of the rock, a material property.

2.3 Coupling between structural and fluid-flow analyses

The cubic relationship between fracture aperture and pressure drop, found in Equation (6), makes the coupled problem highly nonlinear. Two possible methods can be used to solve the system of nonlinear equations formed by Equations (5) and (6).

Interlacing scheme: In this particular implementation, both stress and fluid-flow analyses (taken independently) are linear. Thus, each analysis can be solved alternately, the solution of one determining the boundary conditions of the other.

Quasi-Newton method: With this alternative method, the complete nonlinear system of equations is solved at once. The method is also iterative; it is based on a linearization of the equations, using a Taylor-series expansion. It is found to be somewhat faster and more robust than the interlacing scheme, depending on the original guesses on ϕ and σ.

The rides (relative shear displacement of the sides of the fracture) enter the calculation, resulting in solution matrices that are twice as large as the ones obtained in planar three-dimensional models: to the conventional two unknowns per node (aperture and pressure), one now must add two additional unknowns per node (two perpendicular rides, parallel to the fracture surface).

2.4 Propagation in multilayered formations

The ride DD's have nonzero values when in-situ stresses are not normal to the fracture or when exterior influences, like bimaterial interfaces or existing discontinuities, exist in the rock mass. Multilayered formations can be simulated by explicitly modeling the bimaterial interfaces with DD elements. Instead of imposing given stress distributions, as was done when modeling hydraulic fractures, stress and displacement continuity relationships are imposed across the interfaces [9].

In the case of bimaterial interfaces, the DD values are no longer associated with actual displacements in the solid medium. Hence, a double layer of DD elements is needed on the interfaces because the fictitious displacements that affect one layer are generally not equal to those that affect the neighboring layer (Figure 3). In three dimensions, a pair of associated nodes on the interface yields six unknowns: two normal and four shear DD values. The displacement re-

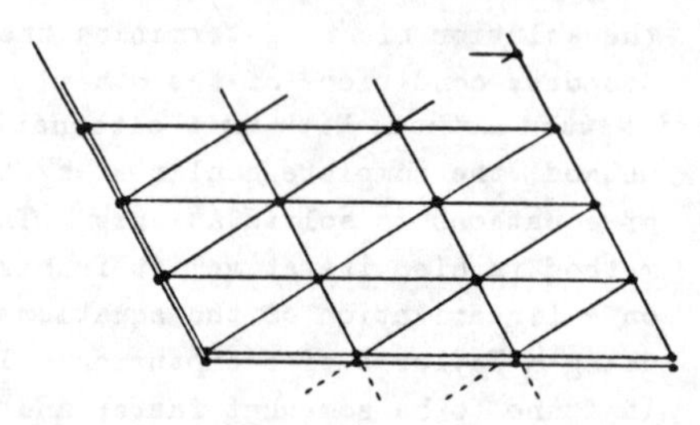

Figure 3: Discretization of bimaterial interfaces.

strictions mentioned above provide three equations, while the normal and shear stress restrictions provide one and two equations, respectively.

In addition to the hydraulic fracture and bimaterial interfaces, the simulator also handles preexisting rock mass discontinuities. The discretization of these discontinuities is similar to that of hydraulic fractures. However, neither fluid-flow nor propagation analysis is provided. The stress distribution on the walls must either be given or follow a frictional Mohr-Coulomb behavior.

Because of the assumptions taken into the fluid-flow analysis part of the model, the hydraulic fracture is always growing. The propagation path is the only unknown that remains to be solved at the end of each time step. Among the several propagation criteria that have been proposed in fracture mechanics [10-12], the maximum circumferential tensile stress criterion was adopted for the following reasons:

- it is easy to implement: a single stress search along the fracture front directly gives the new direction of propagation;
- it is a local criterion, which seems well-suited to the determination of a propagation path;
- it gives results that are in accordance with experimental data; and
- it requires less computational effort than most other criteria.

However, it should be noted that in the case of pure mode III loads, rarely applied to hydraulic fractures, the maximum circumferential tensile stress criterion gives inconsistent results.

3 INTERACTIVE GRAPHICS PROCESSING

3.1 Problem description

Current research focuses on interactive graphics processing: the traditional three-phase program (preprocessor, analysis and postprocessor) is replaced by a single interactive program in which the object being modeled is continuously displayed, as the analysis proceeds. The integration of the graphics phases into the analysis provides a smooth transition from preprocessing, to analysis, to postprocessing functions. It also allows for the display of intermediate results, which give the analyst the option to interactively alter the course of the simulation [13,14].

Since the user is able to view the three-dimensional fracture geometry at each time step, he/she will be able to decide at what point a remeshing becomes necessary, whether the newly generated mesh is acceptable, whether the proposed fracture propagation path is acceptable, and whether some alterations should be considered.

In addition, the analyst can alter such parameters as pumping rates and fluid properties during the simulation in response to the observed fracture behavior, much in the same way that these parameters are actually varied during the treatment. Although such changes are beyond the scope of this paper, it can be said that interactive graphics processing is the first step toward real-time alterations of hydraulic fracturing designs.

In this paper, the following aspects of interactive graphics programming are briefly discussed:

- creation of a boundary element mesh (preprocessor);
- discussion of the data structure and programming patterns used in the analysis part of the model to accommodate the concepts outlined above; and
- interpretation of the results (postprocessor).

3.2 Mesh creation

A hydraulic fracture and any other type of discontinuity in the rock mass can be input as initial geometry. For the sake of simplicity, these initial features are piecewise planar and each planar segment is discretized using a separate mesh. The program can currently generate two types of meshes: elliptical and quadrilateral. Although they are initially generated as planar entities, these meshes are stored as full three-dimensional objects and the program places no planar constraint on the evolution of these objects; e.g., a fracture is not constrained from growing out of its original plane.

A typical mesh-generation session would begin with the selection of a mesh type and a two-dimensional mesh generation. The mapping into the three-dimensional space is done subsequently, through appropriate scaling, rotations and translations.

Elliptical mesh generation is characterized by six parameters supplied by the user. The user must specify the major and minor axes lengths, the number of nodes per quarter of the elliptical arc, two element-size weights, and the portion of the mesh to generate (full mesh, half-mesh or quarter-mesh). The element-size weights are two integer numbers that specify the relative size of the elements at the center and around the edge of the ellipse.

In the case of quadrilateral meshes, the user must specify the coordinates of each of the four corners of the quadrilateral, the number of horizontal and vertical element layers, and the element-size weights along each of the four boundaries.

Figure 4 shows the initial geometry of an elliptical hydraulic fracture contained in a zone limited by two bimaterial interfaces, modeled using quadrilateral meshes.

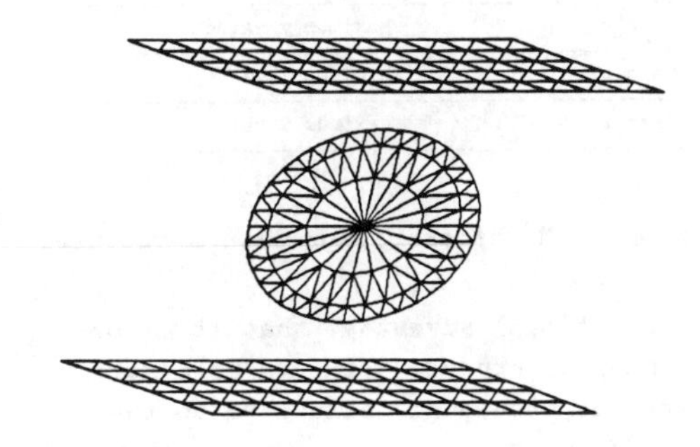

Figure 4: Oblique view of a hydraulic fracture between bimaterial interfaces.

3.3 Data structure

Each discontinuity segment is represented by a separate mesh, and each mesh is defined by a descriptor which contains the following items (Figure 5):

- the size of the mesh segment (number of nodes, elements and edges--an edge is a line segment common to two elements);
- the maximum and minimum nodal coordinates;
- pointers to the data lists for the mesh segment; and
- pointers to other mesh descriptors (doubly linked list; Figure 6).

The fast access to all mesh descriptors, through doubly linked lists, and to the separate lists of all the edges to be drawn is necessary if the meshes are to be continuously displayed and updated during the analysis. The doubly linked list has

desc_type	Descriptor type
fwd_ptr	Forward mesh pointer
back_ptr	Backward mesh pointer
num_nodes	# of nodes
num_elem	# of elements
num_edges	# of edges
coord_ptr	Coordinate list pointer
conn_ptr	Connectivity list ptr
edge_ptr	Edge list pointer
min_x	Minimum x value
max_x	Maximum x value
min_y	Minimum y value
max_y	Maximum y value
min_z	Minimum z value
max_z	Maximum z value

Figure 5: Mesh descriptor data structure.

the additional advantage that it allows any mesh to expand, although no memory space was previously allocated to that specific mesh.

3.4 Programming pattern

The ability to perform smooth display transformations makes three-dimensional geometries like that of Figure 4 easily understood. In this implementation, the display is controlled by interactive functions that allow the user to rotate the picture about its x-, y- or z-axis. Zooming (scaling) and panning (translation) functions also are provided.

It is desirable that these display functions be available at any time during the analysis. This feature is best implemented with an event-driven programming pattern, rather than a control-driven programming pattern; the main difference to the programmer is in the handling of input commands.

With event-driven programs, subroutines are directly invoked by actions of the user, such as a mouse-pick or a keystroke. In a control-driven program, a subroutine is invoked only after having passed through a series of IF-THEN-ELSE struc-

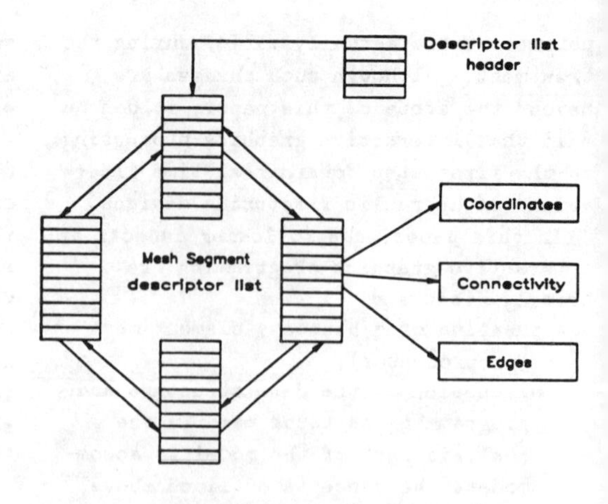

Figure 6: Links between mesh descriptors.

tures. Event-driven programming is particularly important with graphics workstations, where input handling is managed either by a specialized piece of software, or by a portion of the operating system.

The main advantages of the event-driven programming pattern are the following.

- The use of direct CPU-interrupts eliminates the need for a "busy-wait" state, in which computer resources are spent processing the loop which polls the system for various input events.
- No complex control structure is needed, in which input events must be hierarchized to fit in the IF-THEN-ELSE structure. Since each input event is on the same level, it becomes easier to modify or add events in the input routines.

3.5 Result interpretation

In addition to the updated fracture geometry after each time step, quantities like stresses and displacements inside the rock mass, or apertures and pressures inside the fracture, can be displayed either as a two-dimensional graph or as a contour plot. The postprocessor includes the following components:

- line and mesh generation routines (the quadrilateral mesh generation routine described above can be used for this purpose);

- analysis routines that compute results at desired points of the generated line or mesh; and
- display routines.

Unlike the more conventional postprocessors, results can be output at any time in the analysis. The event-driven programming pattern allows the user to call the postprocessing routines with a single mouse-pick. When the main program acknowledges the postprocessing interrupt, a separate window and a child process are immediately created for each postprocessing plot. The spawning of new processes makes possible the immediate update and simultaneous viewing of different plots.

4 CONCLUDING REMARKS AND ACKNOWLEDGMENTS

In this paper, emphasis was brought on the DD method and on interactive graphics techniques, which have made three-dimensional modeling of hydraulic fractures a reality. Selected results from the simulator can be obtained in references [7] and [15].

The authors wish to thank Dowell Schlumberger for its support and the permission to publish this paper.

5 BIBLIOGRAPHY

[1] T.K.Perkins and L.R.Kern, Widths of hydraulic fractures, J.Pet.Tech. 937-949 (1961)

[2] J.Geerstma and F.de Klerk, A rapid method of predicting width and extent of hydraulically induced fractures, J.Pet.Tech. 1571-1581 (1969)

[3] A.Settari, Simulation of hydraulic fracturing processes, Soc.Pet.Eng.J. 219-232 (1980)

[4] M.P.Cleary, R.G.Keck and M.E.Mear, Microcomputer models for the design of hydraulic fractures, Proc. SPE/DOE Symp. on Low-Permeability Gas Reservoirs, Denver, Colorado, 257-279 (1983)

[5] S.L.Crouch and A.M.Starfield, Boundary element methods in solid mechanics, George Allen and Unwin, London, 1983

[6] T.D.Wiles and J.H.Curran, A general 3-D displacement discontinuity method, Proc. 4th Int.Conf.Num.Meth.Geomech., Edmonton, Alberta, 103-111 (1982)

[7] L.Vandamme, A three-dimensional displacement discontinuity model for the analysis of hydraulically propagated fractures, PhD thesis, U.Toronto, Ontario, 1986

[8] J.F.Knott, Fundamentals of fracture mechanics, Halsted Press, New York, 1973

[9] P.K.Banerjee and R.Butterfield, Boundary element methods in engineering science, McGraw-Hill, London, 1981

[10] F.Erdogan and G.C.Sih, On the crack extension in plates under plane loading and transverse shear, ASME J.Bas.Eng. 85:519-527 (1963)

[11] M.A.Hussain, S.L.Pu and J.Underwood, Strain energy release rate for a crack under combined mode I and mode II, Frac.Anal., ASTM STP 560, 2-28 (1974)

[12] G.C.Sih, Strain energy density factor applied to mixed-mode crack problems, Int.J.Frac. 10:305-321 (1974)

[13] A.R.Ingraffea, F.H.Kulhawy and J.F. Abel, Interactive computer graphics for analysis of geotechnical structures, Proc. 1st Int.Conf.Comp.Civ. Eng., 864-875 (1981)

[14] P.A.Wawrzynek and A.R.Ingraffea, Interactive finite element analysis of fracture processes: an integrated approach, Theor.Appl.Frac.Mech, submit. 1986

[15] L.Vandamme, R.G.Jeffrey and J.H.Curran, Effects of three-dimensionalization on a hydraulic fracture pressure profile, 27th US Symp.Rock Mech., Tuscaloosa, Alabama, 580-590 (1986)

Numerical Methods in Geomechanics (Innsbruck 1988), Swoboda (ed.)
© 1988 Balkema, Rotterdam. ISBN 90 6191 809 X

Stress analysis of large structures in hydraulic engineering by weighted residuals method

Chang Yang & Dai Miaoling
Hohai University, Nanjing, People's Republic of China

ABSTRACT: Weighted residuals method (WRM) is well known by its simplicity and generality, especially, collocation method, one of WRM, which is easy to be made into computer programs, and has the advantage of computer time-saving. Despit all this, it is low accuracy and unsteady (the solution is sensitive to the collocation points) for large complex structures. The reason is found and effective treatments put forward in this paper.

1 COLLOCATION METHOD

Collocation method (CM) is one of weighted residuals method (WRM), the principle of which can be written as below:

Take the try function to be

$$\tilde{U} = \sum_{i=1}^{N} C_i U_i \tag{1}$$

for the homogenous problem

$$\left.\begin{aligned} L(U(x)) &= 0 && x \in \Omega \\ G(U) &= g(x) && x \in \partial\Omega_1 \\ H(U) &= h(x) && x \in \partial\Omega_2 \end{aligned}\right\} \tag{2}$$

where the try function terms U_i are predetermined, which satisfy the governing Eq.

Substituting formula (1) into Eqs. (2), we have the inner residual and boundary residuals

$$\left.\begin{aligned} R_i(x) &= L(\tilde{U}(x)) = 0 && x \in \Omega \\ R_g(x) &= G(\tilde{U})-g(x) && x \in \partial\Omega_1 \\ R_h(x) &= H(\tilde{U})-h(x) && x \in \partial\Omega_2 \end{aligned}\right\} \tag{3}$$

respectively.

Let

$$\left.\begin{aligned} &\int_S R_g(X)\,\delta(X-X_i)\,ds = R_g(X_i)=0 \quad X \in \partial\Omega_1 \\ &\qquad (i=1,2,\cdots,m) \\ &\int_S R_h(X)\,\delta(X-X_j)\,ds = R_h(X_j)=0 \quad X \in \partial\Omega_2 \\ &\qquad (i=1,2,\cdots,n) \end{aligned}\right\} \tag{4}$$

where $\delta(X)$ means Dirac-δ function, m+n=N.

Combining with formulas (3) and (1), Eqs. (4) become

$$\left.\begin{aligned} \sum_{i=1}^{m} C_i\, G(u_i) &= g(X_i) \\ \sum_{i=1}^{n} C_i\, H(u_i) &= h(X_j) \end{aligned}\right\} \tag{5}$$

with the hypothesis of G, H being linear operator.

Eqs.(5) is an algebra Eq. system. The coefficients $\{C_i\}$ can be determined by solving it. The approximate solution $\tilde{u}$ is then obtained by formula (1).

In this paper the fundamental solutions will be taken as the try function terms.

2 LINEAR ELASTICITY PROBLEM

The governing Eqs. for linear elasticity problem with the boundary conditions can be written as

$$\left.\begin{aligned} N(U_i(X)) &= 0 && X \in \Omega \\ U_i &= \overline{U}_i(X) && X \in \partial\Omega_1 \\ t_i &= \overline{t}_i(X) && X \in \partial\Omega_2 \end{aligned}\right\} \tag{6}$$

where N is Navier operator, U_i and t_i are displacements and tractions respectively.

The fundamental solutions of displacement and traction for two dimensional problems are:

$$U_{Ki}(X,Y) = \frac{-1}{8\pi\mu(1-\nu)}\left\{(3-4\nu)\,\delta_{Ki}\ln r - r_{,K}r_{,i}\right\} \tag{7}$$

$$T_{ki}(X,Y)=\frac{-1}{4\pi(1-\nu)r}\left\{\left[(1-2\nu)\delta_{ki}+\beta r_{,k}r_{,i}\right]r_{,j}n_j - (1-2\nu)(r_{,k}n_i-r_{,i}n_k)\right\} \quad (8)$$

respectively, where ν, μ are Lame's constants, r and $r_{,i}$ are the distance between the two points X and Y and its derivatives with respect to the coordinate x_i respectively.

We take the try function as

$$\tilde{U}_i = \sum_{j=1}^{N} C_{kj}\, U_{kj}(Y_j, X) \quad (9)$$

or

$$\{\tilde{U}_i\} = \{C\}^T\{U_i\} \quad (10)$$

By CM, the following Eqs.

$$\sum_{j=1}^{N} C_{kj}\, T_{ki}(y_j, x_l) = \bar{t}_i \quad x\in\partial\Omega \quad (11)$$
$$(l=1,2,\cdots,N)$$

hold for the second boundary condition.

Eqs. (11) can be rewritten in matrix form:

$$[T]\{C\} = \{\bar{t}\} \quad (12)$$

from which the coefficients $\{C\}$ are achieved, that is

$$\{C\} = [T]^{-1}\{\bar{t}\} \quad (13)$$

Introducing Eq. (13) into Eq.(10), we obtain the solution

$$\{\tilde{U}_i\} = ([T]^{-1}\{\bar{t}\})^T\{U_i\} = (\{U_i\}^T[T]^{-1}\{\bar{t}\})^T$$

or

$$\{\tilde{U}_i\} = \{U_i\}^T[T]^{-1}\{\bar{t}\} \quad (14)$$

Physically, CM is one of effective-function method. That is:

When the force C_{kj} applied at the point $Y_j \bar{\in} \Omega$ in k direction (j=1, 2,...,N), the displacement at the point $X\in\Omega$ developed by them is $\tilde{U}_i(X)$ in the form of relation (9). If this try function satisfies the boundary conditions at N points, then the try function is the solution of the Eqs. (6).

But the problem is that the method is low accuracy and unsteady. We will discuss the cause so as to find the proper position of the collocation points and the result with high accuracy.

By the collocation method, the boundary conditions are actually unsatisfied in the whole boundary except in the boundary collocation points (BCP). Take the Neumann's problem as an example. $t_i(X)$ in

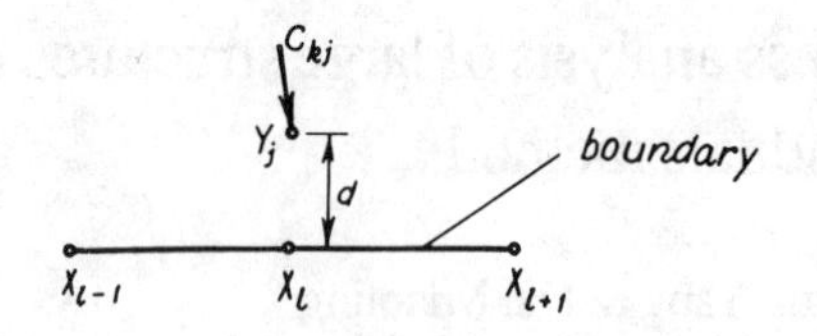

Fig. 1

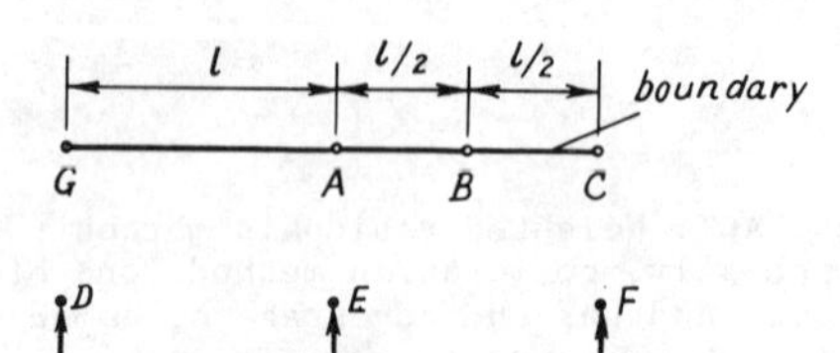

(a)

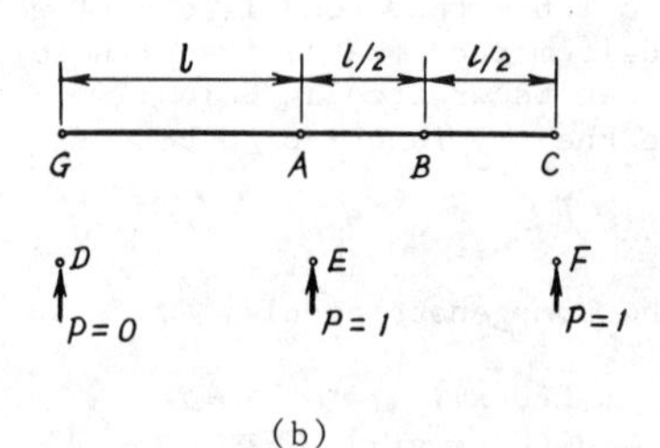

(b)

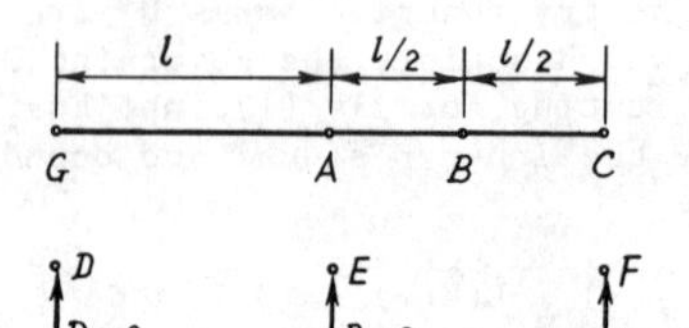

(c)

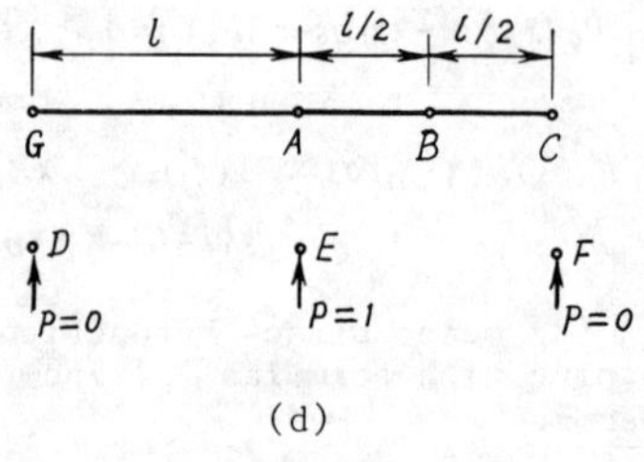

(d)

Fig. 2

formula (11) has the singularity of $1/\gamma$, because T_{ki} has the same singularity. So when the point Y_j (see Fig.1) is very near to the point X_l on the boundary (i.e. d is very small), the traction distribution nearby the point X_l caused by the force C_{Kj} is far away from the linear one. This causes the boundary conditions (11) are not satisfied in the open regions (X_{l-1}, X_l) and (X_l, X_{l+1}), so the results are low accuracy.

On the other hand, if the collocation point (CP) Y_j is taken to be far away from the boundary (i.e. d is very large), then the traction distribution is almost linear and the satisfaction of the boundary condition on the BCP implies the satisfaction in the intervals (X_{l-1},X_l) and (X_l, X_{l+1}). But there is a problem.

Because d is very large, $1/r$ is very small and T_{ki} small either. The variation of the tractions caused by C_{kj} and C_{kj+1} are almost the same, which may cause the matrix [T] in the formula (12) ill-conditioned mathematically. Hence the solution $\{C\}$ is low accuracy and unstable.

In brief, it is most important to take d to be a proper value. Of cause the method solving the Eq. system (12) with high accuracy is necessary. Several typical cases of collocation with the combination of the loads are shown in Fig.2 and the corresponding values of $\Sigma\, 1/r$ are given in the table.

case	d/l	$\Sigma \frac{1}{r_A}$	$\Sigma \frac{1}{r_B}$	$\Sigma \frac{1}{r_C}$	$\varepsilon_B\%$	Suggesting d/l
(a)	0.5	1.8944	1.7304	1.6897	+3.44	0.7
	1.0	2.4142	2.3436	2.1543	-5.69	
	2.0	2.7889	2.7403	2.695	-1.67	
(b)	0.5	1.4472	1.4142	1.4472	+2.28	0.7
	1.0	1.7071	1.7889	1.7071	-4.79	
	2.0	1.8944	1.9403	1.8944	-2.42	
(c)	0.5	0.4472	0.7071	1.0000	+2.28	0.7
	1.0	0.7071	0.8944	1.0000	-4.79	
	2.0	0.8944	0.9701	1.0000	-2.42	
(d)	0.5	1.0000	0.7071	0.4472	+2.28	0.7
	1.0	1.0000	0.8944	0.7071	-4.79	
	2.0	1.0000	0.9701	0.8944	-2.42	

Notice: l is the distance between two BCP nearby, d is the distance between the CP E and BCP A, which has been taken to be unit without losing generality, r_A means the distance between A and the collocation points, P means force and the exact meaning of the error ε_B is:

$$\varepsilon_B = \frac{\Sigma \frac{1}{r_A} + \Sigma \frac{1}{r_C} - 2\Sigma \frac{1}{r_B}}{\Sigma \frac{1}{r_A} + \Sigma \frac{1}{r_C}}$$

From the table we can find that the suitable value of d/l is about 0.7 in all cases. However l_{GA} is not always equal to l_{AC}. When this happens, the l should be taken to be the mean value ($l_{GA} + l_{AC}$)/2. Besides, the limitation $0.6 \leqslant l_{GA}/l_{AC} \leqslant 1.8$ is to be considered.

3 NONHOMOGENOUS EQUALITION

So far, the discussions are in the limitation of homogenous Eqs. For nonhomogenous Eq., CM can not be used directly, because the try function in formula (11) does not satisfies the governing Eq. and the inner residuals are not equal to zero. In order to use CM in this case, we can introduce the special solution of the governing Eq.:

Considered the following problem

$$L(U) + b(X) = 0 \qquad X \in \Omega \tag{15}$$

$$\left.\begin{aligned} G(U) &= g(X) \qquad X \in \partial\Omega_1 \\ H(U) &= h(X) \qquad X \in \partial\Omega_2 \end{aligned}\right\} \tag{16}$$

Let $\quad U = V + U^o \qquad\qquad (17)$

Where V means the special solution of the governing Eq. (15), which satisfies the Eq.

$$L(V) + b(X) = 0 \qquad X \in \Omega \tag{18}$$

If Eqs. (15) and (16) are introduced by Eq. (17), the following Eqs. are derived:

$$\begin{aligned} L(U^o) &= 0 && X \in \Omega \\ G(U^o) &= g(X) - G(V) && X \in \partial\Omega_1 \\ H(U^o) &= h(X) - H(V) && X \in \partial\Omega_2 \end{aligned} \tag{19}$$

This is a homongenous problem, so CM can be used to find the U^o and then the solution of the original problem U can be obtained by the relation (15) immediately.

4 CALCULATION EXAMPLE

The stress analysis for the underground openings of Ertan engineering in China is carried out with 9 seconds in computer M-360 by the collocation method. The geometric shases of the openings are shown in Fig.3. The two openings are exactly the same. The rock foundation is infinite and its elastic modula and Poisson's constant are 2.5×10^5 kg/cm² and 0.2 respectively. The initial stress pattern is:

MAIN STRESS

STRESS--KG/CM**2
COMPRESION IS +
DOUBLE EXCAVATION H
FIST GROUP

10.94 m
17.5 m
13.5 m
5.0 m
22.5 m

Fig. 3

σ_x = 234.50 kg/cm^2 , σ_y = 160.31 kg/cm^2, τ_{xy} = 45.76 kg/cm^2 . The stress results in Fig. 3 show that the large tensile stresses appear in the boundary zone due to the openings.

5 CONCLUSION

From above, the following conclusion can be obtained:

1. Collocation method is one of method with high calculation effects, which can be use for the stress analysis of large complex structures.
2. In order to increase the accuracy of the result, the collocation points should take to be on the normal to the boundary 0.7 l away from the corresponding collocation point (l means the distance between two points nearby). In addition, it will be better to make the l's equal approximately.
3. For two-dimensional linear elasticity problems or Laplace problems with the frist boundary condition, the effectiveness of the collocation method will decrease, because of $ln\ r$ in fundamental solution, which does not decrease with the increase of r. But for three-dimensional problems, the drawback does not exist seemingly.

REFERENCES

P. M, Prenter, SPLINES AND VARIATIONAL METHOD, John & Sons, Inc. 1975.

Shen Jiayin & Chang Yang, Study on Some Problem in Boundary Element Method, Journal of East China Technical University of Water Resources, No.2, 1985.

Numerical Methods in Geomechanics (Innsbruck 1988), Swoboda (ed.)
© 1988 Balkema, Rotterdam. ISBN 90 6191 809 X

An automatic soil parameters identification software

D.Aubry & E.Piccuezzu
Ecole Centrale de Paris, France

ABSTRACT : The need to develop constitutive equations able to follow complex stress-strain path has often led to many parameters to identify a given soil material with respect to such a constitutive model. In an industrial context the time lost by the geotechnical engineer is unacceptable and this gave us the idea to implement an automatic software which would be able to deal with this topic. A discussion of the required steps is presented in this paper.

1. INTRODUCTION

During the last two decades the continous increase in computer possibilities has allowed to use more reliable models in structural engeneering analyses of dams, nuclear power stations or tunnel. This state of affairs entails more and more complex constitutive equations in order to get a good approximation of the real soil behaviour.

But the number of parameters increases as the complexity of constitutive law for soils does. Moreover new kinds of parameters have appeared which are more or less of curve-fitting type. Of course it becomes difficult to determine all those parameters because very often only some of them have a clear mechanical meaning. The latter are very often the easiest to bracket. Obviously the determination of most of those parameters needs an iterative process.

The purpose of this paper is to present an automatic software for the identification of soils parameters and more particularly those of a viscoplastic law which has been developped at Ecole Centrale de Paris during the last years.

First of all the general optimization problem will be presented. Then briefly a classification of the relevant parameters is presented and the tools used in this software namely an optimisation scheme connected to a relationnal database. Finally some results will be discussed.

2. THE OPTIMIZATION PROBLEM

2.1. Experimental path and simulation

The goal of the optimization problem is to minimize a function which represents a measure of the distance between experiments and simulations. First a few notations and definitions will be given.

Let σ' be the effective stresses, ε the strains, u the pore pressure, α the hardening variables and $\mathbf{A}$ the model parameters. Then the constitutive equation in the language of optimal control will be considered as a state equation whereas the parameters represent the control . Broadly speaking this state equation can be formulated by the differential system below :

$$\begin{aligned} d\sigma'/dt &= C^{ep}.(\sigma',\alpha,\mathbf{A}).d\varepsilon/dt \\ d\alpha/dt &= L(\sigma',\alpha,\mathbf{A}) \\ du/dt &= Tr(d\sigma/dt - d\sigma'/dt) \end{aligned} \quad (1)$$

Where C^{ep} stands for a viscoplastic matrix and L for the flow rule of the hardening variables. The following initial conditions :

$$\sigma'(0) = \sigma'_0 \qquad \alpha(0) = \alpha_0 \qquad p(0) = p_0 \quad (2)$$

must be supplied with the state equation.

Along the experimental paths some components of the strain and the stress tensors are imposed by the operator while the others are measured. e.g. with stress controlled or strains controlled tests or more complex path as needed. So a test indexed k can be represented by an equation taking into account the constraints imposed to the components :

$$(\sigma',\varepsilon,p)_k \in C_k \quad (3)$$

In this framework simulating an experimental path k consists in the integration of the state equation (1) satisfying to the constraints (2) and (3).

2.2. The optimization problem's constraint

Another major issue must be considered dealing with the constraints to be included in the formulation of the optimization problem. Firstly the concept of the parameter admissible domain noted Pad is introduced.

Actually the real unknowns of the optimization problem are the set of parameters **A** which controls the state equation (1) and it seems logical to limit the possible parameters values during the optimization process either for obvious physical reasons e.g. positive friction angle or to increase the efficiency of the optimisation process. These remarks lead us to take into account some constraints imposed on the parameters.

These constraints may be expressed usually as linear or non linear inequality which will be summarized by :

$$P_{ad} = \{ \mathbf{A} / f_j(\mathbf{A}) \leq 0 \} \qquad (4)$$

2.3. The Cost Function

Now let J(**A**) be the cost function to minimize, E the set of available tests, T_k the duration of experimental test k and σ'_{dk}, ε_{dk} and p_{dk} the experimental paths.

We define also $w_{\sigma k}$, $w_{\varepsilon k}$ and w_{pk} : The weights affected to a test k. The role of which will be explained shortly. Then the cost function is given by :

$$J(\mathbf{A}) = \frac{1}{2} \cdot \Sigma_{(k \in E)} \{ (w_{\sigma k} \int_0^{T_k} ||\sigma'(t) - \sigma'_{dk}(t)||^2 \, dt + w_{\varepsilon k} \int_0^{T_k} ||\varepsilon(t) - \varepsilon_{dk}(t)||^2 \, dt + w_{pk} \int_0^{T_k} ||u(t) - u_{dk}(t)||^2 \, dt \} \qquad (5)$$

The cost function J is only dependant on the model parameter set A through the state equation (1)

It is believed that this type of norm is general enough to include the basic aspects of most experimental testing conditions either drained or undrained.[3]

Now the optimization problem may be formulated which is to find the optimal set of parameters $A_{opt} \in P_{ad}$ such that :

$$J(A_{opt}) = \mathrm{Inf}_{A \in Pad} \; J(A) \qquad (6)$$

2.4 Control of the optimization process

2.4.1. The role of the weights

The weights used in equation (5) allow to control the optimization process in order to satisfy a pre-defined strategy. Of course the more the constitutive equations are sophisticated the more this strategy is complex and a hierarchy on those weights was developped to combine the possible cases. Practically a weight may be affected to a whole test, to one of its components or to a fraction of the test's duration T_k and the terms w_{ck} are a product of three weights :

$$w_{ck} = w_{test} \cdot w_{compo} \cdot w_{chrono} \qquad (7)$$

w_{test} : weight affected to a whole test.
w_{compo} : weight affected to one component of the stress or the strain tensor.
w_{chrono} : Chronology weight which permit to choose only one part of the test.

This structure allow to eliminate one part of a test which seems doubtful or not interesting for the forecasted application of the simulations. e.g. if we want to apply the finite element method to a static calculation we don't need the cyclic part of a test so we'll affect w_{chrono}=0 after the monotonic loading.

2.4.2. Preliminary elaboration of a strategy

As a preliminary step a sensitivity analysis has been conducted to study which parameter has more influence and on which aspect of the tests. that study allow to build a hierarchy beetween parameters which have been separated into specific classes. Secondly some optimizations have been conducted in order to evaluate the interaction beetween parameters and to find some rules which will be applied to the optimization process.

3. THE VISCOPLASTIC CONSTITUTIVE EQUATION

3.1. Assumptions of the model

A cyclic viscoplastic constitutive equation for soils with kinematical hardening has been used throughout the study but any other constitutive equation would be easily incorporated. That constitutive equation has been developped at Ecole Centrale de Paris in order to describe both cyclic and time effects on clayed soils.

Only a summary of the main features are briefly presented none. The model is built by extending Perzyna's theory to an initial elastoplastic constitutive equation developped by Hujeux which is based on the concept of critical state.[1] [2] [5]

The strain rates are classically divided into an elastic part and a viscoplastic one. The viscoplastic strains results from three elementary plane strains mechanism relating respectively to three ortogonal fixed planes in the physical space and one isotropic mechanism.

3.2. Parameters of the model

The parameters are listed in a type-classification file in order to characterize their nature and consequently to facilitate their identification procedure.

3.2.1. Elastic parameters

Three elastic parameters are thus introduced. K_a : the bulk modulus at atmospheric pressure, G_a : the shear modulus at atmospheric pressure and n_e : the exponent of the non linear elasticity. The identification of the true elastic parameters is rather delicate because the recoverable strains are confined within a limit of 10^{-4}. So the identification of the elastic parameters depends upon the nature of the simulation and the available tests. e.g. if we want to go to geophysical computations the actual values of K_a, K_b and n_e obtained e.g. on a resonant column tests are needed. On the opposite if we simulate a foundation quasi-static problem the average values of these parameters issued from the initial stress-strain tangents of triaxial tests are needed.

These remarks are intended to show that at least in the case of soils the choice of the convenient parameters depends on the applications that the geotchnical enginner has in mind.

3.2.2. Plastic parameters

Eight plastic parameters have to be determined in order to define the yield surfaces, the flow rules and the hardening variable's evolutions :

- ϕ : The internal friction angle.
- β : The compressibility index.
- p_{co} : The initial critical pressure.
- a : The modulus of the monotonous hardening rule for deviatoric mechanism.
- b : Shape parameter of the yield fonction.
- c : The modulus of the monotonous hardening rule for isotropic mechanism.
- d : The distance between isotropic consolidation line and perfect plasticity line.

The plastic parameters can be nicely divided into two major classes :

"State parameters" class : (ϕ, β, p_{co}, d)

The parameters found in this classe are directly identified from classical laboratory or in-sitù tests. More precisely at least two drained or undrained triaxial tests and one isotropic or oedometric test must be available to identify all those parameters.

The determination of β, p_{co} and d is given by the figure below :

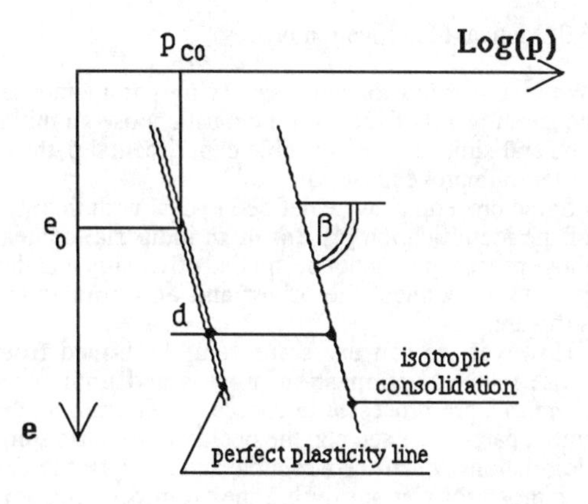

FIGURE 1 . State parameters identification

On the case of clays the positions of the critical state and isotropic lines are explicitely determined by experiments. On the contrary it is very difficult to conduct tests on sands at large strains without any kinematical discontinuities. Consequently difficulties may be expected in the the identification of d and Pco for sands.

"Numerical parameters" class : (a, b, c)

These parameters have to be fitted to experimental curves and their identification requires an iterative process which will be conducted by the optimization software. But there exists some rules which can help us to determine the admissible domain or the initial value for these parameters.

e.g. it has been very found that the parameter b is usually in the range (0.1-0.3) for sands and (0.7-0.9) for clays. This derives from the fact that with b=0 the yield surface of each mechanism is a Mohr-Coulomb criterion which is rather a good criterion for sands. In the opposite way with b=1 the Cam-Clay criterion if obtained which is a better one for clays.

There are a large number of this type of rules and they are included in a pre-processor in charge of defining the initial values and P_{ad}.

3.2.3. Viscous parameters

We need three new parameters to define the viscous part of the constitutive equation : μ^I, the isotropic viscosity, μ^D, the deviatoric viscosity and n_v, the viscous exponent of the Norton-type flow rule. One undrained creep test is required to adjust the value of the viscosity, and the Norton exponent n_v depends on the slope of the experimental line (q_{max},Log(ε)). It was found that n_v has approximately the same value for several clays, about 5.0. This value allow to simulate a sufficiently large interval of strain rates.

3.3. Manual Identification process

When a user has to find a satisfying parameters set concerning a particular soil he must choose an initial one and simulate the available experimental paths in order to improve those values.

Some procedures were defined to deal with this step of the identification.[4] But in an industrial context those procedures cannot be applied efficiently and this lead us to connect the driver and an optimization software.

However we can use some remarks issued from those manual optimization process and implement them in a pre-processor in charge of determining the initial parameters set. e.g. the operator often use some correlations which seem reliable and the software can use those correlations during the pre-processing. e.g. the position of the perfect plasticity line concerning clays is well known. Numerous correlations are also available about the identification of the true elastic parameters.

4. THE TOOLS

This software includes three modules which are the following :

- A relationnal database in which the results of the determination process will be stocked.
- The optimization software in charge of minimizing the cost function inside P_{ad}.
- The driver in charge of the integration of the constitutive equation along the required experimental path.

The essential features of each module is now briefly discussed.

4.1. The relationnal database

The database is an essential module for the whole process. First of all the software is in charge of choosing the initial parameters set. Consequently it must use the database because it is impossible to identify all the parameters from the available tests right at the beginning and also because the engineering judgment and past experience would be very important here. So the software is in charge of looking in the database a previously identified material which has many qualitative features very close to the material to be identified.

Other useful data are recorded into the database for each material. For instance the final value of the cost function at the end of the optimisation and the available testing equipment thus far.

The database is also endowed with a single query language which makes possible to answer to e.g. the question : is there any material with a friction angle larger than 40°.

4.2. The optimization software

The method adopted in this sofware is an augmented lagrangian method which is very robust and efficient for general non-linear constrained optimization problems.[6] The main idea of the penalization concept is to include the constraints (4) into the original cost function (5) in order to transform the constrained optimization problem into an unconstrained one.[3]

Let $\Lambda(\mathbf{A},\lambda,\rho)$ denote the augmented lagrangian function, λ_k the lagrange multiplicators set, ρ the penalty factor and $C_a=\{ j \,/\, f_j(A) = 0\}$ the active constraints set at the current point A. The augmented lagrangian function is given by :

$$\Lambda(\mathbf{A},\lambda_k,\rho) = J(A) - \sum_{j=1}^{m_a} \lambda_j f_j(A) + \frac{\rho}{2} \sum_{j\in C_a} [f_j(A)]^2 \tag{8}$$

If we assume available the initial selection of λ_0, ρ and A_0 then the augmented lagrangian algorithm is given at the point A_k by :

a) check termination criteria

b) with A_k as starting point solve the subproblem :

Minimize $\Lambda(\mathbf{A},\lambda_k,\rho)$ and let A_{k+1} denote the best approximation of A.

We first define a search direction at point A_k from the available informations on the cost function value and its gradients. Then a line search is performed with a parabolic approximation.

c) update C_a and λ_k

d) increase ρ if the constraint violations have not decreased sufficiently from those at A_k

e) k=k+1 and go to a)

At the end of the computation we obtain if available new values of the optimised parameters and also some variables which allow to quantify the optimization quality e.g. the initial and final cost function values.

4.3. The driver

The driver allows to simulate whatever experimental paths, drained or undrained, monotonous or cyclic, taking into account time effects. The model was validated on a large number of experiments concerning sands and clays.[2][4]

The driver is called many times by the optimization module to evaluate the cost function and its gradient. The time spent here must be particularly reduced with automatic time stepping and accurate Runge-Kutta methods.

4.4. Optimization process

The optimization process consists in the following steps:

a) Pre-processing of the data :

Read experimental paths an their quality indices given by the user. Those are grades which give an evaluation of the test reliability.

Choice of the weights from the nature of the previous application and the experimental tests.

b) Choice of the initial parameters set :

From the nature of the available tests, by using some rules, correlations, calculations and by a look-up in the database a pre-processor determines the initial parameters values.

c) Choice of the strategy :

The choice of the strategy consists in the determination of the number of optimization steps and the number and the nature of the parameters to be optimized.

These choices are made in function of the available tests and the nature of the material. In a first approach the above mentionned manual procedure was implemented into the code with some modifications issued from the results of some previously performed optimizations in order to detect the interactions between parameters and the influence of each one on the simulations.

d) Output of results :

All the output is given in a graphic representation with stress strain curves and stress paths at the beginning and the end of the optimisation process.

5. EXAMPLES OF OBTAINED RESULTS

5.1. Consistency results

In this section some results obtained with the optimization of parameters about a clayed material are presented. In a first step we have simulated two drained triaxial tests with a set of parameters close to one determined from tests on Kaolinite clay and chosen as experimental paths. Consequently those paths are very interesting because the true parameters are perfectly known. Then the values of some of the parameters have been modified and we have run the automatic software until we find back the initial parameters set if possible.

We give here the results issued from three optimizations. The first consists in changing only one of the curve fitting parameters. The second in perturbating two of those parameters. And the third consists in changing the values of two curve fitting parameters : a and b, and a state parameter : d. In order to give an idea of the calculation time we have noted that the third optimization with three parameters requires about 200 integrations of the constitutive equation on a drained triaxial path.

The true values of those parameters are a=8E-3, b=.90 and d=1.61 and we can see on the figure 2 that those true values are obtained at the end of the optimization. If we optimize only the parameter a with an initial value of 7.2E-3 or the two parameters a and b with respectively an initial value of 7.2E-3 and .77 we find again the true values.

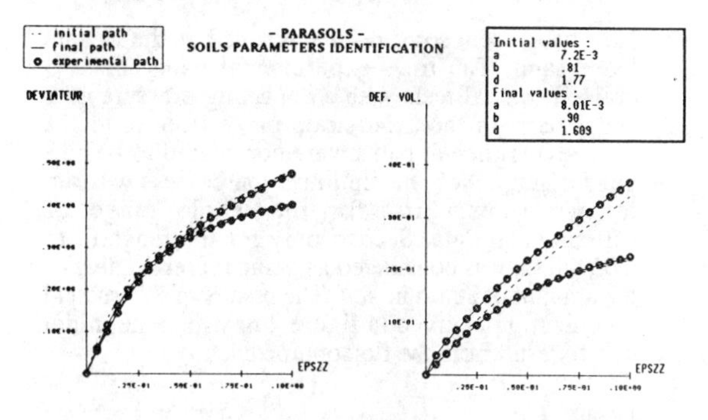

FIGURE 2. Consistency results. Comparison between initial and final parameter values

Several optimizations of this type have been conducted to confirm the consistency of the procedure. It has been found that for a not too large perturbation of the involved parameters the process is stable.

5.2. Decrease rate of the cost function

If we consider the evolution of the cost function which is given in figure 3 we can see that here and also generally it has an hyperbolic shape and that the convergence rate depends on the number of parameters to be optimized.

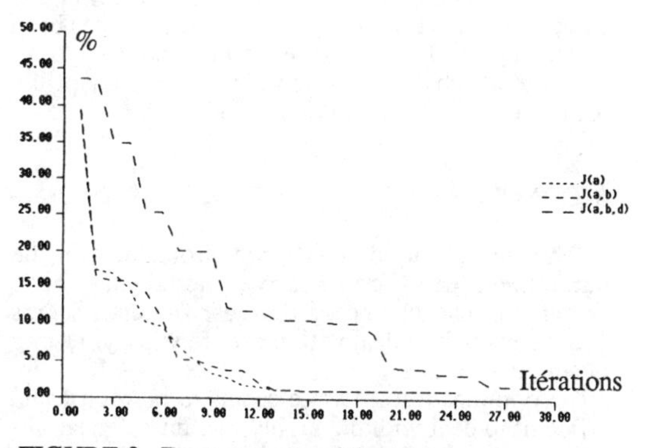

FIGURE 3 . Decrease rate of the cost function

Moreover, we have noted that this rate depends on the nature of the material and the nature of the optimized parameters. In a real situation in wich e.g. the position of the perfect plasticity line and consequently Pco are not well known the identification may become very harder.

For now as a general conclusion it has been found that the optimization process is much faster when dealing with curve fitting parameters than with mechanical parameters which is in fact a nice feature of the constitutive equation.

5.3. Applications oriented optimization

Now we present some results obtained on the Hostun loose sand. The truly experimental paths are two drained triaxial tests with a confining pressure of 1 and 2 bars. In the axial strain range from 0. to .12 those experimental paths were considered to be of a good quality. Now the optimization process was run in two steps. First for the whole range of experimental data. Second only for the range 0. to 0.05 which was considered as being representative of the amount of strain in-sitù. The results of this second optimization is given in figure 4 in which the initial curves result from the first optimization.

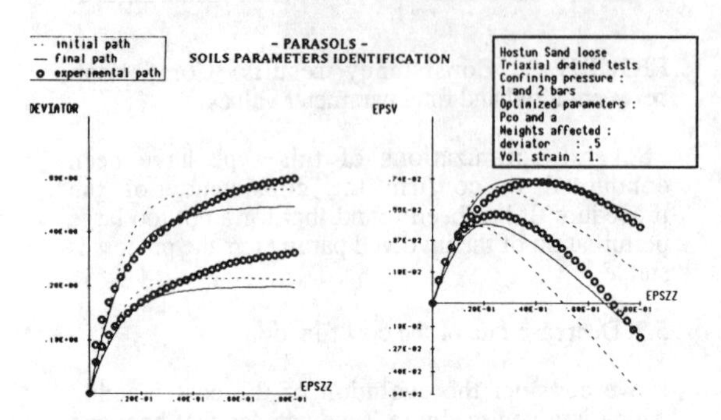

FIGURE 4 . Results of the second step optimization on Hostun loose Sand.

It is very interesting to see that with the weights described above the quality of the simulation has been much improved in the range of interest. That is an example of what can be done in order to adapt the optimization to the forecastable application.

6. CONCLUSIONS

In this paper an identification procedure of the parameters of a constitutive model has been presented. The choice of the cost function seems general enough to deal with most of the usual testing of soil material.

The optimization software seems to be robust and efficient to deal with the highly non linear equations of the model. Finally the parameters have been classified into a relational database which is belevied to be an essential tool either when few data are available for a given soil or as initial values to start the optimization process.

AKNOWLEDGMENTS : This study has been undertaken under contract with Electricité de France / SEPTEN whose support is kindly aknowledged. Several fruitful discussions with Mr Fry from Electricité de France were very helpful.

REFERENCES

1. Aubry D., Kodaissi E., Meimon Y. (1986) A cyclic viscoplastic multimechanism constitutive equation for soils with kinematical hardening. 2nd International Symposium on Numerical Models in Geomechanics. Ghent.
2. Aubry D., Hujeux J.C., Lassoudière F. & Meimon Y. (1982) A double memory model with multiple mechanisms for cyclic soil behaviour. Int. Simp. Numer. Mod. Geomech., Balkema 3-13
3. Gill P., Murray W., Wright M. (1981) Practical Optimization. Academic Press, Inc.
4. Hajal T. (1984) Modelisation elastoplastique des sols par une loi multimecanismes - Application au calcul pressiometrique. These de Docteur-Ingenieur. Ecole Centrale de Paris.
5. Hicher P.Y., (1985) Comportement mécanique des argiles saturées sur divers chemins de sollicitations monotones et cycliques. Application à une modelisation elastoplastique et viscoplastique. These de Doctorat d'Etat. Université Pierre et Marie Curie. Paris.
6. Lowe M., Pierre D. (1975) Mathematical programming via Augmented Lagrangians. Addison-Wesley Publishing Compagny.

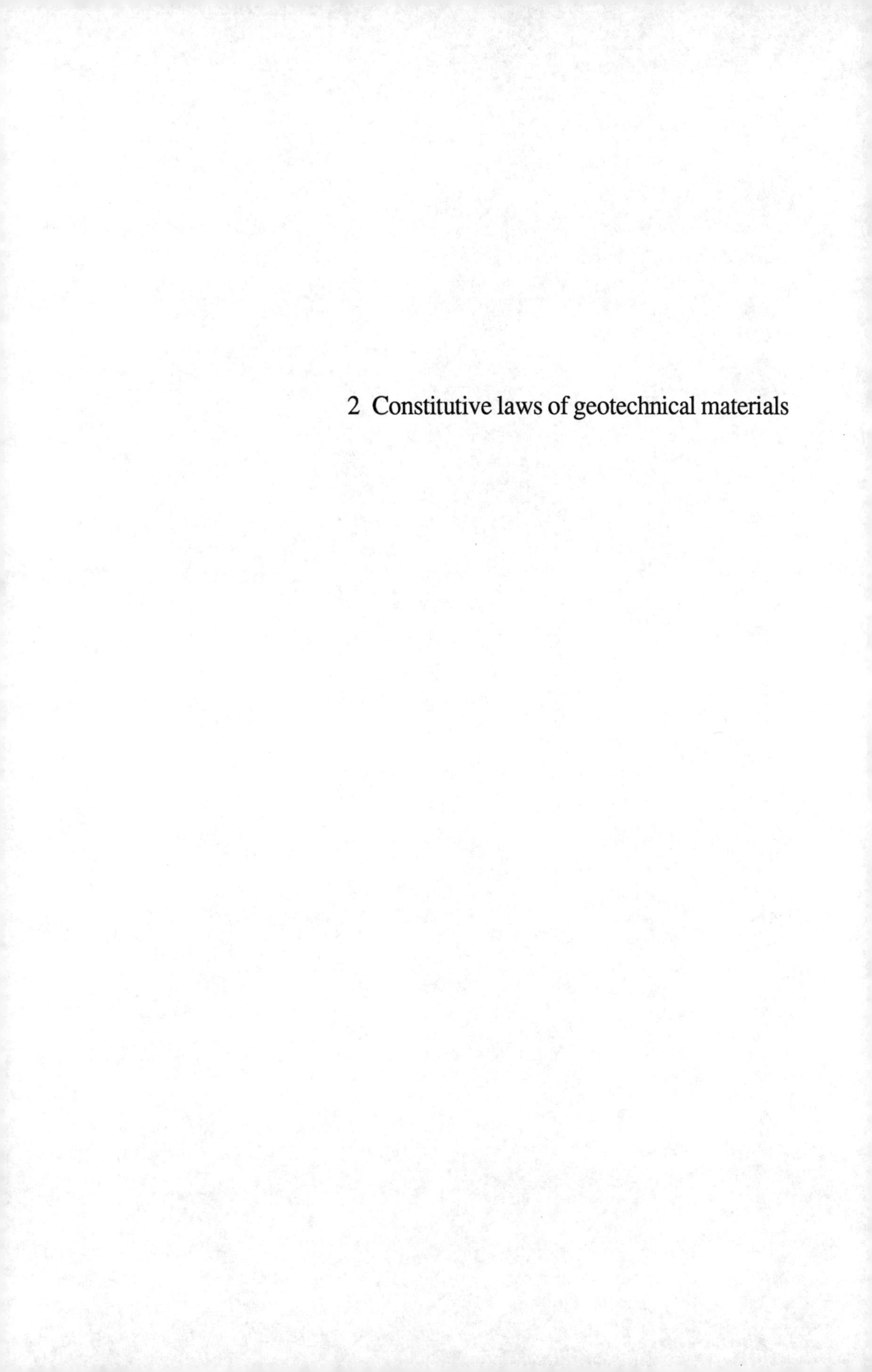

2 Constitutive laws of geotechnical materials

Numerical Methods in Geomechanics (Innsbruck 1988), Swoboda (ed.)
© 1988 Balkema, Rotterdam. ISBN 90 6191 809 X

Generalized constitutive equations of saturated sand

Tej B.S.Pradhan
Institute of Industrial Science, University of Tokyo
Tadanobu Sato & Toru Shibata
Disaster Prevention Research Institute, Kyoto University, Uji, Japan

ABSTRACT: In order to consider the dilatancy and Baushinger effects of soil, the concepts of isotropic and kinematic work hardening are applied to deduce a general non-associate elasto-plastic theory, which gives stress strain relationships for soil under monotonic and cyclic loading conditions. Generalized flow rules have been established based on a newly defined axes system in the stress space. The behaviors of sand are simulated for monotonic and cyclic loading conditions and are compared with experimental data. Eight material parameters are needed which can be determined from one set of conventional triaxial and oedometer tests.

1 INTRODUCTION

Several elasto-plastic models have been proposed to predict the behaviors of soil. There are only a few papers, however, describing the repetitive loading conditions. Mroz/1/ proposed an anisotropic hardening model using the concept of work-hardening moduli to take into account the Baushinger effect. The defect of this model is that the hysteresis loop is completely closed after the application of one full stress cycle. To rectify this shortcoming, Mroz et.al., /2/ proposed a two surface model. Ghaboussi and Momen /3/ proposed a generalized plasticity model based on the assumption of kinematic and isotropic work hardening. Nishi and Esashi /4/ have proposed a model for saturated sand under cyclic loading with the concept of dilatancy limit and equilibrium state. The main object of this paper is to deduce a simpler form of the constitutive equation for granular materials under generalized stress conditions and also to simulate the liquefaction phenomena. A new kinematic strain hardening rule is proposed, which is based on the non-associate flow rule. Simulations are made for different stress paths under drained and undrained conditions and compared with the experimental data.

2 CONCEPT OF NEW AXES SYSTEM

If the stress increment is measured only by a change in octahedral shear stress(τ_{oct}), the effect of its direction change cannot be taken into account because τ_{oct} is a scaler. A new axes system(t-axes system) is defined, such that the line OC (Fig.1), the axis of symmetry of yield surface which has moved in an original stress space depending on a kinematic hardening rule, becomes a new corresponding space diagonal. This is related to the original axes by

$$\{\sigma_{it}'\} = [T]\{\sigma_i'\},\quad (i=1,3) \tag{1}$$

The transformation matrix $[T]$ can be determined from the direction cosine of the line OC. Yield functions and plastic potential functions are also defined in the t-axes system, and flow rules are derived.

3 CONSTITUTIVE EQUATION

The total strain increment ($d\varepsilon_{ij}$) is assumed to be obtained from the sum of the elastic component ($d\varepsilon_{ij}^{e}$) and the plastic component ($d\varepsilon_{ij}^{p}$). The plastic strain increment is calculated based on the non-associated flow rule;

$$d\varepsilon_{ij} = d\varepsilon_{ij}^{e} + d\varepsilon_{ij}^{p} \tag{2}$$

$$d\varepsilon_{ij}^{p} = h\,\frac{\partial g}{\partial \sigma_{ij}}\,df \tag{3}$$

in which h is a strain hardening function, g is a plastic potential function, f is a yield function and σ_{ij} is a stress tensor. $d\varepsilon_{ij}^{e}$ is neglected throughout the remainder of this paper.

(1) Yielding function

According to experimental research by Poorooshasb et.al., /5/ , Frydman et.al., /6/ and Tastuoka et.al., /7/, yielding occurs when the shear stress ratio reaches a critical value. For a lower range of mean effective stress, yield loci can be considered as a family of straight lines radiating outwardly from the origin. Pender/8/ suggested that whenever there is a change in stress ratio(τ_{oct}/p'), plastic yielding occurs.

In this paper the yielding criterion of shear and consolidation are treated separately. The modified Von-Mises type of yield surface is assumed for the yield surface of shear. This surface can expand or contract and can move in stress space keeping the vertex at the origin.

The yield surface for shear corresponding to the t-axes system can be given by

$$f_t(\sigma_{it}') = \sum_{i=1}^{3}\sigma_{it}'^2-(1+\varphi^2)\sum_{i=1}^{3}\alpha_{it}'^2 = 0 \qquad (4)$$

in which σ_{it}',$\alpha_{it}'$and φ represent the coordinates of current stress point, the center of the yield surface and the tangent of half the apex angle, respectively. α_{it}' and φ are parameters measuring the kinematic and isotropic hardenings, respectively.

Yield function for consolidation is expressed by

$$f_c = p'- p_y' = 0 \qquad (5)$$

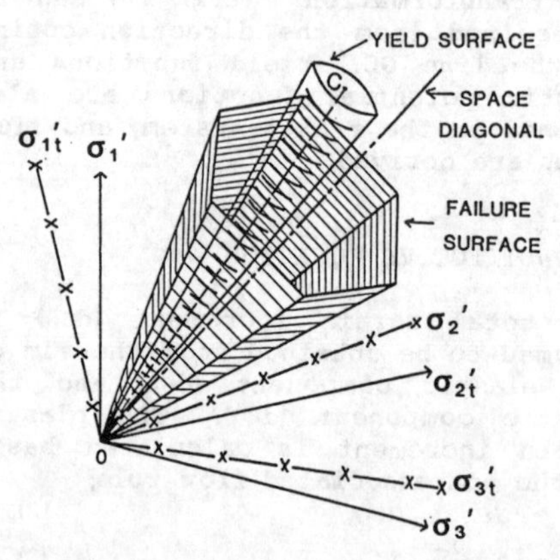

Figure 1. Yield and failure surfaces in stress space.

in which p' is the mean effective principal stress and p_y' is a hardening parameter.

(2) Failure and Phase transformation surfaces

Failure is considered to be a state in which a large amount of plastic strain is generated. According to Bishop's /9/ results, we assume that failure is defined by Mohr-Coulomb's criterion which is expressed by the constant mobilized angle of friction at failure. The shape of the failure surface, which is a deformed hexaconical surface with an apex at its origin, is shown in Figs 1 and 2.

A phase transformation surface is assumed to have a similar shape to that of the failure surface. The implication of this surface is that the dilatancy behavior of the material changes when any arbitrary stress path crosses this surface. It can be defined by parameter φ_m, which is the mobilized angle of friction at the maximum contraction state in the triaxial compression test.

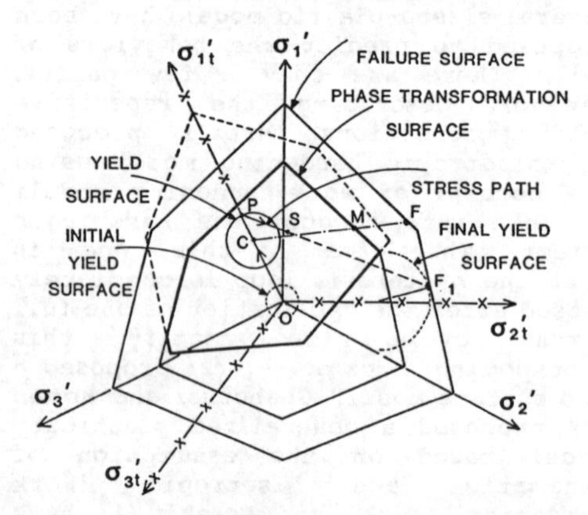

Figure 2. Yield and failure surfaces in octahedral plane.

(3) Stress-dilatancy and plastic potential functions

Plastic potential surfaces are determined from the stress-dilatancy relationship. Fig.3 shows the stress ratio(τ_{oct}/p') vs dilatancy ($dv_d^p/d\gamma_{oct}^p$) relation for Toyoura sand subjected to a cyclic triaxial loading under a constant mean effective stress (p'). The solid line shown in this figure is expressed by

$$\frac{dv_d^p}{d\gamma_{oct}^p} = \frac{3}{2}(M_m-\eta) \qquad (6)$$

in which dv_d^p is the plastic volumetric strain increment due to dilatancy, $d\gamma_{oct}^p$ is the plastic octahedral shear strain increment, $\eta(=\tau_{oct}/p')$ is the stress ratio and M_m is the value of η at the maximum contraction state. A generalized stress dilatancy relationship corresponding to the t-axes system is expressed in a similar form to that in Eq.(6), as follows;

$$\frac{dv_t^p}{d\gamma_t^p} = \frac{3}{2}(\overline{M}_m-\eta_t) \quad (7)$$

in which dv_t^p and $d\gamma_t^p$ are new strain parameters and are related to the original parameters through the transformation matrix $[T]$. $\eta_t(=\tau_t/\sigma_t')$ is a new stress parameter, in which τ_t and σ_t' are the octahedral and mean stress respetively, corresponding to t-axes system. $\overline{M}_m$ is a parameter related to the phase transformation angle.

From Eqs (6) and (7) the derivatives of the plastic potential function (g_s) for shear can be expressed as follows.

$$\frac{\partial g_s}{\partial \tau_t} = 1, \quad \frac{\partial g_s}{\partial \sigma_t'} = (\overline{M}_m-\eta_t) \quad (8)$$

Because no shear strain is assumed to be generated during consolidation , the plastic potential function (g_c) for consolidation can be written as follows:

$$g_c = p'-p_y' = 0 \quad (9)$$

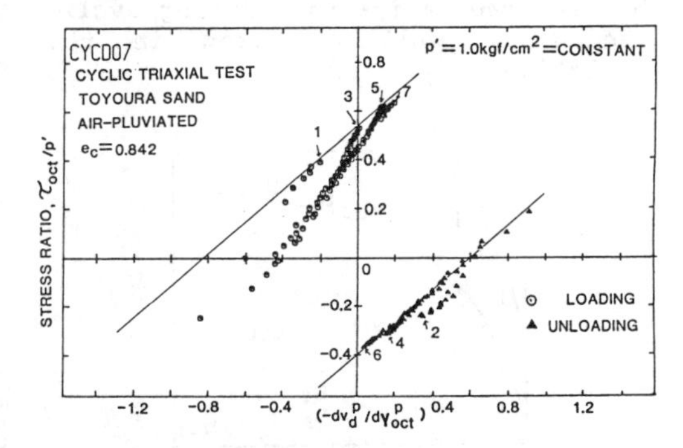

Figure 3. Stress-ratio vs dilatancy relationship.

(4) Hardening functions

The yielding of the material is assumed to occur as a combination of isotropic and kinematic hardening.

(a) Isotropic hardening rule

The measure of isotropic hardening φ is assumed to be a function of plastic strain trajectory $\overline{\gamma}_t^p$ as shown in Fig.4. If the relatioship between φ and $\overline{\gamma}_t^p$ is approximated by a hyperbolic function, its derivative $d\varphi$ is given as follows:

$$d\varphi = \frac{(\varphi_u-\varphi_0)^2\xi_0}{(\varphi_u-\varphi_0+\xi_0\overline{\gamma}_t^p)^2}\, d\overline{\gamma}_t^p \quad (10)$$

in which φ_0 and φ_u represent the size of the initial yield surface and maximum yield surface respectively. ξ_0 represents the degree of isotropic hardening and $\overline{\gamma}_t^p$ is defined by

$$\overline{\gamma}_t^p = \int |d\gamma_t^p| \quad (11)$$

In an undrained condition, the material begins to soften from the onset of the initial liquefaction. As can be seen in many liquefaction tests on sand, a large amount of pore-water pressure is generated if the material is sheared in the opposite direction after the onset of the initial liquefaction. This phenomenon is taken into account in the hardening rule, by considering the shrinkage of the elastic region (Fig.4). ξ_0^* shows the degree of shrinkage of the elastic region and is taken as $5\times\xi_0$.

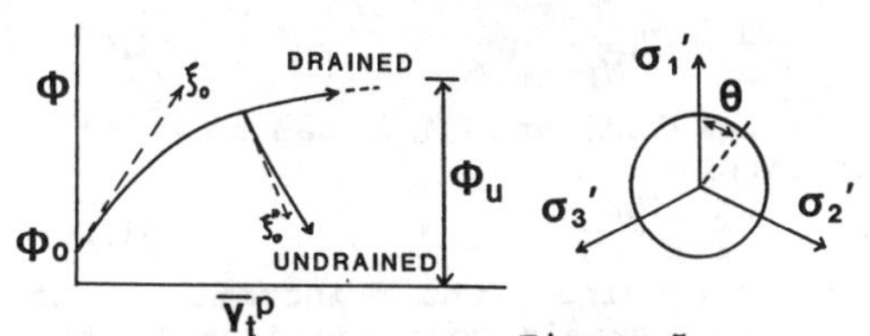

Figure 4. Isotropic hardening function.

Figure 5. Definition of θ in octo-plane.

(b) Kinematic hardening rule

The magnitude of movement of the center of yield surface α_{it}' is considered as a measure of kinematic hardening. The kinematic hardening rule is assumed to be expressed by

$$d\alpha_{it}' = \zeta_i\, d\varepsilon_{it} \quad (i=1,3) \quad (12)$$

in which $\zeta_i\,(i=1,3)$ are kinematic hardening parameters. To determine ζ_i, the following three conditions are required.

(i) Prager's condition of compatibility: This condition must be fulfilled for advancement to plastic deformation and is expressed by

$$df_t = \frac{\partial f_t}{\partial \sigma_{it}'}d\sigma_{it}' + \frac{\partial f_t}{\partial \alpha_{it}'}d\alpha_{it}' + \frac{\partial f_t}{\partial \varphi}d\varphi = 0 \quad (13)$$

(ii) Orthogonality condition: This is a condition in which the center C of the yield surface must move in order to satisfy the orthogonality between vectors CP and OC(Fig.2).

(iii) Non-intersection condition: The yield surface must not intersect the failure surface.

(5) Constitutive relationship

Using Eqs (3),(4) and (8) the strain increment during the shear process can be written as follows:

$$d\varepsilon_{it} = \frac{1}{3} h_s \{(\overline{M}_m - \eta_t) + \frac{(\sigma_{it}' - \sigma_t')}{\tau_t}\} \, df_t \qquad (14)$$

To determine the hardening parameter h_s, we assume a hyperbolic relationship between η_t and γ_t^p, namely,

$$\eta_t - \eta_{to} = \frac{\overline{M}_f G'(\gamma_t^p - \gamma_{to}^p)}{\overline{M}_f + G'(\gamma_t^p - \gamma_{to}^p)} \qquad (15)$$

in which η_{to} and γ_{to}^p are the values of η_t and γ_t^p, respectively, at the start of the plastic deformation. $\overline{M}_f$ is a failure parameter and can be determined from the direction of the stress path and the failure surface. G' is the initial tangent modulus in the η_t- γ_t^p curve and is assumed to be equal to the initial tangent modulus of the $\eta(=\tau_{oct}/p')-\gamma_{oct}^p$ curve. Differentiating Eq.(15) with respect to γ_t^p, we obtain the expression of plastic modulus H;

$$H = \frac{d\eta_t}{d\gamma_t} = G'(\frac{\overline{M}_f - \eta_t}{\overline{M}_f - \eta_{to}})^2 \qquad (16)$$

From Eqs (3),(8), and (14) h_s can be written as follows:

$$h_s = \frac{3}{2} \frac{d\eta_t}{H \, df_t} \qquad (17)$$

To determine the increment of principal strain corresponding to the original axes system, it is asssumed that the direction of principal stress coincides with that of the principal strain increment. Thus, $\{d\varepsilon_i\}$ can be given by

$$\{\varepsilon_i\} = [T]^T \{\varepsilon_{it}\}, \quad (i=1,3) \qquad (18)$$

in which $[T]^T$ denotes the transpose of $[T]$.

The strain increment during consolidation can be written as follows:

$$d\varepsilon_i = \frac{\lambda - \kappa}{(1+e_0)p'} \frac{\partial g_c}{\partial \sigma_i} \, df_c \qquad (19)$$

in which λ and κ are the slopes of the virgin compression line and the swelling line in the e-l_np' plane, respectively, and e_0 is the initial void ratio.

4 SIMULATION OF SAND BEHAVIORS

Simulation results are compared with the experimental stress-stress relation for different sand samples subjected to monotonic and cyclic loading conditions. A list of the parameters used for simulation is given in Table 1.

TABLE 1. PARAMETERS OF SANDS

SANDS	Φ_f^*	Φ_m^*	G_0	Φ_0	Φ_u	ζ_0	λ	κ
TOYOURA	40	29	260	0.04	0.25	10	0.1	0.01
FUJIGAWA	41	33	90	0.04	0.14	12	0.1	0.01

(1) Drained behaviors

Fig.5 shows a octahedral plane. τ_{oct}^* is defined as $\tau_{oct}^* = \tau_{oct} \times \cos\theta$. Shown in Figs 6(a),(b) are the experimental data obtained by Yamada et.al., /10/ in a cubical true triaxial test (p'=1.0kgf/cm^2) of Fujigawa sand for θ= 0° and 30° respectively. The simulated results are shown by the solid line and a good agreement can be seen. When θ=30°, however, the mobilized angle of internal friction at failure seems larger for the experimental data.

Stress-strain curves for a p' constant (=1.0kgf/cm^2) drained cyclic triaxial test on Toyoura sand are shown in Fig.7(a). The experiment gives a larger strain on the extension side, but the general tendency is very well simulated. Fig.7(b) shows the corresponding stress-ratio vs volumetric strain curves. The accumulation of volumetric strain and the decrease in its rate during cyclic loading is well reflected in the simulation.

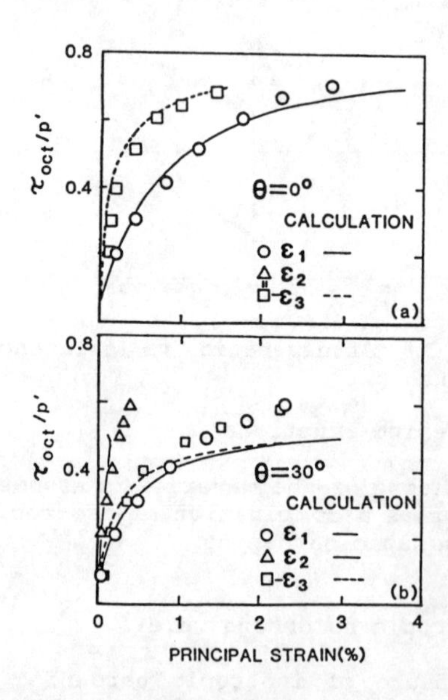

Figures 6(a),(b). Stress-strain relationships of Fujigawa sand.

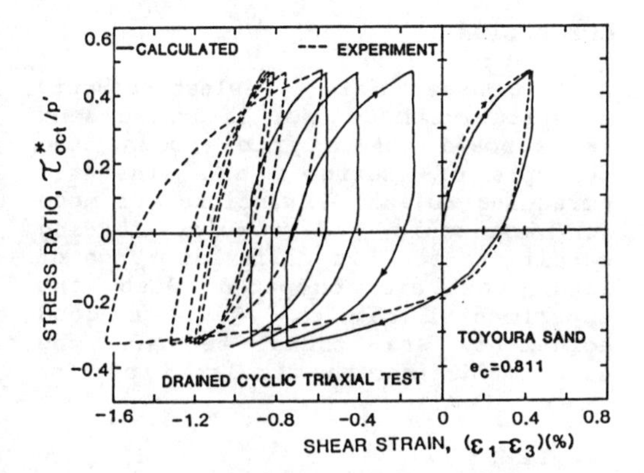

Figure 7(a). Stress-strain relationships of Toyoura sand in drained cyclic triaxial test.

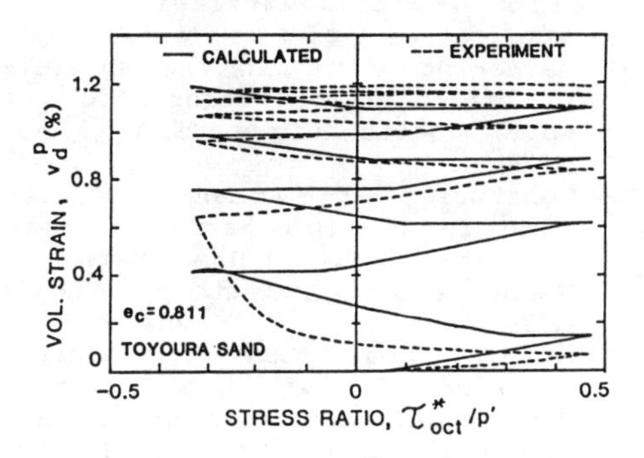

Figure 7(b). Stress-dilatancy relationship of Toyoura sand in drained cyclic triaxial test.

(2) Undrained behaviors

To simulate the undrained condition, the sum of the incremental volumetric strain due to dilatancy (dv_d^p) and consolidation (dv_c^e) is equal to zero. From Eqs (14) and (18), dv_d^p can be written as follows:

$$dv_d^p = \frac{3}{2H}\{T_c\} \quad d\eta_t \tag{20}$$

in which,

$$\{T_c\} = \{\sum_{i=1}^{3}T_{1i}C_1+\sum_{i=1}^{3}T_{2i}C_2+\sum_{i=1}^{3}T_{3i}C_3\} \tag{21}$$

$$C_i = \frac{1}{3}\{(\bar{M}_m-\eta_t)+\frac{(\sigma_{it}'-\sigma_t')}{\tau_t}\},(i=1,3) \tag{22}$$

in which T_{ij} represents a component of the transformation matrix $[T]$.

Calculating dv_c^e from Eq.(19) and using Eq.(20), we can derive the increment in mean effective stress as follows.

$$dp' = -\frac{3(1+e_0)\ \{T_c\}\ p'\ d\eta_t}{2\ \kappa\ H} \tag{23}$$

Shown in Fig.8(a) is the effective stress paths for a constant volume cyclic triaxial test on loose Toyoura sand. The test was performed by controlling the cell pressure of the triaxial cell so as to keep the volume of the sample constant during the strain controlled cyclic loading. This was done by taking into account the volume change due to membrane penetration. The accumulation of excess pore water pressure and the process to liquefaction is well simulated.

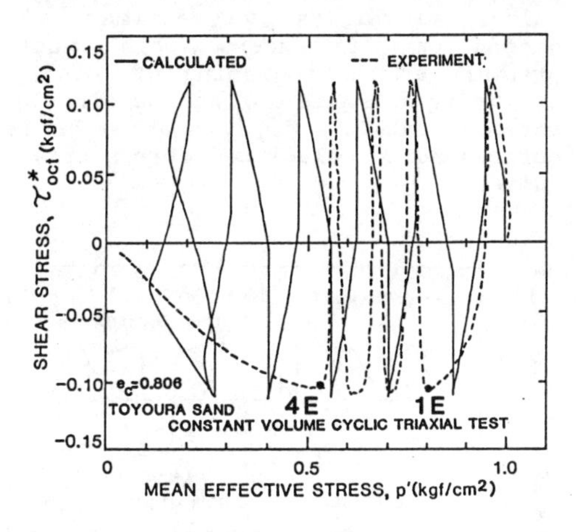

Figure 8(a). Effective stress path of Toyoura sand in constant volume cyclic triaxial test.

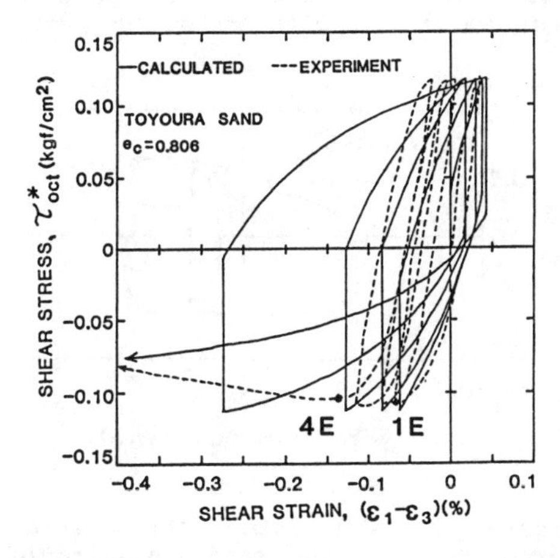

Figure 8(b). Stress-strain relationship of Toyoura sand in constant volume cyclic triaxial test.

For the experimental curve, it should be noted that although the loading direction at point 4E is not changed, the sample is liquefying very quickly. This is a typical phenomenon for loose sand and cannot be reflected in the simulation. Fig.8(b) shows the corresponding stress-strain relationship. The accumulation of strain in the extension side for the constant shear stress amplitude test is well simulated.

Fig.9 shows the stress paths for loose Fujigawa sand subjected to a circular total stress path on the octahedral plane (p'=1.0 kgf/cm^2) as shown in Fig.5. Calculated results over estimate the generation of excess pore water pressure at the beginning of loading, but under estimate during the circular stress path. Fig.10 shows the corresponding calculated stress strain curve.

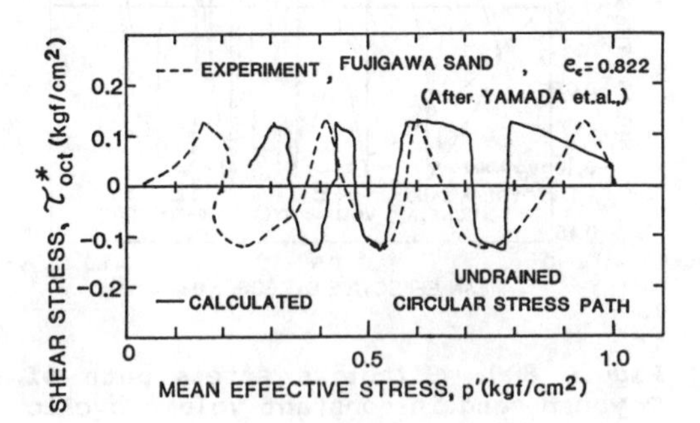

Figure 9. Effective stress path of Fujigawa sand in circular stress path.

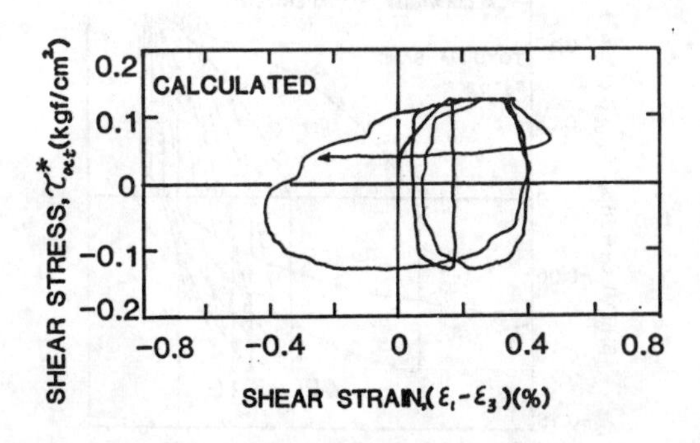

Figure 10. Calculated stress-strain curve of Fujigawa sand in circular stress path.

CONCLUSION

A non-associate elasto-plastic constitutive model for saturated sand is proposed taking into account the concepts of isotropic and kinematic hardening rules. Simulations are made for monotonic and complex loading conditions for two different types of sand, and are compared with the experimental results. Quite a good agreement was found between the experimental and the simulated results.

REFERENCES

/1/ Z.Mroz: On the description of anisotropic work hardening, J. Mech. Physics. Solids,Vol 15,pp 163-175 (1967).

/2/ Z.Mroz,V.A.Norris,Zienkiewicz: Application of an anisotropic hardening model in the analysis of elasto-plastic deformation of soils, Geotechnique 29,No 1,1-34 (1979).

/3/ J.Ghaboussi and H.Momen: Plasticity model for cyclic behavior of sands, 3rd Int. Conf. on Num. Meth. in Geomechanics, Aachen, pp 423-434 (1979).

/4/ K.Nishi and Y.Esashi: Elastic plastic model of fully saturated sand under undrained static and cyclic loading, Proc. 4th Int. Conf. on Num. Meth. in Geomechanics, Edmonton (1982).

/5/ H.B Poorooshasb: Deformation of sand in triaxial compression, Proc. 4th Asian Reg. Conf. on S.M.F.E, Bangkok, Vol.1 pp 63-66 (1971).

/6/ S.Frdman,J.G.Zietlen and I.Alpan: The yielding behavior of particulate media,Can. Geotech. Jour,Vol 10,pp 341-362 (1973).

/7/ F.Tatsuoka and K.Ishihara: Yielding of sand in triaxial compression, S & F, Vol 14,No.2 (1974).

/8/ M.J.Pender: A model for the behaviour of over-consolidated soil,Geotechnique, Vol 28,No 1, pp 1-25 (1978).

/9/ Bishop,A.W: The strength of soils as engineering materials, (1966).

/10/ Y.Yamada and K.Ishihara: Anisotropic deformation characteristics of sands under three dimensional stress conditions, S & F, Vol 19,No 2 (1979).

Numerical Methods in Geomechanics (Innsbruck 1988), Swoboda (ed.)
© 1988 Balkema, Rotterdam. ISBN 90 6191 809 X

A cyclic viscoplastic constitutive model for clay

F.Oka
Gifu University, Japan

ABSTRACT: A cyclic elasto-viscoplastic constitutive model is derived based on the overstress type viscoplasticity theory and the non-associated flow rule. Special attention is concentrated on the random cyclic loadings. The hardening function is extended in order to describe the behavior under cyclic loadings. The proposed theory is applied for the triaxial cyclic tests of silty overconsolidated clay.

1 INTRODUCTION

In relation to the foundation design of offshore structure and the earthquake resistant design , the cyclic plasticity model of clay is an important subject in geotechnical engineering. Especially, in the deep sea bed, soil layers is mainly composed of clay. These clay layer are subjected to the repeated loading due to wave action or earthquakes. The cyclic behavior of clay has been studied experimentally by several researchers(See Wood 1982). Many studies have been performed for the constitutive model for saturated sand in relating to liquefaction analysis. Comparing to the study of cyclic behavior of sand, theoretical study for cyclic behavior of clay is relatively little. This is a motivation of the present work.

In the present paper, a cyclic visco-plasticity model for saturated clay is developed based on the overstress type viscoplasticity theory ,the elasto-plastic constitutive model for overcon-solidated clay proposed by Adachi and Oka(1985) and the general hardening rule for cyclic loading. Using the proposed model, the cyclic triaxial test results for soft silty clays which are normally- and over-consolidated are simulated. The effect of loading rate is discussed. Comparison of predicted and experimental results of stress amplitude constant tests shows that the proposed model is effective for simulating the behavior of silty clay.

2 CYCLIC ELASTO-VISCOPLASTIC CONSTITUTIVE MODEL FOR CLAY

By extending the previous works(Oka 1982, and Adachi and Oka 1986), we have constructed an elasto-viscoplastic constitutive model for saturated clay under general cyclic loading conditions.

2.1 O.C. boundary surface

The O.C. boundary surface is introduced to define the boundary between the normally consolidated region($f_b \geq 0$) and the overconcolidated region($f_b < 0$).

$$f_b = \bar{\eta}^*_{(0)} + M^*_m(\theta)\ln(\sigma'_m/\sigma'_{md}) \tag{1}$$

in which σ'_m is the mean effective stress,

θ is a Lode's angle, M^*_m is a material parameter, σ'_{md} is a material parameter related to deformation history and $\bar{\eta}^*_{(0)}$ is a stress parameter defined as follows:

$$\bar{\eta}_{(0)}^* = [(\eta_{ij}^* - \eta_{ij(0)}^*)(\eta_{ij}^* - \eta_{ij(0)}^*)]^{1/2} \tag{2}$$

$$\text{where } \eta_{ij}^* = s_{ij}/\sigma'_m. \tag{3}$$

In Eq.(2), subscript (0) denotes the value at the end of anisotropic consolidation, and s_{ij} is a deviatoric stress tensor.

Outside the boundary surface(in the normally consolidated region), it is assumed that the behavior of saturated clay is described by the elasto-viscoplastic constitutive equation by

Adachi and Oka 1982). The constitutive model in the overcosolidated region is derived in the followings.

2.2 Plastic yield function

The plastic yield function is given by

$$\bar{\eta}_{(n)}^{*} - \kappa_s = 0 \qquad (4)$$

where κ_s is a hardening parameter and $\bar{\eta}_{(n)}^{*}$ is a relative stress parameter defined as:

$$\bar{\eta}_{(n)}^{*} = [(\eta_{ij}^{*} - \eta_{ij(n)}^{*})(\eta_{ij}^{*} - \eta_{ij(n)}^{*})]^{1/2} \qquad (5)$$

$$\eta_{ij}^{*} = s_{ij}/[c(\sigma'_m/\sigma'_{m0})^d] \qquad (6)$$

in which σ'_{m0} is a unit mean effective stress, c and d are material parameters and $\eta_{ij(n)}^{*}$ denotes the value of η_{ij}^{*} at the nth times turning over state of loading direction. $\eta^{*}_{ij(n)}$ will be updatd when loading direction on the π plane is changed.

2.3 Strain-hardening parameter

In the present paper, the strain-hardening function are generalized for explaining the behavior under the random cyclic loading condition. We define the failure stress components(σ_{1f}, σ_{2f}, σ_{3f}) on the plane, which corresponds to the current direction of stress vector(Fig.1). Using the failure stress components, the relative failure stress ratio is given by

$$\eta^{*}_{(f)} = [(\eta_{ij(n)}^{*} - \eta_{ij(f)})(\eta_{ij(n)}^{*} - \eta_{ij(f)})]^{1/2} \qquad (7)$$

$$\eta_{ij(f)} = s_{ij(f)}/[c(\sigma'_{m(f)}/\sigma'_{m0})^d \qquad (8)$$

where $s_{ij(f)}$ and $\sigma'_{m(f)}$ are the values of s_{ij} and σ'_m derived from the failure stress components.

By use of the relative stress components, the strain hardening function proposed previously(Adachi and Oka 1985) is extended as follows:

$$\gamma^{p*} = \frac{\bar{\eta}^{*}_{(n)}\,\eta^{*}_{(f)}}{G'(\eta^{*}_{(f)} - \bar{\eta}^{*}_{(n)})} \qquad (9)$$

where γ^{p*} is the relative deviatoric strain given as:

$$\gamma^{p*} = [(e_{ij}^{p} - e_{ij(n)}^{p})(e_{ij}^{p} - e_{ij(n)}^{p})]^{1/2} \qquad (10)$$

where $e_{ij(n)}^{p}$ is the value of plastic deviatoric strain tensor at the nth times reversion of loading direction.

2.4 Plastic potential function

The plastic potential function f_p is assumed to be given by

$$f_p = \bar{\eta}^{*}_{(n)} + \tilde{M}^{*}\ln(\sigma'_m/\sigma'_{ma(n)}) = 0 \qquad (11)$$

where the parameter $\tilde{M}^{*}$ is given by

$$\tilde{M}^{*} = \frac{-\eta^{*}}{\ln(\sigma'_m/\sigma'_{mc})} \qquad (12)$$

where η^{*} is a stress parameter defined by

$$\eta^{*} = [\eta_{ij}^{*}/\eta_{ij}^{*}]^{1/2} \qquad (13)$$

$$\eta_{ij}^{*} = s_{ij}/\sigma'_m \qquad (14)$$

In the normally consolidated region, $\tilde{M}^{*}$ is equal to M_m^{*} in Eq.(1).

The rate independent plasticity model for overconsolidated clay can be formulated by use of non-associated flow rule, the yield function Eq.(4), the hardening function Eq.(9) and the plastic potential function Eq.(11).

2.5 Viscoplastic constitutive equations

In this section, an elasto-viscoplastic constitutive model for clay is derived based on the Perzyna's type visco-plasticity theory and the rate independent plasticity theory in the previous section.

The viscoplastic strain rate tensor $\dot{\varepsilon}_{ij}^{vp}$

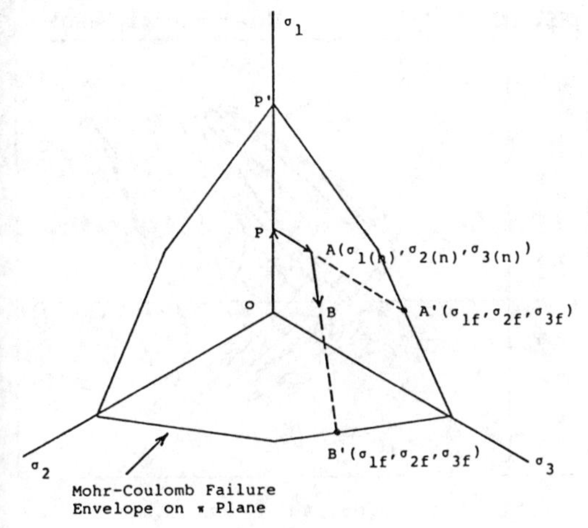

Fig.1 Failure stress components

is given by(Oka 1982)

$$\dot{\varepsilon}_{ij}^{vp} = \langle \Phi_{ijkl}(F) \rangle \frac{\partial f_p}{\partial \sigma_{kl}} \tag{15}$$

$$F = (f - \kappa_s)/\kappa_s \tag{16}$$

$$\langle \Phi_{ijkl}(F) \rangle = 0 \quad (F \leq 0) \tag{17}$$

$$= \Phi_{ijkl}(F) \quad (F>0)$$

Since F=0 expresses the statical yield function, the dynamic yield function f takes the same form of Eq.(4).
In the present theory, $\Phi_{ijkl}(F)$ is assumed to be 4th order isotropic tensor expressed by

$$\Phi_{ijkl}(F) = C_{ijkl}\Phi'(F) \tag{18}$$

$$C_{ijkl} = A\delta_{ij}\delta_{kl} + B(\delta_{ik}\delta_{jl} + \delta_{il}\delta_{jk}) \tag{19}$$

where δ_{ij} is Kronecker's delta, $\Phi'(F)$ is a material function, and A and B are material parameters.

Considering the elastic strain rate tensor, the total strain rate is obtained by adding the elastic strain rate to viscoplastic one.

$$\dot{\varepsilon}_{ij}^{e} = \dot{s}_{ij}/(2G) + \frac{\kappa}{(1+e)\sigma'_m}\dot{\sigma}'_m\delta_{ij} \tag{20}$$

where κ is a swelling index and G is a elastic shear modulus.

As for the shape of the funtion $\Phi'(F)$ in Eq.(18), we have adopted the following form(Oka 1982).

$$\Phi'(F) = \sigma'_m \exp(m'_o(\bar{\eta}_{(n)}^{*} - \kappa_s)) \tag{21}$$

where m'_0 is a material parameter.

3 VERIFICATION OF THE MODEL UNDER CYCLIC LOADING CONDITION

In this section, the proposed cyclic viscoplastic model for clay are verified for the cyclic triaxial tests of over-consolidated silty clay whose O.C.R. is 2.0. In the experiemnts, the reconsolidated Fukakusa Clay are used. I_p is 20.1. The material parameters and the test conditions are listed on Tables 1 and 2. Tests were carried out under the condition of constant deviator stress rate with constant amplitude. Fig.2 presents the experimental result for higher stress rate and Fig.3 shows the result for low stress rate. Comparing two results, it is seen that the amount of the decrease of mean effective stress is larger for the case of low stress rate. This fact is due to the effect of stress rate, e.g., the effect of viscoplastic property of clay. Figs. 4 and 5 show the predicted results by the proposed model. The stress paths are in agreement with the experimental results, however, the amount of strain induced during the cyclic loading are underestimated.

4 CONCLUSION

The present work is the extention of the previous elasto-plastic constitutive model for overconsolidated clay by considering the general hardening rule and the modified yield function. Further research is necessary to imporve the prediction about the stress-strain relations and more verification should be done for the complicated loading conditions.

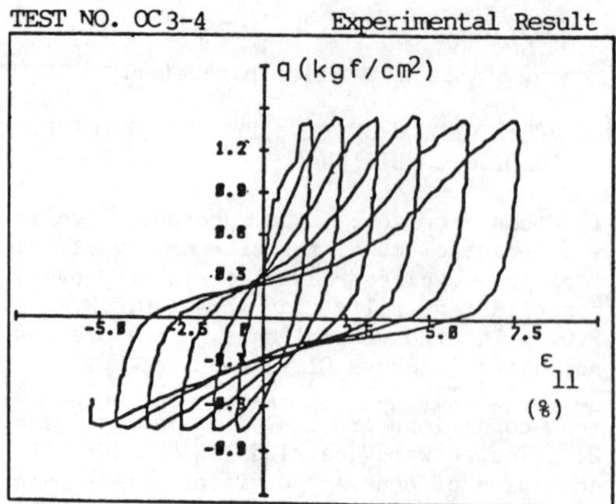

Fig.2(a) Stress-strain curve

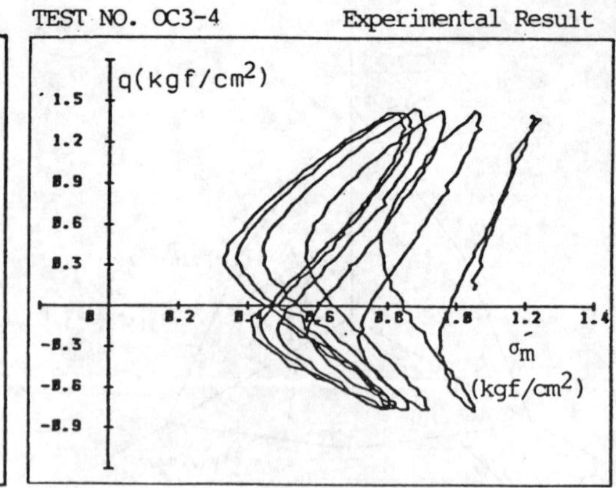

Fig.2(b) Effective stress path

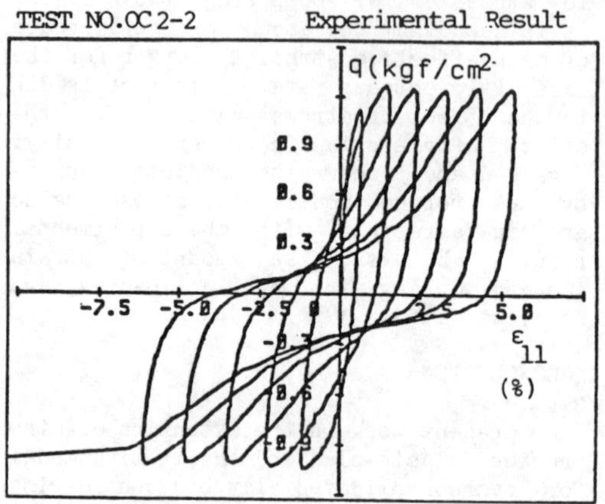

Fig.3(a) Stress-strain curve

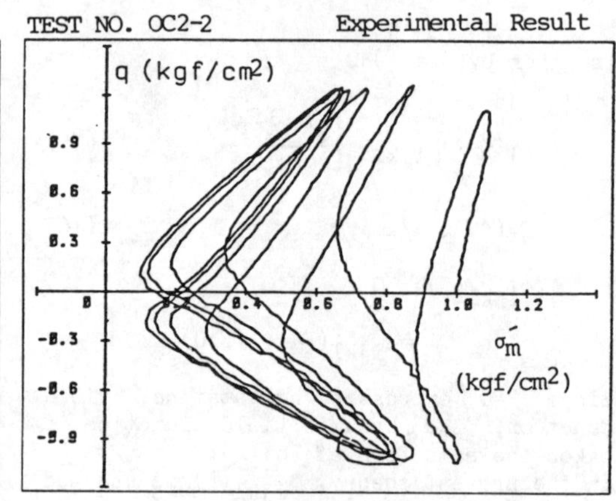

Fig.3(b) Effective stress path

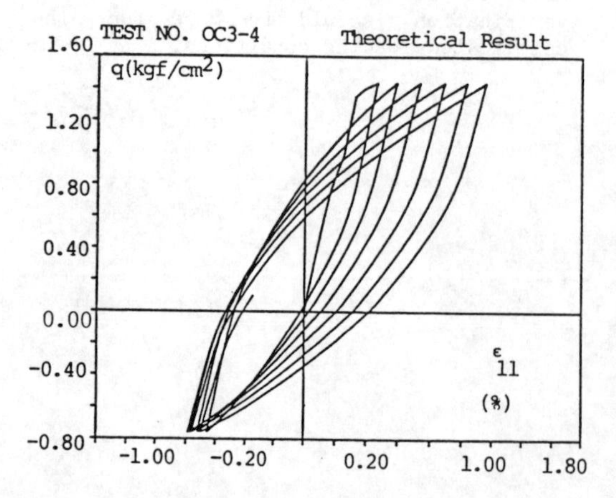

Fig.4(a) Stress-strain curve

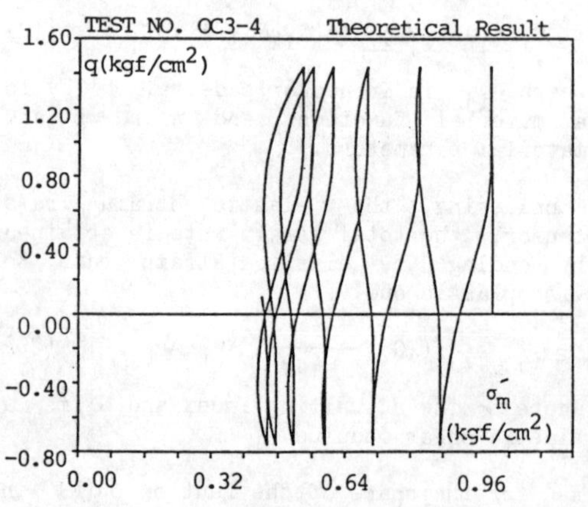

Fig.4(b) Effective stress path

Table 1 Test conditions

Test No.	Initial mean effective stress σ'_{m0} (kgf/cm^2)	q_{max} (kgf/cm^2)	q_{min} (kgf/cm^2)	e_0	Deviator stress rate (kgf/cm^2/sec)
OC2-2	2.0	1.27	-1.05	0.903	2.4×10^{-4}
OC3-4	2.0	1.20	-1.01	0.910	2.4×10^{-3}

Table 2 Material parameters

Elastic Young's modulus (kgf/cm^2)	Swelling index κ	Consolidation index λ	$\eta^*_{(f)}$ (compression)	$\eta^*_{(f)}$ (extension)
700	0.0087	0.091	1.71	1.14

G'	m'_o	C_1 (1/sec)	C_2 (1/sec)	C	D
240	10	2.0×10^{-9}	2.0×10^{-10}	1	0.4

$C_1 = 2B$

$C_2 = 3A + 2B$

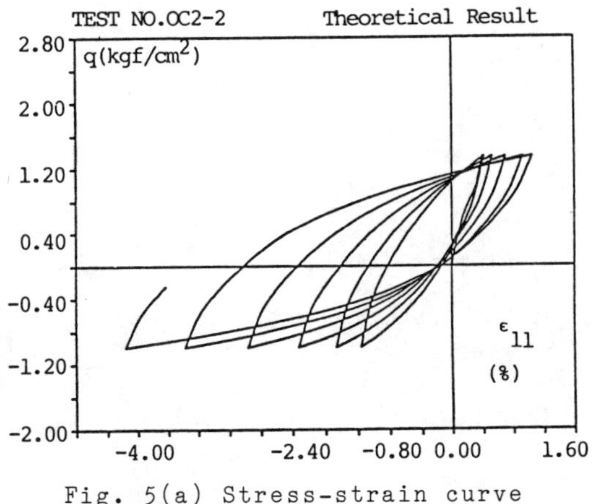

Fig. 5(a) Stress-strain curve

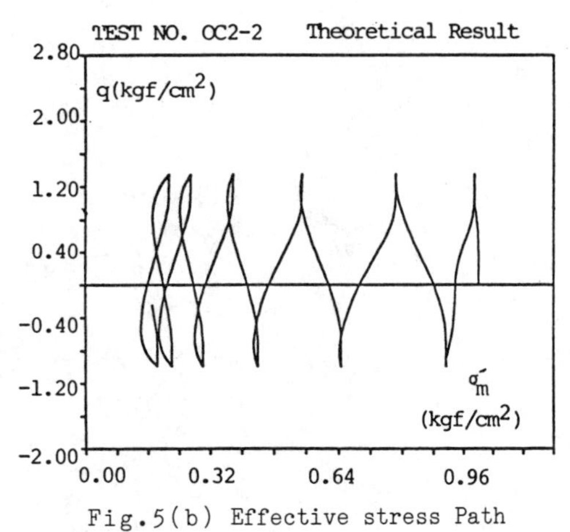

Fig.5(b) Effective stress Path

REFERENCES

(1) D.M.Wood, Laboratory investigations of the behaviour of soils under cyclic loading, Soil Mechanics-Transient and Cyclic Loads,edt.by G.N.Pande and O.C. Zienkiewicz),513-582(1982)

(2) T.Adachi and F.Oka,Constitutive equations for sands and overconsolidated clays, and assigned works for sand, Results of the Int. workshop on Constitutive Relations for soils, Grenoble, edt. by G.Gudehus et al.,141 -157(1986)

(3) F.Oka,Elasto-viscoplastic constitutive equation for overconsolidated clay, Proc. of Int. Symposium on Numerical models in Geomechanics, Zurich,147-156 (1982)

(4) T.Adachi and F.Oka,Constitutive equations for normally consolidated clay based on elasto-viscoplasticity,Soils & Foundations,Vol.22,No.4,57-70(1982)

(5) P.Perzyna,The constitutive equations for work hardening and rate sensitive plastic materials,Proc. Vibrational problems,Warsaw,Vol.4,No.3,281-290 (1963)

Numerical Methods in Geomechanics (Innsbruck 1988), Swoboda (ed.)
© 1988 Balkema, Rotterdam. ISBN 90 6191 809 X

Coupled elasto-plastic deformation-flow finite element analysis using imaginary viscosity procedure

M.Shoji, T.Matsumoto & S.Morikawa
Kajima Corporation, Tokyo, Japan
H.Ohta
Kanazawa University, Japan
A.Iizuka
Kyoto University, Japan

ABSTRACT : Coupled elasto-plastic deformation-flow finite element analysis has been frequently used for the prediction of displacements, stresses and pore water pressures in geotechnical problems. However, the analysis is not suitable for the stability problems since its numerical results often become unstable after stresses of elements arrive at the critical state.
The analytical method including " imaginary viscosity method " which is a usual visco-plasticity computation scheme is developed to analyse the behavior of soil from the initial stress stage to the failure state.
In order to assess the validity and the applicability of this method, a problem of bearing capacity and an actual banking are analysed.

1 INTRODUCTION

When constructing a soil structure like a banking on soft ground, it is important to predict its deformation and stability.

Coupled elasto-plastic deformation-flow finite element analysis (hereinafter called coupled elasto-plastic analysis) is one of the most powerful and useful techniques for analysing the deformation and stress since the coupled problem of pore water and a soil skeleton with an elasto-plastic constitutive relation can be treated easily.

However, even by this method, it is very difficult to obtain both stable and precise solutions for the behavior of soil elements which come to the perfect plastic region such as the failure state, since its incremental strain in the perfect plastic region becomes infinity. Therefore, a usual finite element analysis would give either unstable stresses or overestimated stresses.

The purpose of this paper is to employ a new technique to obtain a stable and precise solution not only for deformation problems but also for stability problems. In order to imple-ment this purpose, an "imaginary viscosity method" proposed by Zienkiewicz and Cormeau (1974) is introduced into coupled elasto-plastic analysis.

Deformation and stress of ground can be analysed stably from an initial loading state to a failure state by the proposed method.

The elasto-plastic constitutive model proposed by Sekiguchi and Ohta (1977) is used to take the anisotropic strength of K_0-consolidated clays into account.

2 ANALYTICAL PROCEDURE BY IMAGINARY VISCOSITY METHOD

2.1 Concept of imaginary viscosity method

Fig.1 schematically illustrates the stress path and the stress-strain relationship in the analysis using imaginary viscosity method. Point A in Fig.1(a) and Point A′ in Fig.1(b) show the initial stress state of normally K_0-consolidated soil. The stress state which is initially at Point A or Point A′ moves with loading and eventually comes to Point B (B′) in the critical state.

Further loading may cause the stresses to overshoot Point B′ at a level of the shear strength in Fig.1(b) , which is not appropri-ate. In order to correct such a misconduct, the imaginary viscosity method is applied to the elements of which stresses overshoot Point B′ . Repeating the imaginary viscosity com-putation, stress relaxation will occur in the element and the stresses come to the steady state (Point D(D′)). During imaginary viscosity computation, the real time is suspended temporarily.

Therefore, line A′−B′−C′ in Fig.1(b) is the stress-strain curve in real time and the line

C′−D′ is the stress-strain curve in imaginary time in viscosity computation while real time does not elapse.

The coupled finite element formulation is same as Iizuka and Ohta (1987). It is basically similar to that proposed by Christian and Boehmer (1970) , but the backward difference in this study is used so as to ensure better stability in computation.

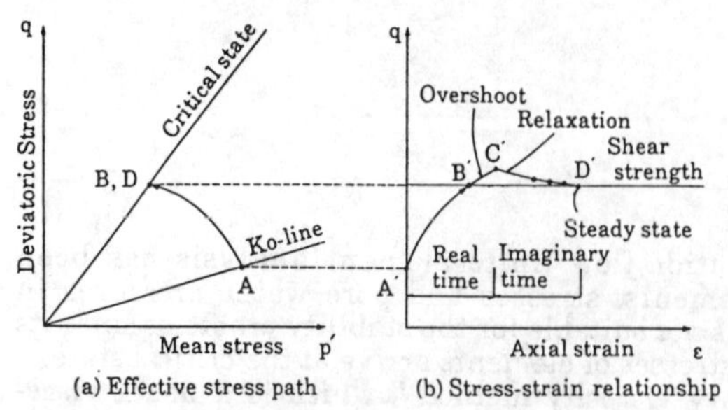

Fig.1 Stress path and stress-strain relationship

2.2 Finite element analysis using imaginary viscosity method

First of all, assuming that the total strain vector $\{\varepsilon\}$, is composed of elastic component $\{\varepsilon^e\}$ and visco-plastic component $\{\varepsilon^{vp}\}$ when the stress point of an element exceeds the initial yield surface, the total strain vector is expressed as follows :

$$\{\varepsilon\}=\{\varepsilon^e\}+\{\varepsilon^{vp}\} \tag{1}$$

By applying associated flow rule, visco-plastic strain velocity vector $\{\dot{\varepsilon}^{vp}\}$ is expressed as :

$$\{\dot{\varepsilon}^{vp}\}=\beta<\frac{f}{f_0}>\{\frac{\partial f}{\partial \sigma}\} \tag{2}$$

in which,

$$\text{at} f<0, <\frac{f}{f_0}>=0,\ \text{at} f\geqq 0, <\frac{f}{f_0}>=\frac{f}{f_0} \tag{3}$$

where, f is yield function for viscosity; f_0' is yield function value at a certain point of time for normalizing f. $\{\sigma\}$ is stress vector, and β is constant related to viscosity.

Backward difference approximation of visco-plastic strain increment related to the time interval $(\Delta\bar{t})$ from a certain time (imaginary time) of $\bar{t}=\bar{t}_0$ to $\bar{t}=\bar{t}_0+\Delta\bar{t}$ can be expressed as follows [Kobayashi (1984)] :

$$\{\Delta\varepsilon^{vp}\}=\Delta\bar{t}\{\dot{\varepsilon}^{vp}\}_{\bar{t}=\bar{t}_0+\Delta\bar{t}} \tag{4}$$

Expanding Eq.(4) in Tayler series, the following is obtained by neglecting the terms of higher order.

$$\{\Delta\varepsilon^{vp}\}=\Delta\bar{t}\left(\{\dot{\varepsilon}^{vp}\}_{\bar{t}=\bar{t}_0}+\left[\frac{\partial\dot{\varepsilon}^{vp}}{\partial\sigma}\right]_{\bar{t}=\bar{t}_0}\{\Delta\sigma\}\right) \tag{5}$$

Expressing the elastic stress-strain matrix as $[D^e]$, then

$$\{\Delta\sigma\}=[D^e]\{\Delta\varepsilon^e\}=[D^e](\{\Delta\varepsilon\}-\{\Delta\varepsilon^{vp}\}) \tag{6}$$

Stress increment vector $\{\Delta\sigma\}$ is obtained by substituting Eq.(5) into Eq.(6),

$$\{\Delta\sigma\}=[\bar{D}]\left(\{\Delta\varepsilon\}-\Delta\bar{t}\{\dot{\varepsilon}^{vp}\}_{\bar{t}=\bar{t}_0}\right) \tag{7}$$

where,

$$[\bar{D}]=\left([D^e]^{-1}+\Delta\bar{t}\left[\frac{\partial\dot{\varepsilon}^{vp}}{\partial\sigma}\right]_{\bar{t}=\bar{t}_0}\right)^{-1} \tag{8}$$

On the other hand, when the following relation is established between strain $\{\varepsilon\}$ and the nodal displacement $\{u\}$:

$$\{\varepsilon\}=[B]\{u\} \tag{9}$$

From the principle of virtual work, the following relational expression is obtained :

$$\int_V [B]^T\{\sigma\}dV-\{L\}=0 \tag{10}$$

where, $\{L\}$ is external nodal force equivalent to surface force and body force. If Eq.(10) is rewritten in the incremental form and Eq.(7) is substituted into the above, the following equation.is obtained.

$$[K]\{\Delta u\}-\{\Delta L^{vp}\}-\{\Delta L\}=0 \tag{11}$$

where,

$$[K]=\int_V [B]^T[\bar{D}][B]dV \tag{12}$$

$$[\Delta L^{vp}]=\Delta\bar{t}\int_V [B]^T[\bar{D}]\{\dot{\varepsilon}^{vp}\}_{\bar{t}=\bar{t}_0}dV \tag{13}$$

Eq.(13) expresses the external nodal force equivalent to the visco-plastic strain specified in Eq.(5). The solution of Eq.(11) can be obtained in a time stepping manner and the relaxation stress is given by Eq.(7).

In regard of computational scheme of $\Delta\bar{t}$, the reader may refer to Zienkiewicz and Cormeau (1974).

2.3 Constitutive model and critical state

The yield function (F) of model by Sekiguchi and Ohta (1977) is given by the following :

$$F=\frac{\lambda-\kappa}{1+e_0}ln\frac{p'}{p_0'}+D\eta^*-v^p \tag{14}$$

where, λ and κ are compression index and swelling index, respectively. e_0, p_0 are void ratio and effective mean principal stress at the end of pre-consolidation respectively. D is dilatancy coefficient.

The effective mean principal stress p and relative shear stress η^* are expressed by the following equation.

$$p' = \frac{1}{3}\sigma'_{ij}\,\delta_{ij}\,,\ s_{ij}=\sigma'_{ij}-p'\,\delta_{ij}$$

$$\eta_{ij}=\frac{s_{ij}}{p'}\,,\ \eta^{*}=\sqrt{\frac{3}{2}(\eta_{ij}-\eta_{ijo})(\eta_{ij}-\eta_{ijo})} \qquad (15)$$

where, σ'_{ij} is effective stress tensor, δ_{ij} is Kronecker's delta, and η_{ijo} is value of η_{ij} in the pre-consolidated state .

Assuming that the failure state of clay as such state where plastic volumetric strain increment (dv^p) becomes zero, the equation of the critical state obtained from Eq.(14) is given by the following.

$$M-\frac{3}{2\eta}+\eta_{ij}(\eta_{ij}-\eta_{ijo})=0 \qquad (16)$$

where,

$$M=\frac{\lambda-\kappa}{D(1+e_0)} \qquad (17)$$

2.4 Treatment of yield function used for imaginary viscosity method

The imaginary yield function for applying imaginary viscosity method should be defined by Eq.(16) as follows :

$$f=\frac{3}{2\eta^{*}}\eta_{ij}(\eta_{ij}-\eta_{ijo})-M=0 \qquad (18)$$

Using Eq.(14), the stress-strain relationship before failure in soil element is defined. And, Eq.(18) expresses the yield function at the critical state in the imaginary viscosity method. This yield function f will not cause any strain hardening.

In order to satisfy the condition of $dv^p=0$ at the critical state, we takes the function g given by the following expression instead the function f as the visco-plastic potential function in the imaginary viscosity method.

$$g=\left\{\frac{3}{2\eta^{*}}\eta_{ij}(\eta_{ij}-\eta_{ijo})\right\}_{p'=p'_f}-M=0 \qquad (19)$$

where, p'_f is the effective mean principal stress at the critical state of each element.

In this paper, in order to avoid complicating the computational algorithm due to non-associated flow rule, Eq.(19) is regarded as the yield function for applying imaginary viscosity method. Eq.(18) is not used in the computations.

2.5 Coupled elasto-plastic analysis by introducing imaginary viscosity method

In the coupled elasto-plastic analysis by imaginary viscosity method, it is necessary to clearly classify the real time governing to water flow and the imaginary time used in the imaginary viscosity method. The former is taken as t and the latter is taken as $\bar{t}$. Fig.2 shows the computational algorithm for this analysis method.

Undrained condition should be prescribed during imaginary visco-plasticity computation since no real time step progresses.

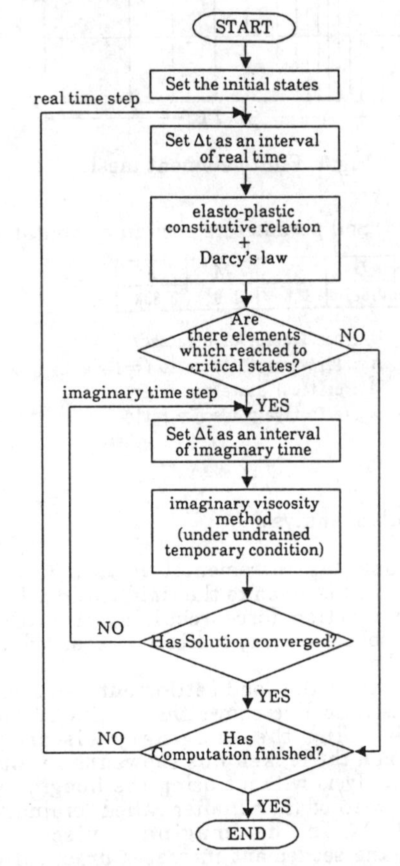

Fig.2 Computational algorithm

3 ANALYSIS OF BEARING CAPACITY

3.1 Analytical model

As an example of analysis, the punching of the rigid smooth footing under undrained condition on an uniform clay ground is computed. Fig.3 shows the finite element mesh. The initial stress is uniform in the whole area, assuming isotropically and normally consolidated conditions of $p'=p_0'=1.0\text{tf/m}^2$. Soil parameters are given in Table 1.

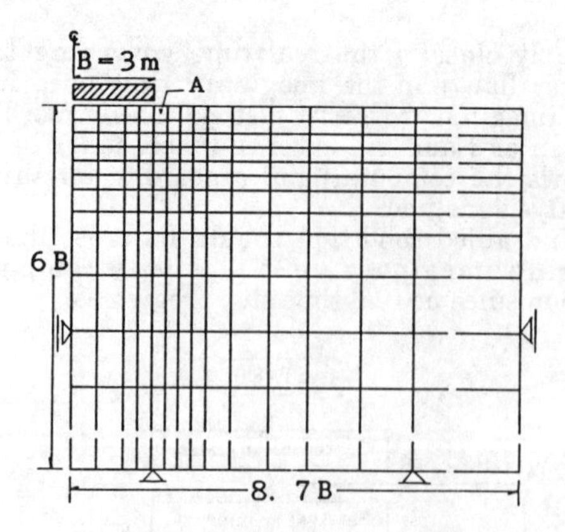

Fig.3 Finite element mesh

Table.1 Soil parameters used in computation

D	Λ	M	ν'
0.086	0.6	1.0	0.333

D : coefficient of dilatancy
Λ : irreversibility ratio ($=1-\kappa/\lambda$)
M : critical state parameter
ν': effective poisson's ratio

3.2 Result of analysis

In this analysis, incremental vertical forced displacement is given to the rigid footing. The vertical reaction force resulting from the above displacement is, therefore, regarded as load in Fig.4.

Fig.4 shows the load-settlement relationship. The solid line shows the results by the analysis using the imaginary viscosity method and the broken line shows the results by the analysis without using the imaginary viscosity method (hereinafter called "ordinary method") . In the imaginary viscosity method, the settlement increases drastically when the load comes to 1.94tf/m^2. On the other hand, the ordinary method gives the load much higher than the imaginary viscosity method and with increase in settlement. The load has a tendency to increase continuously. In Fig.4, the chain-dotted line shows Prandtl's solution (5.14*c*; where, *c* is the undrained shear strength of isotropically and normally consolidated clay sheared under plane strain condition obtained by Ohta, Nishihara and Morita (1985)).

Fig.5 shows effective stress paths in element A in Fig.4. The dotted line illustrates theoretical undrained effective stress paths obtained from Eq.(14). The imaginary viscosity method gives *p* and *q* remaining unchanged, when the stress reaches the critical state. On the other hand, the ordinary method presents that the stress diverging.

These results suggest that the present method is much more stable and better than the ordinary method.

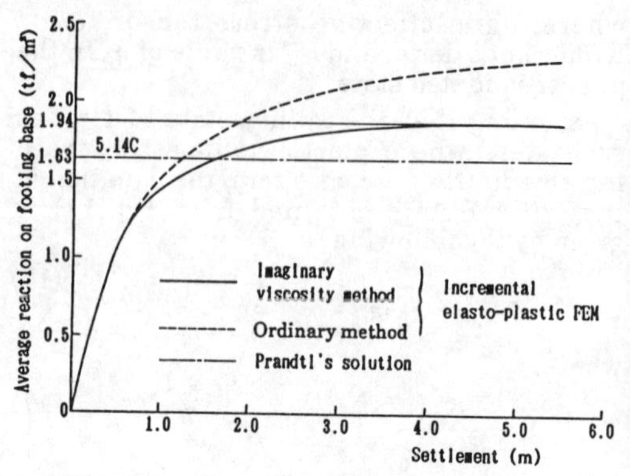

Fig.4 Load-settlement relationship

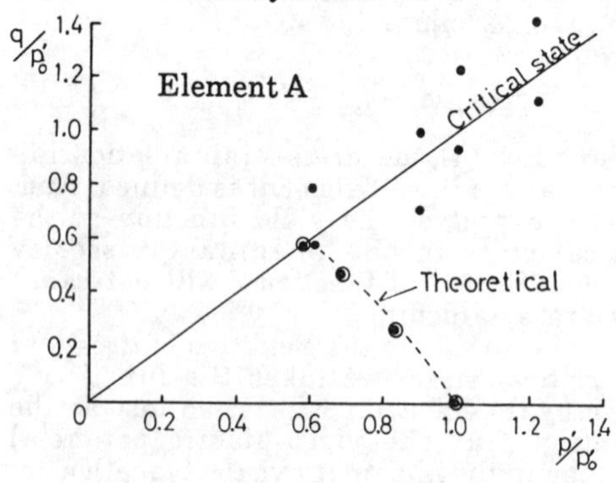

Fig.5 Comparison of stress path

4 ANALYSIS OF PORTSMOUTH TEST BANKING

4.1 Outline of Portsmouth test banking

Fig.6 shows the section of Portsmouth test banking [Ladd (1972)]. The ground consists of soft sensitive marine clay laid down to approx. 10m from the ground level, under which sand silt, clay layer (having such characteristic as marine clay of the upper layer), and Till with relatively high permeability are accumulated in 1m to 2m thick, respectively.

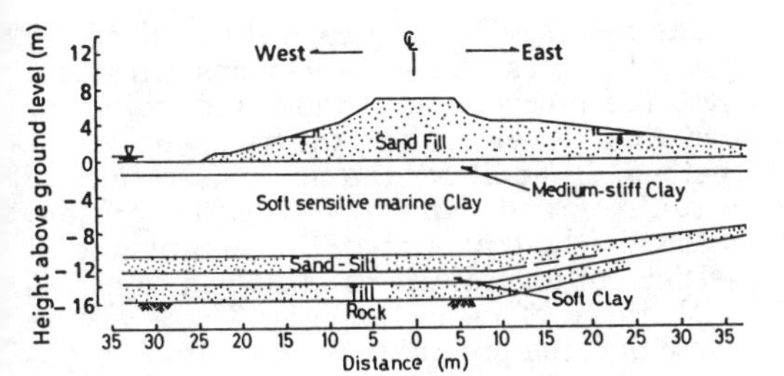

FIg.6 Section of Portsmouth test banking (after Ladd,1972)

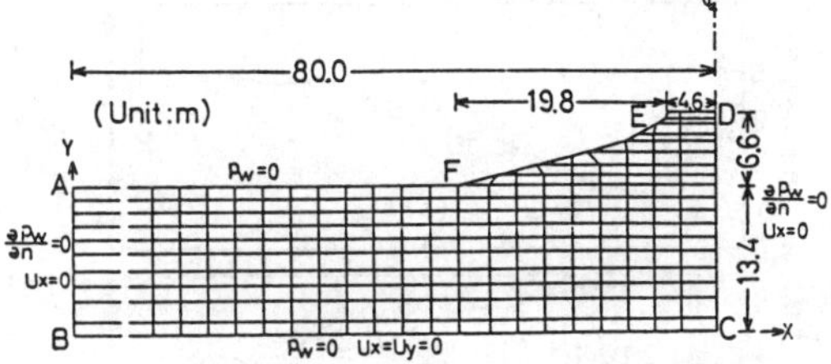

Fig.7 Finite element mesh and boundary conditions

4.2 Analytical model and soil parameters

As shown in Fig.7, the west side half where the banking has failed is analysed. This figure also shows the finite element mesh and the boundary conditions.

The foundation soil parameters are shown in Table.2.

Ladd (1972) has reported about the effective internal friction angle (ϕ'), the compression ratio (λ / $(1+e_0)$) and the swelling ratio (κ $(1+e_0)$) of the marine clay. Therefore, parameters M, D, Λ, are determined by ϕ', $\lambda/(1+e_0)$, $\kappa/(1+e_0)$. The coefficient of earth pressure at rest (K_0'), the coefficient of in-situ earth pressure at rest (K_i') and ν' are obtained from plasticity index (PI) following a chart given by Iizuka and Ohta (1987). The coefficient of permeability (k) of marine clay is also assumed from PI and k of sand silt is selected 10^{-5} cm/sec.

The banking materials are approximated as elastic for simplification, assuming that the Young's modulus (E') is 500tf/m^2 , ν' is 0.33 and k is 0.1cm/sec.

4.3 Results of analysis

Fig.8 shows the comparison of the measured values and calculated values of the settlement at the center of the banking. The solid line in Fig.8(a) shows actual banking construction process and the broken line shows its simplified banking process used in the analysis. The failure occurred actually when the banking height reached 6.6m.

The settlement-time relation is shown in Fig.8(b). Considering that input parameters are determined partly based on assumptions, it may be said that the calculated settlement and measured value agree well to each other.

Fig.9 shows the failure zone at point A, B and C shown in Fig.8(b).

First, the failure zone appears under the center of the banking and on the ground surface around the toe of slope (Fig.9 (a)).

With sharp increase in the settlement, the failure zone under the center of the banking starts expanding. (Fig.9(b)). At point C where the banking height reaches 7.2m in the computation, it is found that the failure zone at the center of the banking touches the failure zone around the toe of slope (Fig.9(c)). At this stage, a kind of sliding line is formed, along which the banking could be considered to fail. The broken line in Fig.9 (c) shows the measured failure arc. The calculated failure zone agrees well with it.

Table.2 Foundation soil properties

	Depth (m)	PI (%)	M	Λ	D	ν'	σ'vo (tf/m²)	K'o	σ'vi (tf/m²)	K'i	k (m/day)	E (tf/m²)
Medium Clay	0~1.15	24	0.81	0.93	0.10	0.35	13.8	0.54	0.51	2.33	0.0004	-
	1.15~2.3						13.8		1.54	1.43	0.00018	-
Soft Clay	2.3~3.53	18	0.81	0.93	0.10	0.34	7.26	0.52	2.50	0.83	0.00034	-
	3.53~4.76						5.80		3.41	0.66	0.00043	-
	4.76~6.0						6.92		4.33	0.64	0.00036	-
	6.0~7.6						8.07		5.38	0.62	0.00031	-
Soft Clay	7.6~8.95	15	0.81	0.93	0.10	0.33	9.23	0.50	6.59	0.58	0.0003	-
	8.95~10.3						10.2		7.83	0.56	0.00023	-
Sand-Silt	10.3~12.2	-	-	-	-	0.33	-	-	9.48	0.50	0.00086	450.0
Soft Clay	12.2~13.4	15	0.81	0.93	0.10	0.33	13.3	0.50	11.1	0.54	0.00021	-

in which the unit weight of sand-fill γ_t is 1.56 tf / m^3
Sand-Silt layer is elastic material.

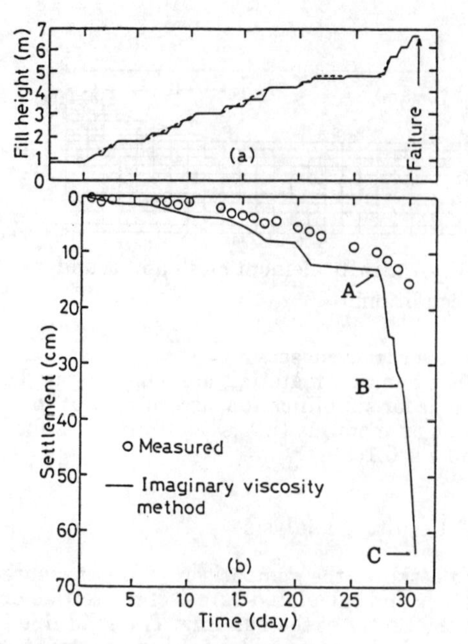

Fig.8 Banking height-settlement relationship

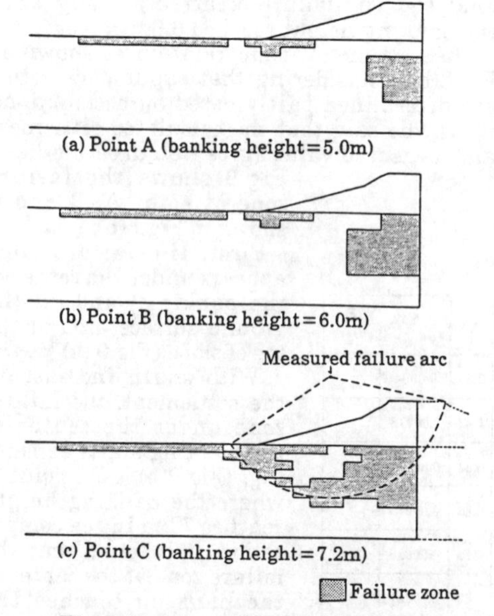

Fig.9 Progress of failure zone

5 CONCLUSION

In this paper, we proposed a coupled elasto-plastic analysis using an imaginary viscosity method.

The result of bearing capacity analysis by using the present method demonstrates that both the precision and stability of solutions are in a level higher than the ordinary method. In addition, the analysis results of actually failed banking using the present method illustrates that the magnitude of settlement and failure patterns had coincided with those of measurements. These results show that the present method makes it possible to simultaneously carry out both deformation and stability analyses for grounds with more realistic condition considering elasto-plastic characteristics and pore water characteristics including its partial draining effect.

REFERENCES

Zienkiewicz, O. C. and Cormeau, I. C. 1974 : Visco-plasticity-Plasticity and Creep in Elastic Solids-A Unified Numerical Solution Approach, International Journal for Numerical Methods in Engineering, Vol.8, No.4, pp.821~845

Kobayashi, M 1984 : Stability Analysis of Geotechnical Structures by Finite Elements, Report of the Port and Harbour Research Institute, Vol.23, No.1, pp.83~101, (in Japanese)

Christian, J. T. and Boehmer, J. W. 1970 : Plane Strain Consolidation by Finite Elments, JSMFD Proc. ASCE, Vol.96, SM4, pp.1435~1457

Sekiguchi, H. and Ohta, H. 1977 : Induced Anisotropy and Time Dependency in Clays, Constitutive equation of Soils, Proc. Speciality Session 9, 9th Int. Conf. Soil Mechanics & Foundation Engineering, Tokyo. pp.229~238

Iizuka, A. and Ohta, H. 1987 : A Determination Procedure of Input Parameters in Elasto-Viscoplastic Finite Element Analysis, Soils and Foundations, Vol.27, No.3, pp.71~87

Ohta, H. Nishihara, A. and Morita, Y. 1985 : Undrained Stability of Ko-consolidated Clays Proc. of 11th ICSMFE, Vol.2, pp.613 ~616

Ladd, C. C. 1972 : Test Embankment on Sensitive Clay, Proc. ASCE, Spec. Conf. on Performance of Earth and Earth-Supported Structures, Vol.1, pp.101~128

Numerical Methods in Geomechanics (Innsbruck 1988), Swoboda (ed.)
© 1988 Balkema, Rotterdam. ISBN 90 6191 809 X

Strength behavior of granular materials using discrete numerical modelling

John M.Ting & Brent T.Corkum
University of Toronto, Canada

ABSTRACT: A discrete particle model based on two dimensional disk-shaped elements is used to model granular soil behavior numerically. Simulations of standard geotechnical laboratory tests such as direct simple shear, triaxial compression and extension and one dimensional compression indicate that realistic nonlinear, stress-history dependent behavior can be obtained from the numerical model. However, use of circular-shaped particles allows particle rolling to dominate the observed behavior. Inhibiting this particle rotation results in strength, deformation and volume change behavior more appropriate for "real" soil.

1 INTRODUCTION

The Distinct (or Discrete) Element Method (DEM) models soil numerically as an assemblage of particles which can freely make and break contacts with its neighbors. The DEM can model static and dynamic, nonlinear, large deformation, bifurcation and local yield behavior in granular materials with relative ease. Since all normal and shear contact forces are monitored, average internal stresses can be measured inobstrusively. Sample reproducibility can be assured. External stresses and stress paths can be controlled precisely. As a result, this method is well suited for investigating the fundamental strength and deformation behavior of granular materials and validating constitutive relations for soil.

The DEM was pioneered in geomechanics by Cundall [3], who used it to analyze the stability of rock slopes, and was extended to soil using two dimensional disk elements [1,2,9,10,11] and three dimensional spheres [3]. In Cundall and Strack's [2] formulation, quasi-rigid disks interact with each other through deformable contacts. These contacts consist of a spring and viscous dashpot in the normal and tangential directions, with an upper limit on the tangential contact force before sliding occurs. Local force equilibrium is used to compute the acceleration, velocity and displacement of each particle in time utilizing a constant timestep, second order-accurate leapfrog integration algorithm which is generally regarded as the best overall for accuracy, stability and efficiency in many-body simulations [5]. Because of the enormous computational requirements of the DEM, most implementations have not progressed significantly past the two dimensional stage with simple disk shapes.

Appropriate values of the DEM parameters such as contact stiffnesses, dampings and strengths, and particle masses, moments of inertia and gradations must be chosen. Because the DEM is a particulate model, one cannot specify desired aggregate properties at the outset such as the Young's modulus E, Poisson ratio μ, Mohr-Coulomb cohesion intercept c and angle of internal friction ϕ. Instead, simulations of standard geotechnical laboratory tests need to be conducted for a given set of DEM properties in order to assess the aggregate strength and deformation behavior.

This Paper presents the results of simulated laboratory testing using a two dimensional disk-based implementation of the DEM using program "DISC", developed and coded by the authors using Cundall and Strack's [2] algorithm. These tests include one dimensional compression, direct simple shear (DSS) and triaxial compression and extension tests and indicate that two dimensional disk systems exhibit excessive rolling compared with real three dimensional soil systems, resulting in a decreased observed shear resistance.

2 RESULTS

All the tests described in this Paper were conducted on samples composed of 562 two dimensional disks of three different radii sizes in the ratio 1:1.5:2 numbering in the ratio 6:3:1. Actual radius magnitude was shown to not affect the overall normalized behavior. Gravitational acceleration was set to zero for each test. The normal contact stiffness k_n was varied between 100 MN/m/m (chosen for expediency) and 10,000 MN/m/m (the analytical value obtained for elastically-loaded granite cylinders).

Each sample was first randomly generated in a specified region to the required gradation, then formed by rapidly moving the walls inward until a partial soil skeleton developed, whereupon either stress-controlled or deformation-controlled walls were applied gradually. For the one dimensional compression tests (reported in Ting et al. [9]), the top wall was then lowered, raised, and re-lowered while the rigid side walls remained in place.

2.1 Direct Simple Shear Tests

Simulations of the direct simple shear test were conducted by rotating the side walls and translating the top wall at various levels of consolidation stress and with different soil properties. Because the conventional DSS test is conducted "fully drained" at constant volume, the dry DEM simulation should replicate the DSS exactly (except that the DEM is two dimensional). These test results, reported fully in [9], are summarized in Table 1.

In each test, it was observed that the $½(\sigma_1+\sigma_3)$ - $½(\sigma_1-\sigma_3)$ (p-q) stress paths eventually bend to the right, indicative of an increase in effective (intergranular) stress during shear. This suggests that while the tests are carried out at constant volume, there is a strongly dilatant tendency, with particles trying to override each other during shear. This tendency decreases with increasing confining stress, as one would expect.

It can be seen that there is an unusually low measured aggregate friction angle ϕ compared with the contact friction angle ϕ_μ (tests A,B). As well there is a lack of an apparent system cohesion intercept c during undrained loading even when contact cohesion C_s is present (tests C-G). Note, however, that even when effectively no contact slip occurs (test F, $C_s = 10^{28}$ MN/m), elastic shear contact deformations can still exist.

TABLE 1. Summary of Direct Simple Shear test results

Test Series	Model Parameters C_s (MN/m)	ϕ_μ (°)	I_o (kg·m^2)	Factor (I_o/I_o*)	Measured c (kPa)	ϕ (°)
A	0	0	$6.7 \cdot 10^4$	1	0	5
* B	0	25	$6.7 \cdot 10^4$	1	0	19
C	0.5	25	$6.7 \cdot 10^4$	1	0	23
D	1.25	25	$6.7 \cdot 10^4$	1	0	25
E	10	0	$6.7 \cdot 10^4$	1	0	31
F	10^{28}	0	$6.7 \cdot 10^4$	1	0	31
G	0.0025	25	$6.7 \cdot 10^{-9}$	1	0	26
H	0	25	$1 \cdot 10^5$	10	0	19
I	0	25	$1 \cdot 10^6$	10^2	0	19
J	0	25	$1 \cdot 10^8$	10^4	0	30
K	0	25	$1 \cdot 10^{10}$	10^6	0	39
L	0	25	$1 \cdot 10^{25}$	10^{21}	0	39
M	1.25	25	$1 \cdot 10^{25}$	10^{21}	250	39
* N	0	25	$6.7 \cdot 10^{-12}$	1	0	20
O	0	25	$1 \cdot 10^{-9}$	10^3	0	23
P	0	25	$1 \cdot 10^{-8}$	10^4	0	30
Q	0	25	$1 \cdot 10^{-7}$	10^5	0	38
R	0	25	$1 \cdot 10^{-6}$	10^6	0	42
S	0	25	$1 \cdot 10^{10}$	10^{22}	0	42
T	0	25	$1 \cdot 10^{10}$	10^{22}	0	49
U	0	25	$1 \cdot 10^{-7}$	10^5	0	40

k_n = 100 MN/m^2 for A-M; 1000 for N-S; 10000 for T; 300 for U
A-F, H-M large radii; N-U small radii
I_o = polar moment of inertia of largest particle

Based on these observations, it was hypothesized that rolling between particles was controlling the deformation and strength of the aggregate. This conjecture is reinforced by the high measured lateral stress ratio K_o (σ_3/σ_1) values in the one dimensional compression tests and the observation of low shear stresses in zones of high shear deformations in the triaxial tests described later. While DISC models the two dimensional disk system correctly, it appears that the circular particle shape does not inhibit rotation between particles to the same degree as "real" angular or subangular soil particles. Consequently, it was decided to investigate the effect of inhibiting the rolling of each particle. This was achieved by increasing the polar moment of inertia I_o of each disk above the normal value I_o* computed from geometry and mass density, effectively decreasing the rotational acceleration at each timestep.

These tests (H-M,O-U) are also summarized in Table 1, and indicate that the samples in which rotation is totally inhibited ($[I_o/I_o^*] > 10^{21}$) possess a far greater (and more realistic) measured aggregate angle of friction ϕ, as seen in Figure 1. An apparent cohesion intercept c now exists when contact cohesion C_s is input, as indicated in Figure 2. All tests had similar gradation, but tests A-F and G-M were conducted on large radius particles, test G had radii 1000 times smaller, and tests N-U had radii 10,000 times smaller than A. The p-q stress path for reduced rotation test series "Q", with $[I_o/I_o^*] = 10^5$, is plotted in Figure 3, together with the locus of pole locations for each test. Note that nearly coincident stress path and pole locus plots indicate that the maximum shear stress is being applied on the horizontal and vertical planes.

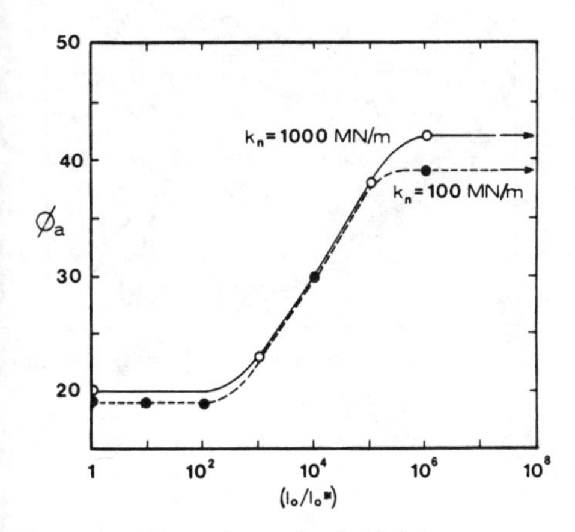

Figure 1. Effect of rotation inhibition on observed aggregate ϕ.

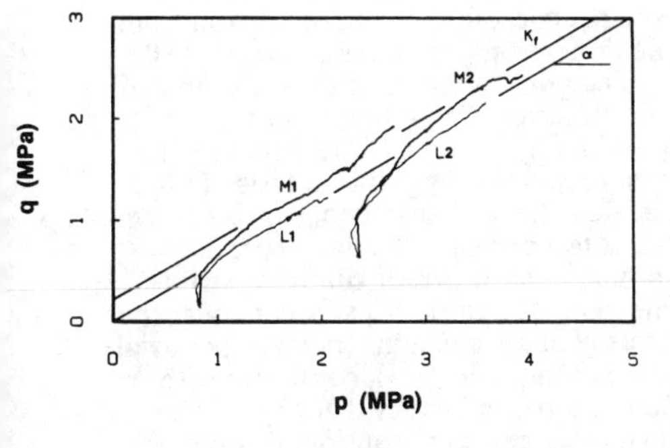

Figure 2. Effect of contact cohesion on p-q stress paths.

These results confirm that excessive rolling appears to occur in the two dimensional disk model when rotation is not inhibited. While this rolling is appropriate for an aggregate composed of two dimensional disks, it does not realistically reflect the behavior of a three dimensional assemblage of non-spherical particles. When the rolling is inhibited using an increased polar moment of inertia (preferably with $[I_o/I_o^*] \sim 10^5$), realistic soil behavior is achieved, with more appropriate aggregate friction angles and the presence of apparent aggregate cohesion when contact cohesion exists.

Another way to inhibit this rolling is to use non-circular particles, such as polygons. While this is algorithmically possible, it is not computationally feasible at present. Since realistic behavior can be obtained using disk particles with high I_o with no additional computational expense, it was decided not to investigate the use of polygonal elements at the present time.

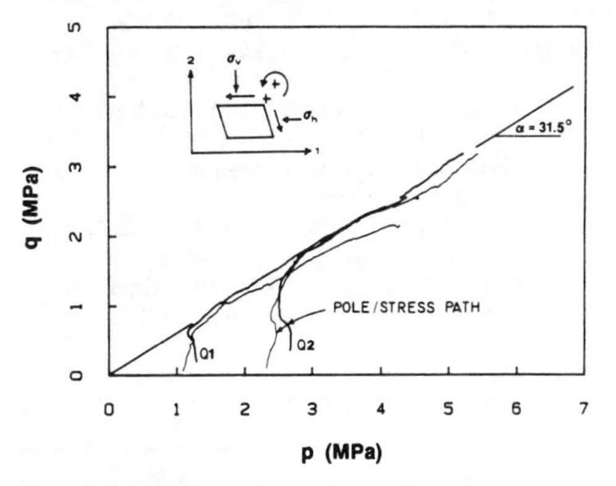

Figure 3. Stress paths and pole locations for DSS series Q, reduced rotation, $k_n = 1000$ MN/m/m.

2.2 Triaxial Tests

Triaxial (actually biaxial) tests were conducted using stress-controlled walls. A simple stress-controlled boundary was first implemented by varying the position of a rigid-geometry wall to achieve an appropriate total confining force. A more refined flexible-geometry wall apportions the total applied force along a wall to all the particles along the stress-controlled boundary, as if a membrane of finite flexibility were present. The triaxial tests are detailed in Ting et al. [9] and summarized in Table 2.

TABLE 2. Summary of 562 disk triaxial tests

Test No.	initial σ_3 (MPa)	failure $\frac{1}{2}(\sigma_1-\sigma_3)$ (MPa)	initial void ratio	final void ratio	E (MPa)
TR.05	0.057	0.040	0.269	0.256	1.7
TR.27	0.27	0.23	0.262	0.259	14
TR1.6	1.6	1.06	0.211	0.168	50
TR3.2	3.2	1.94	0.161	0.111	72
TR3.9	4.5	2.75	0.149	0.069	90
TR4.5	4.5	2.82	0.140	0.070	92
TF.05a	0.05	0.04±	0.263	0.259	2.3
TF.05b*	0.05	0.04±	0.256	0.256	2.5
TF.05c*	0.05	-----	0.256	-----	2.5
TF.05d**	0.05	0.045	0.225	0.241	3.1
TF.17**	0.17	0.13	0.225	0.225	7.8
TF3.2a	3.2	1.5	0.158	0.120	51
TF3.9	3.9	2.3	0.149	0.101	39
TFR.50	0.50	1.20	0.256	0.272	34
TFR1.1	1.11	1.71	0.250	0.258	63
TFR2.0	2.0	2.26	0.244	0.226	75
TFR.50U	0.50	0.20	0.256	0.256	22
TFR1.1U	1.11	0.43	0.250	0.250	38
TFR2.0U	2.0	0.75	0.244	0.244	72

TR rigid wall compression loading, k_n=100 MN/m^2
TF flexible wall compression loading, k_n=100 MN/m^2
TFR flexible wall compression loading, reduced rotation, k_n=300 MN/m^2
TFRxxU flexible wall extension unloading, reduced rotation, k_n=300 MN/m^2

ϕ_μ=25°, C_s=0 for all tests; * shaken once
E computed at 2% strain; ** shaken 5 times

Triaxial test series "TR" (rigid wall) and "TF" (flexible wall) were conducted with contact strengths $\phi_\mu=25°$ and $C_s=0$ and rotation not inhibited. The observed aggregate Mohr-Coulomb parameters for the triaxial and DSS series "B" tests are c=0 and $\phi=19°$. Figures 4 and 5 plot the normalized deviator stress $(\sigma_1-\sigma_3)/\sigma_3$ and volumetric strain for the "TF" series, respectively. Large fluctuations in the deviator stress, up to ±20% of peak, can be observed at low confining stress levels. This effect diminishes when rotation is inhibited, suggesting that it is caused by structural instability inherent with the rolling mechanism of deformation. Comparison of the deviator stress and volumetric strain plots indicates that these tests do not conform with Rowe's [6] stress-dilatancy theory, which in any case does not account for interparticle rolling [8].

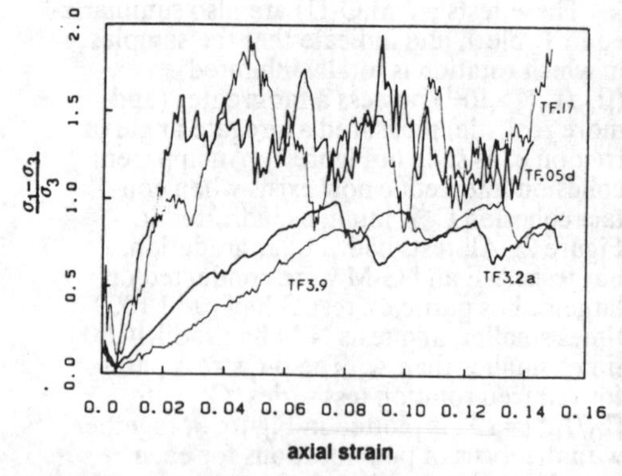

Figure 4. Normalized deviator stress for triaxial test series TF.

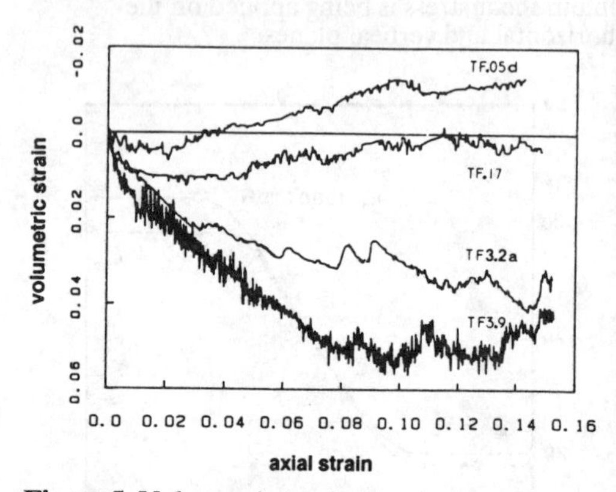

Figure 5. Volumetric strain for triaxial test series TF.

Flexible boundary triaxial tests "TFR" and "TFRxxU" had rotations inhibited using $[I_0/I_0^*]=10^5$. A contact stiffness $k_n=300$ MN/m/m and contact strengths $C_s=0$ and $\phi_\mu=25°$ were used, similar to DSS series "U". Each triaxial test in "TFR" and "TFRxxU" used K_0-consolidated samples taken from one dimensional compression test "1D4".

The normalized deviator stress and volumetric strain are plotted against axial strain in Figures 6 and 7, respectively, for compression loading triaxial tests "TFR". Classic volume change behavior is present in this test series, with dilatancy more prominent at the lower confining stresses. The peak deviatoric stress occurs near the location of peak volume increase per axial strain change $(\partial\epsilon_v/\partial\epsilon_1)$, consistent with Rowe's stress-dilatancy theory [6]. The deviator stresses are considerably less erratic in this test series than the previous test series "TF". The final disk locations for test TFR.50 are shown in Figure 8.

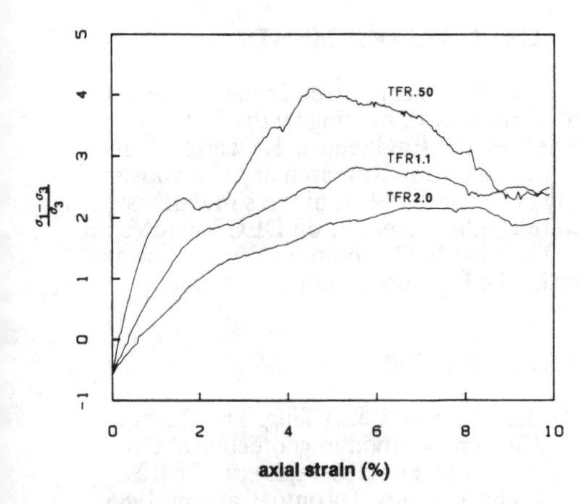

Figure 6. Normalized deviator stress for reduced rotation triaxial series TFR.

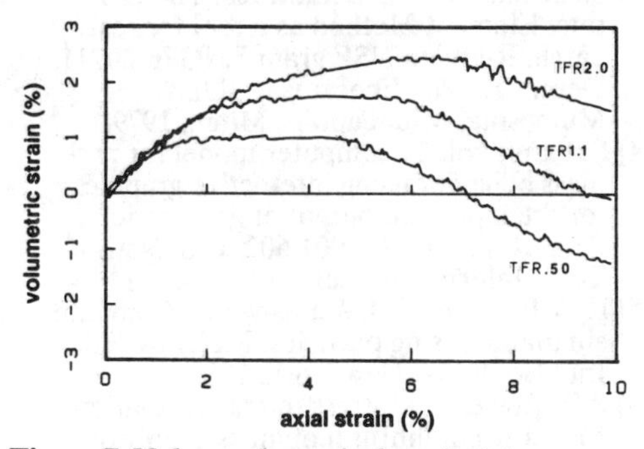

Figure 7. Volumetric strain for triaxial test series TFR.

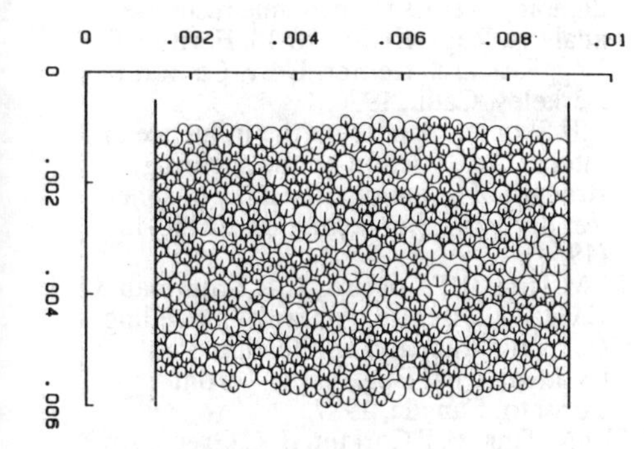

Figure 8. Final disk locations, test TFR.50. Rays initially vertical.

To evaluate the low confining stress triaxial test behavior of this material, extension unloading triaxial tests "TFRxxU" were conducted using the same samples as the "TFR" series, and were sheared by unloading the rigid end walls. The p-q stress paths for the triaxial and DSS tests are shown in Figure 9, based on average internal stresses ascertained using the statistical technique described in Cundall [3]. For the loading triaxial tests, the initial stress state is plotted with negative q, indicative of lateral normal stresses greater than axial normal stresses. The triaxial unloading tests are plotted in the positive (rather than the negative) q quadrant for convenience. This Figure shows that the strength results from the different tests are consistent with each other, as one linear Mohr-Coulomb yield envelope with $c=0$ and $\phi=40°$ ($\alpha=33°$) describes the peak behavior of all tests with mean stresses p at yield of 2 MPa or less. For mean stresses greater than 2 MPa at yield, the yield envelope appears to bend downward.

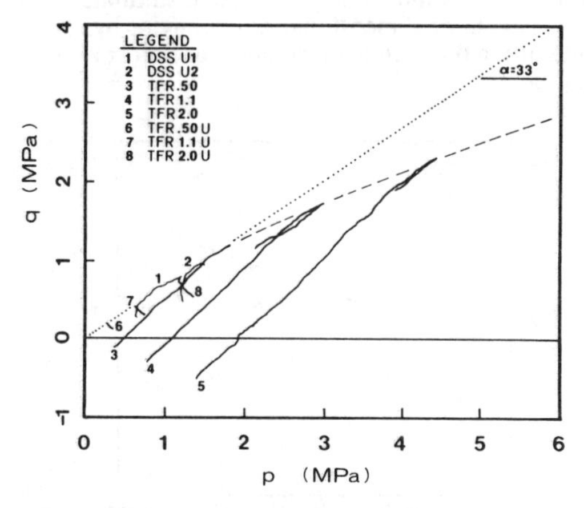

Figure 9. Summary of all reduced rotation strength tests, $k_n=300$ MN/m/m.

It is possible to compare the measured compressibilities from the different test types. Figures 10 and 11 plot the computed shear moduli G for each test for low stiffness particles ($k_n=100$ MN/m/m) with freely rolling particles and medium stiffness particles ($k_n=300$ MN/m/m) with rotation inhibited, respectively, together with correlations for round sand from Seed and Idriss [7], evaluated at 1% shear strain. These Figures indicate that the reduced rotation material exhibits compressibility values and stress-dependencies similar to "real" three dimensional sand, with reasonably similar estimates of the shear modulus obtained from different test methods. The slight variations in modulus values observed are not unexpected, owing to the different degree and nature of shearing in each case.

3 CONCLUSIONS

The Distinct Element Method has been shown to be a powerful numerical technique for modelling the mechanical behavior of granular materials. Based on DEM simulations of standard geotechnical laboratory tests such as triaxial, direct simple shear and one dimensional compression tests, appropriate and consistent stress-strain and strength behavior can be observed.

It has been shown that use of two dimensional disk-shaped particles in the DEM may overestimate the effect of individual particle rolling in an assemblage of non-spherical soil particles. This rolling leads to a lower and more erratic aggregate shear resistance, higher lateral stresses and lower system cohesion than would be expected for an equivalently graded subangular three dimensional soil system. However, the two dimensional disk model may be used to simulate "proper" three dimensional soil behavior by inhibiting this rolling mechanism of deformation.

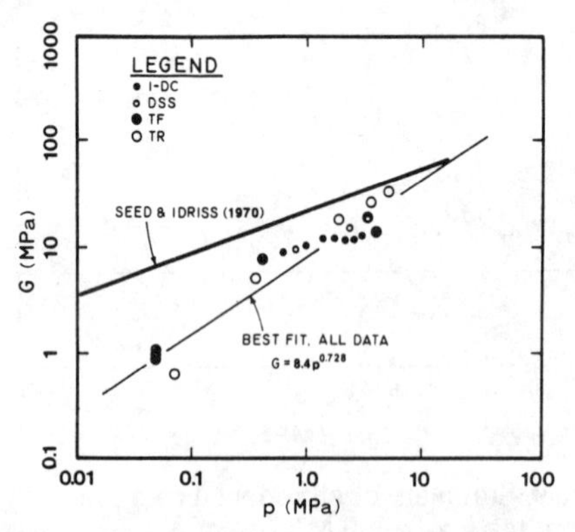

Figure 10. Shear modulus G as a function of mean confining stress, rotation not inhibited.

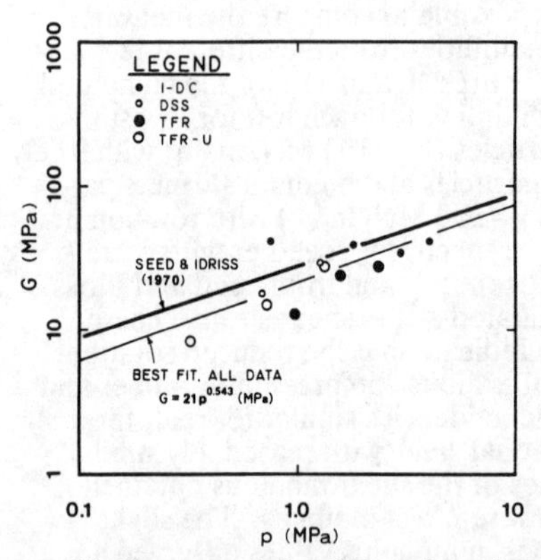

Figure 11. Shear modulus G as a function of mean confining stress, rotation inhibited.

4 ACKNOWLEDGEMENTS

This work was conducted at the University of Toronto with funding by the Natural Sciences and Engineering Research Council of Canada, Cray Research and the University of Toronto. Most of the simulations were conducted either on DEC MicroVax II or Cray X-MP/22 computers funded primarily by the Province of Ontario, Canada.

5 REFERENCES

[1] B.T.Corkum & J.M.Ting. The Discrete Element Method in geotechnical engineering. Publ. 86-11, Dept. Civil Eng., Univ. Toronto, Toronto, Canada, 1986.

[2] P.A.Cundall & O.D.L.Strack. A discrete numerical model for granular assemblies. Geotechnique 29:47-65 (1979).

[3] P.A.Cundall. & O.D.L.Strack. The Distinct Element Method as a tool for research. Rept. on NSF grant ENG76-20711, Dept. Civil & Mineral Eng., Univ. Minnesota, Minneapolis, Minn., 1979.

[4] P.A.Cundall. A computer model for rock-mass behavior using interactive graphics for the input and output of geometric data. Rept. AD/A-001 602, U.S. Natl. Tech. Information Service, 1974.

[5] R.W.Hockney & J.W.Eastwood. Computer simulation using particles. McGraw-Hill Int'l Book Co., New York, 1981.

[6] P.W.Rowe. The stress dilatancy relation for static equilibrium of an assembly of particles in contact. Proc. Royal Society of London A269: 500-527 (1962).

[7] H.B.Seed & I.M.Idriss. Soil moduli and damping factors for dynamic response analysis. Rept. EERC 70-10, Earthquake Eng. Research Center, Univ. California, Berkeley, Calif., 1970.

[8] A.E.Skinner. A note on the influence of interparticle friction on the shearing strength of a random assembly of spherical particles. Geotechnique 19:150-157 (1969).

[9] J.M.Ting, B.T.Corkum, C.R.Kauffman & C.Greco. Discrete numerical modelling of soil: validation and application. Publ. 87-03, Dept. Civil Eng., Univ. Toronto, Toronto, Canada, 1987.

[10] J.M.Ting, B.T.Corkum & C.Greco. Application of the Distinct Element Method in geotechnical engineering. Proc. 2nd Int'l Symp. Numerical Models in Geomechanics, Ghent, Belgium, 1986.

[11] Y.Zhang & P.A.Cundall. Numerical simulation of slow deformations. Proc. Symp. Mechanics of Particulate Media, 10th Nat'l Congress Applied Mechanics, Austin, Texas, 1986.

Numerical Methods in Geomechanics (Innsbruck 1988), Swoboda (ed.)
© 1988 Balkema, Rotterdam. ISBN 90 6191 809 X

A complete constitutive law for soil structure interfaces

M.Boulon, N.Hoteit & P.Marchina
Institut de Mécanique de Grenoble, France

ABSTRACT : The bases of the formulation of directionnaly-dependent three-dimensional interface law, including possible changes in the direction of loading and the effect of progressive grain crushing on large relative tangential displacements, are presented and discussed. Direct shear tests at constant normal sress and at constant volume were used as identifying tests. Numerical simulations of direct shear paths at imposed normal stiffness are carried out and compared to corresponding tests.

Practical applications of this model are numerous in civil engineering : estimation of the capacity of piles and anchorings, reinforcement of soft material embankments or rock faces using "nails", drilling and driving problems.

1. INTRODUCTION

Interface laws have been used for a long time in rock mechanics (|11|, |12|, |10|, |14|) but much more recently in soil mechanics (|8|, |9|, |3|) despite their obvious advantages in problems of contact between solids. In the case of piles, anchorings or nailing, strain localization visualised by Davis and Plumelle (|7|), Habib (|13|) but also using stereophotogrammetry on laboratory shear tests, leads to include in design calculations an interface behaviour as rheological model of the thin transition layer between the structure (concrete or steel) and the soil represented by an equivalent continuum. Directionnal dependance or the coupling between tangential and normal phenomena is a fundamental characteristics and yet ignored in commonly used interface laws due to its small influence for normal stresses of medium intensity and for grain crushing resistant granular materials (thus excluding calcareous sand).

The first part of this paper deals with the presentation and comparison of various formulations of directionaly dependent interface laws. The second part is concerned with the description of the direct shear paths at normal imposed stiffness which are the most relevant to soil-structure interaction problems. In the third part we descibe the laboratory equipment allowing shear tests along such paths. Finally we present some results obtained from tests and from numerical simulations.

2. POSSIBLE CHOICES FOR THE FORMULATION OF A DIRECTIONNALY DEPENDENT INTERFACE LAW

The interface behaviour which is a surface behaviour, actually represents that of a very thin layer. Besides the specific weight within this layer, the significant variables are the stress vector $\underline{\sigma}$ acting on the idealised surface and $\underline{u}$ the relative displacement vector between the two faces of this layer. Laboratory tests involving truly three dimensionnal $\underline{\sigma}$ and $\underline{u}$ vectors being very delicate to carry out we limited our study to the two dimensional case. We also assume the constitutive equation is directionaly dependent, i.e. a first degree homogenous function of the derivative of the loading ($\underline{u}$ or $\underline{\sigma}$) with respect to time).

Identification tests for an interface law always include plane direct shear tests where the tangential relative displacement w(t) (t is the time) is imposed. History of another variable or a relationship between two or more variables and the time may also be imposed.

The most commonly used laboratory tests that are readily exploitable are :

- direct shear tests at constant normal stress ($\dot{w}$ = ct, $\dot{\sigma}_n = 0$)
- direct shear tests at constant volume ($\dot{w}$ = ct, $\dot{u} = 0$)
- oedometric tests ($\dot{w} = 0$, $\dot{\sigma}_n$ = ct).

The so-called creep test ($\underline{\sigma}$ = ct) is sometimes used.

with :
- w tangential component of $\underline{u}$
- u normal component of $\underline{u}$
- τ tangential component of $\underline{\sigma}$
- σ_n normal component of $\underline{\sigma}$

Two choices are possible as to the definition of the loading and of the response.

- loading $\dot{\underline{\sigma}}$, response $\dot{\underline{u}}$

The advantage of this formulation is that the directions of loadings corresponding to different paths are clearly different before a rheological peak. But they may be identical past that peak (see Fig. 1).

Furthermore certain loadings may be impossible to represent, when a limit state has been reached for example.

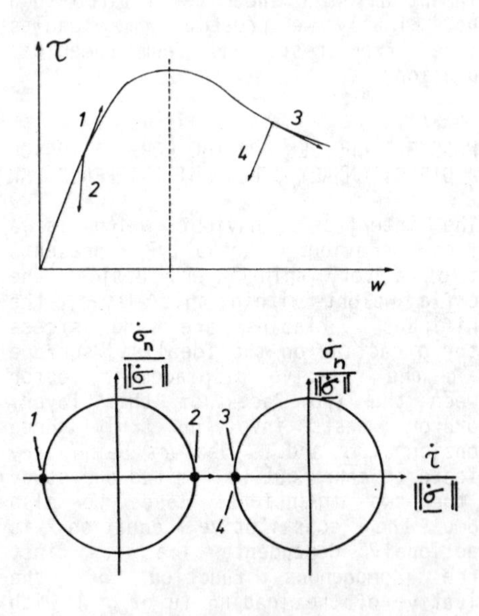

Fig 1 : Example of a path at constant normal stress

- loading $\dot{\underline{u}}$, response $\dot{\underline{\sigma}}$

This formulation overcomes the drawbacks of the first one but raises problems, the main one beeing the closeness of loading on some parts of the identification paths. (see Fig. 2) Nevertheless, this formulation was chosen to modelize the interface behaviour.

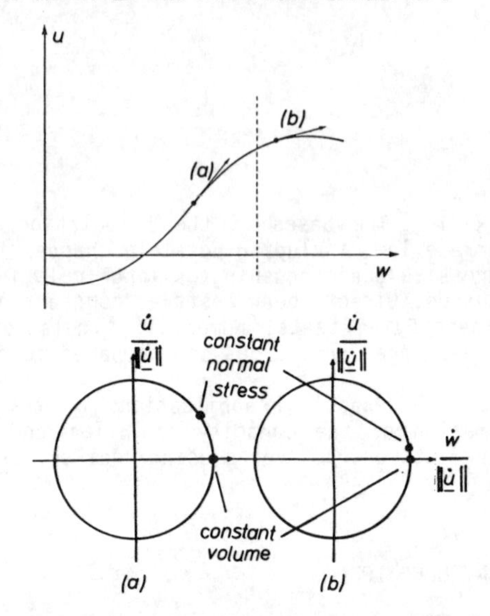

Fig. 2 : Example of directionnal closeness between constant normal stress and constant volume paths.

Paths different from the identification paths are generated by interpolation. The incidence of the proximity of two loadings is important since in this case the interpolation fonctions may take high values positive or negative, physically meaningless.

3. IMPOSED NORMAL STIFFNESS DIRECT SHEAR PATHS.

Among all the different direct shear paths, the imposed normal stiffness paths ($\frac{\Delta\sigma}{\Delta u}$ = cte) are of particular relevance to the modelling of rigid inclusions (metallic or composite concrete-metal) within a soil. Let us consider an inclusion, lateraly rigid, of radius Ro, submitted to axial loading, anchored in a soil of pressiometric modulus E_p. Let e be the soil-inclusion interface thickness (e $\ll$ Ro). The kinemanic and static soil-inclusion compatibility implies (assuming axial symmetry)

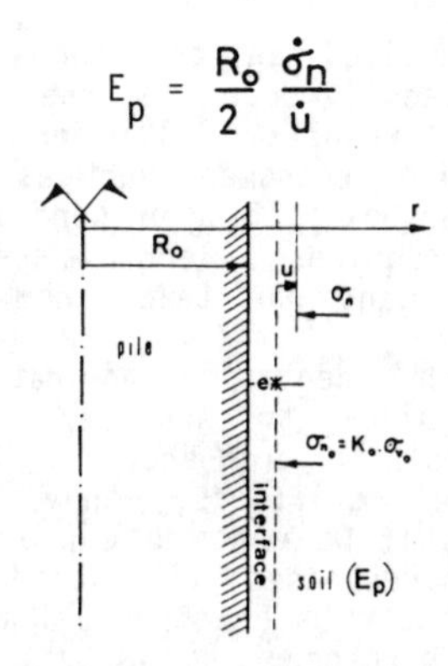

Fig. 3 : Simplified scheme of soil-structure interaction

Hence the expression of the lateral rigidity imposed by the surrounding soil to the soil inclusion interface :

$$k = \frac{\dot{\sigma}_n}{\dot{u}} = \frac{2\,E_p}{R_o}$$

We can see that this rigidity is a function of a mechanical parameter (Ep) but also of a geometrical one (Ro) which induces a scale effect (|2|), however somewhat concealed by soil hetereogenities and experimental inaccuracies. Nevertheless this simplified scheme clearly shows that axial loading of an inclusion induces variations in the normal stress, which overrules the use of constant normal stress shear tests as soil-inclusion friction characterization tests. These tests, fairly numerous in the litterature (|17|, |5|, |4|, |19|), can only give a soil structure angle of friction which is not sufficient in certain cases.

4. IMPOSED NORMAL STIFFNESS DIRECT SHEAR TESTS AND THEIR NUMERICAL SIMULATION.

4.1. Principle of the tests

Fig. 4 shows the principle of these tests whose importance has been reported (|3|, |1|). The weight in the conventional shear box apparatus is replaced by a system of adequate rigidity. The first tests were performed using springs (|18|). Current tests are carried out using a microcomputer driven regulation system. Similar tests have also been performed on rocks (|18|, |15|).

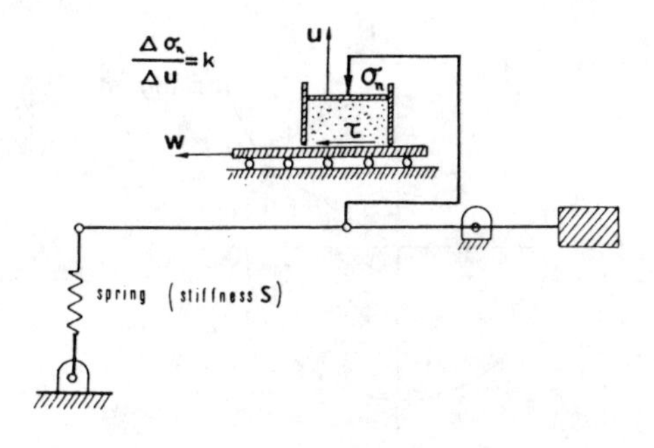

Fig. 4 : Principle of the shear test at imposed normal stiffness

4.2. The numerical simulation

The N identification paths are characterized by :

- The state of interface material $(\sigma_{n_i}, \tau_i, u_i)$.
- The loading $(\dot{w}_i, \dot{u}_i)$.
- The response $(\dot{\tau}_i, \dot{\sigma}_{ni})$.

(i = 1 to N)

We used 4 paths : loading and unloading at constant normal stress, loading and unloading at constant volume.

In order to compare paths between them we need to norm the space of loadings

$$\begin{cases} ||\dot{\underline{u}}_i|| = (\dot{w}_i^2 + \dot{u}_i^2)^{1/2} \\ \text{Normed loading} \quad \lambda_i = \frac{\dot{w}_i}{||\dot{\underline{u}}_i||} \quad \mu_i = \frac{\dot{u}_i}{||\dot{\underline{u}}_i||} \\ \text{Normed response} \quad \xi_i = \frac{\dot{\tau}_i}{||\dot{\underline{u}}_i||} \quad \eta_i = \frac{\dot{\sigma}_n}{||\dot{\underline{u}}_i||} \end{cases}$$

Any path can be interpolated from the N basic normed paths :

$$\begin{cases} ||\dot{\underline{u}}|| = (\dot{w}^2 + \dot{u}^2)^{1/2} \\ \text{Normed loading} \quad \lambda = \frac{\dot{w}}{||\dot{\underline{u}}||} \quad \mu = \frac{\dot{u}}{||\dot{\underline{u}}||} \\ \text{Normed response} \quad \xi = \sum_{i=1}^{N} W_i \xi_i \quad \eta = \sum_{i=1}^{N} W_i \eta_i \end{cases}$$

The interpolation functions must satisfy the following equations :

$$\begin{cases} \sum_{i=1}^{N} W_i = 1 \\ W_i\,(\Theta_j) = \delta_{ij} \end{cases}$$

where δ_{ij} is the Kroenecker symbol and Θ_j represents the trigonometric angle of loading n° j in the space of loadings $(\Theta_j \in [-\pi, \pi])$.

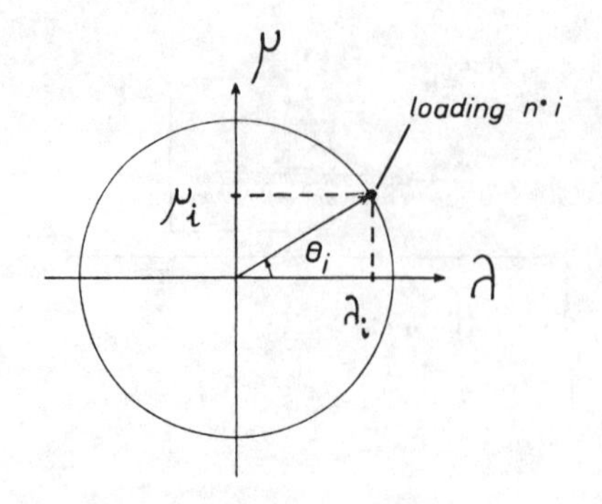

Fig. 5 : The space of loadings

The Wi functions are built using a method inspired from the spline functions interpolation. The $[-\pi, \pi]$ interval is divided into N $[\Theta_i, \Theta_{i+1}]$ intervals ($\Theta_1 = -\pi$ $\Theta_{N+1} = +\pi$).

On each of these intervals Wj is taken as a polynomial satisfying the boundary conditions $Wj(\Theta_i) = \delta_{ij}$. In order to ensure the derivability of Wj on $[-\pi, \pi]$ we must add conditions on the value of the derivatives of the polynomials at the interval boundaries which leads to polynomials of the third degree. We then norm each Wj dividing it by $\sum W_i$.

This type of functions which may not be the simplest to use has the great advantage of giving sensible results even with very small intervals (# 0.01 rad), as may occur (see Fig. 2).

4.3. Results

We used a quartz sand at a high density. The inclusion was made of a metal plate on which sand of the same nature had been glued. Tests were performed at both low and high initial normal stresses (see Figs. 6, 7).

In both cases two identification paths are shown (constant normal stress and constant volume paths) as well as low (5 000 kPa/mm) and high (40 000 kPa/mm) imposed normal rigidity tests. A stepwise numerical simulation is presented for imposed normal stiffness tests. The step of integration Δw is of 0.1 mm.

From Figs. 6 and 7 it can be seen that the simulation gives better results at high initial normal stress. At low initial normal stress, the peak angle of friction is generally underestimated and so is the normal stress.

In order to appraise the representativity of these simulations we need a statistical study of all the parameters involved in the formulation which can be readily done from the set of experimental results. However the quantification of phenomena such as grain crushing, shear box tilting or sand losses during shear requires a lengthy specific study if we want to take them into account.

If the four identification paths we have chosen allow to achieve a fair numerical simulation of shear tests at imposed normal stiffness, they prove unsatisfactory if we want to explore the whole space of responses. The introduction of oedometric tests - or other laboratory tests - is then needed.

Some problems remain to be solved to further improve the validity of this law such as its extension to very high and very low levels of initial stress since such extreme values may be encountered in the simulation of the behaviour of axially loaded piles for instance.

5. CONCLUSION

The results obtained show that the concept of angle of friction, although important, does not entirely characterize soil-structure friction when the normal stress acting on the structure is unknown beforehand, which is generally the case. It is also of obvious interest to be able to predict by a numerical simulation based on simple identification tests, the mobilizable friction between a given inclusion anchored in a given soil and this soil. The above presented law does allow such a prediction. However this law would require reliable truly three dimensional identification tests to get a wider field of applications. Also, problems related to the mode of installation of the inclusion (nails, in situ cast piles ...) have not been resolved yet neither that of the presence of water in the soil.

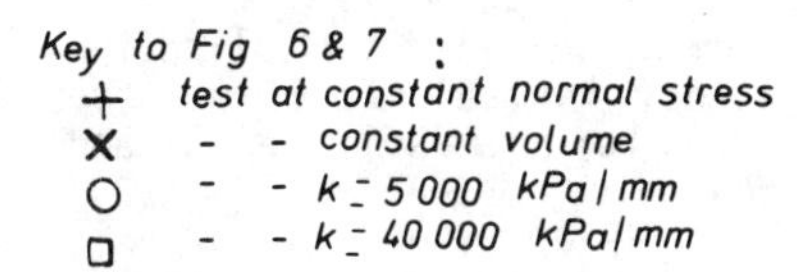

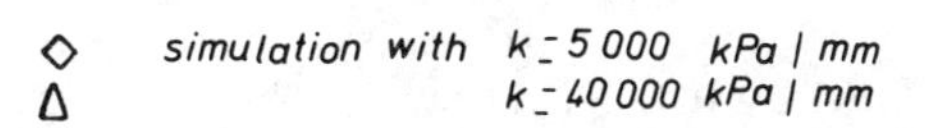

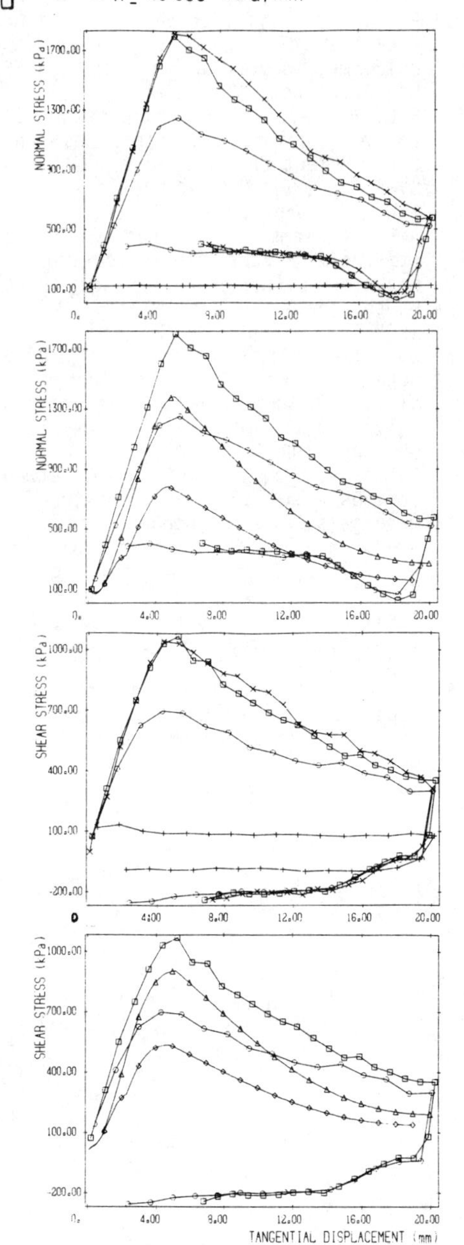

Fig. 6 : Direct shear tests along various paths between a quartz sand and a rough plate.
High density ; initial normal stress 124 kPa

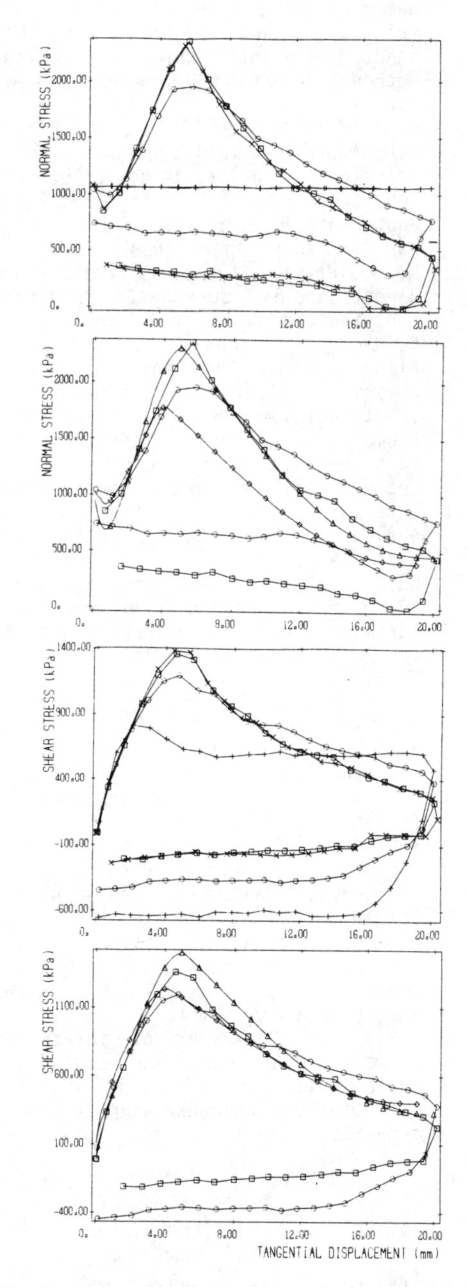

Fig. 7 : Direct shear tests along various paths between a quartz sand and a rough plate.
High density ; initial normal stress 1061 kPa

REFERENCES

|1| M. Boulon, P. Foray ; Physical and numerical simulation of lateral shaft friction along offshore piles in sand. IIIrd Int. Conf. on numerical methods in offshore piling, Nantes, France, 1986.

|2| M. Boulon, P. Le Tirant, P. Foray, J.F. Nauroy ; similitude and lateral friction of piles : scale effect, submitted to Int. Conf. on Geotechnical Centrifuge Modelling, Paris, France, April 1988.

|3| M. Boulon, C. Plytas, Soil-structure directionnally dependent interface constitutive equation ; application to the prediction of shaft friction along piles. 2nd int. Conf. on numerical models in Geomechanics, Ghent, Belgium, pp 43-54, March 1986.

|4| Brumunds, Leonards, Experimental study of static and dynamic friction between sand and typical construction material, Jnl. of testing and evaluation, vol 1, n° 2, pp 162-165, March 1975.

|5| Butterfield, Andrawes, on the angle of friction between sand and plane surfaces, Jnl. of Terramechanics, vol 8, n° 4, pp 15-23, 1972.

|6| W. Cichy, M. Boulon, J.Desrues, Etude expérimentale stéréophotogrammétrique des interfaces sol-fondation à la boîte de cisaillement direct, 4e colloque franco-polonais de mécanique des sols appliquée, Grenoble, France, Septembre 1987.

|7| A.G. Davis, C. Plumelle, Comportement des tirants d'ancrage dans un sable fin, Revue Française de Géotechnique, Vol 10, 1979.

|8| C.S. Desai, Numerical design analysis for piles in sand, Jnl. of the Geotech. Eng. Div. A.S.C.E., GT 6, 100, pp 613-635, 1974.

|9| C.S. Desai Behaviour of interfaces between structural and geological media, Int. Conf. on recent advances in Geotech. earthquake eng. and soil dynamics, Saint Louis, USA, 2, pp 619-638, 1981.

|10| J. Ghaboussi, E.L. Wilson, J. Isenberg, Finite elements for rock joints and interfaces, Jnl. of the soil mech. and found. eng. div., A.S.C.E. SM 10, 99, pp 833-848, 1973.

|11| R.E. Goodman, R.L. Taylor, T.L. Brekke, A model for the mechanics of jointed rock, Jnl. of the soil mech. and found. eng. div., A.S.C.E. SM 3, 94, pp 637-659, 1968.

|12| R.E. Goodman, J. Dubois, Duplication of dilatancy in analysis of jointed rocks, Jnl. of the soil mech. and found eng. div., A.S.C.E. SM 4, 98, pp 399-422, 1972.

|13| P. Habib Thermographie au cours d'essais cycliques en champ non uniforme, proc. of the XI ICSMFE, San Francisco, USA, 1987.

|14| F.E. Heuze, T.G. Barbour, New models for rock joints and interfaces, Jnl. of the Geotech. eng. div., A.S.C.E. GT 5, 108, pp 757-776, 1982.

|15| I.W. Johnston, T.S.K. Lam, A.F. Williams, constant normal stiffness direct shear testing for socketed pile design in weak rock, Geotechnique 37, n° 1, pp 83-89, 1987.

|16| W. Leichnitz, Mechanishe Eigen schaften von Felstrenn-Flächen in direkten Sherversuch, Dissertation Universität Fridericana in Karlsruke, Germany, 1981.

|17| Potyondy, skin friction between various soils and construction materials, Geotechnique, vol. 24, n° 3, pp 339-353, December 1981.

|18| M. Valin, Etude experimentale des interfaces sol-pieu, essais de cisaillement direct à rigidité imposée, Mémoire de DEA, Grenoble, 1985.

|19| E. Wernick, A "true direct shear apparatus" to measure soil parameters of shear bands, Design parameters in geotechnical eng. BGS, London, Vol 2, pp 175-182, 1979.

Numerical Methods in Geomechanics (Innsbruck 1988), Swoboda (ed.)
© 1988 Balkema, Rotterdam. ISBN 90 6191 809 X

Numerical investigations on discharging silos

J.Eibl & G.Rombach
Institut für Massivbau und Baustofftechnologie, University of Karlsruhe, FR Germany

Abstract: In this paper a constitutive viscoplastic law for cohesionsless granular materials is presented ranging steadily from regions with low strain rates resp. static conditions up to the free flow of material with high strain and shear rates. This law is used in conjunction with a nonlinear Finite Element Program to simulate discharging processes in silos.

1. INTRODUCTION - PROBLEM

As a demand for economical storage systems silo structures are growing larger and larger, primarely higher. Nowadays silos are built with a height of more than 80 meters. With growing dimensions it was stated that the commonly used analaytical theory of Janssen [1] for the calculation of wall and bottom loads is not satisfying and leading to unsafe design. World-wide great failures and heavy damages have occured during the last decades. Therefore great effort has been made to develop better theories but many of them are again insufficient as due to their underlying assumptions and their incompleteness they are restricted to simple geometries and static conditions while from experiments and measurements it is known that the highest pressures occur not under conditions at rest but at discharing.

Therefore the authors and co-workers have developed a more rigorous method for the calculation of velocity- and stress fields in silos [2] which is presented in the following. This theory takes into considertation a realistic nonlinear law for the bulk material, the interaction between bulk material and structure and the movement of the material.

2. BASIC EQUATIONS

Starting with the classical theory of continuum mechanics the motion of a solid is characterized by the following set of nonlinear equations:

equilibrium equ.:

$$\nabla T + \rho (b_v + \dot{v}) = 0 \qquad (1)$$

continuity equ.:

$$\partial\rho/\partial t + \nabla (\rho v) = 0 \qquad (2)$$

kinematic equ.:

$$d = 0.5 (v\nabla + \nabla v) \qquad (3a)$$

$$w = 0.5 (v\nabla - \nabla v) \qquad (3b)$$

constitutive equ.:

$$\overset{\circ}{T} = \overset{\circ}{T} (T, d, \overset{\circ}{d}, c_i \ldots) \qquad (4)$$

In these equations ρ is the bulk density, b_v the body force per unit volume, d the strain tensor, w the spin tensor and T the Cauchy stress tensor. Equation (4) stands for the constitutive law which is described in the following chapter and wich is the principal problem. Both geometric and physical nonlinearities are included in the setting of equation (1)-(4).

Due to the high velocities and great strains in the transient flow regimes of the material the calculations have to be done in a spatial fixed cartesian coordinate system, whereby the local derivative of the velocity vector is given as follows:

$$v = \partial v/\partial t + \nabla v \, \dot{v} \qquad (5)$$

This Eulerian frame of reference causes difficulties in so far as the history of all path dependent qualities like e.g. yield stresses or hardening parameters need to be traced throughout the whole space of geometry and time.

3. CONSTITIVE LAW - CALCULATION OF STRESSES

Besides the numerical difficulties the essential problem lies in the formulation of a constitutive law for the granular bulk material. In silos the filling material occupies an intermediate condition between solid and fluid. Under conditions at rest the bulk material may be regarded as solid whereas during discharging it acts like a fluid. This complicated behaviour can not be described by simple models like e.g. the commonly used Mohr Coulomb yielding criterium. Constitutive laws have to be formulated which are able to decribe both the static deformation at low strain rates as well as the free flowing conditions at high strain rates. For both of these regions seperatly several laws are already available but not for intermediate conditions. To circumvent these difficulties the well-know co-rotational stress rate due to Jaumann $\overset{\circ}{T}$ is divided into a rate independant static part $\overset{\circ}{T}_s$ and a rate dependent viscous part $\overset{\circ}{T}_v$.

$$\overset{\circ}{T} = \overset{\circ}{T}_s + \overset{\circ}{T}_v = H\ (T_s, d, c_i, ..)\ d + G\ (\ \mu, d, ...\ )\ \overset{\circ}{d} \qquad (6)$$

The so-called static part describes the behavior of the material mainly during at rest conditions whereas the second part dominates during discharging.

For the strain acceleration $\overset{\circ}{d}$ an objective rate similar to the Jaumann stress rate is introduced:

$$\overset{\circ}{d} = \dot{d} + d\ w - w\ d \qquad (7)$$

Regarding equation (6) - (7) the local stress rate in a spatial fixed Eulerian frame of reference can be calculated by:

$$\frac{\partial T}{\partial t} = H\ d + G\ \frac{\partial d}{\partial t} - (Tw - wT + \nabla T\ v) + G\ (dw - wd + \nabla d\ v) \qquad (8)$$

Herein the physical properties of the granular material are represented by the two fourth order tensors H and G.

The geometric nonlinearities result in the third and fourth part of the right hand side of equation (8). The time derivatives $\partial d/\partial t$and the spatial gradients ∇T are approximated by finite difference techniques. Furtheron in computing stresses it has to be considered that no tensile stresses are allowed and that the limit condition must be fullfilled.

Several constitutive laws for the calculation of the rate independent tensor H have been studied. Best results have been obtained by the viscoplastic law of Lade [3] and the rate type formulation of Kolymbas [4]. In order to check these different formulations several comparison calculations have been carried out whereby significant differences have been observed only when the material reaches the limit state condition resp. the zero stress level.

From the numerical point of view the mayor disadvantage of the elastoplastic formulation due to Lade comes from the great amount of computer time that has to be spent checking the yielding criterias and updating the material dependent parameters especially in a fixed Eulerian frame of reference.

For the calculation of free-flowing marial a viscous part $\overset{\circ}{T}_v$ is introduced for which despite of its practical importance for granular flow, no general mechanical theory is available. Most so far developed material models are of a very theoretical nature and their results differ quite significantly. Therefore a new practical approach first proposed by Buggisch [5] is used for modelling free flowing material. This model is based on equilibrium conditions on a microscopic level analogous to the kinematic gas theory. The formulation results in a rate dependent tensor G which is similar to the deviatoric part of an incompressible Newtonian fluid whereby the viscosity depends on the trace of the strain rate.

$$\overset{\circ}{T}_v = G\ \overset{\circ}{d}$$

$$G_{ijs} = \sqrt{tr\ d^2}\ \ (\delta_{ir}\delta_{js} - 1/3\ \delta_{ij}\delta_{rs}) \qquad (9)$$

The viscosity parameter is gained from experiments and comparison calculations and is assumed to be 1.0 KN sec/m².

4. NUMERICAL SOLUTION METHOD

Equation (1) - (9) results in a set of 19 nonlinear equations which are solved by the Finite Element Method in terms of the velocity as the primary variable.

Time discretisation is done by finite difference techniques resp. the Euler Forward Method. The resulting nonlinear equations are solved by the modified Newton Raphson scheme.

The bulk material is discretized by constant strain triangle elements for plane strain conditions as well as by isoparametric elements for two and three dimensional calculations. In case of cylindrical geometries a thin shell element is used for the silo walls.

The solid elements and the shell elements are coupled by a compatible thin layer 6-node isoparametric contact element (Fig. 1).

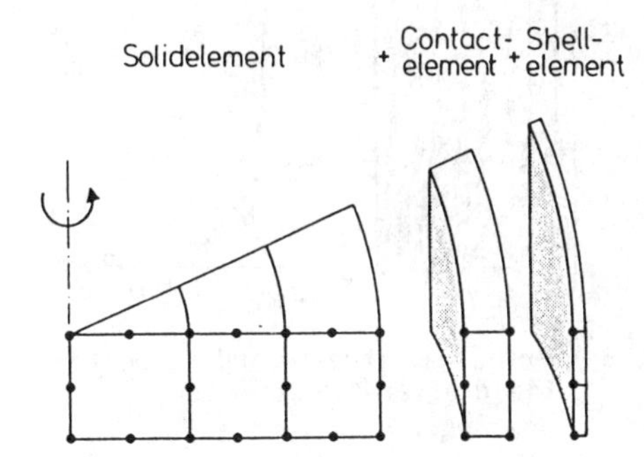

Fig. 1 Isoparametric solid-, contact and shell element

According to the Coulomb friction law which is used to model the interface behavior only normal and shear stresses are transfered to the wall. The friction angle is assumed to be constant. Besides debonding deformations normal to the wall are excluded by this element.

5. RESULTS

With the aforementioned theory and the developed Finite Element Program serval quantities like the influence of silo geometrie especially the location of the openings, the interaction between the filling material and the walls and the influence of bulk material parameters can be studied systematically.

Only one example a axissymmetric silo bin is presented. The filling material is dry sand. The rate type formulation due to Kolymbas is used for the calculation of the static material tensor H.

The silo geometry and the parameters for the bulk material are given in figure 2. Due to the symmetry only one half of the silo is discretized.

To simulate filling of the bin the silo is incrementally loaded by the self weight of the bulk material. Due to the low strain rates viscous and mass terms can be omitted. All bottom nodes are fixed in vertical direction.

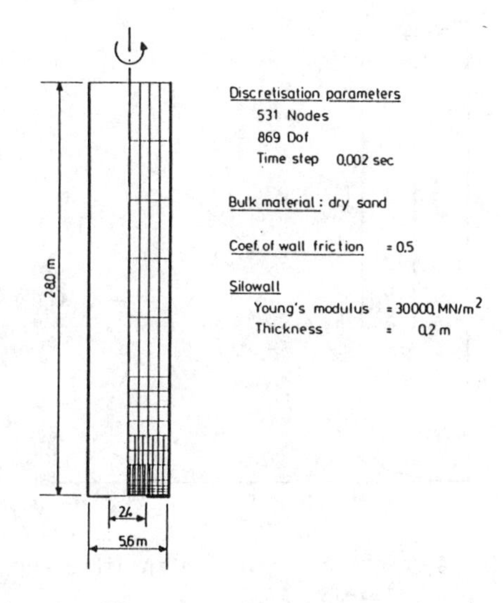

Fig. 2 Element mesh and discretisation parameters

The principal stress field and the wall pressure distribution are shown in Fig. 3 As the material being at rest the larger principal stresses act in vertical direction.

The discharing phase starts at opening of the bin. This is done by changing the kinematic restrictions of all nodes at the outlet region.

As a result the vertical stresses decrease to zero at a free outflow of the bulk material. A zone of low stresses but high velocities expands gradually into the bin until a stable condition is reached.

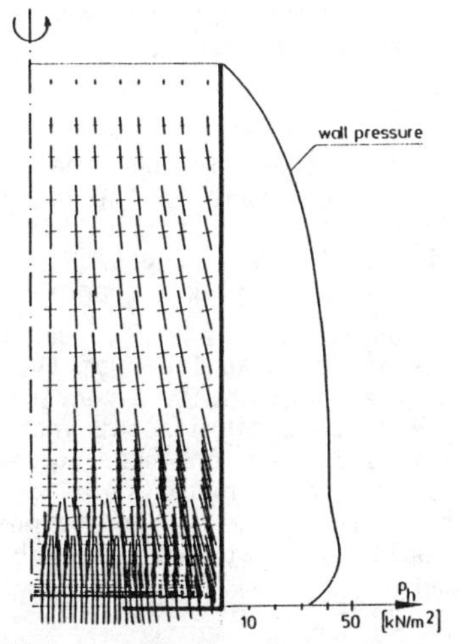

Fig. 3 Principal stress field after filling

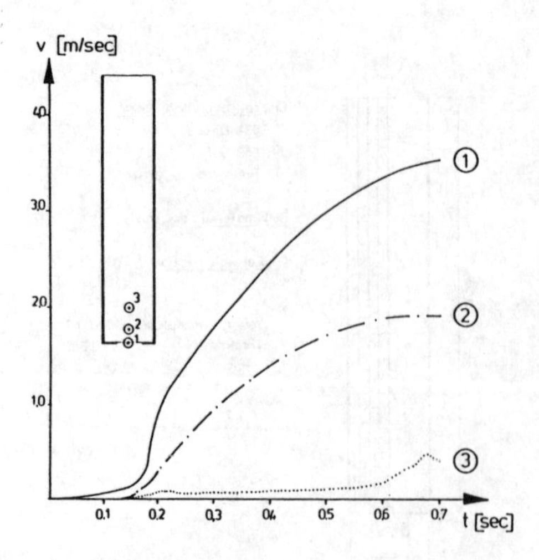

Fig. 4 Vertical velocity in different levels

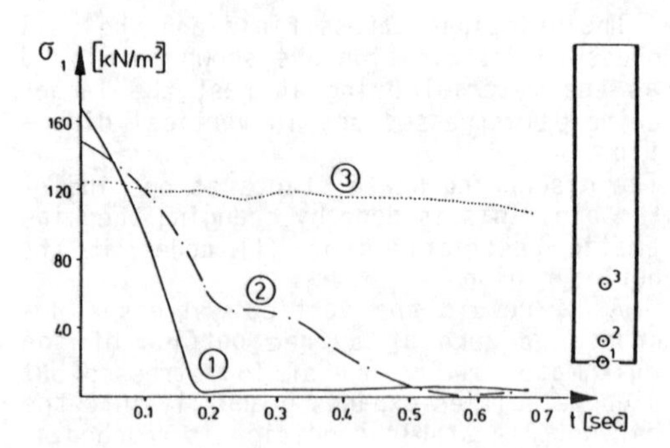

Fig. 5 Max. principal stresses in different levels.

In contrast to the flow zone the stresses remain nearly constant in the upper part of the bin. The material in this region is not in limit state condition as it is assumed by most analytical theories.

In the transition zone the larger principal stresses are changing from the vertical to the horizontal direction. This change leads to a significant increase in wall pressure. For engineering practice the resulting patch loads are of great importance as they lead to bending moments in bin walls so that eventually buckling may occur.

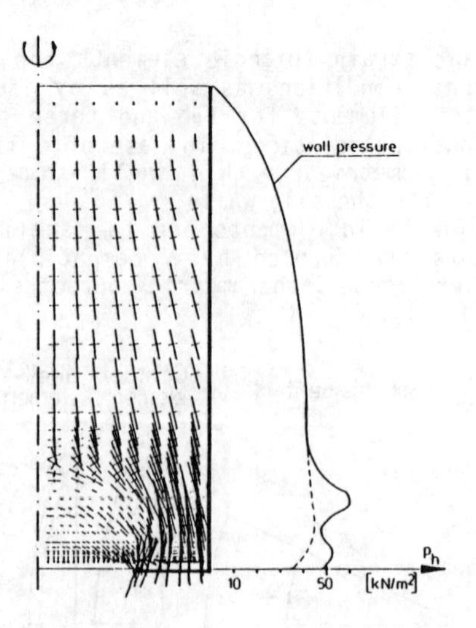

Fig. 6 Principal stress- and velocity field after t = 0.68 sec

6. CONCLUSIONS

A consistent silo theory has been developed which does not neglect significant influences with regard to simplifications at the computing shape.
It will help to understand the behavior of outflowing bulk material in silos better than in the past. Its aim is not to reject the use of simple design formulae which should be developed at the end of the ongoing investigations.

REFERENCES

[1] Janssen, H. A.
"Versuche über Getreidedruck in Silozellen",
VDI-Zeitschrift, 1985

[2] Eibl, J. et al
"Zur Frage des Silodruckes",
Beton und Stahlbetonbau 4, 1982

[3] Lade, P. V.
"Elasto-Plastic Stress Strain Theory for Cohe sionsless Soils with Curved Yield Surfaces",
Int. J. of Solids and Structures 13, 1977

[4] Kolymbas, D.
"A Constitutive Law of Rate Type for Soils and other Granular Materials",
Num. Methods in Geomech., Kosice 1987

[5] Buggisch, W.
"Rheologie of Rapid Flowing Granular Powders",
Int. Fine Part. Research Inst, Brüssel 1986

Numerical Methods in Geomechanics (Innsbruck 1988), Swoboda (ed.)
© 1988 Balkema, Rotterdam. ISBN 90 6191 809 X

A generalization of Hvorslev's equivalent stress

D.Kolymbas & M.Topolnicki
Institute for Soil and Rock Mechanics, Karlsruhe University, FR Germany

ABSTRACT: A tensorial memory parameter is introduced into a constitutive equation in order to describe anisotropic cohesion. Plane strain tests with clay in a kinematically controlled biaxial apparatus are used in order to calibrate and corroborate this model.

1 THEORETICAL CONSIDERATIONS

The most outstanding feature of clay is its capability to memorize previous deformations. This property is expressed by the fact that preloaded clay behaves differently than a "virgin" one, the main difference being the fact that preloaded clay possess cohesion. It is generally accepted that cohesion originates from the physicochemical interaction between the clay particles. This interaction depends strongly on the mutual placement of the particles and is only perceptible if they are very close to each other. Many efforts have been undertaken to derive the macroscopic mechanical properties from microscopic considerations. However, the results are rather meagre and the authors believe that — for the time being — the history dependent behaviour of clay should be investigated only in terms of macroscopic variables. Obviously, the various strain measures are macroscopic parameters that express the mutual placement of the individual clay particles. Among them the volumetric strain or, equally, the density seems to be the most important one. In fact, extensive experimental investigations with clay can be summarized by the statement that the shear strength of clay does not depend in a unique way on the actual normal stress. Instead, a unique relation between the actual density and the actual strength can be established for a limited variety of test types. Thus, the density, which in its part depends on the normal stress history, seems to carry the loading memory that is relevant to the actual mechanical behaviour. Note that in clay mechanics the density is usually expressed by the void ratio e. For saturated clays, the water content w serves also as an appropriate density measure. For several engineering applications it seems convenient to express the density in terms of stress. For that reason, HVORSLEV introduced the so-called equivalent stress which is nothing but a stress-like parameter, which expresses the density of the material. The equivalent stress is defined as the stress on the virgin consolidation line which corresponds to the actual density irrespective of wether the actual state is virgin or overconsolidated (s. Fig. 1). Thus the equivalent stress is a general sort of fictitious stress, i.e. it needs not to prevail in reality and it does not enter the equilibrium equations.

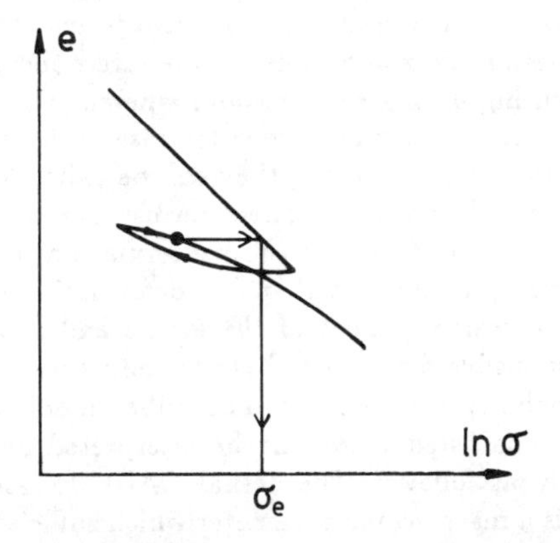

Fig. 1: Compression diagram for clay. Definition of equivalent stress.

It should be noted that HVORSLEV's equivalent stress is a scalar quantity and, thus, it expresses the loading history merely in an isotropic way. In other words, it neglects all sorts of anisotropy which might have been imposed by the previous loading. Of course, it is in fact reasonable to consider as a first step the *isotropic* part of the material memory. However, the next step should be the investigation of the anisotropy, as it is felt that cohesion, being a strength property, is in general anisotropic and cannot be sufficiently expressed by the scalar quantity c.

Now let us consider, how these preliminary remarks can be incorporated into a realistic constitutive equation. Denoting the actual stress by $\mathbf{T}$, the stretching by $\mathbf{D}$ and the co-rotated (or JAUMANN) stress rate by $\mathring{\mathbf{T}}$ (i.e. $\mathring{\mathbf{T}} = \dot{\mathbf{T}} - \mathbf{WT} + \mathbf{TW}$, where $\mathbf{W}$ is the spin tensor) KOLYMBAS [2] has shown that a simple generalized-hypoelastic constitutive equation of the form

$$\mathring{\mathbf{T}} = \mathbf{h}(\mathbf{T}, \mathbf{D}) \tag{1}$$

is capable of describing many aspects of soil behaviour, such as limit condition, dilatancy, irreversibility, stress-dependency of stiffnesses. In order to do this, terms which are non-linear in $\mathbf{D}$ have to be introduced, a fact which deviates from the originally introduced notion of hypoelasticity. A particular equation of the above form proposed for sand reads:

$$\begin{aligned}\mathring{\mathbf{T}} &= C_1 \frac{1}{2}(\mathbf{TD} + \mathbf{DT}) + C_2 \operatorname{tr}(\mathbf{TD})\,\mathbf{1} \\ &+ C_3 \mathbf{T}\sqrt{\operatorname{tr}\mathbf{D}^2} + C_4 \frac{\mathbf{T}^2}{\operatorname{tr}\mathbf{T}} \cdot \sqrt{\operatorname{tr}\mathbf{D}^2} \,.\end{aligned} \tag{2}$$

C_1 to C_4 are 4 material constants describing a particular material. The first two terms of this equation are responsible for the stress increase occuring during deformation, whereas the last two terms count for stress decrease with deformation. Consequently, they can be called 'constructive' and 'destructive' terms respectively. The action of the destructive terms can be attributed to the fact that any deformation causes a rearrangement of the grains and is thus responsible for a partial stress reduction. The capability of equation 2 to describe the soil phenomena listed above can be interpreted physically as follows: The actual CAUCHY stress $\mathbf{T}$ is a macroscopic parameter which sufficiently expresses the material behaviour as this results from the mutual grain position, the contact forces distribution etc.. On the other hand, this equation cannot describe cohesion and some other history dependent effects, since it contains only the *actual* load $\mathbf{T}$. Therefore, the list of parameters in equation 1 has to be enlarged in order to contain a futher tensorial parameter $\mathbf{S}$. $\mathbf{S}$ expresses the internal (probably electrochemical) attraction between the grains and is a stress-like memory parameter, which depends on the loading history. This may be expressed by an appropriate evolution equation expressing the rate of $\mathbf{S}$ as a tensor valued function of $\mathbf{D}$, $\mathbf{T}$ and $\mathbf{S}$:

$$\mathring{\mathbf{S}} = \mathbf{g}(\mathbf{T}, \mathbf{S}, \mathbf{D}) \ . \tag{3}$$

$\mathbf{S}$ is not the real stress and does not enter the equilibrium equation. It is a fictitious stress and can also be called an 'internal stress' or a 'back stress' etc. By association with HVORSLEV's equivalent stress it can also be called the *equivalent stress tensor.*

More precisely, the evolution equation 3 is proposed as follows:

$$\mathring{\mathbf{S}} = \mu_1 \mathbf{T}\operatorname{tr}\mathbf{D} + \mu_2 \frac{1}{2}(\mathbf{TD} + \mathbf{DT}) \ . \tag{4}$$

The first term of equ. 4 is the straightforward generalization of HVORSLEV's equivalent stress, since $\operatorname{tr}\mathbf{D}$ is the volume change rate. However, due to the proportionality to $\mathbf{T}$, this term is not isotropic. According to this term, $\mathbf{S}$ is affected only by volume changes. As it seems reasonable that also isochoric motions affect $\mathbf{S}$, the second term, found by trial and error, has been introduced. μ_1 and μ_2 are material constants.

The action of $\mathbf{S}$ is manifested by the fact that $\mathbf{S}$ appears in the constitutive equation of the considered material (e.g. clay) and influences thus its mechanical behaviour. $\mathbf{S}$ can be built in into the constitutive equation 2 as follows:

$$\begin{aligned}\mathring{\mathbf{T}} &= \frac{C_1}{2}\left[(\mathbf{T} + \mathbf{S})\mathbf{D} + \mathbf{D}(\mathbf{T} + \mathbf{S})\right] \\ &+ C_2 \cdot \operatorname{tr}\left[(\mathbf{T} + \mathbf{S})\mathbf{D}\right] \mathbf{1} + \\ &+ C_3 (\mathbf{T} - \mathbf{S}) \cdot \sqrt{\operatorname{tr}\mathbf{D}^2} \\ &+ C_4 \frac{(\mathbf{T} - \mathbf{S})^2}{\operatorname{tr}(\mathbf{T} + \mathbf{S})} \cdot \sqrt{\operatorname{tr}\mathbf{D}^2} \,.\end{aligned} \tag{5}$$

This equation takes into account that the action of the 'constructive' terms is enhanced by the internal stress $\mathbf{S}$, whereas the action of the 'destructive' terms is reduced. Consequently, $\mathbf{S}$

has been added to **T** in the first two terms and subtracted in the last two. The normalization occuring in the forth term has been accomplished by tr(**T**+**S**).

The determination of the material constants $C_1, C_2, C_3, C_4, \mu_1, \mu_2$, is rather complicated, as **S** is a non-observable internal variable which cannot be directly measured. In addition, **S** appears here via the differential equation 4. Thus, the material constants can only be obtained by a lengthy procedure which is briefly outlined in the third part of the present paper. Here, the material constants are taken as follows:

$$C_1 = -15.84 \quad C_2 = -8.71$$
$$C_3 = 28.77 \quad C_4 = -147.66$$
$$\mu_1 = -1.0 \quad \mu_2 = 0.05 \ .$$

The introduction of **S** makes possible a more detailed reference to the soil history, than this is the case by the mere statement that a particular soil is 'normally consolidated' or 'overconsolidated'. Reference to the so-called OverConsolidatio Ratio (OCR) makes use of the inequality $\sigma_p > \sigma_a$, where σ_p and σ_a are the previous and the actual stresses respectively. As it takes into account only scalar stress components (which are not further specified) it serves only as a rough characterization of the loading history and is insufficient for several detailed considerations. As soon as the full **S**-tensor is specified, any reference to the OCR is superfluous.

The system of differential equations given by equations 4 and 5 may be incorporated to any numerical or analytical procedure intended to solve particular initial-boundary-value problems. If the considered deformation process is an element test (i.e. a test with homogeneous sample deformation) the integration is very simple, since a spatially variable field needs not be calculated (by means of, say, finite elements). Only a time integration is needed. This can be accomplished by any usual procedure such as EULER or RUNGE-KUTTA integration.

It is clear to the reader that the present constitutive theory departs completely from elastoplasticity: It does not decompose the strain in elastic and plastic parts nor does it use any sort of yield or subyield or bounding surfaces. It appears, therefore, to be more straightforward and simpler than the elastoplastic equations. Furthermore, it provides realistic results referring to anisotropic consolidation. This cannot be checked by the usual laboratory equipment, which allows either oedometric or isotropic consolidation. These two sorts of consolidation are rather similar and, therefore, they do not produce markedly different loading histories.

2 EXPERIMENTAL INVESTIGATIONS

In order to overcome the limitation stated above, tests executed with a HAMBLY type plain strain device (in the sequel called 'biaxial apparatus') have been executed. This apparatus (s. Fig. 2) enables to control the strains independently in 1- and 2- directions, respectively. Thus, a large variety of plain consolidation paths can be applied.

TOPOLNICKI has conducted a series of tests where the direction of the consolidation paths has been systematically varied. The consolidation paths have been carried out as proportional strain paths, i.e. the stretching **D** remained constant during this part of loading (in other words: the ratios of the strain rates remained constant), and the volume of the sample decreased continuously. In usual laboratory practice only isotropic, i.e. $\mathbf{D} = -\delta_{ij}$, or oedometric consolidations, i.e. $(\mathbf{D})_{11} = \mathbf{-1}$, else $(\mathbf{D})_{ij} = \mathbf{0}$, are possible. Subsequently to the consolidations identical isochoric deformations up to the limit (failure) state have been applied. It has been observed that the isochoric paths depend markedly on the previous consolidation, which thus turns out to be of anisotropic nature. Especially, the obtained maximum deviatoric stress (i.e. the shear strength) has been found to depend on the previous consolidation path. Usually, this dependence is attributed only to the isotropic deformation history and is expressed by the fact that the undrained shear strength is recognized to depend on the water content (i.e. on the density). The detected dependence on the deformation anisotropy is not marked if isotropic and oedometric consolidations are compared to eachother, but it is increasingly pronounced as the principal stresses change at the transition from the consolidation to the isochoric motion. Especially an interchange of the maximum and minimum principal stress directions (i.e. a rotation by 90°) on this transition clearly reveals the influence of anisotropic consolidation as shown in Fig. 3

Fig. 2: Apparatus with top plate removed:
1 - movable platen, 2 - space for a sample to be pumped into rubber membrane, 3 - ram with load cell, 4 - gearing, 5 - translatory sliding guide, 6 - base plate, 7 - sliding bearing, 8 - displacement transducer, 9 - hole for bottom sealing disc containing drainage stone and hypodermic needle.

3 CALIBRATION

The material constants appearing in equations 4 and 5 can be determinated by the calibration procedure to be stated below. This procedure assumes that the values μ_1 and μ_2 are given and provides a method of obtaining the constants C_1, C_2, C_3, C_4. The resulting set of material constants C_1, C_2, C_3, C_4, μ_1, μ_2 is subsequently tested by means of numerical simulation of triaxial tests. If necessary, i.e. if the calculated results are not satisfactory, the determination of the constants C_i should be repeated with a new choice of μ_1 and μ_2. This trial and error iterative procedure, which can be carried out either manually or automatically, is surely laborious but it appears, for the time being, inevitable in view of the immense difficulty of calibrating a set of equations containing internal variables.

Consider an oedometric consolidated clay sample which is subsequently subject to isochoric compression and isochoric extension respectively. The isochoric motion (in soil mechanics laboratory practice usually obtained under so-called undrained conditions) can be carried out in a triaxial apparatus. The stretching tensor reads then:

$$\mathbf{D} = \begin{pmatrix} -1 & 0 & 0 \\ 0 & 0.5 & 0 \\ 0 & 0 & 0.5 \end{pmatrix} \cdot I \qquad (6)$$

where I equals 1 for compression and -1 for

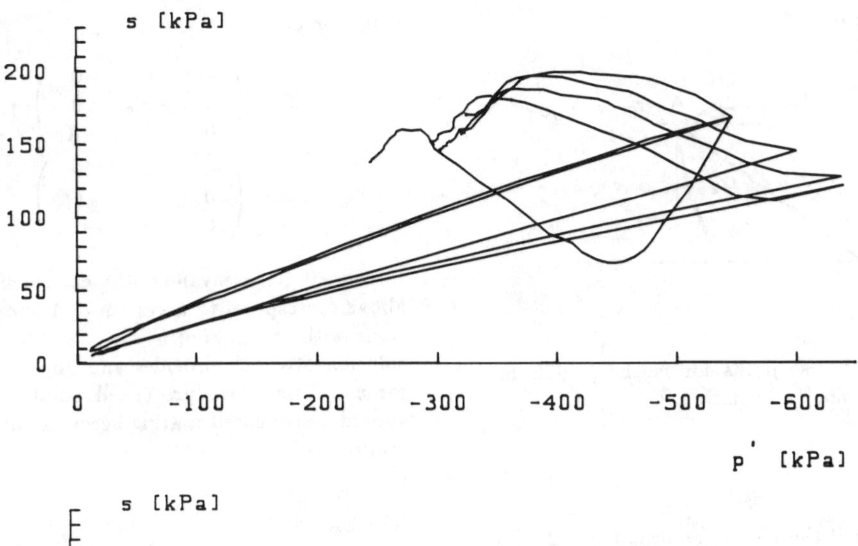

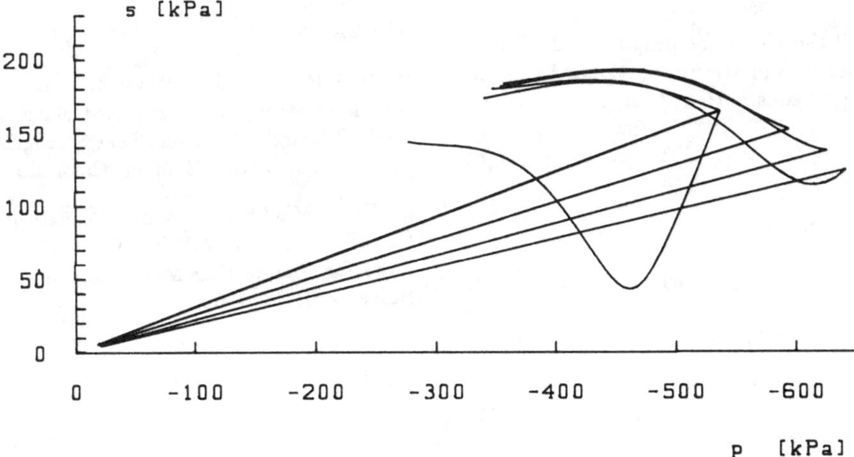

Fig. 3: Stress path representations.
s =stress deviator := $\sqrt{\frac{2}{3}(\sigma_1'^2 + \sigma_2'^2 + \sigma_3'^2 - \sigma_1'\sigma_2' - \sigma_1'\sigma_3' - \sigma_2'\sigma_3')}$, p' = mean stress := $\frac{1}{3}(\sigma_1' + \sigma_2' + \sigma_3')$, (a) experimental, (b) theoretical.

extension. The corresponding stress paths are shown in Fig. 4. From an experimentally obtained plot of these paths the following quantities, refering to the same point A, can be picked off: K_0, C_c, α_1, α_2 (K_0 =coefficient of earth pressure at rest, C_c =compression coefficient).

At the point A, the — known — stress components are

$$\mathbf{T}_A = \sigma_{1A} \begin{pmatrix} 1 & 0 & 0 \\ 0 & K_0 & 0 \\ 0 & 0 & K_0 \end{pmatrix} \tag{7}$$

and the — unknown — components of $\mathbf{S}$ are:

$$\mathbf{S}_A = s_{1A} \begin{pmatrix} 1 & 0 & 0 \\ 0 & K_s & 0 \\ 0 & 0 & K_s \end{pmatrix} \tag{8}$$

The tensors $\mathbf{T}_A$ and $\mathbf{S}_A$ are axisymmetric since the preceding oedometric deformation is also axisymmetric. K_s can be obtained by inserting $\mathbf{T}_A$ and $\mathbf{D}$ into equ. (4). Thus, the only quantity, which is still unknown, is s_{1A}. The ratios of corresponding components of $\mathbf{T}$ and $\mathbf{S}$ are

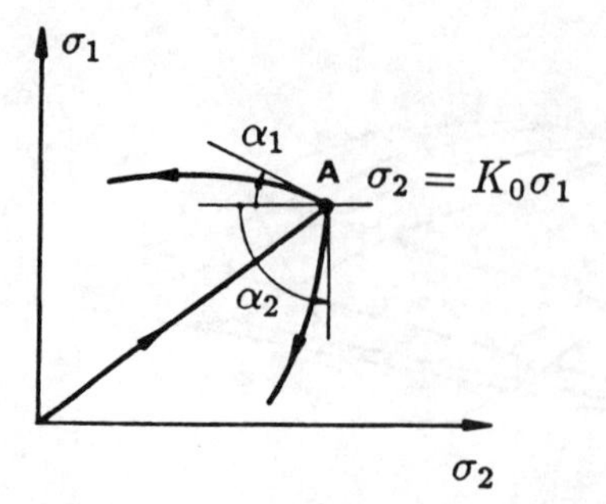

Fig. 4: Stress paths for isochoric deformation (schematic).

constant if the paths are proportional. Consequently, for oedometric proportional paths the ratio s_1/σ_1 is constant:

$$\psi := \frac{s_{1A}}{\sigma_{1A}} \tag{9}$$

It then follows:

$$\dot{s}_1 = \psi\dot{\sigma}_1 \tag{10}$$

On the other hand, from equ. (5 and 4) it follows

$$\begin{aligned} \dot{\sigma}_1 &= h(\mathbf{T},\mathbf{S},\mathbf{D}) \ , \\ \dot{s}_1 &= g(\mathbf{T},\mathbf{D}) \ . \end{aligned} \tag{11}$$

Insertion of (11) into equ. (10) gives an algebraic equation for the determination of ψ. In order to solve this equation, the constants C_1, C_2, C_3, C_4 must be known. For a fixed ψ-value they can be easily determined by solving a system of 4 linear equations. These equations fit the constitutive equation to the experimentally obtained values $K_0, C_c, \alpha_1, \alpha_2$. They read:

$$\dot{\sigma}_2 - K_0\dot{\sigma}_1 = 0 \ , \tag{12}$$

$$\dot{\sigma}_1 C_c + \sigma_1(1+e)\cdot \mathrm{tr}\mathbf{D} = 0 \ , \tag{13}$$

$$\tan\alpha_1 \cdot \dot{\sigma}_1 + \sqrt{2}\dot{\sigma}_2 = 0 \ ,(I=1) \ , \tag{14}$$

$$\tan\alpha_2 \cdot \dot{\sigma}_1 - \sqrt{2}\dot{\sigma}_2 = 0 \ ,(I=-1) \ . \tag{15}$$

In the equations (12),(13),(14) and (15) the rates $\dot{\sigma}_1$ and $\dot{\sigma}_2$ should be expressed by means of the constitutive equation (5), the $\mathbf{T}$ and $\mathbf{S}$ tensors read

$$\mathbf{T} = \begin{pmatrix} -1 & 0 & 0 \\ 0 & -K_0 & 0 \\ 0 & 0 & -K_0 \end{pmatrix} ; \quad \mathbf{S} = \psi \begin{pmatrix} -1 & 0 & 0 \\ 0 & -K_s & 0 \\ 0 & 0 & -K_s \end{pmatrix} . \tag{16}$$

Introducing these values, the equations stated above correspond to a system of 4 linear equations with the unknowns C_1, C_2, C_3, C_4, the solution of which provides and improved value for ψ. This procedure (fixed point iteration) should be repeated until ψ becomes sufficiently stationary.

REFERENCES

[1] HVORSLEV, J. : Physical components of the shear strength of saturated clays- In Proceed. Research Conf. on Shear Strength of Cohesive Soils, ASCE, Boulder, Colorado, 1960.

[2] KOLYMBAS D. : A novel constitutive law for soils. Proceed. 2nd Int. Conf. Constitutive Laws for Engineering Materials, Tucson, 1987. Balkema, Rotterdam.

Numerical Methods in Geomechanics (Innsbruck 1988), Swoboda (ed.)
© 1988 Balkema, Rotterdam. ISBN 90 6191 809 X

Incremental theory of elasticity and plasticity under cyclic loading

Y.Ichikawa, T.Kyoya & T.Kawamoto
Nagoya University, Japan

ABSTRACT: The flow theory does not provide a sufficient framework in describing the behaviour of geomaterials, since the volumetric response cannot be treated independently from the shearing one, and the Drucker's stability postulate is not usually satisfied by the flow theory. The multi-response theory proposed by Ichikawa, Kyoya and Kawamoto (1985) is a natural extension of the incremental plasticity theory. The model is here applied for the behaviour of geomaterials under cyclic loading with nonlinear elastic and plastic responses.

1 INTRODUCTION

Geomaterials exhibit strong nonlinearity and dilatancy characteristics. The incremental plasticity theory based on the flow rule is commonly used in modelling these behaviours (Drucker and Prager 1952, Roscoe, Schofield and Thurairajah 1963, Lade and Duncan 1975, Sekiguchi and Ohta 1979). The flow theory defines the yielding or failure by introducing a hypersurface in the surface space (Prager 1949), and the post-yielded incremental plastic strain is determined by the stress gradient of a plastic potential function. This procedure is called the flow theory. The common shortcoming of the flow theory is that the Drucker's stability postulate is not always satisfied for the dilatant materials. This is the crucial point in the framework of the flow theory, and it is caused by introducing a scalar function for determining the strain increment. That is, as long as the scalar potential function is used, the volumetric response cannot be represented independently from the shearing one (Ichikawa, Kyoya and Kawamoto 1985). The authors have proposed a new concept of the incremental plasticity theory, called the multi-response theory which defines the plastic response functions both in the shearing and volumetric behaviours (Ichikawa, Kyoya and Kawamoto 1985), and the theory gives the complete solution for ambiguities involved in the flow theory. Comprehensive results applied for some rocks were also shown.

The endochronic theory is seemed to give a very similar framework as the multi-response theory for the response of dilatant materials (Varanis and Read 1982). It is, however, essentially different because the multi-response theory occupies a more fundamental position in the sequence of the incremental plasticity theory. The Prager's consistency condition plays an important role in this sequence. The response functions of the multi-response theory are represented by a Laplace transformation, and its discrete form is determined by a spectral approximation method of the Laplace transformation.

Similarly, the elasto-plastic behaviours under cyclic loading have been represented in the framework of the flow theory, then the problem is how to describe the hardening rule. For the soil response, either the multi-surface model (Mroz 1967, Prevost 1977,1978) or the two-surface model (Mroz, Norris and Zienkiewicz 1978, Dafalias and Herrmann 1982) is commonly used. However, these models still involve no versality in describing the shearing and volumetric responses independently.

We here extend the multi-response theory to elasto-plastic behaviours of geomaterials under cyclic loading, then it is applied and verified for some rock materials.

Notations used henceforth are as follows: $\boldsymbol{\sigma}$ is the stress tensor, $\bar{\boldsymbol{\sigma}} = tr(\boldsymbol{\sigma})\boldsymbol{I}/3$ the mean stress tensor where $\boldsymbol{I}^{\sigma}$ is the unit tensor in the stress space, $\bar{\sigma} = tr(\ \boldsymbol{\sigma})/\sqrt{3}$ the mean stress, $\boldsymbol{s} = \boldsymbol{\sigma} - \bar{\boldsymbol{\sigma}}$ the deviatoric stress tensor, $s = \|\boldsymbol{s}\| = (\boldsymbol{s}\cdot\boldsymbol{s})^{1/2}$ the norm of the deviatoric stress tensor, $\theta^{\sigma} = \frac{1}{3}\cos^{-1}\{3\sqrt{3}J_3^{\sigma}/2(J_2^{\sigma})^{3/2}\}$ the stress Lode angle where $J_2^{\sigma} = \boldsymbol{s}\cdot\boldsymbol{s}/2$ is the second invariant and $J_3^{\sigma} = det(\boldsymbol{s})$ the third invariant of the deviatoric stress, respectively; $\boldsymbol{\varepsilon}$ is the strain tensor, $\bar{\boldsymbol{\varepsilon}} = tr(\boldsymbol{\varepsilon})\boldsymbol{I}/3$ the mean strain tensor where $\boldsymbol{I}^{\varepsilon}$ is the unit tensor in the strain space, $\bar{\varepsilon} = tr(\ \boldsymbol{\varepsilon})/\sqrt{3}$ the mean strain, $\boldsymbol{e} = \boldsymbol{\varepsilon} - \bar{\boldsymbol{\varepsilon}}$ the deviatoric strain tensor,

$e = \|e\| = (e \cdot e)^{1/2}$ the norm of the deviatoric strain tensor, $\theta^{\varepsilon} = \frac{1}{3}\cos^{-1}\{3\sqrt{3}J_3^{\varepsilon}/2(J_2^{\varepsilon})^{3/2}\}$ the strain Lode angle where $J_2^{\varepsilon} = e \cdot e/2$ is the second invariant and $J_3^{\varepsilon} = det(e)$ the third invariant of the deviatoric strain, respectively.

Note that in the principal stress space $(\sigma_1, \sigma_2, \sigma_3)$, the mean stress $\bar{\sigma}$, the deviatoric stress s and the stress Lode angle θ^{σ} are represented as shown in Figure 1. The axis $\sigma_1 = \sigma_2 = \sigma_3$ is called the hydrostatic axis.

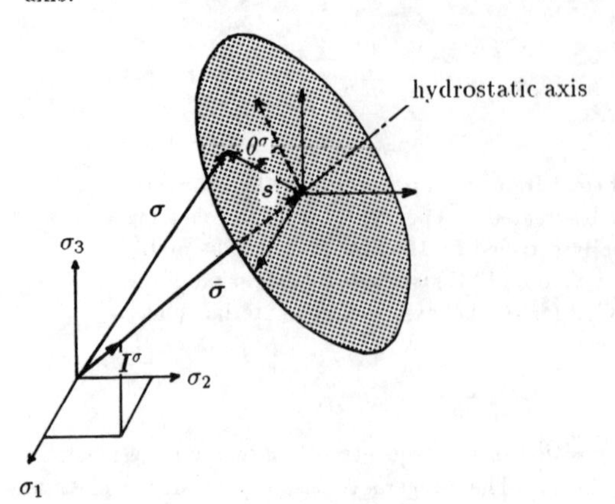

Fig.1 Mean stress σ, deviatoric stress s and stress Lode angle θ^{σ} in the principal stress space.

2 A BRIEF REVIEW OF THE MULTI-RESPONSE PLASTICITY THEORY

Let $f(\sigma, \varepsilon^p)$ be a generic form of the yield function where the superscript p denotes the plastic component. The Prager's consistency condition

$$df = \frac{\partial f}{\partial \sigma} \cdot d\sigma + \frac{\partial f}{\partial \varepsilon^p} \cdot d\varepsilon^p = 0 \tag{1}$$

gives a condition for the subsequent yield surface.

If an incremental relation is now given as

$$d\varepsilon^p = C^p d\sigma \quad \text{or} \quad d\sigma = D^p d\varepsilon^p, \tag{2}$$

the consistency can be written as

$$\frac{\partial f}{\partial \sigma} + C^{p*}\frac{\partial f}{\partial \varepsilon^p} = \mathbf{o} \quad \text{or} \quad D^{p*}\frac{\partial f}{\partial \sigma} + \frac{\partial f}{\partial \varepsilon^p} = \mathbf{o} \tag{3}$$

where C^{p*} and D^{p*} are the adjoint tensors of C^p and D^p, respectively. The differential equation (3), called the yield surface equation, gives the relation between the yield function f and the tensor C^p or D^p.

There are several ways to determine the relation (2); the endochronic theory (Varanis and Read 1982), the multi-response theory (Ichikawa, Kyoya and Kawamoto 1985), and the flow theory which is given by

$$d\varepsilon^p = \lambda \frac{\partial g}{\partial \sigma} \tag{4}$$

where g is a plastic potential function. With the aid of the consistency condition (1), the relation (4) is written as

$$d\varepsilon^p = \frac{1}{h}(\frac{\partial g}{\partial \sigma} \odot \frac{\partial f}{\partial \sigma})d\sigma \quad \text{(Melan's formula)} \tag{5}$$

where $h = -(\partial f/\partial \varepsilon^p)\cdot(\partial g/\partial \sigma)$, so that we have $C^p = \frac{1}{h}(\partial g/\partial \sigma) \odot (\partial f/\partial \sigma)$ for the flow theory. The endochronic theory introduces an intrinsic time and the response is represented by a convolution integral. The multi-response theory directly uses a Laplace transformation; details are shown in the following section.

Let us observe closely the yield surface equation (3). If the yield function f is a scalar one, the equation (3) cannot be solved, so that in the flow theory the terms in f with respect to σ should be known *a priori*. Thus, by experiments we determine the yield function such as the type of Mises, Mohr-Coulomb, Drucker-Prager or Cam-clay type. On the other hand, if f is a vector function which has six components in the most general case, we are free from the above restriction, and by introducing a vector form of the response function between σ and ε^p, called the multi-response function, which is an alternative of the yield function f, the yield surface equation (3) is automatically satisfied.

The response may be represented by some integral transformation method, and we use the Laplace transformation such that

$$\begin{aligned} \sigma_i &= \hat{\phi}_i(\varepsilon^p) \\ &=< \phi_i(\boldsymbol{\xi}), e^{-\varepsilon^p_j \xi_j} > \\ &= \int_0^{\infty} \cdots \int_0^{\infty} \phi_i(\boldsymbol{\xi}) e^{-\varepsilon^p_j \xi_j} d\xi_1 d\xi_2 \cdots d\xi_6; \\ &\qquad i, j = 1, 2, \cdots, 6 \end{aligned} \tag{6}$$

where the stress and plastic strain tensor is rewritten in vector form $(\sigma_1 = \sigma_{xx}, \sigma_2 = \sigma_{yy}, \cdots, \sigma_4 = \sigma_{xy}, \cdots; \varepsilon^p_1 = \varepsilon^p_{xx}, \varepsilon^p_2 = \varepsilon^p_{yy}, \cdots, \varepsilon^p_4 = \varepsilon^p_{xy}, \cdots)$. The function $\phi_i(\boldsymbol{\xi})$ is called the plasticity spectrum. It is easy to see that the vector yield function is given by

$$f_i = \sigma_i - \hat{\phi}_i(\varepsilon^p) = 0. \tag{7}$$

If the material is isotropic, the response is given between $(\bar{\sigma}, s, \theta^{\sigma})$ and $(\bar{\varepsilon^p}, e^p, \theta^p)$. Furthermore, if the response is hydrostatically symmetric (that is, the response is similar with respect to the hydrostatic axis of the stress and strain), the response is restricted for the relation between $(\bar{\sigma}, s)$ and $(\bar{\varepsilon^p}, e^p)$.

The incremental relation (2) is then easily obtained by differentiating the response function.

The Drucker's stability postulate

$$d\sigma \cdot d\varepsilon^p \geq 0 \qquad \text{(stability in local)}$$

requires in the flow theory to coincide the yield function f with the potential function g, while in the multi-response theory, it requires only the nonnegative definiteness of $\boldsymbol{C}^p$ or $\boldsymbol{D}^p$:

$$d\boldsymbol{\sigma} \cdot \boldsymbol{C}^p d\boldsymbol{\sigma} \geq 0 \qquad \text{or} \qquad \boldsymbol{D}^p d\boldsymbol{\varepsilon}^p \cdot d\boldsymbol{\varepsilon}^p \geq 0 \quad (8)$$

It is commonly observed that in the actual response of geomaterials, the volume change $d\bar{\varepsilon}^p$ is not entirely subjected to the shearing strain de^p. However, as long as the flow rule is used in which the tensor $\boldsymbol{C}^p$ is represented by a tensor product of a form of Eqn(5), $d\bar{\varepsilon}^p$ is dependent to de^p even if the non-associated flow rule ($f \neq g$) is applied; see in details Ichikawa, Kyoya and Kawamoto (1985). The multi-response theory avoids this ambiguity and the internal friction angle is distinguished from the dilatancy factor.

3 MULTI-RESPONSE THEORY UNDER CYCLIC LOADING

The material treated here is assumed to be isotropic and hydrostatically symmetric, which implies that the response is given between $(\bar{\sigma}, s)$ and $(\bar{\varepsilon}, e)$. This is conventional but useful because the conventional triaxial test using cylindrical specimens is commonly employed for soil and rock tests. It is not difficult to extend this theory to the general isotropic material (Tsuruhara, Kyoya, Ichikawa and Kawamoto 1985).

3.1 Strains under cyclic loading and response curves

If we unload at, for example, a pont A in Figure 2, and reload again at a point B, the actual loading path for geomaterials may trace the dotted lines. For simplicity, we here assume a straight unloading and reloading path (shown as a solid line in Figure 2), and the strain is directly separated into elastic and plastic components:

$$\boldsymbol{\varepsilon} = \boldsymbol{\varepsilon}^e + \boldsymbol{\varepsilon}^p \qquad (9)$$

In the incremental form, it is written as

$$d\boldsymbol{\varepsilon} = d\boldsymbol{\varepsilon}^e + d\boldsymbol{\varepsilon}^p \qquad (10)$$

One-dimensional elastic and plastic responses are schematically shown in Figure 3.

In addition, the elastic and plastic strains are composed by the deviatoric and mean components, respectively. Thus, we obtain responses of dilatant materials as shown in Figure 4.

These response curves basically define the upper loading surface, while the lower loading surface of the deiviatoric stress s is zero because $\sigma_1 = \sigma_2$ for the conventional triaxial test at a completely unloaded stage, and similarly the lower surface of the mean stress is

$$\bar{\sigma} = \sqrt{3}\,\sigma_3 \;:\; \text{constant}.$$

It should be noted that in order to obtain the elastic and plastic responses of the mean stress $\bar{\sigma}$, we must perform loading and unloading tests for each σ_3 =constant stage.

3.2 Plastic response

The plastic response for the hydrostatically symmetric material is written in general as

Fig.2 Definition of strains (one dimensional case).

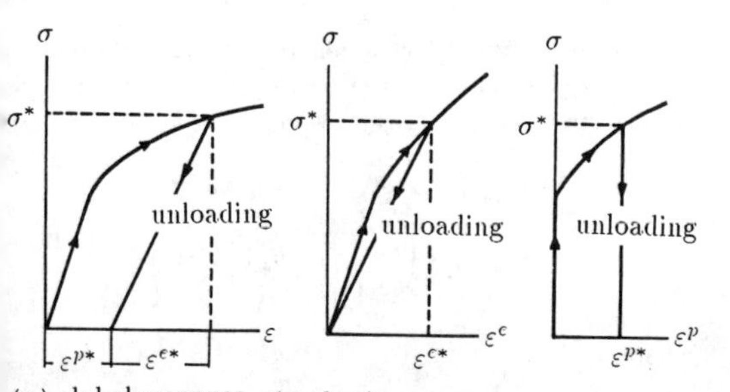

(a) global response (b) elestic response (c) plastic response

Fig.3 Schematic loading-unloading curves.

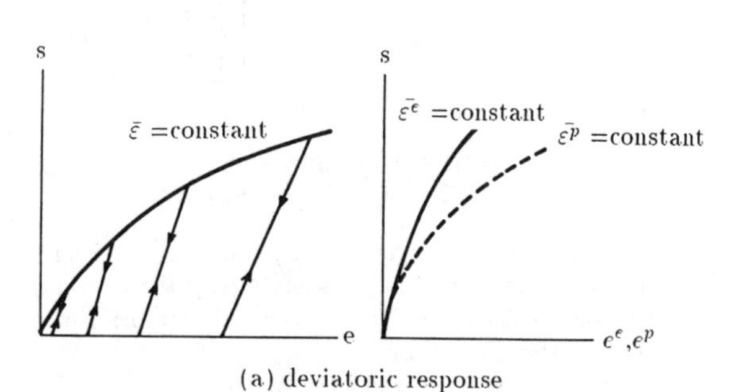

(a) deviatoric response

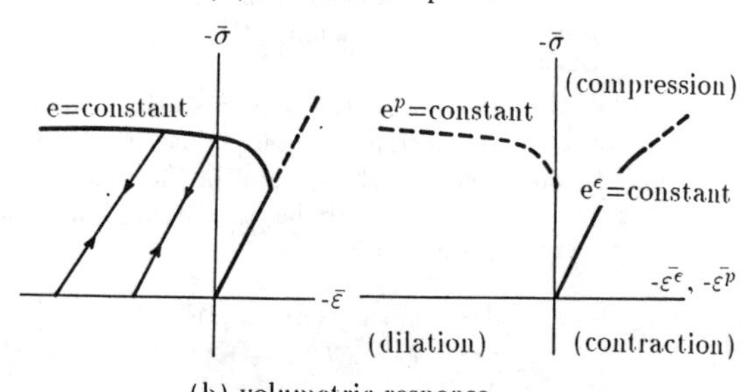

(b) volumetric response

Fig.4 Typical elasto-plastic response of dilatant materials.

$$\begin{aligned} s &= \hat{\phi}^p(e^p, \bar{\varepsilon}^p) \\ \bar{\sigma} &= \hat{\psi}^p(\bar{\varepsilon}^p, e^p) \end{aligned} \tag{11}$$

The vector form of the yield function (8) is then

$$\boldsymbol{f} = \left\{ \begin{array}{c} f_1 \\ f_2 \end{array} \right\}, \qquad \begin{aligned} f_1 &= s - \hat{\phi}^p(e^p, \bar{\varepsilon}^p) \\ f_2 &= -\bar{\sigma} + \hat{\psi}^p(\bar{\varepsilon}^p, e^p) \end{aligned} \tag{12}$$

If we consider that the softening is not a material characteristics but a structural phenomenon caused by a boundary condition of experiments, the response becomes monotone, and we represent it as a Laplace transformation (see Eqn(6)):

$$\begin{aligned} s = s_0 &+ \int_0^\infty \phi_1^p(\xi)\Lambda(\bar{\varepsilon}^p;\xi)d\xi + \int_0^\infty \phi_2^p(\eta)\Lambda(e^p;\eta)d\eta \\ &+ \int_0^\infty \int_0^\infty \phi_3^p(\xi,\eta)\Lambda(\bar{\varepsilon}^p;\xi)d\xi\Lambda(e^p;\eta)d\eta \\ \bar{\sigma} = \bar{\sigma}_0 &- \int_0^\infty \psi_1^p(\eta)\Lambda(e^p;\eta)d\eta + \int_0^\infty \psi_2^p(\xi)\Lambda(\bar{\varepsilon}^p;\xi)d\xi \\ &+ \int_0^\infty \int_0^\infty \psi_3^p(\eta,\xi)\Lambda(e^p;\eta)d\eta\Lambda(\bar{\varepsilon}^p;\xi)d\xi \end{aligned} \tag{13}$$

where $\Lambda(a;\alpha) = 1-\exp(-a\alpha)$, and $\phi_1^p(\xi), \phi_2^p(\eta), \phi_3^p(\xi,\eta)$ and $\psi_1^p(\eta), \psi_2^p(\xi), \psi_3^p(\eta,\xi)$ are called the shearing and volumetric plasticity spectra, respectively.

The incremental response is then obviously obtained as

$$\left\{ \begin{array}{c} ds \\ d\bar{\sigma} \end{array} \right\} = \left[\begin{array}{cc} G_1^p & G_2^p \\ K_2^p & K_1^p \end{array} \right] \left\{ \begin{array}{c} de^p \\ d\bar{\varepsilon}^p \end{array} \right\} \tag{14}$$

where

$$\begin{aligned} G_1^p &= \partial\hat{\phi}^p/\partial e^p, \quad G_2^p = \partial\hat{\phi}^p/\partial\bar{\varepsilon}^p, \\ K_1^p &= \partial\hat{\psi}^p/\partial\bar{\varepsilon}^p, \quad K_2^p = \partial\hat{\psi}^p/\partial e^p. \end{aligned}$$

The inverse relation is

$$\left\{ \begin{array}{c} de^p \\ d\bar{\varepsilon}^p \end{array} \right\} = \left[\begin{array}{cc} 1/h_s & \mu/h_s \\ \beta/h_v & 1/h_v \end{array} \right] \left\{ \begin{array}{c} ds \\ d\bar{\sigma} \end{array} \right\} \tag{15}$$

where

$$\begin{aligned} h_s &= G_1^p + \mu K_2^p, \quad \mu = -G_2^p/K_1^p, \\ h_v &= K_1^p + \beta G_2^p, \quad \beta = -K_2^p/G_1^p. \end{aligned}$$

Here h_s and h_v are the hardening parameters for shearing and volume changing, respectively, μ the internal friction coefficient, and β the dilatancy factor. Note that in the flow theory the internal friction coefficient cannot be distinguished from the dilatancy factor, while in the multi-response theory, they are distinguished.

Let the direction of the plastic strain increment be assumed to coincide with the global stress. This may be an experimental fact, and in the flow theory it is basically assumed. Under this assumption, the relation (15) is now written as

$$\begin{aligned} d\boldsymbol{e}^p &= \frac{1}{h_s}\boldsymbol{m} \otimes (\boldsymbol{m} + \mu\boldsymbol{n})d\boldsymbol{\sigma} \\ d\bar{\boldsymbol{\varepsilon}}^p &= \frac{1}{h_v}\boldsymbol{n} \otimes (\beta\boldsymbol{m} + \boldsymbol{n})d\boldsymbol{\sigma} \end{aligned} \tag{16}$$

where

$$\boldsymbol{m} = \frac{\partial s}{\partial \boldsymbol{s}} = \frac{\boldsymbol{s}}{s} \qquad \boldsymbol{n} = \frac{\partial\bar{\sigma}}{\partial\bar{\boldsymbol{\sigma}}} = \frac{\bar{\boldsymbol{\sigma}}}{\bar{\sigma}} = \boldsymbol{I}^\sigma/\sqrt{3} \tag{17}$$

The total incremental response is then given as

$$d\boldsymbol{\varepsilon}^p = d\boldsymbol{e}^p + d\bar{\boldsymbol{\varepsilon}}^p = \boldsymbol{C}^p d\boldsymbol{\sigma} \tag{18}$$

where

$$\boldsymbol{C}^p = \frac{1}{h_s}\boldsymbol{m} \otimes (\boldsymbol{m} + \mu\boldsymbol{n}) + \frac{1}{h_v}\boldsymbol{n} \otimes (\beta\boldsymbol{m} + \boldsymbol{n}).$$

In order to get an asymptotic expansion of the response (13), the variables are changed as $\tau = 1/\eta$ and $\omega = 1/\xi$, then

$$\begin{aligned} s = s_0 &+ \int_0^\infty \mu_1^p(\omega)\Gamma(\bar{\varepsilon}^p;\omega)d(\ln\omega) \\ &+ \int_0^\infty \mu_2^p(\omega)\Gamma(e^p;\tau)d(\ln\tau) \\ &+ \int_0^\infty \int_0^\infty \mu_3^p(\omega,\tau)\Gamma(\bar{\varepsilon}^p;\omega)d(\ln\omega)\Gamma(e^p;\tau)d(\ln\tau) \\ \bar{\sigma} = \bar{\sigma}_0 &- \int_0^\infty \chi_1^p(\tau)\Gamma(e^p;\tau)d(\ln\tau) \\ &+ \int_0^\infty \chi_2^p(\omega)\Gamma(\bar{\varepsilon}^p;\omega)d(\ln\omega) \\ &+ \int_0^\infty \int_0^\infty \chi_3^p(\tau,\omega)\Gamma(e^p;\tau)d(\ln\tau)\Gamma(\bar{\varepsilon}^p;\omega)d(\ln\omega) \end{aligned} \tag{19}$$

where

$$\begin{aligned} &\mu_1^p(\omega) = \phi_1^p(1/\omega)/\omega, \qquad \mu_2^p(\tau) = \phi_2^p(1/\tau)/\tau, \\ &\mu_3^p(\omega,\tau) = \phi_2^p(1/\omega,1/\tau)/\omega\tau, \\ &\chi_1^p(\tau) = \psi_1^p(1/\tau)/\tau, \qquad \chi_2^p(\omega) = \psi_2^p(1/\omega)/\omega, \\ &\chi_3^p(\tau,\omega) = \psi_2^p(1/\tau,1/\omega)/\tau\omega, \end{aligned}$$

and $\Gamma(a;\beta) = 1 - \exp(-a/\beta) = \Lambda(a;1/\beta)$.

Differentiating Eqn(19), the plasticity spectra can be asymptotically written as

$$\begin{aligned} \mu_{23}^p(\tau;\bar{\varepsilon}^p) &= \lim_{k\to\infty} \frac{(-1)^{k+1}}{(k-1)!}\tau^k \frac{\partial^k s(k\tau,\bar{\varepsilon}^p)}{\partial\tau^k} \\ \chi_{23}^p(\omega;e^p) &= \lim_{k\to\infty} \frac{(-1)^{k+1}}{(k-1)!}\omega^k \frac{\partial^k \bar{\sigma}(k\omega,e^p)}{\partial\omega^k} \end{aligned} \tag{20}$$

where

$$\begin{aligned} \mu_{23}^p(\tau;\bar{\varepsilon}^p) &= \mu_2^p(\tau) + \int_0^\infty \mu_3^p(\omega,\tau)\Gamma(\bar{\varepsilon}^p;\omega)d(\ln\omega) \\ \chi_{23}^p(\omega;e^p) &= \chi_2^p(\omega) + \int_0^\infty \mu_3^p(\tau;\omega)\Gamma(e^p;\tau)d(\ln\tau) \end{aligned}$$

The first order approximation for Eqn(20) gives

$$\begin{aligned} \mu_{23}^p(e^p;\bar{\varepsilon}^p = \text{constant}) &= e^p \frac{\partial s(e^p;\bar{\varepsilon}^p=\text{constant})}{\partial e^p} \\ \chi_{23}^p(\bar{\varepsilon}^p;e^p = \text{constant}) &= \bar{\varepsilon}^p \frac{\partial\bar{\sigma}(\bar{\varepsilon}^p;e^p=\text{constant})}{\partial\bar{\varepsilon}^p} \end{aligned} \tag{21}$$

and the discretized spectral points (τ_i, ω_i) are obtained as to the peak points given by the curves (21) of triaxial tests. The discretized response is then

$$\begin{aligned} s = s_0 + \textstyle\sum_i \{ &a_{1i}\Gamma(\bar{\varepsilon}^p;\omega_i) + a_{2i}\Gamma(e^p;\tau_i) \\ &+ a_{3ii}\Gamma(\bar{\varepsilon}^p;\omega_i)\Gamma(e^p;\tau_i)\} \\ \bar{\sigma} = \bar{\sigma}_0 - \textstyle\sum_i \{ &b_{1i}\Gamma(e^p;\tau_i) - b_{2i}\Gamma(\bar{\varepsilon}^p;\omega_i) \\ &- b_{3ii}\Gamma(e^p;\tau_i)\Gamma(\bar{\varepsilon}^p;\omega_i)\} \end{aligned} \tag{22}$$

where

$$a_{1i} = \mu_1^p(\omega_i)\Delta\omega_i/\omega_i, \qquad a_{2i} = \mu_2^p(\tau_i)\Delta\tau_i/\tau_i,$$
$$a_{3ii} = \mu_3^p(\omega_i,\tau_i)\Delta\omega_i\Delta\tau_i/\omega_i\tau_i,$$
$$b_{1i} = \lambda_1^p(\tau_i)\Delta\tau_i/\tau_i, \qquad b_{2i} = \lambda_1^p(\omega_i)\Delta\omega_i/\omega_i,$$
$$b_{3ii} = \lambda_3^p(\tau_i,\omega_i)\Delta\tau_i\Delta\omega_i/\tau_i\omega_i.$$

Here the coefficients $a_{1i}, a_{2i}, a_{3ii}; b_{1i}, b_{2i}, b_{3ii}$ are determined by the least square method. The full procedure is shown in Ichikawa, Kyoya and Kawamoto (1985).

3.3 Elastic response

Most rocks exhibit the plastic strain at the very beginning stage of cyclic loading. The elastic response is also non-linear. Then, the elastic response of the hydrostatically symmetric material is given by the similar form of the plastic response (13):

$$\begin{aligned} s &= \hat{\phi}^e(e^e,\bar{\varepsilon}^e) \\ &= s_0 + \int_0^\infty \phi_1^e(\xi)\Lambda(\bar{\varepsilon}^e;\xi)d\xi + \int_0^\infty \phi_2^e(\eta)\Lambda(e^e;\eta)d\eta \\ &\quad + \int_0^\infty\int_0^\infty \phi_3^e(\xi,\eta)\Lambda(\bar{\varepsilon}^e;\xi)d\xi\Lambda(e^e;\eta)d\eta \\ &= s_0 + \int_0^\infty \mu_1^e(\omega)\Gamma(\bar{\varepsilon}^e;\omega)d(\ln\omega) \\ &\quad + \int_0^\infty \mu_2^e(\tau)\Gamma(e^e;\tau)d(\ln\tau) \\ &\quad + \int_0^\infty\int_0^\infty \mu_3^e(\omega,\tau)\Gamma(\bar{\varepsilon}^e;\omega)d(\ln\omega)\Gamma(e^e;\tau)d(\ln\omega) \end{aligned}$$

$$\begin{aligned} \bar{\sigma} &= \hat{\psi}^e(\bar{\varepsilon}^e,e^e) \\ &= \bar{\sigma}_0 - \int_0^\infty \psi_1^e(\eta)\Lambda(e^e;\eta)d\eta + \int_0^\infty \psi_2^e(\xi)\Lambda(\bar{\varepsilon}^e;\xi)d\xi \\ &\quad + \int_0^\infty\int_0^\infty \psi_3^e(\eta,\xi)\Lambda(e^e;\eta)d\eta\Lambda(\bar{\varepsilon}^e;\xi)d\xi \\ &= \bar{\sigma}_0 - \int_0^\infty \lambda_1^e(\tau)\Gamma(e^e;\tau)d(\ln\tau) \\ &\quad + \int_0^\infty \lambda_2^e(\omega)\Gamma(\bar{\varepsilon}^e;\omega)d(\ln\omega) \\ &\quad + \int_0^\infty\int_0^\infty \lambda_3^e(\tau,\omega)\Gamma(e^e;\tau)d(\ln\tau)\Gamma(\bar{\varepsilon}^e;\omega)d(\ln\omega) \end{aligned} \tag{23}$$

where $\phi_1^e(\xi), \phi_2^e(\eta), \phi_3^e(\xi,\eta)$ and $\psi_1^e(\eta), \psi_2^e(\xi), \psi_3^e(\eta,\xi)$ are the elasticity spectra.

By differentiating these, we get the incremental relation

$$\begin{Bmatrix} ds \\ d\bar{\sigma} \end{Bmatrix} = \begin{bmatrix} G_1^e & G_2^e \\ K_2^e & K_1^e \end{bmatrix} \begin{Bmatrix} de^e \\ d\bar{\varepsilon}^e \end{Bmatrix} \tag{24}$$

where

$$G_1^e = \partial\hat{\phi}^e/\partial e^e, \quad G_2^e = \partial\hat{\phi}^e/\partial\bar{\varepsilon}^e,$$
$$K_1^e = \partial\hat{\psi}^e/\partial\bar{\varepsilon}^e, \quad K_2^e = \partial\hat{\psi}^e/\partial e^p.$$

or conversly

$$\begin{Bmatrix} de^e \\ d\bar{\varepsilon}^e \end{Bmatrix} = \begin{bmatrix} 1/g_s & \lambda/g_s \\ \alpha/g_v & 1/g_v \end{bmatrix} \begin{Bmatrix} ds \\ d\bar{\sigma} \end{Bmatrix} \tag{25}$$

where

$$g_s = G_1^e + \lambda K_2^e, \quad \lambda = -G_2^e/K_1^e,$$
$$g_v = K_1^e + \alpha G_2^e, \quad \alpha = -K_2^e/G_1^e.$$

The direction of the elastic strain increment is assumed to coincide with the stress increment. This implies that

$$\begin{aligned} de^e &= \tfrac{1}{g_s}\boldsymbol{m}^* \odot (\boldsymbol{m} + \lambda\boldsymbol{n})d\sigma \\ d\bar{\varepsilon}^e &= \tfrac{1}{g_v}\boldsymbol{n}^* \odot (\alpha\boldsymbol{m} + \boldsymbol{n})d\sigma \end{aligned} \tag{26}$$

where

$$\boldsymbol{m}^* = \frac{ds}{\|ds\|} = \boldsymbol{m}^{-1}, \qquad \boldsymbol{n}^* = \frac{d\bar{\sigma}}{\|d\bar{\sigma}\|} = \boldsymbol{n}^{-1}.$$

The procedure to get the discretized response functions are entirely same as the plasticity case, which is omitted here.

4 TRIAXIAL TESTS UNDER CYCLIC LOADING FOR ROCKS

We have performed cyclic triaxial tests for many rocks under naturally dried condition. Lateral deformations are measured by O-type gauges attached at the surface of the specimen. Details of experiemnts are found in Ichikawa, Kyoya, Nitta and Kawamoto (1985). We here only present the results for Funyu-tuff whose physical properties are shown in Table 1.

Table 1 Physical properties of Funyu-tuff.

unit weight	γ_t(g/cm^3)	2.08
moisture content	ω (%)	6.59
void ratio	e	0.33
porosity	n(%)	24.9
degree of saturation	S_r (%)	41.3

The results of the conventional triaxial tests (σ_3 =constant) are shown in Figure 5. We arrange these for the relation between $(\bar{\sigma}, s)$ and $(\bar{\varepsilon}^p, e^p)$, and $(\bar{\sigma}, s)$ and $(\bar{\varepsilon}^e, e^e)$ in Figure 7, and we get the following discrete spectral points:

deviatoric plastic response;
$\omega_{11} = 0.00140, \tau_{11} = 0.0020,$
$\omega_{12} = 0.00203, \tau_{12} = 0.0050,$
volumetric plastic response,
$\omega_{21} = 0.00120, \tau_{21} = 0.0020,$
$\omega_{22} = 0.00148, \tau_{22} = 0.0040,$
deviatoric elastic response;
$\omega_{31} = 0.00248, \tau_{31} = 0.0050,$
$\omega_{32} = 0.00325, \tau_{32} = 0.0060,$
volumetric elastic reponse;
$\omega_{41} = 0.00352, \tau_{41} = 0.0050,$
$\omega_{42} = 0.00340, \tau_{42} = 0.0060.$

The discretized plastic response functions (22) are then determined by the least square method as

$$\begin{aligned} s = \; & a_1\Gamma(e^p;\tau_{11}) + a_2\Gamma(\bar{\varepsilon}^p;\omega_{11}) + a_3\Gamma(e^p;\tau_{11})\Gamma(\bar{\varepsilon}^p;\omega_{11}) \\ & + a_4\Gamma(e^p;\tau_{12}) + a_5\Gamma(\bar{\varepsilon}^p;\omega_{12}) + a_6\Gamma(e^p;\tau_{12})\Gamma(\bar{\varepsilon}^p;\omega_{12} \\ \bar{\sigma} = \; & b_1\Gamma(e^p;\tau_{21}) + b_2\Gamma(\bar{\varepsilon}^p;\omega_{21}) + b_3\Gamma(e^p;\tau_{21})\Gamma(\bar{\varepsilon}^p;\omega_{21}) \\ & + b_4\Gamma(e^p;\tau_{22}) + b_5\Gamma(\bar{\varepsilon}^p;\omega_{22}) + b_6\Gamma(e^p;\tau_{22})\Gamma(\bar{\varepsilon}^p;\omega_{22}) \end{aligned}$$

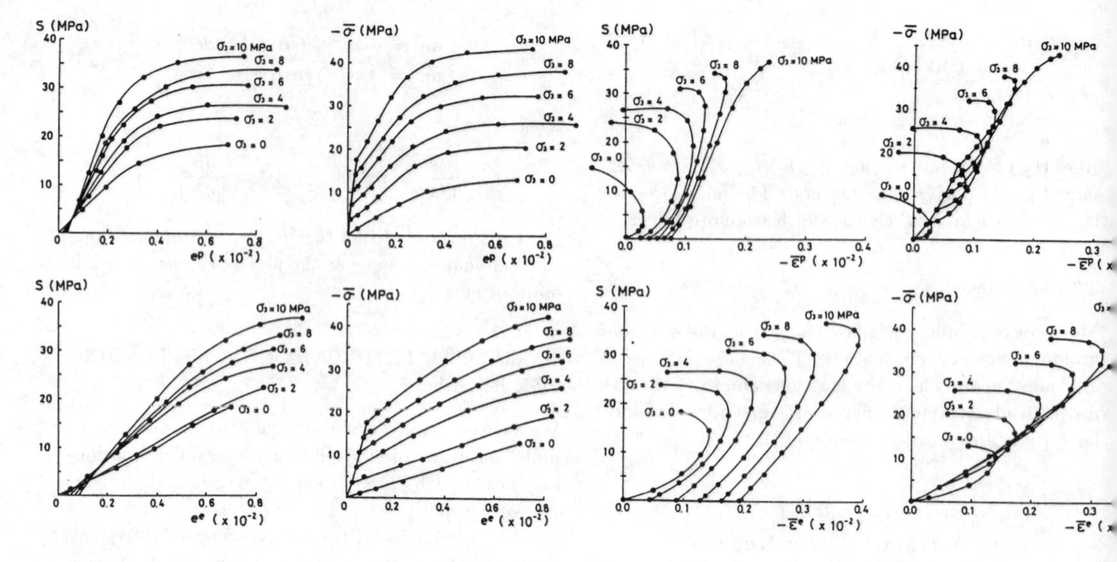

Fig.5 Curves for σ_3=constant of Funyu-tuff.

where

$$
\begin{array}{lll}
a_1 = -17.327 & a_2 = -53.059 & a_3 = 74.139 \\
a_4 = 51.727 & a_5 = 58.642 & a_6 = -80.756 \\
b_1 = -106.747 & b_2 = -131.708 & b_3 = 119.261 \\
b_4 = 146.965 & b_5 = 188.899 & b_6 = -172.312
\end{array}
$$

Similarly, the elastic ones are

$$
\begin{aligned}
s = \; & c_1\Gamma(e^e;\tau_{11}) + c_2\Gamma(\bar{\varepsilon}^e;\omega_{11}) + c_3\Gamma(e^e;\tau_{11})\Gamma(\bar{\varepsilon}^e;\omega_{11}) \\
& + c_4\Gamma(e^e;\tau_{12}) + c_5\Gamma(\bar{\varepsilon}^e;\omega_{12}) + c_6\Gamma(e^e;\tau_{12})\Gamma(\bar{\varepsilon}^e;\omega_{12}) \\
\bar{\sigma} = \; & d_1\Gamma(e^e;\tau_{21}) + d_2\Gamma(\bar{\varepsilon}^e;\omega_{21}) + d_3\Gamma(e^e;\tau_{21})\Gamma(\bar{\varepsilon}^e;\omega_{21}) \\
& + d_4\Gamma(e^e;\tau_{22}) + d_5\Gamma(\bar{\varepsilon}^e;\omega_{22}) + d_6\Gamma(e^e;\tau_{22})\Gamma(\bar{\varepsilon}^e;\omega_{22})
\end{aligned}
$$

where

$$
\begin{array}{lll}
c_1 = -182.541 & c_2 = -42.498 & c_3 = 133.490 \\
c_4 = 204.464 & c_5 = 48.687 & c_6 = -106.899 \\
d_1 = -788.448 & d_2 = -912.203 & d_3 = 972.646 \\
d_4 = 854.348 & d_5 = 942.740 & d_6 = -1015.69.
\end{array}
$$

The response functions obtained are shown in Figure 8.

5 CONCLUDING REMARKS

We have presented a generic theory of incremental plasticity for dilatant materials. The elasto-plastic responses are formulated by introducing a Laplace transformation method. The procedure is a natural extension of the classical incremental theory.

We have discussed on the Drucker's stability postulate, however, for the obtained responses, the condition is not discussed fully. Results for some geomaterials have been obtained.

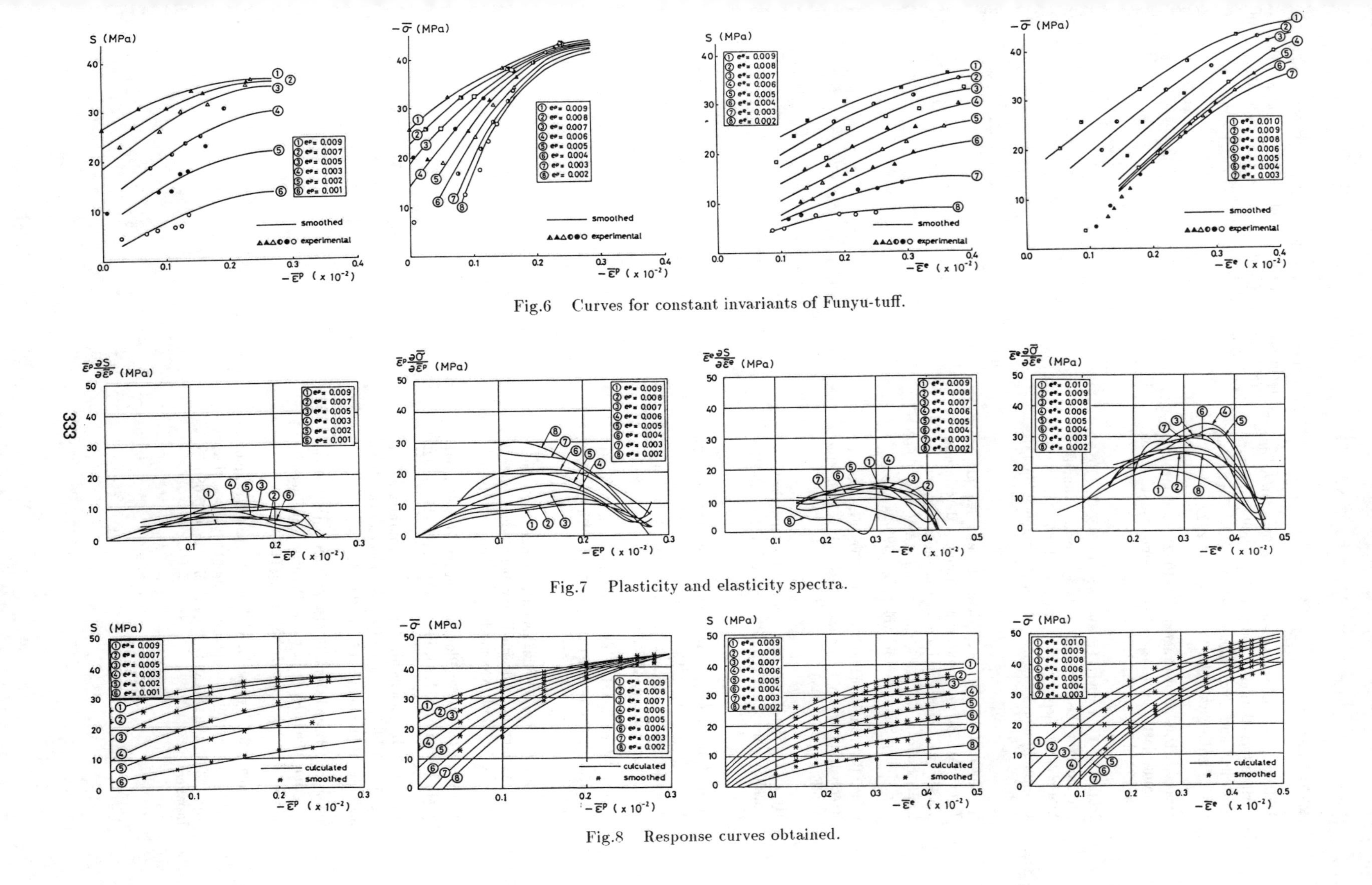

Fig.6 Curves for constant invariants of Funyu-tuff.

Fig.7 Plasticity and elasticity spectra.

Fig.8 Response curves obtained.

REFERENCES

Dafalias, Y.F., and Herrmann, L.R. (1982): Bounding surface formulation of soil plasticity, Soil Mechanics-Transient and Cyclic Loads, ed. by G.N. Pande and O.C. Zienkiewicz, John Wiley, pp.253-282.

Drucker, D.C., and Prager, W. (1952): Soil mechanics and plastic analysis or limit design, Quat. Appl. Math., vol.10, pp.157-165.

Ichikawa, Y., Kyoya, T., and Kawamoto, T. (1985): Incremental theory of plasticity for rocks, Numerical Methods in Geomechanics Nagoya 1985, ed. by T. Kawamoto and Y. Ichikawa, vol.1, Balkema, pp.451-462.

Ichikawa, Y., Kyoya, T., Nitta, H., and Kawamoto, T. (1985): Microcomputer measurement determining plasticity spectrum in incremental theory, Proc. Int. Conf. Education, Practice and Promotion Comp. Meth. Eng. Using Small Comp., vol.3B, pp.329-341.

Lade, P.V., and Duncan, J.M. (1975): Elastoplastic stress-strain theory for cohesionless soil, GT10, Proc. ASCE, pp.1037-1053.

Mroz, Z. (1967): On the description on anisotropic workhardening, J. Mech. Phy. Solids, Vol.15, pp.163-175.

Mroz, Z., Norris, V.A., and Zienkiewicz, O.C. (1978): An anistropic hardening model for soils and its application to cyclic loading, Int. J. Num. Anly. Meth. Geomech., Vol.2, pp.203-221.

Prager, W. (1949): Recent developments in the mathematical theory of plasticity, J. Appl. Phy., vol.20, pp.235-241.

Prevost, J.H. (1977): Mathematical modelling of monotonic and cyclic undrained clay behaviour, Int. J. Num. Anly. Meth. Geomech., Vol.1, pp.159-216.

Prevost, J.H. (1978): Plasticity theory for soil stress-strain behavior, EM104, Proc. ASCE, pp.1177-1194.

Roscoe, K.H., Schofield, A.N., and Thurairajah, A. (1963): Yielding of clays in states of wetter than critical, Geotechnique, vol.13, pp.211-240.

Sekiguchi, H., and Ohta, H. (1977): Induced anisotropy and time dependency in clays, Preprints of Specialty Session 9, 9th Int. Conf. SMFE, pp.229-238.

Tsuruhara, T., Kyoya, T., Ichikawa, Y., and Kawamoto, T. (1985): On anisotropic fracturing of rock materials, Proc. 40th Annual Conf. JSCE, vol.3, pp.685-686 (in Japanese).

Valanis, K.C., and Read, H.E. (1982): A new endochronic plasticity model for soils, Soil Mechanics-Transient and Cyclic Loads, ed. by G.N. Pande and O.C. Zienkiewicz, John Wiley, pp.375-417.

Numerical Methods in Geomechanics (Innsbruck 1988), Swoboda (ed.)
© 1988 Balkema, Rotterdam. ISBN 90 6191 809 X

Numerical simulation of the behaviour of saturated sand

M.J.Prabucki & W.Wunderlich
Ruhr University Bochum, FR Germany

ABSTRACT: The paper deals with the geometrically and physically nonlinear description and the numerical simulation of the behavior of watersaturated sand under cyclic loading. Based on the theory of mixtures a Lagrangean formulation of the equations of motion and continuity is given for the two-phase system. To describe the nonlinear behavior of sand under loading and unloading an elasto-plastic material law is employed. The material parameters are determined by corresponding experimental investigations.

1 INTRODUCTION

In the simulation of the ultimate behavior of structures under extreme loading conditions - such as earthquake or explosion - nonlinear effects are crucial. A dynamic analyis of the structure itself can be performed by standard finite element programs in most cases.

It is much more difficult, however, to include the influences of soil-structure-interaction. This is necessary in many cases, especially in problems in which effects of liquefaction or shear failure play an important role (Cramer, Wunderlich 1981, Strauß et al. 1984,1986). The substitution of the problem by simpler models (like spring-damper models) very often turns out to be too crude, and only a more detailed nonlinear investigation of the coupled problem with inclusion of pore water (Biot 1956) and a more sophisticated material description lead to reliable results.

In the paper we are dealing with the nonlinear theory of mixtures (Bowen 1976) with two constituents (skeleton of the sand and water) to model the above mentioned problem more realistically. The basic equations in local form are formulated in the Lagrangian reference frame including nonlinear terms in the damping and mass matrices. On the material side the nonlinear behavior of sand under dynamic loading and unloading is included by a kind of bounding-surface-model (Dafalias/Popov 1978) in the framework of a theory of elasto-plasticity. The effects due to the change of the hydrostatic stress parts on the one hand, and those due to change of the deviatoric parts on the other are first included in seperate descriptions which are then combined in a consistent manner. The determination of the corresponding material parameters is a crucial point, and exemplary it is indicated how to obtain them by experimental data.

2 A THEORY OF MIXTURES FOR SAND

For soils composed of solid, fluid and gaseous parts the same balance equations as

Table 1 Definitions

Mixture element:
fluid (f) / skeleton (s) $dV = dV^s + dV^f$
$n^\alpha = dV^\alpha / dV$
γ^α = density of constituent α referred to dV^α
$\rho^\alpha = n^\alpha \gamma^\alpha$
$\rho = \sum_\alpha \rho^\alpha$
v_i^α = velocity of constituent α
$v_i = 1/\rho \sum_\alpha \rho^\alpha v_i^\alpha$
q_i^α = interaction force of constituent α
T_{ij}^α = Cauchy stress of constituent α
$T_{ij} = \sum_\alpha T_{ij}^\alpha$ + (neglected terms)

for homogeneous material apply. However, these principles have to be satisfied not only for the mixture as a whole but also for every constituent taking into account the mass, momentum and energy coupling between the particles. For the two-phase system (skeleton-fluid) we assume that thermal effects are neglected, mass changes due to chemical processes are excluded, and all terms, given in Table 1, are continuous.

Then, for each spatial point x_i at time t under the condition of mass conservation the continuity equation for the mixture (1a), the skeleton (1b) and the fluid (1c) can be written as

$$\frac{n^s}{\gamma^s}\frac{D^s}{Dt}\gamma^s + \frac{n^f}{\gamma^f}\frac{D^f}{Dt}\gamma^f + (v_i^s + n^f\bar{v}_i)_{,i} = 0, \tag{1a}$$

$$\frac{1}{n^s}\frac{D^s}{Dt}n^s + \frac{1}{\gamma^s}\frac{D^s}{Dt}\gamma^s + v^s_{i,i} = 0, \tag{1b}$$

$$\frac{1}{n^f}\frac{D^f}{Dt}n^f + \frac{1}{\gamma^f}\frac{D^f}{Dt}\gamma^f + (v_i^s + \bar{v}_i)_{,i} = 0. \tag{1c}$$

D^α/Dt denotes the material time derivative referred to the motion of the constituent $\alpha = f,s$ and $\bar{v}_i = v_i^f - v_i^s$ is the velocity of the fluid relative to that of the skeleton. For imcompressible grains these equations can be simplified with $D^s/Dt\ \gamma^s = 0$.

In the equations of motion derived from principle of momentum balance the internal forces appear as the result of interaction between each part of the mixture:

$$\rho\frac{D}{Dt}v_i = T_{ij,j} + \rho b_i, \tag{2a}$$

$$\rho^s\frac{D^s}{Dt}v_i^s = T^s_{ij,j} + \rho^s b_i + \bar{q}_i^s, \tag{2b}$$

$$\rho^f\frac{D^f}{Dt}(v_i^s + \bar{v}_i) = T^f_{ij,j} + \rho^f b_i + \bar{q}_i^f. \tag{2c}$$

The vector $\bar{q}_i^\alpha$ describes the forces due to the interaction. Since these forces are balanced for the mixture $\bar{q}_i^s = \bar{q}_i^f$ holds. Splitting up the total stresses into partial stresses can be achieved only if the stresses caused by diffusion are neglected. Notice, that these partial stresses of the constituents are not identical with effective stresses in the usual concepts of soil mechanics:

$$T^s_{ij} = T^{eff}_{ij} - n^s p\,\delta_{ij}, \tag{3a}$$

$$T^f_{ij} = -n^f p\,\delta_{ij}. \tag{3b}$$

The interaction forces arise through friction between skeleton and fluid, moving relative to the structure, and are described by the constitutive relations

$$q_i^s = -q_i^f = \Xi_{ij}\,\bar{v}_j, \tag{4}$$

$$\Xi_{ij} = n^f\gamma^f g\,k^{-1}_{ij}. \tag{5}$$

The material tensor Ξ_{ij} contains the Darcy-coefficients which can be determined by experiment.

3 EQUATIONS IN THE REFERENCE STATE

Expressions (1) and (2) are referred to an actual unknown spatial configuration (index i). They have to be referred to a known configuration (index I) in order to clearly determine the action of external forces and to describe the pathdependent constitutive material law. In case of the two-phase system being composed of structure and fluid material coordinates of the skeleton X_I^s are used for the definition of the reference configuration:

$$x_i = x_i(X_I^s, t). \tag{6}$$

As fluid and structure particles follow different deformation pathes, the motion of the fluid has to be referred to the motion of the skeleton. Thus, all material derivatives must be related to the movement of the structure.

With the assumption of incompressibility of the grains, it follows from the continuity equations (1b) and $J^s v_{i,i} = D^s/Dt\ J^s$ (J^s is the Jacobian of the skeleton) that

$$\frac{D^s}{Dt}(J^s n^s) = 0, \tag{7a}$$

$$J^s n^s = n^{so}, \tag{7b}$$

$$J^s n^f = (J_s - n^{so}). \tag{8}$$

In deriving the continuity equation for the fluid in the reference state it is taken into account that

- changes of density of an elastic fluid is proportional to the change of pressure

$$\frac{1}{\gamma^f}\frac{D^f}{Dt}\gamma^f = -\frac{1}{k^f}\left(\frac{D^s}{Dt}p + \bar{v}_i\, p_{,i}\right),$$

- the change of volume fraction n^f is given by

$$\frac{D^f}{Dt}n^f = \frac{n^{so}}{J^s}\, v^s_{i,i} + \bar{v}_i\, n^f_{,i}.$$

This leads to:

$$J^s(x_{I,i}\, v^s_{i,I}) + (J^s - n^{so})(x_{I,i}\, \bar{v}_{i,I})$$

$$+ J^s(\bar{v}_i\, x_{I,i}\, n^f_{,I}) \tag{9}$$

$$- \frac{J^s - n^{so}}{k^f}\left(\frac{D^s}{Dt}p + \bar{v}_i\, x_{I,i}\, p_{,I}\right) = 0.$$

The equations of motion in the reference state are expressed in terms of the second Piola Kirchhoff stress tensor $\hat{T}_{ij}$. This quantity and also its conjugate variable, the Green's deformation tensor, as well as their rates are objective, an important feature for the elastoplastic material law.

With the transformation of the acceleration of the fluid and the stresses the equations of motion for the constituents in the reference configuration read:

Skeleton: $\tilde{\rho}^s = n^{so}\,\gamma^{so}$

$$\tilde{\rho}^s \frac{D^s}{Dt} v^s_i = (x_{i,I}\,\hat{T}^{eff}_{IJ})_{,J} + x_{I,i}(n^{so}\,p)_{,I} \tag{10}$$

$$+ x_{I,i}\,\Xi_{IJ}\,x_{J,j}\,\bar{v}_j + \tilde{\rho}^s\, b^o_i,$$

Fluid: $\tilde{\rho}^f = \gamma^f(J^s - n^{so})$

$$\tilde{\rho}^f \frac{D^s}{Dt}(v^s_i + \bar{v}_i) + \tilde{\rho}^f((v^s_i + \bar{v}_i)_{,J}\, x_{J,j}\, \bar{v}_j)$$

$$= x_{I,i}((J^s - n^{so})p)_{,I} \tag{11}$$

$$- x_{I,i}\,\Xi_{IJ}\,x_{J,j}\,\bar{v}_j + \tilde{\rho}^f\, b^o_i.$$

4 MATERIAL LAW FOR SAND

For the description of the reversible and irreversible deformations under loading and unloading a rate independent, elasto-plastic material law is used. We assume with small elastic strains that the total strains can be written as the sum of the elastic and plastic parts:

$$\dot{\varepsilon}_{ij} = \dot{\varepsilon}^{el}_{ij} + \dot{\varepsilon}^{pl,iso}_{ij} + \dot{\varepsilon}^{pl,dev}_{ij}, \tag{12}$$

$\dot{\varepsilon}^{pl,iso}_{ij}$ = strain rates due to hydrostatic stresses,

$\dot{\varepsilon}^{pl,dev}_{ij}$ = strain rates due to deviatoric stresses.

Each component is related to the corresponding stresses by a separate law without violating restrictions due to consistency. For the elastic part a nonlinear law, using a variable bulk modulus and a constant Poisson's ratio, is assumed. Generally, the bulk modulus is a function of the first isotropic and the second deviatoric stress invariant and is defined by (13) with the material parameters k^{el} and n

$$K = k^{el}\left(\frac{1}{3}I_1^2 + \frac{2}{3}\frac{(1+\nu)}{(1-2\nu)}J_2^2\right)^{n/2}, \tag{13}$$

$$I_1 = \frac{1}{\sqrt{3}}\sigma_{ij}\,\delta_{ij} \quad ; \quad J_2 = \frac{1}{2}S_{ij}\,S_{ij} \quad ;$$

$$S_{ij} = \sigma_{ij} - \frac{1}{\sqrt{3}}I_1\,\delta_{ij}.$$

This relation satisfies requirements of path independence and energy conservation in closed cycles of stress.

Plastic deformations induced by shear loading are modelled using a modified form of a constitutive law proposed recently by Poorooshasb-Pietruszczak (1986). In order to describe the material behavior under various pathes the stress space is subdivided in different regions utilizing a three-surface-model (Figure 1) based on bounding plasticity concepts.

The failure of the material is defined by a rigid limit surface F_F. The memory surface F_M, which expands isotropically, describes the maximum stress state reached during loading history. The yield surface F_Y which in un- and reloading moves kinematically within the memory surface seperates

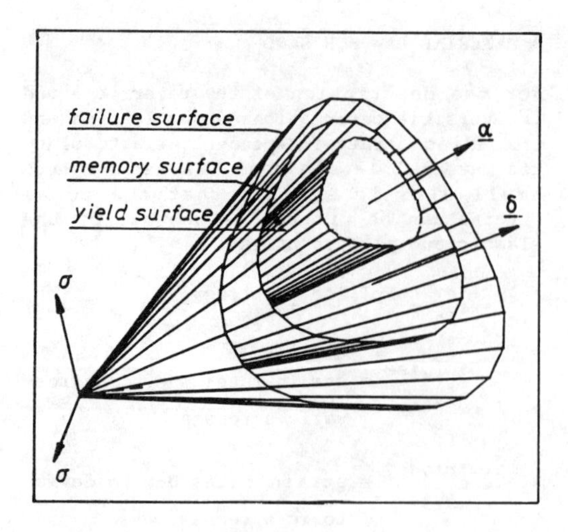

Figure 1. Different surfaces in stress space

the elastic and plastic regions. In addition, it is expanding under primary loading:

$$F_F(\sigma_{ij}) = \sqrt{(2J_2)} + \eta_F\, g(\theta)\, I_1, \quad (14a)$$

$$F_M(\sigma_{ij}, \eta_M) = \sqrt{(2J_2)} + \eta_M\, g(\theta)\, I_1, \quad (14b)$$

$$F_Y(\sigma_{ij}, \alpha_{ij}, \eta_Y) = \sqrt{(2\bar{J}_2)} + \eta_Y\, g(\bar{\theta})\, \bar{I}_1. \quad (14c)$$

With the direction vector of the yield surface α_{ij} and $\bar{S}_{ij} = \sigma_{ij} - (\sigma_{kl}\,\alpha_{kl})\,\alpha_{ij}$ the anisotropic invariants are obtained:

$$\bar{I}_1 = \sigma_{ij}\,\alpha_{ij} \quad , \quad \bar{J}_2 = \frac{1}{2}\,\bar{S}_{ij}\,\bar{S}_{ij} \quad .$$

The shape of the surfaces in the Π-plane is described with function $g(\theta)$ (original form is due to Stutz 1972). θ is the Lode angle and γ a material parameter connected with different angles of friction in triaxial compression and extension:

$$g(\theta) = \gamma\left(\frac{2}{(1-\sin 3\theta) + \gamma^4 (1+\sin 3\theta)}\right)^{.25}, \quad (15)$$

$$\eta_F = \frac{2\,\sqrt{2}\,\sin\phi}{3 - \sin\phi}\,. \quad (16)$$

Under primary loading both the memory and the yield surface are expanding. The isotropic hardening parameters are described by a hyperbolic hardening law (Momen 1980) as a function of the accumulated plastic deviatoric strains:

$$\eta_M = \eta_F\,\frac{\xi}{a + \xi}, \quad (17)$$

$$\eta_Y = \eta_F\,\frac{b\,\xi}{a + \xi}\,, \quad (18)$$

$$\xi = \sqrt{(2J_2^{\varepsilon,pl})}. \quad (19)$$

The parameter b describes the growth of the elastic region and usually ranges up to 0.5, the parameter a defines the shape of hyperbolic relation.

In unloading the yield surface moves unchanged within the memory surface. The magnitude of the plastic strain rates depends on the distance of the two surfaces, on the elastic moduli, on the tangent modulus of the memory surface, and on the material parameter d (Mroz et al. 1979).

The directions of plastic strain rates are defined using a potential formulation corresponding to the critical state concept:

$$G(\sigma_{ij}) = I_1\,\exp\left(\frac{\sqrt{(2J_2)}}{\eta_c\, g(\theta)\, I_1}\right) - I_0. \quad (20)$$

The potential function belonging to the yield surface is determined by geometric considerations (see Figure 2).

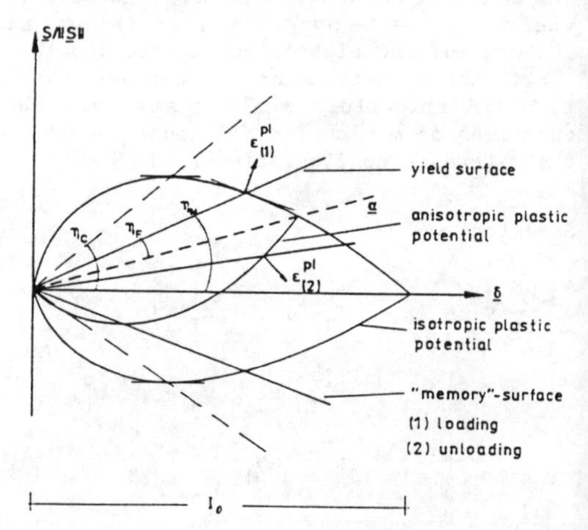

Figure 2. Isotropic and anisotropic plastic potentials

The last material relation gives the plastic deformations due to hydrostatic loading and it is assumed that the material behaves elastically during unloading. The rates of the plastic strains which arise from initial loading are determined by a cap placed upon the yield cone, its directions are given by an associated flow rule:

$$F_{\kappa}(\sigma_{ij},\alpha_{ij},\kappa) = \bar{I}_1 + \kappa = 0, \qquad (21)$$

$$\dot{\varepsilon}^{pl,iso}_{ij} = \frac{1}{H_k} \overset{\circ}{\bar{I}}_1 \alpha_{ij}, \qquad (22)$$

$$H_k = k^{pl} (\bar{I}_1)^m. \qquad (23)$$

k^{pl} and m are material parameters.

According to Table 2 all material parameters should be determined from conventional experiments in soil mechanics. Because of various phenomena and restrictions in experimental investigations it is appropriate to use different test procedures. For experimental investigations a fine sand (ρ_{max} = 1.70 g/cm^3, ρ_{min} = 1.41. g/cm^3) with ρ_d = 1.63 g/cm^3 and n^s = 0.387 was used.

Table 2. Material parameters

Material Law	Param.	Test
nonlinear elastic	k^{el}	isotropic compression
	n	-> un-/ reloading
isotropic plastic	k^{pl}	isotropic compression
	m	-> loading
deviatoric plastic	ν	static triaxial compresssion/ extension
	η_F	
	η_C	
	γ	
	a	
	b	cyclic triaxial shear test
	d	

The parameters for the nonlinear elastic and isotropic plastic law are calculated from cyclic compression tests assuming that unloading is defined by the elastic law. Figure 3 shows comparisons between experiment and numerical simulation.

Informations about deformation and failure behavior under shear loading are obtained from triaxial compression tests in view of known elastic and isotropic plastic parts. Poisson's ratio is evaluated from small unloading cycles defining stress dependent elastic shear modulus. The relation between stress ratio $\eta = \sqrt{2}\, q/3p$ and plastic deviatoric strain allows to determine the coefficient a by matching experimental data and theoretical expression (17), while the relation between η and the plastic volumetric strains shows the volume changes under shear and continuous transition from contractant to dilatant behavior.

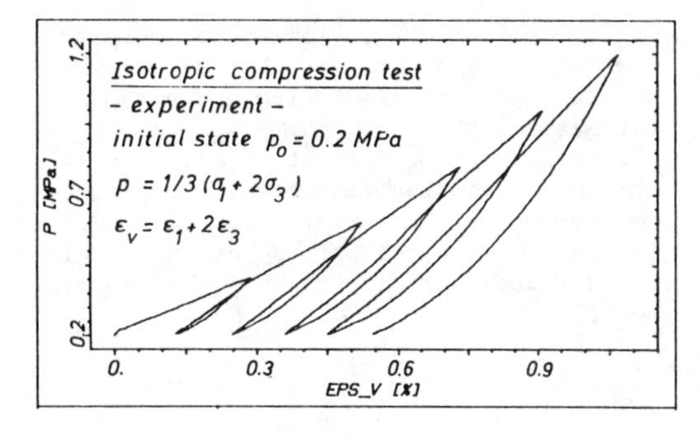

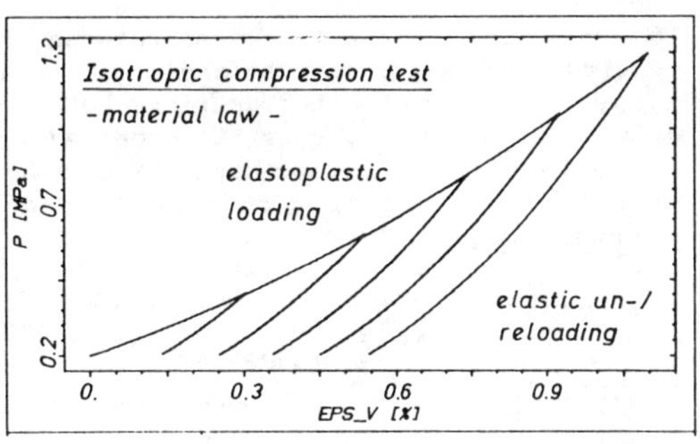

Figure 3. Isotropic compression - experiment and numerical simulation

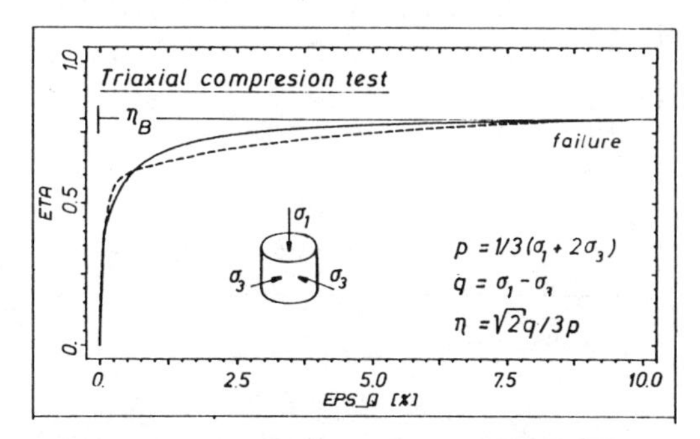

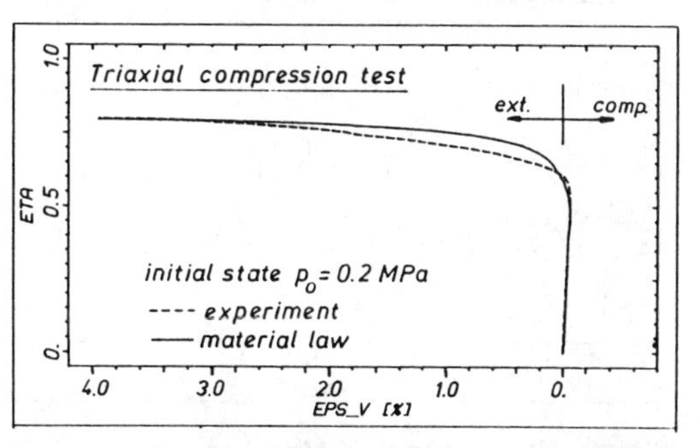

Figure 4. Triaxial compression - experiment and numerical simulation

According to the concept of the "characteristic line" (Luong 1982) the condition $\dot{\varepsilon}_v^{pl} = 0$ renders the coefficient η_c of the plastic potential. Following this procedure numerical simulation (Figure 4) demonstrates good agreement with the experimental data.

Finally the coefficients b and d defining the unloading behavior have to be calculated from cyclic triaxial shear tests. This is done by statistical operations in order to guarentee optimal curve fit of experimental data.

5 NUMERICAL RESULTS

To demonstrate the differences between the results of a linear and the nonlinear analysis described in the paper the behavior of a layer of sand (Figure 5a) under earthquake loading is investigated. The system is subjected initially to dead load and constant load p_o.

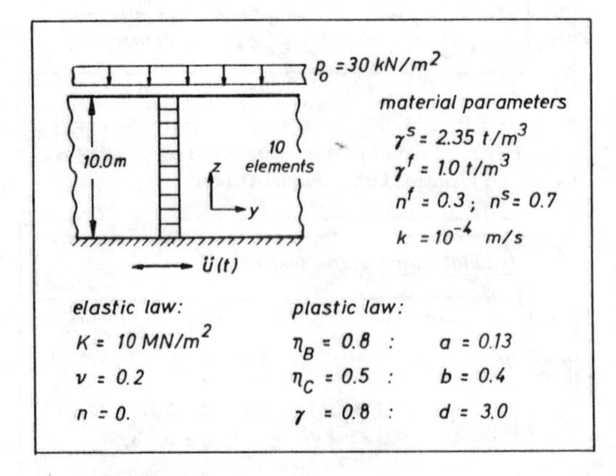

Figure 5a. System

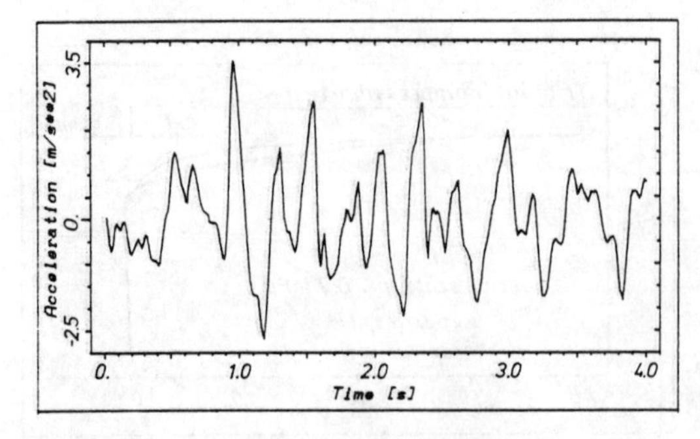

Figure 5b. Earthquake time history

The time history of the loading (taken from the earthquake of Friaul, Italy) is given in Figure 5b and is applied as the horizontal component of the deflection at the foot. As resulting response the corresponding vibrations on the surface are calculated, the reactions of the fluid being characterized by a fictitious deflection $\bar{u}_i = \int_t \bar{v}_i\, dt$.

Example 1: Saturated sand

In the first analysis only the geometrically nonlinear terms are taken into account. Compared with the pure linear problem the horizontal deflections first show smaller but later considerably larger amplitudes than the linear case (Figure 6a,b). This is

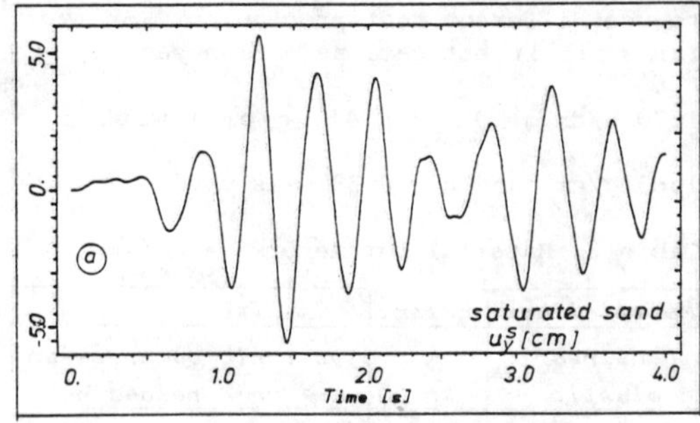

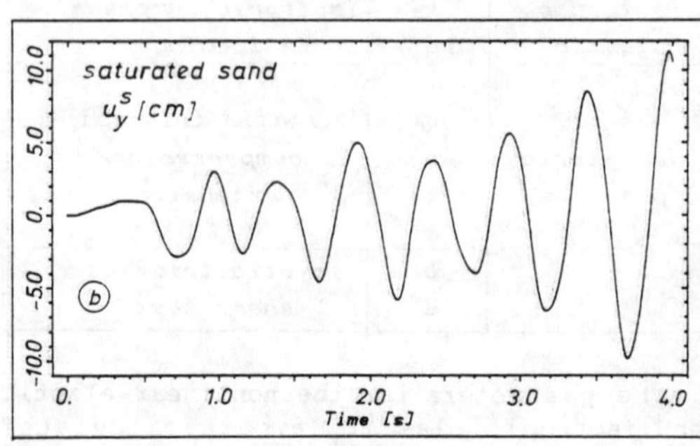

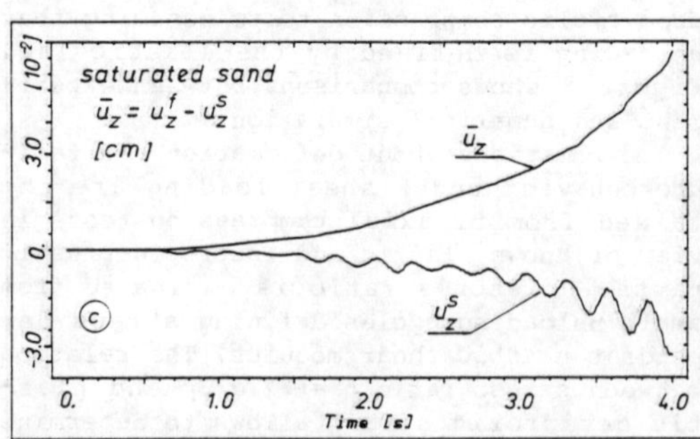

Figure 6. Saturated sand - geom. linear (a) and geom. nonlinear (b and c) description

caused by the nonlinear coupling of the volumetric and deviatoric parts of the deflection. The material first behaves stiffer due to the fluid which is nearly incompressible compared with the skeleton. The resulting pressure of the pore water then decreases caused by its flow and leads to a smaller stiffness of the whole soil (Figure 6c). This reduction in turn gives rise to higher compaction in the skeleton resulting in increasing vertical deflections of the solid. The whole process can be modelled only by its geometric nonlinear description and gives important information in connection with the material relations.

Example 2: Dry sand

In this analysis the geometric nonlinear and physically nonlinear terms are taken into account. The system initially reacts elastically but is damped later by the material nonlinearities (Figure 7a). This damping effect is important for the coupled soil-structure-interaction as it reduces the stresses and deflections of the structure. Also the soil undergoes permanent deflections. They follow from the cyclic shear load and lead to accumulating volumetric compaction (Figure 7b). This effect may also be observed in experiments and can be described realistically only if the plastic strains under un- and reloading are considered. An isotropic hardening rule is therefore not sufficient to model the behavior under cyclic loading.

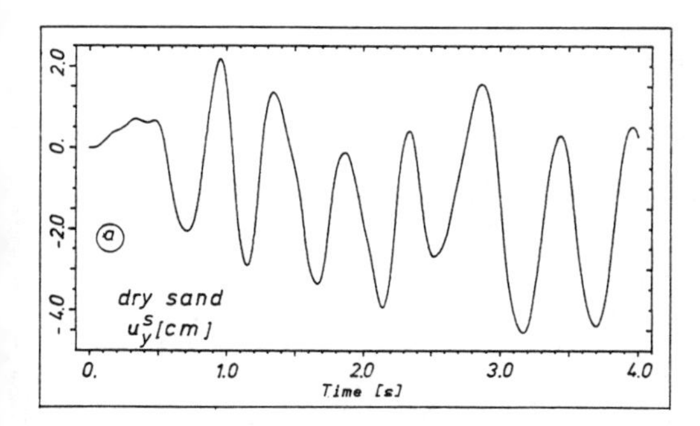

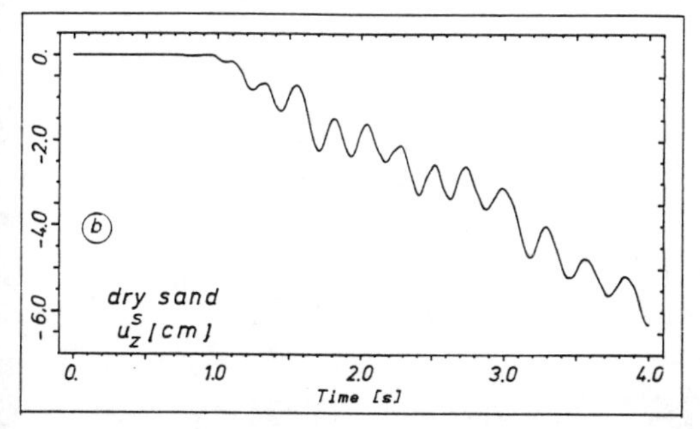

Figure 7. Dry sand

Example 3: Saturated sand

The last example also deals with both the geometric and physical nonlinear description of the soil behavior but includes the influences of the fluid. In difference to dry soils, we observe pore pressure build up due to cyclic loading thus reducing the isotropic effective stresses. Shear stresses which are of the same magnitude as in Example 2 now give rise to higher plastic strains with the actual stress state approaching the failure. Indeed, the horizontal displacements (shown in Figure 8a) are much more pronounced as in case of the dry soil. Furthermore strong damping of excitation response indicates high energy dissipation due to plastic flow. Both, the pronounced deformation as well as the reduction of shear strength are important for stability analysis of the structure. Thus, the higher effort in numerical calculation is justified by a more realistically modelling of the soil behavior.

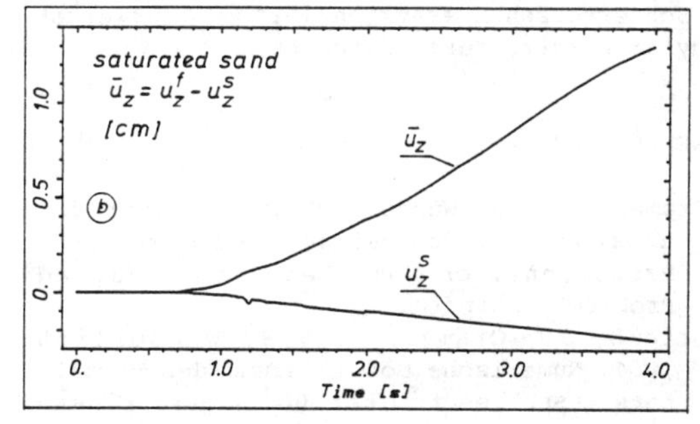

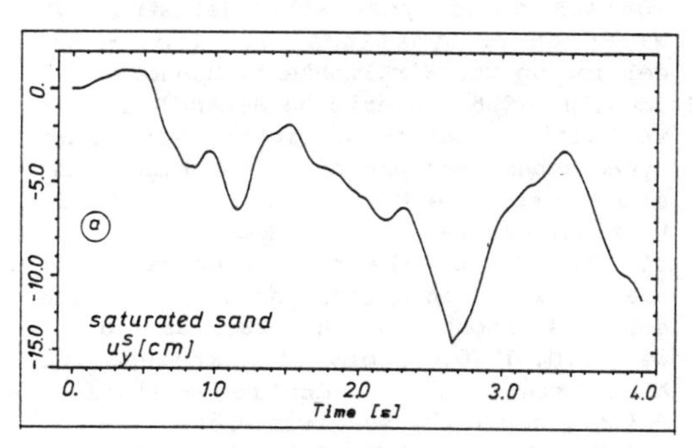

Figure 8. Saturated sand - geom. and phys. nonlinear description

6 CONCLUSIONS

The procedure presented in the paper is well suited for the nonlinear problem of soil-structure interaction, and it allows to extend existing finite element programs by the two-phase model. It is applicable for dry as well as for water-saturated sand under drained and under undrained conditions. It gives satisfactory results for monotic and for cyclic loading and describes most of the phenomena which are observed in experiments.

For dry sand the model describes continuous compaction of the skeleton under cyclic shear. In the case of water-saturated sand it correctly simulates the increase in the pore pressure leading to a decrease in effective stress of the skeleton and to reduced stiffness of the soil. In the numerical simulations of the behavior of the soil this allows to include the imporant case of liquification under dynamic loading in addition to the description of shear failure.

ACKNOWLEDGEMENT

The authors greatfully acknowledge the financial support of the Deutsche Forschungsgemeinschaft through project C2 of the "Sonderforschungsbereich 151" on Tragwerksdynamik (structural dynamics).

REFERENCES

Cramer, H. & W. Wunderlich 1981. Numerical treatment of rock-structure interaction. Proc. Conf. on Num. Meth. for Coupled Problems, Swansea.

Strauß, J., Cramer, H. & W. Wunderlich 1984. Numerische Modellierung des Verhaltens wassergesättigter Böden unter harmonischer und dynamischer Belastung. In Natke (ed.): Dynamische Probleme - Modellierung und Wirklichkeit, Hannover.

Strauß, J. 1986. Numerische Behandlung des Verhaltens wassergesättigter Böden unter dynamischer Beanspruchung. Technisch wissenschaftliche Mitteilungen Nr. 86-2, Ruhr-Universität Bochum, Bochum.

Biot, R.M. 1956. Theory of propagation of elastic waves in a fluid saturated porous solid. J. Acous. Soc. Am. 28: 168-178.

Bowen, R.M. 1976. Theory of mixtures. In: A.C. Eringen (ed.), Continuum Physics, Vol. 3, New York, Academic Press.

Dafalias, Y.F. & E.P. Popov 1975. A model of nonlinearly hardening materials for complex loading. Acta Mechanica 21: 173-192.

Poorooshasb, H.B. & S. Pietruszczak 1986. A generalized flow theory for sand. Soils and Foundations 26: 1-15.

Stutz, P. 1972. Comportement élasto plastique des milieux granulaires. Proc. Int. Symp. Foundation of Plasticity, Warscew.

Momen, H. 1980. Modelling and analysis of cyclic behavior of sands. PH.D. Thesis, University of Illinois, Urbana.

Mroz, Z., Norris, V.A. & O.C. Zienkiewicz 1979. Application of an anisotropic hardening model in the analysis of elastoplastic deformation of soils. Geotechnique 29: 1-34

Schofield, A. & P. Wroth 1968. Critical state soil mechanics. McGraw-Hill, London.

Luong, M.P. 1982. Mechanical aspects and thermal effects of cohesionsless soils under cyclic and transient loading. In P.A. Vermeer and H.J. Luger (eds.), Deformation and Failure of Granular Material, Wiley.

Numerical Methods in Geomechanics (Innsbruck 1988), Swoboda (ed.)
© 1988 Balkema, Rotterdam. ISBN 90 6191 809 X

A constitutive model of soils for evaluating principal stress rotation

H.Matsuoka & Y.Suzuki
Nagoya Institute of Technology, Japan
K.Sakakibara
Japan Highway Public Corporation

ABSTRACT: A constitutive model for soils by which the general strain increments ($d\varepsilon_x$, $d\varepsilon_y$ and $d\gamma_{xy}$) are directly related to the general stress increments ($d\sigma_x$, $d\sigma_y$ and $d\tau_{xy}$) is proposed. The stress-strain matrix between the strain increments and the stress increments is expressed in general coordinates. In order to evaluate the influence of rotation of the principal stress axes on strains, "principal stress rotation tests", in which the stress path is the circumference of a Mohr's circle, have been carried out by a "two-dimensional general stress apparatus" using a stack of aluminium rods. Simple shear tests and liquefaction tests due to principal stress rotation are also analyzed by the proposed model. The model is extended to a three-dimensional one.

1 INTRODUCTION

It is natural to consider that the change in stresses on an arbitrary plane causes strains in every kind of material. In the theory of plasticity, however, such an idea is not satisfied because the constitutive relation is usually formulated by the stress invariants. For example, the strains caused by the "principal stress rotation test" along a Mohr's stress circle under constant principal stresses cannot be explained by the usual theory of plasticity. In the present paper, a constitutive model for soils by which the general strain increments are directly related to the general stress increments is presented. The model predicts the strains not only due to "shear" and "consolidation", but also due to "principal stress rotation".

2 A CONSTITUTIVE MODEL FOR SOILS EXPRESSED IN GENERAL COORDINATES

The following equation has been derived from assuming the hyperbolic relationship between the shear-normal stress ratio (τ_{xy}/σ_x or τ_{xy}/σ_y) and the shear strain (γ_{xy}) expressed in general coordinates (Matsuoka et al., 1986; 1987).

$$\gamma_{xy} = \frac{1}{G_o} \cdot \frac{\sin\phi \cdot \sin\phi_{mo} \cdot \sin 2\alpha}{\sin\phi - \sin\phi_{mo}} \tag{1}$$

where G_o is the gradient of the initial tangent of the hyperbolic relationship, ϕ is the internal friction angle, ϕ_{mo} is the mobilized internal friction angle ($\sin\phi_{mo}=(\sigma_1-\sigma_3)/(\sigma_1+\sigma_3)$) and α is the angle of the principal stress direction. As $1/G_o$ ($= k_s$) is approximately proportional to the logarithm of the mean principal stress σ_m, γ_{xy} in Eq.(1) is considerd to be the function of ϕ_{mo}, α and σ_m. By totally differentiating γ_{xy} with respect to ϕ_{mo}, α and σ_m, [1]$d\gamma_{xy}^s$: shear strain increment due to shear ($d\phi_{mo}$), [2]$d\gamma_{xy}^r$: shear strain increment due to principal stress rotation ($d\alpha$) and [3]$d\gamma_{xy}^{ac}$: shear strain increment due to anisotropic consolidation ($d\sigma_m$) have been derived. It is considered in $d\gamma_{xy}^r$ that the principal stress direction deviates by δ from the principal strain increment direction. Combining the above three shear strain increments $d\gamma_{xy}^s$, $d\gamma_{xy}^r$ and $d\gamma_{xy}^{ac}$ with the stress ratio vs. strain increment ratio relation, which corresponds to the flow rule in the theory of plasticity, the normal strain increments $d\varepsilon_x$ and $d\varepsilon_y$ for each shear strain increment can be obtained. In addition to the above-mentioned strain increments, [4]$d\varepsilon_x^{ic}=d\varepsilon_y^{ic}$: normal strain increments due to isotropic consolidation should be taken into account. As $d\phi_{mo}$, $d\alpha$ and $d\sigma_m$ can be expressed by the general stress increments ($d\sigma_x$, $d\sigma_y$ and $d\tau_{xy}$), the general strain increments can be directly related to the general stress increments as follows (Matsuoka et al., 1986; 1987):

$$\{d\varepsilon_x, d\varepsilon_y, d\gamma_{xy}\}^T = [D]^{-1}\cdot\{d\sigma_x, d\sigma_y, d\tau_{xy}\}^T \quad (2)$$

where [D] is the stress-strain matrix.

3 COMPONENTS OF THE STRESS-STRAIN MATRIX

The components of the matrix $[D]^{-1}$ can be written in the four domains (see Fig.1) as follows (Matsuoka et al., 1986; 1987):

$$[D]^{-1} = \begin{bmatrix} D_{11} & D_{12} & D_{13} \\ D_{21} & D_{22} & D_{23} \\ D_{31} & D_{32} & D_{33} \end{bmatrix} \quad (3)$$

I. $(d\phi_{mo} \geq 0,\ d\sigma_m \geq 0)$

$D_{11} = FF \times SS \times A1 + FR \times RR \times B1 + FF \times CA \times C1 + CCI \times C1$
$D_{12} = FF \times SS \times A2 + FR \times RR \times B2 + FF \times CA \times C1 + CCI \times C1$
$D_{13} = FF \times SS \times A3 + FR \times RR \times B3$
$D_{21} = GG \times SS \times A1 + GR \times RR \times B1 + GG \times CA \times C1 + CCI \times C1$
$D_{22} = GG \times SS \times A2 + GR \times RR \times B2 + GG \times CA \times C1 + CCI \times C1$
$D_{23} = GG \times SS \times A3 + GR \times RR \times B3$
$D_{31} = HH \times SS \times A1 + HR \times RR \times B1 + HH \times CA \times C1$
$D_{32} = HH \times SS \times A2 + HR \times RR \times B2 + HH \times CA \times C1$
$D_{33} = HH \times SS \times A3 + HR \times RR \times B3$

II. $(d\phi_{mo} \geq 0,\ d\sigma_m < 0)$

$D_{11} = FF \times SS \times A1 + FR \times RR \times B1 + CSI \times C1$
$D_{12} = FF \times SS \times A2 + FR \times RR \times B2 + CSI \times C1$
$D_{13} = FF \times SS \times A3 + FR \times RR \times B3$
$D_{21} = GG \times SS \times A1 + GR \times RR \times B1 + CSI \times C1$
$D_{22} = GG \times SS \times A2 + GR \times RR \times B2 + CSI \times C1$
$D_{23} = GG \times SS \times A3 + GR \times RR \times B3$
$D_{31} = HH \times SS \times A1 + HR \times RR \times B1$
$D_{32} = HH \times SS \times A2 + HR \times RR \times B2$
$D_{33} = HH \times SS \times A3 + HR \times RR \times B3$

III. $(d\phi_{mo} < 0,\ d\sigma_m < 0)$

$D_{11} = FFU \times SSU \times A1 + FR \times RR \times B1 + CSI \times C1$
$D_{12} = FFU \times SSU \times A2 + FR \times RR \times B2 + CSI \times C1$
$D_{13} = FFU \times SSU \times A3 + FR \times RR \times B3$
$D_{21} = GGU \times SSU \times A1 + GR \times RR \times B1 + CSI \times C1$
$D_{22} = GGU \times SSU \times A2 + GR \times RR \times B2 + CSI \times C1$
$D_{23} = GGU \times SSU \times A3 + GR \times RR \times B3$
$D_{31} = HH \times SSU \times A1 + HR \times RR \times B1$
$D_{32} = HH \times SSU \times A2 + HR \times RR \times B2$
$D_{33} = HH \times SSU \times A3 + HR \times RR \times B3$

IV. $(d\phi_{mo} < 0,\ d\sigma_m \geq 0)$

$D_{11} = FFU \times SSU \times A1 + FR \times RR \times B1 + CCI \times C1$
$D_{12} = FFU \times SSU \times A2 + FR \times RR \times B2 + CCI \times C1$
$D_{13} = FFU \times SSU \times A3 + FR \times RR \times B3$
$D_{21} = GGU \times SSU \times A1 + GR \times RR \times B1 + CCI \times C1$
$D_{22} = GGU \times SSU \times A2 + GR \times RR \times B2 + CCI \times C1$
$D_{23} = GGU \times SSU \times A3 + GR \times RR \times B3$
$D_{31} = HH \times SSU \times A1 + HR \times RR \times B1$
$D_{32} = HH \times SSU \times A2 + HR \times RR \times B2$
$D_{33} = HH \times SSU \times A3 + HR \times RR \times B3$

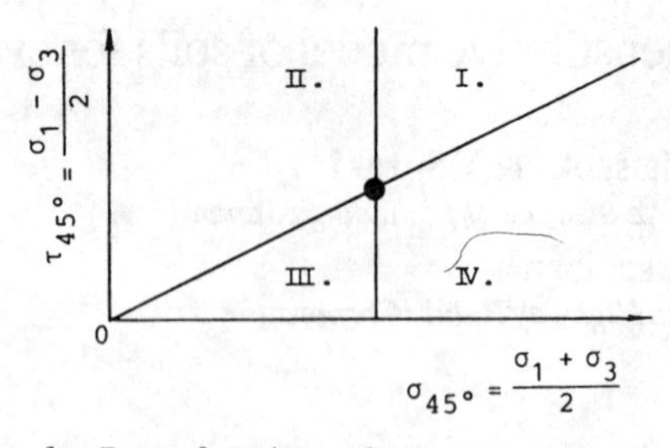

Fig. 1. Four domains of stress changes

where

$$FF = \frac{1}{2}\{2\frac{\mu - \tan\phi_{mo}}{\lambda}\cdot\cos\phi_{mo} + \sin\phi_{mo} + \cos 2\alpha\}$$

$$GG = \frac{1}{2}\{2\frac{\mu - \tan\phi_{mo}}{\lambda}\cdot\cos\phi_{mo} + \sin\phi_{mo} - \cos 2\alpha\}$$

$$FFU = \frac{1}{2}\{2\frac{-\mu - \tan\phi_{mo}}{\lambda}\cdot\cos\phi_{mo} + \sin\phi_{mo} + \cos 2\alpha\}$$

$$GGU = \frac{1}{2}\{2\frac{-\mu - \tan\phi_{mo}}{\lambda}\cdot\cos\phi_{mo} + \sin\phi_{mo} - \cos 2\alpha\}$$

$$HH = \sin 2\alpha$$

$$FR = \frac{1}{2}\{2\frac{\mu - \tan\phi_{mo}}{\lambda}\cdot\cos\phi_{mo} + \sin\phi_{mo} + \cos 2(\alpha + \delta)\}$$

$$GR = \frac{1}{2}\{2\frac{\mu - \tan\phi_{mo}}{\lambda}\cdot\cos\phi_{mo} + \sin\phi_{mo} - \cos 2(\alpha + \delta)\}$$

$$HR = \sin 2(\alpha + \delta)$$

$$SS = k_s\cdot\frac{\sin^2\phi\cdot\cos\phi_{mo}}{(\sin\phi - \sin\phi_{mo})^2}$$

$$SSU = k_s\cdot\frac{\sin^2\phi\cdot\cos\phi_{mo}}{(\sin\phi + \sin\phi_{mo})^2}$$

$$RR = 2k_s\cdot\frac{\sin\phi\cdot\sin\phi_{mo}}{\sin\phi - \sin\phi_{mo}}$$

$$CA = k_c\cdot\frac{\sin\phi\cdot\sin\phi_{mo}}{\sin\phi - \sin\phi_{mo}}$$

$$CCI = \frac{0.434}{2}\cdot\frac{C_c}{1 + e_o} \qquad CSI = \frac{0.434}{2}\cdot\frac{C_s}{1 + e_o}$$

$$A_1 = \frac{I_1/I_2}{8\tan\phi_{mo}}\cdot\cos^2\phi_{mo}\cdot\{2 - (I_1/I_2)\cdot\sigma_y\}$$

$$A_2 = \frac{I_1/I_2}{8\tan\phi_{mo}}\cdot\cos^2\phi_{mo}\cdot\{2 - (I_1/I_2)\cdot\sigma_x\}$$

$$A_3 = \frac{I_1/I_2}{8\tan\phi_{mo}}\cdot\cos^2\phi_{mo}\cdot\{2(I_1/I_2)\cdot\tau_{xy}\}$$

$$B_1 = \cos^2 2\alpha\cdot\frac{(-\tau_{xy})}{(\sigma_x - \sigma_y)^2} \qquad B_2 = \cos^2 2\alpha\cdot\frac{\tau_{xy}}{(\sigma_x - \sigma_y)^2}$$

$$B_3 = \cos^2 2\alpha\cdot\frac{1}{(\sigma_x - \sigma_y)} \qquad C_1 = \frac{1}{(\sigma_x + \sigma_y)}$$

$I_1 = \sigma_x + \sigma_y \qquad I_2 = \sigma_x\sigma_y - \tau_{xy}^2$

In the above equations the letter U means the unloading process. The stress-strain parameters in this constitutive model are k_s, k_c, λ, μ, ϕ, $C_c/(1+e_o)$ and $C_s/(1+e_o)$. ϕ denotes the internal friction angle in terms of effective stresses, C_c the compression index, C_s the swell index and e_o the initial void ratio. k_s is a parameter to be found in Eq.(1) ($1/G_o=k_s$) or in SS, SSU and RR, which decides the magnitude of the shear strains due to shear and principal stress rotation. It is seen from Eq.(1) that k_s is the gradient of the initial tangent of the relation between $(\varepsilon_1-\varepsilon_3)$ and $(\sigma_1-\sigma_3)/(\sigma_1+\sigma_3)=\sin\phi_{mo}$ (in the case of $\alpha=45°$) under a constant mean principal stress σ_m. k_c is a parameter in CA, which decides the magnitude of the shear strain due to anisotropic consolidation. k_c can be calculated from the K_o compression condition as follows:

$$k_c = -\frac{0.434C_c}{1+e_o} \Big/ [\{2\cdot\frac{\mu-\tan\phi_{mo}}{\lambda}\cdot\cos\phi_{mo} + \sin\phi_{mo} - 1\} \cdot\frac{\sin\phi\cdot\sin\phi_{mo}}{\sin\phi-\sin\phi_{mo}}] \qquad (4)$$

where $\phi_{mo} = \sin^{-1}(\frac{1-K_o}{1+K_o})$, $\quad K_o = 1 - \sin\phi$

k_c is also approximately estimated by the following simple equation.

$$k_c = 0.44\frac{C_c}{1+e_o} \qquad (5)$$

λ and μ are parameters in FF, GG, FFU, GGU, FR and GR, which correspond to the gradient and the ordinate intersection of the linear relation between the shear-normal stress ratio and the normal-shear strain increment ratio on the mobilized plane respectively (see Fig.6).

4 EXTENSION TO THREE DIMENSIONS

The constitutive relation mentioned here is a two-dimensional one. In order to extend it to three dimensions, the superposition of the "two-dimensional" principal strain increments is assumed as follows (Matsuoka, 1974; 1982):

$$\left.\begin{aligned} d\varepsilon_1 &= d\varepsilon_{1(12)} + d\varepsilon_{1(13)} \\ d\varepsilon_2 &= d\varepsilon_{2(12)} + d\varepsilon_{2(23)} \\ d\varepsilon_3 &= d\varepsilon_{3(23)} + d\varepsilon_{3(13)} \end{aligned}\right\} \qquad (6)$$

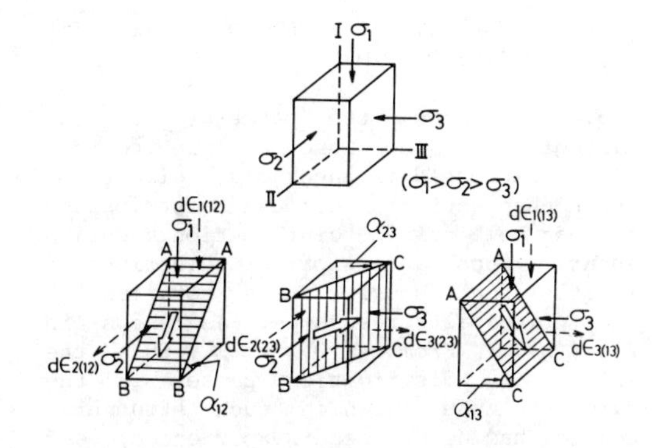

Fig. 2. Superposition of "two-dimensional" principal strain increments under respective pairs of principal stresses

where $d\varepsilon_{1(12)}$, for example, denotes the "two-dimensional" major principal strain increment caused under the two principal stresses σ_1 and σ_2 (see Fig.2). The terms in the right sides of Eq.(6) mean such "two-dimensional" principal strain increments as obtained by Eq.(2).

5 TWO-DIMENSIONAL GENERAL STRESS APPARATUS

A general stress apparatus which can apply the general stresses (σ_x, σ_y and τ_{xy}) independently has been made for a stack of aluminium rods (mixture of ϕ1.6mm and ϕ3.0mm, 50mm in length) as a two-dimensional model of granular materials. The normal stress σ_x in the vertical direction and the normal stress σ_y in the horizontal direction are applied by the wires through pulleys, and the shear stress τ_{xy} is applied by the loading plate on the sample. These three stresses σ_x, σ_y and τ_{xy} can be measured by the load cells independently (see Fig.3). The general strains ε_x, ε_y and γ_{xy} can be also measured by dial gauges independently.

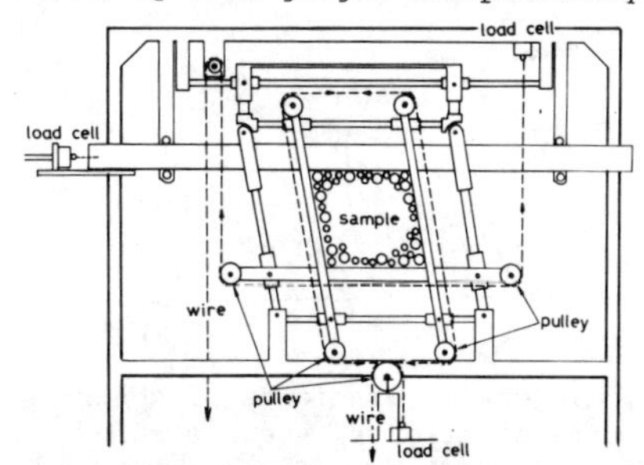

Fig. 3. Main part of two-dimensional general stress apparatus for rod mass

6 PRINCIPAL STRESS ROTATION TESTS WITH SEVERAL ROTATION TIMES

Fig.4(a) shows the pattern of the histogram of the interparticle contact angles expressed in the radial direction, which changes from a circular distribution to an elliptical distribution during shear. Such an elliptical distribution rotates continuously during rotation of the principal stress axes, as shown in Fig.4(b). From $\alpha=0°$ to $\alpha=180°$ the elliptical distribution passes in the "virgin" zone in which such structural change has never been experienced, and after $\alpha=180°$ the experienced structural change will be repeated. Paying attention to the fact that the shear strain increment due to principal stress rotation $d\gamma_{xy}^{r}$ has the same parameter k_s as the shear strain increment due to shear $d\gamma_{xy}^{s}$, the value of k_s is reduced in the experienced structure by measuring the ratio of the tangential gradient Ⓐ to Ⓑ of the cyclic stress-strain relationship at the stress ratio $\sin\phi_a=0.25$ which is the same stress ratio in the principal stress rotation, as shwon in Fig.5. This is based on the similarity of the change in structures during cyclic shearing and principal stress rotation (see Fig.4).

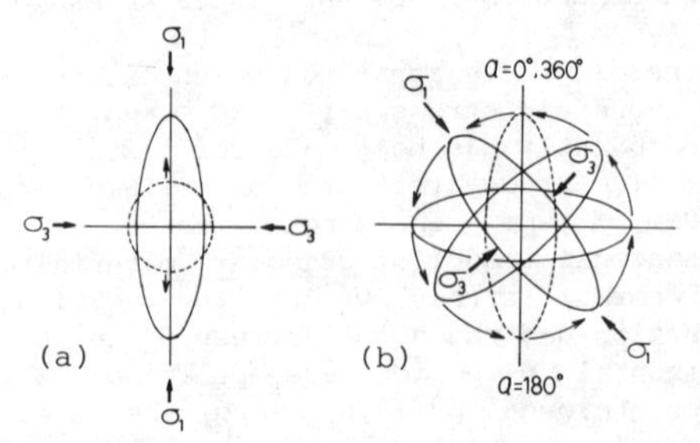

Fig. 4. Pattern of histogram of inter-particle contact angles expressed in radial direction (a) under shear and (b) under principal stress rotation

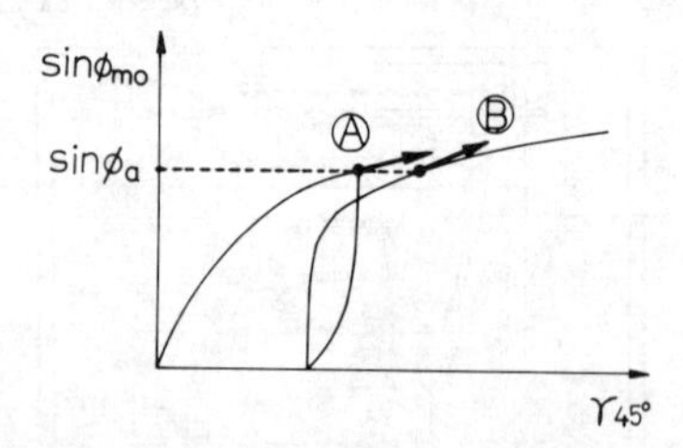

Fig. 5. Tangential gradients of shear-normal stress ratio vs. shear strain relation on 45° plane under virgin and cyclic loadings

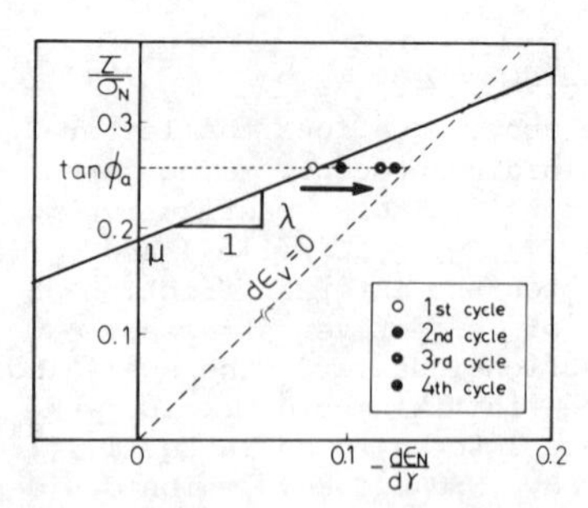

Fig. 6. Shear-normal stress ratio vs. normal-shear strain increment ratio relation on mobilized plane under four principal stress rotations

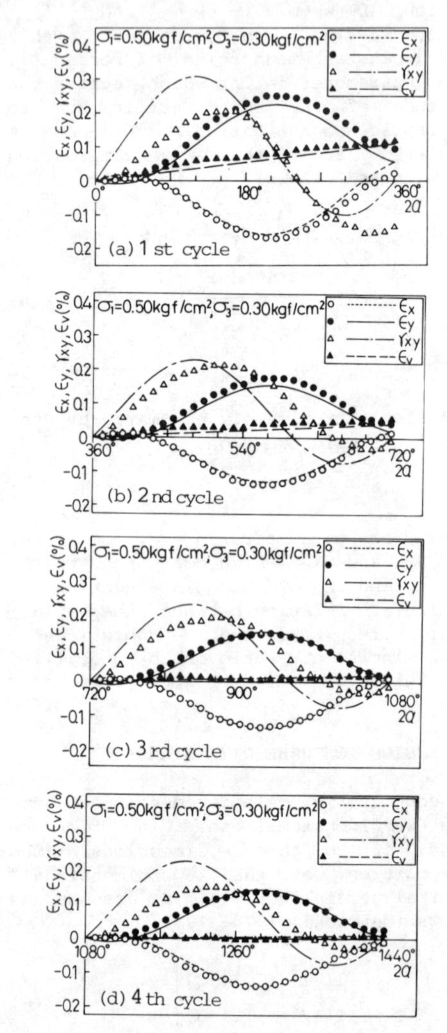

Fig. 7. Comparison between measured strains of principal stress rotation test on rod mass and calculated strains

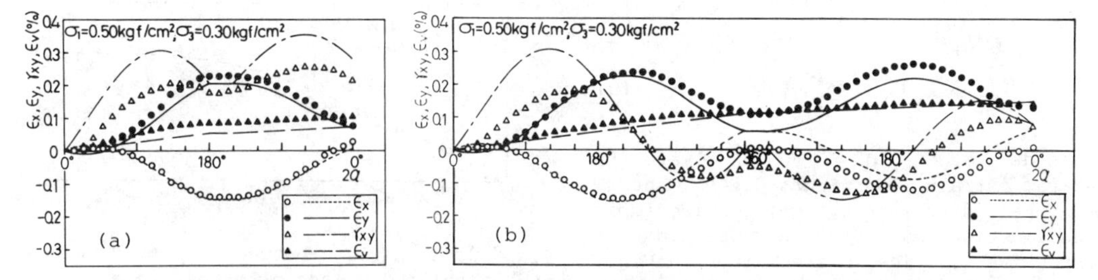

Fig. 8. Comparison between measured strains of principal stress rotation test on rod mass and calculated strains, when rotated reversely at (a)$2\alpha=180°$ and (b)$2\alpha=360°$

Fig.6 shows the shear-normal stress ratio vs. normal-shear strain increment ratio relationship on the mobilized plane ($\alpha=45°+\phi_{mo}/2$) when the stress state circles four times along a Mohr's circle. The broken line in the figure denotes the line of no volumetric strain increment ($d\varepsilon_v=0$). It is seen from Fig.6 that the measured values converge to the line of no volumetric strain increment with the number of rotation times. Therefore, in the analysis the value of the strain increment ratio is assumed to become the intermediate value between the present value and the value at $d\varepsilon_v=0$ of the strain increment ratio.

Fig.7(a)~(d) shows the results of the principal stress rotation test on the stack of aluminium rods in which the stress state circles four times along a Mohr's circle, and the analytical values by the proposed model. It is seen from this figure that the measured strains in the 2nd to 4th cycles are smaller than in the 1st cycle, and the measured volumetric strain ε_v becomes smaller with the number of rotation times. The analytical values based on the above-mentioned ideas explains well such tendency of the measured values.

Fig.8(a) and (b) shows the results of the principal stress rotation tests on the same sample in which the stress states circle from $2\alpha=0°$ to $2\alpha=180°$ and $360°$ in the same direction and then the stress states circle reversely to $2\alpha=0°$, and the analytical values by the proposed model. It is assumed in this analysis that the strains are not produced during $\Delta(2\alpha)=30°$ after the reverse rotation of the principal stress direction, because the measured strains do not change sensitively just after the reverse rotation.

7 SIMPLE SHEAR TESTS

Fig.9(a), (b) and (c) shows the comparison between the measured values of a simple shear test on the stack of aluminium rods and the calculated values. The calculated values are obtained from the following conditions of no vertical stress increment and no lateral strain increment.

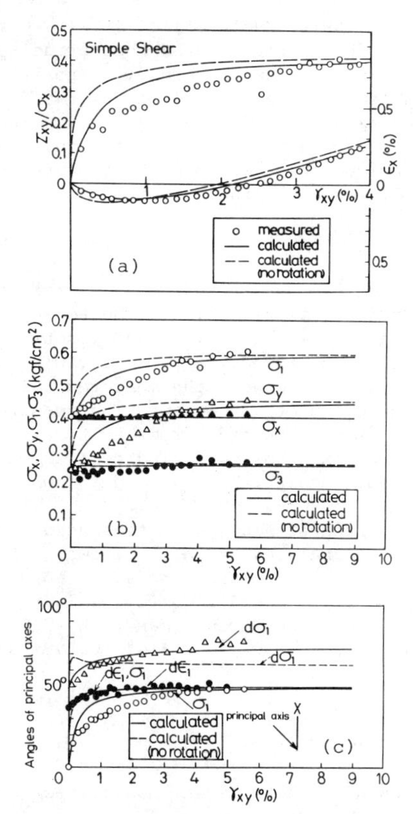

Fig. 9. Comparison between measured values of simple shear test on rod mass and calculated values

$$d\sigma_x = 0 \qquad (\sigma_x = \text{const.}) \tag{7}$$

$$d\varepsilon_y = d\varepsilon_y^s + d\varepsilon_y^r + d\varepsilon_y^{ac} + d\varepsilon_y^{ic} = 0 \tag{8}$$

where $d\varepsilon_y^s$, $d\varepsilon_y^r$, $d\varepsilon_y^{ac}$ and $d\varepsilon_y^{ic}$ are the lateral strain increments due to shear, principal stress rotation, anisotropic consolidation and isotropic consolidation respectively. The broken lines in Fig.9 are the calculated values neglecting the lateral strain increment due to principal stress rotation ($d\varepsilon_y^r$=0). Fig.9(b) shows the measured and calculated values of the changes in stresses (σ_x, σ_y, σ_1 and σ_3) during simple shear. Fig.9(c) also represents the measured and calculated values of the changes in angles of the principal axes (σ_1, $d\sigma_1$ and $d\varepsilon_1$) during simple shear.

8 EXCESS PORE PRESSURE CAUSED BY PRINCIPAL STRESS ROTATION

As compressive volumetric strains are produced during rotation of the principal stress axes as shown in Fig.7, a positive excess pore pressure will be generated if the undrained condition is maintained during the principal stress rotation. We can analyze such excess pore pressure by the following undrained condition.

$$d\varepsilon_v = d\varepsilon_v^s + d\varepsilon_v^r + d\varepsilon_v^{ic} = 0 \tag{9}$$

where $d\varepsilon_v^s$, $d\varepsilon_v^r$ and $d\varepsilon_v^{ic}$ are the volumetric strain increments due to shear, principal stress rotation and isotropic consolidation respectively. Fig.10 shows the comparison between the measured values and the calculated values. The measured values were obtained from a torsional shear test on a hollow-cylindrical Toyoura sand specimen (Ishihara and Towhata, 1983).

The stress-strain parameters used for all these analyses are as follows. For the stack of aluminium rods: k_s=0.31%, λ=0.8, μ=0.19, ϕ=24°, $C_c/(1+e_o)$=1.2% and δ=30°. For the Toyoura sand: k_s=0.23%, λ=1.2, μ=0.20, ϕ=40°, $C_s/(1+e_o)$=0.578% and δ=30°.

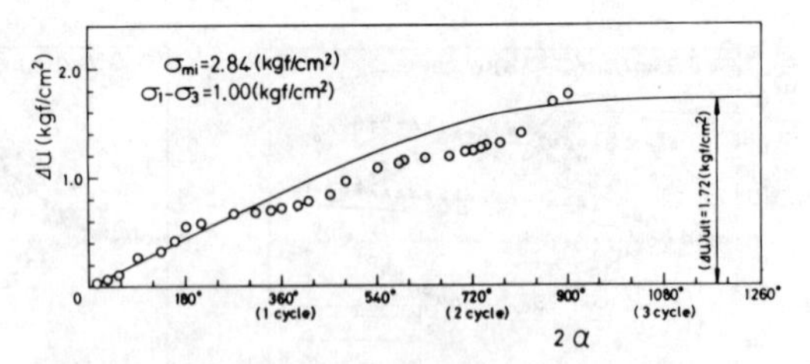

Fig. 10. Comparison between measured values and calculated values of excess pore pressure caused by principal stress rotation

9 CONCLUSIONS

The main results are summarized as follows:

1) A constitutive model for soils by which the general strain increments ($d\varepsilon_x$, $d\varepsilon_y$ and $d\gamma_{xy}$) are directly related to the general stress increments ($d\sigma_x$, $d\sigma_y$ and $d\tau_{xy}$) is proposed. The stress-strain matrix is expressed in general coordinates. The model predicts the strains not only due to "shear" and "consolidation", but also due to "principal stress rotation".
2) The model explains well the results of "principal stress rotation tests" with several rotation times, along a Mohr's stress circle under constant principal stresses. The influence of rotation of the principal stress axes on strains is evaluated by the proposed model.
3) The stress-strain behaviour under simple shear and the excess pore pressure caused by rotation of the principal stress axes are also analyzed by the proposed model.

ACKNOWLEDGEMENTS

The authors wish to thank Prof. T. Nakai at Nagoya Institute of Technology for helpful discussions. They are also indebted to Messrs. Y. Hara and J. Nozoe for their assistance.

REFERENCES

Ishihara, K. & Towhata, I. 1983. Soils and Foundations (23), 4, 11-26.

Matsuoka, H. 1974. Soils and Foundations (14), 2, 47-61.

Matsuoka, H. 1982. Proc. of 4th Int. Conf. on Numerical Methods in Geomechanics, 1, 223-233.

Matsuoka, H., Iwata, Y. & Sakakibara, K. 1986. Proc. of 2nd Int. Symp. on Numerical Models in Geomechanics, Ghent, 67-78.

Matsuoka, H., Iwata, Y., Sakakibara, K. & Hara, Y. 1987. Proc. of 2nd Int. Conf. on Constitutive Laws for Engineering Materials, Theory and Applications, 1, 413-420.

Numerical Methods in Geomechanics (Innsbruck 1988), Swoboda (ed.)
© 1988 Balkema, Rotterdam. ISBN 90 6191 809 X

Assessment of a new class of implicit integration schemes for a cone-cap plasticity model

K.Runesson
Chalmers University of Technology, Gothenburg, Sweden
(Presently at University of Colorado, Boulder, USA)
E.Pramono & S.Sture
University of Colorado, Boulder, USA
K.Axelsson
University of Luleå, Sweden

ABSTRACT: A generalization of the closest point projection algorithm is developed for the integration of constitutive relations in incremental elastic-plastic theory. The algorithm is applied to a two-invariant cone-cap model and treats corner solutions in a straight-forward manner.

1 INTRODUCTION

Finite element algorithms for calculating elastic-plastic deformations of soils are predominantly based on explicit integration of the constitutive relations. Usually the matrix of tangent stiffness moduli is employed in the analysis, or the "elastic" (trial) stress is returned back to the yield surface while assuming that the plastic strain increment direction is known a priori. In the former approach the "plastic" portion of the stress increment becomes tangential to the yield surface in the case of perfect plasticity, and as a consequence the stress has to be corrected in order to satisfy the yield criterion. This correction has often been implemented in a rather arbitrary fashion in many nonlinear finite element codes. A variety of procedures have been proposed during the past two decades such as "back-scaling", "geometric projection", "sub-incrementation of strain"; etc. which have been described by Owen and Hinton [1], Potts and Gens [2].

Algorithms that are based on implicit integration have gained in popularity recently. The most widely used is, undoubtedly, the Backward Euler scheme (BE), which is also known as the Closest Point Projection Method (CPM). For instance, for purely cohesive materials the "elastic" stress is projected onto the yield surface in a geometric sense. The Radial Return Method (RRM), originally proposed by Wilkins [3], is obtained in the special case of J_2-plasticity.

The development of implicit algorithms seems to have followed two main courses: On one hand, the rate equation defining the flow rule is taken as the basis for the straightforward application of, for example, the Generalized Midpoint Rule (GMR). This approach has been followed by Ortiz and Popov [4], Simo and Taylor [5]. On the other hand, the basic constitutive variational inequality may be considered as the basis for the application of a modified GMR leading to the Generalized Closest Point Projection Method (GCPPM) which is the one advocated in this paper. The theory is based on convex analysis as described by Johnson [6]. Numerical algorithms, mainly for the application of the BE method, have been suggested by Nguyen [7], Fröier and Samuelsson [8], Runesson and Booker [9], Runesson et al. [10]. These developments considered algorithms for cohesive materials.

In soil plasticity a non-associated flow rule is usually necessary for realistically modeling dilatancy and other features that are inherent in the drained behavior of soils. This case and algorithmic requirements have been addressed by Runesson [11]. Cone-cap models are often used for soils to account for different failure mechanisms in shear and hydrostatic compression. In such multi-surface models the corners require special attention and, consequently, this problem has been addressed extensively; e.g. Sandler and Rubin [12], Pramono [13], Simo et al. [14]. In the present paper it is shown how the GCPPM can be conveniently used to treat corners and thereby obtain a generally applicable tool.

2 CONSTITUTIVE RELATIONS

2.1 Preliminaries

In this paper we shall focus on a simple analytical modelling problem. To this end it will be sufficient to consider the case of perfect plasticity and we shall in this context use the (convex) set B of admissible stresses $\underset{\sim}{\sigma}$:

$$B = \{\underset{\sim}{\sigma} \mid F(\underset{\sim}{\sigma}) \leq 0\} \tag{1}$$

where $F(\underset{\sim}{\sigma})$ is the yield function. The yield surface ∂B may in the general case be only piecewise smooth; that is a finite number of corners are permitted. We shall then make use of the space of subdifferentials $F_a(\underset{\sim}{\sigma})$ to $F(\underset{\sim}{\sigma})$ that is defined by

$$F_a(\underset{\sim}{\sigma}) = \{\underset{\sim}{a} \mid (\underset{\sim}{\sigma}-\underset{\sim}{\tau}):\underset{\sim}{a} \geq 0,\ \forall\, \underset{\sim}{\tau} \in B\} \tag{2}$$

This space can be said to represent a "fan of admissible normals" located at each corner of the yield surface, where the fan is bounded by the (unique) normal $\partial F/\partial\underset{\sim}{\sigma}$ at each adjacent smooth surface. As F_a is defined in (2), it follows that $\underset{\sim}{a} = 0$ when $\underset{\sim}{\sigma} \in \mathrm{int}(B)$ while $\underset{\sim}{a} \neq 0$ when $\underset{\sim}{\sigma} \in \partial B$. It is clear that $\underset{\sim}{a} = k\ \partial F/\partial\underset{\sim}{\sigma}$, where k is a constant, is the unique direction on a smooth part of ∂B. Subsequently we shall set k = 1 and, hence, denote $\underset{\sim}{a} = \partial F/\partial\underset{\sim}{\sigma}$. This notation will also be used in the non-smooth case by which it is understood that $\partial F/\partial\underset{\sim}{\sigma}$ is not unique but belongs to $F_a(\underset{\sim}{\sigma})$.

The constitutive relations of incremental plasticity can be expressed in terms of three equivalent formulations, that are henceforth denoted (a) the Rate Formulation, (b) the Inequality Formulation and (c) the Tangent Stiffness Formulation. Whereas the latter is frequently adopted for finite element analysis of plasticity problems, Owen and Hinton [1], the two first-mentioned formulations will be adopted in this paper. Small deformation theory and a restricted form of non-associated flow rule will also be assumed. However, in this subsection we shall, for the sake of simplicity, retain the notation pertaining to an associated flow rule. The following expressions relate to the two categories considered.

Rate Formulation:

$$\dot{\underset{\sim}{\sigma}} = \underset{\sim}{D} : (\dot{\underset{\sim}{\varepsilon}} - \dot{\underset{\sim}{\varepsilon}}^P) \tag{3a}$$

$$\dot{\underset{\sim}{\varepsilon}}^P = \lambda\underset{\sim}{a},\quad \underset{\sim}{a} \in F_a \tag{3b}$$

where

$$\lambda \geq 0 \text{ if } \Phi^E(\underset{\sim}{a},\dot{\underset{\sim}{\varepsilon}}) \geq 0, \text{ for some } \underset{\sim}{a} \in F_a;$$

$$\lambda = 0 \text{ otherwise}$$

$$\Phi^E \equiv \underset{\sim}{a} : \dot{\underset{\sim}{\sigma}}^E,\quad \dot{\underset{\sim}{\sigma}}^E = \underset{\sim}{D} : \dot{\underset{\sim}{\varepsilon}} \tag{4}$$

Inequality Formulation:

$$\dot{\underset{\sim}{\sigma}} = \underset{\sim}{D} : (\dot{\underset{\sim}{\varepsilon}} - \dot{\underset{\sim}{\varepsilon}}^P) \tag{5a}$$

$$(\underset{\sim}{\sigma} - \underset{\sim}{\tau}) : \dot{\underset{\sim}{\varepsilon}}^P \geq 0 \quad \forall\underset{\sim}{\tau} \in B \tag{5b}$$

The following notation is used: $\underset{\sim}{D}$ is the tensor of elastic moduli, $\underset{\sim}{\varepsilon}$ is the strain ($\underset{\sim}{\varepsilon}^P$ is the plastic part), and $\dot{\underset{\sim}{\sigma}}^E$ is the "elastic" stress rate. A dot denotes rate. Plastic loading is defined by $\underset{\sim}{\sigma} \in \partial B$ and $\Phi^E > 0$, where Φ^E is the scalar loading function given in (4).

2.2 Model problem: Cone-cap Model

We shall consider a two-surface yield locus, which comprises features that are inherent in existing models for soil and rock-like materials: It consists of a straight cone in the low-confinement region representing cohesive-frictional behavior and an elliptic cap controlling the high-confinement behavior as illustrated in Fig. 1. The locus is defined by five parameters (p_0, p_1, p_2, M_1, M_2) in the p- and q-invariant space, where

$$p = \frac{1}{3}\sigma_{kk},\quad q = \sqrt{\frac{3}{2}}\,|\underset{\sim}{s}|,\quad \underset{\sim}{s} = \underset{\sim}{\sigma} - p\underset{\sim}{\delta} \tag{6}$$

The functional representation is as follows:

$$\text{Cone:}\quad F = q - M_1(p - p_0) \tag{7}$$

$$\text{Cap:}\quad F = \left(\frac{p - p_m}{p_R}\right)^2 + \left(\frac{q}{q_R}\right)^2 - 1 \tag{8}$$

$$p_m = p_2 - p_R$$

$$p_R = \begin{cases} \dfrac{\Delta p_{21}(\Delta p_{21} + \frac{1}{m}\Delta p_{10})}{2\Delta p_{21} + \frac{1}{m}\Delta p_{10}}, & m \neq 0 \\ \Delta p_{21}, & m = 0 \end{cases}$$

$$q_R = \begin{cases} p_R M_1 \sqrt{\dfrac{m\Delta p_{10}}{\Delta p_{21} - p_R}} & , m \neq 0 \\ M_1 \Delta p_{10} & , m = 0 \end{cases}$$

$$m = \frac{M_2}{M_1} \geq 0, \ \Delta p_{10} = p_1 - p_0 > 0,$$

$$\Delta p_{21} = p_2 - p_1 > 0$$

It is clear that there is a corner at the intersection of the cone and the cap when $M_1 \neq M_2$. The top of the cap is given by the coordinates (p_m, q_m), where $q_m = q_R$.

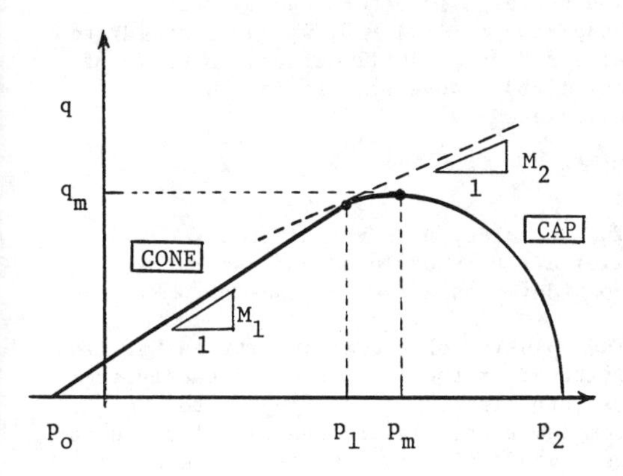

Fig. 1 Cone-Cap Model

3 INCREMENTAL RELATIONS

3.1 Elastic and elastic limit solutions

At the updated time level $t_{n+1} = t_n + \Delta t$ the "elastic" stress ${}^{n+1}\sigma^E$ is obtained by integrating Eq. (3a) in conjunction with $\dot{\varepsilon}^P = 0$; thus

$$^{n+1}\sigma^E = {}^n\sigma + \Delta\sigma^E, \ \Delta\sigma^E = D : \Delta\varepsilon \quad (9)$$

where $\Delta\sigma^E$ is the incremental elastic stress.

The two possibilities involving either elastic (un)loading (E) or plastic loading (P) are defined by

$$^{n+1}\sigma^E \in B \quad (E), \qquad {}^{n+1}\sigma^E \notin B \quad (P)$$

In the case of plastic loading the elastic limit stress ${}^n\hat{\sigma} \in B$ is defined by

$$^n\hat{\sigma} = {}^n\sigma + k\Delta\sigma^E, \ F({}^n\hat{\sigma}) = 0 \quad (10)$$

and the scalar $k \geq 0$ is determined from ${}^n\hat{\sigma} \in B$. When a generalized midpoint rule is adopted, it will be apparent that ${}^n\hat{\sigma}$ will be used except in the limiting case of fully implicit integration.

3.2 Generalized Midpoint Rule (GMR)

The conventional GMR for integration of the rate equation (3b) has been discussed elsewhere, Ortiz and Popov [4], but it will be stated here for the sake of completeness:

$$\begin{cases} ^{n+1}\sigma = {}^{n+1}\sigma^E - \lambda D : {}^{n+\alpha}a, \quad 0 < \alpha < 1 & (11a) \\ F({}^{n+1}\sigma) = 0 & (11b) \end{cases}$$

where the intermediate variables are

$$^{n+\alpha}a = a({}^{n+\alpha}\sigma) \quad (12a)$$

$$^{n+\alpha}\sigma = \alpha \, {}^{n+1}\sigma + (1-\alpha) \, {}^n\hat{\sigma} \quad (12b)$$

The increment of plastic strain which is used in hardening and softening laws is obtained in terms of the solution ${}^{n+1}\sigma$ in Eqs. (11a,b):

$$\Delta\varepsilon^P = \lambda \, {}^{n+\sigma}a = D^{-1} : ({}^{n+1}\sigma^E - {}^{n+1}\sigma) \quad (13)$$

The stress solution is shown schematically in Fig. 2 under the (unrealistic) assumption that $D = r\,I$, where r is a scalar, for the purpose of illustration. This means that the scaling of ${}^{n+1}\sigma^E$ along ${}^{n+\alpha}a$ onto ∂B will take place in the Euclidean metric. Clearly, when $\alpha=1$ the "Closest Point Mapping" algorithm will be retained as illustrated in Fig. 2.

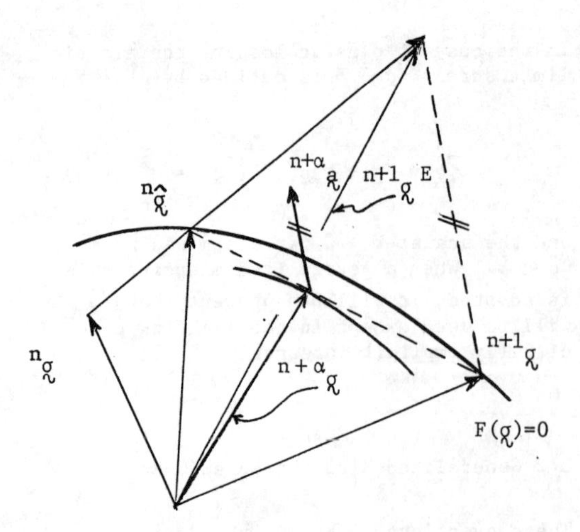

Fig. 2 Stress calculation for GMR ($D = r\ I$)

There are essentially two ways of solving Eqs. (11a,b) for $^{n+1}\sigma,\lambda$:

In the Stress Scaling algorithm (SS) λ is solved in an iterative manner for a given $^{n+\alpha}a$, which is calculated from Eqs. (12a,b) via an initial estimate of $^{n+1}\sigma$. The obtained λ-value is then used for the next higher iteration level to produce a new $^{n+1}\sigma$ from Eq. (1a) etc.

In the Stress Projection algorithm (SP) $^{n+1}\sigma$ and λ are solved simultaneously from Eqs. (11a,b) and $^{n+\alpha}a$ is not assumed to be known a priori. While a scalar equation is solved in SS (although a number of times) the dimension of the stress space generally determines the number of variables that must be solved simultaneously in the SP-algorithm; e.g. 4 variables in plane stress (3 stresses and the scalar λ). However, for simple yield criteria a reduction may be carried out analytically, as will become clear in the discussion of the model problem considered in the following.

3.3 Generalized Closest Point Projection Method (GCPPM)

A generalisation of the Backward Euler rule for application to the inequality formulation expressed in Eqs. (4a,b) is defined by letting $\dot{\sigma} = (^{n+1}\sigma - {^n\sigma})/\Delta t$ and $\sigma = {^{n+\alpha}\hat{\sigma}}$ in Eqs. (4a,b), which results in

$$(^{n+\alpha}\hat{\sigma} - \tau) : D^{-1} : (^{n+\alpha}\hat{\sigma}^E - {^{n+\alpha}\hat{\sigma}}) \geq 0$$

$$\forall \tau \in B \tag{14}$$

where the intermediate stresses are defined as

$$^{n+\alpha}\sigma^E = \alpha\ {^{n+1}\sigma^E} + (1-\alpha)\ {^n\hat{\sigma}} \tag{15}$$

$$^{n+\alpha}\hat{\sigma}^E = \frac{1}{c}\ {^{n+\alpha}\sigma^E},\quad {^{n+\alpha}\hat{\sigma}} = \frac{1}{c}\ {^{n+\alpha}\sigma} \tag{16a,b}$$

The scalar c is determined by the condition $F(^{n+1}\sigma) = 0$, while intermediate stress $^{n+\alpha}\sigma$ is still defined in terms of Eq. (12b). However, $\Delta\varepsilon^P$ is now calculated as

$$\Delta\varepsilon^P = \frac{c}{\alpha} D^{-1} : (^{n+\alpha}\hat{\sigma}^E - {^{n+\alpha}\hat{\sigma}}) \tag{17}$$

The formulation is valid for $0 < \alpha \leq 1$, i.e. $\alpha = 0$ is excluded and can rather be considered as a special case of GMR.

The solution of the inequality in Eq. (14) is found as the projection of the intermediate elastic stress $^{n+\alpha}\sigma^E$ onto B in complementary elastic energy. The equivalent projection in the special case $\alpha = 1$ has been described elsewhere, e.g. Runesson [10]. Thus, $^{n+\alpha}\hat{\sigma}$ is the solution of the minimization problem

$$\min E(\tau),\quad E(\tau) = \frac{1}{2} \left| {^{n+\alpha}\hat{\sigma}^E} - \tau \right|_C^2$$

$$\tau \in B \tag{18}$$

$$|\tau|_C^2 = \tau : D^{-1} : \tau$$

The stress projection has a unique solution even if ∂B is non-smooth. However, let us for the moment assume that ∂B is smooth and let us consider the case with corners later, whereby the projection algorithm for given $^{n+\alpha}\hat{\sigma}$ may be formulated as follows:

Seek the stationary value of L defined by

$$L(\tau,\mu) = E(\tau) + \alpha\mu F(\tau) \tag{19}$$

which has the extremum condition

$$\begin{cases} {}^{n+\alpha}\hat{\underset{\sim}{\sigma}} = {}^{n+\alpha}\hat{\underset{\sim}{\sigma}}^E - \alpha\lambda \underset{\sim}{D} : {}^{n+\alpha}\hat{\underset{\sim}{a}}, & (20a) \\ \qquad 0 < \alpha \leq 1 & \\ F({}^{n+\alpha}\hat{\underset{\sim}{\sigma}}) = 0 & (20b) \end{cases}$$

where $\quad {}^{n+\alpha}\hat{\underset{\sim}{a}} = \underset{\sim}{a}({}^{n+\alpha}\hat{\underset{\sim}{\sigma}})$

The introduction of $\alpha\lambda$ as the Lagrangian multiplier is made for convenience. When ${}^{n+\alpha}\hat{\underset{\sim}{\sigma}}$ has been found it is possible to calculate ${}^{n+1}\underset{\sim}{\sigma}$ by combining the Eqs. (12b) and (16b) to obtain

$${}^{n+1}\underset{\sim}{\sigma} = \frac{c}{\alpha}\, {}^{n+\alpha}\hat{\underset{\sim}{\sigma}} - \frac{1-\alpha}{\alpha}\, {}^{n}\underset{\sim}{\sigma}, \quad 0 < \alpha \leq 1 \qquad (21)$$

The stress solution is depicted schematically in Fig. 3.

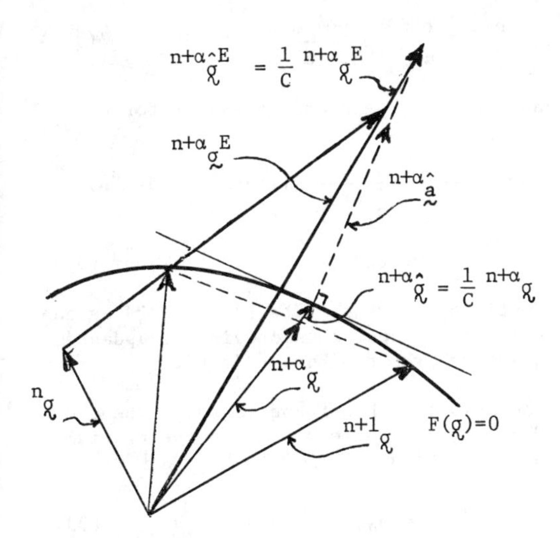

Fig. 3 Stress calculation for GCPPM ($\underset{\sim}{D} = r\,\underset{\sim}{I}$)

In a manner that is analogous to GMR, the updated stress ${}^{n+1}\underset{\sim}{\sigma}$ will be obtained in an iterative procedure; the main additional difficulty being that in order to find ${}^{n+\alpha}\hat{\underset{\sim}{\sigma}}^E$ we must calculate the scalar c from $F({}^{n+1}\underset{\sim}{\sigma}) = 0$.

Remark 1: The problem expressed in Eqs. (20a,b) resembles that stated in Eqs. (11a,b) defining the GMR. When $\alpha=1$ the two methods are clearly identical and we have $c=1$.

Remark 2: The "Closest Point Mapping" algorithm is retained for the projection of ${}^{n+\alpha}\hat{\underset{\sim}{\sigma}}^E$ onto B.

3.4 Two-invariant models, non-associated volumetric strains

When isotropic elasticity is assumed, defined by the (constant) shear modulus G and bulk modulus K, and when non-associated volumetric strains are incorporated via the scalar β ($\beta=0$ represents an associated flow rule), Runesson [11], then the (adjusted) complementary elastic energy may be restated as

$$\left|\underset{\sim}{\sigma}\right|_C^2 = \frac{1}{3G} q^2 + \frac{1+\beta}{K} p^2 \qquad (22)$$

For the two-invariant models defined by $F = F(p,q)$ it may be concluded that the stress solution $\hat{\underset{\sim}{\sigma}} = \pi_C \hat{\sigma}^E$ for a given stress $\hat{\underset{\sim}{\sigma}}^E$ is obtained as

$$\hat{\underset{\sim}{\sigma}} = \frac{\hat{q}}{\hat{q}^E} \hat{\underset{\sim}{s}}^E + \hat{p}\underset{\sim}{\delta} \qquad (23a)$$

i.e. the deviator stress $\hat{\underset{\sim}{s}}$ is proportional to $\hat{\underset{\sim}{s}}^E$; which means "radial return" in the deviator plane. The unknowns (p,q) are then solved from

$$\begin{cases} \min E(p',\, q'), \\ F(p',q') = 0 \end{cases} \qquad (23b)$$

where

$$2E = \frac{1}{3G} (\hat{q}^E - q')^2 + \frac{1+\beta}{K} (p^E - p')^2$$

The solution may be found, analogously to (19) and (20a,b), from

$$\hat{p} = \hat{p}^E - \alpha\mu \frac{K}{1+\beta} \frac{\partial F}{\partial p} (\hat{p},\hat{q})$$

$$\hat{q} = \hat{q}^E - \alpha\mu\, 3G \frac{\partial F}{\partial q} (\hat{p},\hat{q}) \qquad (24)$$

$$F(\hat{p},\hat{q}) = 0$$

When the stress solution (p,q) is known, the plastic strain increments may be obtained in terms of the invariants

$$\Delta e^{p} = \frac{c}{K\alpha} (\hat{p}^{E} - \hat{p})$$

$$\Delta\gamma^{p} = \frac{c}{3G\alpha} (\hat{q}^{E} - \hat{q}) \tag{25}$$

where

$$\Delta e = \Delta\varepsilon_{kk} , \quad \Delta\gamma = \sqrt{\frac{2}{3}} \left| \Delta\underset{\sim}{\gamma} \right| , \tag{26}$$

$$\underset{\sim}{\gamma} = \underset{\sim}{\varepsilon} - \frac{1}{3} e \underset{\sim}{\delta}$$

In conjunction with the developments above, we should note the following:

Remark 1: Without loss of generality we could in the present case have resorted to the triaxial conditions at the onset, for which we have (in matrix notation)

$$\underset{\sim}{\sigma} = \begin{bmatrix} p \\ q \end{bmatrix} , \quad \underset{\sim}{D} = \begin{bmatrix} K & 0 \\ 0 & 3G \end{bmatrix} , \quad \underset{\sim}{\varepsilon} = \begin{bmatrix} e \\ \gamma \end{bmatrix}$$

If we, for the time being, let $\beta = 0$, we see that Eq. (22) is obtained.

4 ANALYSIS OF CONE-CAP MODEL

4.1 Stress solution for the Cone

The projection solution for the intermediate stress $({}^{n+\alpha}\hat{p}, {}^{n+\alpha}\hat{q})$ is immediately obtained via Eq. (24) in the form:

$${}^{n+\alpha}\hat{p} = \frac{1}{1 + \phi M_1^2} \left({}^{n+\alpha}\hat{p}^{E} + \phi M_1 \, {}^{n+\alpha}\hat{q}^{E} + \phi M_1^2 \, p_0 \right) \tag{28a}$$

$${}^{n+\alpha}\hat{q} = M_1 \left({}^{n+\alpha}\hat{p} - p_0 \right) \tag{28b}$$

whereby it is realized that the solution is strongly dependent on the nondimensional parameter ϕ, where

$$\phi = \frac{K}{3(1+\beta)G} = \frac{2(1+\nu)}{9(1+\beta)(1-2\nu)} \tag{29}$$

However, the as yet unknown scalar c, which is inherent in ${}^{n+\alpha}\hat{p}^{E}$ and ${}^{n+\alpha}\hat{q}^{E}$ according to Eq. (16a), must be determined in order to completely define the solution. At this point it is extremely important to observe that it is not known in general whether the final solution is a Cone-,Corner-or a Cap-solution even if the intermediate solution was found to be a Cone solution. The same argument applies when the intermediate stress solution is in the Corner or on the Cap. Let us, therefore, consider the case when the updated stress solution is on the Cone, which gives the condition

$$F({}^{n+1}p, {}^{n+1}q) \equiv {}^{n+1}q - M_1 \left({}^{n+1}p - p_0 \right) = 0 \tag{30}$$

where the updated solution is given as

$${}^{n+1}p = \frac{c}{\alpha} \, {}^{n+\alpha}\hat{p} - \frac{1-\alpha}{\alpha} \, {}^{n}\hat{p} \tag{31a}$$

$${}^{n+1}q = \frac{c}{\alpha} \, {}^{n+\alpha}\hat{q} - \frac{1-\alpha}{\alpha} \, {}^{n}\hat{q} \tag{31b}$$

and we may obtain an expression for c:

$$c = \frac{1}{{}^{n+\alpha}\hat{q} - M_1 \, {}^{n+\alpha}\hat{p}} \left[(1-\alpha)\left({}^{n}\hat{q} - M_1 \, {}^{n}\hat{p}\right) - \alpha M_1 p_0 \right] \tag{32}$$

Combining the Eqs. (28a,b) and (32) we can calculate c and, thus obtain the updated solution from the Eqs. (31a,b).

The solution thus found is valid whenever ${}^{n+1}p \geq p_0$, otherwise we obtain the trivial apex solution

$${}^{n+1}p = p_0 , \quad {}^{n+1}q = 0 \tag{33}$$

In order for the Cone solution to be physically valid we also need to consider the restriction ${}^{n+1}p \leq p_1$, otherwise we obtain either a Corner solution or a Cap-solution. Here we shall make the following observations:

Remark 1: The Euclidean projection in the (p,q)- space is obtained from Eqs. (28a,b) when $\phi = 1$. In the case of associated plasticity (β=0) this corresponds to ν=0.35, which is not an unrealistic value for soil-and rock-like materials. It should be noted, however, that such an orthogonality property is entirely dependent on the stress space in which the yield criterion is represented.

Remark 2: When both the intermediate stress solution $({}^{n+\alpha}p, {}^{n+\alpha}q)$ and the initial stress $({}^{n}p, {}^{n}q)$ are on the cone, then we obtain c=1 from (32). In this special case the updated solution is independent of α and is simply obtained from (28a,b) by replacing all intermediate values with the updated values at the time level t_{n+1}. Furthermore, this solution is the exact solution for the given strain increment provided the strain path during the time interval is straight.

4.2 Stress solution for the Cap

From (24) we obtain the projection solution for the intermediate stress $({}^{n+\alpha}\hat{p}, {}^{n+\alpha}\hat{q})$:

$$ {}^{n+\alpha}\hat{p} = p_m + \frac{1}{1+\frac{k\phi}{p_R^2}} \left({}^{n+\alpha}\hat{p}^E - p_m \right) \qquad (34a) $$

$$ {}^{n+\alpha}\hat{q} = \frac{1}{1+\frac{k}{q_R^2}} \, {}^{n+\alpha}\hat{q}^E \qquad (34b) $$

where it is observed that ${}^{n+\alpha}\hat{p}$, ${}^{n+\alpha}\hat{q}$ are functions of the two unknowns $k = 3G\alpha\lambda$ and c. For a given c we can obtain k from the condition

$$ F({}^{n+\alpha}\hat{p}, {}^{n+\alpha}\hat{q}) = 0 \qquad (35) $$

However, as for the cone-solution, we also know that the updated solution given by (31a,b) must satisfy the yield criterion, which gives the second necessary condition to determine (k,c):

$$ F({}^{n+1}p, {}^{n+1}q) = 0 \qquad (36) $$

Let us assume that Eq. (36) defines a cap-solution. As both Eqs. (35) and (36) result in simple scalar equations, they are solved iteratively and in an interactive fashion. The converged solution is then valid whenever ${}^{n+1}p \geq p_1$. The algorithm implementation will be described briefly in the next subsection.

Remark 1: Fully implicit integration (α=1) means that c=1 and, consequently, (3a,b) will become trivial identities. Only the scalar Eq. (35) remains for the calculation of k via Eqs. (34a,b).

4.3 Algorithm implementation

The implementation and use of the GCPPM are relatively straightforward except at corners, which in the present model appear at the apex of the Cone ($p = p_0$) and at the Cone-Cap intersection ($p = p_1$). The actual procedure to deal with these corners is, however, much simplified since only two different surfaces are involved. Furthermore, the apex solution is easily detected as discussed in Subsection 4.1.

The algorithm outlined in the following considers the Cone-Cap intersection. The strategy for a more complicated case of multisurface intersection will, in principle, be similar although the computational procedure will become more tedious.

A comment on the terminology seems appropriate: Firstly, the term 'corner' is used as a synonym for 'intersection' of yield surfaces, although such a corner may well be smooth and the gradient of the yield function thus be unique. Secondly, each analytical expression of the yield function is associated with a separate yield surface (Cone and Cap in the present model).

In the algorithm outlined below it is convenient to characterize the solution $\underset{\sim}{g} = (p,q)$ projected from a given elastic stress $\underset{\sim}{g}^E = (p^E,q^E)$ as either an Apex, Cone, Corner or Cap solution, as shown in Fig. 4. The pertinent regions, to which $\underset{\sim}{g}^E$ belongs, are limited by the lines L_A, L_{C1} and L_{C2} defined by the (transformed) gradients $\underset{\sim}{a}_A$, $\underset{\sim}{a}_{C_1}$ and $\underset{\sim}{a}_{C_2}$ and given in parameter form:

$$ L_A : \underset{\sim}{g}(s) = \underset{\sim}{a}_A^* s, \quad \underset{\sim}{a}_A^* = (-\phi M_1, 1) \qquad (37a) $$

$$ L_{C1} : \underset{\sim}{g}(s) = \underset{\sim}{g}_C + \underset{\sim}{a}_{C1}^* s, $$
$$ \underset{\sim}{g}_C = (p_1, M_1 \Delta p_{10}), \quad \underset{\sim}{a}_{C1}^* = \underset{\sim}{a}_A^* \qquad (37b) $$

$$ L_{C2} : \underset{\sim}{g}(s) = \underset{\sim}{g}_C + \underset{\sim}{a}_{C2}^*, \quad \underset{\sim}{a}_{C2}^* = (-\phi M_2, 1) \qquad (37c) $$

Points on these lines with given $q = q^E$ are defined by the p-values p_A, p_{C1} and p_{C2}, respectively. It is now a trivial matter to assess the character of the solution.

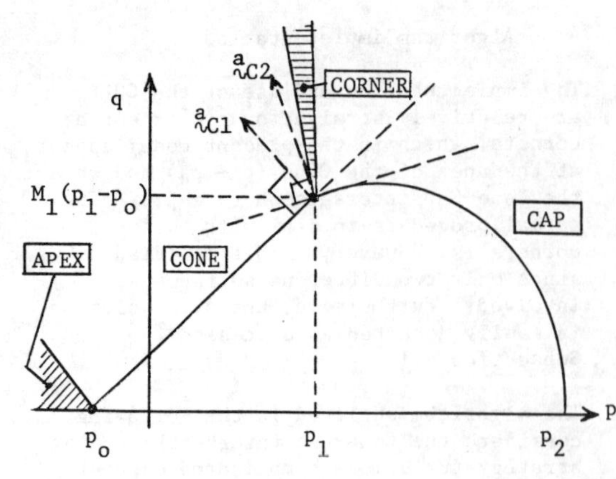

Fig. 4 Characteristic solution regions

Algorithm

1. Check the character of the solution:

a) ${}^{n+1}p^E \leq {}^{n+1}p_A$: Apex solution (33) accepted. Stop.

b) ${}^{n+1}p_A \leq {}^{n+1}p^E$: Either Cone, Corner or Cap solution. Go to 2.

2. Iterate for the solution. Assume ${}^{n+1}\underset{\sim}{g}_i$, c_i given in the ith iteration. Start values: ${}^{n+1}\underset{\sim}{g}_0 = {}^{n}\hat{\underset{\sim}{g}}$, $c_0 = 1$.

3. Calculate ${}^{n+\alpha}\hat{\underset{\sim}{g}}^E_{i+1} \dfrac{1}{c_i}\, {}^{n+\alpha}\underset{\sim}{g}^E$

4. Calculate ${}^{n+\alpha}\hat{\underset{\sim}{g}}$ by projecting ${}^{n+\alpha}\underset{\sim}{g}^E$ on the appropriate yield surface.

Projection on Cone: Use Eqs. (28a,b) to obtain ${}^{n+\alpha}\hat{p}_{i+1}$, ${}^{n+\alpha}\hat{q}_{i+1}$ from given ${}^{n+\alpha}\hat{p}^E_{i+1}$, ${}^{n+\alpha}\hat{q}^E_{i+1}$

Projection on Cap: Calculate ${}^{n+\alpha}\hat{p}_{i+1}$, ${}^{n+\alpha}\hat{q}_{i+1}$ from (34a,b) by first solving k from (35) using Newton type iterations.

Procedure: It is assumed that ${}^{n}\hat{\underset{\sim}{g}}$ is on surface S_1 which may be either the cone-or the cap-surface. The following cases can be directly distinguished with the aid of ${}^{n+\alpha}\hat{\underset{\sim}{g}}^E_{i+1}$:

a) ${}^{n+\alpha}\hat{\underset{\sim}{g}}_{i+1} \in S_1$

b) ${}^{n+\alpha}\hat{\underset{\sim}{g}}_{i+1} \in S_1 \cap S_2$ (corner solution)

c) ${}^{n+\alpha}\hat{\underset{\sim}{g}}_{i+1} \in S_2$

5. Calculate the updated stress ${}^{n+1}\underset{\sim}{g}_{i+1}$ on the appropriate yield surface:

Cone: Find ${}^{n+1}p_{i+1}$, ${}^{n+1}q_{i+1}$ directly from Eqs. (31a,b) with c_{i+1} calculated from Eq. (32).

Cap: Calculate c_{i+1} from (36) using (31). After convergence of the iterations in (36), then ${}^{n+1}p_{i+1}$, ${}^{n+1}q_{i+1}$ are obtained from (31).

Procedure: The following cases are distinguished:

a) ${}^{n+\alpha}\hat{\underset{\sim}{g}}_{i+1} \in S_1$: Assume ${}^{n+1}\underset{\sim}{g}_{i+1} \in S_1$

b) ${}^{n+\alpha}\hat{\underset{\sim}{g}}_{i+1} \in S_1 \cap S_2$: ${}^{n+1}\underset{\sim}{g}_{i+1} \in S_2$

c) ${}^{n+\alpha}\hat{\underset{\sim}{g}}_{i+1} \in S_2$: ${}^{n+1}\underset{\sim}{g}_{i+1} \in S_2$

Check that $F_1 \leq 0$, $F_2 \leq 0$ in the case a. If the wrong surface was assumed to be active then switch and repeat this step.

6. Check convergence of ${}^{n+1}p_{i+1}$, ${}^{n+1}q_{i+1}$ and c_{i+1}:

$$\frac{\left|{}^{n+1}p_{i+1} - {}^{n+1}p_i\right|}{p_1} \leq \delta_\sigma, \quad \frac{\left|{}^{n+1}q_{i+1} - {}^{n+1}q_i\right|}{p_1} < \delta_\sigma, \quad \left|c_{i+1} - c_i\right| \leq \delta_c$$

where δ_σ and δ_c are tolerances.

Remark 1: Fully implicit integration ($\alpha = 1$) means that $c = 1$ and the above iteration procedure will not be necessary. Only step 4 which involves the projection step remains, whereby ${}^{n+1}p^E$, ${}^{n+1}q^E$ will directly give the updated solution ${}^{n+1}p$, ${}^{n+1}q$.

4.4. Accuracy analysis

The accuracy of the proposed algorithm will be assessed for given strain increments $(\Delta e, \Delta\gamma)$ under the assumption of continuous plastic loading. The incremental strain is defined by the parameters ρ and Ψ:

$$\rho = \left|\Delta\underset{\sim}{x}\right|, \quad \Delta\underset{\sim}{x} = \left(\frac{\Delta e}{e_y}, \frac{\Delta\gamma}{\gamma_y}\right) = \left(\frac{\Delta p^E}{p_m}, \frac{\Delta q^E}{q_m}\right) \quad (38a)$$

$$\Psi = \cos^{-1}\frac{\underset{\sim}{a}:\Delta\underset{\sim}{x}}{|\underset{\sim}{a}|\rho}, \quad \underset{\sim}{a} = \left(\frac{\partial F}{\partial p}, \frac{\partial F}{\partial q}\right) \tag{38b}$$

where Eq. (36a) was obtained by assuming that the strains at yield are $e_y = p_m/K$, $\gamma_y = q_m/3G$ and by using the identities $\Delta p^E = K\Delta e$, $\Delta q^E = 3G\Delta\gamma$.

The response is assessed for

$$0 < \rho \leq 10 \quad \text{and} \quad -\pi/2 \leq \Psi \leq \pi/2$$

when the strain increments are applied at certain characteristic points P_i at the yield surface as illustrated in Fig. 5. At the point P_4 we choose $\underset{\sim}{a}$ as the vector that bisects the "fan" of admissible normals.

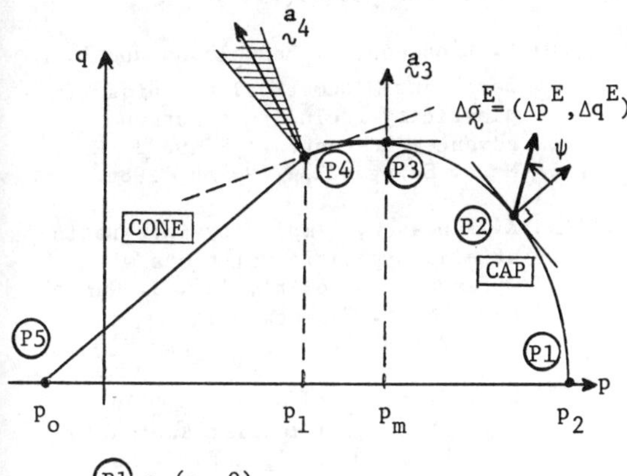

Fig. 5 Accuracy assessment points

The influence of varying α, the corner angle at P_4 and the shape of the cap were investigated. The "exact" solution for given (ρ,Ψ) was obtained in a straightforward fashion by subincrementing (Δe,Δγ). Normally, Δρ = 0.01 was sufficient to achieve the convergent solution ${}^{n+1}\underset{\sim}{g}^* = ({}^{n+1}p^*, {}^{n+1}q^*)$. The error is measured as

$$e = \frac{|{}^{n+1}\underset{\sim}{g} - {}^{n+1}\underset{\sim}{g}^*|}{|{}^{n+1}\underset{\sim}{g}^*|} \tag{39}$$

where ${}^{n+1}\underset{\sim}{g} = ({}^{n+1}p, {}^{n+1}q)$ is the approximate solution.

Results are shown in Tables 1 and 2 for the points P_1 and P_4 respectively for the choice $M_1 = 0.8$, $M_2 = 0.2$, $P_0/P_1 = -0.25$ and $P_2/P_1 = 2$.

Table 1. Errors (%) in GCPPM for Point 1.

Point 1		ρ				
α	Ψ°	0.5	1	2	5	10
0.5	0	0	0	0	0	0
	30	0.1	0.6	3.2	13.2	21.2
	60	0.2	1.6	9.9	38.9	51.4
	90	0.3	3.2	36.8	59.9	64.4
1	0	0	0	0	0	0
	30	1.2	3.3	6.1	5.8	3.3
	60	1.7	5.5	10.8	7.2	3.6
	90	0.6	4.2	11.7	6.6	3.1

Table 2. Errors (%) in GCPPM for Point 4. (1) Corner solution (2) Cone solution (3) Apex solution

Point 4		ρ				
α	Ψ°	0.5	1	2	5	10
0.5	-90	1.3	5.0	12.9	29.2	40.6
	-60	0.9	3.5	9.2	19.1	24.6
	-30	0.3	1.4	3.8	7.5	9.4
	0	0(1)	0(1)	0(1)	0(1)	0(1)
	30	0(2)	0(2)	0(2)	0(2)	0(3)
	60	0(2)	0(2)	0(3)	0(3)	0(3)
	90	0(2)	0(2)	0(3)	0(3)	0(3)
1	-90	3.1	6.3	6.3	6.6	3.8
	-60	3.1	5.2	5.2	3.1	1.7
	-30	1.4	2.2	2.2	1.1	0.6
	0	0	0	0	0	0
	30	0	0	0	0	0
	60	0	0	0	0	0
	90	0	0	0	0	0

5 CONCLUSIONS

The GCPM method proposed in this paper generalizes the concept of "closest point projection" in complementary elastic energy in the sense that an intermediate trial stress is projected. For smooth yield criteria the solution will be similar to that of GMR. Non-associated volumetric strains and sharp corners are included in the analysis. It is considered to be a major advantage that the interpretation of the solution is quite transparent in the presence of corners.

A numerical algorithm has been developed for a Cone-Cap model and has proved to perform very well; even for the corner solutions. The numerical results confirm previous experience with the GMR-method that BE (α = 1) is most accurate for large strain increments while midpoint projection (α = 1/2) is preferred for small and moderate strain increments.

ACKNOWLEDGEMENTS

The authors wish to acknowledge the support provided by NASA Marshall Space Flight Center contract NAS8-35668 at the University of Colorado.

REFERENCES

[1] D.R. Owen and E. Hinton, Finite elements in plasticity, McGraw-Hill, 1980.

[2] D.M. Potts and A. Gens, 'A critical assessment of methods of correcting for drift from the yield surface in elasto-plastic finite element analysis', Int. J. Num. Anal. Meth. Geomech. 9, 149-159(1985).

[3] Wilkins, "Calculation of elastic-plastic flow', Methods of Computational Physics, Vol. 3 (Eds Alder et al.), Academic Press, 1964.

[4] M. Ortiz and E.P. Popov, 'Accuracy and stability of integration algorithms for elastoplastic constitutive relations', Int. J. Num. Meth. Engng., 21, 1561-1576(1985).

[5] J.C. Simo and R.L. Taylor, 'A return mapping algorithm for plane stress elastoplasticity', Int. J. Num. Meth. Engng., 22, 649-670(1986).

[6] C. Johnson, 'A mixed finite element method for plasticity problems with hardening', SIAM J. Num. An. 14, 575-584(1977.

[7] Q.S. Nguyen, 'On the elastic plastic initial-boundary value problem and its numerical integration', Int. J. Num. Meth. Engng. 11, 817(1977).

[8] M. Fröier and A. Samuelsson, 'Vairational inequalities in plasticity, Recent Developments', Finite Elements in Nonlinear Mechanics, Vol. 1 (Ed. Bergan et al.), Tapir Publishers, Trondheim, 1978.

[9] K. Runesson and J.R. Booker, 'On mixed and displacement finite element methods in perfect elasto-plasticity', Proc. Fourth Int. Conf. in Australia on Finite Element Methods, (Ed. Hoadley), Melbourne, 85-89(1982).

[10] K. Runesson, A. Samuelsson and L. Bernspång, 'Numerical technique in plasticity including solution advancement control', Int. J. Num. Meth. Engng. 22, 769-788, 1986.

[11] K. Runesson, 'Implicity integration of elasto-plastic relations with reference to soils', Int. J. Num. Anal. Meth. Geomech. 11, 315-321(1987).

[12] I.S. Sandler and D. Rubin, 'An algorithm and a modular subroutine for the cap model', Int. J. Num. Anal. Meth. Geomech. 3, 173-186(1979).

[13] E. Pramono, Progressive failure analysis in concrete structures, Ph.D. Thesis, University of Colorado (in preparation).

[14] J.C. Simo, J.W. Wu and R.L. Taylor, 'Softening response, completess conditions, and numerical algorithms for the cap model', submitted to Int. J. Num. Meth. Engng.

Numerical Methods in Geomechanics (Innsbruck 1988), Swoboda (ed.)
© 1988 Balkema, Rotterdam. ISBN 90 6191 809 X

Model and parameters for the elastic behaviour of soils

Poul V.Lade
Department of Civil Engineering, University of California, Los Angeles, USA

ABSTRACT: A nonlinear, isotropic model for the elastic behavior of soils has been developed on the basis of theoretical considerations involving the principle of conservation of energy. Energy is therefore neither generated nor dissipated in a closed-loop stress path or strain path. The framework for the model consists of Hooke's law in which Poisson's ratio is constant and Young's modulus is expressed as a power function involving the first invariant of the stress tensor and the second invariant of the deviatoric stress tensor. The characteristics of the model are described, and the accuracy is evaluated by comparison with experimental results from triaxial tests. Parameter determination from unloading-reloading cycles in conventional triaxial compression tests is demonstrated, and parameter values are given for a variety of soils.

1 INTRODUCTION

The models employed for the elastic behavior most often assume the soil to behave as an isotropic material that obeys Hooke's law with a constant Poisson's ratio. Experimental data tend to confirm these assumptions, but the types of dependence on stress state assumed for Young's modulus, the shear modulus and the bulk modulus have shortcomings.

Presented here is an isotropic model for the nonlinear elastic behavior of soils. The model is based on a theoretical development which guaranties lack of energy generation or dissipation for any closed-loop stress path or strain path. The characteristics of the model are described, and determination of parameters from simple tests is demonstrated. Test data for fifteen different soils have been analyzed, and the parameter values are listed and evaluated.

2 COMMON ELASTIC MODELS AND THEIR SHORTCOMINGS

The two most well-known types of expressions for the elastic moduli are those given below. The first expression gives Young's modulus E as a function of the minor principal stress $\sigma_3 = \sigma_{min}$ [1,2]:

$$E = K \cdot p_a \cdot \left(\frac{\sigma_3}{p_a}\right)^n \qquad (1)$$

in which p_a is atmospheric pressure in the same units as E and σ_3. The two material parameters K and n are constant for a soil at a given void ratio. Conditions for which the minor principal stress is constant, thus producing the same value of Young's modulus, are illustrated in Figure 1. Young's modulus is constant along three perpendicular planes crossing the hydrostatic axis at the common value of minor principal stress. The expression in equation (1) has been used with constant and with variable Poisson's ratios.

For an isotropic elastic material, Young's modulus and Poisson's ratio may be replaced by the shear modulus G and the bulk modulus B in Hooke's law. These moduli have often been expressed in terms of the mean normal stress σ_0 [3,4]:

$$G \text{ and/or } B = A \cdot F(e,OCR) \cdot \sigma_0^n \qquad (2)$$

The material parameters incorporated in this class of expressions have occasionally included variations due

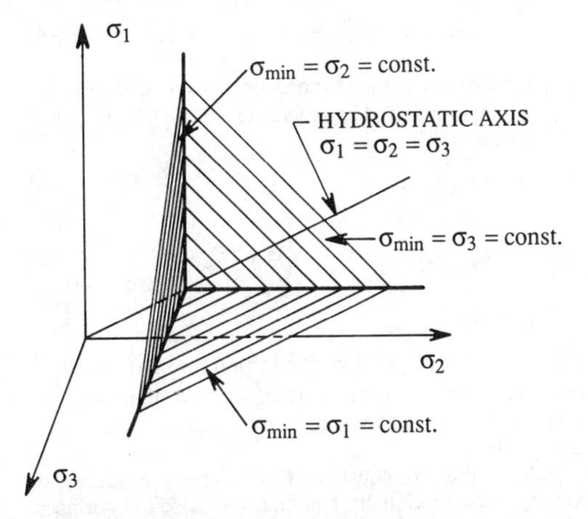

Figure 1. Three Perpendicular Planes in Which Elastic Modulus is Constant According to Equation (1).

to factors such as void ratio e and overconsolidation ratio OCR [4] as indicated in equation (2). For a soil with given e and OCR the values of A,F and n are constant. The mean normal stress and consequently the moduli given by equation (2) are constant in an octahedral plane as illustrated in Figure 2.

Theoretical reasons exist for the variation of elastic moduli with stress state [e.g. 5]. The simple formulations indicated above are empirical in nature, i.e., they have been devised on the bases of results from available, simple tests such as isotropic compression and triaxial compression. The expressions are widely used, because they appear to capture the elastic behavior observed in such tests, and because the material parameters can be determined from the results of conventional triaxial tests.

However, these models for the variation of elastic moduli result in violation of the principle of conservation of energy [6]. Thus, depending on the direction of a closed stress loop, these models will generate or dissipate energy in violation of the basic premise of elastic behavior. According to elasticity theory, a material exhibits true elastic behavior if energy is conserved at the end of a closed stress loop or strain loop.

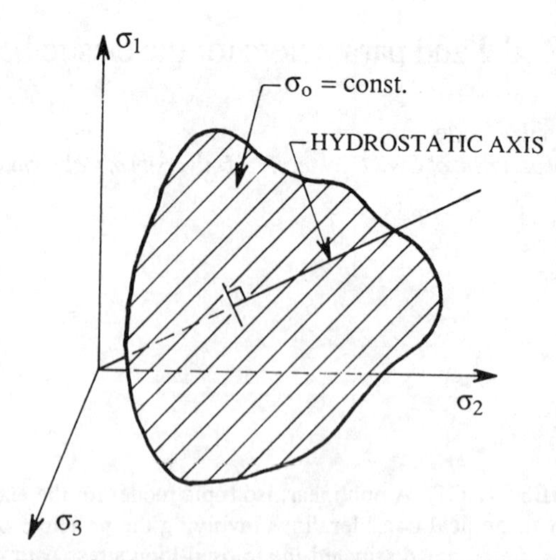

Figure 2. Octahedral Plane in Which Elastic Modulus is Constant According to Equation (2).

3 TRUE ELASTIC MODEL FOR SOILS

An isotropic model for the nonlinear elastic behavior of soils was developed by Lade and Nelson [7]. This model is based on Hooke's law in which Poisson's ratio ν = constant. The expression for Young's modulus was derived from theoretical considerations based on the principle of conservation of energy. According to this derivation, Young's modulus E can be expressed in terms of a power law involving nondimensional material constants and stress functions as follows:

$$E = M \cdot p_a \cdot \left[\left(\frac{I_1}{p_a} \right)^2 + R \cdot \frac{J_2'}{p_a^2} \right]^\lambda \qquad (3)$$

in which

$$R = 6 \cdot \frac{1+\nu}{1-2\nu} \qquad (4)$$

I_1 is the first invariant of the stress tensor, and J_2' is the second invariant of the deviatoric stress tensor, given as follows:

$$I_1 = \sigma_x + \sigma_y + \sigma_z \qquad (5)$$

$$J_2' = \frac{1}{6} \left[(\sigma_x - \sigma_y)^2 + (\sigma_y - \sigma_z)^2 + (\sigma_z - \sigma_x)^2 \right] + \tau_{xy}^2 + \tau_{yz}^2 + \tau_{zx}^2 \qquad (6)$$

The parameter p_a is atmospheric pressure expressed in the same units as E, I_1, and $\sqrt{J_2'}$, and the modulus number M and the exponent λ are constant, dimensionless numbers.

According to equation (3) Young's modulus is constant along rationally symmetric ellipsoidal surfaces whose long axis coincides with the hydrostatic axis and whose center is located at the origin of the principal stress space. The magnitude of Poisson's ratio determines the shape of the ellipsoidal surface. For $\nu = 0$ the value of $R = 6$ and the surface becomes spherical, whereas for $\nu = 0.5$ the value of $R = \infty$ and the surface degenerates into a line coinciding with the hydrostatic axis.

Figure 3 shows cross-sections of the ellipsoidal surfaces in triaxial and octahedral planes for different values of Poisson's ratio. The cross-sections in the octahedral plane are always circular, whereas the cross-sections in the triaxial plane are shaped as ellipses whose aspect ratio depends on the value of Poisson's ratio. The locations of typical failure surfaces for soils with friction angles from 30° to 50° are shown for triaxial compression and extension. Only the portions of the ellipsoids located inside the conical failure surface can be reached by stress states in soils, and only these portions are of interest.

Figure 4 shows a three-dimensional ellipsoidal surface along which Young's modulus is constant according to equation (3). Comparison of this surface with those in Figures 1 and 2 indicates that the conditions for which Young's modulus are constant are quite different for the three expressions. As shown in Figure 1, the model expressed by equation (1) produces surfaces of constant moduli which are convex with regard to the origin, whereas the truly elastic model described by equation (3) results in surfaces that are concave with regard to the origin. The model described by equation (2) produces planes of constant moduli that are neither convex nor concave with regard to the origin. None of the two previous models are contained as special cases within the truly elastic model described here.

4 DETERMINATION OF PARAMETER VALUES

The material parameters M, λ, and ν required to describe the elastic behavior of soils may be deter-

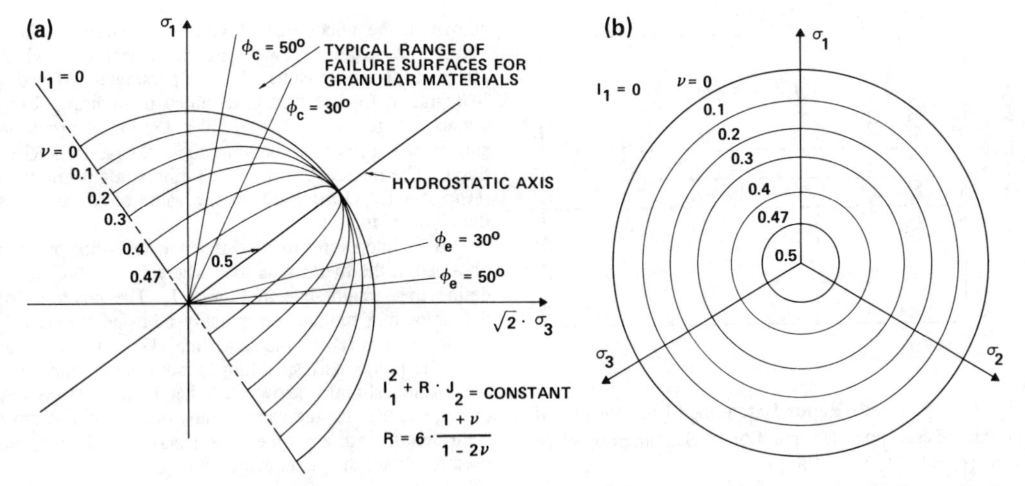

Figure 3. Contours of Constant Young's Modulus Shown in (a) Triaxial Plane and in (b) Octahedral Plane.

mined from conventional drained triaxial compression tests which include unloading-reloading cycles.

4.1 Possion's ratio

The best estimate of Poisson's ratio is obtained from the slope of the volume change curve associated with reloading initiated at or near the hydrostatic axis according to .

$$\nu = -\frac{\dot{\varepsilon}_3}{\dot{\varepsilon}_1} = \frac{1}{2} \cdot \left[1 - \frac{\dot{\varepsilon}_v}{\dot{\varepsilon}_1}\right] \qquad (7)$$

in which $(\dot{\varepsilon}_v/\dot{\varepsilon}_1)$ is the slope of the volume change curve immediately after stress reversal, as indicated in the lower part of Figure 5.

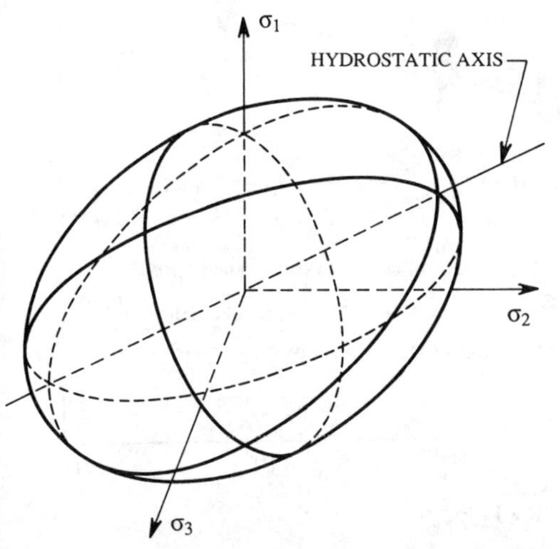

Figure 4. Ellipsoidal Surface Along Which Young's Modulus is Constant According to Equation (3).

It is preferable to obtain an average value of Poisson's ratio from several tests. Figure 6 shows values determined from equation (7) for medium dense Sacramento River Sand plotted against I_1/p_a. Although there is some scatter in the data, there is no discernible

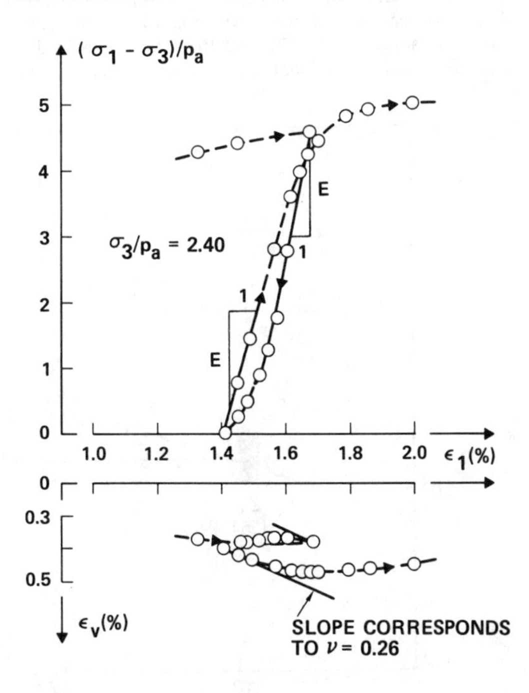

Figure 5. Young's Moduli and Slopes of Volume Change Curve for Determination of Poisson's Ratio from Unloading-Reloading Cycle in Conventional Triaxial Compression Test on Loose Santa Monica Beach Sand.

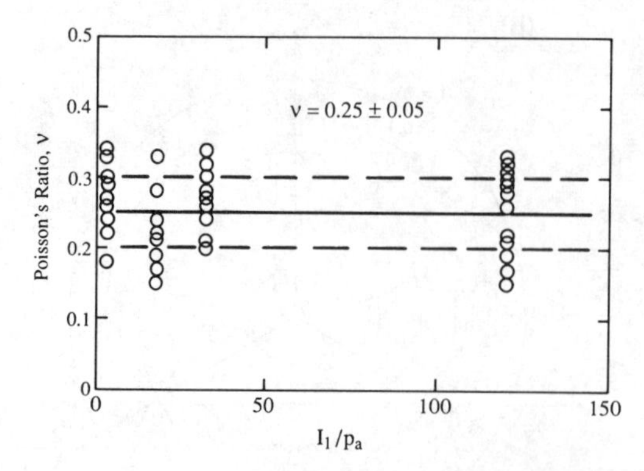

Figure 6. Possion's Ratios Determined from Results of Triaxial Tests on Medium Dense Sacramento River Sand ($e_i = 0.71$, $D_r = 78\%$)

effect of mean normal stress on Poisson's ratio. Similar results have also been obtained for several other soils [e.g. 7,8]. The average value and the standard deviation for Poisson's ratio are indicated on Figure 6.

4.2 Young's modulus

Using the average value of Poisson's ratio, the value of R can be determined from equation (4). Individual modulus values may be obtained from the slopes of the unloading and the reloading stress-strain relations as shown in the upper part of Figure 5. These modulus values are associated with the stress states at which stress reversals are initiated. The parameters M and λ in equation (3) are then determined by plotting (E/p_a) versus $[(I_1/p_a)^2 + R\cdot (J_2'/p_a^2)]$ in a log-log diagram as shown in Figure 7 for medium dense Sacramento River Sand. The intercept of the best fitting straight line with $[(I_1/p_a)^2 + R \cdot (J_2'/p_a^2)] = 1$ is the value of M, and λ is the slope of the line.

It is interesting to note that the reloading moduli obtained at the hydrostatic axis where $J_2' = 0$ essentially define the straight line in Figure 7. The positions of the unloading moduli are suitably adjusted through J_2' and R so as to fall on the same line. For this particular case, the points corresponding to isotropic compression are located slightly below the other points. However, this is usually not the case. Thus, once ν and R have been determined, any type of test may, in principle, be used to obtain the parameter M and λ.

5 ELASTIC PARAMETER VALUES

Seventeen sets of data from fifteen different soils have been analyzed to study the range of parameter values obtained. Descriptions of soils and material parameter values are given in Table 1. The average Poisson's ratios were obtained from reloading, as explained above, as well as from unloading at constant confining pressure and from K_0-unloading. Considerable scatter in the values of Poisson's ratio were obtained as indicated by the standard deviations. However, it was

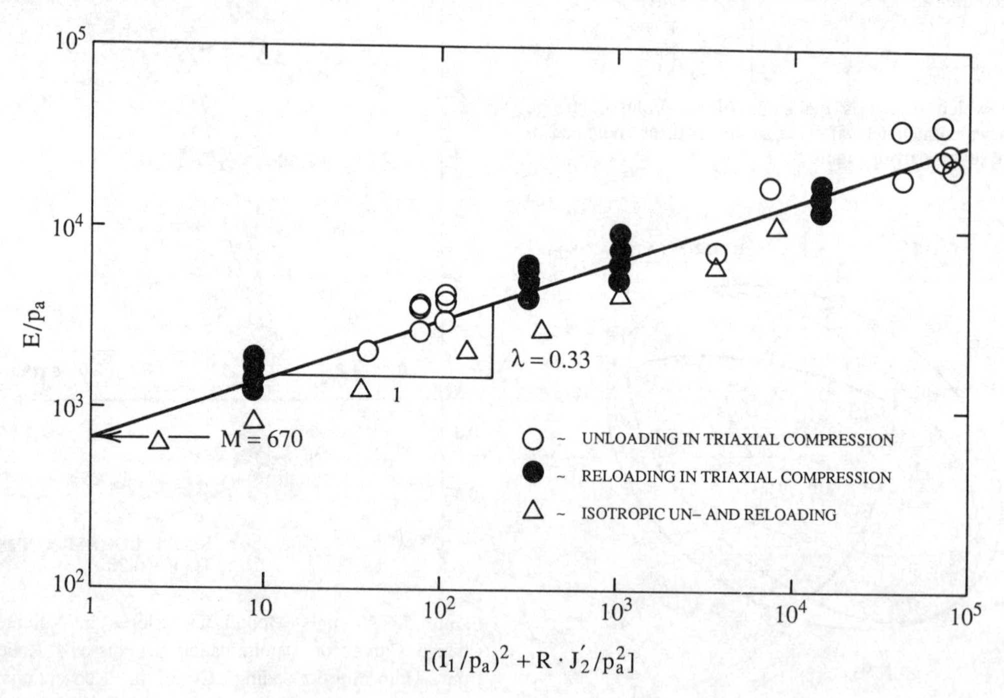

Figure 7. Determination of Parameters for Young's Modulus for Medium Dense Sacramento River Sand ($e_i = 0.71$, $D_r = 78\%$)

Table 1. Description and Elastic Parameter Values for 15 Different Soils

Soil (Reference*)	Description				Poisson's Ratio		Young's Modulus		
	D_{50}(mm)	C_u	e	D_r (%)	ν_{avg}	s	M	λ	Coeff. of Correlation r
Santa Monica Beach Sand (Lade and Nelson [7])	0.27	1.58	0.61	90	0.15	0.08	1270	0.23	0.878
			0.81	20	0.26	0.06	600	0.27	0.792
Montery No. 0 Sand Lade and Duncan [9])	0.43	1.53	0.57	98	0.17	0.14	1120	0.33	0.849
			0.78	27	0.17	0.05	800	0.26	0.659
Fine Silica Sand	0.18	2.0	0.76	30	0.27	0.08	440	0.22	0.673
Mohawk Model Soil	3.5	6.2	0.64	73	0.25	0.13	630	0.27	0.782
Niagara-Type A Material	4.8	1.52	0.71	95	0.23	0.12	670	0.26	0.872
Niagara-Type B Material	1.2	8.6	0.52	95	0.24	0.11	570	0.29	0.788
Dense Silica Sand (Duncan and Chang [2])	0.25	1.8	0.50	100	0.16	-	750	0.21	0.960
Ottawa Sand (Yong and Ko, [10])	0.42	1.3	0.51	87	0.13	0.04	800	0.18	0.429
Dense Sand (Goldscheider [11])	0.42	1.8	0.56	100	0.14	-	280	0.38	0.970
Hostun Sand (Saada [12])	0.34	1.7	0.62	98	0.22	0.04	400	0.32	0.914
Reid Bedford Sand (Saada [12])	0.24	1.7	0.68	51	0.11	-	550	0.33	0.859
Intact Clayey Sand	0.20	200	0.51	65	0.25	0.03	400	0.36	0.963
Remolded Clayey Sand	0.20	200	0.51	65	0.25	0.03	390	0.36	0.987
Compacted Silt	0.013	U.S. Class. ML	0.42	RC = 95% Mod. Proctor	0.27	0.06	340	0.20	0.762
Sacramento River Sand (Lee [13])	0.21	1.5	0.71	78	0.25	0.05	670	0.33	0.975

* Those without reference are from the author's files.

shown in [7] that the average values of ν may well be within ranges in which M and λ are relatively insensitive to Poisson's ratio. Table 1 shows that Poisson's ratio varies from 0.11 to 0.27 with an average value for all soils (mostly granular) of 0.21.

The modulus number M varies from 280 to 1270, whereas the exponent λ varies from 0.18 to 0.38. The coefficient of correlation r is also listed in Table 1 to indicate the quality of fit with the model for each data set. In view of the number of data sets that have been evaluated, the magnitudes and ranges of M, λ, and ν listed in Table 1 are considered to be representative for granular materials.

6 CONCLUSION

An isotropic model for the nonlinear elastic behavior of soils has been presented. The model is based on Hooke's law in which Poisson's ratio is constant, and Yong's modulus is expressed in terms of stress invariants. The expression for Young's modulus has been derived from theoretical considerations based on the principle of conservation of energy. Material parameters may be determined from unloading-reloading cycles in conventional triaxial compression tests. Parameters were determined and studied for fifteen different soils.

REFERENCES

[1] N. Janbu, Soil Compressibility as Determined by Odometer and Triaxial Tests, Proc. Europ. Conf. Soil Mech. Found. Eng., Wiesbaden, 1, 19-25 (1963)

[2] J.M. Duncan and C.-Y. Chang, Nonlinear Analysis of Stress and Strain in Soils, J. Soil Mech. Found. Div., 96, 1629-1653 (1970)

[3] B.O. Hardin and W.L. Black, Closure to Vibration Modulus of Normally Consolidated Clay, J. Soil Mech. Found. Div., 95, 1531-1537 (1969)

[4] B.O. Hardin, The Nature of Stress-Strain Behavior of Soils, Proc. Spec. Conf. Earthquake Eng. Soil Dyn., Pasadena, 1, 3-90 (1978)

[5] R.D. Mindlin, Compliance of Elastic Bodies in Contact, J. Appl. Mech., 71, A-259-268 (1949)

[6] M. Zytynski, M.F. Randolph, R. Nova and C.P. Wroth, On Modeling the Unloading-Reloading Behavior of Soils, Int. J. Anal. Methods Geomech., 2, 87-94 (1978)

[7] P.V. Lade and R.B. Nelson, Modeling the Elastic Behavior of Granular Materials, Int. J. Anal. Methods Geomech., 11 (1987)

[8] P.W. Rowe, Theoretical Meaning and Observed Values of Deformation Parameters for Soil, Proc. Roscoe Memorial Symp. Stress-Strain Behavior of Soils, 143-194 (1971)

[9] P.V. Lade and J.M. Duncan, Cubical Triaxial Tests on Cohesionless Soils, J. Soil Mech. Found. Div., 99, 793-812 (1973)

[10] R.N. Yong and H.-Y. Ko, Pre-Workshop Package, Proc. Workshop Limit Equilibrium, Plasticity and Generalized Stress-Strain in Geotechnical Engineering, Montreal (1980)

[11] M. Goldscheider, True Triaxial Tests on Dense Sand, Constitutive Relations for Soils, Grenoble (1982), Balkema, Rotterdam, 11-54 (1984)

[12] A.S. Saada, Information Package, Proc. Int. Workshop on Constitutive Equations for Granular Non-Cohesive Soils, Cleveland (1987)

[13] K.L. Lee, Triaxial Compressive Strength of Saturated Sand Under Seismic Loading Conditions, Ph.D.-thesis, Univ. of California, Berkeley (1965)

Numerical Methods in Geomechanics (Innsbruck 1988), Swoboda (ed.)
© 1988 Balkema, Rotterdam. ISBN 90 6191 809 X

Discontinuous numerical model for partially saturated soils at low saturation

J.A.Gili & E.E.Alonso
Technical University of Catalunya, Barcelona, Spain

ABSTRACT: Liquid and gaseous phases have been incorporated into a discontinuous particulate system in order to analyze the coupled flow-mechanical behaviour of a granular silt-like soil at low water contents.
The following basic phenomena are represented in the model : surface tension at the gas-liquid interface (which is governed by suction), the stiffnening effect of the menisci at the contacts between particles; the phase changes of both water and air and their associated water content redistribution. The model is used to predict the evolution of water content and pore size distribution, contact forces and deformations for a given state of saturation, geometrical arrangement and externally applied loads or changing water conditions. A number of illustrative numerical examples are described.

1 INTRODUCTION

The mechanical behaviour of partially saturated soils is difficult to characterize due to the presence of interparticle forces which are a direct consequence of the three-phase nature of these soils. Water suction is often selected as an (internal) stress variable representing the interparticle forces induced by capillary actions. However, the attempts to use this suction as an isotropic reference stress in an effective stress principle, have met very limited success. This approach seems to work only in saturated or quasi-saturated states under water tension. In dryer states a separate consideration of total stresses and suction, as an independent stress variable, seems to be necessary to describe the mechanical behaviour of partially saturated soils.
Constitutive modelling of partially saturated soils is still at its beginnings. Reasons for this are the difficulty in understanding and modelling the effect of internal forces between particles and, on the other hand, the relatively involved experimental work often required. It was therefore decided to explore the possibilities of devising a discontinuous particulate model in which the internal forces induced by capillary action at the particle contacts could be adequately modelled using basic principles involving the motion and mutual influence of solid, liquid and gas phases. A model of this kind is described in this paper.
The solid particles have been modelled as quasi-rigid spheres. In order to model the mechanical interaction between them, ideas already established in the formulation of the Distinct Element Method [1] have been followed. In addition, the void space and the water menisci have also been modelled as discrete items. The transfer of liquid among these items has also been considered. In this way coupled flow-deformation phenomena can be analyzed. Some examples are included in the paper to illustrate the capabilities of the model developed.
A model of this kind may be a useful tool to investigate the behaviour of partially saturated granular materials at relatively low degrees of saturation. Hopefully, it could also be used to relate macroscopic constitutive parameters to more fundamental physical constants and geometrical configurations.

2 GEOMETRICAL CONFIGURATION OF A PARTIALLY SATURATED IDEAL GRANULAR MATERIAL

The different phases and species which can be identified in a partially saturated soil are schematically represented in Fig.1.
At relatively low water contents the free

water, forced by surface tension effects, will adopt a minimum air-water interface surface. This can be achieved by means of menisci of water around the contacts between solid particles. In addition a water film of small thickness will remain adhered to the particle surfaces as a result of liquid-solid adhesion forces. The remaining pore volume will be occupied by air mixed with some amount of water vapour. This air may move rather freely through the network of pores.

Based on these ideas, a simplified model contituted by highly undeformable spherical particles, annular water menisci at their contacts and pores has been conceived (Fig. 2).

In order to develop the model, the full three-dimensioned geometry has been simplified in the sense that the center of the spherical particles lie always in the same plane. This leads essentially to a two-dimensional model as far as the kinematics of the particles is concerned. However menisci are truly three-dimensional. They are limited by the adjacent solid particles and a sector of a toroidal surface. A computer generated view of an array of particles, their connectivities and the water menisci at the contacts is shown in Fig. 3.

The angle between the liquid-gas interface and the particle surface has been assumed to be zero. This interface has a double curvature. Therefore the liquid-gas pressure difference, Δp , will be given by the following Laplace equation (Fig.4)

$$\Delta p = \sigma \left(\frac{1}{r_1} + \frac{1}{r_2} \right) \qquad (1)$$

where σ is the surface tension of the liquid-gas interface. Note also the different sign of the two curvature radii r_1 and r_2 .

In a general case, the two spherical

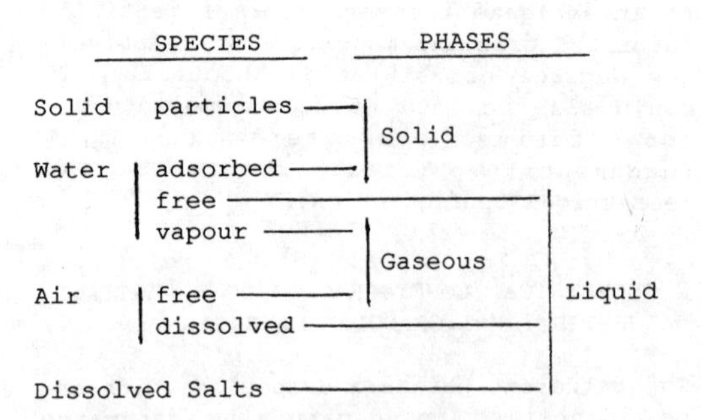

Fig.1. Species and phases in a partially saturated soil.

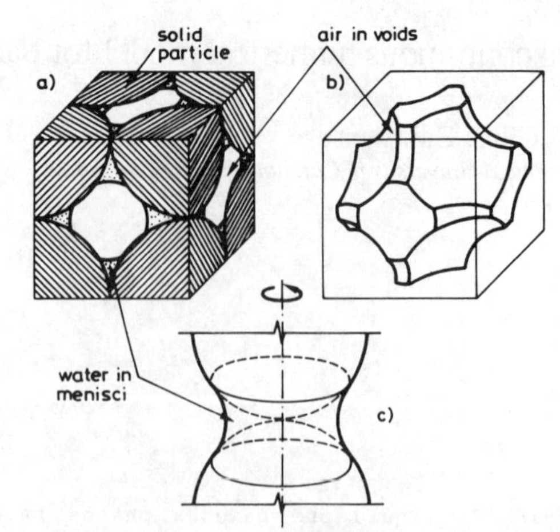

Fig.2 a) Element of unsaturated soil.
b) Approximate shape of a pore. A regular cubic array has been selected for illustration purposes.
c) Geometry of the water menisci at a particle to particle contact.

particles may be in strict contact (δ=0), overcontact (δ<0) or no contact (δ>0), which is the case represented in Fig.4a. Given a relative position of two spherical particles (R_1, R_2, δ) and the θ angle, which is a measure of the wetted area of the particle surfaces (Fig.4), a relationship between the meniscus volume, its curvature radii and the pressure difference, Δp , can be found. For these values, the force exerted by the liquid meniscus on the adjacent particles can also be found (Fig.4b). The variation of

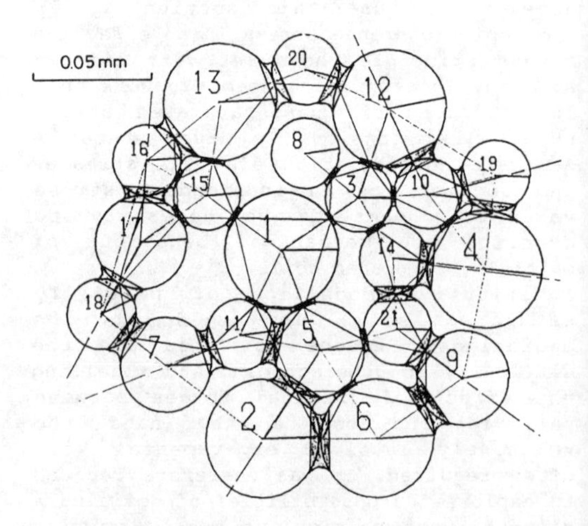

Fig.3 Example of a computer generated array of particles, pores and water menisci.

dimensionless pressure difference Δp and total force exerted by a meniscus on the particles, F, for two identical spherical particles (radius R) are shown in Fig.5 as a function of the angle θ and the contact distance δ.

For high values of θ the absolute value of the curvature radii of the meniscus may be such that the liquid inside the meniscus reaches a pressure higher than the surrounding air. However before this situation is reached, the water menisci will probably overlap (this is, for instance, the case in regular cubic packing whenever $\theta > 45°$) and the conditions for validity of the present model are lost in a strict sense.

In the model, a pore is represented by an enclosure bounded by a closed chain of particles and menisci and, occasionally, by the boundary itself.

A program module to simulate the geometry of a partially saturated porous medium has been developed. It generates a random configuration of particles in a circular domain. Initially the particles are generated in strict contact. In a subsequent stage external loads and prescribed conditions of air and water pressure may be applied to the boundary. An example of a soil configuration is given in Fig.6a. The size of the particles corresponds in this case to a fine silt. At any stage of the computational process graphs showing the state of the soil in terms of the pore size distribution (Fig.6b) or suction in the menisci (Fig.8) may be obtained.

3 FLUID FLOW AND PHASE CHANGES

At any transient state of the model, the pressure differences (of water in the menisci, of the air and water vapour in the pores) or the concentration gradients (of dissolved air in the menisci) will induce "flow" of the different species. In some cases these "flows" constitute actually a phase change. This will be the case of water vaporization or the solution of air into water. A schematic representation of the various flows considered in the model is given in Fig.7. The main characteristics of these mass transfer phenomena are indicated in Table 1.

The flow equations used, at the microscopic level, define the mass of a given species which will flow between two neighboughring pores or menisci. In general, this mass will be proportional to

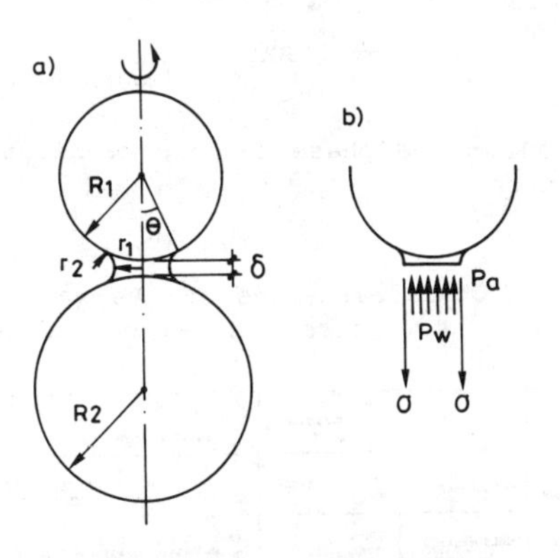

Fig. 4 a) Geometry of a water meniscus at a particle contact. b) Forces through an equatorial section of the meniscus.

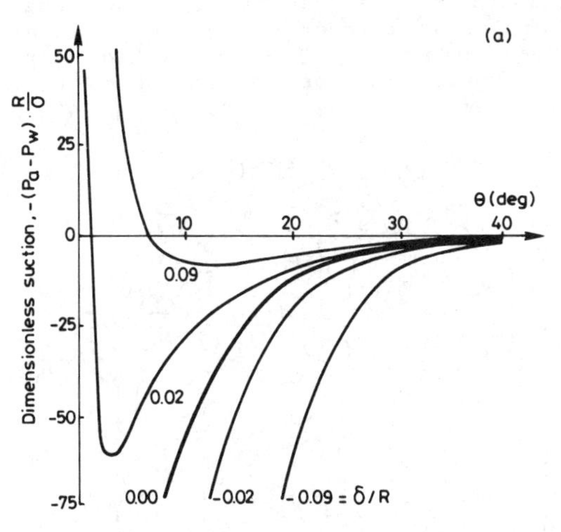

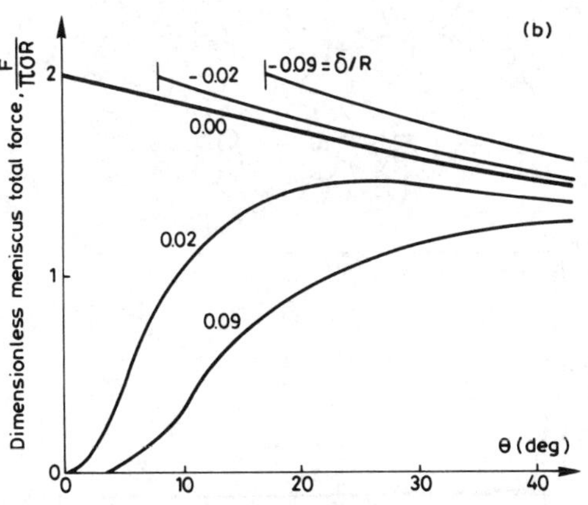

Fig.5 a) Influence of meniscus geometry, as given by the angle θ, on pressure difference at the liquid gas interface.
b) Force exerted on adjacent particles.

the relevant pressure differences and will also be a function of the geometry involved (i.e. film dimension of water adhered to particles or cross sectional area and leght of conduits). The film thickness may be related, through the Van der Waals forces, to the existing suction through a 6th order hyperbolic relationship [3].
A numerical scheme to handle these flows has been developed. It uses a relaxation-type finite difference scheme explicit in time. Similarly to the D.E.M. the flow between two given domains (pores or menisci), at a given computational cycle, is considered to be independent of the remaining domains. This condition requires a sufficiently small time step.
Space limitations prevent a detailed presentation of the flow equations. However two examples are given to illustrate the capabilities of the formulation to model flow phenomena. The first example (Fig.8) shows the evolution in time of suction associated to the menisci of the particle arrangement shown in the previous figure 6. The initial state (Fig. 8a) shows a nonhomogeneous distribution of suction which corresponds to a given constant volume of water for all the menisci. In subsequent instants of time the water flows from and to the menisci until a constant final suction, equal to the boundary imposed suction (0.9 kp/cm^2), is reached at all the particle contacts (Fig. 8c).
In the second example (Fig.9) a certain mass of solute is injected in a point of a column of soil whose two boundaries (L=0 and L=1) are maintained at two given water suctions. Fig.9 shows the propagation of the solute concentration due to convection effects (in the direction of flow) and to molecular diffusion (which induces a symmetric dispersion).

0.1mm

a)

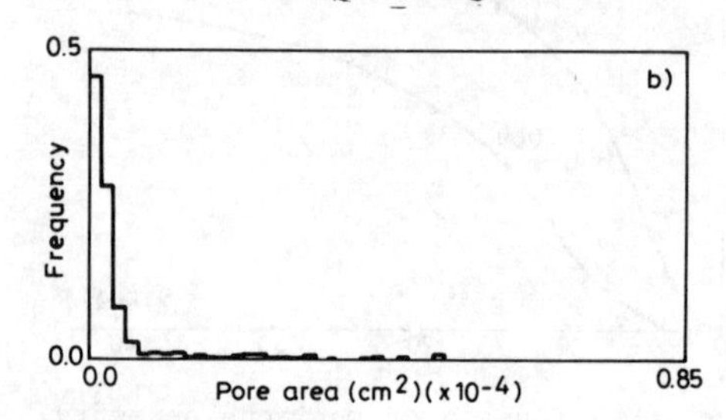

Fig.6 a) Example of generation of a particle configuration. b) Pore size distribution.

4 MECHANICAL MODELLING

Some basic features of the mechanical modelling have been taken from the work of Cundall and Strack [1]. The main differences between a dry and a moist particulate system lies in the nature of the interparticle forces. Accordingly, they will be summarized in this section.
Particles interact with each other through mechanical contacts and through the action of water menisci. Particles are assumed to be highly undeformable in the sense that they maintain always its spherical shape. Some overlapping is allowed and it is

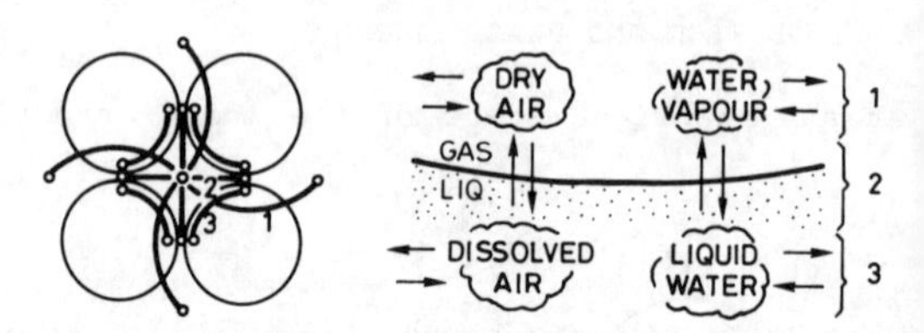

Fig.7. Flows and phase changes considered in the model.

TABLE 1. Characteristics of the various fluid flows and phase changes considered in the model.

SPECIES INVOLVED	ORIGIN AND DESTINATION OF FLOW	CONDUCTION ZONE	PHYSICAL PRINCIPLE
AIR TO AIR (G) VAPOUR TO VAPOUR	1 PORES-PORES	CONDUITS BETWEEN PORES	AIR CONVECTION PROPORT. TO GRADIENT MOLECULAR DIFFUSION (FICK'S LAW)
AIR TO AIR (D) VAPOUR TO WATER	2 PORES-MENISCI	MENISCI INTERFACES	HENRY'S LAW (E) PSYCHROMETRIC RELATIONSHIP (E)
WATER TO WATER AIR (D) TO AIR (D)	3 MENISCI-MENISCI	ADHERED FILMS	WATER CONVECTION PROPORT. TO GRADIENT MOLECULAR DIFFUSION (FICK'S LAW)

(G): Gas; (D): Dissolved; (E) For the state of equilibrium

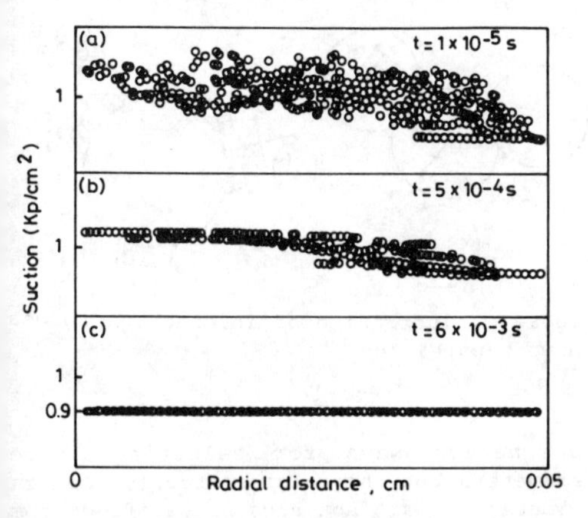

Fig.8. Evolution of suction of water menisci around particle contacts in Fig.6. Contact position is defined by its distance to the center of circular array

assumed to be proportional to the contact forces.

The rheological modelling of the contacts is schematically indicated in Fig.10. Some features of the laws adopted are:

a) The normal stiffness has been described by a constant coefficient, K_n, in loading states. K_n gives an average stiffness to the nonlinear hertzian type of contact. Upon unloading, a different coefficient, K_D, has been used.

b) Normal forces, N_δ, are obtained by multiplying the stiffness coefficients by the overlapping between (undeformed) particles (Fig.11). The value of δmax is used to know the history of the contact and to select the appropriate stiffness value.

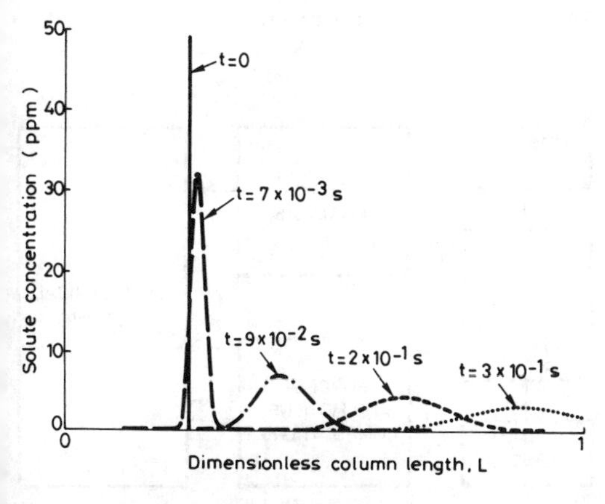

Fig.9. Evolution of solute concentration in a column of particles.

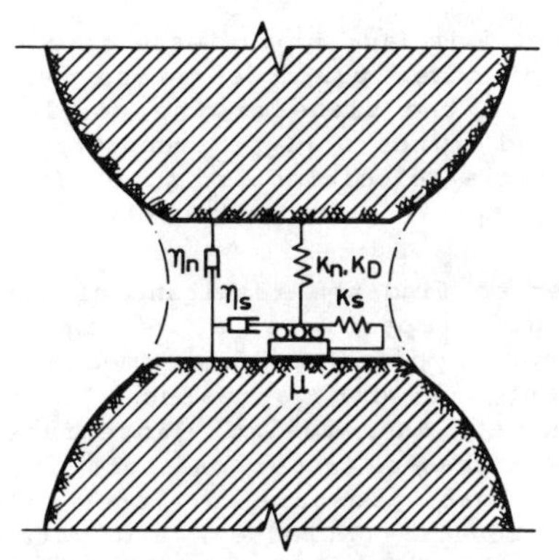

Fig.10. Schematic representation of the rheological behaviour of particle mechanical contacts.

c) In a similar way, a tangential stiffness, K_s, provides the force, T_δ, induced by a relative tangential displacement. T_δ is bounded by the frictional force μN_δ.

d) In a state of no mechanical contact ($\delta \leq o$), both N_δ and T_δ are zero. However $\delta \leq 0$ does not necessarily implies the breaking of the water meniscus and its associated interaction force.

e) Linear damping, controlled by the relative normal and tangential velocities, is also assumed. It has been suggested that this damping may take into account energy losses due to impact and delayed response of the contact (due for instance, to the delayed plastification of contact asperities). In our case a third mechanism is probably more relevant: the resistance of the water within the menisci to accommodate to a new geometric configuration. This effect will be present also (in a normal and tangential sense) even if the mechanical contact is lost but the water meniscus is maintained. These damping forces associated to the contacts will be denoted as T_η and N_η.

In addition, some global damping of the particles themselves may be proposed to

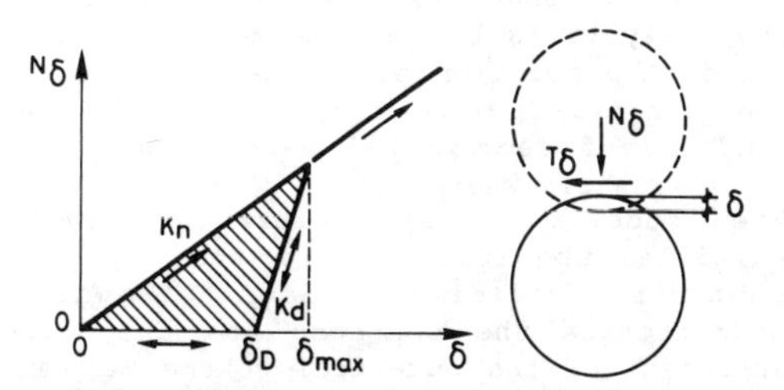

Fig.11. Description of normal mechanical interaction forces.

represent, in an approximate way, the action of other particles whose centers are not in the plane of the model. The associated forces will be proportional to the particle velocities ($\dot{x}$, $\dot{y}$, $\dot{\theta}$) (forces $F_{x\eta}$, $F_{y\eta}$, M_{η}).

In order to find the resultant of forces on a given particle at every time increment, point forces (mechanical contacts), linear stresses (surface tension at the menisci interfaces), surface stresses (water pressure in the menisci, gas pressure in the pores, applied boundary stresses) and particle self weight, should be appropriately added. Fig. 12 shows these forces in a simplified two-dimensional representation (gas pressure has been omitted). If the gas pressure is similar in all the pores, it can be shown that the normal and tangential forces at each contact or meniscus are given by (see Fig.12)

$$\left(N_{\delta}+N_{\eta}+N_{\omega}\right)\left(-\bar{s}_{i}\right) \quad (2)$$

and

$$\left(T_{\delta}+T_{\eta}\right)\left(-\bar{t}_{i}\right) \quad (3)$$

where

$$N_{\omega}=(p_{\omega}-p_{a})(\pi r_{1i}^{2}-\pi r_{8i}^{2})-p_{ai}.\pi r_{8i}^{2}-\sigma .2\pi r_{1i} \quad (4)$$

Adding the forces at the contacts and the (global) damping induced forces gives the resultant force vector (F_x, F_y, M) for every particle. Single and double integrations of Newton's second law provide the updated velocities and displacements for a single time step.

5 SOLVED EXAMPLES

5.1. Preliminary Remarks

Computations have been organized in the way indicated in Fig.13. It must be pointed out that the geometrical part of the analysis is by far the more involved one in the solution procedure. Details of the numerical techniques used and the handling of geometrical aspects will be dealt with elsewhere [2].

When modelling the interaction between particles the selection of the physical contacts (stiffness, damping) which characterize the contacts and the time increment of the numerical scheme becomes a relevant and often delicate decision. Several references ([4],[5],[6]) discuss

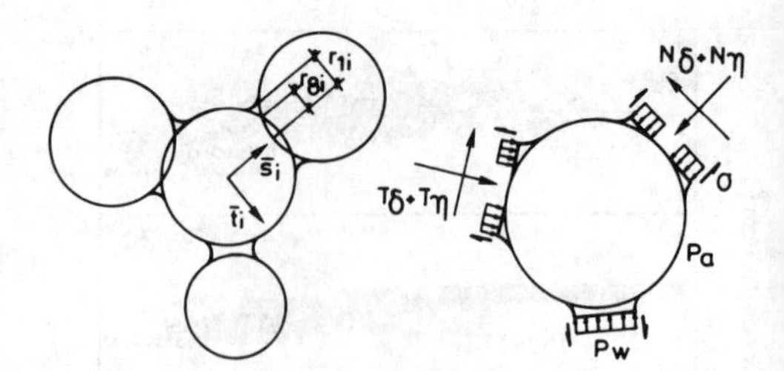

Fig.12. Illustration of interaction forces between particles.

alternative ways for selecting these parameters but the values actually used in computations are not always justified from a physical point of view. Their choice is often governed by computational reasons (stability of the numerical scheme).

An effort has been made in this work to base the contact parameters in a physical background. They are indicated here to allow for discussion and comparison with similar work elsewhere. For an average grain size of $4x10^{-3}$ cm the following constants were adopted for grain to grain contacts: Loading normal striffness, K_n = 10 Kp/cm; unloading normal stiffnes, K_d = 20 Kp/cm; shear stiffness, K_s = 2.5 Kp/cm; friction coefficient, μ = 0.5 (=tan 26.6°) ;normal damping, η_n = $9x10^{-7}$ Kp/cm/s; shear damping, η_s = $4.5x10^{-7}$ Kp/cm/s. The global damping used in computations was $1x10^{-7}$ Kp/cm/s.

The K_n value represents an average stiffness for the nonlinear hertzian relationship. Once this value was fixed, the remaining stiffnesses (K_d,K_s) were derived from theoretical relationships of the Hertzian contact theory.

The friction coefficient is an acceptable value for several rock minerals.

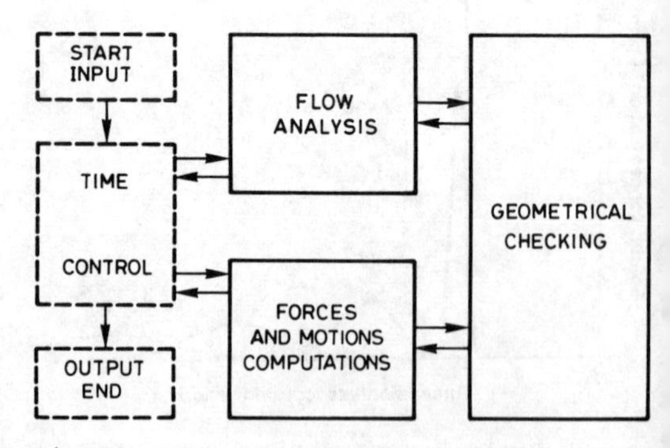

Fig.13. Scheme of the computational procedure.

The normal damping of the contacts was approximated from the well known critical damping for a single degree of freedom system (mass : m, stiffness : k):

$$\eta_{crit} = 2\sqrt{m\,k} \qquad (5)$$

It is know that a necessary condition for numerical stability of an undamped single degree of freedom dynamic system is to use a time step $\Delta t < 2\sqrt{m/k}$ which correponds to the half period of the eigenfrequency ($2x10^{-7}$ sec in our case).However, the existence of damping requires consider ably smaller values of Δt. Given the relatively high value of the damping used in our case (the critical value), the time step finally selected for computations did not exceed 10^{-8} sec (for motions and forces computations).

The flow analysis did not require such a small value of Δt. The numerical scheme used guaranteed stability for any time step and reasonable "exact" values were obtained for $\Delta t + 10^{-6}$ sec.

5.2. Simple cases of equilibrium.

In a first example two particles are linked by a single meniscus. Initially the particles are in strict contact. The forces induced by the meniscus on the two particles will close slightly the distance between them. Some results of the computational process are shown in Fig.14. The overlapping distance between the two particles, initially zero, converges eventually towards an equilibrium value of δ (approximately $0.62x10^{-7}$ cms). The figure shows also the evolution of contact forces. The force induced by the meniscus, N_w, changes slightly. The mechanical force N_δ, proportional to the overlapping, increases monotonically towards the equilibrium value. The damping force, N_η, proportional to the relative velocity of the particles, firts increases and subsequently decreases to a nil value at equilibrium. In a second simple example (Fig.15) a particle A is held in contact with two fixed particles by two water menisci. A modification of the boundary suction induces and increase in water content at the menisci, a parallel reduction of linking forces and the instability of the lower particle when its weight exceeds the holding forces. A sequence of computer simulated successive states is shown in Fig.15

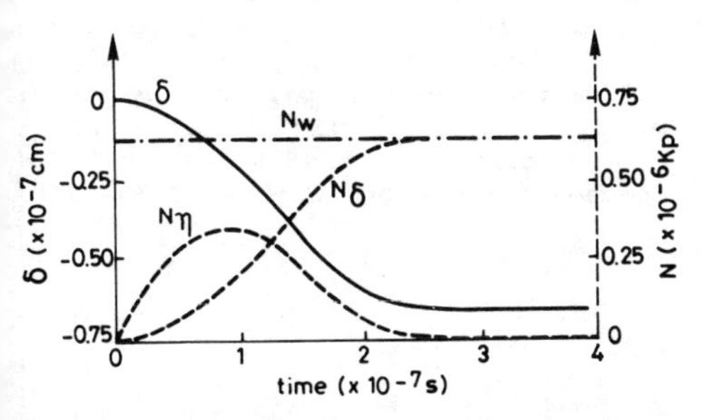

Fig.14. Evolution of relative displacements and forces between two particles.

5.3. Collapse, upon wetting, of an array of particles.

The array of 21 particles shown in Fig.3, with an initial suction of 0.9 Kp/cm^2 is first subjected to an all-round pressure of 0.1 Kp/cm^2. Subsequently the array of particles is saturated, maintaining the external pressure. The changes in the position of the particles is shown in Fig.16. The new arrangement of particles shows a definite collapse of the structure. This is better described with the aid of Fig.17 which indicates the changes in accumulated pore size distribution when the particle system is saturated. The influence of shear stresses on collapse may also be analyzed by the model. Fig.18 and 19 show the collapse rearrangements induced in the same particle array when it is subjected to a mean pressure of 0.1 Kp/cm^2 and a deviatoric stress of 0.1 Kp/cm^2. A slighly larger collapse (in terms of areal porosity) was computed in this case. The relatively small number of particles used in this example prevents a comparison of the computed results with real collapse tests on granular soils. It serves however to illustrate the capabilities of the model developed to simulate "tests" on partially saturated particulate soils.

6 CONCLUDING REMARKS

A discrete model in which particles, pores and menisci are considered as individual interacting items has been developed to represent the behaviour of partially saturated particulate materials. Transient flow conditions involving relevant mass transfer mechanisms of water and air may

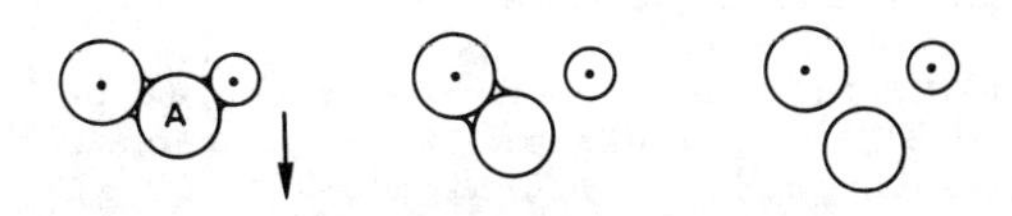

Fig.15. Instability of a particle induced by water content changes.

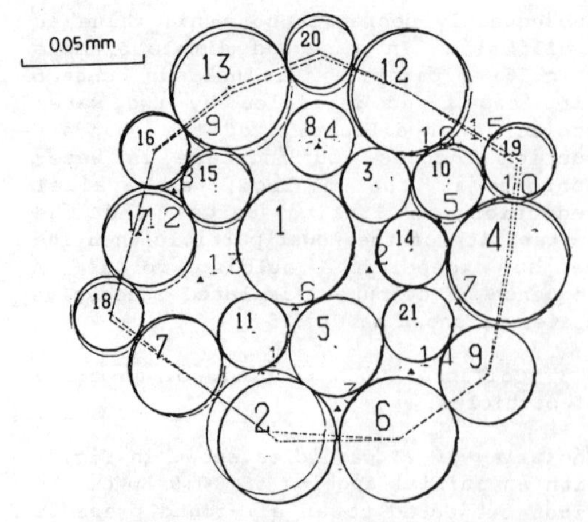

Fig.16. Collapse of a grain structure. Isotropic external stress

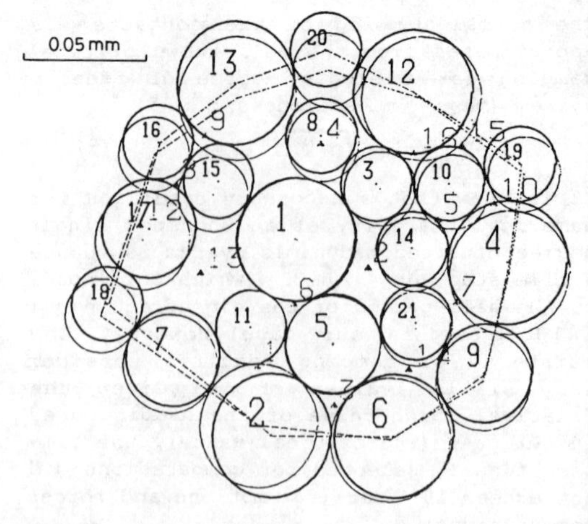

Fig.18. Collapse of a grain structure. Anisotropic external stress.

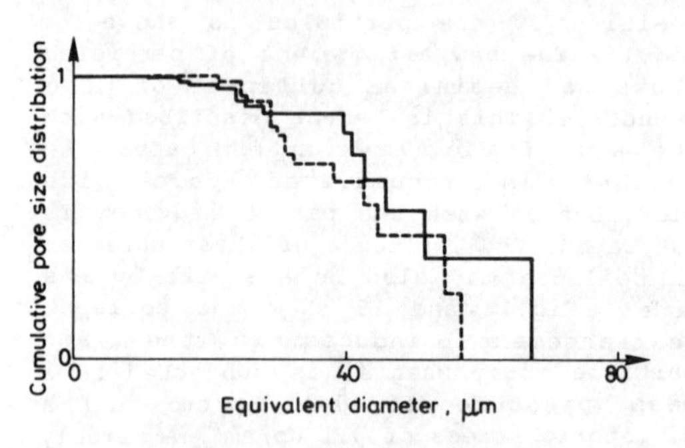

Fig.17. Change in cumulative pore size distribution. Isotropic external stress.

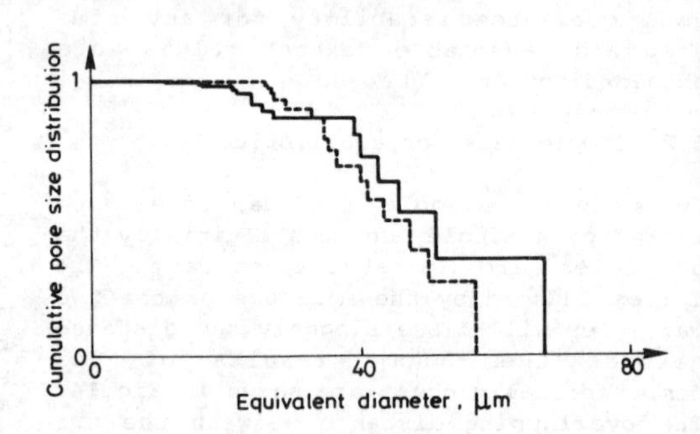

Fig.19. Change in cumulative pore size distribution. Anisotropic external stress.

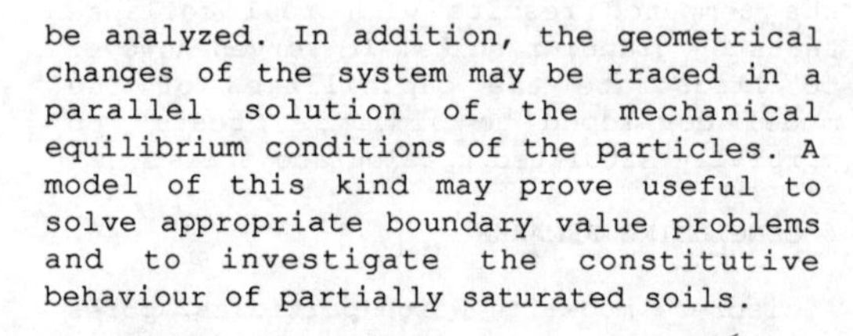

be analyzed. In addition, the geometrical changes of the system may be traced in a parallel solution of the mechanical equilibrium conditions of the particles. A model of this kind may prove useful to solve appropriate boundary value problems and to investigate the constitutive behaviour of partially saturated soils.

7 REFERENCES

[1] P.A.Cundall and O.D.L.Strack. A discrete numerical model for granular assemblies. Géotechnique 29, No.1, 47-65 (1979)

[2] J.A.Gili. Modelo microestructural de medios granulares no saturados. Ph.D.Thesis. Technical Univ. of Catalunya (1988)

[3] G.Kovacs. Seepage Hidraulics. Elsevier, Amsterdam (1981)

[4] Y.Ohnishi, T.Mimuro, N.Takewaki and J.Yoshida. Verification of input parameters for distinct element analysis of jointed rock mass. Proc.Int.Symp.on Fundamentals of Rock Joints. Björkliden, 205-214 (1985)

[5] O.D.L.Strack and P.A.Cundall. The distinct element method as a tool for research in granular media. Report to N.S.F. Univ.of Minnesota.U.S.A. (1978)

[6] O.R.Walton. Explicit particle dynamics model for granular materials. Proc.Fourth Int.Conf.Num.Meth. in Geomech., Edmonton, Canada, 1261-1268 (1982)

Numerical Methods in Geomechanics (Innsbruck 1988), Swoboda (ed.)
© 1988 Balkema, Rotterdam. ISBN 90 6191 809 X

Modelling the development of rupture surfaces using displacement-type finite element methods

R.G.Wan, D.H.Chan & N.R.Morgenstern
University of Alberta, Canada

ABSTRACT: The ability to model the development of a rupture surface as a shear band in a soil medium is of considerable interest in geotechnical engineering. The approach adopted here is the use of a modified principle of virtual work equation which accounts for discontinuities within a given domain of interest. The total displacement field is expressed in terms of continuous and discontinuous components so that degrees of freedom are now global and relative displacements respectively. The strategy adopted is the development of a special two-dimensional element with an implicit internal discontinuity in its interpolation functions, thereby obviating the need to update the finite element mesh. An interface relationship for the rupture is incorporated in the analysis, while the surrounding material can follow any conventional constitutive law. Finally the effectiveness of the formulation is demonstrated by means of a numerical example.

1 INTRODUCTION

The issues of rationally understanding and realistically modelling progressive failure in geomaterials have been the object of intense activity in the last decade [1]. Past approaches adopted have been based mainly upon continuum features of an unfaulted medium, with analyses carried out through the classical finite element method enhanced by an induced local anisotropy in order to account for localization. Among these formulations have been incremental hardening plasticity with negative modulus [2,3] and a stress transfer procedure [4,5]. More recent works [6] attempt to improve the analysis by means of a modified finite element formulation of the stiffness matrix, which accounts for the existence of a shear band within an element through a sub-element approach. This concept, though having the merit of eliminating mesh sensitivity with respect to shear band topology, has the weakness of not considering band position within an element. Also, due to the smeared technique, it does not represent localized deformation well. Obara et al. [7] subsequently developed a cracked triangular element following a similar approach.

The observation of a transition to localized deformation in geological bodies under initially perfectly homogeneous conditions has led some researchers [8,9] to adopt a constitutive stability analysis through bifurcation theory. de Borst [10] probably presented the first numerical simulation of shear bands in sand bodies by means of bifurcation theory and successfully traced the post peak regime by a combination of an incremental-iterative loading procedure and an eigenvalue analysis of the tangential stiffness matrix. The analysis is very involved and relies upon a good numerical strategy to ensure proper convergence to an equilibrium point. Most recently Ortiz et al. [11] introduced suitably defined shape functions in order to reproduce the localized modes of deformation. A bifurcation analysis was also performed to define the onset of localization. However, though the methodology is quite original, resulting shape functions lead to an incompatible element not satisfying C^{o} interelement continuity.

In the present study a conceptually different approach is adopted which represents the actual physical problem in a better manner. In precise terms, the essence of the model is as follows: failure occurs by the propagation of a shear band, here treated as a discontinuity surface, upon which relative sliding of the adjacent surfaces may lead to an ongoing reduction of the material body effective strength. The tip of the band, where the surface gives way to still intact material, advances in space with time and loading history. Initially it might appear logical to use an adaptive mesh to simulate the phenomenon by FEM, i.e., the location of the shear band has to be made to coincide with the edges of the elements. However, the generation of meshes to fit the element topology with a moving band would impose a considerable penalty on the solution technique, if not make it intractable.

The approach adopted here circumvents the problem by writing a modified principle of virtual work expression for a material body containing an internal discontinuity surface as known in classical continuum mechanics [12] and used for soil-pile interaction [13]. A special element whose associated interpolation (shape) functions explicitly cater to discontinuities has been developed, obviating the need to update the mesh topology. The direction of the shear band will be predicted by a Mohr-Coulomb type of criterion, but can also be governed by a bifurcation based equation [14].

2 FORMULATION OF THE PROBLEM

2.1 Virtual work for displacement discontinuities

In the light of the derivation of the incremental equilibrium equations for a material body traversed by a surface of displacement discontinuity, the classical principle of virtual work (PVW) is amended and conveniently recast as follows:

$$\int_{\Omega} \sigma_{ij}\delta\varepsilon_{ij} d\Omega = \int_{\Omega} f_i \delta u_i d\Omega + \int_{\Gamma_\sigma} \overline{T}_i \delta u_i d\Gamma + \int_{\Gamma_s} t_i^1 \delta g_i d\Gamma \qquad (1)$$

where σ_{ij} are Cauchy stresses, δu_i virtual displacements, $\delta\varepsilon_{ij}$ associated virtual strains, f_i body forces, $\overline{T}_i$ prescribed boundary tractions along the surface Γ_σ, t_i^1 surface tractions at internal discontinuity surface Γ_s and δg_i associated virtual displacement jumps or slips. Note that t_i^1 is an internal traction force to be subsequently determined through a contact constitutive relationship. The last term introduced in equation (1) refers to virtual frictional work along Γ_s.

2.2 Incremental virtual work equations

The foregoing equations of equilibrium are formulated in rate form in order to capture the evolution (kinematics) of the rupture surface over time [0,t[. An updated Lagrangian description is chosen and the evolution considered as a sequence of equilibrium states assumed infinitesimally small so as to preclude effects of inertia.

Using the modified PVW expression, the equilibrium of the body in configuration $C^{(1)}$ requires that, see Figure 1:

$$\int_{\Omega^t} \sigma_{ij}\delta(\Delta\varepsilon_{ij})\, d\Omega = \int_{\Omega^t} f_i \delta(\Delta u_i)\, d\Omega + \int_{\Gamma_\sigma} \overline{T}_i\, \delta(\Delta u_i) d\Gamma + \int_{\Gamma_s} t_i^1\, \delta(\Delta g_i)\, d\Gamma \qquad (2)$$

At time t_2, equilibrium of configuration $C^{(2)}$ is implied by

$$\int_{\Omega^t} (\sigma_{ij} + \Delta S_{ij})\, \delta(\Delta E_{ij})\, d\Omega = \int_{\Omega^t} (f_i + \Delta f_i)\, \delta(\Delta u_i)\, d\Omega + \int_{\Gamma_\sigma} (\overline{T}_i + \Delta\overline{T}_i)\, \delta(\Delta u_i)\, d\Gamma + \int_{\Gamma_s} (t_i^1 + \Delta t_i^1)\, \delta(\Delta g_i)\, d\Gamma \qquad (3)$$

The components of Green's strain tensor due to increments in displacements are given by:

$$\Delta E_{ij} = \frac{1}{2}(\Delta u_{i,j} + \Delta u_{j,i}) + \frac{1}{2}\Delta u_{k,i}\Delta u_{k,j} = \Delta\varepsilon_{ij} + \Delta\varepsilon''_{ij} \qquad (4)$$

A comma followed by a subscript indicates a partial derivative with respect to current material coordinates (x_1, x_2, x_3). Furthermore incremental deformations are considered small ($\Delta S_{ij} \simeq \Delta\sigma_{ij}$) and with piecewise linear material behaviour, i.e.,

$$\Delta\sigma_{ij} = C_{ijkl}\, \Delta\varepsilon_{kl}, \qquad \Delta t_i^1 = -\eta_{ij}\Delta g_j, \qquad (5)$$

where C_{ijkl} is a fourth order material constitutive tensor and η_{ij} a second order tensor describing a contact relationship along the discontinuity surface Γ_s relating normal and tangential tractions

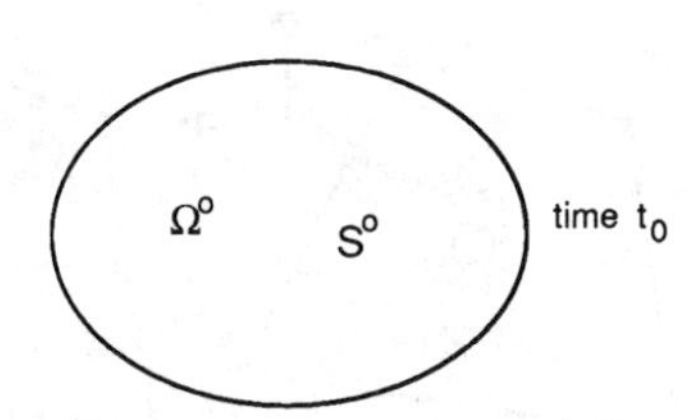

reference configuration $c^{(0)}$

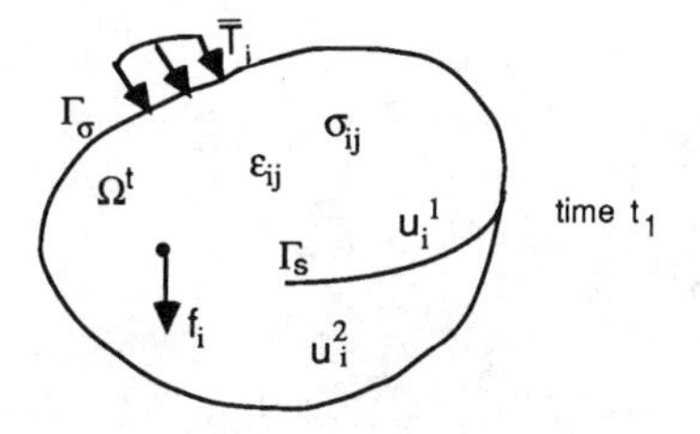

updated configuration $c^{(1)}$

$\delta g_i = \delta u_i^1 - \delta u_i^2$

S_{ij} : 2nd. Piola-Kirchhoff stress tensor

E_{ij} : Green's strain tensor

$\bar{T}_i + \Delta\bar{T}_i$

$\sigma_{ij} + \Delta s_{ij}$

$f_i + \Delta f_i$

$\varepsilon_{ij} + \Delta E_{ij}$

$u_i^1 + \Delta u_i^1$ Γ_s

Ω^{t+dt} time $t_2 = t_1 + \Delta t$

$u_i^2 + \Delta u_i^2$

current configuration $c^{(2)}$

Figure 1. Description of motion

to conjugate displacement discontinuities. For instance, as a simple case, the Coulomb friction law can be used. The incremental form of the modified PVW arises when equations (4) and (5) are substituted into equation (3) and equation (2) subtracted from the latter neglecting higher order terms. The final expression recast into finite element form reads:

$$\begin{aligned}\int_\Omega C_{ijkl}\Delta\varepsilon_{kl}\,\delta(\Delta\varepsilon_{ij})\,d\Omega &= \int_\Omega \Delta f_i\,\delta(\Delta u_i)\,d\Omega \\ &+ \int_{\Gamma_\sigma} \Delta\bar{T}_i\,\delta(\Delta u_i)\,d\Gamma - \int_{\Gamma_s} \eta_{ij}\Delta g_j\,\delta(\Delta g_i)\,d\Gamma \\ &+ \Big\{ \int_\Omega f_i\,\delta(\Delta u_i)\,d\Omega + \int_{\Gamma_\sigma} \bar{T}_i\,\delta(\Delta u_i)\,d\Gamma \\ &- \int_{\Gamma_s} t_i^1\,\delta(\Delta g_i)\,d\Gamma - \int_\Omega \sigma_{ij}\,\delta(\Delta\varepsilon_{ij})\,d\Omega \Big\}\end{aligned} \qquad (6)$$

It should be noted that the expression in brackets on the right hand side of equation (6) would be zero if the initial stress distribution satisfied the equilibrium equation (2).

3 INCREMENTAL FINITE ELEMENT FORMULATION

Let the continuum be divided into an arbitrary number of finite elements, say M, joined together at N global nodes and NL localized nodes where are defined displacement and slip displacement degrees of freedom respectively. Furthermore, let the displacement at a generic material point be interpolated in terms of continuous and discontinuous components. Thus in compact symbolic notation, the relationship in incremental form is given by:

$$\Delta\underset{\sim}{u} = \left[\underline{\underline{N}}^{(\alpha)} \;\middle|\; \underline{\underline{N}}^{(\beta)}\right] \left\{\frac{\Delta\bar{\underset{\sim}{u}}}{\Delta\underset{\sim}{g}}\right\} \qquad (7)$$

in which $\Delta\bar{\underset{\sim}{u}}$ is the vector of incremental nodal displacements, $\Delta\underset{\sim}{g}$ the vector of incremental slip displacement at 'dual' nodes, and $\underline{\underline{N}}^{(\alpha)}$ and $\underline{\underline{N}}^{(\beta)}$ associated matrices of interpolation polynomials. Superindices (α) and (β) refer to continuous and discontinuous components respectively. The incremental strain can be formally computed from kinematic equations as follows:

$$\Delta\underset{\sim}{\varepsilon} = \left[\underline{\underline{B}}^{(\alpha)} \;\middle|\; \underline{\underline{B}}^{(\beta)}\right] \left\{\frac{\Delta\bar{\underset{\sim}{u}}}{\Delta\underset{\sim}{g}}\right\} \qquad (8)$$

with $\underline{\underline{B}}^{(\alpha)}$ and $\underline{\underline{B}}^{(\beta)}$ as compatibility matrices.

Substituting equations (7) and (8) into the modified PVW equation (6) finally yields a set of linear matricial equations with $\Delta\bar{\underset{\sim}{u}}$ and $\Delta\underset{\sim}{g}$ as principal unknowns. This, after some algebraic manipulations is:

$$\left[\frac{\underline{\underline{K}}_{\Omega\Omega}}{\underline{\underline{K}}_{s\Omega}} \;\middle|\; \frac{\underline{\underline{K}}_{\Omega s}}{\underline{\underline{K}}_{ss}}\right] \left\{\frac{\Delta\bar{\underset{\sim}{u}}}{\Delta\underset{\sim}{g}}\right\} = \left\{\frac{\underset{\sim}{f}_\Omega}{\underset{\sim}{f}_s}\right\} + \left\{\frac{\Delta\underset{\sim}{f}_\Omega}{\Delta\underset{\sim}{f}_s}\right\} - \left\{\frac{\underset{\sim}{Q}_\Omega}{\underset{\sim}{Q}_s}\right\} \qquad (9)$$

where

$$\underline{\underline{K}}_{\Omega\Omega} = \sum_{e=1}^{M} \int_{\Omega^e} \underline{\underline{B}}^{(\alpha)^T} : \underline{\underline{C}} : \underline{\underline{B}}^{(\alpha)} d\Omega,$$

$$\underline{\underline{K}}_{\Omega s} = \sum_{e=1}^{M} \int_{\Omega^e} \underline{\underline{B}}^{(\alpha)^T} : \underline{\underline{C}} : \underline{\underline{B}}^{(\beta)} d\Omega,$$

$$\underline{\underline{K}}_{s\Omega} = \sum_{e=1}^{M} \int_{\Omega^e} \underline{\underline{B}}^{(\beta)^T} : \underline{\underline{C}} : \underline{\underline{B}}^{(\alpha)} d\Omega,$$

$$\underline{\underline{K}}_{ss} = \sum_{e=1}^{M} \int_{\Omega^e} \underline{\underline{B}}^{(\beta)^T} : \underline{\underline{C}} : \underline{\underline{B}}^{(\beta)} d\Omega + \sum_{e\epsilon\Gamma_s} \int_{\Gamma_s^e} \underline{\underline{N}}^{(\beta)^T} : \underline{\underline{\eta}} : \underline{\underline{N}}^{(\beta)} d\Gamma,$$

$$\underset{\sim}{f}_{\Omega} = \sum_{e=1}^{M} \int_{\Omega^e} \underline{\underline{N}}^{(\alpha)^T} \cdot \underset{\sim}{f}^e d\Omega + \sum_{e\epsilon\Gamma_\sigma} \int_{\Gamma_\sigma^e} \underline{\underline{N}}^{(\alpha)^T} \cdot \overline{\underset{\sim}{T}}^e d\Gamma,$$

$$\underset{\sim}{f}_{s} = \sum_{e=1}^{M} \int_{\Omega^e} \underline{\underline{N}}^{(\beta)^T} \cdot \underset{\sim}{f}^e d\Omega + \sum_{e\epsilon\Gamma_\sigma} \int_{\Gamma_\sigma^e} \underline{\underline{N}}^{(\beta)^T} \cdot \overline{\underset{\sim}{T}}^e d\Gamma,$$

$\Delta\underset{\sim}{f}_{\Omega}$, $\Delta\underset{\sim}{f}_{s}$ are incremental forms of $\underset{\sim}{f}_{\Omega}$ and $\underset{\sim}{f}_{s}$ respectively.

$$\underset{\approx}{Q}_{\Omega} = \sum_{e=1}^{M} \int_{\Omega^e} \underline{\underline{B}}^{(\alpha)^T} : \underline{\underline{\sigma}} \, d\Omega,$$

$$\underset{\approx}{Q}_{s} = \sum_{e=1}^{M} \int_{\Omega^e} \underline{\underline{B}}^{(\beta)^T} : \underline{\underline{\sigma}} \, d\Omega - \sum_{e\epsilon\Gamma_s} \int_{\Gamma_s^e} \underline{\underline{N}}^{(\beta)^T} \cdot \underset{\sim}{t}^1 d\Gamma .$$

4 4-NODE ISOPARAMETRIC DISCONTINUOUS ELEMENT, QD4

Figure 2 illustrates the two master configurations which account for all possible geometric locations in a typical element. Any other configuration is simply obtained by local cyclic permutation of node numbers.

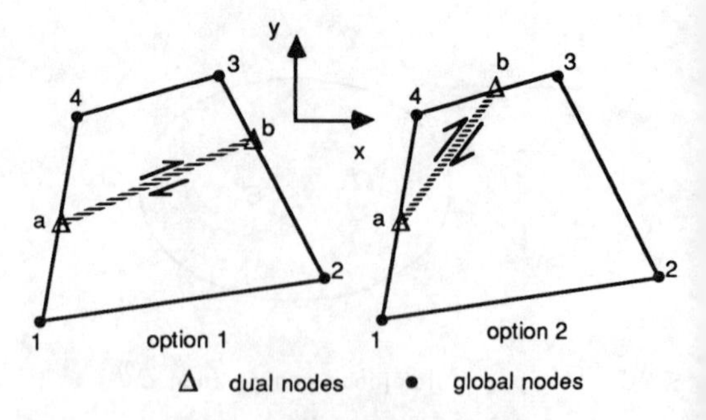

Figure 2. QD4 elements with two options on discontinuity location

The derivation of interpolation functions in the above situations has been performed in the following way. Functions suffering a discontinuity along element boundaries crossed by a discontinuity line are found and interpolated linearly for the interior by means of blending techniques [15]. A comprehensive description of the underlying procedure and derivations of discontinuous interpolation functions are given in [16].

5 NUMERICAL EXAMPLE

Many practical examples of application, for instance the one illustrated in Figure 3 after [17], become amenable to analysis by means of the foregoing formulation.

However in order to pursue an investigation of fundamentals, the case examined herein is restricted to a fairly elementary configuration (see Figure 4), namely an initially homogeneously stressed region which succumbs to the localization phenomenon by artificial generation of shear bands within it. A uniform distributed load is applied on element edge denoted by nodes 4 and 5. The medium is considered elastic and all geometric and material non linearity is assumed concentrated in the shear band.

Shear bands are artificially made to branch from different sites in the mesh, involving elements 1, 3, 2 and 4. Initiation and orientation of the shear band are dictated by a Mohr-Coulomb type of criterion based upon the principle of maximum stress obliquity. Furthermore, in the event of conjugate directions, the orientation compatible with prevailing kinematic constraints is so chosen. In order to establish compatibility between discontinuous (shear band) elements and adjacent continuum elements, the tip of the band is prescribed as a 'dual' node of

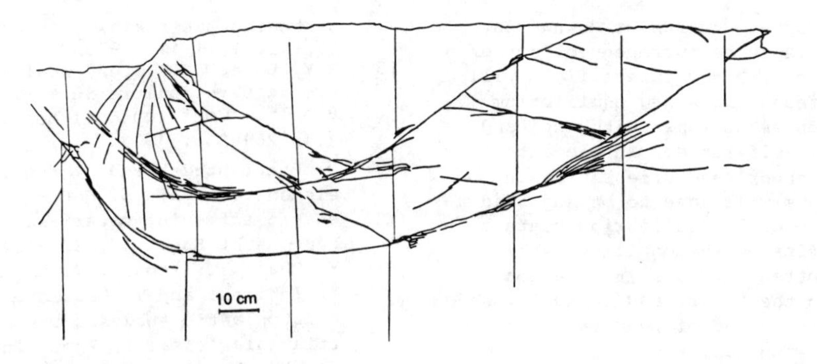

Fig. 3 Slip surfaces beneath a foundation, after [17]

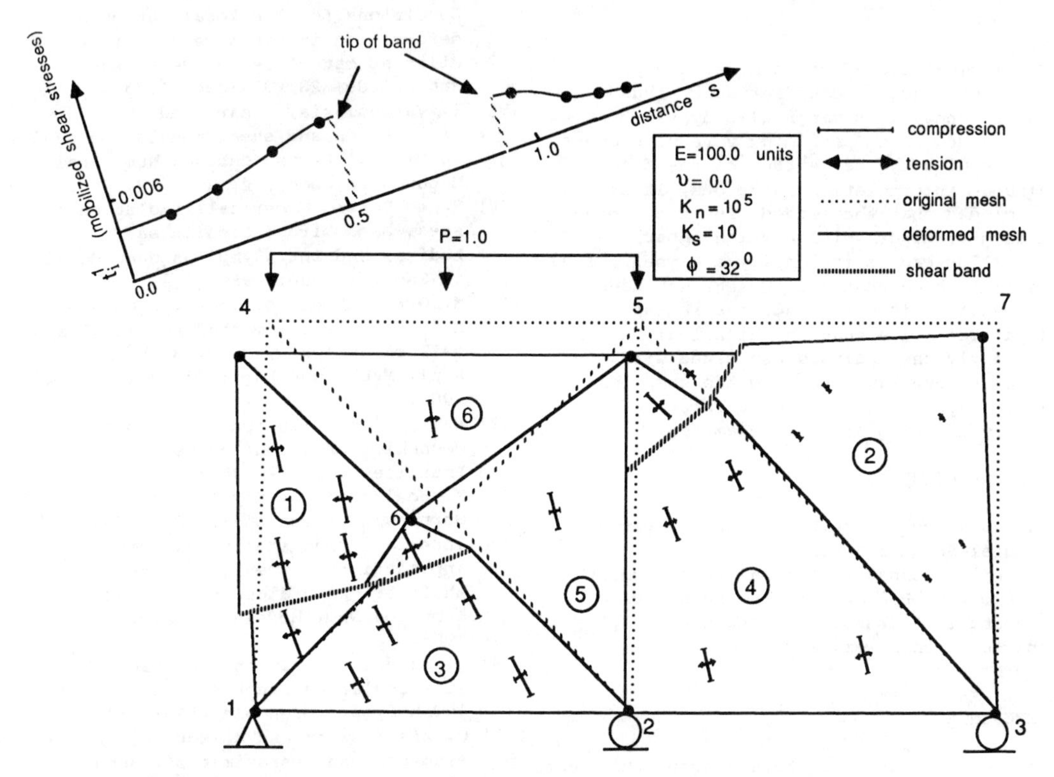

Figure 4. Finite element schematization of slippage along a shear band

zero slip in the computations. The mobilized slippage tendency indicates that most of the movement occurs in the vicinity of the applied loads. Also, the stress distribution with the shear band in place reveals a substantial difference from the one expected for the unfaulted case, i.e., stresses are released and principal stress rotations reduced in regions outside the band. For example the rotation of principal stresses adjacent to line 1-4 in Figure 4 may be noted. This suggests that loads tend to localize along the band leaving the surrounding areas in an unloading state. Moreover, stresses in element number 2 are very small in comparison with others while some tension develops in elements 1 and 4 as one segment of corresponding shear band elements tries to detach itself from the rest of the elements. Here the normal stiffness of the shear band has been prescribed at a high enough value that tensile opening or separation is prohibited although the shear band curves upwards or downwards. If the load is to be incremented further in the analysis, a new stress field will have to be found and a new equilibrium configuration established with the band nucleating itself further in element 5. However the shear band orientation in other elements will have to be adjusted until a new overall equilibrium state is achieved. Finally the mobilized shear stresses plotted at Gauss integration points along the shear band typically show the well known trend of peak values at the leading tip with lesser values at sites where most of the movement has taken place.

6 CONCLUSIONS

A numerical procedure to model the presence of discontinuity surfaces, thus shear bands, in a material body has been presented. From preliminary results concerned with the net effect of the discontinuous interpolation functions, it can be concluded that the method provides a promising framework within which shear band modelling can be rationally analysed. It is also shown that the present enhanced formulation is more conducive in capturing localized modes than the classical FEM is. Presently the analysis concerned with the constitutive law governing the shear band and its inception through bifurcation theory is being actively examined.

ACKNOWLEDGEMENTS

This research is being supported by the Natural Science and Engineering Research Council of Canada. The valuable comments of Prof. M. Boulon from the Mechanics Institute of Grenoble during the early stages of this work are deeply appreciated.

REFERENCES

[1] M.P. Cleary, Fracture discontinuities and structural analysis in resource recovery endeavours, Journal of Pressure Vessel Technology, ASME, 10, 2-12, 1978.

[2] K. Hoeg, Finite-element analysis of strain softening clay, Proc. ASCE, 98, SM1, 43-58, 1972.

[3] R.H. Gates, Progressive failure model for clay shale, Proc. Sym. on Finite Element Method in Geotechnical Engineering, Vicksburg, Mississippi, 327-347, 1972.

[4] K.Y. Lo and C.F. Lee, Stress analysis and slope stability in strain softening materials, Geotechnique, 23, No. 1, 1-11, 1973.

[5] K.Y. Lo and C.F. Lee, Analysis of progressive failure in soils, Proc. 8th Int. Conf. on Soil Mech., Moscow, 1.1, 251-258, 1973.

[6] S. Pietruszczak and Z. Mroz, Finite element analysis of deformation of strain softening materials, Int. J. Num. Meth. Eng., 17, 327-334, 1981.

[7] Y. Obara, T. Yamabe, Y. Shimizu, Y. Ichikawa and T. Kawamoto, Elastoplastic analysis by cracked triangular element, Proc. Int. Conf. on FEM, Beijing, China, 756-760, 1982.

[8] J.W. Rudnicki and J.R. Rice, Conditions for the localization of deformation in pressure sensitive dilatant materials, J. Mech. Phys. and solids, 23, 371-394, 1975.

[9] I. Vardoulakis, Shear band inclination and shear modulus of sand in biaxial tests, Int. J. Num. Anal. Methods in Geom., 4, 103-119, 1980.

[10] R. de Borst, Numerical simulation of shear-band bifurcation in sand bodies, 2nd Int. Symp. on Num. Models in Geom., 91-98, 1986.

[11] M. Ortiz, Y. Leroy and A. Needleman, A finite element method for localized failure analysis, Comp. Meth. in Appl. Mech. and Eng., 61, 2, 189-214, 1987.

[12] L.E. Malvern, Introduction to the Mechanics of a Continuous Medium, Prentice-Hall, New Jersey, 1969.

[13] M. Boulon and C. Plytas, Soil Structure directionally dependent interface constitutive equation-application to the prediction of shaft friction along piles, 2nd Int. Symp. on Num. Models in Geom., 43-54, 1986.

[14] J. Mandel, Conditions de stabilité et Postulat de Drucker, Proc. of the IUTAM Symp., Grenoble, 58-68, 1966.

[15] O. Zienkiewicz and K. Morgan, Finite elements and approximation, John Wiley & Sons, New York, 1982.

[16] R. Wan, Finite element simulation of shear band development in geologic media, Ph.D. thesis, under preparation, University of Alberta, Canada, 1988.

[17] C. Chazy and P. Habib, Les piles du quai de Floride, 5th Congres Int. de Mec. des Sols, Paris, Com. 6/27, 669, 1961.

Numerical Methods in Geomechanics (Innsbruck 1988), Swoboda (ed.)
© 1988 Balkema, Rotterdam. ISBN 90 6191 809 X

Constitutive subroutine to simulate alternating loading of granular materials

F. Molenkamp
Delft Geotechnics, Netherlands

ABSTRACT: From experiments with alternating loading on granular materials it is known that densification and deformation occur. Depending on the water content and the rate of loading this material behaviour can lead to liquefaction and cyclic mobility. To simulate these properties in a finite element analysis a numerical code of a constitutive model has been developed. The constitutive model is of the elasto-plastic, kinematic hardening type. Its major component simulates the behaviour under cyclic deviatoric loading. In the paper the structure of the constitutive subroutine and its implementation in a finite element code are discussed.

1. TYPE OF CONSITUTIVE MODEL

The constitutive model (Molenkamp, 1982) is intended to simulate the behaviour of soil under alternating deviatoric loading (see fig. 1). It is based on the concept of kinematic yield surfaces (see eg. Iwan 1967, Mroz, Norris, Zienkiewicz, 1979) to describe the distribution of the plastic stiffness in the stress space.

The elasticity is taken as nonlinear in accordance with experimental data (eg. Lade, 1977, Vermeer, 1980).
In partially drained and undrained loading porepressure generation will cause the effective stresses to decrease, possibly leading to fluidisation. Therefore the model must also be able to simulate fluidisation. To this end a simple linear viscous model is used.

For simplicity it is assumed that when fluidisation has occurred it will be maintained whatever the strain and the strain rate.

Also relatively small magnitudes of cohesion c can be accounted for easily by adapting the normal stresses by $\frac{c}{tg\phi}$, in which ϕ is the friction angle at large deformation, before applying the frictional type of material model and by subtracting this additional stress again after the calculation of its response.
The occurrence of cracking of a cohesive material is accounted for in a simple way by giving a limit to the isotropic effective stress.

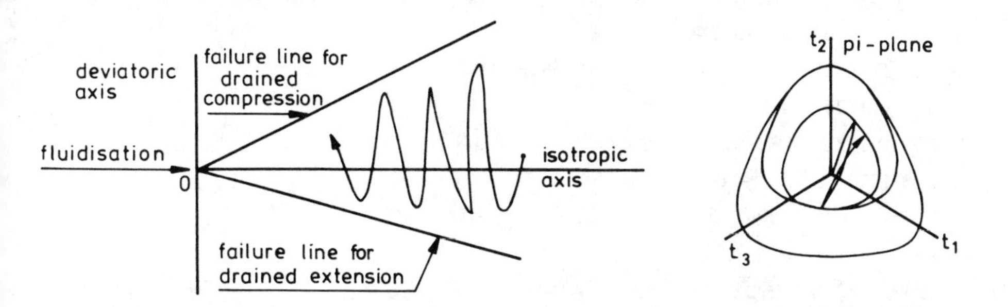

Fig.1 Illustration of the type of stress path for which the material behaviour must be simulated.

Because severely nonlinear material behaviour should be simulated, the increments of both the isotropic stress and the shear stress level should be kept small within the constitutive subroutine in order to maintain accuracy.

2 STRUCTURE OF THE FINITE ELEMENT PROGRAM

The constitutive model can be applied in finite element programs. Next the general structure of such a FEM-program is outlined.

In the finite element program the subroutine package (Smith, 1982) has been used together with the subroutine package LIB3 (Molenkamp, 1986). The mesh generation and the graphical output are generated by means of a graphics package of Delft Geotechnics. The structure of the FEM-program is summarized in table 1. In the program the modified Newton-Raphson method is applied.

First the global pseudo tangent matrix is generated starting from the assumption of elasto-plastic behaviour at every Gauss point. Then this assumption is

Table 1. Structure of FEM-program

```
DECLARATIONS
INPUT OF GEOMETRICAL DATA
FORM NODAL FREEDOM ARRAY
INPUT MATERIAL PARAMETERS (ETC)
IF    FIRST START:
.     DEFINE INITIAL STRESS AND ANISOTROPY
.     CALCULATE INITIAL LOAD
IF    RESTART:
.     READ STRESS ANISOTROPY AND INITIAL LOAD
INPUT OF ENFORCED DISPLACEMENT AND LOAD INCREMENT
START LOOP OF SUCCESSIVE SUBINCREMENTS OF LOAD
.     START LOOP OF COMPOSITION OF TANGENT STIFFNESS
.     .     COMPOSE PSEUDO TANGENT STIFFNESS MATRIX
.     .     COMPOSE LOAD VECTOR
.     .     DECOMPOSE STIFFNESS MATRIX
.     .     SOLVE FOR RESULTING DISPLACEMENT
.     .     CHECK TYPE OF MATERIAL BEHAVIOUR
.     END OF LOOP OF COMPOSITION OF TANGENT STIFFNESS
.     START LOOP OF MODIFIED NEWTON-RAPHSON ITERATION
.     .     CALCULATE LOAD INCREMENT FOR CURRENT ESTIMATE OF DISPLACEMENT INCREMENT
.     .     CALCULATE REMAINING UNBALANCE
.     .     SOLVE FOR RESULTING DISPLACEMENT INCREMENT
.     .     CHECK CONVERGENCE CRITERION
.     END OF LOOP OF MODIFIED NEWTON-RAPHSON ITERATION
.     PREPARE FOR NEXT LOAD INCREMENT
.     COPY OUTPUT DATA TO INPUT DATA
.     PREPARE PRINTED OUTPUT
.     PREPARE DATA FOR GRAPHICAL OUTPUT
.     OUTPUT DATA FOR RESTART
END OF LOOP OF SUCCESSIVE INCREMENTS OF LOAD
GENERATE GRAPHICAL OUTPUT.
```

checked in an iterative scheme until all Gauss points with assumed elasto-plasticity are found to behave elasto-plastically indeed.

Having estimated the pseudo tangent stiffness matrix, the solution is determined iteratively by reducing the remaining unbalance until it becomes acceptably small.

In each iteration step of the modified Newton-Raphson iteration for the load increment the current estimate of the displacement increment must be calculated (see table 1). To this end in fact for each Gauss point for the local strain increment the stress increment must be determined. This calculation is performed by means of a constitutive subroutine, which structure is discussed next.

3 STRUCTURE OF CONSTITUTIVE SUBROUTINE

In table 2 the logic of the constitutive subroutine for strain-controlled loading has been elaborated. The description of the actions starts at the top of this figure.

First a choice is met; it concerns the question whether in the previous increment viscous flow (or cracking) occurred. If so (= Y) the stress will be calculated as viscous flow, this case is indicated as case 7.

If previously no viscous flow (or cracking) occurred (= N) then the following actions are performed. First the effect of cohesion c is taken into account by adapting the normal stresses. Then the co-rotational strain increment and the resulting co-rotational stress increment due to elastic behaviour are calculated. With this elastic stress increment being known, next two choices are considered.

The first choice concerns the occurrence of a stress reversal on the basis of the so-called "differential criterion". This criterion for a stress reversal reads:

$$\text{sign} \left\{ \frac{\partial F^d}{\partial \sigma_{kl}} \Delta\sigma^e_{kl} \right\} < 0 \qquad (1)$$

in which

$$F \{ I(\underset{\approx}{T}) \} - f(\chi) = 0 \qquad (2)$$

is the expression of the kinematic yield surface and

I = indicates invariants

$\underset{\approx}{T}$ = pseudo stress tenson

χ = hardening parameter

In case the criterion of eq. (1) is satisfied (= Y) then it is checked whether the smallest yield surface is sufficiently small, namely:

$$f(\chi) < f_{minimum} \qquad (3)$$

This condition is considered to limit the magnitude of the elastic part of the stress reversal; only within this yield surface elastic reversal will be allowed.

In case both eqs. (1) and (3) are satisfied (= Y) then the choice of the occurrence of a stress reversal on the basis of the so-called "incremental criterion" is considered. This criterion for a stress reversal reads:

$$\text{sign} \{ F^d(\sigma + \Delta\sigma^e) - F^d(\sigma) \} < 0 \qquad (4)$$

If this criterion is satisfied and the resulting stress is in the compressive part of stress space (= Y) then "case 1" of "stress reversal" occurs. The resulting elastic stress could still become too small and viscous flow (or cracking) could result.

In case the "differential criterion" is satisfied and the smallest yield surface is large (see eq. 3) but the "incremental criterion" is not satisfied then the stress path must cross the inner region of the smallest kinematic yield surface and again leave it. This case is called "elastic cross over"; it is indicated by "case 2" (see table 2).

In this case the stress at the intersection of the stress path and the smallest kinematic yield surface is calculated using the assumption of elastic behaviour. In principle this stress could be so small that viscous flow (or cracking) could start. The latter possibility is checked. If viscous flow does not occur then the remaining co-rotational strain increment will be taken into account at a later stage of the subroutine.

In case that the differential criterion is satisfied, but the smallest yield surface is relatively large, then a small elastic reversal is determined to begin with and the resulting new small yield surface is introduced. In the mean time the possible occurrence of liquefaction (or cracking) is checked. This case is called "case 3". The remaining strain increment is calculated. It will be taken into account at a later stage of the subroutine.

At that stage of the calculation only continued loading still has to be considered. This situation can be reached in three ways as shown in table 2.

The calculations for continued loading are different depending on the position of the stress at the start of the increment. Therefore the question arises

Table 2. Structure of constitutive subroutine

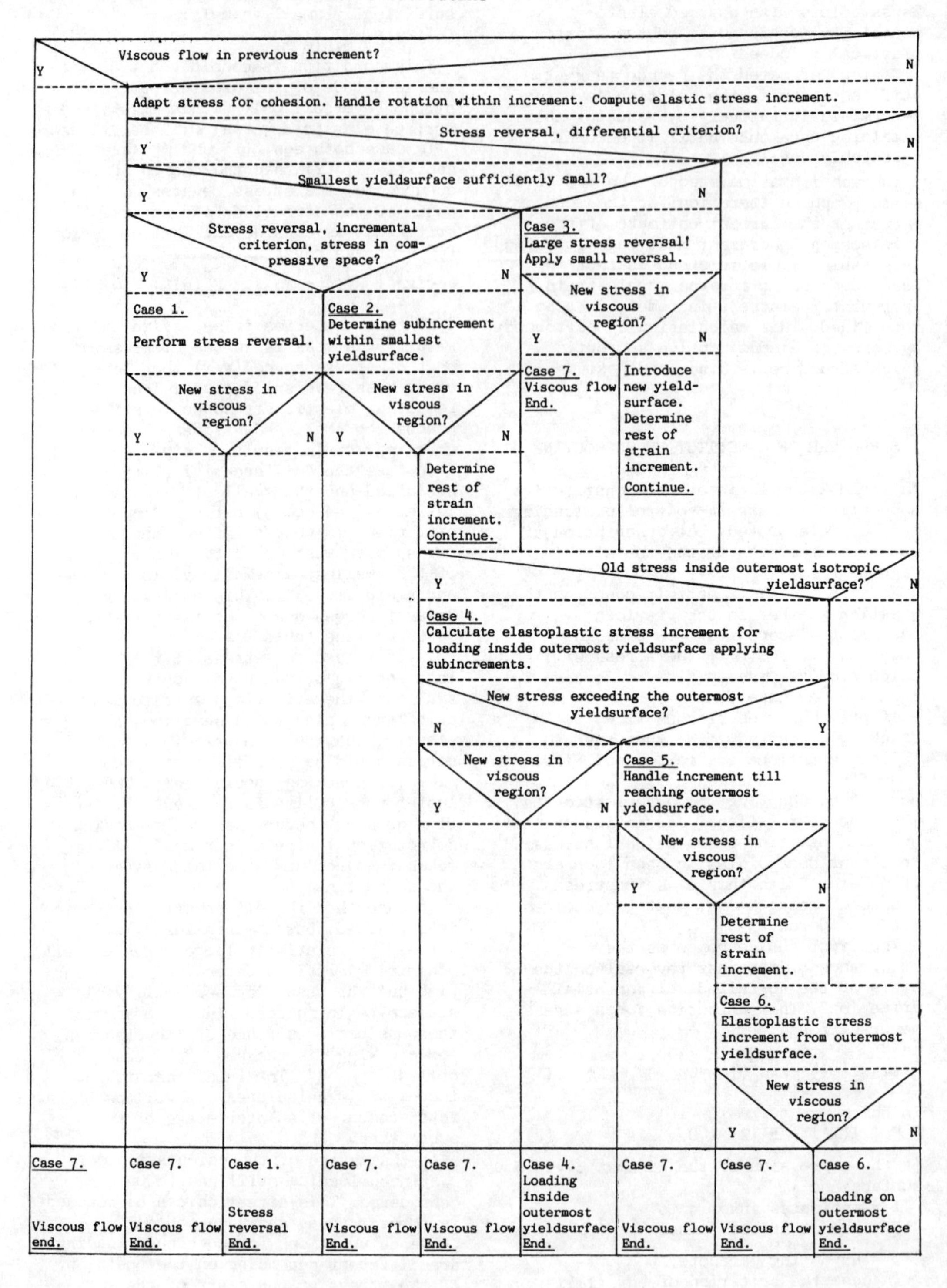

whether the stress at the start of the increment is situated inside or on the outermost isotropic yield surface. If it is situated inside the outermost yield surface (= Y) then "case 4" is arrived at (see table 2). In such a case the elasto-plastic stress increment is calculated. However, the resulting stress can exceed the outermost yield surface in which case it would not be realistic.

This possible occurrence is checked. If the resulting stress is not passing the outermost yield surface (= N) then the possible occurrence of viscous flow (or cracking) is considered.

If the resulting stress exceeds the outermost yield surface (= Y) then "case 5" occurs. In such a case the stress at the intersection of the stress path and the outermost yield surface is determined.

If this stress does not lead to viscous flow (or cracking) then the remaining co-rotational strain increment past this intersection point is determined, to be taken into account at the next stage.

That stage concerns loading on the outermost yield surface; it is indicated in table 2 as "case 6". Also in this case viscous flow or cracking could be initiated.

In the constitutive subroutine the "case 7" of viscous flow or cracking as arrived at before is elaborated at the end.

4 SIMULATION OF DRAINED AND UNDRAINED CYCLIC LOADING

To illustrate the capabilities of the model to simulate the behaviour of sand under both drained and undrained alternating loading the results of a numerical simulation of triaxial tests under one-way cyclic loading at a cell pressure of $\sigma_r = 100 \frac{kN}{m^2}$ are shown.

In the drained case the deviatoric strain increment per cycle is decreasing slightly in consecutive cycles (see figure 2). In the first cycle a permanent deviatoric strain of 0.037% is generated. During cyclic loading the material continues to densify; after 10 cycles of loading the contraction is about 0.1% while the deviatoric strain is of the order of 0.21%.

The simulation of the undrained response is shown in figure 3. For the stiffness of the porewater has been used:

$$\frac{K_w}{n} = \frac{\Delta p}{\Delta \varepsilon_{vol}} = 2 \frac{GN}{m^2}$$

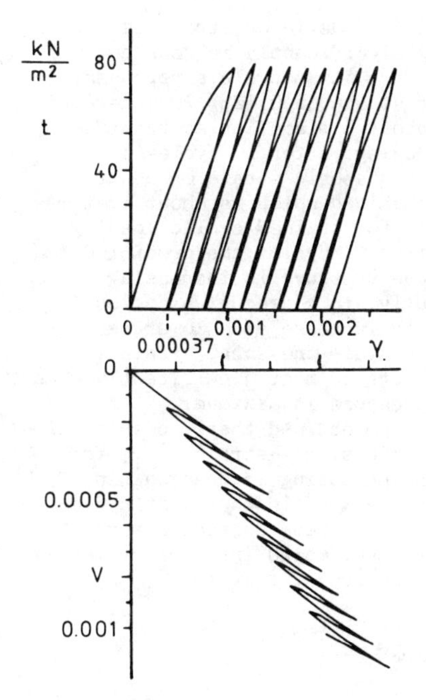

Fig.2 Numerical simulation of drained one-way cyclic loading in triaxial compression

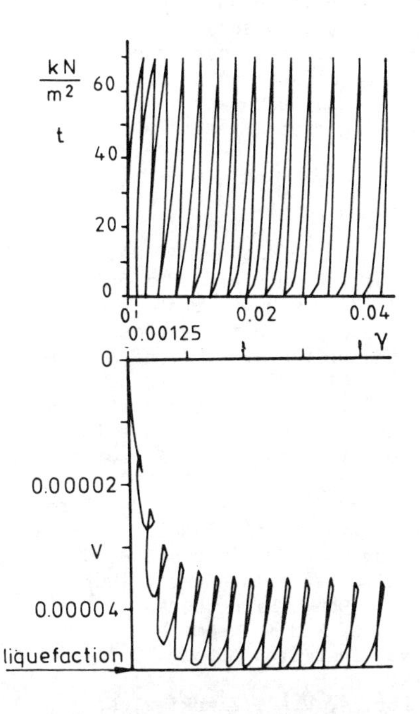

Fig.3 Numerical simulation of undrained one-way cyclic loading in triaxial compression

The predicted deviator strain is increasing significantly in consecutive cycles. In the first cycle a permanent deviatoric strain of 0.125% is generated which is about 3 times larger than for drained loading. After 10 cycles a permanent deviatoric strain of about 2.65% is obtained which is about 12 times larger than for drained cyclic loading.

In the first 4 cycles the predicted average pore pressure is increasing significantly, in consecutive cycles it continues to increase at a lower rate. Within each cycle the shear strain is found to increase most significantly when the pore pressure is maximum.

It should be noticed that the shape of the deviatoric stress-strain curve for loading and unloading is changing in consecutive cycles. In the first cycle the tangent stiffness decreases with increasing load, while in later cycles the opposite occurs.

5 CONCLUSIONS

The constitutive model as considered is able to simulate soil behaviour for alternating loading. Also the occurrence of fluidisation can be accounted for.

Both the function and the logic of the constitutive subroutine have been described. It has been shown that for this severely nonlinear model the accuracy can be maintained by dividing the applied strain increment in small parts, depending on the resulting stress path.

REFERENCES

Iwan, W.D. 1967. On a class of models for the yielding behaviour of continuous and composite systems. Transactions of the ASME, J. of Applied Mechanics, p 612-617.

Lade, P.V., 1977. Elasto-plastic stress-strain theory for cohesionless soil with curved yield surfaces. J. Solids Structures, Vol. 13, p. 1019-1035.

Molenkamp, F. 1982. Kinematic model for alternating loading, ALTERNAT. Laboratorium voor Grondmechanica, CO-218598. Third revision, August 1987.

Molenkamp, F. 1986. Third course on the application of FEM in geomechanics. Dutch Royal Instition of Civil Engineers.

Mroz, Z., Norris, V.A., Zienkiewics, O.C., 1979. Application of an anisotropic hardening model in the analysis of elasto-plastic deformation of soils. Geotechnique 29, 1, p.1-34.

Smith, I.M., 1982. Programming the Finite Element Method with applications to geomechanics. John Wiley & Sons.

Vermeer, P.A. 1980. Formulation and analysis of sand deformation problems. Thesis for the degree of doctor in Technical Science at Delft University of Technology, Netherlands.

Numerical Methods in Geomechanics (Innsbruck 1988), Swoboda (ed.)
© 1988 Balkema, Rotterdam. ISBN 90 6191 809 X

Constitutive equation of tertiary sedimentary mudstone considering strain hardening and strain softening

K.Watanabe, A.Denda, H.Kawasaki & A.Nakazawa
Institute of Technology, Shimizu Construction Co., Ltd, Tokyo, Japan

ABSTRACT : A series of laboratory mechanical tests were performed in order to investigate the strength properties (maximum strength and residual strength) and deformation properties (especially the dilatancy properties of the strain hardening and strain softening process). Based on the test results and the theory of elasto-plasticity, a constitutive equation that considers strain hardening and strain softening was derived. The elasto-plastic behaviors can be simulated with sufficient accuracy, using the constitutive equations proposed in this study.

1 INTRODUCTION

Recently in Japan, there have been yearly increases in the construction of rock engineering structures such as nuclear power plants, the Honshu-Shikoku long-span bridges and the Seikan undersea tunnels (Honshu-Hokkaido) which are based on soft bedrock such as tertiary sedimentary mudstone.

Different from hard rock, the mechanical properties (the strength properties and the deformation properties) of soft rock obtained from laboratory tests are generally considered to be equivalent to those of in-situ rock mass.

It is necessary to simulate the behaviors of soft rock with using the constitutive equation in order to analyze the interaction behaviors between these structure and the soft rock foundation.

A series of laboratory tests such as the uniaxial compression tests, the standard consolidation tests and the triaxial compression tests were conducted in order to investigate the mechanical properties of the tertiary sedimentary mudstone.

The authors[1] have already proposed the elasto-plastic constitutive equations considering strain hardening behaviors which can simulate the behaviors of the specimens up to the maximum strength.

In this research, strain softening behaviors which are characteristic in mudstone are treated and a new constitutive equation is proposed which evaluates both strain hardening and strain softening behaviors.

Many constitutive equations have been proposed to represent the strain softening behaviors. With reference to the researches on elasto-plastic theory, relationship between shear stress and shear strain have been mainly discussed, whereas volumetric strain has not been taken into account very much except for the constitutive equation proposed by Oka and Adachi[2].

In this research, the volumetric behaviors at strain softening are especially studied in detail to report on a new constitutive equation with the simulated test results.

2 TEST RESULTS

2.1 Test procedure and results

Table 1 shows the results of the fundamental test for the physical and mechanical properties of mudstone.

Table 1 Typical properties of mudstone

Item	Value
Specific gravity	2.72
Natural water content	29.3 %
Wet unit weight	18.8kN/m^3
Natural void ratio	0.83
Uniaxial compressive strength	2.17MPa
Tensile strength	0.32MPa
Consolidation yield stress	11.27MPa
Compression index	0.215
Swelling index	0.055

The triaxial compression tests were carried out under constant confining pressure (0.01,0.05,0.10,0.49,0.98,1.96MPa).

All tests were conducted under a strain rate of 0.0025%/min with drained conditions.

The results of the triaxial compression test are shown in Figs.1 and 2.

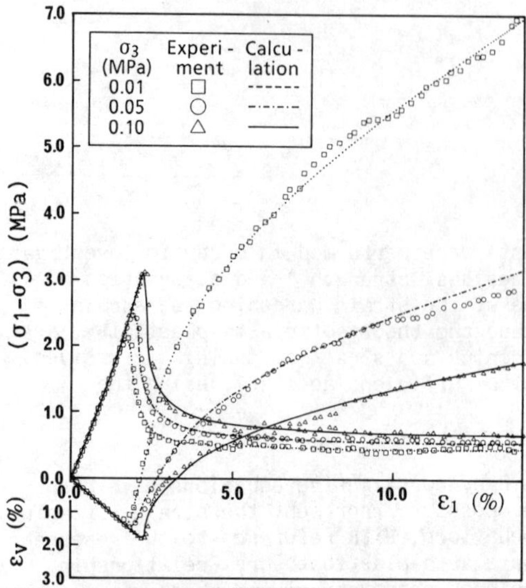

Fig.1 Stress strain curves in the case of confining pressure of 0.01, 0.05, 0.10MPa

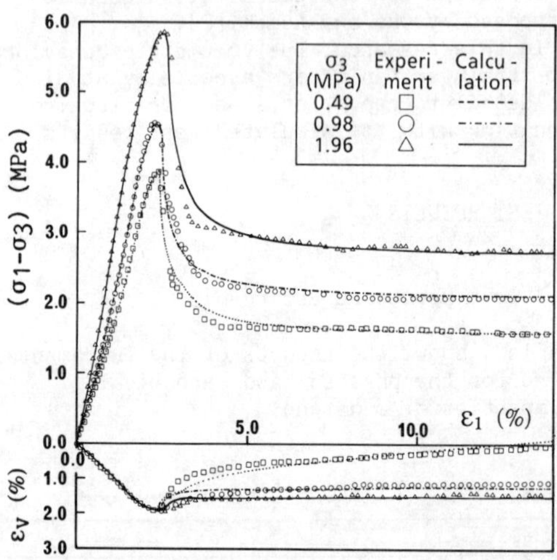

Fig.2 Stress strain curves in the case of confining pressure of 0.49, 0.98, 1.96MPa

According to the stress strain relationship in Fig.1, the specimens behave just like an elastic body up to the maximum strength, followed by clear strain softening and heavy volumetric expansion.

In this case, volumetric strain has not converged even at 15% axial strain.

However, in the stress strain relationship in Fig.2, the specimens show elasto-plastic behaviors with strain hardening up to the maximum strength, followed by clear strain softening but little volumetric expansion. In this case, volumetric strain has almost converged at 15% axial strain.

2.2 Elastic deformation properties

Elastic behaviors are recognized up to the maximum strength in Fig.1 and up to the initial yield stress in Fig.2. From the stress strain relationship in these stress regions, tangential elastic shear modulus G and tangential elastic bulk modulus K are calculated and shown in Figs.3 and 4. From these figures, G and K are expressed as linear functions of σ_m' which are ;

$$G = 46.1 + 36.0 \cdot \sigma_m' \quad \text{(MPa)} \qquad (1)$$
$$K = 44.1 + 21.0 \cdot \sigma_m' \quad \text{(MPa)} \qquad (2)$$

In this paper, elastic strain is estimated by equations (1) and (2).

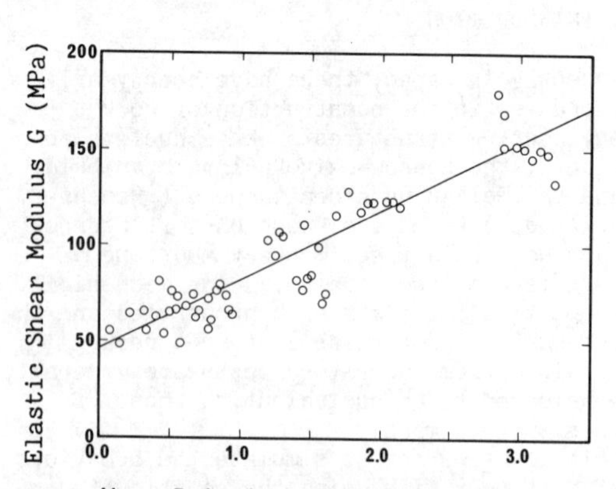

Fig.3 Relationship between G and σ_m'

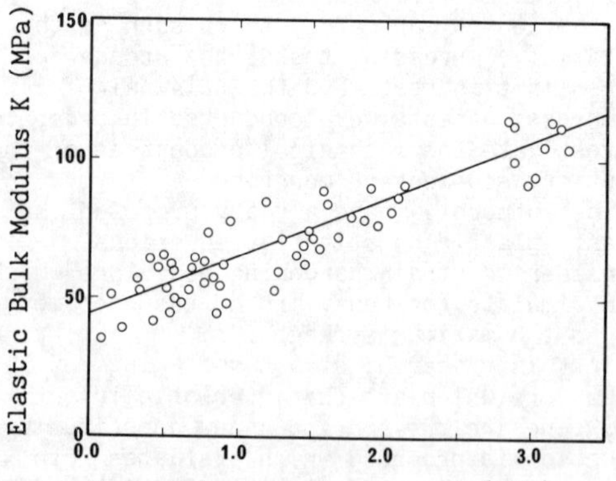

Fig.4 Relationship between K and σ_m'

2.3 Strength properties and plastic deformation properties

Fig.5 shows the maximum strength, the initial yield stress and the vectors of plastic strain increment at strain hardening. The maximum strength is expressed by equation (3) and has a different gradient on the boundary of $\sigma_m'=1.23$MPa.

$$\left(\frac{\tau_{oct}}{\sigma_m'+A^*}\right)_f = M_f \qquad (3)$$

where, M_f is the stress ratio at maximum strength, A^* is the value of σ_m' intersecting the maximum strength line with the mean principal stress and τ_{oct} is the octahedral shear stress. The value of M_f and A^* are respectively 1.12, 0.17 MPa in the range $\sigma_m'<1.23$MPa and 0.434, 2.37MPa in the range $\sigma_m'\geqq 1.23$MPa.

The vectors of the plastic strain increment are almost perpendicular to the σ_m'-axis. That is, plastic volumetric strain was hardly developed.

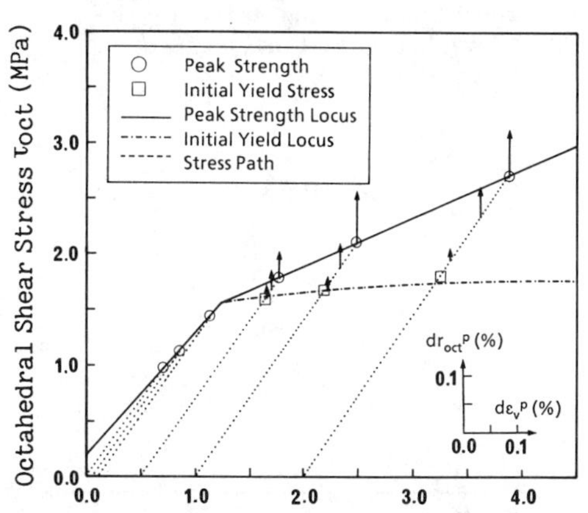

Fig.5 Strength and plastic strain increment vectors before maximum strength

Fig.6 shows maximum strength, residual strength and the vectors of plastic strain increment at strain softening. The residual strength is expressed by equation (4) and has a different gradient on the same boundary value from the case of the maximum strength.

$$\left(\frac{\tau_{oct}}{\sigma_m'+A^*}\right)_r = M_r \qquad (4)$$

where, Mr is the stress ratio at residual strength. The value of A^* is the same value as in the case of maximum strength. The values of Mr are 0.62 in the range $\sigma_m'<1.23$MPa and 0.24 in the range $\sigma_m'\geqq 1.23$MPa.

The vectors of the plastic strain increment are clearly changing direction. That is, the plastic volumetric strain is developed and varies from expansion to compression when the confining stress is varying from a low stress to a high stress.

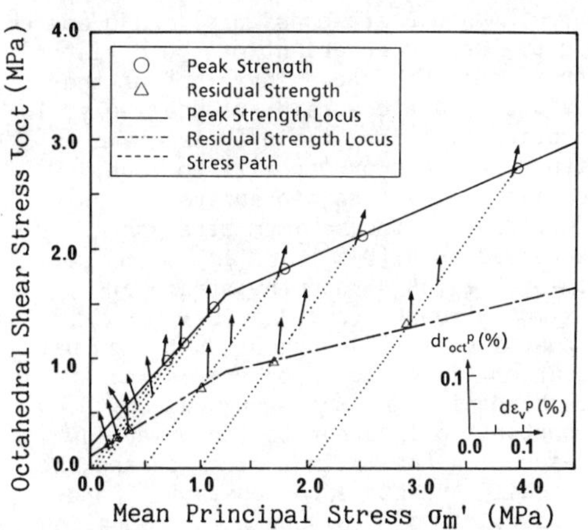

Fig.6 Strength and plastic strain increment vectors after maximum strength

3 CONSTITUTIVE EQUATION AND SIMULATION RESULTS

The yielding condition, the plastic potential function and the hardening and softening rule which justify the elasto-plastic equation are studied to express the behaviors at strain hardening and strain softening. These relationship are described as follows.

3.1 The yield criterion and the initial yield condition

The yield criterion was prescribed by the condition of the stress ratio $\tau_{oct}/(\sigma_m'+A^*)$ to be constant as shown in equation (5).

$$f = \frac{\tau_{oct}}{\sigma_m'+A^*} - \left(\frac{\tau_{oct}}{\sigma_m'+A^*}\right)_y \qquad (5)$$

In order to determine the initial yield stress, the equilibrium state line proposed by Nishi[3] is used as shown in equation (6). The equilibrium state line evaluates the elastic limit of saturated sand in the overconsolidated stress region undergoing a cyclic loading.

$$\left(\frac{\tau_{oct}}{\sigma_m'+A^*}\right)_{yi} = \alpha\, M_f\left(\frac{\sigma_{mc}'+A^*}{\sigma_m'+A^*}-1\right)^{1/2} \qquad (6)$$

where, σ_{mc}' is the consolidation yield stress (P_c), and α is a material constant whose value is 0.6.

3.2 The hardening and softening rule

Prevost et al.[4] proposed equation (7) as the hardening and softening rule.

$$q = A \cdot \frac{B(\bar{\varepsilon}^p)^2 + \bar{\varepsilon}^p}{1 + (\bar{\varepsilon}^p)^2} \qquad (7)$$

where, A and B are constants, and q and $\bar{\varepsilon}^p$ are the second invariant of the deviator stress and the second invariant of the deviator plastic strain, respectively. In equation (7), the strain $\bar{\varepsilon}^p$ is given in either per cent or per mill to model realistic stress strain curves. By using equation (7), the maximum strength q_f is expressed as $A \cdot (B+\sqrt{1+B^2})/2$, the residual strength q_r as AB and the strain at maximum strength $(\bar{\varepsilon}^p)_f$ as $(B+\sqrt{1+B^2})$.

Adachi et al.[5] suggest that when using equation (7) proposed by Prevost et al. as the hardening and softening rule, it is important to incorporate the power law prescribing the maximum strength and the residual strength into constants A and B.

According to this suggestion, equation(7) is modified to equation (8) as the hardening and softening rule.

$$\frac{\tau_{oct}}{\sigma_m' + A^*} = A \frac{B(\xi)^2 + C\xi}{C + (\xi)^2} \qquad (8)$$

In equation (8), the stress ratio $\tau_{oct}/(\sigma_m'+A^*)$ is used instead of q, and the normalized strain ξ instead of $\bar{\varepsilon}^p$. The constant C is newly added in addition to A and B. In this case, the stress ratio at maximum strength M_f is expressed as $A \cdot (B+\sqrt{c+B^2})/2$, the stress ratio at residual strength M_r is expressed as AB and the normalized strain ξ_f as $(B+\sqrt{c+B^2})$.

Here, the value of ξ_f is defined as 1.0. Constants A, B and C are defined by M_f and M_r as follows ; $A=2M_f$, $B=M_r/A=M_r/2M_f$ and $C=1-2B=1-M_r/M_f$. The value of A,B and C are obtained as follows ; 2.24,0.2765, 0.447 in the range $\sigma_m' < 1.23$MPa, respectively, 0.868,0.2765, 0.447 in the range $\sigma_m' \geqq 1.23$MPa, respectively.

Figs.7 and 8 show schematically the hardening and softening rule proposed by Prevost et al. and the newly proposed rule in this study, respectively.

ξ is the normalized strain, and is expressed by equation (9) before maximum strength and by equation (10) after maximum strength.

$$\xi = 1 + (1-\xi_0) \frac{r_{oct}^p - (r_{oct}^p)_f}{(r_{oct}^p)_f} \qquad (9)$$

$$\xi = \left(1 + \frac{r_{oct}^p - (r_{oct}^p)_f}{\lambda}\right)^n \qquad (10)$$

In equation (9), ξ_0 is the value of ξ at initial yield and becomes equation (11) by solving equation (8) for ξ.

$$\xi_0 = \frac{AC - (A^2C^2 + 4ABC\eta_i - 4C\eta_i^2)^{1/2}}{2(\eta_i - AB)} \qquad (11)$$

where, η_i is the stress ratio at initial yield.

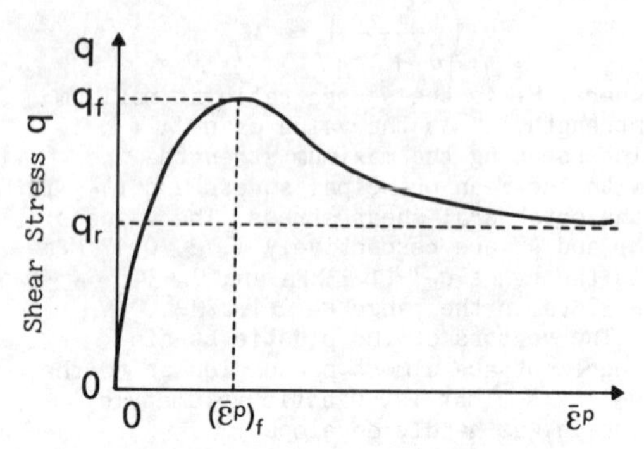

Plastic Shear Strain $\bar{\varepsilon}^p$

Fig.7 The hardening and softening rule proposed by Prevost et al.

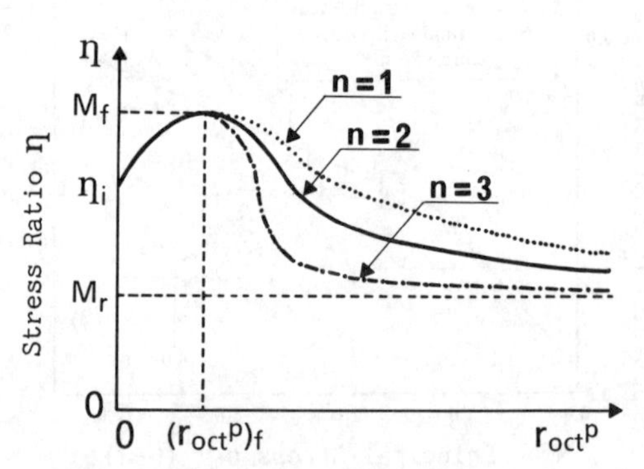

Plastic Shear Strain r_{oct}^p

Fig.8 The hardening and softening rule proposed in this study

Moreover, to prescribe the normalized strain ξ in equation (9) and equation (10), it is necessary to evaluate $(r_{oct}^p)_f$ which is the octahedral plastic shear strain at maximum strength. Fig.9 shows $(r_{oct}^p)_f$ against mean principal stress at initial yield $(\sigma_m')_i$ for σ_3 =0.49, 0.98, 1.96MPa.

As is shown in Fig.9, there is a linear relationship between $(r_{oct}^p)_f$ and $(\sigma_m')_i$ which can be expressed as equation (12).

$$(r_{oct}^p)_f = a_1 + a_2 \cdot (\sigma_m')_i \qquad (12)$$

where, a_1 and a_2 are constants, calculated as -2.91×10^{-3} and 3.70×10^{-3}, respectively.

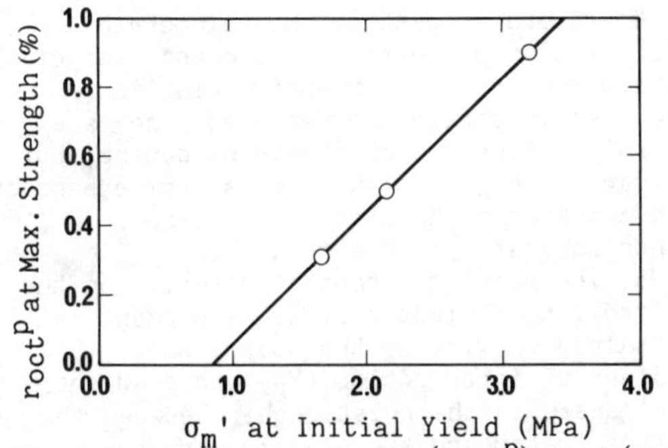

Fig.9 Relationship between $(r_{oct}^p)_f$ and $(\sigma_m')_i$

λ and n are the shape factors which determine the degree of decrease in strength after the maximum strength. Decrease in strength becomes bigger, as n becomes bigger with constant λ, or as λ becomes smaller with constant n. The value of λ is defined with respect to the value of mean principal stress at maximum strength $(\sigma_m')_f$ as ; λ is 0.030 in the range $(\sigma_m')_f < 1.23$MPa, and 0.012 in the range $(\sigma_m')_f \geqq 1.23$MPa. The value of n is 2.0.

When the newly proposed hardening and softening rule is used, strain is unnecessary to be multiplied by 100 or 1000, and, compared with the original model, it is possible to control the degree of decrease in strength freely.

3.3 Plastic potential

The plastic potential before the maximum strength is determined by Von-Mises criterion expressed as equation (13) since plastic volumetric strain is hardly observed in Fig.5.

$$g = \tau_{oct} - k \tag{13}$$

The plastic potential after the maximum strength is defined as follows. Adachi et al.[6] proposed equation (16) as the plastic potential for the estimation of the elasto-plastic behaviors of tertiary sedimentary tuff called Oya stone. They derived the plastic potential using the relationship between the stress ratio $\tau_{oct}/(\sigma_m'+A^*)$ and the plastic strain increment ratio $(-d\varepsilon_v^p/dr_{oct}^p)$ in equation (14), and the normality rule of equation (15). Therefore, the plastic potential is given by determining the value of intersection (Mg) and gradient (α^*) used in equation (14).

$$\frac{\tau_{oct}}{\sigma_m'+A^*} = M_g + \alpha^*\left(-\frac{2}{3}\cdot\frac{d\varepsilon_v^p}{dr_{oct}^p}\right) \tag{14}$$

$$\frac{d\tau_{oct}}{d(\sigma_m'+A^*)} = -\frac{2}{3}\frac{d\varepsilon_v^p}{dr_{oct}^p} \tag{15}$$

$$g = \tau_{oct} - \frac{M_g}{1-\alpha^*}(\sigma_m'+A^*)\left\{1-\left(\frac{\sigma_m'+A^*}{\sigma_{mys}'+A^*}\right)^{\frac{1-\alpha^*}{\alpha^*}}\right\} \tag{16}$$

Equation (14), calculated for the entire confining pressure, is expressed as shown in Figs.10 and 11.

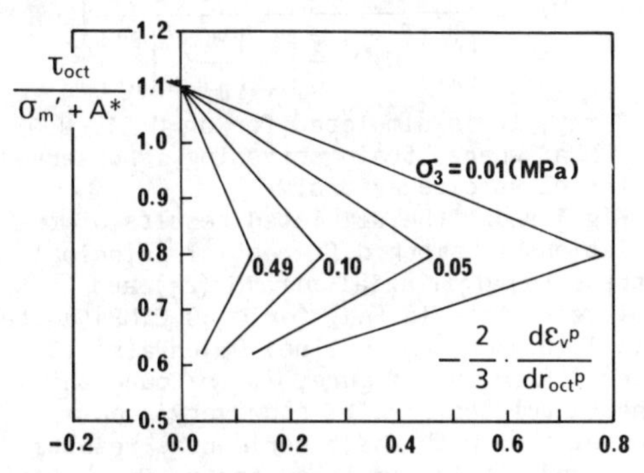

Fig.10 Relationship between stress ratio and plastic strain increment ratio ($\sigma_3 \leqq 0.49$MPa)

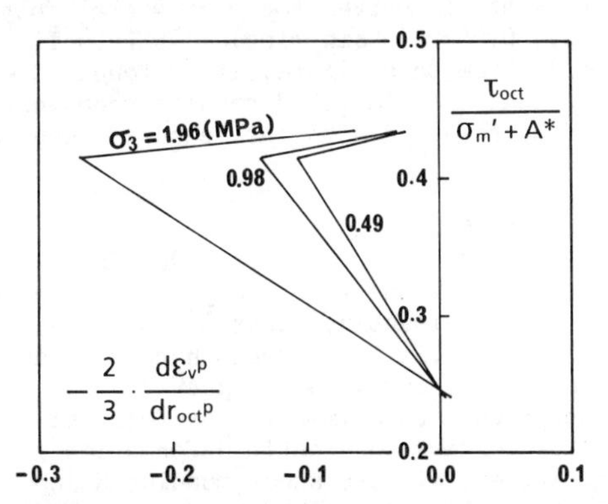

Fig.11 Relationship between stress ratio and plastic strain increment ratio ($\sigma_3 \geqq 0.49$MPa)

Fig.10 and Fig.11 show that, M_g is a constant value, whereas α^* is stress-dependent and is expressed by mean principal stress at maximum strength $(\sigma_m')_f$ as ;

$$\alpha^* = b_1 + b_2\cdot(\sigma_m')_f \tag{17}$$

The values of M_g, b_1 and b_2 are given in Table 2.

Table 2 The value of M_g, b_1 and b_2

Stress condition	M_g	b_1	b_2
$(\sigma_m')_f < 1.23$MPa $\eta \geqq 0.8$	1.10	0.90	-1.84
$(\sigma_m')_f < 1.23$MPa $\eta < 0.8$	0.55	-0.75	1.53
$(\sigma_m')_f \geqq 1.23$MPa $\eta \geqq 0.415$	0.44	0.35	-0.07
$(\sigma_m')_f \geqq 1.23$MPa $\eta < 0.415$	0.25	-2.59	0.49

3.4 Results of the simulation

The simulation is carried out by the constitutive equation (18) derived, using the relationships given above.

$$\{d\sigma\}=\left[[D^e]-\frac{[D^e]\left\{\frac{\partial g}{\partial \sigma}\right\}\left\{\frac{\partial f}{\partial \sigma}\right\}^T[D^e]}{\left\{\frac{\partial f}{\partial \sigma}\right\}^T[D^e]\left\{\frac{\partial g}{\partial \sigma}\right\}-\left\{\frac{\partial f}{\partial \varepsilon^p}\right\}^T\left\{\frac{\partial g}{\partial \sigma}\right\}}\right]\{d\varepsilon\} \quad (18)$$

First, it is simulated for σ_3=0.01, 0.05, 0.10MPa, where strain softening is observed after elastic behaviors.

Fig.1 shows the simulated results of the relationship among differential principal stress (σ_1-σ_3), axial strain (ε_1) and volumetric strain (ε_v) for σ_3=0.01MPa(dotted line), 0.05MPa(chain line) 0.10MPa(solid line). From this figure, it is found that the calculated results show very good correlation with the experimental results.

Second, it is simulated for σ_3=0.49, 0.98, 1.96MPa, where strain softening is observed after strain hardening.

Fig.2 shows the simulated results of the stress strain curves for σ_3=0.49MPa (dotted line), 0.98MPa(chain line) 1.96MPa(solid line). From this figure, it is found in each case that the simulated results show very good correlation with the measured results.

4 CONCLUSIONS

By conducting the laboratory tests with tertiary sedimentary mudstone, the elasto-plastic behaviors at strain hardening and strain softening were researched. The conclusions are summarized as follows:

1. According to elastic deformation properties, elastic shear modulus G and elastic bulk modulus K depend on confining stresses and can be expressed by linear functions of the mean principal stress σ_m'.

2. The strength properties and plastic deformation properties change at the boundary value of the confining stress, σ_m'=1.23MPa. Therefore, maximum strength and residual strength are closely approximated by a bi-linear line on the boundary of this stress. Also, deformation properties after yielding are determined for the each stress regions.

3. According to the hardening and softening rule determining the relationship between shear stress and shear strain at strain hardening and strain softening, it is possible to control the degree of decrease in strength by modifying the original relationship proposed by Prevost et al. to the newly proposed relationship between stress ratio and normalized strain.

4. In order to study the volumetric deformation properties, the quantities of the compressive or expansive plastic volumetric strain are measured. They are hardly recognized at strain hardening, but clearly recognized and varies from expansion to compression as confining pressure increases at strain softening.

5. The plastic potential determining the direction of plastic strain increment is obtained in view of the state, where the values of intersection (M_g) and gradient (α^*) used in the relationship between the stress ratio and the plastic strain increment ratio are dependent on stresses. From these relationships, it is possible to estimate the volumetric deformation properties at strain hardening and strain softening.

According to the relationships mentioned above, it is evident that the behaviors of tertiary sedimentary mudstone at strain hardening and strain softening can be reasonably simulated by the elasto-plastic constitutive equations expressed as equation (18) within the limits of this experiment.

5 ACKNOWLEDGEMENT

The authors wish to thank Professor Toshihisa Adachi of Kyoto University for his profound guidance and valuable advices.

REFERENCES

[1] A.Denda and K.Watanabe, Elasto-plastic Behavior of Tertiary Sedimentary Mudstone, Proc. of 8th ARCSMFE Vol.1, 159-162 (1987)
[2] F.Oka and T.Adachi, A Constitutive Model of Soft Rock with Strain-softening, Report of ISSMFE subcommittee on constitutive laws of soils and Proc. of Discussion session X I ICSMFE, 152-155 (1985)
[3] K.Nishi, Elasto-plastic Behaviors of Saturated Sand under Undrained Cyclic Loading and its Constitutive Equation, Proc. of JSCE Vol.319, 115-128 (In Japanese) (1982)
[4] J.H.Prevost and K.Hoeg, Soil mechanics and plasticity analysis of strain softening, Geotech. Vol.25, 279-297 (1975)
[5] T.Adachi and T.Ogawa, Mechanical Properties and Failure Criterion of Sedimentary Soft Rock, Proc. of JSCE Vol.295, 51-63 (In Japanese) (1977)
[6] K.Akai, T.Adachi and K.Fujimoto, Constitutive Equations for Geomechanical Materials based on Elasto-viscoplasticity, 9th ICSMFE, 1-10 (1977)

Numerical Methods in Geomechanics (Innsbruck 1988), Swoboda (ed.)
© 1988 Balkema, Rotterdam. ISBN 90 6191 809 X

Theoretical evaluation of the frictional damping of rocks

Giovanni Alpa
Politecnico di Torino, Italy
Luigi Gambarotta
Università di Genova, Italy

ABSTRACT: Rock like materials have been modelled as elastic solids containing frictional flaws in order to perform a numerical evaluation of the damping effect in dynamical processes. Hysteretic loops have been analised for various cyclic loading paths. The results show a strong influence of the stress state on the energy loss and a maximum damping for relatively low values of the friction coefficient.

1 INTRODUCTION

In numerical simulation of the mechanical behaviour of rocks, expecially when the attention is focused on the basic properties of a nominally homogeneus portion of material rather than to the structural complexity of rock masses, the model of elastic solid with internal flaws has been shown useful to analyse some material features [1,2]. Such a model allows to account for the strong dependence of the material response on the stress state, considering that compressive stresses tend to close the flaws and induce a non-linear, anisotropic, path-dependent behaviour, mainly due to the asperities of the flaw faces and to the consequent frictional sliding.

It is implicit that in the mentioned model the friction force acting on the flaw faces memorizes the stress history. Then the constitutive equations cannot be directly established in terms of stress and strain tensors. It needs to establish the stress-strain relationships starting from a microscopic point of view and accounting for the sliding due to any orientation of flaws, i.e. to any plane on which the resolved stresses act.

This idea has been developed in a previous paper in which an incremental form of the constitutive equations for microcracked solids under compressive stresses has been obtained and applied to some cyclic load histories [3]. Particular phenomena such as the modification of the subsequent elastic domains, the hysteretic loops and the associated damping effect in dynamic processes were analysed.

In the present paper the constitutive equations for microcracked solids, in a more general form including tensile stresses, are employed for a numerical evaluation of the damping in rocks. The aim is mainly to obtain a knowledge about the dependence of the damping on the stress and on the friction coefficient. Obviously, the model here employed give detailed informations only about the damping caused by the frictional sliding, providing that stress state and friction simulate a real situation in rock masses, while any other source of damping is excluded. It has been thought to be convenient to consider other sources of damping, such as viscoplasticity, only after a detailed analysis of the purely frictional system has been developed.

In order to perform a step by step procedure, the incremental stress-strain relations have been explicity integrated under the assumption of proportional loading for any step. Since these incremental constitutive equations give the strain increments as functions of the actual stresses, strains, friction forces and stress increments (see eqn.10), their invertion has required an iterative procedures.

Some dynamic processes for a single degree of freedom system have been analysed in the case of free vibrations. Cases have been considered which simulate real situations in rock masses, where shearing stresses, varying from an initial amplitude, are superimposed to constant normal stresses (i.e. geostatic pressure). The results show a relevant dependence of the damping on the stress state. The influence of the coefficient of internal friction has also been analysed: a maximum for relatively low values of this parameter has been found.

2 CONSTITUTIVE EQUATIONS

The constitutive equations for microcracked solids have been deduced under the following hypotheses:

a) elastic solid containing a random distribution of plane flaws (microcracks);

b) asperities of crack faces causing Coulomb friction forces;

c) crack interaction is considered irrelevant so that sliding of each crack depends only on the resolved stresses acting on the crack planes.

The strain rate tensor $\dot{\varepsilon}_{ij}$ is considered as the sum of two contribution, i.e. the part ... related to the stress rate $\dot{\sigma}_{ij}$ by means of the elastic compliance tensor C_{ijlm} (associated to the uncracked solid) and the contribution of crack opening and sliding $\dot{\varepsilon}_{ij}^*$:

$$\dot{\varepsilon}_{ij} = C_{ijlm}\,\dot{\sigma}_{lm} + \dot{\varepsilon}_{ij}^* . \tag{1}$$

Let us consider all the cracks having unit vector $\mathbf{n}$ normal to the crack plane inside an infinitesimal solid angle $d\Omega$; their opening contribution is a normal strain $d\varepsilon_n$ given by

$$d\varepsilon_n = h\,(\sigma_n - p_n)\,d\Omega, \tag{2}$$

where h is a constant depending on the elastic properties of the solid and on the crack density; σ_n is the resolved normal stress; p_n is a reaction force, associated to the unilateral character of the crack, which increment $\dot{p}_n$ is given by:

$$- \dot{p}_n = 0 \quad \text{for } \sigma_n > 0 \quad \text{or } \sigma_n = 0,\ \dot{\sigma}_n \geq 0 \tag{3}$$

$$- \dot{p}_n = \dot{\sigma}_n \quad \text{for } \sigma_n < 0 \quad \text{or } \sigma_n = 0,\ \dot{\sigma}_n < 0 . \tag{4}$$

More complex is the evaluation of the sliding contribution. The shear strain $d\boldsymbol{\gamma}_n$ can be put in the form:

$$d\boldsymbol{\gamma}_n = k\,(\boldsymbol{\tau}_n - \mathbf{f}_n) \tag{5}$$

where k is still a constant depending on the elastic matrix and on the crack density; $\boldsymbol{\tau}_n$ is the resolved shear stress and $\mathbf{f}_n$ is the friction force (which memorizes the load history). The complexity of the shear strain-stress relationship is due to the increment $\dot{\mathbf{f}}_n$ of the friction force which is given by the following relations (see ref. [3]):

$$- \dot{\mathbf{f}}_n = 0; \tag{6}$$

for $\sigma_n > 0$;

or $\sigma_n = 0,\quad \dot{\sigma}_n \geq 0$

$$- \dot{\mathbf{f}}_n = \dot{\boldsymbol{\tau}}_n \tag{7}$$

for $\sigma_n = 0, \dot{\sigma}_n < 0, |\dot{\boldsymbol{\tau}}_n| + \mu\dot{\sigma}_n < 0$;

or $\sigma_n < 0, |\mathbf{f}_n| < -\mu\sigma_n$;

or $\sigma_n < 0, |\mathbf{f}_n| = -\mu\sigma_n,\ \dot{\boldsymbol{\tau}}_n \dfrac{\mathbf{f}_n}{|\mathbf{f}_n|} + \mu\dot{\sigma}_n < 0$

$$- \dot{\mathbf{f}}_n = -\mu\dot{\sigma}_n \frac{\dot{\boldsymbol{\tau}}_n}{|\dot{\boldsymbol{\tau}}_n|} \tag{8}$$

for $\sigma_n = 0,\ \dot{\sigma}_n < 0, |\dot{\boldsymbol{\tau}}_n| + \mu\dot{\sigma}_n \geq 0$

$$- \dot{\mathbf{f}}_n = -\mu\dot{\sigma}_n \frac{\mathbf{f}_n}{|\mathbf{f}_n|} + \left(\dot{\boldsymbol{\tau}}_n \cdot \mathbf{n}\times\frac{\mathbf{f}_n}{|\mathbf{f}_n|}\right)\left(\mathbf{n}\times\frac{\mathbf{f}_n}{|\mathbf{f}_n|}\right) \tag{9}$$

for $\sigma_n < 0, |\mathbf{f}_n| = -\mu\sigma_n,\ \dot{\boldsymbol{\tau}}_n \cdot \dfrac{|\mathbf{f}_n|}{|\mathbf{f}_n|} + \mu\dot{\sigma}_n \geq 0$

in which μ is the friction coefficient.

The strain rate is then:

$$\dot{\varepsilon}_{ij} = C_{ijlm}\,\dot{\sigma}_{lm} + h\int_\Omega(\dot{\sigma}_n - \dot{p}_n)\,d\Omega + \tag{10}$$
$$+ k\int_\Omega |\dot{\boldsymbol{\tau}}_n - \dot{\mathbf{f}}|\tfrac{1}{2}(n_i t_j + n_j t_i)\,d\Omega,$$

where n_i, n_j and t_i, t_j are respectively the components of $\mathbf{n}$ and of the unit vector of $d\boldsymbol{\gamma}_n$; Ω is the semisphere representing all the orientations.

Under the assumption of proportional loading the above relations may be easily integrated, obtaining in such a way the finite increment constitutive equations which can be employed in a step by step analysis.

Aiming to analyse dynamic processes it needs to invert at each step the finite increment constitutive equations in order to get the stress rate as functions of the strain rate. This requires the use of an iterative numerical procedure of the 'initial strain' type [4,5].

For rocks or rock-like materials it can be observed that reasonable values of h range in the large interval between 10/E and 1/10E (E Young modulus), depending on the extent and density of the internal fessures. This may be argued observing that ratios between uncracked and cracked solid Young moduli of .1 and .8 correspond to the above mentioned limits. The constant k is related to h because both are consequence of the same cause: the presence of cracks. From the analysis of an elastic solid containing penny shaped cracks [6], it can be deduced that $k/h = 2/(2-\nu)$, where ν is the Poisson ratio which can be assumed approximately constant and equal to that of the uncracked solid.

Before to proceed to dynamic applications, it is interesting to show some features of the proposed model in the cases of static response for cyclic loads.

Let us consider a homogeneous body homogeneously stressed by an isotropic compression $\sigma_1=\sigma_2=\sigma_3=-\sigma$ ($\sigma_1,\sigma_2,\sigma_3$ principal stresses) and successively consider a superimposed pure shear stress $\Delta\sigma_1=-\Delta\sigma_2=\tau$.

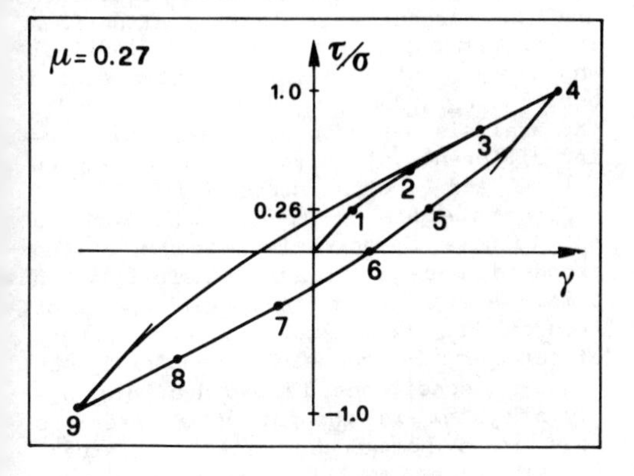

Figure 1. τ vs. γ diagram for low values of the friction coefficient.

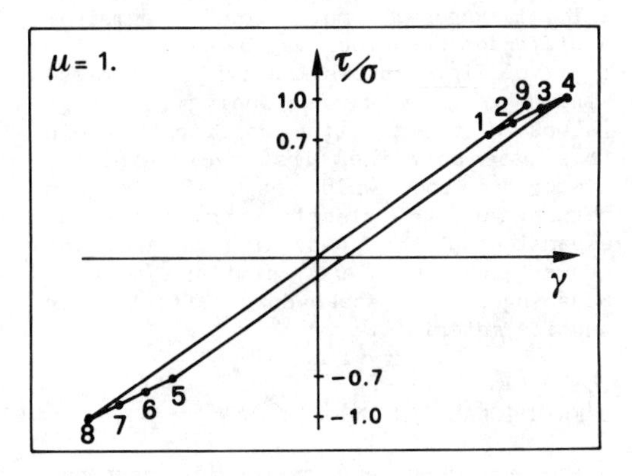

Figure 2. τ vs. γ diagram for high values of the friction coefficient.

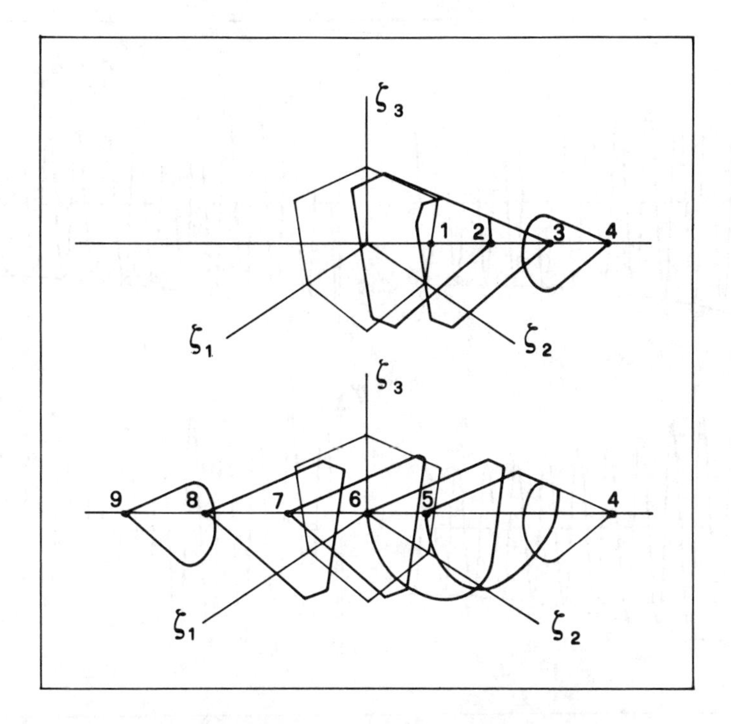

Figure 3. Dragging and deformation of the no sliding domain for low values of the friction coefficient.

A cyclic loading path with τ varying between σ and $-\sigma$ has been analysed for two values of μ (μ= 0.3 and μ= 1.) The results corresponding to μ= 0.3 are shown in figs. 1,3. In fig. 1 the stress-strain diagram is plotted which shows the hysteretic behaviour of the material. The hysteretic loop is repeated with no change as τ cyclically varies. Fig. 3 shows the dragging and deformation of the no-sliding domain (in the deviatoric plane) when the stress point goes out of initial exagonal domain (Coulomb domain). A different behaviour may be observed for μ = 1.0. Fig. 2 shows the typical stress-strain diagram corresponding to high values of the friction coefficients; in this case only the first cycle exhibits hysteretic loop which is followed by a return into the elastic range due to an expansion of the no-sliding domain. The above phenomena are similar to the Baushinger and shake-down effects for ductile materials.

3 FRICTIONAL DAMPING

The nonlinear and hysteretic response of the proposed mechanical model leads to energy dissipation i.e. to a damping effect in dynamic processes.

Since the constitutive equations show a strong dependence of the material response on the friction coefficient and on the stress state, some dynamic processes corresponding to different situations have been analysed. Different values of the friction coefficient have been considered in order to quantify the influence of this material parameter on the damping. The stress states considered are the current ones in many situations in rock masses, i.e. constant vertical and lateral stresses and a shear stress oscillating from an initial amplitude. The analysis has been performed considering different values of the ratio between lateral and vertical stress.

Since the aim of the present work is to evaluate the dynamic response of the proposed model and its feasibility, a simple system having a single degree of freedom has been analysed so that any effect due to geometric complexity or boundary conditions is avoided. The dynamic system is thought as a masseless body in a homogeneous state of stress (constant and oscillating) due to constant external applied forces and to a

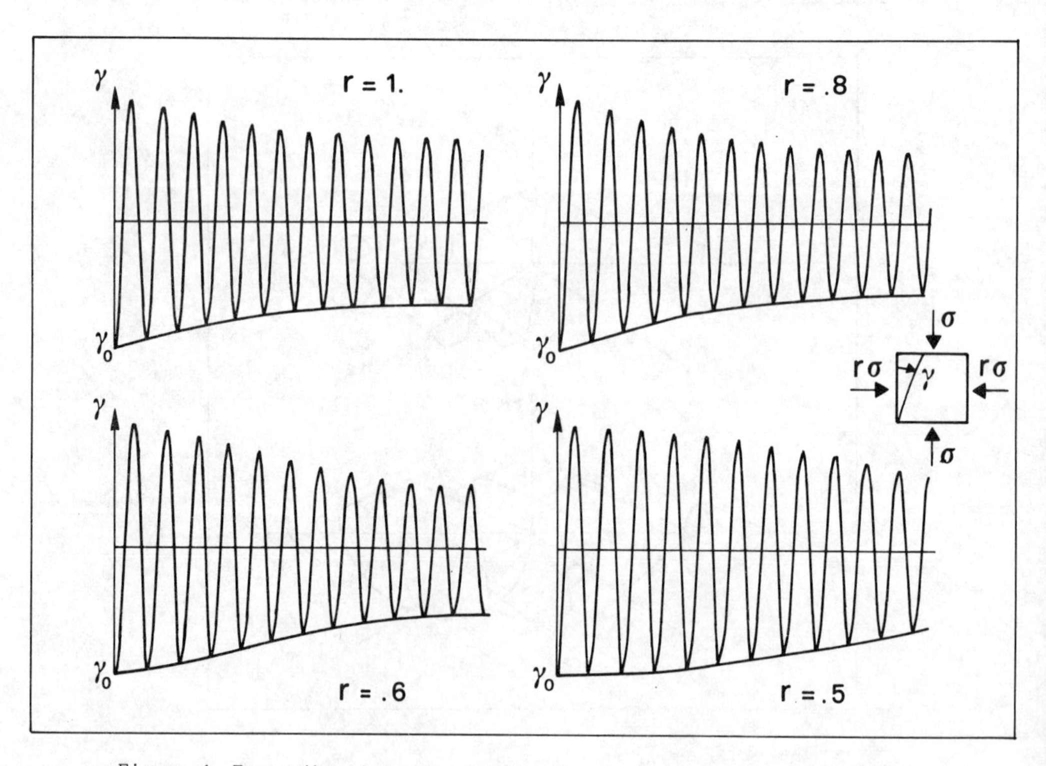

Figure 4. Free vibrations for different values of the lateral stress.

rigid external oscillating mass.

Referring to fig. 4 for the mean of the simbols, the equations of the motion are

$$c\,\ddot{\gamma} = \tau \qquad (11)$$
$$\dot{\gamma} = f\,(\sigma, \tau, r, \dot{\tau})$$

where c is a constant and f (·) summarizes the constitutive equations. The free vibration of the system has been obtained by means of the Runge-Kutta method starting at rest from an initial value γ_0, obtained by a preliminary application of a shear stress τ from 0 to $.75\sigma$. It has been assumed $h = 10/E$, $\nu = 0.2$.

The influence of the lateral stress on free vibrations is shown in Fig. 4, in which the dynamic material response, corresponding to values of the ratios between lateral stress and vertical stress varying from 0.5 to 1, is plotted. From these diagrams it can be drawn that:

a) the frequency does not practically depend on the amplitude (as is commonly observed when the non-linearity of the system is not too great);

b) a small dependence of frequency on the constant stress state may be observed;

c) applied constant stresses have a great influence on the decay rate so that both concave or convex trends are possible.

To obtain further informations for a more wide range of variations of the parameters the energy dissipation per cycle has been evaluated as a function of the shear stress amplitude for different values of the constant stresses and of the friction coefficient.

The diagrams of fig. 5 show maximum values of the damping ratio ξ, i.e. the ratio between the energy loss per cycle and 4π times the maximum elastic energy, depending on the lateral stresses. The maxima shift toward greater values as the ratio $r = \sigma_H/\sigma_V$ diminishes. Furthermore it can be seen that the trend itself is strongly affected by μ and r. The portion of curve on the right of each maximum reminds to the characteristic response of the classic simple oscillator consisting of a spring and a Coulomb frictional damper. If one considers that the decrement in amplitude per cycle is approximately proportional to the product $\xi\tau$ (constant in the classic simple oscillator [7]), the following relation between the trend in decay and the curves of fig. 5 may be obtained:

a) if $-\frac{d}{d\tau}(\xi\tau) = -\tau\frac{d\xi}{d\tau} - \xi > 0$

the trend in decay is convex;

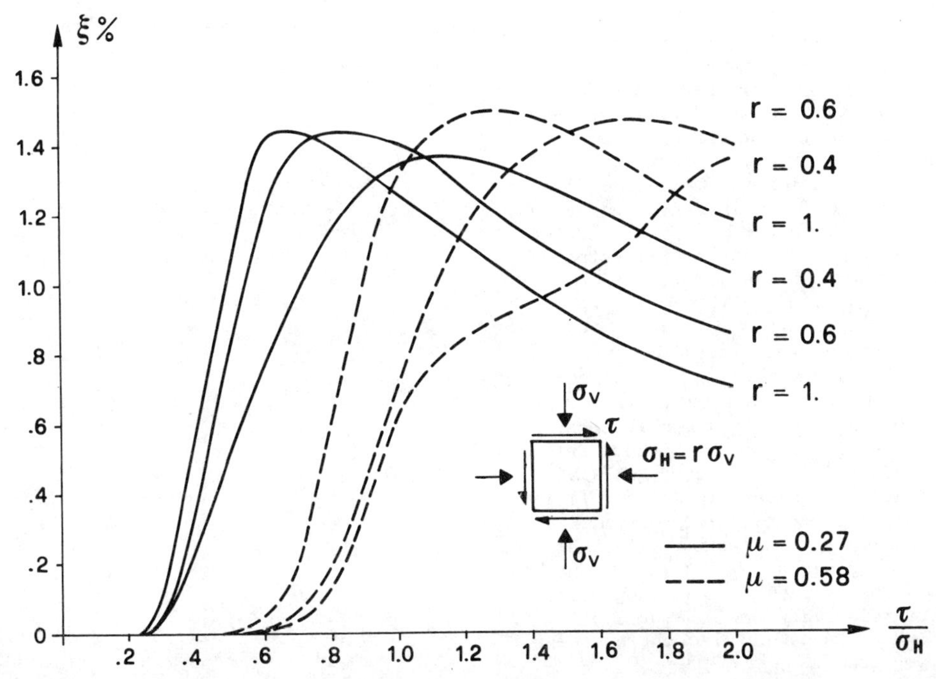

Figure 5. Damping ratio vs. shear stress amplitude.

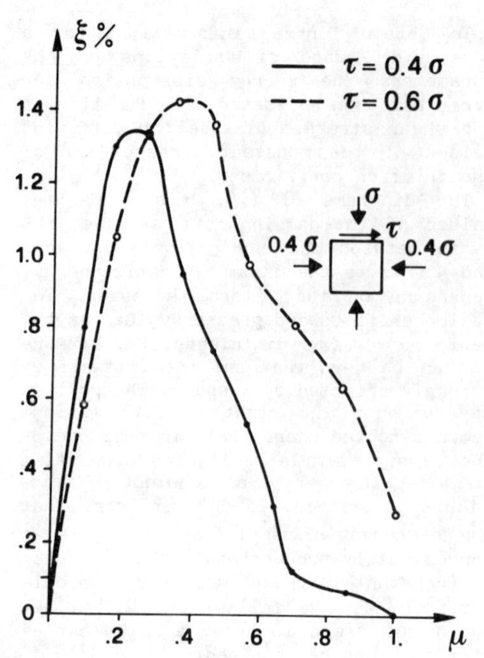

Figure 6. Damping ratio vs. friction coefficient diagram.

b) in the opposite case the trend is concave.

Then from an examination of the curves of fig. 5 it may be seen that, after a possible initial convex trend, the decrement in amplitude becomes always concave i.e. similar to that occurring in viscous damping. As the amplitude diminishes the material behaviour tends to become elastic because the internal flaws jam. Such a phenomena correspond to the vanishing of the hysteretic loop as illustrated in fig. 2 and it depends on the ratio τ/σ_H as illustrated in fig. 5.

The dependence of the damping ratio on the friction coefficient in shown in fig. 6. As is expected, the influence of the friction coefficient in very strong. For $\mu = 0$ the material behaves as a cracked elastic solid without friction and then $\xi = 0$. For high values of μ the friction forces lock the flaws and no sliding occurs; in this case the material behaves as an uncracked elastic solid. The diagram of fig. 6 shows that maximum values of damping ratio occur for relatively low values of μ ($\mu = 0.2 \div 0.5$) depending also on the shear stress amplitude.

REFERENCES

[1] J.B. Walsh: The effect of cracks on the uniaxial elastic compression of rocks, Geophys. Res., Vol. 70 (1965), pp. 399-411.

[2] M.L. Kachanov: A microcrack model of rock inelasticity - part. I: frictional sliding on microcracks, Mech. of Materials, Vol. 1 (1982), pp.19-27.

[3] G. Alpa, L. Gambarotta and A. Tafanelli: Effetto dell'attrito interno nei solidi con microfessure, Atti VIII Congresso Nazionale AIMETA, Vol. 1 (1986), pp. 165-169.

[4] G. Alpa, L. Gambarotta: Numerical evaluation of the influence of frictional sliding in the behaviour of microcracked solids, Proceedings of the Int. Conf. on Computational Engineering Mechanics, Beijing, 1987.

[5] O.C. Zienkiewicz and G.N. Pande : Time - dependent multilaminate model of rocks - a numerical study of deformation and failure of rock masses, Int. Journ. Num. Anal. Methods in Geomechanics, Vol. 1 (1977), pp. 219-247.

[6] H. Horoii and S. Nemat - Nasser: Overall moduli of solids with microcracks: loads - induced anisotropy, J. Mech; Phys. Solids, Vol. 31 (1983), pp. 155-171.

[7] J.P. Den Hartog: Mechanical vibrations, Mc. Graw Hill, New York 1956.

Numerical Methods in Geomechanics (Innsbruck 1988), Swoboda (ed.)
© 1988 Balkema, Rotterdam. ISBN 90 6191 809 X

Analysis of the cone penetration test by the strain path method

C.I.Teh & G.T.Houlsby
Department of Engineering Science, Oxford University, UK

ABSTRACT: A method of analysis for the penetration of a cone penetrometer into clay is presented. The analysis is based on the Strain Path Method, with the initial estimate of the velocity field provided by the solution to the equations of fluid flow. Corrections to account for the equilibrium equations are achieved by the use of large strain finite element analysis. The calculations demonstrate the primary importance of horizontal stress and rigidity index in controlling cone resistance.

Keywords: analysis, clays, cone factor, finite elements, piezocone, strain path

1 INTRODUCTION

The purpose of this paper is to present in outline the method of analysis used to study the penetration of a piezocone into clay. Whilst some of the results are presented here, the intention is to concentrate on the analytical techniques. Full details of the analysis are given by Teh (1987) and a discussion of the results and their implications for the interpretation of piezocone data given by Houlsby and Teh (1988).

The full analysis consists of two stages. In the first the undrained penetration of a CPT into clay is analysed, with the clay idealised as an incompressible elastic-perfectly plastic (Von Mises) material. Values of the cone factor N_{kt} (defined as $(q_t - \sigma_{vo})/s_u$) are derived, as well as stresses and displacements around the cone. In the second stage the consolidation around a static cone is analysed using uncoupled Terzaghi-Rendulic consolidation theory. Since this second stage involves a straightforward solution to the diffusion equations in axial symmetry, and has been solved using conventional Alternating Direction Implicit techniques, only the first stage of the analysis is described in this paper. A more sophisticated coupled analysis of the consolidation phase is planned as a future research project.

2 METHODS OF ANALYSIS FOR CONE PENETRATION

Approximate derivations of N_{kt} factors have in the past been based mainly on either bearing capacity theory or cavity expansion theory. Modification of the results of plane strain bearing capacity calculations for application to axially symmetric problems is often carried out empirically, although numerical solutions to axisymmetric problems can be obtained using plasticity theory (e.g. Cox et al. (1961), Houlsby and Wroth (1983)). Adaptation of bearing capacity theory to deep penetration problems is much more difficult, and either involves adoption of boundary conditions which are not appropriate for real cone penetrometers (e.g. Koumoto and Kaku (1982)) or results in N_{kt} factors which increase indefinitely with increased penetration (Houlsby and Wroth, 1982). In practice empirical depth factors are applied to shallow bearing capacity solutions.

The problem of the transition between shallow and deep penetration is readily explained by the fact that shallow penetration involves a mechanism in which displaced material moves entirely outwards to the free surface, whilst in deep penetration the displaced material is accommodated by elastic deformation of the soil. Thus cavity expansion theories are applied to deep penetration problems (Vesic, 1972). It is widely held that the limit solution for the cylindrical cavity expansion pressure ψ_c is applicable to estimation of the radial stress on the shaft of a penetrometer. For the von Mises yield criterion:

$$(\sigma_r-\sigma_\theta)^2 + (\sigma_\theta-\sigma_z)^2 + (\sigma_z-\sigma_r)^2 + 6\tau_{rz}^2 = 8s_u^2 \quad ....(1)$$

the solution for cylindrical cavity expansion is:

$$\psi_c = \sigma_{ro} + \frac{2s_u}{\sqrt{3}}\left(1 + \ln\left[\frac{\sqrt{3}I_r}{2}\right]\right) \quad ....(2)$$

where I_r is the rigidity index G/s_u.

Note that s_u is the undrained strength in triaxial compression. The von Mises criterion implies a shear strength in plane strain of $2s_u/\sqrt{3}$.

The spherical cavity expansion pressure ψ_s is often considered as an estimate of the pressure at the cone tip:

$$\psi_s = \sigma_{ro} + \frac{4s_u}{3}\left(1 + \ln\left[I_r\right]\right) \quad ....(3)$$

Note that both of these expressions demonstrate the important influence of the rigidity index through the terms in $\ln(I_r)$.

Baligh (1986) has pointed out that an inconsistency in the application of cavity expansion theory to cone penetration problems is that it does not model correctly the strain paths followed by soil elements. He suggests the use of the Strain Path Method for problems of deep penetration, and this method has been used in this research.

By changing the frame of reference, the penetration of a cone in a homogeneous material can be viewed as a steady state flow of soil past a static penetrometer. An initial estimate of the streamlines must be made, for instance by using fluid mechanics theory. The flow pattern will of course be slightly in error since an incorrect constitutive law has been used, but because the problem is heavily constrained kinematically (by the incompressibility condition and the boundary conditions on the cone) the estimate represents a reasonable first approximation. From this flow pattern the complete history of strain for each soil element may be determined. Using appropriate initial stresses (which may be anisotropic) far ahead of the cone, the deviatoric stresses may be determined by integration of the appropriate constitutive laws up the streamlines. In this study a simple linear elastic-perfectly plastic model with a von Mises yield surface was used, involving only two material parameters, the shear modulus G and the shear strength in triaxial compression s_u. The appropriate incremental constitutive equations for the steady flow problem (Houlsby et al., 1985) are given by:

$$\dot{\varepsilon}_{rr} = \frac{\partial u}{\partial r} = \frac{1}{2G}\dot{\sigma}'_{rr} + 6\lambda\sigma'_{rr} \quad ....(4)$$

$$\dot{\varepsilon}_{zz} = \frac{\partial v}{\partial z} = \frac{1}{2G}\dot{\sigma}'_{zz} + 6\lambda\sigma'_{zz} \quad ....(5)$$

$$\dot{\varepsilon}_{\theta\theta} = \frac{u}{r} = \frac{1}{2G}\dot{\sigma}'_{\theta\theta} + 6\lambda\sigma'_{\theta\theta} \quad ....(6)$$

$$\dot{\gamma}_{rz} = \frac{\partial v}{\partial r} + \frac{\partial u}{\partial z} = \frac{1}{G}\dot{\tau}_{rz} + 12\lambda\tau_{rz} \quad ....(7)$$

Where u and v are the velocities in the r (radial) and z (axial) directions respectively and λ is the plastic multiplier. A superposed dot indicates the time differential and a dash indicates the deviatoric component. Note of course that the velocities must be such that the incompressibility condition is satisfied. This is most conveniently achieved by obtaining the velocities by the differentiation of a potential function:

$$u = \frac{1}{r}\frac{\partial\psi}{\partial z}, \quad v = -\frac{1}{r}\frac{\partial\psi}{\partial r} \quad ....(8)$$

The stress rates in steady flow are given by convective equations:

$$\dot{\sigma}'_{rr} = u\frac{\partial\sigma'_{rr}}{\partial r} + v\frac{\partial\sigma'_{rr}}{\partial z} \quad ....(9)$$

and three similar equations for $\dot{\sigma}'_{zz}$, $\dot{\sigma}'_{\theta\theta}$ and $\dot{\tau}'_{rz}$.

The above equations are supplemented by either $\lambda = 0$ for elastic behaviour if the stresses are within the yield surface or equation (1) in the case of plasticity. In addition the equilibrium equations:

$$\frac{\partial\sigma_{rr}}{\partial r} + \frac{\partial\tau_{rz}}{\partial z} + \frac{\sigma_{rr} - \sigma_{\theta\theta}}{r} = 0 \quad ...(10)$$

$$\frac{\partial\tau_{rz}}{\partial r} + \frac{\partial\sigma_{zz}}{\partial z} + \frac{\tau_{rz}}{r} = 0 \quad ...(11)$$

must be satisfied.

Alternative models for the deviatoric stress-strain behaviour of the soil could readily be included in the method. Note that for an incompressible material the mean stress is not determined by the constitutive relations but appears as a reaction which is determined solely by the equilibrium equations. Thus, having calculated the deviatoric stresses, the mean normal stress may be determined by using one of the equilibrium equations, radial or axial, and integration from the outer boundary (which must be set at some point sufficiently distant from the cone). It is then found that the stresses do not exactly obey the other equilibrium relationship, with the discrepancy reflecting the error in the initial flow field. In principle the other relationship could be used to correct the flow field, and the following schemes have been adopted to attempt this correction.

3 NEWTON-RAPHSON CORRECTION

By cross differentiating the equilibrium equations they may be combined to eliminate the mean stress:

$$\frac{\partial}{\partial z}\left[\frac{\partial \sigma'_{rr}}{\partial r} + \frac{\partial \tau_{rz}}{\partial z} + \frac{\sigma'_{rr} - \sigma'_{\theta\theta}}{r}\right] + \frac{\partial}{\partial r}\left[\frac{\partial \tau_{rz}}{\partial r} + \frac{\partial \sigma'_{zz}}{\partial z} + \frac{\tau_{rz}}{r}\right] = H$$

where $H = 0$ for a set of stresses in equilibrium, and H is taken as a measure of the error for a set of stresses which are not in equilibrium.

An initial set of values of the stream function ψ_o is estimated and the corresponding values H_o determined. At each grid point ψ is then perturbed by a small amount, and the resulting effect on H determined at every other grid point. The changes are confined to a region downstream and extending a small distance to either side of the perturbed point. After perturbing every grid point in turn an NxN matrix (where N is the number of grid points) of "partial derivatives" of H with respect to ψ can be calculated. By multiplying $-H_o$ by the inverse of this matrix the appropriate correction to ψ may be determined to eliminate the error approximately (exactly in the case of a linear problem). The calculation is then repeated with the new ψ values until the error is reduced to a satisfactory value.

Testing of this approach with some simple problems showed that as long as the perturbed flow does not include any regions in which plastic deformation occurs, then the method converges very rapidly on the correct solution. This is because the equations are purely linear in these cases. If, however, there is any plastic deformation, as is the case for all realistic cone penetration problems, then the method does not converge (even when acceleration or deceleration techniques are also adopted).

4 PSEUDO-DYNAMIC CORRECTION

The second correction method is based on an unsteady flow approach. In this case the equilibrium equations are considered for a material with density ρ:

$$\frac{\partial \sigma_{rr}}{\partial r} + \frac{\partial \tau_{rz}}{\partial z} + \frac{\sigma_{rr} - \sigma_{\theta\theta}}{r} = \rho\left[\frac{\partial u}{\partial t} + u\frac{\partial u}{\partial r} + v\frac{\partial u}{\partial z}\right] \qquad ...(13)$$

and a similar equation in the vertical direction.

Introducing the vorticity $\xi = \frac{\partial u}{\partial z} - \frac{\partial v}{\partial r}$ and cross differentiating the modified equilibrium equations again to eliminate the mean stress leads to the equation:

$$\frac{\partial \xi}{\partial t} = \frac{H}{\rho} - \left[u\frac{\partial \xi}{\partial r} + v\frac{\partial \xi}{\partial z} - \frac{u\xi}{r}\right] \qquad ...(14)$$

which is a form of the vorticity transport equation.

A choice of a sufficiently small value of ρ can always be made so that the bracketed term in the above equation is small relative to H/ρ. Considering a small time step the value of H may be used to calculate an increment of ξ, which may in turn be used as a source term in the equation:

$$\xi = -\frac{1}{r}\frac{\partial^2 \psi}{\partial r^2} - \frac{1}{r}\frac{\partial^2 \psi}{\partial z^2} - \frac{1}{r^2}\frac{\partial \psi}{\partial r} \qquad ...(15)$$

(which arises from the definition of ξ and equation (8)) to determine a new field of ψ values. This last equation is solved using successive-over-relaxation techniques.

Whilst this method might be expected to be robust, it has been found so far that, in spite of the choice of a wide variety of densities and time steps, the method reduces major errors but does not eliminate the inequilibrium completely.

5 FINITE ELEMENT CORRECTION

The third scheme considered involved the use of finite element analysis. The stresses from the strain path method were passed to a finite element program and the equivalent out-of-balance forces calculated. These forces were then negated in a conventional incremental analysis. The corresponding displacements were considered as indicating the shift in the streamline pattern, and the updated streamlines were then used to compute a new strain path solution. Whilst not based on any rigorous interpretation, it was considered that this approach, which involves arriving at an equilibrium solution, albeit by an "incorrect" stress path, could lead to a workable correction scheme. The scheme did not in fact work, but this is thought to be at least in part due to difficulties in interpolation between the rectangular mesh used for the strain path method and the Gauss points in the triangular mesh used in the finite element analysis (this choice was for reasons beyond the scope of this paper). Even when no correction was carried out the two stages of interpolation resulted in significant stress changes.

It remains to be seen whether the method might be successful with more closely matched grids in the two numerical schemes.

6 STRAIN PATH FINITE ELEMENT METHOD

Instead of attempting to carry out iterative corrections to the strain path calculation, a different approach has been adopted, and has proved successful. The method involves combining the merits of the strain path method, which correctly accounts for steady flow but results in an error in equilibrium, with the finite element method which satisfies equilibrium correctly.

The problem of cone penetration has previously been analysed by finite element methods (e.g. de Borst and Vermeer (1984), Kiousis et al. (1987)). However, in these analyses the cone has effectively been introduced into a pre-bored hole, with the surrounding soil still in its in situ stress state. An incremental plastic collapse calculation is then carried out and the collapse value identified as the indentation pressure.

This interpretation is not entirely correct, however. During the real penetration of the cone, very high lateral and vertical stresses develop adjacent to the shaft of the cone. By influencing the stresses around the cone tip, these result in higher cone penetration pressures than are predicted by the analysis of a cone in a pre-bored hole. A careful distinction is necessary between a plastic collapse solution and a steady state penetration pressure.

In an attempt to solve the inequilibrium problem of the strain path method, whilst still accounting for the effects of continuous penetration, the following finite element analysis has been carried out. The solution from the strain path method is taken as the initial stress state. These stresses are not quite in equilibrium, with the inequilibrium being represented by a set of out of balance body forces. The cone is then held fixed while the out of balance forces are eliminated by incrementally applying equal and opposite forces. After the inequilibrium has been eliminated the cone is penetrated further until a steady load is reached. Because of the substantial displacements involved in this operation, a large strain formulation of the finite element method is used. Fifteen noded triangular elements are used, and the mesh is shown in Fig. 1.

Fig. 2 shows a comparison of three analyses using the finite element method, in which cone resistance is plotted against penetration. Curve AB is a conventional analysis starting from the in situ stresses, curve CD an analysis as described above, and EF a similar analysis but without correction of the inequilibrium. It can be seen that the equilibrium correction in fact has a very small effect on the ultimate resistance value. The lower resistance given by the analysis of a cone in a pre-bored hole is clearly apparent.

The in situ horizontal stress may be varied in the calculation. The variation of the cone factor with a measure of the horizontal stress $\Delta = (\sigma_{vo} - \sigma_{ho})/2s_u$ where $-1 \leq \Delta \leq 1$) has been examined. The calculations indicate that $(q_t - \sigma_{ho})/s_u$ is almost constant rather than $(q_t - \sigma_{vo})/s_u$. Since, however, σ_{ho} is not usually known accurately, it seems more practicable to retain the use of the factor $N_{kt} = (q_t - \sigma_{vo})/s_u$, but to recognise the effect of the horizontal stress by using a cone factor which depends on Δ.

Finite element analyses (for the pre-bored hole case) have been made of cones with smooth and rough shafts; the results are shown in Fig. 3. The cones on rough shafts show a slightly lower tip resistance since part of the vertical load is carried by friction on the shaft,

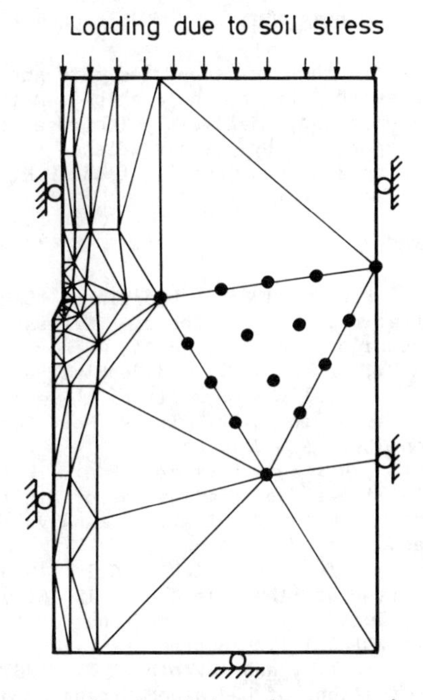

Fig. 1 Finite element mesh

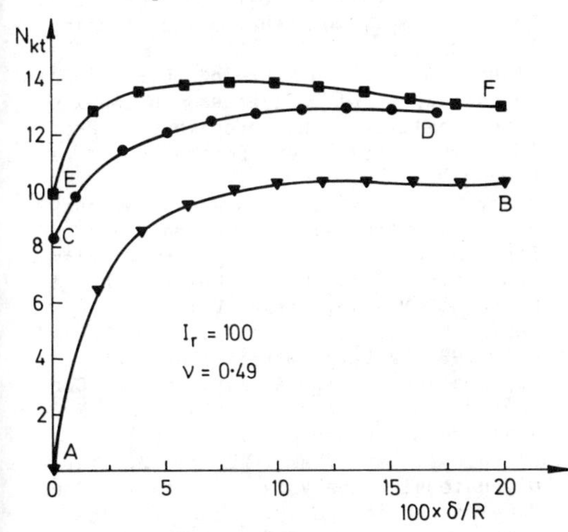

Fig. 2 Load-penetration curves from finite element analysis

although it is difficult to quantify the results from these analyses. Unfortunately the roughness on the inclined cone face could not be varied in this study (in which all finite element analyses were for rough faced cones). More quantitative information on the effects of cone and shaft roughness has been obtained from a study of cones near the surface of a clay using the method of characteristics (Houlsby and Wroth, 1982). This study indicates an increase in mean stress on the cone face due to surface roughness, defined as $\alpha_f = \tau_f\sqrt{3}/2s_u$ where τ_f is the shear stress on the cone face. There is a slight reduction due to the shaft roughness factor α_s (defined similarly to α_f).

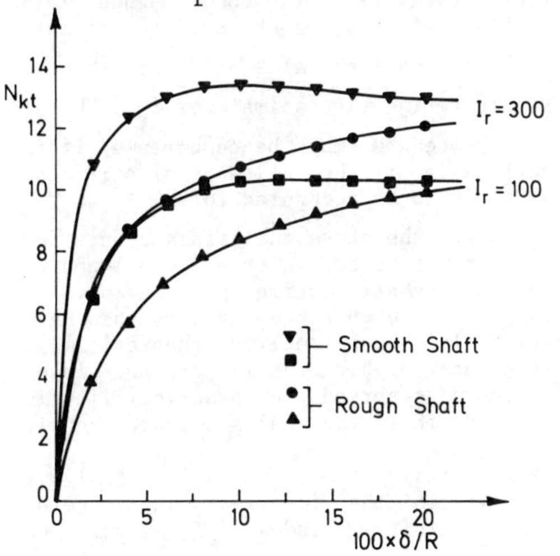

Fig. 3 Finite element analyses of rough cones with smooth and rough shafts

The finite element analyses do not result in a precisely linear variation of N_{kt} with $\ln(I_r)$, but it has been found that they can be very satisfactorily fitted by applying a simple factor to the solution for spherical cavity expansion. The final result is an approximate expression for N_{kt} which includes the effects of rigidity index, cone roughness and in situ stresses:

$$N_{kt} = N_s\left(1.25 + \frac{I_r}{2000}\right) + 2.4\alpha_f - 0.2\alpha_s - 1.8\Delta \qquad \ldots(16)$$

where N_s is the factor which appears in the spherical cavity expansion expression, i.e. $N_s = \frac{4}{3}(1 + \ln\left[I_r\right])$. Eq.16 applies for the following ranges:

I_r 50 to 500 (hence N_s 6.55 to 9.62)
Δ -1 to 1
α_f 0 to 1
α_s 0 to 1

In extreme combinations this results in N_{kt} factors from 6.4 to 18.6, but for more typical cases gives N_{kt} factors in the range 9 to 17. Although it is not possible to test the full range of roughness cases, the above expression approximates the finite element calculations (13 runs) to better than 2% in every case.

Note that Eq. 16 requires estimates of the rigidity index and in situ horizontal stress (as well as the cone roughness) to be made before q_t can be used to estimate s_u. The derived value of s_u may then be used to refine the estimate of I_r. Whilst this procedure may be cumbersome, it is unavoidable if the influence of stiffness on N_{kt} is to be accounted for.

Because the above analysis accounts for the high stresses which are developed on the cone shaft during penetration, it results in higher cone factors than have previously been derived theoretically. Since these higher cone factors are closer to those observed in practice, it is believed that the analysis is reasonably realistic.

Note that the penetration analysis demonstrates the important influence of the rigidity index I_r on the interpretation of the cone. Whilst this may be inconvenient, it cannot be ignored. If the CPT is to be rationally interpreted then information about soil stiffness is required.

7 CONCLUSIONS

The penetration of a cone into clay under undrained conditions has been analysed using the strain path method. The resulting stress field is not quite in equilibrium, and various methods have been explored to achieve a correction to the flow field which would eliminate the equilibrium. None of the methods attempted was entirely successful. An alternative approach in which the results of the strain path analysis are used as the starting point for a finite element analysis has been successful.

For a cone penetrating an elastic-perfectly plastic material theoretical values of cone factors have been derived. The cone factor depends on:

a. The rigidity index
b. The horizontal stress
c. Roughness of cone and shaft.

An approximate expression for the cone factor in terms of these parameters has been proposed (Eq. 16).

ACKNOWLEDGEMENTS

The first author was supported throughout this research by a scholarship from the Kuok Foundation, Malaysia. The research was supported by the Science and Engineering Research Council of the U.K.

REFERENCES

Baligh M.M. 1986. Undrained Deep Penetration, I: Shear Stresses. Geotechnique, Vol. 36, No. 4

Cox, A.D., Eason, G. and Hopkins, H.G. 1961. Axially Symmetric Plastic Deformation in Soils. Trans. Roy. Soc. London, Ser. A., Vol. 25

de Borst, R. and Vermeer, P.A. 1984. Possibilities and Limitations of Finite Element for Limit Analysis. Geotechnique, Vol. 34, No. 2

Houlsby, G.T. and Teh, C.I. 1988. Analysis of the Piezocone in Clay. Proc. 1st Int. Symp. on Penetrometer Testing, ISOPT-1. In press

Houlsby, G.T. and Wroth C.P. 1982. Determination of Undrained Strengths by Cone Penetration Tests. Proc. 2nd. Euro. Symp. on Penetration Testing, Amsterdam, Vol. 2

Houlsby G.T. and Wroth C.P. 1983. Calculation of Stresses on Shallow Penetrometers and Footings. Proc. IUTAM/IUGG Symp. on Seabed Mechanics, Newcastle, U.K.

Houlsby G.T., Wheeler, A.A. and Norbury, J. 1985. Analysis of Undrained Cone Penetration as a Steady Flow Problem. Proc 5th. Int. Conf. Num. Meth. in Geomech., Nagoya, Japan, Vol. 4

Kiousis, P.D., Voyiadjis, G.Z. and Tumay, M.T. 1987. A Large Strain Theory and its Application in the Analysis of the Cone Penetration Mechanism. Int. J. Num. Anal. Meth. in Geomech., In Press

Koumoto, T. and Kaku, K. 1982. Three Dimensional Analysis of Static Cone Penetration into Clay. Proc. 2nd. Euro. Symp. on Penetration Testing, Amsterdam, Vol. 2

Teh, C.I. 1987. An Analytical Study of the Cone Penetration Test. D.Phil. Thesis, Oxford University, U.K.

Vesic, A.S. 1972. Expansion of Cavities in Infinite Soil Mass. Proc. ASCE, J. Soil Mech. Found. Eng. Div., Vol. 96, SM1

Numerical Methods in Geomechanics (Innsbruck 1988), Swoboda (ed.)
© 1988 Balkema, Rotterdam. ISBN 90 6191 809 X

Constitutive laws including kinematic hardening for clay with pore water pressure and for sand

H.Duddeck
Institut für Statik, Technical University of Braunschweig, FR Germany
D.Winselmann
Duddeck und Partner, Braunschweig, FR Germany
F.T.König
Ph. Holzmann AG, Hannover, FR Germany

ABSTRACT: Two constitutive laws for elasto-plastic finite element analyses are proposed. Covering kinematic as well as isotropic hardening they are valid for general three dimensional stress paths including slow cyclic loading. The flow rules are non-associated allowing a better description of volumetric deformations.

1. INTRODUCTION

Constitutive laws used in finite-element-analyses have to be valid for general three dimensional stress paths. The ratios of the principal stresses are different within the continuum and change even during monotonous loading continuously. The analysed structure may have zones of loading as well as unloading. The applied constitutive laws should cover these changes in loading and the corresponding stress history. They should describe the stress-strain behaviour in all phases phenomenologically correct. The analysis should yield displacements as well as stresses correctly since in boundary problems both are related to each other.

For saturated soil the consideration of time dependent pore water pressure is of decisive importance because even slow step by step loading with interim consolidation may cause cyclic effects on stresses. The accurate numerical analysis of pore water pressure requires adequate consideration of the dilatant and contractant soil properties.

2. FUNDAMENTAL SOIL PROPERTIES

In describing the soil properties, the hydrostatic and deviatoric stress components have to be considered separately. Hydrostatic stresses increase the contact forces between soil particles. They reduce the pore volume. Deviatoric stresses are causing shear deformations up to failure. According to the density or the over-consolidation-ratio volumetric deformations may be dilatant or contractant. From the very beginning of loading there are reversible and irreversible parts of ground deformations. Significant unloading yields hysteretic effects because of opposite irreversible deformations. Small unloading remains almost completely elastic.

For sand this behaviour is more pronounced for deviatoric loading than for hydrostatic parts. Failure will occur only by deviatoric failure criteria. Hydrostatic loading yields small deformations by reducing the pore volume causing progressive hardening. Hydrostatic failure only occurs by the crushing of soil particles and is therefore not relevant in the usual range of loading.

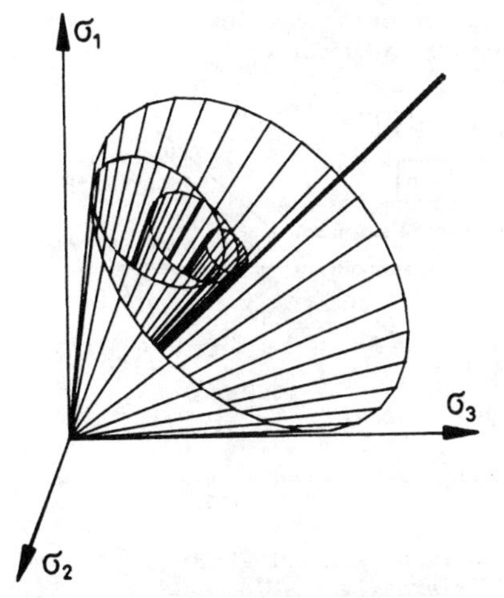

Figure 1. Combined kinematic and isotropic hardening shown in the stress space (cap not plotted).

The properties of sand can be described by constitutive laws consisting of an isotropic hardening rule for hydrostatic loading and a combined isotropic and kinematic hardening rule for deviatoric loading.

For cohesive soil the excess pore water pressure is causing fundamental differences in comparison to sand. Total stresses and effective stresses have to be considered separately.

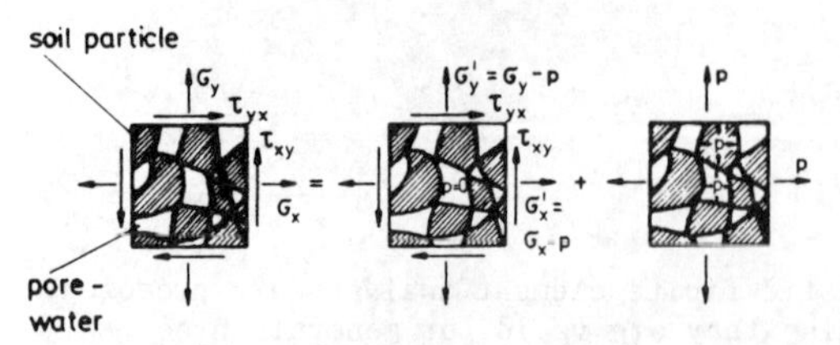

Figure 2. Definition of stresses in saturated soil

The pore water pressure participates in taking the hydrostatic stresses and is therefore reducing the stresses between the soil particles which are decisive for the load carrying capacity. This increases the ratio of deviatoric to hydrostatic stresses and the capacity of failure.

The pore water pressure is time dependent, increasing during loading and dissipating during the following consolidation. The time needed for complete consolidation depends on the permeability of the soil and the drainage conditions.

The pore water pressure and the volumetric deformations of the ground are coupled by the pore volume. Therefore constitutive laws for cohesive soils have to give special account to volumetric strains during hydrostatic loading.

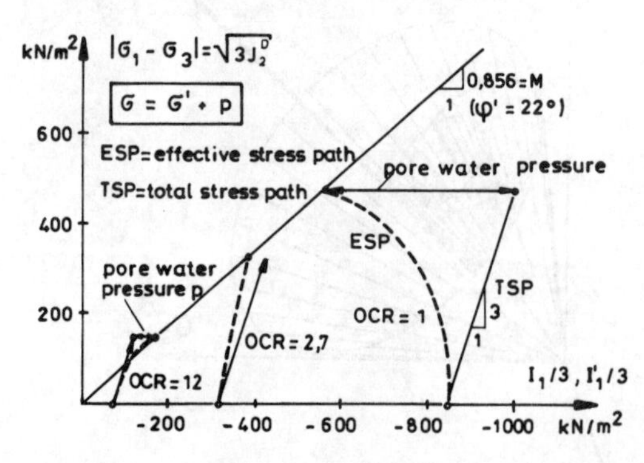

Figure 3. Stress paths of "Weald Clay" in undrained triaxial tests /6/

Preloading of the ground is given by the over-consolidation-ratio (OCR). It characterizes volumetric deformation properties. In triaxial tests normally consolidated clay exhibits contractant behaviour whereas overconsolidated clay yields dilatant volumetric strains. Normally consolidated clay behaves similar to loose sand, overconsolidated clay similar to dense sand.

Contrary to sand - where the stress-strain behaviour depends on the stress-level - the behaviour of clay may be normalized by the consolidation stress.

The behaviour of cohesive soil can be described by constitutive models consisting of an elliptical yield surface and a combination of isotropic and kinematic hardening. Regarding for transient pore pressure effects the coupled stress-strain-flow problem has to be solved. Rules for kinematic hardening are required because monotonous loading causes non-monotonous stress paths due to consolidation.

3. CONSTITUTIVE MODEL FOR SAND INCLUDING CYCLIC LOADING

3.1 Isotropic-kinematic cone model with cap

The constitutive law for sand, proposed here, consists of two yield criteria one for deviatoric and one for hydrostatic loading. Both criteria have isotropic hardening rules. The yield criterion for deviatoric stresses includes an additional kinematic hardening rule and a failure criterion. In the three dimensional principal stress space this model can be visualized by a cone-cap model.

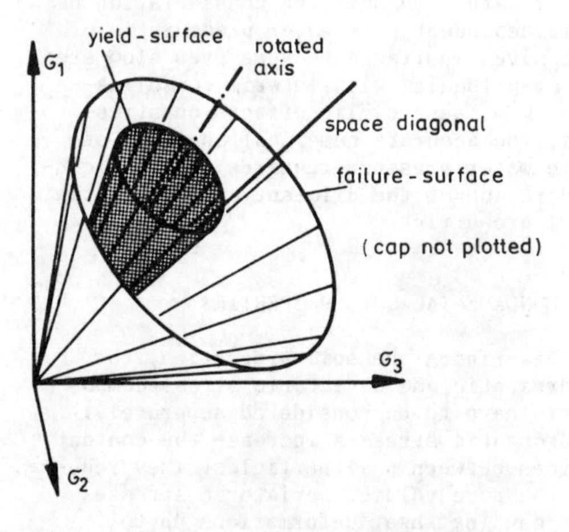

Figure 4. The kinematic cone model (cap not plotted)

In case of monotonous loading the actual state of stress is a point of both yield surfaces. By following the stress path the inner cone, which represents the deviatoric yield surface, expands from zero to a maximum opening angle (isotropic hardening). Simultaneously it changes its position by a rotation around the origin (kinematic hardening).

In the direction of the space diagonal the yield cone is closed by a cap. Considering the small influence of hydrostatic stresses no kinematic hardening rule was applied for the cap. Diversing of the direction of the stress path yields also a change of the position of the yield cone. The region enclosed by the yield cone and the cap is the elastic region. Moving into this region after a stress reversal, the actual state of stress yields only elastic displacements and the yield surfaces do not alter. The constitutive law is valid for general three dimensional stress paths. The possible positions of the yield cone are limited by the failure surface, which divides allowable and not allowable stress conditions.

The constitutive law is an extension and a generalization of the model proposed by Ghaboussi and Momen /3/ for the analysis of special triaxial tests .

The hardening rules have been derived from tests published by Arslan /1/ and Früchtenicht /2/ by standardizing and by generalizing the properties for three dimensional stress conditions.

3.2 Deviatoric yield criteria

The expansion and the rotation of the conical yield surface is limited by the following isotropic failure criterion. The index ° notes that the expressions are related to the space diagonal and not to the rotated cone axis.

$$F^B_{(\sigma_{ij})} = s^\circ_{ij} \cdot s^\circ_{ij} - \varkappa_B^2 \cdot (I^\circ)^2 = 0 \quad (1)$$

$$\alpha^\circ_{ij} = \frac{1}{\sqrt{3}} \cdot \delta_{ij} \; ; \; (\alpha_{ij} \cdot \alpha_{ij})^{1/2} = 1{,}0 \quad (2)$$

$$I^\circ = \sigma_{ij} \cdot \alpha^\circ_{ij} \; ; \; s^\circ_{ij} = \sigma_{ij} - I \cdot \alpha^\circ_{ij}$$

$\varkappa(B)$ specifies the opening angle of the failure cone.

$$\tan \gamma_B = \varkappa_B \quad (3)$$

The yield surface is described in analogy to the failure surface.

$$F_{(\sigma_{ij}, \alpha_{ij}, \varkappa)} = s_{ij} \cdot s_{ij} - \varkappa_F^2 \cdot I^2 = 0 \quad (4)$$

$\varkappa(F)$ specifies the variable opening angle of the yield cone. This angle is limited by max $\varkappa(F)$. The ratio B of the opening angles of the failure and the yield cone can be evaluated from triaxial tests with loading and unloading cycles. It is shown in Fig. 5. $\alpha(ij)$ specifies the direction of the rotating cone axis.

A non-associated flow rule is used to describe the volumetric deformations. The flow rule can be fitted to dilatant and contractant behaviour by the factor a of the plastic potential.

$$\Delta\varepsilon^p_{ij} = \lambda \cdot \frac{\partial Q}{\partial \sigma_{ij}} \quad . \quad (5)$$

$$Q = s_{ij} \cdot s_{ij} - a \cdot \varkappa_F^2 \cdot I^2 \quad (6)$$

The factor a may be evaluated from triaxial tests. It depends on the parameter B. For the tested sand a is given by

$$a = -0.80 + 3.20 * \varkappa_F^2 \qquad B=1.0 \quad (7)$$

$$a = -1.85 + 1.45 * \varkappa_F^2 \qquad B=2.0 \quad (8)$$

Since the yield criterion must be fulfilled within each increment the condition of consistency yields:

$$\Delta F = \frac{\partial F}{\partial \sigma_{ij}} \cdot \Delta\sigma_{ij} + \frac{\partial F}{\partial \alpha_{ij}} \cdot \Delta\alpha_{ij} + \frac{\partial F}{\partial \varkappa_F} \cdot \Delta\varkappa_F = 0 \quad (9)$$

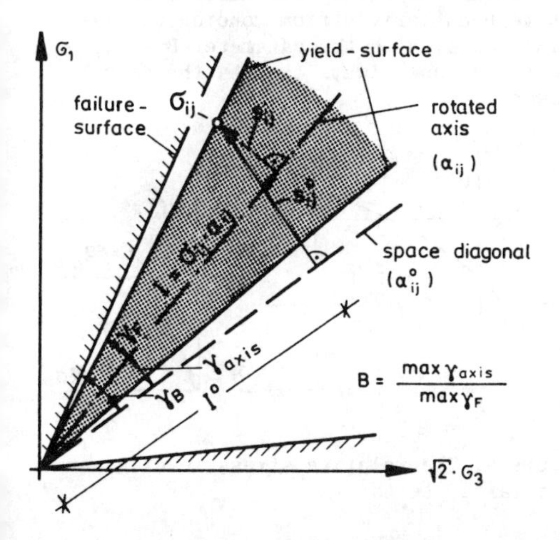

Figure 5. The kinematic cone-model in the triaxial plane

$$\Delta\alpha_{ij} = \beta_{ij} \cdot c_{(w_p)} \cdot \Delta w_p \tag{10}$$

$$\Delta\varkappa_F = g_{(w_p)} \cdot \Delta w_p$$

$$\Delta w_p = \sigma_{ij} \cdot \Delta\varepsilon_{ij}^{p} \tag{11}$$

The variation of the direction of the cone axis and the opening angle is determined by the amount of plastic work during plastic deformation.

For deriving the elasto plastic matrix D(ep) Eq. (9) is transformed. A is determined by matrix multiplication.

$$\frac{\partial F}{\partial \sigma_{ij}} \cdot \Delta\sigma_{ij} - A \cdot \lambda = 0 \tag{12}$$

$$A = -\left[\frac{\partial F}{\partial \alpha_{ij}} \cdot \beta_{ij} \cdot c_{(w_p)} + \frac{\partial F}{\partial \varkappa_F} \cdot g_{(w_p)} \right] \cdot \frac{\partial Q}{\partial \sigma_{ij}} \cdot \sigma_{ij} \tag{13}$$

For several yield criteria the matrix of the elasto plastic material law is derived by eliminating the proportionality constants λ in Eq. (14), Winselmann /11/.

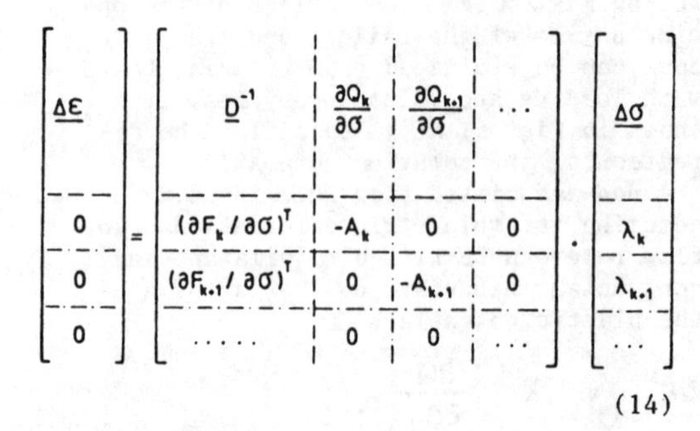

$$\begin{bmatrix} \underline{\Delta\varepsilon} \\ 0 \\ 0 \\ 0 \end{bmatrix} = \begin{bmatrix} \underline{D}^{-1} & \frac{\partial Q_k}{\partial \sigma} & \frac{\partial Q_{k+1}}{\partial \sigma} & \cdots \\ (\partial F_k / \partial \sigma)^T & -A_k & 0 & 0 \\ (\partial F_{k+1} / \partial \sigma)^T & 0 & -A_{k+1} & 0 \\ \cdots & 0 & 0 & \cdots \end{bmatrix} \cdot \begin{bmatrix} \underline{\Delta\sigma} \\ \lambda_k \\ \lambda_{k+1} \\ \cdots \end{bmatrix} \tag{14}$$

The tensor $\beta(ij)$ in Eq.(10) specifies the moving of the variable cone axis. If there is an additional failure surface - as usually for sand - no intersection of the yield and the failure surface is allowed. With a 'moving-direction' as shown in Fig. 6b the yield surface contacts the failure surface tangentially and does not intersect.

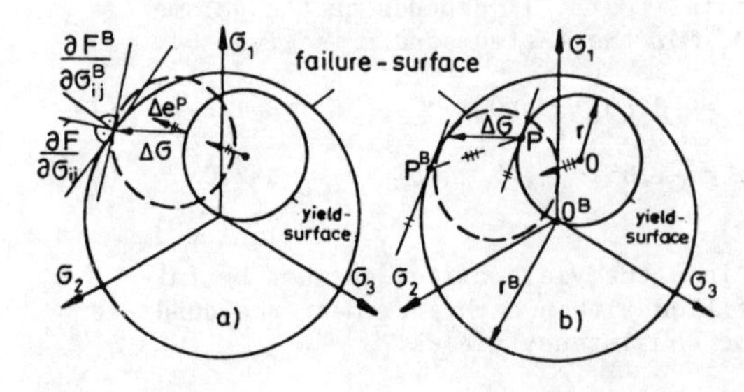

Figure 6. The translation of the yield surface in the deviatoric plane

$$\beta_{ij}^{*} = \sigma_{ij}^{B} - \sigma_{ij}$$

$$\beta_{ij} = \beta_{ij}^{*} - \beta_{kl}^{*} \cdot \alpha_{kl} \cdot \alpha_{ij} \tag{15}$$

The isotropic part of the hardening function is shown in Fig. 7. It may be

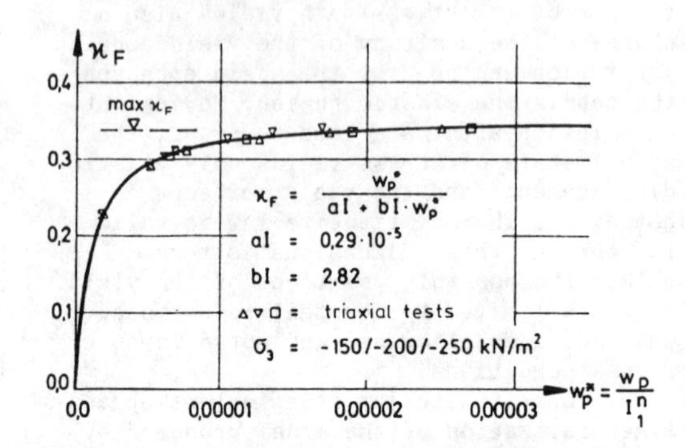

Figure 7. The invariant hardening function for deviatoric loading, from triaxial test results

determined from triaxial tests by plotting the stress-level $\varkappa(F)$ versus the normalized plastic work $w(p)^*$. For the analysis of three dimensional stress states the test parameter $\sigma(3)$ has to be eliminated, Winselmann /11/. The test results in Fig. 7 have been derived from monotonous triaxial tests. With the parameter B = 1.0 the yield cone always touches the space diagonal.

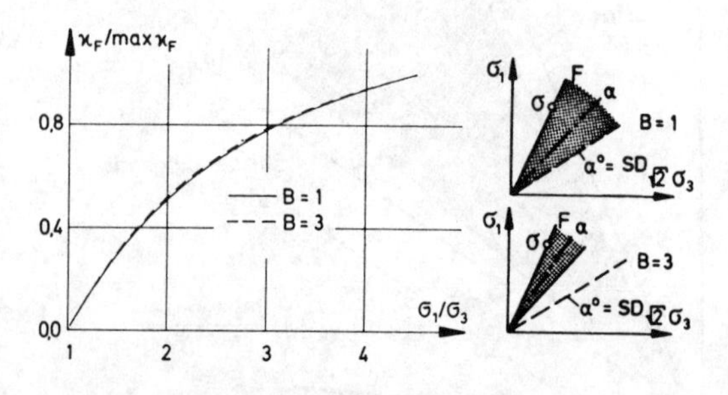

Figure 8. The relative stress-level given by triaxial tests

The relative stress-levels $\varkappa(F)$ / max $\varkappa(F)$ are nearly equal for different values of the parameter B. Therefore, the hardening function can be generalized.

$$g_{(w_p)} = \frac{\Delta\varkappa_{F(B)}}{\Delta w_{p(B=1)}} \cdot \frac{\Delta\varkappa_{F(B=1)}}{\Delta w_p^*} \cdot \frac{\Delta w_p^*}{\Delta w_p} \qquad (16)$$

The kinematic part of the hardening function yields:

$$c_{(w_p)} = \frac{\sqrt{\Delta\alpha_{ij} \cdot \Delta\alpha_{ij}}}{\Delta w_p} \cdot \frac{1}{\sqrt{\beta_{kl} \cdot \beta_{kl}}} \qquad (17)$$

With given B the kinematic part c(wp) can also be expressed by the isotropic part g(wp):

$$c_{(w_p)} = \frac{B \cdot \frac{\partial F}{\partial \varkappa}}{\frac{\partial F}{\partial \alpha_{ij}} \cdot \beta_{ij}} \cdot g_{(w_p)} \qquad (18)$$

The stress-level $\varkappa$(F) in Fig. 7 is valid only for initial loading. After the first stress reversal it is replaced by the stress-level $\varkappa$(U), which gives the length of the stress path up to the failure surface.

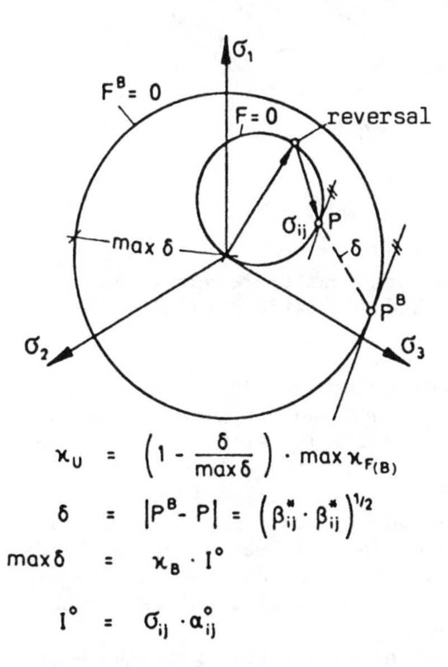

Figure 9. The stress-level $\varkappa$ (U)

As fictious end of the stress path the conjugate point P(B) is chosen, see also Fig. 6, which has the same tangent in the deviatoric stress plane. For small unloading with following reloading $\varkappa$ (U) has to be equal to $\varkappa$ (F). Unloading and reloading is expressed by the hardening functions of initial loading, yielding:

$$c_{(w_p)} = \frac{\frac{\partial F}{\partial \varkappa_F}}{\frac{\partial F}{\partial \alpha_{ij}} \cdot \beta_{ij}} \left[(1+B) \cdot g_{(\varkappa_U)} - g_{(w_p)}\right] \qquad (19)$$

$$g_{(\varkappa_U)} = \left[1 - bI \cdot \varkappa_U \cdot \frac{\max\varkappa_{F(B=1)}}{\max\varkappa_{F(B)}}\right]^2 \cdot \frac{1}{aI} \cdot \frac{\Delta w_p^*}{\Delta w_p} \cdot \frac{\Delta\varkappa_{F(B)}}{\Delta\varkappa_{F(B=1)}} \qquad (20)$$

3.3 Hydrostatic loading

The yield criterion of isotropic hardening for hydrostatic stresses corresponds to a variable cap on top of the yield cone. It may be a section of a sphere, see Lade /7/, or a flat ellipsoid, see Arslan /1/, as proposed here:

$$F_2 = f_{2(\sigma)} - f_{2(w_{p2})} = 0 \qquad (21)$$

$$f_{2(\sigma)} = \frac{2I_1^2 - I_2}{p_a^2}, \quad f_{2(w_{p2})} = \frac{M \cdot w_{p2}}{p_a - L \cdot w_{p2}}$$

The hardening parameter in Eq. (12) is:

$$A_2 = \frac{\partial f_{2(w_{p2})}}{\partial w_{p2}} \cdot \underline{\sigma} \cdot \frac{\partial Q_2}{\partial \sigma} \qquad (22)$$

3.4 Results

In Fig. 10 and 11 results are shown, applying the kinematic cone model.

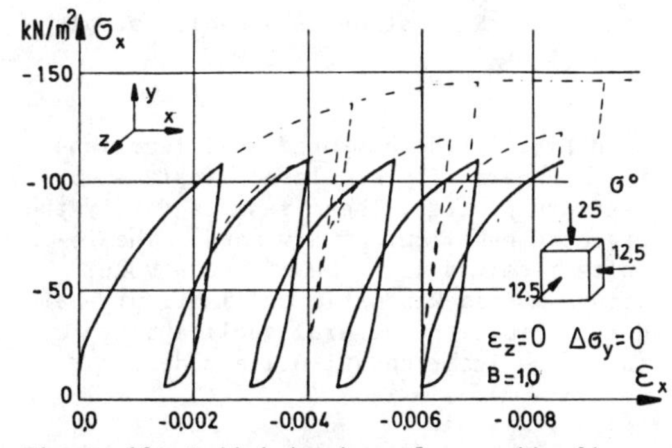

Figure 10. Soil behaviour for cyclic displacements

In Fig. 10 the computed stress-strain relation for a soil cube is plotted. Beginning with the in situ state of stress the cube was cyclically expanded and compressed by inducing displacements in direction of the x-axis. As in plane strain problems the displacement in z-direction was set to zero. The stress in vertical direction was kept constant. This is similar to the action on the ground due to mining subsidence. With larger strains the hysteretic effects increase. For small cyclic strains the reaction is nearly elastic, the stress path remains within the yield-cone.

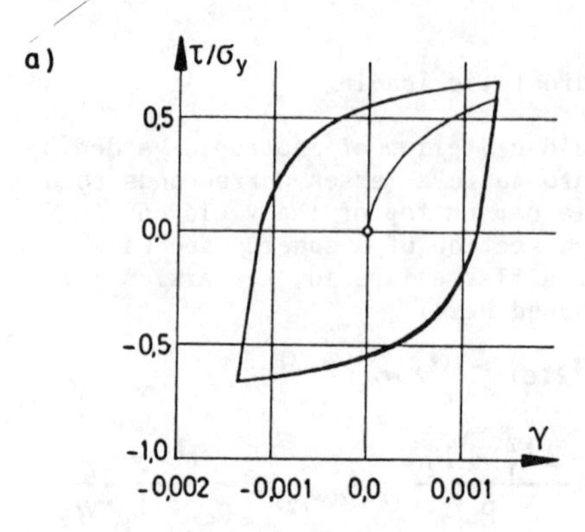

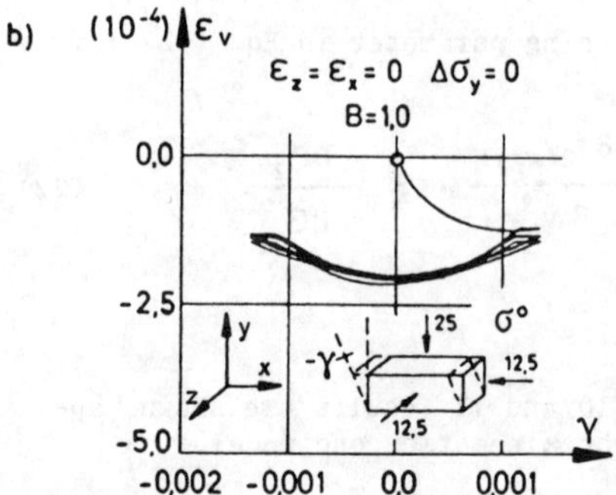

Figure 11. Simulation of a cyclic simple shear test

In Fig. 11 the computed soil reactions for a numerically simulated simple shear test are plotted. These results verify the expected behaviour of dry sand. The results received with the cone-cap model with combined kinematic and isotropic hardening prove the general applicability of this non-linear constitutive model.

4. CONSTITUTIVE MODEL FOR CLAY INCLUDING CYCLIC LOADING

4.1 A critical state model with isotropic and kinematic hardening

Granular materials are usually sufficiently permeable to avoid the build-up of excess pore water pressures during slow loading. For cohesive soils however, pore pressure generation and dissipation are of fundamental importance. Furthermore, the behaviour of clays is much more affected by volumetric strains due to hydrostatic loading than that of sands.

Considering the specific aspects of clay behaviour a modified critical state model is presented. The model consists of an isotropic bounding surface and an inner yield surface with isotropic and kinematic properties.

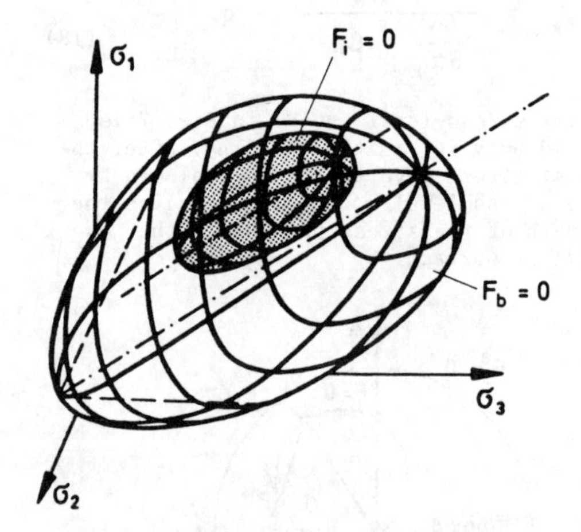

Figure 12. Critical state model with isotropic and kinematic hardening

This two-surface-model is most suited to describe drained and undrained behaviour, including the development of excess pore pressures due to volumetric strains, stress history, and stress induced anisotropy. The outer bounding surface serves as a kind of initial loading or consolidation surface. It expands or contracts isotropically according to a volumetric hardening rule.

In the principal stress space the bounding surface forms an ellipsoid around the space diagonal. The shape of the ellipsoid has been adapted to failure criteria valid for general three dimensional stress states. Traces of the ellipsoid in the deviatoric plane are shown in Fig. 13.

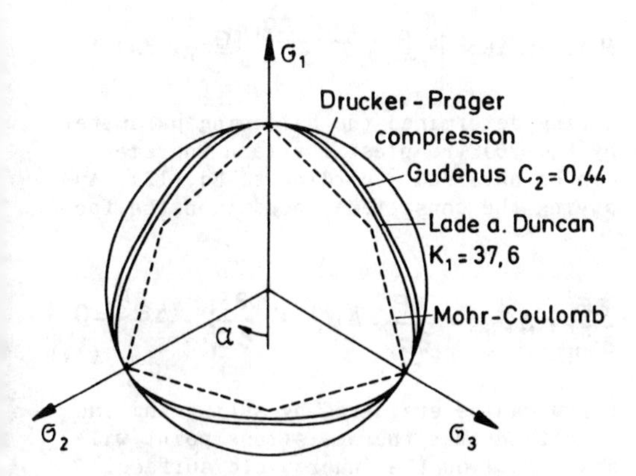

Figure 13. Comparison of different failure criteria in the deviatoric plane

The Gudehus criterion /4/ has been used in the following analyses. It is very adaptable by means of the parameter c2 and well suited for mathematical handling.

The inner yield ellipsoid encloses the elastic domain. Following the stress path it translates within the bounding surface, simulating plastic deformations and hysteretic phenomena during loading cycles.

Size and position of the inner yield surface represent the memory of the soil for stress history events within the initial loading surface. The expansion or contraction of the yield surface is assumed to be proportional to that of the bounding surface.

The model is a further development and generalization of the two-surface model presented by Mroz et al /8/, /9/ for triaxial tests.

4.2 Yield surfaces in the triaxial plane

The yield surfaces are first presented with circular traces in the deviatoric plane. They will be generalized in section 4.3. In Fig. 14 it is assumed that the

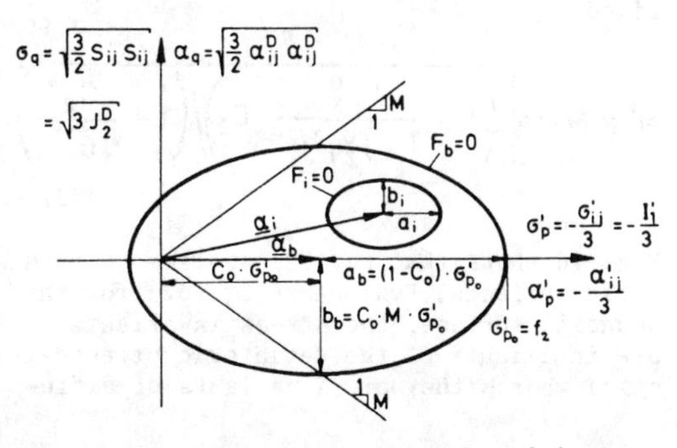

Figure 14. Bounding surface Fb and inner yield surface Fi in the triaxial plane

axis of the inner yield ellipsoid coincides with the triaxial plane. The stress level f2 (maximum consolidation pressure) determines the size of the bounding surface. The ratio of the two semi-axes of the inner and outer ellipses are equal. This ratio can be modified by means of parameter Co. This allows the model to be calibrated to different soil characteristics.

The equation of the inner yield surface

$$F_i = \left(\frac{M \cdot C_o}{1 - C_o}\right)^2 \cdot \left[\left(\left(-\frac{\sigma'_{ii}}{3}\right)_i - \left(-\frac{\alpha'_{ii}}{3}\right)_i\right)^2 - a_i^2\right]$$

$$+ \frac{3}{2}\left(S_{ij} - \alpha^D_{ij}\right)_i \left(S_{ij} - \alpha^D_{ij}\right)_i = 0 \qquad (23)$$

applies correspondingly to the bounding surface, for which the deviatoric components of the direction tensor α(ij) vanish.

The function of the plastic potential Q is also expressed by Eq.(23). Introducing a non-associated flow rule requires the parameter Co to be changed in order to obtain a different ratio of the semi-axes.

The hardening parameter of the bounding surface A(b), which is required in the consistency condition Eq.(12) can be determined according to Eq.(9) using the volumetric strain hardening relationship of the Cam Clay model.

$$A^* = \frac{\partial F}{\partial \sigma'_{p_o}} \cdot \frac{1 + e_o}{\lambda_c - \varkappa_s} \cdot \frac{\sigma'_{p_o} \cdot \underline{m}^T \cdot \underline{n}_Q}{\|[\partial F / \partial \sigma]\|} \qquad (24)$$

It is denoted A(b)*, since normalized derivatives of the yield condition and the plastic potential are used for this model. This leads to a proportionality factor λ* which is identical with the plastic volumetric strain increment. The bounding surface defines the state of initial loading by an equivalent hydrostatic stress level f2=σ(po'). Once the stress path reaches the bounding surface it will become the active surface. While expanding with the stress path it governs the stress strain relationship according to its hardening parameter A(b)*. Having solved the system equations, the isotropic expansion of the

$$\Delta a_b = (1 - C_o) \cdot \Delta \sigma'_{p_o} \; , \qquad (25a)$$

$$\underline{\Delta \alpha_b} = \delta_{ij} \cdot C_o \cdot \Delta \sigma'_{p_o} \qquad (25b)$$

$$\Delta \sigma'_{p_o} = -\left(\frac{1 + e_o}{\lambda_c - \varkappa_s}\right) \cdot \sigma'_{p_o} \cdot \Delta \varepsilon^p_{ii} \qquad (25c)$$

bounding surface follows from the consistency condition.

As long as the bounding surface is active, the inner yield surface remains in tangent contact with the outer surface. The size of the inner yield surface is specified by a constant ratio between the semi-axes of the inner and the outer surface.

The hardening parameter A(i)* of the inner surface follows from an interpolation rule in connection with a conjugate point which has to be specified on the bounding surface in analogy to Fig. 6.

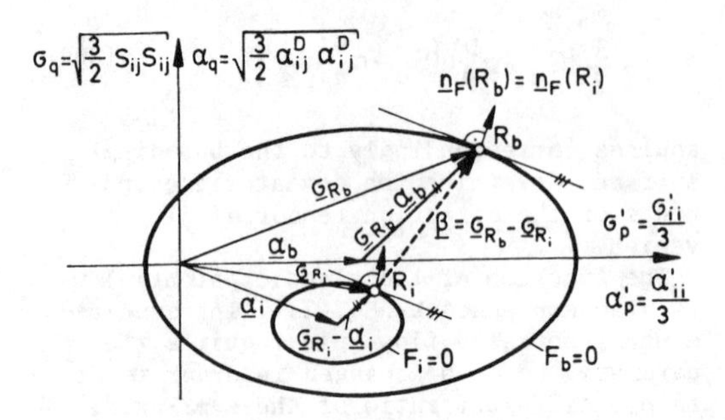

Figure 15. Stress point and conjugate point

Knowing the hardening parameter of the conjugate point A(b) and given an initial value A(oi)* the parameter A(i)* can be determined.

$$A^*_i = A^*_b + (A^*_{oi} - A^*_b)\cdot\left(\frac{\delta}{\delta_o}\right)^{\gamma} . \quad (26)$$

$$\delta = |\underline{\beta}| = \sqrt{\beta_{ij}\cdot\beta_{ij}} . \quad (27)$$

$$\delta_o = \sqrt{3}\cdot(2\cdot a_b - 2\cdot a_i) . \quad (28)$$

The term δ(o) is a measure for the maximum possible distance between inner and outer yield surface. Calibration calculations have shown that a smooth transition from the elastic domain to the elasto-plastic domain between inner and outer yield surface can be achieved by fixing the parameter A(oi)* at approximately ten times the amount of the elastic modulus.

The translation rule of the inner ellipsoid follows from the postulate that the surfaces must not intersect, but engage each other along a common normal, Prevost /10/.

$$\underline{\Delta\sigma}_{R_i} - \underline{\Delta\sigma}_{R_b} = \mu\cdot\underline{\beta} . \quad (29)$$

Assuming combined isotropic and kinematic hardening yields:

$$\underline{\Delta\alpha}_i - \underline{\Delta\alpha}_b = \mu\cdot\underline{\beta} + \frac{\Delta a_b - \Delta a_i}{a_i}(\underline{\sigma}_{R_i} - \underline{\alpha}_i) \quad (30)$$

Having determined the hardening parameter by Eq.(26) the plastic strain increments can be obtained according to Eq.(12). Applying the consistency condition the factor

$$\frac{\partial F_i}{\partial \sigma_{ij}}\cdot\Delta\sigma_{ij} + \frac{\partial F_i}{\partial \alpha_{ij}}\cdot\Delta\alpha_{ij} + \frac{\partial F_i}{\partial \varepsilon^p_{ii}}\cdot\Delta\varepsilon^p_{ii} = 0 \quad (31)$$

tor μ can be evaluated by satisfying the condition that the new stress point will again fall on the inner yield surface. Therefore, the tensor Δα(ij) in Eq.(31) has to be substituted by $\underline{\Delta\alpha(i)}$ of Eq.(30).

Restricting the model to kinematic hardening within the bounding surface the last term of Eq.(31) vanishes. This means that the sizes of the yield surfaces do not change any further, once the stress path moves within the bounding surface.

Because of the affinity of the inner and outer yield surface the same non-associated flow rule is applicable to both surfaces. The normalized derivatives of the plastic potential Q for the conjugate point on the bounding surface can be taken as the unit vector components normal to the inner yield surface.

4.3 Shape of yield surfaces in the deviatoric plane

Yield surfaces in a generalized model for three dimensional stress paths have to account for generally valid yield criteria. Therefore the slope M of the critical state line (Fig.14) is continuously adjusted according to its position in the deviatoric plane. Using the Gudehus criterion the following function for the variable slope M* of the critical state line is derived:

$$M^* = M_K\cdot\sqrt{\left(1 - \frac{3J^D_3}{(2J^D_2)^{3/2}}\cdot C_2\right)\Big/\left(1 + \frac{C_2}{\sqrt{6}}\right)} \quad (32)$$

Fig. 16 shows that Eq.(32) applies to both yield surfaces. Evaluating Eq.(32) for the bounding surface, the stress invariants are invariants of the deviatoric stresses s(ij) wheras they are invariants of diffe-

$$M^*(F_b) \Longrightarrow J^D_2,\ J^D_3 = f(S_{ij}) \quad (33)$$

rential stresses s(ij)-α(ij) in case of the inner yield surface.

$$M^*(F_i) \Longrightarrow J_2^D, \; J_3^D = f(S_{ij} - \alpha_{ij}^D) \; . \qquad (34)$$

The functional expression for M^* has to be inserted in the yield and plastic potential functions, which affects all partial stress derivatives.

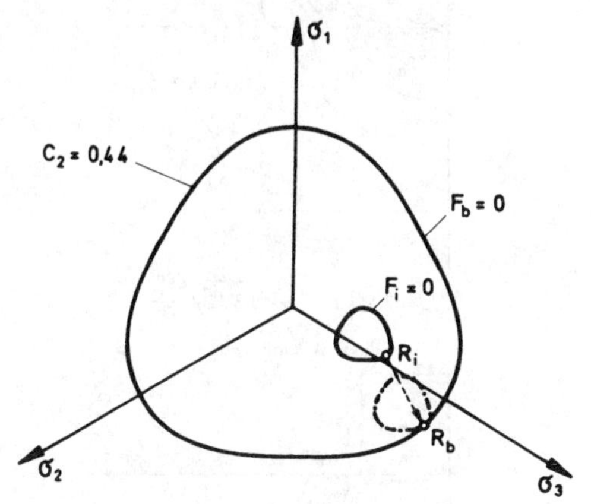

Figure 16. Traces of generalized inner and outer yield surfaces (deviatoric plane)

4.4 Numerical modelling of transient pore water pressures

In order to account for the time dependent consolidation the transient pore water pressures have to be introduced as nodal unknowns into the finite element discretization. For the discretization scheme in the time domain a linear interpolation along each time step is chosen. Application of the Galerkin integration scheme yields $\xi = 2/3$ as characterizing coefficient.

Using the displacement method the following set of matrix equations is derived:

$$\begin{bmatrix} \underline{K} & \beta \cdot \underline{C} \\ \beta \cdot \underline{C}^T & -\beta^2 (\frac{2}{3} \Delta t \cdot \underline{W}_F + \underline{W}_S) \end{bmatrix} \cdot \begin{bmatrix} \Delta \underline{\hat{u}}_{(\Delta t)} \\ \Delta \underline{\hat{p}}^*_{(\Delta t)} \end{bmatrix} = \qquad (35)$$

$$\begin{bmatrix} 0 & 0 \\ 0 & \beta \cdot \Delta t \cdot \underline{W}_F \end{bmatrix} \cdot \begin{bmatrix} \underline{\hat{u}}_{(t)} \\ \underline{\hat{p}}_{(t)} \end{bmatrix} + \begin{bmatrix} \Delta \underline{F}_{(\Delta t)} \\ \beta \cdot \Delta t (\underline{Q}_{(t)} + \frac{2}{3} \Delta \underline{Q}_{(\Delta t)}) \end{bmatrix}$$

$\underline{\hat{u}}$ and $\underline{\hat{p}}$ are the unknown nodal displacements and pore pressures respectively. $\underline{K}$ is the elasto plastic element stiffness matrix, describing the stress strain behaviour of the soil skeleton. $\underline{W}_F$ is the flow matrix governing the fluid flow through porous media. $\underline{W}_S$ is a water storage matrix. It vanishes if the pore water is assumed to be incompressible. The transposed matrix $\underline{C}^T$ couples the volumetric deformation of the soil with the pore water volume change, while $\underline{C}$ couples the pore water pressures and the effective stresses with the total stresses.

Large differences in the main diagonal terms because of large stiffness coefficients and small permeability values, may cause numerical difficulties especially at the start of the consolidation phase, König /5/. Introducing the factor β the conditioning of the matrix equations can be improved.

4.5 Results

Oil storage tanks often have to be founded on soft ground in coastal areas. Therefore a tank foundation problem is chosen to demonstrate the model's ability to simulate the pore pressure development which is crucial for the tanks stability and its settlement prediction.

When filled, a typical tank with 15 m radius and a storage capacity of about 10000 m3 gives rise to a uniform foundation load of 150 kN/m2.

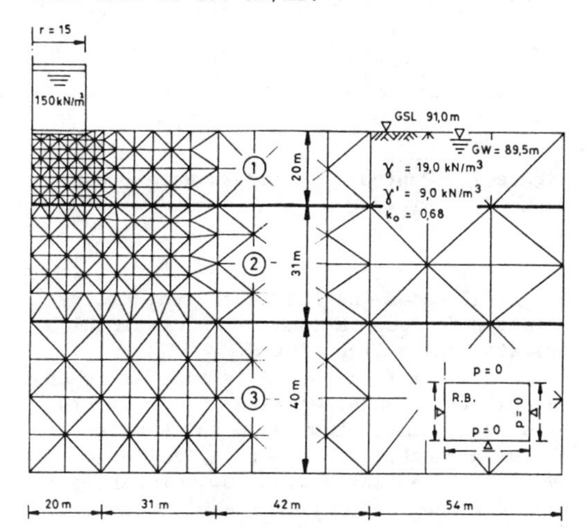

Figure 17. Finite element discretization of tank on layered ground

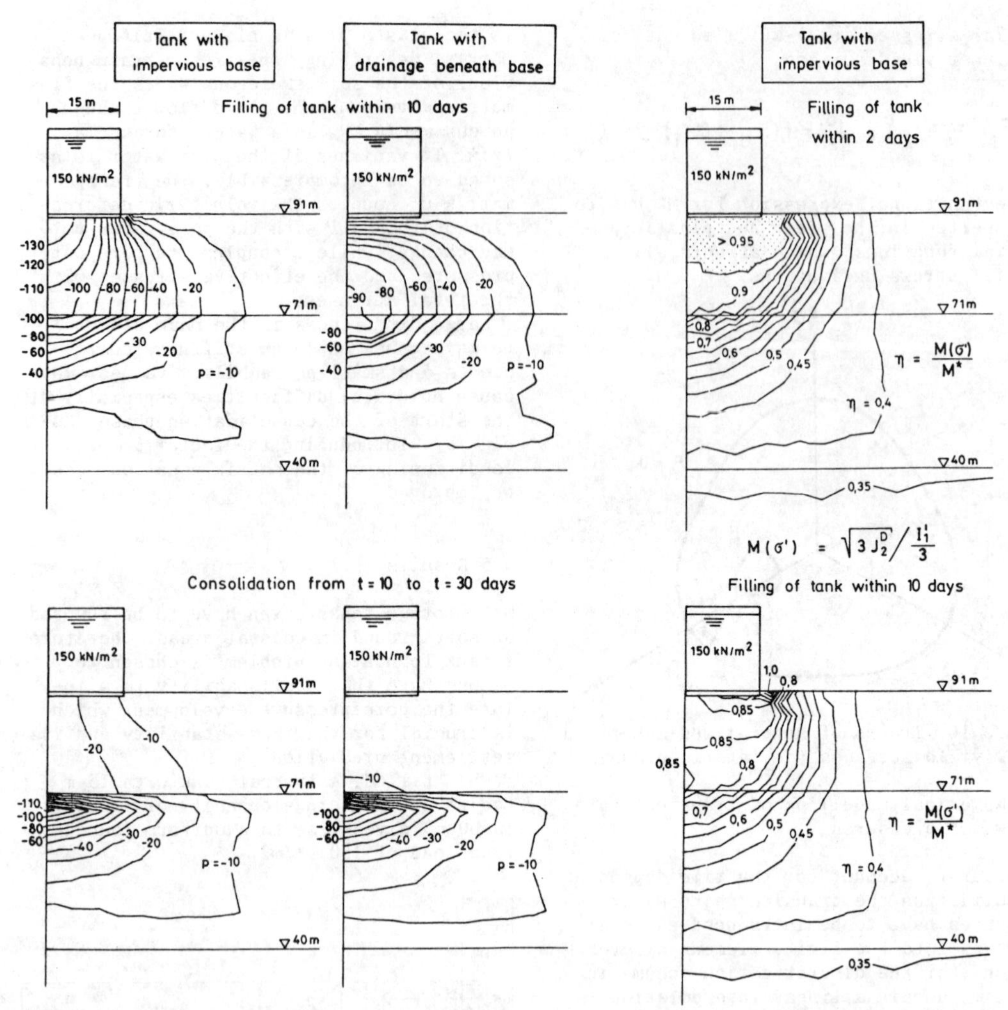

Figure 18. Contours of excess pore pressure under tank

Figure 19. Contours of mobilized shear strength

Fig. 17 shows the finite element discretization for such a tank with rotational symmetry on layered ground.

The soil profile, which is similar to an example from Lambe/Whitmann /6/, consists of a sandy silt, about 20 m thick with medium permeability (Layer 1), followed by a soft clay layer with low permeability (Layer 2) and a highly permeable sand layer (Layer 3).

During a rapid first filling of the tank large excess pore pressures develop beneath the impervious tank base. The pore pressure contours for a tank filled within ten days (Fig. 18) show that the excess pore pressure beneath the centre line is almost equal to the applied load.

As a consequence large plastic zones with fully mobilized shear strength (Fig. 19) give rise to large deformations and eventually may lead to a tank failure.

The drainage conditions beneath the tank can be improved by means of a drainage layer. Fig. 18 shows that this leads to a significant reduction of the resulting pore pressure with the maximum pore pressure contour appearing close to the low permeable clay layer.

After allowing the soil to consolidate under the tank load for about 20 days (Fig. 18, bottom), the excess pore pressure in the upper layer is almost completely dissipated, whereas pore pressure spreading combined with stress redistribution due to consolidation leads to even increased pore pressures in the less permeable clay layer. This demonstrates that settlements will occur over a long time span until complete pore pressure dissipation is achieved.

Filling and discharging of tanks give rise to continuous loading and unloading cycles. With each cycle new pore pressures will be generated. In soft cohesive soils a continuous pore pressure build-up due to rapid loading cycles with short intermediate consolidation times eventually may lead to a failure by liquefaction.

By means of the coupled stress -strain-flow model with combined isotropic and kinematic hardening the effect of e.g. drains or preloading on the soil behaviour can be investigated.

Allowable loading rates and required consolidation times can be determined.

5. SUMMARY

Two constitutive laws with combined isotropic and kinematic hardening have been presented for sand and for clay. The models are valid for general three dimensional stress paths including slow cyclic loading. Accounting for volumetric deformation characteristics, non-associated flow rules are used.

For cohesive soils special regard was given to model undrained behaviour. In order to predict transient pore pressures during loading and consolidation a numerical solution of the coupled stress-strain-flow problem by means of an incremental finite element formulation has been presented. Examples demonstrate the great potential of the proposed models to simulate actual soil behaviour.

REFERENCES

/1/ M.U. Arslan, Beitrag zum Spannungs-Verformungsverhalten der Böden, Ph.D Thesis, TU Darmstadt (1980)

/2/ H. Früchtenicht, Das Verhalten nicht bindigen Bodens beim Aushub tiefer Baugruben, Ph.D Thesis, TU Braunschweig (1984)

/3/ J.Ghaboussi & H. Momen, Plasticity Model for cyclic behaviour of sand, Third Int. Conference on Numerical Methods in Geomechanics, Rotterdam, Balkema (1979)

/4/ G. Gudehus, Elastoplastische Stoffgleichungen für trockenen Sand, Ing. Archiv 42, 151-169 (1973)

/5/ F.T. König, Stoffmodelle für isotrop kinematisch verfestigende Böden bei nichtmonotoner Belastung und instationären Porenwasserdrücken, Report 85-46, Institut für Statik, TU Braunschweig (1985)

/6/ T.W. Lambe & R.V. Whitmann, Soil Mechanics, Wiley & Sons, New York (1969)

/7/ P.V. Lade, Elasto-Plastic Stress-Strain Theory for Cohesionless Soil, Journal of the Geotechnical Engineering Division 1037-1053 (1977)

/8/ Z. Mroz, V.A. Norris & O.C. Zienkiewics, An anisotropic hardening model for soils and its application to cyclic loading, Int. Journal Numerical & Analytical Methods in Geomechanics, Vol. 2, 203-221 (1978)

/9/ Z. Mroz, V.A. Norris & O.C. Zienkiewics, Application of an isotropic hardening model in the analysis of elasto-plastic deformations of soils, Geotechnique 29, No. 1, 1-34 (1979)

/10/ J.H. Prevost, Plasticity Theory for Soil Stress-Strain Behavior, Journal of the Engineering Mechanics Division, Oct., 1177-1194 (1978)

/11/ D. Winselmann, Stoffgesetze mit isotroper und kinematischer Verfestigung sowie deren Anwendung auf Sand, Report 84-44, Institut für Statik, TU Braunschweig (1984)

Numerical Methods in Geomechanics (Innsbruck 1988), Swoboda (ed.)
© 1988 Balkema, Rotterdam. ISBN 90 6191 809 X

An anisotropic hardening model for the mechanical behaviour of clay

Hiroyoshi Hirai
Department of Civil Engineering, Kumamoto University, Japan

ABSTRACT: A model introduced in the present paper is capable of describing the mechanical behaviour of anisotropically consolidated clay reasonably well. This model provides the non-associative flow rule based on combined isotropic-kinematic hardening. The good agreement is observed between simulation and experimental data.

1 INTRODUCTION

Various anisotropic behaviours of clay can be observed in many fields. Experiments of clay under anisotropically consolidated stress have showed that the stress-strain relationships are influenced by the initial degree of anisotropic consolidation. Specimens trimmed at different angle for the clay deposit consolidated one-dimensionally exhibit different undrained failure strength.

Various constitutive models of clay have been developed to incorporate the influence of anisotropic initial condition on deformation and strength. Pietruszczak et al.(1983) simulated the anisotropic behaviour on K_0-consolidated clays by using multisurface model based on combined isotropic-kinematic hardening. The constitutive models to account for the initial anisotropy and subsequent evolution of induced anisotropy of clay are presented by Banerjee et al. (1984) and Anandarajah et al.(1986) on the basis of the boundary surface plasticity. An anisotropic hardening model is proposed by Hirai(1987) in order to describe the induced anisotropy of sand. This model can be easily applied to the simulation of clay behaviour.

The objective of this paper is to present a constitutive model that is capable of describing the mechanical behaviour of clay subjected to shear deformation subsequent to anisotropically consolidated stress. The initial yield surface rotates from isotropic stress line when subjected to anisotropic consolidation and the subsequent yield surface expands and translates as the shear stresses are applied to a sample. The clay behaviour is assumed to be simulated by combination of isotropic and kinematic hardening for which the expansion of yield surface is prescribed by conventional isotropic hardening and the translation of yield surface is specified by the evolution of kinematic hardening. The model simulations are compared with some experimental data obtained under undrained triaxial tests of clay.

2 FAILURE FUNCTION

The general failure surface of clay has been proposed in a variety of stress forms. It is indicated from available experimental data that the failure surface takes a conical shape with smooth curved meridians in the principal stress space and round cornered hexagon of Mohr-Coulomb type on the octahedral plane. The failure surface employed in the present paper is of a conical shape proposed by Willam and Warnke(1975). The failure surface is given by the following form:

$$F = J_2^{1/2} + \omega R(\theta) I_1 = 0 \tag{1}$$

where ω is a failure material constant and I_1 and J_2 are the first invariant of stress σ_{ij} and the second invariant of deviatoric stress s_{ij} respectively as follows:

$$I_1 = \sigma_{kk} \tag{2}$$

$$J_2 = s_{ij}s_{ij}/2 \tag{3}$$

$$s_{ij} = \sigma_{ij} - I_1/3\,\delta_{ij} \tag{4}$$

The variable θ denotes the angle similar to the Lode angle on the octahedral plane, defined by the following equation:

$$\cos(3\theta) = -3\cdot 3^{1/2} J_3/(2J_2^{3/2}) \tag{5}$$

where J_3 is the third invariant of deviatoric stress:

$$J_3 = s_{ij}s_{jk}s_{ki}/3 \tag{6}$$

The function $R(\theta)$ determines the shape of the cross section of the failure surface on the octahedral plane and takes the form:

$$R(\theta) = U(\theta)/V(\theta) \tag{7}$$

where

$$U(\theta) = 2(1-\bar{R}^2)\cos\theta + (2\bar{R}-1)\times \{4(1-\bar{R}^2)\cos^2\theta + 5\bar{R}^2 - 4\bar{R}\}^{1/2} \tag{8}$$

$$V(\theta) = 4(1-\bar{R}^2)\cos^2\theta + (1-2\bar{R})^2 \tag{9}$$

in which $\bar{R}$ is a material constant related to the distance from the origin to failure surface on the octahedral plane.

3 YIELD SURFACE AND PLASTIC POTENTIAL

An important feature of cyclic behaviour of clay is induced anisotropy due to plastic deformation. The deformation-induced plastic anisotropy is described by the combination of isotropic and kinematic hardening. The yield surface in general takes the following form:

$$f(\sigma_{ij}, \alpha_{ij}, \beta_{ij}, k) = 0 \tag{10}$$

where α_{ij} and k denote kinematic hardening and isotropic hardening respectively and β_{ij} indicates the anisotropic tensor to represent the rotation of yield surface.

The specific shape of the yield surface used in the present paper is of a generalized form of the Modified Cam clay model. The yield surface which translates with expansion is given by the following equation:

$$f = \bar{J}_2/r(\bar{\theta})^2 + m\bar{I}_1^{\,2} + k\bar{I}_1 = 0 \tag{11}$$

where m is a material constant and the invariants are

$$\bar{I}_1 = \bar{\sigma}_{kk} \tag{12}$$

$$\bar{J}_2 = \bar{s}_{ij}\bar{s}_{ij}/2 \tag{13}$$

$$\bar{J}_3 = \bar{s}_{ij}\bar{s}_{jk}\bar{s}_{ki}/3 \tag{14}$$

$$\cos(3\bar{\theta}) = -3\cdot 3^{1/2}\bar{J}_3/(2\bar{J}_2^{\,3/2}) \tag{15}$$

where

$$\bar{\sigma}_{ij} = \sigma_{ij} - \alpha_{ij} - \beta_{ij} I_1/3 \tag{16}$$

$$\bar{s}_{ij} = s_{ij} - \alpha'_{ij} - \beta'_{ij} I_1/3 \tag{17}$$

$$\alpha'_{ij} = \alpha_{ij} - \alpha_{kk}\,\delta_{ij}/3 \tag{18}$$

$$\beta'_{ij} = \beta_{ij} - \beta_{kk}\,\delta_{ij}/3 \tag{19}$$

The variable $r(\bar{\theta})$ in eq.(11) is given by replacing θ and $\bar{R}$ in eqs.(7) to (9) with $\bar{\theta}$ and $\bar{r}$ respectively. Since the tensor α_{ij} represents translation of yield surface, the yield function given by eq.(11) can expand with isotropic hardening and simultaneously translate with kinematic hardening in stress space, as shown in Figure 1.

The direction of incremental plastic strain is defined by specifying the shape of the plastic potential surface. As it has been suggested that plastic potential surface is usually different from yield surface, the non-associative flow rule is essentially to be employed in the incremental plasticity theory. The plastic potential used in the present paper is a generalized form of the Cam clay model and is given by the following form:

$$g = J_2^{1/2}/(R(\theta)I_1) - \eta\ln(I_1/h_1) = 0 \tag{20}$$

where η is a material constant and h_1 is unnecessary to define explicitly.

4 HARDENING FUNCTIONS

It is generally considered that the hardening process is composed of isotropic and kinematic hardening due to plastic deformation. It is necessary to provide the evolutional rules of each hardening. If the isotropic hardening may depend on not only the plastic work related with the change in shape but also that related with the change in volume, the evolution of isotropic hardening is expressed in the following form :

$$\dot{k} = \phi\,(\sigma_{ii}\dot{e}_{jj}^{(p)}/3 + n s_{ij}\dot{e}_{ij}^{\prime(p)}) \tag{21}$$

where ϕ and n denote material constants and $\dot{e}_{ij}^{(p)}$ and $\dot{e}_{ij}^{\prime(p)}$ are the plastic strain rate and the deviatoric plastic strain rate respectively.

The kinematic hardening used in the present paper is modified form of Ziegler's law. This evolution is expressed as

$$\dot{\alpha}_{ij} = C\bar{s}_{ij}\dot{e}_p \tag{22}$$

where

$$C = D\, J_2^{1/2} r(\bar{\theta})^2 / (2\bar{J}_2 (1+(W(\theta)/R(\theta))^2)^{1/2} \quad (23)$$

where D is a material constant, $W(\theta)= \partial R(\theta) / \partial \theta$ and

$$\dot{e}_p = (\dot{e}'^{(p)}_{mn} \dot{e}'^{(p)}_{mn})^{1/2} \quad (24)$$

5 STRESS-STRAIN RELATIONSHIPS

A general elastoplastic formulation is based on the assumption that the total strain response consists of elastic and plastic strain responses. This is represented as follows:

$$\dot{e}_{ij} = \dot{e}_{ij}^{(e)} + \dot{e}_{ij}^{(p)} \quad (25)$$

where e_{ij} and $e_{ij}^{(e)}$ are the total strain and elastic strain respectively.

If the principal axes of plastic strain rate tensor conincide with the principal stress axes, this condition is satisfied with sufficient generality by the following equation:

$$\dot{e}_{ij}^{(p)} = \Lambda\, \partial g / \partial \sigma_{ij} \quad (26)$$

where Λ is a scalar function dependent on stress and strain, and g is the plastic potential.

The function Λ is determined by use of the consistency condition which is expressed as

$$\partial f / \partial \sigma_{ij} \dot{\sigma}_{ij} + \partial f / \partial \alpha_{ij} \dot{\alpha}_{ij} + \partial f / \partial k\, \dot{k} = 0 \quad (27)$$

The constitutive relation in elastic range is given by the following equation:

$$\dot{e}_{ij}^{(e)} = (\dot{\sigma}_{ij} - \dot{p}\,\delta_{ij})/(2\mu p) + 0.435 C_s \dot{p}\, \delta_{ij} / \{3(1+e_0)p\} \quad (28)$$

where μ is an elastic material constant, e_0 is the initial void ratio, C_s is the rebound compressibility of clay, $p=I_1/3$ and μ is given as follows:

$$\mu = \frac{3(1-2\nu)(1+e_0)}{2(1+\nu)\cdot 0.435 C_s} \quad (29)$$

where ν is the Poisson's ratio.

6 SIMULATION OF UNDRAINED TRIAXIAL TESTS OF ANISOTROPICALLY CONSOLIDATED CLAY

In order to investigate the strength and deformation of saturated clay, undrained shear tests have been carried out under anisotropic consolidation or K_0-consolidation. The proposed model is applied to the simulation of mechanical behaviour of anisotropically consolidated clay. The predicted results are compared with experimental data reported by Stipho(1978). The physical property and material parameters used in prediction are as follows: e_0=1.0, ν=0.3, C_s=0.115, m=0.0133, $\bar{r}$=0.95, η= -0.204, $\bar{R}$=0.72, n=1.0, ϕ =-0.57, D=342.

Figures 2-5 show the comparison between simulation and experimental results for effective stress paths, stress-strain curves and pore water pressure responses when a sample is subjected to isotropic consolidation. It is found that the model simulation is capable of describing experimental results of triaxial compression and extension tests for every value of OCR. The prediction performed by Banerjee et al.(1984) seems to be not close to experimental data of triaxial compression test for OCR=1.0. This may be because the associative flow rule employed by Banerjee et al.(1984) is not adequate for the model prediction. On the other hand, the present simulation using the non-associative flow rule is successful in predicting experimental data of triaxial compression tests for OCR=1.0.

Figures 6-10 show the comparison between model simulation and experimental results in compression and extension tests for anisotropically consolidated clay (N= σ_1/σ_3). The proposed model is capable of predicting the effective stress path, stress-strain curve and pore water pressure obtained in experiments.

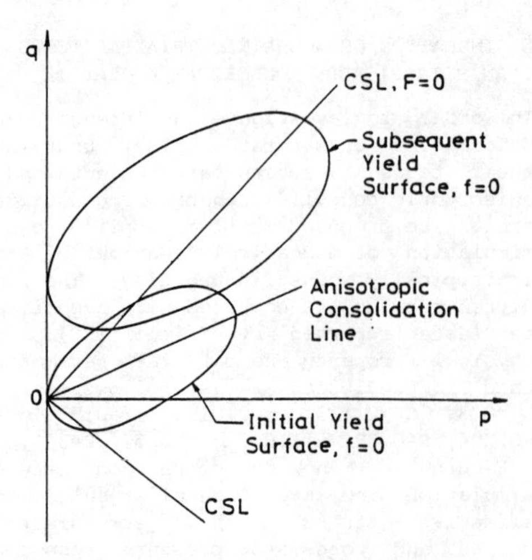

Fig.1 Yield surfaces of anisotropically consolidated clay

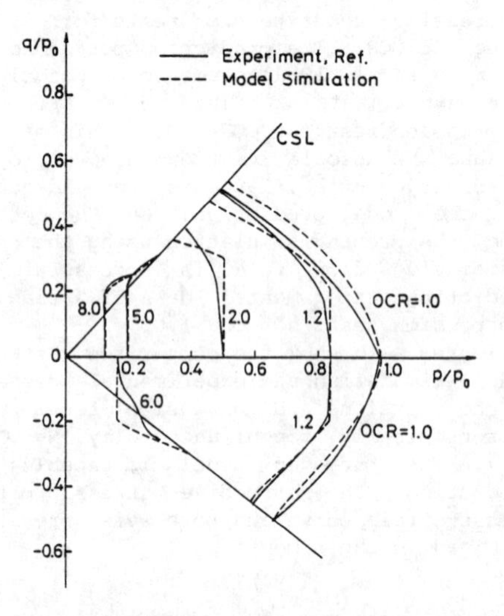

Fig.2 Comparison of model simulation with triaxial compression and extension tests of isotropically consolidated clay

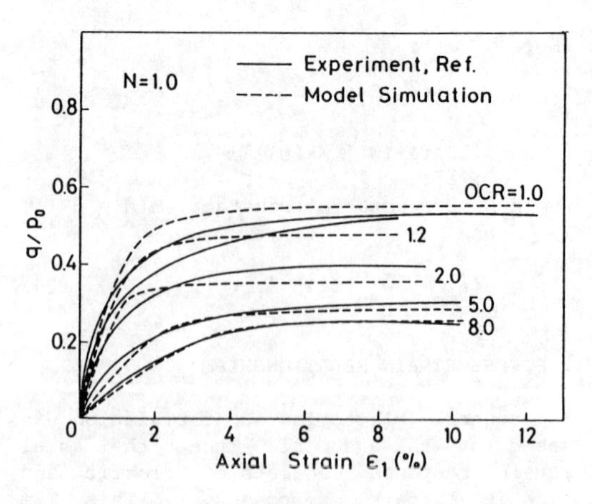

Fig.3 Relationships between stress and strain in triaxial compression tests

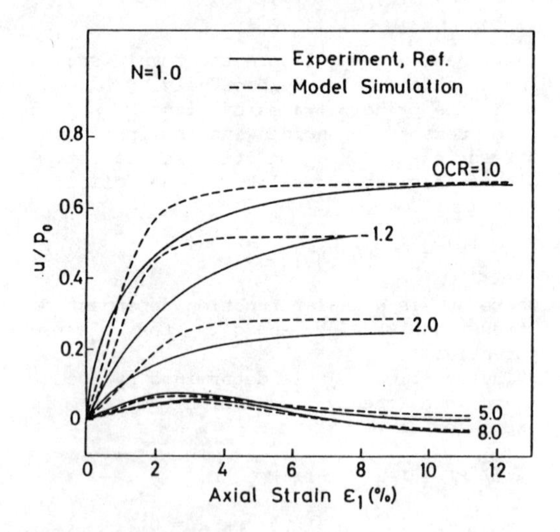

Fig.4 Relationships between strain and pore water pressure in triaxial compression tests

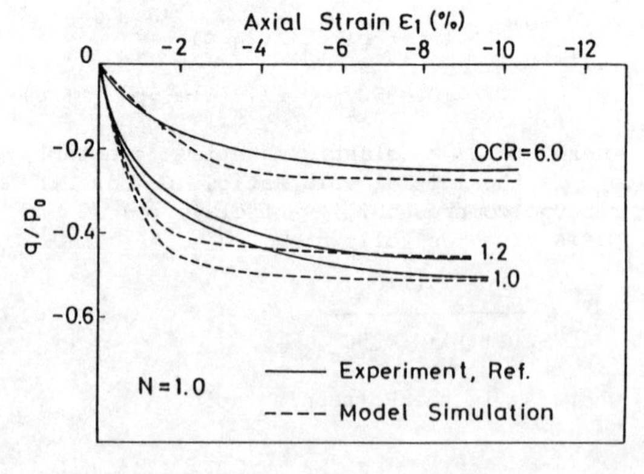

Fig.5 Relationships between stress and strain in triaxial extension tests

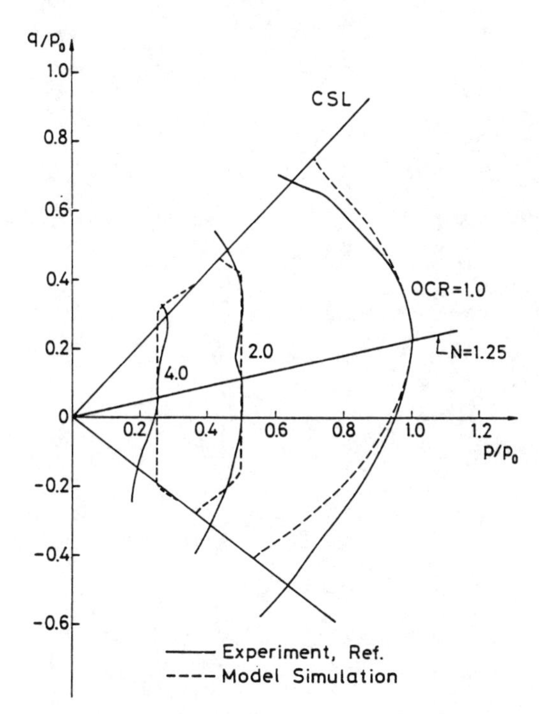

Fig.6 Effective stress paths in undrained test of anisotropically consolidated clay

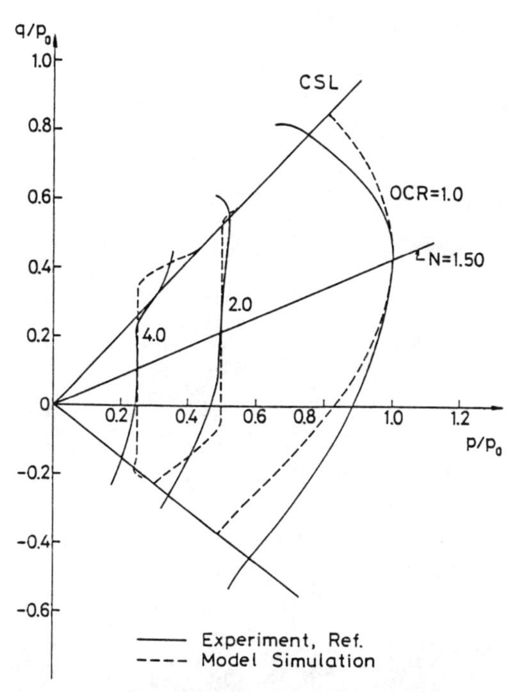

Fig.7 Effective stress paths in undrained test of anisotropically consolidated clay

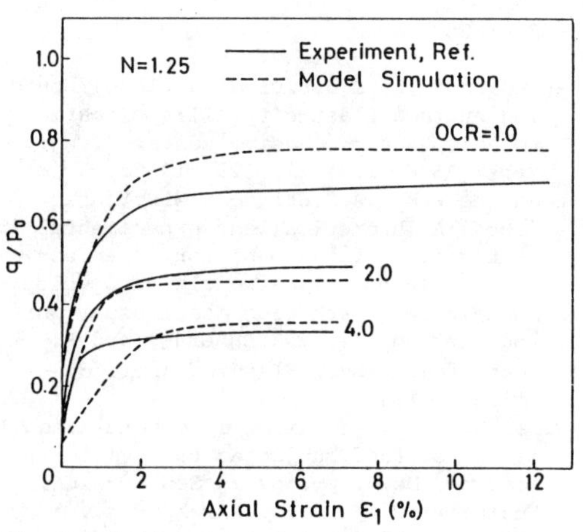

Fig.8 Relationships between stress and strain in triaxial compression tests

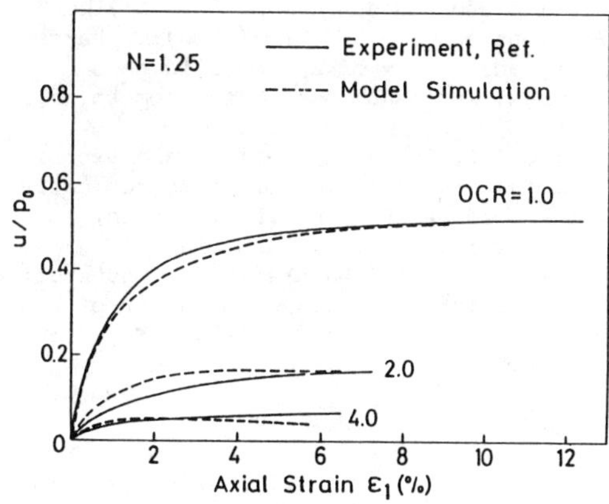

Fig.9 Relationships between strain and pore water pressure in triaxial compression tests

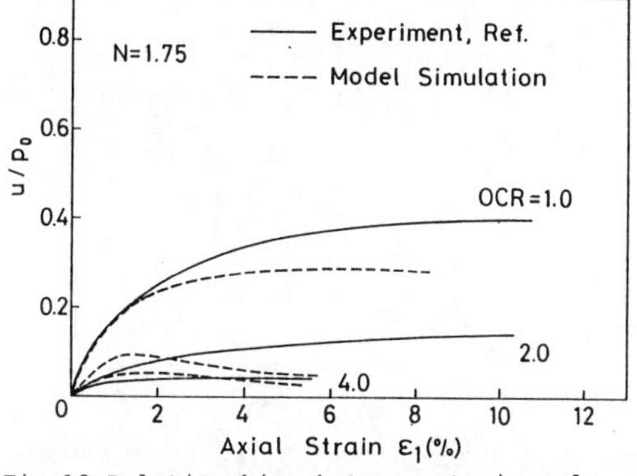

Fig.10 Relationships between strain and pore water pressure in triaxial compression tests

REFERENCES

Anandarajah,A. & Y.F.Dafalias(1986):Bounding surface plasticity.III:application to anisotropic cohesive soils, J.Eng. Mech. ASCE, 112(12), 1292-1318.

Banerjee,P.K., A.S.Stipho & N.B.Yousif (1984):A theoretical and experimental investigation of the behavior of anisotropically consolidated clays, Development in Soil Mechanics and Foundation Engineering, II, P.K.Banerjee and R. Butterfield, Eds.,Elsevier Applied Science, 1-41.

Hirai,H.(1987):An anisotropic hardening model for sand subjected to cyclic loading, Developments in Geotechnical Engineering, 42, A.S.Cakmak, Eds., Elsevier and Computational mechanics publications, 53-67.

Hirai,H.(1987):Modelling of cyclic behaviour of sand with combined hardening, Soils and Foundations, 27(2), 1-11.

Pietruszczak,ST. & Z.Mroz(1983):On hardening anisotropy of K_0-consolidated clays, Int.J.Num.Anal.Meth.Geomech. 7, 19-38.

Stipho,A.S.(1978):Theoretical and experimental investigation of anisotropically consolidated Kaolin, Ph.D. Thesis, University College, Cardiff.

Willam,K.J. & E.P.Warnke(1975):Constitutive models for the triaxial behavior of conrete, Int.Assoc.Bridge Struct.Eng., 19, 1-30.

Numerical Methods in Geomechanics (Innsbruck 1988), Swoboda (ed.)
© 1988 Balkema, Rotterdam. ISBN 90 6191 809 X

Three-dimensional shearing deformation of sand

N.Moroto
Hachinohe Institute of Technology, Japan

ABSTRACT: The author focused on dissipative energy during shearing process of granular material and has proposed the state function S_s called 'entropy of granular material'. In this paper, first he carried out the true triaxial test on Toyoura sand and then using the parameter S_s he made up a shearing deformation model for three dimensional stress system.

1 INTRODUCTION

The author focused on dissipative energy during shearing process of granular material such as sand and has proposed the state parameter S_s defined by

$$dS_s = \frac{\text{(increment of plastic work done)}}{\text{(confining pressure)}} \quad (1)$$

which can be regarded as an interesting measure of procession of the shearing deformation and called 'entropy of granular material' by the author.

The reason why the author introduced this parameter is as follows:

The large stress path dependency of the plastic work done due to shear is attributed to the fundamental nature of granular materials which are characterized by the deformation rules given by
(a) plastic shearing strain is strongly related to the stress ratio, not to the shear stress,
(b) plastic volumetric strain takes place due to dilatancy.
Thus, the author has emphasized the importance of using the parameter S_s given by Eq.(1) for the study of shearing deformation of sand(Moroto(1974),Moroto(1975),Moroto(1976),Moroto(1980),Moroto(1983),Moroto(1985),Moroto(1987)).

In this paper, he tries to make up a three dimensional deformation model by using the parameter S_s.

2 ENERGY EQUATION

The work done to the soil sample by external forces from without may be largely dissipated and stored to a limited extent:

$$dW = dU + dW^P$$

dW: increment of work done
dU: increment of recoverable elastic energy stored
dW^P: increment of plastic work done

The plastic work done may be separated into two components, dW^p_s and dW^p_c :

$$dW^p = dW^p_s + dW^p_c$$

dW^p_s: increment of the plastic work done due to shear
dW^p_c: increment of the plastic work done due to consolidation

3 S_s IN GENERAL THREE DIMENSIONAL SYSTEM

In general three dimensional stress system, the author's parameter S_s is defined as

$$dS_s = dW^p_s \,/\, p \quad (2)$$

Then,

$$dS_s = dv_d + \frac{3}{2}\,\alpha\frac{\tau_{oct}}{p}\,d\gamma_{oct} \quad (3)$$

$$dS_s = dv_d + \alpha\eta d\varepsilon \quad (4)$$

$$\alpha = (3+\mu\nu)/\sqrt{(3+\mu^2)(3+\nu^2)} \quad (5)$$

$$\mu = (2\sigma_2-\sigma_3-\sigma_1)/(\sigma_1-\sigma_3) \quad (6)$$

$$\nu = (2d\varepsilon_2-d\varepsilon_3-d\varepsilon_1)/(d\varepsilon_1-d\varepsilon_3) \quad (7)$$

$p=(\sigma_1+\sigma_2+\sigma_3)/3$

dv_d: increment of volumetric strain due to dilatancy (plastic)

τ_{oct}: octahedral shear stress

$d\gamma_{oct}$: octahedral shear strain increment (plastic)

$\eta = q/p$

$q = \frac{3}{\sqrt{2}}\tau_{oct}$

$d\varepsilon = \frac{1}{\sqrt{2}} d\gamma_{oct}$

The stresses are all the effective stress.

4 TEST RESULTS

The author and Okamoto carried out the true triaxial test on Toyoura sand under drained condition in keeping the mean principal stress constant($p=1.0\mathrm{kgf/cm^2}$). The sand was tested in air dried condition. The phisical properties of the material are as follows;

Specific gravity of grains $G_s=2.65$
Uniformity coefficient $U_c=1.5$
Minimum void ratio $e_{min}=0.60$
Maximum void ratio $e_{max}=0.96$

The test was run for the three cases of the initial relative densities:

DENSE : $D_r = 0.84$
MEDIUM : $D_r = 0.55$
LOOSE : $D_r = 0.18$

The normal pressures were applied to each face of cuboidal specimen(10x10x10cm) in stress controled manner(Okamoto,1978).

4.1 Experimental results on ν and μ

For each μ- constant tests(-1 ∿1 in μ), the average value of ν was caluculated for all the deformational stages except the initial and the results are shown in Fig.1. From these values,one can easily know that the coefficient α given by Eq.(5) remains in the range of 1.00-0.98. Thus, one can reasonably put that

$$\alpha = 1.0 \tag{8}$$

empirically

Mathematically, the coefficient α becomes unity in the case $\nu = \mu$, however,in the frictional material like sand, the value of ν deviates from that of μ . Neverthless, the test result indicates the avilability of Eq.(8). From this evidence, Eqs.(3) and (4) can be simply rewritten as

$$dS_s = dv_d + \frac{3}{2}\frac{\tau_{oct}}{p} d\gamma_{oct} \tag{8}$$

$$dS_s = dv_d + \eta d\varepsilon \tag{9}$$

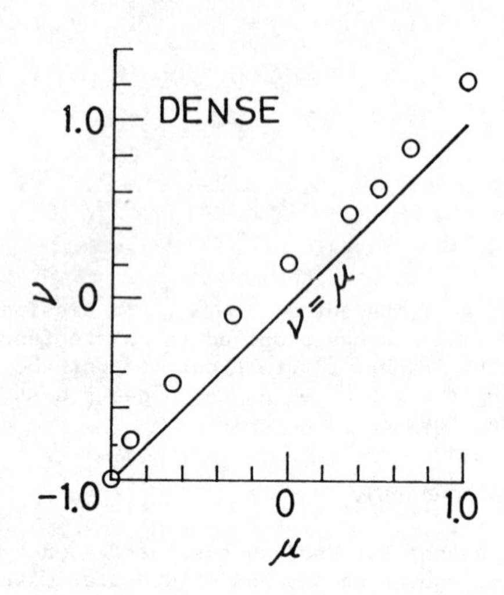

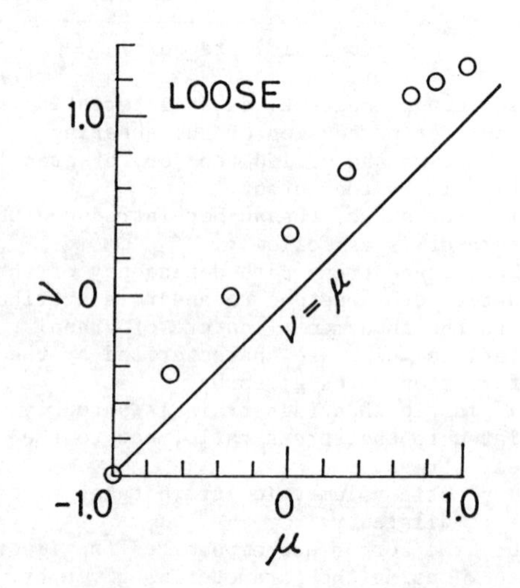

Fig.1 Relationship between ν and μ

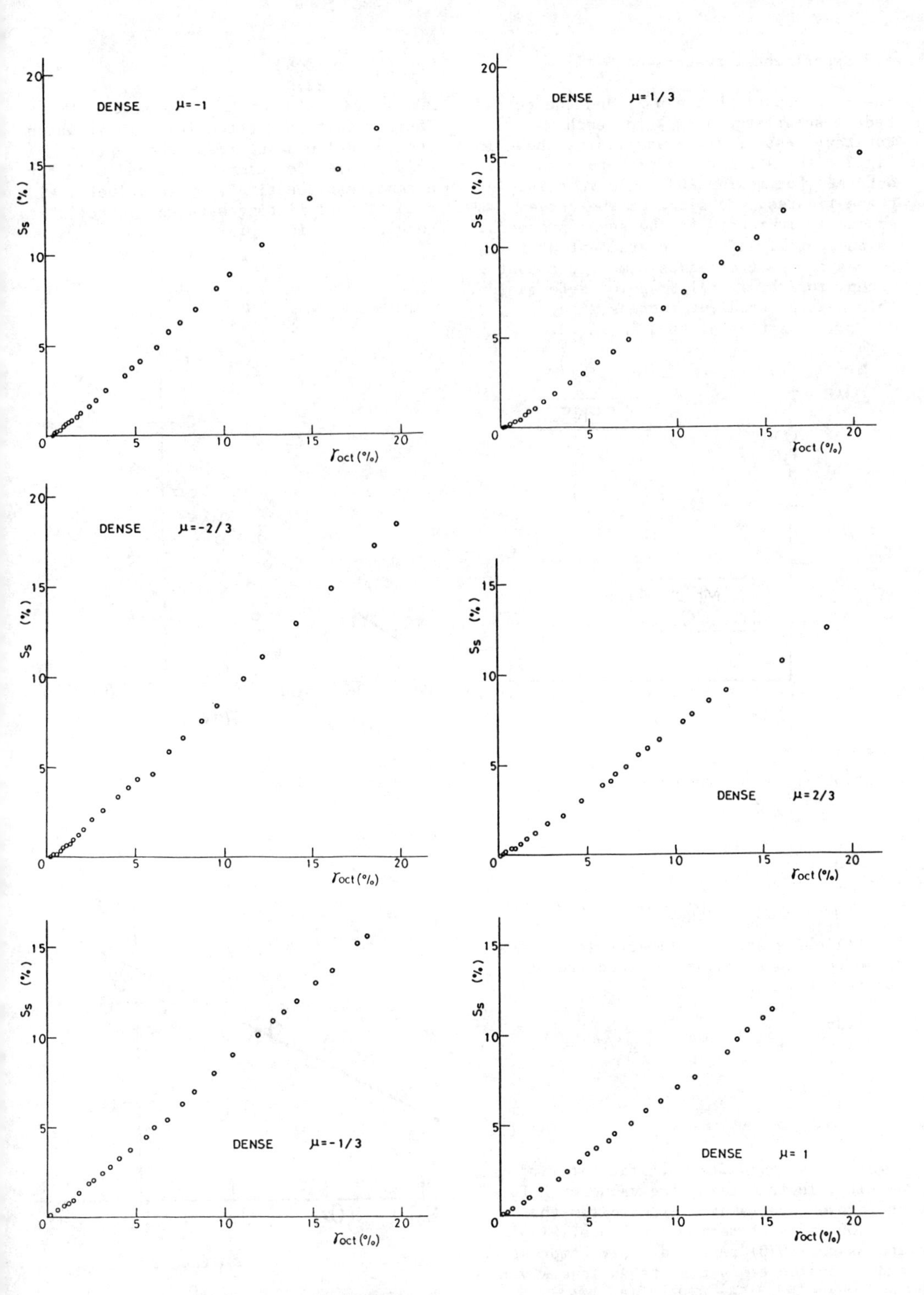

Fig.2 Relationship between S_s and γ_{oct}

4.2 Experimental results on S_s

The relationship between S_s and the octahedral shear strain γ_{oct} for each μ - constant test of dense samples is shown in Fig.2. The similar figures are also obtained for medium and loose samples. In these figures, the slope of the curves gradually increases as the shear proceeds. As an approximation, the gradient of the curves can be treated as almost constant except for the intial stage of deformation. This average gradient expressed by $dS_s/d\gamma_{oct}$ vs μ is plotted in Fig.3.

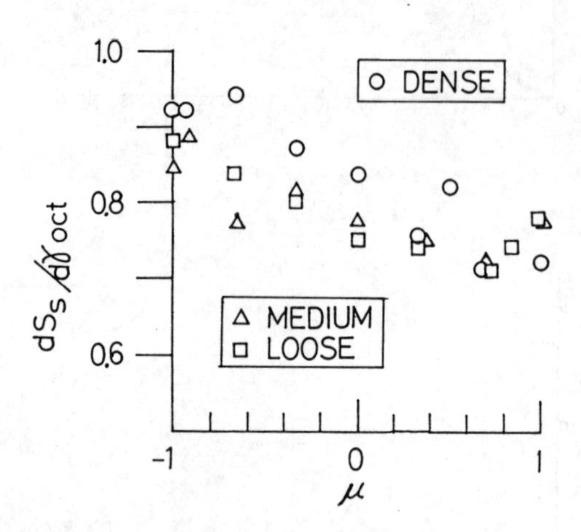

Fig.3 Relationship between $dS_s/d\gamma_{oct}$ and μ

If we write

$$M/\sqrt{2} = dS_s / d\gamma_{oct}$$

the following stress ratio-strain increment ratio equations can be obtained from Eqs. (8) and (9):

$$\frac{dv_d}{d\gamma_{oct}} + \frac{3}{2}\frac{\tau_{oct}}{p} = M/\sqrt{2} \qquad (10)$$

$$\frac{dv_d}{d\varepsilon} + \eta = M \qquad (11)$$

where $M=M(\varepsilon,\mu)$ for specified initial density. In this test, the value of M for the dense case takes a bit greater than for the cases of medium and loose. Nishi and Esashi(1978) reported their compression and extension tests results on Tone River sand whose physical properties are:

G_s : 2.70
e_{max} :0.99
e_{min} :0.63
U_c :2.0

,the initial density is in a medium dense. Their result is plotted in Fig.4 in which the solid lines intercept the vertical axes at 0.61 for compression and 0.48 for extension,respectively.These values yield a sufficient fitting with the author's result shown in Fig.3

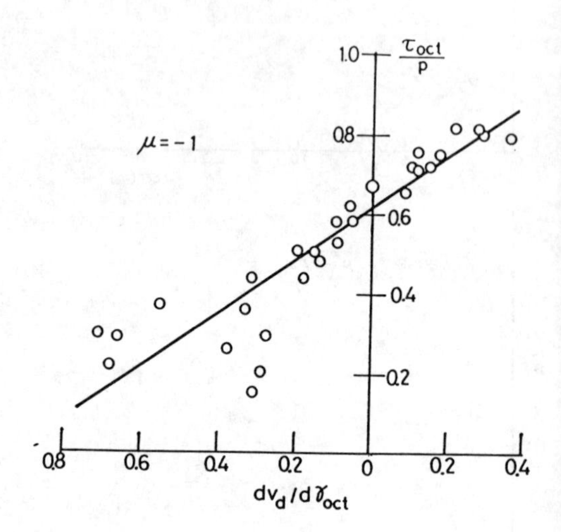

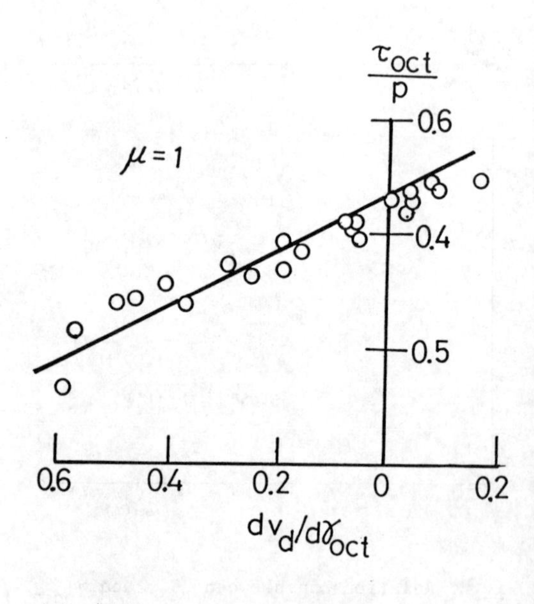

Fig.4 Stress ratio and dilatancy rate plot (Nishi and Esashi(1978))

4.3 Equi-S_s line

If sand specimens subjected to shear arrive at an equi-S_s line along different deformation paths, they have been accompanied by an equal amount of 'entropy production'. In Fig.5, this equi-S_s line is illustrated.

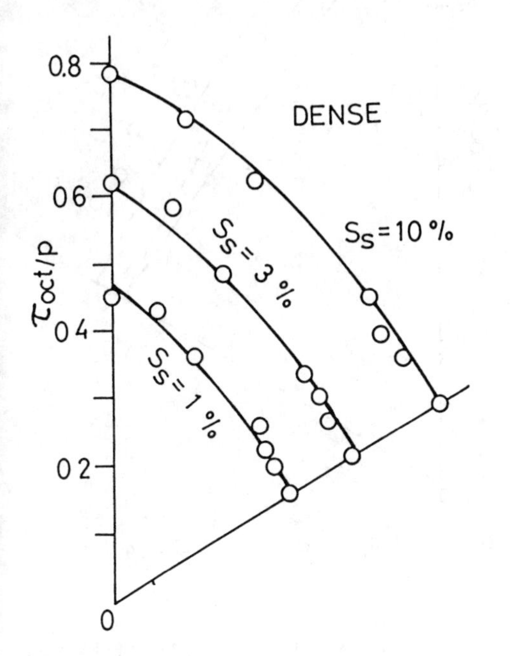

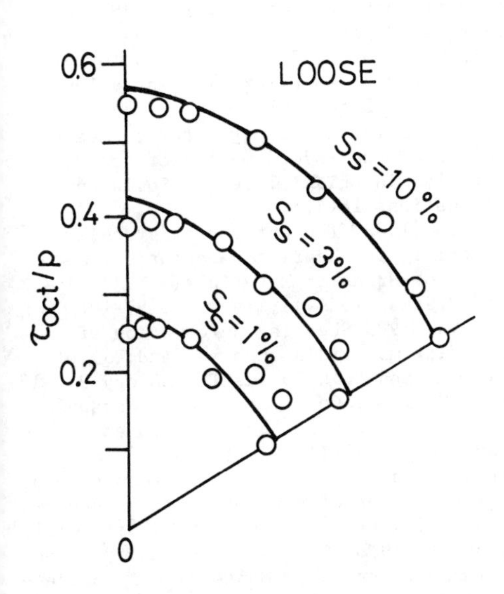

Fig.5 Equi-S_s lines

5 CONSTITUTIVE MODEL

For a given μ, from Eq.(11) one can write

$$\frac{dv_d}{d\varepsilon} + \eta = M(\varepsilon) \tag{12}$$

In the case that the constant $M(\varepsilon)$ can be written only by the stress ratio η, the plastic potential ψ is obtained as

$$\psi = \int \frac{d\eta}{M(\eta)} + \log_e p = \text{const.} \tag{13}$$

which is identical with Cambridge's one (Schofield and Wroth,1968) in form. The second order derivative of plastic work done can be written as

$$dW_s^2 /p = dS_s \, d\psi \tag{14}$$

Thus, the deformation model can be made up by

$$d\varepsilon_i = p \frac{\partial \psi}{\partial \sigma_i} dS_s. \tag{15}$$

in which S_s plays a role of yield function. The material stability is governed by the sign of $d\psi$. It is interesting to note that:

(a) The flow rule expressed by Eq.(15) is similar to the one in non-associate theory of soil plasticity,

(b) The material stability is determined simply by Eq.(14)

(c) The yield condition is specified with dS_s=0 and the yield loci are shown in Fig. 5. These yield loci are compared with those proposed by Lade and Duncan(1975),Matsuoka and Nakai(1974) and also of Mohr-Coulomb in Fig.6. This figure shows that the inherent anisotropy in an early stage of deformation and the induced anisotropy thereafter may not be overlooked.

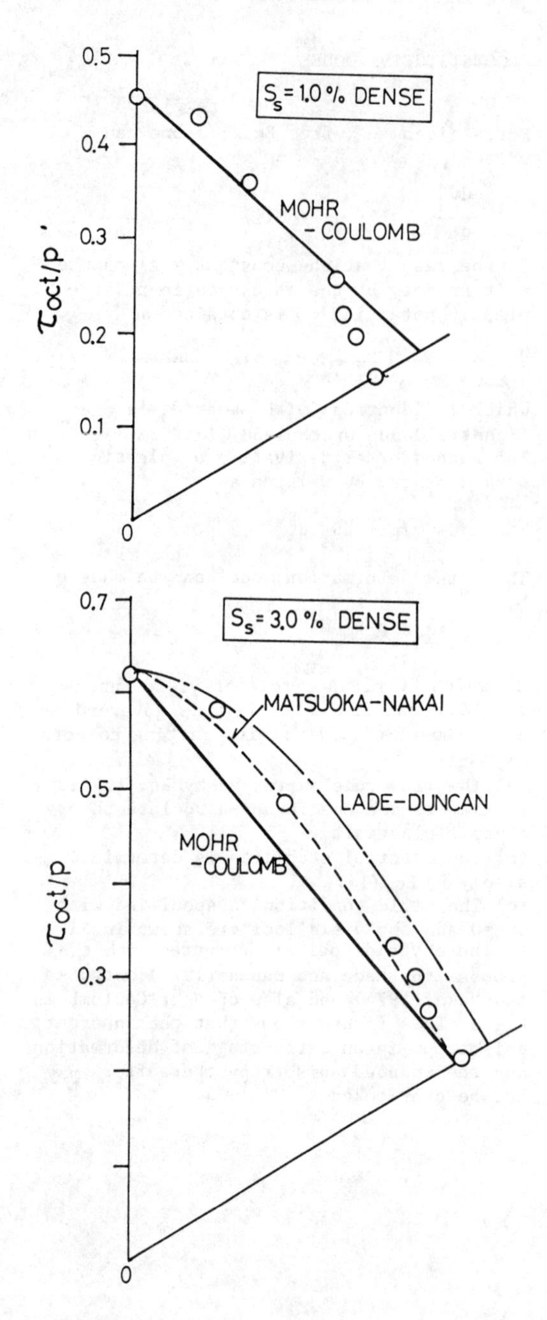

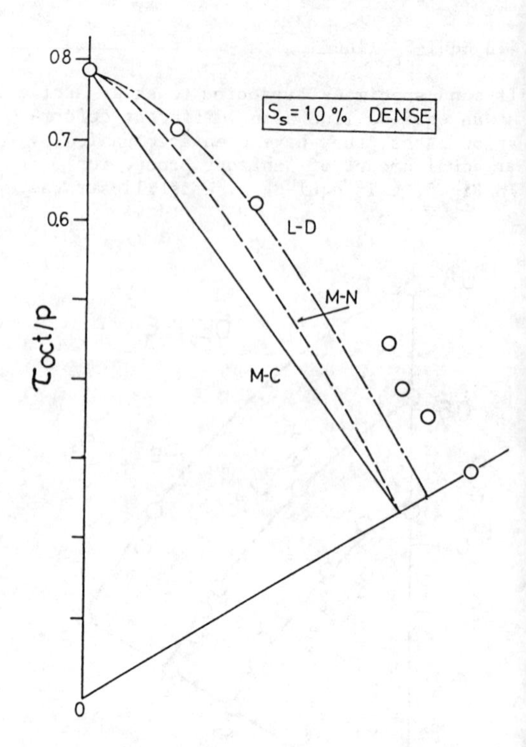

Fig.6 Equi-S_s and other yield loci

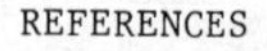

REFERENCES

Lade,P.V. and Duncan,J.M. 1975. Elasto-plastic stress-strain theory for cohesionless soil. Proc.ASCE, GT Div.Vol.101, No.10: 1037-1053

Matsuoka,H. and Nakai,T. 1974. Stress-deformation and strength characteristics of soil under three different principal stress. Proc.JSCE 232:59-70

Moroto,N. and Kawakami,F. 1974. State function of sand deformation. Proc.JSCE 229: 77-85(in Japanese)

Moroto,N. et al. 1975. Deformation of granular material in triaxial compression. Proc.JSCE 239:47-56

Moroto,N. 1976. A new parameter to measure degree of shear deformation of granular material in triaxial tests, Soils and Foundations 16(4):1-9

Moroto,N. 1983. The entropy of granular materials in shearing deformation, Mechanics of Granular Materials(J.T.Jenkins and M.Satake(eds.)),Elsevier:187-194

Moroto,N. 1985.Shearing deformation of granular materials such as sand. J. of Powder and Bulk Solids Technology 9:7-18

Moroto,N. 1987. On deformation of granular material in simple shear. Soils and Foundations 27(1): 77-85

Nishi,K. and Esashi,Y. 1978. Stress-strain relationships of sand based on elasto-plasticity theory, Proc.JSCE 280:111-121

Okamoto,T. 1978. Shear of sand under three principal stresses. Master Thesis presented to Tohoku University

Schofield,A.N. and Wroth,C.P. 1968. Critical State Soil Mechanics. MaGraw Hill.

Numerical Methods in Geomechanics (Innsbruck 1988), Swoboda (ed.)
© 1988 Balkema, Rotterdam. ISBN 90 6191 809 X

Model of sand behaviour towards shearing and compressibility in three-dimensional conditions of stress and strain

S.Chaffois
Laboratoire de Géotechnique INSA, Lyon, France
J.Monnet
Grenoble University, France

ABSTRACT : This theoretical model concerns sand behaviour when true triaxial shearing, and compressibility occur. The main interest is the low number of parameters which always have a physical meaning. This model uses only 8 constant parameters for over-consolidated soil, and 6 for normaly consolidated soil. The behaviour of the sand is assumed to be elastic in an area limited by an elliptical cap ; beyond this cap, "standard" plasticity occurs in the compressible area, and "non-standard" plasticity in the dilatancy area. Theoritical results from the conventional triaxial test with a constant stress ratio $\sigma 1/\sigma 3$ are compared with the experimental results.

1) INTRODUCTION

The computer technology is improving more and more. It allows to conduct large computations for sheap price. Engineers who are in charge of a construction will be able to use large computers to predict the behaviour of buildings and fondations. It is necessary to built constitutive equations for the soil behaviour, which must be used in finite element programs.

The theoritical model must handle only physical parameters which can be fit by engineers. This model is built from energitic assumptions and uses only 8 physical constants for the overconsolidated soil.

2) Theoretical modelling

2-1) Hypothesis

- the soil is non-cohesive and isotropic, normally consolidated.
- the soil is compressible in an area limited in the stress space by a conical surface : Beyond this surface, dilatancy occurs, governed by the Chaffois and Monnet (1985) law.
- there is no volume variation on the three-dimensional characteristic surface.
- a three-dimensional elliptical cap separates the elastic region from the plastic region in the compressible area.
- in the compressible area there is a linear relation between the void ratio and the logarithm of main stress.
- the three dimensional conical surface which governs the failure, uses a quite similar formulation than the characteristic surface.
- stress are negative in compression.

2.2) Compression theory

2.2.1) Three-dimensional surface limiting the compressive area

Chaffois and Monnet (1985) have shown that the three-dimensional surface where no volume occurs, and which limits the dilatancy region and compressive region, can be fitted by the LADE (1972) formula :

$$f(\sigma) = -(\sigma_1+\sigma_2+\sigma_3)^3 + L_1.\sigma_1.\sigma_2.\sigma_3 \quad (1)$$

where L1 is a constant. (See figure 1).

Monnet and Gielly (1979), have shown that in conventional triaxial conditions ($\sigma 2 = \sigma 3$), in compression the shearing ratio is related to the interparticulate angle of friction :

$$-\frac{\tau_{oct}}{\sigma_{oct}} = tg\ \phi_\mu \quad for \quad \frac{\Delta \dot{v}}{v} = 0$$

and is equal to the "characteristic state" of Luong (1978).

There is a single mathematical relation between the value of ϕ_μ and L1, such that if ϕ_μ is known, L1 can be computed.

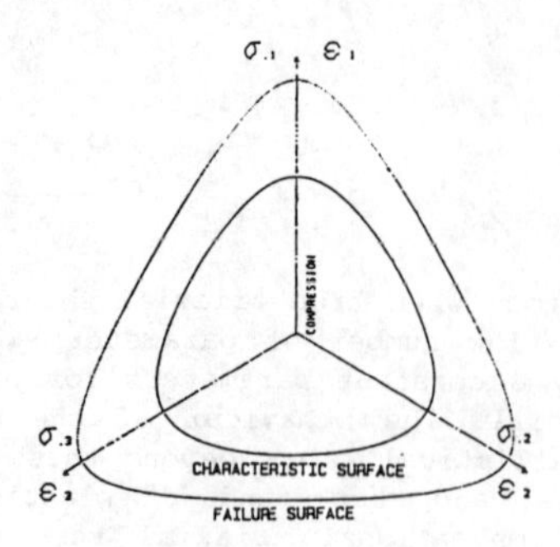

Fig. 1 Section in the π plane of the Lade surface.

2.2.2.) Void ratio variation and Young's modulus determination.

It was assumed that the void ratio variation in terms of the logarithm of $-\sigma_{oct}$ is a straight line of slope $-$ Cs in the elastic region and a straight line of slope $-$ Cc in the plastic region.

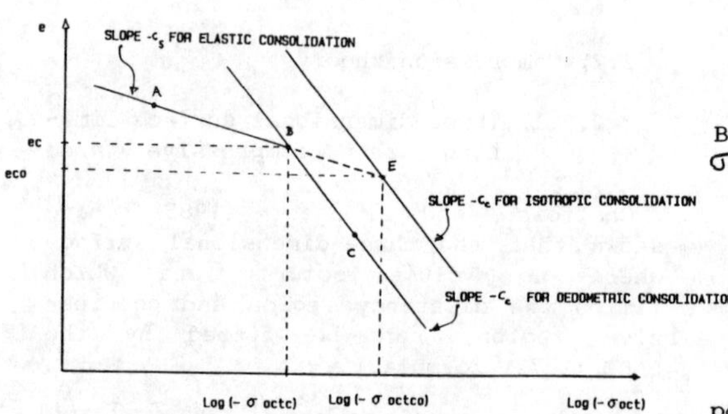

Fig. 2 The CAM CLAY concept in the consolidation area.

The cap model that limits the elastic region is an ellipse (figure 3)

When the soil is in an elastic condition, the point in the : e-log($-\sigma_{oct}$) plane is in A on a straight

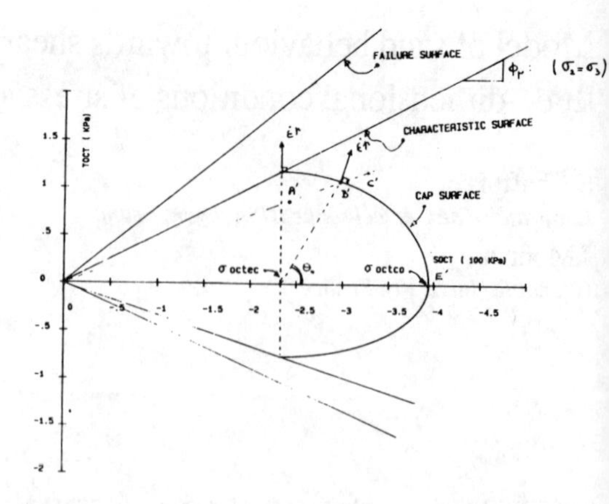

Fig. 3 The cap model used.

line of slope -Cs. When plasticity begin, the point is on the ellipse in B' on the τ_{oct} σ_{oct} plane, and at point B on the e-log ($-\sigma_{oct}$) plane. The plastic potential is an iso work three-dimensional curve. With a single large diameter of σ_{octco} and a small diameters function of τ_{oct}, σ_{oct} and L1.

When plasticity occurs, the ellipse grows until point C' is in the τ_{oct}-σ_{oct} plane, and the point is displaced along the line of slope -Cc in C.

It is assumed that Hooke's law of elasticity applies in this area

$$E = -\frac{3(1-2\nu)(1+e)\sigma_{oct}}{C_s \log(1+\Delta\sigma_{oct}/\sigma_{oct})} \qquad (2)$$

The elastic variation between B and C is also a function of Cs, σ_{oct} and can be computed by (2).

2.2.3) Plastic potential of compression

In the compressive area, the plastic potential is a three-dimensional surface of revolution. This surface, cut by the τ_{oct} - σ_{oct} plane, gives 2 straight lines. (See fig. 3).

Cut by the π plane, this surface gives the inner curve of fig. 1, corresponding to equation (1) where L1 is related to ϕ_μ.

The upper ellipse in figure 3 has its maximum axis at σ_{oct} co, and minimum axis at τ_{oct}, σ_{oct} ec.

To determine the parameters of the ellipse, conventional triaxial conditions are assumed, that is $\sigma_2 = \sigma_3$.

The slope of the upper line is then ϕ_p and :

$$-\frac{\tau_{oct}}{\sigma_{oct}} = tg\,\phi_p$$

At point D', there is no volume variation, and $\dot{\varepsilon}^p$ is normal to the axis σoct, and σoct ec is the center of the ellipse.

The equations of the ellipse in cylindrical coordinates are :

$$\sigma_{oct} = (\sigma_{oct\,co} - \sigma_{oct\,ec})\cos\theta + \sigma_{oct\,ec} \quad (3)$$

$$\tau_{oct} = \sigma_{oct\,ec} \cdot tg\,\phi_{ec} \cdot \sin\theta \quad (4)$$

and the following are obtained :

$$-\frac{\tau_{oct\,ec}}{\sigma_{oct\,ec}} = tg\,\phi_{ec} \quad (5)$$

For E' point on figure 3 we have :

$$\sigma_{oct\,co} = \alpha \cdot \sigma_{oct\,ec}\,;\ \alpha = 1 + \frac{\sqrt{2}\,tg\,\phi_{ec}}{tg\,\theta_0} \quad (6)$$

Equation (6) gives the relation between two vary important parameters of the ellipse : σoct co and σoct ec

Equations allows the determination of θ_0 in the oedometric conditions by the equation :

$$\sin\theta_0 = \frac{\sqrt{2}(1-2\nu) + \sqrt{2(1-2\nu)^2 + 24(1-\nu)(1-2\nu)tg^2\phi_{ec}}}{6(1-\nu)\,tg\,\phi_{ec}} \quad (7)$$

The cartesian equation of the plastic potential is a function of τoct, σoct, τoct ec, σoct ec and we find σoct ec by :

$$\sigma_{oct\,ec} = \frac{\sigma_{oct} + \sqrt{\sigma_{oct}^2 - \alpha(2-\alpha)\left(\sigma_{oct}^2 + \frac{(\alpha-1)^2\tau_{oct}^2}{tg^2\phi_{ec}}\right)}}{\alpha(2-\alpha)} \quad (8)$$

Along the plastic potential defined by σoct ec, σoct co and τoct ec (equations 5, 6, 8) plastic work is the same at any point. Considering a current point defined by τoct, σoct and the isotropic consolidation point define by σoct co, the following may be written :

$$3\sigma_{oct} \cdot \varepsilon_{oct} + 3\,\tau_{oct} \cdot \varepsilon'_{oct} = 3\,\sigma_{oct\,co}\,\varepsilon_{oct\,co}$$

Or

$$g(\sigma, \varepsilon) = -\sigma_{oct} \cdot \varepsilon_{oct} - \tau_{oct} \cdot \varepsilon'_{oct} + \sigma_{oct\,co}\,\varepsilon_{oct\,co} \quad (9)$$

$g(\sigma, \varepsilon)$ is the plastic potential ; $g(\sigma, \varepsilon)$ is negative in elasticity and equal to zero in plasticity.

2.2.4) Flow rule of compression

The normality condition apply to the elleptical can gives the flow rule :

$$\frac{\dot{\varepsilon}'^p_{oct}}{\dot{\varepsilon}^p_{oct}} = \frac{(\alpha-1)\sqrt{\alpha\,\sigma_{oct\,ec}^2(\alpha-2) + \sigma_{oct}(2\sigma_{oct\,ec} - \sigma_{oct})}}{tg\,\phi_{ec}(\sigma_{oct} - \sigma_{oct\,ec})} \quad (10)$$

2.3) Theory of shearing and dilatancy

2.3.1) Flow rule

When the stress path goes beyon the characteristic surface, dilatancy occurs. The increment of plastic work gives :

$$3\sigma_{oct} \cdot \dot{\varepsilon}^p_{oct} + 3 \cdot \tau_{oct} \cdot \dot{\varepsilon}'^p_{oct} \cos(\gamma - \gamma') = -3\sigma_{oct} \cdot \dot{\varepsilon}'^p_{oct}\,tg\,\phi_{ec} \cdot \cos(\gamma - \gamma') \quad (11)$$

where γ is the lode angle of stress and γ' is the lode angle of plastic strain. The equality between the real increment of plastic work and the assumed one's gives the flow rule :

$$\frac{\dot{\varepsilon}^p_{oct}}{\dot{\varepsilon}'^p_{oct}\cos(\gamma - \gamma')} = -\frac{\tau_{oct}}{\sigma_{oct}} + tg\,\phi_{ec} \quad (12)$$

2.3.2.) Yield finition

The total plastic work is assumed to be due to a sliding of particulate in the octaedrical plane, so :

$$3\sigma_{oct} \cdot \varepsilon^p_{oct} + 3 \cdot \tau_{oct} \cdot \varepsilon'^p_{oct} = -3\,\sigma_{oct}\,\varepsilon'^p_{oct}\,tg\,\phi_{ec}$$

This relation rewrited, gives the yield function :

$$h(\sigma, \varepsilon) = \tau_{oct} + \sigma_{oct}\left(tg\,\phi_{ec} + \frac{\varepsilon^p_{oct}}{\varepsilon'^p_{oct}}\right) \quad (13)$$

3) Comparison between theory and experiment for the consolidation of sand

Vuez et al. (1980) have conducted a lot of experimental tests on a Leucate medium sand of unit weight 17 KN/m3.

We extract of the experimental programme a series of 5 tests which have been made with controlled consolidation. The device, was a conventional triaxial test, but a computer was used to adjust the strain. The consolidation process was done with a constant ratio $\sigma'1/\sigma'3$ from zero, until $\sigma'3$ reached the imposed value of 153 KPa. Then the water pressure $\sigma'3$ was kept constant and $\sigma'1$ was increased to failure point of the sample.

For computations, the theoritical model neads only 8 geotechnical constants that are :

- eo Initial void ratio
- σ_{co} overconsolidation stress
- ec the void ratio which corresponds to σ_{co}
- Cs the slope of the "elastic" part of the e-logσoct graph
- Cc the slope of the "plastic" part of the e-logσoct graph

All these 5 parameters are shown in figure 10 :

- ν Poisson's ratio
- ϕ the internal of friction
- ϕ_μ the interparticulate angle of friction

The parameters ν and ϕ are well known, ϕ_μ is shown in figure 3.

We have written a Pascal programme for computing these parameters from conventional or true triaxial tests, and a special set of parameters for each test or mean parameters can be found.

Table 1 : The set of parameters used for computation

Test numbers	eo	σ_{co} 100 kPa	ec	Cs	Cc	ν	ϕ	ϕ_μ
	Special sets of parameters							
12.1	0,546	-0,6	0,538	0,018	0,051	0,36		
12.2	0,563	-0,6	0,554	0,029	0,067	0,37		
12.3	0,513	-0,7	0,509	0,014	0,058	0,42	39,8°	30,2°
12.7	0,557	-1	0,549	0,023	0,093	0,0419		
12.9	0,514	-0,6	0,499	0,011	0,011	0,392		
CREDS 5	0,538	-5,05	0,529	0,005	0,023	0,315	43°	30,9°
	Mean parameters							
From	eo	σ_{co} 100 kPa	ec	Cs	Cc	ν	ϕ	ϕ_μ
12.1 to 12.7	0,554	-0,6	0,5396	0,020	0,066	0,392	39,8	30,2

The experimental programme of Vuez et al. is computed with our theoritical model and results are shown in figures 4 to 11. If we except test 12.9, differences of 3 % on strain between the theoritical results and the experimental ones can be observed when the mean parameters are used, and of less than 2 % when the special set of parameters are used.

On examining the stress strain curves, a change of shape of the curve cannot be found when the stress path goes from the constant ratio consolidation to the triaxial shearing path ($\sigma 3$ close to 200 or 300 kPa). But for all the tests an increasing of the curvature can be found near the maximum shearing value. Note that for all tests a correct curve shape along the constant ratio consolidation can be found on examining the volume variation, it is found that theory and experiment correspond well. The correct volume dilatation for all tests is observed. In tests 12.1, 12.2 and 12.3 the volume compaction is computed very well, but for tests 12.7 and 12.9 there is a difference of 2 % in the evaluation or the volume variation.

An oedometric test made on a true triaxial apparatus for the Aussois meeting (1984) has also been computed.

The determination of parameters Cs, Cc eo and ec, $\sigma_{c\sigma}$ is shown in figure 10, and a comparison between experiment and calculus in figure 11 has been made. A close correspondance between the two curves, with differences that are less than 0,2 % of strain may be observed.

4) CONCLUSION

This paper has presented a new theoretical model which uses only simple geotechnical parameters, easily measurable by classic laboratory tests. These are only 8 constant parameters :

- ν is related to the elastic properties of the soil and can be measured with the conventional triaxial test or by Ko determination.
- ϕ the internal angle of friction, and ϕ_μ, the interparticulate angle of friction, are related to the shearing characteristics of the soil. They can be quantified with the conventional triaxial test.

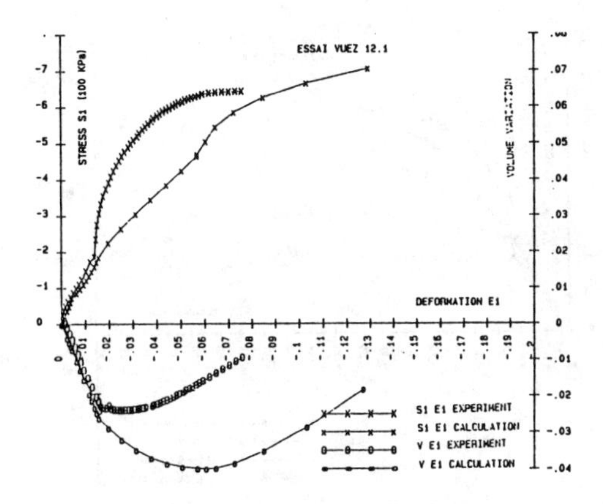

Figure 4 Test Vuez 12.1 with mean parameters

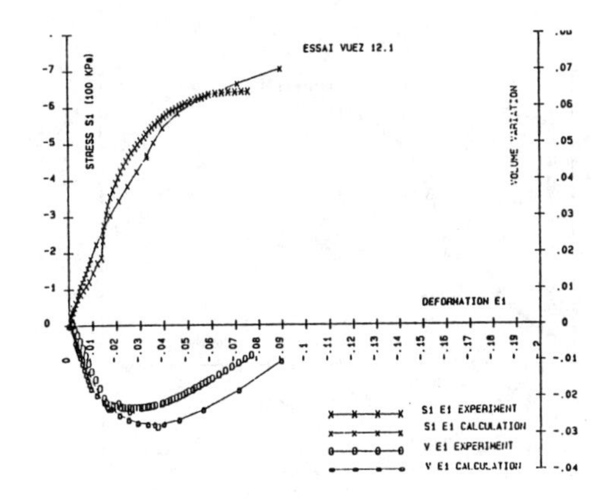

Figure 5 Test Vuez 12.1 with special parameters

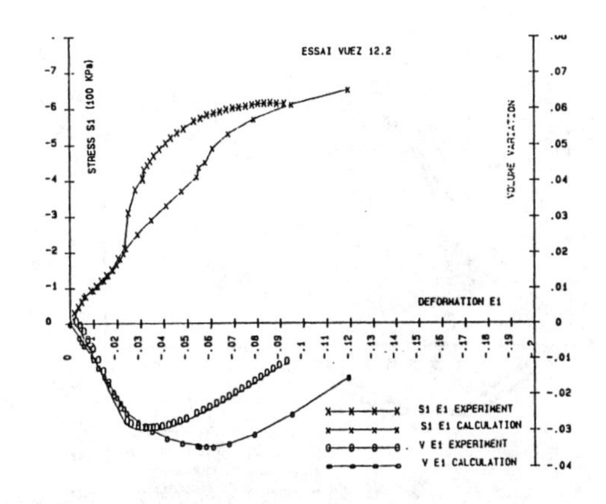

Figure 6 Test Vuez 12.2 with mean parameters

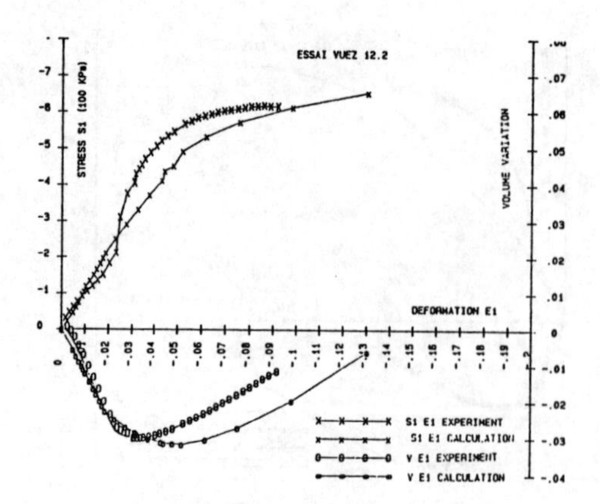

Figure 7 Test Vuez 12.2
with special parameters

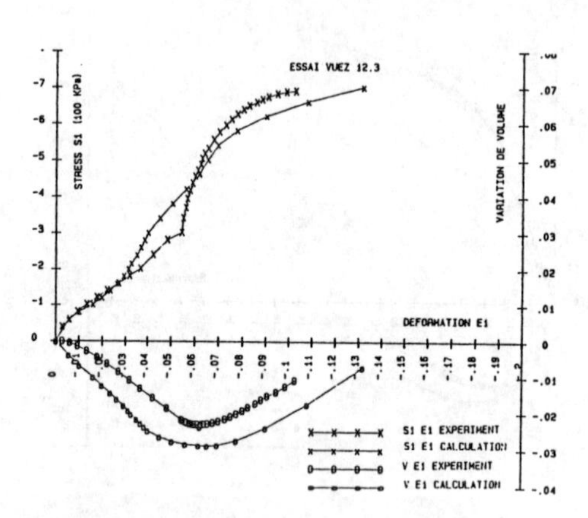

Figure 8 Test Vuez 12.3
with mean parameters

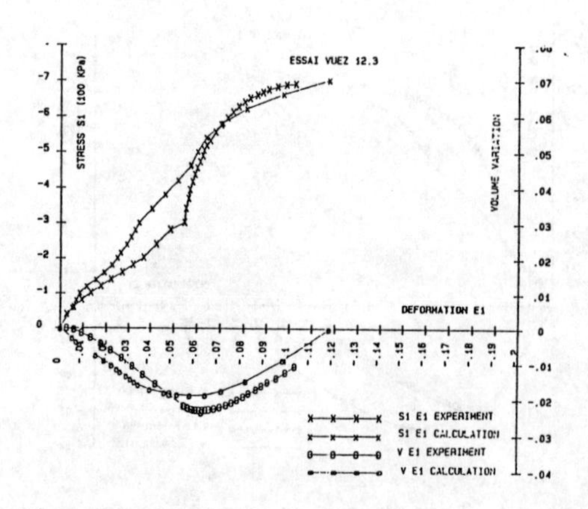

Figure 9 Test Vuez 12.3
with special parameters

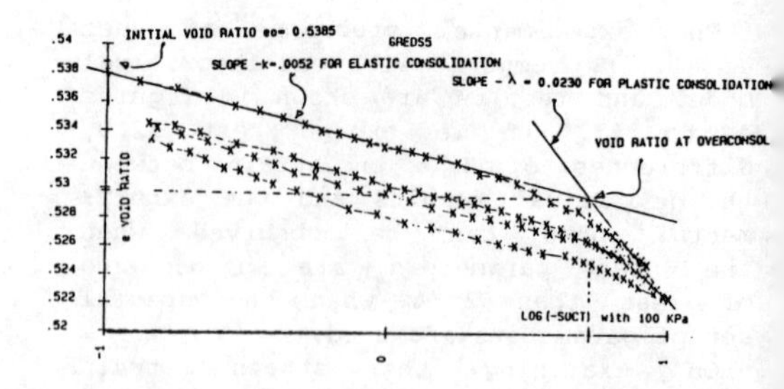

Figure 10 Determination of 5 parameters for the oedometric tree dimensional test.

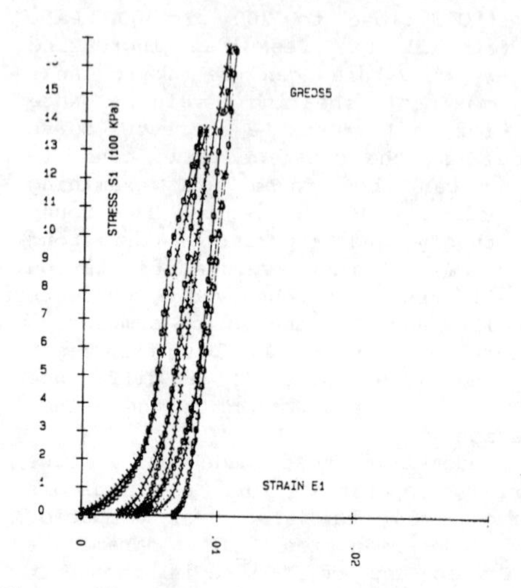

Figure 11 Comparison between experience and calculation for the oedometric tree dimensional test.

- e_σ the initial void ratio, Cc and Cs the slopes of the : e-log($-\sigma_{oct}$) straight line in plastic and elastic contitions. $\sigma_{c\sigma}$ the over consolidation pressure and ec the void ratio at the overconsolidated pressure, can all be determined easily by an oedometer test, or by a conventional triaxial device with isotropic consolidation.

This model is able to compute the behaviour of the non-cohesive soil along a three-dimensional stress path which gives either consolidation or dilatation.

Consolidation plasticity is modelled by the normality condition with an elliptical plastic potential, where as dilatancy is modelled by a "non-standard" plasticity. The elliptical plastic potential is normal to the plastic strain along the oedometric stress path, and gives no volume variation along the three dimensional surface which separates, the compressive area from the dilatancy area. In the τ_{oct}, σ_{oct} plane of the conventional triaxial test, this surface gives a straight line, which is the "characteristic line" of Luong (1978).

This new formulation provides close agreement between theoritical and experimental results for straight stress paths, but needs some improvement to take into account the change of direction of the stress path.

BIBLIOGRAPHY

Aussois, 1984. Ecole d'hivers Rhéologie des Géomatériaux - CNRS - IMG Grenoble.

Burland, 1965. The yielding and dilatation of clay Geotechnique n° 2 p. 211-214.

Burland, 1969. Deformation of soft clay beneath loaded areas. Int. Conf. Mexico Vl p. 55-63.

Chaffois, Monnet, 1985. Modèle de comportement du sable au cisaillement dans un état tridimensionnel de contrainte et de déformation. Revue Française de Géotechnique n° 32 p. 59-69.

Chaffois, Monnet, 1986. A theoritical model using a few number of parameters. 2 Int. Symposium on Num. Models in Geom. Ghent. p. 99-104.

Lade, 1972. The stress strain characteristics of coherionless soils. Phi. theses Berkeley.

Luong, 1978. Etat caractéristique du sol. C.R. Acad. Sciences. Paris 287 n° 15 p. 305-307.

Monnet, 1982. Calcul au cisaillement du sable sollicité en déformation plane. Revue Française de Géotechnique n° 7 p. 41-55.

Monnet, Gielly, 1979. Determination d'une loi de comportement pour le cisaillement du sol pulvérulent. revue Française de Géotechnique n° 7 p. 45-66.

Ohmaki, 1979. Strenght and deformation characteristics of overconsolidated cohesive soil. 3 Int. Conf. Num. Meth. Aachen Vl p. 465-474.

Nova, Wood, 1979. A constitutive model for sand in triaxial compression. Int. J. Num. Meth. Geom. Vol. 3 1979 p. 255-278.

Roscoe, Schofield, Wroth, 1958. On the yielding of soil - Geotechnique 8 - n° 1 p. 22-58.

Sture, Desai, Janardhanam, 1979. Development of a constitutive law for artificial soil. 3 Int. Conf. Num. Meth. Geom. Aachen 1979. Vol.1 p. 309-317.

Vuez, Deveaux, Amoras, Monnet, 1980. Renforcement de matériaux dilatants en compression triaxiale. Journée Universitaire Génie Civil. INSA de Lyon p. 58-78.

Numerical Methods in Geomechanics (Innsbruck 1988), Swoboda (ed.)
© 1988 Balkema, Rotterdam. ISBN 90 6191 809 X

Elasto-plastic anisotropic hardening model for sand in a wide stress region

H.Murata & N.Yasufuku
Yamaguchi University, Ube, Japan

ABSTRACT: The elasto-plastic constitutive model, taking account of anisotropy effects induced by the initial consolidation and the following shear processes, is established in order to describe the stress-strain behavior of an anisotropically consolidated sand in a wide stress region. The proposed model consists of a set of the yield function, plastic potential and hardening modulus which are expressed based on the theoritical considerations and experimental evidences. The resultant equations are not only a simple form enough for use but also can give good qualitative predictions for the axisymmetric triaxial test results of sand.

1 INTRODUCTION

It is generally recognized that both deformation and strength behaviors of sand in a wide stress region are remarkably influenced by a confining pressure and the anisotropy effects induced by the initial consolidation and loading processes. Therefore, in order to analyze and precisely predict the stress-strain behavior of sand as a boundary value problem, it is essential to establish the adequate constitutive model which can evaluate these complicated characteristics of sand.

The aim of this paper is to present an elasto-plastic constitutive model, taking account of both isotropic behaviors of sand in a wide stress region and anisotropy effects of sand induced by the consolidation and shear processes. This will be mainly done by extending the isotropic hardening model proposed by Murata, Hyodo and Yasufuku (1987a) for describing the stress-strain behaviors of normally or over-consolidated sand for static loading in a wide stress region.

The proposed model consists of a set of the yield function, plastic potential and hardening modulus involving the newly assumed two internal variables. These functions are expressed based on a few basic assumptions and experimental evidences obtained from axisymmetric triaxial tests with various stress paths for an anisotropically consolidated sand.

The predicted behaviors are compared with available experimental data for sand.

2 GENERAL STRESS-STRAIN INCREMENT FOR PROPOSED MODEL

During the application of the stress increment, it is assumed that the total strain increment $d\varepsilon_{ij}$ can be divided into elastic and plastic parts as follows:

$$d\varepsilon_{ij} = d\varepsilon_{ij}^{e} + d\varepsilon_{ij}^{p} \tag{1}$$

where superscripts e and p denote elastic and plastic components, respectively. In this study, the elastic strain increment is expressed by

$$d\varepsilon_{ij} = \frac{1}{3} dv^{e}\delta_{ij} \tag{2}$$

where dv^e is elastic volumetric strain increment which is formulated by Eq.21 and δ_{ij} denotes Kronecker's delta.

Based on the non-associated flow rule, the plastic strain increment is derived as follows:

$$d\varepsilon_{ij}^{p} = \Lambda\frac{\partial g}{\partial\sigma_{ij}} = \frac{1}{H}\frac{\partial g}{\partial\sigma_{ij}}\left(\frac{\partial f}{\partial\sigma_{kl}}d\sigma_{kl}\right) \tag{3}$$

where σ_{ij} is the stress component, Λ is a proportional factor, H is the hardening modulus, g is the plastic potential function and f is the yield function which depends on stress states and a set of internal variables.

To simplify the model as much as possible, we assume the yield function involving only two internal variables such as:

$$f(\sigma_{ij}, k, \alpha) = 0 \tag{4}$$

where, k and α are internal variables which represent the changes of the size of yield function and represent the rotation of the yield function around a origin of a stress space, respectively. These two internal variables depend on the plastic strain undergone by sand.

In the following sections, our analysis and experimental considerations are restricted to the case of axisymmetric triaxial tests. Then, the following stress and strain increment parameters are used,

$$p = (\sigma_a + 2\sigma_r)/3 \ ; \quad q = \sigma_a - \sigma_r \tag{5}$$

$$dv = d\varepsilon_a + 2d\varepsilon_r \quad ; \quad d\varepsilon = 2(d\varepsilon_a - d\varepsilon_r)/3 \tag{6}$$

and also the stress ratio η is defined as follows:

$$\eta = q/p \tag{7}$$

where σ_a and σ_r are the effective axial and radial stresses, $d\varepsilon_a$ and $d\varepsilon_r$ are the axial and radial strain increments, dv and $d\varepsilon$ are the volumetric and shear strain incremens, respectively. The compressive stresses and strains are taken as positive

3 YIELD AND PLASTIC POTENTIAL FUNCTION

3.1 Yield function

We have investigated the yield characteristics of isotropically consolidated sand in detail using the multi-stage stress path method under confining pressure from 0.1MPa to 10MPa. Based on the experimental results, we proposed the isotropic yield function as follows:

$$f = \eta^2 + n\,\ell n\, p/p_o = 0 \tag{8}$$

where n is an experimental parameter and p_o is the equivalent isotropic consolidation pressure. The detail discussion about the foundamental experimental finding which lead to Eq.8 has been already reported by Murata, Hyodo and Yasufuku (1987b). In this section, the yield function which can evaluate the stress-induced anisotoropy of sand is proposed by extending the isotropic yield function given by Eq.8.

The typical yield curve for the anisotropically consolidated sand corresponding to point A is shown in Fig.1, which are obtained from the tests with nine stress paths carried out after being consolidated to point A and then being unloaded to point B, as shown on Fig.1. From this figure, it can be seen that the cap-type yield curve rotates around a origin of stress space. Figure 2 shows the relation between $(\eta-\alpha)^2$ and $\ell n\, p/p_o$ re-

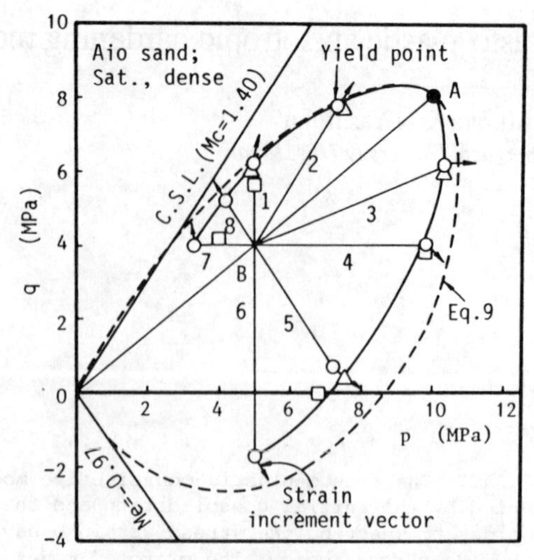

Fig.1 Comparison of the predicted and experimental yield curve for anisotropically consolidated sand.

plotted using the yield curve presented in Fig.1. From this result, it can be considered that the relation between $(\eta-\alpha)^2$ and $\ell n\, p/p_o$ is approximated by a unique straight line. Therefore, we propose the yield function which can evaluate the experimental yield characteristics of an anisotropically consolidated sand such as:

$$f = (\eta - \alpha)^2 + n\,\ell n\, p/p_o = 0 \tag{9}$$

where, α is an anisotropic variable accounting for the degree of the rotation of the yield curve which is defined by the value of η at which the gradient of the yield curve, $dq/dp=\infty$, p_o is the value

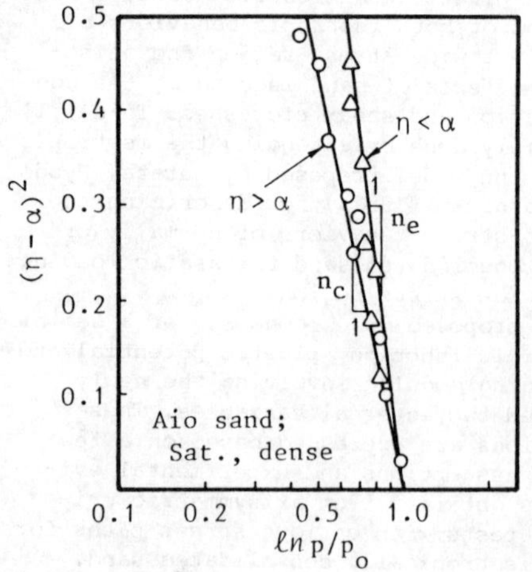

Fig.2 $(\eta-\alpha)^2 - \ell n p/p_o$ characteristics of yield curve represented in Fig.1.

of p at $\eta=\alpha$, and n is an experimental parameter, that is, the slope of linear relation between $(\eta-\alpha)^2$ and $\ell n\, p/p_o$.

Based on the Eq.8, the gradient of yield curve, dq/dp is given by

$$\frac{dq}{dp}=\frac{2(\eta-\alpha)\eta-n}{2(\eta-\alpha)} \qquad (10)$$

From Eq.10, we get the value of η at dq/dp=0 (= N), that is

$$\eta|_{dq/dp=0}=\frac{\alpha\pm\sqrt{\alpha^2+2n}}{2}=N \qquad (11)$$

where, N is a experimental parameter relating n with α. Here, based on the experimental evidences, N is approximated by 0.7M, where M is the value of η at $dv^p/d\varepsilon^p=0$. The yield curve depicted using Eq.11 is shown in Fig.1. It is recognized from this figure that the predicted yield curve can sucessfully express the characteristics of the experimental yield curve.

3.2 Failure locus

Performing a series of conventional triaxial tests, it is well known that the stress-strain relations such as $\eta-\varepsilon$ curves show peaks. In this study, the locus of these peaks are considered the failure locus. Figure 3 shows the relation between peaks of η,η_p and the relevant values of p, p_p obtained from the triaxial drained compression tests with various stress paths. It is indicated that the value of η decreses exponentially with the increse of p and then reaches the constant value, i.e. $\eta=M$. Based on this experimental evidence, we assume that the failure locus is expressed as follows:

$$\eta_p=|M|-D\,\ell n\, p/p_\ell \qquad (12)$$

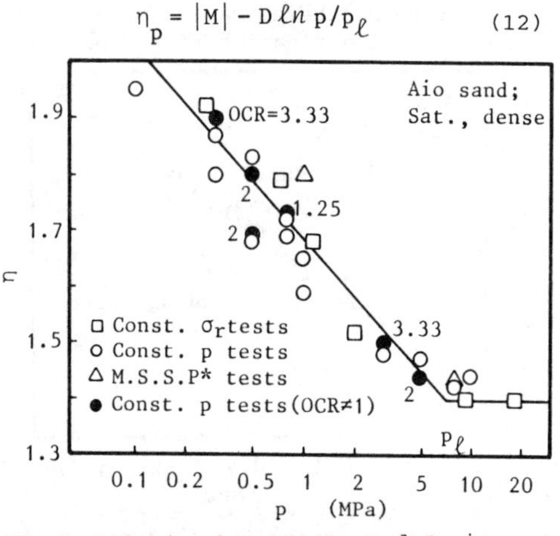

Fig.3 Relation between η_p and p_p in triaxial compression tests under wide stresses.

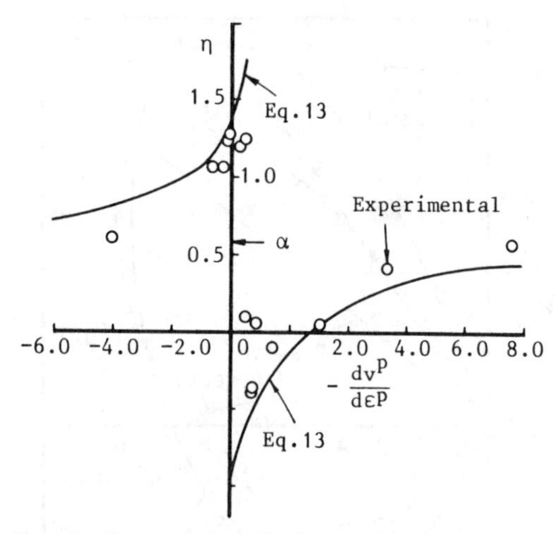

Fig.4 Comparison of the predicted and experimental stress-dilatancy relations of anisotropically consolidated sand.

where D and p_ℓ are the experimental parameters. D is defined by the slope of η_p-p_p relation, and when $p>p_p$,D=0.

3.3 Plastic potential

Figure 4 indicates the relation between and plastic strain increment vector $dv^p/d\varepsilon^p$ at each yield point depicted in Fig.1, in which dilatancy is defined as $dv^p/d\varepsilon^p$. From this figure, we suppose that the stress-dilatancy relation for an anisotropically consolidated sand can be expressed as

$$\frac{dv^p}{d\varepsilon^p}=\frac{m-2(\eta-\alpha)\eta}{2(\eta-\alpha)} \qquad (13\text{-a})$$

$$m=\frac{|2M-\alpha|^2-\alpha^2}{2} \qquad (13\text{-b})$$

where m is a experimental parameter relating α with M. When m is equal to n in Eq.9, Eq.13 gives the plastic potential equivalent to the yield function defined by Eq.9. Eq.13 can represent that when η is equal to α, no plastic shear strain occurs, and when η is equal to M, no plastic volumetric strain increment occurs. Furthermore, when α is equal to 0, Eq.13 reduces to the following equation which was already prposed by Murata, Hyodo and Yasufuku (1987a) for isotropic hardening model.

$$\frac{dv^p}{d\varepsilon^p}=\frac{M^2-\eta^2}{\eta} \qquad (14)$$

Now, applying the normality rule to Eq.13, the plastic potential function is derived as

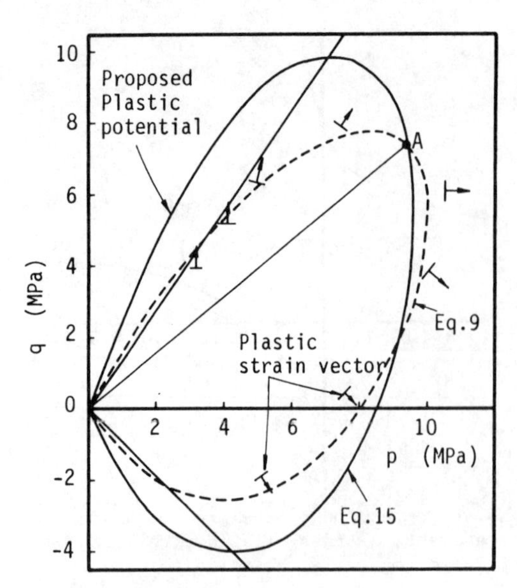

Fig.5 Comparison of the proposed plastic potential and yield curves.

$$g = (\eta - \alpha)^2 + m \ell n\, p = \text{const.} \qquad (15)$$

The picture of the plastic potential is shown in Fig.5, together with the predicted yield curve and the directions of the plastic strain increment vector obtained experimentaly.

4 EVALUATION OF HARDENING MODULUS

In order to complete the proposed model, it is necessary to evaluate two internal variables k and α entering the yield function in Eq.4, concretely. Then, we firstly assume that the evolution equation for isotropic hardening variable k ,which describe the changes of the size of the yield curve in a wide stress region, can be expressed as

$$dk = \Lambda\{\frac{\partial g}{\partial p} + A(1 - \frac{p}{p_\ell})\eta\frac{\partial g}{\partial q}\} = \Lambda\bar{k} \qquad (16)$$

where A = 1 when $0 \leq \eta \leq \eta_p$ (hardening process) and A = 0 when $\eta_p \geq \eta \geq M$ (softening process) or $p > p_\ell$. The similar idea has been reported by Nova and Wood (1979). Equation 16 indicates that the development of the hardening behavior depends on not only changes of dv^p but also changes of $d\varepsilon^p$. Both strain increments contribute to the development of the hardening behavior. The term $(1-p/p_\ell)$ in Eq.16 is introduced to evaluate the contributing rate of $d\varepsilon^p$ to dv^p with p. This term yields 1 at p=0 and 0 at $p=p_\ell$.

Here, when p=0, Eq.16 reduces to the state parameter proposed by Moroto (1976) and when $p \geq p_\ell$, Eq.16 reduces to the hardening variable for the Cam-clay model (Schofield and Wroth (1968)).

Secondly, we assume to use the evolution equation for the internal variable α as follows:

$$d\alpha = \Lambda\frac{a}{p_o}(\sqrt{A\,|\frac{\partial g}{\partial p}|\,|\frac{\partial g}{\partial q}|})(q - \eta_{in}p) = \Lambda\bar{\alpha} \quad (17)$$

where η_{in} is the value of η at the end of the anisotropic consolidation. When $\eta=\eta_{in}$ or $\eta_p \geq \eta \geq M$, from the difinition of $d\alpha$,$d\alpha=0$. The constant a is a experimental parameter controling the degree of the anisotropy development. Term $|\partial g/\partial p|\,|\partial g/\partial q|$ is introduced to evaluate that the development of the anisotropy is produced by the coupling of dv^p and $d\varepsilon^p$.

Now, appling the consistency condition to Eq.4, the yield condition yields

$$df = \frac{\partial f}{\partial \sigma_{kl}}d\sigma_{kl} + \frac{\partial f}{\partial k}dk + \frac{\partial f}{\partial \alpha}d\alpha = 0 \qquad (18)$$

Substituting Eqs.16 and 17 as the internal variables k and α respectively into Eq.18 and then rearranging Eq.18, the hardening modulus is given by

$$H = -(\frac{\partial f}{\partial k}\bar{k} + \frac{\partial f}{\partial \alpha}\bar{\alpha}) \qquad (19\text{-a})$$

$$\frac{\partial f}{\partial k} = \frac{\partial f}{\partial p_o}\frac{\partial p_o}{\partial v^p}\frac{\partial v^p}{\partial k} \qquad (19\text{-b})$$

where the definition of k and α in Eq.19 are defined by Eqs.17 and 18, respectively. Because the yield function and plastic potential are concretely given by Eqs.9 and 15, respectively, and $\partial v^p/\partial k$ in Eq.19-b is evident from the definition of dk in Eq.16 (i.e. $\partial v^p/\partial k = 1$), the rest work to determine the hardening modulus is the experimental evaluation of the partial

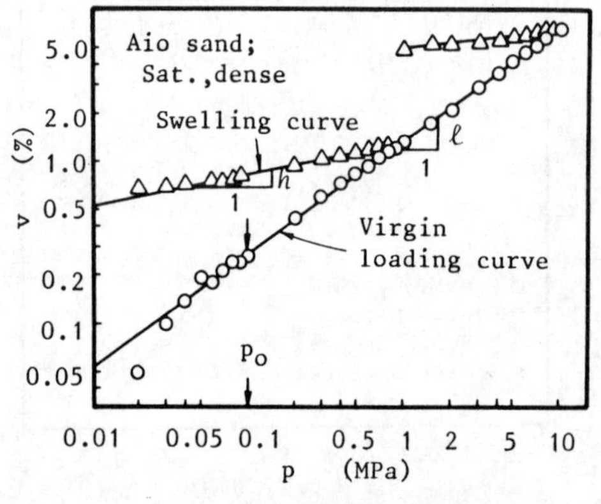

Fig.6 $\ell n\,v$-$\ell n\,p$ relation for isotropic consolidated test up to 10MPa confining pressure.

derivative, $\partial p_o/\partial v^p$. Figure 6 shows the relation between $\ell n\, v$ and $\ell n\, p$ obtained from the isotropic consolidation test up to 10MPa confining pressure. This figure indicate the $\ell n\, v$-$\ell n\, p$ relation for the virgin loading and swelling or reloading under consolidation process. The $\ell n\, v$-$\ell n\, p$ relation can be approximated by a straight line at least in the stress region up to 10MPa. Based on this linear relation, the total and elastic volumetric strain increments under isotropic consolidation process can be expressed as follows, respectively:

$$dv|_{\eta=0} = \ell v_a \left(\frac{p_c}{p_a}\right)^{\ell} \frac{dp_c}{p_c} \tag{20}$$

$$dv^e = h v_a \left(\frac{p}{p_a}\right)^{\ell} \frac{dp}{p} \tag{21}$$

where p_c is the equivalent isotropic consolidation pressure which is the intersection of f=0 with the p-axis. v_a is the value of v at $p=p_a$, the unit pressure is chosen for p_a. ℓ and h are the experimental parameters which are slopes of the $\ell n\, v$-$\ell n\, p$ virgin loading and swelling or reloading line, respectively.

Using Eqs.20 and 21, the plastic volumetric strain increment under the isotropic consolidation can be expresed as

$$dv^p|_{\eta=0} = (\ell-h)\, v_a \left(\frac{p_c}{p_a}\right)^{\ell} \frac{dp_c}{p_c} \tag{22}$$

Because we regard $d\varepsilon^p$ under the isotropic consolidation process as zero, Eq.22 can be rewritten as

$$\frac{\partial p_c}{\partial v^p} = \frac{1}{(\ell-h)v_a} \left(\frac{p_a}{p_c}\right)^{\ell} p_c \tag{23-a}$$

$$\frac{\partial p_o}{\partial v^p} = \frac{\partial p_o}{\partial p_c} \frac{\partial p_c}{\partial v^p} \tag{23-b}$$

Then, from Eqs.23-a and 23-b, we finally get the following equation.

$$\frac{\partial p_o}{\partial v^p} = \frac{e^{\alpha^2/n}}{(\ell-h)v_a} \left(\frac{p_a}{p_c}\right)^{\ell} p_c \tag{24-a}$$

$$p_c = p_o e^{-\alpha^2/n} \tag{24-b}$$

The resultant expressions for the strain increments is determined from Eqs. 9, 12, 15, 16, 17, 19 and 24, accurately.

Total volumetric and shear strain increments written in matric form becomes

$$\begin{vmatrix} dv \\ d\varepsilon \end{vmatrix} = \frac{1}{H} \begin{vmatrix} \frac{H}{K} + \frac{\partial f}{\partial p}\frac{\partial g}{\partial p} & \frac{\partial f}{\partial q}\frac{\partial g}{\partial p} \\ \frac{\partial f}{\partial p}\frac{\partial g}{\partial q} & \frac{\partial f}{\partial q}\frac{\partial g}{\partial q} \end{vmatrix} \begin{vmatrix} dp \\ dq \end{vmatrix} \tag{25-a}$$

$$\frac{1}{K} = h v_a \left(\frac{p}{p_a}\right)^{\ell} \tag{25-b}$$

where, Eq.25-b is derived from Eq.21.

5 PREDICTION OF STRESS-STRAIN RELATION

The applicability of the proposed model is investigated by using the results of con-

Table 1. Values of experimental parameters for Aio sand used in analysis.

Parameters				
ℓ	0.699	M	M_c	1.40
h	0.206		M_e	-0.95
v_a	0.0027	N	N_c	0.98
D	0.148		N_e	-0.67
p_ℓ	7.0 (MPa)			
α	0.58			
a	10.0			

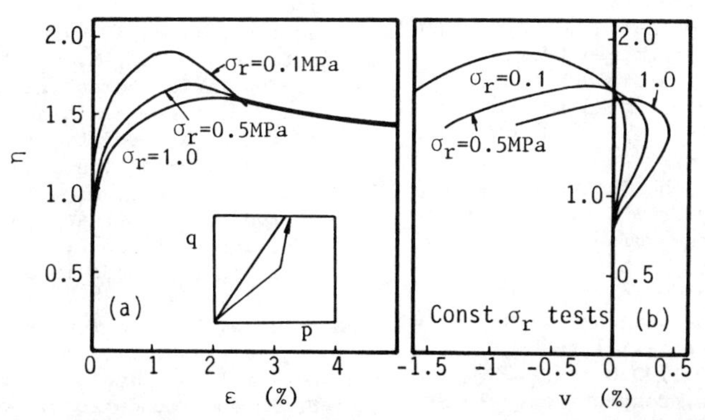

Fig.7 Prediction of stress-strain curves of anisotropically consolidated sand for constant σ_r tests under various confining pressures; (a) η-ε curves, (b) η-v curves.

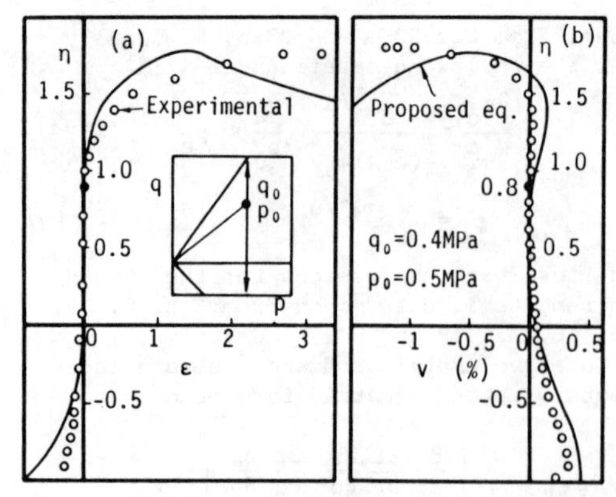

Fig.8 Prediction of stress-strain curves of anisotropically consolidated sand for constant p compression and extension tests; (a) η-ε curve, (b) η-v curve.

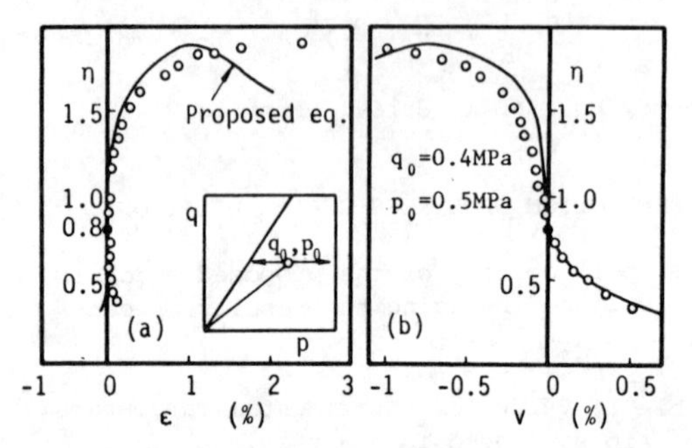

Fig.9 Prediction of stress-strain curves of anisotropically consolidated sand for constant q tests; (a) η-ε curve, (b) η-v curve.

ventional triaxial tests with various stress paths. All of the experimental parameters for "Aio" sand are listed in Table I. Here, ℓ, h and v_a are determined from an isotropic consolidation test, D, P_ℓ and M are determined from a few conventional triaxial compression tests. In general, N and α are determined from the characteristics of the yield curve for the anisotropically consolidated sand as shown in Fig.1. We however adopt 0.7M for N and $0.75\eta_{in}$ for α approximately, based on the experimental evidences. The values of n and m are estimated from Eqs.10 and 13-b respectively, and the value of a is conveniently estimated to be 10.

Figure 7(a),(b) show the plots of η against the deviator strain and volumetric strain for drained triaxial tests at constant σ_r under 0.1, 0.5 and 1.0MPa confining pressures, respectively. This result reasonably describes the strain hardening-softening behavior of sand and the dependency of stress-strain curve on the confining pressures. Figures 8 and 9 show the predicted and measured results of drained triaxial tests at constant p and constant q for an anisotropicaly consolidated sand, respectively. From these figures, the satisfactory agreements in the stress-strain curves can be noted for an anisotropicaly consolidated sand by using the proposed model.

6 CONCLUSIONS

In order to evaluate the machanical behaviors of an anisotropically consolidated sand in a wide stress region, the elasto-plastic constitutive model, taking account of the stress-induced anisotropy, was derived by intoducing two new internal variables. Two internal variables can describe the changes of size of yield surface in a wide stress region and control the degree of the anisotropic developments,respectively.

It is shown that the resulting model can reasonably explain the stress-strain behaviors of an anisotropically consolidated sand under the axisymmetric stress condition.

ACKNOWLEDGEMENTS

The authors wish to thank Prof. N. Miura of Saga University and Prof. M. Hyodo of Yamaguchi University for their helps and advices.

REFERENCES

Murata, H., Hyodo, M. and Yasufuku, N. 1987a. Elasto-plastic constitutive model for sand in wide stress region. 2nd Int. Conf. on Constitutive Laws for Engineering Materials: Theory and Applications, Vol.1, Tucson: 421-428.

Murata, H., Hyodo, M. and Yasufuku, N. 1987b. Yield characteristics of dense sand under low and high pressure. Proc. of the Japan Society of Civil Engineers, No.382: 183-191 (in Japanese).

Nova, R. and Wood, D.M. 1979. A constitutive model for sand in triaxial compression. Int. Jour. for Numerical and Analytical Methods in Geomechanics, Vol.3: 255-278.

Moroto, N. 1976. New parameter to measure degree of shear deformation of granular material in triaxial compression tests. Soils and Foundations, Vol.16, No.4: 1-9

Schofield, A.N. and Wroth, C.P. 1968. Critical State Soil Mechanics. McGrow-Hill, London.

Numerical Methods in Geomechanics (Innsbruck 1988), Swoboda (ed.)
© 1988 Balkema, Rotterdam. ISBN 90 6191 809 X

Cap parameters for clayey soils

D.N.Humphrey
University of Maine, Orono, USA
R.D.Holtz
Purdue University, West Lafayette, Ind., USA

ABSTRACT: A work hardening cap model with a variable aspect ratio end cap is described. Simple equations to obtain the model parameters from readily available soil properties are reviewed. Parameters for 52 clayey soils are presented and correlations with index properties are examined. For many soils the aspect ratio from the cap model was significantly different than from the Cam clay and modified Cam clay models. However, the average aspect ratio from the cap and modified Cam clay models were about the same. The aspect ratio for soils with initial hydrostatic and nonhydrostatic consolidation are shown to be different.

1 INTRODUCTION

Work hardening cap models with a cone shaped Drucker-Prager [1] ultimate failure surface and an elliptical cap are frequently used to represent the undrained behavior of normally and lightly overconsolidated clays. A feature of this model is that the aspect ratio of the elliptical cap, defined as the ratio of the major and minor radii of the ellipse, can be adjusted to best represent soil behavior. The model parameters can be determined from readily available soil properties using simple equations [2,3]. The equations are summarized in a subsequent section.

The purpose of this paper is to examine the range of cap model parameters for clayey soils. Parameters for 52 clayey soils are presented and correlations with soil index properties are examined. The aspect ratio for clayey soils is compared with the aspect ratio of the Cam clay and modified Cam clay models. Finally, parameters for initial hydrostatic and nonhydrostatic conditions are compared.

2 DESCRIPTION OF CAP MODEL

The form of the cap model used herein was proposed by DiMaggio and Sandler [4] and Sandler, et al. [5]. It has a cone shaped Drucker-Prager ultimate failure surface [1] and an elliptical work hardening cap (Fig. 1). They are expressed in terms of

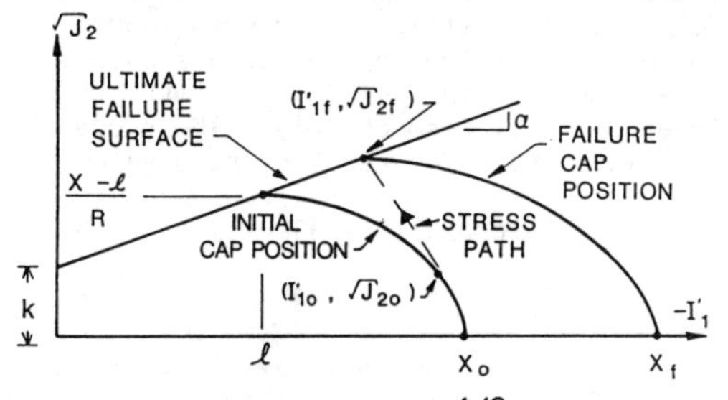

Fig. 1. Cap model in $I_1'-J_2^{1/2}$ space.

effective stresses by the first invariant of the stress tensor I_1' and the second invariant of the stress deviator tensor J_2. Compressive stresses and strains are taken as negative. The ultimate failure surface has a slope α and an intercept with the $J_2^{1/2}$ axis of κ (Fig. 1). Its equation is

$$\alpha I_1' + J_2^{1/2} - \kappa = 0 \quad (1)$$

The cap intersects the I_1' axis at x and the ultimate failure surface at coordinates (ℓ, $(x-\ell)/R$), where R is the aspect ratio (Fig. 1). The cap's shape is described by the equation of an ellipse

$$(I_1' - \ell)^2 + R^2 J_2 - (x - \ell)^2 = 0 \quad (2)$$

The position of the cap is coupled to the plastic volumetric strain ε_v^p

$$\varepsilon_v^P = W[\exp(Dx) - 1] \quad (3)$$

where W and D are curve fitting parameters [2,3]. A normal flow rule is assumed. An increment of loading on the cap causes the cap to expand (Fig. 1) and results in the soil undergoing negative plastic volumetric strain (decrease in void ratio) and plastic shear strain.

3 PROCEDURE TO DETERMINE CAP PARAMETERS

A detailed procedure to determine the cap parameters was given by Humphrey and Holtz [2,3] and is reviewed below.

The Drucker-Prager ultimate failure surface was matched to the hexagonal cross section of the Mohr-Coulomb criterion for triaxial compression ($\sigma_1 = \sigma_3$) resulting in the following equations

$$\alpha = \frac{2\sin\phi'}{\sqrt{3}(3-\sin\phi')} \qquad \kappa = \frac{6c'\cos\phi'}{\sqrt{3}(3-\sin\phi')} \quad (4)$$

where ϕ' and c' are the effective Mohr-Coulomb parameters.

The hardening parameters, D and W in Eq. 3, are determined from hydrostatic consolidation test results. The virgin loading and unloading-reloading curves are assumed to be linear on an ε_v-$\ln(I_1'/3)$ plot with slopes a and b, respectively. The slopes are related to oedometer test results by

$$a = C_c/[2.303(1+e_o)] \quad (5)$$

$$b \simeq C_r/[2.303(1+e_o)] \quad (6)$$

where: C_c = compression index, C_r = recompression index, and e_o = initial void ratio. Eq. 6 is only an approximation since the lateral earth pressure changes during unloading [2]. Volumetric strain is separated into its elastic and plastic components; then Eq. 3 is matched to the plastic component at three points resulting in

$$Dx_o = \frac{2\ln[(1-P)/P]}{(x_f/x_o - 1)}; \; P = \frac{\ln[(1+x_f/x_o)/2]}{\ln[x_f/x_o]} \quad (7)$$

$$\frac{W}{(a-b)} = \frac{-\ln(x_f/x_o)}{\exp[(Dx_o)(x_f/x_o)] - \exp[Dx_o]} \quad (8)$$

This shows that Dx_o and $W/(a-b)$ are functions only of the ratio of failure to initial cap positions, x_f/x_o.

The ratio x_f/x_o is found from

$$x_f/x_o = \exp\left[\frac{-b}{a-b}\ln\left[\frac{\kappa/\sigma'_{vo} - J_{2f}^{1/2}/\sigma'_{vo}}{\alpha(1+2K_o)}\right]\right] \quad (9)$$

where K_o is the normally consolidated coefficient of lateral earth pressure at rest. The ratio, $J_{2f}^{1/2}/\sigma'_{vo}$, is introduced as a normalized measure of the shear stress at failure, where σ'_{vo} is the initial vertical effective stress. For undrained triaxial tests

$$J_{2f}^{1/2}/\sigma'_{vo} = \frac{2}{\sqrt{3}}(s_u/\sigma'_{vo}) \quad (10)$$

where s_u is the undrained shear strength.

The cap aspect ratio R is determined from undrained shear tests on normally consolidated soil where the loading path causes the cap to expand (Fig. 1). It is shown in [2,3] that R is determined by trial and error solution of

$$x_f/x_o = \frac{(\kappa/\sigma'_{vo})(1-\alpha R) - (J_{2f}^{1/2}/\sigma'_{vo})(1-\alpha^2 R^2)}{[-R(\kappa/\sigma'_{vo}) + (1+2K_o) - R(H)^{1/2}]\alpha} \quad (11)$$

$$\text{where: } H = (1-K_o)^2(\alpha^2R^2-1)/3 + \alpha^2(1+2K_o)^2 + (\kappa/\sigma'_{vo})^2 - 2\alpha(\kappa/\sigma'_{vo})(1+2K_o) \quad (12)$$

For hydrostatic initial conditions (K_o = 1) and zero cohesion (κ = 0) Eqs. 11 and 12 are solved directly for R

$$R = -3(x_f/x_o)/(J_{2f}^{1/2}/\sigma'_{vo}) - \frac{1}{\alpha} \quad (13)$$

4 CAP PARAMETERS FOR CLAYEY SOILS

4.1 Hydrostatic Initial Consolidation

The parameters Dx_o, $W/(a-b)$, and R were computed from consolidation and consolidated hydrostatic (isotropic) undrained (CIU) triaxial test data for 52 normally consolidated soils. The soil data was obtained primarily from a summary prepared by Mayne and Swanson [6,7]. The data and calculated parameters are summarized in Table 1. The parameter κ was assumed to be 0 which is reasonable for normally consolidated clayey soils.

The volume change parameter Dx_o ranged from -0.969 to -0.730 and $W/(a-b)$ ranged from 2.719 to 2.757. A statistical analysis was made and no significant correlation was found between either parameter and the soil index properties.

The computed aspect ratios ranged from 0.57 to 12.80. There was no significant correlation of R with Atterberg limits, ϕ' or consolidation properties. R tended to

Table 1. Cap parameters from CIU tests on normally consolidated clayey soils.

SOIL	LL	PI	ϕ'	C_c	C_r	s_u/σ'_{vo}	Dx_o	$\frac{W}{(a-b)}$	R CAP	R CC	R MCC	REFERENCE
WEALD (REMOLDED)	43	25	22.0	0.214	0.081	0.279	-0.874	2.726	6.02	10.44	6.07	[8]*
SPESTONE KAOLIN	72	31	23.5	0.573	0.170	0.313	-0.921	2.721	4.11	9.72	5.65	[9]*
SAN FRANCISCO BAY MUD	88	45	35.2	0.665	0.221	0.432	-0.879	2.725	4.08	6.26	3.64	[10]*
LONDON CLAY (REMOLDED)	78	52	18.4	0.371	0.143	0.250	-0.895	2.724	5.53	12.67	7.36	[8]*
KEUPER MARL	32	14	25.9	0.272	0.058	0.256	-0.909	2.722	7.16	8.74	5.08	[11]*
ILLITE	57	31	24.6	0.431	0.138	0.341	-0.920	2.721	3.60	9.25	5.38	[12]*
RESED. BOSTON BLUE CL.	41	21	29.5	0.338	0.051	0.285	-0.936	2.720	5.96	7.59	4.41	[13]*
OSLO	39	18	27.0	0.175	0.035	0.380	-0.957	2.719	2.59	8.35	4.86	[14]*
BRADWELL	95	65	20.0	0.714	0.152	0.300	-0.964	2.719	2.54	11.58	6.73	[15]*
SHELLHAVEN	110	80	23.0	0.937	0.051	0.205	-0.969	2.719	7.48	9.95	5.78	[16]*
CALCIUM MONTMORILL.	203	169	12.5	0.979	0.440	0.221	-0.969	2.719	1.15	19.16	11.14	[17]*
NEWFIELD	28	10	30.5	0.210	0.018	0.475	-0.957	2.720	1.35	7.32	4.25	[18]*
VIRGINIA COASTAL	54	27	28.5	0.444	0.055	0.409	-0.969	2.719	2.07	7.88	4.58	[19]*
SPESTONE KAOLIN	72	32	22.6	0.693	0.069	0.210	-0.959	2.719	7.53	10.14	5.89	[20]*
JAPANESE	64	37	33.7	0.403	0.078	0.400	-0.937	2.720	3.56	6.56	3.82	[21]*
ATCHAFALAYA	95	75	23.5	1.040	0.104	0.250	-0.964	2.719	5.47	9.72	5.65	[13]*
MILAZZO	61	33	23.0	0.431	0.111	0.333	-0.949	2.720	2.87	9.95	5.78	[22]*
CALCIUM ILLITE	85	48	24.2	0.758	0.274	0.250	-0.827	2.733	9.47	9.41	5.47	[23]*
DRAMMEN	33	15	28.0	0.265	0.055	0.310	-0.925	2.721	5.10	8.03	4.67	[24]*
BACKSWAMP	70	40	22.2	0.583	0.163	0.280	-0.918	2.721	4.97	10.34	6.01	[25]*
KARS LEDA	45	21	28.3	1.121	0.021	0.265	-0.940	2.722	5.33	7.94	4.61	[26]*
PLASITC HOLOCENE	65	38	32.9	0.640	0.087	0.335	-0.947	2.720	4.72	6.74	3.92	[27]*
GHANA	42	29	20.8	0.182	0.058	0.380	-0.961	2.719	0.58	11.09	6.45	[28]*
HALLOYSITE	62	26	34.3	0.304	0.053	0.418	-0.947	2.720	3.17	6.44	3.74	[29]*
SIMPLE CLAY	NA	NA	23.1	0.207	0.083	0.290	-0.858	2.728	6.29	9.90	5.76	[30]*
SEATTLE	52	26	28.8	0.260	0.081	0.372	-0.904	2.723	3.97	7.79	4.53	[31]*
SODIUM ILLITE	79	44	20.7	0.362	0.200	0.340	-0.901	2.723	2.87	11.15	6.48	[32]*
GRUNDITE	55	29	32.3	0.359	0.177	0.356	-0.730	2.757	9.11	6.87	4.00	[33]*
TERRA ROXA	43	22	29.2	0.216	0.046	0.313	-0.918	2.721	5.36	7.67	4.46	[34]*
SCOTT	34	12	33.4	0.138	0.037	0.231	-0.814	2.736	12.80	6.63	3.85	[35]*
TOLEDO	42	23	20.0	0.288	0.067	0.200	-0.903	2.723	9.13	11.58	6.73	[36]*
KAWASAKI	65	31	35.9	0.668	0.058	0.370	-0.965	2.719	3.92	6.13	3.56	[37]*
CONCORD BLUE	32	10	24.8	0.163	0.039	0.355	-0.950	2.719	2.76	9.16	5.33	[38]*
LAGUNILLAS	61	37	26.5	0.771	0.127	0.305	-0.947	2.720	4.52	8.53	4.96	[37]*
DRAMMEN	NA	NA	30.7	0.599	0.127	0.285	-0.899	2.723	6.99	7.26	4.22	[39]*
LISKEARD	56	33	26.1	1.073	0.035	0.298	-0.949	2.721	3.84	8.67	5.04	[40]*
LILLA EDET	61	32	24.3	0.860	0.140	0.290	-0.952	2.719	4.42	9.37	5.45	[41]*
HOKKAIDO SILT A	52	21	37.2	0.299	0.058	0.420	-0.930	2.721	3.70	5.90	3.43	[42]*
HOKKAIDO SILT B	51	21	35.1	0.193	0.030	0.362	-0.939	2.720	4.47	6.28	3.65	[42]*
HOKKAIDO CLAY	72	32	36.1	0.412	0.058	0.410	-0.953	2.719	3.43	6.09	3.54	[42]*
KANPUR CLAY	38	18	29.0	0.284	0.060	0.295	-0.912	2.722	6.05	7.73	4.49	[43]*
RANN OF KUTCH	91	49	26.0	0.610	0.264	0.326	-0.835	2.732	6.21	8.70	5.06	[43]*
LANSISALMI	78	46	19.5	1.059	0.401	0.215	-0.838	2.731	10.06	11.90	6.92	[44]*
SAULT STE MARIE	55	32	28.9	0.165	0.050	0.327	-0.881	2.725	5.65	7.76	4.51	[45]*
BATH KAOLINITE	48	15	24.5	0.189	0.045	0.425	-0.968	2.719	0.96	9.29	5.40	[46]*
BANGALORE MONTMORILL.	580	495	12.5	2.260	1.350	0.211	-0.927	2.721	3.15	19.16	11.14	[47]*
LONG ISLAND COASTAL	64	34	22.8	0.222	0.068	0.271	-0.894	2.724	6.10	10.04	5.84	[19]*
HACKENSACK VARVED	65	35	19.0	0.481	0.068	0.158	-0.932	2.720	11.76	12.24	7.11	[48]*
KODIAK ISLAND	30	14	42.2	0.124	0.056	0.510	-0.795	2.740	4.87	5.16	3.00	[49]*
WINNIPEG CLAY	94	60	12.8	0.996	0.240	0.211	-0.968	2.719	1.95	18.68	10.86	[50]*
REMOLDED BOSTON BLUE	33	15	27.5	0.338	0.051	0.300	-0.947	2.720	4.87	8.19	4.76	[51]
PORTSMOUTH	35	15	21.1	0.711	0.035	0.253	-0.958	2.720	4.18	10.92	6.35	[52]

Notes: *From data summarized by Mayne and Swanson [6,7]; CC = Cam clay; MCC = Modified Cam clay

decrease as s_u/σ'_{vo} increased as shown in Fig. 2. The equation of the best fit straight line through the data is

$$R = 10.9 - 18.8(s_u/\sigma'_{vo}) \qquad (14)$$

The r^2 was 0.306 which indicates only poor correlation and that R must be determined from test results.

4.2 Comparison with Cam clay and Modified Cam clay Models

The shape of the elliptical cap for the Cam clay [53] and modified Cam clay [54] models are fixed. Their aspect ratios are related to α by

Cam clay $\quad R = 1.72/\alpha \qquad (15)$

Modified Cam clay $\quad R = 1/\alpha \qquad (16)$

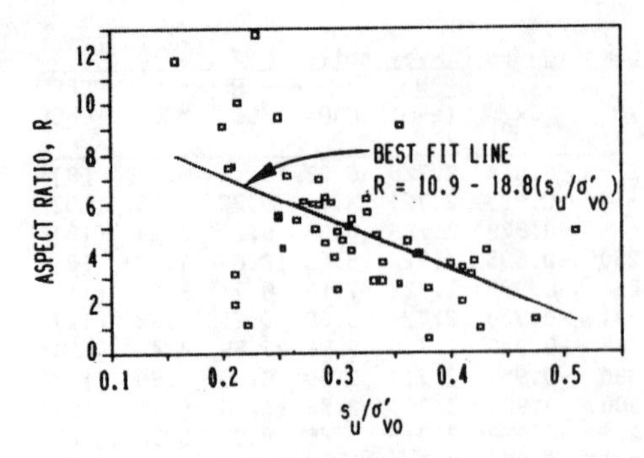

Fig. 2. Aspect ratio R vs. s_u/σ'_{vo}

The aspect ratios for these models is shown in Table 1. It is seen that for many soils they differ considerably from the R for the cap model. The average R for the cap model is 5.0, while for Cam clay it is 9.3. However, the average R for modified Cam clay is 5.4 which agrees quite well with the cap model average. Nonetheless, the cap model offers the considerable advantage of a variable aspect ratio which can be adjusted to model observed soil behavior.

4.3 Nonhydrostatic Initial Consolidation

The ϕ', s_u/σ'_{vo}, and R from nonhydrostatic and hydrostatic consolidated tests are generally different as shown in Table 2. The same was true for R computed from hydrostatic strength parameters but with the nonhydrostatic K_o. This emphasizes the importance, for at least some soils, of using strength parameters obtained from tests with nonhydrostatic consolidation.

5 CONCLUSIONS

(1) Cap parameters for 52 clayey soils are presented. This will provide a useful library of typical cap parameters. (2) The aspect ratio R tended to decrease as the ratio s_u/σ'_{vo} increased. (3) For many soils, R from Cam clay and modified Cam clay models was significantly different than from the cap model; thus, the cap model has the advantage of a variable aspect ratio which can be adjusted to match observed soil behavior. (4) The R for soils with initial hydrostatic and nonhydrostatic consolidation differed considerably for many soils. This emphasizes the importance of obtaining soil parameters from tests which model in situ stress conditions.

6 ACKNOWLEDGEMENTS

Financial support for this research was provided by the Indiana Department of Highways and the Federal Highway Administration, through the Joint Highway Research Project at Purdue University. This support is gratefully acknowledged.

Table 2. Comparison of tests with hydrostatic and nonhydrostatic consolidation.

Soil	K_o	ϕ' I*	ϕ' A*	s_u/σ'_{vo} I	s_u/σ'_{vo} A	R I	R A	R **	Reference
Resedimented Boston Blue	0.51	29.5	26.5	0.285	0.330	5.96	0.62	5.11	[13]
Resedimented Boston Blue***	0.52	29.5	29.0	0.285	0.365	5.96	0.81	4.87	[13]
Atchafalaya	0.71	23.5	21.0	0.250	0.240	5.47	3.41	3.91	[13]
Mexico Volcanic	0.36	----	47.0	-----	0.430	----	2.56	----	[55]
Agnew	0.58	----	25.0	-----	0.308	----	1.70	----	[38]
Hokkaido silt A	0.45	37.2	35.1	0.420	0.400	3.70	0.84	0.85	[42]
Hokkaido silt B	0.45	35.1	34.9	0.362	0.361	4.47	1.75	1.76	[42]
Hokkaido clay	0.47	36.1	34.0	0.410	0.360	3.43	1.64	0.88	[42]
Weald clay	0.59	25.9	25.0	0.323	0.256	5.56	19.5	1.78	[56]
Drammen clay	0.50	----	30.7	-----	0.280	----	5.68	----	[57]
Remolded Boston Blue	0.54	27.5	26.5	0.300	0.330	4.87	0.85	2.49	[51]
Connecticut varved	0.68	----	20.3	-----	0.234	----	3.85	----	[58]
Portsmouth	0.63	21.1	24.2	0.253	0.245	4.18	4.19	2.17	[53]
Middleton fibrous peat	0.47	----	57.4	-----	0.740	----	0.06	----	[59]
Remolded weald clay	0.61	26.0	26.0	0.320	0.270	5.77	8.10	2.19	[60]

*I = hydrostatic tests; A = nonhydrostatic tests consolidated with given K_o
**R computed for anisotropic conditions using strength parameters from CIU tests
***K_o consolidated plane strain active test

7 REFERENCES

Abbreviations:
ICSMFE = Int.Conf.Soil Mech.Found.Eng.
JGED = J.Geot.Eng.Div.
JSMFD = J.Soil Mech.Found.Div.
PCSMFE = Panam.Conf.Soil Mech.Found.Eng.

[1] D.C.Drucker and W.Prager, Soil mechanics and plastic analysis or limit design,Q.Appl.Math. 10,157-165 (1952)

[2] D.N.Humphrey, Design of reinforced embankments, Joint Highway Research Project, School of Civil Eng., Purdue Univ., W. Lafayette, Indiana, Report No. FHWA/IN/JHRP-86/17 (1986)

[3] D.N.Humphrey and R.D.Holtz, A procedure to determine cap model parameters, Second Int.Conf.Constitutive Laws for Eng.Materials, Tucson, Ariz. II, 1225-1232 (1987)

[4] F.L.DiMaggio and I.S.Sandler, Material model for granular soils, J.Eng. Mech.Div., ASCE 97, 935-950 (1971)

[5] I.S.Sandler, F.L.DiMaggio and G.Y.Baladi, Generalized CAP model for geological materials, JGED, ASCE 102, 683-699 (1976)

[6] P.W.Mayne, Cam-clay predictions of undrained strength, JGED, ASCE 106, 1219-1242 (1980)

[7] P.W.Mayne and P.G.Swanson, The critical-state pore pressure parameter from consolidated undrained shear tests, Shear Strength of Soil, ASTM STP 740, 410-430 (1980)

[8] D.J.Henkel, The shear strength of saturated remoulded clays, Shear Strength of Cohesive Soils, ASCE, Univ.of Colorado, 533-554 (1960)

[9] S.F.Amerasinghe and R.H.Parry, Anisotropy in heavily overconsolidated kaolin, JGED, ASCE 101, 1277-1293 (1975)

[10] J.K.Mitchell, Fundamentals of Soil Behavior, John Wiley and Sons, Inc., New York, N.Y., 1976

[11] S.F.Brown, A.Lashine and A.Hyde, Repeated load triaxial testing of a silty clay, Geotechnique 25, 95-114 (1975)

[12] J.France and D.A.Sangrey, Effects of drainage in repeated loading of clays, JGED, ASCE 103, 769-785 (1977)

[13] C.C.Ladd and L.Edgers, Consolidated-undrained direct-simple shear tests on saturated clays, Soils Publication 284, Dept.of Civil Eng., Massachusetts Institute of Technology (1972)

[14] N.E.Simons, The effect of overconsolidation on the strength characteristics of an undisturbed Oslo clay, Shear Strength of Cohesive Soils, ASCE, Univ.of Colorado,747-763 (1960)

[15] A.W.Skempton, Horizontal stresses in an overconsolidated Eocene clay, Fifth ICSMFE, Paris I, 351-357 (1961)

[16] A.W.Skempton and D.J.Henkel, Post glacial clays of the Thames Estuary at Tilbury and Shellhaven, Third ICSMFE, Zurich I, 302-312 (1953)

[17] G.Mesri and R.Olson, Shear strength of montmorillonite, Geotechnique 20, 261-270 (1970)

[18] D.A.Sangrey, D.J.Henkel and M.I.Esrig, Effective stress response of saturated clay soil to repeated loading, Can.Geot.J. 6,241-252 (1969)

[19] P.G.Swanson and R.E.Brown, Triaxial and consolidation testing of cores from the 1976 Atlantic margin coring project, Report OF78-124, U.S.Geol. Survey, Law Eng.Testing Co. (1977)

[20] R.H.Parry and V.Nadarajah, Observations on laboratory prepared, lightly overconsolidated specimens of kaolin, Geotechnique 24, 345-358 (1973)

[21] T.Shibata and D.Karube, Creep rate and creep strength of clays, Seventh ICSMFE, Mexico City I, 361-368 (1969)

[22] A.Croce, R.Japelli, A.Pellegrino and C.Viggiani, Compressibility and strength of intact clays, Seventh ICSMFE, Mexico City I, 81-89 (1969)

[23] R.Olson, The shear strength properties of calcium illite, Geotechnique 12, 23-43 (1962)

[24] N.E.Simons, Comprehensive investigations of the shear strength of an undisturbed Drammen Clay, Shear Strength of Cohesive Soils, ASCE, Univ.of Colorado, 727-745 (1960)

[25] R.V.Whitman, Some considerations and data regarding the shear strength of clays, Shear Strength of Cohesive Soils, ASCE, Univ. of Colorado, 581-614 (1960)

[26] G.P.Raymond, Kars Leda Clay, Performance of Earth and Earth-Supported Structures, ASCE, Purdue Univ. I, 319-340 (1972)

[27] D.Koutsoftas and J.Fischer, In-situ undrained shear strength of two marine clays, JGED, ASCE 103, 989-1005 (1976)

[28] J.W.S.deGraft-Johnson, H.S.Bhatia and D.M.Gidigasu, The strength characteristics of residual micaceous soils, Seventh ICSMFE, Mexico City I, 165-172 (1969)

[29] P.Taylor and D.Bacchus, Dynamic cyclic strain tests on a clay, Seventh ICSMFE, Mexico City I, 401-409 (1969)

[30] C.C.Ladd, Stress-strain behavior of saturated clay and basic strength principles, Rpt.No. R64-17 to U.S.

Army Waterways Exp.Sta., Vicksburg, Miss., Apr., 1964
[31] M.A.Sherif, M.J.Wu, and R.C.Bostrum, Reduction in soil strength due to dynamic loading, Microzonation Conf., National Science Foundation and ASCE II, 439-454 (1972)
[32] R.Olson and J.Hardin, Shearing properties of remolded sodium illite, Second PCSMFE, Mexico I,204-218(1963)
[33] W.H.Perloff and J.O.Osterberg, The effect of strain rate on the undrained shear strength of cohesive soils, Second PCSMFE,Mexico I,103-128 (1963)
[34] P.T.daCruz, Shear strength characteristics of some residual compacted clays, Second PCSMFE, Mexico I, 73-102 (1963)
[35] B.Ladanyi, et al., Some factors controlling the predictability of stress-strain behavior of clay, Can.Geot.J. 2, 60-89 (1965)
[36] T.H.Wu, N.Chang and E.M.Ali, Consolidation and strength properties of a clay, JGED, ASCE 104, 889-905 (1978)
[37] C.C.Ladd and T.W.Lambe, The strength of an undisturbed clay determined from undrained tests, Laboratory Shear Testing of Soils, ASTM STP 361, 342-371 (1963)
[38] J.A.Egan, A critical state model for the cyclic loading pore pressure response of soils, M.S.Thesis, Cornell Univ., Ithaca, N. Y., June (1977)
[39] H.Van Eekelen and D.M.Potts, The behavior of Drammen Clay under cyclic loading, Geotechnique 28, 173-196 (1978)
[40] G.P.Raymond, Foundation failure of New Liskeard Embankment, Highway Research Board Bull. 463, 1-17 (1973)
[41] L.Bjerrum and N.Simons, Comparison of shear strength characteristics of normally-consolidated clays, Shear Strength of Cohesive Soils, ASCE, Univ.of Colorado, 711-726 (1960)
[42] T.Mitachi and S.Kitago, Change in undrained strength characteristics of a saturated remolded clay due to swelling, Soils and Found., JSSMFE 16, 45-58 (1976)
[43] Yudhbir and A.Varadarajan, Undrained behavior of overconsolidated saturated clays during shear, Soils and Found., JSSMFE 14, 1-12 (1974)
[44] K.H.Korhonen, Stresses and strains in undrained tests, Ninth ICSMFE, Tokyo, I, 165-168 (1977)
[45] T.H.Wu, A.Douglas and R.Goughnour, Friction and cohesion of saturated clays, JSMFD, ASCE 90, 1-32 (1962)
[46] B.Broms and M.Ratnam, Shear strength of an anisotropically-consolidated clay, JSMFD, ASCE 89, 1-26 (1963)
[47] A.Sridharan, S.Rao and G.Rao, Shear strength characteristics of saturated montmorillonite and kaolin clays, Soils and Found., JSSMFE 11, 1-22 (1971)
[48] S.Saxena, J.Hedberg and C.C.Ladd, Geotechnical properties of Hackensack Valley varved clays of New Jersey, Geot.Testing J.,ASTM 1,148-161 (1978)
[49] R.W.Sparrow, P.G.Swanson and R.E.Brown, Report of laboratory testing of Gulf of Alaska Cores, Open File Report, U.S.Geol.Survey, Law Eng.Testing Co. (1979)
[50] C.Crawford, Some characteristics of Winnepeg Clay, Can.Geot.J. 1, 227-235 (1964)
[51] C.C.Ladd, Stress-strain behaviour of anisotropically consolidated clays during undrained shear, Sixth ICSMFE, Montreal I, 282-286 (1965)
[52] C.C.Ladd, Test embankment on sensitive clay, Performance of Earth and Earth-Supported Structures, ASCE, Purdue Univ. I, Part 1,101-128 (1972)
[53] A.N.Schofield, and C.P.Wroth, Critical State Soil Mechanics, McGraw-Hill Book Co., London, 1968
[54] K.H.Roscoe and J.B.Burland, On the generalized stress-strain behavior of 'wet' clay, Engineering Plasticity, J.Heyman and F.A.Leckie, eds., Cambridge Univ.Press, 535-609 (1968)
[55] K.Y.Lo, Shear strength properties of a sample of volcanic material of the Valley of Mexico, Geotechnique 12, 303-318 (1962)
[56] D.J.Henkel and V.A.Sowa, The influence of stress history on stress paths in undrained triaxial tests on clay, Laboratory Testing of Soils, ASTM STP 361, 280-291 (1964)
[57] K.H.Andersen, J.H.Pool, S.F.Brown and W.F.Rosenbrand, Cyclic and static laboratory tests on Drammen Clay, JGED, ASCE, 106, 499-530 (1980)
[58] C.C.Ladd, Foundation design of embankments constructed on Connecticut Valley varved clay, Research Report R75-7, Dept.of Civil Eng., Mass.Inst.of Technology (1975)
[59] A.W.Dhowian and T.B.Edil, Consolidation behavior of peats, Geot.Testing J., ASTM 3, 105-114 (1980)
[60] A.W.Skempton and V.A.Sowa, The behavior of saturated clays during sampling and testing, Geotechnique, 13, 269-290 (1963)

Numerical Methods in Geomechanics (Innsbruck 1988), Swoboda (ed.)
© 1988 Balkema, Rotterdam. ISBN 90 6191 809 X

A constitutive model with two yield surfaces for soils

Z.Z.Yin
Hohai University, Nanjing, People's Republic of China

ABSTRACT: A new constitutive model for soils with elliptic and parabolic yield surfaces was developed. It can reflect shear dilation. Comparison of calculated results using the model with the measured ones shows that it can fit various loading conditions. There are only 8 parameters in the model easy to be determined. It is convenient to be used.

1 INTRODUCTION

The 'cap' models with single yield surface, such as the Cam Clay model [7], the model proposed by Khosla and Wu [5], the model of Hwang [3], can get satisfactory results for general loading conditions. But they have some limitations. If a triaxial test sample is subjected to certain stresses p and q, and then the deviator stress q keeps constant while the mean normal stress p decreases, these models will regard the state as elastic rebounding since the stress state drops inside the 'cap' yield surface. As a matter of fact, the shear strain can be measured in the test. When the cell pressure continues to decrease, the sample even achieves to failure, and the shear strain may be very large. This shear strain is not elastic. The change of p can not result in elastic shear strain. The shear strain must be a kind of plastic strain. Therefore, inside the 'cap', there still exists yield. The second yield surface is necessary.

There are many models with two yield surfaces, such as those proposed by Lade [4], by prevest and Hoeg [6], by Vermeer [8], by the author and Duncan [10], etc.. Each has its own merits and shortcomings. This paper developed a new model with a elliptic yield surface and a parabolic yield surface.

2 IRRECOVERABLE DEFORMATION OF SOILS

The plastic deformation of soils is due to the mutual slide between particles. The microscopic slide between particles exists even for the plastic volumetric deformation. An isotropic compressive stress p applied on a sample only causes volumetric deformation. Fig.1(a) shows the particle arrangement before p is applied, and (b) is after loading. It can be seen that some of the particles dropped into original pores due to compression, and the mutual slide of particles existed.

For soils subjected to shear stress, the particle slide is more remarkable. From the macroscopic point of view, the slide is along the shear direction, but from the microscopic point of view, the slide is random since the particle shape and arrangement are irregular. In Fig.2, the shear force acts horizontally. but the particle A slides upwards and the particle B slides

(a)

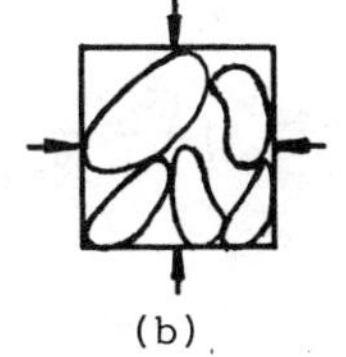

(b)

Fig.1 Change of particle arrangement
(a) Before compression
(b) After compresion

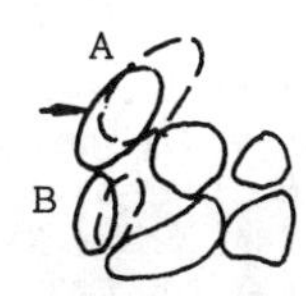

Fig,2 Slide between particles

downwards. Particles like A tend to be lifted resulting in dilation, while particles like B have a tendency to compress the soil. Both slides, causing dilation and causing compression, exist simultaneously in a given soil mass during the same loading process. The soil dilates if the former is dominant, otherwise the soil compresses.

It is reasonable to assume that the plastic strain consists of two parts: One reflects mainly the behavior of the particles which tend to cause compression, designated by ε_1^p, and the other reflects what cause dilation, designated by ε_2^p. They make up the total plastic strain.

$$d\varepsilon^p = d\varepsilon_1^p + d\varepsilon_2^p \tag{1}$$

It is necessary to use different kinds of yield surfaces and hardening laws to express these two parts of strain.

3 YIELD CRITERIA

3.1 The yield in connection with compression

The Modified Cam Clay Model provided a good yield criterion for the plastic strain relating to compression. Duncan et al.[2] revised it to be used for the soils with cohesion. However, the shape of elliptical locus in the Modified Cam Clay Model, namely, the ratio of the major axis to the minor axis, only depends on the strength parameter M. (M is the slope of the failure line q_f-P), and does not vary with the deformation characteristics of soils. For many soils it does not fit practice well. In this paper, the yield criterion for compression is based on the Modified Cam Clay Model and some revisions are made. The first Yield surface f_1 is expressed as

$$p + \frac{q^2}{M_1^2(p+p_r)} = P_0 = F(\varepsilon_v^p) \tag{2}$$

where

$$P = (\sigma_1 + \sigma_2 + \sigma_3)/3 \tag{3}$$

$$q = \frac{1}{\sqrt{2}}\sqrt{(\sigma_1-\sigma_2)^2+(\sigma_2-\sigma_3)^2+(\sigma_3-\sigma_1)^2} \tag{4}$$

p_r is the intercept of failure line q_f-p with p axis, M_1 is a parameter slightly greater than the value of M. The value of M_1 relates to the type of test σ-ε curves.

The yield surface locus and the meaning of parameters are shown in Fig.3.

The value of p_0 is the abscissa of the point where the yield locus intersects the p axis. In the Modified Cam Clay Model, a straight e-ln p relation is assumed. The void ratio e will be infinite when p=0 according to the assumption. In fact, it is not a straight line when p is small, especially for compacted soils. It was discovered that the $p_0 - \varepsilon_v$ or $p_0 - \varepsilon_v^p$ relation can be expressed as a hyperbola, as shown in Fig.4. Similarly, Wei has suggested the hyperbolic relation for the loading - compression strain relationship

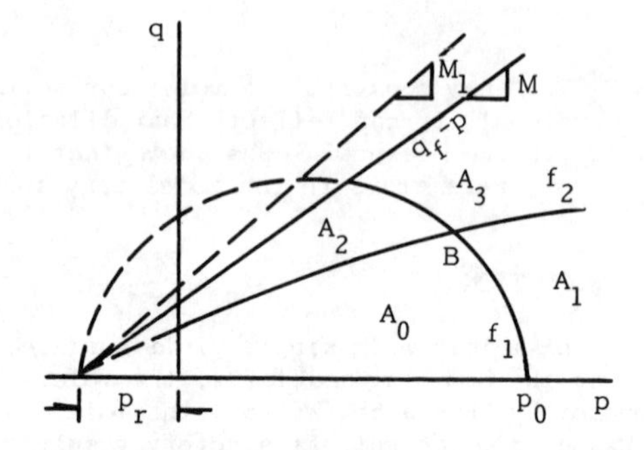

Fig.3 The yield surfaces

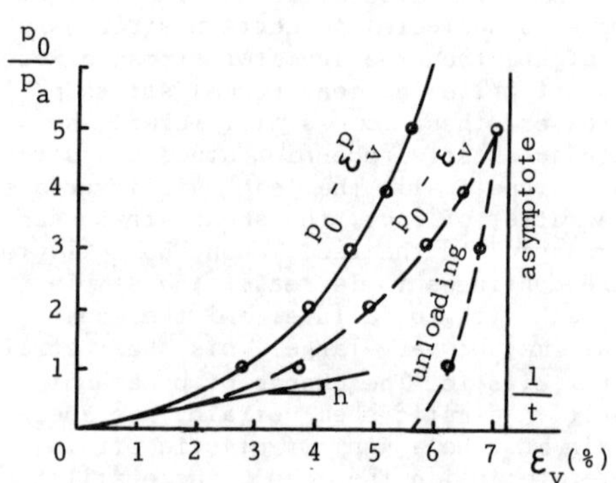

Fig.4 $p_0/p_a - \varepsilon_v$ relationship (Xiao Langdi clay)

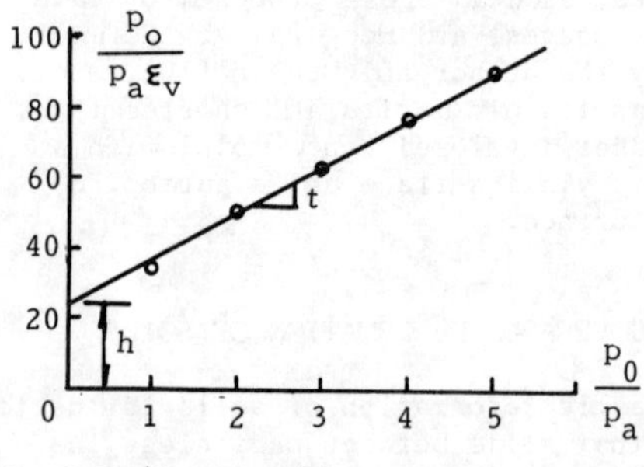

Fig.5 $P_0/\varepsilon_v - p_0$ line

[9]. Plotting $p_0/p_a\varepsilon_v$ vs. p_o/p_a as in Fig.5, results in a straight line with intercept h and slope t. Here p_a is the atomospharic pressure. The linear relationship has the equation

$$\frac{p_o}{p_a \varepsilon_v^p} = h + t \frac{p_0}{p_a} \quad (5)$$

Then,

$$p_0 = \frac{h \varepsilon_v^p}{1-t\varepsilon_v^p} p_a \quad (6)$$

This is the hardening law. Substitute it into Eq. (2)

$$p + \frac{q^2}{M_1^2 (p+p_r)} = \frac{h \varepsilon_v^p}{1-t\varepsilon_v^p} p_a \quad (7)$$

3.2 The yield in connection with dilation

There are two kinds of dilation. One is the elastic dilation, resulting from the reduction of the mean normal stress. The other is plastic dilation, which results from shearing. The yield of dilation is only in connection with shear stress and shear strain except tensile stress is applied. It is reasonable to take the plastic deviator strain ε_s^p as the hardening parameter of the second yield criterion,

$$\varepsilon_s^p = \int \frac{\sqrt{2}}{3} \sqrt{(d\varepsilon_1^p - d\varepsilon_2^p)^2 + (d\varepsilon_2^p - d\varepsilon_3^p)^2 + (d\varepsilon_3^p - d\varepsilon_1^p)^2} \quad (8)$$

The second yield criterion was developed according to the analysis of several triaxial test results.

$$\frac{a\,q}{G} \sqrt{\frac{q}{M_2(p+p_r)-q}} = \varepsilon_s^p \quad (9)$$

Where G is the elastic shear modulus, M_2 is the parameter slightly greater than M, the slope of q_f - p line, and 'a' is the parameter which mainly reflects the dilation or compression. Eq.(9) represents a parabola in the p - q plane, as is shown by curve f_2 in Fig.3.

Associated flow rule is assumed for both kinds of the yield surfaces.

In Fig.3. point B represents the current stress state . Yield surfaces f_1 and f_2 divide the p - q plane into four areas. A_0 is elastic area. A_1 is the plastic area with only first kind of yield. A_2 has only the second kind of yield. There exist both kinds of yield in area A_3. If q keeps constant and p decreases, the stress state falls into area A_2. There will be plastic shear strain and expansive volumetric strain according to the proposed model. If p increases, the stress state will be in area A_1, and the volumetric strain will be compressive. On the other hand, if p=constant and q increases, or both p and q increase like in the triaxial compression test, the stress state will drop in area A_3. At low stress level, yield locus f_2 is flat, the volumetric component of plastic strain is small according to associated flow rule, which is dilative; while yield locus f_1 is steep, and the compressive volumetric strain is great. Therefore, the total volumetric strain is compressive. At high stress level, the dilative strain relating to f_2 and the compressive strain relating to f_1 are competitive. Which will be dominant depends on the parameters, especially the parameter 'a'. If 'a' is high, yield surface f_2 dominates, and the volumetric strain is dilative. All these rules predicted by the new model are in accordance with the experiment results.

4 DETERMINING THE PARAMETERS

The parameters in the proposed model can be determined from conventional triaxial test results.

4.1 Elastic parameters G and ν

The Poisson's Ratio ν is assumed to be 0.3 for all soils. Since the elastic strain is a small portion of the total strain, the constant Poisson's Ratio does not result in great error.

The elastic shear modulus G can calculated from Young's Modulus E and Poisson's Ratio ν. Duncan et al. suggested that E varies from $1.2E_i$ to $3.0E_i$, where E_i is the initial tangent modulus (1), and can be determined as following

$$E_i = K\, p_a \left(\frac{p}{p_a}\right)^n \quad (10)$$

Approximately, let $E = 2.0E_i$, then the shear modulus

$$G = \frac{E}{2(1-\nu)} = \frac{1}{1.3} K p_a \left(\frac{p}{p_a}\right)^n \quad (11)$$

The parameters K and n can be determined according to Duncan's hyperbolic model.

Let

$$K_G = K/1.3 \quad (12)$$

Then

$$G = K_G\, p_a (p/p_a)^n \quad (13)$$

G varies with p. Two parameters K_G and n are used to determine G.

An alternative method for evaluating parameters K_G and n is based on q - ε_s curves. The initial slope of q - ε_s curve is $3G_i$. Assume $G = 2G_i$ .plotting G/p_a vs. p/p_a on log-log sheet, a straight line is achieved. Its intercept is K_G and slope is n.

4.2 M_1, M_2, and P_r

The values of p_r and M can be derived from c and φ from triaxial compressive tests using the equations:

$$p_r = c \cot\varphi \quad (14)$$

$$M = \frac{6 \sin\varphi}{3 - \sin\varphi} \quad (15)$$

They can also be determined from q_f - p line, as shown in Fig.3.

The value of M_1 varies from 1.0M to 1.5M. It relates to the ratio of $\varepsilon_v / \varepsilon_s$. To a certain extent, it influences dilation or compression. Generally, the ratio of M_1/M is 1.02 - 1.20. For soils of high compression, it may be greater than 1.20.An experiential equation for M_1 is as following

$$M_1 = (1+0.25\beta^2)M \quad (16)$$

where β is the average value of $\varepsilon_{v75}/\varepsilon_{a75}$. Here, ε_{v75} and ε_{a75} are the volumetric strain and axial strain respectively at stress level s = 75% for triaxial drained tests.

The value of M_2 influences the shape of calculated q - ε_s curves. It varies between 1.02M to 1.15M, and can be calculated using the following experiential equation,

$$M_2 = M/R_f^{0.25} \quad (17)$$

where R_f is failure ratio as in Duncan's model. $R_f = q_f/q_{ult}$. and q_{ult} is the asymptotic value of q in the hyperbolic q - ε_s curve.

4.3 h and t

The method shown in Fig.5 can be used to evaluate h and t if the isotropic compressive and rebounding test results are available.

Another method is to make use of triaxial

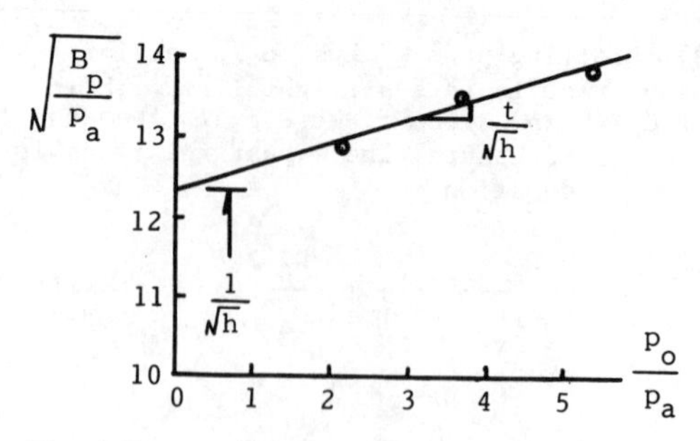

Fig.6 Determining h and t

drained test results. From Eq.(5),

$$\varepsilon_v^p = \frac{p_0}{hp_a + tp_0} \quad (18)$$

Differentiate it,

$$d\varepsilon_v^p = \frac{hp_a dp_0}{(hp_a + tp_o)^2} \quad (19)$$

Let

$$B_p = \frac{dp_0}{d\varepsilon_v^p} \quad (20)$$

Then,

$$B_p = \frac{p_a}{h}\left(h + t\frac{p_0}{p_a}\right)^2 \quad (21)$$

or

$$\sqrt{\frac{B_p}{p_a}} = \sqrt{h} + \frac{t\, p_0}{\sqrt{h}\, p_a} \quad (22)$$

At stress level s = 50%, the dilative strain caused by f_2 is small, and approximately compensates for the elastic compression. The measured volumetric strain ε_{v50} can be taken as $\Delta\varepsilon_{v1}^p$, the plastic volumetric strain associated with f_1. Calculate the value of p_0 from Eq.(2) according to the corresponding stresses p_{50} and q_{50}, and let $\Delta p_0 = p_0 - \sigma_3$, then

$$B_p = \frac{\Delta p_0}{\Delta\varepsilon_{v1}^p} \quad (23)$$

For different curves, value of B_p is calculated. The corresponding average p_0 is

$$\bar{p}_0 = (p_0 + \sigma_3)/2 \quad (24)$$

Plotting $\sqrt{B_p/p_a}$ vs. $\bar{p}_0/p_a$ results in a straight line with intercept $\sqrt{h}$ and slope $t/\sqrt{h}$, as shown in Fig.6. Then parameters h and t can be derived.

4.4 'a'

The parameter 'a' reflects what role the

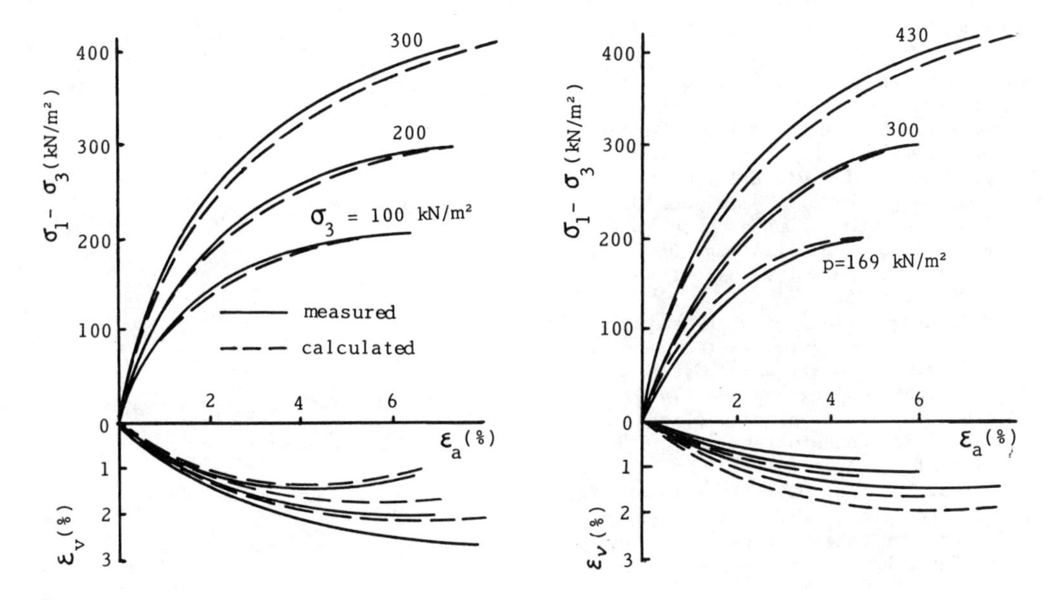

(a) σ_3 = constant (b) p = constant

Fig. 7 Triaxial test results and calculated results (Teton Dam silt)

yield surface f_2 plays in total deformation. The parameter 'a' varies from 0.1 to 0.5. A higher value of 'a' corresponds to dilation, while a lower value corresponds to compression. It can be estimated using the equation

$$a = 0.25 - 0.15\,d \tag{25}$$

where d is the average slope of the last part of ε_v - ε_a curves. The last section represents the part where the stress level varies from 75% to 95%.

Above experiential equations may not fit all soils. Adjustment may need for some soils.

All together, there are 8 parameters K_G, n, p_r, M_1, M_2, h, t, and 'a'. Their meaning is clear, and the method for determining them is simple.

5 EXAMINING THE MODEL

Several soils in different types have been studied to examine the proposed model. Good agreement was achieved between the calculated stress-strain curves and the measured ones for all these soils. The results of different stress paths, σ_3= constant and p = constant, are also good. As an example, the results of Teton Dam silt (see Ref.10) is shown in Fig.7, in which (a) is σ_3 = constant, and (b) is p= constant. The solid lines represent measured results, and the dotted lines represent the calculated results. The same parameters are used in the calculation for both the cases, which are derived from σ_3 = constant data. They are: K_G = 75, n = 0.21, p_r = 80 kN/m, M_1 = 0.89, M_2 = 0.89, h = 154, t = 3.5, a = 0.33. It was observed that calculated relults and the measured ones are quite close to each other.

The model was used to analyse the Stranstead Abbotts reinforced embankment in England by finite element method. The calculated results are close to the meassured ones. Reference [11] gives the detail.

The proposed model is based on the modified Cam Clay Model, and the concept of Duncan's hyperbolic model is used to determine some parameters. It has the merits of both the models. It overcomes the limitations of the single yield surface model, can fit various loading conditions well, and can reflect dilation. The preliminary examination analysis gives the satisfactory results. Besides, its equations are simple. All parameters can be readily determined from triaxial drained tests. Consequently, it is a elasto-plastic model good for application.

REFFERENCES

[1] J.M.Duncan and C.Y.Chang, Non-linear analysis of stress and strain in soils, proc. ASCE, Journal of Soil Mechanics and Foundation Division, 96, SM5 (1970)
[2] J.M.Duncan, T.B.D'Orazio, C-S. Chang, K.S. Wong, and L.I.Namiq, Con2D: A finite element computer program for analysis of consolidation, Univ. of California, Berkeley, Report No. UCB/GT/81-1. (1981)
[3] W.X.Huang, Effect of work hardening rules on the elasto-plastic matrix, Int. Symposium on Soils under Cyclic and Transient Loading,Swansea (1980)
[4] P.V.Lade, Elasto-plastic stress strain theory for cohesionless soil with curv - ed yield surfaces, Int. J.Solids structures, Vol. 13, NO11, 1019-1035(1977)
[5] K.V.Khosla and T.H.Wu, Stress-strain theory for cohesionless soil, Journal of Geotechnical Engineering Division, ASCE, 102, GT4, 303-321 (1976)
[6] J.H.Prevost and K.Hoeg, Effective stress strain strength model for soils, J.of Geotechnical division, ASCE,101, GT3 (1975)
[7] K.H.Roscoe and J.B.Burland,On the generalized stress-strain behavior of 'wet' clay, Engineering plasticity(ed. J.Heyman and F.A.Leckie) Cambridge University, Press. 535-609 (1968)
[8] P.A.Vermeer, a double hardening model for sand, Geotechnique, 28,No.4, 413-433 (1983)
[9] R.L.Wei, A new method for analysing the compression test results, J. of Nanjing Hydraulic Research Institute, No.3 (1980)
[10] Z.Z.Yin and J.M.Duncan, A stress-strain relationship for dilative and non-dilative soils, University of califonia, Berkeley, Report No. UCB/GT/83-1 (1983)
[11] Z.Z.Yin and D.A.Sun, An elasto-plastic analysis of the Stranstead Abbotts embankment, under publication (1987)

Numerical Methods in Geomechanics (Innsbruck 1988), Swoboda (ed.)
© 1988 Balkema, Rotterdam. ISBN 90 6191 809 X

Verification of elasto-viscoplastic model of normally consolidated clays in undrained creep

T.Matsui & N.Abe
Osaka University, Suita, Japan

ABSTRACT: In this paper, the performance of the authors' elasto-viscoplastic model for normally consolidated clays to predict the undrained creep behavior is verified. First the undrained creep theory under triaxial stress condition is derived from our proposed model, then it is examined by the available test results of undrained triaxial creep. As the result, it is demonstrated that the proposed elasto-viscoplastic model can describe the creep behavior of natural undisturbed clays quite well.

1 INTRODUCTION

The clay is a material of plasticity and time-dependency. To represent these characteristics many elasto-viscoplastic models of clays have been proposed. However, the ability of these models to predict the realistic time-dependent behavior is not necessarily satisfactory.

For example, it is essential to mechanically understand the undrained creep characteristics of clays, which is a typical example of the time-dependency. Since Murayama and Shibata (1961) pointed out that the upper yield strength existed in the undrained creep behavior of clay, many experimental data showed that the upper yield strength, at which the undrained creep mode changed, was a very important attribute of undrained creep characteristics of clays. By the available elasto-viscoplastic models, however, it is impossible to predict the change of undrained creep mode depending on the creep stress level. In other words, each available elasto-viscoplastic model can only predict either the failure creep mode in which the creep failure always occurs or the stationary creep mode in which the creep failure never occurs.

The authors' elasto-viscoplastic model (Matsui and Abe, 1985) of normally consolidated clays being based on the flow surface theory can theoretically predict the upper yield strength and represent the change of undrained creep mode depending on the creep stress level.

In the present paper, first the theoretical background of the authers' model is described, and then the ability of our model to predict the undrained creep behavior including the change of the creep mode is verified with the some available data of the undrained triaxial creep tests.

2 ELASTO-VISCOPLASTIC MODEL BASED ON FLOW SURFACE THEORY

Matsui and Abe (1985) have proposed the following elasto-viscoplastic model of normally consolidated clays based on flow surface theory of viscoplasticity, in which the viscoplastic volumetric strain v^{vp} was assumed to be the strain hardening parameter

$$\dot{\varepsilon}_{ij} = H_{ijkl}\dot{\sigma}_{kl} + \left[\frac{\dfrac{\partial F}{\partial \sigma_{mn}}\dot{\sigma}_{mn} + \dfrac{\partial F}{\partial t}}{\dfrac{\partial F}{\partial p}} \right] \frac{\partial F}{\partial \sigma_{ij}} \quad (1)$$

$$F = \mu \ln\left[\frac{1}{\delta}\left[\left\{1-\exp\left(-\frac{\delta \cdot v}{\mu}{}_r t\right)\right\}\exp\left(\frac{f}{\mu}\right) + \delta\exp\left(-\frac{\delta \cdot v}{\mu}{}_r t\right)\right]\right] - v^{vp} = 0 \quad (2)$$

in which, $\dot{\varepsilon}_{ij}$ is the strain rate tensor, $\dot{\sigma}_{ij}$ the stress rate tensor, H_{ijkl} the elastic tensor, p the mean effective stress, t the elapsed time, v^{vp} the viscoplastic volumetric strain and μ, δ, $\dot{v}^v_r$ the viscoplastic material parameters. F is the viscoplastic loading function being called flow function. f in Eq.(2) is the volumetric strain in the reference state which is a function of stress. The flow function F is derived being based on the following basic assumptions.

(1) The deformation of clays is decomposed into the deformation in the reference state and the rate-dependent deformation.

(2) The reference state assumed in the model is not the static equilibrium state, but the viscous state specified with a finite volumetric strain rate.

(3) To introduce both the logarithmic creep law and the concept of creep equilibrium state into the model, the following characteristic equation is assumed

$$v^v = -\mu \ln\left(\frac{\dot{v}^v}{\dot{v}^v_r} + \delta\right) \qquad (3)$$

in which, $\dot{v}^v$ is the viscous volumetric strain rate and $\dot{v}^v_r$ the reference viscous volumetric strain rate (material parameter).

(4) The instantaneous viscoplastic response of clays does not exist. Therefore, clays always respond elastically at the moment of loading.

3 UNDRAINED CREEP THEORY UNDER TRIAXIAL STRESS CONDITION

Analytical expressions for undrained creep behavior under triaxial stress condition can be derived from the proposed elasto-viscoplastic model.

3.1 UNDRAINED CREEP EQUILIBRIUM STATE SURFACE AND UPPER YIELD STRENGTH

The volumetric strain f in Eq.(2) is assumed to be evaluated by original Cam-clay model. In other words, it is assumed that the mechanical behavior of normally consolidated clays in the reference state is described by original Cam-clay model. f is given by the following equation

$$f = \frac{\lambda-\kappa}{1+e_o}\left[\ln\left(\frac{p}{p_o}\right) + \frac{\eta}{M}\right] \qquad (4)$$

in which, λ is the compression index, κ the swelling index, e_0 the initial value of the void ratio e, p_0 the initial value of p, η the stress ratio and M the value of η in the critical state.

The following equation is derived from Eqs.(2) and (4) under the undrained condition.

$$\frac{\kappa}{1+e_o}\ln\left(\frac{p}{p_o}\right) + \mu\ln\left[\frac{1}{\delta}\left[\left\{1-\exp\left(-\frac{\delta}{\mu}\dot{v}^v_r t\right)\right\}\exp\left[\frac{\lambda-\kappa}{\mu(1+e_o)}\left\{\ln\left(\frac{p}{p_o}\right) + \frac{\eta}{M}\right\}\right] + \delta\exp\left(-\frac{\delta}{\mu}\dot{v}^v_r t\right)\right]\right] = 0 \qquad (5)$$

Assumed the time to be infinite, the following equation at the the undrained creep equilibrium stress state is obtained, in which the asterisk signifies the undrained creep equilibrium state.

$$\lambda\ln\left(\frac{p^*}{p^*_o}\right) + (\lambda-\kappa)\frac{\eta^*}{M} = 0$$
$$p^*_o = p_o\delta^{\frac{\mu(1+e_o)}{\lambda}} \qquad (6)$$

The undrained creep equilibrium stress line of Eq.(6) is similar to the undrained effective stress path in the reference state. Consequently, the following undrained creep equilibrium state surface can be generally obtained.

$$e^*_o - e^* = \lambda\ln\left(\frac{p^*}{p^*_o}\right) + (\lambda-\kappa)\frac{\eta^*}{M} \qquad (7)$$

Let us consider the undrained creep of a constant deviator stress. Assuming that the failure criterion of clays is not rate-dependent, the creep rupture occurs when the stress ratio η reaches M. Therefore, the deviator stress q* at the intersection of the undrained creep equilibrium stress line and the failure line (η=M) becomes the upper yield strength as shown in Fig.1. That is, the upper yield strength q_Y is theoretically obtained as follows:

$$q_Y = Mp_o\exp\left[\frac{\mu(1+e_o)}{\lambda}\ln\delta - \left(\frac{\lambda-\kappa}{\lambda}\right)\right] \qquad (8)$$

3.2 CREEP RUPTURE LIFE

Solving Eq.(5) for the elapsed time t under a constant creep stress (deviator stress) $\bar{q}$, the following equation is obtained.

$$t = \frac{\mu}{\delta \dot{v}_r^v} \ln \left[\frac{1-\delta \exp[-\frac{\lambda-\kappa}{\mu(1+e_0)}\{\ln(\frac{p}{p_0})+\frac{\eta}{M}\}]}{1-\delta \exp[-\frac{\lambda-\kappa}{\mu(1+e_0)}\{(\frac{\lambda}{\lambda-\kappa})\ln(\frac{p}{p_0})+\frac{\eta}{M}\}]} \right] \quad (9)$$

Eq.(9) represents the migration of effective stress state of undrained creep, that is, the excess pore water pressure changes with the elapsed time during undrained creep.
As the creep rupture occurs when the stress ratio η becomes M, the creep rupture life t_R is given by the following equation.

$$t_R = \frac{\mu}{\delta \dot{v}_r^v} \ln \left[\frac{1-\delta \exp[-\frac{\lambda-\kappa}{\mu(1+e_0)}\{\ln(\frac{\bar{q}}{Mp_0})+1\}]}{1-\delta \exp[-\frac{\lambda-\kappa}{\mu(1+e_0)}\{(\frac{\lambda}{\lambda-\kappa})\ln(\frac{\bar{q}}{Mp_0})+1\}]} \right] \quad (10)$$

3.3 CREEP STRAIN AND CREEP STRAIN RATE

In the undrained triaxial creep of a constant creep stress $\bar{q}$, the creep strain (axial strain) ε and the creep strain rate (axial strain rate) $\dot{\varepsilon}$ are theoretically given by the following equation.

$$\varepsilon = \frac{\bar{q}-q_0}{3G} + \frac{\kappa}{M(1+e_0)} \ln[\frac{Mp_0-\bar{q}}{Mp-\bar{q}}] \quad (11)$$

$$\dot{\varepsilon} = \frac{\dot{v}_r^v(\delta A-B)(B-\delta)}{[(\delta A-B)+(1-A)B\{1+(\frac{\lambda-\kappa}{\lambda})(1-\frac{\eta}{M})\}](M-\eta)} \quad (12)$$

in which

$$A = (\frac{p}{p_0})^{-\frac{\kappa}{\mu(1+e_0)}}$$

$$B = \exp[\frac{\lambda-\kappa}{\mu(1+e_0)}\{\ln(\frac{p}{p_0})+\frac{\eta}{M}\}]$$

4 EXPERIMENTAL VERIFICATION

In this section the availability of the triaxial undrained creep theory being previously described is verified by comparing with the two available triaxial undrained creep test results.

Vaid and Campanella (1977) carried out a series of triaxial undrained creep test for an undisturbed Haney clay. All specimens were normally-consolidated by an

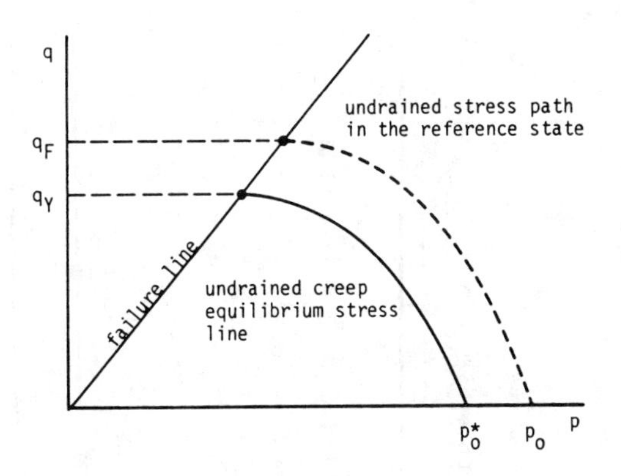

Fig.1 Undrained creep equilibrium stress line and upper yield strength

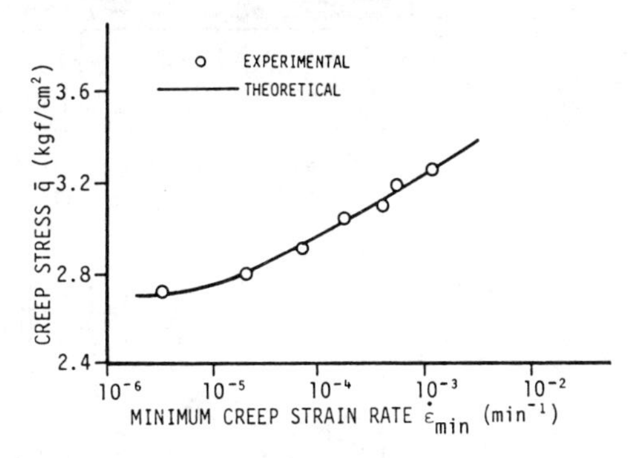

Fig.2 Relationship between creep stress and minimum creep strain rate for Haney clay

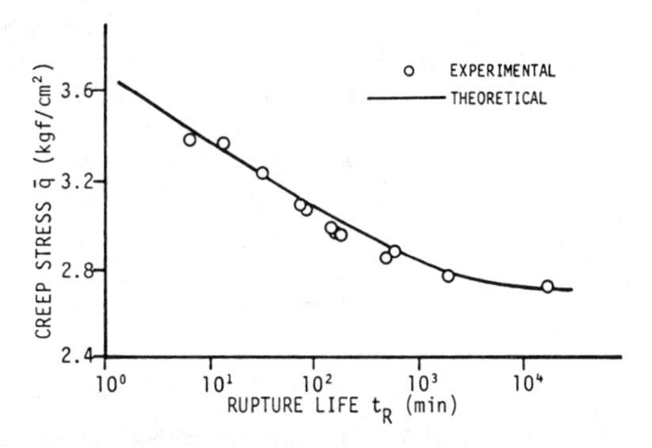

Fig.3 Relationship between creep stress and creep rupture life for Haney clay

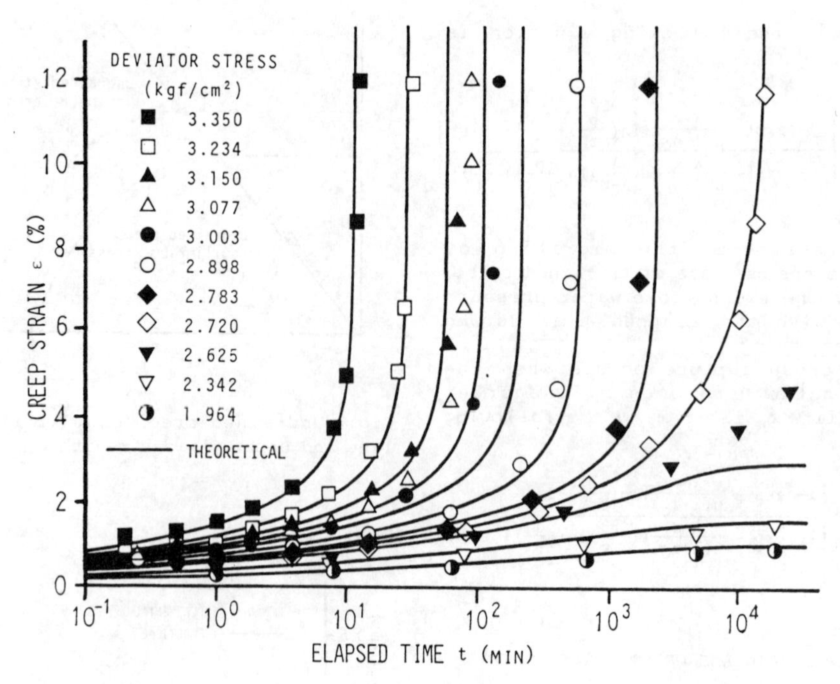

Fig.4 Developments of creep strain with elapsed time for Haney clay

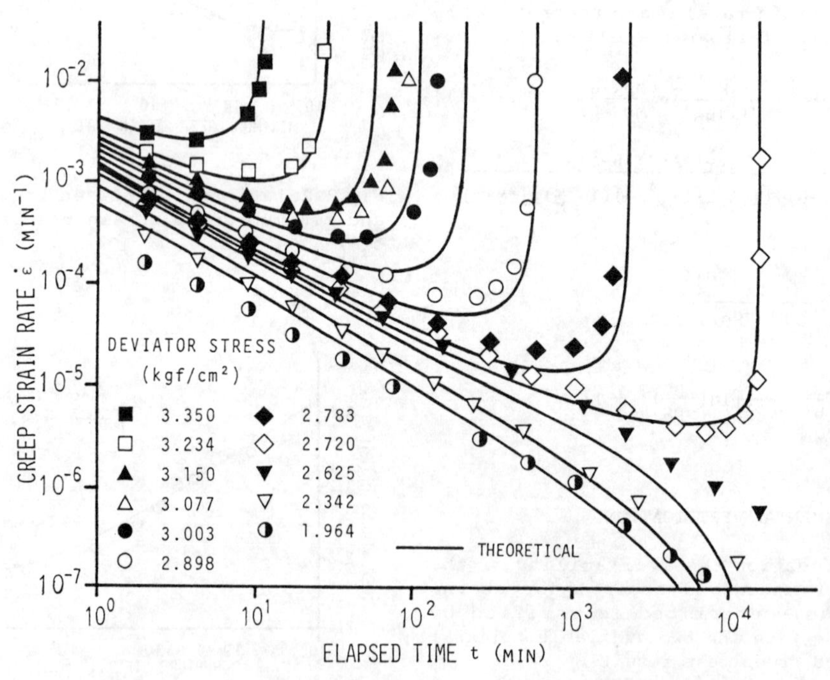

Fig.5 Variations of creep strain rate with elapsed time for Haney clay

isotropic pressure of 5.25 kgf/cm² for 36 hours. After then, keeping them in the undrained condition for 12 hours, the creep loading was applied to them. The reader can refer the detail of the test to their published paper.

Our elasto-viscoplastic model has eight material parameters. The five material parameters λ, κ, e_0, G and M for Haney clay are decided form the experimental results of the consolidation test and the undrained triaxial compression test, while the three viscoplastic parameters μ, δ and $\dot{v}_r^v$ are decided by the curve fitting for the test results in the relation between the creep stress and the minimum creep strain rate, as shown in Fig.2. All material parameters for Haney clay are summarized in Table 1.

Fig.3 shows the relation between the creep stress and the creep rupture life. It is seen from this figure that theoretical curve agrees well with the experimental data points and approaches asymptotically to a certain value of the deviator stress as decreasing the creep stress. This suggests the existence of the upper yield strength. The similar trend can be recognized in Fig.2, which suggests that the minimum strain rate in the undrained creep do not come out for the creep stresses of less than the upper yield strength.

Figs.4 and 5 show the comparison between the theoretical curves and the experimental data points for the variation of the creep strain and the creep strain rate with time, respectively. The experimental data points show that specimens for creep stress of more than 2.72 kgf/cm² go through states of the minimum strain rate and then accelerating creep states, and finally reach failure. Whereas, in specimens for creep stresses of less than 2.63 kgf/cm², their strain rates decrease with the elapsed time and never reach failure in 2 or 3 weeks.
The theoretical upper yield strength is obtained by Eq.(8) as 2.719 kgf/cm² (p_0=5.25 kgf/cm²) for Haney clay.

The theoretical curves in Figs.4 and 5 show that the failure creep mode appears in creep stress of more than the yield value, while the stationary creep mode in creep stress of less than that. Those creep mode characteristics agree quite well with the experimental results both qualitatively and quantitatively.

Murayama et al. (1970) carried out a series of triaxial undrained creep tests for an undisturbed Osaka clay. All specimens were normally-consolidated by an isotropic pressure of 3.0 kgf/cm² for 24 hours. Then, the creep loading was applied

Table 1 Material parameters for Haney clay

λ	κ	e_0	M	G	μ	δ	$\dot{v}_r^v$
0.20	0.031	0.896	1.29	600 [kgf/cm²]	0.004	0.168	8.0×10^{-6} [min⁻¹]

Table 2 Material parameters for Osaka clay

λ	κ	e_0	M	G	μ	δ	$\dot{v}_r^v$
0.343	0.105	1.303	1.47	128 [kgf/cm²]	0.00595	0.006335	4.13×10^{-6} [min⁻¹]

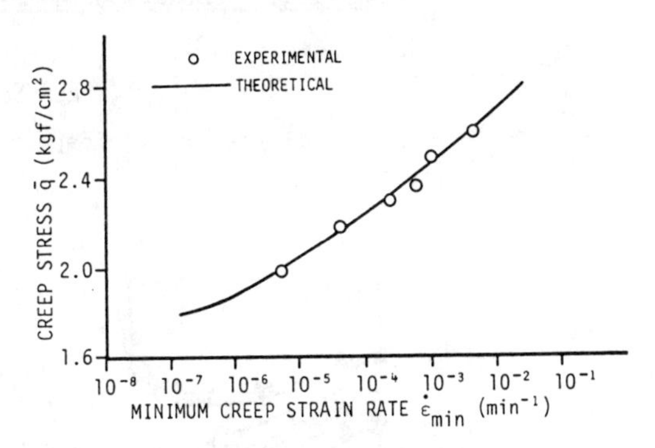

Fig.6 Relationship between creep stress and minimum creep strain rate for Osaka clay

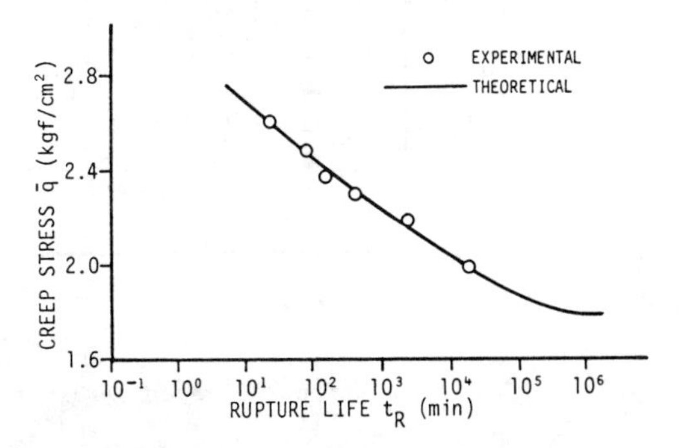

Fig.7 Relationship between creep stress and creep rupture life for Osaka clay

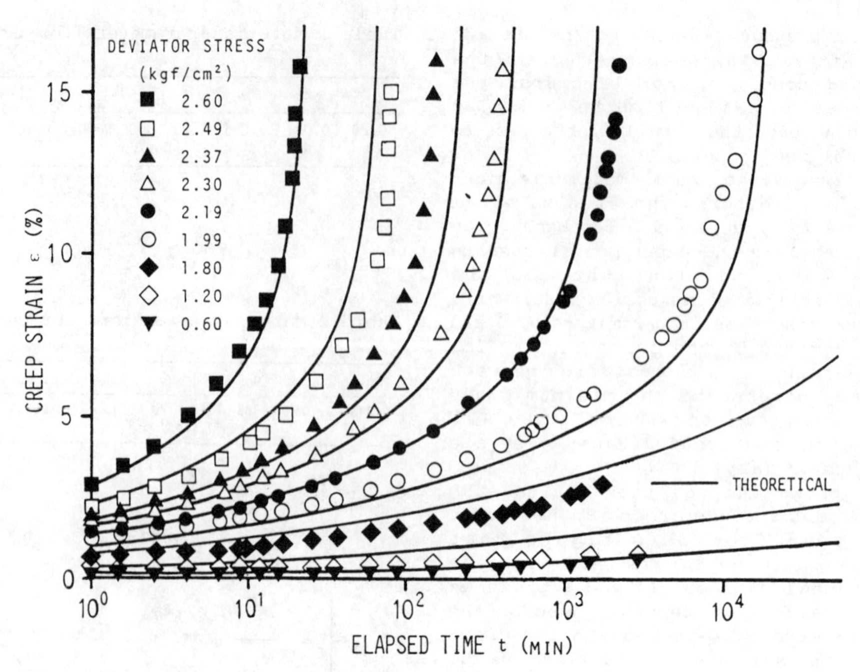

Fig.8 Developments of creep strain with elapsed time for Osaka clay

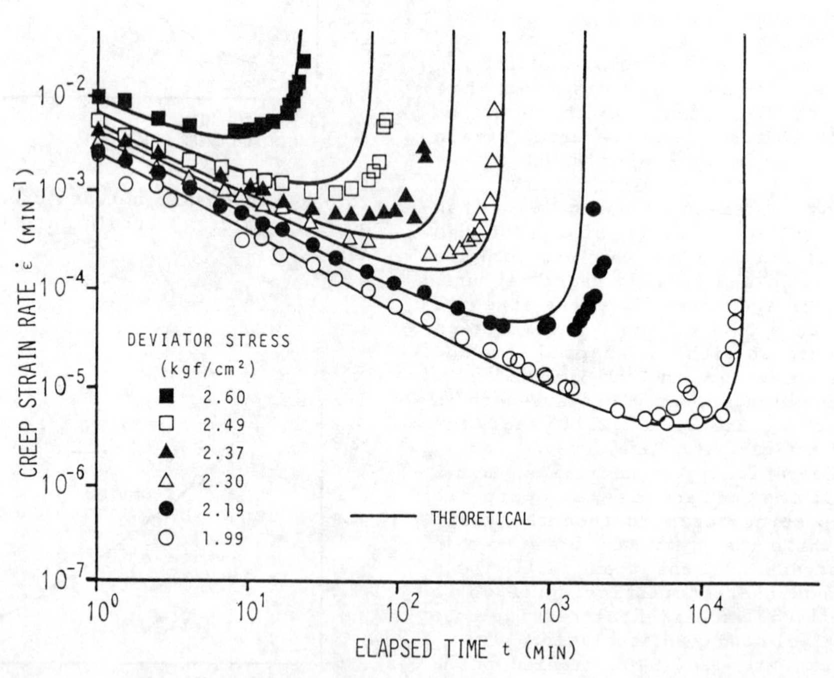

Fig.9 Variations of creep strain rate with elapsed time for Osaka clay

to them. The material parameters were decided by Sekiguchi (1984) except a viscoplastic parameter δ, which was decided in this paper by assuming that the upper yield strength is 1.80 kgf/cm² being based on the experimental result. All material parameters for Osaka clay are summarized in Table 2.

Figs.6 to 9 show the comparison between the theoretical curves and the experimental data points for Osaka clay. It is seen from these figures that the proposed creep theory being based on authors' elasto-viscoplastic model can successfully predict the undrained triaxial creep behavior for Osaka clay.

5 CONCLUSION

In the present paper, the ability of the authors' elasto-viscoplastic model of normally consolidated clays to predict the undrained creep behavior has been verified. That is, the triaxial undrained creep theory being derived from our proposed model was examined by comparing with the available test results of the undrained creep. As the result, it was demonstrated that the creep theory could adequately estimate the undrained creep mode depending on the creep stress level. Also, the theoretical creep behavior agreed quite well with the experimental behavior both qualitatively and quantitatively.

REFERENCE

Abe,N 1987. Elasto-viscoplastic model of clays including creep rupture criteria. Proc. 8th ARC, Vol.1, Kyoto : 1-4.

Matsui,T.& N.Abe 1985. Elasto/viscoplastic constitutive equation of normally consolidated clays based flow surface theory. Proc. 5th Int. Conf. Num. Methods Geomech., Vol.1, Nagoya:407-413.

Matsui,T. & N.Abe 1985. Undrained creep characteristics of normally consolidated clay based on flow surface model. Proc. 11th ICSMFE, Discussion Session IA, San Francisco : 140-143.

Matsui,T. & N.Abe 1986. Flow surface model of Viscoplasticity for normally consolidated clay. Proc. 2nd Int. Symp. Num. Models Geomech., Ghent : 157-164.

Murayama,S. & T.Shibata 1961. Rheological properties of clays. Proc. 5th ICSMFE, Vol.1, Paris : 269-273.

Murayama,S., N,Kurihara & H.Sekiguchi 1970. On creep rupture of normally consolidated clays. Annuals, Disaster Prevention Research Institute, Kyoto University, No.13B : 525-541 (in Japanese).

Sekiguchi,H. 1984. Theory of Undrained creep rupture of normally consolidated clay based on elasto-viscoplasticity. Soils and Foundations, 24-1 : 129-147.

Vaid,Y.P. & R.G.Campanella 1977. Time-dependent behavior of undrained clay. J. GT Div., ASCE, No. GT7 : 639-709.

Numerical Methods in Geomechanics (Innsbruck 1988), Swoboda (ed.)
© 1988 Balkema, Rotterdam. ISBN 90 6191 809 X

A solution method for continua with continuously varying stiffnesses

M.A.Koenders & A.F.Williams
Kingston Polytechnic, UK

ABSTRACT: A solution method is put forward to solve the displacement field for a situation where stiffness variations are confined to a prescribed area. A perturbation method is employed which is akin to the one used to obtain estimates for overall stiffnesses for heterogeneous materials.

As an example a two dimensional half space is deformed by a spatially periodic point load. An anisotropic material model is used. Fluctuations in stiffness are imposed on the part of the medium adjacent to the loaded boundary. An error analysis is demonstrated.

1 GENERAL PRINCIPLES

The problem addressed in this paper is to obtain estimates for a displacement field $\underline{u}(\underline{x})$ defined by the static equilibrium equations with position dependent stiffnesses $\underline{A}(\underline{x})$:

$$\frac{\partial}{\partial x_j}\left(A_{ijkl}\frac{\partial u_k}{\partial x_l}\right) = 0 \qquad (1)$$

In equation (1), as in the rest of this paper, Einstein's summation convention is employed: subscripts occurring twice are summed over. A further assumption is that all deformations are sufficiently small to allow the approximation of geometrical lenearisation (see Becker (1975)).

The idea is to use a perturbation theory type of approach. Such an approach has been employed in the past by Kneer (1965) and Kroner (1967) for the case of a heterogeneous material where the heterogeneity is more or less random. A first order estimate for the distortion is available from the overall strain. No boundary conditions need to be considered, just the average field is required.

Here the approach is similar but the first order approximation which is needed for subsequent perturbations is determined by boundary conditions. Therefore no constant first order solution can be applied. The method outlined in detail below relies on Fourier transformation as the main vehicle to arrive at a solution. As a result the solution can be written in convolution form which is very suitable for numerical processing.

2 OUTLINE OF THE METHOD

The fluctuations in stiffnesses are assumed to be confined to a region. This region need not be the same as the region of the whole. It is necessary to define an average value of the stiffnesses. This is done by integration over a suitable volume V:

$$\bar{A}_{ijkl} = \int_V dv\, A_{ijkl}(\underline{x}) \qquad (2)$$

The fluctuations are now:

$$B_{ijkl}(\underline{x}) = A_{ijkl}(\underline{x}) - \bar{A}_{ijkl} \qquad (3)$$

In what follows a two dimensional approach is applied. This does not affect the generality of the problem.

A first order estimate for the displacement field $\underline{u}$ follows from:

$$\bar{A}_{ijkl}\frac{\partial^2 \bar{u}_k}{\partial x_l \partial x_j} = 0 \qquad (4)$$

supplemented with boundary conditions. This estimate will be called the zero solution, the existence of which is assumed. The displacement field is written as the zero solution plus perturbations $\underline{v}(\underline{x})$:

$$\underline{u}(\underline{x}) = \overline{\underline{u}}(\underline{x}) + \underline{v}(\underline{x}) \quad (5)$$

Using (3) and (5) the equilibrium equations (1) become:

$$\overline{A}_{ijkl} \frac{\partial^2 \overline{u}_k}{\partial x_l \partial x_j} + \frac{\partial}{\partial x_j} (B_{ijkl} \frac{\partial \overline{u}_k}{\partial x_l}) + + \overline{A}_{ijkl} \frac{\partial^2 v_k}{\partial x_l \partial x_j} + \frac{\partial}{\partial x_j} (B_{ijkl} \frac{\partial v_k}{\partial x_l}) = 0 \quad (6)$$

The first term in (6) vanishes due to (4). The last term is a product of two fluctuating quantities and is assumed to be of an order less than the other terms. Thus:

$$\overline{A}_{ijkl} \frac{\partial^2 v_k}{\partial x_l \partial x_j} = - \frac{\partial}{\partial x_j} (B_{ijkl} \frac{\partial \overline{u}_k}{\partial x_l}) \quad (7)$$

A particular solution of (7) is found using Fourier transformation. The homogeneous equation serves to adjust the boundary conditions which have been changed by the particular solution:

$$\overline{A}_{ijkl} \frac{\partial^2 v_k}{\partial x_l \partial x_j} = 0 \quad (8)$$

supplemented with boundary conditions. (8) has the same form as (4). The method is particularly efficient when (4) can be solved under more or less general boundary conditions so that the result is easily adapted to suit (8).

To derive a particular solution the Fourier transform $\tilde{\underline{v}}(\underline{\omega})$ is introduced:

$$v_k(\underline{x}) = \frac{1}{(2\pi)^2} \int d_2\omega \, \tilde{v}_k(\underline{\omega}) \, e^{i\underline{\omega}.\underline{x}} \quad (9)$$

Similarly:

$$\frac{\partial}{\partial x_j} (B_{ijkl} \frac{\partial \overline{u}_k}{\partial x_l}) = \frac{i}{(2\pi)^2} \int d_2\omega \omega_j \widetilde{(B_{ijkl} \frac{\partial \overline{u}_k}{\partial x_l})}(\underline{\omega}) \, e^{i\underline{\omega}.\underline{x}} \quad (10)$$

So that the solution can be written as a convolution:

$$v_a(\underline{x}) = \int d_2y \; F^{-1}\{P^{-1}_{ap} i\omega_j\}(\underline{x}+\underline{y}) B_{pjkl} \frac{\partial \overline{u}_k}{\partial y_l} \quad (11)$$

where:

$$P_{ik} = \overline{A}_{ijkl}\omega_j\omega_l \quad (12)$$

The accuracy of the method can be read from the magnitude of the neglected term in (6). A scalar measure is:

$$\varepsilon = \left| \frac{B_{iikl} \frac{\partial v_k}{\partial x_l}}{\overline{A}_{iikl} \frac{\partial \overline{u}_k}{\partial x_l}} \right| \quad (13)$$

The expression $F^{-1}\{P^{-1}_{ap} i\omega_j\}$ in (11) is important. It is easily verified that to evaluate it one needs integrals of the following form:

$$I_p = i \int d_2\omega \frac{\omega_1^{\,p} \omega_2^{\,3-p} \, e^{i\underline{\omega}.}}{a\omega_1^{\,4} + b\omega_1^{\,2}\omega_2^{\,2} + c\omega_2^{\,4}} \quad (p=0..3) \quad (14)$$

where a, b and c are defined as:

$$a = \overline{A}_{1111}\overline{A}_{2121} \; ; \; c = \overline{A}_{2222}\overline{A}_{1212}$$

$$b = \overline{A}_{1111}\overline{A}_{2222} + \overline{A}_{1212}\overline{A}_{2121} + - (\overline{A}_{2112}+\overline{A}_{2211})(\overline{A}_{1122}+\overline{A}_{1221}) \quad (15)$$

The integrals (14) are evaluated. The result depends on the location of the $\frac{a}{c}\omega_1^{\,4} + \frac{b}{c}\omega_1^{\,2}\omega_2^{\,2} + \omega_2^{\,4} = 0$. If $b^2-4ac<0$ $\omega_2 = (\pm\alpha \pm i\beta)\omega_1$; if $b^2-4ac>0$ then $\omega_2 = \pm i\gamma\omega_1, \pm i\delta\omega_1$. The result of the integration is given here:

$b^2-4ac<0$:

$$I_p = \frac{\pi}{2\alpha\beta c} \left[\frac{J_p^+}{\beta^2 x_2^{\,2} + (x_1+\alpha x_2)^2} + \frac{J_p^-}{\beta^2 x_2^{\,2} + (x_1-\alpha x_2)^2} \right] \quad (16)$$

where:

$$J_0^{\pm} = \pm(\beta^2-\alpha^2)(x_1 \pm \alpha x_2) - 2\alpha\beta^2 x_2$$

$$J_1^{\pm} = -\alpha(x_1 \pm \alpha x_2) \mp \beta^2 x_2$$

$$J_2^{\pm} = \pm(-x_1 \pm \alpha x_2)$$

$$J_3^{\pm} = -\alpha(x_1 \pm \alpha x_2) \pm \beta^2 x_2$$

and:

$b^2-4ac>0$:

$$I_p = \frac{2\pi}{c(\delta^2-\gamma^2)} \left(\frac{F_p}{\delta^2 x_2^2+x_1^2} - \frac{G_p}{\gamma^2 x_2^2+x_1^2} \right) \tag{17}$$

where:

$F_0 = \delta^3 x_2$; $G_0 = \gamma^3 x_2$

$F_1 = -\delta x_1$; $G_1 = -\gamma x_1$

$F_2 = \delta x_2$; $G_2 = \gamma x_2$

$F_3 = -x_1$; $G_3 = -x_1$

Note that in the above $\underline{\overline{A}}$ is given transverse isotropic properties with the axes coinciding with the x_1-x_2 axes. For the average properties this is not a great restriction. No conditions are imposed on the variations $\underline{B}$.

The reconstruction of the vectors $\underline{v}$ from the I_p's goes as follows. Introduce the quantity $D(\underline{x})$:

$$D_{qj}^{(p)} = \frac{1}{(2\pi)^2} \int d_2y \; I_p B_{qjkl} \frac{\partial \overline{u}_k}{\partial y_l} \tag{18}$$

Now:

$$v_1(\underline{x}) = \overline{A}_{2121}D_{11}^{(3)} + \overline{A}_{2222}D_{11}^{(1)} + \overline{A}_{2121}D_{12}^{(2)} + \\ + \overline{A}_{2222}D_{12}^{(0)} - (\overline{A}_{2112}+\overline{A}_{2211})(D_{21}^{(2)}+D_{22}^{(1)}) \tag{19}$$

$$v_2(\underline{x}) = \overline{A}_{1111}D_{21}^{(3)} + \overline{A}_{1212}D_{21}^{(1)} + \overline{A}_{1111}D_{22}^{(2)} + \\ + \overline{A}_{1212}D_{22}^{(0)} - (\overline{A}_{1122}+\overline{A}_{1221})(D_{11}^{(2)}+D_{12}^{(1)}) \tag{20}$$

From the form of the I_p's it is seen that the method does not work if $\beta \to 0$ or if δ, $\gamma \to 0$. This is equivalent to the requirement of ellipticity for the average problem:

$$\det(\overline{A}_{ijkl}\omega_j\omega_l) \neq 0 \tag{21}$$

The implications for a, b and c and the resulting conditions on the moduli are discussed in the literature (Koenders, Arthur and Dunstan 1987).

Higher order approximations for the problem are found by substituting the first order approximation back into the equilibrium equations. This is not carried out here. Kroner (1967) has shown how it can be done for a random material.

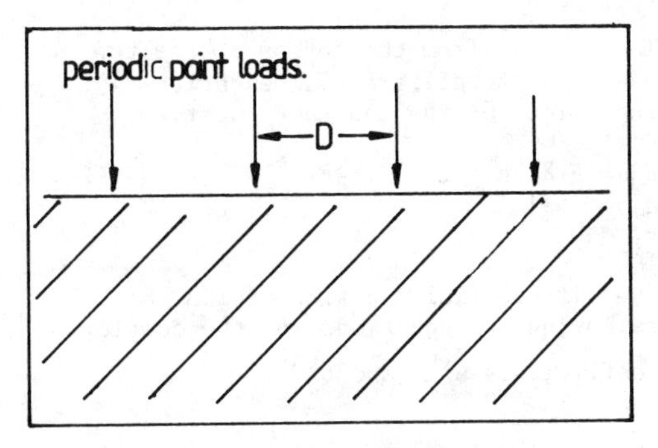

Fig 1. Illustration of the geometry.

3 A PERIODIC POINT LOAD ON A HALF SPACE

An example is worked out to demonstrate the method. A periodic point load with spacing D works on a half space. The properties of the half space will be assumed to be anisotropic. The space is homogeneous, but where the deformation varies, that is near the boundary the constitutive properties will undergo slight changes. This will model to a degree the stress-dependence of a soil.

Due to the periodic nature of the problem the variations in stiffness are also periodic.

The solution of eqation (4) has the form:

$$\overline{u}_k = a_k^{(0)}x_k + \sum_{n=1}^{\infty} \exp(-\beta n\omega x_2) \\ \mathrm{Re}(a_k^{(n)})\cos(n\omega(x_1+\alpha x_2)) + \\ - \mathrm{Im}(a_k^{(n)})\sin(n\omega(x_1+\alpha x_2)) + \\ + \mathrm{Re}(b_k^{(n)})\cos(n\omega(x_1-\alpha x_2)) + \\ - \mathrm{Im}(b_k^{(n)})\sin(n\omega(x_1-\alpha x_2)) \tag{22}$$

where:

$$\omega = 2\pi/D \tag{23}$$

Note that the solution (22) is not correct when the material has isotropic average properties. When that is the case solutions of the form:

$\exp(-\beta n\omega x_2)x_2\cos(n\omega x_1)$ and

$\exp(-\beta n\omega x_2)x_2\sin(n\omega x_1)$

must be taken into account. For isotropic materials $\alpha \to 0$. Alternatively the limit can be taken in formula (22).

The coefficients $\mathrm{Re}(a_k)$, $\mathrm{Im}(a_k)$, $\mathrm{Re}(b_k)$ and $\mathrm{Im}(b_k)$ for all n are now found.

They follow from the forces at the top and from the differential equation (4). The forces at the top take the form:

$$F_2 = F \sum_{n=1}^{\infty} \cos(n\omega x_1) + \tfrac{1}{2}F \qquad (24)$$

$$F_1 = 0$$

The differential equation yields the following two equations for the complex coefficients $a_k^{(n)}$ and $b_k^{(n)}$:

$$(\bar{A}_{1111} + (\alpha + i\beta)^2\bar{A}_{1212})a_1^{(n)} +$$
$$+ (\bar{A}_{1122} + \bar{A}_{1221})(\alpha + i\beta)a_2^{(n)} = 0 \qquad (25)$$

$$(\bar{A}_{1111} + (\alpha - i\beta)^2\bar{A}_{1212})b_1^{(n)} +$$
$$(\bar{A}_{1122} + \bar{A}_{1221})(\alpha - i\beta)b_2^{(n)} = 0 \qquad (26)$$

Thus there are 8 equations for 8 unknowns which is solved numerically.

It pays to set up the routine to solve this in a general way. Instead of (24) the forces at the top can be written as:

$$F_2 = \sum_{n=1}^{\infty} F_2^{(n)}\cos(n\omega x_1) + G_2^{(n)}\sin(n\omega x_1) + F_2^{(0)} \qquad (27)$$

$$F_1 = \sum_{n=1}^{\infty} F_1^{(n)}\cos(n\omega x_1) + G_1^{(n)}\sin(n\omega x_1) + G_1^{(0)} \qquad (28)$$

In so doing the solution for equation (8) is found with very little extra effort.

From the form of (22) it is seen that the coefficients a_k and b_k depend on n as 1/n. Due to the presence of the factor $\exp(-\beta n\omega x_2)$ the series converges always for $x_2>0$. It is therefore justified to replace the infinite sum by a finite one. All the time it is assumed that $\beta>0$, which is one of the key assumptions for the whole method to work.

An alternative method to arrive at the zero solution is to compute the deformation field of one point load and to convolve this solution with the Dirac comb (24). This is not pursued here because that method is not so easily generalised to include general conditions as considered in equations (27) and (28).

The term with n=0 in equation (22) is not found from (24)...(26), but more directly:

$$a_1^{(0)} = 0 \quad ; \quad a_2^{(0)} = \tfrac{1}{2}F/\bar{A}_{2222}/D \qquad (29)$$

All other terms in (22) decay rapidly with increasing x_2. With (29) the displacement increases. The strain however is a constant and this quantity is important for the evaluation of the correction (formula (11)).

The fluctuations are assumed to be of the form:

$$B_{ijkl} = \tfrac{1}{2}\bar{A}_{ijkl}\cos(n\omega x_1)\exp(-\beta\omega x_2) \qquad (30)$$

The integration (11) is done numerically using the trapezium rule in two dimensions. Note that due to the form of the influence function (formulas (16) and (17)) only the contribution directly around the point of evaluation needs to be taken into account. The amount of computation is therefore small. The method is suitable for a microcomputer. In this case the calculation is particularly efficient because the quantities needed for the calculation of the I_p's are also part of the zero solution.

After the calculation of $\underline{v}(\underline{x})$ a correction is introduced to readjust the boundary conditions. The gradients needed for this are also computed numerically. The traction has the form:

$$A_{i2kl}\frac{\partial v_k}{\partial x_1} + B_{i2kl}\frac{\partial \bar{u}_k}{\partial x_1} \quad (x_2=0) \qquad (31)$$

Numerical Fourier transformation makes the correction suitable for formula (27).

The errors in the calculation may be due to the following factors:

a. Numerical errors in the integration of (11) or the determination of the numerical derivatives for (31) and their subsequent Fourier transformation.
b. Turning the series in (22) into a finite series rather than an infinite one.
c. Neglecting the last term of (6).

Factors a. and b. can be made arbitrarily small by refining the mesh of the calculation and by adding more terms in the Fourier series. Using formula (13) the error in c. can be investigated. A pilot calculation is done for an anisotropic medium with stress/strain relation:

$$\begin{bmatrix}\sigma_{11}\\ \sigma_{12}\\ \sigma_{21}\\ \sigma_{22}\end{bmatrix} = k \begin{bmatrix}1. & 0. & 0. & .5\\ 0. & .7 & .7 & 0.\\ 0. & .7 & .7 & 0.\\ .5 & 0. & 0. & .5\end{bmatrix} \begin{bmatrix}e_{11}\\ e_{12}\\ e_{21}\\ e_{22}\end{bmatrix} \qquad (32)$$

where k is a scale factor. The value of β is .62. The values of ε are plotted in figure 4. It is seen that the maximum value never exceeds 15%.

In figure 2 the displacement field of the zero solution is shown. Figure 3

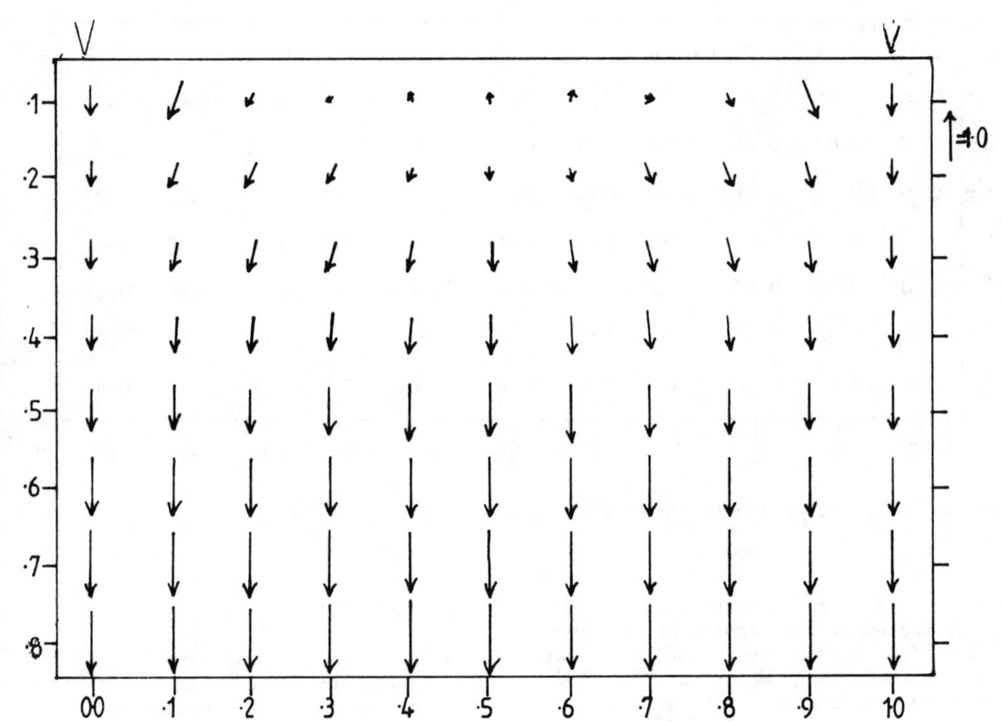

Fig 2. Displacement field of the zero solution. The point loads apply at the points marked V .

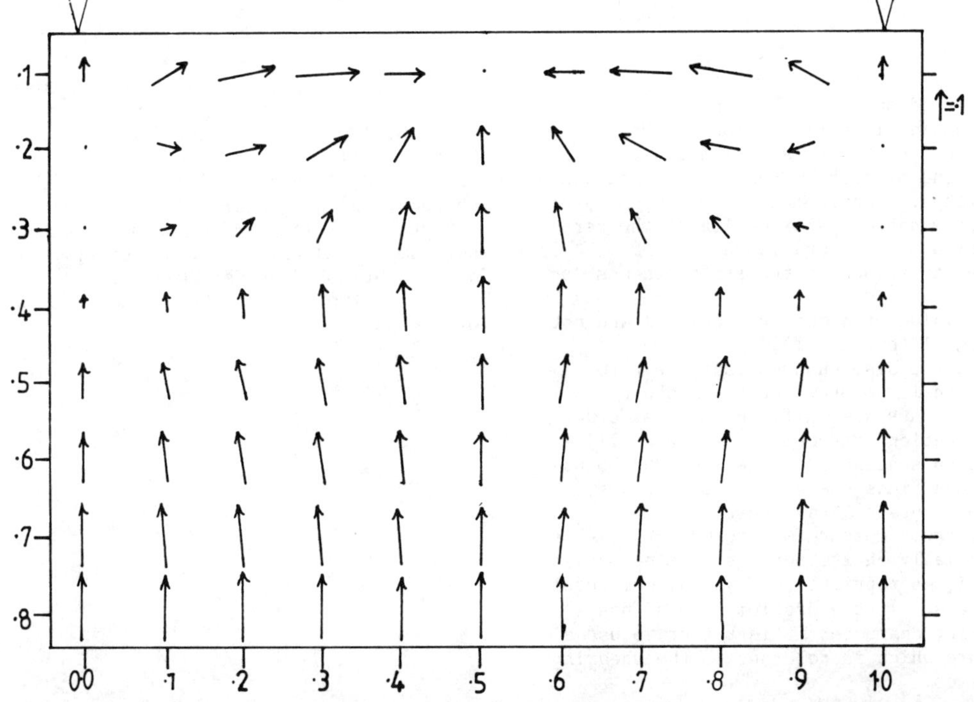

Figure 3. Displacement field of the first order perturbation; the point loads apply at the points marked V .

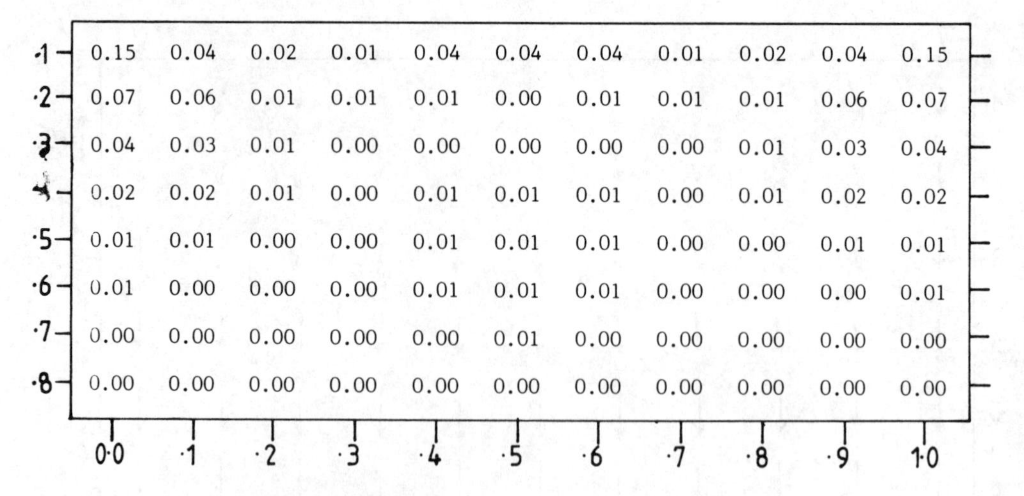

Figure 4. Errors according to formula (13).

shows the first order correction $\underline{v}(\underline{x})$. Note that the scales of the figures are different. The figures show one period only. The weakness introduced by equation (30) attracts extra displacement. The greater stiffness under the point loads causes an overall correction opposing the zero solution.

4 DISCUSSION

The method outlined works well for the kind of example given in section 3. Here it is easy to find a zero solution and it is for this kind of problem that the method was developed. Requirements for its use are:
- A reasonable approximation of the zero solution must be available.
- The variations in the stiffness must be smooth.
- The value of β must be positive and not too small (e.g. $\beta > .1$).

The latter also ensures that the zero solution is smooth. When the above requirements are fulfilled a first order perturbation correction - formula (11) - and a subsequent adjustment of the boundary conditions - equation (8) - gives quite accurate, fast results.

The error measure ε - formula (13) - is essentially an isotropic stress comparison. This is appropriate for typical compression problems. If the problem to hand has a shearing character it is better to use a measure which is relevant to the shearing stresses.

5 ACKNOWLEDGEMENTS

The work was carried out with financial support from the Dutch Government section Civil Engineering of Rijkswaterstaat. The project title is "Settlements associated with Unstable Filters".

6 REFERENCES

Becker, E. & Burger, W. 1975. Kontinuumsmechanik. Teubner. Stuttgart.
Kneer, G. 1965. Phys. Stat. Solids 9, pp 825-838.
Koenders, M, Arthur, J, Dunstan, T. 1987. The behaviour of granular materials at peak stress. IUTAM/ICM Symposium on Yielding, Damage and Failure of Anisotropic Solids. Grenoble August 24-28.
Kroner, E. 1967. Journ. Mech. Phys. Solids. Vol 15, pp 319-329.

Numerical Methods in Geomechanics (Innsbruck 1988), Swoboda (ed.)
© 1988 Balkema, Rotterdam. ISBN 90 6191 809 X

Mathematically consistent formulation of elastoplastic constitutive equations

K.Hashiguchi
Department of Agricultural Engineering, Kyushu University, Fukuoka, Japan

ABSTRACT: The original subloading surface model proposed by the author (or the bounding surface model with a radial mapping by Dafalias) does not assume a yield surface enclosing a purely elastic domain, and instead it assumes a subloading surface which always passes through a current stress point even in the unloading state. Thus, it describes a continuous stress rate-plastic strain rate relation (smooth elastic-plastic transition) and its loading criterion does not require the judgement whether a current stress lies on a yield surface or not. It can not, however, describe reasonably an induced anisotropy and a hysteresis effect, since the center of similarity of the normal-yield or the subloading surfaces is fixed or is not formulated reasonably. Recently, the author presented the mathematically exact formulation of the subloading surface model. In this paper, this extended subloading surface model is applied to granular materials, and it is compared with some test data of sand.

1 INTRODUCTION

The classical theory of plasticity is concerned only with a description of the remarkable plastic deformation in the yield state, ignoring a plastic deformation due to a stress change within the yield surface by assuming its interior to be a purely elastic domain. In order to extend the classical theory so as to describe also the plastic deformation due to a stress change within the yield surface, the author proposed the "subloading surface model" and refined it mathematically (Hashiguchi and Ueno, 1977a; Hashiguchi, 1978, 1980a). In this extension the yield state in which a stress lies on the classical yield surface and the state within that surface are called a "normal-yield state" and a "subyield state", respectively, and the classical yield surface is called a "normal-yield surface". The classical yield surface is also called a "bounding surface" (Dafalias and Popov, 1975) or a "distinct-yield surface" (Hashiguchi, 1980a), while it is doubtful to replace the term "yield" surface used historically in the theory of plasticity by the term "bounding" surface expressing a geometrical meaning rather than a physical one. The salient feature of the subloading model is the assumption of the "subloading surface" which expands/contracts passing always through a current stress point even in the unloading state and retaining a similarity to the normal-yield surface, and the plastic modulus depends on the ratio of the size of the subloading surface to that of the normal-yield surface. Since a purely elastic domain is not assumed in this model, a continuous stress rate-strain rate relation is described bringing about a smooth elastic-plastic transition and the loading criterion does not require the judgement whether a stress lies on the yield surface or not. Later on, Dafalias and Herrmann (1980) presented a similar idea in which a plastic modulus depends on the ratio of the magnitude of the current stress to that of the conjugate stress on the normal-yield surface whereas the subloading surface is not utilized explicitly in their formulation. They call it a "bounding surface model (with a radial mapping)" as well as the earlier formulations (Dafalias and Popov, 1975, 1976, 1977) which fall within the framework of the so-called "two-surface model". The two surface model assumes a small yield surface, called a "subyield surface" (Hashiguchi, 1981), which encloses a purely elastic domain and moves with a plastic deformation within the normal-yield surface, keeping its size constant relatively to the size of the normal-yield surface, while this surface moves without a plastic deformation for the stress change along the normal-yield surface. Then, the two surface model as well as the multi surface model (model of a field of hardening moduli) proposed by Mroz (1966, 1967) and Iwan (1967) is regarded to be an extension of the kinematic hardening model (Edelman and Drucker, 1951; Ishlinski 1954; Prager, 1956; Hodge, 1957) to the subyield state. Eventually, the subloading surface model has a different structure from the two surface model. The term "subloading surface model" would express concisely the physical feature of this model which is an extension of the classical theory to the subyield state by assuming the subloading surface within the normal-yield surface, while both two and subloading surface models are called a "bounding surface model" by Dafalias (1986) though they have different structures and the bounding surface is just a yield surface in the classical theory as was described above. The subloading or bounding surface model has been applied widely to the prediction of irreversible deformation of soils, metals and concretes. In the past, however, the center of similarity of the normal-yield and the subyield surfaces is fixed in the origin of stress space or on the central axis of the normal-yield surface. Then, an open hysteresis loop is predicted for the partial unloading-reloading cycle of stress as was criticized by Mroz and Norris (1982) and Mroz and Zienkiewicz (1984) to the author's papers (Hashiguchi, 1978, 1980a), and also Masing effect (Masing, 1926) cannot be described. Then, the author so extended it that the center of similarity translates with a plastic deformation (Hashiguchi, 1986a, b). In this paper, the subloading surface model is applied to soils and is compared with some test data of sand.

2 BASIC CONCEPT OF THE SUBLOADING SURFACE MODEL

Prior to description of constitutive equation of soils, let a qualitative nature of the subloading surface model be explained briefly.

The normal-yield and the subloading surfaces assumed in this model are described as

$$f(\hat{\boldsymbol{\sigma}}) - F(H) = 0 \tag{1}$$

$$f(\bar{\boldsymbol{\sigma}}) = \mathscr{f} \quad (\leqq F) \tag{2}$$

where

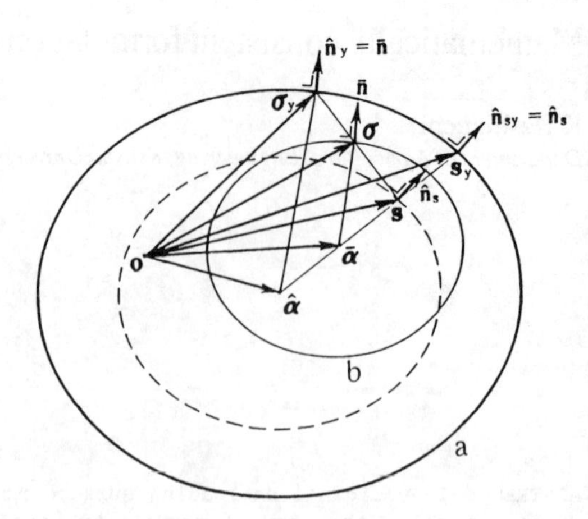

Fig.1 Normal yield surface: a and subloading surface: b.

$$\hat{\boldsymbol{\sigma}} \equiv \boldsymbol{\sigma} - \hat{\boldsymbol{\alpha}} \tag{3}$$

$$\bar{\boldsymbol{\sigma}} \equiv \boldsymbol{\sigma} - \bar{\boldsymbol{\alpha}} \tag{4}$$

$\boldsymbol{\sigma}$: stress
$\hat{\boldsymbol{\alpha}}$: center of normal-yield surface
$\bar{\boldsymbol{\alpha}}$: center of subloading surface
H : hardening/softening parameter
$\mathscr{f}$: value of $f(\bar{\boldsymbol{\sigma}})$ calculated by substituting a current value of $\bar{\boldsymbol{\sigma}}$.

These surfaces are assumed to retain a similarity and to be similar to each other with respect to a center of similarity $\mathbf{s}$ which translates with a plastic deformation*. They are illustrated schematically in Fig.1, in which

$$\bar{\mathbf{n}} \equiv \frac{\partial f(\bar{\boldsymbol{\sigma}})}{\partial \bar{\boldsymbol{\sigma}}} \Big/ \left\| \frac{\partial f(\bar{\boldsymbol{\sigma}})}{\partial \bar{\boldsymbol{\sigma}}} \right\| \tag{5}$$

$\boldsymbol{\sigma}_y$: conjugate stress on the normal-yield surface

---

* A similarity of two figures F and F' means that there exists a one to one correspondence between points on F and those on F' and that a ratio of a length of line segment PQ connecting two points on F and that of a corresponding line segment P'Q' on F' is always constant. Further, if PQ and P'Q' are paralell to each other, we define that F and F' are similar and locate in a similar position to each other. In this case, the line-segment PP' meets always at one point which is called a "center of similarity". Letting a center of similarity be denoted by S, a ratio of the length of PS and that of P'S is constant always. A function describing similar figures must be homogeneous.

Further, it is assumed that a ratio R ($\equiv f(\bar{\sigma})/F$) of the size of the subloading surface to that of the normal-yield surface increases and decreases in the loading and the unloading state, respectively.

In Fig.2, a uniaxial loading behaviour predicted by the subloading surface model is schematically illustrated for the simplest case assuming a perfectly-plastic normal-yield surface, where σ_1-ε_1 and s_1-ε_1 (subscript 1 corresponds a axial component) are depicted by solid and dashed lines, respectively. In the initial loading illustrated in Fig.2(a), the similarity-center $\boldsymbol{s}$ translates with a plastic deformation, following a current stress $\boldsymbol{\sigma}$. In the reverse loading illustrated in Fig.2(b), the subloading surface first contracts and shrinks to a point when the current stress returns to the similarity-center. In this process, only an elastic deformation proceeds. After that, the subloading surface again expands, and thus a plastic deformation occurs diverging from the elastic curve. In the reloading illustrated in Fig.2(c) which is magnified for clarity, an elastic deformation proceeds until the current stress reaches the similarity-center. Thereafter, a plastic deformation occurs again. A closed hysteresis loop is obtained in such a simple way.

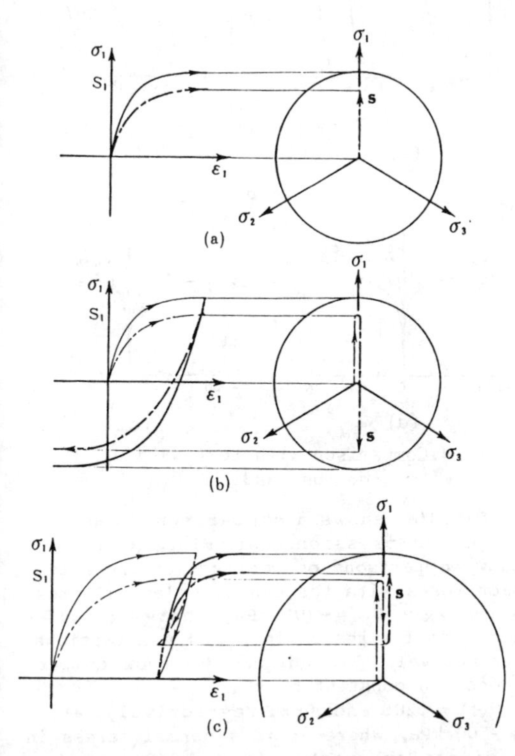

Fig.2 Uniaxial loading behavior predicted by the subloading surfcae model: (a) initial loading, (b) reverse loading, (c) reloading.

3 CONSTITUTIVE EQUATIONS OF SOILS

Constitutive equations of soils based on the subloading surface model (Hashiguchi, 1986a, b) are presented below.

Let the normal-yield surface (see Fig.3) be given as

$$f(\hat{\boldsymbol{\sigma}}) = \sqrt{\hat{\sigma}_m^{\,2} + (\hat{\sigma}'/m)^2} \tag{6}$$

$$m = 2\sqrt{6}\,\sin\phi_t / \{ \tfrac{1}{2}(A^4 + B^4) - \tfrac{1}{2}(A^4 - B^4)\sin 3\theta \}^{1/4} \tag{7}$$

$$\left.\begin{matrix}\Phi_a\\ \Phi_b\end{matrix}\right\} \equiv 3 \pm \sin\phi_t \quad (\Phi_a : +,\ \Phi_b : -) \tag{8}$$

$$F = F_0 \exp\{ -H/(\rho - \gamma) \} \tag{9}$$

$$\dot{H} \equiv \dot{\varepsilon}_v^p + \mu\,(V^p - V_c^p)\,\dot{\varepsilon}^{p\prime\,\nu} \tag{10}$$

where

$$\hat{\sigma}_m \equiv \tfrac{1}{3}\mathrm{tr}\hat{\boldsymbol{\sigma}},\quad \hat{\boldsymbol{\sigma}}' \equiv \hat{\boldsymbol{\sigma}} - \hat{\sigma}_m \boldsymbol{I},\quad \hat{\sigma}' \equiv \|\hat{\boldsymbol{\sigma}}'\| \tag{11}$$

$$\dot{\varepsilon}_v^p \equiv \mathrm{tr}\dot{\boldsymbol{\varepsilon}}^p,\quad \dot{\boldsymbol{\varepsilon}}^{p\prime} \equiv \dot{\boldsymbol{\varepsilon}}^p - \tfrac{1}{3}\dot{\varepsilon}_v^p \boldsymbol{I},\quad \dot{\varepsilon}^{p\prime} \equiv \|\dot{\boldsymbol{\varepsilon}}^{p\prime}\| \tag{12}$$

$$V^p = V_0^p \exp(\varepsilon_v^p) \tag{13}$$

- V_c^p: material constant - specific volume limiting between a densification and a liquefaction
- ϕ_t: angle of internal friction in axisymmetric stress state
- ρ, γ: material constants - slopes of elastoplastic and elastic lines, respectively, in lnP-lnV (P: pressure, V: specific volume) diagram of isotropic consolidation
- μ, ν: material constants - regulating a deviatoric strain hardening
- F_0: initial value of F

Let a translation rule of normal-yield surface be given as

$$\hat{\boldsymbol{\alpha}} = -F\boldsymbol{I} \tag{14}$$

which means that a normal-yield surface

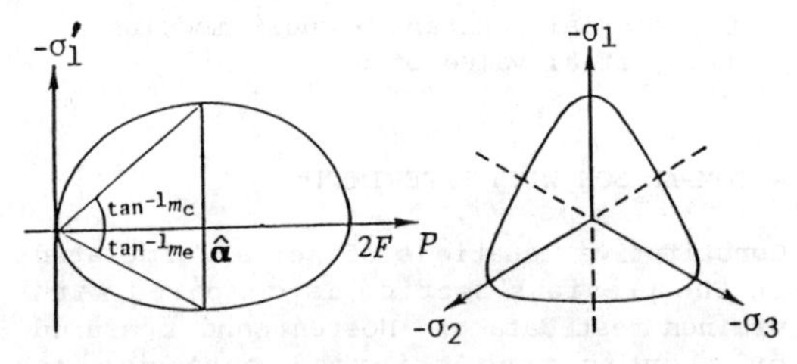

Fig.3 Normal-yield surface for soils.

translates along a hydrostatic axis and passes through an origin of stress space.

The meridian section of the normal-yield surface is an ellipsoid, and the π-section is a smooth convex curve coinciding with Coulomb-Mohr failure condition in axisymmetric stress state.

Let a translation rate of similarity-center be given as

$$\dot{\mathbf{s}} = \dot{\hat{\boldsymbol{\alpha}}} + \frac{\dot{F}}{nF}\hat{\mathbf{s}} + c\,\dot{\varepsilon}^p\left(\frac{\bar{\boldsymbol{\sigma}}}{R} - \frac{\hat{\mathbf{s}}}{\chi}\right) \quad (15)$$

where

$$\dot{\varepsilon}^p \equiv \| \dot{\boldsymbol{\varepsilon}}^p \| \quad (16)$$

c : material constant - regulating a rate of translation of similarity-center

χ : material constant - maximum value of $f(\hat{\mathbf{s}})/F$

Let an expansion rule of subloading surface be given as

$$\dot{R} \equiv \zeta\,(1-R)^{\upsilon}\,\dot{\varepsilon}^p / \{ f\,(\boldsymbol{\sigma}-\mathbf{s})\,/\,F \} \quad \text{for } \dot{\boldsymbol{\varepsilon}}^p \neq \mathbf{0} \quad (17)$$

ζ, υ: material constants - regulating a rate of expansion of subloading surface

A plastic strain rate is derived from the above equations and the associated flow rule as follows:

$$\dot{\boldsymbol{\varepsilon}}^p = \mathrm{tr}(\bar{\mathbf{n}}\dot{\boldsymbol{\sigma}}) / \mathrm{tr}[\,\bar{\mathbf{n}}\{\frac{F'}{F}\bar{h}\,\hat{\boldsymbol{\sigma}} + \bar{\mathbf{a}} + c(1-R)(\frac{\bar{\boldsymbol{\sigma}}}{R} - \frac{\hat{\mathbf{s}}}{\chi}) + \frac{U}{R}\tilde{\boldsymbol{\sigma}}\}\,]\,\bar{\mathbf{n}} \quad (18)$$

where

$$F' \equiv dF/dH = -F/(\rho-\gamma) \quad (19)$$

$$\bar{h} \equiv \{ (1+\mu\varepsilon^{p\prime\,\nu}V^p)\,\bar{\sigma}_m + \nu\,\mu\,(V^p-V^p_c)\,\varepsilon^{p\prime\,\nu-1}\bar{\sigma}' / m^2 \} / \sqrt{\{\tfrac{1}{3}\bar{\sigma}_m^2 + (\bar{\sigma}'/m)^2\}} \quad (20)$$

$$\bar{\mathbf{a}} = -F'\,\bar{h}\,\mathbf{I} \quad (21)$$

$$\bar{\sigma}_m \equiv \tfrac{1}{3}\mathrm{tr}\,\bar{\boldsymbol{\sigma}}, \quad \bar{\boldsymbol{\sigma}}' \equiv \bar{\boldsymbol{\sigma}} - \bar{\sigma}_m\mathbf{I} \quad (22)$$

Let an elastic strain be given as

$$\boldsymbol{\varepsilon}^e = \gamma \ln\left(\frac{p}{p_0}\right)\mathbf{I} + \frac{1}{2G}\boldsymbol{\sigma}' \quad (23)$$

where

$$p = -\sigma_m, \quad \boldsymbol{\sigma}' \equiv \boldsymbol{\sigma} - \sigma_m\mathbf{I} \quad (24)$$

G: material constant - shear modulus

p_0 : initial value of p

4 COMPARISON WITH EXPERIMENT

Constitutive equations of soils formulated in the previous section is compared with drained test data of Hostun sand measured by a cubic true triaxial test device (Saada, 1987). The material constants and initial values used in calculation are as follows:

ϕ_t = 28°, ρ = .008, γ = .004, μ = 4, V^p_c = 1.85, ν = 1.5, c = 30, χ = .95, ζ = 10,000, υ = 15, G = 200,000 kPa,
F_0 = 500 kPa, V^p_0 = 1.616, $\mathbf{s}_0$ = $-50\,\mathbf{I}$ kPa.

where $\mathbf{s}_0$ is the initial state of $\mathbf{s}$. Theoretical curves are depicted from the results calculated from the same isotropic stress state σ_m = -100 kPa.

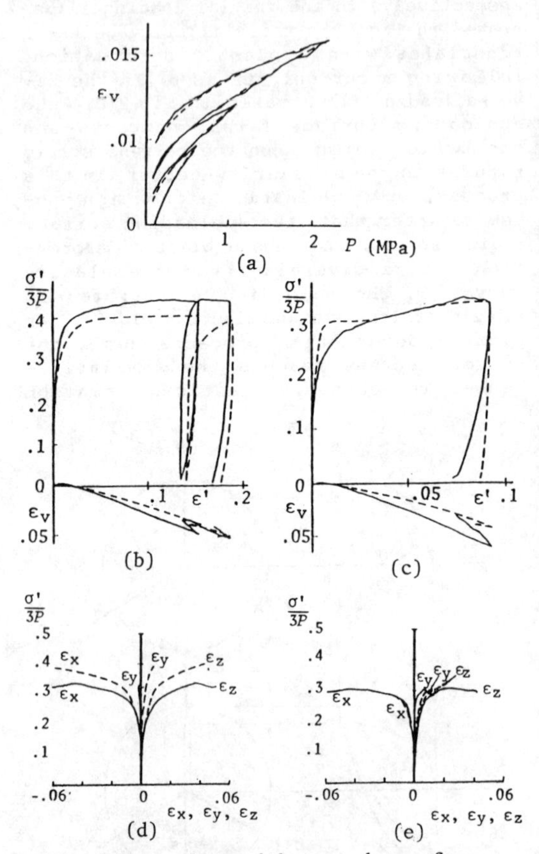

Fig.4 Comparison with test data of Hostun sand.

Fig.4(a) shows a comparison on an isotropic compression test. Fig.4 (b) and (c) show comparisons on compression and extension tests with the constant lateral pressure: $\sigma_x = \sigma_y = -200$ kPa, where σ_x and σ_y are normal stresses in x and y directions, respectively. Fig.4(d) and (e) show comparisons on constant b (= $(2\sigma_z - \sigma_x - \sigma_y)/(\sigma_x - \sigma_y)$) = .286 and .666, respectively, at σ_m = -500 kPa, where σ_z is a normal stress in z direction) tests. In Fig.5(d) and (e), ε_x, ε_y and ε_z are normal strains in x, y and z directions, respectively, and

$$\dot{\varepsilon}_v = \mathrm{tr}\,\dot{\boldsymbol{\varepsilon}},\ \dot{\boldsymbol{\varepsilon}}' = \dot{\boldsymbol{\varepsilon}} - \frac{1}{3}\dot{\varepsilon}_v \mathbf{I},\ \dot{\varepsilon}' = \|\dot{\boldsymbol{\varepsilon}}'\| \quad (25)$$

5 DISCUSSIONS - OTHER EXTENSIONS OF THE SUBLOADING SURFACE MODEL

In the original subloading (or bounding) surface model, the subloading surface expands/contracts under the restraint that the similarity-center of the normal-yield and the subyield surfaces is fixed in the origin of stress space. Accordingly, the state in which a current stress returns to the origin of stress space is regarded to be purely elastic. In order to modify it so as to describe the plastic deformation in this state, some ideas have been proposed other than the translation of the similarity-center formulated by the author.

On the other hand, in stead of making the similarity-center translate, many workers (Aboim and Wroth, 1982; Pande and Pietruszczak, 1982; Zienkiewicz and Mroz, 1984; Zienkiewicz et al., 1985; Pietruszczak, 1986) assume that a plastic deformation is produced even when the stress increment has an inward direction of the subloading surface and postulate the following equations

$$\dot{\boldsymbol{\varepsilon}}^p = \frac{\mathrm{tr}(\bar{\boldsymbol{n}}\dot{\boldsymbol{\sigma}})}{D_l}\boldsymbol{m} \quad \text{for loading } (\mathrm{tr}(\bar{\boldsymbol{n}}\dot{\boldsymbol{\sigma}}) \geqq 0)$$

$$\dot{\boldsymbol{\varepsilon}}^p = -\frac{\mathrm{tr}(\bar{\boldsymbol{n}}\dot{\boldsymbol{\sigma}})}{D_u}\boldsymbol{m} \quad \text{for unloading } (\mathrm{tr}(\bar{\boldsymbol{n}}\dot{\boldsymbol{\sigma}}) \leqq 0) \quad (26)$$

where

?: normalized tensor having an outer-normal direction to the plastic potential surface

D_l, D_u: plastic moduli in the states $\mathrm{tr}(\bar{\boldsymbol{n}}\dot{\boldsymbol{\sigma}}) \geqq 0$ and $\mathrm{tr}(\bar{\boldsymbol{n}}\dot{\boldsymbol{\sigma}}) \leqq 0$, respectively.

According to (26), variations of plastic strain rate for the change of direction of the stress rate in Fig.5(a) are predicted as in Fig.5(b) for the case putting $\boldsymbol{m} = \bar{\boldsymbol{n}}$.

On the other hand, the traditional theory of plasticity with the associated flow rule predicts them as in Fig.5(c).

In Fig.5(b), the smallest plastic strain rate does not occur for the stress rate having an inward direction of the subloading surface but occurs by the stress rate having a tangential direction. In other words, the plastic strain rate once decreases and again increases as the stress increment rotates from the outward to the inward direction of the loading surface. Thus, the plastic strain rate predicted by (26) conflicts with a real material response illustrated in Fig.5(d). This unreal property causes the prediction of a reverse loading curve having an inflection point, since a purely elastic state occurs in the loading curve having an inflection point as illustrated by the point b in Fig.6,

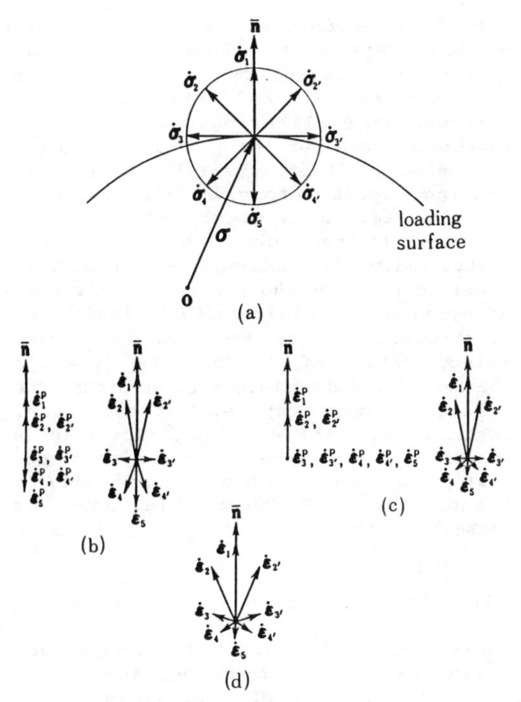

Fig.5 Plastic strain rates and strain rates predicted by Eq.(26) with $\boldsymbol{m} = \bar{\boldsymbol{n}}$: (b) and the traditional theory of plasticity: (c), and the strain rate response of real materials: (d) for the change of direction of the stress rate: (a).

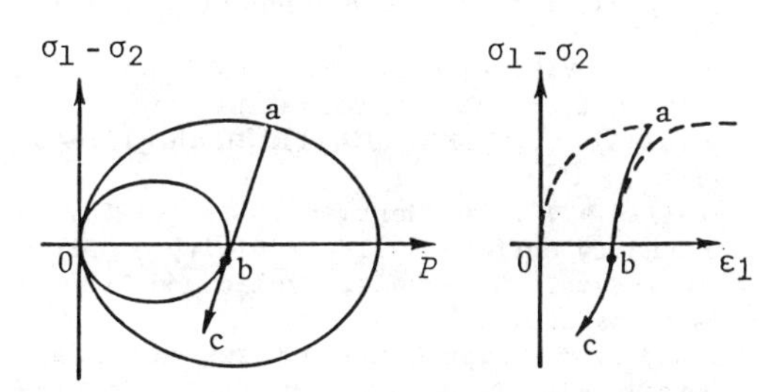

Fig.6 Stress-strain curve with an inflection ponit and an open hysteresis loop predicted by Eq.(26).

since a purely elastic state occurs in the moment that a stress rate has a tangential direction to the subloading surface. Further, it leads to the prediction of an open hysteresis loop for the stress cycle of a partial unloading-reloading.

Naylor (1985) proposed to shift the similarity-center from the origin of stress space to the center of the ellipsoidal normal-yield surface for soils. While the original subloading surface model predicts an open hysteresis loop, this modification offers an advantages that it predicts a closed hysteresis loop for a large unloading-reloading cycle of stress in isotropic as well as deviatoric stress changes. It still brings about an open hysteresis loop for the partial unloading-reloading cycle of stress in which a reloading occurs before an elastoplastic state is realized by the unloading.

The above-mentioned approaches are restricted to the isotropic deformation. In order to describe the induced anisotropy, Anandrajah and Dafalias (1985, 1986) proposed to incorporate the rotational hardening (Hashiguchi, 1977b, 1979) by which the normal-yield surface rotates around the origin of stress space. The rotational hardening with the associated flow rule is unacceptable in general as known from supposing that it brings about an unreality in the pressure-independent (viz, Mises or Tresca) materials.

REFERENCES

Aboim, C.A. and Wroth, C.P. 1982. Bounding-surface-plasticity-theory applied to cyclic loading of sand. Proc. 2nd Int. Symp. Numer. Models in Geomech.: 65-72.

Anandrajah, A. and Dafalias, Y.F. 1986. Bounding surface plasticity.-III. ASCE J. Eng. Mech. 112: 1292-1318.

Dafalias, Y.F. & Popov, E.P. 1975. A model of nonlinearly hardening materials for complex loading. Acta Mechanica. 21: 173 -192.

Dafalias, Y.F. & Popov, E.P. 1976. Plastic internal variables formalism of cyclic plasticity. ASME J.Apll. Mech. 43:645-651.

Dafalias, Y.F. & Herrmann, L.R. 1980. A bounding surface soil plasticity. Proc. Int. Symp. Soil under Cyclic and Transient Loading: 335-345.

Dafalias,Y.F.1986.Bounding surface plasticity.-I. ASCE J.Eng. Mech. 112: 966-987.

Edelman, F. & Drucker, D.C. 1951. Some extensions of elementary plasticity theory. J. Franklin Inst. 251: 581-605.

Hashiguchi, K. & Ueno, M. 1977a. Elastoplastic constitutive equations of soils: Proc. 9th ICSMFE - Spec. Sess. 9: 73-82.

Hashiguchi, K. 1977b. An expression of anisotropy. ibid.: 302-305.

Hashiguchi, K. 1978. Plastic constitutive equations of granular materials. Proc. US -Japan Seminar on Continuum & Stat. Appr. Mech. of Granular Materials: 321-329.

Hashiguchi, K. 1979. Constitutive equations of granular media. Proc. 3rd ICONMIG: 435 -439.

Hashiguchi, K. 1980a. Constitutive equations of elastoplastic materials with elastic-plastic transition. ASME J. Appl. Mech. 47: 266-272.

Hashiguchi, K. 1980b. Anisotropic hardening model for granular media. Proc. Int. Symp. Soils Cyclic Trans. Load.: 469-474.

Hasiguchi, K. 1981. Constitutive equations of elastoplastic materials with anisotropic hardening and elastic-plastic transition. J. Appl. Mech. 48: 297-301.

Hashiguchi, K. 1986a. A mathematical description of elastoplastic deformation in normal-yield and sub-yield states. Proc. 2nd Int. Conf. Numer. Models in Geomech.: 17-24.

Hashiguchi, K. 1986b. Elastoplastic constitutive model with a subloading surface. Proc. Int. Conf. Computational Mech. 1: IV-65-70.

Ishlinski, I.U. 1954. Generalized theory of plasticity with linear strain hardening. Ukr. Mat. Zh. 6: 314-324.

Iwan, W.D. 1967. On a class of models for the yielding behavior of continuous and composite systems. ASME J. Appl. Mech. 34: 612-617.

Masing, G. 1926. Eigenspannungen und Verfestigung heim Messing. Proc. 2nd Int. Cong. Appl. Mech.: 332-335.

Mroz, Z. 1967. On the description of anisotropic workhardening. J. Mech. Phys. Solids. 15: 163-175.

Mroz, Z. & Norris, V.A. 1982. Elastoplastic and viscoplastic constitutive models for soils with anisotropic loading. Soil Mech.-Transient & Cyclic Loads: 305-320.

Mroz, Z. & Zienkiewicz, O.C. 1984. Uniform formalization of constitutive equations for clays and sands. Mechanics of Engineering Materials: 415-450.

Naylor, D.J. 1985. A continuous plasticity version of the critical state model. Int. J. Numer. Meth. in Engng. 21: 1187-1204.

Pande, G.N. and Pietruszczak, St. 1982. Reflecting surface model for soils. Proc. Int. Conf. Numer. Models Geomech.: 50-64.

Pietruszczak, St. 1986. A flow theory of soil: Concept of multiple neutral loading surface. Comput. & Geotech. 2: 185-203.

Prager, W. 1956. A new methods of analyzing stresses and strains in workhardening plastic solids. ASME J. Appl. Mech. 23: 493-496.

Saada, A.S. 1987. Information Package for Int. Workshop on Constitutive Equations for Non-Cohesive Soils. Cleveland.

Zienkiewicz, O.C. 1984. Generalized plasticity formulation and its applications to geomechanics. Mechanics of Engineering Materials. John Wiley & Sons: 665-679.

Zienkiewicz, O.C., Leung, K.H. and Pastor, M. 1985. Simple model for transient soil loading in earthquake analysis.-I. Int. J. Numer. Anal. Meth. Geomech. 9:453-476.

Numerical Methods in Geomechanics (Innsbruck 1988), Swoboda (ed.)
© 1988 Balkema, Rotterdam. ISBN 90 6191 809 X

On errors involved in the experimental-numerical investigation in triaxial compression tests

Janusz M.Dluzewski & Lech A.Winnicki
Department of Civil Engineering, Warsaw University of Technology, Poland

ABSTRACT: Authors' experimental investigations by means of triaxial compression tests on sand are reported. Dense sand was tested and the strain softening and dilatation were observed. All the measured variables were related to the current configuration i.e. the total Eulerian description was employed. The experiments were simulated numerically with the help of the finite element method. A model for the granular medium, proposed by the authors in (2), was employed. The large deformation problem was analyzed in the total Eulerian description leading to nonsymmetrical secant stiffness matrix. Finite element formulation for strong nonlinearities, both physical and geometrical, was also considered in the total Eulerian description and the iterative procedures were used. The study of the errors involved in the whole process starting from experiment to the final numerical results is reported.

1 INTRODUCTION

Two types of experiments are mainly performed to study behaviour of soils: plane strain test and triaxial test. The landslide, failures of retaining walls, strip footings are all problems in plane strain, therefore the deformations are not well represented by data from the triaxial test. However, plane strain tests are difficult to perform while the triaxial test combines simplicity with versatility. As it was found from experiments on sand and on saturated clays (1), the strength determined by plane strain test is usually greater than that determined by triaxial test. From the designer's point of view its use is on the safe side. Therefore, the triaxial test has been commonly accepted for evaluation of soil parameters for engineering problems as well as for research studies of soil behaviour. To obtain soil strength data such as cohesion or friction angle from the standard triaxial test equipment is comparatively simple. Difficulties arise when deformations or strain data are to be measured for situations in which large strains and large displacements are involved. The nature of these difficulties, which may or may not create errors, are as follow:

i - Special care must be paid to the specimen preparation process, lubrication of the platens ends and coaxiality of the applied load with the sample in order to obtain the highest homogeneity of the whole sample and thus create stress and strain states as close to uniform as possible.

ii - Definition of proper pairs of stress and large strain tensors should be suitably chosen (2), (5), to avoid misunderstandings while interpreting the results obtained from experiments.

iii - Strain-rate control type of loading apparatus should be used to study post-critical behaviour of dense sand.

2 EXPERIMENTAL INVESTIGATION

Experimental results (2),(3),(7) have shown that during tests on sand in the triaxial compression apparatus both the dilatation and strain softening effects are observed. These phenomena appear when the strains exceed 5% and it is when the strain softening begins to exert an influence on the ultimate load and on local unloading in soils under foundations. The dilatation affects the magnitude of settlements and deformations of the subsoil.

The dilatation and softening phenomena have been investigated by the authors. All tests were performed on dry dense sand from the locality of OSOWIEC. The height to diameter ratio of the sample was 2:1 (height 76mm, diameter 38 mm). The results of the sieve analysis are shown in Fig.1. All samples were taken from very dense sand, $e = e_{max}$ (for maximum compaction: void ratio e_{max} = 0,38 bulk unit weight ρ_{max} = 19 KN/m^2; for loose sand: void ratio e_{min} = 0,60 , bulk unit weight ρ_{min} = 16 KN/m^2). The sample was

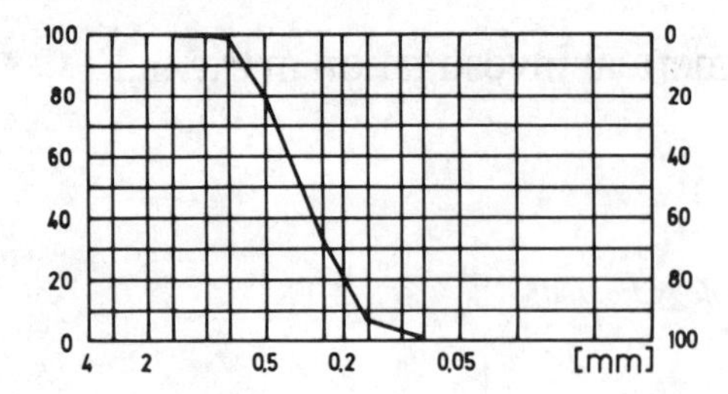

Figure 1. The results of the sieve analysis

prepared by dividing 165 g of sand into ten equal portions which were subsequently poured and compacted. The strains and stresses were related to the current deformed configuration of the sample, thus the elements of the Cauchy stress tensor and the linear parts of the Almansi strain tensor were used. All calculated variables were related to the whole sample. Loading was applied in kinematic way, thus the strain softening could be observed. The volumetric changes were measured for the whole specimen $\varepsilon_V = \Delta V/V$, where the current volume V was obtained from the shadows of the sample with the help of the optical method. To minimize the friction between platens and sand, platens were polished and lubricated. The friction causes nonhomogeneity of the sample and constitutes the main problem in the triaxial compression test. The following two damage mechanisms of the specimen were observed; first, shearing and second, bulging of the sample, Fig.2. The shearing is caused by either initial nonhomogeneity of the sample or initial nonaxiality of the piston in triaxial apparatus. In the experiments, the coaxiality of the piston and sample as well as verticality of the apparatus were checked with the use of the levelling instrument. Provided the sample is homogeneous and the apparatus is properly set the second damage mechanism is observed. During the experiments the confining pressure was being varied from 0 to 1 MP. The strains varied from 0 to 40%.

The experimental evidence for standard compression tests are shown in Fig.3. Two other results for different loading paths are shown in Fig.4 and 5. These experiments are required to supply the necessary results which can be used to describe the behaviour of the medium with the help of theoretical model. The obtained results of triaxial compression tests can be suitable for such modeling, provided the user is aware of errors involved during the whole experiment. The errors due to initial nonhomogeneity, friction between platens and sand are accumulated and the total result is the average of the behaviour of

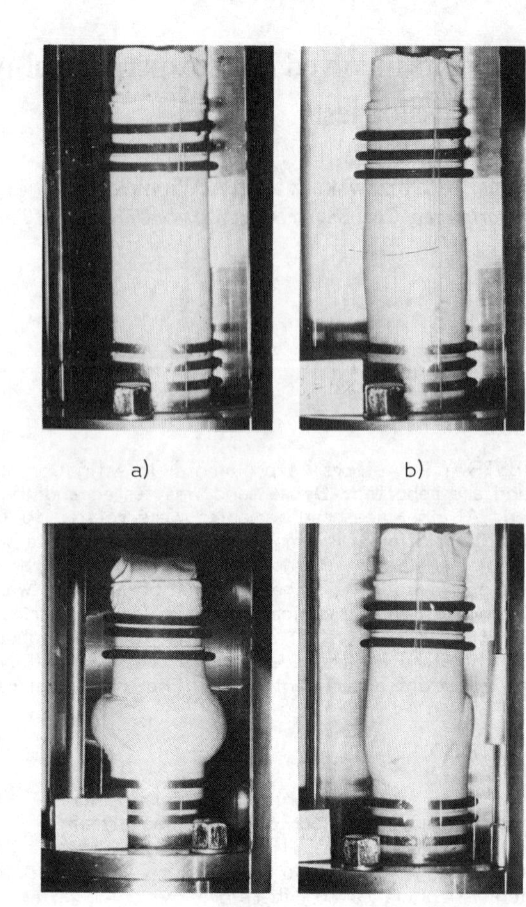

Figure 2. The side views of the sampless:
a) before compression b) during experiment
c) bulging - the fist damage mechanism
d) shearing - the second damage mechanism

sample and not the medium alone. To show the distributions of strains in the sample a few diagrams from (4) are shown, Fig.6. Here the measurements were made with the help of the lead balls embedded into specimen of dense sand which was then X-rayed. The strains were calculated from X-ray photos on the basis of differences of the initial and current positions of the lead balls.
The known conclusion has been confirmed that, for a given vertical strain $\bar{\varepsilon}_1$, the volumetric changes increase with increasing initial density, attaining as much as 18 % in the middle of the sample, as shown in the right-hand side of Fig. 6.

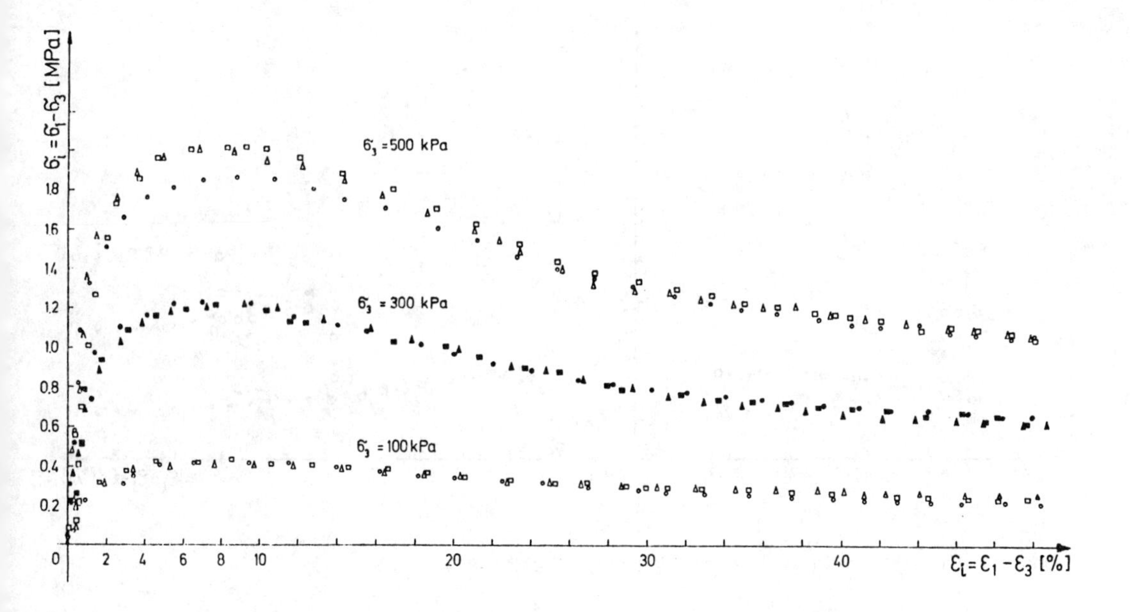

Figure 3.a. The experimental results of the standard compression tests.

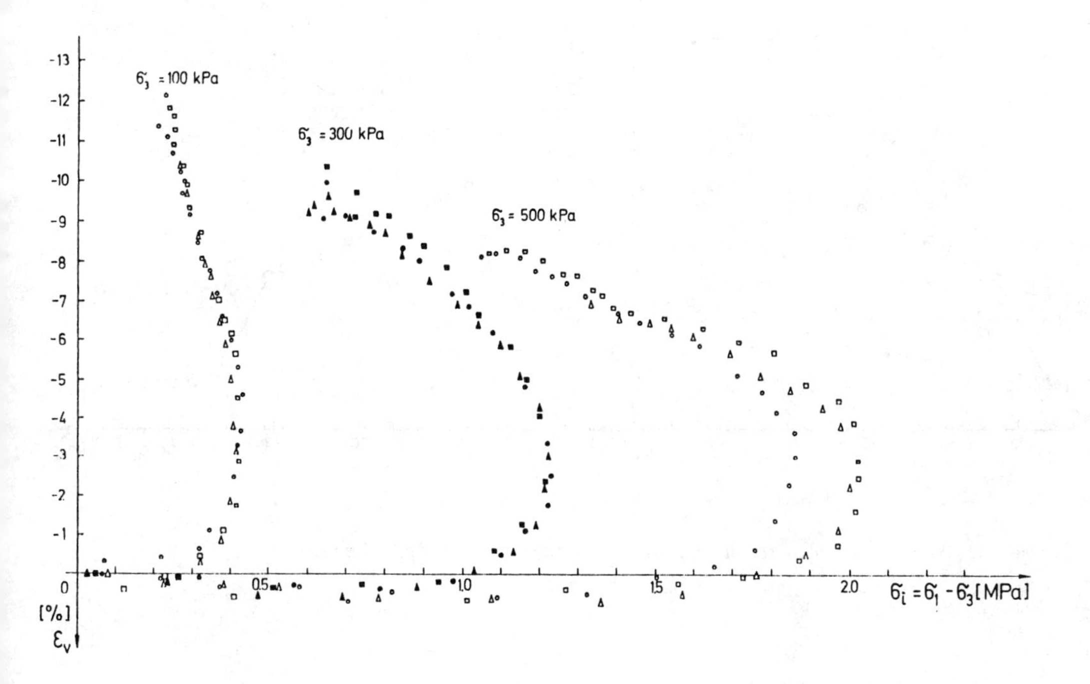

Figure 3.b. The experimental results of the standard compression tests.

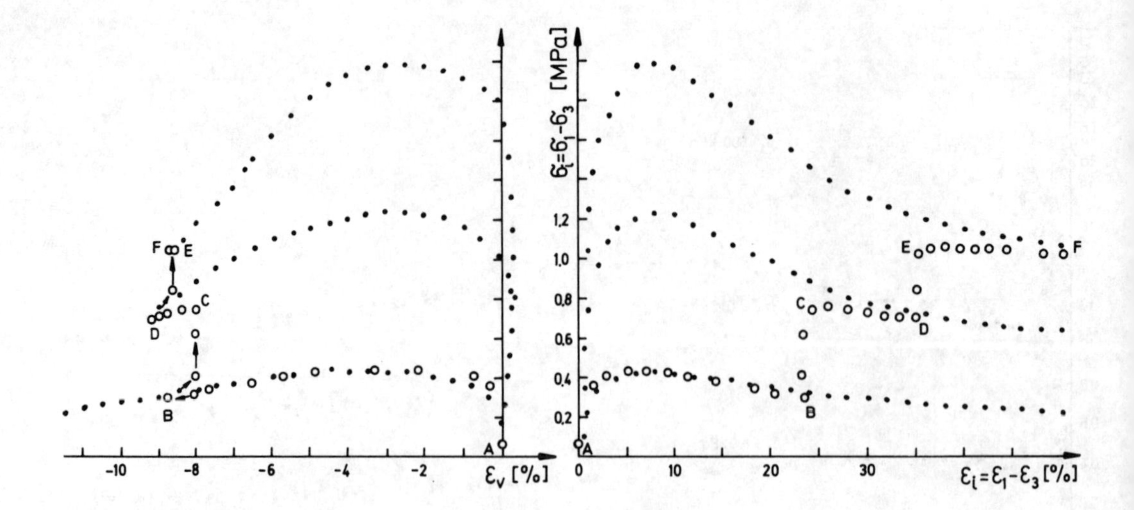

Figure 4. The tests results for increasing confining pressure

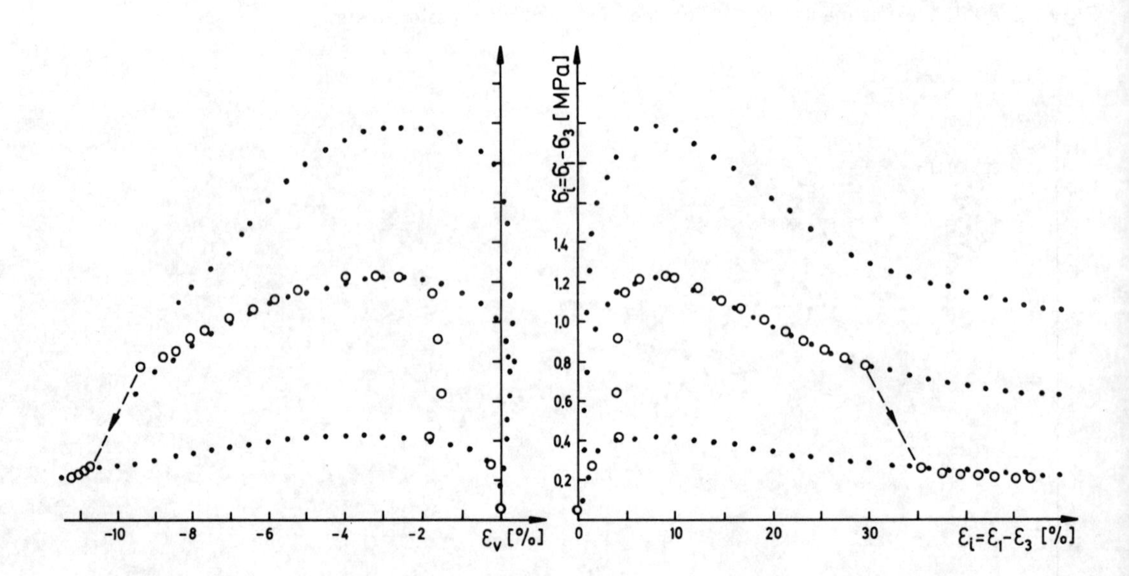

Figure 5. The tests results for increasing and decreasing confining pressure

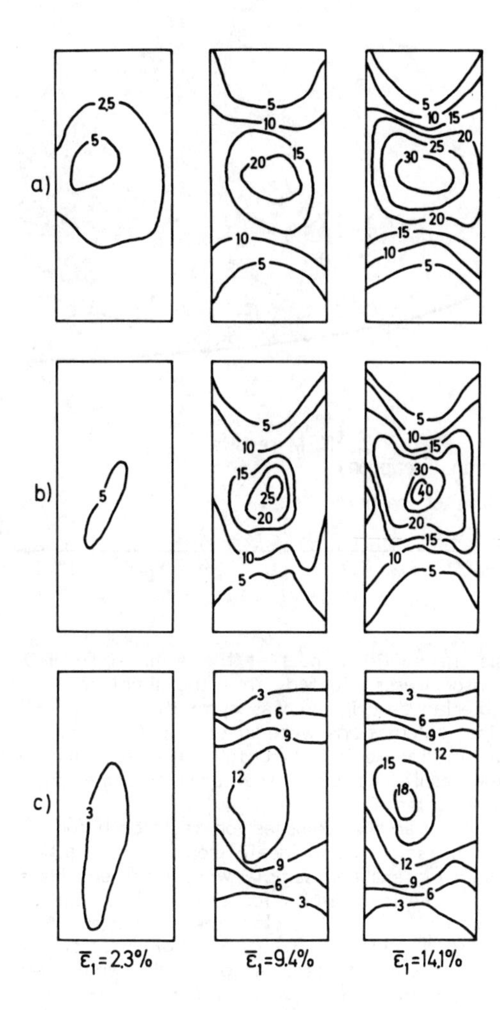

Figure 6. The strains distributions obtained by Nitecki (4) a) ε_3 - horizontal b) ε_1 - vertical c) ε_v - volumetric changes

3 THEORETICAL MODEL FOR SAND

The theoretical model for sand proposed in (2) is adopted. The equations of the model stem from the deformation theory of plasticity for loading process; in other words the nonlinear elastic model is engaged. The theoretical model is described by two pairs of equations; first for prepeak (precritical) state, second for postpeak (postcritical) behaviour. For precritical behaviour the equations have the following form:

$$\varepsilon_i = \frac{c\,T}{\alpha - T}, \qquad (3.1)$$

$$\varepsilon_v = \vartheta \left(\frac{\sigma_m}{p_\alpha}\right)^m - \frac{c'T}{\alpha - T}, \qquad (3.2)$$

where

$$T = \frac{\sigma_i}{\sigma_m} \qquad (3.3)$$

$\varepsilon_i = \sqrt{I_1^2 - 3I_2}$ is intensity of the strain tensor, $\varepsilon_v = \Delta V/V$ is the volumetric change, $\sigma_i = \sqrt{J_1^2 - 3J_2}$ is the intensity of the stress tensor, $\sigma_m = J_1/3 = (\sigma_1 + \sigma_2 + \sigma_3)/3$ is the mean stress, c, α, ϑ, m, p_α, c' are the parameters of the model.

The postcritical behaviour of sand is described by the following equations:

$$\varepsilon_i = \frac{a\,(2\overline{T} - T)}{T - b}, \qquad (3.4)$$

$$\varepsilon_v = \vartheta \left(\frac{\sigma_m}{p_\alpha}\right)^m - K + \frac{c''T}{\beta - T}, \qquad (3.5)$$

where $\overline{T}$, a, c'', β, K, b are the parameters $\overline{T}$, c'', β, a depend on the remaining ones. The detailed description of the model and the way of determining the necessary parameters is shown in (2).

4 NUMERICAL ANALYSIS

The finite element formulation was applied for both physically and geometrically nonlinear problem. The total Eulerian description and the secant stiffness matrix method were used. No history of the process was taken into account. The principle of virtual work for large strains in the total Eulerian description leads to the following equations of the nodal equivalence:

$$\int_V B_o^T D(\varepsilon)\,(B_o - \frac{1}{2} B_L)\,dv\; q = \int_V N^T f\,dv + \int_S N^T t\,dv, \qquad (4.1)$$

where B_o, B_L are the linear and nonlinear geometrical terms due to Almansi strain tensor, $D(\varepsilon)$ is the secant stiffness matrix, q is the nodal displacements vector, N is the matrix of shape functions, f and t are the volume and surface forces, respectively. The equation (4.1) is defined in the unknown current configuration of the body. The configuration, strains and stresses were sought for in the iterative process (6). In the numerical analy-

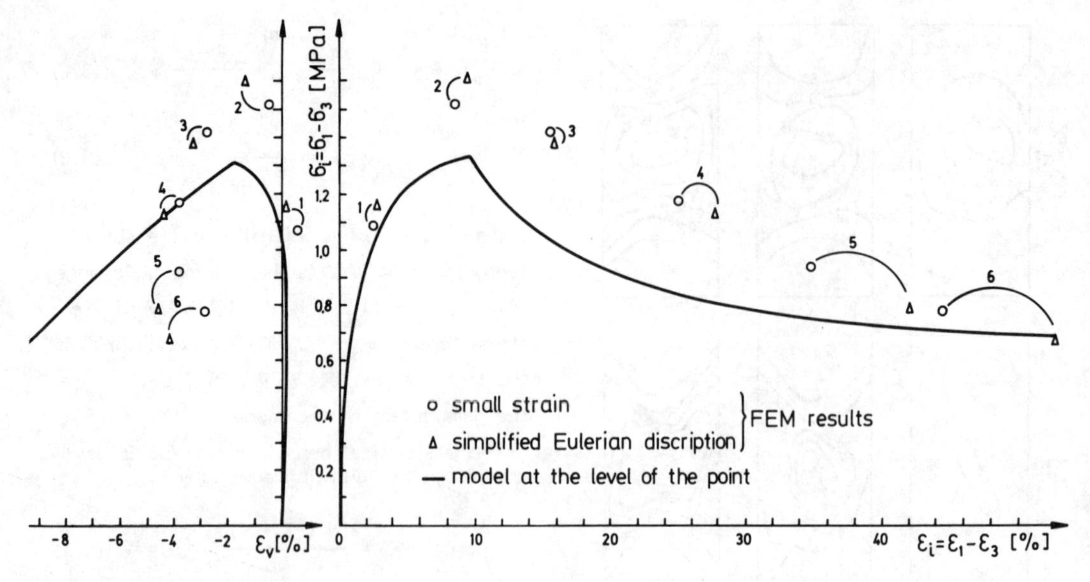

Figure 7. Finite element results for the whole sample

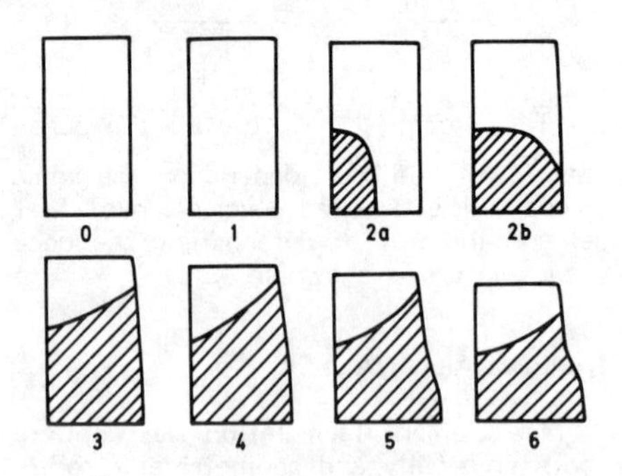

Figure 8. FEM results. Evolution of the post-critical zones, right hand side upper quarter of the sample

sis the nonlinear part of the Almansi strain tensor was neglected, thus the simplified Eulerian description was applied. The numerical calculations were also done for small strains in the initial configuration to compare the results for small and large deformation description.

The standard compression tests both for ideally smooth and ideally rough contacts between platens and sample were modeled. Here, the numerical results for the rough contact, which is much closer to experiment, are shown in Figs. 7, 8, 9. The values of strains and stresses are calculated for the whole sample, similarly as it was done in the experiments. The differences between theoretical predictions and numerical results are caused by nonhomogeneity of the sample in the FEM modeling.

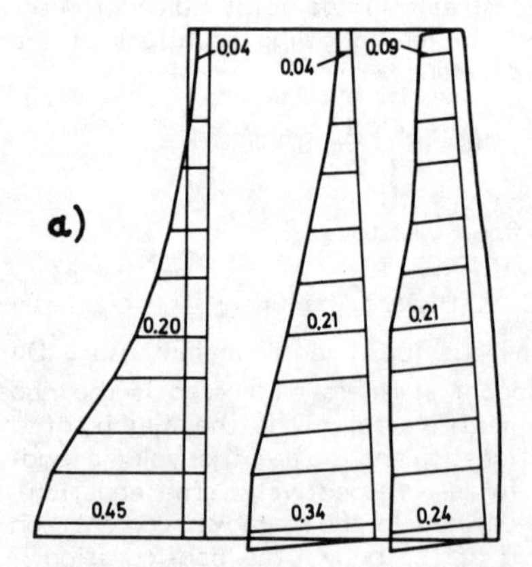

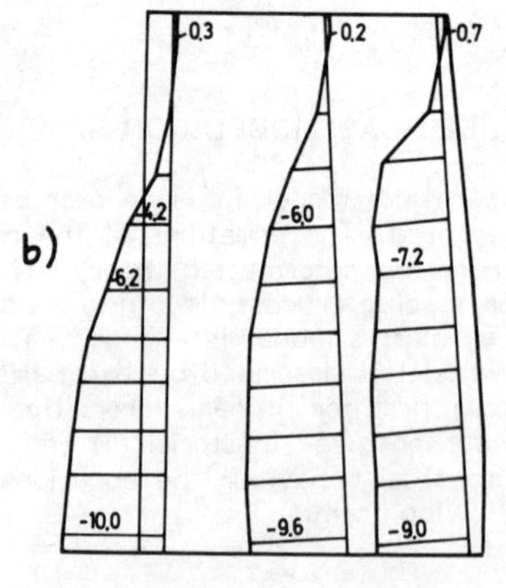

Figure 9. FEM results. The distribution of: a) ε_1 - vertical strain b) ε_v - volumetric changes, (right hand side upper quarter of the sample)

5 CONCLUSIONS

For some geotechnical failure problems it is the knowledge about postcritical behaviour of dense sand masses that is important. Therefore the foregoing study was undertaken as containing not only the laboratory triaxial tests, but also a proposal of the theoretical model for sand and its finite element implementation.

At each stage of investigations certain errors can be present, due to imperfections in tests and/or the employed kind of interpretation, transfer of data from tests to the proposed theoretical models and numerical procedures used.

The most frequently encountered errors are:

i - lack of homogeneity of samples (initial anisotropy) and nonaxiality of sample with loading that can lead to shearing failure,

ii - the fact that the whole sample behaves as a specific structure (i.e. represents a boundary value problem and not a material point),

iii - more or less suitable choice of stress and strain tensors in the theoretical description and in the numerical analysis.

Moreover, some unloading zones can appear in the sample which are not reflected in the proposed model.

It seems desirable to measure and interpret test data referring to the central part of the sample instead of the whole volume (to avoid the disturbances caused by the known phenomena of friction and stiffness of platens).

For future investigation another philosophy of determining model parameters can be proposed. The nonuniformity of the sample due to friction between platen and soil is known to have the main influence on the experimental errors. These errors enter the whole subsequent process of analysis, i.e. the determination of the model parameters and numerical calculations. The following way of determining the model parameters can be suggested. The iterative process, whose starting point is supplied by the model parameters obtained from standard compression test, can be activated. The boundary value problem for triaxial compression test will then be solved and model parameters will be sought for in the iterative process. The consistency of the numerical results for the whole sample with the $\sigma - \varepsilon$ curves obtained directly from the experiments will be the criterion for the termination of the iterative process.

ACKNOWLEDGMENTS

Great appreciation is expressed to prof. M. Kwieciński for encouragement to perform this study and for help in improving the manuscript.

The authors are also grateful to prof. M. Gryczmański for making it possible to perform tests on sand in his laboratory. Special thanks go to prof. C.S.Desai for his invitation to submit this paper.

Financial assistance of the Polish Academy of Sciences under Grant No 501/072/5 is dully acknowledged.

REFERENCES

(1) K.L.Lee, Comparison of plane strain and triaxial tests on sand, J. of Soil Mech. and Found.Div. ASCE, SM3, 901-923, (1970).

(2) J.M.Dłużewski, L.A.Winnicki, Loading path dependent behaviour of sand, Proc. 2-nd Int. Conf. on Num. Models in Geomech., Gent, 105-113, (1986).

(3) M.Mohkam, Contribution a l'etude experimentale et theorique du comportement des sables sous chargements cycliques, PhD theses, L'Institut National Politechnique de Grenoble, (1983)

(4) T.Nitecki, The use of X-rays for the determining of soil deformations in the triaxial test, Proc.8-th National Conf. on Soil and Found.Mech., Wrocław, 93-98, (1987), (in Polish).

(5) J.M.Dłużewski, Geometrically nonlinear model for cohesionless soil in terms of deformation theory of plasticity, PhD thesis, Warsaw University of Technology, (1986), (in Polish).

(6) J.M.Dłużewski, Total formulation for large strains in soils, Computers and Geotechnics, Int. J., (submitted).

(7) R.Lungren, J.K.Mitchell and J.H.Wilson, Effects of loading method on triaxial test results, J.of Soil Mech. and Found. Div. ASCE, SM2, 407-419, (1968).

Numerical Methods in Geomechanics (Innsbruck 1988), Swoboda (ed.)
© 1988 Balkema, Rotterdam. ISBN 90 6191 809 X

Non-associated flow rules in computational plasticity

Martin B.Reed & J.Cassie
Brunel University, Uxbridge, UK

ABSTRACT: A non-associated flow rule is proposed for materials with low plastic dilation, in which a Mohr-Coulomb yield surface is combined with a plastic potential of Drucker-Prager type. It is simpler to implement than the alternative approach of using a modified Mohr-Coulomb plastic potential, and does not have the tendency to attract stress states towards the corners of the yield surface. Results from the two flow rules are compared for an axisymmetric tunnel excavation problem, using a special one-dimensional finite element formulation.

1 INTRODUCTION

The classical theory of plasticity [1] assumes an associated flow rule; that is, the plastic potential $Q(\underset{\sim}{\sigma})$ whose gradient defines the direction of plastic straining, is identical to the yield function $F(\underset{\sim}{\sigma})$. In this case, Drucker's postulate is satisfied, and this guarantees stability and uniqueness of the stress solution under prescribed boundary conditions. However, when such a flow rule is applied in problems involving soil or rock, an unrealistically high degree of plastic dilation is in general predicted. It is therefore necessary to employ a non-associated flow rule, where the yield function and plastic potential are defined separately. Mroz [2] has shown that Drucker's postulate is a sufficient but not a necessary condition for uniqueness, and that uniqueness can be proved for some non-associated flow rules; in particular, he considered a Tresca yield surface with a Von Mises plastic potential. General uniqueness theorems for non-associated rules are so far lacking, however.

Perhaps the most commonly-used yield criterion in geotechnical analyses is that of Mohr-Coulomb. In three-dimensional principal stress space the yield surface $F(\underset{\sim}{\sigma}) = 0$ is a cone with irregular hexagonal cross-section in the deviatoric plane (Fig. 1). Davis [3] proposed a non-associated flow rule to be used with this surface, based on conditions of plane strain, with the intermediate principal stress in the out-of-plane direction, and neglecting elastic strains. The result of this is that the plastic potential is of the same form as the yield function but with the angle of friction ϕ replaced by an angle of dilation ψ , $0 \leqslant \psi \leqslant \phi$. The two extremes of $\psi = \phi$ and $\psi = 0$ correspond to associated flow and flow with zero plastic dilation, respectively. The determination of ψ for a particular material is not a practical proposition, and numerical applications of the model - [4], for example - tend to use both $\psi = 0$ and $\psi = \phi$, with the former producing the more realistic displacement field.

In an algorithm using this flow rule, special techniques are required for handling stress states lying on or close to the sharp corners of the yield surface, where the gradients of $F(\underset{\sim}{\sigma})$ and $Q(\underset{\sim}{\sigma})$ are discontinuous, and this problem is worsened by the fact that during the deformation, stresses in the plastic region are attracted to precisely these points. Another drawback is that $Q(\underset{\sim}{\sigma})$ is a complicated function when written in terms of stress invariants, necessitating substantial calculation to evaluate the matrix of second partial derivatives of Q , required if the implicit elasto-viscoplasticity algorithm is used.

In this paper an alternative non-associated flow rule is proposed, which also incorporates a parameter controlling the degree of plastic dilation. The Mohr-Coulomb yield function is combined with a plastic potential of Drucker-Prager type;

this does not involve the third stress invariant, so that the task of evaluating derivatives is greatly simplified.

The implementation of the two flow rules is considered in the context of the implicit elasto-viscoplasticity algorithm. After presenting the flow rules and their implementation, they are compared by applying them to a standard plane strain problem, namely the excavation of a cylindrical cavity in a pre-stressed elastic-brittle plastic rock mass. It is seen that with the proposed flow rule, stress states do not gravitate towards the singularities of the yield surface, and a more realistic distribution of the out-of-plane stress is obtained. The analyses are carried out for the case of zero plastic dilation, and there is very little difference between the predictions of cavity wall displacements from the two flow rules. It is concluded that the proposed flow rule provides a simple and useful model for geotechnical materials.

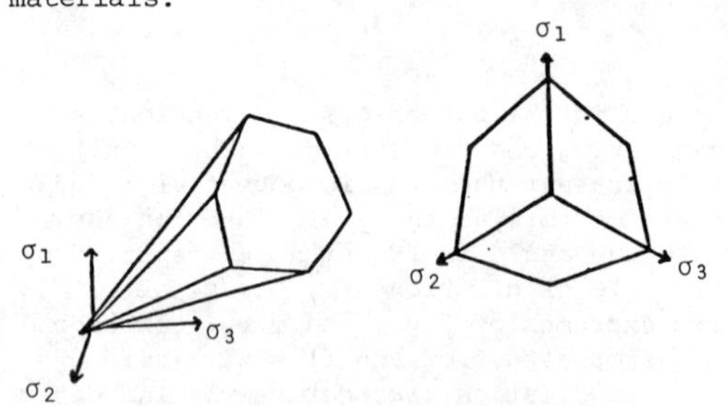

Figure 1. Mohr-coulomb surface.

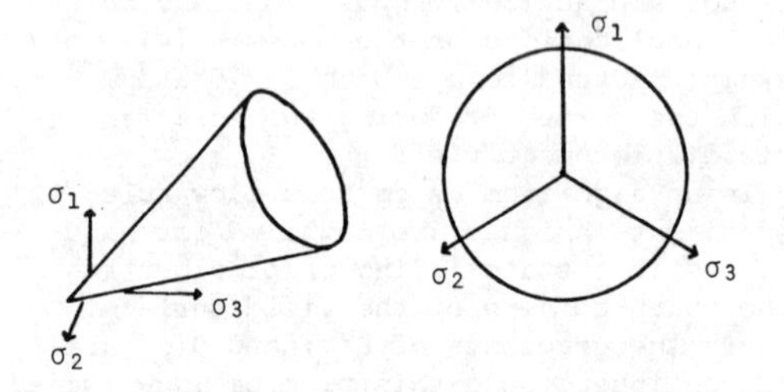

Figure 2. Drucker-Prager surface.

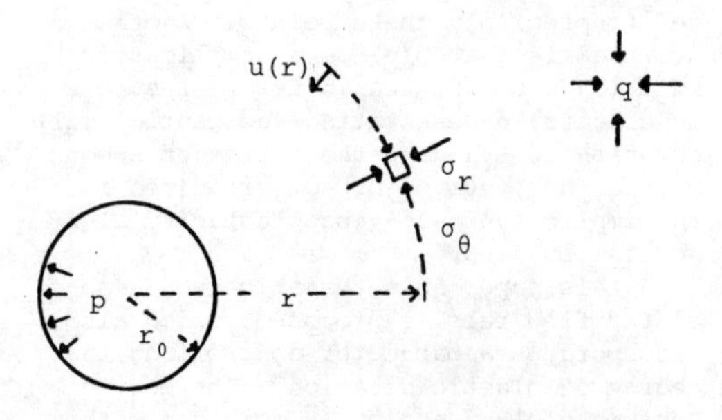

Figure 3. Axisymmetric tunnel problem

2. PLASTICITY ALGORITHM

Plastic straining of an element of material is initiated when its state of stress $\underset{\sim}{\sigma}$ reaches the yield surface $F(\underset{\sim}{\sigma}) = 0$; at points within the yield surface $F(\underset{\sim}{\sigma}) < 0$ the deformation is purely elastic. Plastic strain rates are governed by the flow rule [1]

$$\dot{\underset{\sim}{\varepsilon}}^{p} = d\lambda \frac{\partial Q}{\partial \underset{\sim}{\sigma}} \tag{1}$$

where $d\lambda$ is a constant of proportionality and $Q(\underset{\sim}{\sigma})$ is the plastic potential function.

In the elasto-viscoplasticity algorithm stress states exceeding the yield surface are permitted, and the viscoplastic strain rates at such a point are defined [5] by the flow rule

$$\dot{\underset{\sim}{\varepsilon}}_{vp} = \gamma\ \Phi(F) \frac{\partial Q}{\partial \underset{\sim}{\sigma}} . \tag{2}$$

Here, γ is a fluidity parameter, and $\Phi(F)$ a monotonically increasing function of F. If $F(\sigma)$ is a convex function of the form

$$F(\underset{\sim}{\sigma}) = f(\underset{\sim}{\sigma}) - f_0 \tag{3}$$

where $f(t\underset{\sim}{\sigma}) > f(\underset{\sim}{\sigma})$ for any $t > 1$, so that f_0 indicates the strength of the material, then a convenient choice of Φ is

$$\Phi(F) = \frac{f(\underset{\sim}{\sigma}) - f_0}{f_0} . \tag{4}$$

In fact, if the viscoplasticity algorithm is to be used, then $F(\underset{\sim}{\sigma})$ must have the form (3), so that the value of $F(\underset{\sim}{\sigma})$ at any point outside the yield surface is a measure of the distance from that point to the surface. (That is, the point lies on an expanded yield surface $F^*(\underset{\sim}{\sigma}) = f(\underset{\sim}{\sigma}) - f_0^* = 0$ with an increased strength f_0^*.). This requirement is particularly important if an elastic-brittle plastic material model is used, in which stress states are allowed to reach an initial yield surface $F^*(\underset{\sim}{\sigma}) = 0$ before plastic straining begins, and thereafter relaxed over time to the smaller residual yield surface $F(\underset{\sim}{\sigma}) = 0$. There is then a minimum relaxation distance from one surface to the other which is not affected by reducing the size of the loading increments. Some of the recently-proposed yield surfaces for soil and rock [6.7] are not appropriate in this context.

In the viscoplasticity algorithm the viscoplastic strain rate is integrated over the timestep $[t_n, t_{n+1}]$ by a linear approximation using a time discretization parameter θ lying between 0 and 1. The algorithm (full details of which are in reference [8]) leads to an element stiffness matrix

$$K^n = \int_\Omega B^T \hat{D}^n B d\Omega \qquad (5)$$

where $\hat{D}^n = (D^{-1} + \theta \Delta t H^n)^{-1}$, D being the elastic constitutive matrix. The evaluation of H requires the Hessian matrix of second partial derivatives of $Q(\underset{\sim}{\sigma})$, but only the gradient vector of first partial derivatives of $F(\underset{\sim}{\sigma})$. For an unconditionally stable algorithm, $\frac{1}{2} \leq \theta \leq 1$, although if an explicit algorithm ($\theta = 0$) is used the need to evaluate H is avoided. In the results presented below, an implicit algorithm with $\theta = \frac{1}{2}$ was used.

3. FLOW RULES

The Mohr-Coulomb yield surface is clearly of the form (3), being written

$$F(\sigma) = \sigma_1 - k\sigma_3 - \sigma_c \qquad (6)$$

where σ_1, σ_3 are the major and minor principal stresses (using the 'compression positive' sign convention), k the triaxial stress factor and σ_c the unconfined compressive strength. The parameters k and σ_c may be expressed in terms of the cohesion c and angle of friction ϕ.

The shape of the yield surface $F(\underset{\sim}{\sigma}) = 0$ in three-dimensional principal stress space is shown in Fig.1.

The flow rule proposed by Davis [3] takes as its plastic potential

$$Q(\underset{\sim}{\sigma}) = \sigma_1 - \ell\sigma_3 - \sigma_c \qquad (7)$$

where $\ell = \dfrac{1+\sin\psi}{1-\sin\psi}$, ψ being an angle of dilation. Then $\phi = \psi$ implies $Q \equiv F$, i.e. associated flow.

The volumetric plastic strain rate

$$\dot{\varepsilon}^p_v \equiv \dot{\varepsilon}^p_1 + \dot{\varepsilon}^p_2 + \dot{\varepsilon}^p_3 \qquad (8)$$

is equal to $d\lambda(1-\ell)$ by the flow rule (1), so that $\psi = 0$ implies zero plastic dilation. The shape of $Q(\underset{\sim}{\sigma})$ is that of $F(\underset{\sim}{\sigma})$, but with a smaller apex angle for $0 < \psi < \phi$.

In terms of the stress invariants I_1, J_2 and Lode angle Θ, the Mohr-Coulomb yield function (6) becomes

$$F(\sigma) = \sqrt{J_2}\,[\sqrt{3}(k-1)\sin\Theta + (k+1)\cos\Theta] - \frac{1}{3}(k-1)I_1 - \sigma_c \qquad (9)$$

with the plastic potential $Q(\underset{\sim}{\sigma})$ being written similarly. The prospect of finding second partial derivatives of this function with respect to the Cartesian stresses is a daunting one, and the complexity seems unwarranted when we recall that the flow rule was proposed as a simple linearized approximation to reality. As remarked in the Introduction, this flow rule was intended for plane strain situations where the out-of-plane stress remains strictly intermediate between σ_1 and σ_3. Since $Q(\underset{\sim}{\sigma})$ is independent of the intermediate stress, it is possible to perform a two-dimensional elasto-plastic analysis with this flow rule omitting any consideration of σ_2. However, the intermediacy of σ_2 is by no means guaranteed. Potts and Gens [9] have shown that, with this flow rule, stress states at yielded sampling points are attracted towards the corners of the yield surface, where either $\sigma_1 > \sigma_2 = \sigma_3$ or $\sigma_1 = \sigma_2 > \sigma_3$. Since the gradient of $Q(\underset{\sim}{\sigma})$ is discontinuous on these corners, it is necessary to either 'round off' the corners in a smooth way [8,10] or to invoke a special return algorithm [11,12].

The alternative approach to be considered in this Paper is to remove the dependence of $Q(\underset{\sim}{\sigma})$ on the Lode angle Θ by defining the plastic potential as

$$Q(\underset{\sim}{\sigma}) = \sqrt{J_2} - \alpha I_1. \qquad (10)$$

The surface $Q(\underset{\sim}{\sigma})$ = constant is then a Drucker-Prager cone about the hydrostatic axis, with circular cross-section in the deviatoric plane (Fig.2). The Mohr-Coulomb yield function (6,9) is used as before. The degree of plastic dilation is controlled by the parameter α. Since use of (10) in the flow rule (1) leads to

$$\dot{\varepsilon}^p_v = -3\alpha d\lambda \qquad (11)$$

there will be zero plastic dilation when $\alpha = 0$. There is no value of α for which associated flow will occur, but an upper limit on α is given by

$$\alpha_{max} = \frac{2\sin\phi}{\sqrt{3}(3-\sin\phi)} = \frac{k-1}{\sqrt{3}(k+2)} \qquad (12)$$

in which case $Q(\underset{\sim}{\sigma})$ generates a family of cones with the same apex angle as the yield surface; in the deviatoric plane the Drucker-Prager circle coincides with the outer apices of the Mohr-Coulomb hexagon [Ref.8, p.220].

Now by the flow rule (1), the plastic strain increment in the out-of-plane direction for a plane strain problem is

$$\dot{\varepsilon}_2^p = d\lambda\left\{\frac{1}{6\sqrt{J_2}}(2\sigma_2 - \sigma_1 - \sigma_3) - \frac{\alpha}{3}\right\}. \quad (13)$$

If the elastic strains after yielding has occurred are ignored, then the plane strain condition implies that $\Delta\varepsilon_2^p = 0$, so that for zero plastic dilation

$$\Delta\sigma_2 = \tfrac{1}{2}(\Delta\sigma_1 + \Delta\sigma_3). \quad (14)$$

That is, we can expect the out-of-plane stress to remain intermediate when using this flow rule. By contrast, the absence of plastic out-of-plane strains with flow rule (7) leads to

$$\Delta\sigma_2 = \nu(\Delta\sigma_1 + \Delta\sigma_3) \quad (15)$$

from the elasticity equations, so that if Poisson's ratio is zero σ_z will remain constant while the other stresses drop in the plastic zone, until σ_z equals the major principal stress - that is, a corner of the yield surface is reached. This situation will be demonstrated for a standard rock mechanics problem in the following sections.

4. AXISYMMETRIC TUNNEL PROBLEM

The performance of the two flow rules will be compared on a classic problem of rock mechanics, that of the excavation of a cylindrical cavity in a prestressed rock continuum. The problem is illustrated in Fig.3 : an elastic-brittle plastic rock mass is under a hydrostatic in situ stress q , and a tunnel of circular cross-section, radius r_0, is excavated in it (represented by reduction of the tunnel support pressure p from $p = q$ to $p = 0$). Because of the axial symmetry of the problem, it may be analysed in the radial direction only, for the radial and tangential stresses σ_r, σ_θ and the tunnel wall displacement $u(r)$. If the out-of-plane stress σ_z is assumed to be the intermediate principal stress, then an analytic solution for the stresses is well-known ([13], for example) for brittle Mohr-Coulomb rock. A zone of plastic rock develops around the tunnel, with a discontinuity in σ_θ at the elastic/plastic interface. The Author has extended this solution to displacements [14] using the flow rule (7), and has shown that it is possible for the stresses in the plastic region to drop until an inner plastic zone develops in which $\sigma_\theta = \sigma_z > \sigma_r$ [15], i.e. the stress state lies on a corner of the yield surface.

As an analytic solution using the proposed flow rule (10) is not available, a one-dimensional finite element method is used to compare the two rules. In this method [14,16], each element represents an annulus of rock around the tunnel.

Since the analysis is only one-dimensional, it is possible to use a fine mesh and small timesteps and loading increments, to avoid the need for special techniques at the yield surface corner.

5. RESULTS

The method of the previous section was applied to the following three sets of data for brittle Mohr-Coulomb rock with initial unconfined compressive strength σ_c, residual strength σ_c', triaxial stress factor k , Young's modulus E and Poisson's ratio ν :

(i) E=20GPa ν=0.4 σ_c=40MPa σ_c'=20MPa

k=3 q=30MPa

(ii) As (i), except that Poisson's ratio was reduced to ν=0.0 .

(iii) E=40GPa ν=0.2 σ_c=142MPa σ_c'=5.8MPa

k=5 q=108MPa

In each case, the tunnel radius r_0 = 4 metres, and the final tunnel support pressure $p = 0$. The flow rules (7) and (10) were used with zero plastic dilation, so that the plastic potential may be written

(a) $Q(\sigma) = \sigma_1 - \sigma_3$ from flow rule (7) .

(b) $Q(\sigma) = \sqrt{J_2}$ from flow rule (10) .

The yield function was given by (6) in each case. Fig.4 shows a cross-section in the deviatoric plane of the yield surface, plastic potential (respectively a Tresca and a Von Mises cylinder) and plastic strain increment vectors, for the two rules.

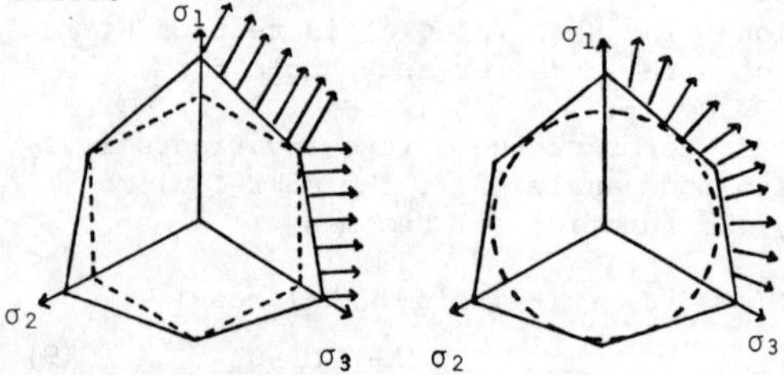

Figure 4. Tresca and Von Mises flow rules.

Results for the tunnel wall inward displacement u , in mm are given in Table 1. It is seen that flow rule (b) predicts slightly larger displacements but there is no significant difference.

Table 1. Tunnel wall displacements (in mm)

	(a) $Q=\sigma_1-\sigma_3$	(b) $Q=\sqrt{J_2}$
Problem (i)	10.891	10.898
Problem (ii)	9.012	9.371
Problem (iii)	49.014	49.950

In Figs. 5-7 are graphs of principal stresses at the Gauss points for problems (i) - (iii), using flow rules (a) and (b) in each case. Also shown are the analytic solutions for σ_r and σ_θ. In problem (i) σ_z remains intermediate, and there is little difference between the flow rules. In problem (ii), because of the lower value of ν , an inner plastic zone with $\sigma_z = \sigma_\theta > \sigma_r$ develops with flow rule (a), while with rule (b) all three stresses drop simultaneously. In problem (iii), with a much greater drop in strength at yield, $\sigma_z = \sigma_\theta$ throughout the plastic

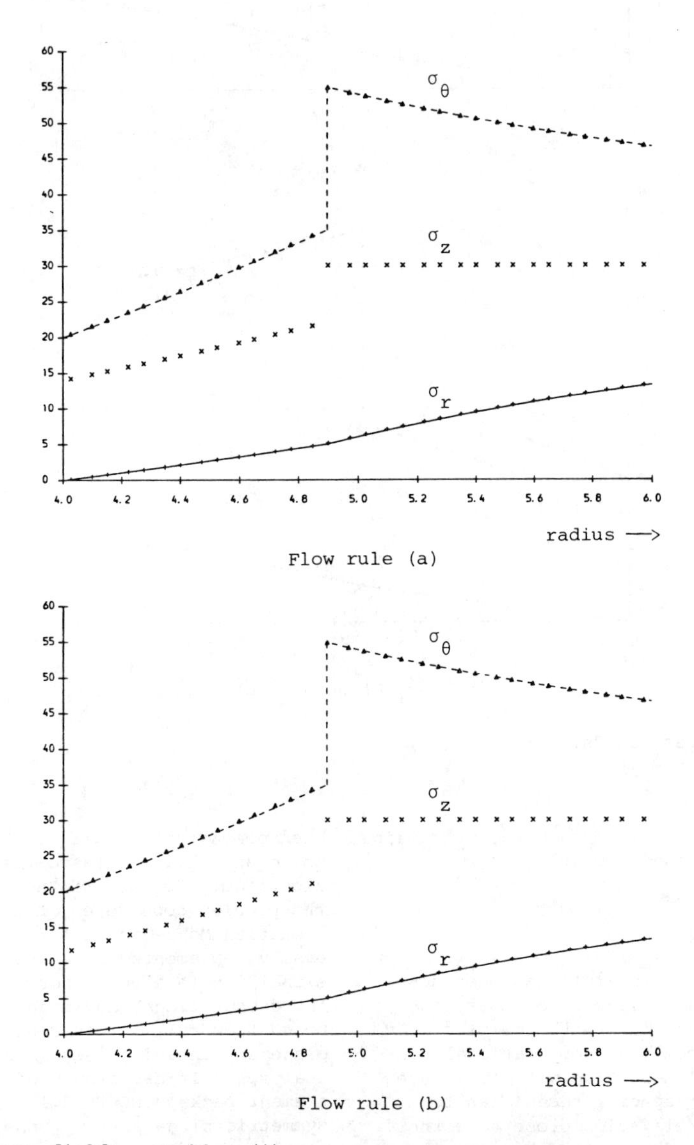

Figure 5. Stress fields: problem (i)

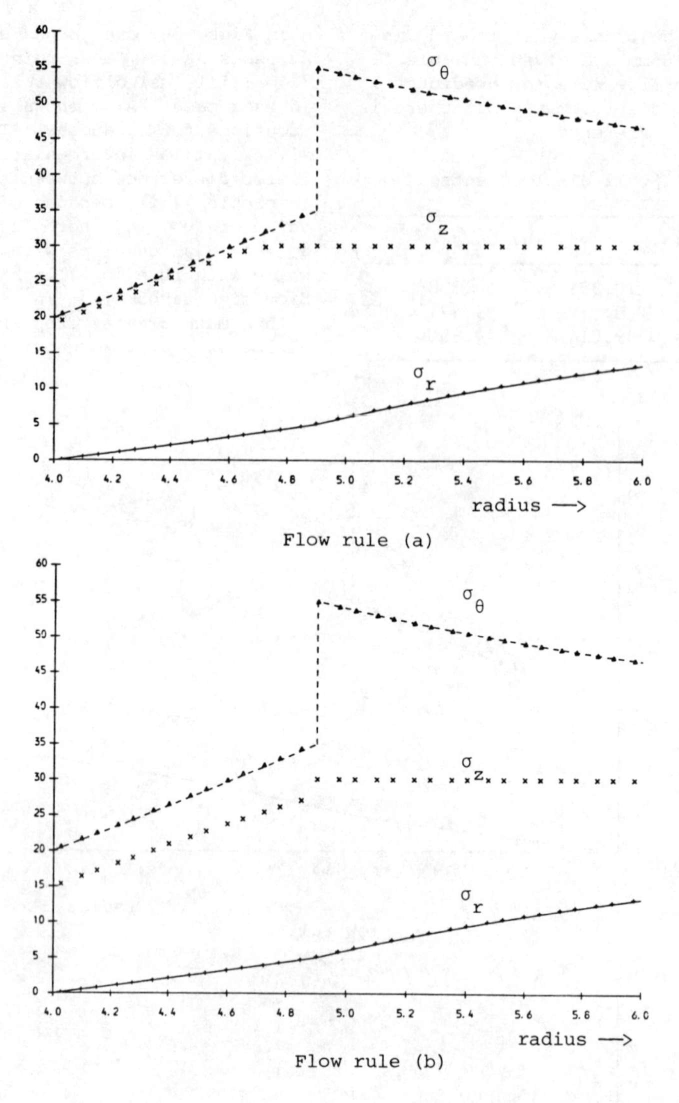

Figure 6. Stress fields: problem (ii)

region with rule (a); even here, σ_z remains intermediate using rule (b).

6. CONCLUSIONS

A non-associated flow rule has been described, which is suitable for modelling geotechnical materials with low plastic dilation. It combines computational simplicity in implementation with avoidance of the need for special techniques to handle stress relaxation close to corners of the yield surface. Its performance has been demonstrated on a rock tunnel problem using an implicit elasto-viscoplasticity algorithm. Because of the axisymmetry of the problem considered, the alternative 'modified Mohr-Coulomb' flow rule was easily implemented in a special one-dimensional finite element formulation, but the computational advantages of the proposed flow rule are evident for general plane strain and three-dimensional analyses. It has been used in a finite element package FESTER [17] for non-axisymmetric plane strain tunnelling problems.

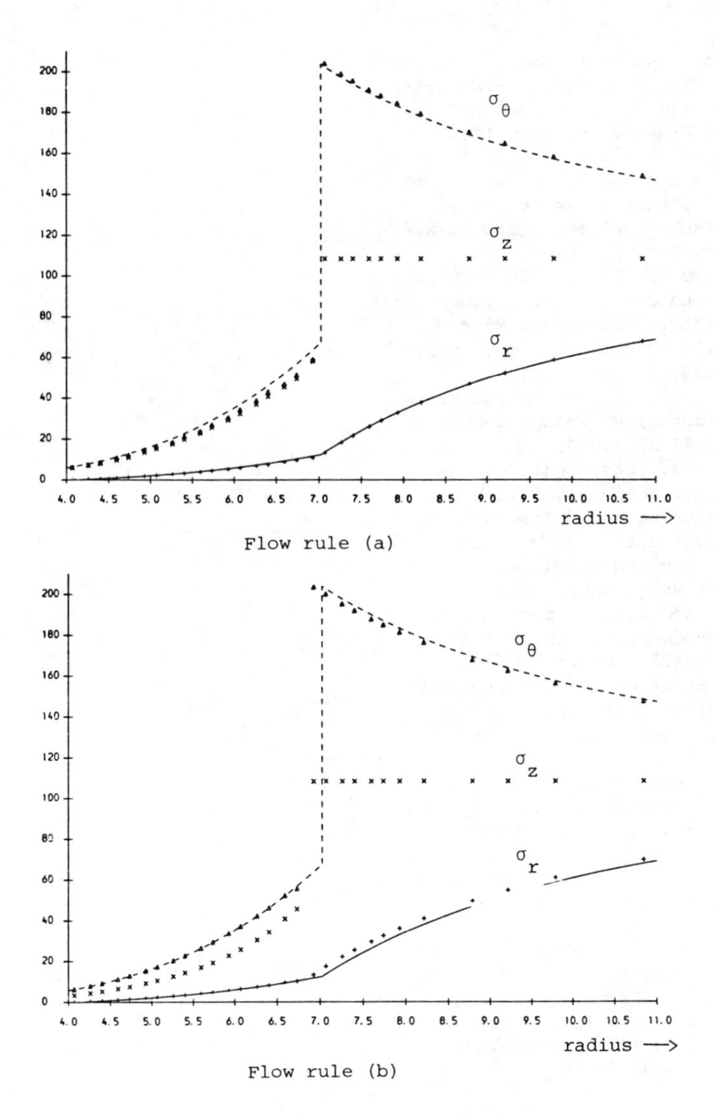

Figure 7. Stress fields: problem (iii)

REFERENCES

[1] R. Hill, Mathematical Theory of Plasticity, Oxford University Press, Oxford, 1983.

[2] Z. Mroz, Non-associated flow laws in plasticity, J. de Mecanique, 2, 21-42 (1963).

[3] E.H. Davis, Theories of plasticity and the failure of soil masses. In: Soil Mechanics - Selected Topics, Ed. by I.K. Lee, Butterworths, London, 341-380 (1968).

[4] O.C. Zienkiewicz, C. Humpheson and R.W. Lewis, Associated and non-associated visco-plasticity and plasticity in soil mechanics, Geotechnique, 25, 671-689 (1975).

[5] O.C. Zienkiewicz, and I.C. Cormeau, Visco-plasticity - plasticity and creep in elastic solids - a unified numerical solution approach, Inter. J. Numer. Methods Engng. 8, 821-845 (1974).

[6] H. Matsuoka and T. Nakai, Stress-Deformation and strength characteristics of soil under three different principal stresses, Proc. J.S.C.E., 32, 59-70 (1974).

[7] M.K. Kim and P.V. Lade, Modelling rock

strength in three dimensions, Int. J. Rock Mech. Min. Sci., 21, 21-33 (1984).

[8] D.R.J. Owen and E. Hinton, Finite Elements in Plasticity, Pineridge Press, Swansea, 1980.

[9] D.M. Potts and A. Gens, The effect of the plastic potential in boundary value problems involving plane strain deformation, Int. J. Numer. Anal. Methods Geomech., 8, 259-286 (1984).

[10] S.W. Sloan and J.R. Booker, Removal of singularities in Tresca and Mohr-Coulomb yield functions, Comm. Appl. Numer Methods, 2, 176-179 (1986).

[11] J.M.M.C. Marques, Stress computations in elastoplasticity, Engng. Computations 11, 42-51 (1984)

[12] M.A. Crisfield, Consistent Schemes for plasticity computation with the Newton-Raphson method. In: Computational Plasticity, Ed. by D.R.J. Owen, E. Hinton and E. Oñate, Pineridge Press, Swansea, 1987, 133-160

[13] J.C. Jaeger and N.G.W. Cook, Fundamentals of Rock Mechanics (3rd Ed.) Chapman and Hall, London, 1979.

[14] M.B. Reed, Stresses and displacements around a cylindrical cavity in soft rock, IMA J. Appl. Math 36, 223-245 (1986).

[15] M.B. Reed, The influence of out-of-plane stress on a plane strain problem in rock mechanics, Int. J. Numer. Analytic Methods Geomech. (to appear)

[16] J.P. Carter and S.K. Yeung, Analysis of cylindrical cavity expansion in a strain-weakening material, Comp. Geotech, 1, 161-180 (1985).

[17] M.B. Reed, A viscoplastic model for soft rock. In: Computational Plasticity, Ed. by D.R.J. Owen, E. Hinton and E. Oñate, Pineridge Press, Swansea, 1677-1689 (1987).

ACKNOWLEDGEMENTS

The work described in this paper was conducted through a project sponsored by the Science and Engineering Research Council, and British Coal.

Numerical Methods in Geomechanics (Innsbruck 1988), Swoboda (ed.)
© 1988 Balkema, Rotterdam. ISBN 90 6191 809 X

Constitutive relations with general Masing rule under multi-dimensional stress condition

K.Nishi & M.Kanatani
Central Research Institute of Electric Power Industry, Abiko, Japan

ABSTRACT : Constitutive relations to adequately evaluate the deformation behaviour of granular materials under rotation of principal stress axis and multi-dimensional stress strain field during cyclic loading are proposed. It was found through comparison between experimental and computational results that these relations can predict with high accuracy the strain softening due to generation of excess pore water pressure and undrained strength for fully saturated sand under cyclic shear.

1 INTRODUCTION

While some valuable constitutive relations have been proposed to evaluate the mechanical behavior of soils under cyclic loading like earthquake and sea wave motion untill now from various stands, it is not yet sufficient for explaining the effect of the rotation of principal stress axis to deformation behavior and for giving the reference about the judgement of loading and unloading under multi-dimensional stress field.

Here, first, the general Masing rule, which has been frequently used for one dimensional problems on response of ground during earthquake, are developed so as to make possible its application for two and three dimensional dynamic ploblems. Elasto-plastic costitutive relations are proposed by paying attention to stress-dilatancy relations obtaind from drained cyclic loading tests for saturated sand by triaxial torsional shear apparatus and by adopting the above mentioned general Masing rule as the hardening function.

To verify the usefulness of newly proposed constitutive relations, computations for simple shear tests under undrained cyclic loading are conducted.

2 CONSTITUTIVE RELATIONS BASED ON ELASTO-PLASTICITY THEORY

2.1 Generalized Masing Rule

Stress parameter η^* defined by Sekiguchi and Ohta (1977) is introduced in order to express the plastie deformation behavior during cyclic loading under general stress condition.

$$\eta^* = \sqrt{\left(\frac{S_{ij}}{\sigma_m'} - \frac{S_{ijo}}{\sigma_{mo}'}\right)\left(\frac{S_{ij}}{\sigma_m'} - \frac{S_{ijo}}{\sigma_{mo}'}\right)} \qquad (1)$$

where S_{ij} and σ_m' are deviatoric stress tensor and effective mean stress, respectively. Suffix 0 means initial stress state before cyclic loading. η^* is originally defined to formulate the deformation behavior of anisotropic consolidated soils under monotonic loading. Here, η_r^* is newly introduced to enlarge the adoptability of η^* to cyclic loading.

$$\eta_r^* = \sqrt{\left(\frac{S_{ij}}{\sigma_m'} - \frac{S_{ijr}}{\sigma_{mr}'}\right)\left(\frac{S_{ij}}{\sigma_m'} - \frac{S_{ijr}}{\sigma_{mr}'}\right)} \qquad (2)$$

where S_{ijr} and σ_{mr}' are S_{ij} and σ_m' at time judged as reverse.

Also, reverse is defined as the stress states that the increment of η^* or η_r^* shows negative value. By adopting such relative stress ratio, skeleton and histeretic curves are given as follows.

$$\text{skeleton curve}: \sqrt{2I_2} = \frac{d\eta^*}{Go^*(1-\eta^*/M_f)^2} \qquad (3)$$

$$\text{histeretic curve}: \sqrt{2I_2} = \frac{d\eta_r^*}{Go^*(1-\eta_r^*/2M_f)^2} \qquad (4)$$

where $\sqrt{2I_2}\,(=\sqrt{de_{ij}^p\,de_{ij}^p})$ is second invariant of increment of plastic deviatoric strain tensor, G_0^* is initial tangent modulus at the relation between η^* and $\int\sqrt{2I_2}$, M_f is relative stress ratio at failure. Eq. (3) means that the relation between η^* and $\int\sqrt{2I_2}$ is given as hyperbolic formula by Nishi and Esashi (1978). Also, eq.(4) is formulated by adoption of so called Masing rule. Both equations are adopted as the hardening functions.

2.2 Yield functions and stress-dilatancy equations

It is postualted that the plastic strain continuously generates when stress parameter η^* or η_r^* changes, that is $d\eta^*, d\eta_r^* \neq 0$. In this case, yield function f is given as

$$\eta^* \geq \eta^*_{max} : f=\eta^* \quad (5)$$

$$\eta^* < \eta^*_{max} : f=\eta_r^* \quad (6)$$

where η^*_{max} is the maximum value of η^* previously experienced.

Stress-dilatancy equations are adopted as follows.

$$\eta^* \geq \eta^*_{max} : \frac{dV_d}{\sqrt{2I_2}} = M_m - \eta^* \quad (7)$$

$$\eta^* < \eta^*_{max} : \frac{dV_d}{\sqrt{2I_2}} = \alpha \cdot (M_m + \eta^{*r} - \eta_r^*) \quad (8)$$

where V_d is volumetric strain due to dilatancy, M_m is the value of η^* showing maximum contraction, that is $dV_d=0$, η^{*r} is η^* at the reverse point, and α is parameter expressing the extent of accumulation of dilatancy and is given as follows.

$$\alpha = 1 - \frac{V_{dr}}{V_{df}} \quad (9)$$

$$V_{df} = m^* \cdot \sigma_{mr}' \cdot (\eta^{*r})^n \quad (10)$$

where V_{dr} is volumetric strain accumulating at the reverse point and V_{df} is maximum volumetric strain brought about by the drained cyclic loading under $\eta^*(=\eta^{*r})$ =constant condition. Eq.(10) represents the relation proposed by Yagi(1974) and Ohoka(1976) under general stress condition. According to the triaxial torsion shear tests, m^* is given as the function of relative density Dr and n nearly equals to 5.0. Eq. (7) was previously proposed by Nishi and Esashi(1978), and plastic potential function was induced based on it. Eq.(8) is newly formulated based on the test results by triaxial

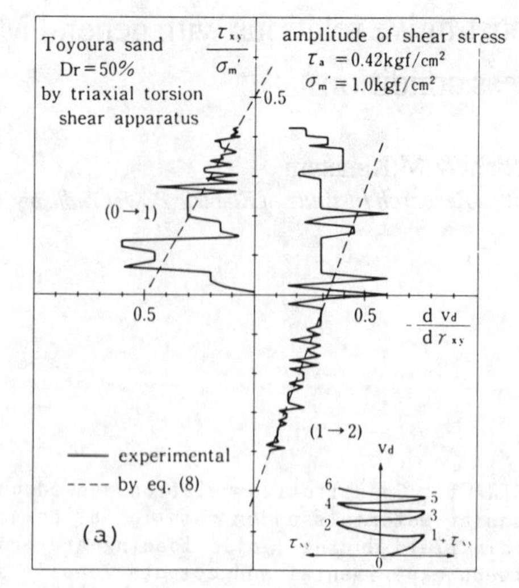

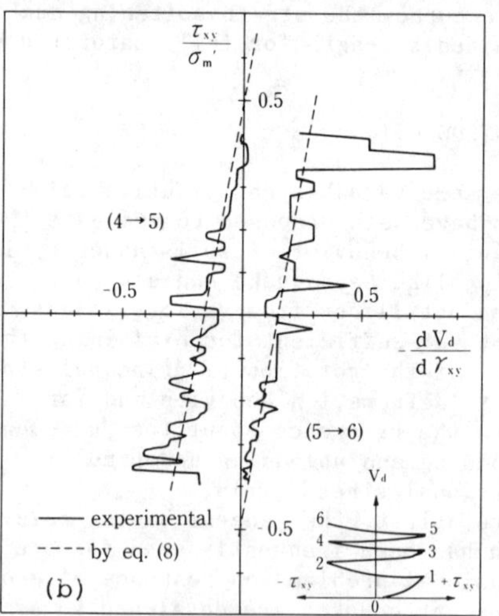

Fig.1 Relations between stress ratio and strain incremental ratio by drained cyclic shear test

torsion shear apparatus. Fig.1(a),(b) show the relations between stress ratio τ_{xy}/σ_m' and strain incremental ratio $-dV_d/d\gamma_{xy}$ obtained by drained cyclic shear test under isotropic consolidation and constant amplitude of shear stress τ_{xy} shown in figure. It is noticed that τ_{xy}/σ_m' and $d\gamma_{xy}$ are equivalent to η^* and $\sqrt{2I_2}$ respectively, in the case of stress condition such as $\sigma_x'=\sigma_y'$ and $\tau_{xy_o}=0$. It is clearly demonstrated in Fig.1 that 1) the relation between τ_{xy}/σ_m' and $-dV_d/d\gamma_{xy}$ is approximately given as liner one under cyclic loading, 2) α

nearly equals to 1.0 in the case of virgin loading shown as 0→1 in Fig.1 (a), and 3) accompanying with the increase in number of cycles, α gradually decreases.

While volumetric strain V_d accumulates by drained cyclic shear loading, increment of V_d becomes approximately 0 after some ten cycles. Consequently it is concluded that α in eq.(8) approaches 0 with the increase in number of cycles . Paying attention to these test results, eq.(8) is set up (Nishi, Tohma and Kanatani (1985)). On the other hand, the relations between strain ratio divided by $(M_m - \tau_{xy}/\sigma_m')$ and $1-V_{dr}/V_{df}$ are shown in Fig.2. It is known from this figure that the relation between parameter α and accumulated volumetric strain due to dilatancy is approximately given by simple formulation such as eq.(9).

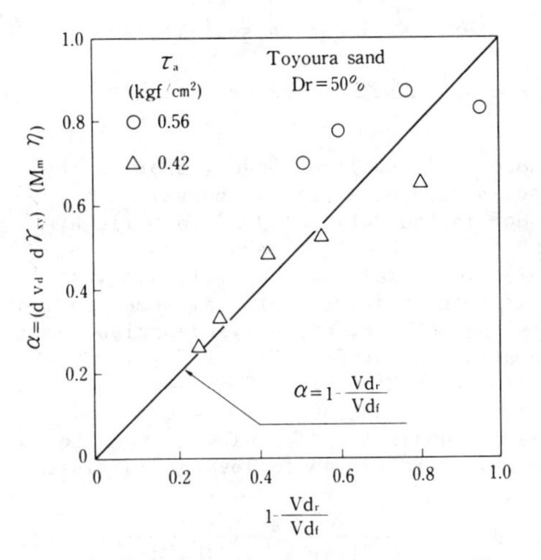

Fig.2 Relations between parameter α and accumlated volumetric strain V_{dr} by dilatancy

2.3 Constitutive relations

From the above mentioned relations, the following constitutive relations are given.

$$d\varepsilon_{ij} = \frac{1+\nu}{E} dS_{ij} + \frac{1-2\nu}{E} d\sigma_m' \delta_{ij} + h \frac{\partial g}{\partial \sigma_{ij}} df \tag{11}$$

where η_{ij} is S_{ij}/σ_m', E is Young's modulus and ν is Poisson's ratio.

$\underline{\eta^* \geq \eta^*_{max}}$

$$h = \frac{M_m \cdot \sigma_m'}{Go^*} \left(1 - \frac{\eta^*}{M_f}\right)^{-2} \tag{12}$$

$$\frac{\partial g}{\partial \sigma_{ij}} = \frac{1}{M_m \sigma_m'} \left\{ \frac{\eta^*_{ij} - \eta^*_{ijo}}{\eta^*} + (M_m - \eta^*) \frac{\delta_{ij}}{3} \right\} \tag{13}$$

$$df = \frac{1}{\sigma_m'} \left\{ \frac{\eta^*_{ij} - \eta^*_{ijo}}{\eta^*} - \frac{S_{ij}(\eta^*_{ij} - \eta^*_{ijo})}{\sigma_m' \eta^*} \frac{\delta_{ij}}{3} \right\} d\sigma_{ij} \tag{14}$$

$\underline{\eta^* < \eta^*_{max}}$

$$h = \frac{\sigma_m' \{\alpha(M_m + \eta^{*r} - \eta_r^*) + \eta_r^*\}}{Go^*} \times \left(1 - \frac{\eta_r^*}{2M_f}\right)^{-2} \tag{15}$$

$$\frac{\partial g}{\partial \sigma_{ij}} = \frac{1}{\sigma_m' \{\alpha(M_m + \eta^{*r} - \eta_r^*) + \eta_r^*\}} \left\{ \frac{\eta^*_{ij} - \eta^*_{ijr}}{\eta_r^*} + \alpha(M_m + \eta^{*r} - \eta_r^*) \frac{\delta_{ij}}{3} \right\} \tag{16}$$

$$df = \frac{1}{\sigma_m'} \left\{ \frac{\eta^*_{ij} - \eta^*_{ijr}}{\eta_r^*} - \frac{S_{ij}(\eta^*_{ij} - \eta^*_{ijr})}{\sigma_m' \eta_r^*} \frac{\delta_{ij}}{3} \right\} d\sigma_{ij} \tag{17}$$

3 MATERIAL CONSTANTS

It is ideal that constitutive relations are simple and material constants involved in those are as few as possible and these physical meaning is clear. A number of material constants in constitutive relations proposed here is 7, that is M_f, M_m, m^*, n, Go^* to describe the plastic behavior and E, ν for elastic behavior. Adopting Mohr-Coulomb's failure criterion, M_f is given as

$$M_f = \sqrt{2S_x^2 + S_y^2 + [\{\pm(\sigma_x' + \sigma_y')\sin^2\phi - (\sigma_x' - \sigma_y')\}/2\sigma_m' - S_{xyo}/\sigma_{mo}']^2} \tag{18}$$

where ϕ is the angle of internal friction at failure state. Also, sign of + and - in above equation corresponds to $dS_{xy} > 0$ and $dS_{xy} < 0$, respectively. M_m is computed from following equation by using M_f, ϕ, and ϕ_m that is the angle of internal friction mobilized at the stress state of $dV_d=0$.

$$M_m = M_f \frac{\sin\phi_m}{\sin\phi} \tag{19}$$

Young's modulus is given as follows by assuming that Poisson's ratio ν is constant ($\nu=0.0$) and void ratio $e - \sigma_m'$ relation at swelling by isotropic compression test is linear on semi-log scale.

$$E = \frac{3(1-2\nu)(1+e)\sigma_m'}{\kappa} \tag{20}$$

where κ is swelling index, that is the inclination of $e - \ln\sigma_m'$ curve.

Go* is the ratio of $d\eta^*$ to $\sqrt{2I_2}$ at $\eta^*=0$. On the other hand, shear modulus Go at very small strain amplitude was sufficiently investigated by some kinds of testing methods. Here, we describe the relations among Go*, Go and G given by E and ν ($G=E/2(1+\nu)$). Consider the stress field that shear stress τ_{xy} cyclicaly changes under $\sigma_x'=\sigma_y'=\sigma_z'$. Then, Eq.(3) is rewritten by the following equation.

$$d\gamma^p_{xy} = \frac{2d(\tau_{xy}/\sigma_m')}{Go^*(1-\sqrt{2}\tau_{xy}/\sigma_m' M_f)} \tag{21}$$

When $\tau_{xy} \rightarrow 0$ and $\sigma_m' \rightarrow \sigma_{mo}'$, the above equation becomes

$$d\gamma^p_{xy} = \frac{2d\tau_{xy}}{Go^*\sigma_{mo}'} \tag{22}$$

Also, as total shear strain increment $d\gamma_{xy}$ is summention of elastic shear strain increment $d\gamma^e_{xy}$ and $d\gamma^p_{xy}$, and $d\gamma^e_{xy}$ is given as $d\tau_{xy}/G$, after all Go* is connected with G and Go as follows.

$$Go^* = \frac{2G \cdot Go}{\sigma_{mo}'(G-Go)} \tag{23}$$

Namely, Go* is obtained by material constants κ and Go(or shear wave velocity). m* and n are determined by drained cyclic shear tests under constant amplitude of shear stress. As before mentioned, n is approximately 5.0 due to triaxial torsion shear tests and m* is given as the function of relative density Dr such as shown in Fig.3.

It is concluded that material constants in proposed constitutive relations are easily obtained from popular laboratory tests and are decided with clear physical meaning.

4 SIMULATION OF DEFORMATION BEHAVIOR DURING UNDRAINED CYCLIC LOADING

4.1 Shear stress reversal effect to deformation behavior

By introducing stress parameter η^* and η_r^*, shear stress reversal effect to deformation behavior can be evaluated by proposed constitutive relation. Here,

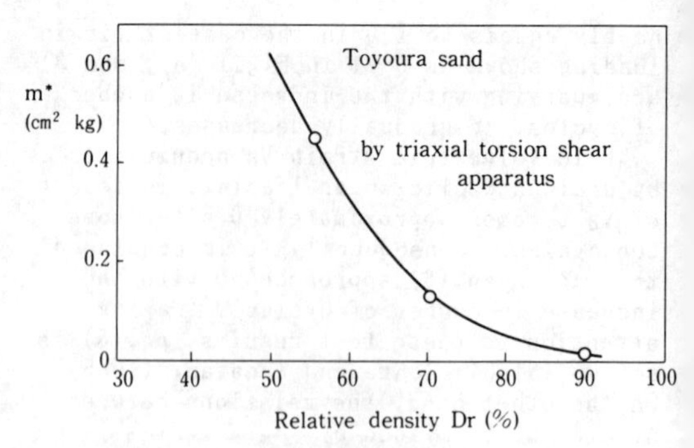

Fig.3 Relations between material constants m* and relative density

we concretely show the adoptability of constitutive relations by analysing the deformation behavior of soil element under undrained cyclic shear stress and plane strain field. Briefly, consider the stress condition that shear stress changes only. In this case, $d\varepsilon_x=0$ by lateral strain constraint condition and $d\varepsilon_y=0$ by undrained condition. Computational conditions are 1) initial stress : $\sigma_{xo}'=1.0$, $\sigma_{yo}'=0.5$, $\sigma_{zo}'=0.5$, $\tau_{xyo}'=0$ kgf/cm², 2) cyclic shear stress : case 1 is $\tau_{xy}=+0.13 \sim -0.13$kgf/cm² (two way cyclic loading), case 2 is $\tau_{xy}=0 \sim +0.13$kgf/cm² (one way cyclic loading). Used material constants are Go*=2456, $\phi=36°$, $\phi_m=30°$, m*=0.86cm²/kgf, n=5, $\kappa=0.0015$, which are obtained by triaxial torsional shear tests for Toyoura sand.

Fig.4(a),(b) and Fig.5 show the computational results on the relation between shear stress τ_{xy} and effective mean stress σ_m', relation between second invariant of deviator stress tensor $\sqrt{2J_2}$ and σ_m', respectively. Also, relation between τ_{xy} and shear strain γ_{xy} is shown in Fig.6(a),(b). It is found from Fig.4(a) that perfect liquefaction takes place under a number of cycles of 6 in the case of two-way cyclic loading. However, for one way cyclic loading excess pore water pressure built up in computation is about 50% under same number of cycles. Also, accumlated shear strain is very small comparing with the case of two-way cyclic loaing. Of course, these results are experimentaly observed as well known. It is shown in Fig.5 that the decrease in $\sqrt{2J_2}$ is clearly computed accompanying with the decrease in effective mean stress, which is phenomenon brought about by lateral strain constraint condition, that is real deformation behavior observed in

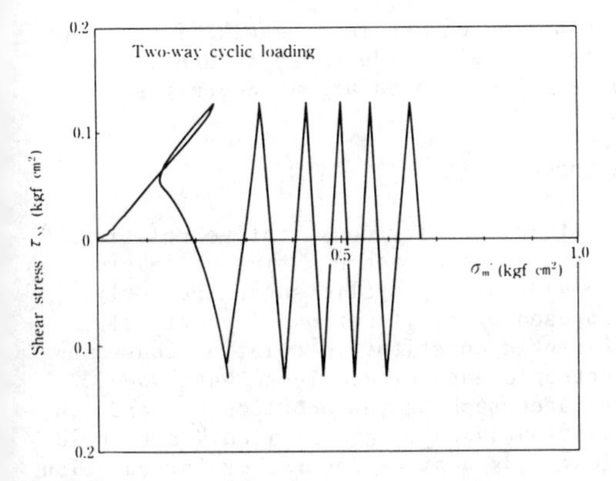

Fig.4(a) Effective stress path under two way cyclic loading computed by constitutive relations

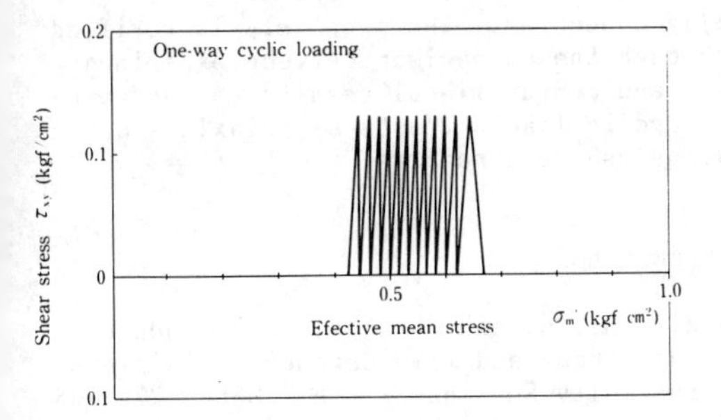

Fig.4(b) Effective stress path under one way cyclic loading computed by constitutive relations

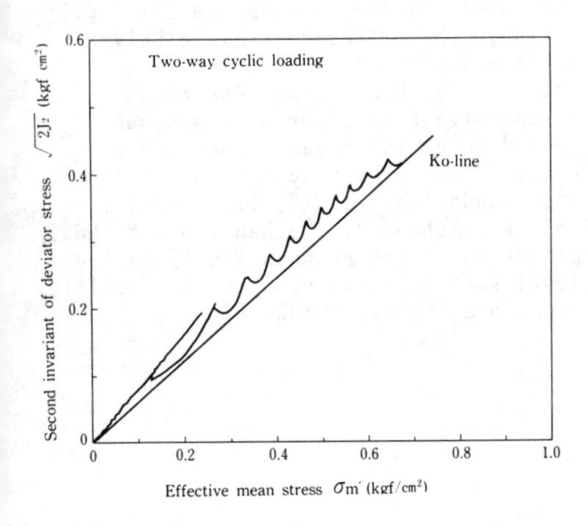

Fig.5 Relations between $\sqrt{2J_2}$ and σ_m' during two way cyclic loading computed by constitutive relations

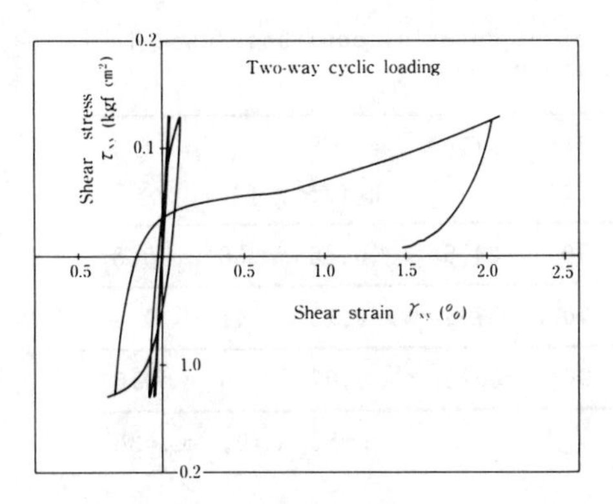

Fig.6(a) Shear stress and shear strain relations during two way cyclic loading computed by constitutive relations

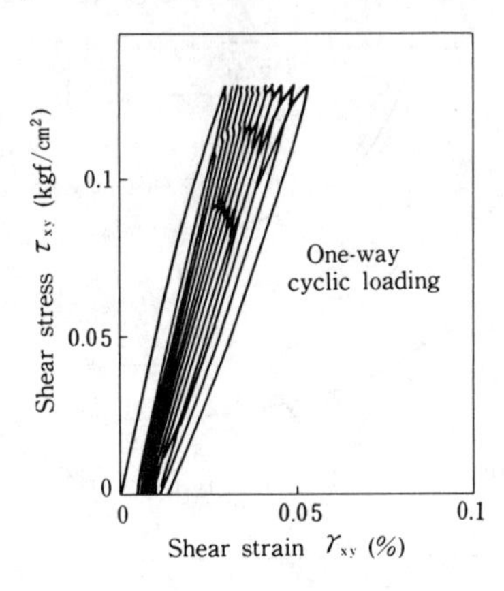

Fig.6(b) Shear stress and shear strain relations during one way cyclic loading computed by constitutive relations

the way of perfect liquefaction is fully demonstrated by proposed constitutive relations.

4.2 Prediction of undrained shear strength under cyclic loading

Here, comparison between experimental and computational results on undrained shear strength is demonstrated to show the propriety of proposed constitutive relations. Experiments were performed by triaxial torsion shear apparatus for hollow cylindrical specimen with relative density of Dr=50, 70 and 90%. Toyoura sand was used as sample. All tests were executed under undrained condition by the application of

Table-1 Material constants used in computation

Dr (%)	Go*	m* (kgf/cm²)	ϕ (°)	eo
50	2456	0.86	36	0.8
70	3734	0.15	41	0.7
90	6615	0.02	46	0.6

$n=5,\ \nu=0,\ \phi_m=30°$

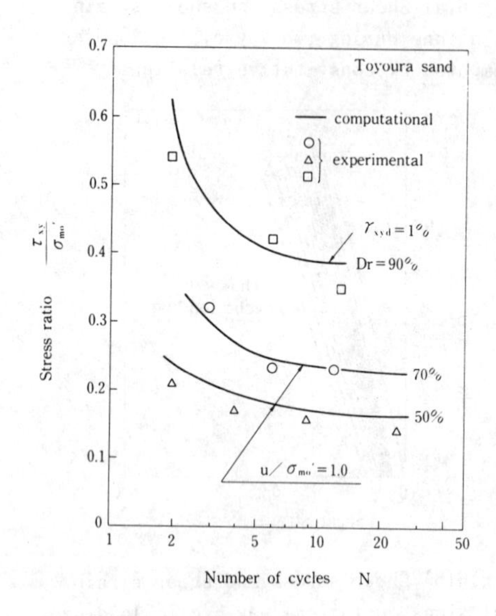

Fig.7 Comparison between computed results and experimental results on dynamic strength vs. number of cycles relations

torsion shear after isotropic consolidation. Material constants are shown in Table.1 Fig.7 shows the relations between stress ratio defined as the ratio of undrained shear strength to initial effective mean stress and nubmer of cycles N. For both of experiments and computation these are defined as states where effective stress shows 0 for the specimens with Dr of 50 and 70%, on the other hand, double amplitude of shear shrain attains to 1% for the specimens with Dr of 90%. It is clearly shown that good correlations between experimental and computational results are existed. Namely, proposed constitutive relations give the real evaluation on cyclic strength of granular materials with wide range of density if material constants are adequately set up.

5 CONCLUSIONS

Elasto-plasticity constitutive relations to evaluate the non-elastic deformation behavior during cyclic shear are newly proposed by modification of previously presented constitutive relations based on isotropic hardening. These have some features such as the adoption of hardening function based on generalized Masing rule, clear judgement of reverse by introduction of stress parameter defined by Sekiguchi· Ohta(1977) under multi-dimensional stress-strain field, and so on. Material constants in constitutive relations are easily determined by laboratory tests generally conducted. The propriety is verified through the comparison between experimental and computational results for undrained cyclic loading tests by triaxial torsional shear apparatus.

REFERENCES

Sekiguchi, H. and Ohta, H. 1977. Induced anisotropy and time dependency in clays. Specialty Session 9, 9th ICSMEF : 229-238

Nishi, K. and Esashi, Y. 1978. Stress-strain relationships of sand based on elasto-plasticity theory. Proc. of JSCE, Vol.280 : 111-122

Yagi, N. (1974). Mechanical properties of sand under cyclic loading and its application. Ph. D Thesis in Kyoto University(in Japanese)

Oh-oka (1976). Drained and undrained stress-strain behavior of sands subjected to cyclic shear stress under nearly plane strain condition. Soils and Foundation, Vol.16, №3 : 19-31.

Nishi, K., Tohma, J. and Kanatani, M.1985. Effective stress response analysis for level sand deposits with cyclic mobility. Proc. 5th ICONMIG : 389-397

Numerical Methods in Geomechanics (Innsbruck 1988), Swoboda (ed.)
© 1988 Balkema, Rotterdam. ISBN 90 6191 809 X

Some constitutive laws for creeping soil and for rate-dependent sliding at interfaces

L.Vulliet
University of Arizona, Tucson, USA
K.Hutter
Technische Hochschule, Darmstadt, FR Germany

ABSTRACT: Soil under creeping deformations is modeled either as an incompressible viscous fluid or a binary mixture of a viscous fluid matrix filled with an inviscid ideal fluid. Explicit expressions for the constitutive relation of stress are deduced and phenomenological coefficients are determined by means of laboratory and field data. The rate dependent constitutive models are compared with corresponding relations deduced from viscoplasticity. Finally, rate dependent sliding laws are presented.

1 INTRODUCTION

Soil cannot in general be considered as a material with a single unique material response; the constitutive behavior is related not only to the characteristics of the soil, but also to the nature of the physical processes under consideration.

For landslide problems we know that a stability analysis can be successfully achieved by assuming that the soil mass behaves as a rigid-perfectly plastic body. However, such a procedure provides no information on the displacement field and consequently does not cover cases like progressive failure or soil-structure interaction. For these an elasto-plastic approach is needed. Creep flows require yet a different constitutive approach. Depending on the time-scales of the physical processes either elasto-visco-plastic, visco-plastic or purely viscous constitutive laws will apply. In this paper, we focus on the long term behavior of slowly creeping soils and neglect the starting phase of transient creep.

We use methods of continuum mechanics to deduce physically objective forms of the constitutive relations of the stress tensor. We model the soil either as a non-Newtonian (very) viscous fluid or a saturated mixture of an inviscid interstitial fluid and a non-Newtonian (very) viscous fluid matrix. In both cases the stress tensor of the "solid" phase is of Reiner-Rivlin type, but additional assumptions are invoked to simplify the stress-stretching relationship. We derive the most general form of such reduced constitutive relationships. Laboratory and in-situ creep data are used in the identification of constitutive parameters. They demonstrate the usefulness of the approach. We briefly show how our approach may be related to the classical visco-plastic theory, and how some proposed response functions may be used in this context.

Sliding along a basal surface is also studied. Viscous-type relationships between the shear traction and the sliding velocity are postulated and functional relationships are suggested which may adequately account for the slip along the basal surface. For two sites (in Switzerland and in France), the usefulness of the scheme is demonstrated.

2 CONSTITUTIVE LAWS FOR DIFFERENTIAL CREEP

As mentioned in the introduction the complexity of the constitutive class will delimit the number of phenomena that can be described by the model. We exclude anisotropy and assume that time scales of the transient processes and the creeping motion are distant enough that interaction of the two can be ignored. Thus, we omit elasticity effects in comparison to viscous creep. Two classes of soil models, henceforth simply called "soils" will be analysed: (i) an incompressible viscous (one-component) fluid, which is representative of a soil whose moisture content hardly varies and thus does not explicitly affect its creep response, and (ii) a binary mixture of a solid (matrix) and an (interstitial) fluid, which is appropriate for saturated conditions.

2.1 Incompressible viscous model

Let $\underset{\sim}{t}$ be the Cauchy stress and $\underset{\sim}{D}$ the stretching tensor, $D_{ij} = 1/2(v_{i,j} + v_{j,i})$, where $\underset{\sim}{v}$ is the velocity field. It follows from general priciples of continuum mechanics (see e.g. Truesdell, 1977, 78, Müller, 1973, or others) that the most general isotropic stress-stretching (strain rate) relation is given by

$$\underset{\sim}{D} = a\underset{\sim}{1} + b\underset{\sim}{t} + c\underset{\sim}{t}^2 \qquad (1)$$

in which $\underset{\sim}{1}$ is the identity tensor and a, b, c are scalar valued functions of the stress invariants

$$a = a(t_I, t_{II}, t_{III}), \; \ldots$$

$$t_I = tr(\underset{\sim}{t}),$$

$$t_{II} = \frac{1}{2}\, tr(\underset{\sim}{t}^2), \qquad (2)$$

$$t_{III} = \frac{1}{6}\, tr(\underset{\sim}{t}^3),$$

where tr(.) is the trace operator. The class (1) of constitutive relations is known as a Reiner-Rivlin fluid.

Incompressibility permits to assume that $\underset{\sim}{t}$ is a deviator, so $tr(\underset{\sim}{t}) = t_I = 0$. Then, since $tr(\underset{\sim}{D}) = 0$, equation (1) reduces to

$$\underset{\sim}{D} = b\underset{\sim}{t} + c\,(\underset{\sim}{t}^2 - \frac{2}{3}\, t_{II}\, \underset{\sim}{1}),$$

$$b = b(t_{II}, t_{III}), \qquad (3)$$

$$c = c(t_{II}, t_{III}).$$

Well-known creep laws such as those of Newton, Norton or Prandtl (Findley et al. 1976) are still of a more restrictive class than is equation (3). This simpler class is obtained by the following two postulates: (i) the stress tensor is collinear to the stretching tensor, and (ii) the coefficient functions b or c depend on the stress only through the invariant t_{II}. With these

$$\underset{\sim}{D} = b(t_{II})\, \underset{\sim}{t}, \qquad (4)$$

where b(.) is an as yet arbitrary function of its argument which must be determined from experiments. Table 1 shows the form of this creep response function when some popular creep laws are considered.

These laws (though not in the generalized form of Table 1) were applied to soil mechanics problems by different authors, see Vulliet (1986), Vulliet and Hutter (1987a). Orders of magnitude of some parameters entering these laws are also collected in Table 1; they correspond to a vatiety of clayey soils.

Table 1. Generalized creep response functions for various different creep laws in terms of stress invariants, and order of magnitude of parameters.

flow law	$b(t_{II})$	remarks	order of magnitude
Newton	$1/\mu$		$\mu = 10^9$ kN·m^{-2} s
Bingham	$\dfrac{1 - \sqrt{\dfrac{t_{II_0}}{t_{II}}}}{\mu}$	applicable if $t_{II} > t_{II_0}$	$\mu = 10^8$-10^{10} kN·m^{-2} s
Norton	$A\, t_{II}^{\frac{m-1}{2}}$	also called power law	A depends on m. m = 3 - 10, with frequent value of 5
Prandtl-Eyring	$\dfrac{A \sinh(B\sqrt{t_{II}})}{\sqrt{t_{II}}}$	$\rightarrow$ AB if $t_{II} \rightarrow 0$	-
Prandtl	$\dfrac{A \exp(B\sqrt{t_{II}})}{\sqrt{t_{II}}}$	$\rightarrow \dfrac{A}{\sqrt{t_{II}}}$ if $t_{II} \rightarrow 0$	-

2.2 Constitutive laws for saturated compressible soils

We regard the soil now as a saturated mixture of a very viscous fluid and an inviscid ideal fluid. Thus the peculiar stresses are: for the water a simple pressure, p_w, called pore water pressure, corrected by the void-ratio e or porosity n, and for the "solid" the complement of this pressure plus an effective stress tensor σ', governed by constitutive quantities, viz:

$$\sigma^w = -n\, p_w \mathbf{1}, \qquad \sigma^s = -(1-n)\, p_w \mathbf{1} + \sigma', \tag{5}$$

where
$$n = \frac{e}{1+e}, \quad e = \frac{n}{1-n}. \tag{6}$$

Adding the two yields the (approximate) mixture stress,

$$\sigma = \sigma^w + \sigma^s = -p_w \mathbf{1} + \sigma'. \tag{7}$$

The constitutive stress σ' will be taken from the class $\sigma'(n,D)$, where n is the porosity as before and D is the stretching tensor of the deformation field of the solid phase.

Consider the decomposition

$$\sigma'(n,D) = \sigma^e(n) + \sigma^v(n,D). \tag{8}$$

Following common usage we shall call $\sigma^e(n)$ the elastic stress (terms like "pseudo-elastic", "quasi-elastic" and "elasto-plastic" would be more appropriate) and σ^v the viscous stress.

Explicit expressions for (8) are given in Vulliet and Hutter (1987a) where the following assumptions were made: a) the "elastic" stress is isotropic ($\sigma^e(n) = -p^e(n)\mathbf{1}$), b) an oedometric law applies for the "elastic" pressure (i.e. linear relation between the logarithm of p^e and the void ratio e), c) the higher-order stress effects are neglected, d) the effect of the third stress-invariant is not considered. Two simple cases arise when further assumptions are made.

Case A

If the bulk viscosity is small, i.e. if secondary consolidation effects are negligible, the constitutive equation reads:

$$\mathrm{tr}\, D = -0.435\, C_c (1-n) \frac{\dot{p}'}{p'}, \qquad D^d = D - \frac{1}{3}\mathrm{tr}(D)\mathbf{1} = b(t_{II}, n)\, t, \tag{9}$$

where D is the stretching tensor related to the solid phase, D^d its deviatoric part, $p'\mathbf{1}$ and t the volumetric, resp. deviatotic part of the effective stress tensor σ', and C_c is the compression index. $(9)_1$ shows that the rate of volume change is governed by the time rate of change of the effective pressure. $(9)_2$ describes the deviatoric behavior. The creep response function b(.) which has the dimension of an inverse viscosity depends on both, t_{II} and n. The porosity (or void ratio) most likely influences the viscous behavior; this is not so for incompressible materials like metals, where (4) may apply as an approximation.

Case B

The other extreme applies for very compressible soils, where volumetric creep prevails. The constitutive equation reduces to:

$$\mathrm{tr}\, D = 3(\tilde{a} - b)\, p', \qquad D^d = D - \frac{1}{3}\mathrm{tr}(D)\mathbf{1} = b\, t, \tag{10}$$

where
$$(\tilde{a}, b) = (\tilde{a}, b)(p', t_{II}, n). \tag{11}$$

Here, two creep response functions are needed. In $(10)_1$, $(\tilde{a}-b)$ is a bulk viscosity. As this viscosity is a function of stress and porosity, $(10)_1$ does not imply a constant rate of volumetric strain for a constant effective pressure p' !

3 DETERMINATION OF RESPONSE FUNCTIONS AND PARAMETERS

There exists a large number of laboratory experiments under pure shear, torsion and compression from which the materially dependent constitutive functions could be determined. We demonstrate here two cases, a triaxial compression test and the shear cell test "Géonor H12" (Bjerrum & Landva, 1966). Moreover, we show how in-situ inclinometer measurements may be equally used, as the stress-strain state in the field is often close to simple shear.

3.1 Incompressible soil

a) triaxial compression

In the usual laboratory coordinate system all stresses and strain rate matrices have diagonal forms. For our one-component model equation (4) applies; writing it down for any component of t and D yields $b(t_{II}) = D_{ij}/t_{ij}$, or

$$b(t_{II}) = \frac{3\dot{\varepsilon}_1}{2(\sigma_1 - \sigma_3)}$$
$$t_{II} = \frac{1}{3}(\sigma_1 - \sigma_3)^2; \qquad (12)$$

σ_1 and $\dot{\varepsilon}_1$ are the axial stress and strain rate, and σ_3 the radial stress (or confining pressure). Evidently, the interstitial pore water pressure has disappeared from these formulas; p_w does not affect the constitutive relation explicitly. Measuring $\dot{\varepsilon}_1$ and the stresses σ_1, σ_3 determines $b(t_{II})$. To this end, we plot $3\dot{\varepsilon}_1/[2(\sigma_1 - \sigma_3)]$ against $1/3\,(\sigma_1 - \sigma_3)^2$ and then find b by postulating a mathematical relationship that fits the data points.

b) Simple shear test

The Géonor H12 device permits application of a combined shear and compression. A cylindrical specimen is subjected to a vertical pressure p_z and an horizontal shear stress τ. As the specimen is laterally confined by a reinforced menbrane, the radial strain rates $\dot{\varepsilon}_x = \dot{\varepsilon}_y$ are null and a radial stress $p_x = p_y$ develops.

Here, because of incompressibility, $\dot{\varepsilon}_z = 0$, so $\underset{\sim}{D}^d = \underset{\sim}{D}$, and from (4) we obtain $\underset{\sim}{D}^2 = b^2(t_{II})\,\underset{\sim}{t}^2$, from which we deduce by taking the trace

$$b(t_{II}) = (D^d_{II}/t_{II})^{1/2}, \qquad (13)$$

which finally yields

$$b(t_{II}) = \left[\frac{\frac{1}{4}\dot{\gamma}^2}{\frac{1}{3}(p_z - p_x)^2 + \tau^2}\right]^{1/2}, \qquad (14)$$

$$t_{II} = \frac{1}{3}(p_z - p_x)^2 + \tau^2,$$

where $\dot{\gamma}$ is the measured shear strain rate. (14) equally applies to a ring shear creep test, or to in-situ inclinometer measurements in long and shallow creeping slopes. In the field, by assuming $p_z = p_x$ (Tan-Tjong-kie, 1981, Kovazanjian and Mitchell, 1984), (14) reduces to

$$b(t_{II}) = \frac{1}{2}\frac{\dot{\gamma}}{\tau}, \qquad t_{II} = \tau^2. \qquad (15)$$

This simplified version of (14) was used for practical cases and the in-situ shear stress was estimated through the infinite slope assumption (Vulliet, 1986).

c) An expression for $b(_{II})$

Vulliet (1986) showed that the finite viscosity law (Hutter, 1983)

$$b(t_{II}) = \frac{1}{\mu} + A\, t_{II}^{\frac{m-1}{2}} \qquad (16)$$

is numerically convenient and offers a satisfactory data fit. Values of the material parameters μ, A and m are collected in Table 2 for different soils and various experimental procedures.

3.2 Compressible saturated mixture

a) Triaxial compression

For case A, i.e. small bulk viscosity, equation (9) applies. In an undrained triaxial experiment the interstitial pressure is kept constant, $\dot{p} = 0$, so that $\mathrm{tr}(\underset{\sim}{D}) = 0$, only the deviatoric part of the stress-stretching relation remains to be determined. Thus

Table 2. Values of parameters entering in equation (16) for three different soils.

Soil	μ $[kN\cdot m^{-2}\cdot s]$	m	A $[(kN\cdot m^{-2})^{-x}\cdot s^{-1}]$	device, reference
San Francisco Bay Mud	∞	3.5	$1.85\cdot 10^{-10}$	triax Singh & Mitchell (1968) Vulliet & Hutter (1987a)
Remoulded illite clay	$5\cdot 10^9$	9.2	$2.16\cdot 10^{-25}$	"
Villarbeney clay	∞ $1.4\cdot 10^{10}$	5.3 9.6	$6.7\cdot 10^{-17}$ $1.2\cdot 10^{-23}$	in situ measurements Vulliet (1986) Vulliet & Hutter (1987a)

$$b(t_{II},n) = b(\frac{1}{3}(\sigma_1-\sigma_3)^2,n) = \frac{3\,\dot\varepsilon_1}{2(\sigma_3-\sigma_3)}. \quad (17)$$

The creep response function b depends now on two variables: undrained triaxial experiments need be perfomed for various degrees of consolidation. Strictly, a three-dimensional representation (t_{II},n,b) ought to be determined. This is a formidable endeaver which to our knwoledge has not been undertaken in this generality.

In case B, when the compressibility of the soil is large, equations (10) apply. For the conventional (axisymmetric) stress-strain state we get:

$$\begin{aligned} b(p',t_{II},n) &= \frac{\dot\varepsilon_1-\dot\varepsilon_3}{\sigma_1-\sigma_3}, \\ \tilde a(p',t_{II},n) &= \frac{\dot\varepsilon_1-\dot\varepsilon_3}{\sigma_1-\sigma_3} - \frac{\dot\varepsilon_1+2\dot\varepsilon_3}{\sigma_1'+2\sigma_3'}. \end{aligned} \quad (18)$$

In the laboratory p_w, σ_1, σ_3, $\dot\varepsilon_1$ and $\dot\varepsilon_3$ need be controlled and measured to obtain explicit functional relationships for $\tilde a$ and b.

b) simple shear test

For case A and performance of the shear experiment without drainage we have again $\dot\varepsilon_z = 0$, $\dot p = 0$. Thus, (14) applies again, but b is now also a function of n:

$$b(t_{II},n) = \left[\frac{\frac{1}{4}\dot\gamma^2}{\frac{1}{3}(p_z-p_x)^2+\tau^2}\right]^{1/2}. \quad (19)$$

For case B, i.e. large compressibility, equations (10) apply. From these, it follows that

$$\begin{aligned} b(p',t_{II},n) &= \left(\frac{D^d_{II}}{t_{II}}\right)^{1/2} \\ \tilde a(p',t_{II},n) &= \frac{1}{3}\,\frac{\mathrm{tr}\,\underset{\sim}{D}}{p'} + b(p',t_{II},n); \end{aligned} \quad (20)$$

upon consideration of the previously described stress strain rate state in simple shear these become

$$\begin{aligned} b(p',t_{II},n) &= \left[\frac{\frac{1}{3}\dot\varepsilon_z^2+\frac{1}{4}\dot\gamma^2}{\frac{1}{3}(p_z'-p_x')^2+\tau^2}\right]^{1/2} \\ \tilde a(p',t_{II},n) &= b(p',t_{II},n) - \frac{\dot\varepsilon_z}{2\,p_x'+p_z'}. \end{aligned} \quad (21)$$

An experimental determination of these trivariate functions is still missing.

c) An expression for $b(_{II},n)$ (case A)

We challenge experimental soil mechanicians to perform the experiments leading to (17) and (19) and conjecture that

$$b(t_{II},n) = A(n)\ t_{II}^{\frac{m-1}{2}}, \quad (22)$$

in which A is a power or an exponential law, is a suitable candidate to match such experiments.

An alternative procedure is obtained as follows: Assume that the soil is normally consolidated and that there exists a one-to-one correspondence between porosity n and pressure p'. Then, we may write

$$b = b(t_{II},p'). \quad (23)$$

The pressure dependence is introduced through a Drucker-Prager yield criterion. For a power law we thus propose

$$b(t_{II},p') = A\,\frac{t_{II}^{\frac{1-m}{2}}}{(k'+3\alpha' p')^m}, \quad (24)$$

where

$$\begin{aligned} k' &= \frac{6\,c'\cos\phi'}{(3-\sin\phi')\sqrt{3}}, \\ \alpha' &= \frac{2\sin\phi'}{(3-\sin\phi')\sqrt{3}}. \end{aligned} \quad (25)$$

c' is the cohesion and ϕ' the internal angle of friction. A has the dimension $[s^{-1}]$. This law expresses that for a fixed deviatoric stress the strain rate increases with decreasing spherical stress. This agrees qualitatively with experimental results, obtained among others by Ter-Stepanian (1975). For two soils values of material parameters A and m are collected in Table 3.

Table 3. Values of parameters entering in equation (24) for two different soils.

Soil	m	A $[s^{-1}]$	device references
San Francisco Bay Mud	4.7	$2.6\cdot10^{-5}$	triax Singh-Mitchell (1968) Vulliet-Hutter (1987a)
remoulded Villarbeney clay	4.8	$5.85\cdot10^{-9}$	Géonor H12 Vulliet (1986)

4 APPLICATION OF THE MODEL IN THE CONTEXT OF VISCO-PLASTICITY

So far it was assumed that creep occurs at all stress levels; one exception was the Bingham law in Table 1, where the strain rate remains zero up to a "yield" stress t_{II0}. A simple way to incorporate such plastic effects is to use an extension of Perzyna's theory of viscoplasticity (Perzyna, 1966, Cormeau, 1976) in which the strain rate $\underset{\sim}{D}$ is decomposed into an elastic part $\underset{\sim}{D}^{e}$ and a viscoplastic part $\underset{\sim}{D}^{vp}$:

$$\underset{\sim}{D} = \underset{\sim}{D}^{e} + \underset{\sim}{D}^{vp} \tag{26}$$

and where the elastic strain rate is related to the objective stress derivative (Jaumann derivative)

$$\omega_{ij} = \frac{1}{2}(v_{i,j} - v_{j,i}), \quad \overset{\nabla}{\underset{\sim}{\sigma}}{}' = \dot{\underset{\sim}{\sigma}}{}' + \underset{\sim}{\omega}\,\underset{\sim}{\sigma}' - \underset{\sim}{\sigma}'\,\underset{\sim}{\omega}, \tag{27}$$

for instance via the linear relation

$$\overset{\nabla}{\underset{\sim}{\sigma}}{}' = \underset{\sim}{C}_e\, \underset{\sim}{D}^{e} \tag{28a}$$

or, in view of isotropy, and with $\underset{\sim}{\sigma}' = -p\,\underset{\sim}{1} + \underset{\sim}{t}$

$$\underset{\sim}{D}^{e} = \frac{1}{3K}\dot{p}'\,\underset{\sim}{1} + \frac{1}{2G}\overset{\nabla}{\underset{\sim}{t}}. \tag{28b}$$

$\underset{\sim}{C}_e$ is the elastic constitutive matrix and G & K are the shear and bulk moduli, respectively; $\underset{\sim}{t}$ is the stress deviator. The viscoplastic strain rate is given by

$$\underset{\sim}{D}^{vp} = \gamma < \phi(F) > \frac{\partial Q}{\partial \underset{\sim}{\sigma}'}, \tag{29}$$

where $F(\underset{\sim}{\sigma}')$ is a scalar valued yield function, Q a scalar viscoplastic potential, γ a so called fluidity parameter which may be a function of stress and

$$< \phi(F(\sigma')) > = \begin{cases} \phi(F), & \text{for } F > 0, \\ 0, & \text{for } F \le 0, \end{cases} \tag{30}$$

i.e. the viscoplastic strain rate equals zero for a state of stress inside the yield surface (in the stress space).

4.1 Material without yield

If we choose

$$Q \equiv F \equiv \sqrt{t_{II}} \tag{31}$$

viscoplastic strain rate occurs at any stress level. In the special case of pure creep ($\overset{\nabla}{\sigma} = 0$) the elastic strain rate tensor is null; further, by substituting (31) in (29) and then (27) gives

$$\underset{\sim}{D} = \gamma < \phi(\sqrt{t_{II}}) > \frac{1}{2\sqrt{t_{II}}}\,\underset{\sim}{t}, \tag{32}$$

which is identical to (4) for the viscous incompressible body provided that

$$b(t_{II}) = \frac{\gamma < \phi\,\sqrt{t_{II}}) >}{2\sqrt{t_{II}}}. \tag{33}$$

4.2 Material with yield and non-associative viscoplasticity

For any yield function $F(\underset{\sim}{t})$ and a von-Mises-type viscoplastic potential $Q = t_{II} - c^2$, c = const., we find a constitutive law close to that given in (9). With

$$\frac{\partial Q}{\partial \underset{\sim}{\sigma}'} = \frac{\partial Q}{\partial t_{II}}\,\frac{\partial t_{II}}{\partial \underset{\sim}{\sigma}'} = 1 \cdot \underset{\sim}{\sigma}', \tag{34}$$

(29) implies

$$\underset{\sim}{D}^{vp} = \underset{\sim}{D}_d^{vp} = \gamma < \phi(F) > \underset{\sim}{\sigma}', \tag{35}$$

where $\underset{\sim}{D}_d^{vp}$ is the deviatoric part of $\underset{\sim}{D}^{vp}$.

With the elastic components as given in (28b) the complete stress strain rate relationship becomes

$$\frac{1}{3}\operatorname{tr}\underset{\sim}{D} = \frac{1}{3K}\dot{p}',$$

$$\underset{\sim}{D}_d = \frac{1}{2G}\overset{\nabla}{\underset{\sim}{\sigma}} + \gamma < \phi(F(\underset{\sim}{t}))\,\underset{\sim}{\sigma}', \tag{36}$$

which should be compared with (9) (saturated compressible soil, case A). The two laws are similar but not equivalent for constant stress creep tests ($\overset{\nabla}{\sigma}' = 0$) or for a material with a very large shear modulus ($G \to \infty$). However, in (9), the creep response function b is expressed as a function of the porosity n, and not only of the stress state; moreover, the volumetric response in (9) is non linear.

Finally, to complete the comparison, we may propose to use the rhs of (24) as fluidity parameter γ, and to put $\phi(F) = 1$ in (29). However, this will create a jump in velocity at $F = 0$. To avoid it, let

$$\gamma < \phi(F) > = A\,\frac{F^{m-1}}{(\sqrt{t_{II}} - F)^m} \tag{37}$$

with, for example, $F = \sqrt{t_{II}} - k' - 3\alpha' p'$ for a Drucker-Prager yield criterion. The main difference between this expression (37) and the often proposed expression (Cormeau, 1976)

$$\gamma < \phi(F) > = A\left(\frac{F}{F_0}\right)^n \tag{38}$$

with F_0 as a constant, is that the strain rate contours in the stress space will not be parallel to the yield surface. In other words, equation (37) predicts that two different points in the stress space, situated at the same distance from the yield surface F = 0 will not generate the same strain rate, except if identical mean pressures are applied.

5 SLIDING LAWS

There exist numerous situations where the deformations occur in a narrow zone or layer, with small deformations within the adjacent bodies. Such conditions can be described as "contact problems" and for landslides, arise as "sliding surfaces". Here, we present viscous-type sliding laws which relate the (relative) sliding velocity at the interface to the state of stress at this surface.

We propose the law

$$\underset{\sim}{v}_B = - f(.)\, \underset{\sim}{\tau}_B \qquad (39)$$

where $\underset{\sim}{v}_B$ is the (relative) velocity at the base (or interface), $\underset{\sim}{\tau}_B$ is the shear traction within the sliding surface and f(.) a sliding coefficient which may depend on the state of stress at the sliding surface and other internal variables. Several sliding laws are collected in Table 4 with values of parameters determined for two landslides in Switzerland (Le Day, La Frasse) and one landslide in France (Sallèdes). In each case, inclinometer measurements were used to determine the location of the sliding surface and the value of the sliding velocities. The state of stress was estimated through the infinite-slope assumption. The power sliding law with dependence on the effective pressure seemed to be very convenient, however in some cases, a yield stress will be needed: In effect sliding does not occur until a given stress level is reached. This stress level will be the limit between no-slip (f(.) = 0) and sliding (f(.) > 0) Some of the laws proposed in Table 4 allow for these considerations. For more details, the reader may consult Vulliet & Hutter (1987c).

6 CONCLUSION

Continuum mechanical concepts were used to derive rate dependent constitutive relations for soil models that are either treated as a viscous fluid or a saturated binary mixture of two fluid. The proposed explicit constitutive models prove to reasonably match laboratory and in situ creep data.

Table 4. Some expressions for the sliding law f(.) in equation (39) and related values of parameters for three different landslides.

Type	$f(\sigma'_B, \tau_B)$	value of parameters	references
Newton	$1/\mu$		Vulliet (1986)
Bingham	$\left[1 - \frac{c + \sigma'_B \, tg\,\phi}{\tau_B}\right]/\mu$, $\tau_B > c + \sigma'_B \, tg\,\phi$		Vulliet (1986)
Weertman	$B(\tau_B)^{n-1}$, $n \geq 1$		Weertman (1964)
Bingham, non linear	$B\left[1 - \frac{c + \sigma'_B \, tg\,\phi}{\tau_B}\right]\tau_B^{n-1}$, $\tau_B > c + \sigma'_B \, tg\,\phi$		
power law with stress ratio	$B \frac{\tau_B^{n-1}}{(c + \sigma'_B \, tg\,\phi)^n}$	$n = 1$, $c = 0$, $B/tg\,\phi = 1.27 \cdot 10^{-9}\, m \cdot s^{-1}$ $n = 2$, $c = 0$ $B/tg\,\phi = 2.6 \cdot 10^{-6}\, m \cdot s^{-1}$ $n = 15$, $c = 0$, $B/tg\,\phi = 9.56 \cdot 10^{3}\, m \cdot s^{-1}$	Le Day, Switzerl. La Frasse, " Sallèdes, France
linear, with yield	$\frac{B}{c + \sigma'_B \, tg\,\phi} - \frac{A}{\tau_B}$	$A = 1.46 \cdot 10^{6}\, m \cdot s^{-1}$, $B = 1.54 \cdot 10^{-6}\, m \cdot s^{-1}$, $c = 0$, $\phi = 9.5^{o}$	Sallèdes, France
hyperbolic, with yield	$\frac{C}{c + \sigma'_B \, tg\,\phi - \tau_B} - \frac{C \cdot D}{\tau_B}$	$C = 1.22 \cdot 10^{-9}\, m \cdot s^{-1}$, $D = 5$, $c = 0$, $\phi = 9.5^{o}$	Sallèdes, France

The models are very similar to some visco-plastic formulations and permit determination of the flow in creeping landslide motions. Such computations have been presented in Vulliet (1986) and Vulliet & Hutter (1987b,c) and thus prove to be a useful tool for the prediction of the explicit motion of unstable landslides.

REFERENCES

Bjerrum, L. & A. Landva 1966. Direct simple shear test on a Norwegian quick clay. Géotechnique, 16/1:1-20.

Cormeau, I.C. 1976. Visco plasticity and plasticity in the finite element method. Thesis University of Wales, Swansea.

Findley, W.N., Lai, J.S. & K. Onaran 1976. Creep and relaxation of nonlinear visco-elastic materials. North-Holland Pub. Co., Amsterdam.

Hutter, K. 1983. Theoretical Glaciology, D. Reidel, Boston.

Kavazanjian, E. & J.K. Mitchell 1980. Time-dependent deformation behavior of clays. J. of Geot. Eng. Div. ASCE, GT6:611-630.

Mueller, I. 1973. Thermodynamik. Die Grundlagen der Materialtheorie. Bertelsmann Universitätsverlag, Düsseldorf.

Perzyna, P. 1966. Fundamental problems in viscoplasticity. Adv. in Appl. Mech. 9: 243-377.

Singh, A. & J.K. Mitchell 1968. General stress-strain-time function for soils. J. Soil Mech. and Found. Div., ASCE. 94:SMI: 21-46.

Tan Tjong-kie 1981. Time dependent lateral pressures and Poisson's ratio measurement. Proc. 10th ICSMFE, Stockholm, Vol.1: 797-800.

Ter-Stepanian, G. 1975. Creep of a clay during shear and its rheological model. Géotechnique. 25/2:299-320.

Truesdell, C. 1977,78. Rational continuum mechanics, Vol. I and II, Academic Press, New York.

Vulliet, L. 1986. Modélisation des pentes naturelles en mouvement. Thèse EPFL No 635, Lausanne.

Vulliet, L. & K. Hutter 1987a. A set of constitutive models for soils under slow movement. Submitted to J. of Geotech. Eng., ASCE.

Vulliet, L. & K. Hutter 1987b. A continuum model for natural slopes in slow movement. Géotechnique, London. (Submitted)

Vulliet, L. & K. Hutter 1987c. Viscous-type sliding laws for landslides. Canadian Geotechnical Journal (submitted).

Weertman, J. 1964. The theory of glacier sliding. J. of Glaciology, 5:287-303.

Numerical Methods in Geomechanics (Innsbruck 1988), Swoboda (ed.)
© 1988 Balkema, Rotterdam. ISBN 90 6191 809 X

Model with multiple mechanisms for anisotropic behaviours of sands

D.Pradel
University of Tokyo, Japan

ABSTRACT: In this paper an elastoplastic constitutive model based on the concept of multiple plane strain mechanisms is briefly described. The numerical predictions explain well the yielding and flow of anisotropic sands observed from element tests on true triaxial and hollow cylindrical specimens.

INTRODUCTION

Several experimental investigations have shown that not only the magnitude but also the orientation of the three principal stresses affect the deformation of sand (Arthur and Menzies 1972, Yamada and Ishihara 1979, Miura et al. 1986,...). The concept of multiple plane mechanisms does not ignore the influence of the direction of principal stress axes and is therefore well suited to predict inherent anisotropic behaviors.

SEPARATION INTO MECHANISMS

This model is based on the postulate that plastic strains develop in plane strain mechanisms (Aubry et al. 1982). The number of mechanisms can be infinite however for computational reasons a limited number is required. For usual purposes six mechanisms defined in three planes (Fig.1) are sufficient for accurate predictions.

The total plastic strains $\tilde{\varepsilon}^p$ are equal to the sum of the contributions from all the mechanisms

$$\tilde{\varepsilon}^p = \sum_{k=1}^{3} \tilde{\varepsilon}^{pk} = \sum_{k=1}^{3} (\tilde{\varepsilon}^{pdk} + \tilde{\varepsilon}^{pvk}) \quad (1)$$

where $\tilde{\varepsilon}^{pdk}$ and $\tilde{\varepsilon}^{pvk}$ are the contributions from the deviatoric and volumetric mechanisms in plane Ω_k (k = 1,2,3).

The expression of the stress variables are derived from the equations of the Mohr circle of stress. For a small element in plane Ω_k (Fig.2) p_k q_k and x_k represent respectively the abscissa of the center, the radius, and the cosine of the stress point in the Mohr circle; with $i \neq j \neq k$

$$p_k = \frac{1}{2}.(\sigma_{ii} + \sigma_{jj}) \quad (2)$$

$$q_k = \{ (\frac{\sigma_{ii} - \sigma_{jj}}{2})^2 + \sigma_{ij}^2 \}^{0.5} \quad (3)$$

$$x_k = \frac{\sigma_{ii} - \sigma_{jj}}{2.q_k} \quad (4)$$

Because of the plane strain formulation the mechanisms in Ω_k contain effectively only three strain components ε_{ii}^{pk} ε_{jj}^{pk} and ε_{ij}^{pk} (with $i \neq j \neq k$), and all the other contributions are null. Hence

$$\text{for } i = k \text{ or } j = k \qquad \varepsilon_{ij}^{pk} = 0. \quad (5)$$

The volumetric and deviatoric variables of strain are obtained as follows

$$i \neq j \neq k \qquad \varepsilon_v^{pk} = \varepsilon_{ii}^{pk} + \varepsilon_{jj}^{pk} \quad (6)$$

$$\bar{\varepsilon}^{pk} = \{ (\varepsilon_{ii}^{pk} - \varepsilon_{jj}^{pk})^2 + (2.\bar{c}.\varepsilon_{ij}^{pk})^2 \}^{0.5} \quad (7)$$

where $\bar{c}$ is a material function justified by considerations of energy dissipation. If E^{pk} is the plastic work, according to (5)

$$E^{pk} = \sigma_{ii}.\varepsilon_{ii}^{pk} + \sigma_{jj}.\varepsilon_{jj}^{pk} + 2.\sigma_{ij}.\varepsilon_{ij}^{pk} \quad (8)$$

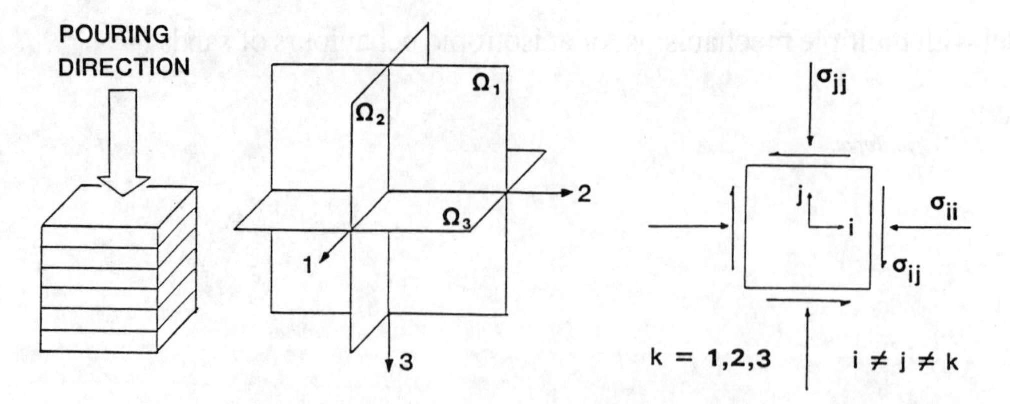

Fig.1 The planes of the mechanisms.

Fig.2 State of stress of an element in Ω_k

$$\text{for } \bar{c} = 1 : \quad E^{pk} \leq (p_k.\varepsilon_v^{pk} + q_k.\bar{\varepsilon}^{pk}) \quad (9)$$

When the directions of the stresses and the plastic strain increments are coincident the two terms in (9) are equal. However, experimental studies have shown that coaxiality is usually limited to triaxial type of loading (Miura et al. 1986). Therefore in order not to dissipate to much energy values of $\bar{c}$ smaller than the unity are required ($\bar{c} < 1$). Experiments on hollow cylindrical specimens (Symes et al. 1982, Pradel 1987) show that for constant directions of the principal stress axes the directions of the strain increments do not vary considerably. Hence, it is reasonable to consider $\bar{c}$ as a material constant.

ELASTIC STRAINS

The elastic strains are assumed to be isotropical and defined with the following incremental equation

$$\dot{\varepsilon}_{ij}^{e} = \frac{3.\kappa'}{2(1+e).p}.\dot{s}_{ij} + \frac{\kappa}{3(1+e)}.\frac{\dot{p}}{p}.\delta_{ij} \quad (10)$$

where e is the void ratio, p the mean pressure, δ_{ij} the Kroneker's operator, κ and κ' two elastic constants and s_{ij} the deviatoric stress tensor ($s_{ij} = \sigma_{ij} - p.\delta_{ij}$).

VOLUMETRIC MECHANISMS

Associated yield functions are adopted for the volumetric mechanisms

$$f_k^v = \beta_k.\varepsilon_v^{pvk} - \mathrm{Log}(\frac{p_k}{p_{ko}}) \quad (11)$$

where p_{ko} is a value of reference for p_k and represents the pressure for which the volumetric component ε_v^{pvk} is null, and where β_k is a set of material constants determined for a given density, because of orthotropy $\beta_1 = \beta_2$.

DEVIATORIC MECHANISMS

In order to accurately predict the stress-strain behaviors in both compression and extension tests the use of the stress ratio R_k has been justified (Pradel and Ishihara 1987). This stress parameter is defined as follows

$$R_k = \frac{(1 + c_r).q_k}{p_k + c_r.p} \quad (12)$$

where c_r is a material constant.

The use of R_k gives a unique flow rule (Fig.3) for triaxial compression and extension tests. The experimental data is approximated with the expression in equation (12), where m α_o α_n and n are model constants common for the three deviatoric mechanisms.

$$\frac{\dot{\varepsilon}_v^{pdk}}{\dot{\bar{\varepsilon}}^{pdk}} = \alpha_o.(m - R_k) + \alpha_n.(m - R_k) \quad (13)$$

The direction of the plastic strain increments follows the definition of the deviatoric component of strain (7).

$$\dot{\varepsilon}_{ii}^{pdk} - \dot{\varepsilon}_{jj}^{pdk} = \lambda.\bar{c}.(\sigma_{ii} - \sigma_{jj}) \quad (14a)$$

$$\dot{\varepsilon}_{ij}^{pdk} = \lambda.\sigma_{ij} \quad (14b)$$

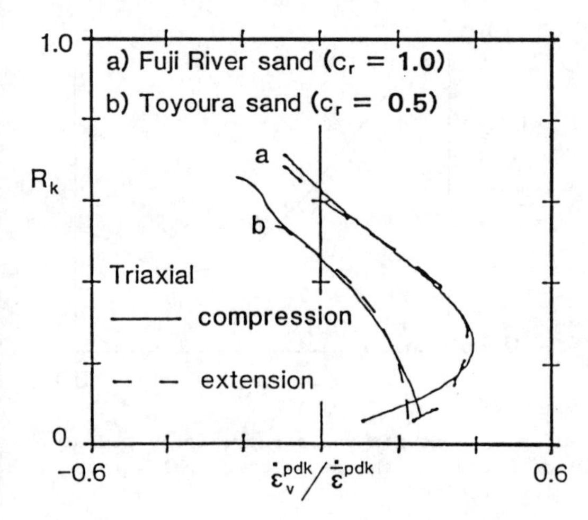

Fig.3 Experimental flow rule for two sands.

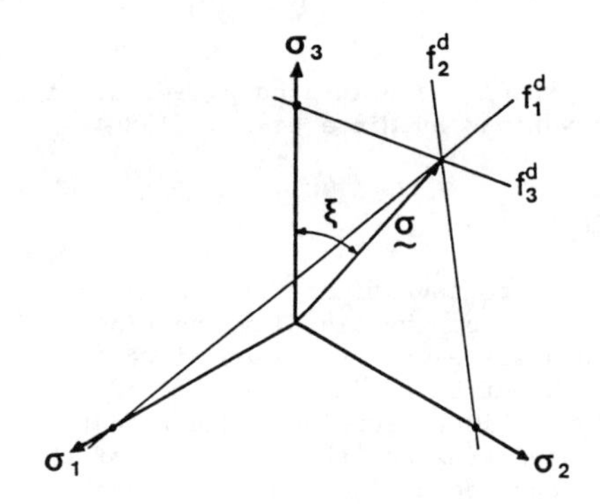

Fig.4 Yield surfaces of the deviatoric mechanisms in the octahedral plane.

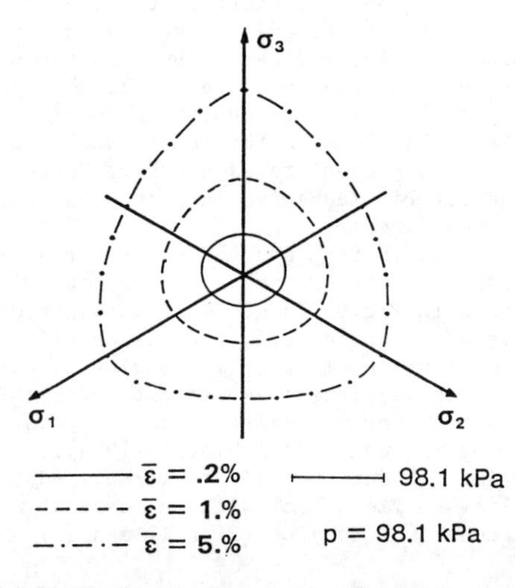

Fig.5 Shape of the predicted equal shear strain contours in the octahedral plane.

It is assumed that for a given density the yield functions are entirely defined in terms of the deviatoric component of strain $\bar{\varepsilon}^{pdk}$ and of the stress variables. The following expression is proposed

$$f_k^d = R_k - \frac{\bar{\varepsilon}^{pdk}}{h_k(\bar{\varepsilon}^{pdk}) + x_k . h_k'(\bar{\varepsilon}^{pdk})} . L_k(p) \quad (15)$$

where h_k h_k' and L_k are polynomial functions to be chosen by the modeler depending on the required precision. The following expressions were adopted for Fuji River and Toyoura sands (Pradel 1987, Pradel and Ishihara 1987)

$$h_k(\bar{\varepsilon}^{pdk}) = a_k . \bar{\varepsilon}^{pdk} + b_k . (\bar{\varepsilon}^{pdk})^s \quad (16a)$$

$$h_k'(\bar{\varepsilon}^{pdk}) = c_k . \bar{\varepsilon}^{pdk} + d_k . (\bar{\varepsilon}^{pdk})^s \quad (16b)$$

$$L_k(p) = 1.0 \quad (16c)$$

TRUE TRIAXIAL PREDICTIONS

It can be shown that the yield surfaces given by equation (15) can be represented with straight lines in the octahedral plane (Fig.4). This is apparently in contradiction with the experimental and theoretical evidence by Lade (1972) which revealed that the yielding of sand can be correctly approximated with a smooth cone. However the predictions in Fig.5 show that at large strains there is a good agreement between the model and Lade's experimental results.

The predictions in Figs.5 and 6 were obtained using the data from the true

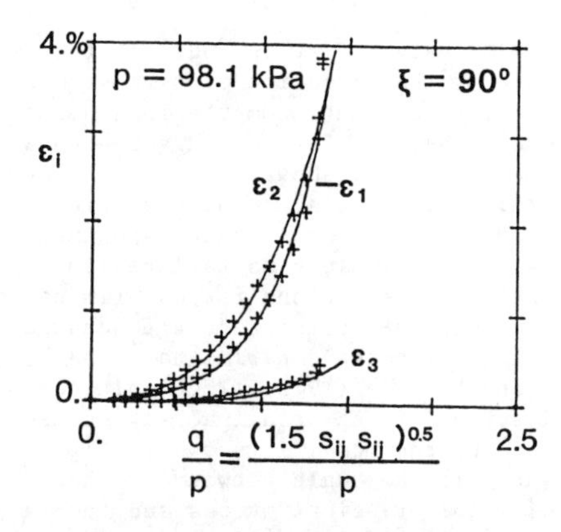

Fig.6 Comparison between measured and predicted strains for a true triaxial experiment (+ : experiment ——— : model).

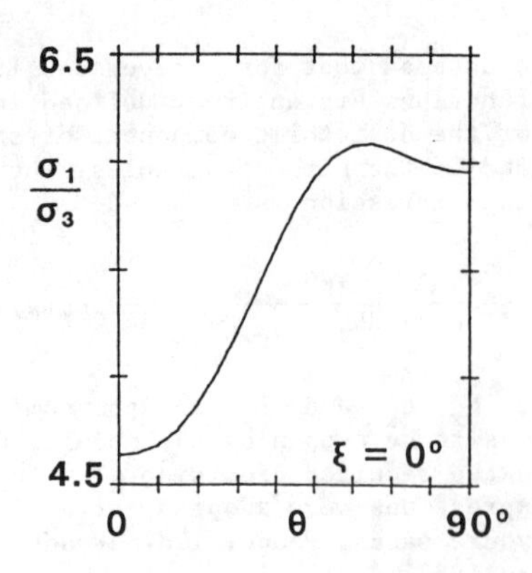

Fig.7 Strength anisotropy, behavior of tilted samples (prediction).

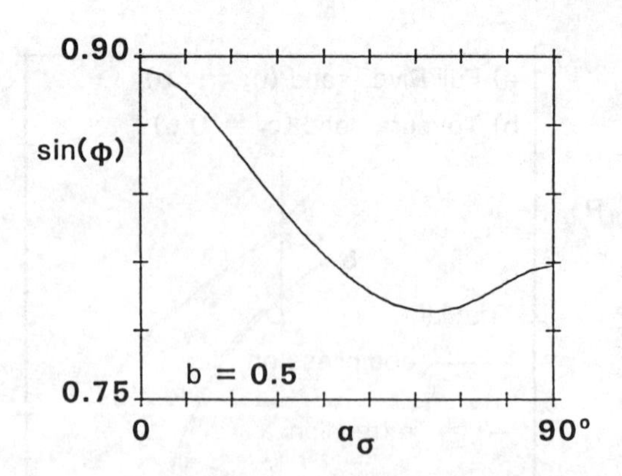

Fig.8 Strength anisotropy, influence of the direction of the major principal stress (prediction)

triaxial experiments performed on loose samples of Fuji River Sand (e = 0.83 ~ 0.85) by Yamada and Ishihara (1979). The details concerning the material constants of the model have been described in a prior publication (Pradel and Ishihara 1987). The strain $\bar{\varepsilon}$ is defined as

$$\bar{\varepsilon} = \{\frac{2}{3}\cdot(\varepsilon_{ij} - \frac{1}{3}\varepsilon_v\cdot\delta_{ij}).(\varepsilon_{ij} - \frac{1}{3}\varepsilon_v\cdot\delta_{ij})\}^{0.5} \quad (17)$$

STRENGTH ANISOTROPY

The compression triaxial tests performed by Arthur and Menzies (1972) on tilted samples showed that the strength of sand is greatly influenced by the angle of tilt θ . Similarly experimental investigations by Symes et al. (1982) and Miura et al. (1986) using hollow cylindrical specimens revealed that the influence of the direction of the principal stress axes on the strength is significant.

The main feature of this model is that the influence of the direction of stresses can be taken into account. Computations using the model constants for Fuji River sand were performed until the shear strain $\bar{\varepsilon}$ reached 5% at this level the slope of the stress-strain curves is so small that experimentally it would be consider as failure. In Fig.7 the numerical predictions for experiments on tilted samples are presented, the behavior agrees well with the experimental data by Arthur and Menzies (1972). The influence of the direction of the principal stress axes predicted by the model is shown in Fig.8, where α_σ is the angle between the direction of major principal stress and the vertical direction. The numerical values were obtained for tests with p = 98.1 kPa and

$$\sigma_1 \geqq \sigma_2 \geqq \sigma_3 \qquad b = \frac{\sigma_2 - \sigma_3}{\sigma_1 - \sigma_3} = 0.5 \qquad (18)$$

The results are in good agreement with the observations by Miura et al. (1986).

YIELD LOCI

In order to investigate the influence of anisotropy on the yielding and plastic flow of sand series of experimental tests were carried out by Pradel et al. (1987) using a hollow cylinder torsional shear test device. This apparatus has the advantage of the individual control of four components of stress, i.e., the axial σ_z , the radial σ_r , the tangential σ_θ , and the circumferential stress $\sigma_{z\theta}$; the respective strains ε_z ε_r ε_θ and $\varepsilon_{z\theta}$ where calculated from measurements of the axial displacement, the torsional angle, and the volume changes of the sample and the inner cell. All the tests were performed with a constant mean pressure p = 98.1 kPa and with the radial and tangential stresses equal to each other. Hence, the deviatoric mechanism in plane Ω_3 was never mobilized.

The stress paths involved cycles of stress reversal and stress axes rotation, such as the ones in Fig.9. These tests were performed in order to identify the states of stress at which plastic flow begins to take place (Fig.10) (these experiments are similar to the conventional triaxial compression tests by Tatsuoka and Ishihara (1974)). The experimental data revealed well defined families of yield loci with shapes that are independent of density (Figs.11 and 12).

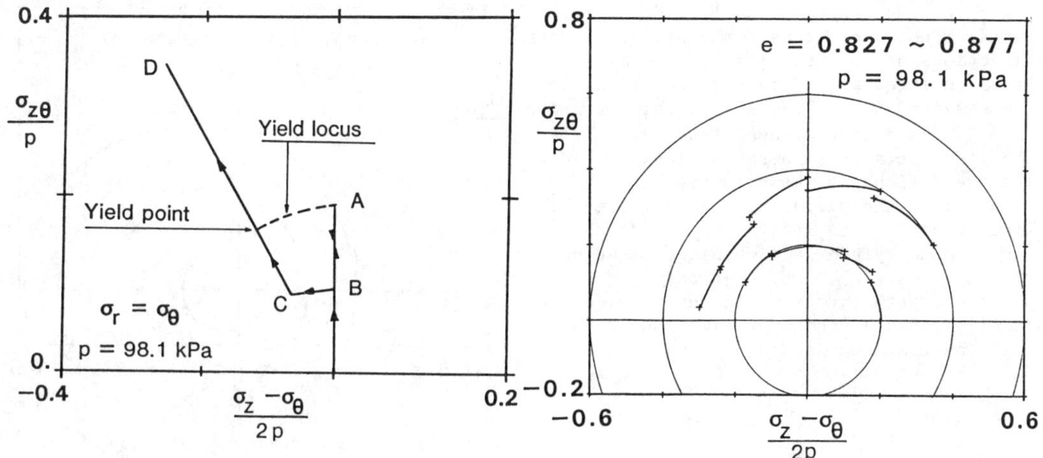

Fig.9 Stress path used for the determination of a yield locus.

Fig.12 Yield loci of loose Toyoura sand from the $\sigma_{z\theta}$ vs. $\varepsilon_{z\theta}$ stress-strain curves.

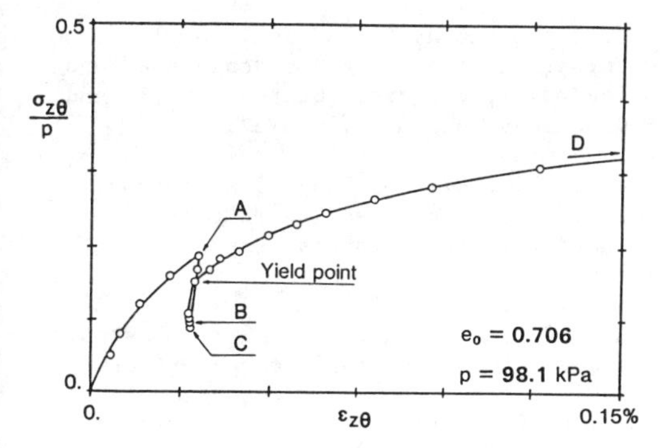

Fig.10 Stress-strain curve used for the determination of a yield locus.

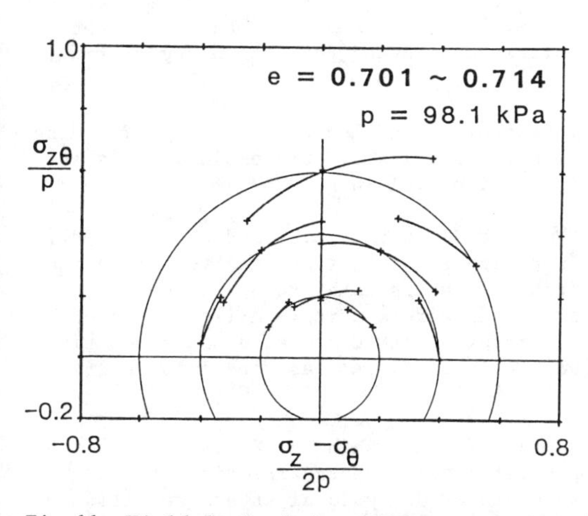

Fig.11 Yield loci of dense Toyoura sand based on the $\sigma_{z\theta}$ vs. $\varepsilon_{z\theta}$ stress-strain curves.

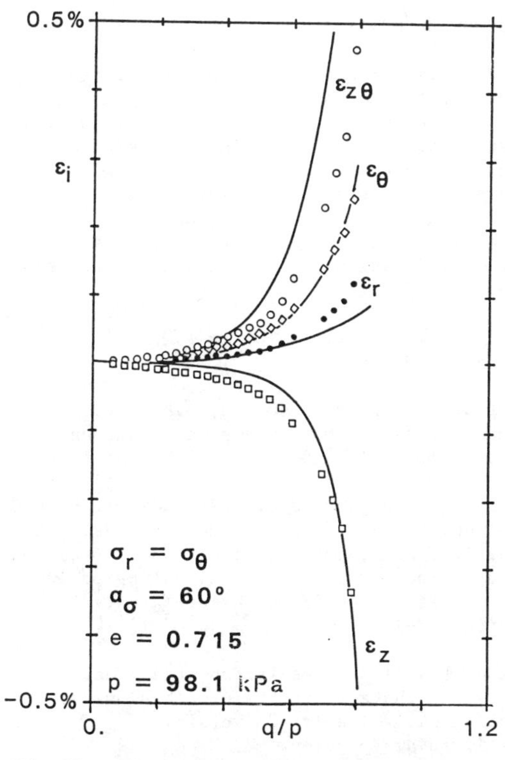

Fig.13 Comparison between measured and predicted strains for a torsional shear experiment on a hollow cylindrical sample.

The testing program also included some radial monotonic stress paths with constant directions of the principal stress axes. Based on three of the tests (one triaxial compression $\alpha_\sigma = 0^0$, one triaxial extension $\alpha_\sigma = 90^0$, and a pure shear test $\alpha_\sigma = 45^0$) the model parameters were obtained; one of the predictions is shown in Fig.13.

In Fig.14 the yield surface of the mechanism producing the strain $\varepsilon_{z\theta}$ and the experimental yield loci obtained based on the $\sigma_{z\theta}$ vs. $\varepsilon_{z\theta}$ stress-strain curves are plotted together; the yield function defined in (15) gives a good approximation of the yielding characteristics of sand.

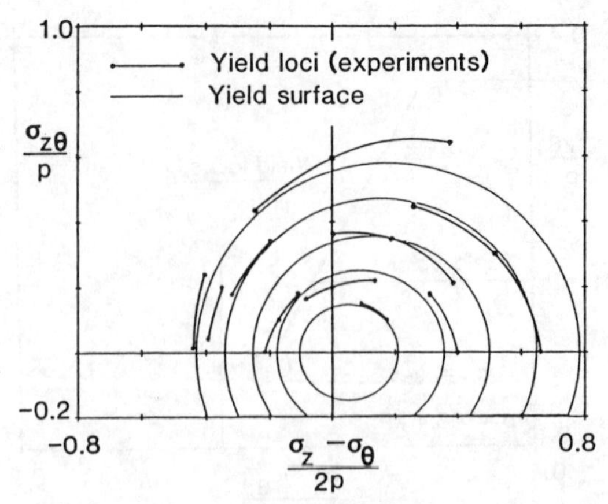

Fig.14 Comparison between the yield loci of dense sand and the yield surface of a mechanism.

CONCLUSIONS

Based on the concept of multiple plane strain mechanisms a model that does not ignore the influence of the direction of the principal stress axes was developed.

The model predictions are consistent with experimental evidence showing the influence of inherent anisotropy on the yielding and plastic flow of sands.

ACKNOWLEDGMENTS

The research described here was performed as part of the author's Doctor of Engineering work under the supervision of Professor K. Ishihara. The author is indebted with him and also with Mr. M. Gutierrez for his invaluable assistance during the experimental stage. This study was possible thanks to the financial support of the Japanese Ministry of Education and of the Swiss National Research Fund.

REFERENCES

Arthur, J.R.F. and Menzies, B.K. 1972. Inherent anisotropy in a sand. Geotechnique Vol.22 No.1 115-128

Aubry, D., Hujeux, J.C., Lassoudiere, F. and Meimon, Y. 1982. A double memory model with multiple mechanisms for cyclic soil behavior. Proc. of the International Symposium on Numerical Models in Geomechanics, Zurich: 3-13.

Lade, P.V. 1972. The stress-strain and strength characteristics of cohesionless soils. Ph.D. Thesis, University of California, Berkeley.

Miura, K., Miura, S. and Toki, S. 1986. Deformation of anisotropic dense sand under principal stress axes rotation. Soils and Foundations Vol.26 No.1 36-52.

Pradel, D. 1987. Modeling of anisotropic behaviors of sand based on multiple plane mechanisms. D.Eng. Thesis, University of Tokyo.

Pradel,D. and Ishihara, K. 1987. Elastoplastic model for anisotropic behaviors of sands. Proc. of the 2nd International Conference on Constitutive Laws for Engineering Materials, Tucson: 631-638

Pradel, D., Ishihara, K. and Gutierrez, M. 1987. On modeling of inherent anisotropy of sands. The 22th Japan National Conference on Soil Mechanics and Foundation Engineering, Niigata: 359-362

Tatsuoka, F. and Ishihara,K. 1974. Yielding of sand in triaxial compression. Soils and Foundations Vol.14 No.2 63-76

Symes, M.J., Hight, D.W. and Gens,A. 1982. Investigating anisotropy and the effects of principal stress axes rotation and of principal stress using a hollow cylinder apparatus. IUTAM Conference on Deformation and Failure of Granular Materials, Delft: 441-449

Yamada, Y. and Ishihara, K. 1979. Anisotropic deformation characteristics of sand under three dimensional stress conditions. Soils and Foundations Vol.19 No.2 79-94

Numerical Methods in Geomechanics (Innsbruck 1988), Swoboda (ed.)
© 1988 Balkema, Rotterdam. ISBN 90 6191 809 X

On the constitutive equations of the chalk

R.Charlier & J.P.Radu
University of Liège, Belgium

ABSTRACT: Chalk layers outcrop near Liège and exist in depth in some North Sea oil reservoirs. A lot of mechanical experiments have been realized in many laboratories on test-pieces extracted from these fields. The results show a large scatter , but some significant trends exist. Two elastoplastic isotropic constitutive laws are proposed, following the Cam-Clay and Cap model ideas. One of them is used to model a classical loading. Some characteristics of this kind of plastic laws are pointed out.

1 INTRODUCTION

Chalk layers outcrop near Liège and exist in some North Sea oil reservoir. The hereafter described research was developed in order to model by the finite elements method the compaction of a chalk oil reservoir and the induced subsidence of the sea bottom. A lot of mechanical experiments have been realized on test-pieces extracted from the above-mentioned fields by research centers and petroleum industry laboratories [2], [6]. Unfortunately, the experimental conditions were very dissimilar : some samples were cleaned with toluene or other solvent which modify the chalk structure and decrease its strength. The test-pieces were then saturated during the experiments by different liquids : oil, distilled water, sea water, chemicaly equilibrated water,... The last factor to be noted is the silica content influence on the strength. All these effects could not be quantified. They induce some scattering of the experimental results. However, we have used all the results to develop and fit the mechanical model. Typical results are presented at the figure 1.

2. ELASTICITY

We assume here that chalk is isotropic. At the beginning of the loading, its mechanical behaviour is approximately linear elastic. The elasticity parameters are measured on the experimental diagram (fig. 1) and related to the pore volume. Young's modulus decreases when the porosity increases, and Poisson's ratio remains roughly constant (fig. 2).

3 PLASTICITY

Especially in the North Sea oil reservoirs, irreversible strains seem to appear. Therefore, an elastoplastic incremental constitutive law will be developed. First, one defines the shape of the initial "apparent" yield locus (fig. 1).

All the experimental points are plotted in a $(I_\sigma, \sqrt{II_{\hat{\sigma}}})$ diagram (fig. 3), where

$I_\sigma = \sigma_{ii}$ is the first invariant of the Cauchy stress tensor,

$\hat{\sigma}_{ij} = \sigma_{ij} - 1/3\ I_\sigma\ \delta_{ij}$ is the deviator of the Cauchy stress tensor.

$II_{\hat{\sigma}} = 1/2\ \hat{\sigma}_{ij}\ \hat{\sigma}_{ij}$ is the second invariant of the deviator of the Cauchy stress tensor.

Despite the scatter of the results, some trends appear immediately :

- a pure volumic compression ($I_\sigma > 0$, $II_{\hat{\sigma}} = 0$) can induce plastic strains ,
- the maximum shear strength appears for a mean volumic stress ,
- when the volumic stress decreases to zero, the deviatoric strength also decreases to zero ,
- few informations are available for weakly stressed situations.

Therefore, the yield locus looks like the Cam-Clay and the Cap model. Two shapes have been tried. First, one has chosen to represent the yield locus by means of an ellipse in the $(I_\sigma, \sqrt{II_{\hat{\sigma}}})$ space (fig. 4). The plasticity conditions are then the following :

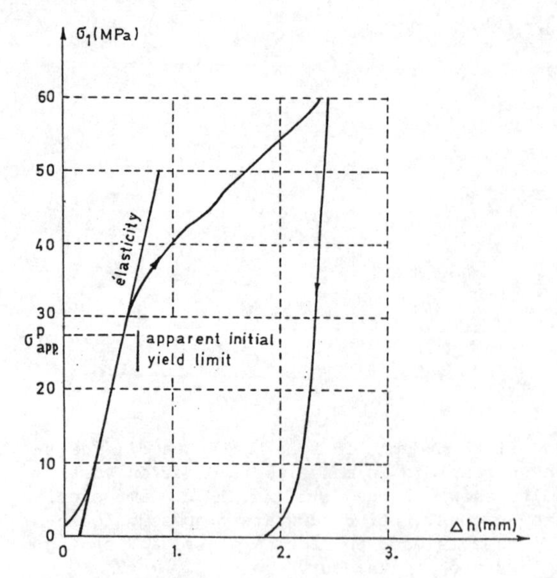

vertical shortening axial stress relation.

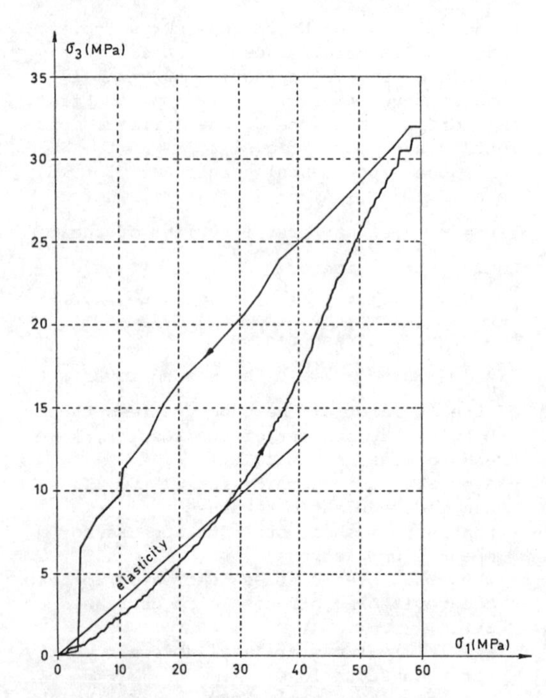

axial stresses radial stress relation.

σ_1 = axial stress

σ_3 = radial stress

Δh = height variation

Figure 1. Typical experimental results for a triaxial test (without radial strain).

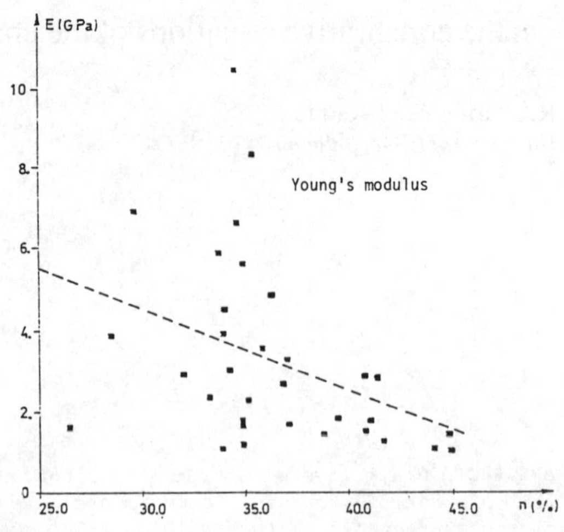

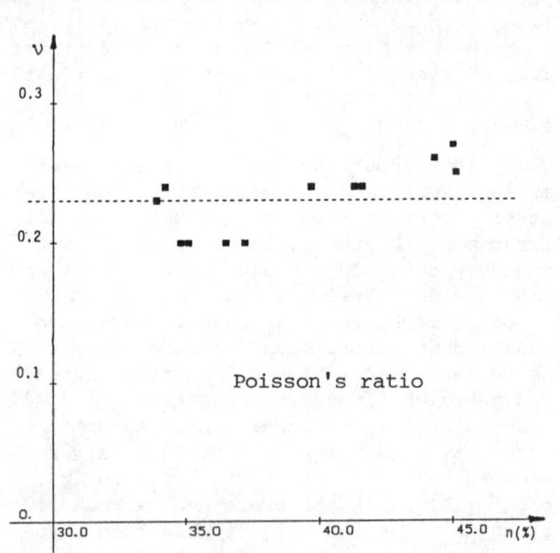

Figure 2. Variation of Young's modulus E and Poisson's ratio ν with respect to the porosity n.

$$f = II_{\hat{\sigma}} + m^2 I_\sigma (2a - I_\sigma)$$

$f < 0 \rightarrow$ elasticity

$f = 0 \rightarrow$ elastoplasticity

$f > 0 \rightarrow$ impossible

Cohesion is not represented.

One assumes normality of the plastic strain rate $\dot{\varepsilon}^p_{ij}$ to the yield surface :

$$\dot{\varepsilon}^p_{ij} = \dot{\lambda} \frac{\partial f}{\partial \sigma_{ij}}$$

and scalar isotropic hardening

$$\frac{da}{dt} = B \dot{\lambda}$$

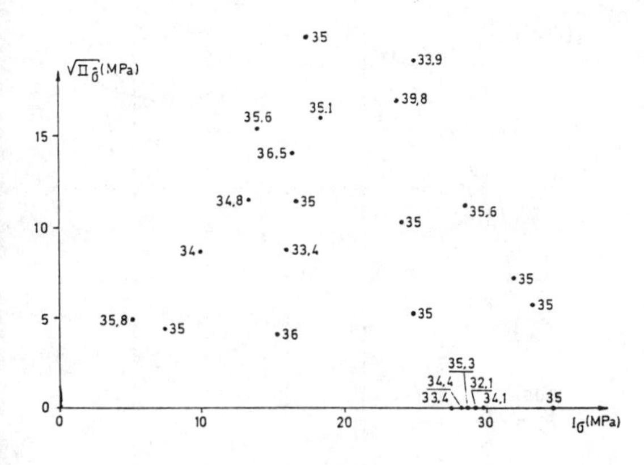

(porosities around 35 %)

Figure 3 - Experimental data : initial apparent yield locus.

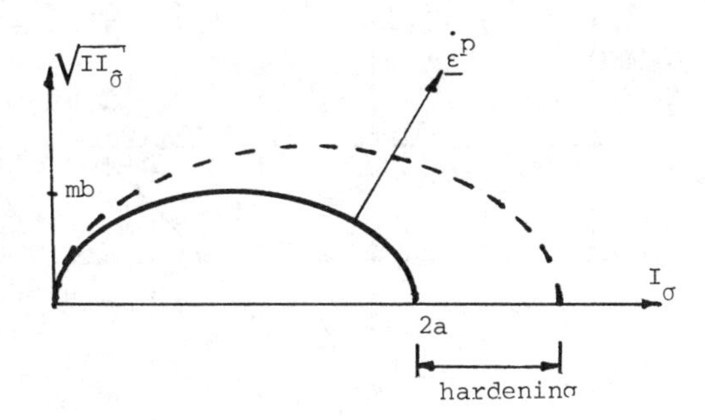

Figure 4 - Elliptic yield locus.

The second yield locus is composed of a Coulomb's straight line and a spherical cap (fig. 5). The plasticity conditions are now the following :

if $I_\sigma \leqslant R/\sqrt{1+m^2} \rightarrow f = \sqrt{II_{\hat{\sigma}}} - m\, I_\sigma$

if $I_\sigma \geqslant R/\sqrt{1+m^2} \rightarrow f = II_{\hat{\sigma}} + I_\sigma^2 - R^2$

$f < 0 \rightarrow$ elasticity

$f = 0 \rightarrow$ elastoplasticity

$f > 0 \rightarrow$ impossible

Cohesion is not represented. One assumes normality of the plastic strain rate to the yield surface

$$\underline{\dot{\varepsilon}}^p = \dot{\lambda}\, \frac{\partial f}{\partial \sigma_{ij}}$$

Scalar isotropic hardening modifies only the spherical cap :

$$\frac{dR}{dt} = B\, \dot{\lambda}$$

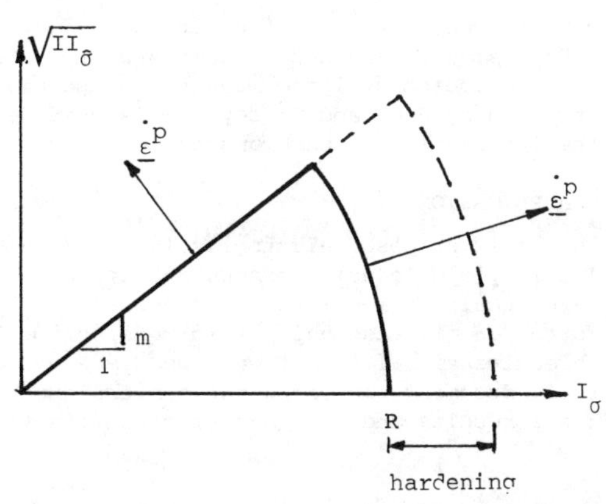

Figure 5 - Chalk cap model

The parameters describing the yield locus are (m, a) or (m, R). They have been evaluated for some discrete porosity values using a modified least squares method : If one point of the figure 3 is on the yield surface, then f = 0. Otherwise, f ≠ 0. Therefore f is a measure of the distance between the experimental point and the theoretical law. The least squares method minimizes the cumulated errors

$$S = \sum_{i=1}^{N} f_i^2$$

where N is the member of experimental points. The derivation of S with respect to the parameters gives nonlinear relations between the parameters which can easily be solved.

The above described method neglects the influence of porosity on the yield locus. Therefore we modify the least squares method using the porosity weights W_i

$$W_i = 1/(|\phi_{ref} - \phi_i|),\ W_i \leqslant 10.$$

where ϕ_{ref} is the reference porosity of the adjusted yield locus and ϕ_i is the sample porosity. We minimize now the cumulated weighted errors

$$S = \sum_{i=1}^{N} W_i\, f_i^2$$

and we obtain the parameters values depending on the porosity.

4 NUMERICAL IMPLEMENTATION

The chalk constitutive laws have been implemented in the finite elements code LAGAMINE developed in the MSM department, Liège University [1]. Large strains, geometrical non linearities and unilateral contact with friction can be modeled, together with the material non linearities.

Each step is divided in some subintervals In each subinterval the mean normal schema proposed by Rice and Tracey [8] is used for the integration of the constitutive law.

5 APPLICATION

The first proposed elastoplastic law (elliptic yield locus) has been applied at a large scale to model the subsidence of a North Sea oil reservoir. Bidimensional and threedimensional meshes were used. A very good agreement was found between the numerical results and the in-situ measurements.

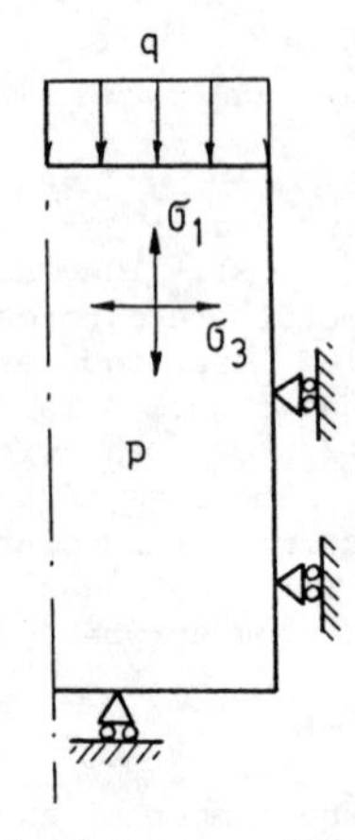

Figure 6 - Triaxial test without radial strain.

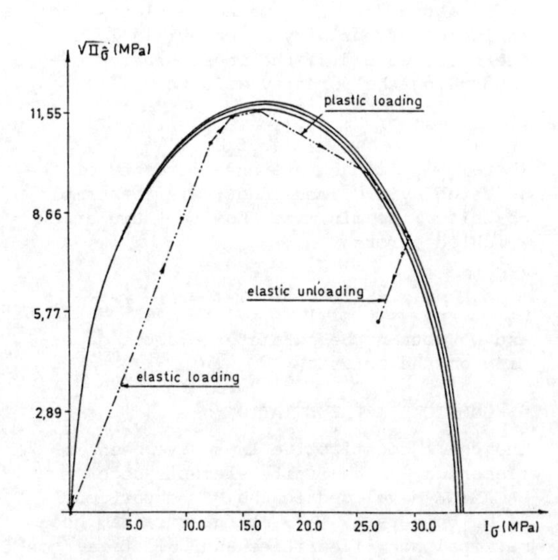

Figure 7 - Numerical model : stresses path for a triaxial test without radial strain.

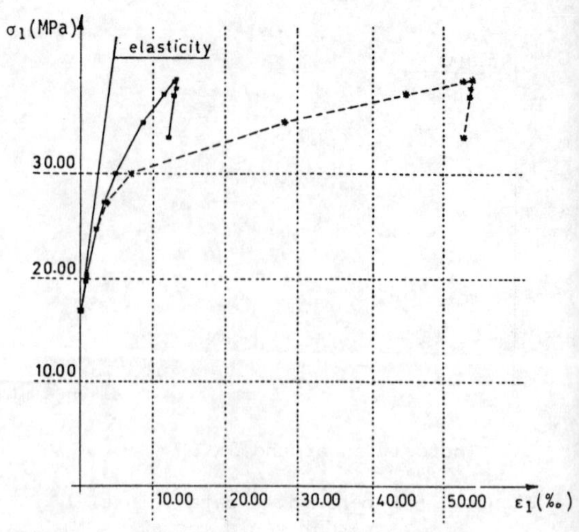

axial strain vs axial stress relation

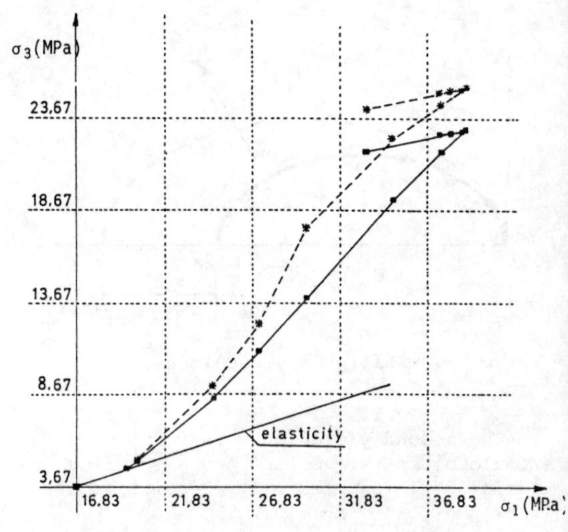

axial stress vs radial stress relation

—■— high hardening

- -✱- - - low hardening

Figure 8 - Numerical model. Results for a triaxial test without radial strain.

Details can be found in [1, 3, 7]. Here we will show the numerical modeling of a triaxial test where the radial strain remains zero (fig.6).

First a vertical load q and an initial pore pressure p are applied. Then q remains constant and the pore pressure p is decreased. The stress path is first linear.

After the contact with the yield surface, the path follows it, and some isotropic hardening appears. The pore pressure increase corresponds to an elastic unloading. When the first elliptic law is used, the stress path is shown on the figure 7 in the case of a small hardening. The numerical results show a good qualitative and quantitative agreement with the experiments, as it appears from the comparison of figures 8 and 1.

6 CONCLUSIONS

A large number of mechanical experimental tests on high porosity chalk have been analysed. Two elastoplastic isotropic constitutive laws are proposed. One of them is applied to the modeling of a simple triaxial test and of a North Sea oil reservoir subsidence. The numerical results are in good agreement with the experiments.

REFERENCES

[1] Charlier, R. 1987. Approche unifiée de quelques problèmes non linéaires de mécanique des milieux continus par la méthode des éléments finis (grandes déformations des métaux et des sols, contact unilatéral de solides, conduction thermique et écoulements en milieux poreux). Thèse de doctorat. Université de Liège, Faculté des Sciences Appliquées.

[2] Da Silva, F., Sarda,J.P. et Schroeder, Ch. 1985. Mechanical Behaviour of Chalks. The Chalk Research Program Symposium, paper 2/4, Stavanger,Norway.

[3] Monjoie, A., Schroeder, Ch., Fonder,G. Charlier, R., Radu, J.P. 1985-1987. Ekofisk. Subsidence-compaction. Modèle éléments finis. Rapports de recherches, Université de Liège, Faculté des Sciences Appliquées, MSM-LGIH.

[4] Naylor, D.J. et Pande, G.N. 1981. Finite Elements in Geotechnical Engineering. Pineridge Press, Swansea,B.

[5] Vutukuri, V.F. and Lama, R.D. 1978. Handbook on Mechanical Properties of Rocks. Transtech. Publications.

[6] Shao, J.F. 1987. Etude du comportement d'une craie blanche très poreuse et modélisation. Thèse de 3e cycle. Université des Sciences et Techniques de Lille - Flandres - Artois.

[7] Charlier, R., Radu, J.P. and Da Silva, F. To appear. Finite elements model of a North-Sea oil field.

[8] Rice, J.R. and Tracey, D.M. 1973. Computational fracture mechanics. Proc. Symp. on Num. and Comput. Meth. in Struct. Mech. Ed. S.J. Fenves, N. Perrone, A.R. Robinson and W.C. Schnobrick. Academic Press.

Numerical Methods in Geomechanics (Innsbruck 1988), Swoboda (ed.)
© 1988 Balkema, Rotterdam. ISBN 90 6191 809 X

Mathematical modelling and prognose dimensions of karst sinkholes in multilayered soil bases of buildings and structures

E.A.Sorochan & G.M.Troitzky
Research Institute of Bases and Underground Structures, Moscow, USSR
P.D.Grigoruk
State University, Gomel, USSR
V.V.Tolmachyov
Research Institute for Engineering Site Investigation, Moscow, USSR

ABSTRACT: According to the requirements of the USSR Building Code (SNiP) for the designing of the antikarst protection it is necessary to know the design dimensions of the likely sinkholes. The authors worked out a probability-statistical model of karst deformations for both: a nonhomogeneous multilayered massive of soils overlaying the karst zone and the development of sinkholes on soil surface and within soil-bases of structures, as well as a procedure of numerical implementation on computers. The program pack is used for the evaluation of the karst danger when working out the recommendations on designing the constructional antikarst protection for buildings and structures as well as on their location at the site in conditions of a covered sulphate or carbonate karst including chalk.

Various engineering arrangements for the protection of buildings and structures from karst deformation danger are employed(Ref.1). Sinkholes are the most dangerous affects for the majority of buildings and structures. Usually the sulphate and carbonate rocks are bedded at more than 20-30 m depth and the most reasonable antikarst protection is a special foundation designing.

The USSR Building Code for soil-base and footing design in karst dangerous regions was commited in 1986. In accordance to this Code sinkhole parameters within footing subsoil are to be calculated by probability-statistical and/or by analitical methods, the last resting on geo-engineering conditions with regard to their possible variations, soil deformation regularities, constructive pecularities of buildings, grade of responsibility and the maintenance period. Those requirements suggest a special theory (engineering karstology), being substantially formalized to solve the design problems. Its methodology supports the idea of solving all the problems within a singular geotechnical system, conventially identified as 'karst-structure'. This takes to the concepts and terminology formation, to special task statements while examining karst process regularities, their interpretation, karst parameter selection, calculation scheme, etc.

Thus, adhering the position of the system under consideration, the main parameters of sinkholes are: the probability of sinkhole formation within the structure soil-base while maintaining its t_n, the design sinkhole span, being taken as an initial data for the designing of the antikarst protection, taking into account a stable sinkhole slope beneath the footing, etc. Let's consider they way of such calculation.

The probability of structure damage is based on the sinkhole affection regularities. In 1968 certain conditions had been proved (Ref.3), under which definite conditions distribution of the sinkhole in time for a given territory of an area F, approach the Poisson Law with the distribution parameters λFt, where λ is an evarage annual number of sinkholes per area unit, t -the design time due to the consideration in practice, as, for example, the structure maintenance time t_n. The observations that had been done in a number of karst regions in the Soviet Union confirm this regularity. Recently this conclusion was supported by American investigators (Ref.4).

The λ-parameter for most of the territories is found by the observation of natural sinkholes, and as for separate area under consideration it is found by calculation methods using the theory of information, correlation and dispersion analysis, the theory of qualititative features, the theory of point processes. As an areal technogeneous affect is tested, for example, the rise or fall of an underground water level, the λFt_n-parameter in conformity with the concrete situation may be evaluated as

$$\lambda Ft_n = \lambda \sum_{i=1}^{n} \sum_{j=1}^{m} K_i F_i \Delta t_j$$

where K_i is a coefficient of the λ parameter changing under the influence of the technogeneous factors over F_i (the area) per time interval t_j found in the observations on the karst deformation mechanism.

The sinkhole diameter value distribution over the soil surface is proved to be approaching log-normal law (Ref.2). If the stochastic regularities of the sinkhole parameters are to be found, one may calculate the risk levelfor different probility of the structure damage (loss of stability, failure in some of its parts or complete damage of the structure, etc.). This suggests the change or the correction of the sinkhole diameter distribution law with due regard for the depth of the footing soil-base and the affecting loading at the soil surface. Hence, the antikarst protection means are to be effective and economically advantageous if there are maps of the sinkhole diameter distribution.

To achieve the trustworthy of the forecasted data concerning sinkhole diameters on the karst territories it is only possible in case of mathematic and probability-statistic models, that are computer-aided. In our paper the karst danger is consi dered as a possibility to definate the size of the sinkhole emergance at any of a city building plan with regard for the surface loading affection. Thus the territory regioning consists in mapping the curves of equal probability for the sinkhole diameters.

In this case the mathematical formulation of the problem has the form: 'Let under certain conditions in certain points of the soil surface and within structure soil-base the values of sinkhole diameters d_i ($i=1,2,...,n$), with centers in points of coordinates (x_i,y_i) be known In addition to that, the sinkhole diameters could be found both by calculations with regard for the design load of the located buildings, and by the recording of natural sinkholes at the area under consideration. In the last case the geo-engineering conditions and the surface loading define the sinkhole diameters d indirectly. The said initial data (d_i,x_i,y_i) necessitate to find the values $d(x,y)$ in an arbitrary set point (x,y) of the site surface under consideration'.

The formulated problem corresponds to the interpolation or the approximation of a function. An interpolating (approximating) function may be any function which values are equal to the d_i-values of are approaching them. Thus, each finite set of points (x_i,y_i) ($i=1,2,...,n$) has a corresponding infinite interpolating or approximating function $d(x,y)$. To pick up any representative output of the whole set, the interpolating (apprpximating) function $d(x,y)$ should be restricted. Or, for example, may be belonging to $d(x,y)$ function of a certain fixed class of functions and only one single function of the whole set either:

(1) satisfies certain of the values proximity $d(x_i,y_i)$ and

$|d(x_i,y_i)-d_i| \leq \delta$ (δ=const>0); or

(2) satisfies the equations

$d_i=d(x_i,y_i)$ ($i=1,2,...,n$)

Thus, to select a certain function $d(x,y)$ out of the infinite set of functions one should put forward a hyposesis on the pecularities of unknown interpolating (approximating) surface. In this paper the procedure of selecting the parametrizing class of functions is based on probability-statistical initial data. One should first find a certain statistical sampling (i.e. a set of surfaces) for which there is a complete information as for as their characteristics are concerned. Then the whole set of surfaces is described in terms of a random field theory. Having definated those field probability characteristics one may assume that within the site where the point discreted system is set ($d_i(x_i, y_i)$) the actual surface $d(x,y)$ is described by one of the random field implementations. Thus, the function $d(x,y)$ restoring task discrete system of points is formulated as a task directed on a random field $D(x,y)$ characteristics: definition by measurements in points (x_i,y_i). This presupposes additional information on field possible values in points (x,y) in the form of a probability distribution and a reciprocal correlative ratio while the function $d(x,y)$ is derived. The said mathematical model is valid if d_i in points (x_i,y_i) $(i=1,2,\dots,n)$ are defined within the error equal to Δd_i.

According to Ref.2 provided quasi-homogeneous geo-engineering conditions the sinkhole diameters at the soil surface could be related to the Gaussian processes, characterizing the mathematical expectation and autocorrelating functions. The mathematical expectation $\mu D(x,y)$ at the point (x,y) determines the most probable random variable value. Having plotted the function $\mu D(x,y)$ by given random variables at the points (x_i,y_i) $(i=1,2,\dots,n)$ one is to obtain the most probable approximating (interpolating) surface.

The homogeneous process of the correlation function turn on $x^{(2)}-x^{(1)}$ difference where x^i $(i=1,2)$ in common case is a vector. It means that the function is

$$K(x,y)=r^f \qquad (1)$$

where $r=\text{const}$, $|r|<1$, $f=(x^2+y^2)^k$, $0<k\leq 1$.

Let the number of points under consideration be equal to n and let the values of the karst sinkhole diameters d_i of a random field $D(x,y)$ be known in points (x_i,y_i). For any set of the random values given in the points (x_i,y_i) $(i=1,2,\dots,n)$ by aid of the correlation function the covariation matrix being built $K(x,y) = (K(x_i,y_i,x_jy_j))$ $(i,j=1,2,\dots,n)$.

This matrix is known to be determined non-negative and due to arbitrariness (randomness) (x_i,y_i) it is considered as posively defined. Thus, it follows that $\det(x,y)\neq 0$.

The mode is chosen for the evaluation of the field $D(x,y)$, i.e. the mean $\mu_y D(x,y)$. So for the Gaussian field the evaluation of the mean values may be defined by

$$\det(A)=0 \qquad (2)$$

where

$$A=\begin{pmatrix} & K(x_1,y_1;\ x,y) \\ & K(x_2,y_2;\ x,y) \\ K(x,y) & \cdot\ \cdot\ \cdot \\ & K(x_m,y_m;\ x,y) \\ d_1\ d_2\ \cdot\ d_n & {}_xD(x,y) \end{pmatrix}$$

For explicit function $\mu_y D(x,y)$ expression (1) should convert the matrix $K(x,y)$. After that one receives the form $\mu_y D(x,y)$. For this purpose let's introduce the matrix B.

$$B=\begin{pmatrix} & & & 0 \\ & K^{-1}(x,y) & & 0 \\ & & & 0 \\ 0 & 0 & \dots & 1 \end{pmatrix}$$

The result of the product of the matrices A and B from the right tacking into account (2) and that $\det K^{-1}\neq 0$ is

$$\mu_y D(x,y)=\sum_{i=1}^{n} d_i \sum_{j=1}^{m} r_{ij} K(x_j,y_j;\ x,y) = \sum_{i=1}^{n}\sum_{j=1}^{m} d_i r_{ij} K(x_j,y_j;\ x,y) \qquad (3)$$

Here r_{ij} $(i,j=1,2,\dots,n)$ are the elements of the converse matrix K^{-1}.

It should be mentiones that when there is large number of data (for example, when n>200) the correlation matrix conversion is a difficult task, for in most cases the initial data $d_i(x_i,y_i)$ $(i=1,2,\dots,n)$ are arranged on an irregular network. Correlation matrix measure can be increased to several thousands, while the initial data are given in the nodes (x_i,y_i) of a regular network.

Thus, the formula (3) allows to find the d-values in any point of the region under consideration. For building site regioning by the values of the karst danger the knowledge of the sinkhole probability diameters in the nodes of a network is advisible. Using formula (3) we may find the diameters in the nodes of a regular network with a given step of the nodes. And for the calculation time shortening the procedure is not done by all of the data, but only by the data of the environmental nodes, which values are defined by the field autocorrelation function $D(x,y)$ behaviour. The network step is found provided by the sinkhole forecasted diameters in the nodes of a regular network, different from the given dimensions not more then by one unit of the mean-square-root of the latter. Below we draw on a conclusion of the independence in calculation on a regular network with any chosen step of the network.

If the plotted network has m horizontal and n vertical lines, then the K-matrix in (1) contains (m.n) elements. Now the K-matrix may be represented in the form of a tensor product of matrices $K^{(m)}$ and $K^{(n)}$, i.e. $\mathbf{K}=K^{(m)}\otimes K^{(n)}$. By the force of the tensor product of characteristics there is an equility

$K^{-1}=(K^{(m)})^{-1}\otimes(K^{(n)})^{-1}$ for the matrix.

As matrices $K^{(m)}$ and $K^{(n)}$ order is less than the order of the K-matrix, they may be converted using standard computer software. If the matrices size is still large ($200\leq(m,n)<400$) the square-root method is applied for their conversion. By force of the matrices symmetry and with need of computer storage economy, we form and store only the upper triangle part of the matrix into the computer memory.

So let $r_{ij}^{(m)}$ and $r_{ij}^{(n)}$ correspond to the elements of the converse matrices $(K^{(m)})^{-1}$ and $(K^{(n)})^{-1}$. Then the elements r_{ij} $(i,j=1,2,\ldots,m.n)$ of the converse matrix K^{-1} are found by the matrices elements $r_{ql}^{(m)}$ $(q,l=1,2,\ldots,m)$ and $r_{sp}^{(n)}$ $(s,p=1,2,\ldots,n)$ of the matrices $(K^{(m)})^{-1}$ and $(K^{(n)})^{-1}$ by the formula $r_{ij}=r_{ql}^{(m)}\cdot r_{sp}^{(n)}$ respectively, where indices q,l,s and p are found as follows:

$$\begin{cases} q=\nu+1 \\ l=\tau+1 \end{cases} \quad \text{and} \quad \begin{cases} s=i-\nu n \\ p=j-\tau n \end{cases}$$

and $\nu=((2i-1)/2n)$,
$\tau=((2j-1)/2n)$.

By (2) the random variable $d(x,y)$ of the field $D(x,y)$ in an arbitrary point (x,y) of the site under consideration within a regular network is found from the equation

$$d(x,y)=\sum_{j=1}^{m}\sum_{i=1}^{n}\hat{d}_{ij}K(x_i,y_i;x,y) \quad (4)$$

The elements $\hat{d}_{ij}$ $(i=1,2,\ldots,n;\ j=1,2,\ldots,m)$ are found from the formula $\hat{d}_{ij}=\sum_{s=1}^{m.n}d_{ql}\cdot r_{sp}$ where $q=\sigma+1$, $l=s-\sigma m$, $p=j+(i+1)m$ and $\sigma=((2s-1)/2m)$

The solution (4) allows regional mapping by a large number of points and takes into account the field heterogeneousness, varying from site to site.

The authors worked out a SOKOS-program pack in FORTRAN. The pack is realized in a dialogue regime on the computer CM-4. For regioning the karst danger on the building site or large territories based on the results of geo-engineering investigations and the data characteristics of the buildings (i,e. the length, width and depth of the footings and their location at the site, the affect of the soil surface loading, etc.) there is a need of calculations using the PROWAL-program. This program calculates the probable sinkhole diameters in a given point of coordinates with regard for the above mentioned statements. Then the PROGNOZ-program with the d_i-data in points (x_i,y_i) $(i=1,2,\ldots,n)$ finds the sinkhole forecasted diameters over the surface of the karst territory at the nodes of a network.

The solution results by the PROGNOZ-program are printed either in the form of a table of sinkhole diameters in the network nodes under consideration, as shown below in the example of calculation given on the figure, or in the form of curves using the graphic features of the computer.

The next stage comprises the calculations of the probable footing

*** INITIAL DATA ***
NUMBER OF BOREHOLES: N=11

X	Y	D
77.0	118.5	14.90
57.0	145.0	14.40
55.0	57.5	14.70
80.0	50.0	14.30
100.0	30.0	15.40
77.0	111.0	14.70
90.0	90.0	14.20
110.0	143.0	15.50
122.0	38.0	16.00
147.0	63.0	12.40
165.0	65.0	15.00

Fig.1 Table of results after PROWAL-program calculations - initial for the next step calculations

*** RESULTS ***
STEP H=10.0 M

	55.0					85.0					135.0	
	I	.	.	.	.	I	.	.	.	.	I	.
30	-	-	-	-	7	6	-	-	-	-	-	-
40	-	-	6	6	7	7	7	7	9	8	-	-
50	-	7	6	6	5	5	7	8	7	8	-	-
60	6	7	6	6	6	7	8	9	8	7	7	-
70	7	7	5	6	6	7	7	9	8	7	6	-
80	5	5	6	5	7	7	8	8	7	7	5	6
90	6	5	5	6	7	8	8	7	7	6	6	6
100	-	5	5	7	7	7	7	7	6	5	6	-
110	-	5	6	7	5	7	7	5	6	6	6	-
120	-	5	5	7	8	8	7	6	6	6	-	-
130	-	5	6	7	5	7	7	7	7	-	-	-
140	-	-	6	7	7	7	7	8	7	-	-	-

Fig.2 Result of regioning of the basin footing (radii of predeterminated sinkhole dimensions)

span dimensions satisfying the given safety factor for the buildings. This calculation is done by the PROLET-program. The main principles for the sinkhole dimensions and design spans for the antikarst footing of structures are given in Ref.5

As an example the results of regioning under a basin of a power station cooling tower are represented. The figure gives an outline of sinkhole radii in nodes of a regular network in the plan of the basin footing. The initial data for those calculations were the coordinates and the subsoil conditions of 11 boreholes at the building site. The sinkhole diameters were previously calculated using the PROWAL-program. The regioning of the site was restricted by the coordinates of the boundary boreholes.

The analysis of several numeric experiments have shown that the SOKOS-program pack is efficacious for karst deformation forecasting while designing or maintaining the structures. It may facilitate for cost minimization of the antikarst measures taking into account the damage losses due to the influence of karst deformations on the buildings while designing their location in the plan of the site.

The SOKOS-program pack was used in designing antikarst protection means in conditions of covered sulphate and carbonate karst (including chalk) in the Ukraine, the Volga regions, the Bashkiria and other parts of the USSR.

Bibliography

(1) E.A.Sorochan, G.M.Troitzky, V.V. Tolmachyov. Kompleksnye zaschitnye meropriiatia pri stroitel'stve na zakarstovannykh territoriakh. J.:Osnovania, fundamenty i mekhanika gruntov, 4, 16-19 (1982)

(2) V.V.Tolmachyov, G.M.Troitzky, V.P.Khomenko. Ingenerno-stroitel'noie osvoienie zakarstovannykh territorii. Stroiizdat. 177. Moscow. USSR. 1986

(3) V.V.Tolmachyov. Ob otzenke effektivnosti protivokarsto ykh meropriatii na geleznykh dorogakh. Proc.MIIT, 319. Moscow, (1968)

(4) D.Raghu, C.Tiedeman. Sinkhole risk analysis for a selected area in Warran County, New Jersey. Sinkholes: their geology, engineering & environmental impact. Balkema, Rotterdam. 167-169 (1984)

(5) E.A.Sorochan, G.M.Troitzky, V.V.Tolmachyov, V.P.Khomenko, S.N.Klepikov, N.S.Metelyuk, P.D.Grigoruk. Antikarst protection for buildings and structures. Proc. XI Int.Conf.on SM & FE. San Francisco/12-16 Aug. 1985/. Balkema, Rotterdam, 1985

(6) V.Kh.Kivelidi, M.E.Starobinetz, V.M.Eskin. Veroiatnostnye metody v seismorazvedke. Nedra, 347, Moscow, USSR, 1982

Numerical Methods in Geomechanics (Innsbruck 1988), Swoboda (ed.)
© 1988 Balkema, Rotterdam. ISBN 90 6191 809 X

A model for predicting the viscoplastic stress-strain behaviour of clay in three-dimensional stresses

T.Nakai
Nagoya Institute of Technology, Japan
K.Tsuzuki
Shimizu Construction Co., Ltd, Japan

ABSTRACT: An elasto-viscoplastic model for clay is presented. This model can describe uniquely various time-dependent deformation and strength characteristics of normally consolidated clay in three-dimensional stresses. The influence of intermediate principal stress is taken into consideration by using the concept of the mechanical quantity t_{ij}, and the influence of stress path on the direction of viscoplastic flow is taken into consideration by dividing the viscoplastic strain rate into two components. The time-dependent behaviour is expressed by introducing the non-stationary flow surface theory developed by Sekiguchi.

1 INTRODUCTION

Although the well-known Cam-clay model (Roscoe, Schofield and Thurairajah, 1963; Schofield and Wroth, 1968) is a simple and fine model for clay, it is not able to describe sufficiently the stress-strain behaviour under general stress conditions, e.g., (1)influence of intermediate principal stress on the deformation and strength characteristics, (2)influence of stress path on the direction of plastic flow, (3)rheological behaviour, (4)behaviour under cyclic loading, (5)inherent anisotropy and others. Many studies have then been done to construct more general constitutive models for clay.

Nakai and Matsuoka (1986) developed an elastoplastic model for clay (named t_{ij}-clay model) on the basis of the concept of mechanical quantity t_{ij} (Nakai and Mihara, 1984). This model can in particular describe precisely the influence of intermediate principal stress and the influence of stress path. On the other hand, Sekiguchi (1977) proposed an elasto-viscoplastic model based on a non-stationary flow surface theory to explain the various rheological behaviour of clay.

In the present paper, the authors extend the t_{ij}-clay model to an elasto-viscoplastic model which can describe uniquely the various behaviour of clay including the time-dependent behaviour, by introducing the viscoplastic theory by Sekiguchi. As another type of viscoplastic model applicable in three-dimensional stresses, Oka (1986) developed a model by combining the concept of t_{ij} mentioned above and Adachi and Oka's model (1982) based on the overstress theory.

2 ELASTO-VISCOPLASTIC MODEL FOR CLAY

An elastoplastic model for clay which particularly takes into consideration the influence of intermediate principal stress and stress paths—named t_{ij}-clay model — has been presented (Nakai and Matsuoka, 1986). This model is here extended to an elasto-viscoplastic model, with reference to the theory of non-stationary flow surface proposed by Sekiguchi (1977).

2.1 Stress parameters used

The mechanical quantity t_{ij} is a symmetric tensor given by

$$t_{ij} = \sigma_{ik} \cdot a_{kj} \qquad (1)$$

in which σ_{ik} is the stress tensor, and a_{kj} is the tensor which has the direction cosines (a_1, a_2 and a_3) of the normal to the SMP as its principal values (Nakai and Mihara, 1984; Nakai and Matsuoka, 1986). Here, the SMP is the plane ABC in Fig.1, and its direction cosines (a_1, a_2 and a_3) are expressed as

$$a_i = \sqrt{\frac{J_3}{\sigma_i \cdot J_2}} \qquad (i = 1, 2, 3) \qquad (2)$$

where J_1, J_2 and J_3 are the first, second and third effective stress invariants (Matsuoka and Nakai, 1974, 1977).

$$\left.\begin{aligned} J_1 &= \sigma_1 + \sigma_2 + \sigma_3 \\ J_2 &= \sigma_1\sigma_2 + \sigma_2\sigma_3 + \sigma_3\sigma_1 \\ J_3 &= \sigma_1\sigma_2\sigma_3 \end{aligned}\right\} \quad (3)$$

Then, the stress parameters (t_N and t_S) used in the present model are defined by

$$t_N = t_1a_1 + t_2a_2 + t_3a_3 = t_{ij}a_{ij} \quad (4)$$

$$t_S = \sqrt{(t_1^2 + t_2^2 + t_3^2) - t_N^2} = \sqrt{t_{ij}t_{ij} - (t_{ij}a_{ij})^2} \quad (5)$$

For reference, the stress parameters (p and q) used in most of models such as Cam-clay model are

$$p = \frac{1}{3}(\sigma_1 + \sigma_2 + \sigma_3) = \frac{1}{3}J_1 \quad (6)$$

$$q = \frac{1}{\sqrt{2}}\sqrt{(\sigma_1 - \sigma_2)^2 + (\sigma_2 - \sigma_3)^2 + (\sigma_3 - \sigma_1)^2} = \sqrt{J_1^2 - 3J_2} \quad (7)$$

According to the t_{ij}-concept, if we make a yield function and/or a plastic potential function using the stress parameters (t_N and t_S) instead of (p and q) and assume the flow rule in t_{ij}-space instead of σ_{ij}-space, the stress-strain behaviour in three-dimensional stresses can be described with unified soil parameters.

2.2 Derivation of model

According to the viscoplastic theory based on the concept of non-stationary flow surface by Sekiguchi (1977), the viscoplastic potential F is specified as

$$F = C_\varepsilon \cdot \ln\left[\frac{\dot{\varepsilon}_{v0} \cdot t}{C_\varepsilon}\exp\left(\frac{f}{C_\varepsilon}\right) + 1\right] = \varepsilon_v^p \quad (8)$$

in which C_ε is the secondary compression index, $\dot{\varepsilon}_{vo}$ is the reference volumetric strain rate, t is the elapsed time, ε_v^p is the viscoplastic volumetric strain which is used as the strain hardening parameter and f is the scalar function of stress. In Sekiguchi's model, employed is the scalar function which represents the shape of the yield surface at a constant plastic volumetric strain of Cam-clay model (Schofield and Wroth, 1968) or Ohta's model (Ohta, 1971).

$$f = \frac{\lambda - \kappa}{1 + e_0}\left[\ln\frac{p}{p_0} + \frac{1}{M}\cdot\frac{q}{p}\right] \quad (9)$$

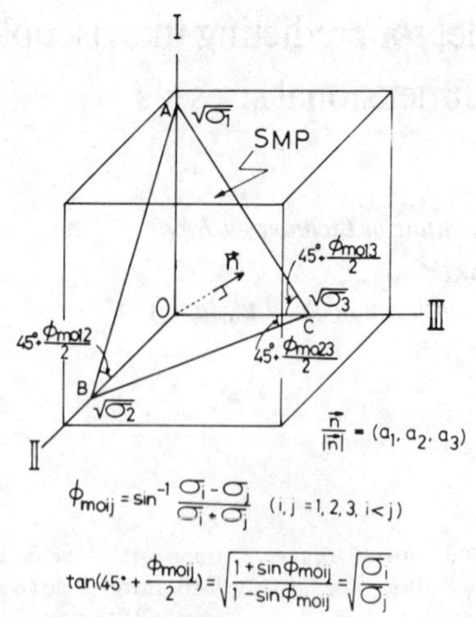

Fig. 1. Cubical soil element and Spatially Mobilized Plane (SMP).

where λ and κ is the compression and swelling indices, and M is the value of q/p at critical state. In the present model, we employ the following function which represents the shape of the yield surface of the t_{ij}-clay model (Nakai and Matsuoka, 1986) in order to take into consideration the influence of intermediate principal stress:

$$f = \frac{\lambda - \kappa}{1 + e_0}\left[\ln\frac{t_N}{t_{N0}} + \frac{-\alpha}{1 - \alpha}\ln\left|1 - (1 - \alpha)\frac{X}{M^*}\right|\right] \quad (10)$$

where $X \equiv t_S/t_N$ and M^* is expressed as follows using the stress ratio X_f and strain increment ratio Y_f at critical state:

$$M^* = X_f + \alpha Y_f \quad (11)$$

The values of X_f and Y_f are determined using the principal stress ratio at critical state under triaxial compression R_f (Nakai and Matsuoka, 1986):

$$X_f = \frac{\sqrt{2}}{3}\left(\sqrt{R_f} - \sqrt{1/R_f}\right) \quad (12)$$

$$Y_f = \frac{1 - \sqrt{R_f}}{2(R_f + 1/2)} \quad (13)$$

where

$$R_f \equiv (\sigma_1/\sigma_3)_{f(comp.)} = \frac{1 + \sin\phi'_{(comp.)}}{1 - \sin\phi'_{(comp.)}} \quad (14)$$

Next, in order to take into consideration the influence of stress path on the direction of viscoplastic flow, we assume that the viscoplastic strain rate $\dot{\varepsilon}_{ij}^p$ is divided into the component satisfying the associated flow rule in t_{ij}-space $\dot{\varepsilon}_{ij}^{p(AF)}$ and the component compressive isotropically $\dot{\varepsilon}_{ij}^{p(IC)}$ in the same way as the plastic strain increments in the elastoplastic t_{ij}-clay model. As a result, total strain rate is given by the following equation as the summation of these viscoplastic components and the elastic component:

$$\dot{\varepsilon}_{ij} = \dot{\varepsilon}^{e}_{ij} + \dot{\varepsilon}^{p}_{ij} = \dot{\varepsilon}^{e}_{ij} + \dot{\varepsilon}^{p(AF)}_{ij} + \dot{\varepsilon}^{p(IC)}_{ij} \quad (15)$$

Elastic strain rate

Using the well-known relation for isotropic elastic materials, $\dot{\varepsilon}^{e}_{ij}$ is given by

$$\dot{\varepsilon}^{e}_{ij} = \frac{1+\nu_e}{E_e}\dot{\sigma}_{ij} - \frac{\nu_e}{E_e}\dot{\sigma}_{kk}\delta_{ij} \quad (16)$$

where

$$E_e = 3(1-2\nu_e)(1+e_0)\frac{p}{\kappa} \quad (17)$$

Viscoplastic strain rate compressive isotropically

The isotropic component $\dot{\varepsilon}^{p(IC)}_{ij}$ of the viscoplastic strain rate is given by the following equation with reference to the formulation of the isotropic plastic component in the elastoplastic t_{ij}-clay model (Nakai and Matsuoka, 1986):

$$\dot{\varepsilon}^{p(IC)}_{ij} = K\langle \dot{t}_N \rangle \frac{\delta_{ij}}{3} \quad (18)$$

where the bracket ⟨ ⟩ implies that ⟨A⟩=A if A>0 and ⟨A⟩=0 if A≦0, and the coefficient K is

$$K = \frac{\lambda - \kappa}{1+e_0}\cdot\frac{1}{t_{N1}}\cdot\frac{\partial F}{\partial f} \quad (19)$$

Here, t_{N1} is the value of t_N where the surface of $f=\varepsilon_v^{\,p}$ intersect the t_N-axis and is given by

$$t_{N1} = t_N \cdot \left|1 - (1-\alpha)\frac{X}{M^*}\right|^{\frac{-\alpha}{1-\alpha}} \quad (20)$$

Viscoplastic strain rate satisfying associated flow rule

Since the direction of $\dot{\varepsilon}^{p(AF)}_{ij}$ is normal to the viscoplastic potential in t_{ij}-space, $\dot{\varepsilon}^{p(AF)}_{ij}$ is defined as

$$\dot{\varepsilon}^{p(AF)}_{ij} = \Lambda\frac{\partial F}{\partial t_{ij}} \quad (21)$$

The proportionality coefficient Λ is determined as follows from the condition of consistency $(\dot{F}=(\partial F/\partial\sigma_{k\ell})\dot{\sigma}_{k\ell} + (\partial F/\partial t)=\dot{\varepsilon}_v^{\,p})$ and Eqs.(15), (18) and (21):

$$\Lambda = \frac{\dfrac{\partial F}{\partial \sigma_{ij}}\dot{\sigma}_{ij} + \dfrac{\partial F}{\partial t} - K\langle \dot{t}_N \rangle}{\dfrac{\partial F}{\partial t_{kk}}} \quad (22)$$

We can finally calculate the total strain rate $\dot{\varepsilon}_{ij}$ using Eq.(15) and these three components expressed by Eqs.(16), (18) and (21). The model proposed here is named VP t_{ij}-clay model.

Now then, replacing the scalar function f of Eq.(10) by that of Eq.(9), assuming the viscoplastic flow in σ_{ij}-space $(\dot{\varepsilon}_{ij}^{\,p(AF)} = \Lambda(\partial F/\partial\sigma_{ij}))$ instead of t_{ij}-space and not taking into consideration the isotropic component given by Eq.(18), we obtain Sekiguchi's model.

3 EXPERIMENTAL ASSESSMENT

3.1 Determination of soil parameters

Saturated remoulded normally consolidated Fujinomori clay was used in the experiments. Its physical properties are that the liquid limit w_L=41%, the plastic limit w_p=23% and the specific gravity G_s=2.67.

The values of soil parameters of Fujinomori clay are listed in Table 1. Here, $\lambda/(1+e_0)$ and $\kappa/(1+e_0)$ are determined from a consolidation test (see Fig.2(a)), and $\phi'_{(comp.)}$ is the angle of shear resistance in terms of effective stress under triaxial compression condition (see Fig.2(b)). The value of α is determined in the following way. From the condition that the ratio p_c/p_o in undrained triaxial compression test (p_c and p_o: mean principal stresses at critical state and initial state as shown in Fig.2(b)) calculated by t_{ij}-clay model is equal to that by Cam-clay model, the relation between p_c/p_o and $\phi'_{(comp.)}$ shown in Fig.2(c) is obtained. We can therefore estimate α from $\phi'_{(comp.)}$ using this figure. The parameters C_ε and $\dot{\varepsilon}_{vo}$, which are concerned with viscous properties, can be determined from the gradient of ε_v-ln t relation in duration of secondary compression in a consolidation test and the period of consolidation t_c before experiment (see Fig.2(d)). The Poisson ratio ν_e in the elastic component is usually assumed to be zero, since the

Table 1. Values of soil parameters for Fujinomori clay.

$\lambda/(1+e_0)$	5.08×10^{-2}
$\kappa/(1+e_0)$	1.12×10^{-2}
$\phi'_{(comp.)}$	33.7°
α	0.74
ν_e	0.0
C_ε	0.001
$\dot{\varepsilon}_{v0}$ (%/min)	7.0×10^{-5}

elastic shear strain is relatively small. Thus, all of the soil parameters are determined from the results of a consolidation test and the shear strength in terms of effective stress in the same way as those of Sekiguchi's model.

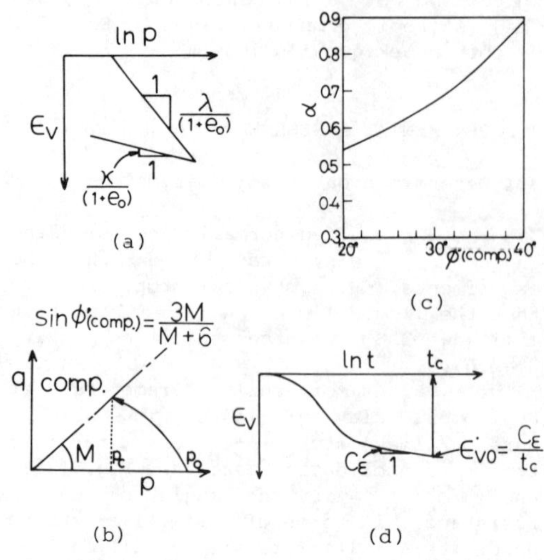

Fig. 2. Explanation of soil parameters
(a)ε_v~ln p relation in consolidation test
(b)q~p relation in undrained test
(c)α~$\phi'_{(comp.)}$ relation
(d)ε_v~ln t relation in consolidation test

3.2 Undrained constant strain rate tests

Fig.3 shows the analytical effective stress paths (curves) and the observed values (dots) in undrained triaxial compression and extension tests at two kinds of axial strain rates ($\dot{\varepsilon}_a$=5.5x10^{-2} (%/min.) and $\dot{\varepsilon}_a$=5.5x10^{-4} (%/min.)), in terms of the relation between q/p_0 and p/p_0 (p_0: initial consolidation pressure =196kN/m^2). Figs.4(a) and (b) also show the relation between shear stress and axial strain in the same tests. Here, axial strain ε_a coincides with the deviatoric strain ε_d=(2/3)(ε_a-ε_r) under undrained triaxial conditions. The analytical curves in Figs.3 and 4 describe well the observed strain rate effect on undrained stress-strain behaviour. In addition, the analytical results explain the differences of the effective stress paths and the stress-strain relations between triaxial compression and extension conditions, which are not described by Sekiguchi's model. In Fig.5 the analytical and experimental results in Fig.3 are rearranged in terms of the

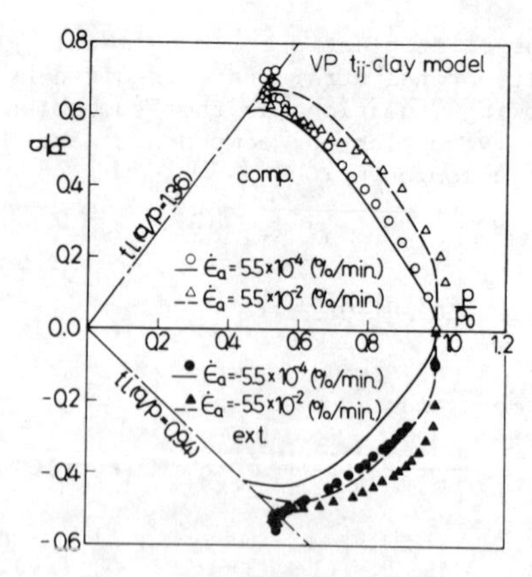

Fig. 3. Effective stress paths in (p, q) space in undrained constant strain rate tests

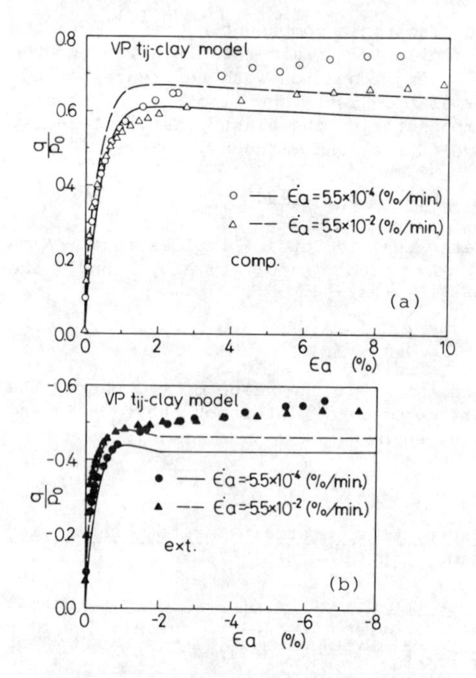

Fig. 4. Stress-strain relations in undrained constant strain rate tests
(a)triaxial compression
(b)triaxial extension

relations between t_S/t_{NO} and t_N/t_{NO}. Here, t_{NO} is the initial value of t_N and is equal to p_0 under the isotropic stress condition. It is noticed from this figure that the stress-strain behaviour of clay including the strain rate effect in three-dimensional stresses is uniquely arranged not in (p, q) space but in (t_N, t_S) space.

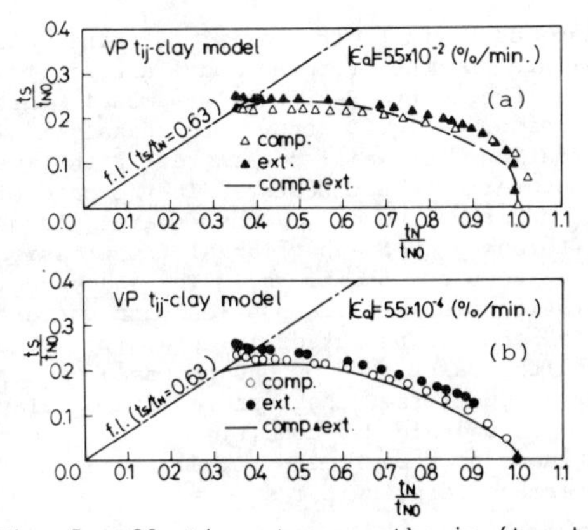

Fig. 5. Effective stress paths in (t_N, t_S) space in undrained constant strain rate tests
(a) $\dot{\varepsilon}_a = 5.5 \times 10^{-2}$ %/min.
(b) $\dot{\varepsilon}_a = 5.5 \times 10^{-4}$ %/min.

3.3 Undrained stress relaxation tests

The undrained stress relaxation tests were performed after being sheared up to the stress level of q/p_0=0.4 at the strain rate of $\dot{\varepsilon}_a=5.5\times10^{-2}$ (%/min.) under undrained condition. The specimen were then resheared at the same strain rate after stress relaxation for two days. Fig.6 shows the effective stress paths in (p, q) space during all these processes. Here, the analytical curves represented by broken line at the period of reshearing are obtained using the value of soil parameter $\dot{\varepsilon}_{v0}=7.0\times10^{-5}$ (%/min.) which is the same as that during the primary shearing and the relaxation periods. As is shown in this figure, these broken curves underestimate the shear stress after stress relaxation and do not correspond to the experimental results. Therefore, in the present analysis the analytical curves (solid lines) after stress relaxation are calculated by replacing the values of $\dot{\varepsilon}_{v0}$ with the value of the viscoplastic strain rate $\dot{\varepsilon}_v^p$ calculated at the end of stress relaxation. Fig.7 shows the relation between shear stress and axial strain, and Fig.8 shows the variations in shear stress with time on a log scale during the stress relaxation period. As is seen from Fig.7, the present analytical stress-strain curves agree well with the experimental results under triaxial compression and extension conditions. It is also seen from Fig.8 that the analytical results explain the observed undrained stress relaxation phenomena though the calculated shear stresses are a little larger.

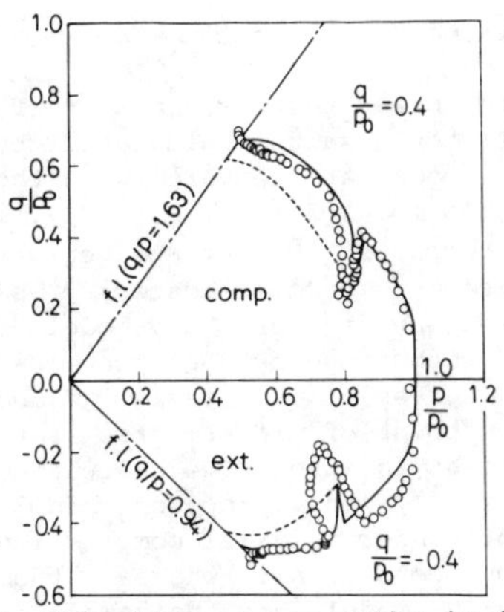

Fig. 6. Effective stress paths in (p, q) space in undrained constant strain rate-stress relaxation-constant strain rate tests.

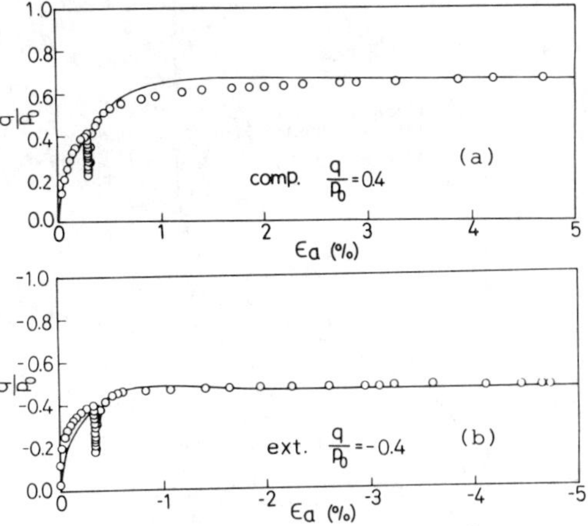

Fig. 7. Stress-strain relations in undrained constant strain rate-stress relaxation-constant strain rate tests.

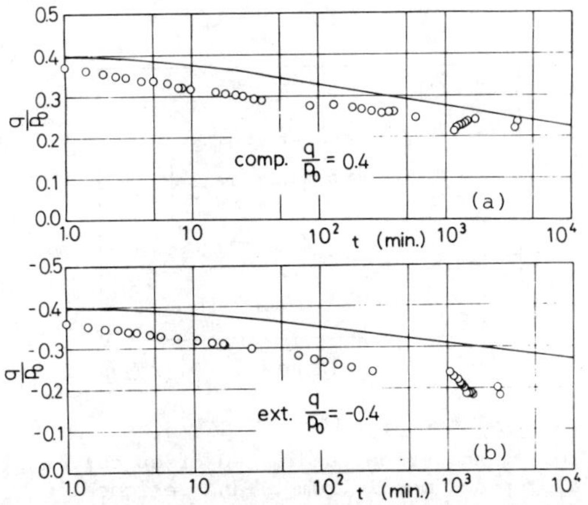

Fig. 8. Variations in shear stress with time in stress relaxation tests.

3.4 Undrained creep tests

Undrained creep tests are not performed, so that the results calculated by the proposed model are compared with those by Sekiguchi's model(Sekiguchi, 1977, 1984). Fig.9 compares the creep curves of isotropically consolidated samples calculated by the proposed model (solid curves) with those by Sekiguchi's model (broken curves). The marks "O" and "×" show the point of minimum creep rate and the point at failure. The calculated ε_a-t relations by the proposed model are different under triaxial compression and extension conditions, whereas those by Sekiguchi's model are the same. It is also seen that the axial strains to failure by the proposed model are finite, but those by Sekiguchi's model are infinite. Fig.10 shows the analytical variations of axial strain rate with time in undrained creep tests. In the case of the proposed model, there are large differences between triaxial compression and extension conditions. The analytical relations between creep stress and rupture life are illustrated in Fig.11. The results calculated by the proposed model may suggest that the rupture life of clay under undrained condition is much influenced by the magnitude of intermediate principal stress.

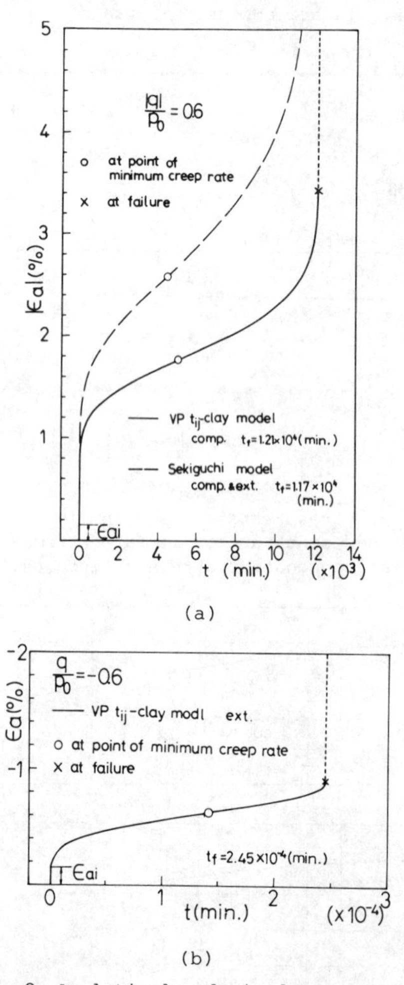

Fig. 9. Analytical undrained creep curves
(a)proposed model (comp.) and Sekiguchi's model (comp. & ext.)
(b)proposed model (ext.)

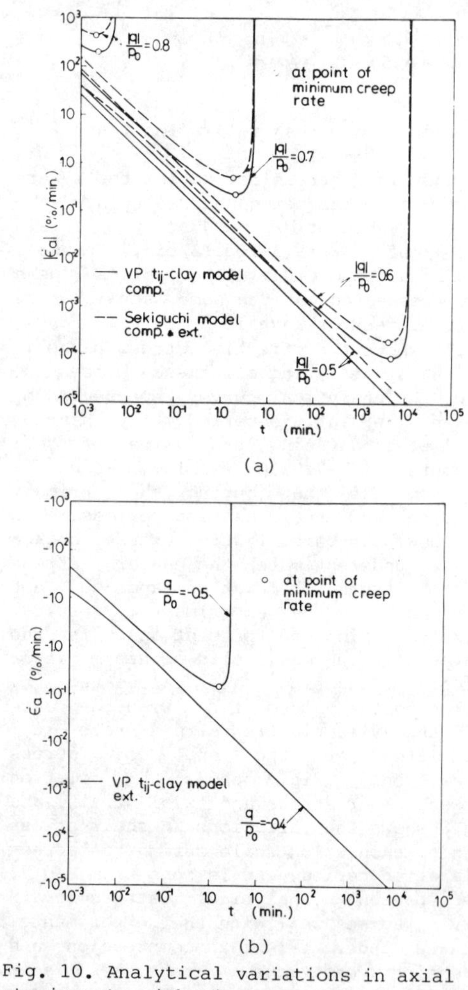

Fig. 10. Analytical variations in axial strain rate with time
(a)proposed model (comp.) and Sekiguchi's model (comp. & ext.)
(b)proposed model (ext.)

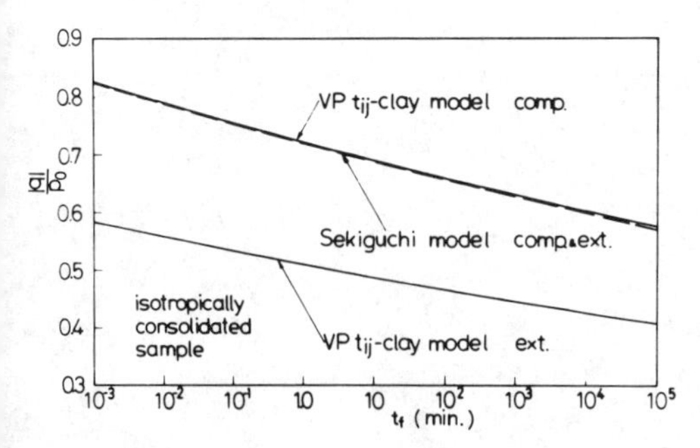

Fig. 11. Analytical creep stress-rupture life relations.

4 CONCLUSIONS

The main results of this paper are summarized as follows:

(1) By introducing the non-stationary flow surface theory by Sekiguchi the authors extend the elastoplastic model for clay (t_{ij}-clay model) proposed before to an elasto-viscoplastic model. The model presented here can in particular describe uniquely the influence of intermediate principal stress and the influence of stress path, together with the time-dependent behaviour, regardless that its soil parameters can easily be determined.

(2) Undrained constant strain rate tests and undrained stress relaxation tests on Fujinomori clay under triaxial compression and extension conditions are performed. The results calculated by the proposed model agree well with these test results and describes the behaviour of undrained creep tests including the difference between triaxial compression and extension conditions. Although the validities of the proposed model are checked here under undrained triaxial conditions alone, the model is applicable in general three-dimensional stress conditions including drained condition.

ACKNOWLEDGEMENTS

The authors wish to thank Prof. H.Matsuoka at Nagoya Institute of Technology for his helpful supports and discussions, and Prof. H.Sekiguchi at Kyoto University for his kind advice on the viscoplastic theory. They are also indebted to Messrs. K.Ishikawa and M.Miyake for their experimental assistance. This study was done by the financial support of Grant-in-Aid for Scientific Research (No.61550355) from the Ministry of Education.

REFERENCES

Adachi, T. & F.Oka 1982. Constitutive equations for normally consolidated clays based on elasto-viscoplasticity. Soils & Foundations, Vol.22, No.4:57-70.

Ohta, H. 1971. Analysis of deformations of soils based on the theory of plasticity and its application to settlement of embankments. Dr. Eng. Thesis, Kyoto Univ.

Oka, F. 1986. A viscoplastic constitutive model of normally consolidated clay under three-dimensional stress conditions. Proc. 2nd Int. Symp. on Numerical Models in Geomechanics, Ghent:165-170.

Matsuoka, H. & T.Nakai 1974. Stress-deformation and strength characteristics of soil under three different principal stresses. Proc. JSCE, No.232:59-70.

Matsuoka, H. & T.Nakai 1977. Stress-strain relationship of soil based on the SMP. Proc. Specialty Session 9, 9th ICSMFE:153-162.

Nakai, T. & H.Matsuoka 1986. A generalized elastoplastic constitutive model for clay in three-dimensional stresses. Soils & Foundations, Vol.26, No.3:81-98.

Nakai, T. & Y.Mihara 1984. A new mechanical quantity for soils and its application to elastoplastic constitutive models. Soils & Foundations, Vol.24, No.2:82-94.

Roscoe, K.H., Schofield, A.N. & Thurairajah, A. 1963. Yielding of clays in states wetter than critical. Géotechnique, Vol.13, No.3:211-240.

Schofield, A.N. & Wroth, C.P. 1968. Critical State Soil Mechanics, McGraw-Hill, London.

Sekiguchi, H. 1977. Rheological characteristics of clays, Proc. 9th ICSMFE, Vol.1:289-292.

Sekiguchi, H. 1984. Theory of undrained creep rupture of normally consolidated clay based on elasto-viscoplasticity. Soils & Foundations, Vol.24, No.1:129-147.

Numerical Methods in Geomechanics (Innsbruck 1988), Swoboda (ed.)
© 1988 Balkema, Rotterdam. ISBN 90 6191 809 X

Effect of soil strength on numerical simulation

Shad M.Sargand
Ohio University, Athens, USA
R.Janardhanam
University of North Carolina, Charlotte, USA

ABSTRACT: Loess behaves distinctly different after a threshold stress state. Introduction of this concept in the constitutive law improves the agreement between the numerical analysis and the actual field test data. Two case studies are examined and the results reported.

1 INTRODUCTION

A need for investigation exists to describe the action of loess under load so as to substantiate present foundation design practices. If found necessary, new design practices need to be developed. Much of the present knowledge related to the performance of loess as a foundation supporting medium has been provided by the U. S. Bureaû of Reclamation. The complex structural composition of loess has a predominent effect on its strength properties. Numerical simulation of the soil structure interaction problems to determine the behavior of foundations in loess will not be meaningful if a realistic constitutive law for soil is not incorporated. Therefore, a comprehensive series of laboratory tests on specimens of loess was conducted to understand the effect of soil fabric and moisture content on strength properties of loess.

Shear strength of loess greatly depends on its (1) mineralogical composition, (2) moisture content, and (3) consolidation pressure. Loess has been observed to behave distinctly different after a threshold stress state. This threshold stress concept is unique to loess. It is introduced in the constitutive law developed in order to improve the agreement between the numerical analysis and the actual field test data.

2 OBJECTIVE

Loess has been identified as a problem soil for foundation use. Its properties are very sensitive to disturbance of its structure. Two factors presumably contribute to sample disturbance which destroy the cemented clay bond that constitutes the shear strength of loess; one during sample preparation and the other due to high confining pressure. In addition, only a small increase of moisture quickens the destruction of the loess structure. For realistic design of a foundation, a meaningful analysis of soil-structure interaction is essential. A reliable soil-structure interaction study can be accomplished only if a realistic constitutive law is available. These considerations lead to three primary objectives for this work.

(1) Conduct a series of tests to determine the effect of soil fabric, moisture content, and consolidation pressure of the strength properties of loess.

(2) Develop a constitutive law for the soil under study, and

(3) Refine the model in order to make a good agreement of predicted foundation capacity values with field test data.

3 SOIL TESTED

Soil samples tested in this investigation were taken from the deposits of loess in eastern Nebraska. It is made up of relatively uniform silt-size particles which are bonded together. This bond is the result of their clay coatings which surround the soil particles. The silt particle-size fraction is found to be composed of rock-forming minerals like quartz and feldspar whereas clay minerals of montmoillonite and Illite are present in substantial amount in fine fractions. Unit weight varies from 70 pcf to a maximum of 100 pcf. High density materials over about 90 pcf (with void ratio of less than about 0.85) show no indication of collapse upon wetting at any load. They do on the other hand, exhibit substantial swelling or heave at low loads. For low density loess, moisture contents below about 15 percent result in firm loess of high strength, whereas weathered loess having about 20 percent moisture content results in a soft, easily consolidated material of very low shearing resistance. The apparent cohesion varies from 0.1 to 0.4 tsf and the angle of internal friction varies from 23 to 33 degrees.

4 TESTING

To develop constitutive relations for loess at its natural moisture content (w = 19 to 24%), loess samples are tested in undrained triaxial compression. In each test at least one unloading-reloading cycle is included. A typical stress-strength response curve in triaxial compression for a confining pressure of 1.5 tsf is shown in Fig. 1. A series of conventional triaxial compression (CTC) tests have been run on undisturbed cylindrical samples. A set of typical stress-strain curves is shown in Fig. 2. Since the peak deviator stress is not apparent, an arbitrary strain level has been chosen to define failure. Figure 3 shows Mohr's envelop for 10% strain and 18% strain. For 20% strain, the friction angle is 29 degrees and cohesion 0.4 tsf.

5 ANALYSIS OF TEST RESULTS

The results obtained show that the mechanical behavior of macroporous loess depends to a large extent on the values of the consolidation pressure. For the soil investigated (with moisture content w = 18-22%), a consolidation pressure (σ_3) of 1 tsf has been found to be threshold state of stress separating the two loesses - stiff and plastic. Such a behavior is a consequence of failure in the internal loess structure exposed to a corresponding stress state. This value of 1 tsf is obtained by interpolating stress-strain triaxial curves. In the case of $\sigma_3 \leq 1$ tsf, it behaves like a stiff material while for $\sigma_3 > 1$ tsf its behavior is plastic.

Examination of this unique bahavior suggests that during loading there is a partial, local or complete break of internal links. The attached particles move away from each other, sliding, rolling, rotating and better arranging themselves, thus densifying the entire mass. The original coherent medium changes into a granular mass as consolidation pressure approaches the threshold stress state. Under these conditions the inherent, primary cohesion is highly weakened, and practically exhausted. By a further increase of the stress state (over the threshold stress state), the hydrocolloidal links are restored again. But their effect is completely different. Its behavior is marked plastic now and the strength grows with the loading intensity. This phenomenal behavior of loess makes it difficult to develop a simple constitutive law.

6 CONSTITUTIVE LAW

The stress-strain responses observed in CTC tests suggest the material to behave elasto-plastically from the start of loading. Two types of constitutive models have been developed: (1) cap model and (2) hyperbolic model. Details of the two models are given. The hyperbolic model is then used in the illustrated soil-structure interaction problem.

7 THE CAP MODEL

The cap model consists of an upper yield surface $F_1 = F_1(J_1, J_{2D})$ and a moving yield surface $F_2 = F_2(J_1, J_{2D}, \epsilon_v^p)$. Here J_1 is the first invariant of the stress tensor, J_{2D} the second invariant of the stress deviator tensor and ϵ_v^p the accumulated volumetric plastic strain. Motion of the cap is governed by the hardening rule, $X = X(\epsilon_v^p)$. The perfectly plastic upper yield surface is described considering the condition of 15% axial strain as the failure criterion in a triaxial compression test. The Drucker-Prager yield surface fits well with the test data, with $\alpha = 0.22$ and $\beta = 0.7$ tsf in the form

$$F_1 = (J_{2D})^{1/2} + \alpha J_1 + \beta = 1 \quad (1)$$

The moving cap has the form, Fig. 4.

$$F_2 = \frac{1}{(X-C)^2}\,[R^2 J_{2D} + (J_1 - C)^2] = 1 \quad (2)$$

for isotropic hardening. The parameters X and C represent the J_1 distances to the cap intersections with the axis and the upper yield surface respectively. R is the aspect ratio of the cap. Loess exhibited a more pronounced strain hardening effect with increased confining pressure. To account for this, R is written as R = R(C) to increase with C. A second order polynomial is chosen to define R as

$$R = R_1 + R_2C + R_3C^2 \tag{3}$$

with constants evaluated as R_1 = 0.279, R_2 = -0.010 tsf^{-1} and R_3 = -0.034 tsf^{-2}. The parameters in the hardening rule

$$X = \frac{1}{D} \ln\left(1 + \frac{(\epsilon_v^P)}{W}\right) + Z \tag{4}$$

are then determined by curve fitting. The tension cut off is where J_1 = 0.77 tsf. The model developed is verified for prediction of static laboratory test results by numerically integrating the model over the triaxial compression and the comparison of predicted and observed test results are shown in Fig. 5. Very good agreement is seen between the model and experiment.

8 HYPERBOLIC MODEL

A typical hyperbolic model is shown in Fig. 6. The stress-strain responses observed in CTC tests, Fig. 2 suggest fitting a hyperbolic model type of constitutive law to loess. The deviator-stress at failure $(\sigma_1 - \sigma_3)_f$, determined using Mohr's circle is related to the asymtote of the stress strain curve, shown in Fig. 6, as

$$(\sigma_1 - \sigma_3)_f = R_f(\sigma_1 - \sigma_3)_u \tag{5}$$

$$\text{where } (\sigma_1 - \sigma_3)_f = \frac{2c \cos\emptyset + 2\sigma_3 \sin\emptyset}{1 + \sin\emptyset} \tag{6}$$

The tangent modulus E_t can then be expressed in terms of initial modulus E_i as

$$E_t = E_i(1 - \lambda)^2 \tag{7}$$

where

$$\lambda = \frac{R_f(\sigma_1 - \sigma_3)(1-\sin\emptyset)}{2(c \cos\emptyset + \sigma_3\sin\emptyset)} \tag{8}$$

and

$$E_i = KP_a(\sigma_3/P_a)^n \tag{9}$$

Here P_a = atmospheric pressure, n = exponent determining the relation of rate of variation of E_i with σ_3 and k = material constant.

From the transformed hyperbolic radial strain relationship, the tangent poisson's ratio is determined as

$$\gamma_t = \frac{(G - F \log(\sigma_3/P_a)}{(1 - A)^2} \tag{10}$$

where

$$A = \frac{(\sigma_1 - \sigma_3)^D}{E_i(1 - \lambda)^2} \tag{11}$$

$$D = \frac{(\epsilon_3/\epsilon_1) - f}{\epsilon_3} \tag{12}$$

and f = γ_i = initial poisson's ratio.

Parameters for the hyperbolic model are determined from the laboratory test data as

R_F = 0.85   n = 0.566
C = 0.45 tsf   G = 0.311
P_a = 1.058 tsf   F = 0.358
Ø = 29°   D = 1.755
K = 315.5

9 SHEAR STIFFNESS

The shear stiffness, K_{ss}, is determined using the direct-shear test with half of the box filled with soil material and the other half pile material. K_{ss} is defined as the derivative of shear stress with respect to shear strain and can be expressed as

$$K_{ss} = (1-\xi)^2K_i \tag{13}$$

where K_i = initial shear stiffness expressed as

$$K_i = K_j\gamma_w(\sigma_n/P_a)^n \tag{14}$$

Here K_i = K_j = parameter of hyperbolic shear stiffness, γ_w = unit weight of water and n = exponent determining the relation of rate of variation of K_i with σ_n and

$$\xi = \frac{R_f\tau_a}{C_a + \sigma_n \tan\emptyset_a} \tag{15}$$

The parameters are determined from the laboratory test results as

$C_a = 0.255$ tsf $n = 0.01$
$\emptyset = 29°$ $R_f = 0.95$
$K_j = 7800$

10 APPLICATION

An illustrative example like determining the ultimate axial load capacity of a pile, using the hyperbolic model developed is presented. There are several methods to determine the ultimate axial load capacity of a pile. The finite element method is one of them. This method mostly overcomes the various limitations other methods experience. A four node quadralateral isoparametric element, Fig. 7, is used in the finite element displacement approach. This numerical method permits considering the smear zone differently from the unsmeared soil around the pile, as shown in Fig. 8. The element equilibrium equation can be expressed as

$$[K]\ \{q\} = \{Q\}_x + \{Q\}_T \tag{16}$$

where [K] = element stiffness matrix, {q} = element nodal displacement vector due to body forces and $\{Q\}_T$ = element modal vector due to prescribed surface tractions.

When a pile foundation is axially loaded, the pile-soil interaction is accounted for by introducing an interface finite element. The element equilibrium equation can be written in the form

$$[K]\ \{q\} = 0 \tag{17}$$

where [K] = element stiffness matrix. It can be expressed as

$$[K] = \begin{bmatrix} k_{ss} & k_{sn} \\ k_{ns} & k_{nn} \end{bmatrix} \tag{18}$$

Here, k_{ss} = stiffness coefficient in shear direction, k_{nn} = stiffness coefficient in normal direction, $k_{sn} = k_{ns}$ = stiffness coefficient in cross-direction usually assumed to be zero.

The element stiffness matrix is assembled by adding the contribution of the solid element and the interface element using the direct stiffness approach, into the global equation and modified by the prescribed boundary conditions. The final equation is in the form

$$[\bar{k}]\ \{\bar{r}\} = \{\bar{R}\} \tag{19}$$

where $[\bar{k}]$ = modified assemblage stiffness matrix, $\{\bar{r}\}$ = modified assemblage nodal displacements vector and $\{\bar{R}\}$ = modified assemblage load vector. The solution of Eqn. 19 gives the displacements at nodal points of the discretized body.

11 PILE CAPACITY

A typical boundary used for finite element calculation is shown in Fig. 8. To account for the effect of pile installation on the soil around the pile (the smear zone), a different value of K_s is assigned to the soil surrounding the pile. K_{s1}, coefficient of horizontal pressure of the unsmeared soil is taken equal to $k_o = (1-\sin\emptyset)$. K_{s2} for the smeared zone is varied to reflect different degrees of disturbance. Experimental and numerically simulated load settlement curves for two cases PT1 and PT2 are shown in Figs. 9 and 10. The experimental values are high in both cases. They are not in agreement with the predicted values. This results in describing the constitutive model with two hyperbolic models.

12 MODIFIED HYPERBOLIC MODEL

Since the threshold stress state for soil tested is about 1.0 tsf, the soil is divided into two: (1) soil with horizontal in-situ stress less than or equal to 1.0 tsf. (2) soil with horizontal in-situ stress greater than 1.0 tsf.

The hyperbolic model parameters are evaluated accordingly. From a parametric study, it is found that two factors mainly control the model K and R_F. K controls the steepness of the curve particularly at initial strain and R_F controls the deviator-stress at failure. Keeping K_{s1} constant = 0.5, K_{s2} and K are varied to get a good approximation of load settlement curve. Load-settlement curve at $K_{s2} = 0.945$ and K = 1500 is found to be in good agreement with PT1 experimental curve. For comparison, results assuming uniform "stiff" soil are analyzed which indicates piles fail faster than when using two type stress-strain behavior.

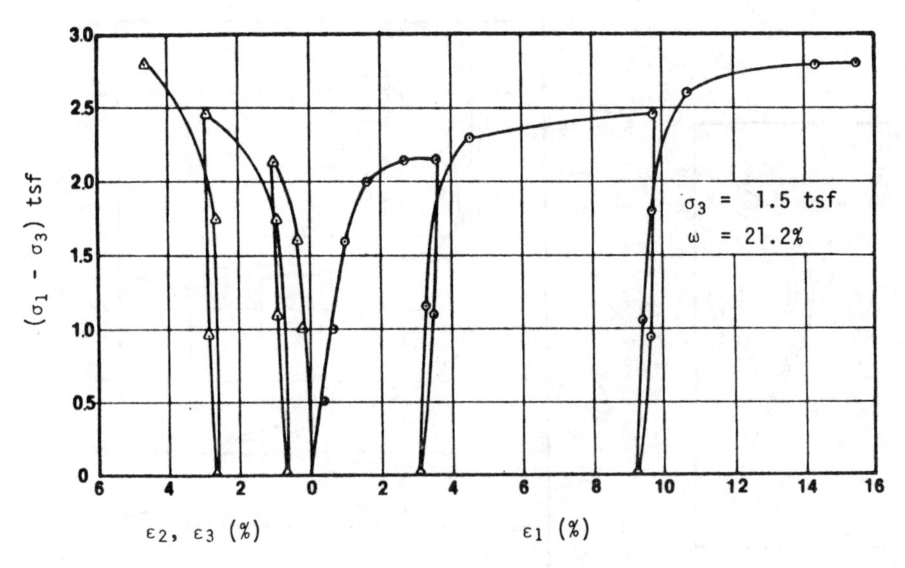

Figure 1. A typical stress-strain curve for CU triaxial compression(σ_3=1.5 tsf)

13 CONCLUSIONS

The mechanical properties of loess are governed mostly by consolidation pressure. The critical state of stress separating the two "loesses," namely stiff and plastic, is about 1 tsf. Shear strength of loess is greatly dependent on moisture content. The finite element method gives results comparable with experimental values. The introduction of interface element has enhanced the quality of results. However, the choice of a proper constitutive law dictates the perfection and reliability of the numerical method. Selection of a constitutive law appropriate to the field conditions is greatly recommended.

14 REFERENCES

Bozininovic, D. and et al. 1985. Stress-strain strength characteristics of Macroporous loess in Belgrade. Proceedings of the International Conference on Soil Mechanics and Foundation Engineering. San Francisco, California.

Desai, C.S. 1974. Numerical design analysis for piles in sands. Journal of Geotechnical Engineering, ASCE, Vol. 100, No. GT6.

Duncan, J.M. and C.Y. Chang 1970. Nonlinear analysis of stress and strain in soils. Journal of Soil Mechanics and Foundation Engineering, ASCE, Vol. 96, No. 5.

Indra, S.H.Harahap 1984. Calculation of ultimate capacity of an axially loaded single pile in loess. M.S. thesis, Ohio University. USA.

Goodman, R.E., R.L. Taylor and T.L. Brekke 1968. Model for the mechanics of jointed rock. Journal of Soil Mechanics and Foundation Engineering, ASCE, Vol. 94, No. SM3.

Kuppasamy, T. and C.S. Desai 1978. A computer code for axially and laterally loaded piles. Virginia Tech Report 78-2.

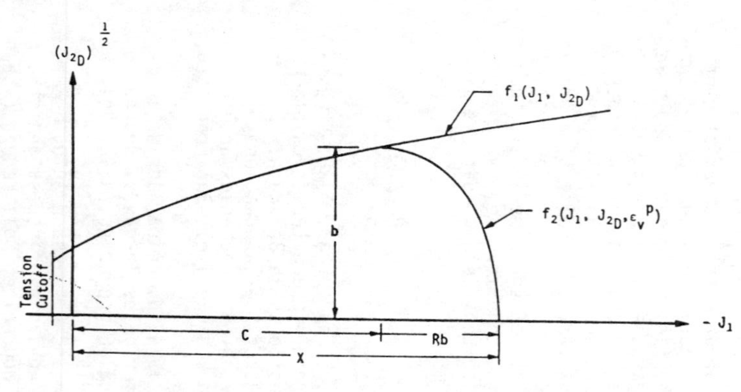

Figure 2. A set of stress-strain curves for CTC.

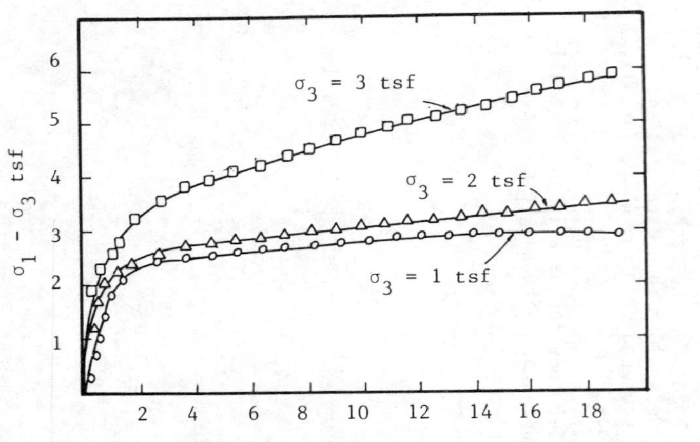

Figure 4. Details of a cap model.

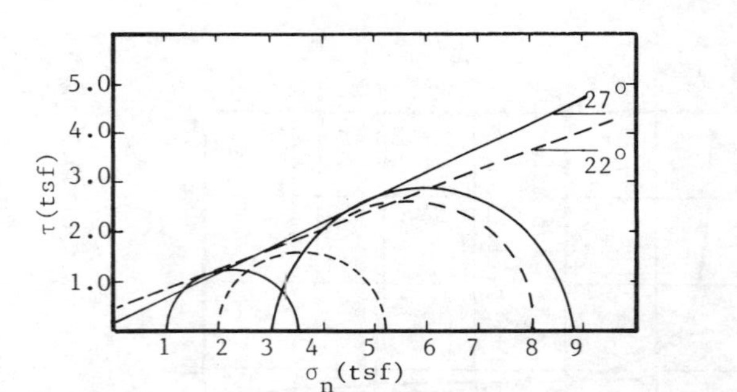

Figure 3. Mohr's envelops for 10 and 18% strain.

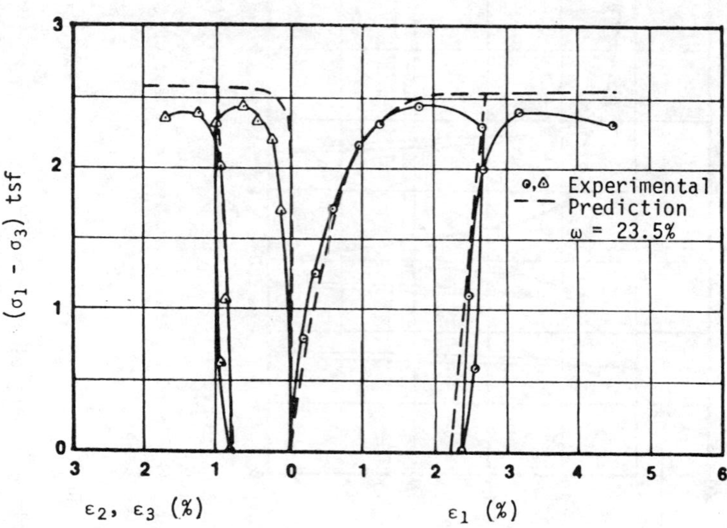

Figure 5. Comparison of observed and predicted stress-strain responses for CTC test.

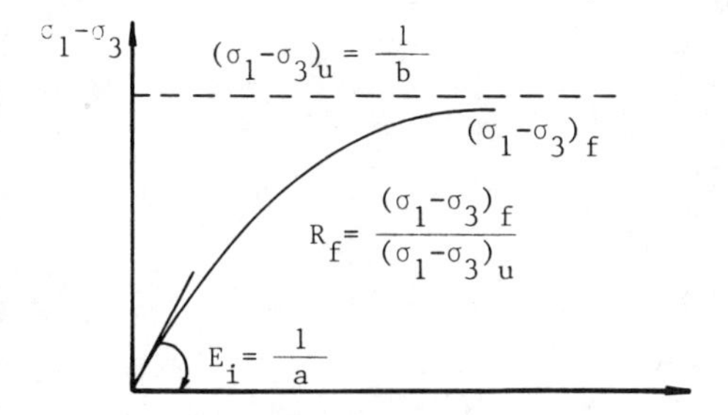

Figure 6. Details of a hyperbolic model.

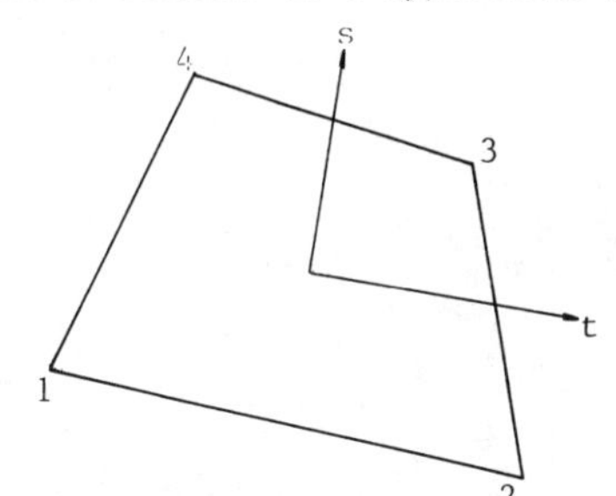

Figure 7. Quadrilateral Isoparametric element.

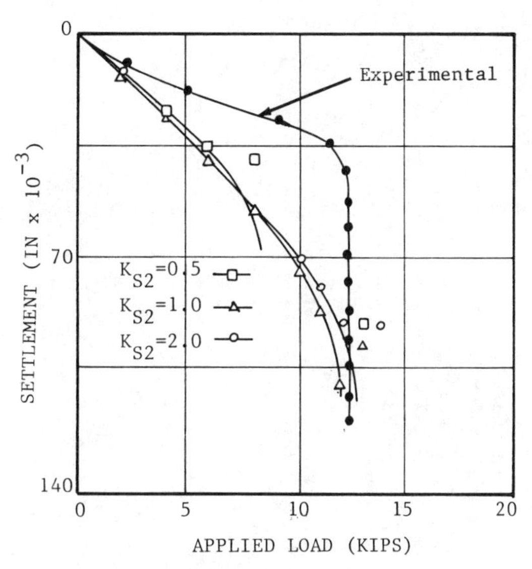

Figure 9. Comparison of load-settlement curves between experimental and finite element method. Pile #1

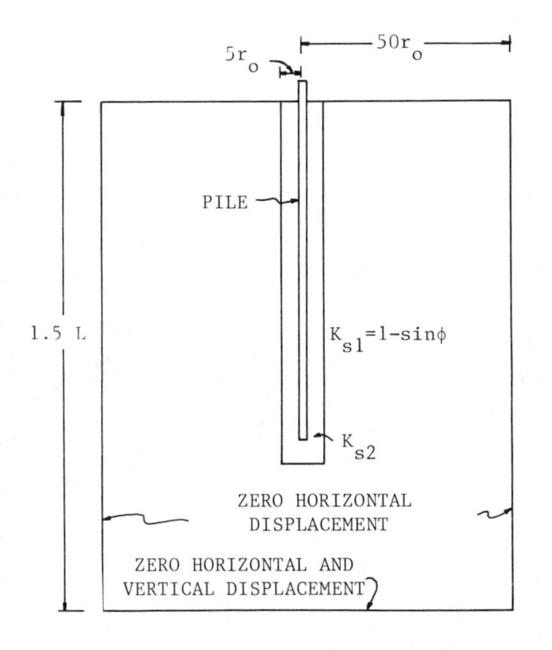

Figure 8. Typical boundary for finite element analysis.

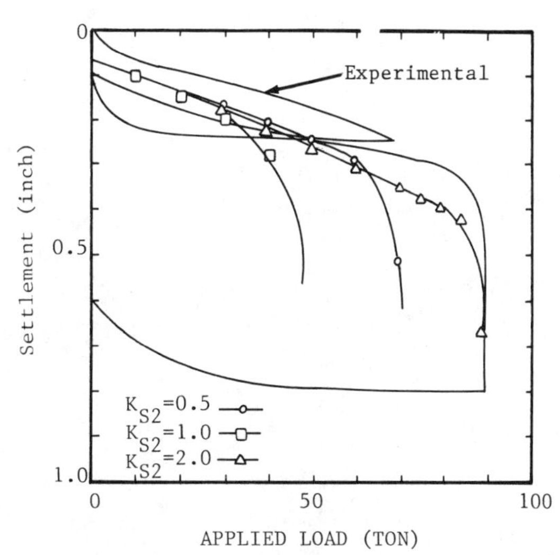

Figure 10. Comparison of load settlement curves between experimental and finite element method. Pile #2. (PT2)

Numerical Methods in Geomechanics (Innsbruck 1988), Swoboda (ed.)
© 1988 Balkema, Rotterdam. ISBN 90 6191 809 X

Constitutive models constrained by the entropy maximum principle

P.Scharle
Ministry for Building and Urban Development, Budapest, Hungary

ABSTRACT: The thermodynamic approach used recently for modelling the behaviour of the granular media has brought into focus the concepts of thermic and statistic entropy. The purpose of the paper is to present a short overview of the essential ideas and to outline the concept of mesoentropy. Having made a distinction between the microentropy and the mesoentropy the principle of maximum entropy proves to be a constitutive condition. The idea finds its precedent in the theory of radial stress distribution postulated and elaborated by Fröhlich (1934).

1. THERMODYNAMIC CONSIDERATIONS

The use of concepts first introduced in thermostatics and interpreted later in the theory of particulate materials has a history of decades. For twenty years the research on this field broadens continuously. Among the concepts taken over by analogy the entropy has got a key role. Many efforts have resulted in different definitions (both of structural, e.g. Mogami, 1965, and of phenomenologic character, e.g. Moroto, 1976). The connections between these definitions, the possible interpretations of the traditional thermic entropy and that of the statistic entropy taken from the information theory became a common subject of wide investigations (Gudehus, 1968). Many results proved to be useful even from practical point of view, too, however no general attitude about the merits and advantages of the thermodynamic approach has been established.

Scrutinies have been made, for example, to find (or define) an appropriate intensive field, dual to the entropy (which is of extensive nature). A short overview (Scharle, 1987) shows, that the idea of the geotechnical variables (as introduced by Nova and Hueckel, 1980), the concept of the internal variables (applied by Houlsby, 1982) and other steps were essential but have not resulted in a comprehensive field theory up to now. Several authors have arrived at a deeper understanding of some well known phenomena in soil mechanics using the concept of statistic entropy and its maximum principle (Feda, 1971, Jowitt, Munro, 1975, Brown, 1978, Lőrincz, 1986). However, it remained a question, which variable of the possible ones like void ratio, coordination number, grain size, contact normal density or any other, having a probabilistic distribution over the domain investigated should be selected as fundamental for an overall field theory in the frames of mechanics. Relationships among the different entropies were elaborated (Moroto, 1980, Satake, 1984) but the analogy to the thermic entropy remained somewhat ambiguous.

Some of the difficulties referred to up to now can be solved by separating two kinds of entropies. To find reasons for the idea we have to remind the reader of the characteristic feature of the granular materials. For this sort of materials a common convention is put on when defining continuous fields for analytic models: the postulation of distinction between the micro- and mesodomain (Axelrad, 1978, Kestin, Bataille, 1980). It seems to be possible to extend this distinction for the concept of the entropy. The microentropy is of thermic nature, strictly connected with the internal energy states at the atomic level and, by this way with the statistic entropy in the traditional (Boltzmann) sense. The mesoentropy is to be defined over the mesodomain, it refers to the randomness of the granules. Making distinction between these entropies is a decision of constitutive character, emphasizing the intrinsic connection between the structure of the material considered and the field variable chosen to describe the mechanical behaviour.

2. THE USE OF THE ENTROPY MAXIMUM PRINCIPLE

Having specified the micro- and mesoentropies we have lost the direct analogy between the plastic work and the heat loss expressed by the increment of entropy and temperature. At the same time we are set free from looking for sophisticated arguments about the analogy between the dissipation of energy and the uncertainty of particle configuration. Transport of heat is allowed without the necessity of rearrangements in the grain configuration (while the microentropy may increase). The random motion of the granules can be connected with the overall stress field inhomogeneities rather than with the heat loss (the magnitude of which being often unknown or neglected).

The use of statistic entropy as a "bulk parameter" for any probabilistic distribution defined over the mesodomain, rather than a kind of extensive variable, does not impel to give up the idea of introducing new dual variables into the field equations. Dynamic and kinematic fields corresponding with the mechanisms like rolling, sliding, break of bonds etc. are to be defined. Simultaneously, there arise the lesson to interpret the maximum entropy principle. This principle was formulated by Jaynes as follows: "the least prejudiced or biased assignement of the probabilities is that which maximizes S (called entropy) subject to the given information" (Tribus, 1961). In other words, this principle makes possible to chose the least biased distribution from those allowed by the prescribed conditions. From this point of view, another interpretation of the constitutive character of the mesoentropy can be given: in the frames of mechanics the maximum entropy principle can be used to find a structural field among all those satisfying the equilibrium and compatibility equations.

3. FORMULATION OF THE FIELD PROBLEM

The introduction of mesoentropy brings into focus the statistic variables defined over the mesodomain (e.g. coordination number, void ratio, fabric tensors, contact normal distribution—Gudehus, 1968, Oda, 1978, Satake, 1983) when we try to formulate the governing equations of a geomechanical field problem. To be more descriptive consider briefly the process of deformation from this point of view.

The configuration of the granular material in the initial (equilibrium) state of stress can be characterized by a probabilistic distribution of elementary cells, which are known and classified for regular spheres (Filep, 1937), and can be described by simplifying assumptions in more general cases as well. The frequency of the cells may be very different (the assumption of the equal initial probability, made by several workers, seems to be unrealistic), however their distribution can be estimated using the maximum principle applied jointly with the condition of equilibrium for gravitational forces. Known statistic averages of void ratio, coordination number etc. may constitute further conditions.

The initial configuration is able to bear some changes of the stress field without rearrangement; if we neglect the elastic deformation of the granules it behaves like a rigid body. The threshold level of the change in the stress state is determined by the resistance of the contact microregions. Exceeding this level rearrangement begins, resulting in changes of all distributions considered before. The region of the rearrangement depends on the stress increment related to the rigid body threshold level. Thus, the field problem adequately formulated for particulate materials should have a characteristic feature: moving boundaries and regions of variable range are to be considered. The resulting configuration must be quasirigid again with respect to the new stress field and the entropy maximum principle with the simultaneous conditions of equilibrium and compatibility is to be satisfied.

In this formulation we must not count with the traditional definitions of stress and strain. In accordance with the concept of mesoentropy, statistic averages of stress and strain are to be interpreted over the mesodomain. Among several known possibilities (Christoffersen et al., 1981, Kanatani, 1983) the stress and strain tensors defined by Satake (1983) using statistic and graph approach foreshadow and make probable an appropriate selection of dual variables. These tensors reflect the key role of the contact normals and characterize both the induced anisotropy and porosity of the grain configuration. Other interpretations may prove to be appropriate, too. For the case of quasistatic deformation of granular soils the separation of the dynamic internal variables (like the resolution of the macro-stress tensor) may prove to be expedient (Cundall, Strack, 1983). The components of the forces acting on a single particle can be resolved into four subsets, first of them corresponding to the average hydrostatic stress in the particle, the second to the anisotropic stress distribution due to fabric, the third corresponding to the tendency of contacts to slide (thus relating to the dissipation) and the fourth to excentricity of forces causing the buckling of particle chains. This resolution has been justified with numerical experimental data (Cundall, Drescher, Strack, 1982), and stimulated further considerations (Papadopoulos, 1986). It seems to be possible to select appropriate dual (kinematic) fields and for the quasiconservative and dissipative parts free energy and dissipation potentials can be constructed.

Having had selected the dual extensive and intensive fields arises the next problem of describing their relationship. This constitutive relationship will be (must be) less deterministic than that of the classic stress and strain, because the maximum entropy principle was implemented into the model.

4. REMARKS AND CONCLUSIONS

The implementation of the maximum entropy principle as a constitutive condition may seem to be unusual. Nevertheless, we can refer to an interesting precedent, connected with the stress distribution problem in soil masses. The radial stress distribution theory elaborated by Fröhlich (1934) and others (Ohde, 1939, Holl, 1940) results in stress fields satisfying the field and boundary equilibrium equations. In this theory the compatibility condition can not be satisfied but giving up the assumption of homogeneity. Keeping constant the Poisson's ratio different distributions of the Young modulus are required, and these distributions depend on the "concentration index" (Harr, 1966). The condition of radial distribution, as introduced by Fröhlich, is of constitutive character: it reflects the lack of tensile strength. Consequently, some of the traditional constitutive assumptions are to be abandoned (in Fröhlich's theory the distribution of the modulus of elasticity is determined by the concentration index and Poisson's ratio; it is clear that other sets of conditions can be derived, too).

All those comments objecting the considerations outlined up to now are right and kept in mind which draw the attention to the threatening problems in defining new material parameters, contriving new experimental methods (both laboratory and field ones) and numerical techniques. The number of parameters determining the initial state increases step by step loading process is to be considered, nonlinear programming procedures are to be used, etc. On the other hand, however, elaboration of the statistic models can result in more rigorous description of the mechanical behaviour. For example, in the statistic-thermodynamic models one can interpret better

- the development and motion of rigid regions and those of, where particle rearrangement occurs (generalized "limit depth" formulations can be established);
- the anisotropic hardening, shake down phenomena and their dependence on the change of the stress state;
- the effect of the internal kinematic constraints (inlets, reinforcements) increasing the strength of the soil mass.

Thus, the difficulties seem to be comparable to the advantages. Elaboration of the statistic models can decrease the gap between the coarse physical assumptions and delicate computational techniques often compounded inconsistently in our days.

5. ACKNOWLEDGEMENTS

The considerations presented have been developed in the frames of a broader research work about the behaviour of multiphase dissipative systems. The author is indebted to I. Bojtár, G. Páti, A. Szijártó and V. Szabó A. both for their kind remarks and critics given during the preparation of this paper.

REFERENCES

Axelrad, D. R.: Micromechanics of solids, Elsevier, 1978

Brown, C. B.: The use of maximum entropy in the characterization of granular materials, in Continuum Mechanical and Statistical Approaches in the Mechanics of Granular Materials (ed. Cowin, S. C., Satake, M.), Gakujutsu Bunken Fukyu-kai, 1978, 98–108

Cundall, P. A., Drescher, A., Strack, O. D. L.: Numerical experiments on granular assemblies; measurements and observations, in Deformation and failure of granular materials (ed. by Vermeer, P. A., Luger, H. J.), Balkema, 1982, 355–370

Cundall, P. A., Strack, O. D. L.: Modelling of microscopic mechanisms in granular materials, in Mechanics of granular materials (ed. by Jenkins, J. T., Satake, M.), Elsevier, 1983

Christoffersen, J., Mehrabadi, M. M., Nemat-Nasser, S.: A micromechanical description of granular material behavior, J. Appl. Mech., 1981, 339–344

Feda, J.: Struktura partikulárních látek a Boltzmannův princip, Stavebnícky casopis SAV, 1971, 310–330

Filep L.: Egyenlő gömbökből álló halmazok, Vízügyi Közlemények 19, 1937, 128–144

Fröhlich, O. K.: Druckverteilung im Baugrunde, Springer, 1934

Gudehus, G.: Gedanken zur statistischen Boden mechanik, Der Bauingenieur, 43 (1968), 9, 320–326

Harr, M. E.: Foundations of theoretical soil mechanics, McGraw-Hill, 1966

Holl, D. L.: Stress transmission in earths, Proc. Highway Research Board, 1940, 20

Houlsby, G. T.: A derivation of the small-strain incremental theory of plasticity from thermomechanics, in Deformation and failure of granular materials, (ed. by Vermeer, P. A., Luger, H. J.), Balkema, 1982. 109–118

Jowitt, P. W., Munro, J.: The influence of void distribution and entropy on the engineering properties of granular media, in Applications of Statistics and Probability in Soil and Structural Engineering, Aachen, 1975, 365–385

Kanatani, K.: Mechanical properties of ideal granular materials, in Mechanics of granular materials, (ed. by Jenkins, J. T., Satake, M.), Elsevier, 1983, 137–149

Kestin, J., Bataille, K.: Thermodynamics of solids, in Continuum models of discrete systems, University of Waterloo, 1980, 99–147

Lőrincz J.: A talajok szétosztályozódásáról, MTA Műszaki Mechanikai Tanszéki Kutatócsoport IV. Tudományos Ülésszaka, Tanulmányok, 1986, 60–68

Mogami, T.: A statistical theory of mechanics of granular materials, Journ. Fac. Eng., University of Tokyo, 1965, B, 28, 2, 65–79

Moroto, N.: A new parameter to measure degree of shear deformation of granular material in triaxial compression tests, Soils and Foundations, 1976, 4, 1–10

Moroto, N.: Shearing deformation of granular materials such as sand, Hachinohe Institute of Technology, Aomori, Japan, 1980

Nova, R., Hueckel, T.: A "geotechnical" stress variables approach to cyclic behaviour of soils, in Soils under cyclic and transient loading (ed. Pande, G. N., Zienkiewicz, O. C.), Balkema, 1980, 301–314

Oda, M.: Significance of fabric in granular mechanics, in Continuum Mechanical and Statistical Approaches in the Mechanics of Granular Materials (ed. Cowin, S. C., Satake, M.), Gakujutsu Bunken Fukyu-kai, 1978, 7–26

Ohde, J.: Zur Theorie der Druckverteilung im Baugrund, Der Bauingenieur, 1939, 33/34

Papadopoulos, J. M.: Incremental deformation of an irregular assembly of particles in compressive contact, M. I. T, 1986 Ph.D. Thesis

Satake, M.: Fundamental quantities in the graph approach to granular materials, in Mechanics of granular materials, (ed. by Jenkins, J. T., Satake, M.), Elsevier, 1983, 9–19

Satake, M.: A new entropy theory for granular materials, 8th Congr. of Chemical Engineering (CHISA), Praha, 1984, F5.2

Scharle, P.: Thermodynamic considerations for some modelling problems in soil mechanics, in Results and prospects of joint research in the Academies of Sciences of socialist countries, Prague, 1987

Tribus, M.: Thermostatics and thermodynamics, Van Nostrand, 1961

Numerical Methods in Geomechanics (Innsbruck 1988), Swoboda (ed.)
© 1988 Balkema, Rotterdam. ISBN 90 6191 809 X

Geomechanical applications of fully coupled, transient thermoelasticity

J.P.Carter & J.R.Booker
University of Sydney, Australia

ABSTRACT: In this paper the equations for fully coupled, transient thermoelastic deformations are presented and a numerical procedure is suggested for the approximate solution of these equations. Solutions are obtained for several thermomechanical problems in geomechanics and it is concluded that the application of a fully coupled theory provides solutions which, for most materials, are not very different to those obtained from the more traditional, semi-coupled theory.

1. INTRODUCTION

When a solid body is heated or cooled at its boundaries the interior will undergo time-dependent changes of temperature. These temperature changes induce thermal strains and if the body is restrained then transient stress changes will also be induced. When the same body is strained by the application of boundary loadings or body forces, or by thermally induced stresses, the possibility also arises of the strains inducing temperature changes in the body. The first of these effects, viz the influence of temperature on strains and stresses, has received much attention in the literature, e.g. Nowacki (1962), while the latter has had only scant attention, e.g. Smith, Carter and Booker, (1987); Carter and Booker, (1987).

It has been traditional for thermoelastic analyses to be carried out in two stages. First, the temperature distribution is determined by solving either the Laplace equation for steady state conditions, or the diffusion equation for the transient heat flow problem. In the second stage of the analysis the computed temperature distribution is used to calculate the response of the elastic body to the imposed thermal gradients and any other applied forces. It has generally been considered that this approximate, semi-coupled approach is satisfactory when applied to materials like metals.

In geotechnical engineering most analysis is restricted to isothermal conditions, however, a number of important problems are encountered that require an analysis of thermal effects. Examples include the disposal of radioactive waste and "hot rock" mining. For these cases knowledge of the thermal properties of the geological materials is less precise than for metals, and at this stage it is difficult to be sure whether the traditional, semi-coupled approach to thermal analysis will be adequate. For this reason a fully coupled approach to the problem has been adopted in this paper. The governing equations for the fully coupled problem are developed and a finite element solution scheme is suggested. Numerical results are evaluated for one problem for which it is possible to obtain a closed form solution and the accuracy obtainable with the numerical approach is demonstrated. Results are also obtained for a problem typical of those encountered in the storage of hot radioactive waste in a long borehole.

2. GOVERNING EQUATIONS

The physical process of coupled thermoelastic deformation is governed by the following set of equations.

2.1 Equilibrium

In the absence of body forces, equlibrium of the body is expressed, in a cartesian cordinate system, as

$$\partial^T \sigma = 0 \tag{1}$$

in which $\sigma = (\sigma_x, \sigma_y, \sigma_z, \tau_{xy}, \tau_{yz}, \tau_{zx})^T$

and

$$\partial^T = \begin{bmatrix} \partial/\partial x & 0 & 0 & \partial/\partial y & 0 & \partial/\partial z \\ 0 & \partial/\partial y & 0 & \partial/\partial x & \partial/\partial z & 0 \\ 0 & 0 & \partial/\partial z & 0 & \partial/\partial y & \partial/\partial x \end{bmatrix}$$

σ is the vector of stress components, with tensile normal stress regarded as positive. These quantities represent the increase over the initial state of stress due to the applied loading and the temperature change.

2.2 Strain-displacement Relations

These relations may be expressed in matrix form as,

$$\epsilon = \partial u \qquad (2)$$

where

$\epsilon^T = (\epsilon_x, \epsilon_y, \epsilon_z, \gamma_{xy}, \gamma_{yz}, \gamma_{zx})$ is the vector of strain components, and

$u^T = (u_x, u_y, u_z)$ is the vector of displacement components.

2.3 Constitutive Law

For the case of thermoelastic deformations, Hooke's law for an isotropic material may be written as

$$\sigma = -\beta\theta a + D\epsilon \qquad (3)$$

in which $\theta = T - T_0$,

$a^T = (1,1,1,0,0,0)$.

T_0 and T represent the initial and current absolute temperatures, respectively, θ represents the temperature change and D is a matrix of elastic constants, given by

$$D = \begin{bmatrix} \lambda+2G & \lambda & \lambda & 0 & 0 & 0 \\ & \lambda+2G & \lambda & 0 & 0 & 0 \\ & & \lambda+2G & 0 & 0 & 0 \\ & & & G & 0 & 0 \\ & & & 0 & G & 0 \\ \text{symmetric} & & & 0 & 0 & G \end{bmatrix}$$

with λ and G the Lamé modulus and elastic shear modulus of the material, respectively. The thermal stress modulus, β, is given by

$$\beta = E\alpha/(1-2\nu) \qquad (4)$$

where E and ν are Young's modulus and Poisson's ratio of the material and α is the coefficient of linear thermal expansion.

2.4 Conservation of Energy

If complete coupling of the elastic and thermal processes is considered, then the condition for conservation of mechanical and thermal energy at any point can be written in integral form as

$$-\int_0^t \nabla^T h \, dt = \rho c_v \theta + T\beta\epsilon_v \qquad (5)$$

in which

$h^T = (h_x, h_y, h_z)$ = the heat flux,

ρ = the mass density,

c_v = the specific heat (at constant volume),

ϵ_v = the volume strain,

$\nabla^T = (\partial/\partial x, \partial/\partial y, \partial/\partial z)$ = the gradient operator,

and where the thermoelastic deformations of interest occur during the time interval 0 to t.

The term $T\beta\epsilon_v$ accounts for the mechanical energy involved in the coupled process, in the more conventional uncoupled analysis this term is ignored. In the present case the temperature changes considered will be small compared to the ambient absolute temperature ($\theta << T_0$), and thus to sufficient accuracy T may be approximated by the constant T_0 in equation (5).

2.5 Heat Flow

The conduction of heat through the solid body is governed by the Fourier law, viz.

$$h = -k\nabla\theta \qquad (6)$$

in which k is the thermal conductivity.

3. VIRTUAL WORK

In the absence of any increase in body force the equation of virtual work, relating the internal strain energy to the work done by the surface tractions F, can be written as,

$$\int \delta\epsilon^T\sigma \, dV - \int \delta u^T F \, dS = 0 \qquad (7)$$

where δu represents a virtual displacement field consistent with the displacement boundary conditions and $\delta\epsilon$ represents the associated virtual strains.

If the boundary points are either insulated or subjected to a specified temperature, and the virtual temperature field, $\delta\theta$, is consistent with these conditions, then from (5) the following equation must be satisfied:

$$\int \delta\theta[\rho c_v \theta + \beta T_0 \epsilon_v - \int_0^t \nabla^T h \, dt] \, dV = 0. \qquad (8)$$

Substitution of equation (6) into (8) gives

$$\int [\delta\theta (\frac{c_v \rho}{T_0}) \theta + \delta\theta\beta\epsilon_v + \int_0^t (\nabla^T \delta\theta (\frac{k}{T_0}) \nabla\theta \, dt]dV = 0 \quad (9)$$

The incorporation of specified flux or convection boundary conditions into equation (9) is relatively straightforward but is not pursued here.

4. FINITE ELEMENT SOLUTION

An approximate solution of the governing equations presented above may be obtained by application of the finite element method of spatial discretisation.

Suppose that the continuous values of **u** and θ can be represented adequately by their values at selected nodes, i.e.,

$$u = N\delta \quad (10)$$

and

$$\theta = X\varphi \quad (11)$$

where δ and φ are the vectors containing the nodal values and N and X contain the shape functions of the displacements and temperature, respectively.

The strain components now may be written as

$$\epsilon = B\delta \quad (12)$$

where $B = \partial N$,

and the temperature gradients can be expressed as

$$\nabla\theta = Y\varphi \quad (13)$$

where $Y = \partial^T \mathbf{a} X$.

When equations (10) to (12) and the extended form of Hooke's law (3) are substituted into equation (7), it is found that

$$K\delta - L\varphi = \mathbf{b} \quad (14)$$

where

$$K = \int B^T DB \, dV,$$

$$L = \int Z^T \beta X \, dV,$$

$$Z = \mathbf{a}^T B,$$

and **b** is a vector of nodal forces corresponding to the applied surface tractions.

Substitution of equations (10) to (13) into (9) gives

$$- L^T\delta - M\varphi - \int_0^t \Phi\varphi \, dt = 0 \quad (15)$$

where

$$\Phi = \int Y^T (\frac{k}{T_0}) Y \, dV$$

and

$$M = \int X^T (\frac{\rho c_v}{T_0}) X \, dV$$

The solution of equations (14) and (15) is obtained using a time-marching procedure as follows. For a thermoelastic material, the material properties are independent of stress level, temperature and time, so that matrices K, L, M and Φ contain only constant terms. Equation (15) may be written incrementally and for the time interval $t\text{-}\Delta t$ to t, it takes the form,

$$- L^T\Delta\delta - M\Delta\varphi - \int_{t-\Delta t}^{t} \Phi\varphi(t) \, dt = 0 \quad (16)$$

where

$$\Delta\delta = \delta(t) - \delta(t-\Delta t),$$

$$\Delta\varphi = \varphi(t) - \varphi(t-\Delta t).$$

To sufficient accuracy the time integral may be evaluated numerically, so that (16) may be written as

$$- L^T\Delta\delta - M\Delta\varphi - f\Delta t\Phi\Delta\varphi = \Delta t\Phi\varphi(t-\Delta t) \quad (17)$$

where $0 \leqslant f \leqslant 1$. Therefore the incremental solution is determined from the system of equations

$$\begin{bmatrix} K & -L \\ -L^T & -(M + f\Delta t\Phi) \end{bmatrix} \cdot \begin{bmatrix} \Delta\delta \\ \Delta\varphi \end{bmatrix} = \begin{bmatrix} \Delta b \\ \Delta t\Phi\varphi(t-\Delta t) \end{bmatrix} \quad (18)$$

The solution over the full time range of interest is obtained by incrementing the time and successively solving equations of the type given in (18). For stability of the time marching scheme f must be greater than or equal to one-half (Booker and Small, 1975), and often it is convenient to select f = 1.

5. RESULTS

The formulation presented in the previous sections has been encoded in a general purpose finite element program called AFENA (Carter, 1986). This program was used on a microcomputer to calculate the results for the following test problems.

5.1 One-dimensional Transient Heat Flow

This problem is defined in Fig. 1. A layer of thermoelastic material of thickness h is initially at a uniform absolute temperature of T_0 throughout. The base of the layer (z

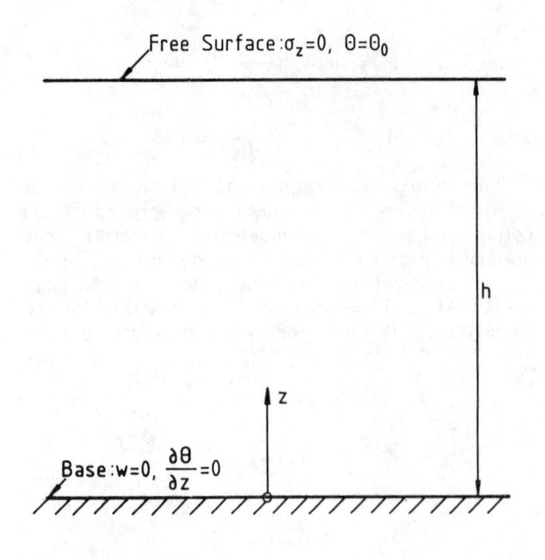

Fig. 1 One-dimensional Problem

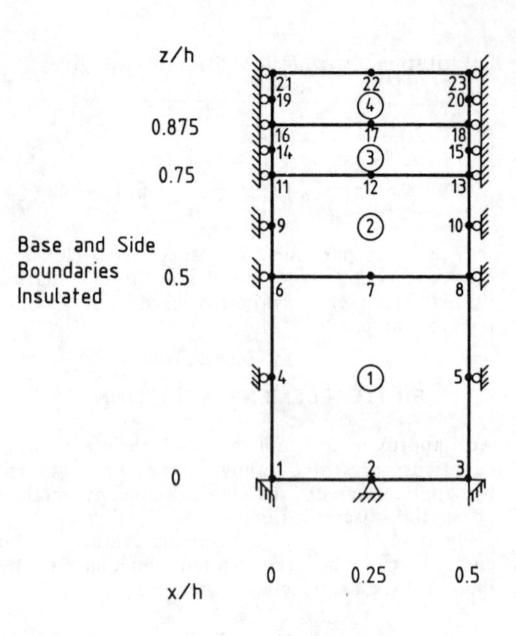

Fig. 2 F.E. Mesh used for One-dimensional Problem

= 0) is rigid (w = 0) and insulated ($\partial\theta/\partial z$ = 0) and for t > 0 the temperature of the unstressed surface is maintained at an absolute value of $T = T_0 + \theta_0$; i.e. the temperature increase at the surface is θ_0. The time-dependent response of the layer has been calculated both analytically and with AFENA, using the mesh depicted in Fig. 2.

It can be shown that the exact solution for this problem is

$$\bar{\theta} = \left(\frac{\theta_0}{s}\right) \frac{\cosh[\mu z]}{\cosh[\mu h]} \tag{19}$$

and

$$\bar{w} = \left(\frac{\theta_0}{\mu s}\right)\left(\frac{\beta}{M}\right)\frac{\sinh[\mu z]}{\cosh[\mu h]} \tag{20}$$

where

$$\mu = \sqrt{(sA/\kappa M)} \tag{21}$$

In the above equations $\bar{\theta}$ and $\bar{w}$ are the Laplace transforms of the temperature change and vertical displacement, respectively. A and M are the constrained moduli corresponding to adiabatic and isothermal conditions, κ is the thermal diffusivity, and s is the Laplace transform varible. For isothermal conditions M is related to Young's modulus, E, and Poisson's ratio, ν, by

$$M = \frac{E(1-\nu)}{(1+\nu)(1-2\nu)} \ . \tag{22}$$

It may be shown that the adiabatic modulus, A, is given by

$$A = M + \beta^2 T_0/(\rho c_V) \tag{23}$$

where β, c_V and ρ have been defined previously. The thermal diffusivity is related to the other material properties by

$$\kappa = k/(\rho c_V) \ . \tag{24}$$

To obtain the quantities θ and w, the transforms $\bar{\theta}$ and $\bar{w}$ must be inverted. The inversion process is relatively easy when carried out numerically using the efficient scheme developed by Talbot (1979). In addition, the case corresponding to A = M is simply one-dimensional diffusion, the solution for which is well known (e.g. Carslaw and Jaeger, 1986).

Analytical solutions for the isochrones of temperature change and for the variation of the surface heave are plotted in Figs. 3 and 4. These curves demonstrate the influence of the parameter A/M on the coupled heat flow; the case where A = M corresponds to uncoupled thermoelastic behaviour. It can be seen that the coupling of the thermal and elastic processes is significant for a material with A/M = 1.5. This is likely to be an upper limit on the modulus ratio for many materials and thus for many practical problems the semi-coupled approach is likely to give answers of sufficient accuracy.

The accuracy of the finite element formulation and coding was validated by comparing the analytical and numerical solutions for a material with A/M = 2. Although this is likely to be an unrealistic choice of the modulus ratio, it does allow a comparison for a case where coupling effects are particularly significant. As can be seen

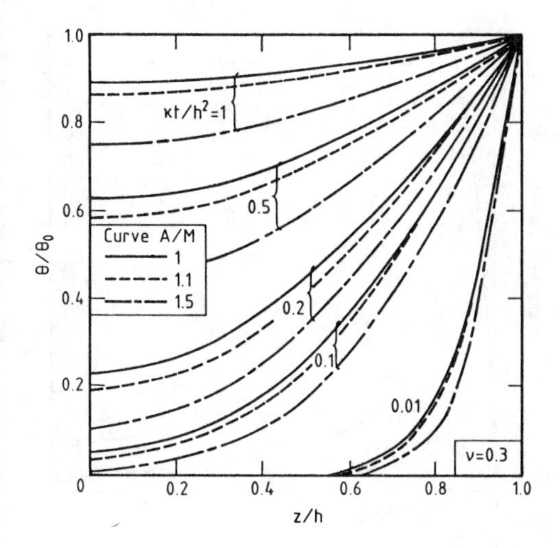

Fig. 3 Temperature Isochrones for One-dimensional Heat Flow

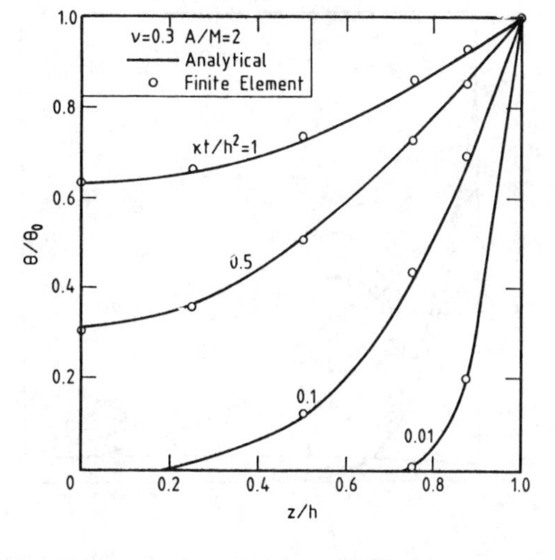

Fig. 5 Comparison of Predicted Temperature Isochrones

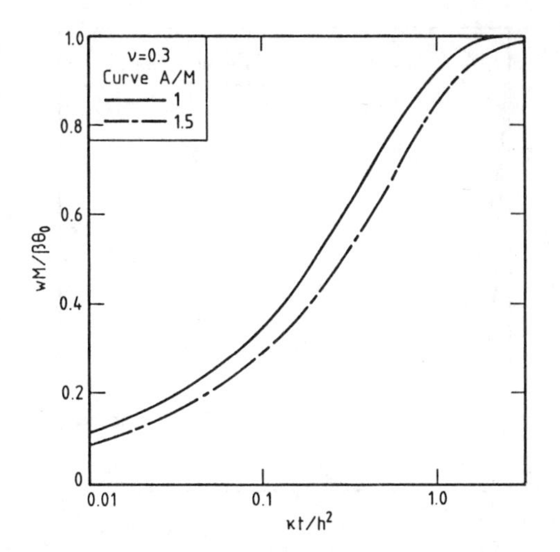

Fig. 4 Surface Heave for One-dimensional Heat Flow

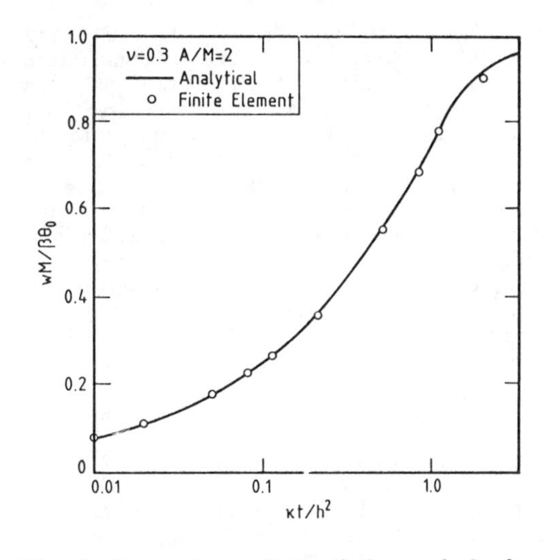

Fig. 6 Comparison of Predictions of Surface Heave

from Figs. 5 and 6, the agreement between the analytical and numerical solutions is very satisfactory.

5.2 Heat Flow from a Borehole

The problem of the disposal of radioactive wastes is currently being tackled by the nuclear industry. Various schemes for the long term storage of these wastes have been proposed and one of these involves the placement of cannisters of waste in deep boreholes in stable geological deposits.

An important aspect of the storage problem concerns the generation of heat by the decaying radioactive waste materials and the effect of this heat on the surrounding rock masses. Heat will be generated for a significant period of time after initial placement and thus predictions of the transient thermal response of the soil or rock masses are required.

A typical problem of the heat generation within a deep cylindrical borehole is considered here. As the hole is deep, conditions of plane strain are assumed and thus the field quantities will vary only in the radial direction. Finite element computations have been made of the transient response of a rock mass for which

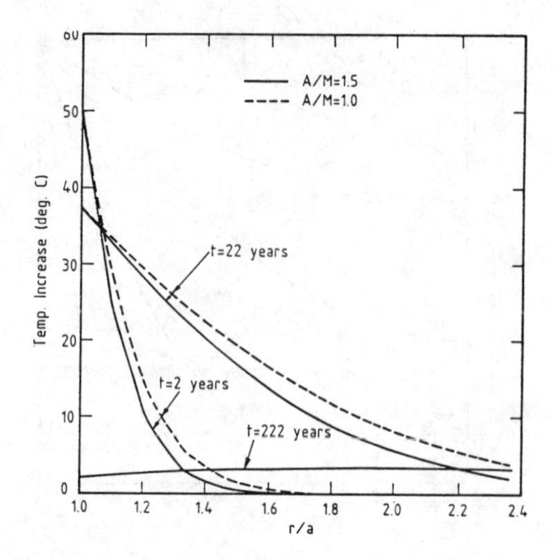

Fig. 7 Temperature Isochrones for Borehole Problem

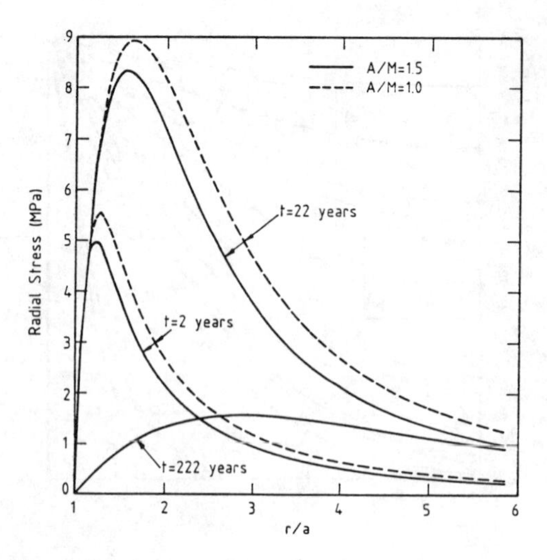

Fig. 8 Isochrones of Radial Stress Change for Borehole Problem

$E = 80{,}000$ MPa, $\nu = 0.3$, $\alpha = 10^{-5}\,°C^{-1}$ and $\kappa = 0.02$ m^2/year. The temperature boundary condition at the wall of the borehole (r = a) is assumed to vary with time. In particular, the temperature is considered to increase initially linearly with time, until it reaches a maximum increase of 50°C after 2 years. Thereafter, the boundary temperature reduces exponentially with time, so that 50 years after initial placement of the waste it will drop to an overall increase of 25°C. These boundary conditions at r = a, can be summarised as follows:

$$\theta_a = 25t, \qquad 0 \leqslant t \leqslant 2 \text{ years}$$

$$= 50 \exp(0.0288 - 0.0144t), \quad t \geqslant 2 \text{ years}$$

Predictions of the transient thermoelastic response of the rock mass are summarised in Figs 7 to 9. Fig. 7 shows the temperature distributions at selected times, viz t = 2, 22 and 222 years after initial placement of the heat source. Isochrones at 2 and 22 years have been plotted for rock masses for which A/M = 1 and 1.5, i.e. predictions using the semi-coupled and fully coupled theories have been plotted. It can be seen that the fully coupled theory predicts a slightly slower flow of heat away from the borehole and into the rock mass. Only one curve has been plotted for t = 222 years, because the differences between predictions for A/M = 1 and 1.5 cannot be detected at the scales used in Figs 7 to 9.

The transient stress changes are shown in Figs 8 and 9, with compression plotted as positive. Fig. 8 indicates that the radial stress changes are entirely compressive and it is interesting that these stresses continue to increase for some time after t = 2 years, even though the boundary temperature is decreasing for t > 2 years. These stress increases occur as the heat is redistributed within the rock mass. Eventually the radial stresses close to the hole decrease as can be seen, for example, at t = 222 years.

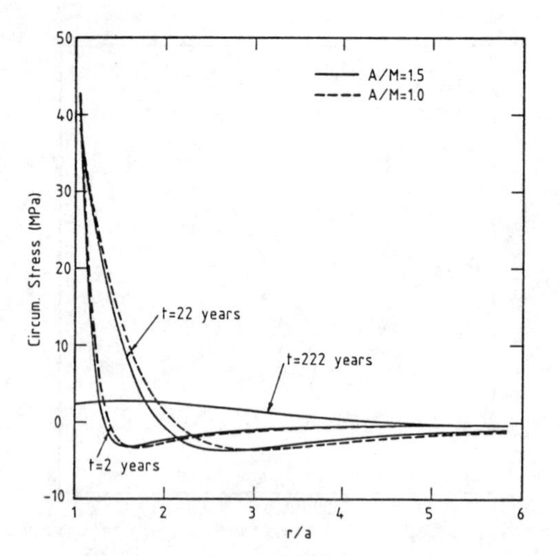

Fig. 9 Isochrones of Circumferential Stress Change for Borehole Problem

The stress changes in the circumferential direction are plotted in Fig. 9 and it is notable that very high compressive stress changes (in excess of 30 MPa) are induced in the rock close to the borehole wall. At larger radii, e.g. for r/a $\geqslant$ 1.5, significant tensile stress changes are induced in the circumferential direction at t = 2 and 22 years. At the later time t = 222 years the circumferential stress changes are relatively small and compressive. Of course, at large times the predicted stress changes become infinitesimal.

It is also interesting to note that the fully coupled theory predicts significant but not large differences in the stress distributions, compared to those obtained using the semi-coupled approach. Indeed, it is likely that for most geological materials A/M < 1.5, and so the semi-coupled theory should adequately predict the thermoelastic response.

6. CONCLUSIONS

The theory of fully coupled thermoelasticity has been presented, together with a numerical solution scheme based on the use of the finite element technique with a time-marching strategy. This procedure has allowed the solution for the full transient response in thermoelastic problems. Solutions obtained numerically have been compared with analytical solutions and the agreement was found to be satisfactory. An examination was made of the significance of full coupling on the predicted response of the thermoelastic material for two initial value problems, viz. a one-dimensional problem involving the raising of the temperature of the surface of a thermoelastic layer, and the raising of the surface temperature of a borehole. It has been demonstrated for both problems that coupling has a significant influence on the rate of heat diffusion throughout the material whenever the adiabatic and isothermal moduli are sufficiently different, say by about 25 per cent or more. It is expected that many natural materials will have ratios of A to M less than 1.25 and so the traditional, semi-coupled theory may provide sufficient accuracy in most practical problems.

ACKNOWLEDGEMENTS

The authors wish to acknowledge the value of discussions with D.W. Smith during the preparation of this paper.

REFERENCES

Booker, J.R. and Small, J.C. (1975), "An Investigation of the Stability of Numerical Solutions of Biot's Equations of Consolidation", Int. J. Solids Structures, Vol. 11, pp. 907-917.

Carslaw, H.S. and Jaeger, J.C. (1986), Conduction of Heat in Solids, (2nd Edn - Paperback), Clarendon Press, Oxford, U.K.

Carter, J.P. (1986), "AFENA - A Finite Element Numerical Algorithm - Users' Manual", School of Civil and Mining Engineering, University of Sydney, Australia.

Carter, J.P. and Booker, J.R. (1987), "Finite Element Analysis of Coupled Thermoelasticity", Proc. 5th Int. Conf. in Australia on Finite Elements, pp. 340-345.

Nowacki, A. (1962), Thermoelasticity, Pergamon Press Limited. International Series of Monographs in Aeronautics and Astronautics.

Smith, D.W., Carter, J.P. and Booker, J.R. (1987), "Numerical Analysis of Linear, Quasi-Static, Coupled, Transient Thermoelasticity", Proc. CTAC'87, Sydney, North-Holland, to appear.

Talbot, A. (1979). "The Accurate Numerical Inversion of Laplace Transforms", J. Instit. Maths. Applics., Vol. 23, pp. 97-120.

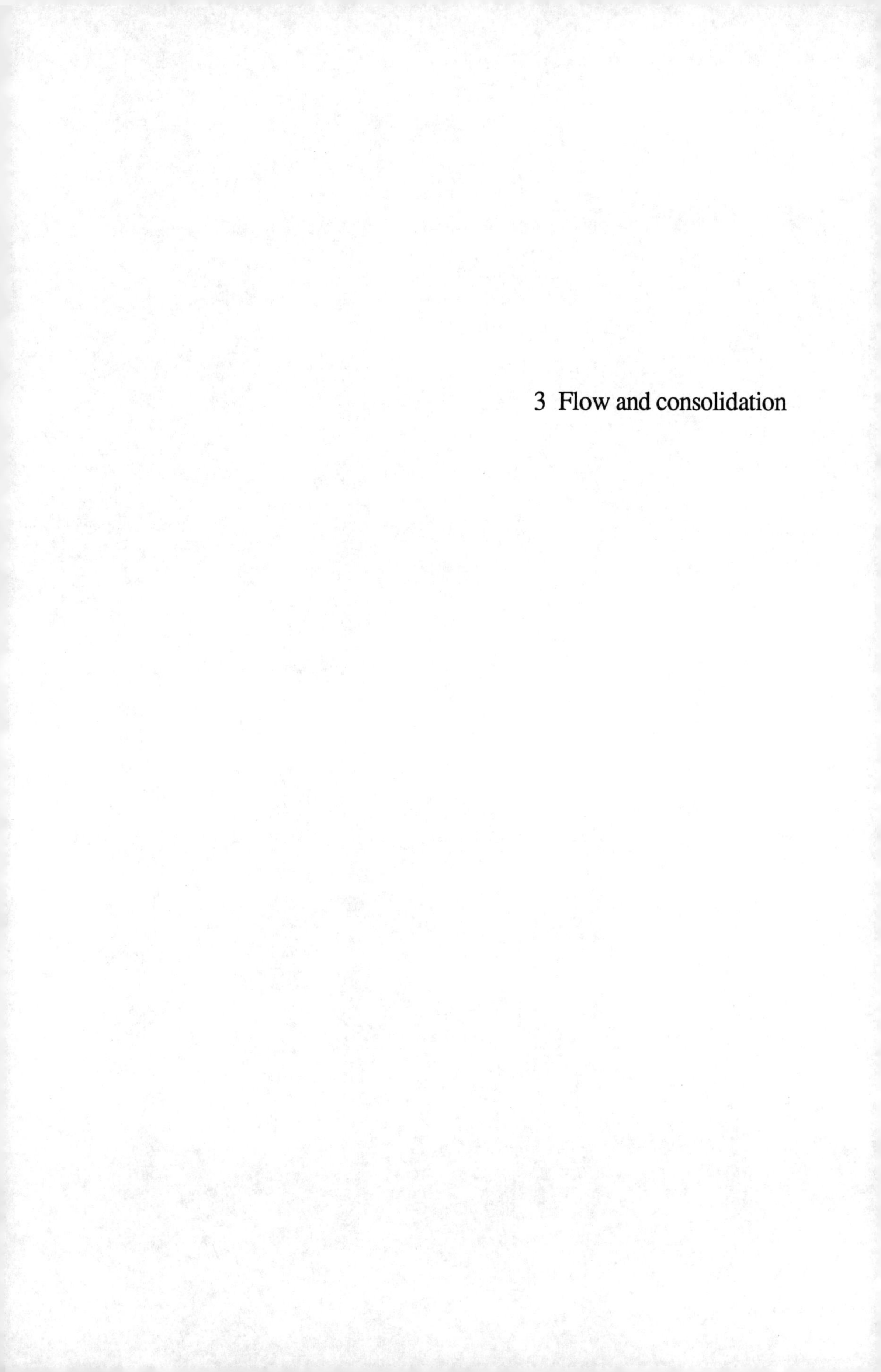

3 Flow and consolidation

Numerical Methods in Geomechanics (Innsbruck 1988), Swoboda (ed.)
© 1988 Balkema, Rotterdam. ISBN 90 6191 809 X

Numerical analysis of soil stability around deep wellbores

H.H.Vaziri
Golder Associates, Seattle, Wash., USA

ABSTRACT: The mechanics of soil failure around deep wellbores are described. Based upon coupled flow and deformation formulations developed by Biot, a methodology has been developed for predicting the extent of soil failure, both in tensile and shear modes, due to the stress relief caused by well drilling and subsequent fluid production. Application of the numerical model to simulate the recovery processes conventionally employed in oil production has revealed that, for specific geometry and formation characteristics, there exists a critical fluid pressure gradient which if exceeded, would result in soil collapse around the well face. The increase in permeability that occurs in the collapsed zone causes a corresponding reduction in fluid pressure gradient to subcritical levels. This results in a temporary stabilization of the soil failure until the fluid pressure gradient exceeds a new critical level commensurate with the modified wellbore geometry and formation properties. The analyses also demonstrate that evolution of gases in an occluded form can result in an appreciable increase in fluid production.

1. INTRODUCTION

The stability of subsurface formations in the vicinity of a borehole is a problem of continuing interest to the petroleum industry. From the earliest days of exploration and production, petroleum engineers have been confronted with the difficult task of maintaining the stability of deep boreholes. Greater production depths, exploitation of marginal reserves, horizontal drilling for production from thin oil zones, increased use of strongly deviated wells from expensive deep-sea gravity platforms, and the use of enhanced recovery techniques such as steam injection and hydrofracturing in heavy oil-sand deposits, have been responsible for the recent sharp increase of interest in re-examining the mechanics of borehole stability. The consequences of wellbore instability, including increased drilling time, stuck pipe, side tracking, shearing of well casings, severe channeling in water floods, direct communication between offset wells, and significant loss of heat during secondary recovery processes, are believed to form an appreciable component of drilling and production capital expenditure. With regard to such problems as sand production accompanying the exploitation of a hydrocarbon reservoir, accurate prediction of the response of a formation to the local pore pressure and stress environment is prerequisite to proper anticipation and prevention of costly downhole problems. Reliable recovery predictions must take account of geometric modifications that develop as a result of sand production, especially the possible enlargement of the cavity around the borehole, and the enhancement of permeability in the zone affected by the removal of sand.

As a vertical borehole is drilled, the radial stress is relieved and the load is transferred around the circumference as a hoop stress. The radial component decreases from the far-field horizontal stress level and becomes equal to the pressure drop across the mud cake, or to zero if there is no fluid overpressure. Wellbore failure results from the mechanical incapacity of the wall material to sustain the loss of radial support and the redistributed stresses. In uncemented formations, the potential for sand production develops once fluid starts flowing towards the wellface; however, sand production will only occur once the fluid-pressure gradient exceeds a certain critical value. The fluid-pressure gradient depends on soil strength properties, permeability, and fluid characteristics.

The initiation of sand production requires the fluid-pressure gradient, and consequently the fluid flowrate, to exceed a certain critical value. Risnes et al. [1], developed formulations for the stresses around a wellbore that account for the development of arching and the subsequent failure of arches when subjected to seepage forces. Assuming an incompressible fluid, homogeneous and isotropic soil having an elastic/perfectly-plastic behavior, uniform and constant permeability, and steady-state conditions, the maximum flowrate, q_c, the formation can support if the well is uncased is given by:

$$q_c = 4\pi \frac{k}{\mu} c'D \tan\left(45 + \frac{\phi'}{2}\right) \tag{1}$$

Where D is the thickness of the formation, k is the intrinsic permeability, c' is the cohesion, ϕ' is the angle of friction, and μ is the fluid dynamic viscosity. Flows in excess of this critical rate will, according to Risnes et al. [1], result in collapse of the formation.

The practical application of equation (1) is quite limited, however, because the assumption of steady-state flow generally is not met. Initially, fluid-pressure gradients near the wellbore may be greatly in excess of the long-term steady-state values, resulting in unpredictable early instability. As collapse of the surrounding material occurs, permeability in those zones must increase and, assuming no other changes, the pressure gradient should be reduced. When the pressure gradient is reduced to subcritical levels, sand production ceases. By the time steady-state conditions have been reached, the well geometry and formation properties may have changed significantly.

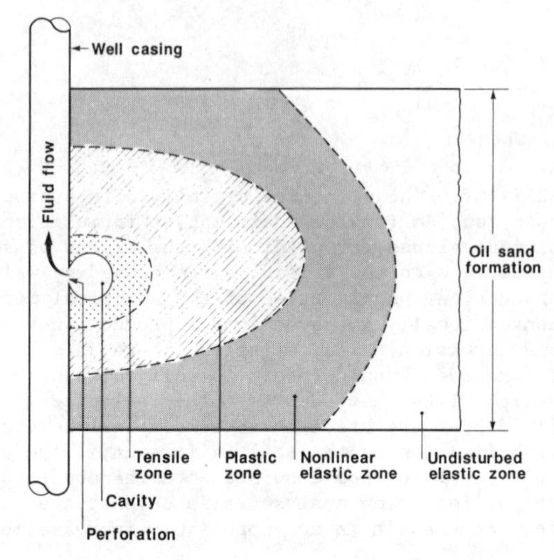

Figure 1. Schematic view of sand failure around a single perforation

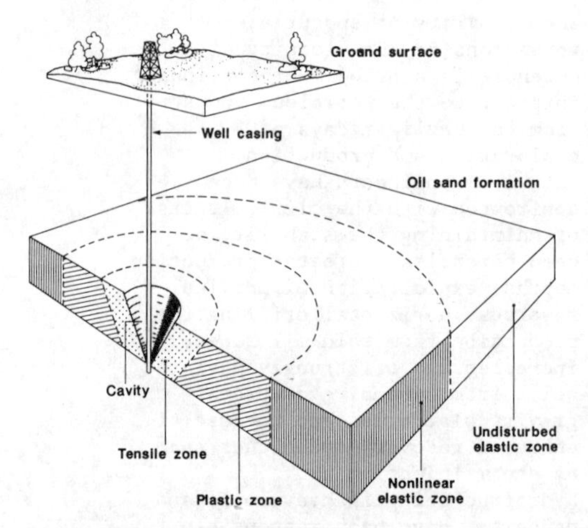

Figure 2. Global view of sand failure around multi-perforated well casing

Figure 1 is a schematic representation of the possible changes in geometry and material characteristics that can develop locally around a single perforation in a producing well. Figure 2 is a global illustration of the effects of failure that may occur under more practical conditions when the producing section of the well is left uncased or blasted to create a number of closely-spaced perforations.

Another limitation of Risnes' model is that it does not allow for the possibility of gas evolution resulting from pressure reduction. Bitumen generally contains large quantities of dissolved gases, as can also be found in many reservoirs under high confining pressures. Pressure relief induced around wellbores can lead to the evolution of dissolved gases which, even in small quantities, can significantly increase the probability of shear or tensile failure around the openings. This is because gas evolution results in pore fluid pressures that remain at higher levels and for longer periods than fully saturated conditions (Byrne and Vaziri [2]).

This paper provides an approach for analyzing the significant effects of sand failure and sand production on fluid flow and the mechanical behavior of the formation. Using the effective stress principle, fully coupled flow and deformation solutions developed by Biot [3] have been employed and modified to allow for the effects of a compressible fluid phase (which might develop through the evolution of dissolved gases) and nonlinear soil behavior. Numerical schemes have been devised for tracking the soil behavior at failure and under liquefied conditions. The results deduced from applying a finite-element code embodying the above mentioned features to model the primary recovery process in a single wellbore are discussed. Emphasis is placed on identifying the factors which have a predominant influence on sand and fluid production. The mechanisms initiating and terminating sand instability and their consequences on flow behavior are also described.

2. DERIVATION OF COUPLED FLOW AND STRESS EQUATIONS

Using the conventional cartesian tensor notation, the following equations can be written:

(a) Equilibrium equation:

$$\sigma_{ij,j} - F_i = 0 \tag{2}$$

where σ_{ij} is the total stress tensor and F_i is the body force vector.

(b) Effective stress strain relationship:

$$\sigma'_{ij} = \sigma_{ij} - u\delta_{ij} = D_{ijk\ell}\,\epsilon_{k\ell} \tag{3}$$

where u is the pore pressure, δ_{ij} is Kronecker's delta, $\epsilon_{k\ell}$ is the strain vector, $D_{ijk\ell}$ is made up of tangent stiffness moduli E_t and B_t whose variations are described by:

$$E_t = k_E P_a \left(\frac{\sigma'_m}{P_a}\right)^{n_E} \left[1 - \frac{R_f(1-\sin\phi')(\sigma'_1-\sigma'_3)}{(2c'\cos\phi'-2\sigma'_3\sin\phi')}\right]^2 \tag{4}$$

$$B_t = k_B P_a \left(\frac{\sigma'_m}{P_a}\right)^{n_B} \tag{5}$$

$$\phi' = \phi'_1 - \Delta\phi' \log\left(\frac{\sigma'_3}{P_a}\right)^{n_B} \tag{6}$$

where k_E, k_B, n_B, n_E and R_f are the hyperbolic parameters, ϕ'_1 is the angle of friction at one atmosphere.

The effective stress equation has been shown (Sparks [4]) to be also valid for partially saturated soils, provided the pore gas remains in an occluded state.

(c) Continuity condition for a compressible pore fluid:

$$-v_{i,i} + \dot{w}_{i,i} - \dot{u}/B_e = 0 \tag{7}$$

where v_i is the superficial velocity, w_i is the displacement vector and B_e is the equivalent bulk modulus of the compressible phases that constitute the soil skeleton:

$$\frac{1}{B_e} = \frac{n}{B_f} + \frac{1-n}{B_s} \tag{8}$$

The compressibility of the pore fluid, B_f, can be determined using Boyle's and Henry's laws. Assuming negligible surface tension effects, the bulk modulus of the fluid phase (i.e., the mixture of liquid and occluded gas bubbles) can be shown to be (Vaziri [5]):

$$\frac{1}{B_f} = \frac{(e_1 - S_1 e_1 + HS_1 e_1)^2}{e_o(e_o - S_o e_o + HS_o e_o)(u_o + P_a)} + \frac{S_1 e_1}{B_\ell e_o} \quad (9)$$

where B_ℓ is the liquid bulk modulus, e is the void ratio, S is the saturation degree, H is coefficient of gas solubility (Henry's constant), and subscripts $_o$ and $_1$ refer to initial and current states.

(d) Fluid flow behavior in accordance with Darcy's Law.

$$v_i = \frac{k_{ij} u_{,j}}{\mu \gamma_f} \quad (10)$$

where k_{ij} is the intrinsic permeability tensor, and γ_f is the unit weight of fluid. The intrinsic permeability can be expressed using the following empirical relationship which has been found to be representative of fine dense sands (Lambe and Whitman [6]):

$$k = k^* \left[\frac{e_1^3(1 + e_o)S^3}{e_o^3(1 + e_1)}\right] \quad (11)$$

where k^* is the permeability under fully saturated conditions.

Using the above relationships, the equilibrium and continuity equations can be written as:

$$[1/2(w_{i,j} + w_{j,i})D_{ijk\ell}]_{,j} + u_{,j} = F_i \quad (12)$$

$$\frac{(-k_{ij} u_{,j})_{,i}}{\mu \gamma_f} + \dot{w}_{i,i} - \frac{\dot{u}}{B_e} = 0 \quad (13)$$

The finite-element forms of equations (12) and (13) can be developed in a straight forward manner using Galerkin's procedure (Vaziri [7]).

3. METHODOLOGY FOR MODELING FAILURE

A very important aspect of soil behavior, particularly in analyzing problems involving large movements and extensive failure, is the failure criterion. Different failure criteria have been developed for different materials. The Mohr-Coulomb failure criterion appears to best represent the failure conditions in granular soils. Mohr-Coulomb failure criterions states:

$$\sigma_3' = N_\phi \sigma_1' - 2c' N_\phi^{1/2} \quad (14)$$

where

$$N_\phi = \frac{1 + \sin \phi'}{1 - \sin \phi'} \quad (15)$$

A graphical representation of the above criterion is shown in Figure 3.

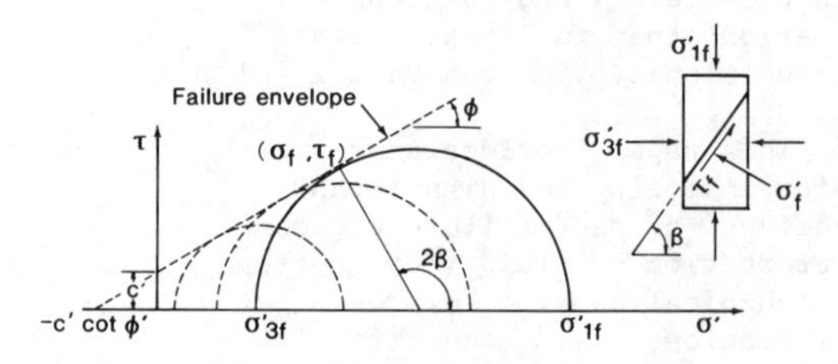

Figure 3. Mohr-Coulomb failure envelope resulting from drained Triaxial tests

For the problems considered herein, i.e., removal of radial support to create open wells, significant changes in the stress state develop that may result in the failure of large regions of soil around the wellbore. The correct simulation of soil behavior at failure is thus of crucial importance in the overall analysis. Two modes of failure generally develop in the soil mass around an open cavity: (1) a loose or tensile zone, existing around the well face, which might have separated from the soil mass by being subjected to large seepage forces; and (2) a plastic zone, existing just outside the tensile zone, which may have undergone some shear deformations, or may have been affected by shear stresses in excess of its shear strength causing it to fail but without becoming detached from the soil mass. The methodology adopted for numerically modeling the behavior at failure is as follows.

With reference to Figure 4, all stress states within the shaded zone are elastic and permissible.

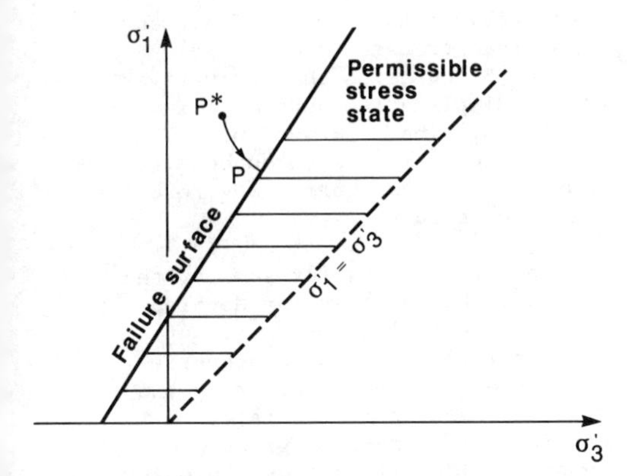

Figure 4. Mohr-Coulomb failure surface

Any stress state lying on the failure surface represents a plastic condition and is permissible also. Stress states outside the region such as $P^*(\sigma_1'^*, \sigma_3'^*)$ are invalid and must be corrected in order to bring them onto the yield surface. The linking path to one such a point, $P(\sigma_1', \sigma_3')$, is shown on this figure. One way of calculating the magnitude and direction of the stress correction is to assume that the correcting path (i.e., P*P) is normal to the yield surface. To mathematically locate such a line,

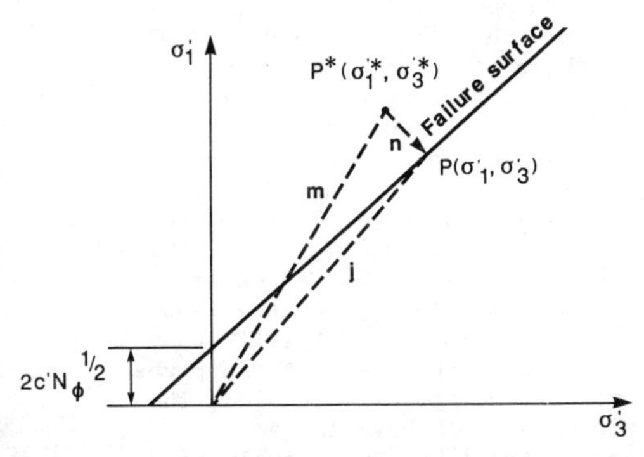

Figure 5. Correction of stresses normal to the yield surface

the state of the corrected stresses (σ_1', σ_3') is denoted by the position vector **j** and the corresponding uncorrected stresses $(\sigma_1'^*, \sigma_3'^*)$ by vector **m**. The correction vector **n** is to be perpendicular to the failure surface and so must be in the direction $(1, -N_\phi)$. These vectors are shown in Figure 5.

Vector **n** can now be expressed as:

$$n = (k, -N_\phi \kappa) \tag{16}$$

where κ is an arbitrary constant.

Since **j** = **m** + **n**, therefore:

$$\begin{aligned}(\sigma_1', \sigma_3') &= (\sigma_1'^*, \sigma_3'^*) + (\kappa, -\kappa N_\phi) \\ &= [(\sigma_1'^* + \kappa), (\sigma_3'^* - \kappa N_\phi)]\end{aligned} \tag{17}$$

The values of σ_1', σ_3' must satisfy the failure criterion, thus:

$$(\sigma_3'^* + \kappa) = N_\phi(\sigma_1' - N_\phi \kappa) - 2c' N_\phi^{1/2} \tag{18}$$

by solving this equation we obtain:

$$\kappa = \frac{N_\phi \sigma_1'^* - 2c' N_\phi^{1/2} - \sigma_3'^*}{1 + N_\phi^2} \tag{19}$$

Hence:

$$\sigma_1' = \sigma_1'^* - N_\phi \kappa \tag{20}$$

$$\sigma_3' = \sigma_3'^* + \kappa \tag{21}$$

To determine the cartesian components of the corrected stress σ_x' and σ_y' in the case of plane strain, or σ_r' and σ_θ' in the case of axial symmetry, the assumption is made that the direction of the principal stresses remains the same both before and after correction. Hence the corrected stress components are:

$$\sigma_x' = S + R \cos \phi' \tag{22}$$

$$\sigma_y' = S - R \cos \phi' \tag{23}$$

$$\tau_{xy} = R \sin \phi' \tag{24}$$

where R and S are the corrected radius and center of the Mohr's circle given by the following equations:

$$S = (\sigma_1' + \sigma_3')/2 \tag{25}$$

$$R = (\sigma_1' + \sigma_3')/2 \quad (26)$$

Also:

$$\theta = \tan^{-1}\left[\frac{2\,|\tau_{xy}{}^*|}{\sigma_x'^* - \sigma_y'^*}\right] \quad (27)$$

where $(\sigma_x'^*, \sigma_y'^*, \tau_{xy}{}^*)$ are the cartesian components of the uncorrected stress state which has principal values $(\sigma_1'^*, \sigma_3'^*)$.

The correction of the stresses is shown in Figure 6 for the plane strain case. For a three-dimensional problem, the correction may be interpreted as a correction normal to the failure plane, keeping the intermediate stress constant.

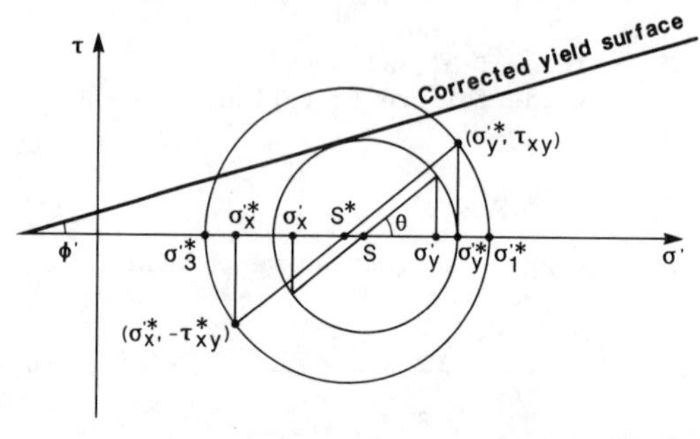

Figure 6. Uncorrect and corrected Mohr circles for the plane strain case

A simple technique is used for implementing the stress correction derived above. The method is to transfer the stresses in excess of those defined by the yield criterion to the adjacent elements, resulting in some additional displacements within an overall stable system. Satisfaction of the yield criterion is therefore achieved by an iterative process. For finite-element analysis, the following order is adopted for activating the stress transfer procedure.

At the termination of every load increment, the stress states which have violated the yield criterion at each integration point (or optionally at elemental centroids) are recorded. The magnitude of the overstress is then determined in accordance with some suitable technique, such as the one described earlier. Denoting the correcting stresses by $\Delta\sigma_x'$, $\Delta\sigma_y'$, τ_{xy}, the next step is to calculate the equivalent nodal loads which balance those stress changes. These are:

$$\{\Delta F\} = \int_v [B]^T \{\Delta\sigma'\} dv \quad (28)$$

where $\{\Delta F\}$ is the vector of element nodal load transfer forces and matrix $[B]$ contains the derivatives of the displacement shape functions. The deformation, strains and stresses throughout the body due to these correcting forces are then computed.

In order to allow for some of the physical changes that may develop when soil experiences tensile or plastic failures, and to properly track the soil state following the failure phase, some special considerations must be given. Tensile failure is assumed to occur when the minor principal stress reduces to below $-c'\cot\phi'$ (see Figure 3). The stresses are then set equal to $-c'\cot\phi'$ and the residual stresses are shed off to the adjoining elements. For the problem of unloading the wellbore, it is assumed that every element which has failed in tension has come apart from the soil mass and entered into a liquefied state. Under these circumstances, those elements can never regain any stiffness or strength; therefore, once an element has experienced tensile failure, its stresses will be maintained at $-c'\cot\phi'$ throughout the rest of the analysis. In order to keep the problem numerically stable, it is assumed that the bulk modulus, B, of these elements will be 10% of the current fluid bulk modulus, B_f, and the shear modulus, G, will be 0.1% of the soil bulk modulus. Since, for gassy soils, B_f may change all the time, B will also change proportionally. Thus, these elements will provide a negligible contribution to sharing the imposed changes in stress, but will behave in a manner that helps to keep the analysis numerically stable. The elements which have failed in tension will be continuously re-examined to ensure that their stresses are maintained at $-c'\cot\phi'$ and the differences are re-distributed (whether positive or negative). The permeability of the elements which have experienced tensile failure is increased by two orders of magnitude to reduce their resistance to drainage and reflect their liquefied state.

Plastic failure is assumed to occur when the shear stress level exceeds the strength level. Under these circumstances, the stresses at variance with the Mohr-Coloumb stress state are redistributed and the shear modulus is reduced to 0.1% of the soil bulk modulus. During the load shedding iterations within each increment, any element which fails plastically will remain plastic throughout the current load shedding cycle (i.e., it will have a very low shear modulus). Furthermore, the stress states of the failed elements are re-examined in each iteration to ensure that they remain on, or very close to, the yield envelope. Since an element of soil which has failed plastically cannot regain strength during the continuation of the loading process that initiated failure, it is assumed that those elements remain plastic and on the yield surface throughout the loading sequence. However, if the stress path is changed (e.g. during consolidation or through change in the direction of loading) then such a restriction is lifted and the elements are allowed to regain their stiffness or move down from the yield surface if they wish to do so.

The preceding formulations and methodologies have been incorporated into a finite-element code, ENHANS, as presented in Vaziri [5].

4. APPLICATION TO SIMULATE PRIMARY FLUID PRODUCTION

Figure 7(a) illustrates the in situ stress state in a typical oil sand formation at approximately 500 m depth. A schematic view of the possible changes in geometry that may develop following the well drilling is shown in Figure 7(b). The problem is modeled as a vertical cylindrical hole through a horizontal layer of oil sand

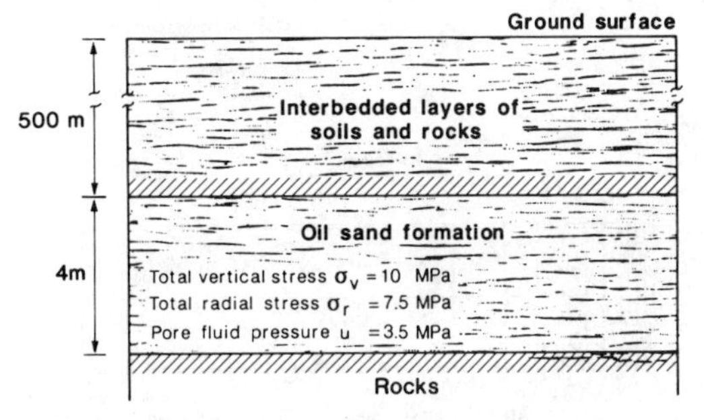

Figure 7(a). In situ stress conditions in the oil-sand formation

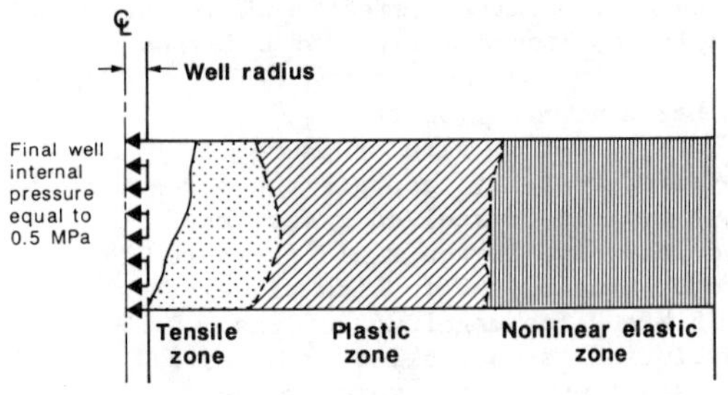

Figure 7(b). Changes resulting from wellbore depressurization

contained between two rigid and impermeable layers of rock. Radial symmetry around the well axis is assumed. The finite-element mesh designed to represent a 0.5-m radius well penetrating a 4-m thick oil-sand formation is shown in Figure 8. Flow is assumed to occur across boundary AC into the well and BD represents the in situ fluid pressure at a distant boundary. Boundaries AB and CD are vertically restrained but free to move laterally. The assumed formation properties shown in Table 1 are typical for oil-sand formations found in the Lloydminister area of

Table 1. Oil-sand formation properties used in the analysis

Formation properties	Initial values	Typical values computed after well-depressurization		
		Nonlinear elastic zone	Plastic zone	Tensile zone
c' (MPa)	0.035	0.035	0.035	0.035
ϕ'	40°	40°	40°	40°
E (MPa)	1000	500 - 1000	0.5 to 1	1E-3
B (MPa)	300	100 to 300	50 to 100	1
B_f (MPa)	2000	1000 to 2000	10 to 1000	10
K (m/s)	1.4E-8	(1.4 to 1.8)E-8	(1.4 to 3)E-8	1E-4
n	0.30	0.3 to 0.31	0.31 to 0.36	>0.5
H	0.3	0.3	0.3	0.3
S (%)	100	99 to 100	90 to 99	<90
σ'_r (MPa)	4.0	1.7 to 4.0	0.0 to 1.7	-0.042
σ'_θ (MPa)	4.0	6.2 to 4.0	0.0 to 6.2	-0.042
u (MPa)	3.5	2.2 to 3.5	0.5 to 2.2	0.5

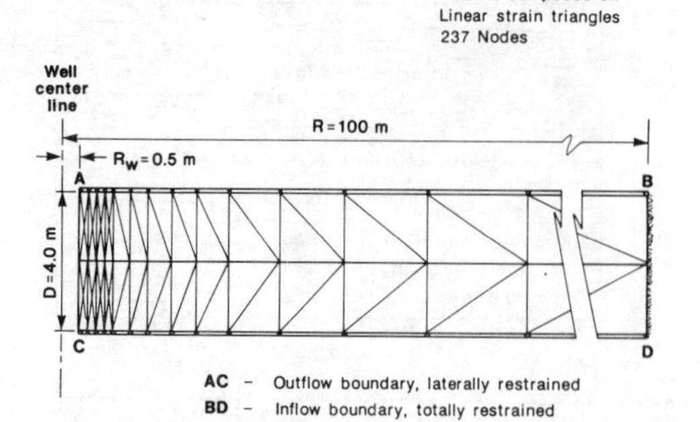

Figure 8. Finite element mesh of the wellbore through a layer of oil-sand

Alberta, Canada (Smith [8]). The fluid, under in situ pressure, is assumed to fully saturate the pores but contains dissolved gases that could evolve if the fluid pressure reduces below the in situ value of 3.5 MPa. The magnitude of gas evolution is controlled by Henry's constant which is assumed to be 0.3.

Pressure reduction induced by the drilling and subsequent production of the wellbore is modeled as a time-dependent process using the finite-element code ENHANS. In the analysis, it is assumed that reduction of the initial internal radial stress of 7.5 MPa and fluid pressure of 3.5 MPa to the final value of 0.5 MPa takes about 50 days to complete. Figure 9(a) depicts the pore-pressure and effective-stress distribution out to a radial distance of 12 m from the well axis at some intermediate stage when the well pressure is 2 MPa. The situation after pressure reduction to its final value of 0.5 MPa is shown in Figure 9(b). Development of the tensile zone and growth of the plastic zone as pressure is reduced can be observed. The fluid pressure in the tensile zone is uniform due to the high permeability developed in this region.

In order to verify the stability analysis of the finite-element model, the Risnes et al. [1] criterion, as described by equation (1) can be examined. Based on this formulation, if cohesion is zero, stability will be violated as soon as the internal well pressure is

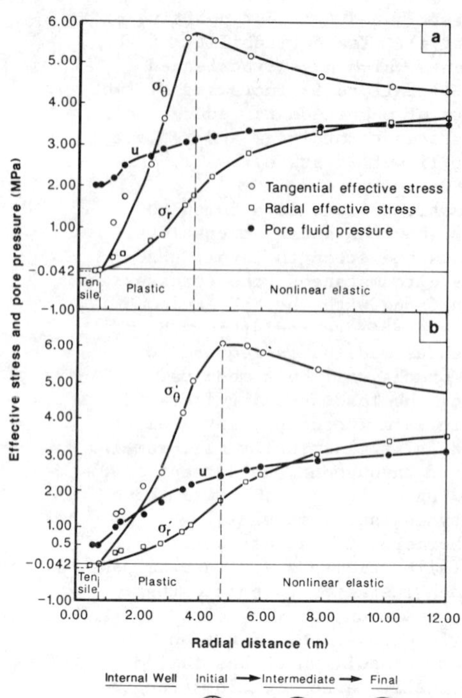

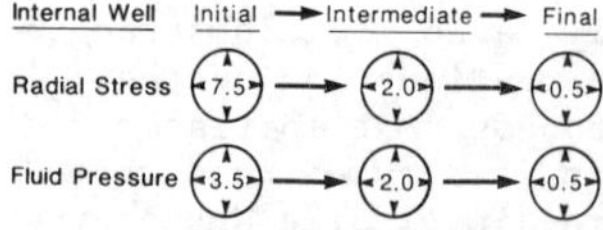

Figure 9. Effective stress and fluid pressure profiles at two different stages (a) Intermediate (b) Final

reduced below the in situ pore pressure, thus causing flow into the well. By simulating such a condition, the analytical solution predicts instability (i.e., progressive increase of the plastic and tensile zones) once the support pressure is reduced below 3.5 MPa. However, at exactly 3.5 MPa, the situation remains stable.

To compare the finite-element and analytical solutions using a cohesion of 0.035 MPa, the same problem is analyzed but with the assumption that the hydraulic conductivity (with respect to bitumen at 25°C) remains constant at 1.4×10^{-8} m/s throughout the analysis. With the parameters chosen, the limit for instability (from equation (1)) is a flow rate of 5.3×10^{-6} m^3/s. Carrying out the finite element-analysis but maintaining a constant value for hydraulic conductivity results in instability at a support pressure less than 2.3 MPa, which corresponds to a persistent flowrate in excess of 5.5 x 10^{-6} m^3/s. For support pressures below 2.3 MPa, the tensile and plastic zones increase at a fast rate and develop an unstable situation with a rapid rise in the flowrate. For the situation when the hydraulic conductivity varies, as is likely to occur in reality, there is no closed-form formulation. Risnes et al. [1], however, state that the conductivity immediately next to the wellbore plays a significant role in the stability. For the problem under consideration, the enhanced conductivity acts to reduce pore pressures close to the well, which in turn increases effective stresses and hence leads to stability. As already mentioned, the formation becomes unstable if the permeability in the tensile zone is not changed.

In order to demonstrate enhancements in fluid production that could be realized by intentionally increasing the compressibility of the fluid phase (e.g., by allowing dissolved gases to evolve in an occluded form), the mesh shown in Figure 8 has been re-analyzed subject to a reduction in fluid pressure only. This is simulated by restraining all the boundary movements, such that any influence due to changes which may have resulted from instability is removed. The flow is simulated by reducing the outflow boundary pressure (along AC) from its initial value of 3.5 MPa to 0.5 MPa. Throughout the analysis, the hydraulic conductivity is kept unchanged from its initial value of 1.4 x 10^{-8} m/s. Two cases are analyzed: (1) assuming the fluid to be free of gas with a constant fluid bulk modulus equal to 2 x 10^3 MPa; and (2) assuming the fluid to contain dissolved gas that can evolve in proportion to Henry's constant equal to 0.1 when the fluid pressure drops below 3.5 MPa. Figure 10 depicts the variation of flowrate with time for these cases. As can be observed, a significant portion of the flow occurs before steady state is reached and, over this period, the average flowrate is two or three times higher for the gassy fluid. It may also be noted that the time to reach steady state is approximately two orders of magnitude greater in the case of a compressible fluid.

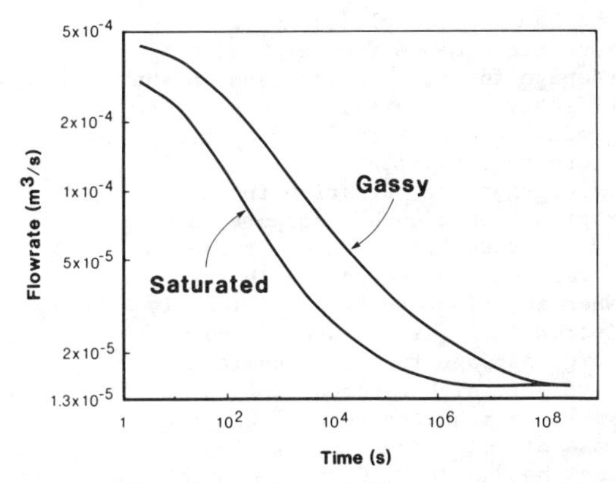

Figure 10. Variation of flowrate with time

5. DISCUSSION

Based on a number of cases studied, it has been found that wellface stability and sand production depend on a number of parameters which affect the flowrate. The soil property which seems to affect stability most of all is the apparent cohesion (c'). In the problem presented herein, a value of c' equal to 0.035 MPa was used. The same problem, however, becomes totally unstable with a c' below 0.025 MPa and exhibits no tensile failure if a c' in excess of 0.1 MPa is employed. Another factor having a major influence on stability is the operational scheme adopted to reduce pressure inside the wellbore. The faster the pressure reduction process is carried out, the steeper will be the fluid pressure gradient at the well face, and hence the greater are the chances of developing instability.

Based on numerous additional studies not presented here, it has been observed that the flowrate during the early stages of pressure reduction in the wellbore has a profound influence on the extent of instability. A sudden initial surge of flow is likely to mobilize an appreciable quantity of sand, particularly within a region close to the wellface where the very high shear stresses have already brought the soil into a state of failure and hence reduced resistance to mobilization. A slow reduction of wellhead pressure, on the other hand, provides an opportunity for the sand to consolidate without much disturbance from excessive seepage forces. An increase in the effective stress state caused by a gradual decrease in fluid pressure close to the wellbore helps to strengthen the formation in that region. Once such a support zone is fully developed, the entire formation can remain stable, even when the flowrate is significantly increased. This is because very large seepage forces concomitant with very high rates of inflow would be required to dislodge the mass of soil around the wellbore that has, by now, developed much higher strength or resistance characteristics compared to its failed state immediately following well completion. Some field operational techniques which can be adopted to control the sand production are presented in Vaziri [9] and [10].

Gradual reduction of the well pressure during early stages of production can have additional benefits in the case of gassy soils, such as oil-sands, since the evolved gas can act to maximize the fluid production. It has been shown that both the flowrate and the total fluid production are greater in situations where gas is kept in an occluded form. A rapid reduction in pressure can result in the evolution of sufficiently large quantities of dissolved gas to cause a phase separation. Under these circumstances, gas forms a separate continuum which can move ahead of the liquid phase, depriving the fluid phase of the additional drive that it could have received from the gas pressure. The maximum yield is likely to be achieved by keeping gas in an occluded form throughout the production cycle.

6. SUMMARY AND CONCLUSIONS

In examining the soil stresses around wellbores, Risnes et al. [1] provided a theoretical formulation for the maximum flowrate that a fluid-bearing formation with certain characteristics can sustain before developing overall instability. The theoretical stability criterion implies that, at sub-critical flowrates, the formation remains totally stable and, once the critical flow rate is exceeded, shells of soil start to spall off into the opening; such a failure mechanism will continue throughout the porous medium. This contradicts the reality of the situation because, in all real systems, there will be geometrical and geomechanical limitations that will inhibit the progression of the collapse.

To overcome some of these limitations and to provide a numerical model that can simulate most of the important processes involved in depleting fully-saturated or gassy reservoirs, a finite-element program (ENHANS) was developed on the basis of Biot's fully coupled flow and deformation formulations. The program takes account of varying compressibility of the pore fluid, as well as varying permeability and nonlinear stress-strain behavior of the soil. Since one of the purposes of this program is to analyze problems that involve significant soil mobilization and failure, a specific procedure is postulated for modeling the soil behavior at failure and a distinction is made between the failure behavior in tension and in shear.

Application of the finite-element program to model the time-dependent response of an oil-sand formation following well drilling and fluid production, has shown that such a process results in the development of a limited tensile zone (in the order of one or two wellbore radii) beyond which the soil is in a nonlinear elastic state, gradually grading into original in situ conditions. Development of the tensile zone and propagation of the extensive plastic zone into the formation are the result of seepage forces that are induced by fluid flow. The numerical analyses performed indicate that the arch formed by the inner surface of the cavity is stable up to some critical value of pressure gradient or flowrate. Once this is exceeded, the walls of the cavity will fall in, leaving behind a new arch and a larger cavity, which in turn can sustain a greater flowrate before it collapses.

Acknowledgements

Support provided by Golder Associates in preparing this paper is gratefully acknowledged. My colleague, Michael Kenrick, provided many constructive comments to improve the general quality and readability of this paper, his interest is deeply appreciated.

BIBLIOGRAPHY

[1] R. Risnes, R. K. Bratli and P. Harsrud, Sand Stresses around a Wellbore, Society of Petroleum Engineers Journal, 883-898, (1982).

[2] P. M. Byrne and H. H. Vaziri, Stress, Deformation and Flow Analysis in Oil Sand Masses, XI Int. Conf. on Soil Mech. and Found. Eng., San Francisco, Numerical Methods Session (1985).

[3] M. A. Biot, General Theory of Three Dimensional Consolidation, J. of Applied Physics, 12, 155-164 (1941).

[4] A. D. W. Sparks, Theoretical Considerations of Stress Equations for Partly Saturated Soils, 3rd Conf. for Africa on Soil Mech. and Found. Eng., Salisburg (1963).

[5] H. H. Vaziri, Nonlinear Temperature and Consolidation Analysis of Gassy Soils, Ph.D. Thesis, University of British Columbia, Vancouver, Canada, 548 p. (1986).

[6] T. W. Lambe and R. V. Whitman, Soil Mechanics, Wiley, New York, 1969.

[7] H. H. Vaziri, Mechanics of Fluid and Sand Production from Oil Sand Reservoirs, Petroleum Society of CIM, 37, 519-535 (1986).

[8] G. E. Smith, Fluid Flow and Sand Production in Heavy Oil Reservoirs under Solution Gas Drive, Society of Petroleum Engineers, Oakland, California, SPE 15094 (1986).

[9] H. H. Vaziri, A Methodology for Predicting Sand Failure around Wellbores, Petroleum Society of CIM, 38, 1011 - 1034 (1987).

[10] H. H. Vaziri, Geomechanics Approach to Reservoir Modeling, 1st Int. Forum on Oil Mining and Gravity Recovery, Dallas, Texas (1987).

Numerical Methods in Geomechanics (Innsbruck 1988), Swoboda (ed.)
© 1988 Balkema, Rotterdam. ISBN 90 6191 809 X

Numerical investigation of hydraulic fracturing in clays

H.K.Tam, H.K.Mhach & R.I.Woods
City University, London, UK

ABSTRACT: To gain insight into the phenomenon of hydraulic fracturing in clay, laboratory tests and complementary numerical studies have been conducted. Cylindrical samples of soil have been fractured in the laboratory by increasing water pressures in a central cavity via a hypodermic probe. This has been simulated with finite elements implementing Biot's consolidation theory together with an elasto-plastic strain hardening constitutive model. Both diametral (axisymmetric) and plan (plane strain) sections have been modelled. Fracturing has been studied with respect to several different failure criteria. The best fit to experimental data is offered by that of zero effective hoop stress in clay elements near the cavity. Rate of cavity pressure increase has also been investigated and is shown to have a strong influence on fracturing pressure.

1 INTRODUCTION

One of the primary causes of failure in embankment dams is believed to be the hydraulic fracturing of the clay core. A program of work is underway in the United Kingdom to assess the susceptibility of a number of older embankment dams to this manner of failure. Part of this research has involved fundamental laboratory studies of the mechanisms involved in hydaulic fracturing. Water pressures are increased at a given rate in the central cavity of a cylindrical clay sample until fracturing occurs, and as there is strong evidence of dependency of fracturing pressure on this rate of increase the problem may be classified as coupled. This paper describes a series of finite element computations designed to simulate the laboratory experiments and to permit further investigation of rate effects.

2 LABORATORY PROCEDURE

Full details of the laboratory experiments and their results are given by Mhach (1987), from which the following points are summarised:

- A cylindrical soil sample is prepared with a central sand-filled cavity
- The sample is set up in an hydraulic triaxial apparatus, with a hypodermic probe inserted in the sand "cell"
- Under constant applied total stress, the water pressure in the cavity p_w is increased until hydraulic fracturing occurs in the clay sample
- In the laboratory, fracturing is evidenced by a sudden increase in flow of water from the cavity
- In numerical studies, fracturing is deemed to have occured as soon as one of the following criteria is satisfied in clay elements near the cavity:
  a) soil reaches critical state,
  b) shear stress achieves maximum value,
  c) effective hoop stress becomes zero.

3 COUPLED LOADING AND CONSOLIDATION

In soil mechanics problems, loading is generally assumed to be applied drained (no change in pore pressures) or undrained (no change in volume). However, it has been shown by Woods (1986) that an important class of problems exists in which the loading and drainage rates are comparable and for which the drained or undrained approximation can be seriously in error.

Under these conditions a fully coupled formulation should be used, and the most common approach is to use Biot's (1941) equation governing excess pore pressures together with the equations of equilibrium and an appropriate constitutive law. Some analytical solutions have been obtained for simple geometries and loadings (Murray, 1978), with numerical methods such as the FEM being used for more general cases.

4 FINITE ELEMENT COMPUTATIONS

The computations described in this paper were performed with the CRISP finite element package, written and developed by the Cambridge Soil Mechanics Group (Britto and Gunn, 1987). The use of CRISP to study various coupled problems has been reported by Carter (1982), Houlsby and Nageswaren (1982), Woods (1986) and Airey (1987).

4.1 Finite Element Mesh

Two different types of mesh have been used

a) Plane strain mesh, Figure 1, and
b) Axisymmetric mesh, Figure 2.

Due to symmetry the plane strain mesh only models one quadrant of a plan cross-section through the soil sample. Both the sand and clay are modelled with 6-noded linear strain triangles (LST) with additional pore pressure degrees of freedom at each vertex node. The mesh is graded in the radial direction, being finer near the cavity. The pore pressure in the sand element nodes adjacent to the probe is increased incrementally over uniform time steps.

In the axisymmetric mesh, one half of a diametral section has been modelled (again due to symmetry). LST elements were used for the clay, sand cell, top platten and hypodermic probe.

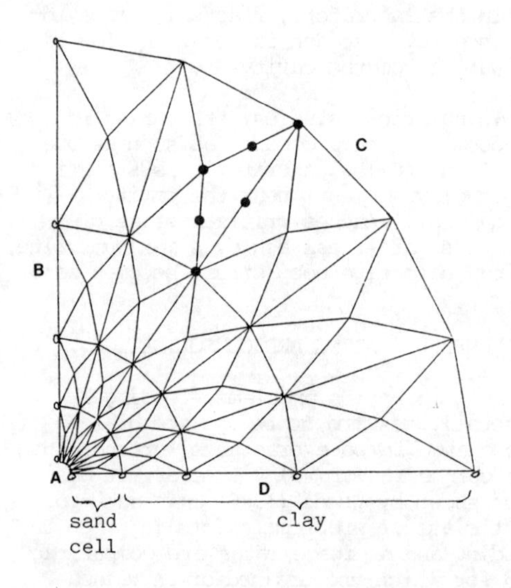

Figure 1. FE Mesh for plane strain analyses

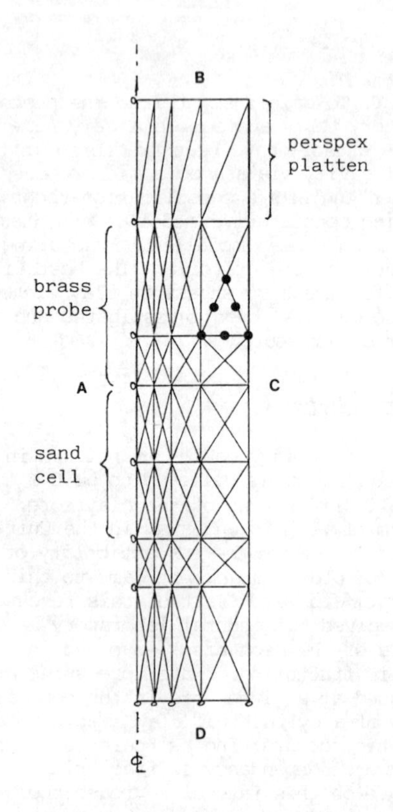

Figure 2. FE Mesh for axisymmetric analyses

4.2 Constitutive Model

Modified Cam-clay was used to model the clay behaviour in both of the meshes described above. An elastic perfectly plastic (Mohr-Coulomb) model was used for the sand, and linear elastic models for both the platten and probe.

The modified Cam-clay model (Roscoe and Burland, 1968) requires 5 basic soil parameters:

κ the slope of the unload/reload line in $v : \ln p'$ space
λ the slope of the virgin compression line in $v : \ln p'$ space
M the slope of the critical state line in $q' : p'$ space
Γ the specific volume on the critical state line at $p' = 1$ kPa
ν' the elastic Poisson's ratio

The initial effective stress state and a measure of overconsolidation must also be specified in order to solve boundary value problems. If consolidation analysis is to be performed, coefficients of permeability in the coordinate directions are required in addition.

Soil parameters used in the analyses described in this paper are for a reconstituted puddle clay:

$\kappa = 0.03$ $\nu' = 0.3$
$\lambda = 0.12$ $k_x = 2.7 \times 10^{-9}$ m/min
$\Gamma = 2.314$ $k_y = 2.7 \times 10^{-9}$ m/min
$M = 1.275$ $\gamma_w = 10.0$ kN/m^3

For the sand cell the following were used:

$E' = 7500$ kN/m^2
$\nu' = 0.3$ $k_x = 2.7 \times 10^{-5}$ m/min
$\phi' = 30°$ $k_y = 2.7 \times 10^{-5}$ m/min

Very stiff linear elastic parameters were used for the top platten and probe.

4.3 Initial Stress State

The samples were isotropically consolidated to a total confining pressure of σ_c = 200 kPa. The total stress was then held constant throughout the remainder of the test.

4.4 Boundary Conditions

In the plane strain case, Figure 1, boundaries B and D were fixed in the tangential direction and free to move in the radial direction. Boundary A simulated the edge of the hypodermic probe and pore pressures were increased incrementally at the vertex nodes of the element along boundary A.

In the axisymmetric case, Figure 2, boundary A was the axis of symmetry of the sample and was fixed horizontally. Pore pressures were increased incrementally at the vertex nodes of the sand elements along boundary A.

4.5 Rate of Loading

The rate at which the cavity water pressure is increased has a significant effect on the fracturing pressure p_f (defined as the water pressure at which the sample begins to crack). The influence of loading rate was studied by using different time intervals for fixed increments of water pressure. Rates between the limits of 1000 kPa/min (effectively undrained) and 0.001 kPa/min (effectively drained) were examined. Table 1 summarises the particular rates chosen for presentation in the next section.

Table 1. Summary of analyses

Rate of pressure increase (kPa/min)	Mesh Type Plane Strain	Axisymmetric
1000	Case A1	Case B1
50	Case A2	Case B2
0.01	Case A3	Case B3

Note : 50 kPa/min is typical of a "rapid" laboratory rate

5 DISCUSSION OF RESULTS

5.1 Group A (Plane Strain)

(a) Changes of stress

Figures 3 to 5 show the variation of stresses in the clay adjacent to the sand cell as the water pressure is increased.

It is clear from these figures that there are reductions in (effective) hoop, radial, and mean stresses, and increases in total mean stress, pore pressure, deviator stress and stress ratio as the cavity water pressure is increased. The criterion which is satisfied first is that of zero effective hoop stress; no peak is observed in the deviator stress and no elements reach critical state (although stress ratios higher than critical are achieved as the stress path for elements near the cavity takes the soil into states "dry" of critical). Incipient fracturing can also be identified by a sharp increase in stress ratio. The rate at which the cavity water pressure is raised produces consistent trends in the data; of great significance is the increasing "lag" in pore pressure behind the cavity pressure at higher rates.

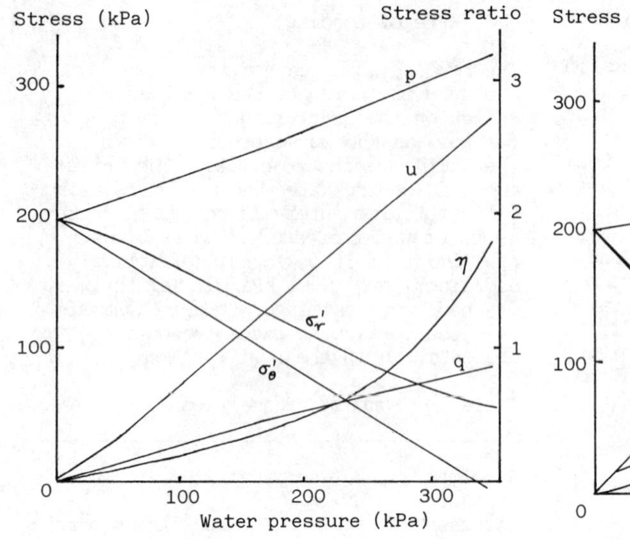

Figure 3. Variation of stresses with cavity water pressure : Case A1

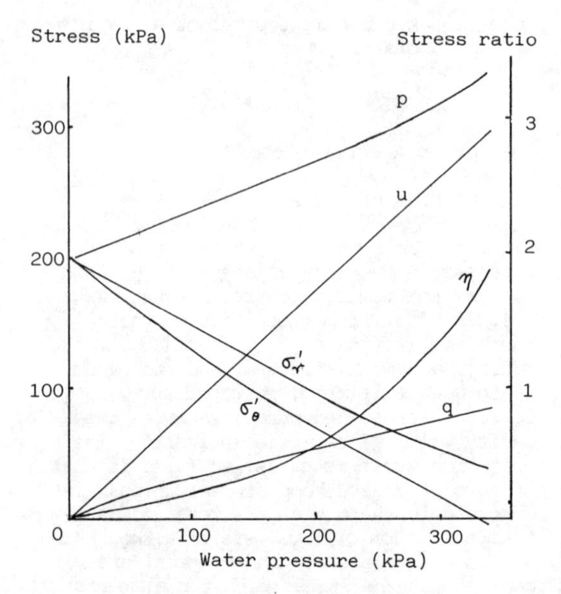

Figure 4. Variation of stresses with cavity water pressure : Case A2

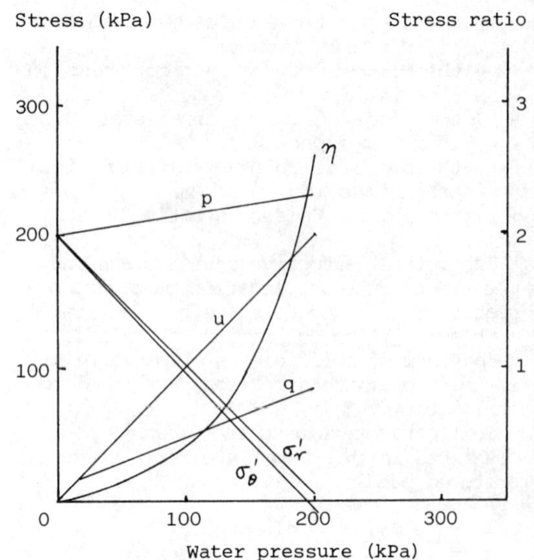

Figure 5. Variation of stresses with cavity water pressure : Case A3

(b) Propogation of Pressure Front

The pore pressure response across the radius of the clay sample was examined at the different rates of cavity pressure increase, Figures 6 to 8. These figures show the radial variation at the point of hydraulic fracturing (ie at $\sigma_\theta' = 0$). Faster rates of increase allow less time for the pore pressure front to traverse the sample radius (as expected), Figure 6. At slower rates the pore pressures become virtually uniform across the sample, as more time has been allowed for drainage, Figure 8.

(c) Rate effects

Having established a suitable criterion for hydraulic fracturing ($\sigma_\theta' = 0$) it is possible to define the fracturing pressure for the various rates of cavity pressure increase and plot these out, Figure 9. A smooth curve is obtained, asymptotic to the drained and undrained limits. This type of curve is typical of coupled (rate dependent) problems and indicates, for example, the range of rates which could produce an effectively undrained response in laboratory tests. Such information is important to the experimentalist, as very little guidance exists on appropriate rates of loading in laboratory (or field) tests.

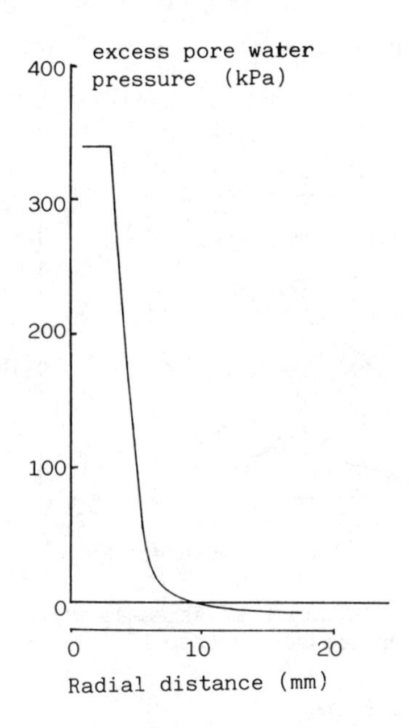

Figure 6. Radial variation in pore pressure at fracture : Case A1

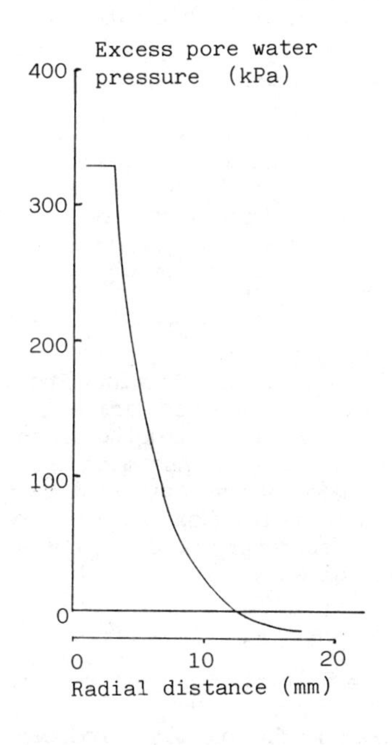

Figure 7. Radial variation in pore pressure at fracture : Case A2

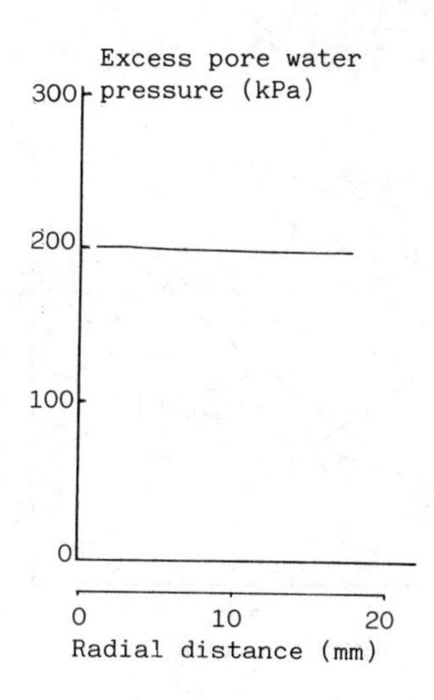

Figure 8. Radial variation in pore pressure at fracture : Case A3

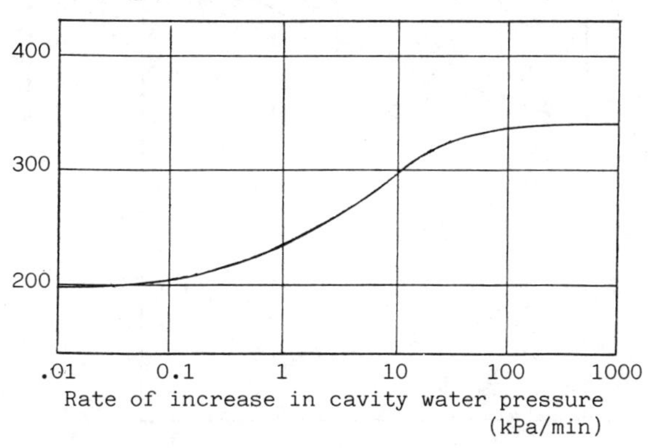

Figure 9. Influence of rate on fracturing pressure for plane strain analyses

5.2 Group B (Axisymmetric)

(a) Change of Stress

Figures 10 to 12 correspond to Figures 3 to 5, showing the variation of stress in the clay with increasing cavity pressure for the axisymmetric analyses. As before, zero effective hoop stress is taken to define the point of fracturing, as neither a peak deviator stress nor the critical state is reached in any element until after $\sigma_\theta' = 0$.

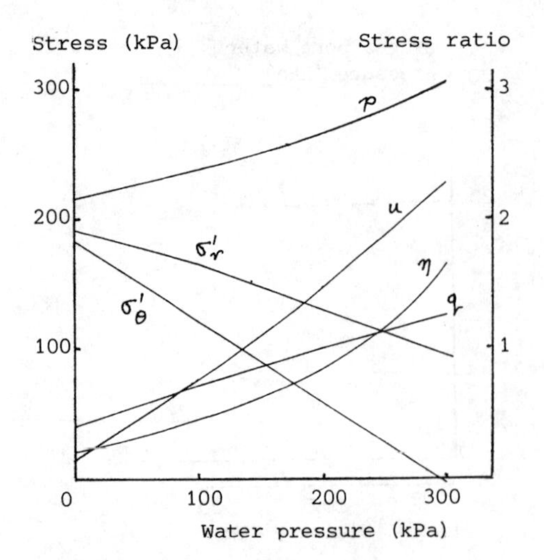

Figure 10. Variation of stresses with cavity water pressure : Case B1

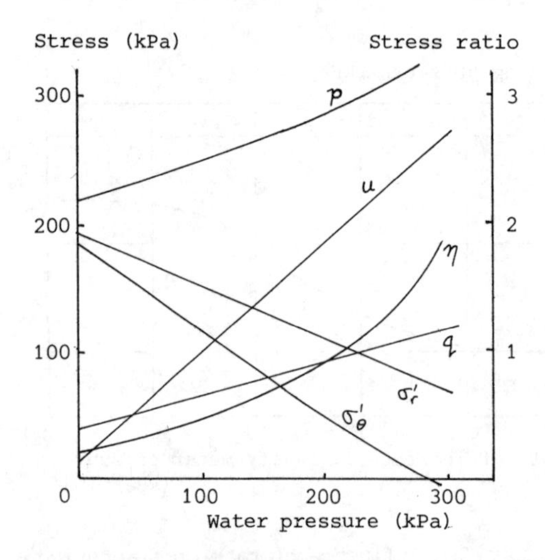

Figure 11. Variation of stresses with cavity water pressure : Case B2

(b) Stress State at Fracture Pressure

Contour plots of effective hoop stress, deviator stress, and excess pore pressure are presented in Figures 13 to 15 respectively, for a rate of 50 kpa/min (typical laboratory rate). The plots were obtained at the cavity pressure causing hydraulic fracture, and it is clear that the locations of minimum effective hoop stress coincide fairly closely with those of maximum deviator stress.

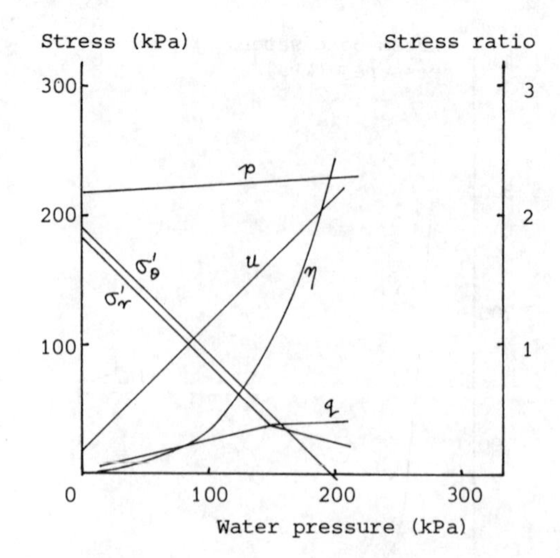

Figure 12. Variation of stresses with cavity water pressure : Case B3

(c) Rate effects

A plot of fracturing pressure against rate of increase of cavity water pressure has been obtained for the axisymmetric analyses, Figure 16. The trend is similar to that shown in Figure 9 for the plane strain analyses, although there is a discrepancy in absolute values which becomes more marked at higher rates.

5.3 Confining pressure

All Group A and B analyses were repeated with different confining pressures of 100, 300 and 400 kPa (to supplement those at 200 kPa). Similar patterns were obtained for both stress changes and pressure front propagation in the clay cample when compared to σ_c = 200 kPa.

In Figure 17, fracturing pressure has been plotted against confining pressure for the single rate of 50 kPa/min. The plane strain and axisymmetric finite element analyses are shown together with the experimental results. Assuming the zero effective hoop stress criterion to be valid, the plane strain f.e. results agree very well with laboratory observations and fall on a line given by

$$p_w - u_o = 1.6 \ \sigma_c \qquad (1)$$

where u_o is the initial steady-state pore pressure in the clay.

The axisymmetric f.e. results produce a best fit line given by

$$p_w - u_o = 1.42 \ \sigma_c \qquad (2)$$

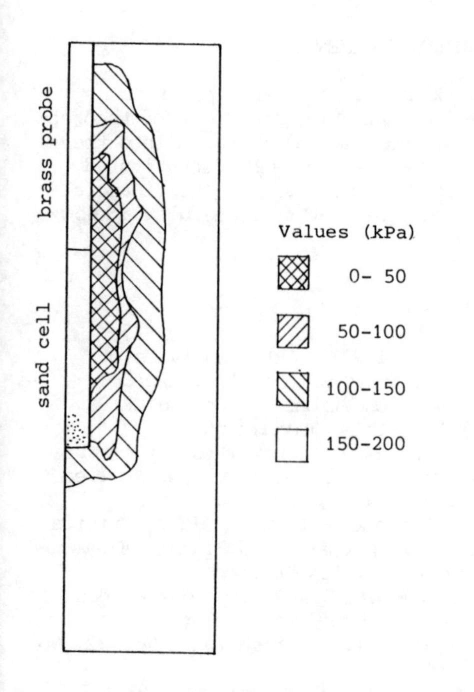

Figure 13. Contours of effective hoop stress at p_w = 310 kPa : Case B2

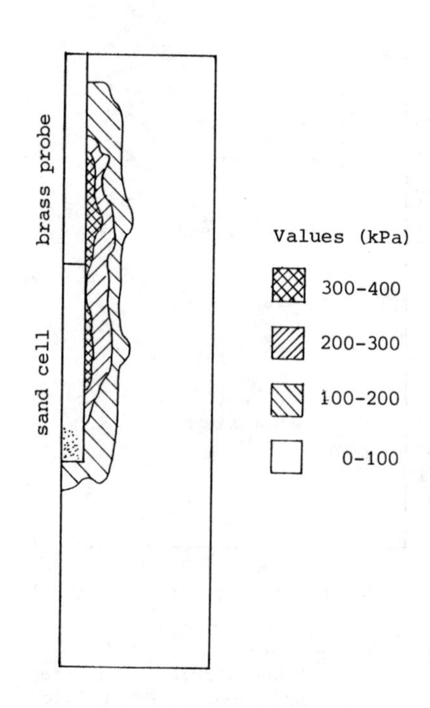

Figure 15. Contours of excess pore pressure at p_w = 310 kPa : Case B2

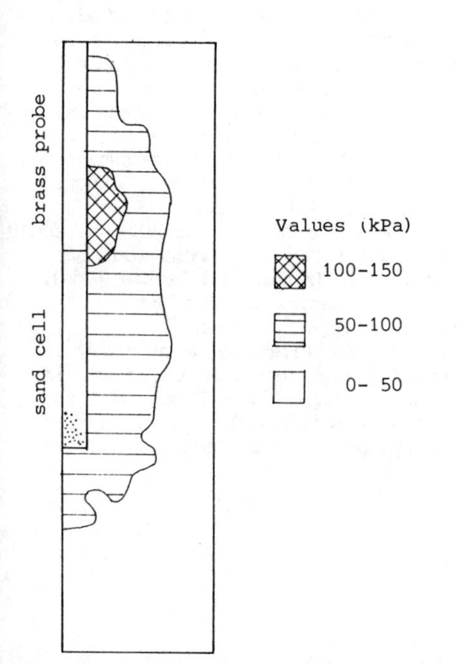

Figure 14. Contours of deviator stress at p_w = 310 kPa : Case B2

Fracturing pressure (kPa)

300

200

.01 0.1 1.0 10 100 1000

Rate of increase of cavity water pressure (kPa/min)

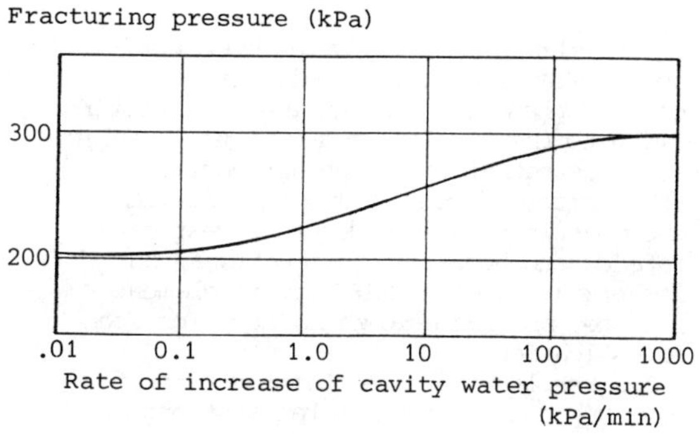

Figure 16. Influence of rate on fracturing pressure for axisymmetric strain analyses

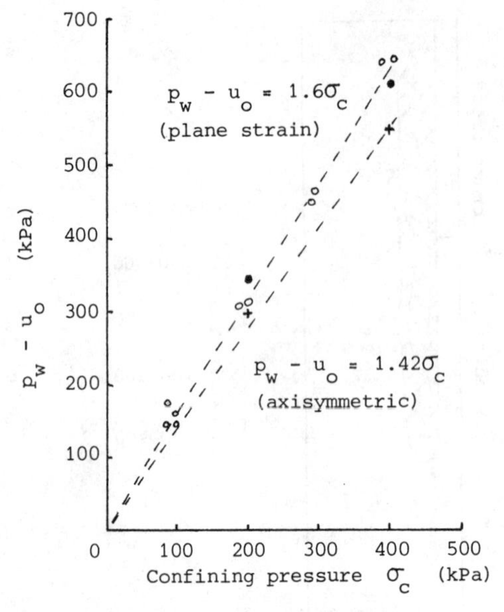

Legend

- • Plane strain FE analyses
- + Axisymmetric FE analyses
- o Experimental data

Figure 17. Variation of fracturing pressure with confining pressure : rate 50 kPa/min

6 CONCLUSIONS

Finite element modelling can usefully complement laboratory studies of the complex phenomenon of hydraulic fracturing. Owing to the coupled nature of the problem, it is essential to use an appropriate formulation of continuity, equilibrium, and constitution in order to investigate rate dependence. Biot's equations, as implemented in the CRISP finite element package, provide a powerful tool for such investigation.

The results indicate that the rate of cavity pressure increase has a strong influence on the pressure at which fracturing occurs. However, rates in excess of 50 kPa/min ensure that the fracturing pressure will be within 3% of the undrained value. A relationship has further been established between the undrained fracturing pressure and the effective confining pressure, indicatng that the plane strain analyses with a zero effective hoop stress criterion provide the closest agreement with experimental results.

7 ACKNOWLEDGEMENTS

The work described in this paper forms part of a continuing programme of theoretical and experimental research into the behaviour of natural and reconstituted soils at City University. The financial support of the SERC is gratefully acknowledged.

REFERENCES

Airey, D W (1987). CRISP analyses of triaxial tests with different end drainage conditions. Proc 3rd CRISP Workshop, City University.

Biot, M A (1941). General theory of three dimensional consolidation. J. Applied Physics, Vol 12, 155-164.

Britto, A M and M J Gunn (1987). Critical State Soil Mechanics via Finite Elements. Chichester, Ellis Horwood.

Carter, J P (1982). Predictions of non-homogenous behaviour of clay in the triaxial test. Geotechnique, Vol 32, No 2, 55-58.

Houlsby, G T and S Nageswaren (1982). A study of consolidation with radial drainage. Workshop on Finite Elements in Critical State Soil Mechanics, Cambridge.

Mhach, H K (1987). An experimental study of undrained hydraulic fracture and erosion tests on puddle clay. Research Report GE/87/8, City University.

Murray, R T (1978). Developments in two- and three-dimensional consolidation theory. in C R Scott (ed.) Developments in Soil Mechanics, pp 103-148, London, Applied Science Publishers.

Roscoe, K H and J B Burland (1968). On the generalised stress-strain behaviour of "wet" clay. in Heyman and Leckie (eds), Engineering Plasticity, pp 535-609, Cambridge University Press.

Woods, R I (1986). Finite element analysis of coupled loading and consolidation. in Pande and Van Impe (eds), Proc 2nd Int Conf Num Models in Geomechanics, Ghent University, pp 701-712. Redruth, Jackson.

Numerical Methods in Geomechanics (Innsbruck 1988), Swoboda (ed.)
© 1988 Balkema, Rotterdam. ISBN 90 6191 809 X

Consolidation analysis of soils by elasto-plastic constitutive models

Y.K.Cheung & Y.Tsui
University of Hong Kong
P.Y.Lu
Guangdong Research Institute of Hydraulic Engineering, Guangzhou, People's Republic of China

ABSTRACT: This paper describes an analysis of a "Slurry-Fall-Fill" Dam in China by finite element technique. The formulation follows the general variational principle of Biot's coupled problem of consolidation with the incorporation of nonlinear elasto-plastic constitutive relationship for soils. The characterization of soil behaviour by four equations relating the elastic bulk modulus K, elastic shear modulus G, strain hardening parameter H, and yield function f as used in the elasto-plastic model is formulated. Finally, the results of the finite element analysis, including the distribution of vertical displacement and pore water pressure in the dam are presented.

1 BIOT'S COUPLED PROBLEM OF CONSOLIDATION

Biot's theory is based upon the following five assumptions:

1. The pores are completely saturated with pore fluid.
2. Fluid are completely incompressible.
3. The equation of motion of the pore fluid obeys Darcy's law.
4. The deformation of soil skeleton is subjected to small displacement theory.
5. The total stress in the soil is equal to the effective stress σ_{ij} plus the pore water pressure p.

By applying the differential method as suggested by Verruijt (1977) to a rectangular Cartesian coordinate system $x_i(i=1,2,3)$, the following four equations can be obtained:

1. Equations of equilibrium:

$$\sigma_{ij,j} + \delta_{ij}p_{,j} + \bar{F}_i = 0 \quad \text{in } V \qquad (1)$$

2. Strain-displacement relations:

$$\varepsilon_{ij} = \tfrac{1}{2}(u_{i,j} + u_{j,i}) \quad \text{in } V \qquad (2)$$

3. Stress-strain relations:

$$\sigma_{ij} = a_{ijkl}\varepsilon_{kl} \quad \text{in } V \qquad (3)$$

4. Equation of continuity:

$$u_{i,i} - u^o_{i,i} + \Delta t\frac{K}{\gamma}(p_{,i})_{,i} = 0 \quad \text{in } V \qquad (4)$$

where σ_{ij}, ε_{ij}, u_i and p are the effective stress tensor, the strain tensor, the displacement vector and the pore water pressure respectively. $\bar{F}_i$ is the volume force vector, a_{ijkl} is elasticity constant tensor, K is the permeability tensor of the porous medium, γ is the specific gravity of pore fluid, δ_{ij} is Kronecker delta, $u^o_{i,i}$ is equal $u_{i,i}$ $(t=t_o)$ and $u_{i,i}$ is equal $u_{i,i}$ $(t=t_o+\Delta t)$.

The boundary conditions for the porous medium are surface tractions, $\bar{T}_i$, and displacements, $\bar{u}_i$, which are prescribed. The boundary conditions for the pore fluid are surface pore pressure, $\bar{p}$, and flux, $\bar{Q}$, which are prescribed too. These boundary conditions are

1. $$(\sigma_{ij} + \delta_{ij}p)l_i = \bar{T}_i \quad \text{on } S\sigma \qquad (5)$$

2. $$u_i = \bar{u}_i \quad \text{on } Su \qquad (6)$$

3. $$\Delta t\frac{K}{\gamma}p_{,i}l_i = \bar{Q} \quad \text{on } S_Q \qquad (7)$$

4. $$p = \bar{p} \quad \text{on } Sp \qquad (8)$$

These boundary conditions in Eqs. (5) to (8) may be considered as the constraint conditions. By following the approach using undetermined Largrange multipliers as suggested by Chien (1980) and the procedures outlined by Lu (1985), then the functional for generalized variational principle of Biot's coupled problem of consolidation can be obtained as follows:

$$\Pi = \int_V \{(\tfrac{1}{2} a_{ijkl}\varepsilon_{kl}\varepsilon_{ij} - \bar{F}_i u_i) - \sigma_{ij}[\varepsilon_{ij} - \tfrac{1}{2}(u_{i,j} + u_{j,i})] + (pu_{i,i} - pu^o_{i,i} - \tfrac{1}{2}\Delta t \frac{K}{\gamma} p_{,i}\, p_{,i})\}dV - \int_{s_\sigma} \bar{T}_i u_i ds - \int_{s_u} (u_i - \bar{u}_i)(\sigma_{ij} + \delta_{ij} p) l_j ds + \int_{s_Q} \bar{Q} p ds + \int_{s_p} (P - \bar{P})\Delta t \frac{K}{\gamma} p_{,i} l_i ds \qquad (9)$$

If the conditions of constraints are satisfied, the approach of the principle of minimum potential energy will yield the following:

$$\Pi_p = \int_V [(\tfrac{1}{2} a_{ijkl}\varepsilon_{kl}\varepsilon_{ij} - \bar{F}_i u_i) + (pu_{i,i} - pu^o_{i,i} - \tfrac{1}{2}\Delta t \frac{K}{\gamma} p_{,i}\, p_{,i})]dv - \int_{s_\sigma} \bar{T}_i u_i ds + \int_{s_Q} \bar{Q} p ds \qquad (10)$$

Equation (10) is analogous to functional presented by Verruijt (1977) as well as Sandhu and Wilson (1969). Equation (10) can also be expressed in tensor and matrix notation for finite element analysis. In plane strain problem with three-noded triangular constant strain element, we can have the following routine standard expressions:

$$u_i = [N_u]\{u_m\}, \quad u_{i,i} = [B_\Delta]\{u_m\}, \ \varepsilon_{ij} = [B_e]\{u_m\}$$

$$\sigma_{ij} = [D]_{ep}[B_e]\{u_m\}, \quad P = [N_p]\{P_m\}, P_{,i} = [B_p]\{P_m\}$$

$$\bar{F}_i = \{\bar{F}\}, \quad \bar{T}_i = [N_u]\{\bar{T}\}, \ \bar{Q} = [N_p]\{\bar{Q}\}$$

Substituting into Eq.(10), functional with constraint conditions can be written as

$$\Pi_p(u_m, P_m) = \sum_{m=1}^{M} \{\int_\Omega \tfrac{1}{2}\{u_m\}^T [B_e]^T [D]_{ep} [B_e]\{u_m\} d\Omega - \int_\Omega \{u_m\}^T [N_u]^T [\bar{F}] d\Omega + \int_\Omega \{u_m\}^T [B_\Delta]^T [N_p]\{P_m\} d\Omega - \int_\Omega \{u^o_m\}^T [B_\Delta]^T [N_p]\{P_m\} d\Omega - \tfrac{1}{2}\frac{\Delta t k}{\gamma}\int_\Omega \{P_m\}^T [B_p]^T [B_p]\{P_m\} d\Omega - \int_{s_\sigma} \{u_m\}^T [N_u]^T [N_u]\{\bar{T}\} ds + \int_{s_Q} \{P_m\}^T [N_p]^T [N_p]\{\bar{Q}\} ds\} \qquad (11)$$

By putting $\delta\Pi_p = 0$ and considering u_m, P_m terms in Eq.(11) as independent variables of variation, Eq.(11) can be reduced to

$$\begin{aligned} [K_1]\{u_m\} + [K_2]\{P_m\} &= \{R_1\} + \{M_1\} \\ [K_2]^T(\{u_m\} - \{u^o_m\}) + [K_3]\{P_m\} &= -\{M_2\} \end{aligned} \qquad (12)$$

where

$$\begin{aligned} [K_1] &= \sum_{m=1}^{M} \int_\Omega [B_e]^T [D]_{ep} [B_e] d\Omega \\ [R_1] &= \sum_{m=1}^{M} \int_\Omega [N_u]^T [\bar{F}]\, d\Omega \\ [K_2] &= \sum_{m=1}^{M} \int_\Omega [B_\Delta]^T [N_p] d\Omega \\ [K_3] &= \sum_{m=1}^{M} -\Delta t \frac{k}{\gamma} \int_\Omega [B_p]^T [B_p] d\Omega \\ [M_1] &= \sum_{m=1}^{N} \int_{s_\sigma} [N_u]^T [N_u]\{\bar{T}\} ds \\ [M_2] &= \sum_{m=1}^{N} \int_{s_Q} [N_p]^T [N_p]\{\bar{Q}\} ds \end{aligned} \qquad (13)$$

If expressed in the incremental form, we then have

$$\begin{bmatrix} [K_1] & [K_2] \\ [K_2]^T & [K_3] \end{bmatrix} \begin{Bmatrix} \Delta u_m \\ P_m \end{Bmatrix} = \begin{Bmatrix} \{R_1\} - [K_1]\{u^o_m\} + \{M_1\} \\ -\{M_2\} \end{Bmatrix} \qquad (14)$$

This is the finite element formulation of Biot's coupled problem of consolidation with constraint condition.

2 ELASTO-PLASTIC CONSTITUTIVE MODEL

In general, the constitutive models describe the relationship between physical quantities such as stress, strain and time. The constitutive model is a central part in setting up numerical techniques to tackle geotechnical problems. The relations arising from plasticity theory usually are incremental; i.e., the stresses and strains are related entirely by their incremental or differential behaviour. The elasto-plastic constitutive relationship presented here is mainly derived from the normality between the yield surface and the direction of plastic strain increments. This key concept leads to two consequences: (a) the yield surface in principal stress space may be plane or curved, if it is curved, it must be convex outward; (b) the plastic strain increment vector must be normal to the yield surface at the yield stress point and proportional to the outward normal. Mathematically, it can be expressed as follows:

$$d\varepsilon^p_{ij} = \frac{\partial f}{\partial \sigma_{ij}} d\lambda \qquad (15)$$

where $d\varepsilon^p_{ij}$ represents the plastic strain increment
f is the yield function
$d\lambda$ is a constant of proportionality

When total strain separated in elastic and plastic components, it can be written in the following matrix form:

$$\{d\varepsilon\} = \{d\varepsilon^e\} + \{d\varepsilon^p\} \qquad (16)$$

where $\{d\varepsilon\} = [D]\ \{d\sigma\}$, i.e. the elastic component is written in the generalized form of Hooke's Law with [D], the elastic stress-strain matrix expressed in terms of elastic bulk modulus K and elastic shear modulus G as $D_{ijkl} = 2G\ \delta_{ik}\ \delta_{jl} + (K - \frac{2}{3}G)\delta_{ij}\delta_{kl}$ and $\{d\varepsilon^p\} = \{\partial f/\partial\sigma\}d\lambda$ as in Equation (15). Once a strain hardening yield criterion f(p,q,H) has been chosen (prescribed or experimentally determined), the derivation of the incremental stress-strain relationship is primarily a matter of mathematical manipulation. Then it can be shown that

$$\{d\sigma\} = ([D] - [D]_p)\{d\varepsilon\}$$
$$= \left[[D] - \frac{[D]\left\{\frac{\partial f}{\partial\sigma}\right\}\left\{\frac{\partial f}{\partial\sigma}\right\}^T [D]}{-\frac{\partial f}{\partial H}\left\{\frac{\partial H}{\partial\varepsilon^p}\right\}^T\left\{\frac{\partial f}{\partial\sigma}\right\} + \left\{\frac{\partial f}{\partial\sigma}\right\}^T [D]\left\{\frac{\partial f}{\partial\sigma}\right\}} \right]\{d\varepsilon\}$$
$$= [D]_{ep}\{d\varepsilon\} \qquad (17)$$

where $[D]_{ep}$ is the generalized Elasto-plastic stress-strain matrix and is expressed in terms of (1) elastic parameters K and G, (2) straining hardening parameter H and (3) yield functin f.

3 DETERMINATION OF PARAMETERS K, G, H AND f OF THE ELASTO-PLASTIC MODEL

The experimental approaches/procedures in obtaining the elasto-plastic model parameters are described in the following subsections.

3.1 Determination of elastic moduli K and G

The expressions for elastic bulk modulus K and shear modulus G can be routinely written as

$$K = K_1 P_a\left(\frac{p}{P_a}\right)^{n_1} \text{ and } G = \frac{G_1}{3} P_a\left(\frac{p - \frac{1}{3}q}{P_a}\right)^{n_2} \qquad (18)$$

where p_a is the atmospheric pressure.

So the parameters K_1 and G_1 can be directly obtained from conventional triaxial tests, say isotropic loading (compression), unloading (swelling) and σ_3=const. shear tests respectively. In lg K vs lg p plot, n_1 is the tangent and K_1 is the value at p=1.0. Likewise, n_2 and G_1 can be determined in the corresponding lg G vs lg σ_3 plot.

3.2 Shape of yield surface, f.

In the theory of plasticity, the yield function f(p,q,H) is generally prescribed, such as indicated in some of the classical treatments by Tresca (1864), Mises (1913). However, another approach, as presented here, is to determine the shape of yield function directly from experimetnal data. This simple concept is best illustrated by an example. From conventional σ_3=const. drained tests, the test data concerning the plastic strains (i.e. ε^p_v and $\bar{\varepsilon}^p$) can be plotted in the relationships σ_3-q-ε^p_v and σ_3-q-$\bar{\varepsilon}^p$ as shown respectively in Figures 1 and 2. Then various intermediate values of ε^p_v and $\bar{\varepsilon}^p$ corresponding to different q and σ_3 values

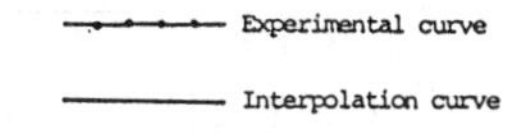

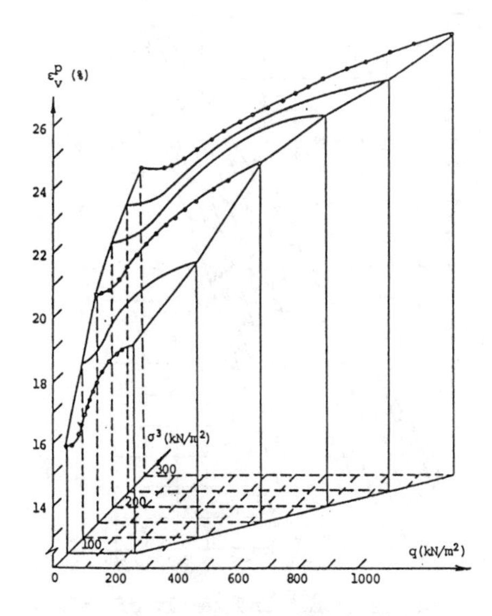

Fig.1 Space plot of σ_3-q-ε^p_v

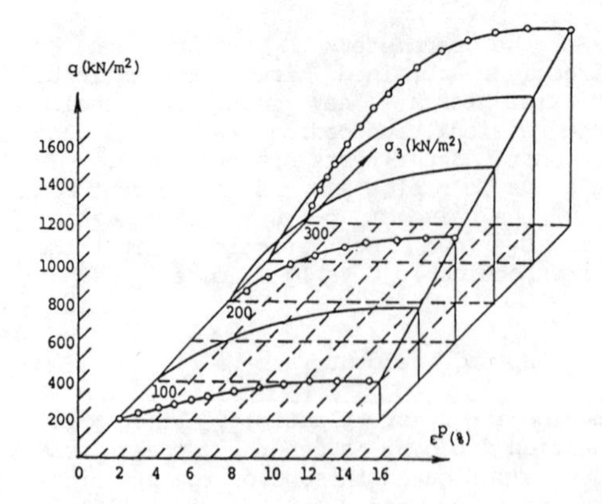

Fig.2 Space plot of σ_3-q-$\bar{\varepsilon}^p$

are interpolated by bicubic spline function (Lu et al, 1982, 1984). Thus the vectors of the incremental plastic strains $\overrightarrow{d\bar{\varepsilon}^p}$ can be obtained from $d\varepsilon_v^p$ and $d\bar{\varepsilon}^p$, and they are scaled accordingly with different stress-paths in a p-q plot as shown in Figure 3. Based on the concept of normality rule and following the procedures outlined by Huang et al (1981),

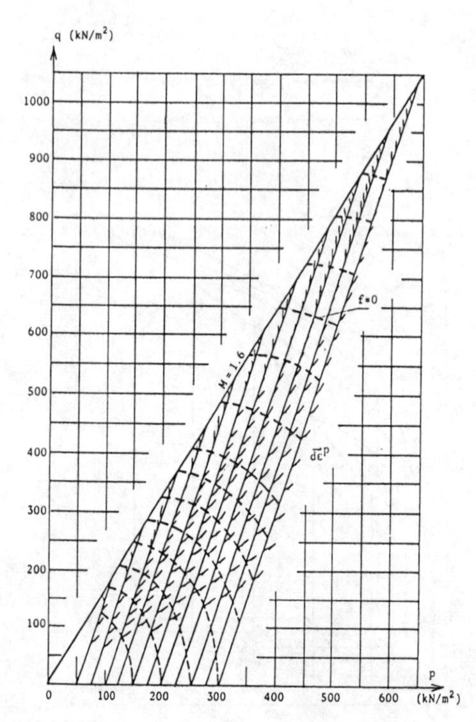

Fig.3 Vectors of $\overrightarrow{d\bar{\varepsilon}^p}$ and family of f=0 curves

family of yield surfaces (f=0) can be determined as shown by dashed curves in Figure 3. At q=0, the curves (f=0) should be normal to p-axis, and the tangents at $q-q_1/p = M$ should be parallel to q-axis (Figure 4). By observation, the shape of family of yield surfaces (f=0) coincides with the expanding elliptical curves. Written in algebraic form, we then have

$$f = q^2 + A_1 p^2 - A_2 p - A_3 p p_o + A_4 \frac{p^2}{p_o} + B_1 \frac{p^2}{p_o^2} - B_2 \frac{p}{p_o} + B_3 p_o + B_4 p_o^2 + B_5 \qquad (19)$$

where
$$A_1 = \left(\frac{\nu M}{\alpha}\right)^2 \qquad B_2 = 2q_1^2 \frac{\nu}{\alpha^2}$$
$$A_2 = 4q_1\left(\frac{\nu}{\alpha}\right)^2 M \qquad B_3 = 2q_1 \nu M F$$
$$A_3 = 2\nu A_1 \qquad B_4 = F(\nu M)^2$$
$$A_4 = 2q_1\left(\frac{\nu M}{\alpha^2}\right) \qquad B_5 = q_1^2 F$$
$$B_1 = \left(\frac{q_1}{\alpha}\right)^2 \qquad F = \left(\frac{\nu}{\alpha}\right)^2 - 1$$

The parameters ν, α, q_1 and M are defined as shown in Figure 4. p_o is the value of p at q=0, and based on experimental data, its expression can be written as:

$$p_o = e^{\ln 10 \, \frac{H-A_o}{n}} \qquad (20)$$

where H is the hardening parameter. A_o and n are determined directly from the plot of H vs log p_o

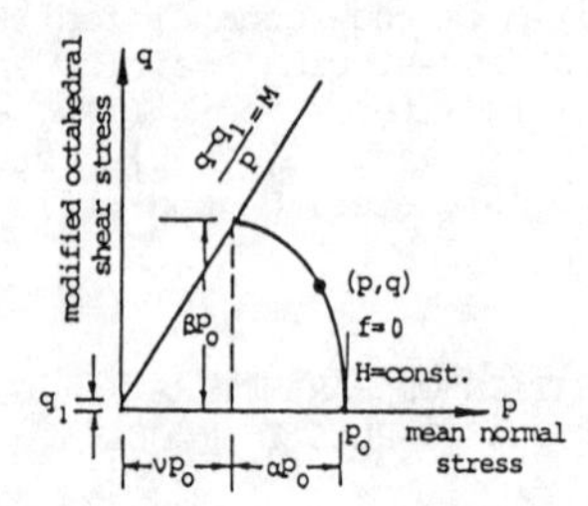

Fig.4 Elliptic cap

3.3 Strain hardening parameter H

Following the concept of Lu et al (1984) and Prevost & Hoeg (1975), the straining hardening parameter H can be chosen as a function of ε_v^p and $\bar{\varepsilon}^p$, $H=H(\varepsilon_v^p, \bar{\varepsilon}^p)$, i.e.,

$$H = (\varepsilon_v^p)^3 + a_1(\varepsilon_v^p)^2\bar{\varepsilon}^p + a_2\varepsilon_v^p(\bar{\varepsilon}^p)^2 + a_3(\bar{\varepsilon}^p)^3 + (\varepsilon_v^p)^2 + a_4\varepsilon_v^p\bar{\varepsilon}^p + a_5(\bar{\varepsilon}^p)^2 + \varepsilon_v^p + a_6\bar{\varepsilon}^p \qquad (21)$$

Where a_1, a_2, ... a_6 are coefficients to be determined. From p-q plot, isovalue lines of ε_v^p and $\bar{\varepsilon}^p$ are drawn to intersect with family curves of f=0 (Figure 5). Using these N points of intersection as N data points, the following expression is formed

$$\sum_{i=1}^{N} [H(\varepsilon_v^p,\bar{\varepsilon}^p)-H_{o_i}]^2 = G(a_1,a_2,...a_6)$$

By means of least square procedure, setting $\frac{\partial G}{\partial a_j} = 0 (j=1,2,...6)$, the values of $a_1,a_2,...a_6$ are evaluated. When back-substituting into Equation 21, the hardening parameter H can be determined.

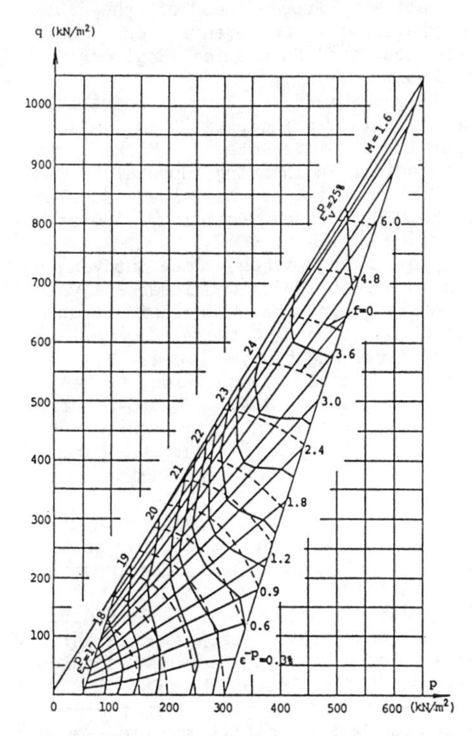

Fig.5 Yield surfaces and isovalue lines of ε_v^p and $\bar{\varepsilon}^p$

4 COMPUTATION AND RESULTS OF ANALYSES

A finite element computer programme based on Equations 14, 17 was coded and simulation computations were performed for a "Slurry-Fall-Fill" dam constructed in South China (Tsui et al, 1987). The dam is 68 m high and is composed of four different kinds of earth materials, namely: rock fill, hand-packed rubble, ridging soil and hydraulic fill (Figure 6). The dam section and the corresponding construction sequences (8 loading stages) are also shown in Figure 6.

4.1 Modelling for media

Rockfill and hand-packed rubble, used as inverted filter in the dam, are assumed to be linear elastic with the following material parameters:

	Rockfill	Hand-pack rubble
elastic bulk modulus K (MN/m^2)	2778	2083
elastic shear modulus G(MN/m^2)	167	125
Coefficient of permeability k (m/day)	43.2	43.2

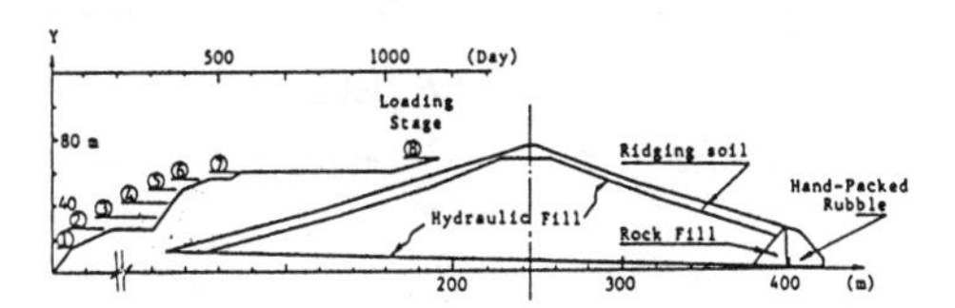

Fig.6 Dam section and construction loading stages

The other two types of earth materials, i.e. hydraulic fill soil and ridging soil, are treated as nonlinear elasto-plastic material. The material properties and the pertinent values as used in the finite element analyses were determined from test results as delineated in Section 3 and are summarized in Table 1.

4.2 Results of finite element analyses

The finite element programme used in the analyses represents plane strain condition and allows for a step by step simulation of the actual construction sequences. Just to illustrate the versatility of the finite element method, few selected results, such as the distribution of pore water pressures and vertical displacements at certain construction stages are presented in Figures 7 and 8.

TABLE 1. Material properties used in finite element Analyses

Pertinent reference in section (3)	Material type	Hydraulic fill	Ridging soil
Eq.(18)	K_1 (MN/m^2)	16.59	16.200
	n_1	0.755	0.910
	G_1 (MN/m^2)	61.03	62.950
	n_2	0.787	0.794
Eq.(19)	α	0.5	0.52
	ν	0.5	0.48
	q (kN/m^2)	6.3	11.40
	M	1.6	1.52
Eq.(20)	A_o	0.2181	0.0262
	n	0.1204	0.0264
Eq.(21)	a_1	-12.6032	294.0455
	a_2	-61.5873	-161.0773
	a_3	-22.4554	78.7212
	a_4	- 5.7422	-22.1030
	a_5	-10.6449	-2.3332
	a_6	2.0644	0.9816
Coefficient of permeability k (m/day)		0.009 to 0.0862	3.382

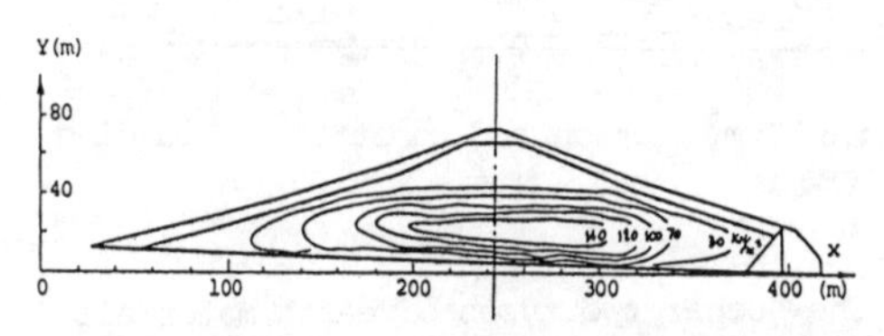

Fig.7 Distribution of pore water pressure at the end of loading stage (5)

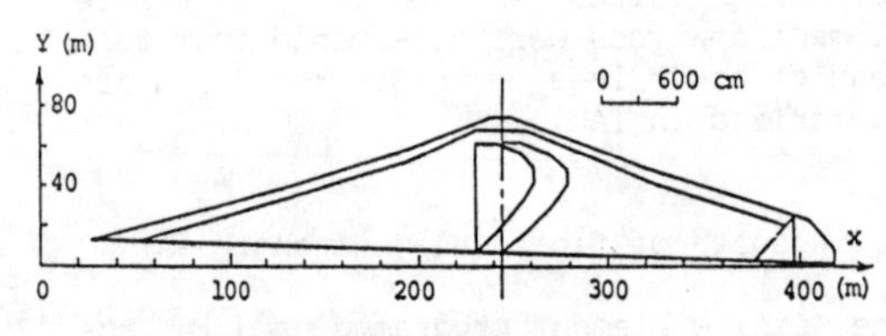

Fig.8 Vertical displacement distribution at the end of loading stage (7)

ACKNOWLEDGMENT

This work has been supported by Haking Wong Foundation under Grant Record No.698. The funding provided during Mr. Lu's tenure as Research Officer at the University of Hong Kong is gratefully acknowledged.

REFERENCES

Biot, M.A. 1941. General theory of three-dimensional consolidation. Journal Appl. Phys., Vol.12.

Chien Weizang 1980. Variational calculus and method of finite element. Sci. Press, Beijing.

Huang Wenxi, Pu Jialiu and Chen Yujiong 1981. Hardening rule and yield function for soils. Proceedings of the 10th International Conference on Soil Mechanics and Foundation Engineering, Stockholm.

Lu Peiyan 1985. Two generalized variational principles in soil mechanics. Selected Works of Geotechnical Engineering, Special volume in Commemoration of the Golden Jubilee of ISSMFE, China Building Industry Press.

Lu Peiyan, Xiong Lizhen, Chen Shaoyong & Chen Mingging 1982. Nonlinear analysis of soils using the cubic and bicubic spline function. Chinese Journal of Geotechnical Engineering, No.1.

Lu Peiyan, Chen Shaoyong, Xiong Lizhen & Chen Mingging 1984. Elasto-plastic constitutive equation for hydraulic-fill soil of slurry-fall dam. Chinese Journal of Geotechnical Engineering, No.2.

Prevost, J.H. & Hoeg K. 1975. Effective stress-strain-strength model for soils. Journal of Geotechnical Engineering Division, ASCE, Vol.101, No.GT3.

Sandhu, R.S. & Wilson E.L. 1969. Finite-element analysis of seepage in elastic media. Journal of Engineering Mechanics Division, ASCE, Vol.95, No. EM3.

Tsui, Y. & Lu, P.Y. 1988. Performance of a 68 m high hydraulic-fill dam in South China. Paper submitted for publication in the proceedings for the ASCE Geotechnical Specialty Conference on Hydraulic Fill Structure. Colorado, USA.

Verruijt, A. 1977. Generation and dissipation of porewater pressure. Ed. by Gudehus, G., Finite Element in Geomechanics.

Numerical Methods in Geomechanics (Innsbruck 1988), Swoboda (ed.)
© 1988 Balkema, Rotterdam. ISBN 90 6191 809 X

Deformation of a sedimentary overburden on a slowly creeping substratum

N.C.Last
Koninklijke Shell Exploratie en Produktie Laboratorium, Rijswijk, Netherlands

ABSTRACT: An explicit finite-difference computer program has been further developed and used to investigate in a geological context the problem of a gently sloping elasto-plastic frictional overburden overlaying a mobile viscous substratum of varying thickness. The computed results compare favourably with a simplified analytical result but show also that a yielding material leads to characteristic features in the overburden which resemble areas of growth faulting (extension) and thrusting (compression) accompanied by large lateral movements. The key parameters which differentiate the styles of deformation are examined and quantified.

1 INTRODUCTION

Creep flow in mobile salt or shale substrata has long been recognised as an important factor in the tectonics of salt basins (Trusheim, 1960) and major deltas (Merki, 1972; Evamy et al., 1978). The available evidence suggests that salt and shale layers may flow in the direction of decreasing overburden load to produce features such as shale waves, salt pillows or ridges, and overburden faulting, which are of particular interest, since any mechanical model capable of explaining their occurrence in a coherent manner will be helpful, for example in quantitative studies of hydrocarbon migration and accumulation patterns.

A simplified theoretical approach was used earlier by Lehner (1977) to predict the evolution of the shape of the interface in an idealised, two-layer salt(or shale)/overburden sequence. The theory provides an estimate of the flow rate within the mobile (viscous) layer and predicts the occurrence of travelling waves (or ridges) and associated trailing depressions (or basins) due to lateral movement within the mobile layer and depressions in the basement. Support for the occurrence of these features can be drawn from field observations and experiments. Moreover, if a numerical model is used, assumptions concerning the velocity profile in the substratum and the behaviour of the overburden can be less restrictive. Indeed, comparisons between the numerical simulations and the predictions of Lehner's model should provide useful quantitative and qualitative data to assess the range of applicability of the simplified analysis. The objective has been therefore to develop a numerical model for investigating this type of geological setting.

2 THE COMPUTER PROGRAM

The two-dimensional, plane strain program GEOMEC (GEOMechanics Explicit Code) is based on the explicit finite-difference method (Wilkins, 1964, Cundall, 1976). Any local disturbance (of equilibrium) is propagated at a materially dependent rate consistent with Newton's Laws of Motion in the form

$$\operatorname{div} \underset{\sim}{\sigma} + \underset{\sim}{F} = \rho \underset{\sim}{a} \qquad (1)$$

where $\underset{\sim}{\sigma}$ is the Cauchy stress tensor, $\underset{\sim}{F}$ the body force and $\underset{\sim}{a}$ the acceleration. Thus the model follows a sequence of locally determined dynamic (d'Alembert) equilibrium states rather than a series of globally determined static equilibrium states. In order for the equilibrium to be assessed locally in this way, the incremental time-step between successive sets of calculations (each representing a state in the evolution process) must be small enough to prevent propagation of information beyond neighbouring calculation points within such a step (the scheme is conditionally stable). Numerically, therefore, many relatively simple calculations are substituted for the fewer, but invariably complex, inversions or multiple iterations that would be required for the solution of the equivalent series of global equilibrium states. The trade-off is that often many thousands of these smaller calculation cycles (each representing a single time-step) may be required to solve a given problem.

2.1 Developments

The development of GEOMEC from the original code NESSI (Cundall et al, 1980) has been motivated by the special requirements of geological modelling. In particular, techniques have had to be implemented to model accurately the slowly evolving, time-dependent processes associated with sedimentation and with creep of the mobile substratum. A brief description of the developments which are of particular relevance to the modelling reported here is included.

2.1.1 Geological Time Scales

The central finite-difference discretisation of equation (1) results in a system of explicit equations which can be solved by progressing incrementally through a large number of time-steps, $n\Delta t$. However, the stability of the scheme depends critically on the local rate of propagation of compressional waves so that typically

$$\Delta t \leq \ell\sqrt{\frac{\rho}{K}} \qquad (2)$$

in which ℓ is a characteristic length of the spatial discretisation, ρ is the material density and K is the appropriate elastic modulus. Using typical numerical meshes and material properties, for geomaterials Δt is usually in the range from 0.001 to 0.1 seconds. Clearly the modelling of events which take place over geological time is not feasible unless a suitable scaling rule can be adopted. Previously an 'adaptive density scaling' technique (Cundall, 1982) was used: essentially the inertial density in equation (2) is scaled to artificially increase Δt provided that the inertial force term remains small in comparison to a reference value, for example the gravitational body forces. In a problem where deformation rates are constrained by slow viscous flow, an alternative approach is possible: the viscosity can be scaled to speed up the process, provided that the inertial forces remain small in comparison to the viscous forces. The Reynolds' number (=kinetic energy per unit volume / viscous stress) is a suitable measure of this ratio,

$$Re = \rho v d/\eta \qquad (3)$$

in which η and ρ are the viscosity coefficient and the density respectively, and v is an average velocity and d a characteristic length of the viscous flow regime. For the scaled system, the coefficient of viscosity becomes

$$\eta^s = \eta/\lambda \qquad (4)$$

Under steady quasi-static conditions, this scaling will simply increase the real velocities v by the same factor so that the computed velocities become

$$\underset{\sim}{v}^s = \lambda \underset{\sim}{v} \qquad (5)$$

and the Reynolds number for the computations becomes

$$Re^s = \lambda^2 Re \qquad (6)$$

Hence the scaling factor λ should be selected such that $Re^s \ll 1$. It is immediately clear that a scaling by λ^2 of the material density in equation (3) would have the same effect on the Reynolds number and would increase the time-step of equation (2) by λ - this is in fact the basis of density scaling. Both approaches have been implemented and successfully used. While adaptive density scaling remains the more generally applicable scheme, viscosity scaling appears to be particularly appropriate for the class of problems studied here. Furthermore, under near steady state conditions, numerical results indicate that the schemes are indeed equivalent.

2.1.2 Sedimentation and Erosion

To model sedimentation, the top zones ('surface zones') of the mesh grow vertically to reflect the specified rate of growth and are treated differently from the underlying regular zones (Figure 1). Within each time-step, the change of stress in the growing zone due to sedimentation corresponds to the addition of a thin slice of frictional material whose stresses are consistent with the stress state in an infinite Rankine slope. The stress state for the whole zone is then evaluated in the usual way by reference to a constitutive model. The only difference is that the surface zone is treated in a step-by-step, semi-Eulerian fashion so that the vertical sides of the zone remain vertical. The surface grid points (Figure 1) are not material points (they have no inertial mass) and their velocity reflects the movement of the sedimentation boundary rather than the local

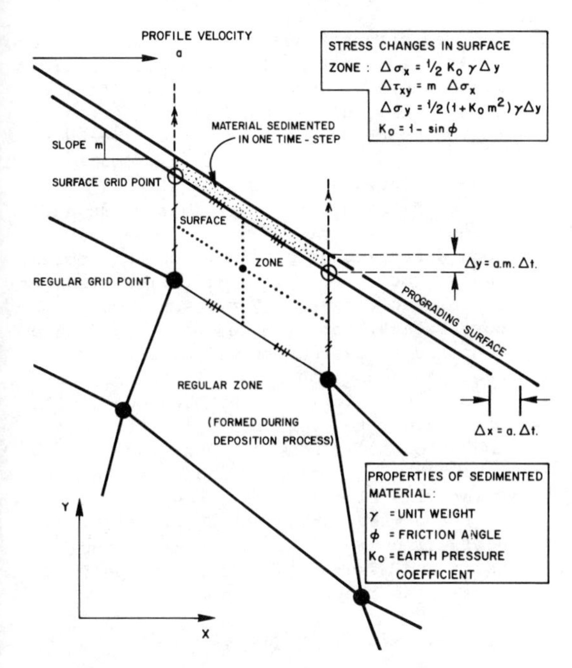

Figure 1. Sedimentation logic.

material velocity. When a surface zone reaches a specified size, it is switched to being treated as a regular zone, and a new surface zone is formed.
The scheme ensures that the growing surface always conforms to a prescribed, time-dependent profile, which implies that locally the effective rate of deposition (or erosion, simply 'negative deposition') varies accordingly.

2.1.3 Material Behaviour

The material behaviour assumed for the frictional overburden is linearly elastic with a limiting Mohr-Coulomb plastic yield surface and a non-associated flow rule (Davis et al., 1974). To model the mobile (salt) substratum, a model of viscous behaviour has been implemented. The model has an elastic volumetric response and portrays Maxwellian behaviour in shear so that the behaviour becomes time-dependent. The viscosity coefficient is prescribed initially (with a spatial variation, if appropriate) and is assumed to remain constant thereafter.

2.2 Validation of GEOMEC for viscous flow

For an ideal incompressible Newtonian fluid, the Navier-Stokes equations can be solved

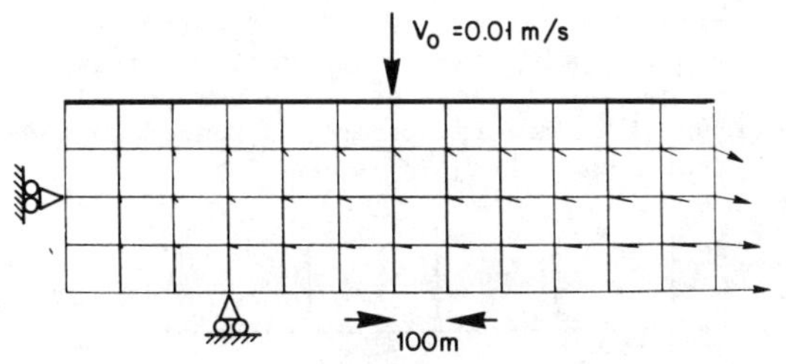

Figure 2. Squeezing of viscous material between parallel plates (steady state velocity vectors superimposed).

for some simple boundary value problems involving plane viscous flow. The analytic solutions apply to an ideal incompressible Newtonian fluid but in the steady state, solutions obtained with the compressible Maxwellian material in GEOMEC should agree. The dimensions and material parameters used in the validations were representative of the values used in the subsequent geological models. The examples of Plane Couette flow, Plane Poiseuille flow and Jaeger's approaching parallel plates (Jaeger, 1956) have been used. In the latter case an excellent test is realised by checking the total force P exerted on the plates (separation 2d, width 2b, velocity of approach $2v_0$),

$$P = 2 \int_0^b \sigma_y(x)\, dx = 2\eta v_0 (b/d)^3 \qquad (7)$$

For the mesh shown in Figure 2, the computed load from GEOMEC is within 1.5% of that given by equation (7).
In fact, for slow rates of flow and steady or near steady state conditions GEOMEC is able to reproduce the analytical predictions in all three cases. Furthermore, the viscosity scaling option was used successfully to obtain these results. These validations encompass the main classes of plane viscous flow that are likely to be encountered in the geological models.

3 DEFORMATION OF SLOPING OVERBURDEN DUE TO MOBILE SUBSTRATUM

The geological setting consists of a gently sloping overburden on a thin mobile substratum of varying thickness. The assumed sequence of events is that an initially uniform overburden (Figure 3a) is subsequently built up by sedimentation to form a wedge-like differential load (Figure 3b) that will cause flow of the substratum in the direction of decreasing load. The time varying deformations in the two-layer sequence and in particular, of the interface, should lead to the features that are of geological interest. The analytical and numerical models attempt to reproduce these features.

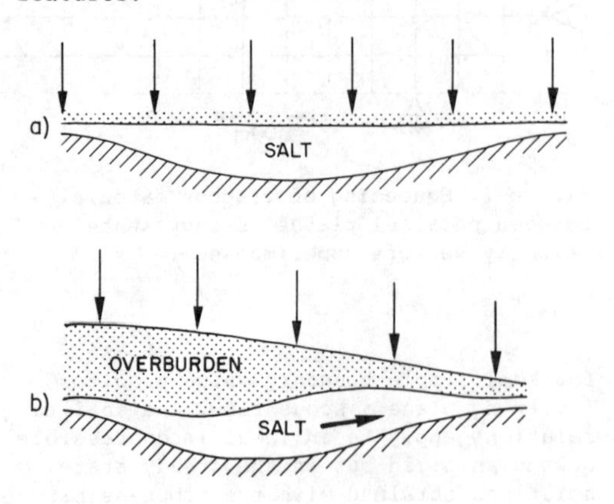

Figure 3. Geological setting.

3.1 Review of the Analytical Model Based on Lubrication Theory

Lehner (1977) examined the deformation of the mobile layer by treating the material as an incompressible, Newtonian viscous fluid. Along the upper and lower surfaces the substratum adheres to the adjacent layers and the upper surface is loaded vertically to reflect the presence of the sloping overburden. Using these basic assumptions, 'Reynolds Equation' can be shown to govern the time-dependent thickness h of the mobile layer (Figure 4) in the form

$$\frac{\partial}{\partial x}(h^3 \frac{\partial p}{\partial x})-6\eta\frac{\partial}{\partial x}[(V_1+V_2)h]-12\eta\frac{\partial h}{\partial t} = 0 \qquad (8)$$

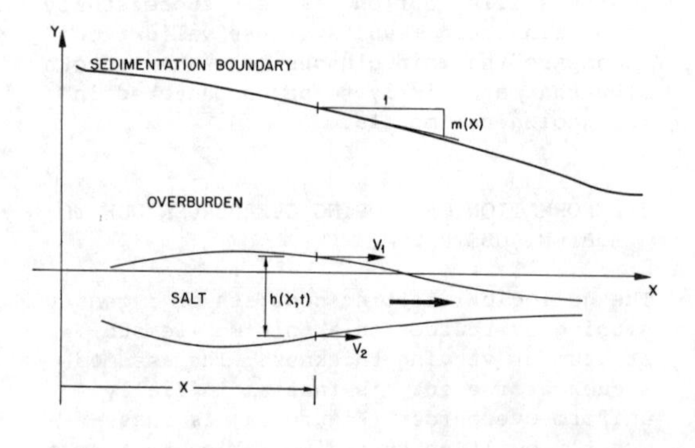

Figure 4. Idealised two-layer sequence used by Lehner (1977).

This equation arises in the theory of hydrodynamic lubrication (for example, Langlois, 1964) where it describes the lubricating effect of a thin viscous layer. In the geological context, to obtain the time-varying shape of the mobile substratum, equation (8) must be integrated. This requires a knowledge of the pressure gradient $\partial p/\partial x$ and the bounding velocities V_1 and V_2. Lehner considered situations in which the mobile layer adheres to a laterally immobile overburden and basement ($V_1 = V_2 = 0$). This corresponds, for example, to a rigid basement and an overburden which suffers negligible horizontal displacements, but is allowed to deform in inhomogeneous simple shear along the vertical. In other words, the second component of flow associated with equation (8) is removed and only the Poiseuille flow remains. This flow is driven by the pressure gradient $\partial p/\partial x$ which, on neglecting the overburden's resistance to shear and assuming a constant slope (m) can be expressed by

$$\partial p/\partial x \simeq -\gamma m \qquad (9)$$

in which γ is the unit weight of the overburden. Equation (8) then readily yields (cf. Lehner, 1977)

$$\left.\frac{dx}{dt}\right|_h = \frac{\gamma m h^2}{4\eta} \qquad (10)$$

which represents the speed at which a vertical section of thickness h propagates horizontally in the downslope direction. Hence, for a given initial geometry, the evolution of the deforming layer can be predicted.

3.2 The Numerical Simulations

The initial geometry and dimensions of the idealised two-layer sequence are outlined in Figure 5 and the principal material properties are given in Table 1.

Table 1. Principal material properties.

Property	Overburden	Salt	Units
Unit Weight	11.0	11.0	KN/m^3
Bulk Modulus	150.0	1500.0	MN/m^2
Shear Modulus	60.0	600.0	MN/m^2
Viscosity	–	10^{17}	Ns/m^2
		(= 10^{18}	poise)
Friction Angle	30.0	–	degrees
Cohesion	0.0	–	MN/m^2
Dilation Angle	0.0	–	degrees

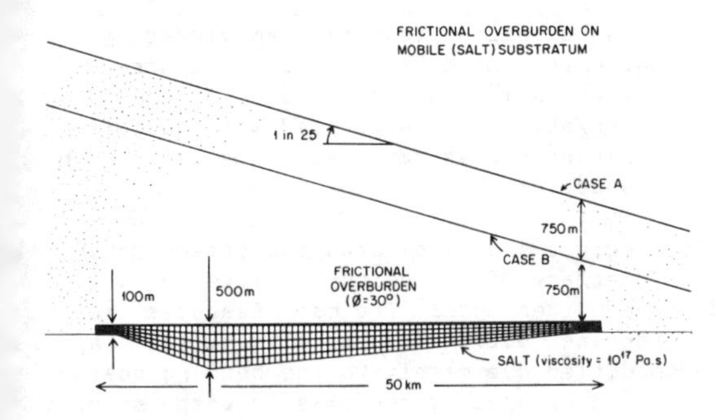

Figure 5. The numerical models.

The overburden was modelled in three ways:

CASE A - A 'thick' frictional layer consisting of elasto-plastic (Mohr-Coulomb) material elements.

CASE B - A 'thin' frictional layer (material elements modelled as in CASE A).

CASE C - A distributed vertical stress that reflects the changing shape of the interface applied directly to the top surface of the substratum.

CASE C corresponds most closely to the lubrication theory model. However, it is not exactly the same because in the numerical simulation no a-priori assumption is made concerning the velocity profile through the viscous layer.

The overburden was deposited during a short period of geological time, with erosion and/or deposition maintaining a fixed sedimentation boundary during subsequent deformation. Figure 6 indicates the deformed numerical grids after about 0.5 million years.

3.3 Interpretation of Computer Simulations

The 'thick' overburden (CASE A, Figure 6a) has remained almost intact but flow has occured in the substratum in the direction of decreasing overburden load. This flow, combined with some lateral movement of the overburden, has created a small depression or basin above the upslope edge of the substratum and a region of compression (with some uplift) above the downslope edge (the 'toe'). The 'thin' overburden (CASE B, Figure 6b) undergoes more severe deformations and a deeper basin (approximately 1 km, Figure 7a) and an anticlinal structure (associated with thrusting, Figure 7b) are produced in the corresponding locations. The horizontal motion of the thick overburden is prevented while the bulk of the thin overburden rides out on the creeping substratum (travelling approximately 2 km after 0.5 million years). The horizontal velocity profile in the substratum exhibit predominantly Poiseuille flow under the thick overburden and Couette flow under the thin overburden. There is no plastic flow in the thick overburden, but in the thin overburden yielding occurs in the regions of concentrated extension (the basin) and compression (the toe) and indicate the areas where faulting might be anticipated. In other words the stresses caused by the tendency for the overburden to move downslope on the lubricating substratum locally exceed the frictional strength of the material.

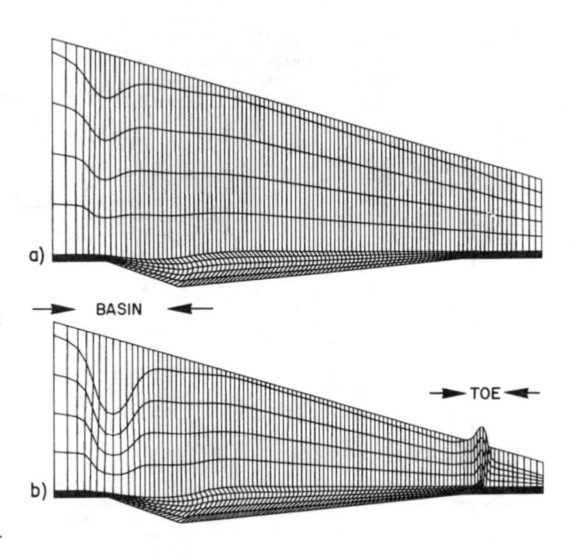

Figure 6. Deformed numerical grids, (a) CASE A and (b) CASE B (not to scale).

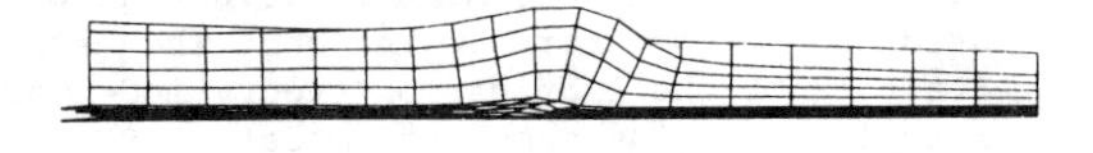

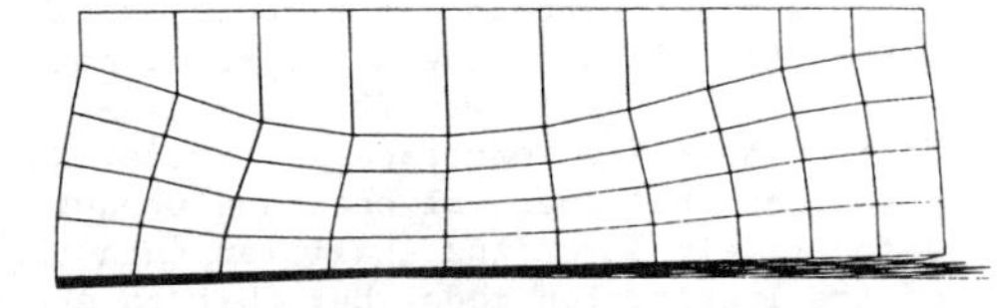

Figure 7. Detail of toe and basin region, CASE B to scale.

This result is clearly significant, because two modes of overall response - yielding and non-yielding - are found to be associated with different overburden thicknesses.

In the simpler case in which the (frictionless) overburden is represented by a distributed load (CASE C), points on the interface are constrained to move vertically and the solution for the interface shape depends only on the overburden gradient (and not on thickness). The velocity profile in the substratum indicates Poiseuille type flow, as assumed in the lubrication theory model. Figure 8 illustrates, on a greatly expanded vertical scale, sections through the 3 models above the deepest part of the substratum.

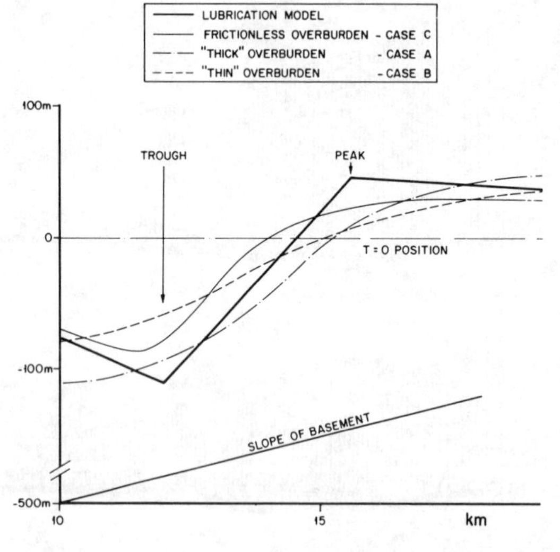

Figure 8. Comparison of interface shapes.

The somewhat unrealistic looking sharp corners of the lubrication theory prediction are simply a reflection of the rather simplistic linear basement profile that was used in the computer simulations. The main features of the lubrication theory model are reproduced by the numerical model (CASE C). Both the position of the maximum layer thickness (the peak) and the trailing depression (the trough) are captured to within the width of one material element (about 500 m), and away from the discontinuities the interfaces coincide. Note also that the numerical peak and trough are respectively lower and shallower than those of the lubrication model but that an overall material balance is maintained (and continuously monitored during simulation). The close agreement between the two approaches is an excellent result; even closer agreement would be possible if a more refined numerical grid was used (the substratum was discretised with just five constant stress material elements through its depth).

Figure 8 also indicates the effect on the interface of including the more realistic overburden model. The main features are retained but, not surprisingly, the shape is modified due mainly to the bending resistance offered by the elastic response of the material and by lateral movement of the overburden.

3.4 Simple Classification Based on the Computed Results

Slope instability will be defined as the conditions which allow significant horizontal motion of the bulk of the overburden (the 'wedge') and will be assumed to occur when the toe and basin regions yield in idealised passive and active Rankine states respectively. Furthermore, for stable slopes the substratum will be treated as a lubricating layer in the manner postulated by Lehner (1977). These assumptions are supported by the results of the computer simulations. Incipiently unstable slopes can now be assessed by examining the overall static equilibrium of the overburden wedge (Figure 9).

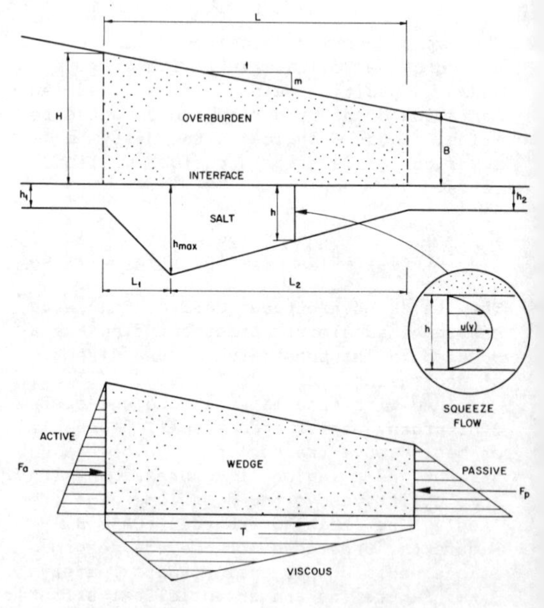

Figure 9. Horizontal equilibrium of the overburden wedge.

In particular, horizontal equilibrium requires

$$F_a + T - F_p = 0 \qquad (11)$$

The active and passive resistances are given through the Rankine earth pressure coefficients (k_a and k_p) by

$$F_a = \frac{1}{2}\gamma H^2 k_a \qquad (12)$$

$$F_p = \frac{1}{2}\gamma B^2 k_p = \frac{1}{2}\gamma B^2 (k_a)^{-1} \qquad (13)$$

Note that the assumption that k_p is simply equal to the inverse of k_a is only strictly true for a horizontal, cohesionless layer. In the numerical model there was no cohesion and the overburden slope was very gentle so that use of this relationship is justified. The shear force T acting along the base of the wedge can be obtained by integrating the shear traction τ at the interface which results from plane Poiseuille flow in a substratum of thickness h. Expressed in terms of the pressure gradient in the substratum, this traction is

$$\tau = -\frac{h}{2}\frac{\partial p}{\partial x} \qquad (14)$$

and by assuming a constant slope,

$$\tau = \frac{h}{2}\gamma m \qquad (15)$$

The thickness h varies gently through the lateral extent of the substratum and can be described by

$$h(x) = h_1 + (h_{max} - h_1)x/L_1 \quad \text{for } x=0 \text{ to } x=L_1$$

$$h(x) = h_{max} - (h_{max} - h_2)(x-L_1)/(L-L_1) \quad \text{for } x=L_1 \text{ to } x=L$$

Hence the shear force can be evaluated as

$$T = \int_0^L \tau(x)\,dx$$

$$= \frac{1}{2}\gamma m \int_0^{L_1} h(x)\,dx + \frac{1}{2}\gamma m \int_{L_1}^{L} h(x)\,dx$$

$$= \frac{1}{4}\gamma m(Lh_{max} + L_1 h_1 + L_2 h_2) = \frac{1}{2}\gamma m A \qquad (16)$$

in which A is the cross-sectional area of the substratum. This simple result holds true only when the overburden slope (and therefore the pressure gradient) is constant. Furthermore, an averaged value $\bar{h}$ for the substratum depth is introduced and defined by

$$A = L\bar{h} \qquad (17)$$

so that substitution of (12), (13), (16) and (17) into equation (11) leads to

$$(H/L)^2(1-k_a^2) - 2m(H/L) + [m^2 - k_a(m\bar{h}/L)] = 0$$

$$\text{for } B/L = H/L - m \geq 0 \qquad (18)$$

in which $\bar{h}/L$ represents a normalised measure of the average depth of the substratum which should be small if the conditions assumed in the analysis are to remain applicable. This result (18) is interesting since it involves neither the coefficient of viscosity nor the position of maximum depth of the substratum. Solutions to this equation are shown in Figure 10 for a range of appropriate values of the geometric and material parameters.

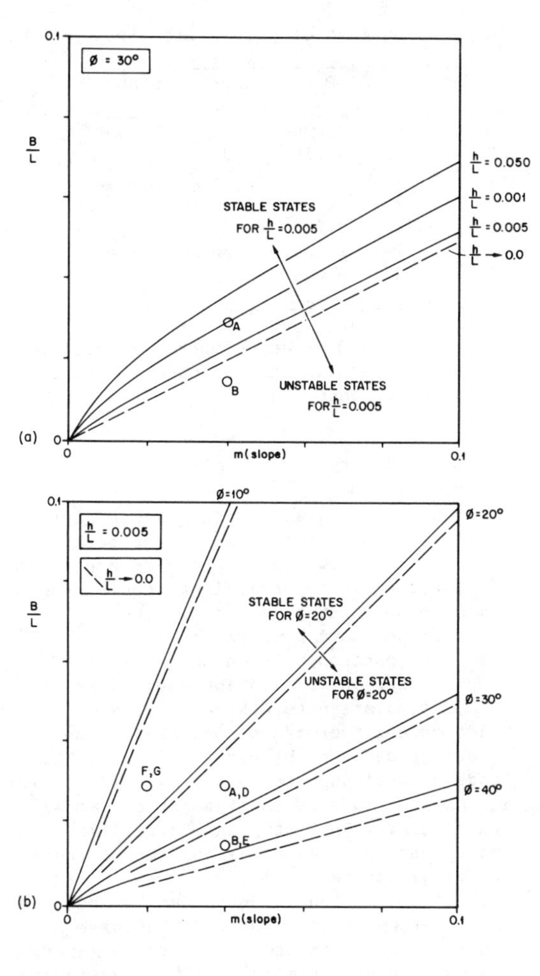

(a) Effect of substratum depth

(b) Effect of overburden strength (Ø = friction angle)

Figure 10. Stable and unstable states for the idealised overburden wedge.

Notice that for relatively shallow substrata ($\bar{h}/L \leq 0.01$ say) the effect of the shear force (17) is quite small. Indeed, if this force is ignored, solutions to equation (18) simplify to

$$H/L = m/(1-k_a) \quad \text{for } B/L = H/L - m \geq 0 \qquad (19)$$

and these are shown as dashed lines in Figure 10.

The curves in Figure 10 separate in a broad sense unstable (yielding) and stable (non-yielding) overburden states. The conditions pertaining to the computer simulations are marked as points A and B. Furthermore, additional simulations have been performed (Table 2) and are included in Figure 10.

Table 2. Summary of parametric study.

CASE	m	B/L	$\phi°$	stable?
A	0.04	0.030	30	yes
B	0.04	0.015	30	no
C	0.04	0.015	-	-
D	0.04	0.030	20	no
E	0.04	0.015	40	yes
F	0.02	0.030	20	yes
G	0.02	0.030	10	no

Clearly the additional computer simulations (CASES D to G) support the general trends indicated by equation (18).

4 CONCLUDING REMARKS

The computer program GEOMEC has been further developed and used to examine the behaviour of a deforming two-layer sequence consisting of a gently sloping frictional overburden on a mobile substratum. The following observations have been made:

1. A comparison between Lehner's (1977) model (based on lubrication theory) and a GEOMEC simulation for the case of a frictionless overburden indicates close agreement between the two treatments. This lends general support to the validity of the assumptions made in the theoretical model (particularly that of plane Poiseuille flow) and serves as a useful validation of the numerical scheme.

2. If a frictional overburden is added to the numerical model, the main features of the lubrication theory model are retained, but the shape of the travelling wave in the overburden/substratum interface is modified, depending on the detailed behaviour of the overburden. For a non-yielding material the interface is smoothed out due to bending resistance and there is negligible horizontal movement of the overburden, while Poiseuille type flow dominates in the substratum. This case conforms most closely with the lubrication theory model. For conditions leading to material yielding and slope instability, large horizontal movements of the overburden are observed and a large component of Couette (simple shear) type flow is present in the substratum.

3. The computer simulations have shown the effect of overburden thickness on a particular two-layer sequence, but suggest a broader classification of behaviour which has been quantified in a simple way by considering horizontal (static) equilibrium for conditions pertaining to incipient slope instability. Further numerical tests need to be performed to establish the general validity of this classification (for example, the sensitivity to salt depth has not been investigated) but the results reported here support the predicted trends.

4. For conditions which promote significant yielding and subsequent movement of the overburden, regions in which faulting might be anticipated are indicated by a deep extensional basin upslope and a ridge structure associated with thrusting near the toe. In particular, growth faulting can be postulated at the back of the basin region with overthrusting at the toe. To explicitly model these features, further development of GEOMEC is required and is under investigation.

5. The general capabilities of GEOMEC to model situations involving coupled creep (viscous) and frictional behaviour together with a sedimentation boundary have been demonstrated.

REFERENCES

Cundall, P.A. 1976. Explicit Finite-Difference Methods in Geomechanics. Int Conf Num Meth in Geom, Vol.1: 132-150, Blacksburg, Virginia.

Cundall, P.A., Hansteen, H., Lacasse, S. and Selnes, P. 1980. NESSI - Soil Structure Interaction Program for Static and Dynamic problems. Norwegian Geotechnical Institute report 51509-8.

Cundall, P.A. 1982. Adaptive Density Scaling for Time-Explicit Calculations. 4th Int Conf Num Meth in Geom, Edmonton, Canada, Vol.1: 23-26.

Davis, E.H., Ring, G.J. and Booker, J.R. 1974. The Significance of the Rate of Plastic Work in Elasto-Plastic analysis. Univ of Sydney, School of Civ Eng, report 242.

Evamy, B., Haremboure, J., Kamerling, P., Knaap, W., Malloy, F. and Rowlands, P. 1978. Hydrocarbon Habitat Of Tertiary Niger Delta. AAPG Bulletin, Vol.62: 1-39.
Jaeger, J.C. 1956. Elasticity, Fracture and Flow with Engineering and Geological Applications. Chapter 3, Methuen, London.
Langlois, W.E. 1964. Slow Viscous Flow. Chapter 9, MacMillan, London.
Lehner, F.K. 1977. A Theory of Substratal Creep Under Varying Overburden With Applications to Tectonics. AGU Spring Meeting, Washington D.C., EOS Abstr, Vol.58: 508.
Merki, P. 1972. Structural Geology of the Cenozoic Niger Delta. Proc 1st Conf African Geol, Ibadan Univ, Nigeria: 635-646.
Trusheim, F. 1960. Mechanism of Salt Migration in Northern Germany. AAPG Bulletin, Vol.44, No.9: 1519-1540.
Wilkins, M.L. 1964. Calculation of Elastic-Plastic Flow. Methods Comput Phys, Vol.3: 211-263.

Numerical Methods in Geomechanics (Innsbruck 1988), Swoboda (ed.)
© 1988 Balkema, Rotterdam. ISBN 90 6191 809 X

Evaluation of a quasi-dynamic algorithm for soil consolidation problems

D.Kovačić
Geotehnika Co., Department Geoexpert, Zagreb, Yugoslavia
A.Szavits-Nossan
Faculty of Civil Engineering, Zagreb University, Yugoslavia

ABSTRACT: The paper deals with quasi-dynamic algorithm for the solution of the equilibrium problem approximated by the finite element method. The proposed algorithm avoids the solution of the system of equations and it can be used without modifications to solve nonlinear problems. The procedure is applied to the consolidation problem. The paper provides the assessment of the efficiency of the proposed algorithm.

1 INTRODUCTION

Use of the finite element method to soil consolidation problems is now well-established. Starting from the first application almost two decade ago /1/, much improvement on theoretical formulations as well as on computational procedures has been done. In one of the latest papers on the subject one can find a comprehensive list of references /2/. Recent advances include dynamic response of saturated soil /3/, /4/, the elasto-viscoplastic constitutive model for clay /5/, non-linearity of soil modulus and coefficient of permeability /6/ as well as infinite boundaries /7/.

The present paper deals with the application of the dynamic relaxation technique /8/ to the solution of two-dimensional FE equations of equilibrium. These equations use relative water displacements as a fundamental variable of the fluid phase in contrast to the usually adopted pore water pressure.

2 THE FIELD EQUATIONS

The model for the two-phase soil medium is based on the following assumptions:

- pores of the deformable skeleton are fully saturated by a compressible fluid,
- due to incompressibility of the soil particles the soil volume change is equal to the pore volume change,
- the flow of the fluid through the soil skeleton is governed by Darcy's law,
- Terzaghi's effective stress principle is satisfied,
- quasi-stationary motion applies, i.e. the acceleration forces are neglected,
- there is no restriction on the type of constitutive equations of the skeleton and the fluid,
- deformations and velocities are assumed to be small.

As already stated the present formulation uses two sets of homogeneous independent variables. The displacement vector for the soil skeleton is denoted by u, the displacement vector for the fluid is denoted by v, and the relative displacement of the fluid with respect to solid is defined as $w = n\,(u - v)$ where n is soil porosity.

The strains are given by

$$\varepsilon_s = -\tfrac{1}{2}\,[\mathrm{grad}\ u + (\mathrm{grad}\ u)^T] \qquad (1)$$

$$\varepsilon_f = -\,(\mathrm{div}\ u + \mathrm{div}\ v)/n \qquad (2)$$

where ε_s is the strain tensor of the soil skeleton and ε_f is the volumetric strain of the fluid.

The constitutive equation for the soil skeleton takes the general form of a functional

$$\sigma' = \sigma'(\varepsilon_s) \qquad (3)$$

σ' is the effective stress tensor which is related to the total stress tensor σ and pore fluid pressure p by

$$\sigma' = \sigma - p \cdot I \qquad (4)$$

and I is the unit tensor.

Similarly the general form of the constitutive equation for the fluid is

$$p = p(\varepsilon_f) \tag{5}$$

In the case of a linear elastic fluid with bulk modulus B_f the relation (5) reduces to

$$p = B_f \cdot \varepsilon_f \tag{6}$$

The equations of equilibrium are

$$-\text{div}(\sigma' + p \cdot I) + b = 0 \tag{7}$$

The generalized Darcy's law is given as

$$\dot{w} = k \cdot (-\text{grad } p + \gamma_f)/\gamma_f \tag{8}$$

where $\dot{w}$ is relative fluid velocity, k is the permeability tensor and γ_f is the fluid unit weight.

Equations (7) and (8) along with the initial and boundary conditions and constitutive relations completely define the mixed boundary value problem.

3 FINITE ELEMENT DISCRETIZATION

3.1 Spatial discretization

The finite element equivalent of equations (7) and (8) is

$$S + W = f \tag{9}$$

$$W + F \cdot \dot{w} = f_f \tag{10}$$

where

$S = S(u)$ ..... vector of skeleton nodal forces
$W = W(u+w)$ ... vector of fluid nodal forces
f ... vector of skeleton nodal load (external forces)
F ... permeability matrix
f_f ... vector of fluid nodal load (external forces)

Substracting (9) from (10) the set of equations becomes

$$S + W = f \tag{9 bis}$$

$$F \cdot \dot{w} = S - (f - f_f) \tag{11}$$

3.2 Time discretization

In order to approximate the variation in time the equations (9) and (11) are integrated over a time step (from t_1 to t_2). Therefore the equations (9) and (11) are transformed into

$$S_2 + W_2 = f_2 - (S_1 + W_1 - f_1)(1 - \theta)/\theta, \quad \theta > 0 \tag{12}$$

$$w_2 = \Delta t \cdot \{[S_1(1 - \theta) + S_1 \cdot \theta] - [f_1(1 - \theta) + f_2\,\theta - f_{f1}(1 - \theta) - f_{f2} \cdot \theta]\}/F + w_1 \tag{13}$$

In equations (12) and (13) Δt denotes time increment and indexes 1 and 2 denote the value of the above defined vector at appropriate time. The value of parameter depends on the type of interpolation used. The system of equilibrium equations (12) ought to be solved at each time step.

The dynamic relaxation procedure is applied to the solution of equilibrium equations. The concept of the dynamic relaxation technique is to extend the static problem to a fictitious dynamic problem by adding a suitably adopted inertial term to equilibrium equations and then extracting the kinetic energy from the system until the static equilibrium state is reached. The kinetic energy extraction is achieved by setting the velocity vector of the oscilating system to zero at each kinetic energy peak.

This method has been already applied to the solution of static equilibrium problem using finite difference scheme for time integration /9/.

4 SOLUTION OF EQUATIONS

4.1 Dynamic relaxation algorithm

The application of the above described procedure to the equations (12) leads to

$$M \cdot \ddot{u}_2 + F_{int} = F_{ext} \tag{14}$$

or

$$\ddot{u}_2 = M^{-1} \cdot R \tag{15}$$

where M is the mass matrix, $\ddot{u}_2$ is the nodal acceleration vector, $F_{int} = S_2 + W_2$, $F_{ext} = f_2 - (S_1 + W_1 - f_1)(1 - \theta)/\theta$ and $R = F_{ext} - F_{int}$ is the vector of the residuals, i.e. the unbalanced nodal forces.

The equation (15) is used with an explicit integration procedure in fictitious time T. The equation is integrated by the Newmark method /10/ with $\beta = 0$ /11/ which gives

$$\dot{u}_{T+\Delta T} = \dot{u}_T + ½\Delta T\cdot(\ddot{u}_{T+\Delta T} + \ddot{u}_T) \quad (16)$$

$$u_{T+\Delta T} = u_T + \Delta T\cdot\dot{u}_T + ½\,\Delta T^2\cdot\ddot{u}_T \quad (17)$$

In these expressions the nodal acceleration vectors are defined by

$$\ddot{u}_T = M^{-1}\cdot R_T \quad (18)$$

$$\ddot{u}_{T+\Delta T} = M^{-1}\cdot R_{T+\Delta T} \quad (19)$$

The lumped, diagonal mass matrix is used. For the sake of simplicity subscript 2 is omitted in expression (16) to (19) and in the further derivation.

The stability limit for the explicit integration scheme is given by /12/

$$\Delta T \le \Delta T_{cr} = T_n/\Pi \quad (20)$$

where ΔT_{cr} is the critical time step and T_n is the smallest natural period of the corresponding eigenvalue of the equivalent linear system.

For 1D linear element

$$\Delta T_{cr} = L\ /\ [(M_c+B_f/n)(1+2\beta\theta)/\rho]^{½} \quad (21)$$

$$\beta = 2\cdot\alpha\cdot c_v\cdot\Delta t\ /\ L^2 \quad (22)$$

$$c_v = k\cdot M_c/\gamma_f \quad (23)$$

$$\alpha = (1 + M_c/B_f)^{-1} \quad (24)$$

where L is the length of the element, M_c is the modulus of compressibility of the soil and ρ is the fictitious mass density.

The dynamic relaxation algorithm was implemented into the finite element computer program taking into account the fact that at kinetic energy peak E_k its first time derivative is equal to zero. This statement is equivalent to

$$R^T\cdot\dot{u} = 0 \quad (25)$$

In the discrete system the condition (25) could be satisfied only approximately. In other words some criteria ought to be established in order to measure the accuracy of the solution. These criteria are defined through nondimensional parameters α_P and α_R.

4.2 Accuracy parameters α_P and α_R

Taking into account that condition (25) should be satisfied approximately and denoting this product by P, the accuracy of P is defined by

$$\frac{|P|}{|P_{max}|} \le \alpha_P \quad (26)$$

P_{max} is the value of the product at maximum value of all $E_{k\ max}$.

The condition that all nodal residuals are approximately equal to zero is defined by

$$\frac{R_{norm}}{F_{norm}} \le \alpha_R \quad (27)$$

where $R_{norm} = (\Sigma R^2/N)^{½}$, N is the number of degrees of freedom, F_{norm} is the sum of all external nodal loads.

5 VERIFICATION AND ACCURACY OF THE ALGORITHM

In order to assess the efficiency of the proposed algorithm and to check its sensitivity to the chosen values of parameters α_P and α_R a series of test cases have been done. The results of 1D linear static equilibrium as well as 1D and 2D consolidation are presented.

5.1 Linear static equilibrium problem

The proposed algorithm has been checked first of all on a simple problem - one-dimensional linear elastic static equilibrium. The test cases with different number of elements have been used. They showed that the value of $\alpha_P < 0.1$ had no significance to the accuracy of the algorithm. On the other hand the same tests showed that the value $\alpha_R = 0.001$ gave safisfactory accuracy, i.e. the error of u_{max} and ε_{max} is less then 1% .

Among a number of tests the results of 10 linear elements model are shown in Table 1. in which K denotes number of fictitious time increment T , NPEAK denotes the number of kinetic energy peaks, u_{max} is the displacement of the top node and ε_{max} is the maximum strain in an element.

Table 1. Accuracy tests for α_R parameter

α_R	K	NPEAK	u_{max}	ε_{max}
0.1	19	2	1.0256	1.1322
0.01	30	6	1.0237	1.0465
0.001	53	12	1.0037	1.0051

(Exact values are: u = 1.000 and ε = 1.000)

Figure 1. shows the graphic presentation of the "velocity damping" for this example.

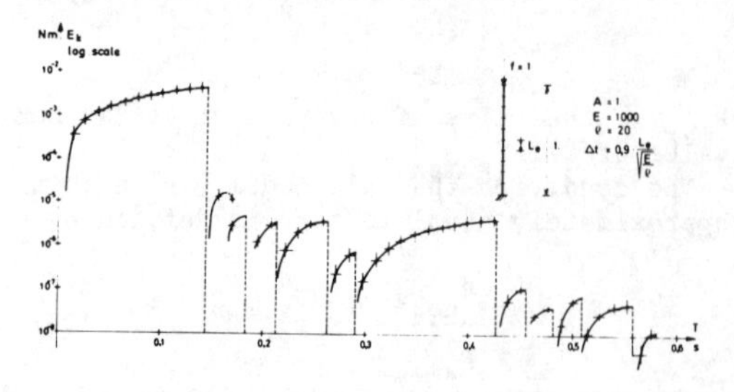

Fig 1. Kinetic energy versus fictitious time plot for 10 elements model

5.2 One-dimensional consolidation

One-dimensional consolidation problem was approximated also with linear elements. The initial excess pore pressure field is assumed to be constant and the process of consolidation was simulated with the ratio $B_f / M_c = 2000$. Backward time interpolation (implicit integration scheme) was applied, $\theta = 1$, with constant time intervals.

The results of the 20 elements model are presented. The behaviour of the dynamic relaxation algorithm is shown by means of the excess pore pressure versus time plot for bottom element (Fig. 2). Taking $\alpha_P = 0.1$ as constant value the variation in α_R shows again that the accuracy of the calculation of residuals influences the state of equilibrium, i.e. the correct distribution of the excess pore pressure. Like in the prevoius example it may be concluded that the value $\alpha_R = 0.001$ gave safistactory accuracy.

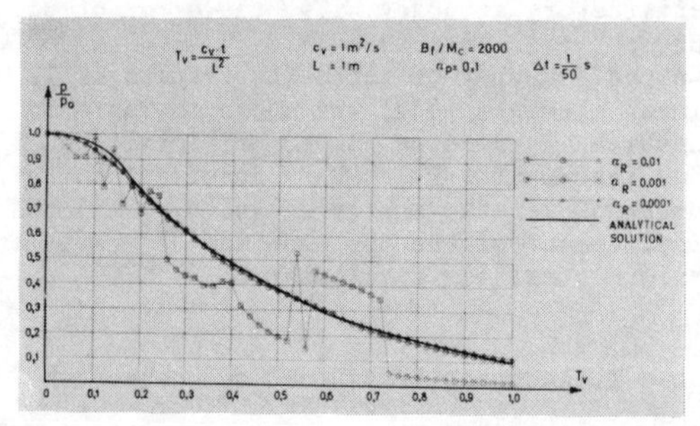

Fig.2 Excess pore pressure versus time plot in bottom element

Table 2. shows that while the accuracy is increased for two orders of magnitude the number of fictitious time steps is almost doubled.

Table 2. Accuracy tests for α_R parameter

α_R	K	NPEAK
0.01	11787	198
0.001	17656	288
0.0001	21153	352

5.3 Two-dimensional consolidation

Layer of finite thickness subject to surface load over a finite area is the usual bench mark problem for two-dimensional consolidation. The example of plain strain consolidation is approximated with four-node quadrilateral elements (Fig. 3). Bilinear shape functions are used to interpolate both u and w inside an element. Vectors of nodal forces are obtained using Gauss rule - 2x2 points (exact integration) for effective stresses and 1 point (reduced integration) for pore pressure /13/.

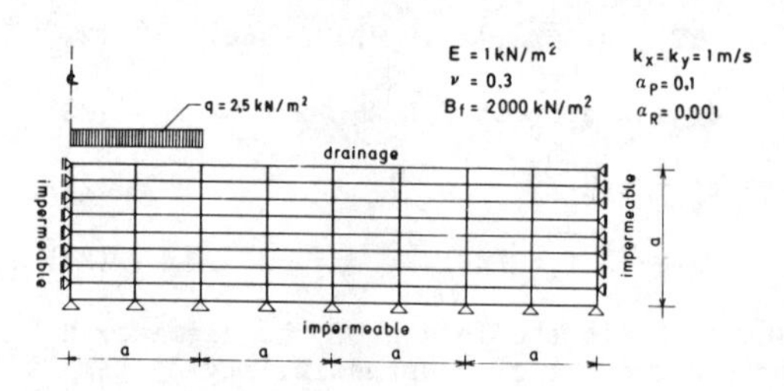

Fig. 3 Finite element mesh for plain strain consolidation

The analysis is carried out in two stages. In the first stage the undrained conditions are simulated by taking $B_f / M_c = 200$. As $w \equiv 0$ only the set of equations (9) remains active.

The effective stress and pore pressure field at the end of this stage of the analysis becomes the initial stress field for the consolidation stage. In the second stage $f = 0$, equations (9) and (11), as there is no additional external load during consolidation. f_f is the excess pore water pressure vector. Implicit time integration scheme has been applied along with logarithmic time increment /14/.

Figure 4. shows the surface displacements distribution. The cumulative displacement curve (undrained stage + consolidation) is in the excellent agreement with the same result of the drained analysis. (Ideally for elastic material and theoretically completed pore pressure dissipation these should be the same.) The drained analysis has been carried out the same way as the undrained analysis but taking $B_f = 0$.

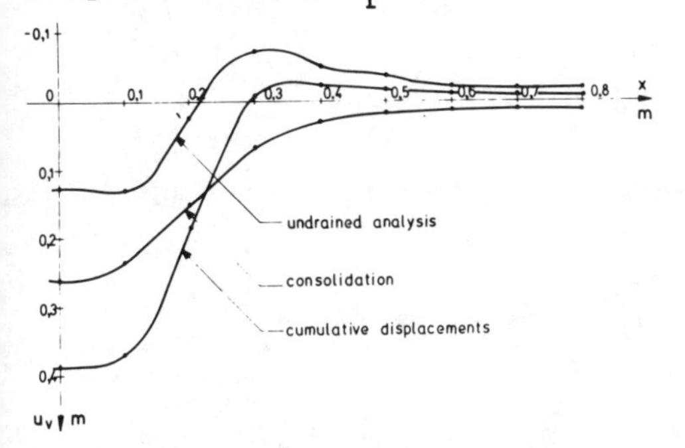

Fig.4 Surface displacements distribution

Table 3. gives an insight into the efficiency of the proposed algorithm in terms of computer time.

Table 3. Efficiency of the algorithm

Type of analysis	K	CPU time
drained	77	36 s
undrained	511	3 min
consolidation	6376	40 min

6 CONCLUSION

Introduction of relative pore water displacement as the field variable instead of the pore water pressure enables the use of the same interpolation functions for the fluid phase as for the soil skeleton. It was shown that the dynamic relaxation algorithm is applicable to consolidation problems. This algorithm bypasses the need for the assembly of the global stiffness matrix since it operates on element level. It is also perfectly suited for easy implementation of nonlinear constitutive equations. The programming of the algorithm results in a simple and tractable structure of the computer program.

BIBLIOGRAPHY

/1/ R.S.Sandhu and E.L.Wilson, Finite element analysis of seepage in elastic media, J. eng. mech. div. ASCE, 95, 641-652 (1969)

/2/ R.S.Sandhu, S.C.Lee and H.The, Special finite elements for analysis of soil consolidation, Int. j. numer. anal. methods geomech. 5, 125-147 (1985)

/3/ B.R.Simon, J.S.-S. Wu, O.C.Zienkiewicz and D.K.Paul, Evaluation of u-w and u-Π finite element methods for the dynamic response of saturated porous media using one-dimensional models, Int. j. numer. anal. methods geomech. 10, 461-482 (1986)

/4/ R.S.Sandhu and S.J.Hong, Dynamics of fluid-saturated soils - variational formulation, Int. j. numer. anal. methods geomech. 11, 241-255 (1987)

/5/ F.Oka, T.Adachi and Y.Okao, Two-dimensional consolidation analysis using an elasto-viscoplastic constitutive equation, Int. j. numer. anal. methods geomech. 10, 1-16 (1986)

/6/ H.R.Thomas and P.S.Chan, Non-linear one-dimensional consolidation analysis on a microcomputer, Microcomputers in engineering (ed. B.A.Schrefler and R.W. Lewis), Pineridge Press, Swansea, 1986

/7/ L.Simoni and B.A.Schrefler, Land subsidence analysis using mapped infinite elements, Microcomputers in engineering (ed. B.A.Schrefler and R.W. Lewis), Pineridge Press, Swansea, 1986

/8/ P.Cundall, Explicit finite difference methods in geomechanics, Numerical methods in geomechanics (ed.C.S.Desai), ASCE, 132-150 (1976)

/9/ A. Szavits-Nossan and D. Kovacic, A relaxation stress-deformation finite element program, Numerical methods in geomechanics Aachen 1979 (ed. W.Wittke), Balkema, Rotterdam, 1979

/10/ N.W.Newmark, Computation of dynamic structural response in range approaching failure, Proceedings of the symposium on earthquake and blast effects on structures, Los Angeles,1952

/11/ T.Belytschko, R.L.Chiapetta and H.D.Bartel, Efficient large scale non-linear transient analysis by finite elements, Int. j. num. meth. eng. 10, 579-596 (1976)

/12/ K.J.Bathe and E.L.Wilson, Numerical methods in finite element analysis, Prentice Hall, Englewood Cliffs, New Jersey, 1976

/13/ D.J.Naylor, Stresses in nearly incompressible materials by finite elements with application to the calculation of excess pore pressures, Int. j. num. meth. eng. 8, 443-460 (1974)

/14/ C.T.Hwang, N.R.Morgenstern and D.W.Murray, On solutions of plain strain consolidation problems by finite element methods, Canadian geotechnical journal 8, 109-118 (1971)

Numerical Methods in Geomechanics (Innsbruck 1988), Swoboda (ed.)
© 1988 Balkema, Rotterdam. ISBN 90 6191 809 X

Consolidation of layered soils under time-dependent loading

J.C.Small & J.R.Booker
School of Civil and Mining Engineering, University of Sydney, Australia

ABSTRACT: A finite layer method is used to compute the behaviour of a layered soil which has anisotropy of permeability and which is subjected to a time dependent loading. The method relies upon obtaining the solution in Laplace transform space and then using numerical inversion to produce the time dependent solution.

Advantages of the method are that it requires very little computer storage and can be run on a microcomputer, it can directly produce a result at any nominated time which has advantages over 'marching' solutions, and that it can be used to solve realistic three dimensional problems involving layered soils with anisotropic permeability.

1. INTRODUCTION

The solution of problems involving the consolidation of a soil layer under an applied loading has long been of interest to engineers. While the simple one-dimensional theory advanced by Terzaghi (1923) is still widely used and may be applicable under some circumstances, it is recognised that for three dimensional loading conditions this simple theory may be grossly in error.

Solutions to problems involving consolidation under three dimensional loading conditions (and which use the three dimensional theory proposed by Biot (1941)) have been obtained by many authors (Gibson and McNamee, 1957; Gibson, Schiffman and and Pu (1970); Booker (1974)) in an attempt to model field behaviour more closely. However, because of the complexity of the problem, true analytic solutions such as those mentioned above are difficult to obtain and therefore have been limited to soils which are homogeneous.

In reality one of the most important three dimensional effects is that of anisotropy of the soil, especially anisotropy of permeability. Many natural soils are also produced by sedimentation which means that the soil consists of different layers of material. Layering and anisotropy greatly complicate the solution process and this has led to the use of numerical techniques. Finite element formulations such as those proposed by Christian and Boehmer (1970), or Sandhu and Wilson (1969) have enjoyed widespread use as they allow anisotropy and material layers of any shape to be dealt with.

While numerical methods of solution allow complex problems to be solved, the amount of effort required in producing the meshes, and the large computer storage required to solve the resultant equations remains a disadvantage. Although full three dimensional problems may be solved, the amount of computer storage required is usually prohibitive and analysis is restricted to the axisymmetric or plane strain cases which involve only two spatial dimensions.

In this paper, the finite layer method is used to obtain solutions to the problem of a layered anisotropic soil subjected to a three dimensional surface loading. Provided that we can assume the soil layers are uniform and horizontal (i.e. they do not dip appreciably and are roughly of uniform thickness), we may greatly simplify the solution process by use of the finite layer method. This approach has been used to consider the consolidation of isotropic layered deposits by Runesson and Booker (1982), and Vardoulakis and Harnpattanapanich (1986). In this paper the approach developed by Booker and Small (1987) is extended to account for anisotropy of the flow properties of the stratified soil.

2. BASIC EQUATIONS

If we consider the horizontally layered soil profile shown in Figure 1a, where a soil layer (i) is considered to lie between z coordinates z_i and z_{i+1}, we may write the

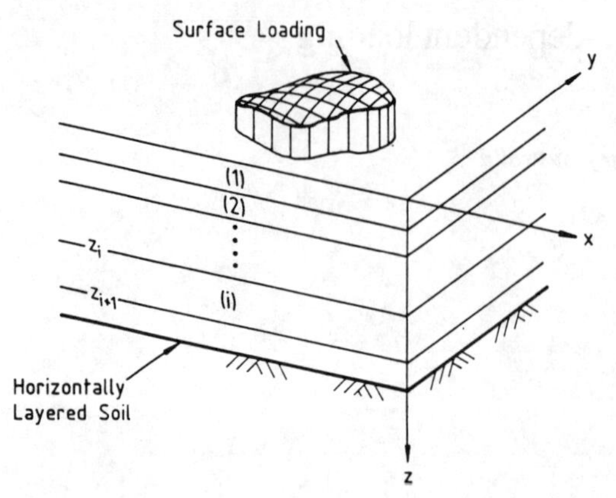

Figure 1a: Surface loading applied to layered soil.

following equations which govern the deformation of the layer.

(a) The equations of equilibrium

$$\sigma_{j\ell,\ell} = 0 \tag{1}$$

where

$\sigma_{j\ell}$ are the components of the total stress increases over the initial stress state. The indices j, ℓ vary over the index set (x, y, z) and the summation convention for repeated indices is adopted. A comma denotes a partial differentiation.

(b) It will be assumed that the soil skeleton is isotropic with respect to its elastic properties so that we may write the stress-strain relationship

$$\sigma'_{j\ell} = \lambda \epsilon_v \delta_{j\ell} + 2G\epsilon_{j\ell} \tag{2}$$

where

$\sigma'_{j\ell} = \sigma_{j\ell} + p\ \delta_{j\ell}$ is the effective stress increase
p is the excess pore water pressure
$\epsilon_{j\ell}$ denotes the strain components
$\epsilon_v = \epsilon_{\ell\ell}$ is the volume strain
and $\delta_{j\ell}$ is the Kronecker delta.

λ and G are the Lamé and shear modulus of the soil respectively.

(c) The strain-displacement relationship. The components of strain are related to the displacements u by:

$$\epsilon_{j\ell} = (u_{j,\ell} + u_{\ell,j})/2 \tag{3}$$

(d) If we assume that the soil particles are incompressible compared to the soil skeleton and that the pore fluid is incompressible it follows that the decrease in volume of a soil element is equal to the volume of water squeezed out of that element, and so:

$$\epsilon_v = -\int_0^t v_{\ell,\ell} dt \tag{4}$$

v_ℓ denotes the components of the superficial velocity of the pore water relative to the soil skeleton.

(e) It will also be assumed that the flow of water in the soil is governed by Darcy's Law and that the permeability in the horizontal (x-y) direction k_h is different to that in the vertical (z) direction k_v so that:

$$v_j = -k_{j\ell} p,_\ell / \gamma_w \tag{5}$$

where $k_{yy} = k_{xx} = k_h$
$k_{zz} = k_v$
and $k_{j\ell} = 0 \quad j \neq \ell$

3. ANALYSIS OF PLANE STRAIN

If a soil deforms under conditions of plane strain, so that deformation is restricted to the x-y plane, we can greatly simplify the equations by applying a Fourier transform eg:

$$(U,W,P) = \frac{1}{2\pi}\int_{\infty}^{+\infty} (iu_x, u_z, p)e^{-i\rho x}dx \tag{6a}$$

with the equivalent inverse transform being

$$(u_x, u_z, p) = \int_{\infty}^{+\infty} (iU, W, P)e^{+i\rho x}dx \tag{6b}$$

For the stresses we may write the transforms as

$$(N,T) = \frac{1}{2\pi}\int_{\infty}^{+\infty} (\sigma_{zz}, i\tau_{xz})e^{-i\rho x}dx \tag{6c}$$

Once this has been done we may write a relationship between the transformed quantities at the top of a layer of material and the bottom of a layer e.g.

$$\begin{bmatrix} A & B \\ B^T & C \end{bmatrix} \begin{bmatrix} \bar{\Delta}_a \\ \bar{\Delta}_b \end{bmatrix} = \begin{bmatrix} -\bar{F}_a \\ \bar{F}_b \end{bmatrix} \tag{7}$$

The matrices A, B, C are defined in full in Appendix A and

$$\Delta = (U, W, P)^T$$
$$F = (T, N, Q)^T \tag{8}$$

Q represents the Fourier transform of the quantity q defined by:

$$q = \int_0^t v_z \, dt$$

The subscripts a, b indicate that the variables are at the top or the bottom of the layer respectively, see Figure 1(b). We therefore have an expression relating the transformed shear stress T, normal stress N and flow Q to the transformed horizontal displacement U, vertical displacement W and pore pressure P for each layer. We need now to propose some conditions which exist on the boundaries between each individual soil layer. Once this has been done we can assemble all of the layer equations (such as Equation 7) to form the overall set of equations for the layered system.

By specifying values of **Δ** or **F** at the boundaries we are able to solve the equations for the quantities

$$\bar{\Delta} = (\bar{U}, \bar{W}, \bar{P})^T$$

at each layer interface.

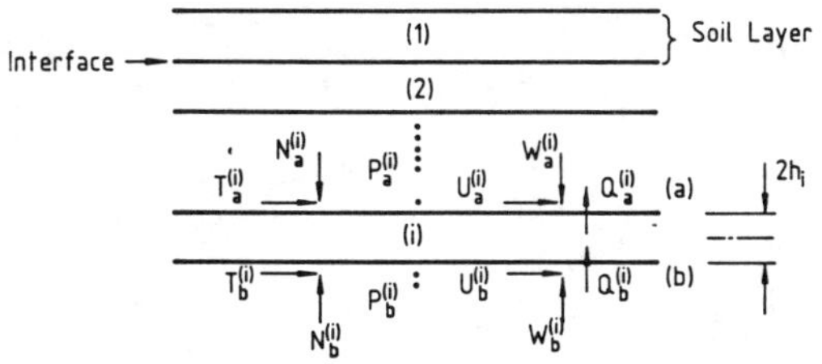

Figure 1b: Transformed quantities at an interface.

4. NUMERICAL INVERSION

Once the transformed field quantities are known it remains to apply an inverse Fourier and an inverse Laplace transform to obtain the final result. This may be done numerically, and in this paper the Fourier inversion was carried out using Gaussian quadrature. To do this the infinite integral such as that of equation (6b) must be truncated at some large but finite value (say A). For example we could evaluate the excess pore pressure from

$$p = \int_{\infty}^{+\infty} Pe^{i\rho x} \approx 2\int_0^A P \cos(\rho x) dx \qquad (9)$$

if P is an even function of x (which would be the case for a symmetric strip loading).

Integrals such as that of equation 9 were numerically determined by use of 80 evaluations at Gauss points in this paper.

In order to invert the Laplace transform the method described by Talbot (1979) was used. This involved evaluation of the transformed field quantities at 20 values of the Laplace transform parameter s. Hence to obtain the final solution the set of equations was solved a total of 80 x 20 = 1600 times to evaluate both the inverse Fourier and Laplace transforms.

5. TRANSFORMED EQUATIONS FOR THREE DIMENSIONAL LOADINGS

For loadings applied over general shaped areas we may evaluate the transformed quantities by applying a double Fourier transform to the loadings and field equations. The resultant set of equations for the layer are identical to equations 7 and so we may readily solve for the transformed variables.

The derivation of equations for the general three dimensional case is not presented here but will be left for a later paper, however it is very similar to that presented by Booker and Small (1982b). An example of a uniform loading applied to the surface of a soil layer over a circular region is provided as an example in the next section.

6. EXAMPLES

Some example problems were analysed to illustrate the preceding theory and these were selected so as to demonstrate the various capabilities of the computer program CONTAL. This program is based on the theory presented in this paper, and is capable of analysing problems which involve different material layers, and loadings applied over strip, circular or rectangular regions which may be uniform or of other distribution or which may be time dependent. Each soil layer may have anisotropy of permeability.

(a) Effect of Anisotropy

In order to examine the effect of anisotropic permeability the problem of a uniform circular loading q applied to the surface z = 0 of a layer of soil over the region $r \leqslant a$ was examined. The problem is shown schematically in the inset to Figure 2. The layer of thickness h = 2a has a horizontal permeability k_h and a vertical permeability k_v. For this example a 'step' loading, which is applied at time $t = 0^+$ and thereafter held constant, was chosen.

Results of the analysis are shown in Figure 2 which shows the central vertical deflection ρ_z plotted in non-dimensional form against the time factor τ where $\tau = c_v t/a^2$ and c_v is the coefficient of consolidation in the vertical direction defined as $c_v = (\lambda + 2G)k_v/\gamma_w$ (λ, G are the Lamé and shear modulus, γ_w is the unit weight of water).

The effect of a higher horizontal permeability is to speed up consolidation considerably; even a ratio $k_h/k_v = 4$ causes a noticeable shift in the time-settlement curve.

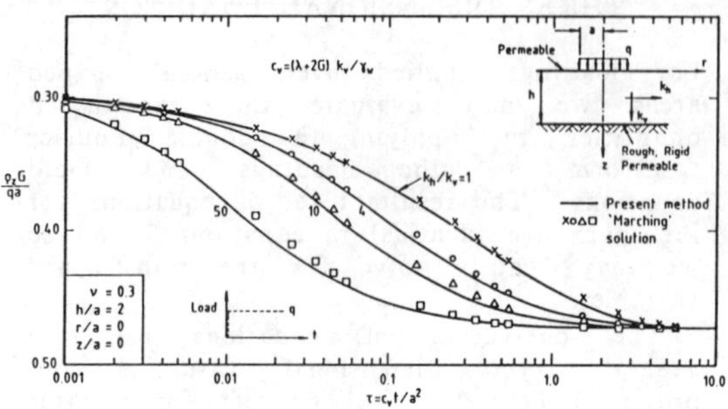

Figure 2: Effect of anisotropic permeability on time-settlement behaviour of circular loading.

(b) Effect of Time-dependent Loading

Loadings are in reality applied over a finite period of time, with the magnitude and perhaps distribution of the loading changing with time. Time-dependent loadings may be dealt with using the method presented here, provided that the solution obtained in Laplace transform space can be numerically inverted.

To demonstrate the effect of a time-dependent loading an 'embankment' load distribution given by

$$q = q_{max} \quad ; \quad |x| \leqslant a$$

$$q = q_{max}\left(2 - \frac{|x|}{a}\right) \quad ; \quad a \leqslant |x| \leqslant 2a.$$

was applied to a layer of soil ($h/a = 2$, $k_h/k_v = 4$), however the load was considered to be applied in two ways. Firstly the load was assumed to be applied in a linear fashion starting from zero at $\tau = 0$ and reaching a maximum value at $\tau = 1$. Secondly the load was considered to be applied as a step loading at $\tau = \frac{1}{2}$ and thereafter held constant.

In order to obtain the time-displacement behaviour of the embankment which is applying a linearly increasing load (which reaches a maximum and then remains constant), the principle of superposition was used. Since the central value of the embankment loading may be written as $q = q_{max}\ \tau - q_{max}(\tau - 1)$ for values of $\tau > 1$, we can obtain the displacements by subtracting the displacements for a loading $q = q_{max} \cdot (\tau - 1)$ from the displacements for a loading $q = q_{max}\tau$ when $\tau > 1$.

A plot of the central deflection of the embankment with time is shown in Figure 3 for the two different loading histories (shown schematically in insets (a) and (b) to the figure). It may be seen that for times greater than $\tau = 1$ the deflection for the 'step' loading gives quite a good approximation to the deflection for a load which increases linearly within the time range $0 < \tau \leqslant 1$. Before this time, as may be expected, the load-deflection curves are quite different.

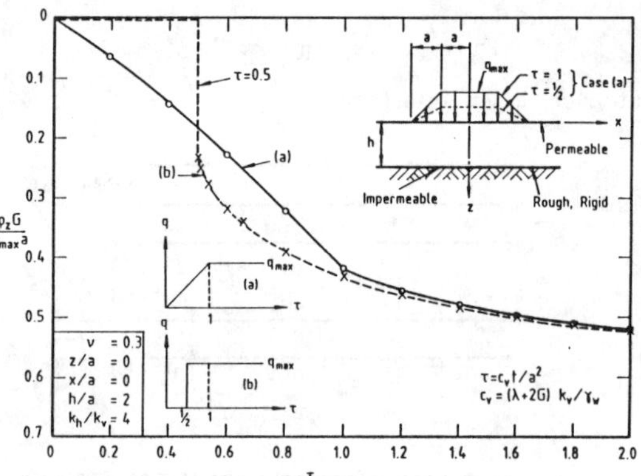

Figure 3: Effect of loading history on time-settlement behaviour of embankment.

7. REFERENCES

Biot, M.A. (1941) "General Theory of Three Dimensional Consolidation", Jl. Appl. Phys., Vol. 12, pp. 155-167.

Booker, J.R. (1974) "The Consolidation of a Finite Layer Subject to Surface Loading", Int. Jl. Solids Structs., Vol. 10, pp. 1053-1065.

Booker, J.R. and Small, J.C. (1982a) "Finite Layer Analysis of Consolidation. I", Int. Jl. Num. Anal. Meth. Geom., Vol. 6, pp. 151-171.

Booker, J.R. and Small, J.C. (1982b) "Finite Layer Analysis of Consolidation. II", Int. Jl. Num. Anal. Meth. Geom., Vol. 6, pp. 173-194.

Booker, J.R. and Small, J.C. (1987) "A Method of Computing the Consolidation

Behaviour of Layered Soils using Direct Numerical Inversion of Laplace Transforms" (to be published in Int. Jl. Num. Anal. Meth. Geom.).
Christian, J.T. and Boehmer, J.W. (1970) "Plane Strain Consolidation by Finite Elements", Jl. Soil Mech. Found. Div., ASCE, Vol. 96, SM4, pp. 1435-1457.
Gibson, R.E. and McNamee, J. (1957) "The Consolidation Settlement of a Load Uniformly Distributed over a Rectangular Area", Proc. 4th Int. Conf. Soil Mech. Found. Eng., Vol. 1, pp. 297-299.
Gibson, R.E., Schiffman, R.L., and Pu, S.L. (1970) "Plane Strain and Axially Symmetric Consolidation of a Clay Layer on a Smooth Impermeable Base", Q. Jl. Mech. Appl. Math., Vol. XXIII, Pt. 4, pp. 505-519.
Runesson, K. and Booker, J.R. (1982) "Exact Finite Layer Method for the Plane Strain Consolidation of Isotropic Elastic Layered Soil", Proc. Int. Conf. Finite Element Meths., Peking, pp. 781-785.
Sandhu, R.S. and Wilson, E.L. (1969) "Finite Element Analysis of Seepage in Elastic Media", Jl. Eng. Mech. Div. ASCE, Vol. 95, EM3, pp. 641-652.
Talbot, A. (1979) "The Accurate Numerical Inversion of Laplace Transforms", J. Inst. Math. Applics., Vol. 23, pp. 97-120.
Terzaghi, K. (1923) "Die Berechnung der Durchlassigkeitziffer des Tones aus dem Verlauf der hydrodynamischen Spannungserscheinungen", Original paper published in 1923 and reprinted in From Theory to Practice in Soil Mechanics, New York, John Wiley & Sons (1960), pp. 133-146.
Vardoularkis, I. and Harnpattanapanich T. (1986) "Numerical Laplace-Fourier Transform Inversion Technique for Layered-Soil Consolidation Problems - I. Fundamental Solutions and Validation", Int. Jl. Num. Anal. Meth. Geom., Vol. 10, No. 4, pp. 347-365.

APPENDIX A

The matrices in equation 7 are defined as follows:

$$A = \begin{bmatrix} \rho G(\Omega_1+\Omega_2) & -\rho G(\Psi_1+\Psi_2) & (\Phi_1+\Phi_2)/2 \\ -\rho G(\Psi_1+\Psi_2) & \rho G(\Lambda_1+\Lambda_2) & -(\Gamma_1+\Gamma_2)/2 \\ (\Phi_1+\Phi_2)/2 & -(\Gamma_1+\Gamma_2)/2 & (\chi_1+\chi_2)/4G\rho \end{bmatrix}$$

$$B = \begin{bmatrix} \rho G(\Omega_1-\Omega_2) & -\rho G(\Psi_1-\Psi_2) & (\Phi_1+\Phi_2)/2 \\ \rho G(\Psi_2-\Psi_1) & \rho G(\Lambda_2-\Lambda_1) & (\Gamma_2-\Gamma_1)/2 \\ (\Phi_1-\Phi_2)/2 & (\Gamma_1-\Gamma_2)/2 & (\chi_1-\chi_2)/4G\rho \end{bmatrix}$$

$$C = \begin{bmatrix} \rho G(\Omega_1+\Omega_2) & \rho G(\Psi_1+\Psi_2) & (\Phi_1+\Phi_2)/2 \\ \rho G(\Psi_1+\Psi_2) & \rho G(\Lambda_1+\Lambda_2) & (\Gamma_1+\Gamma_2)/2 \\ (\Phi_1+\Phi_2)/2 & (\Gamma_1+\Gamma_2)/2 & (\chi_1+\chi_2)/4G\rho \end{bmatrix}$$

where

$$\Omega_1 = N[(1-\delta)C_\mu S_\rho^2]/\Delta_1$$

$$\Psi_1 = N[(1-\delta)C_\mu C_\rho S_\rho]/\Delta_1 - 1$$

$$\Phi_1 = S_\rho \xi_1/\Delta_1$$

$$\Lambda_1 = N[(1-\delta)C_\mu C_\rho^2]/\Delta_1$$

$$\Gamma_1 = C_\rho \xi_1/\Delta_1$$

$$\chi_1 = \frac{c_v\rho}{s}[-\mu S_\mu \zeta_1 + \rho S_\rho \xi_1 / (1-\delta)/N]/\Delta_1$$

with

$$\xi_1 = \frac{c_v\rho}{s}(\rho C_\mu S_\rho - \mu S_\mu C_\rho)$$

$$\zeta_1 = C_\rho S_\rho[(1-\delta) - \frac{\delta}{2N}] - Z[(1-\delta) + \delta/2N](C_\rho^2 - S_\rho^2)$$

$$\Delta_1 = N C_\mu \zeta_1 - C_\rho \xi_1/(1-\delta)$$

$$\delta = (c_v - c_h)\rho^2/s$$

$$N = (\lambda + 2G)/2G \quad \text{and} \quad C_\mu = \cosh\mu h$$

$$Z = \rho h \qquad S_\mu = \sinh\mu h$$

where h is the half layer thickness (see Fig. 1b).

To obtain the expressions for Ω_2, Ψ_2, Φ_2, Λ_2, Γ_2, χ_2, ξ_2, ζ_2, Δ_2 we simply replace sinh by cosh and vice versa.

eg $$\Omega_2 = N S_\mu C_\rho^2/\Delta_2$$

$$\Delta_2 = N S_\mu \zeta_2 - S_\rho \xi_2/(1-\delta)$$

$$\xi_2 = \frac{c_v\rho}{s}(\rho S_\mu C_\rho - \mu C_\mu S_\rho)$$

Numerical Methods in Geomechanics (Innsbruck 1988), Swoboda (ed.)
© 1988 Balkema, Rotterdam. ISBN 90 6191 809 X

Constitutive modelling for anisotropically overconsolidated clay

Y.Kohata & T.Mitachi
Hokkaido University, Sapporo, Japan
M.Kawada
Hokkaido Development Bureau, Japan

ABSTRACT: A series of consolidated drained triaxial test on a saturated remolded clay was performed. Triaxial test specimens were first consolidated and rebounded under K_0 condition to the stress states at which OCR to be 2, 4 and 10. Then drained stress probe tests along six stress paths for each OCR were conducted to investigate the influence of anisotropic stress history and stress path on the stress-strain behavior of overconsolidated clay. Based on the test results, a constitutive model which describes the stress-strain characteristics of anisotropically overconsolidated clay was proposed.

1 INTRODUCTION

Research works on the constitutive model of clays have recently been developed remarkably, and various models have been proposed and discussed. Most of those, however, are derived based on the experimental data on isotropically consolidated clays, and few models focus the attention on anisotropically overconsolidated clays. Based on the assumption of non-associated flow rule, Pender (1978) proposed an elasto-plastic model for overconsolidated clays by assuming the constant stress ratio lines as a yield loci. Mroz et al.(1979) proposed a model for overconsolidated soils based on a bounding surface model. By using exponential function as the bounding surface, Adachi & Oka (1982) proposed a model for sands and overconsolidated clays. These models can describe qualitative stress-strain behavior of overconsolidated clays, but, the coincidence of the calculated stress-strain curves with the experimental ones is not always satisfied.

In this paper, the results of drained stress probe tests on a saturated remolded clay consolidated and rebounded under K_0 condition are shown. Based on the test results, a new constitutive model which describes the stress-strain behavior of anisotropically overconsolidated clay is presented. The method of determination of model parameters is also presented.

2 EXPERIMENTS

2.1 Clay tested and sample preparation

A remolded natural clay, LL and PL of which are 63% and 30 respectively, was sampled in the suburbs of Sapporo, Japan, and was thoroughly mixed with distilled water, sieved through a 420μ size sieve, and stored in the state of slurry. Before making test specimen, the slurry was stirred again in a soil mixer for about two hours and then transferred under a vacuum into a preconsolidation cell. Then the slurry was initially consolidated one-dimensionally under vertical consolidation pressure of 80kPa for about 2 weeks. Cylindrical specimens for triaxial tests, 50mm in diameter and 120mm in height, were trimmed from the preconsolidated sample. the specimen was mounted on the triaxial cell base under water to avoid air entraining, and the drainage during consolidation was forced in the radial direction through the side drain paper wrapped spirally around the specimen. Then, the specimen and side drain were enclosed in a 0.2mm thick rubber membrane which was sealed against the cap and pedestal by O-rings. Lubrication by teflon sheet with silicone grease was applied to top and bottom of the specimen.

2.2 Testing procedure

As shown in Fig.1, triaxial test specimens were first consolidated under K_0 condition by 523kPa effective mean principal stress, then they were rebounded under K_0 condition

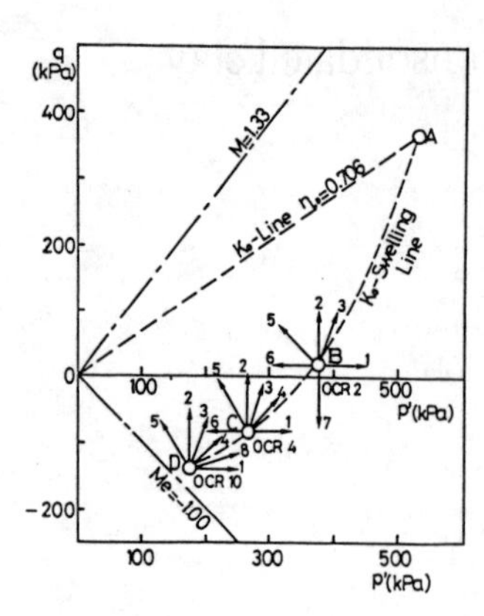

Fig.1 Stress paths in p'-q plane.

to the stress states at which OCR to be 2,4 and 10. In this figure, p' and q denote effective mean principal stress and deviator stress, respectively. Drained stress probe tests along six stress paths for each OCR were performed under following conditions.

(1) CP Test: Constant effective mean stress test
- OCR2; p'=373kPa (Path B2,B7)
- OCR4; p'=263kPa (Path C2)
- OCR10;p'=169kPa (Path D2)

(2) Cq Test: Constant deviator stress test
- OCR2; q= 13kPa (Path B1,B6)
- OCR4; q=-83kPa (Path C1,C6)
- OCR10;q=-143kPa(Path D1)

(3) CIR Test: Constant incremental stress ratio test
- OCR2;$\Delta q/\Delta p'$=3,-1.714 ($-\Delta\sigma_3'/\Delta\sigma_1'$=-1/5) (Path B3,B5)
- OCR4;$\Delta q/\Delta p'$=3,1,-1 (Path C3,C4,C5)
- OCR10;$\Delta q/\Delta p'$=3,1,-1.714,1/3 (Path D3,D4, D5,D8)

For consolidation and rebound stage of the tests mentioned above, both axial and lateral stresses were controlled manually so that the stress paths and final stress points of K_0 consolidation and rebound of all the test specimens with the same OCR to be identical, referring to several preliminary test results of K_0 consolidation and rebound obtained by using automatic K_0 control system (Mitachi & Kitago, 1976). After reaching the final stress point of K_0-rebound, axial and lateral stresses of shear stage were increased manually in several steps up to the final value of stresses with a time interval of 12 hours. Initial back pressure of 100kPa was applied to all specimens. Pore pressure was measured at the bottom of the specimen by pressure transducer, and axial load was measured by load cell set up inside the triaxial cell. The temperature of the laboratory during the test was controlled at 20± 0.5℃.

3 TEST RESULTS AND DISCUSSION

3.1 Shear strain behavior

Typical experimental curves of stress ratio η (=q/p') versus shear strain ε relationship of drained stress probe tests are shown in Fig.2. As can be seen in Figs.2 (a) and (b), the negative shear strains developed during unloading are larger than the positive shear strains developed during loading. This may be explained by the fact that the unloading paths go toward the critical state line. Thus, the shear strain developed on the stress paths toward the critical state line depends strongly on the magnitude of stress ratio η and on the present stress point. On the other hand, the different results from those mentioned above are seen in Figs.2(c) and (d). In Fig.2(c), negative shear strains are developed for both loading and unloading paths and the magnitude of strain for unloading is larger than that for loading. In Fig.2(d), negative shear strain is developed for loading and positive one for unloading. From these test results, it is considered that the shear strain developed in Cq test depends not only on the change of η but also on that of p'.

Fig.3 illustrates the relationship between modulus of rigidity G obtained as the initial tangent modulus of q versus ε curve and preshear effective mean stress p_0'. G/p_0' versus OCR relationship is also shown in the figure. Positive correlation is seen for both relationships. If poisson's ratio and swelling index are assumed to be constant, linear relationship between G and p' is concluded from the theory of elasticity. By applying a linear regression to the G versus p_0' relationship with the assumption that the regression line coincides with the origin, following relationship is obtained.

$$G=112(1+e)p'$$

3.2 Dilatancy behavior

Fig.4 shows the relationship between stress ratio η and volumetric strain v in CP tests for each OCR. Symbols of ◎, ▣ and ◬ in the figure denote the beginning points of each CP tests. In compression tests, remarkable dilatancy is not seen during every stage of shear on OCR2 specimen. For OCR4 and OCR10 specimen, negative dilatancy developed at initial stage of loading changes to positive as the stress path approaching to the critical state. On the other hand, dilatancy behavior of extension test on OCR2 specimen is similar to that of normally consolidated clay, and large negative dilatancy is seen.

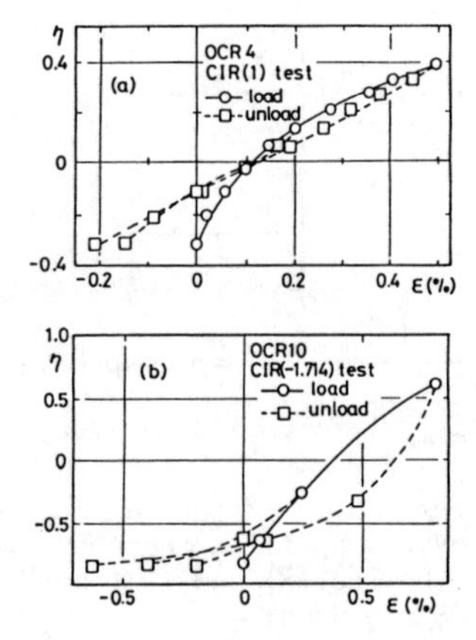

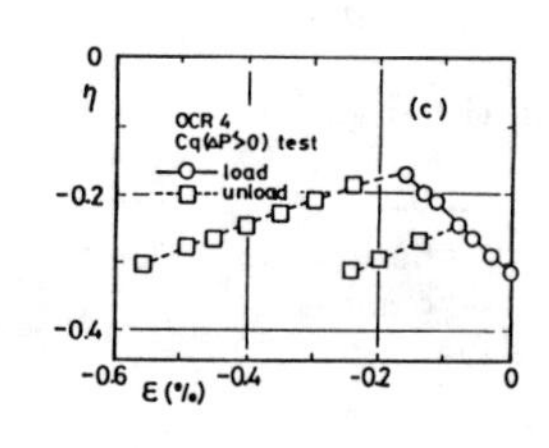

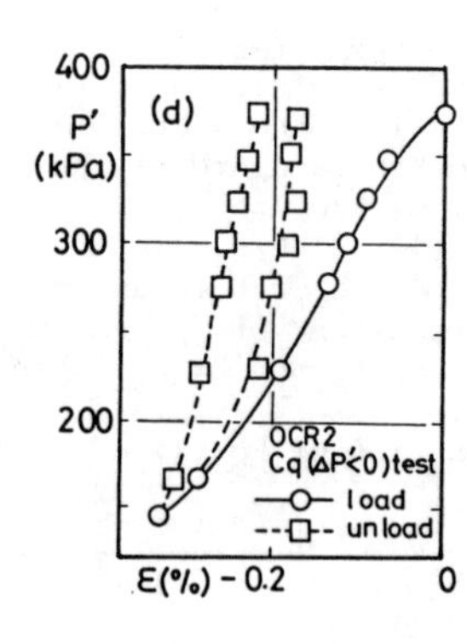

Fig.2 Stress ratio versus shear strain.

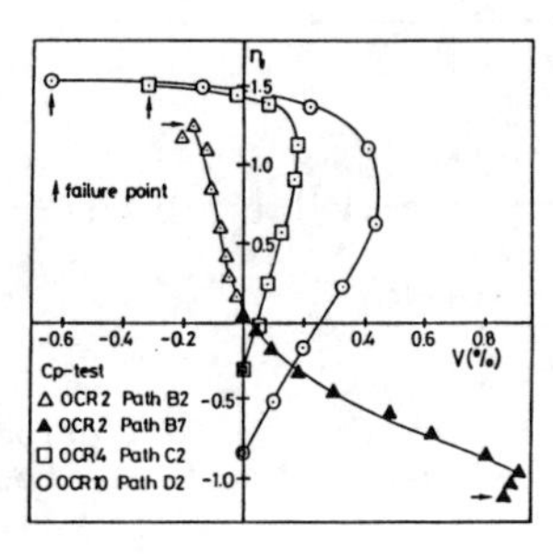

Fig.3 Stress ratio versus volumetric strain.

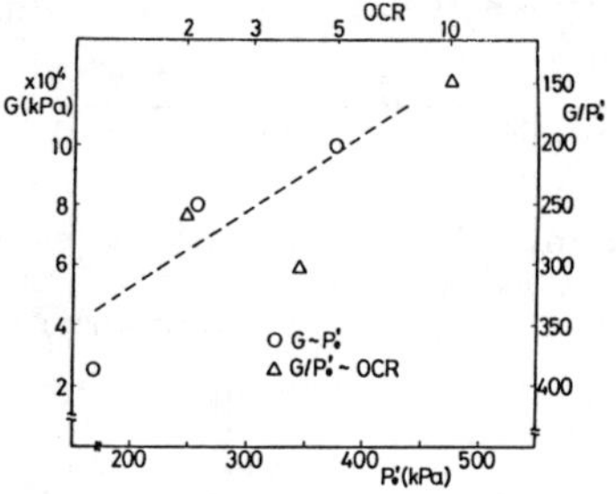

Fig.4 Modulus of rigidity G versus preshear effective stress p_0' and G/p_0' versus OCR relationships.

Fig.5 Schematic representation of proposed model.

4 DERIVATION OF CONSTITUTIVE EQUATION

Schematic representation of proposed model is illustrated in Fig.5. As shown in figure, the model is two-surface model, which consists of boundary surface developed by consolidation and yield surface within the boundary surface. Boundary surface is derived from the basis of the model for anisotropically normally consolidated clays proposed by Ikeura et al. (1987), and yield surface is based on a concept of proposed model by Adachi and Oka (1982).

4.1 Boundary surface

Introducing the idea of kinematic hardening into the basic work equation of modified Cam-clay model by Roscoe and Burland(1968), and a parameter a_0 representing the dependency of plastic strain increment ratio on the stress path during shear, Ikeura and Mitachi(1986) assumed the following plastic strain increment ratio for anisotropically consolidated clay.

$$\frac{d\varepsilon^p}{dv^p} = \frac{a_0\ (\eta - \eta_0)}{(M - \eta_0)^2 - (\eta - \eta_0)^2} \quad --(1)$$

M; stress ratio at the critical state

η_0; stress ratio at the end of anisotropic consolidation

They also modified the normality rule as

$$\frac{d\varepsilon^p}{dv^p} = -\frac{1}{A}\cdot\frac{dp'}{dq}$$

where, parameter A was assumed to change its value as follows, satisfying the conditions A=1 at the start of shearing ($\eta=\eta_0$, $d\varepsilon^p/dv^p=0$) and A=0 at the critical state ($\eta=M$, $d\varepsilon^p/dv^p=\infty$).

$$A=(1-b(d\varepsilon^p/dv^p))^{-1} \qquad \text{-------(2)}$$

where parameter b is assumed as b=0 when the rotation of principal stress occurs during shear and $b=\eta_0$ when it does not occur.

Combination of Eq.(1) with (2) gives

$$\frac{dp'}{p'}+\frac{a_0(\eta-\eta_0)}{(M-\eta_0)^2-(\eta-\eta_0)^2+a_0(\eta-b)(\eta-\eta_0)}d\eta=0 \qquad \text{------ (3)}$$

Integrating Eq.(3) with the condition $\eta=\eta_0$ and $p'=p_0'$, the equation representing the bounding surface is obtained.

Considering that p_0' also is a variable,

$$\frac{dp_0'}{p_0'}=\frac{dp'}{p'}+\frac{a_0(\eta-\eta_0)d\eta}{(M-\eta_0)^2-(\eta-\eta_0)^2+a_0(\eta-b)(\eta-\eta_0)} \qquad \text{------ (4)}$$

By using consolidation and swelling indices λ and κ, respectively, the plastic component of the void ratio change is given as

$$de^p=-(\lambda-\kappa)\cdot\frac{dp_0'}{p_0'} \qquad \text{------ (5)}$$

Combining Eqs.(4) and (5) with the following equation,

$$dv^p=-\frac{de^p}{(1+e)}$$

plastic strain increments are given as

$$dv^p=\frac{\lambda-\kappa}{1+e}\left[\frac{dp'}{p'}+\frac{a_0(\eta-\eta_0)\,d\eta}{(M-\eta_0)^2-(\eta-\eta_0)^2+a_0(\eta-b)(\eta-\eta_0)}\right] \qquad \text{-(6)}$$

$$d\varepsilon^p=\frac{a_0(\eta-\eta_0)}{(M-\eta_0)^2-(\eta-\eta_0)^2}\cdot dv^p \qquad \text{-----(7)}$$

4.2 Yield surface

Yield function is assumed as follows,

$$f=\eta-\eta_{(n)}-k_s=0 \qquad \text{---------(8)}$$

$$k_s=\frac{(M_f-\eta_{(n)})\varepsilon^p}{C(M_f-\eta_{(n)})+\varepsilon^p}$$

M_f; stress ratio at failure

$\eta_{(n)}$; stress ratio at the n th time of change of stress direction

C; a parameter determined as described later in section 4.3

Plastic potential function is assumed as follows,

$$g=|\eta-\eta_{(n)}|+\widetilde{M}\ln(p'/p'_{a(n)}) \qquad \text{---(9)}$$

where

$$d\eta>0;\ p'_{a(n)}=p'\exp\left[\frac{\eta-\eta_{(n)}}{\widetilde{M}}\right]$$

$$d\eta<0;\ p'_{a(n)}=p'\exp\left[\frac{\eta_{(n)}-\eta}{\widetilde{M}}\right]$$

and

$$\widetilde{M}=-\frac{|\eta|}{\ln(p'/p_c')},\ p_c'=p_b'\exp(|\eta_0|/M_m)$$

$$p_b'=p'_{max}\cdot\exp((1+e)/(\lambda-\kappa)\cdot v^p) \qquad \text{-(10)}$$

M_m is a stress ratio at the maximum positive dilatancy. By applying the consistency condition to the Eqs.(8) and (9), hardening function is obtained as,

$$H=\frac{p'C(M_f-\eta_{(n)})^2}{(M_f-\eta)^2}$$

Thus,

$$dv^p=C\cdot\frac{(M_f-\eta_{(n)})^2}{(M_f-\eta)^2}\cdot(|\widetilde{M}|-|\eta|)|d\eta| \qquad \text{----(11)}$$

$$d\varepsilon^p=C\cdot\frac{(M_f-\eta_{(n)})^2}{(M_f-\eta)^2}\cdot d\eta \qquad \text{----(12)}$$

In addition, volumetric yield surface of constant p' was assumed inside the boundary surface.

$$f_v=p'-K_v=0$$

where, K_v is obtained by using plastic hardening modulus K_p proposed by Mroz et al. (1979),

$$dK_v=K_p\cdot dv^p$$

for $dp'>0$

$$K_p=(K_{pr})_{nc}+K_{p0}\left[\frac{p'-p_b'}{p_b'}\right]^{\gamma+1} \qquad \text{----(13)}$$

$$\text{if } p'\geqq p_b',\ K_p=(K_{pr})_{nc}=\frac{(1+e)}{(\lambda-\kappa)}p'$$

for $dp'<0$

$$K_p=(K_{pr})_{oc}+K_{p0}\left[\frac{p'}{p_b'}\right]^{\gamma+1} \qquad \text{---(14)}$$

$$\text{where, } (K_{pr})_{oc}=\frac{(1+e)}{\kappa_{max}}\cdot p'$$

Parameters γ and K_{p0} are determined as described in the following section.

Elastic strain is defined as follows;

$$dv^e=\frac{\kappa}{1+e}\cdot\frac{dp'}{p'},\qquad d\varepsilon^e=\frac{dq}{3G}$$

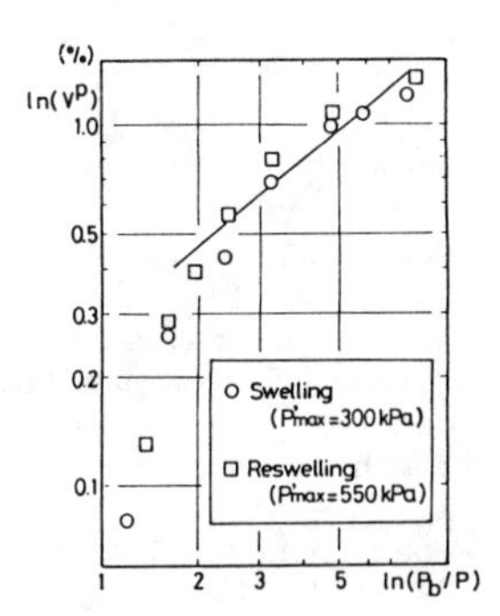

Fig.6 Determination of parameter γ and K_{P0}.

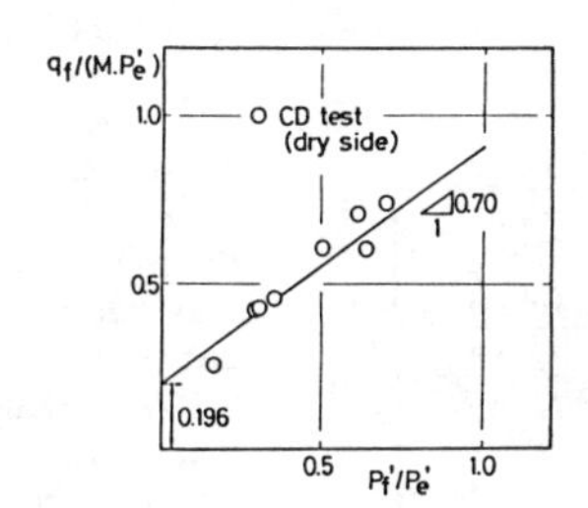

Fig.7 Determination of parameter β.

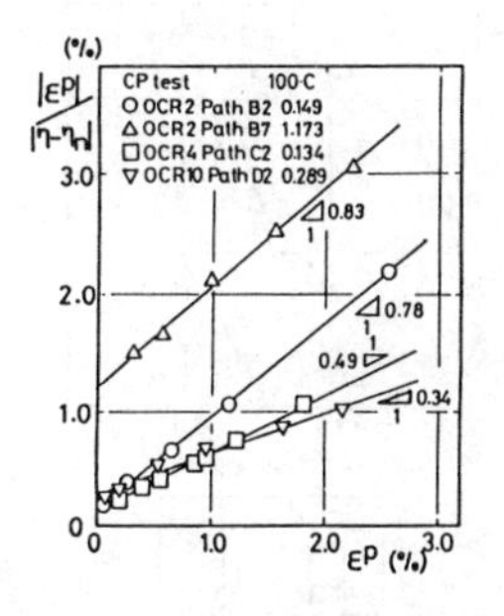

Fig.8 Determination of parameter C.

4.3 Determination of parameters

Parameters required for proposed model are λ, κ, K_{P0}, γ, M_m, M_f, C, a_0 and G. From the results of isotropic consolidation and swelling test, λ=0.126, and κ=0.024 (slope of swelling curve obtained by connecting the point after consolidation with the point of OCR 10), or κ=0.0082 (initial slope of swelling curve) were obtained. K_{P0} and γ can be obtained by Mroz's method. Now, because of $K_{Pr} \ll K_{P0}$, Eqs.(13) and (14) give

$$K_p = K_{P0}\left[\frac{p'-p_b}{p_b'}\right]^{\gamma+1} \quad \text{or} \quad K_p = K_{P0}\left[\frac{p'}{p_b'}\right]^{\gamma+1}$$

If dp'<0,

$$K_{P0}\left[\frac{p'}{p_b'}\right]^{\gamma+1} = \frac{dp'}{dv^p}$$

Integrating with the condition p_b'=constant

$$\frac{p_b'}{\gamma}\cdot\left[1-\left[\frac{p_b'}{p'}\right]^{\gamma}\right] = K_{P0}\cdot v^p$$

Thus,

$$\gamma \ln\frac{p_b'}{p'} = \ln\frac{K_{P0}\cdot\gamma}{p_b'} + \ln\left[\frac{p_b'}{K_{P0}\cdot\gamma} - v^p\right] \quad -(15)$$

If $K_{P0}\gamma/p_b' < 10^4 \sim 10^5$, $\ln(p_b'/K_{P0}\gamma - v^p) \doteqdot \ln(v^p)$. Fig.6 shows the relation between $\ln(p_b'/p')$ and $\ln(v^p)$. By applying least square method to the range of $v_p > 0.5\%$, γ=1.0 and $K_{P0} = 1.58\times10^5$ (kPa) is obtained. Based on the fact that the value of the stress ratio at maximum positive dilatancy is almost equal to that at critical state of undrained shear test, M_m can be determined as 1.26 for compression and -1.0 for extension from a series of consolidated undrained tests. M_f is obtained by drained strength equation proposed by Ohmaki(1984). Ohmaki considered that there exist two critical lines for wet and dry side. And, for dry side, failure strength was defined as

$$q_f = M\ (\ p_f' + \sigma_0\ )$$

where,

$$\sigma_0 = p_f'(1-\beta)((p_{max}'/p_f')^{\Omega}\ \exp(-de_f\,\Omega/(\lambda\cdot\Lambda) - 1) \quad \text{-------}(17)$$

$\Omega = 1-\mu/\lambda$, $\Lambda = 1-\kappa/\lambda$

μ ; slope of e vs. ln p' in dry side

de_f ; vertical distance between C.S.L in wet side and N.C.L

Thus,

$$q_f/(Mp_e') = (1-\beta)\exp(-de_f/\lambda) + \beta\, p_f'/p_e' \quad (18)$$

Fig.7 shows the relation between $q_f/(Mp_e')$ and p_f'/p_e'. From this figure, β=0.7 is obtained, and then from Eqs.(17) and (18), M_f is given as follows;

$$M_f = M(0.231\cdot(p_{max}'/p')^{0.595} + 0.7) \text{ --}(19)$$

Parameter C is a value of an initial slope for $(\eta-\eta_{(n)})\sim\varepsilon^p$ curve. As shown in Fig.8, however, the initial slope changes with OCR. For this reason, $C = C_0\ (p_{max}'/p_0')^{1/2}$ was assumed. Where, p_0' is a value of an initial slope for OCR=1. ($C_0 = 1\times10^{-3}$ for comp., $C_0 = 1.5\times10^{-3}$ for ext.) Parameter a_0 is determined by applying curve fitting to the η versus v_p curve of CP test for normally consolidated clay, and is given as 1.1 for compression and 1.8 for extension.

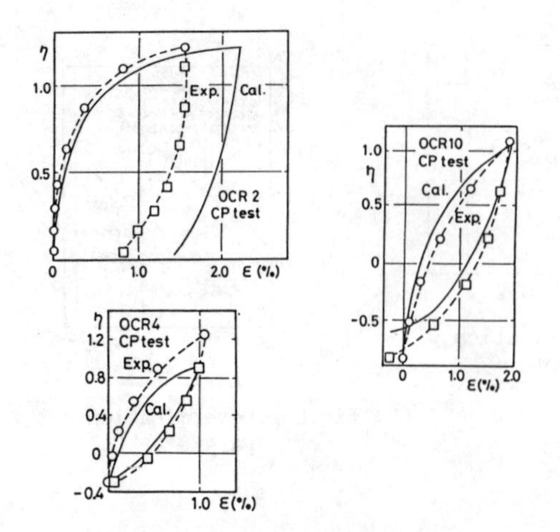

Fig.9 Comparisons of the calculated stress ratio versus shear strain relationships with the observed ones.

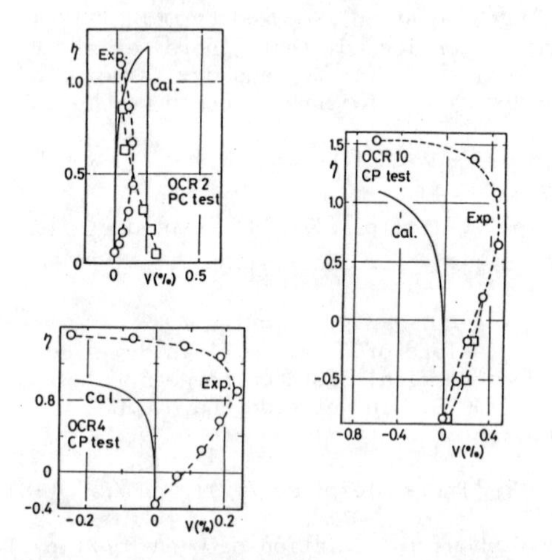

Fig.10 Comparisons of the calculated stress ratio versus volumetric strain relationships with the observed ones.

5 COMPARISON OF CALCULATED STRESS-STRAIN BEHAVIOR WITH OBSERVED ONE

Comparison of the calculated stress ratio and strain relationship with the observed one is shown in Fig.9 for shear strain and in Fig.10 for volumetric strain, respectively. As can be seen from these figures, though the coincidence of the calculated volumetric strain with the experimental one is not so well, the calculated shear strain agrees fairly well with the observed one.

6 CONCLUSIONS

1) Shear strain developing along the stress paths moving towards the critical state line depends on the stress ratio. On the other hand, it depends on both stress ratio and mean effective stress for the stress paths leaving from the critical state line.
2) Dilatancy behavior of overconsolidated clay is influenced by the effect of anisotropic swelling history. And K_0 overconsolidated clay exhibits negative dilatancy in the beginning of loading even in the "dry side", and then significant positive dilatancy is observed on the stress path moving towards the Hvorslev surface in the dry side.
3) Modulus of rigidity of the anisotropically overconsolidated clay is a unique function of effective mean principal stress.
4) Proposed model can well describe the stress-strain, especially for shear strain, behavior of anisotropically overconsolidated clay.

7 ACKNOWLEDGMENTS

The authors gratefully acknowledge the helpful cooperation provided by T.Saito, N.Yamaguchi, A.Kato, and H. Kanazawa.

REFERENCES

Adachi,T. & F.Oka 1982, Constitutive equations for sands and overconsolidated clays, and assigned works for sand, Result of the Int. Workshop on Constitutive Relations for Soils, Grenoble, 141-157

Ikeura,I. & T.Mitachi 1987, Dilatancy behavior of anisotropically consolidated clays, Technical Report, JSSMFE, Hokkaido Branch, No.27, 85-90, (in Japanese)

Ikeura,I. & T.Mitachi 1986, Influence of Stress Path on Stress-Strain Characteristics of Anisotropically Consolidated Clay, Soils and Foundations, Vol.26 No.3, 157-168, (in Japanese)

Mitachi,T. & S.Kitago 1976, Change in undrained shear strength characteristics of saturated remolded clay due to swelling, Soils and Foundations, Vol.16, No.1,45-58

Mroz,Z., V.A.Norris & O.C.Zienkiewicz 1979, Application of an anisotropic hardening model in the analysis of elasto-plastic deformation of soils, Geotechnique 29, No.1, 1-34

Ohmaki,S. 1984, Shear strength characteristics of overconsolidated cohesive soils under axisymmetric stress condition, JSCE No.346, 97-106, (in Japanese)

Pender,M.J. 1978, A model for the behavior of overconsolidated soil, Geotechnique 28, No.1, 1-25

Numerical Methods in Geomechanics (Innsbruck 1988), Swoboda (ed.)
© 1988 Balkema, Rotterdam. ISBN 90 6191 809 X

Pore pressure built up as a result of wave action

S.E.J.Spierenburg
Geotechnical Laboratory, University of Delft, Netherlands

ABSTRACT: Wave-interaction with the seabed results in cyclic shear stress variations. Depending on the compressibility and the permeability, an excess pore pressure may be generated in a saturated seabed during one loading cycle. As a result of storm wave action the cumulated pore pressure can lead to liquefaction. In order to incorporate the cyclic effects into the linear consolidation equation, an uncoupled approach is adopted here. A theoretical relation is proposed for the residual volumetric strain, which includes the effect of preshearing. A comparison is made with results from cyclic loading experiments. Using this method analytical solutions can be derived for the pore pressure buildup in the seabed. The results show that through cyclic loading effects only a limited zone of the seabed is affected. However if liquefaction does occur still a considerable depth is involved.

1 INTRODUCTION

Cyclic loading tends to compact a loose granular material. However, in saturated sand, for example a seabed loaded by waves, the compaction is retarded because pore pressures are generated which have to drain first. Drainage of excess pore pressure is described by the theory of consolidation. The result of one load cycle can be a residual pore pressure which is the difference of the generation and simultaneous drainage. After a long series of successive cycles, for example a seabed under storm wave conditions, the cumulated pore pressure may reduce the effective stress to zero. Such state is called liquefaction. Pore pressure buildup is important for the stability of offshore structures as the bearing capacity is reduced or even failure may occur in case of liquefaction.
A complete description of the simultaneous generation and dissipation is complex. The process of generation is a non-linear type of behaviour because the volume change is always negative regardless of the direction of the shear stress. A general solution procedure for these type of problems is the so-called uncoupled approach, see Finn etal [1], Seed & Rahman, [2].
Different solution procedures have been outlined by Gudehus [3].
In this paper the uncoupled approach will be adopted for the calculation of the pore pressure buildup in a seabed loaded by wind-driven sea-waves. A theoretical expression is proposed for the residual volume after each load cycle. The effect of the stress history, i.e. preshearing, can be included. A comparison is made with results from standard cyclic loading tests. It is concluded that the general behaviour is predicted reasonable well. With the theoretical relation proposed here, analytical solutions can be derived for the pore pressure buildup in the seabed and an evaluation can be made of the liquefaction potential.

2 MATHEMATICAL MODEL

In the so-called uncoupled approach the interrelated processes of the instantaneous wave response and subsequent compaction are separated based on the assumption that the mutual influence of both phenomena is small. It is presumed that the residual pore pressure generated as a result of one load cycle is small compared to the amplitude of the wave load. In order to incorporate the plastic contraction behaviour into a continuum approach a residual volume strain is assumed after each loading cycle. A similar response has been measured by, for instance, Silver & Seed, see [4] , in cyclic simple shear tests on dry sand.

The plastic volume strain ε_c, produced by a series of N cycles of shear stresses may be written in the following form:

$$\varepsilon_c = - D \frac{\hat{\tau}}{\sigma} N \qquad (1)$$

As a first approximation the contraction is related to the ratio of the amplitude of the cyclic shear stress $\hat{\tau}$ and the isotropic effective stress σ. It can be considered that all frictional behaviour of a granular material is expressed in terms of $\hat{\tau}/\sigma$. A simplified method is to suppose linearity with the number of cycles N. Furthermore a material parameter D is introduced.

This model is generally valid for the first part of the process only. Experiments show that densification tends to a certain maximum when N approaches infinity: the maximum density or minimum void ratio. Here a reduction according to a negative exponential function is proposed, see Figure 1.

$$\varepsilon_c = - D \frac{\hat{\tau}}{\sigma} N_t \left(1-\exp\left(- \frac{N}{N_t}\right)\right) \qquad (2)$$

where N_t determines the rate of the reduction with the number of cycles. For large values of N_t the proposed function (2) reduces to the linear model (1), see Figure 1. Note that the volumetric strain per loading cycle is obtained by differentation with respect to N, giving:

$$\frac{\partial \varepsilon_c}{\partial N} = - D \frac{\hat{\tau}}{\sigma} \exp\left(- \frac{N}{N_t}\right) \qquad (3)$$

This formula clearly shows that the strain increment vanishes when N increases.

It will be shown that based on the relation proposed here, analytical solutions can be derived for the pore pressure buildup in a seabed generated by wind-driven sea-waves.

3 PRESHEARING

As shown in experiments, see, for example, Bjerrum [5], previous loaded specimens that are subjected to renewed cylic loading show a different, mostly lower, response than virgin samples. This phenomenon is called preshearing.

In the model the effect of the stress history can be inserted by assuming a certain number of preshearing load cycles N_o The expression for the volumetric compaction then is:

$$\varepsilon_c = - D \frac{\hat{\tau}}{\sigma} N_t (1-\exp(-(N+N_o)/N_t) \qquad (4)$$

The result of a lower response is illustrated by the lower curve in Figure 1.

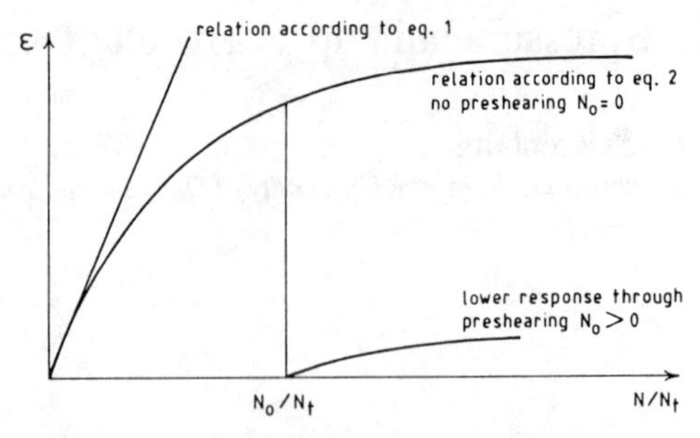

Fig.1. Theoretical model

4 COMPARISON WITH EXPERIMENTAL RESULTS

4.1 Undrained cyclic loading tests

Cyclic loading effects are generally studied in undrained tests. Under undrained conditions the volume of the specimen remains (almost) constant. The plastic strains ε_c induced by the cyclic loading are compensated by an elastic deformation. The latter is determined by Hooke's law. The bulk modulus K of soil depends, more or less linear, on the isotropic stress level. According to critical state soil mechanics [6] we have

$$K = \frac{1+e}{l} \sigma \qquad (5)$$

where l is the slope of the normal consolidation line and e the void ratio.

It can be shown, see [7], that based on the linear model (1), the following solution for the excess pore pressure can be derived.

$$p = D \frac{1+e}{l} \hat{\tau} N \qquad (6)$$

The same response is observed in experiments. Bjerrum [5], for example, reports results of cyclic simple shear tests and proposed the following linear expression for the pore pressure:

$$p = \sigma_i \tan\beta \, N \qquad (7)$$

Here σ_i is the initial vertical effective stress and $\tan\beta$ the slope of the measured linear increasing pore pressure. This equation complies well with the theoretical behaviour. The material parameter D simply follows as:

$$D = \frac{\tan\beta}{\frac{1+e}{l} \hat{\tau}/\sigma_i} \qquad (8)$$

As the measured value of $\tan\beta$ depends on the applied stress level $\hat{\tau}/\sigma$, which conforms to $\hat{\tau}/\sigma_i$, the parameter D, proposed here, seems a more fundamental material parameter than $\tan\beta$.

4.2 Drained cyclic loading tests

When during cyclic loading tests drainage is allowed, instead of pore pressure buildup, the specimen is compacted. In 1972 Youd, see [8], published results of drained strain controlled cyclic simple shear tests (strain ampl. γ_o:0.10-8.45%).

In order to compare the experimental data a conversion of the model is needed from volumetric strain to void ratio. The volumetric strain rate with respect to the number of cycles N is defined as:

$$\frac{d\varepsilon}{dN} = \frac{1}{V}\frac{dV}{dN} \tag{9}$$

The total soil volume V is the sum of the volume of grains and air when dry sand is considered. In the case of saturated samples, pore pressures drain almost instantaneously and therefore the pore fluid can be excluded here. Because the volume of soil grains V_g, determined by the total volume V and the void ratio e, is constant, the derivative with respect to the number of cycles N of V_g is zero.

$$\frac{dV_g}{dN} = \frac{d}{dN}\left(\frac{V}{1+e}\right) = \frac{1}{1+e}\frac{dV}{dN} - \frac{1}{(1+e)^2}\frac{de}{dN}V = 0 \tag{10}$$

With this result the derivative of the volumetric strain can be written as:

$$\frac{d\varepsilon}{dN} = \frac{1}{1+e}\frac{de}{dN} \tag{11}$$

When the proposed theoretical model (2) is substituted in this expression a differential equation for the void ratio is found. With the void ratio e_o at the beginning of the test and the final void ratio e_f, this equation has the following solution:

$$e = (1+e_f)\left(\frac{1+e_o}{1+e_f}\right)^{\exp(-\frac{N}{N_t})} - 1 \tag{12}$$

In Figure 2 the original experimental data from Youd are presented. Predictions for different values of N_t are also given.

It can be concluded that a reasonable prediction can be obtained. A better agreement may be found by introducing a different dependence of the compaction to the number of cycles, for example, by a power function. However, in such case a numerical solution procedure will be necessary, while here, as will turn out in the following sections, analytical solutions can be derived for the pore pressure buildup in the seabed.

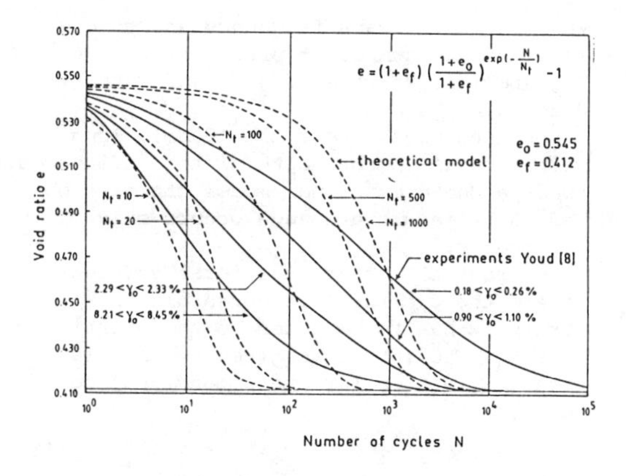

Fig.2 Comparison experimental data and theoretical model

5 CALCULATION OF PORE PRESSURE BUILDUP DUE TO WAVES

5.1 Differential equation

In the case of a homogeneous semi-infinite seabed loaded by harmonical waves the instantaneous wave-induced response, in particular the cyclic shear stress, i.e. the driving force to pore pressure buildup, can be determined analytically, see, for example, Yamamoto et al. [9], or Spierenburg [7].

This solution is rather simplified when the pore fuild is assumed as incompressible, which is reasonable for normal sea conditions. In such case the pore pressure and the effective stresses are given by:

$$\begin{aligned} p &= \hat{p}\exp(-\lambda y)\cos(\omega t-\lambda x) \\ \sigma_{xx}' &= -\sigma_{yy}' = \hat{p}\lambda\exp(-\lambda y)\cos(\omega t-\lambda x) \\ \sigma_{xy}' &= \hat{p}\lambda y\exp(-\lambda y)\sin(\omega t-\lambda x) \end{aligned} \tag{13}$$

The stresses depend mainly on the amplitude of the wave pressure at the seabed $\hat{p}$, which can be calculated from the wave fluctuation at the sea surface using linear wave theory, and the wavenumber λ $(2\pi/L)$.

The ultimate cyclic shear stress $\hat{\tau}$, determined by the deviatoric shear stress, follows as:

$$\hat{\tau} = \hat{p}\lambda y\exp(-\lambda y) \tag{14}$$

The isotropic effective stress changes linearly with depth, while a coefficient of lateral earth pressure K_o determines the relation between the vertical and horizontal effective stress. Substitution into the proposed relation (2) leads to the expression for the residual volume contraction.

$$\varepsilon_c = -D\frac{3\hat{p}\lambda y\exp(-\lambda y)}{(1+2K_o)(\gamma_s-\gamma)}N_t\left(1-\exp\left(-\frac{\omega(t+t_o)}{2\pi N_t}\right)\right) \tag{15}$$

where ω is the wave frequency and t_o determines the number of preshearing waves. The specific weights of soil and water are denoted by γ_s and γ.

For long waves (here L ≈ 100 m), long with respect to the depth of the influenced zone, a one-dimensional model can be adopted. As a result drainage occurs only vertically.

In order to obtain the total volumetric strain the cyclically induced plastic strain ε_c has to be added to the elastic part ε_e. The latter is given by the classical expression for one-dimensional consolidation.

$$\varepsilon = -\alpha\,(q-p) + \varepsilon_c \tag{16}$$

In this analysis only pore pressures due to contraction are considered, therefore the surface load q disappears from the equations. The second equation is the so-called storage equation.

$$\frac{\partial \varepsilon}{\partial t} = \frac{k}{\gamma}\;\frac{\partial^2 p}{\partial y^2} \tag{17}$$

Taking the derivative with respect to time of the volumetric strain (16), the result can be substituted into equation (17), leading to:

$$\frac{\partial p}{\partial t} = c_v\,\frac{\partial^2 p}{\partial y^2} + B\,\exp(-(\lambda y+\omega t/2\pi\,N_t)) \tag{18}$$

The first part of the governing differential equation for the residual pore pressure is the well-known diffusion equation describing one-dimensional consolidation or heat diffusion. The last term represents the cyclic effect and has the character of a source, also referred as the pumping term, see [3].

The consolidation coefficient c_v, determined by the permeability k and the compressibility α, is defined as:

$$c_v = \frac{k/\gamma}{\alpha} \tag{19}$$

The coefficient B is an auxiallary coefficient.

$$B = \frac{D}{\alpha}\,\frac{3}{1+2K_o}\,\frac{\hat{p}\lambda}{\gamma_s-\gamma}\,\frac{\omega}{2\pi}\,\exp(-\omega t_o/2\pi\,N_t) \tag{20}$$

A similar, but simplified, equation can be derived for the linear model.

$$\frac{\partial p}{\partial t} = c_v\,\frac{\partial^2 p}{\partial y^2} + B_1\,\exp(-\lambda y) \tag{21}$$

$$\text{where}\quad B_1 = \frac{D}{\alpha}\;\frac{3}{1+2K_o}\;\frac{\hat{p}\,\lambda}{\gamma_s-\gamma}\;\frac{\omega}{2\pi} \tag{22}$$

5.2 Analytical solutions

The boundary conditions for this problem are that at the surface of the seabed the excess pore pressure is zero. At the start of the wave loading the pressure is also zero and at great depth the fluid velocity must vanish.

$$\begin{aligned} y &= 0 \quad p(0,t) = 0 \\ y &\to \infty \quad \frac{\partial p(y,t)}{\partial y} = 0 \\ t &= 0 \quad p(y,0) = 0 \end{aligned} \tag{23}$$

An analytical solution can be found with the use of an integral transform method. With the Laplace transformation the partial differential equation (18) or (21) can be transformed into an ordinary differential equation in y and s, which can easily be solved. The inverse transformation can be found using standard methods from Laplace transformation analysis, see [7]. The analytical solutions for both models are:

linear densification:

$$\begin{aligned} p_1(y,t) = \frac{B_1}{c_v\lambda^2}\Big(&1-\operatorname{erf}\Big(\frac{y_1}{2\sqrt{t_1}}\Big) - \exp(-y_1) \\ &+ \tfrac{1}{2}\exp(t_1-y_1)\operatorname{erfc}\Big(\sqrt{t_1}-\frac{y_1}{2\sqrt{t_1}}\Big) \\ &- \tfrac{1}{2}\exp(t_1+y_1)\operatorname{erfc}\Big(\sqrt{t_1}+\frac{y_1}{2\sqrt{t_1}}\Big)\Big) \end{aligned} \tag{24}$$

limited densification:

$$\begin{aligned} p(y,t) = \frac{B}{c_v\lambda^2+\omega/2\pi\,N_t}\Big(&-\exp(-(t_2+y_1)) \\ &+ \tfrac{1}{2}\exp(t_1-y_1)\operatorname{erfc}\Big(\sqrt{t_1}-\frac{y_1}{2\sqrt{t_1}}\Big) \\ &- \tfrac{1}{2}\exp(t_1+y_1)\operatorname{erfc}\Big(\sqrt{t_1}+\frac{y_1}{2\sqrt{t_1}}\Big) \\ +\tfrac{1}{2}\exp\Big(-\frac{y_1^2}{4t_1}\Big)\Big\{ &w\Big(\sqrt{t_2}+i\,\frac{y_1}{2\sqrt{t_1}}\Big)+w\Big(-\sqrt{t_2}+i\,\frac{y_1}{2\sqrt{t_1}}\Big)\Big\}\Big) \end{aligned} \tag{25}$$

$$\begin{aligned} \text{where}\quad y_1 &= \lambda y \\ t_1 &= c_v\lambda^2 t \\ t_2 &= t/2\pi\,N_t \\ \operatorname{erf}(z) &= 1-\operatorname{erfc}(z) = \frac{2}{\sqrt{\pi}}\int_o^z \exp(-\zeta^2)\,d\zeta \\ w(z) &= \exp(-z^2)\Big\{1+\frac{2i}{\sqrt{\pi}}\int_o^z \exp(\zeta^2)\,d\zeta\Big\} \\ i &= \sqrt{-1} \end{aligned} \tag{26}$$

5.3 Behaviour at infinite time ($t \rightarrow \infty$)

For large values of t the transient solution for the linear model as given by expression (24) changes into a steady state solution.

$$p_1(y,\infty) = \frac{B_1}{c_v\lambda^2}(1-\exp(-\lambda y)) \qquad (27)$$

This steady state solution itself can be derived rather easily, by integration, from the original differential equation (21) by assuming such behaviour and as a consequence assuming that the derivative with respect to time of the pore pressure tends towards zero for large values of t.

In the model with limited densification the driving force to pore pressure generation gradually disappears because the contraction per load cycle vanishes when the number of cycles increases. As a result all excess pore pressures will be dissipated at infinite time.

6 RESULTS

Numerical values of the error-function and the w-function can be obtained with a polynomal approximation. For large values of the argument up to infinity an asymptotic series can be used. In this way the variation of the excess pore pressure with depth and the generation as a function of time was calculated, see Fig.3 and 4. Realistic values for the soil properties were inserted in these calculations, while the effect of preshearing was excluded here ($t_o = 0$).

The results show that pore pressure buildup according to the linear model tends towards a maximum, steady state, solution while according to the second model all generated pore pressures will be dissipated at large values of t. As presumed in section 2, the model with limited densification reduces to the linear model for large values of N_t.

For the evaluation of the liquefaction potential the maximum pore pressure buildup is important. For the linear model this is given by the steady state solution (27). The maximum for the model with limited densification can be determined numerically. In this paper the maximum was calculated using the Newton-Raphson interpolation technique. In Fig.5 the maximum pore pressure buildup according to both models is shown. In this case a large value for N_t was assumed and the number of preshearing load cycles was set to zero.

It can be concluded that the maximum pore pressure buildup is considerably reduced with the introduction of the limited densi-

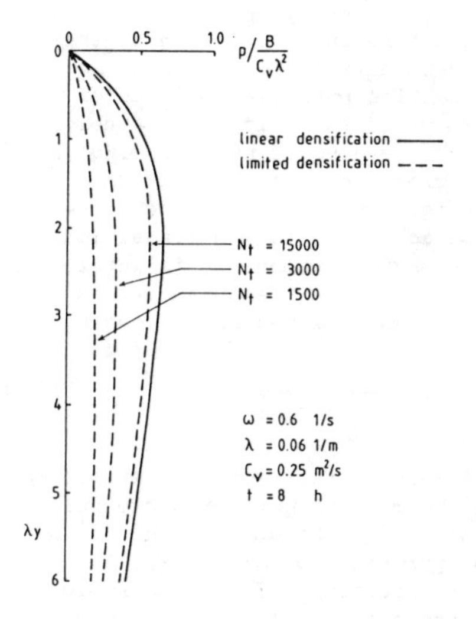

Fig.3 Pore pressure buildup as a function of depth

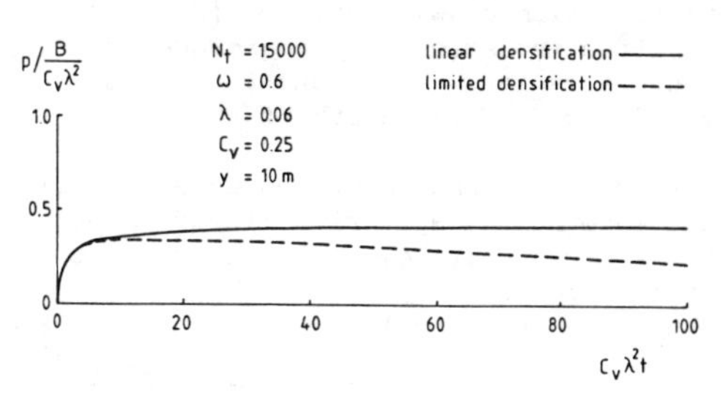

Fig.4 Pore pressure buildup as a function of time

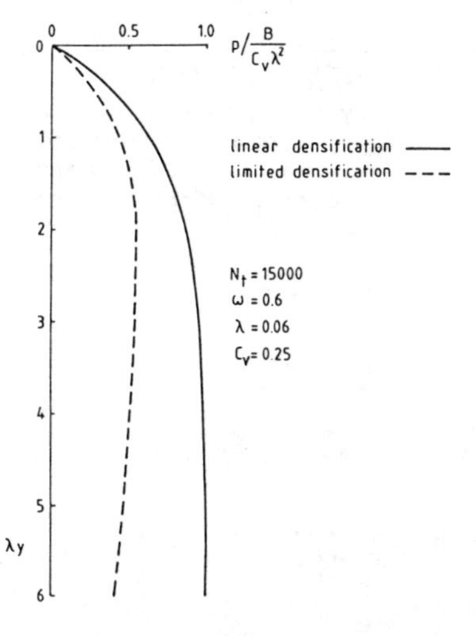

Fig.5 Maximum pore pressure buildup

fication, resulting in a reduction of the liquefaction potential.

The liquefaction potential or rate of maximum pore pressure buildup for a certain location can be evaluated by comparing the excess pore pressure with the initial isotropic effective stress. In such case the solution according to the linear model will yield a maximum value. It is interesting to note that near the surface this solution (27) can be linearized, leading to:

$$\lambda y \ll 1 \quad p_1(y,\infty) = \frac{B_1}{c_v \lambda^2} \lambda y \tag{28}$$

In Figure 6 the initial isotropic effective stress and the maximum pore pressure buildup according to (27) together with the linearized approximation are shown. As both the pore pressure and the isotropic effective stress change linearly with depth, this means that if liquefaction occurs it will occur practically homogeneous over a zone near to the surface. The depth of this zone depends upon the wavelength. In the example, with a wavelength of 100 metres, the effective stress may be reduced to practically zero in a zone for which $\lambda y = 0.1\text{-}0.2$ i.e. a depth of about 1.5-3.0 metres.

A storm can be characterized by a significant wave height H_s. Using linear wave theory the amplitude of the pressure at the seabed $\hat{p}$ follows as:

$$\hat{p} = \gamma H_s / 2\cosh\lambda h \tag{29}$$

where λ is the wavenumber and h the local waterdepth. Using this relation (29)) and (19) and (22), the ratio of the pore pressure and the effective stress can be changed into:

$$\frac{p}{\sigma} = \frac{D}{k} \left(\frac{3}{1+2K_o}\right)^2 \left(\frac{\gamma}{\gamma_s-\gamma}\right)^2 \frac{H_s\lambda}{2\cosh\lambda h} \frac{\omega}{2\pi\lambda} \tag{30}$$

This formula clearly shows that the maximum pore pressure buildup strongly depends on the ratio of the material parameter D and the permeability k together with the wave steepness H_s/L.

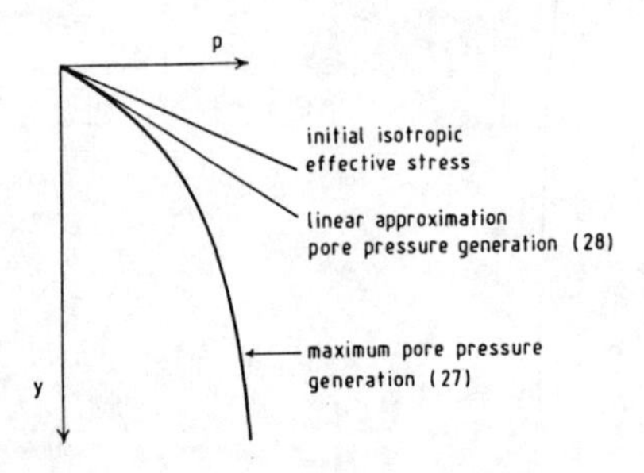

Fig.6 Liquefaction potential

7 CONCLUSIONS

Based on the proposed model describing cyclic effects analytical solutions can be derived for the pore pressure buildup in a semi-infinite seabed. With the results an evaluation can be made of the liquefaction potential. The maximum excess pore pressure is considerably reduced by the effect of the stress history. When included in a numerical method, the model is potentially useful for dealing with problems with a more complex geometry.

ACKNOWLEDGEMENTS

The author is indebted to Prof. A. Verruijt (Dept. of Civil Engng, University of Delft) for discussions and his encouragement during this study. Discussions with Dr. Vermeer are also gratefully acknowledged. These investigations have been supported by the Netherlands Technology Foundation (STW).

REFERENCES

[1] W.D.L. Finn, Lee, K.W. and Martin, G.R. An effective stress model for liquefaction. J. of Geotechnical Engng div. ASCE, 103, GT6, 517-533, 1977.

[2] H.B. Seed and Rahman, M.S. Wave-induced pore pressure in relation to ocean floor stability of cohesionless soils. Mar.Geotechnology, 3,2, 123-150, 1978.

[3] G. Gudehus. Stability of saturated granular bodies under cyclic loading. Proc.Dyn.Meth in Soil and Rock Mech. Vol.2 Plastic and long-term effects, 195-212, 1978.

[4] M.L. Silver and Seed, H.B. Volume changes in sands during cyclic loading, J. of the SMFE div. ASCE, 97, SM9,1971.

[5] L. Bjerrum. Geotechnical problems involved in foundations of structures in the North Sea. Geotechnique, 23, 319-358, 1973.

[6] J.H. Atkinson and Bransby, P.L. The mechanics of soils, An introduction to critical state soil mechanics, McGraw Hill, 1978.

[7] S.E.J. Spierenburg. Seabed Response to Water Waves. PhD-thesis Delft University of Technology, 1987.

[8] T.L. Youd. Compaction of sands by repeated shear straining. J. of the SMFE div. ASCE,98, SM7, 709-725, 1972.

[9] T. Yamamoto, Koning, H.L., Sellmeijer,H. and Hijen, E. van. On the response of a poro-elastic bed to water waves. J. of Fluid Mech., 87, 193-206, 1978.

Numerical Methods in Geomechanics (Innsbruck 1988), Swoboda (ed.)
© 1988 Balkema, Rotterdam. ISBN 90 6191 809 X

Modelling thermal, three-dimensional, three-phase flow in a deforming soil

W.S.Tortike & S.M.Farouq Ali
University of Alberta, Edmonton, Canada

ABSTRACT: A mathematical model for three-dimensional, three-phase fluid flow in a heavy oil reservoir, with heat transfer and spatial displacements, is presented. It is based on the point conservation equations for heat and mass transfer, and the classical equations of solid mechanical equilibrium. Trilinear basis functions in space are used to approximate all the primary variables in a piecewise-continuous solution obtained from applying the method of weighted residuals with Galerkin or Petrov-Galerkin weighting functions. The problem has been discretized in time with a novel method for a spatially integrated general system of equations arising from any non-linear partial differential equation. Some preliminary results will be presented.

1. INTRODUCTION

The application of geomechanics to petroleum reservoir engineering has been limited mainly to the design of hydraulic fractures or the consolidation of partly compacted reservoir sands. In the first case, solutions were obtained by the approximate use of linear elastic fracture mechanics (LEFM) in two-dimensional and some three-dimensional numerical fracture simulators (Mendelsohn [1] and [2]). Although many of these fracture simulators are very sophisticated, they lack the ability to model certain features. One such feature is the effect of an altered effective stress field in the rock or soil as the fracturing fluid leaks off from the walls of the fracture. Consolidation studies have been carried out analytically or by the use of a two-phase flow model coupled to a consolidation model in two dimensions.

These two cases also differ in the type of material considered. In the first case, hydraulic fracturing, the material is a hard rock with finite tensile strength and low absolute permeabilities ($0.001 \leq k \leq 0.1 \mu m^2$). The purpose of fracturing is to stimulate oil production by providing a zone of high permeability in the order of several μm^2. Compaction occurs mainly in reservoir formations that are structurally weak, such as various producing chalks, or that undergo clastic compaction, such as loose sands.

Our problem deals with both a different fluid and a different material from what has just been discussed. We are concerned with the production of bitumen from the oil sands of Alberta. The geomechanical properties of the oil sands in the Athabasca deposit have been described in detail elsewhere (for example, Dusseault and Morgenstern [3]). The sands are an uncemented assemblage of sub-angular quartz grains with less than 10% of other, accessory, elements. The oil sands are absolutely compacted and the undisturbed core samples are very stiff. Irreversible dilation accompanies the failure of oil sand samples, which caused some erroneous estimations of the bulk density and Young's modulus to be made from what were disturbed samples (Dusseault [4]). The bitumen saturating the pore spaces has very high viscosities—from 200 Pa.s in Peace River and Cold Lake, Alberta, to more than 3 000 Pa.s in Athabasca—at the ambient temperature of ca. 15°C. The effective mobility of the bitumen is almost zero because of the high viscosities, even though the extracted oil sands exhibit high absolute permeabilities in the order of several μm^2.

Bitumen recovery schemes rely on the mobilization of the viscous hydrocarbon. The most effective way to do this is to reduce the viscosity by raising the temperature of the bitumen. The viscosity declines exponentially with increasing temperature. However, the only way to heat the bitumen is by conduction, with the heat source extending over the greatest possible area. Typically, this is achieved by injecting steam into the formation through an existing layer of sand permeable to the steam such as one with a mobile water saturation, or by parting the formation with injected hot fluids. Where there is no injectivity the only recourse is to part the oil sand. Gravity plays a major rôle in moving the mobilized bitumen to the wellbore.

The orientation of a parting in the oil sand depends on the local principal stresses, the parting occurring in the plane normal to the least principal stress. Local geologic features such as bedding boundaries and jointing will also influence the orientation of the parting. The orientation is important because a horizontal parting at the base of the bitumenous sands is most effective at heating the bitumen. Thus bitumen mobilization, hence bitumen recovery, is a function of the rate of heating the immobile hydrocarbon. The challenge for any engineer designing an *in situ*bitumen recovery scheme is to determine this rate of mobilization within a given confidence interval. To make this determination a model of steam, water and bitumen flow with heat conduction and convection must be coupled with the mechanical behaviour of the sand under conditions of changing pressures and temperatures. This paper proposes such a model.

2. GENERAL FORMULATION

The fluid flow, energy and consolidation equations are initially developed separately and then coupled through the bulk velocities. The equations are discretized in space using the method of weighted residuals with Galerkin weighting functions. Convective terms in the flow and energy equations use Petrov-Galerkin weighting functions. A method for approximating the solution in time has been developed for spatially integrated non-linear partial differential equations.

The use of finite elements for the discretization of all the fluid flow, energy and consolidation equations enables all the degrees of freedom to be represented at exactly the same mesh points. The finite element method is cumbersome for the fluid flow and energy equations compared to, say, the more common finite difference methods. For that reason the fluid model is first developed separately on the computer. This approach also permits us to test the finite element reservoir simulator against existing case studies before introducing the complications of material displacements. The entire model is being coded on the CDC Cyber 205 vector computer.

2.1 Flow and energy conservation equations

The point conservation equations are taken for the flow of three fluid components: bitumen, water and a non-condensable gas. The general molar balance is written for a mixture where each component can exist in any of the three phases: the oleic, aqueous and gaseous, or vapour phases. The following expression therefore represents three equations:

$$\vec{\nabla} \cdot [\phi(x_{io}\rho_{om}\vec{v}_o S_o + x_{iw}\rho_{wm}\vec{v}_w S_w + y_{iv}\rho_{vm}\vec{v}_v S_v)] + \frac{\partial}{\partial t}[\phi(x_{io}\rho_{om}S_o + x_{iw}\rho_{wm}S_w + y_{iv}\rho_{vm}S_v)] = 0, \quad (1)$$

where the subscript $i = w, b, g$ for the water, bitumen and gas components respectively.In the current formulation the only mixed component phases are the oleic and vapour phases. Bitumen may contain dissolved gas to form the oleic phase, and the non-condensable gas may be mixed with steam to form the vapour phase. A simplified molar balance can then be extracted for each component. The equation for water:

$$\vec{\nabla} \cdot [\phi(x_{ww}\rho_{wm}\vec{v}_w S_w + y_{wv}\rho_{vm}\vec{v}_v S_v)] + \frac{\partial}{\partial t}[\phi(x_{ww}\rho_{wm}S_w + y_{wv}\rho_{vm}S_v)] = 0\,. \quad (2)$$

The equation for oil:

$$\vec{\nabla} \cdot (\phi x_{bo}\rho_{om}\vec{v}_o S_o) + \frac{\partial}{\partial t}(\phi x_{bo}\rho_{om}S_o) = 0\,. \quad (3)$$

The equation for gas:

$$\vec{\nabla} \cdot [\phi(x_{go}\rho_{om}\vec{v}_o S_o + y_{gv}\rho_{vm}\vec{v}_v S_v)] + \frac{\partial}{\partial t}[\phi(x_{go}\rho_{om}S_o + y_{gv}\rho_{vm}S_v)] = 0\,. \quad (4)$$

These equations must be formulated in terms of the principal variables of interest: $S_o$and S_w, the oil and water saturations expressed as pore volume fractions, p_w, the aqueous phase pressure, and T, the temperature. The grains of sand are found to be covered by a thin film of water as they are preferentially wetted by water, so $p_w$is the pore pressure exerted on the grains.

Phase pressures are related through the capillary pressures:

$$\left.\begin{array}{rcl} P_{c_{gw}} & = & p_g - p_w \\ P_{c_{wo}} & = & p_w - p_o \end{array}\right\} \quad (5)$$

It is usually assumed that the steam-water capillary pressure is negligible. Thus $p_g = p_w$. The saturations are constrained by

$$S_v + S_o + S_w = 1 \quad (6)$$

where S_v is the remaining fractional saturation, that of the vapour. The velocities in equations 1,2,3, and 4 are the intrinsic phase velocities, related to the bulk phase velocities through the differential form of Darcy's equation:

$$\vec{v}_i = \frac{-\lambda_i}{\phi S_i}\vec{\vec{k}} \cdot (\nabla p_i + \rho_i \vec{g}) . \quad (7)$$

Non-linear products containing primary variables are approximated and expanded by a first-order Taylor expansion from the previous time step. The three resultant non-linear partial differential equations can now be written exclusively in terms of the four unknowns S_w, S_o, p_w, T. One more equation is required for the solution. This is the point conservation equation for energy. It is assumed that the kinetic energy and kinetic energy changes are small compared to the internal energy and changes in internal energy (Bird *et al*[5]), and that the viscous dissipation term is negligble. Whence:

$$\frac{\partial}{\partial t}(\rho U) + \vec{\nabla}\cdot(\rho\vec{v}U) + \vec{\nabla}\cdot\vec{q} - \rho(\vec{v}\cdot\vec{g}) + \vec{\nabla}\cdot(p\vec{v}) = 0 . \quad (8)$$

The expression must still be posed in terms of S_w, S_o, p_w, T. This is achieved by rewriting the equation in terms of enthalpies, h, volume averaging the terms over the solid and fluid phases, using Fourier's equation for heat conduction and assuming instantaneous thermal equilibrium. Further, the application of the source constraints and relations used with the flow equations results in the fourth equation, of similar form to the other three.

2.2 Flow and energy equations with consolidation

The flow and energy equations have not yet been coupled with the displacement model. Huyakorn and Pinder [6] have shown, using Biot's consolidation theory, how to couple fluid flow with deformation in a porous medium. This can be extended to multiphase flow and the addition of changes in temperature. It is assumed that the solid grains are incompressible, so then the solid density is only a function of temperature. Two coupling terms arise from this analysis. The bulk (Darcy) fluid velocity is now a function of the intrinsic fluid velocity, $\vec{v}_i$, and the solid matrix veclocity, $\vec{v}_m$. The solid phase continuity equation

$$\vec{\nabla}\cdot[(1-\phi)\rho_m\vec{v}_m] + \frac{\partial}{\partial t}[(1-\phi)\rho_m] = 0 , \quad (9)$$

is used to eliminate the time rate of change of porosity in the flow equations as a function of temperature and the spatial derivative of matrix velocity.

The matrix velocities are merely the time rate of change of the component displacements, so the flow and energy equations are now formulated in terms of S_w, S_o, p_w, T, u_1, u_2, and u_3. Whence we obtain the following four non-linear partial differential equations, named after their origins as water, oil, gas, and energy:

The water equation:

$$\begin{aligned} & A_1\frac{\partial S_w}{\partial t} + A_2\frac{\partial S_o}{\partial t} + A_3\frac{\partial p_w}{\partial t} + A_4\frac{\partial T}{\partial t} \\ & + A_5 S_w + A_6 S_o + \vec{\nabla}\cdot(\vec{\vec{A}}_7\cdot\vec{\nabla}p_w) \\ & + \vec{\nabla}\cdot\vec{A}_8 + A_9 + A_{10}\frac{\partial}{\partial t}(\vec{\nabla}\cdot\vec{u}) = 0 . \quad (10) \end{aligned}$$

The oil equation:

$$\begin{aligned} & B_1\frac{\partial S_o}{\partial t} + B_2\frac{\partial p_w}{\partial t} + B_3\frac{\partial T}{\partial t} + B_4 S_o \\ & + \vec{\nabla}\cdot(\vec{\vec{B}}_5\cdot\vec{\nabla}p_w) + \vec{\nabla}\cdot(\vec{\vec{B}}_6\cdot\vec{\nabla}S_o) \\ & + \vec{\nabla}\cdot\vec{B}_7 + B_8 + B_9\frac{\partial}{\partial t}(\vec{\nabla}\cdot\vec{u}) = 0 . \quad (11) \end{aligned}$$

The gas equation:

$$\begin{aligned} & C_1\frac{\partial S_w}{\partial t} + C_2\frac{\partial S_o}{\partial t} + C_3\frac{\partial p_w}{\partial t} \\ & + C_4\frac{\partial T}{\partial t} + C_5 S_w + C_6 S_o + \vec{\nabla}\cdot(\vec{\vec{C}}_7\cdot\vec{\nabla}p_w) \\ & + \vec{\nabla}\cdot(\vec{\vec{C}}_8\cdot\vec{\nabla}S_o) + \vec{\nabla}\cdot\vec{C}_9 + C_{10} \\ & + C_{11}\frac{\partial}{\partial t}(\vec{\nabla}\cdot\vec{u}) = 0 . \quad (12) \end{aligned}$$

The energy equation:

$$D_1\frac{\partial S_o}{\partial t} + D_2\frac{\partial p_w}{\partial t} + D_3\frac{\partial T}{\partial t} + D_4 S_w + D_5 S_o$$

$$+ D_6 p_w + \vec{D}_7 \cdot \vec{\nabla} S_o + \vec{D}_8 \cdot \vec{\nabla} p_w$$
$$+ \vec{D}_9 \cdot \vec{\nabla} T + \vec{\nabla} \cdot (D_{10} \nabla T)$$
$$+ D_{11} \frac{\partial}{\partial t} (\vec{\nabla} \cdot \vec{u}) + \vec{D}_{12} \cdot \frac{\partial \vec{u}}{\partial t}$$
$$+ D_{13} = 0. \qquad (13)$$

The addition of more unknowns — the displacements — stipulates the addition of three displacement equations.

2.3 Displacement equations

The solid model is governed by the three-dimensional equilibrium equations:

$$\frac{\partial \tau_{ij}}{\partial x_j} + F_i = 0. \qquad (14)$$

where $i, j = 1, 2, 3$, varying as the dimensions. The local stress tensor, τ_{ij}, is negative in compression. The governing equation for the soil is developed in terms of the Terzaghi effective stress, the inclusion of the constitutive relationship for a linear elastic isotropic material with initial thermal strains, and a Biot constant of unity, to give the three governing equations for material deformation:

$$\frac{\partial \sigma^0_{ij}}{\partial x_j} + F_i + G\left(\frac{\partial^2 u_i}{\partial x_j \partial x_j} - 2\alpha \frac{\partial T}{\partial x_i}\right)$$
$$+ (\lambda + G)\frac{\partial^2 u_j}{\partial x_i \partial x_j} - 3\lambda\alpha \frac{\partial T}{\partial x_i} - \frac{\partial p_w}{\partial x_i} = 0. \qquad (15)$$

with boundary conditions of prescribed displacements $u_i = U_i$ on boundary portion B_u and prescribed surface tractions $\tau_{ij} n_j = S_{ij}$ on boundary portion B_s. More suitable material behaviour, such as elastoplastic, is introduced after spatial discretization of the above equations.

3. SPATIAL DISCRETIZATION

The primary variables S_w, S_o, p_w, T, and $u_i$are approximated in space by a trial function $N_J(x_j)$. The method of weighted residuals is then applied to all the final equations with the Galerkin choice of weighting function. Convective terms in the flow and heat equations use Petrov-Galerkin weighting functions to give upstream mesh points more influence, as described by Huyakorn and Pinder [7]. Seven different integrands arise from this discretization scheme. The domain is finally divided into finite elements to form a piecewise continuous solution to the the unknown variables. Linear trial functions have been used in the computer model. Two elements are implemented as illustrated in Figure 1: the L8 eight-noded brick and the L6 six-noded triangular prism. The L6 element

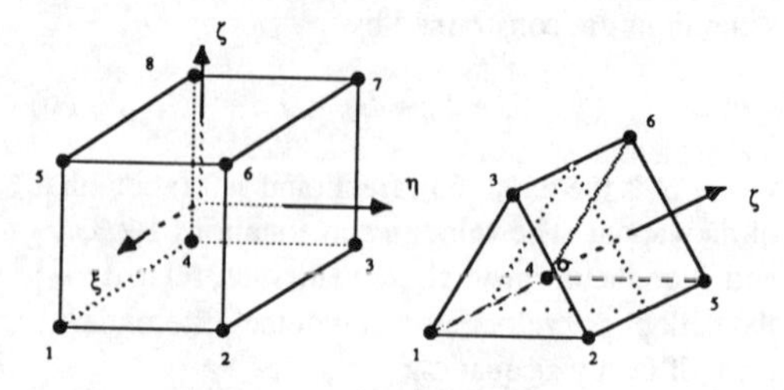

Figure 1: L8 and L6 elements

is required to make the model flexible enough to handle the possible existence of parting planes.

4. TIME DISCRETIZATION AND PROBLEM LINEARIZATION

The purpose of the model is to predict in time the flows and deformations in the problem domain. A general time integration method has been derived for spatially-integrated, non-linear partial differential equations of the form

$$S_{IJ}(\psi, t)\psi_J(t) + D_{IJ}(\psi, t)\frac{d\psi_J(t)}{dt} + F_I(\psi, t) = 0. \qquad (16)$$

The time domain is divided into finite elements with a local coordinate T. The arbitrary variable ψ_j is then approximated in time by a linear trial function. Application of the method of weighted residuals over all the time elements yields the integral expression

$$\int_0^1 \Big\{ D_{ij}(t_n + \Delta t_n T)\frac{\Delta \psi^n_j}{\Delta t_n} + S_{ij}(t_n + \Delta t_n T)$$
$$\cdot [\psi^n_j(1 - T) + \psi^{n+1}_j T] \Big\} W_n \, dT$$
$$+ \int_0^1 \Big\{ F_i(t_n + \Delta t_n T) \Big\} W_n \, dT = 0. \qquad (17)$$

where $\Delta t_n = t_{n+1} - t_n$ and $\Delta \psi^n_j = \psi^{n+1}_j - \psi^n_j$. The coefficients D_{ij}, S_{ij}, and F_i are interpolated by the same trial function as ψ_j. Point collocation at $T = \theta$, for $0 \leq \theta \leq 1$, can be taken for each element. The final equation, discretized in time, has the form

$$(D_{IJ}^n + \theta \Delta D_{IJ}) \frac{\Delta \psi_J}{\Delta t} + (S_{IJ}^n + \theta \Delta S_{IJ})(\psi_J^n + \theta \Delta \psi_J) + (F_I^n + \theta \Delta F_I) = 0, \quad (18)$$

The approximation completes the discretization of the problem. When this is applied to the spatially discretized equations formed earlier, they must still be linearized to find a solution. The Newton-Raphson method generalized to multiple dimensions is applied. If the general discretized non-linear equation has the form

$$\mathcal{M}_{\alpha_I} = 0, \quad (19)$$

where α indicates the origin of the equation (water, bitumen, gas, energy or displacements), then the linearized form is

$$\left[\frac{\partial \mathcal{M}_{\alpha_I}}{\partial \psi_{\beta_K}}\right] \{\Delta \psi_{\beta_K}\} = -\{\mathcal{M}_{\alpha_I}\}, \quad (20)$$

where ψ_β is the solution variable extracted from equation β: S_w, S_o, p_w, T, or u_i. I, K vary as the local node numbers of the element: 1 to 8, or 1 to 6. This equation represents (seven times the number of nodes) equations for a single finite element.

5. COMPUTATION

The problem and our choice of solution are inherently demanding. For that reason an available vector computer was chosen to develop the code on and perform the simulations. This machine is the CDC Cyber 205 at the University of Calgary, Alberta. The Cyber 205 is particularly effective at processing long vectors, so from the beginning of the project the code was designed to force arithmetic operations onto long vectors. Designing this type of software is time-consuming on a batch machine such as the one at the University of Calgary site. It is ultimately a worthwhile venture because of the effectiveness of the code at run-time. However, the benefits of applying the policy of vectorization at the development stage of a simulation project depend on that policy not interfering with the testing and verifying of the code.

The integrand formation and integration routines were fully vectorized before being implemented. The outer products formed at each quadrature point are exceptionally intensive calculations and are considerably speeded up by vectorization. The calculation of the numerous fluid properties has been implemented so that properties at all mesh points are found as one set of vector instructions. The solution of the resulting set of linear equations is made by a general gaussian direct solver for sparse matrices (Tortike and Farouq Ali [8]). The stiffness matrix is asymmetric, and not of regular structure. The linear solver is fully vectorized for the Cyber 205 and is effective even for large problems. A preprocessor inspects the node connectivities and predicts the stiffness matrix structure, forcing it to be structurally symmetric. The symbolic array is then passed to either the Gibbs-King or Gibbs-Poole-Stockmeyer, profile or bandwidth minimization routines, implemented by Lewis [9]. A permutation vector for the old ordering is returned for use with user-supplied nodal information.

No explicit reading and writing is used to manage the computer memory. The Cyber 205 has a very large virtual memory that is most effectively used by large vector operations. The model has been developed in modules which were independently tested with sample problems.

6. PRELIMINARY RESULTS

The model just described is being run only as a thermal flow model at the moment. In the meantime, we have preliminary results from a former student who worked on the same project. Kowalchuk [10] studied the two-dimensional strains associated with the flow of bitumen and hot water in the tar sands. The resulting stresses were used to predict possible parting initiation planes. In this simpler model, the distribution of pore pressure and temperature at time t was predicted in a vertical cross-section extending from an injection well bore. A combined flow and energy model

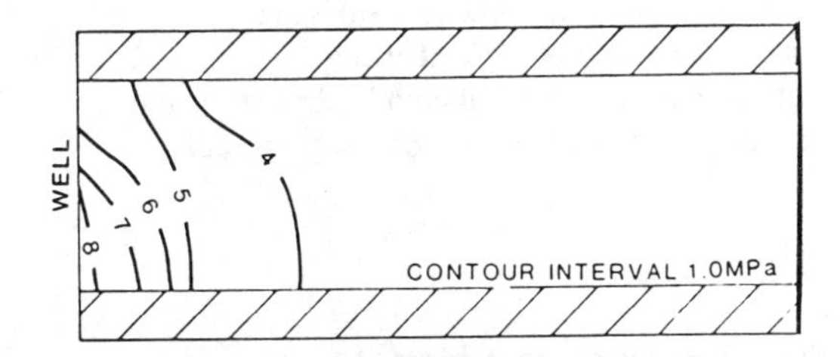

Figure 2: final pore pressure distribution [10]

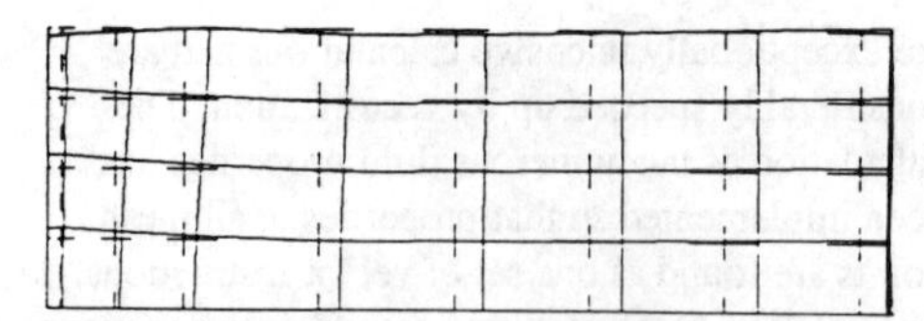

Figure 3: deformation (solid) results compared with original mesh (dashed) for thermo-elastoplastic material at final time [10]

using finite differences and the method of characteristics generated this information. An isoparametric finite element mesh of two-dimensional plain strain elements was superimposed on the finite difference grid so that the mesh nodes coincided with the block boundaries. A free boundary was prescribed at the top of the simulated cross-section, while a rolling support was imposed at the base. This mesh and the pore pressures and temperatures from the thermal flow model were used with the ADINA™ program to investigate the stresses for a thermo-elastoplastic material with a von Mises yield criterion.

The final pore pressure distribution following a period of hot water injection is shown in Figure 2. A zone of plastic deformation appeared around the well bore because of the loading on the sandface (Figure 3). Uplift in the regions where pore pressures had sufficiently increased also occurred (Figure 4). These results have further implications

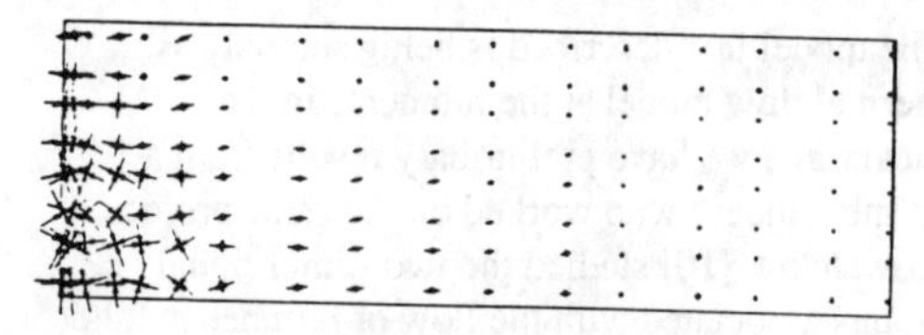

Figure 4: principal stress vectors for the thermo-elastoplastic material at final time [10]

because the Athabasca sand, in particular, demonstrates irreversible dilations when stressed beyond the yield point. The altered material fabric has new flow properties, generally of increased permeability. As the flow properties are instrumental in moving heat energy into the cold bitumen, hence making it mobile, the deformations are of great consequence to recovery operations.

7. CONCLUSIONS

The cohesionless nature of the bituminous sands in Athabasca and elsewhere means that the fabric of the soil can be disrupted with sufficient straining. This is an important attribute in the mobilization and recovery of bitumen by the injection of hot fluids. Preliminary results from a sequential analysis of hot fluid injection into the sands and the stresses and strains created by the changes in pore pressure and temperature support this thesis. The implication is that the geomechanical events associated with hot fluid flow in the tar sands are too important to leave out of the reservoir engineering studies. A quantitative analysis of these effects on bitumen recoveries will be reported upon the successful computer implementation of the solution procedure outlined in this paper.

ACKNOWLEDGEMENTS

The authors would like to thank the Alberta Oil Sands Technology and Research Authority for their continuing generous support of this study.
Mr. Kowalchuk is also thanked for permitting the inclusion of diagrams from his thesis.

NOMENCLATURE

A	non-linear coefficient
B	non-linear coefficient, region boundary
C	non-linear coefficient
D	non-linear coefficient
F_i	body force per unit volume, non-linear coefficient
$\vec{g}$	acceleration due to gravity
G	Lamé constant (shear modulus)
h	enthalpy
$\vec{\vec{k}}$	absolute permeability tensor
$\mathcal{M}$	function value
N	basis function
p	fluid (pore) pressure
P_c	capillary pressure
$\vec{q}$	heat flux
S	liquid saturation (fraction of pore volume)
S_{ij}	prescribed traction, non-linear coefficient
t	time
T	temperature, linear time function
$\vec{u}$	displacement vector with components (u, v, w)
U	internal energy
U_i	prescribed displacement
$\vec{v}$	intrinsic phase velocity
$\vec{V}$	bulk (Darcy) velocity

x	liquid mole fraction, coordinate direction
y	vapour mole fraction, coordinate direction
α	coefficient of volumetric thermal expansion, equation indicator
α'	Biot's constant
β	equation indicator
ε	strain
ε^0	initial strain
κ	volume averaged thermal conductivity
λ	mobility ratio, Lamé constant
ϕ	porosity (fraction)
ψ	any degree of freedom
σ	effective stress
σ^0	initial effective stress
τ	total stress
θ	collocation point

Subscripts

b	bitumen
f	fluid
g	gas
i, j, k	principal direction
I, J, K	node number
m	molar or solid matrix
o	oil
u	displacement component
v	vapour, displacement component
w	water, displacement component

Superscripts

n	time level

REFERENCES

[1] D.A. Mendelsohn, A Review of Hydraulic Fracture Modeling—Part I: General Concepts, 2D Models, Motivation for 3D Modeling, Trans. ASME, J. Energy Res. Tech., 106, 369–376 (1984)

[2] D.A. Mendelsohn, A Review of Hydraulic Fracture Modeling—Part II: 3D Modeling and Vertical Growth in Layered Rock, Trans. ASME, J. Energy Res. Tech., 106, 543–553 (1984)

[3] M.B. Dusseault and N.R. Morgenstern, Shear Strength of Athabasca Oil Sands, Can. Geot. J., 15, 216–238 (1978)

[4] M.B. Dusseault, Sample Disturbance in Athabasca Oil Sand, J. Can. Pet. Tech., 85–92 (April–June, 1980)

[5] R.B. Bird, W.E. Stewart and E.N. Lightfoot, Transport Phenomena, J. Wiley and Sons, 1960

[6] P.S. Huyakorn and G.F. Pinder, Computational Methods in Subsurface Flow, Academic Press, 1983

[7] P.S. Huyakorn and G.F. Pinder, A Pressure-Enthalpy Finite Element Model for Simulating Hydrothermal Reservoirs, in R. Vichnevetsky (ed.), Advances in Computer Methods for Partial DIfferential Equations II, IMACS (AICA) (1977)

[8] W.S. Tortike and S.M. Farouq Ali, STAIRWAY: an effective sparse Gaussian solver for the Cyber 205 vector computer, presented at iciam '87, Paris (1987)

[9] J.G. Lewis, Algorithm 582 The Gibbs-Poole-Stockmeyer and Gibbs-King Algorithms, ACM Transactions on Mathematical Software, 8, 180–189 (1982)

[10] K.K. Kowalchuk, Thermal-Hydraulic Fracture Initiation in Oil Sands, M.Sc. thesis, The University of Alberta (1987)

Numerical Methods in Geomechanics (Innsbruck 1988), Swoboda (ed.)
© 1988 Balkema, Rotterdam. ISBN 90 6191 809 X

Finite difference analysis of consolidation by vertical drains with well resistance

A.Onoue
Shimizu Construction Co., Ltd, Tokyo, Japan

ABSTRACT: A finite difference method which includes consideration of well resistance is presented for consolidation of multilayered anisotropic ground by vertical drains. A new procedure is next outlined for drawing the consolidation curve of multilayered ground using Hansbo's equation concerning single-layered centripetal consolidation. The validity of the numerical analysis is verified through a pair of in-situ experiments. Furthermore, the experimental results also reveal that the proposed procedure yields a sufficiently accurate consolidation curve for any multilayered ground without the need for a computer.

1 INTRODUCTION

When ground in the field is consolidated by vertical drains, the consolidation process tends to be retarded due to the drain well resistance. Yoshikuni & Nakanodo[1] derived a rigorous solution for centripetal consolidation with well resistance under the equal strain condition. Their solution yielded a coefficient of well resistance, L. Hansbo[2] proposed an approximate solution based on a procedure similar to Barron's[3]. One problem with both of these solutions, however, is that their application is confined to single-layered soils only, while alluvial ground generally consists of several layers.

In this paper, an in-situ experiment on the consolidation of an anisotropic multilayered soil by drains is analyzed using the author's finite difference method which takes well resistance into account. Based on the numerical analysis, an additional new procedure is next presented to obtain the consolidation degree for multilayered soils using Hansbo's equation. Finally, the accuracy of this procedure is examined, and the calculated time-settlement curves are compared with the experimental results.

2 ANALYTICAL METHOD

2.1 Basic equation

According to Yoshikuni[4], the consolidation equation for pore water pressure is

$$\dot{u} = c_v \nabla^2 u + \dot{\varphi}\ , \tag{1}$$

where $\nabla^2\varphi = 0$ and φ is the scalar potential.

A soil cylinder dewatered by a drain is illustrated in Fig. 1. Under the boundary conditions of the vertical strain being uniform at the top and bottom of the clay cylinder, the radial strain being zero at the impervious outer boundaries and at the periphery of the drain, and the load applied to the top surface of the cylindrical body being kept constant, the value of $\dot{\varphi}$ equals zero. The consolidation equation for the l-th layer of the ground is then

$$\frac{\partial u_l}{\partial t} = c_h^{(l)}\left(\frac{\partial^2 u_l}{\partial r^2} + \frac{1}{r}\frac{\partial u_l}{\partial r}\right) + c_v^{(l)}\frac{\partial^2 u_l}{\partial z^2}\ , \tag{2}$$

where u_l, $c_h^{(l)}$ and $c_v^{(l)}$ are the excess pore water pressure and horizontal and vertical coefficients of consolidation for the l-th layer. The boundary and initial conditions are given in Table 1.

As the permeability coefficient of the drain material is finite, the pore water flow at the side surface of a drain is given by the following continuity equation:

$$\left(\frac{\partial u}{\partial r}\right)_{r=r_w} + \frac{r_w}{2}\frac{k_w}{k_h}\left(\frac{\partial^2 u}{\partial z^2}\right)_{r=r_w} = 0\ . \tag{3}$$

2.2 Difference equation and dimensionless variables

After dividing the area between $r=r_w$ and $r=r_e$ into M equal segments and H into N equal segments, the segment lengths are expressed as Δr and Δz, respectively. Following the expressions $\lambda = d_e/d_w$, $\mu = H/d_w$, $\eta = r/r_w$, $\zeta = z/H$, $\phi = u/u_o$ and $T = c_h^{(1)}t/d_e^2$ with the horizontal coefficient of consolidation for the first layer, ϕ is replaced with u. Equation (2) then yields the dimensionless equation (after Onoue[5]):

$$u_{i,j}^{k+1} = A\cdot\{u_{i-1,j}^{k},\ u_{i,j-1}^{k},\ u_{i,j}^{k},\ u_{i,j+1}^{k},\ u_{i+1,j}^{k}\}, \quad (4)$$

where

$$A=\left[R_l\left(1-\frac{1}{2P}\right),\ Z_l,\ 1-2R_l-2Z_l,\ Z_l,\ R_l\left(1+\frac{1}{2P}\right)\right], \quad (5)$$

$$R_l=\left(\frac{2M\lambda}{\lambda-1}\right)^2\frac{c_h^{(l)}}{c_h^{(1)}}\Delta T, \ (6) \qquad Z_l=\left(\frac{N\lambda}{\mu}\right)^2\frac{c_v^{(l)}}{c_h^{(1)}}\Delta T, \ (7)$$

and

$$P=\frac{M}{\lambda-1}+i-1 \quad . (8)$$

Figure 2 shows the finite difference scheme.

Continuity Eq.(3) is rewritten along the periphery of the drain well, such that

$$B\cdot\{u_{1,j-1}^{k},\ u_{1,j}^{k},\ u_{1,j+1}^{k}\} = -S_l u_{2,j}^{k} \quad , (15)$$

where $B=[1,\ -(2+S_l),\ 1]$, (16)

$$S_l=\frac{\pi}{4}\frac{ML}{N^2(\lambda-1)}K_h^{(l)}, \ (17) \qquad K_h^{(l)}=k_h^{(l)}/k_h^{(1)}, \ (18)$$

and

$$L=\frac{32}{\pi^2}\frac{k_h^{(1)}}{k_w}\mu^2 \quad . (19)$$

Modifications of vector A and vector B corresponding to other boundary conditions were explained in detail by Onoue[5].

2.3 Boundary conditions at internal layer boundary and at side surface of drain

A model of the internal layer boundary is illustrated in Fig. 3 where $\alpha=\{$(boundary depth)-(relevant grid depth)$\}/\Delta\zeta$. The value of α is positive for the grid intersections on the left side of the layer boundary and negative for those on the right side. Vector Eq.(5) then becomes Eq.(9) and Eq.(10) at the left- and right-hand grids respectively:

$$A=\left[R\left(1-\frac{1}{2P}\right), Z_l, 1-2R-Z_l(2+Q), Z_l(1+Q), R\left(1+\frac{1}{2P}\right)\right], (9)$$

$$A=\left[R\left(1-\frac{1}{2P}\right), Z_{l+1}(1+Q), 1-2R-Z_l(2+Q), Z_{l+1}, R\left(1+\frac{1}{2P}\right)\right], (10)$$

where

$$Q=\frac{(1-|\alpha|)(1-a)}{a+|\alpha|(1-a)}, \ (11) \qquad a=\begin{cases}k_v^{(l)}/k_v^{(l+1)};\ \text{for left side}\\ k_v^{(l+1)}/k_v^{(l)};\ \text{for right side}\end{cases}, \ (12)$$

$$R=\begin{cases}\left.\begin{cases}R_l & \text{(left-hand grid)}\\ R_{l+1} & \text{(right-hand grid)}\end{cases}\right\} & \text{when } |\alpha|\geq\frac{1}{2}\\ \left(\frac{2M\lambda}{\lambda-1}\right)^2\frac{\bar{c}_h}{c_h^{(l)}}\Delta T & \text{when } |\alpha|<\frac{1}{2}\end{cases}, \quad (13)$$

and

$$\bar{c}_h=\frac{\left(\frac{1}{2}+\alpha\right)k_h^{(l)}+\left(\frac{1}{2}-\alpha\right)k_h^{(l+1)}}{\left(\frac{1}{2}+\alpha\right)\frac{k_h^{(l)}}{c_h^{(l)}}+\left(\frac{1}{2}-\alpha\right)\frac{k_h^{(l+1)}}{c_h^{(l+1)}}} \quad . (14)$$

2.4 Various consolidation degrees

The average excess pore water pressure in the radial direction, $u_z(\zeta, T)$, can be found by Eq.(20) using the nondimensional excess pore water pressure, $u(\eta, \zeta, T)$:

$$u_z(\zeta, T)=\int_{1/\lambda}^{1} u(\eta,\zeta,T)\,\eta d\eta \Big/ \int_{1/\lambda}^{1}\eta d\eta \quad . (20)$$

The overall average consolidation degree is also given by

$$U(T)=\frac{\int_0^1\int_{1/\lambda}^1\{1-u(\eta,\zeta,T)\}\,\eta d\eta d\zeta}{\int_0^1\int_{1/\lambda}^1 \eta d\eta d\zeta} \quad . (21)$$

The trapezoidal rule was used in making these numerical calculations.

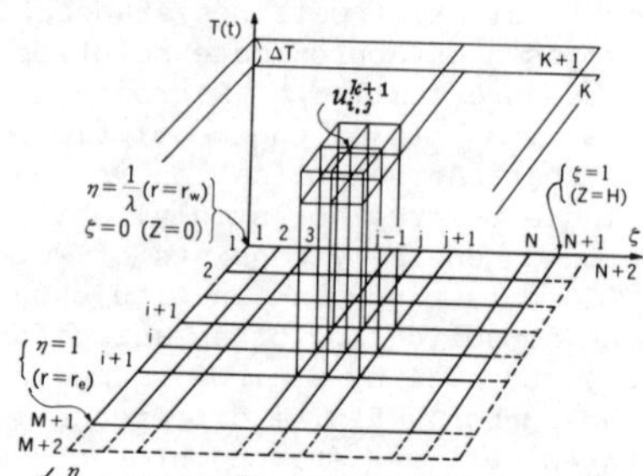

Fig. 2 Finite difference scheme

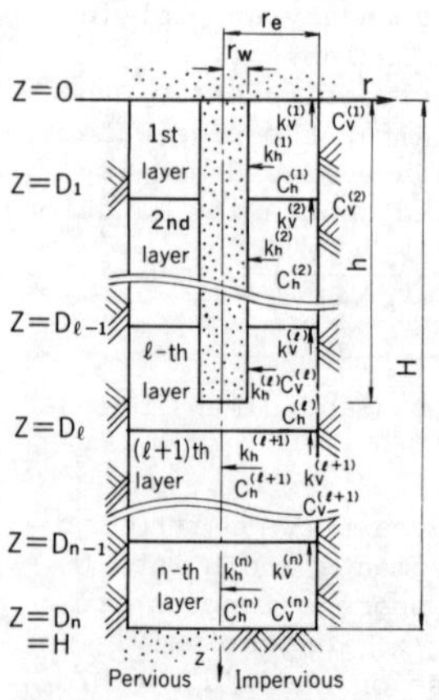

Fig. 1 Drain well in multilayered ground

Table 1 Boundary and initial conditions

Boundary	Condition equations
$z=0$	$u_1=0$
$z=H$	$\frac{\partial u_n}{\partial z}=0$ or $u_n=0$
$z=D_1$	$u_1=u_2,\ k_v^{(1)}\frac{\partial u_1}{\partial z}=k_v^{(2)}\frac{\partial u_2}{\partial z}$
$z=D_2$	$u_2=u_3,\ k_v^{(2)}\frac{\partial u_2}{\partial z}=k_v^{(3)}\frac{\partial u_3}{\partial z}$
⋮	⋮
$z=D_{n-1}$	$u_{n-1}=u_n,\ k_v^{(n-1)}\frac{\partial u_{n-1}}{\partial z}=k_v^{(n)}\frac{\partial u_n}{\partial z}$
$0\leq z\leq h,\ r=r_w$	$\left(\frac{\partial u_l}{\partial z}\right)+\frac{r_w k_w}{2k_h^{(l)}}\left(\frac{\partial^2 u_l}{\partial z^2}\right)=0$
$h\leq z\leq H,\ r=0$	$\frac{\partial u_l}{\partial r}=0$
$t=0$	$u_l=u_0$ (Initial pressure)

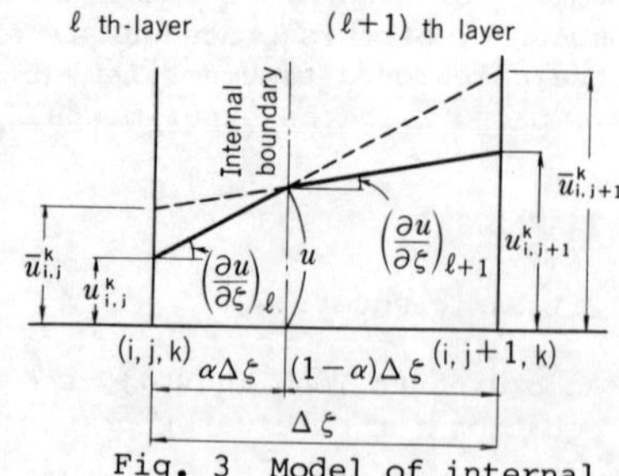

Fig. 3 Model of internal layer boundary

3 NUMERICAL CALCULATIONS

In order to check the accuracy of finite difference analysis, the time histories of the consolidation degree for an isotropic soil were plotted in Fig. 4 with those obtained by Yoshikuni et al.'s rigorous solution[1]. Distributions of u and u_z are also compared in Fig. 5. Good agreement is seen between the finite difference results and those obtained by the rigorous solution.

Depth distributions are given in Fig. 6 for u_z in two-layered soils having equal layer thicknesses. In this figure, Case 1 is the result for the same single-layer soil as shown in Figs. 4 and 5. The upper layers of Cases 2 and 4 are the same soil as in Case 1. In Case 2, the coefficient of volume compressibility, m_v, of the lower layer is one-fourth that of the upper layer; in Case 4, the permeability coefficient, k_h, of the lower layer is four times larger than that of the upper layer. In both Cases 2 and 4, c_h of the lower layer is thus four times that in Case 1. Cases 3 and 5 have reversed layer orders from those of Cases 2 and 4, respectively.

As indicated in Fig. 6, consolidation in the lower layer of Cases 2 and 4 and in the upper layer of Cases 3 and 5 is, therefore, much more accelerated than that of Case 1. However, the pore pressure isochrones in the upper layer of Cases 2 and 4, and in the lower layer of Cases 3 and 5 are very close to those of Case 1. Namely, the pore water pressure isochrone in each layer of a multilayered soil is not greatly influenced by the soil properties of the vicinity layers, except for the area near the layer boundaries. This small influence serves as the background for the subsequent proposal.

4 APPLICATION PROCEDURE FOR MULTILAYERED SOIL BY THE EXISTING SIMPLE EQUATION

4.1 Solution for homogeneous ground

Hansbo[2] proposed Eq.(22) for centripetal consolidation of a homogeneous soil, taking well resistance into account:

$$u_z(\zeta, T) = u_0(\zeta) \cdot e^{-8T/\mu_r} \quad , \quad (22)$$

where

$$\mu_r = \frac{\lambda^2}{\lambda^2-1}\ln\lambda - \frac{3\lambda^2-1}{4\lambda^2} + \frac{\pi^2 L}{8}\left(1-\frac{1}{\lambda^2}\right)\zeta(2-\zeta), \quad (23)$$

and u_0 is initial excess pore water pressure. This equation is based on the assumption that water in the clay layers moves only in the radial direction. It is necessary to use Carrillo's method[6] when taking vertical flow into consideration. The efficiency of consolidation by vertical flow is, however, so slight that it is negligible on site.

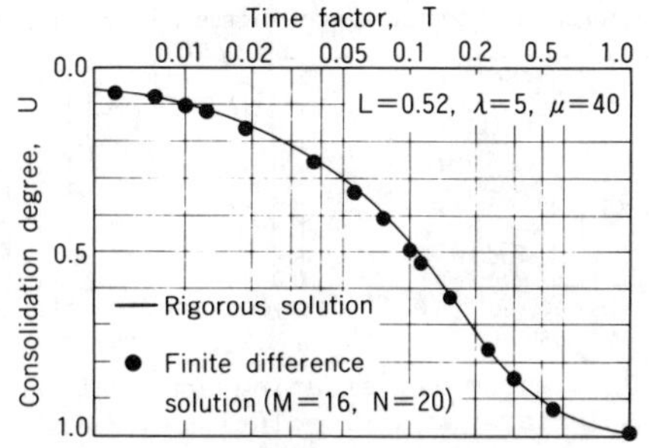

Fig. 4 Accuracy of consolidation degree by finite difference solution

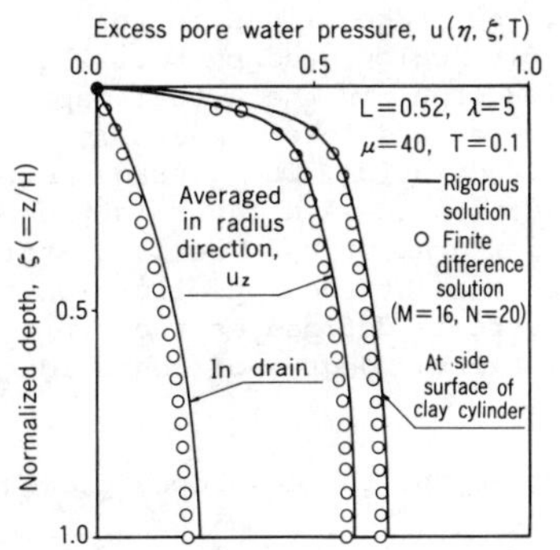

Fig. 5 Accuracy of excess pore water pressure by finite difference solution

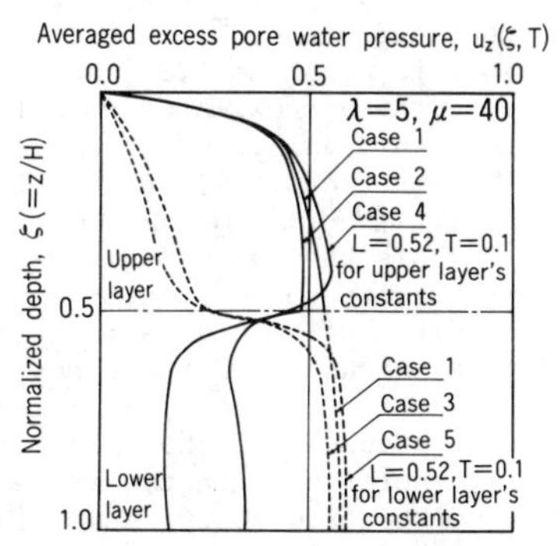

Fig. 6 Excess pore water pressure for various two-layered soil systems

4.2 Application of Hansbo's equation to multilayered soil

The author has developed a procedure for calculating the consolidation degree of multilayered soils by using Eq.(22). This procedure is explained using Fig. 7, where (a) shows the u_z of a two-layered soil system analyzed by the differential method. Let us call this the theoretical result.

The imaginary single-layered soils with a thickness, H, illustrated in (b) and (c)

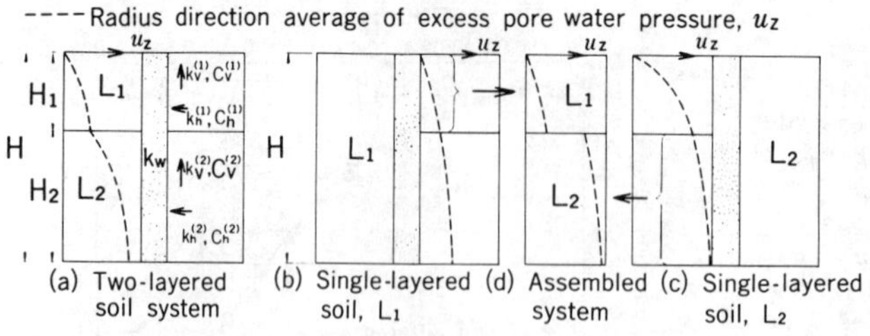

Fig. 7 Diagram illustrating application of single-layered soil solution

have the same property as that of the upper layer and of the lower layer in (a), respectively. The broken lines in (b) and (c) show the results calculated by Eq.(22). The distribution of u_z in the actual depth ranges of the upper and lower layers is extracted and plotted in (d). The average degree of consolidation in each layer is then obtained by the integral of u_z. The layer-thickness-weighted mean of the consolidation degree of both layers is defined as the approximate overall average degree of consolidation.

4.3 Accuracy of the developed procedure

A comparison is shown in Fig. 8, between the approximate consolidation curves and the theoretical ones in the case where the layer-thickness-weighted mean of the well resistance coefficient, $\overline{L}$, of the two-layered soil is equal to 0.52. As is clear from Fig. 8, the approximated consolidation curves are within an allowable range of error. The time factor, T, in the figure is defined $T=(c_h^{(1)}+3c_h^{(2)})t/(4d_e^2)$.

Figure 9 shows the maximum discrepancy of the approximate consolidation degree from the theoretical one as a function of the relative thickness of the upper layer, $r_t(=H_1/H)$. In Fig. 9, the largest approximation error occurs when the value of r_t is 0.5, and it increases with the increase in the degree of heterogeneity, D_h $(=c_h^{(2)}/c_h^{(1)})$.

With the approximation procedure, consolidation is overestimated when $D_h>1$.

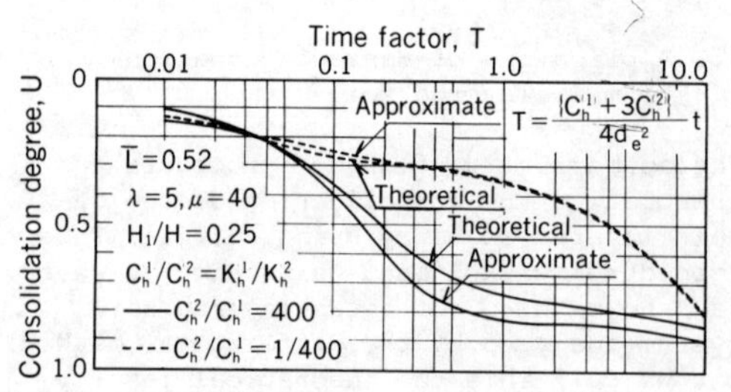

Fig. 8 Approximate vs. theoretical consolidation curves for two-layered soil systems

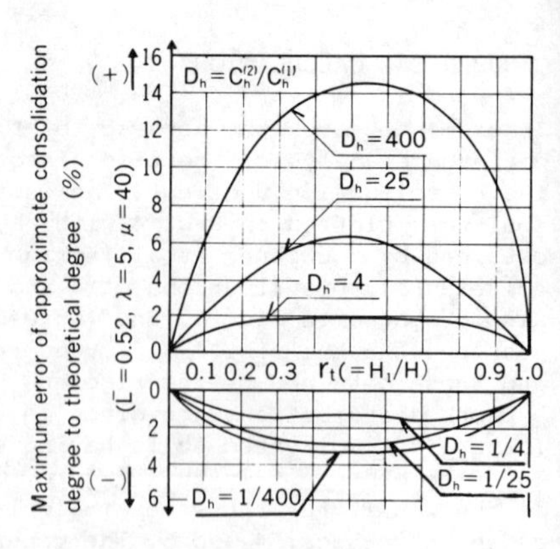

Fig. 9 Maximum error of approximated degree of consolidation

Conversely, it is underestimated when $D_h<1$. Since the actual value of D_h in the field is generally between 10 and 1/10, the developed procedure is accurate to within 4 %.

5 VERIFICATION BY IN-SITU EXPERIMENT

5.1 Ground and experiment

The experiment site was located at the edge of Teganuma(Tega Swamp) in Chiba Prefecture. Figure 10 indicates the plan and cross section of the ground. The experiment consisted of Tests A and B. Test A was started after the conclusion of Test B. All test conditions, for instance, a height of 3.5 m fill (including sand mat), a drain spacing of 1.2 m (square arrangement), the installation method of displacement-type drains using a vibrohammer, were the same in both tests, except for the length and the material of drains. The permeability coefficient of the drain material, k_w, used in Test A was 0.4 cm/sec, and as tabulated later in Table 2, that of the ground was 4.3×10^{-7} cm/sec at most. The coefficient of well resistance, L, was then less than 0.087 and the estimated maximum retardation of the consolidation degree was less than 1.5%. Thus the influence of the well resistance was negligible in Test A. On the other hand, a remarkable influence appeared in Test B where the value of k_w was very small, i.e., 0.02 cm/sec.

The vertical coefficient of consolidation, c_v, through Oed-meter test is also shown in Fig. 10. As is presented in Fig. 10, the alluvial clay ground is considered to be a multilayered ground consisting of six layers. Soil properties of the normally consolidated alluvial clay ground are shown in Fig. 11. Visual inspection of the split continuous piston samples showed no sand seam nor lens.

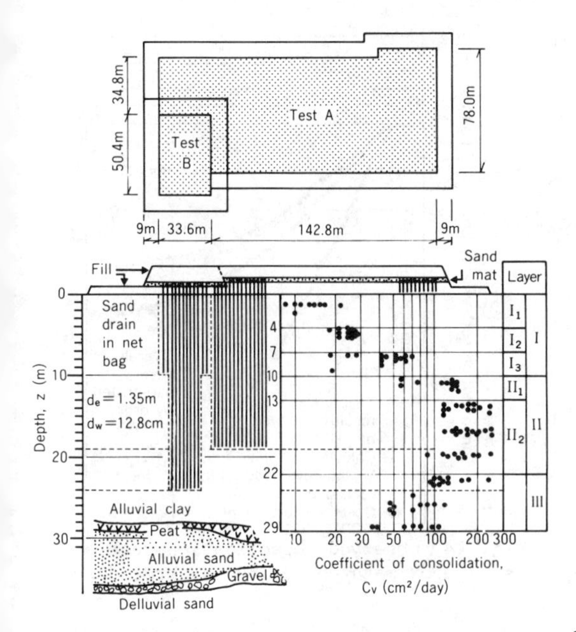

Fig. 10 Plan of experiment site and cross section of ground

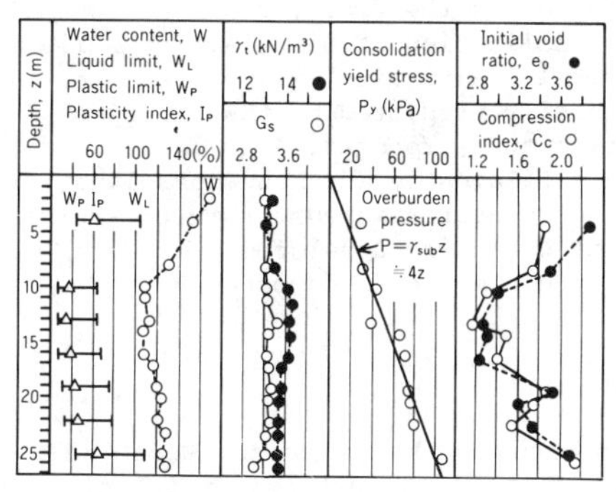

Fig. 11 Soil profile of ground

5.2 Apparent horizontal coefficient of consolidation after drain installation

The observed time-settlement results of five layered groups in Test A are plotted in Fig. 12. The apparent horizontal coefficients of consolidation, $\overline{c_h}$, and of permeability, $\overline{k_h}$, in all layers after drain pile installation are back-analyzed and tabulated in Table 2, together with the Oed-meter test results, c_v and k_v. The value of c_h decreases with the decrease of k_h due to disturbance during drain installation. In this experiment, the value of $\overline{k_h}$ was thus assumed to be equal to that of k_v since the value of $\overline{c_h}$ was nearly equivalent to that of c_v. As seen in Fig. 12, the back-analyzed time-settlement curves well simulate the observed results.

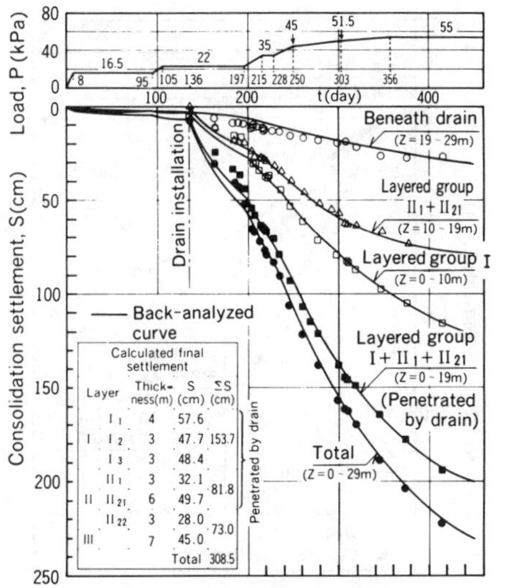

Fig. 12 Observed time-settlement results and back-analyzed curves for Test A

Table 2 Tested and back-analyzed constants

Layer group		Depth (m)	$\overline{C_h}$ (cm²/day)	Tested C_v (cm²/day)	$\overline{K_h}$ (cm/sec)	Tested K_v (cm/sec)
I	I1	0 ~ 4	13	13	0.75×10^{-7}	0.75×10^{-7}
	I2	4 ~ 7	25	25	1.1×10^{-7}	1.1×10^{-7}
	I3	7 ~ 10	60	60	2.2×10^{-7}	2.2×10^{-7}
II	II1	10 ~ 13	70	110	3.3×10^{-7}	3.3×10^{-7}
	II2	13 ~ 22	80	170	4.3×10^{-7}	4.3×10^{-7}
III		22 ~ 29	285	95	1.8×10^{-7}	1.8×10^{-7}

5.3 Observed influence of well resistance

Figure 13 shows the analytic time-settlement curves for Test B using the constants listed in Table 2, except that the unanalyzable value of $\overline{c_h}$ in the depth range z = 22m ~ 24m is assumed to be equivalent to the value of c_v. Those curves accurately coincide with the observed results when well resistance is taken into account. Therefore, the numerical analysis method proposed here is proven to be valid for prediction of consolidation by vertical drains with well resistance. As is obvious from the broken lines in Fig. 13, disregarding well resistance reduces accurate prediction of the consolidation degree.

5.4 Validity of newly developed procedure

Approximate time histories of the consolidation degree for Test B can be obtained by the aforementioned newly developed procedure taking well resistance into account. In Test B, there are six layers which are penetrated by a drain well having a length of 24 m.

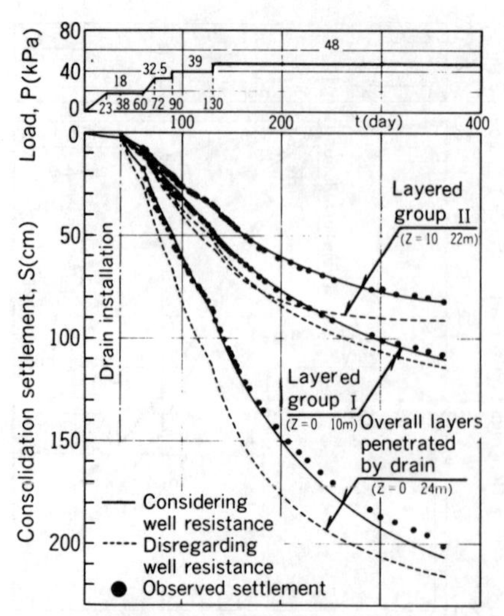

Fig. 13 Observed consolidation settlement vs. finite difference solutions for Test B

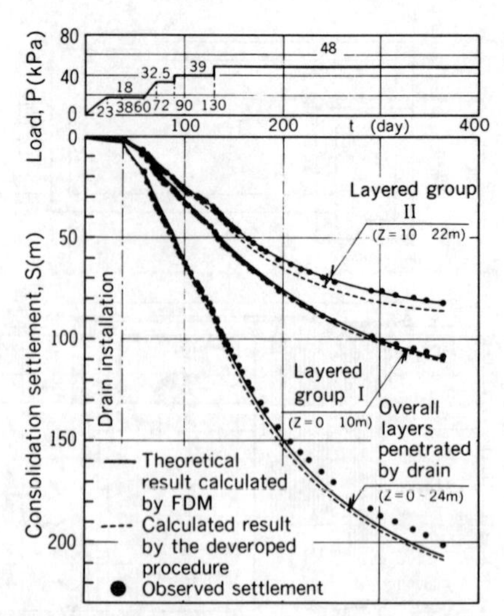

Fig. 14 Observed consolidation settlement vs. approximate curves by new procedure

There are three steps in the procedure.

1. It is presumed that there are six imaginary homogeneous soils each having a thickness of 24 m and an impervious lower boundary. The vertical distribution of u_z in each homogeneous soil is calculated at arbitrary time factors.

2. The distribution of u_z in the depth range where each layer actually exists is extracted from the distributions of u_z in the six imaginary homogeneous soils. The approximate consolidation degree and compression amount of each layer are calculated from each portion of the six extracted pore pressure distributions and the coefficients of volume compressibility.

3. The approximate time-settlement curves of layered groups I and II and the overall ground penetrated by a drain are obtained.

Figure 14 compares the approximate curves and the measured curves, with both curves clearly coinciding well. Therefore, the developed procedure for predicting the consolidation of multilayered ground using existing simple equations, such as Hansbo's, is valid.

In this paper, the finite difference method is used in step 2 for drawing the distribution of u_z because the time history of the effective load in Test B is complicated. In the case of instant loading, Hansbo's Eq.(22) is, however, convenient for calculating the vertical distribution of u_z in each soil without a computer.

6 CONCLUSIONS

The following conclusions can be drawn from the present work.

(1) Well resistance of a vertical drain significantly retards the consolidation rate of clay in the field.

(2) Consolidation of multilayered soils by vertical drains with well resistance can be easily predicted by the numerical analysis method proposed.

(3) A sufficiently accurate consolidation curve can be approximated for multilayered ground by the newly developed procedure without the need for a computer.

REFERENCES

[1] H.Yoshikuni and H.Nakadono, Consolidation of Soils by Vertical Drain Wells with Finite Permeability, Soils and Foundations, 14, 2, 35 - 46 (1974)

[2] S.Hansbo, Consolidation of Fine-Grained Soils by Prefabricated Drains, Proc. 10th ICSMFE, 3, 677 - 682 (1981)

[3] R.A.Barron, Consolidation of Fine-Grained Soils by Drain Wells, Trans. ASCE, 113, No.2346, 718 - 742 (1948)

[4] H.Yoshikuni, A New Development of the Equation of Consolidation Expressed in Terms of Pore Water Pressure (in Japanese), Proc. JSCE, 212, 41 - 50 (1973)

[5] A.Onoue, Numerical Analysis of Consolidation by Drains with Well Resistance (in Japanese), Proc. 1st Symp.Num.Meth. In Geotech.Eng. Union of Japanese Scientists and Engineers, 81 - 88 (1986)

[6] N.Carrillo, Simple Two- and Three-Dimensional Cases in the Theory of Consolidation of Soils, Journ.Math.Phys, 21, 1 - 5, (1942)

Numerical Methods in Geomechanics (Innsbruck 1988), Swoboda (ed.)
© 1988 Balkema, Rotterdam. ISBN 90 6191 809 X

Interpretation of concrete top base foundation behaviour on soft ground by coupled stress flow finite element analysis

Katsuhiko Arai
Fukui University, Japan
Yuzo Ohnishi & Masakuni Horita
Kyoto University, Japan
Ikuo Yasukawa
Fushimi Technical High School, Kyoto, Japan
Shinji Nakaya
Century Research Center Corporation, Osaka, Japan

ABSTRACT: Recently a new foundation method which is called TOP BASE METHOD has attracted engineers attention in Japan. It is used on very soft ground to reduce the consolidation settlement and to increase the bearing capacity of a foundation.

The top base method is to locate top-shape concrete blocks on/in a thin graveled mat laid over very soft ground. In-situ measurements of the behavior of the foundations indicate that consolidation settlement is reduced up to 1/2 - 1/3 and bearing capacity of the foundation increases 50 % - 100 % comparing to the primary non-treated ground.

This paper describes the 1/5 size model tests of the top base foundation performed in a laboratory and the coupled stress-flow(consolidation) finite element analysis to interpret the laboratory measurements. It is found so far that the top base foundation prevents the lateral deformation of soft ground and reduces its negative dilatancy to reduce the surface settlement, and that the the foundation creates rather uniform stress distribution under it to increase its bearing capacity.

1 INTRODUCTION

Most of Japanese cities have been developed on very soft alluvial deposits. With the increasing of population and the lack of land, even a very soft ground has to be used for a regident place. Civil engineers are required to improve such a soft ground for the construction of structures in judging by cost performance, reliability and handling capability. In case that a structure is not very heavy, it can be built safely only with proper improvement (without piles) of the near surface and/or subsurface ground.

Recently a new method has been invented which assembles the top-shaped concrete blocks and places them on the ground. A group of the top blocks can be used as a shallow foundation replacing short piles. It is called "Top Base Foundation".

A schematic view of the top base foundation under construction is shown in Fig. 1, where the top-shaped blocks are placed on crusher-run (gravel) which is spread over the soft ground. To locate the position of individual top block and to reinforce a group of the top blocks, a lattice of iron rods is placed on the crusher-run. It has been reported that several real structures constructed with the new method show drastic effects of reducing settlement and increasing bearing capacity of the structures.[1)5)6)9)] However, few investigation has been carried out so far to clarify the reason why this happens.[1)-10)]

This paper investigates the behavior of the top base foundation on a soft ground. The top base method is believed to modify the overall properties of a soft ground and improve the problems associated with settlement and bearing capacity. Laboratory model tests have been conducted to study in detail the behavior of the foundation ground subjected to the load applied on the top base blocks. For better interpretation of the laboratory test, a coupled stress-flow (negative pore pressure, unsaturated

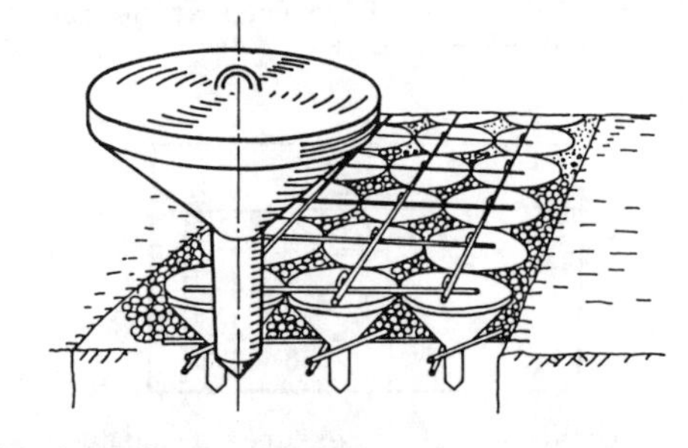

Fig.1 A View of Top Base Foundation

flow included) finite element method has been performed. For the analysis, Sekiguchi-Ohta model[11], a elast-plastic theory, was inplemented for the constitutive model for soft ground. The model can evaluate the effect of dilatancy, anisotropy, creep and so on.

2 LABORATORY MODEL TEST

Some Laboratory model tests of a top base foundation are perform in order to clarify the mechanism of controlling settlement and increasing bearing capacity with top-shaped concrete blocks.

2.1 Axisymetrical plate loading tests

A model ground was prepared with clay in a cylindrical container shown in Fig.2. The clay, which has a liquid limit of 40.4 % and a plastic index of 18.4, was mixed in high water content, twice as high as the liquid limit. Then it was consolidated in the container under a pressure of 0.5 kgf/cm2 after filled in the container. For a foundation, miniature (1/5 of actual scale) top shaped blocks (their diameter is 6 cm) and also scaled down crusher-run (gravel) were set on the ground. Pressure transducers and pore pressure transducers were installed in the clay as shown in Fig.3. The load increment on the foundation was 0.1 kgf/cm2 up to 0.4 kgf/cm2 in a loading interval of 5 minutes, and the load was kept constant at 0.4 kgf/cm2 for a week and later was increased up to 0.8 kgf/cm2.

The ground surface in the soil container settled with time, as shown in Fig.4 for the top base and crusher-run foundations. The immediate settlement is represented at t=0 minute in Fig.4. The top base foundation shows a little more immediate settlement than for the crusher-run foundation because of the compaction of pores among the top-shaped blocks, filling sand and the ground. Fig.4 also shows that the consolidation settlement of the top base foundation is half of that of the crusher-run foundation. It is likely known that the time delayed consolidation settlement is restrained by the existence of the top bases.

Fig.5 for the earth pressure transducer

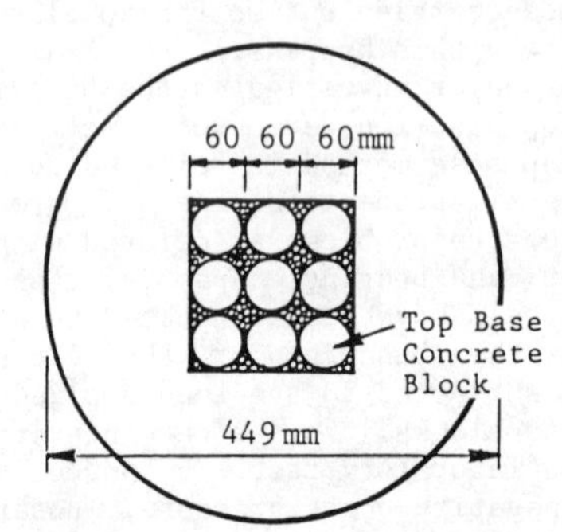

Fig.2 Plan of Top Base Foundation for Laboratory Test

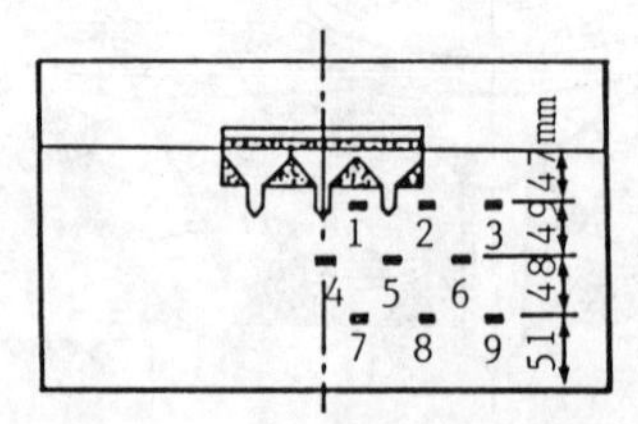

Fig.3 Location of Pressure Transducers and Pore Water pressure Transducers

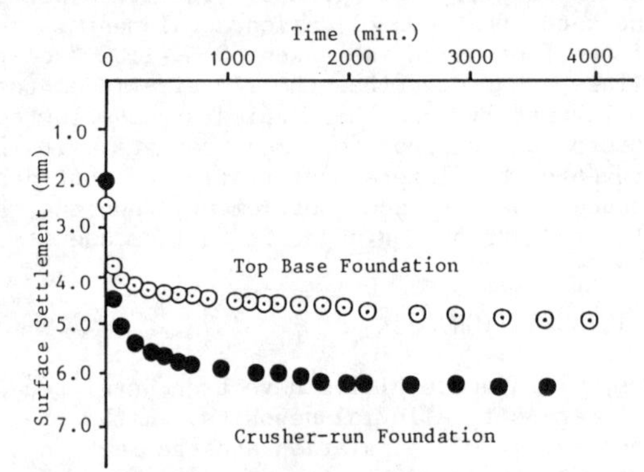

Fig.4 Comparison in Surface Settlement of Laboratory Consolidation tests

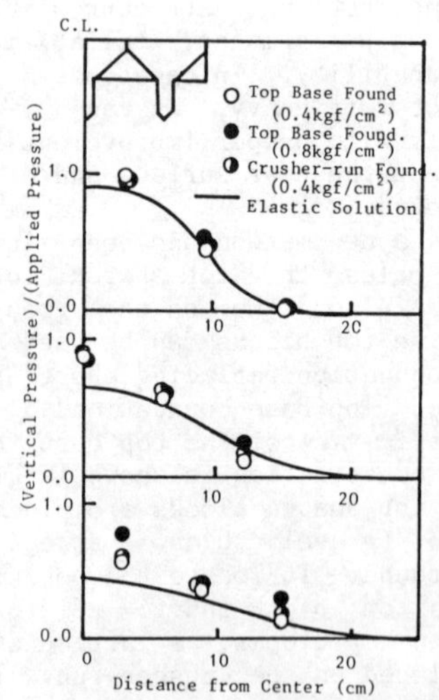

Fig.5 Pressure Distribution Measured in Laboratory Consolidation Tests

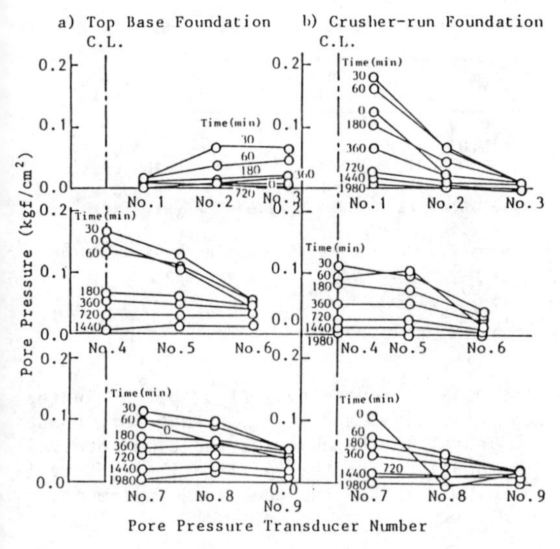

Fig.6 Pore Water Pressure Measured in Loboratory Consolidation Tests

measurements shows only small difference between the top base foundation and the crusher-run foundation. In contrast, Fig.6 clearly shows the difference in distribution of pore pressure between both foundations. The difference in pore pressure at the measurement point No.1 in Fig.3 will be caused by the lateral movement of soil beneath the foundations due to dilatancy. The top base foundation prevents the lateral deformation beneath it with a good interaction between the solid axes of top-shaped blocks and the filling gravel.

2.2 Plane strain plate loading tests

Fig.7 shows the soil container in which some loading tests were performs on the top base foundation in the condition of plane strain. The soil condition and the method installing top blocks are same as the previous axisymetrical tests. In order to clarify the mechanism of failure, many aluminum targets shown in Fig.8 are placed in the consolidated ground to measure the deformation in ground. The load on the plate increases by 0.1 kgf/cm2 in an interval of 5 minutes.

Fig.9 shows the load-settlement curves during plate loading tests. Fig.9 shows that the top base foundation fails with 5.0 tf/m2 while that the primary ground without any top-shaped concrete blocks fails at the load of 3.0 tf/m2. The failure pattern of the top base foundation is imagined from the deformation pattern in Fig.10.

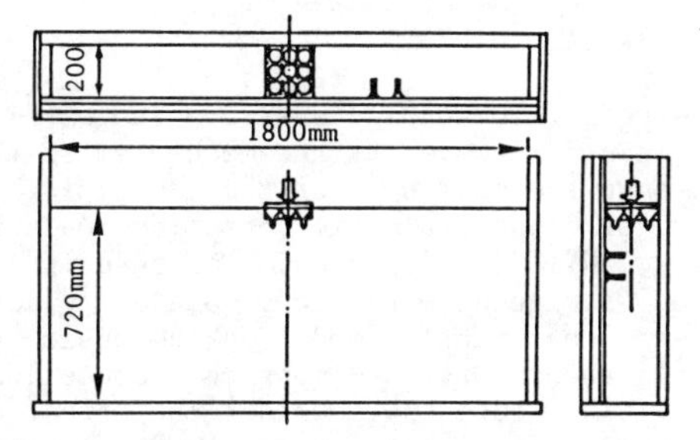

Fig.7 Soil Container for Plane Strain Plate Loading Tests

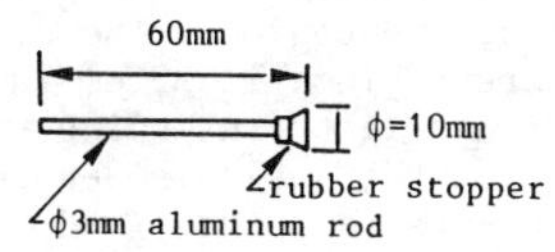

Fig.8 Target for Deformation Measurements

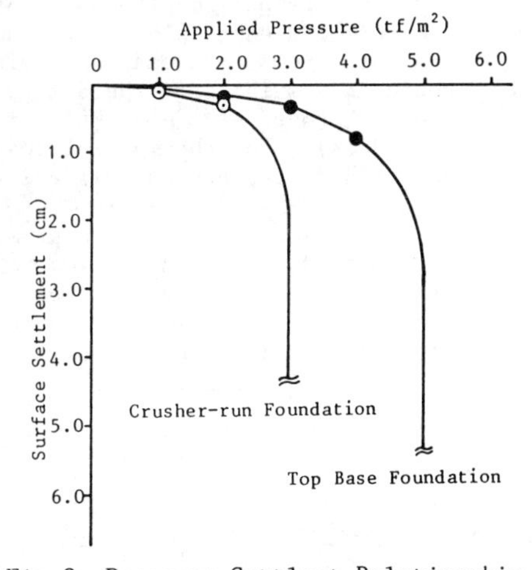

Fig.9 Pressure-Settlent Relationship in Plate Loading Tests

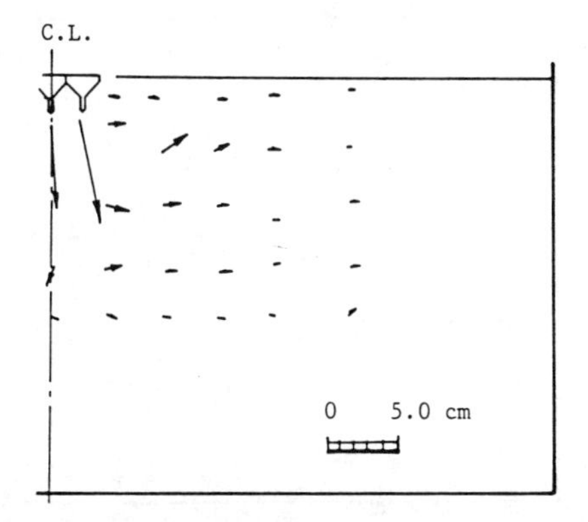

Fig.10 Deformation Pattern at p=5tf/m2 on Top Base Foundation

3 NUMERICAL ANALYSIS

A stress-flow coupled analysis is carried out in order to verify the mechanism of the top base foundation. For the analysis, a code of elastic-plastic consolidation analysis, UNICOUP, is used. The code introduces Sekiguchi-Ohta model as an elast-plastic constitutive model on the basis of Biot's theory of two-dimensional consolidation. The code can consider the effect of dilatancy, anisotropy, creep, and secondary consolidation.[11]

Fig.11 shows the FEM mesh. It assumes the plain strain condition, and analyzes the right half of the region. The size of top-shaped concrete block is 33 cm in diameter and 33 cm in height (including the part of the axis). In the analysis 8-node isoparametric elements are used and the total number of elements is 104 and the total number of nodes is 329. The evaluation of pore water pressure is only on the nodes at the four corners of each elements. The material constants are summarized in Table 1. The ground is simulated by the Sekiguchi-Ohta model (one of elast-plastic constitutive models). For the initial condition of the ground, the pre consolidation pressure is assumed as $\sigma_{yield}=\sigma_0 x2+5$ tf/m² where σ_0 is the effective overburden pressure. The top-shaped block and the loading plate are concrete and are assumed to be elastic. Also the filling gravel is assumed to be elastic for easy interpretation. The load is increased by 1 tf/m2 per 5 minutes up to 5 tf/m2 and kept constant for about 20 years. In the analyses the time interval is doubled with the initial time interval of 0.001 minutes for each load increment.

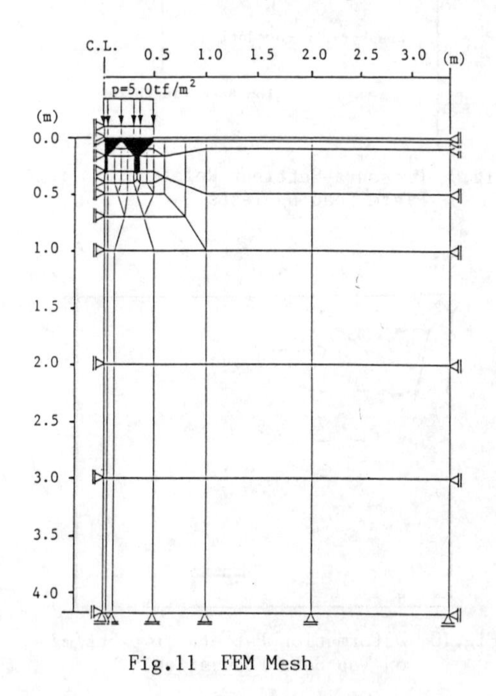

Fig.11 FEM Mesh

Table 1 Material Properties used in FEM

Soil (elast-plastic)	Concrete (elastic)	Gravel (elastic)
$\lambda=0.25$		
$\kappa=0.03$		
$\mu=D=0.0004$	$E=1\times10^4$ tf/m²	$E=200$ tf/m²
$\nu=0.333$	$\nu=0.333$	$\nu=0.333$
$\gamma=1.62$ tf/m²	$\gamma=3.0$ tf/m²	$\gamma=1.8$ tf/m²
$e_0=1.6$	$e_0=0.3$	$e_0=0.7$
$k=6\times10^{-8}$ m/min	$k=6\times10^{-2}$ m/min	$k=6\times10^{-2}$ m/min
$K_0=0.5$	$K_0=0.5$	$K_0=0.5$
$(\sigma_{yield}=\sigma_0 x2+5$ tf/m²$)$		

4 ANALYTICAL RESULTS AND INTERPRETATION OF THE SETTLEMENT CONTROL BY TOP BASE METHOD

Fig.12 shows a comparison in time-settlement curves of primary ground and top base foundation. —○—○— and —△—△— show the surface settlements at the center of loading plate of top base foundation and the primary ground, respectively. —●—●— and —▲—▲— show the settlement of the center points under 1.5 m in depth. The surface settlement of top base foundation is almost one-half of that of the primary ground. Fig.12 emphasizes that the surface settlements of the both foundation have a large difference in the settlement while the points at 1.5 m deep has no difference. It suggests that the mechanism of settlement control by top base foundation is strongly related to the behavior of the ground near the surface.

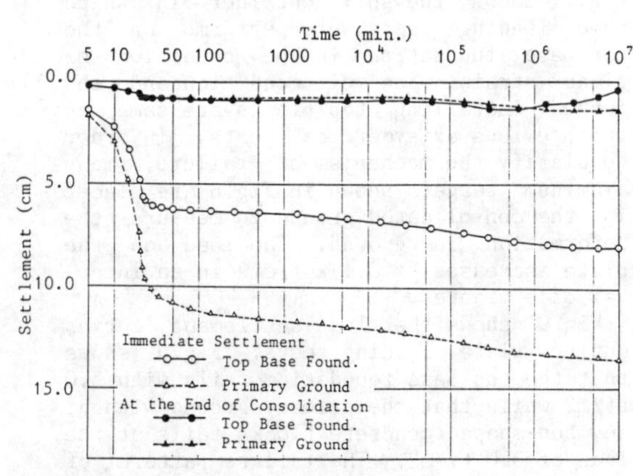

Fig.12 Comparison in Surface Settlements analyzed by FEM

Fig.13 shows the surface deformation. -○-○- and -△-△- show the immediate surface deformations and the top base forndation has 5.0 cm at the center of the foundation while the primary ground has 7.5 cm. At the end of consolidation, the settlement of the primary ground is 13.5 cm, and that of the top base foundation is 7.6 cm. The difference in settlement between those foundations increases with elapsing time. The consolidation settlement of the top base foundation is 50 % of the immediate settlement while that of the primary ground is 80 % of the immediate settlement. From the deformation analysis, the top base foundation is also effective to restrain the consolidation settlement.

Fig.14 show the distribution of lateral deformation along the center line under the loading plate. In the figure, the lateral deformation occurs mainly at the region up to 1 m deep. The immediate lateral deformation of the top base foundation is 2/3 of that of the primary ground. The ratio become less than 1/2 at the end of consolidation. Another important point is that the

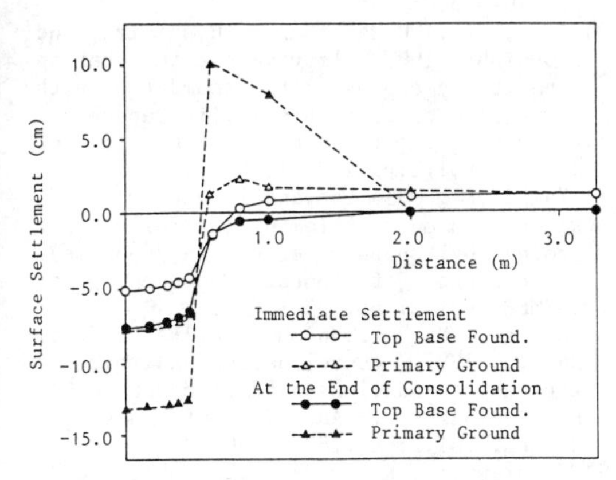

Fig.13 Configulation of Surface during Consolidation

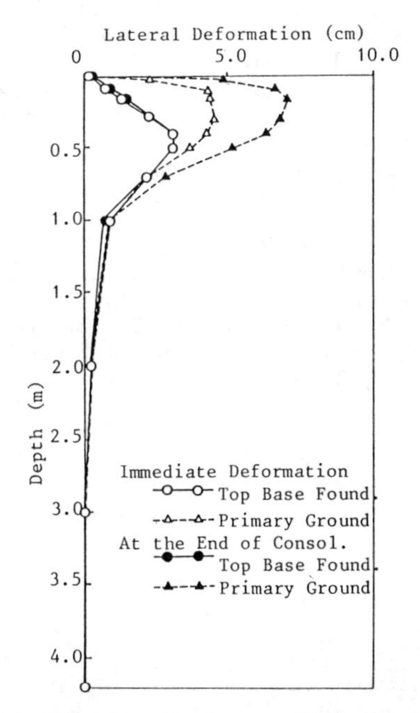

Fig.14 Distribution of Lateral Deformation with Depth

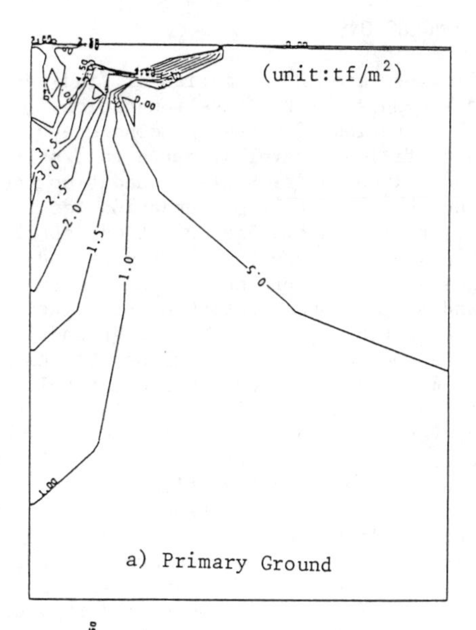

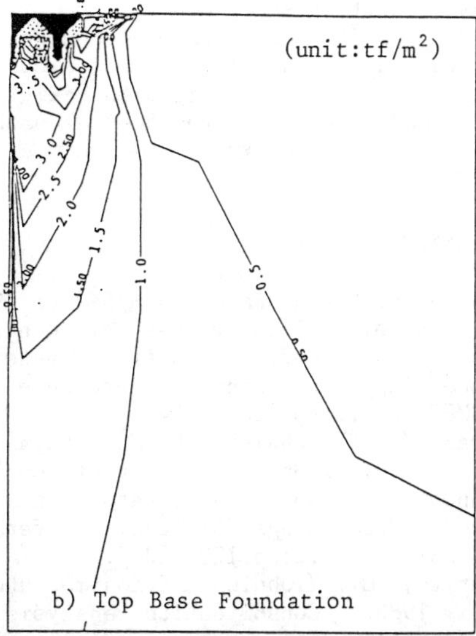

Fig.15 Comparison in Vertical Stress Distribution

primary ground show larger lateral deformation with consolidation than the top base foundation.

Fig.15 shows a comparison in vertical stress distribution between the foundations at the end of consolidation with p=5.0 tf/m2. As seen in the laboratory tests, the distribution in vertical stress under the top base foundation is similar to that of the primary ground though some distaurbance is observed near surface in the primary ground.

5 CONCLUSIONS

Following are the summaries of findings on the mechanism of top base foundation.

1) Combination of top-shape concrete blocks and filling gravel disperse the surface load like a flexible foundation does though the top base foundation consists of rigid assemblage of tied concrete blocks. Therefore the top base foundation has a trend to prevent local deformation and stress concentration and to increase the bearing capacity of soft ground.
2) The combination of top-shaped blocks and filling gravel prevents the lateral deformation and a loading plate, which prevents the negative dilatancy of soft ground due to deformation. It help to control the surface settlement.
3) The top-shaped concrete blocks restrain the filling gravel in it. Therefore the gravel is restrained to decrease, and the above-mentioned effects which are the increase of bearing capacity and the decreasing surface settlement.
4) The mechanism of settlement control by the top base foundation can be simulated well by the FEM analysis with Sekiguchi-Ohta model though it has been a problem in the elastic analysis.

REFERENCES

1) Yamada,K., Y.Yasukawa and M.Saitoh 1986. In-situ plate loading test of concrete top block on soft ground (in Japanese). Proceedings of Annual Conference of JSSMFE. vol.2:pp.1281-1284.
2) Arai, K., Y. Ohnishi, H. Machihara and H. Kokubo, 1986. Model test of top base foundation on soft clay stratum (in Japanese). Proceedings of Annual Conference of JSSMFE. vol.2:pp.1285-1286.
3) Horita, M., Y.Ohnishi, K.Kojima, and K. Arai 1986. Consolidation analysis of foundation with top-shaped concrete blocks (in Japanese). Proceedings of Annual Conference of JSSMFE. vol.2: pp.1287-1290.
4) Arai, K., Y. Ohnishi, H. Machihara, M. Horita, H. Kokubo, and I. Yasukawa 1986. Settlement control of soft ground by top base foundation (in Japanese). Symposium on Lateral Deformation of Ground. JSSMFE: pp.111-114.
5) Arai, K., Y. Ohnishi, M. Horita and I. Yasukawa 1987. Measurement and interpretation of loading test of concrete top block on soft ground. The proceedings of 2nd international Symposium on Field Measurements in Geomechnics, Vol.2 pp.919-926.
6) Arai, K., H.Machihara, H.shimizu, and Y.Ohnishi 1987. Improvement in bearing capacity by employing the foumdation with top-shaped concrete blocks.(in Japanese). Proceedings of Annual Conference of JSSMFE. vol.2:pp.1131-1132.
7) Yamada,K., M.Saitoh,and I.Yasukawa 1987. Discussion on settlement control of soft ground by top base method (in Japanese). Proceedings of Annual Conference of JSSMFE. vol.2: pp.1833-1836.
8) Horita, M., Y.Ohnishi, K.Arai and S. Nakaya 1987. consolidation analysis of top base foundation by FEM.(in Japanese). Proceedings of Annual Conference of JSSMFE. vol.2: pp.1837-1840.
9) Yasukawa,I., K.Yamada, Y.Ohnishi, and M. Saitoh 1987. Settlement control of foundation on soft ground by concrete top blocks (in Japanese). Proc. of 32nd Simposium of JSSMFE on Foundation Procedures without Piles, pp.49-54.
10) Arai,K., H. Machihara, M. Horita, Y. Ohnishi, and I.Yasukawa 1987. Laboratory tests and analysis on settlement and bearing capacity of foundation with top-shaped concrete blocks. Proc. of 32nd Simposium of JSSMFE on Foundation Procedures with-out Piles, pp.55-60.
11) Sekiguchi, H. and Ohta, H. 1984. Induced anisotropy and time dependency in clay. Proceedings of 9th ICSMFE, Specially Session No.9, Constitutive Equation of Soils: pp.229-238.

Numerical Methods in Geomechanics (Innsbruck 1988), Swoboda (ed.)
© 1988 Balkema, Rotterdam. ISBN 90 6191 809 X

On the determination of the matrix of permeability for partially saturated finite elements

S.Gobbi & S.M.Petrazzuoli
Studio Geotecnico S.G.S., Naples, Italy

ABSTRACT: A finite element model for unconfined seepage problems with particular reference to earth dams is developed. It is based on a simple determination of the matrix of permeability that allows to account for the saturated and the unsaturated zone of the elements intersected by the free surface. The proposed model is supposed to improve both the accuracy of the solution and convergence. As the iterative procedure does not require any modification of the initial mesh, superposition of different effects may be easily performed. Obtained results are compared with data published by other authors.

1 INTRODUCTION

In 1966 Zienkiewicz et al. /1/ clearly illustrated for the first time how the Finite Element Method can be successfully used to solve field problems such as two and three dimensional seepage flow in non homogeneus anisotropic porous media.

Since then the FEM has been widely used by many authors to analyze different kind of groundwater seepage problems. Many studies have focused on unconfined flow as this case entails particular difficulties due to presence of a free surface, the location of which is not fully known a priori. This means that the region where seepage occours is initially undetermined, and the correct position of the top flow line, together with its exit point on the downstream face of the considered region, must be derived as part of the solution.

As the boundary conditions to be written for the free surface concern both prescribed total heads and prescribed fluxes, and in general they are not satisfied at the same time, special iterative procedures are required to achieve reasonable results. Up to today the proposed methods to solve unconfined flow can be divided in two main groups: a) methods involving successive modifications of the initial mesh; and b) methods which utilize the same schematization throughout the solution without changes of its geometry.

In an attempt to summarize the principal features which are common to the above mentioned procedures, we could say that the typical approach of the methods pertaining to the first one, like those proposed by Finn /2/, Taylor and Brown /3/, Neuman and Witherspoon /4/ and France et al. /5/, is the most immediate from the physical point of view.

In few words, these models consist in the analysis of the only region lying below the free surface. This is treated as an impervious boundary and shifted along prescribed directions at the end of any iteration untill the condition:
elevation head = total fluid head
is satisfied everywhere on it. In this way, of course, the possibility to account for any contribution from the unsaturated zone is implicitly neglected.

Despite the simplicity of such models, the shifting procedures may imply intermediate solutions extending beyond the geometric or physical boundaries of the analyzed domain, and may also result in lack of convergence. Moreover, the global matrix needs to be rearranged from iteration to iteration as the configuration of the region where flow occurs is modified, so requiring strong computational efforts. Finally, working with a variable mesh makes extremely difficult to superimpose different effects coupled with seepage flow, such as seepage forces or eventual external loads.

On the other hand, the group previously named b) is composed of models utilizing two main different approaches. One of them, proposed by Desai /6/, is the so called "residual flow" procedure.

In opposition to the models synthetized above, here the the free surface is no longer regarded as an impervious boundary, even if it acts so after its right location has been found. If we assume an arbitrary initial position of the free surface and there we impose the cited condition on the total fluid head, in general we will obtain at the end of an iteration non-zero fluxes normal to that surface. Fluxes entering or leaving adjacent elements crossed by the guessed surface are algebraically added together and then transformed into nodal concentrated flows from which a nodal forcing vector that allows to correct the position of the free surface for the next iteration is finally computed. Although fluxes across intermediate positions of the free surface are permitted in this case, this formulation also doesn't allow to consider the presence of the unsaturated zone to determine a more accurate solution.

Later, Li and Desai /7/ modified the residual flow procedure in this direction, while Rodriguez-Roa and Munoz /8/ suggested a new simple model that include saturated-unsaturated flow for both two dimensional and axi-symmetric seepage problems.

According to their model no assumption is made for the initial position of the free surface but for its exit point on the downstream face, while the considered domain is supposed to be fully saturated. At the end of each iteration the pore water pression in the centroid of any element is evaluated and, if somewhere it results negative, which means that the corresponding element is prevailingly above the actual position of the free surface, the coefficients of permeability of that element are changed according to the value of the water content in the centroid of the element. It has been shown in fact that the coefficients of permeability are functions of the degree of saturation and empirical expressions have been proposed to describe this relationship. This procedure is carried on untill the differences in water pressure in two successive iteration are as small as desired. If the so computed exit point coincides with the one assumed at the beginning, the final solution has been obtained, otherwise the position of the exit point is modified and another cycle is performed.

Difficulties in the use of this procedure essentially arise from the cited relations between coefficients of permeability and water content. In fact, these relations don't necessarily imply bi-univocal correspondance depending on the previous degree of saturation history, and their use require the experimental determination of some parameters, what is not always possible to carry out.

However, as at a small decrease of the water content corresponds an immediate decrease of the coefficients of permeability k rapidly approaching to zero, it would seems reasonable to adopt $k = 0$ whenever the pore water pressure in the centroid of an element is negative. Although this assumption shouldn't affect considerably the final result for most engineering purposes, we have noted that in this way convergence is not always achieved, as it will be later discussed.

The present paper is intended to show an improvement of the Rodriguez-Roa and Munoz's model by means of an analytical determination of the matrix of permeability of the elements crossed at any given iteration by the free surface. This is supposed to improve the accuracy of the solution and to achieve convergence even when the previous model doesn't converge.

Two applications of the proposed model relating to earth dams and embankments are reported together with comparisons with results published by other authors.

2 ANALYTICAL FORMULATION OF THE PROPOSED MODEL

For the sake of simplicity, let us consider in the following an orthotropic homogeneus soil in which the local principal directions coincide everywhere with the global ones. Solutions for more complex cases can be handled in the same way discussed below with simple analytical modifications.

If the Darcy's law holds, the general differential equation governing two-dimensional steady state seepage flow in a domain without any source or sink can be written as:

$$\frac{\partial}{\partial x}\left(k_x \frac{\partial \Phi}{\partial x}\right) + \frac{\partial}{\partial y}\left(k_y \frac{\partial \Phi}{\partial y}\right) = 0 \qquad (1)$$

where:

k_x, k_y = coefficients of permeability in the x and y direction respectively;

Φ = total fluid head.

Equation (1), associated with an appropriate set of boundary conditions, holds for both confined and unconfined flow. Referring to Fig. 1, in the proposed model the initial boundary conditions for unconfined flow can be specified as follows:

$$\Phi = H_u \quad \text{on AE} \qquad (2)$$

$$\partial\Phi/\partial n = 0 \quad \text{on AB, FC, CD and DE} \qquad (3)$$

$$\Phi = H_d \quad \text{on BG} \qquad (4)$$

$$\Phi = y \quad \text{on FG} \qquad (5)$$

Mathematically,it can be shown that to solve equation (1) is equivalent to determine a function Φ that not only satisfies the boundary conditions written before but also minimizes the functional:

$$\chi = \int_S \left\{\frac{1}{2}\left[k_x \left(\frac{\partial \Phi}{\partial x}\right)^2 + k_y \left(\frac{\partial \Phi}{\partial y}\right)^2\right]\right\} ds \qquad (6)$$

taken all over the analyzed flow domain S.

If the value of Φ in any internal point of the generical element "e" can be expressed

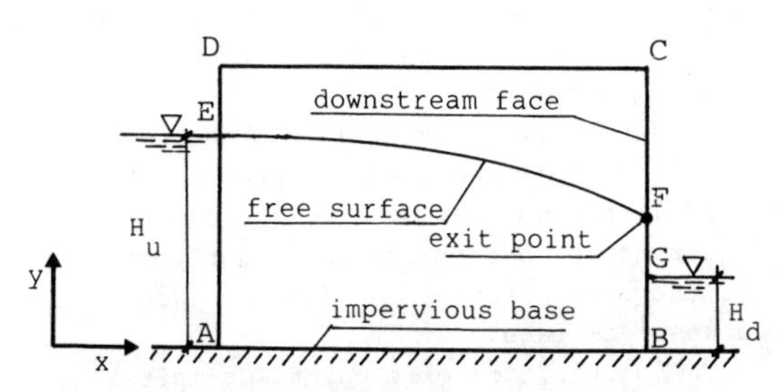

Fig. 1. Unconfined seepage flow diagram.

through the values of the total head at the nodal points of that element as:

$$\Phi^e(x,y) = [N_1, N_2, \ldots, N_n] \{\Phi\}^e \qquad (7)$$

in which $N_1,\ldots,N_n$ are only function of the coordinates of the n nodes of the considered element and $\{\Phi\}^e$ is the list of the corresponding nodal values of Φ , then substitution of equation (7) in (6) yields for element "e" to:

$$\chi^e = \frac{1}{2}\int_{S^e} \{k_x [\frac{\partial}{\partial x}(N_1,\ldots,N_n)\{\Phi\}^e]^2 + k_y [\frac{\partial}{\partial y}(N_1,\ldots,N_n)\{\Phi\}^e]^2\} ds \qquad (8)$$

where S^e is the area of triangle "e".

Minimization of equation (8) implies its differentiation with respect to each nodal value of Φ , that is, in compact form:

$$\left\{\frac{\partial \chi^e}{\partial \Phi^e}\right\} = [h]^e \{\Phi\}^e \qquad (9)$$

where the coefficients of matrix $[h]^e$ are given by:

$$h_{ij}^e = \int_{S^e} \left(k_x \frac{\partial N_i}{\partial x}\frac{\partial N_j}{\partial x} + k_y \frac{\partial N_i}{\partial y}\frac{\partial N_j}{\partial y}\right) ds \qquad (10)$$

As the complete procedure to achieve minimization of equation (6) is well known from the theory of finite element and it has already been clearly illustrated in many of the reported references with emphasis to seepage problems, let us now analyze how equation (10) can be specified in some particular cases.

If we adopt triangular finite elements to discretize the flow domain and r, s, t are the nodal points of element "e", then the functions N_r, N_s and N_t are linear equation of the kind:

$$N_i = (a_i + b_i x + c_i y)/(2S^e) \qquad (11)$$

the coefficients of which being only related to the coordinates of the element's nodes. Hence the partial derivatives contained in the right side of equation (10) just represent costant values, and so it can be simplified as:

$$h_{rs}^e = \frac{1}{(2S^e)^2}\left[b_r b_s \int_{S^e} k_x ds + c_r c_s \int_{S^e} k_y ds\right] \qquad (12)$$

For a fully saturated element, besides, k_x

and k_y are costant within the element and equation (12) can be simplified again, so obtaining:

$$h^e_{rs} = (k_x b_r b_s + k_y c_r c_s)/(4S^e) \quad (13)$$

For a partially saturated element instead, the coefficients of permeability are no longer costant all over the element. In this case as well, we could write an equation similar to (13) by simply assuming for such elements the values of k_x and k_y relative to the computed value of Φ in the centroid of the element or by considering the average values of the coefficients of permeability throughout the same element. Both this procedures, however, seem to imply an oversimplification of the problem according to the rapid variability of k with respect to slight variations of the degree of saturation inside the same element.

Without significant additional computational efforts, then it appears more reasonable to consider separately the saturated and the unsaturated zone, each of them characterized with its own costant or average values of the coefficients of permeability. If indices 1 and 2 respectively denote the saturated and the unsaturated region within the element "e" crossed by the free surface (see Fig. 2), equation (10) now becomes:

$$h^e_{rs} = [(k_{x1} b_r b_s + k_{y1} c_r c_s) S^e_1 + (k_{x2} b_r b_s + k_{y2} c_r c_s) S^e_2]/(2S^e)^2 \quad (14)$$

From a strictly mathematical point of view, the respect of the continuity of flow in each nodal point internal to the flow domain implies that equation (14) holds for any value of k as small as we want, but not equal to zero, even if this last condition has a clear physical meaning. Therefore, if we want to neglect the contribution of the unsaturated zone assuming for this zone $k_2 = 0$, so that:

$$h^e_{rs} = (k_{x1} b_r b_s + k_{y1} c_r c_s) S^e_1/(2S^e)^2 \quad (15)$$

equation (15) should just be considered as the limit of (14) for k_2 approaching zero.

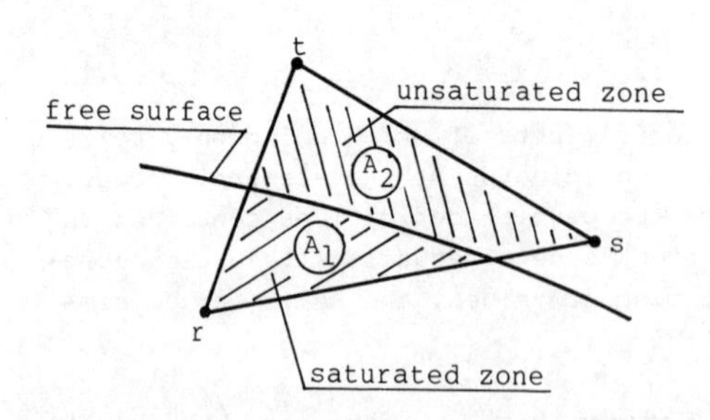

Fig. 2. Triangular finite element crossed by the free surface.

3 APPLICATIONS

To demonstrate the advantages offered by the above illustrated numerical formulation, two applications of seepage analysis through earth dams are discussed herein. The first one concerns the same problem already analyzed in /5/ and /8/, whereas the second one is referred to an actual italian dam cross section.

For each of the studied problems, whose geometrical representations are respectively reported in Fig. 3 and Fig. 4, solutions were sought applying the Rodriguez-Roa and Munoz's procedure and the presented numerical formulation.

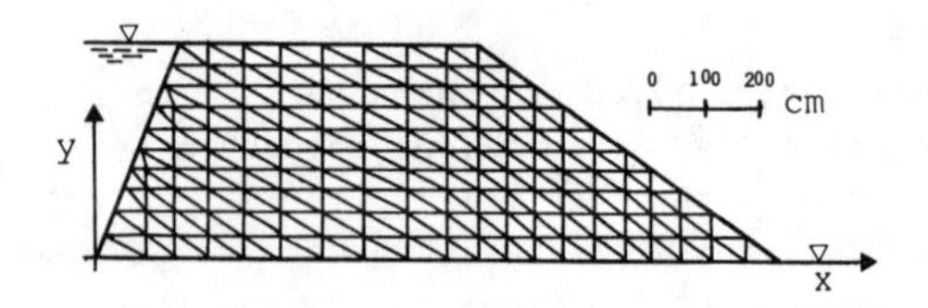

Fig. 3. Mesh employed for the first dam analyzed.

In both cases, using the Rodriguez-Roa and Munoz's procedure with $k_x = k_y = 0$ for the unsaturated zone, we noted that approaching the final solution the results of successive iterations periodically varied among the same two or three solutions, never achieving real convergence.

On the other hand, utilizing the proposed model, we obtained full convergence ($\Delta\Phi/\Phi < 1/100$) in a maximum of 5 iterations only after having determined the right position of the exit point of the free surface on the downstream face.

To do this, the exit point was initially located in correspondance of the last nodal

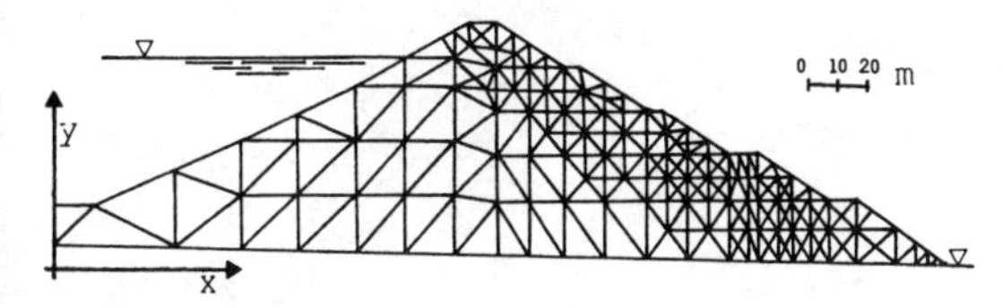

Fig. 4. Mesh employed for the second dam analyzed.

point of the adopted schematizations and then it was moved step by step upward untill the pore water pressure in each node above it resulted negative. In our opinion this seems to be the only way to find the final solution. In fact, if we assume at the beginning of an iteration a position of the exit point that is above its real one, the results of that iteration will not allow to determine its new location for the next iteration.

The so computed free surface profiles are shown in Fig. 5 and in Fig. 6. In particular the solution in Fig. 5 is almost coincident with the ones reported in /5/ and /8/.

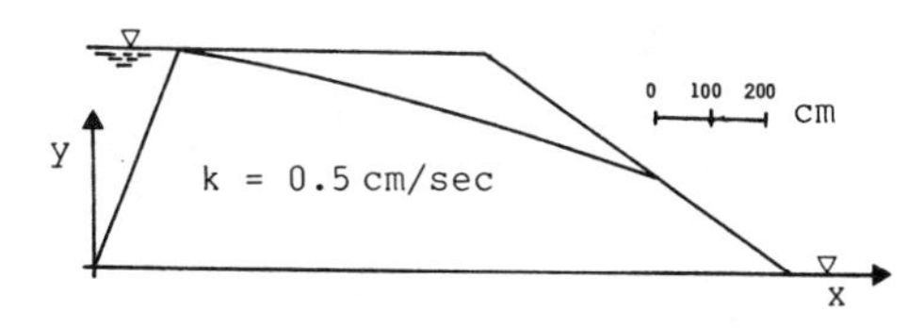

Fig. 5. Free surface profile for the first dam analyzed.

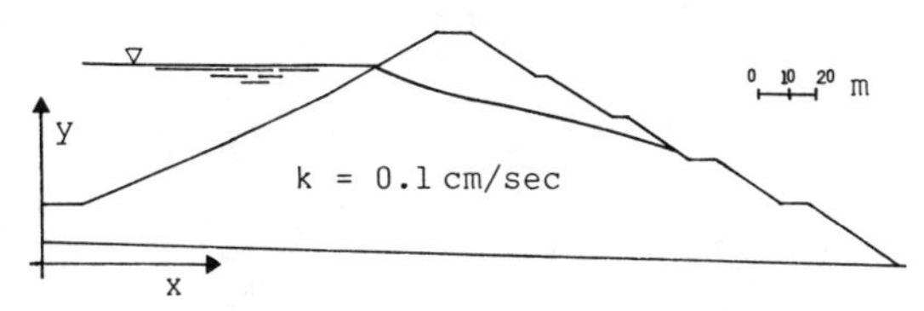

Fig. 6. Free surface profile for the second dam analyzed.

4 CONCLUSIONS

The presented model allows to achieve an accurate solution for unconfined seepage problems considering both saturated and unsaturated regions.

The model involves a simple analytical determination of the matrix of permeability of the partially saturated elements and may be used also with a small personal computer with a reasonable running time.

Convergence is assured also for the special case in which k = 0 is taken for the unsaturated region.

Obtained results are in good agreement with analogous results reported by other authors.

5 REFERENCES

/1/ O.Zienkiewicz, P.Mayer and Y.K.Cheung, Solution of anisotropic seepage by finielements, J.Eng.Mech.Div. ASCE 92 EM1, 111-120, (1966)

/2/ W.D.Liam Finn, Finite-element analysis of seepage through dams, J. Soil Mech. Found. Div. ASCE 93 SM6, 41-48, (1967)

/3/ R.L.Taylor and C.B.Brown, Darcy flow solutions with a free surface, J.Hydr.Div. ASCE 93 HY2, 25-33, (1967)

/4/ S.P.Neuman and P.A.Witherspoon, Finite element method of analyzing steady seepage with a free surface, Water Resources Research 6 no.3, 889-897, (1970)

/5/ P.W.France, C.J.Parekh, J.C.Peters and C.Taylor, Numerical analysis of free surface problems, J.Irr.Drain.Div. ASCE 97 IR1, 165-179, (1971)

/6/ C.S.Desai, Finite element residual schemes for unconfined flow, Int.J.Numer. Method Eng. 10, 1415-1418, (1976)

/7/ G.C.Li and C.S.Desai, Stress and seepage analysis of earth dams, J.Geotech.Eng. 109 no.7, 946-960, (1983)

/8/ F.Rodriguez-Roa and J.F.Munoz, A simplified finite element model for transient two-dimensional and axi-symmetric seepage problems, Proc.Int.Symp.Numer.Models Geomech., Zurich, 475-485, (1982)

Numerical Methods in Geomechanics (Innsbruck 1988), Swoboda (ed.)
© 1988 Balkema, Rotterdam. ISBN 90 6191 809 X

Imperfect underground barriers under transient seepage conditions

R.Jappelli & F.Federico
Università di Roma, Tor Vergata, Italy
C.Valore
Università della Calabria, Arcavacata di Rende, Italy

ABSTRACT : The effectiveness of barriers of assigned thickness and permeability , penetrating in a homogeneous soil layer and keyed into an underlying impervious formation, is analyzed. The plane free-surface seepage domain is defined by two vertical planes where the piezometric head is constant, and a lower horizontal impervious boundary. The unsteady seepage process subsequent to the installation of the barrier is analyzed by the method of successive steady-states. The soil skeleton is considered incompressible. The governing system of differential equations has been integrated by numerical methods; moreover, an analytical solution has been obtained for special classes of the coefficients of the equations. Results of the proposed analysis compare favourably with the few available experimental data.

1. INTRODUCTION

The most important applications of underground barriers in Geotechnical Engineering refer to the control of pore water pressures in the foundations of dams (Casagrande, 1961) , the abatement of leakage from reservoirs, the construction of underground dams for the storage of fluids (Kamon, Aoki, 1987), and recently,to the protection of aquifers from pollution (A.S.T.M.,1984). In the latter case, the barrier mitigates the impact of contaminants on the aquifer and on the underground environment, the control of groundwater velocities playing a key role in the design of protective measures (Barbour, 1983).

In general, the barrier is not watertight owing to the intrinsic, non-zero, hydraulic conductivity of the material of which it is composed, as well as to construction defects.Therefore, the barrier is "imperfect".The effectiveness of the measure depends on the thickness, depth, location and permeability of the barrier, and on the characteristics of the involved ground mass; furthermore, it is influenced by the flow regime,the distribution and the size of possible construction defects.

The response of the ground-barrier system has been so far analyzed under a number of symplifying hypotheses concerning the flow regime (steady, transient), the degree of imperfection of the barrier and its penetration and thickness. Coupled effects of seepage and deformation processes have been generally neglected. Telling and Menzies (1978) consider partial or fully penetrating imperfect cut-off walls of finite thickness under confined steady state seepage conditions. The same problem, but with reference to a perfect (i. e. impervious) barrier -whose thickness is vanishing - has been treated by Polubarinova-Kochina (1962), among others.

As to the unconfined processes, only a few studies are available. The steady state problem has been addressed by Polubarinova-Kochina (1962) in the case of a watertight barrier partially penetrating in a homogeneous half-space, when the thickness of the barrier is negligible.The problem of imperfect barriers fully penetrating a finite layer has been investigated by Outmans (Bear, 1972) and Sato (1982).

The transient free-surface seepage process in presence of fully penetrating imperfect barriers with finite thickness has been dealt with by Sato (1982). Experimental data are reported for the first time by the latter author. The theoretical treatment, however, is in need of further improvement.

In the present paper the analysis of the unsteady free-surface seepage process which takes place after the instantaneous construction of an imperfect fully penetrating barrier of finite thickness, is proposed.

Under the assumptions specified below,a more general system of governing differential equations is derived; this is solved by numerical methods. Moreover, a closed form solution is obtained for special classes of the coefficients of the equations. Finally, theoretical results are compared with available published experimental data.

2. STATEMENT OF THE PROBLEM

The barrier modifies the flow field, both upstream and downstream. Its effects vary with time.Two important and diverse effects of the barrier may be distinguished. The first one consists in the modification of the piezometric head within the composite ground-barrier system when the boundary conditions remain unchanged after the installation of the barrier. The second effect entails variations of the piezometric heads depending upon changes of the boundary conditions due to the presence of the barrier itself, or to external actions .

Both processes may be coupled with consolidation or swelling phenomena.

In this paper the first effect only is considered. The construction of the barrier is supposed to be instantaneous.

The analysis of the two-dimensional unsteady seepage process is developed assuming that the flow domain (fig.1) is defined by a lower impervious horizontal surface and by two vertical surfaces along which the piezometric head is constant and invariable with time. Moreover, the soil is homogeneous and isotropic as far as its coefficient of permeability k_1 is concerned; the barrier is keyed into the lower impervious bedrock; its cross section is rectangular, while its thickness is equal to l_2; the barrier is homogeneous and isotropic, its coefficient of permeability being $k_2 < k_1$. The seepage flow obeys Darcy's law; Dupuit's hypothesis is valid; the fluid and the soil skeleton are incompressible; no capillary phenomena occur.

Under the above assumptions, and when the inertial forces are neglected, the unsteady free surface seepage process is governed by a system of elliptic partial differential equations, with usual conditions - of the Dirichlet or Neumann type - on a part of the boundary .Too many conditions -some being non-linear - must be satysfied on the remaining unknown moving part of the boundary of the flow domain.

Due to the complexity of the problem, the solution can be obtained in general only by numerical methods. Many techniques are available which are based on the discretization of the seepage flow domain or of its boundary (Gioda, Gentile, 1987; Gioda ,Cividini , 1984).

A different approach to the problem was proposed by Baiocchi (1971). According to this method, the domain is fixed and the problem is reconduct to the search of the solution of a quasi-variational inequality, after the introduction of a "non trivial" change of the dependent variable (Craig,1984). However, despite of its brilliant mathematical structure, applications of this method to irregular domains are still very difficult.

A third approach followed for the purposes of the present paper, consists in introducing additional simplifying hypotheses suggested by the actual geometry of the flow domain and by the most important features of the seepage process, in order to formulate and solve simpler, but still significant, engineering problems. Following this last line of thought, some mathematical models that could be usefully employed during the preliminary design stages are derived and discussed.

3. GOVERNING EQUATIONS

It is well known that under the above stated assumptions, the unsteady seepage process is governed by a system of Boussinesq's equations. The initial conditions may be expressed by means of Dupuit's equation for the steady-state free-surface established in the ground prior to the installation of

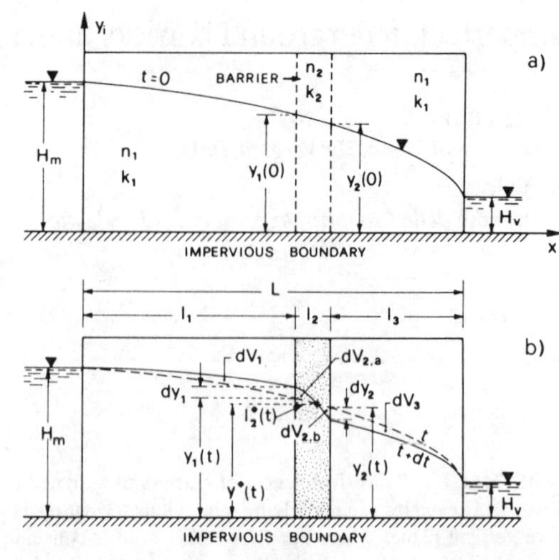

Fig.1 - Reference scheme for the analysis of the problem: a) free-surface before the construction of the barrier ; b) free-surface at times t and t+dt after the construction of the barrier ; dV_1,$dV_{2,a}$,$dV_{2,b}$,dV_3: elemental ground volumes .

the barrier.

The solutions may be obtained by finite differences or finite elements methods,by linearization procedures or by the method of successive steady-states (Bear,1972). According to the latter method, the attention is focused on two significant sections coinciding with the upstream and downstream faces of the barrier. The flow rates $q_1(t)$,$q_2(t)$,$q_3(t)$ through the upstream ground mass, the barrier, and the downstream ground mass respectively, as evaluated by means of the Dupuit's formula, should satisfy the following differential equations:

$$[q_1(t)-q_2(t)]dt=n_1 \cdot dV_1+n_2 \cdot dV_{2,a} \quad (1a)$$

$$[q_2(t)-q_3(t)]dt=n_2 \cdot dV_{2,b}+n_1 \cdot dV_3 \quad (1b)$$

It can be shown that,when the free surface is parabolic and if higher order infinitesimal quantities are neglected,the following system of two ordinary first-order non-linear differential equations is obtained:

$$\left\{\frac{k_1}{l_1}\left[H_m^2 - y_1^2(t)\right] - \frac{k_2}{l_2}\left[y_1^2(t) - y_2^2(t)\right]\right\}dt = 4y_1(t)\, dy_1 \cdot$$

$$\cdot\left[\frac{n_1 l_1}{H_m + y_1(t)} + \frac{n_2 l_2^*(t)}{y_1(t) + y^*(t)}\right] \quad (2a)$$

$$\left\{\frac{k_2}{l_2}\left[y_1^2(t) - y_2^2(t)\right] - \frac{k_1}{l_3}\left[y_2^2(t) - H_v^2\right]\right\}dt = 4y_2(t)\, dy_2 \cdot$$

$$\cdot\left[\frac{n_2(l_2 - l_2^*(t))}{y^*(t) + y_2(t)} + \frac{n_1 l_3}{y_2(t) + H_v}\right] \qquad (2b)$$

in which $y_1(t)$ and $y_2(t)$ are the piezometric heads at upstream and downstream faces of the barrier, respectively, and:

$$l_2^*(t) = l_2 \frac{y_1(t)\, dy_1}{y_1(t)\, dy_1 - y_2(t)\, dy_2} \qquad (3)$$

$$y^*(t) = \sqrt{y_1^2(t) - [y_1^2(t) - y_2^2(t)] \cdot \frac{l_2^*(t)}{l_2}} \qquad (4)$$

The system (2) cannot, in general, be written in explicit form. However, the elemental volumes $dV_{2,a}$ and $dV_{2,b}$ (see fig.1) may be neglected in many engineering situations.As a consequence, it is possible to assume $l_2 << l_1, l_3$.

For the purpose of the present analysis two additional hypotheses have been formulated. According to the first, (A), the volumes $dV_{2,a}$ and $dV_{2,b}$ are neglected; this implies that the effective porosity n_2 of the barrier is no longer considered. Alternatively, the values of $l_2^*(t)$ and $y^*(t)$ may be expressed by a suitably chosen function of l_2 , $y_1(t)$ and $y_2(t)$. In the following, (B), $l_2^*(t)=l_2/2$ and $y^*(t)=[y_1(t)+y_2(t)]/2$ have been considered.

The resulting systems of differential equations in dimensionless form are given in table I, where the equations applying when the free surface is assumed to be straight (Casagrande,Shannon,1952) are also shown .All the systems are autonomous in the variables $r_1(t)=y_1(t)/H_m$ and $r_2(t)=y_2(t)/H_m$. The first variable represents the effectiveness of the barrier,when expressed in terms of piezometric heads.For the cases of straight free surface, since $\psi_1(y_1,y_2)$ and $\psi_2(y_1,y_2)$ are simple polynomial functions, the solution may be easily found by linearization procedures.

The initial and the steady state conditions, in dimensionless form, can be found by the Dupuit's expression and the continuity conditions.

The function ψ_1 and ψ_2 (see table I) vanish when :

$$r^2_{1,c}=(\lambda_1 r^2_2 + \beta\lambda_2)/(\lambda_1+ \beta\lambda_2) \qquad (5a)$$

$$r^2_{2,c}=(\lambda_3 r^2_1 + \beta\mu^2\lambda_2)/(\lambda_3+\beta\lambda_2) \qquad (5b)$$

respectively; $r_1(\tau)$ may reach a maximum and $r_2(\tau)$ a minimum for special combinations of the values of parameters λ_1, λ_2, β, μ. The implications of these last results are currently being investigated.

Table I - Governing systems of differential equations in dimensionless form.

FORM OF FREE SURFACE	(A) $dV_{2,a} = dV_{2,b} = 0$	(B) $l_2^* = l_2/2$; $y^* = (y_1+y_2)/2$
PARABOLIC	$\frac{dr_1}{d\tau} = \frac{1}{4n_1\lambda_1}\left(1+\frac{1}{r_1}\right)\psi_1(r_1,r_2)$ $\frac{dr_2}{d\tau} = \frac{1}{4n_4(1-\lambda_1-\lambda_2)}\left(1+\frac{\mu}{r_2}\right)\psi_2(r_1,r_2)$	$\frac{dr_1}{d\tau} = \frac{1}{4r_1}\frac{\psi_1(r_1,r_2)}{\left(\frac{n_1\lambda_1}{1+r_1}+\frac{n_2\lambda_2}{3r_1+r_2}\right)}$ $\frac{dr_2}{d\tau} = \frac{1}{4r_2}\frac{\psi_2(r_1,r_2)}{\left[\frac{n_2\lambda_2}{3r_2+r_1}+\frac{n_4(1-\lambda_1-\lambda_2)}{r_2+\mu}\right]}$
STRAIGHT	$\frac{dr_1}{d\tau} = \frac{\psi_1(r_1,r_2)}{n_1\lambda_1}$ $\frac{dr_2}{d\tau} = \frac{\psi_2(r_1,r_2)}{n_4(1-\lambda_1-\lambda_2)}$	$\frac{dr_1}{d\tau} = \frac{\psi_1(r_1,r_2)}{n_1\lambda_1+n_2\lambda_2/2}$ $\frac{dr_2}{d\tau} = \frac{\psi_2(r_1,r_2)}{n_4(1-\lambda_1-\lambda_2)+n_2\lambda_2/2}$

$L=l_1+l_2+l_3$; $\lambda_1=l_1/L$; $\lambda_2=l_2/L$; $\mu=H_v/H_m$; $r_1=y_1/H_m$;
$r_2=y_2/H_m$; $\beta=k_1/k_2$; $\tau=\frac{t \cdot K_2}{H_m}$;

$$\psi_1(r_1,r_2)=\delta^2\left[\frac{\beta}{\lambda_1}(1-r_1^2)-\frac{1}{\lambda_2}(r_1^2-r_2^2)\right]$$

$$\psi_2(r_1,r_2)=\delta^2\left[\frac{1}{\lambda_2}(r_1^2-r_2^2)-\frac{\beta}{(1-\lambda_1-\lambda_2)}(r_2^2-\mu^2)\right]$$

4. SOLUTIONS

The four systems of equations have been integrated by the Runge-Kutta method of fourth order, using the Richardson extrapolation technique for better accuracy (Ledermann, 1981).

When the barrier is located in the central zone of the aquifer, $l_1 \cong l_3$ (see fig.1) , an approximate closed - form solution of the system (2), case A, when the free surface is straight, may be obtained. Due to space limitations,the detailed steps of the derivation will be reported elsewhere. It may be shown that the solution of the system is given by:

$$y_1(t) = \frac{(\sqrt{M}-p) - (\sqrt{M}+p)\, F \cdot \exp\left(\frac{\sqrt{M}\cdot A \cdot t}{N}\right)}{2\left[1+F\cdot\exp\left(-\frac{\sqrt{M}\cdot A\cdot t}{N}\right)\right]} \qquad (6a)$$

$$y_2(t) = H^* - y_1(t) \qquad (6b)$$

where $A = k_1/l_1$; $N = n_1 l_1$; $p = DH^*/A$; $D = k_2/l_2$; $H^*=y_1(0)+ y_2(0)$; $M = p^2-4s$; $s = D\,(H^{*2})/A+Hm^2$;

$$F = \left[\sqrt{M}-p-2y_1(0)\right]/\left[\sqrt{M}+p+2y_1(0)\right]$$

The assumptions for the validity of the analytical solution imply that $H^*=y_1(t)+y_2(t)=y_1(0)+y_2(0)$ is approximatively constant.

5. DISCUSSION OF RESULTS

Some results computed using the proposed

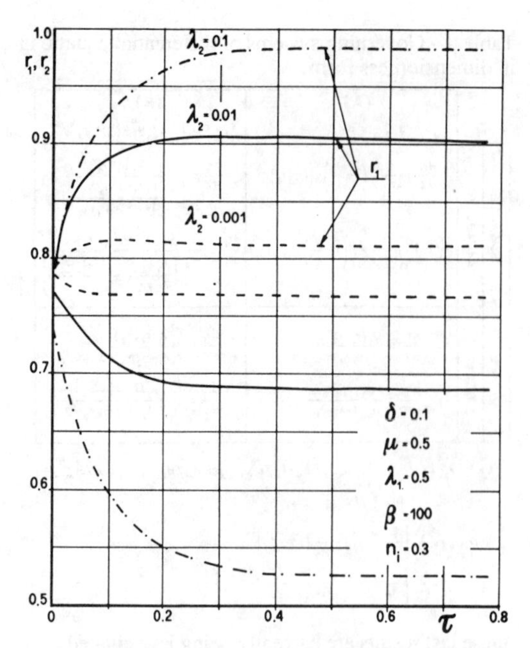

Fig.2 -Influence of the time factor τ and of the thickness ratio $\lambda_2=l_2/H_m$ on the effectiveness of the barrier $r_1=y_1(t)/H_m$ and on $r_2=y_2(t)/H_m$.

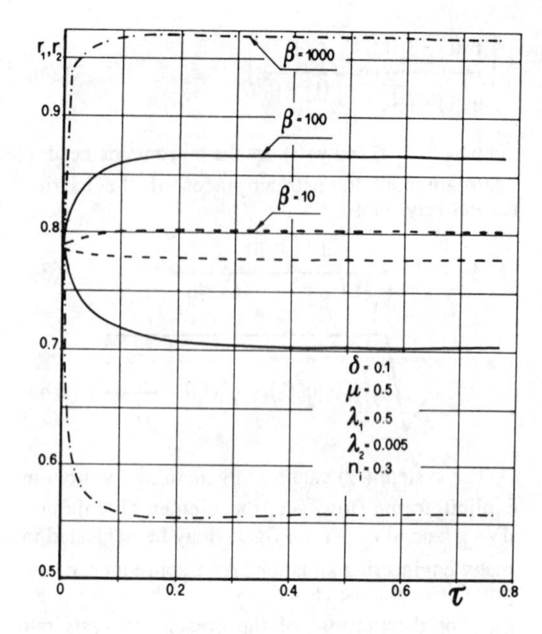

Fig.3 - Influence of the time factor τ and of the permeability ratio $\beta=k_1/k_2$ on the effectiveness of the barrier $r_1=y_1(t)/H_m$ and on $r_2=y_2(t)/H_m$.

mathematical models, are shown in figures 2, 3 and 4.The results represented in figures 2 and 3 have been calculated by the numerical solution of the system (2), for parabolic free - surface , case B. Other results, not reported in this paper and evaluated for the corresponding straight free-surface cases, point out that the influence of the form of free-surface is generally slight , provided that δ is small. In the initial phase of the process, values of r_1 obtained for straight surfaces are higher than those for parabolic free surfaces, for both cases A, B. The values of r_2 are, on the contrary, lower. However, in the initial phase the maximum difference does not exceed 5 percent; afterwards, it is less than 1 percent or one thousandth when λ_2 is lower than 0.01.

Under transient flow conditions, the effectiveness of the barrier is also influenced by the dimensionless ratio $\delta=H_m/L$, and by the porosity n_2 of the barrier; it is markedly dependent upon the time factor τ. Other factors being equal, the influence of n_2 on the effectiveness is not marked; when n_2 decreases, however, the free - surface configuration changes more rapidly. When μ decreases, the rate of variation of $y_1(t)$ and $y_2(t)$ increases. When δ decreases, the absolute value of the derivatives $dr_1/d\tau$ and $dr_2/d\tau$ decreases and the seepage process slows down, as expected.

The influence of β and λ_2 may be easily inferred by comparing figures 2 and 3, from which appears that the most influential factor is β.

The permeability coefficient of the barrier k_2 greatly affects the progress of the seepage process. In fact, an assigned modification of y_i $(i = 1, 2)$ may be induced after a time interval t inversely proportional to k_2. As a consequence, the reduction of k_2 (β increases) permits a diminishing of the thickness of the barrier, but involves a marked increase of the time necessary to obtain the same effect on y_i. Geometrically identical geotechnical systems, with the same value of β, will present the same initial and final configuration of the free-surface, but the progress of the phenomenon in the case of the system characterized by lower value of k_2 will be slower.Values of $r_1$and r_2 approach those pertaining to the steady -state condition (Bear, 1972; Sato, 1982; Federico,Valore, 1987) when the time factor τ is greater than 0.3.

As far as the authors know,very few full scale investigations of the transient seepage problem under consideration have been carried out. The first published results have been obtained on analogic Hele - Shaw models by Sato (1982) and are shown in fig. 4, where the results of the proposed analyses are also represented. Due to the high value of k_1, about 1

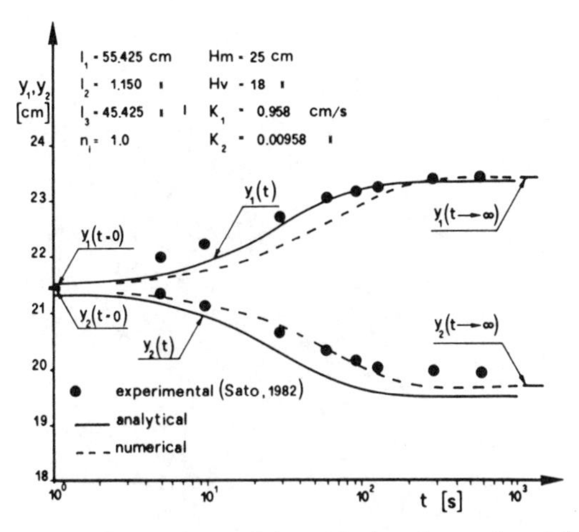

Fig.4 - Comparison of theoretical and experimental results from Hele-Shaw analogic model tests.

cm/s, the steady state condition is reached after about 200 seconds.

The agreement of analytical and numerical results is satisfactory.Both solutions, in turn, agree with experimental data. The maximum difference is never higher than 3 percent.

It may be noted that analytical and experimental results compare favourably even if the barrier is not close to the center of the system.

This result could have been predicted on the basis of verification of the condition on H*, which shows that the analytical solution is still applicable when the barrier is located at appreciable distances from the center ($\lambda_2 = 0.4 \div 0.6$), provided that the time factor τ is greater than 0.1.

6. CONCLUDING REMARKS

The process of transient seepage flow which takes place in a homogeneous isotropic porous medium resting on an impervious bedrock after the instantaneous construction of an "imperfect", fully penetrating, barrier has been analyzed.

A governing system of two first - order, non - linear differential equations has been derived under simplifying assumptions; the system has been solved by numerical methods. Moreover, a closed-form solution has been found for special classes of the coefficients of the equations. The analytical solution is useful for checking the numerical computations (Gibson,1974).

The analytical and numerical solutions compare favourably with the few available experimental data and confirm that the effectiveness of the barrier varies appreciably in function of the time elapsed from its installation.

The proposed analysis enables an investigation of the influence of the involved geotechnical factors and indicates the relevance of the flow regime for the design of the barrier.

ACKNOWLEDGEMENTS

This research has been carried out with the partial financial support of CNR, National Research Council of Italy, within the framework of the Research Project Difesa Catastrofi Idrogeologiche, LR 4, Contract nº 86-00053-42.
The authors are equally responsible for the paper.

REFERENCES

A.S.T.M., 1984 Hydraulic barriers in soil and rock. Special Technical Publication , nº 874.

Baiocchi, C. 1971. Su un problema di frontiera libera connesso a questioni di idraulica.Ann.Mat.Pura Appl., (4), 92, pp.102-127.

Barbour, S.L. and Krahn, J. 1983. Contaminant transport in soils and its significance in the design of waste management facilities - 7th Pan Am. Conf. SMFE, Vol. 5, pp. 525-538.

Bear, J. 1972. Dynamics of fluids in porous media. American Elsevier, New York.

Casagrande, A. 1961. Control of seepage through foundations and abutments. Geotechnique, Vol.11 n. 3, pp. 161 - 181.

Casagrande, A. and Shannon, W. L. 1952. Base course drainage for airport pavements. Trans. ASCE, Vol. 117, pp. 792 - 820.

Craig, A. W. 1984. Quasi - variational inequalities and unconfined flow through a non homogeneous, anisotropic dam. Int. J. of Engng. Sci., Vol. 22, n. 7, pp. 891 - 971.

Federico, F. and Valore, C. 1987. Le barriere sotterranee "imperfette" per il controllo delle pressioni neutre e per la protezione degli acquiferi. Atti Riunione del Gruppo CNR di Ingegneria Geotecnica, Roma.

Gibson R.E.,1974. The Analytical Method in Soil Mechanics. Rankine Lecture, Geotechnique, Vol.24, pp. 115-140

Gioda, G. and Cividini, A. 1984. An approximate F.E. analysis of seepage with a free surface. Int. J. for Num. and Anal. Met. in Geomech., Vol. 8, pp. 549 - 566.

Gioda, A. and Gentile, C. 1987. A non - linear programming analysis of unconfined steady - state seepage. Int. J. for Num. and Anal. Meth. in Geomech., Vol. 11, pp. 283 - 305.

Kamon, M. and Aoki, K. 1987. Groundwater control by Tsunegami underground dam. IX E.C.S.M.F.E., Vol. 1, pp. 175-178, Dublin.

Ledermann, W. ed. 1981. Numerical Methods. In Handbook of Applicable Mathematics, Vol. III.

Polubarinova - Kochina, P.M. 1962. Theory of ground water movement. Princeton Univ. Press.

Sato, K. 1982. Hydro - dynamic behavior of ground water in confined and unconfined layers with cut - off wall. Soils and Found., 22, n. 1, pp.14-22.

Telling, R.M., Menzies, B.K. and Simons, N.E. 1978. A design method for assessing the effectiveness of partially penetrating cut - off walls. Ground Engineering, November.

Numerical Methods in Geomechanics (Innsbruck 1988), Swoboda (ed.)
© 1988 Balkema, Rotterdam. ISBN 90 6191 809 X

Prediction of consolidation of elastoplastic subsoils

A.L.Goldin, A.A.Gotlif & V.S.Prokopovich
B.E.Vedeneev All-Union Research Institute of Hydraulic Engineering (VNIIG), Leningrad, USSR

ABSTRACT: A problem of clay bed consolidation is formulated and solved by the finite element method. Used as constitutive for the soil skeleton are the nonassociated plasticity equations suggested by V.N.Nikolaevsky. Stabilized displacements and the extent of plastic zones are shown to increase if the pore pressure is accounted for.

Typical of the present-day power engineering is the construction of foundations exerting rather high loads on subsoil. To make a realistic prediction of foundation settlements and heelings the design calculations shall allow for plastic properties of soil skeleton and the consolidation phenomenon. This approach enables one to refine the magnitudes of stabilized foundation settlements and estimate stresses and strains in the subsoil at different stages of structure construction and operation.

For mathematic simulation of subsoil a model of the two-phase medium consisting of soil skeleton and pore liquid is adopted. According to the M.Bio theory of dynamic consolidation the two-phase soil deformation process is described by the following set of equations (a two-dimensional problem):

$$\frac{\partial \sigma_{ij}}{\partial x_j} - \delta_{ij}\frac{\partial p}{\partial x_j} = f_i , \qquad (1)$$

$$\frac{\partial}{\partial x_i}\left(\frac{K_\varphi}{\gamma_w}\frac{\partial p}{\partial x_i}\right) = \frac{\partial \theta}{\partial t}, \quad i,j=1,2 \qquad (2)$$

where σ_{ij} are the stress tensor components, p is the pore pressure, f_i are the components of external volume forces, θ is the volume strain of the soil skeleton, K_φ is the coefficient of permeability, γ_w is the volumetric weight of water, δ_{ij} is Kronecker's symbol.

The soil skeleton is described by the dilatant model of non-associated plasticity with hardening suggested by V.N.Nikolaevsky and applied earlier for calculating stabilized strains in foundation subsoils (Nikolaevsky et al., 1975; Goldin et al., 1980). Consider briefly the fundamentals of this model.

The loading surface used is of the form

$$T + \alpha(q)\sigma - Y(q) = 0 \qquad (3)$$

where T is the intensity of tangential stresses, σ is the average normal stress, α and Y are the hardening functions, q is the hardening parameter (here the plastic volume strain). In the pre-limit state (when the loading surface is reached) under active loading the plastic component

$$d\varepsilon_{ij}^{p} = [\sigma_{ij} + (\frac{Q}{3} - \sigma)\delta_{ij}]d\lambda, \quad i=1,2 \qquad (4)$$

is added to the linear strains. In this relationship $Q = 2\Lambda(q)(Y - \alpha\sigma)$; the dilatancy rate Λ is the assigned hardening function.

$d\lambda$ is determined from (3), plastic strain of a volume e_p and that of shear Γ_p turn out to be related to each other by the dilatancy relationship

$$de_p = \Lambda \, d\Gamma_p \qquad (5)$$

The above equations together with the Cauchy relationships constitute a closed system for determining the condition of both soil components. Besides, the formulation of the problem includes initial and boundary conditions of these components.

The solution algorithm is based on the finite element method with the piecewise-linear approximation of the displacement field and pore pressures in the triangular element net. The computational algorithm relationships are obtained from the basic equation set by the Galerkin method.

The matrix analogue of the set of equations (1)-(2) is integrated according to the implicit difference scheme. The plastic flow relationships are allowed for by the method of additional loads. Iterations are performed at each time step. The solution procedure makes it possible to construct and invert just one matrix for all intervals with the constant time step.

The algorithm described is realized in the system of programs which enable the long-term settlements and heelings of foundations to be predicted within the framework of the two-dimensional consolidation problem and the dimensions of plastic strain zones and stress and pore pressure distributions at different time moments to be defined. Taken as initial data are the geometry of the foundation subsoil region, its boundary conditions, deformation, strength and permeability characteristics, the time relationship of external loads.

CALCULATION EXAMPLE

The interaction between a 70 m wide and 3 m thick concrete foundation and a 30 x 190 m soil layer was considered. The soil material was clay with the deformation modulus of 36 MPa, Poisson's ratio of 0.36, the angle of internal friction of 19°, the cohesion of 0.046 MPa and the permeability coefficient of $4 \cdot 10^{-5}$ m/day. The soil skeleton was assumed to be an ideal elastoplastic material. Upper and lower boundaries of the clay layer were adopted to be drained. The intensity of the uniformly distributed load applied to the foundation has been growing linearly from 0 to 0.5 MPa during two years. Allowance was made for the linear distribution of natural stresses over the depth at the lateral pressure coefficient of 0.6 (γ_{sb} = = 10 kN/m).

All in all three design alternatives were considered. The first one calculated the clay layer with the linearly deformable skeleton. The second one took into account the soil skeleton plasticity and neglected the pore pressure. The last one allowed for the plasticity and consolidation of clay. Figure 1a depicts the variation of settlements of the clay layer upper boundary. Figure 1b shows the excess pore pressure curves. Figure 2 demonstrates the development of plasticity zones at different loading stages for the second and third design alternatives.

The calculations reveal that 2 years after cessation of loading the excess pore pressure falls about tenfold and the consolidation process practically stops. The third alternative gives the maximum settlement of the foundation. The solution of the elastoplastic problem yields less (by about 5%) settlement which is explained by the influence of the loading path. In this solution the plasticity zones are localized under the foundation corners and reach the depth of about 14 m whereas with the consolidation taken into account they are twice as deep and join each other. For the first alternative (neglecting the plasticity) the stabilized settlement is 20% less than that with the elastoplastic soil layer consolidation taken into account.

The program developed has been used for calculating long-term settlements and heeling of the power engineering structure foundations.

REFERENCES

Nikolaevsky, V.N., Syrnikov, N.M., Shefter, G.M. 1975. Dynamics of elastoplastic dilating media. In: "Progress in Mechanics of Deformable Media", Moscow: Nauka, p.397-413.

Goldin, A.A., Prokopovich, V.S. 1980. Determination of the bearing capacity of structure foundations using the non-associated law of solifluction. Izvestia VNIIG imeni B.E.Vedeneeva: Sbornik nauchnykh trudov, v.137, p.3-7.

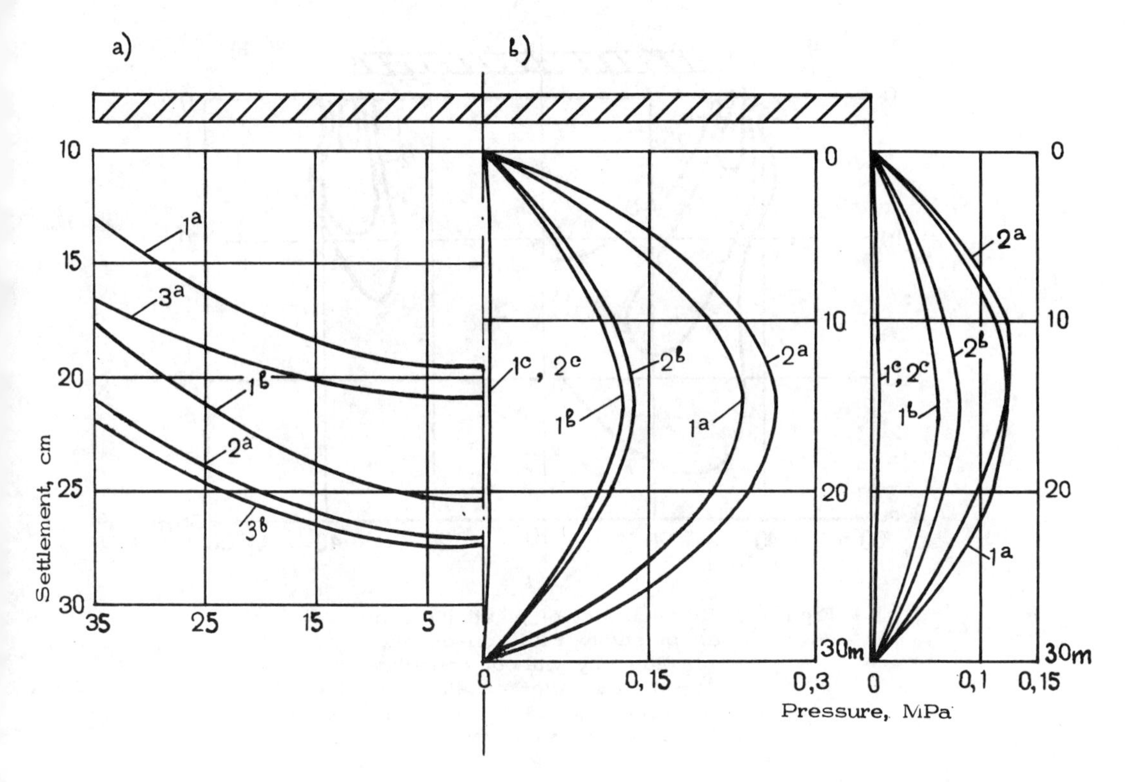

Figure 1. Settlements and pressures
a) Curves of settlements of soil layer surface:
1 - consolidation with linear deformation;
2 - plasticity; 3 - consolidation with plasticity;
a - 2 years after construction start;
b - 6 years after construction start
b) Curves of excess pore pressure variations in the soil layer under the centre and corner of stamp:
1 - consolidation with linear deformation;
2 - consolidation with plasticity;
a - 2 years after construction start;
b - 2.5 years after construction start;
c - 4 years after construction start

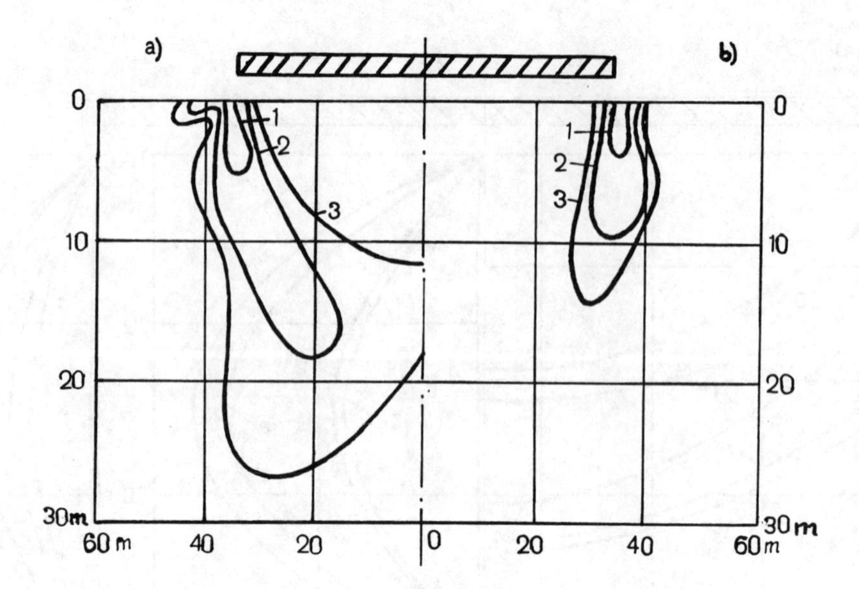

Figure 2. Development of plasticity zones:
a) plasticity with consolidation;
b) plasticity without consolidation;
1 – 1 year after construction start;
2 – 1.5 year after construction start;
3 – 2 years after construction start

Numerical Methods in Geomechanics (Innsbruck 1988), Swoboda (ed.)
© 1988 Balkema, Rotterdam. ISBN 90 6191 809 X

Effects of test procedure on constant rate of strain pressuremeter tests in clay

I.C.Pyrah, W.F.Anderson & L.S.Pang
University of Sheffield, UK

ABSTRACT: To examine strain rate effects in pressuremeter tests in clay a combined laboratory and numerical study has been carried out. In the laboratory, specimens have been consolidated around a cylindrical cavity to simulate 'perfect' insertion of a self boring pressuremeter, and the cavity has then been expanded at differing rates of strain. A complementary numerical study using the Cam clay model and including both consolidation and deviatoric creep has allowed the separate influences of these time dependent phenomena on the cavity expansion to be studied.
Results from both investigations indicate that varying the rate of expansion influences the test results, and hence derived parameters.

1. INTRODUCTION

The Menard pressuremeter [1] consists of a cylindrical cell which can be inflated in a pre-drilled borehole to test the deformation and strength properties of soil. The borehole expansion technique with this type of instrument is usually stress-controlled in that the pressure is applied in increments, each increment being held for a certain period of time. However, the use of a pre-bored hole means that the soil is subjected to stress relief and mechanical disturbance prior to the test. The development of the self-boring pressuremeter [2, 3] enables the instrument to be inserted into the ground with minimum disturbance. In addition, with current equipment tests can be carried out using a constant rate of strain technique similar to that used in most other field and laboratory tests. In fine grain soil the pressuremeter test results may be analysed to give the undrained shear strength and shear modulus of the clay, but inevitably there will be some consolidation and creep of the soil around the expanding probe. The assumption of undrained behaviour therefore, is questionable.

The effects of consolidation and creep on stress controlled pressuremeter tests have been studied both experimentally and numerically and reported elsewhere [4, 5]

These studies showed that the effect of both consolidation and creep is to reduce the deduced modulus values, but that consolidation of the soil around the probe tends to produce higher deduced undrained shear strengths, while creep tends to have an opposite effect, i.e. give a lower deduced strength. Soil parameters derived from a stress-controlled test are thus dependent on the relative effect of consolidation and creep and for these stress-controlled tests the effect of creep appears to be more influential.

The present study investigates the relative importance of consolidation and creep in strain-controlled pressuremeter tests. Laboratory simulations of the test have been carried out and the results compared with numerical predictions which model the influence of these time-dependent phenomena.

2. LABORATORY SIMULATION

A large diameter triaxial cell was modified so that hollow cylindrical specimens 150mm outside diameter, 25mm inside diameter and approximately 150mm high could be subjected to internal and external pressures independently. The soil specimen was taken from a clay bed prepared with known stress history. It was then re-consolidated inside the modified triaxial cell to a higher mean stress with equal internal and external pressures so as to simulate the 'perfect' insertion of a self-boring pressuremeter in an isotropic soil mass. The cavity was subsequently expanded at a constant rate of radial strain by raising the internal pressure using a Cambridge In-situ self boring pressuremeter control system, and the cavity size continuously monitored by a system of three strain-gauged spring feelers inside the cavity.

The main objective of the work was to examine the influence of rate of testing on soil behaviour during strain-controlled cavity expansion. Series of tests were carried out on two clay types (kaolin and pottery clay) with radial strain rates ranging from 0.2% per minute to 4% per minute. The results show similar trends of behaviour for both clays and this paper concentrates on the behaviour of pottery clay. The test results are compared with numerical simulations in order to assess the relative significance of consolidation and creep.

3. NUMERICAL SIMULATION

The numerical simulation had to take into account both the effects of consolidation and creep during the expansion of the cavity. A finite element computer program CAMFE developed by Carter [6] includes consolidation in the analysis of the plane-strain expansion of a cylindrical cavity. The program uses a Biot type consolidation analysis with the flow of pore-fluid through the soil governed by Darcy's law. The present analysis is based on the modified Cam Clay model with solutions obtained using an incremental time-marching technique which can deal with both material and finite deformation non-linearity.

The effects of creep in a stress-controlled expansion have been studied by incorporating the Singh & Mitchell [7] creep model into the CAMFE program as reported by Pyrah et al [5]. The model accounts only for undrained deviatoric creep and requires the determination of three creep parameters from undrained triaxial creep tests at constant but different, deviator stress levels.

Unlike stress-controlled tests in which the cavity pressure is held for a certain period of time, the cavity pressure changes throughout a strain-controlled test to achieve the required rate of increase of radial strain. To evaluate creep strain using the Singh and Mitchell model, the creep time is taken from the beginning of the test and this corresponds to the lower bound of creep proposed by Pyrah et al [5]. However, this method may also over-predict creep strains because the current deviator stress is assumed to have been acting over the whole period of the test, but the deviator stress has been lower prior to the current time. Despite this uncertainty, it is possible to assess the influence of creep on the test results and compare the predictions with the laboratory results.

The computer program allows the loading at the inner and outer boundaries of the thick hollow cylinder to be defined in terms of stress or displacement increments. A strain-controlled test can easily be simulated by adopting the displacement defined type of loading at the inner boundary. Constant stress is assumed at the outer boundary simulating the in situ horizontal stress in a pressuremeter test.

The hollow cylinder (inner radius 12.5mm, outer radius 75mm) was represented by a finite element mesh consisting of 73 nodes, i.e. 36 elements, and the total cavity radial displacement was divided into small equal steps, each associated with a time step depending on the strain rate. A suitable number of displacement steps was established by trial and error so that no numerical instability occurred for the range of strain rates considered. This number was 250 and the associated time steps ranged from 0.02 minute to 0.4 minute for strain rates ranging from 4% to 0.2% per minute. Maximum expansion of the cavity was taken as 20% radial strain.

The modified Cam Clay soil parameters were obtained from conventional tests (oedometer and triaxial) and the Singh-Mitchell parameters were measured in undrained creep tests carried out at constant deviator stress levels of 40%, 60%, 75% and 80% of the failure value.

Table 1 Soil Parameters used in analysis

Critical State Parameters		Creep Parameters	Initial Stresses
κ = 0.025	λ = 0.101	$A = 1.5 \times 10^{-3}$ %/min	$\sigma'_r = 200$ kN/m^2
e_{cs} = 1.265	M = 1.1	$\bar{\alpha}$ = 5.7	$\sigma'_\theta = 200$ kN/m^2
G = 6300 kN/m^2	$k = 2.77 \times 10^{-10}$ m/s	m = 1.0	$\sigma'_z = 200$ kN/m^2

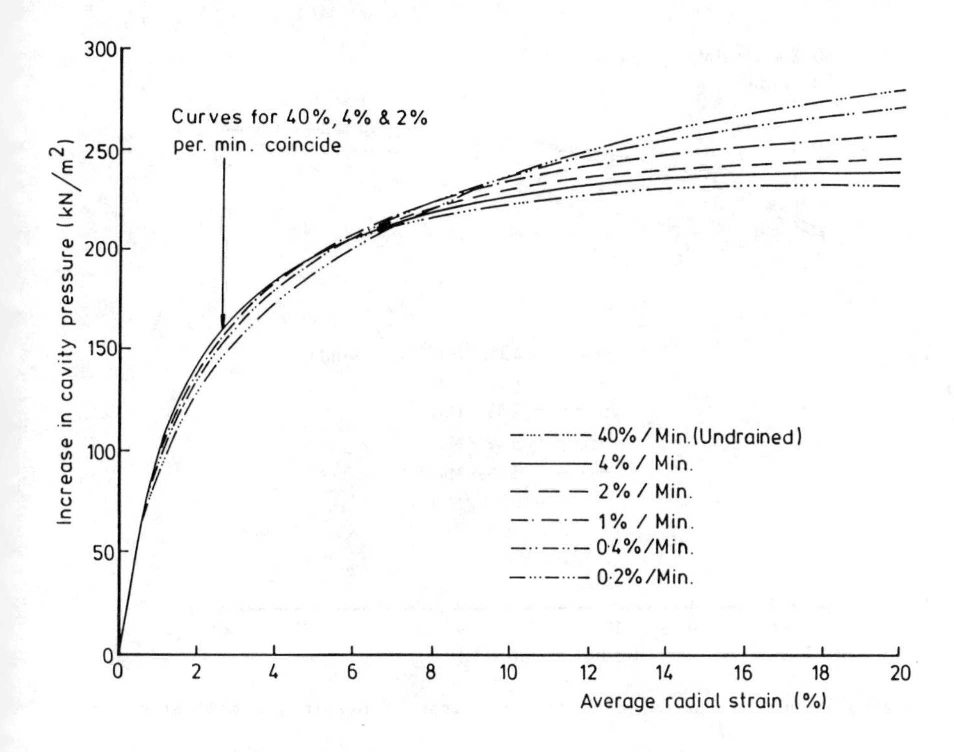

Figure 1. Simulated expansion curves for different rates of strain (consolidation only)

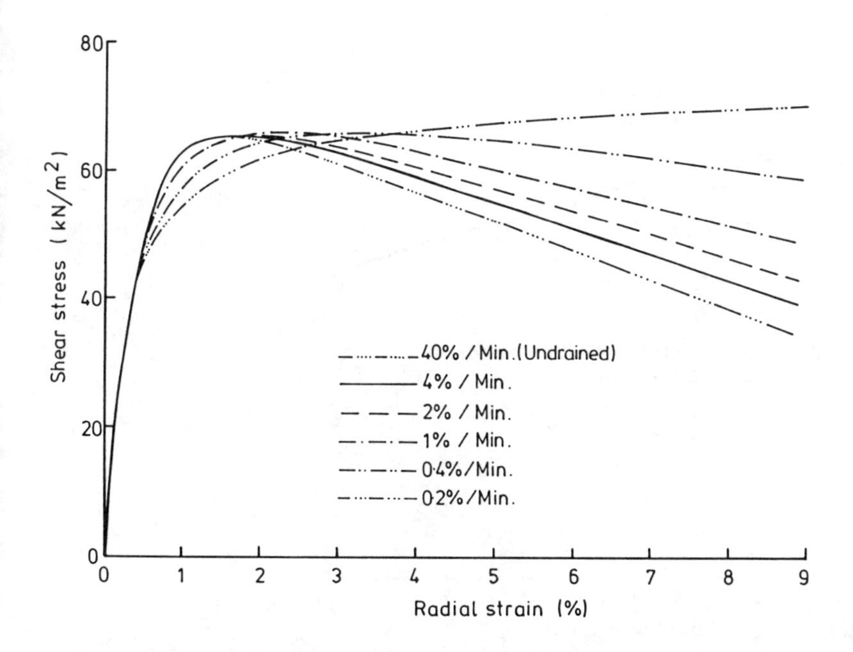

Figure 2. Derived stress-strain curves for different rates of strain (consolidation only)

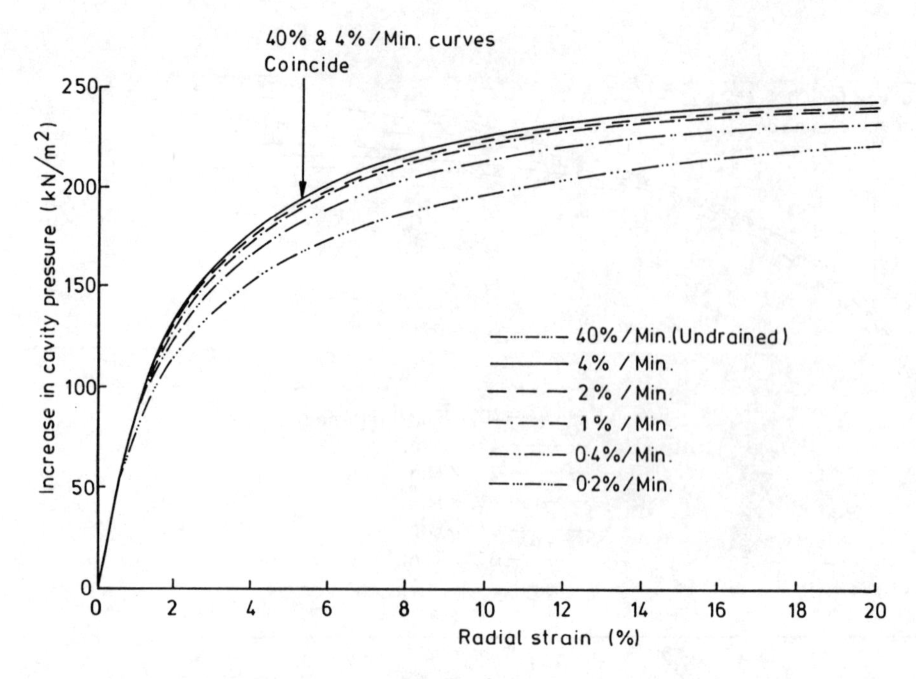

Figure 3. Simulated expansion curves for different rates of strain (consolidation and creep)

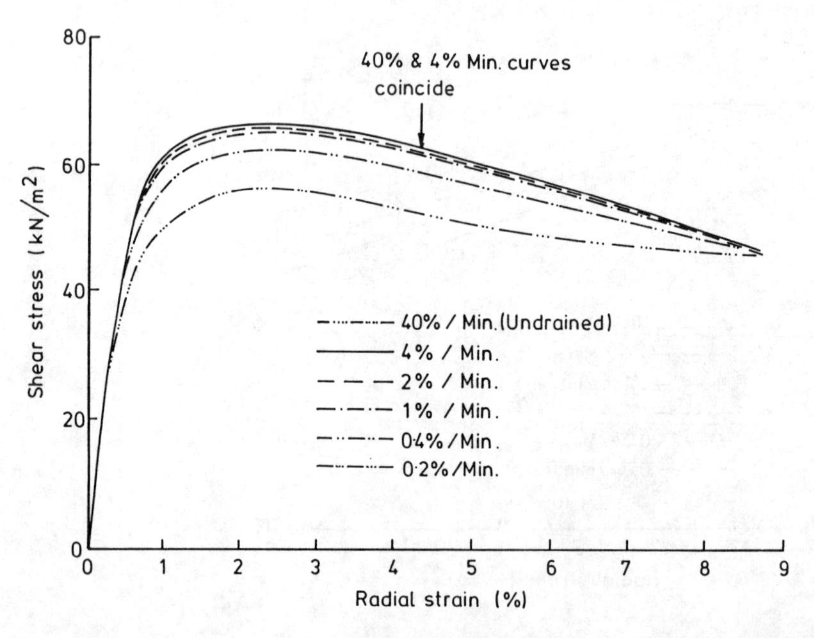

Figure 4. Derived stress-strain curves for different rates of strain (consolidation and creep)

4. RESULTS AND DISCUSSION

By considering consolidation as the sole time-dependent phenomenon in the analysis (consolidation only), the simulated expansion curves for tests at different strain rates are given in fig. 1. For comparison purposes, an undrained test was also simulated using a very fast strain rate (40% per minute) so that little time is allowed for consolidation. The effects of consolidation are:

1. During the early part of the test, the radial strain for a given pressure increases as the strain decreases.
2. As the test proceeds, the soil close to the cavity gains in strength as consolidation takes place, resulting in a higher limit pressure.

The expansion curves were interpreted using the Palmer (1972) method to generate the entire stress-strain curves from the expansion data. The deduced stress-strain curves for tests at different strain rates are given in fig. 2. The curves reflect the effect of consolidation in that the deduced shear strengths remain unchanged, or show a slight increase for the very slow test. The strains at which these strengths occur increase with decreasing strain rate.

With creep included in the analysis (consolidation and creep), the simulated expansion curves and the derived stress-strain curves are given in figs. 3 and 4. Although the initial increase in cavity radial strain as the strain rate decreases is clearly seen, there is no increase in the limit pressure nor in the deduced shear strength. In contrast, the slower tests give lower deduced strengths.

This reduction in the deduced shear strength cannot be observed in the laboratory tests (fig. 5). Apart from the fastest test (4% per minute), which shows some soil strain-rate effect causing higher values of limit pressure and deduced strength, the other tests show that the limit pressure and the deduced shear strength increase with decreasing strain rate. The strains associated with the deduced strengths also increase. This trend can only be due to consolidation and not to creep. This is in marked contrast to the results of the stress-controlled cavity expansion tests [8] which showed the predominant influence of creep causing apparent reduction of deduced strengths in the slower tests.

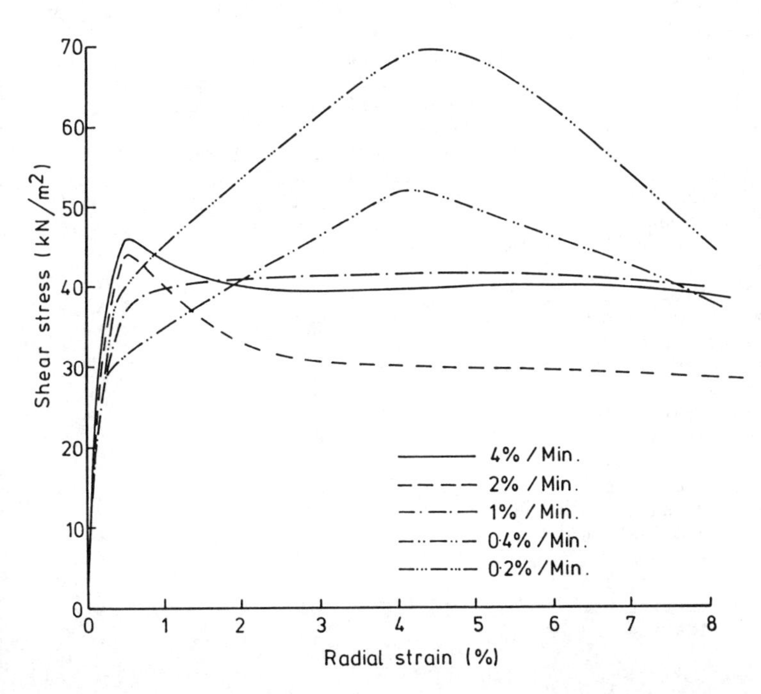

Figure 5. Derived stress-strain curves from laboratory hollow cylinder tests

Since consolidation is dependent on the length of the drainage path and creep is not explicitly related to it in the Singh-Mitchell model, it is difficult to extrapolate the findings of this study to a full scale test. The relative importance of consolidation and creep in such a test may be different from the tests at laboratory scale.

Another factor not taken into account in the interpretation of the results is the effect of the ratio of the outer to inner diameter of the hollow cylinder. The effect of a finite, rather than an infinite, diameter ratio (6 in this study) is to under-estimate the deduced shear stress as the radial strain increases. This needs further investigation; nevertheless, the trend from this study is clear.

5. CONCLUSIONS

1. For the numerical simulations considering consolidation only, the effect of reducing the strain rate for constant rate of strain cavity expansion is to increase the limit pressure. When creep is included in the analysis, the trend is reversed, i.e. giving lower limit pressure with decreasing strain rate.

2. As the strain rate decreases, more consolidation occurs and the deduced shear strengths increase, as do their associated strain values. With creep included in the analysis, the deduced shear strength decreases with decreasing strain rate.

3. In a strain-controlled test, the deduced modulus is insensitive to the testing strain rate. However creep of the clay results in lower deduced modulus values compared with those deduced from analyses considering consolidation only.

4. Comparing the experimental results with the numerical simulations shows that in a stress-controlled test, creep is the more significant time-dependent phenomenon, while in a strain-controlled test creep is of minor importance and consolidation is predominant.

5. The present study is restricted to cavity expansion of a hollow cylinder of 25mm inner diameter. Since consolidation depends on the length of the drainage path, it will be useful to carry out full scale testing to confirm the relative importance of creep and consolidation as indicated in this study.

REFERENCES

[1] L.F. Menard, An apparatus for measuring the strength of soils in place, MSc Thesis, University of Illinois, (1956).

[2] F. Baguelin, J.F., Jezequel and A. Le Mehaute, Expansion of cylindrical cavities in cohesive soils, proc. ASCE, 98, SM11, 1129-1142, (1972).

[3] C.P. Worth and J.M.O. Hughes, An instrument for in-situ measurement of the properties of soft clay, Report CUED/D/SOIL/TR13, University of Cambridge, (1972).

[4] W.F. Anderson, I.C. Pyrah and F. Haji-Ali, Pressuremeter testing of normally consolidated clays : the effects of varying test technique, Site Investigation Practice : Assessing BS 5930, Geological Society Engineering Geology Special Publication, No. 2, 125-132, (1986).

[5] I.C. Pyrah, W.F. Anderson and F. Haji-Ali, The interpretation of pressuremeter tests - time effects for fine grained soils, Proc. 5th Int. Conf. on Numerical Methods in Geomechanics, Nagoya, 1629-1636, (1985)

[6] J.P. Carter, CAMFE, A computer program for the analysis of a cylindrical cavity expansion in soil, Report CUED/SOILS/TR52, University of Cambridge, (1978)

[7] A. Singh and J.K. Mitchell, General stress-strain-time function for soils, Proc. ASCE, 94, SM1, 21-46, (1968).

[8] W.F. Anderson, I.C. Pyrah and F. Haji-Ali, Rate effects in pressuremeter tests in clay, Proc. ASCE, 113, GT11, (1987).

Numerical Methods in Geomechanics (Innsbruck 1988), Swoboda (ed.)
© 1988 Balkema, Rotterdam. ISBN 90 6191 809 X

Analytical procedure for evaluating pore-water pressure and deformation of saturated clay ground subjected to traffic loads

M.Hyodo
Yamaguchi University, Ube, Japan

K.Yasuhara
Nishinippon Institute of Technology, Kanda, Fukuoka-ken, Japan

ABSTRACT: This paper presents a procedure to calculate the settlements of low embankments built on the soft clays subjected to traffic-induced transient loads. The vehicle running tests were performed on the trial embankment and vertical earth pressures were measured. The results were compared with those calculated by both Boussinesq's equation and dynamic finite element method in which the traffic loads were treated as impulsive loads. Then the cyclic triaxial compression tests were carried out and the empirical formulations, the variations of pore-water pressure and shear strain with time, were developed. Combining the stresses calculated by the finite element method at the elements of ground and the cyclic responses of clay obtained by cyclic triaxial tests, deformation and settlement of the low embankment and clay ground were estimated.

1 INTRODUCTION

It has become one of the important problem for the maintenance of highways in Japan that the low embankment for highways constructed on the soft clays may induce harmful settlement affected by traffic loads. This type of highway is often constructed because of stability and for economical reason, and is designed by means of Boussinesq's equation or currently available method of static calculation to require a minimum height in which the subgrades will not be under the influence of traffic loads. We, however, have encountered several cases that these highways suffer from abnormal settlements after the beginning of travelling of motor vehicles. Therefore, it is supposed that the applied forces other than the static loads may act on the grounds.

This kind of problems should be involved in the category of soil dynamics. In order to explain this phenomenon, the following two models must be developed: one for stress propagation due to traffic loads and the other to account for responses of clays subjected to cyclic loads by travelling of vehicles.

This paper aims at proposing an analytical method for evaluating the state of stresses and estimating the settlements when the traffic loads are prevailed. In order to carry out development the method of analysis and its verification, vehicle running tests were performed on the trial embankment. In addition, to clarify the behavior of elements in clays, cyclic triaxial compression tests were carried out. It is shown in this paper how to evaluate the pore-water pressure and the deformation of soft clay grounds beneath the low embankments subjected to traffic loads, combining the calculated stresses and the results of cyclic triaxial tests.

2 ANALYSIS PROCEDURE

When the grounds are subjected to such long-term cyclic loading as the case of traffic loads, the soil element under this condition is supposed to be in partial-drained condition with generation and dissipation of pore-water pressure. It is considered that the settlement during the partial-drained cyclic loading is divided into two parts. One is due to shear deformation in the undrained cyclic shear and the other is due to the volume change in the partial-drained cyclic shear process. The schematic diagram for explaining the behavior is shown in Fig.1. The total settlement is evaluated by the combination of these two factors.

The total flow of the proposed procedure for evaluating the settlement of ground suffered by traffic-induced cyclic load is presented in Fig.2. At first, the stress increments due to traffic loadings are evaluated by the dynamic finite element analysis and the stresses obtained by the analysis are assumed to be invariable in all the cycles. Then the route is divided into two ways: one for evaluating the factor of undrained cyclic shear deformation by using an empirical formulation developed by the result of cyclic triaxial test, and the other to obtain the volume decrease due to partial drainage of generated pore-water pressure by means of finite element analysis. The detail procedures will be described in the following sections.

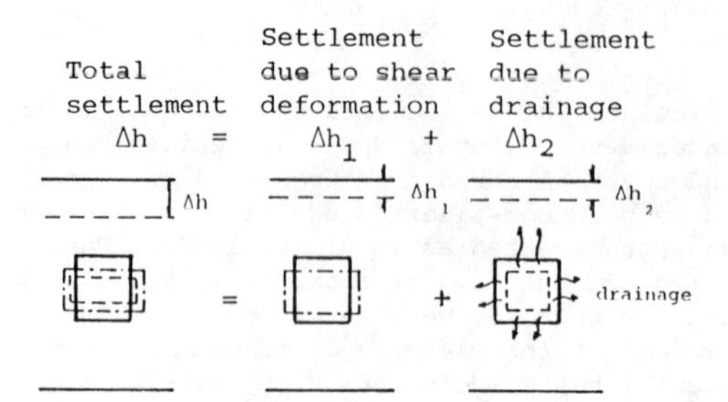

Fig.1 Schematic diagram for explaining the settlement of ground due to long-termed cyclic loading

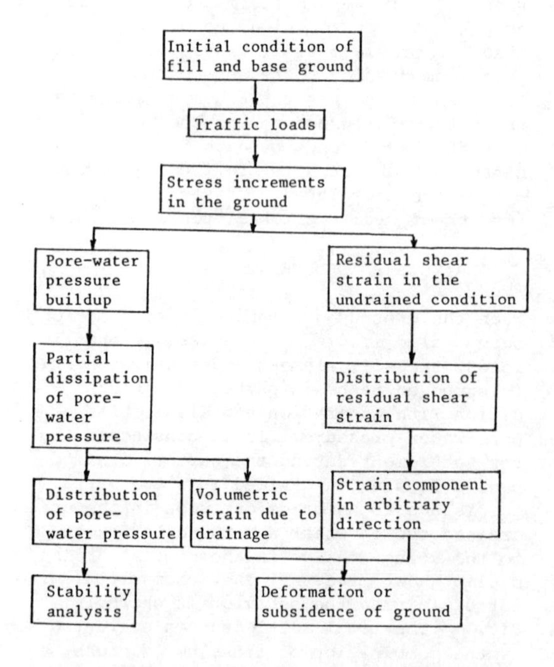

Fig.2 The flow diagrm of the proposed procedure for evaluating the settlement of ground subjected to traffic load

3 STRESS INCREMENTS DUE TO TRAFFIC LOADS

There are few data about the stresses which are observed in the soft clay grounds induced by traffic loads because it is not easy to measure them at fields. Recently, we have had an opportunity to measure the vertical earth pressure induced by traffic loads in the saturated silty sand deposit with low bearing capacity. A trial embankment with approximate 2.0m height was built on the deposit and vehicle running tests were performed by travelling of 10 ton truck on it.

The traffic-induced vertical earth pressure and excess pore-water pressure with depth were measured by the pressure transducers burried in the ground as shown in Fig.3 in accordance with the different truck speed of 0, 10, 20, 30 and 35km/h, respectively. Each wheel load was measured to be 2.8tf of front wheel and 5.2 tf of middle and rear wheels, respectively. The observed distributions of vertical pressures are summarized in Fig.4 in which measured results are compared by Boussinesq's equation. It is recognized in this figure that the earth pressures measured during the running of truck are about three times as large as ones measured at a standstill of it. The calculated results by the Boussinesq's equation are close to the static ones but are different from those of vehicle running tests.

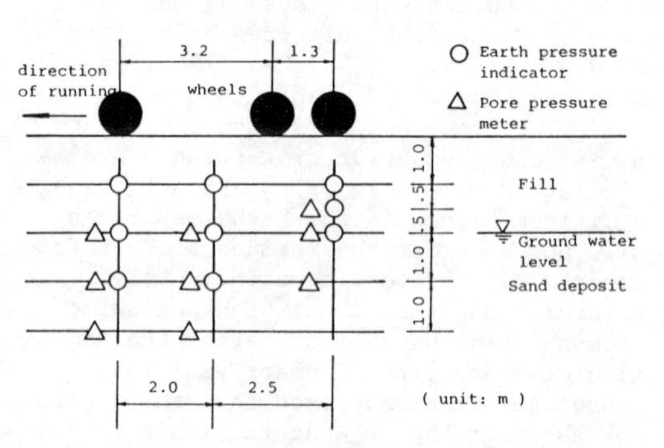

Fig.3 Earth pressures and pore-water pressure transducers burried in the embankment and ground

Therefore, a dynamic finite element method, which enables the traffic load to be regarded as an dynamic impulsive one, is employed to evaluate these stresses. The measurement of vertical earth pressure shows that the wave of traffic load is of triangular shape as shown in Fig.5. The duration of impulsive load is correlated with the speed of truck as presented in Fig.6.

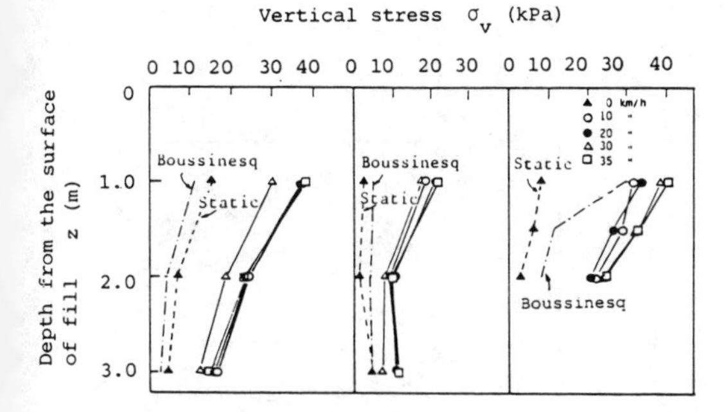

Fig.4 Observed distributions of vertical earth pressure

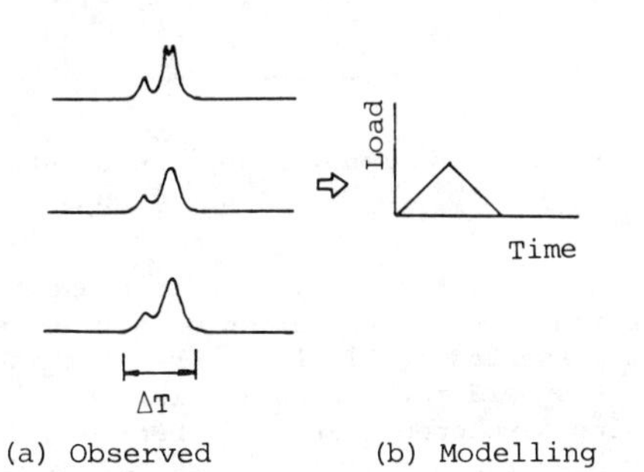

Fig.5 Observed wave form of vertical earth pressure

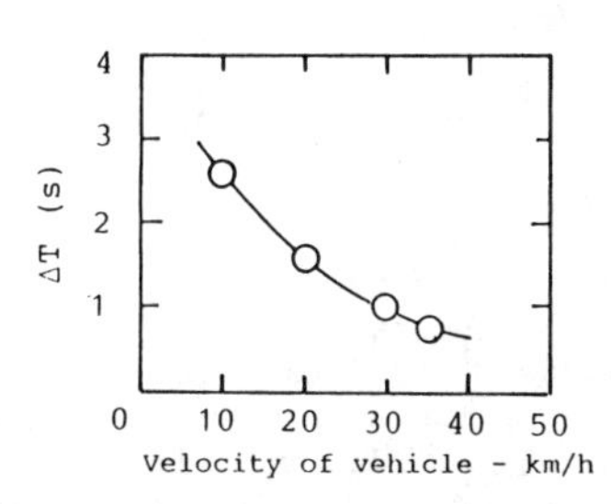

Fig.6 Relationship between duration of earth pressure and velocity of vehivle

The governing equation of motion in this case is as follows:

$$[M]\{\ddot{u}\} + [C]\{\dot{u}\} + [K]\{u\} = \{R(t)\} \quad (1)$$

in which [M],[C] and [K] are mass,damping and stiffness matrices, and $\{\ddot{u}\}$,$\{\dot{u}\}$,$\{u\}$ and $\{R(t)\}$ are the acceleration, velocity, displacement and vector of external force, respectively.

The observed and calculated earth pressures are compared with each other in Fig.7, in which the case 1 means the pressures under the wheel and the case 2 means ones under the midpoint of right and left wheels. In this figure, it is recognized that the analysed and observed results are in good correspondence with each other in both cases. From this results, it is satisfactory to consider the problem of stress propagation by traffic load as dynamic one. Therefore, we are going to apply the procedure to the case of clay ground.

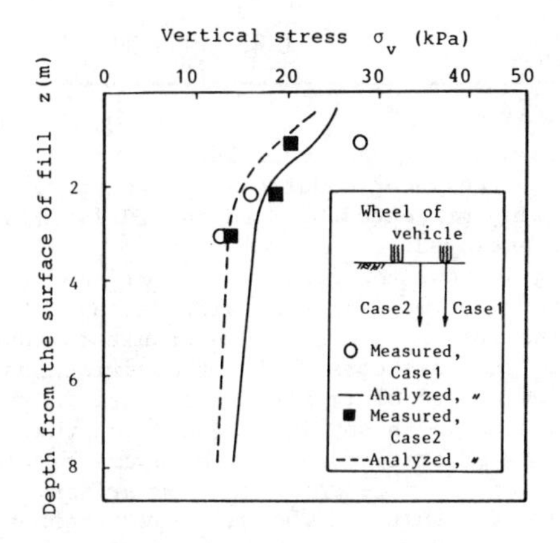

Fig.7 Observed and calculated vertical earth pressures in the embankment and ground

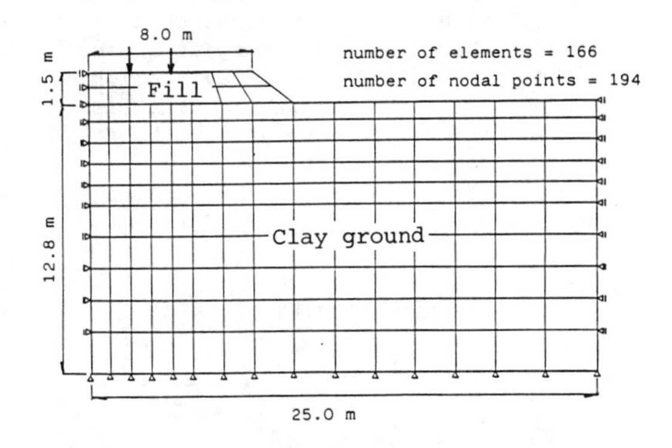

Fig.8 Finite element model of low embankment and clay ground

In order to illustrate the numerical procedure presented in this paper, a hypothetical low embankment and clay ground was analysed. The cross section considered which is presented as a finite element model is illustrated in Fig.8. The assumed material properties are listed in Table 1.

In the dynamic analysis, the equivalent linear method was carried out for the ground to satisfy the dependency of shear moduli and damping ratios on shearstrains. On the other hand, for the embankment, assuming it as an elastic body, the elastic analysis was

Table 1 Material properties used in the analysis

Material	E(static) (kPa)	ν	G_0(dynamic)	ρ (t/m^3)
Fill	10,000	0.35	76,000	1.95
Clay ground	7,000	0.45	46,000	1.58

performed. In the equivalent linear analysis, the shear modulus and damping ratio for clay presented by Seed and Idriss [1] were employed.

Fig.9 shows the distribution of analysed vertical normal stresses with depth of the embankment and ground. The results of dynamic analysis considering the variations of the period of impulsive load are presented in these figures in which the results of static analysis by Boussinesq's equation are also shown for comparison. In every results, although large stresses are produced in the embankment under the point of traffic load, they become fairly smaller in the ground. However, the vertical stresses due to dynamic analysis become several times as large as the Boussinesq's solution, which is of same tendency as observed in Fig.4. Furthermore it is found that the difference due to the variation of load's period is not remarkable, that is also same tendency as observed in the field test.

The maximum shear stress was normarized by the effective mean principal stress

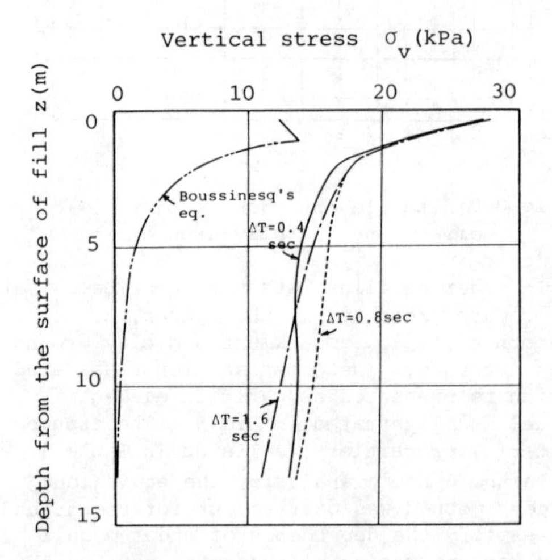

Fig.9 Distribution of vertical stress with depth

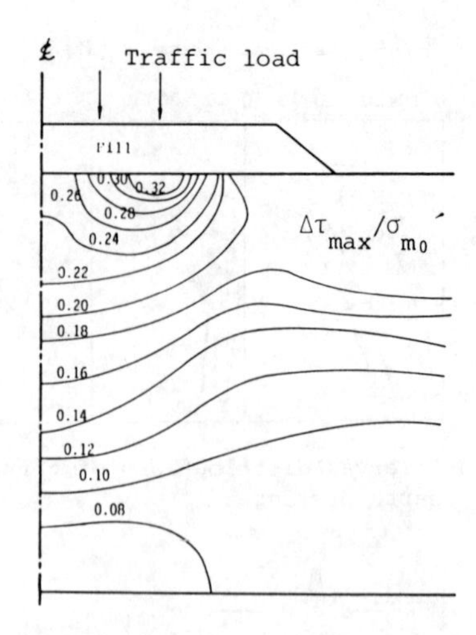

Fig.10 Distribution of stress ratio $\Delta\tau_{max}/\sigma_{m0}'$ in the ground

obtained by the static finite element solution. The distribution of stress ratio is presented in Fig.10 in which it is found that stress ratio with a relatively large value developed near the embankment decreases rapidly with increasing the depth. The stress ratio calculated is applied to the evaluation of strain in the ground which will be treated in the later part of the paper.

4 CYCLIC TRIAXIAL TESTS

One-way cyclic triaxial compression tests with the constant lateral pressure were performed assuming the variation of incremental stresses in the ground due to traffic loads. The detail of tests' procedure and results were presented in the other paper [2]. Although, in practice the clay grounds are considered to be consolidated under the anisotropical stress circumstance, the cyclic tests were carried out using the specimens consolidated isotropically for approximation.

The pore-water pressure and axial strain were measured in the undrained cyclic triaxial tests. They depended on the confining pressure, the amplitude of cyclic axial stress and the number of load cycles. Regression analyses for these formulations were performed and the following relations were obtained.

$$\varepsilon_a = 0.096(\Delta\sigma_a/\sigma_c)^{1.745}(\log 10N)^{1.763} \quad (2)$$

$$u/\sigma_c = 0.064(\Delta\sigma_a/\sigma_c)^{1.418}(\log 10N)^{1.535} \quad (3)$$

The coefficients of correlation resulted from the regression analyses were about

0.93~0.94, which is considered to be fairly high value.

The next procedure is to evaluate the pore water pressure and strain of the elements of the ground, combining the analysed stresses and results of cyclic triaxial tests. In this study, we are going to correspond each maximum shear stress in the plane with inclined angle of 45° to the principal plane as shown in Fig.11. Then eqs.(2) and (3) are rewritten into the function of maximum shear stress $\Delta\tau_{max}$ and number of load cycles. Further, tha axial strain ε_a is converted into the maximum shear strain γ_{max}, and cyclic axial stress $\Delta\sigma_a$ into the maxumum shear stress as shown in the following equation:

$$\Delta\sigma_a = 2\Delta\tau_{max} \tag{4}$$

$$\varepsilon_a = 4/3\ \gamma_{max} \tag{5}$$

Substituting above equations into eqs.(2) and (3), the following equations are obtained.

$$\gamma_{max} = 0.241(\Delta\tau_{max}/\sigma_c)^{1.745}(\log 10N)^{1.763} \tag{6}$$

$$u/\sigma_c = 0.171(\Delta\tau_{max}/\sigma_c)^{1.418}(\log 10N)^{1.535} \tag{7}$$

The correlation of strain state between plane strain and axisymmetrical condition in undrained cyclic triaxial test is presented by the Mohr's strain circles as shown in Fig.11. In the undrained condition, it is satisfied to keep following relations: $\varepsilon_1 + \varepsilon_3 = 0$, in the plane strain, and $\varepsilon_1 + 2\varepsilon_3 = 0$, in the axisymmetrical condition, respectively. If it is assumed that the maximum shear stresses in the analyses and triaxial tests can be corresponded each other, the maximum shear strain is calculated using eq.(6) in the analyses. Then the major principal strain is obtained because there is a relation of $\varepsilon_1 = \gamma_{max}$ in the Mohr's circle. Further, an arbitrary component of strain can be calculated if the angle of objective plane from the principal plane would be known. Thus, taking the angle between the vertical direction and the major principal axis to be α, the vertical strain is calculated as

$$\varepsilon_v = \varepsilon_1 \cos 2\alpha \tag{8}$$

Then the behavior of clay in partial-drained condition which is, as mentioned above, one of the factor of settlement was investigated. Drained cyclic triaxial tests were carried out in order to make clear the behavior. In the drained cyclic triaxial test, the generated pore-water pressure does not dissipate perfectly during a cycle because the clay is of very low permeability. A part of generated pore-water pressure dissipates through the boundary surface, in which the pore-water pressure is always zero, and the pore pressure is supposed to be distributed in the specimen.

Then the model for presenting the partial drainedbehavior of clay was developed as shown in Fig.12. In partial drained condition, the time dependent pore-water pressure traces the curve with a peak and that of volumetric strain traces the monotonically increasing curve like the figure. From a certain time t to the time after small increment t+Δt, the path of pore-water pressure assume to trace the points A, C and B, that is, the pore-water pressure increment during Δt is considered to be added to the value at time t and then drainage occurs until the point C. The

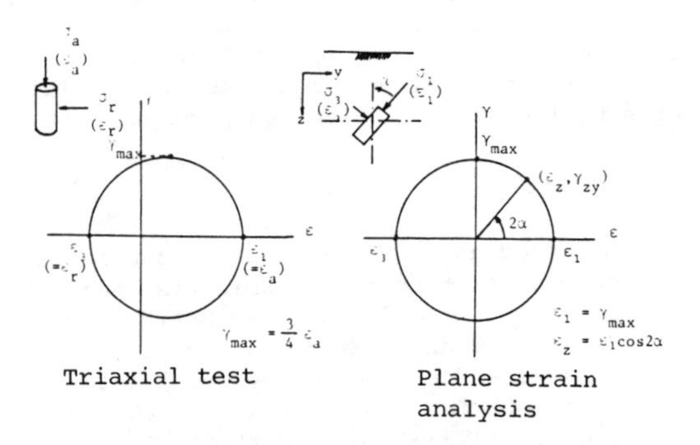

Fig.11 Correlation of strain states between axisymmetric and plane strain condition in undrained cyclic loading

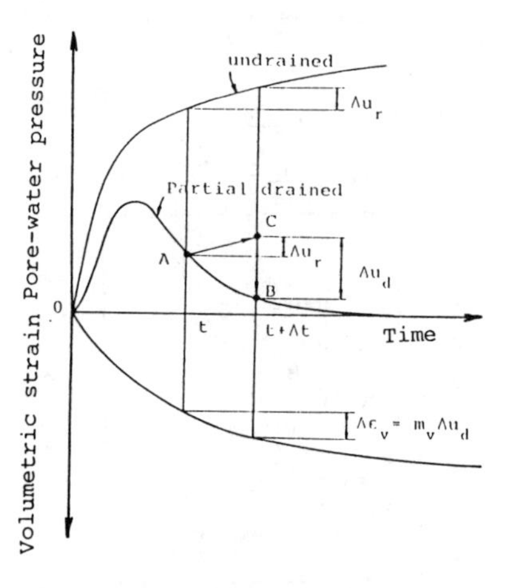

Fig.12 Schematic diagram for presenting the behavior of pore-water pressure and volumetric strain in partial drained condition

volumetric strain is calculated by the change of effective stress in the dissipation process of pore-water pressure.

The following equation which was originally introduced by Booker et al [3] was employed as the governing equation for the partial drained behavior of clay.

$$\{\nabla\}^T[K]\{\nabla \frac{u}{\gamma_w}\} = m_v(\frac{\partial u}{\partial t} - \frac{\partial u_g}{\partial t}) \quad (9)$$

in which [K]: permeability matrix, u_g: generated pore-water pressure. The parameters contained are coefficients of permeability and compressibility. They were obtained by the post-cyclic reconsolidation tests [2]. The coefficients of permeability and compressibility for tested clay were obtained by averaging the tests' results as 1.20×10^{-8}cm/s and 0.0126cm²/kgf, respectively. To investigate the adjustment between the model and the practical behavior of clay, some analyses were performed about the triaxial clay specimen in the partial drained condition using finite element method. After several case studies, it was confirmed that the tendency of test and numerical results correspond well each other [2].

5 SETTLEMENT OF GROUND SUBJECTED TO TRAFFIC LOAD

Settlement of the clay ground subjected to traffic loads was calculated by the analytical procedure mentioned above. The object of analysis is the low-embankment and clay ground of which incremental stresses were calculated in section 3. In the partial-drained analysis, the same finite element meshes as shown in Fig.8 were used.

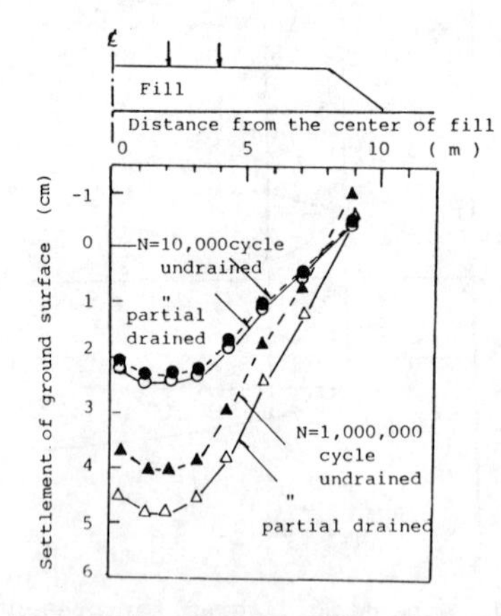

Fig.13 Calculated settlement of ground in undrained and partial-drained condition

The clay ground is assumed to be constituted by the same soil as the specimens used in the cyclic triaxial tests. The traffic load that was treated in section 3 are supposed to act once per a minute, that is, 1,440 cycles/day cyclic loads are taken.

Fig.13 shows the calculated settlements after the 10,000 and 1,000,000 cycles of cyclic loads are acted. In the figure, there are the results of analyses under the undrained and partial-drained condition. It is recognized that settlement is predomonant under the points of traffic loads, on the other hand,uplift of ground appears at a place apart from the loading points. The effect of partial-drainage does not appear at the 10,000 cycles but it does at the 1,000,000 cycles at the rate of about 20% of the total settlement. However, the magnitude of settlement seems to be relatively smaller than that of reality. That is considered to be due to underestimation of shear deformatin because here, we used the cyclic triaxial tests'data under isotropically consolidated condition. In reality, the ground is consolidated anisotropically and so it is necessary to use the tests'data performed under such condition. The analytical procedure presented here is possible to be applied to such data and also to the other problems such as wave loading and the other cyclic loading.

6 CONCLUSIONS

An analytical procedure for evaluating the settlement of low-embankment highway subjected to traffic loads was presented. At first, it was made clear that the stress increments in the ground due to that load could be obtained by the dynamic stress analysis which assumed the traffic loads to be impulsive ones. Then the settlement was estimated by combining the analysed stress increments and results of cyclic triaxial tests. The quantity of settlement was represented by the sum of two factors: one for undrained shear and the other for partial dissipation of cyclic-induced pore pressure.

REFERENCES

[1] H.B.Seed and I.M.Idriss:Soil moduli and damping factors for dynamic response analysis, EERC 70-10, Univ.Calif. (1970)

[2] K.Yasuhara and M.Hyodo:Partial-drained behavior of clay under cyclic loading, Proc. 6th Int. Conf. on Num. Meth. in Geomech. (1988) in submitted

[3] J.R.Booker, M.S.Rahman and H.B.Seed: GADFLEA- A computer program, EERC 76-24, Univ.Calif. (1976)

Numerical Methods in Geomechanics (Innsbruck 1988), Swoboda (ed.)
© 1988 Balkema, Rotterdam. ISBN 90 6191 809 X

Partial-drained behaviour of clay under cyclic loading

Kazuya Yasuhara & Kazutoshi Hirao
Nishinippon Institute of Technology, Fukuoka, Japan
Masayuki Hyodo
Yamaguchi University, Ube, Japan

ABSTRACT : An analytical model with simplicity is presented for evaluating the volumetric change and dissipation of pore pressure during drained cyclic triaxial tests on a highly plastic marine clay. It is suggested from the analytical results that the model proposed should be a clue to predict the deformation, settlement and stability of soft clay under long-term cyclic loading.

1 Introduction

There are some possibilities that clay deposits are placed in the partially drained conditions in which excess pore pressure is generated and partly dissipated when they are subjected to long-term cyclic loading from traffics and waves. It is considered that settlements of clay in these cases consist of shear deformation during undrained cycling and recompressive volumetric change due to dissipation of cyclically induced pore pressure. It is not clear whether deformation of clay under partial-drained cyclic loading is equivalent to the one under cyclic loading followed by drainage. If both behaviours are different each other and interrelationship between both is found out, it will greatly contribute to the evaluation of long-term settlements of clay under cyclic loading.

The main scope of the present study is to investigate the clay behaviour of long-term drained and undrained cyclic loading through triaxial tests on a highly plastic Ariake clay. At the same time, an analytical model was pursued for explaining the behaviour of deformation and pore pressure of clay under drained cyclic triaxial tests by means of the results from undrained cyclic triaxial tests followed by drainage. In other words, drained cyclic triaxial tests were analyzed as a kind of model test while undrained cyclic triaxial tests with drainage was regarded as a element test. Consequently, if this would be successful,we would obtain a clue for estimating the deformation and settlement of clay subjected to long-term cyclic loading.

2 Mechanism of clay deformation under long-term cyclic loading

Clay deformation under long-term cyclic loading with inclusion of drainage may constitute :

1) Shear deformation under undrained cyclic loading.
2) Volumetric change due to dissipation of cyclically induced pore pressure.

Settlements under cyclic loading are, in general, contributed by these two components.

In the actual fields, clay may sometimes be situated under partial-drained cyclic loading conditions. The interrelation between undrained cyclic loading with drainage and partial-drained cyclic loading is compared in Fig. 1 by the e - log β plots.

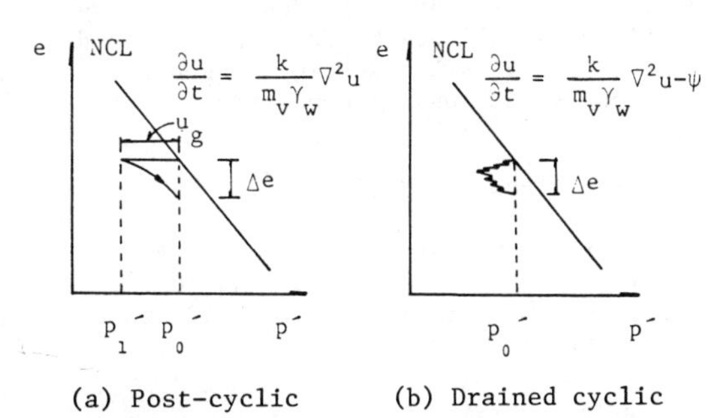

Fig. 1 State paths in post-cyclic recompression and drained cycling

The deformation model proposed in the present paper was obtained by superposition of deformation due to both undrained cyclic loading and dissipation of cyclically induced pore pressure. The variations of pore pressure with time during partial-drained cyclic loading in clay are modelled in Fig. 2.

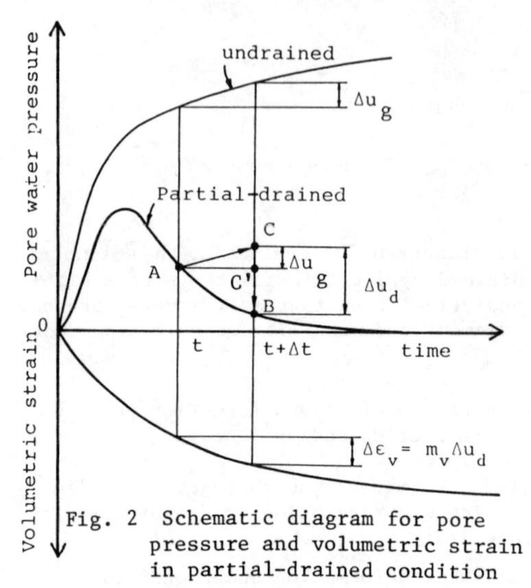

Fig. 2 Schematic diagram for pore pressure and volumetric strain in partial-drained condition

In the heavy curve corresponding to the partial-drained cyclic loading, pore pressure generated during an interval from t to t + Δt will dissipate from A to B in Fig. 2. In the analytical model, it is assumed that the incremental pore pressure produced by undrained cyclic loading during a time interval Δt should be superimposed on the above-mentioned pore pressure. Then, a part of this superimposed pore pressure dissipates as a soil element follows the A→C→B path as shown in Fig. 2. It is also postulated in the current paper that this amount of pore pressure during partial-drained cyclic loading should be predicted by the results from undrained cyclic triaxial tests followed by drainage. Accordingly, pore pressure developed under drained-cyclic loading conditions is given by

$$\Delta u_d = \Delta u_g + (\Delta u_d - \Delta u_g) \qquad (1)$$

Each term of the right side of Eq. (1), Δu_g and $(\Delta u_d - \Delta u_g)$, corresponds to the pore pressure components during $\vec{AB}$ and $\vec{BC'}$, respectively.

3 Cyclic triaxial tests on reconstituted marine clay

To pursue the generation of pore pressure under undrained cycliling and to construct the analytical model for behaviour under partial-drained cycling, three kinds of cyclic triaxial tests were carried out as follows:

1) Undrained cyclic triaxial tests without drainage (type-A).
Results from this test were used to estimate the residual pore pressure and shear strain.

2) Undrained cyclic triaxial tests followed by drainage.
In this test drainage is permitted after a certain number of load cycles is applied to the specimen. Since cyclically induced pore pressure is then dissipated with time, the specimen is recompressed. In this period, therefore, the coefficients of volume compressibility and permeability were determined from measurement of volumetric strain due to dissipation of cyclically induced pore pressure. Also, the effect of cyclic loading history on these coefficients was investigated from a family of tests of this kind.

3) Drained cyclic triaxial tests (type-AD).
The cyclic load is applied to a specimen under drained conditions by opening the drainage system. Throughout the cylindrical specimen (3.5 cm in diameter and 8.75 cm in height) in triaxal cell, distribution of pore pressure is not homogeneous during drained cycling. In this sense, this type of cyclic test can be classified as a partial-drained shear test.

Reconstituted highly plastic clay called Ariake clay was used in these tests. Index properties and some mechanical parameters are : G_s = 2.65, w_L = 123%, I_p = 69, C_c = 0.700, C_s = 0.163 and ϕ' = 39°. The clay cylinder was trimmed as a specimen for every triaxial test from the clay block which was fully preconsolidated under 59 kPa of the vertical pressure in the large consolidation vessel. The water content was 90 to 95% on the average.
Pore pressure was measured through the porous stone with 3 mm in diameter buried in the center of lower pedestal. Drainage was done from the side wall around the lower pedestal through the drain paper surrounded by the specimen. The conditions of cyclic triaxial tests were summarized in Table 1. The cyclindrical specimen for the triaxial test was isotropically preconsolidated for 24 hrs under the confining pressures of 100, 200 and 300 kPa. Then, cyclic loading starts after isotropic consolidation. The number of load cycles was 3600 for most of tests to 172800 as the maximum.

Table 1 Conditions of cyclic triaxial tests on Ariake clay

Test No.	σ_c (kPa)	$\Delta\sigma_a$ (kPa)	N (cycle)	W_i (%)	e_c^*
A-1	200	40	3600	95.6	2.021
A-2	200	80	3600	93.3	1.982
A-3	200	100	3600	90.6	1.953
A-4	200	120	3600	93.8	1.960
A-5	200	140	3600	90.5	1.871
A-7	200	100	172800	92.5	1.934
A-10	200	120	172800	93.2	1.893
A-15	100	36	3600	93.8	2.086
A-16	100	64	3600	94.5	2.071
A-17	100	86	3600	93.3	2.084
A-18	300	76	3600	94.1	1.822
A-19	300	137	3600	95.6	1.861
C-1	200	100	3600	90.9	1.895
C-2	200	120	172800	91.7	1.896
C-3	200	140	3600	92.7	1.895
C-4	200	80	172800	91.6	1.950
C-5	200	80	3600	92.2	1.915
C-6	200	80	17280	91.4	1.850
C-7	200	80	360	93.3	1.952
C-10	200	80	5000	91.9	1.972
C-11	200	80	87000	92.4	1.869
C-12	200	80	3600	91.2	1.894
AD-1	100	40	3600	94.3	2.013
AD-2	100	60	3600	93.1	2.000
AD-3	100	80	3600	93.0	2.039
AD-4	200	80	3600	93.1	1.895
AD-5	200	120	3600	93.5	1.850
AD-6	200	160	3600	92.5	1.861
AD-8	300	120	3600	92.1	1.708
AD-9	300	180	3600	92.7	1.688
AD-10	300	240	3600	93.2	1.741

Remarks:
σ_c: confining pressure
*void ratio at the end of consolidation of each specimen

4 Formulation of cyclically induced pore pressure and shear strain in clay

The pore pressure and axial stain were measured in undrained cyclic triaxial tests. Both parameters mainly depend upon the confining pressure, the cyclic load intensity and the number of load cycles. It is, therefore, assumed that both the axial strain and the pore pressure ratio can be formulated as a function of the stress ratio and the number of load cycles. Regression analyses for this formulation were performed and the following relations were obtained.

$$\varepsilon_a = 0.096(\Delta\sigma_a/\sigma_c)^{1.745}(\log_{10}10N)^{1.763} \quad (2)$$

$$u_g/\sigma_c = 0.064(\Delta\sigma_a/\sigma_c)^{1.412}(\log_{10}10N)^{1.535} \quad (3)$$

Figs. 3 and 4 show the variations of residual strain and pore pressure ratio with the number of load cycles, respectively. All of the observed values are compared

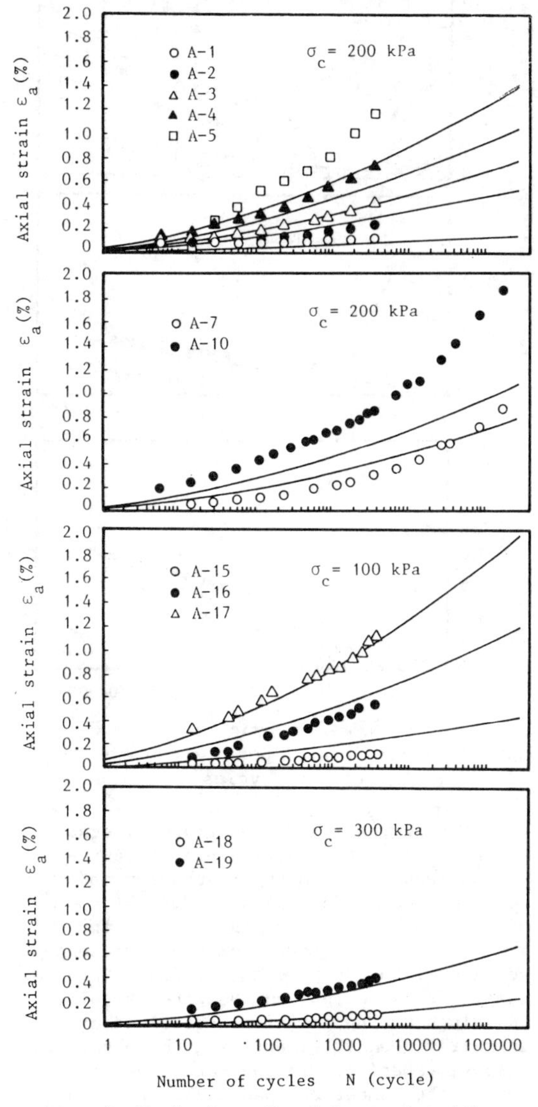

Fig. 3 Variation of axial strain with number of load cycles (comparison of observed with clculated results)

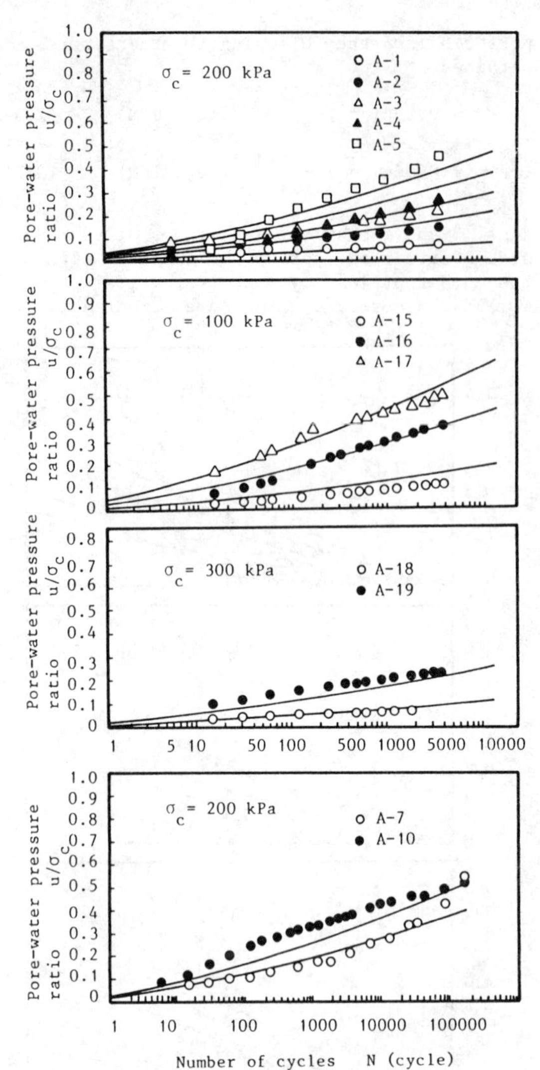

Fig. 4 Variations of pore pressure with number of load cycles

with the regression curves given by eqs. (2) and (3). The coefficients of correlation resulted from the regression analyses are about 0.93~0.94 which may be considred to be a fairly high value. In both kinds of figures, the results are presented classifying them by the confining pressure. It is recognized that the calculated values by the empirical equations are in fairly good agreement with the observed values, especially in the pore pressure versus the number of load cycles relations. Judging from this fact, it can be said that eqs. (2) and (3) are available for describing the behaviour of normally consolidated clay under long-term cyclic loading.

5 Analytical model

As was mentioned previously, an element of clay grounds under long-term cyclic loading may sometimes be situated under partial-drained conditions in which the sequence of pore pressure build-up and its dissipation is repeated. According to the research works on the effect of partial drainage on sand liquefaction, subsequent undrained shear is affected by the process of partial drainage in case where a significant amount of pore pressure is dissipated during each cycle. It has been concluded that the effect of partial drainage should be considered in the drainage boundary and its neibourhood in sand (Umehara and Zen : 1985). In comparison with sand, the effect of dissipation of cyclically induced pore pressure on clay behaviour should be investigated under long-term cyclic loading conditions while, because of low permeability in clay, it is regarded as undrained in case of short-term cyclic loading conditions such as at earthquake.

Following the model illustrated in Fig. 2, volumetric strain, $\Delta\varepsilon_v$, during path $\vec{BC}$ under the partial-drained condtion is written as

$$\Delta\varepsilon_v = m_v \Delta u_d \qquad (4)$$

where m_v : coefficient of volume compressibility for partial drainage. Numerical analysis of pore pressure build-up and its partial dissipation was carried out by extension of the Booker et al.'s method which was originally introduced in the liquefaction problems in sand (Booker et al. : 1976). The governing equation in this case is

$$\{\nabla\}^T[K]\{\nabla\frac{u}{\gamma_w}\} = m_v\{\frac{\partial u}{\partial t} - \frac{\partial u_g}{\partial t}\} \qquad (5)$$

in which [K] : permeability matrix, u_g : cyclically induced pore pressure. The parameters contained in the governing equation are the coefficients of permeability and volume compressibility. Those were determined by the results from post-cyclic recompression tests (so called type-C test). The observed values of those coefficients are illustrated in Figs. 5 and 6, respectively.

Although the number of data is limited and the plots are somewhat scattering, it can be said that those two coefficients are kept constant independet of the cyclically induced pore pressure ratio. Those constant values are as follows:

$$k = 1.93 \times 10^{-8} \text{ cm/s}$$

$$m_v = 0.0126 \text{ cm}^2/\text{kgf}$$

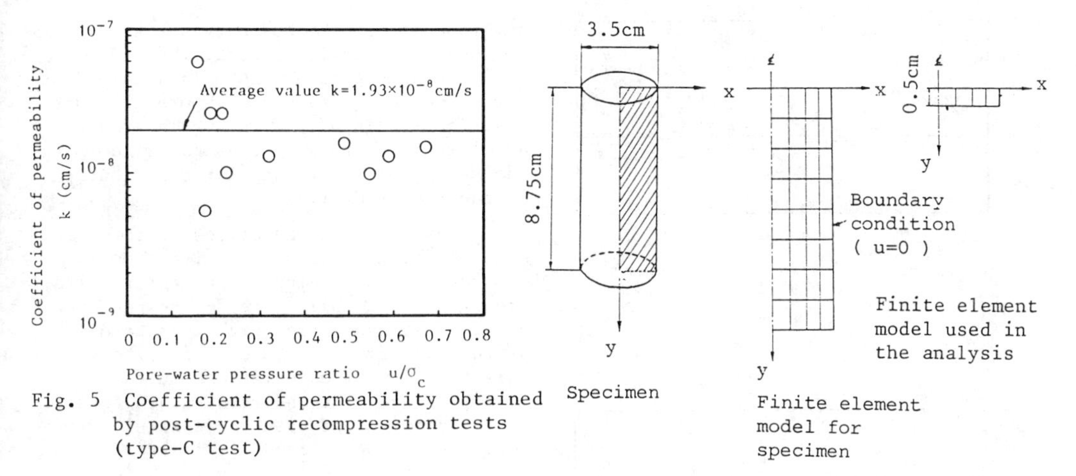

Fig. 5 Coefficient of permeability obtained by post-cyclic recompression tests (type-C test)

Fig. 7 Finite element model for specimen in triaxial test

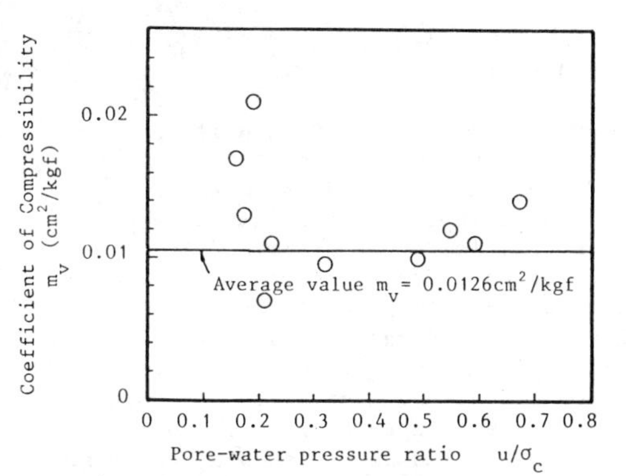

Fig. 6 Coefficient of volume compressibility obtained by post-cyclic recompression tests (type-C test)

6 Results of analysis

Numerical analysis using the finite element method were conducted for the clay cylinder in triaxial tests under partial-drained condtions. A sketch of specimen and its finte element model are presented in Fig. 7. The boundary condition which is to be zero pore pressure is given in the side boundary of specimen. Since only the radial flow is permitted in this case, the finite element model can be replaced by the simpler one which is shown in Fig. 7c.

Fig. 8 illustrates the computed results of distribution of pore pressure build-up and dissipation in a specimen under drained cyclic loading. The results in Fig. 8 point out that the effect of drainage gradually propagates from surface to inside of a specimen while the pore pressure rises up homogeneously except the

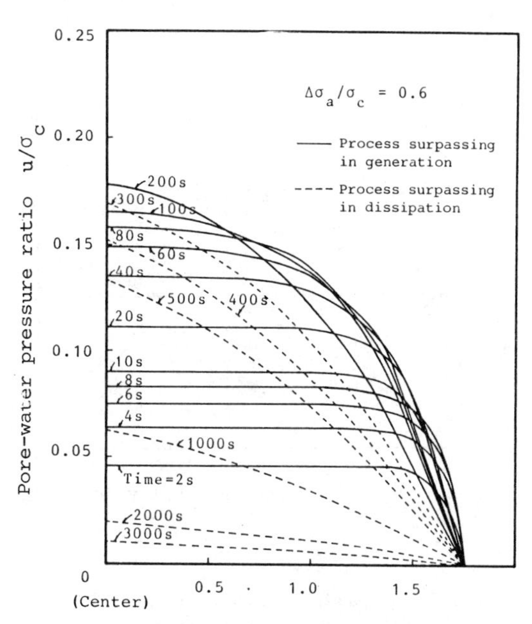

Fig. 8 Generation and dissipation of pore pressure in specimen under partial-drained cyclic loading

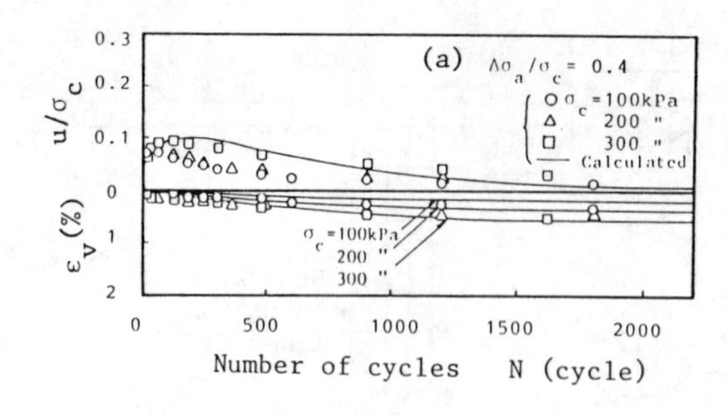

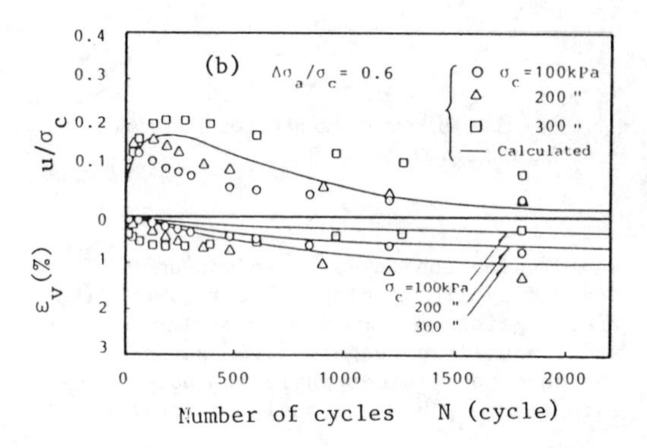

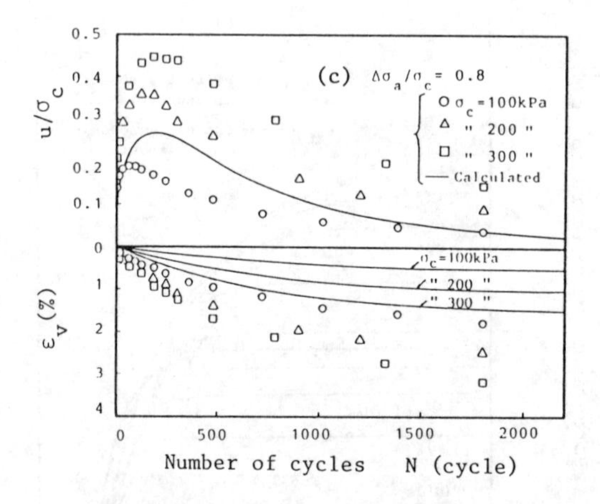

Fig. 9 Comparison between observed and calculated pore pressure and volumetric strain versus number of load cycles under partial drained cyclic loading on Ariake clay

neighbouhood of the boundary surface at the beginning of cyclic loading. Thereafter, when the peak pore pressure is attained at the center of the specimen, pore pressure distribution is of the triangular shape and then pore pressure dissipates keeeping the same shape of triangular distribution by the end of drainage.

Fig. 9 shows the variations of pore pressure and volumetric strain with time in drained cyclic loading. The test results are shown by a solid line for respective value of stress ratio which is regardless of the confining pressure. In spite of this tendency in calculated results of pore pressure, the observed values tend to increase dependent upon the confining pressure. On the other hand, it is seen that the analytical results of volumetric strain fit fairly well with the observed plots. Consequently, it is concluded from this comparison that the proposed model is appropriated for analysis of the results from drained cyclic triaxial tests on clay.

7 Conclusion

The Analytical model proposed in the current paper is aimed at estimating the combined undrained shear deformation and partial drained volumetric strain under long-term cyclic loading. Undrained deformation was detected as an exponential function obtained by undrained cyclic triaxial tests. The volumetric strain in partial-drained cyclic loading was given by combination of the consolidation theory with the model which is possible to express the pore pressure build-up and dissipation at the same time. The finite element analysis by means of this model was carried out on clay behaviour in drained cyclic triaxial tests. Analytical results for variations of pore pressure and volumetric strain with time are in good correspondence with test results. This model is also featured by the possible minimizing of error and easy treatment in numerical computation.

REFERENCES

Booker,J.R. et al. (1976) : GADFLEA. A computer program for the analysis of pore presure generation and dissipation during cyclic or earthquake loading, EERC 76-24, Univ. of California.
Umehara, Y. et al. (1985) : Evaluation of soil liquefaction potentials in partial drained conditions, Soils and Foundations, Vol. 25, No. 2, pp. 57-72.

Numerical Methods in Geomechanics (Innsbruck 1988), Swoboda (ed.)
© 1988 Balkema, Rotterdam. ISBN 90 6191 809 X

Consolidation settlements of interacting structural foundations

A.P.S.Selvadurai
Department of Civil Engineering, Carleton University, Ottawa, Ontario, Canada
K.R.Gopal
Department of Civil Engineering, Royal Military College, Kingston, Ontario, Canada

ABSTRACT: The present paper examines the consolidation behaviour of interacting strip footings located on a clay stratum of finite thickness and underlain by a smooth impervious rigid base. The problem is formulated within the context of Biot's classical theory for consolidation and solved by using a finite element technique. The numerical results presented in the paper indicate the manner in which the consolidation settlement response of the interacting foundations is influenced by the spacing between them.

1 INTRODUCTION

The category of problems which examine the performance of structural foundations which exert a mutual influence is of importance to geotechnical engineering practice. Results of such analysis have been extensively employed in the treatment of adjacent footings, tower foundation structures and tripod shaped offshore structures. The study of pile and anchor groups are of course a very common situation where interaction effects need to be properly assessed to interpret clearly the group response. The literature in this area have primarily focussed on the evaluation of the time independent response of interacting foundations which relate to the evaluation of time independent elastic settlement and bearing capacity. Some examples of these studies are given by Poulos and Davis (1975, 1980), Selvadurai (1979, 1982, 1983) and Zienkiewicz et al. (1980). The objective of this study is the evaluate the manner in which time-dependent phenomena manifest in such group interaction problems. The time dependent phenomena in soil-like materials can occur either due to processes such as viscoelastic or viscoplastic creep, due to soil-consolidation or due to a combination of these material responses. In this paper attention is restricted to the modelling of time-dependent response of the soil by appeal to Biot's (1941) theory for soil consolidation. To illustrate the influence of the interactive processes attention is focussed on the plane strain problem of two impervious interacting rigid strip footings which rest on a layer of finite thickness which is underlain by a smooth impervious rigid stratum. The field equations governing the consolidation response is solved by appeal a finite element scheme. In particular an infinite element is used to model the extended infinite region of the consolidating layer. The numerical results presented in the paper illustrates the manner in which the mutual interactions influence the rate of settlement, the rotation, etc. of the respective foundations.

2 FIELD EQUATIONS

The general theory of consolidation for a fluid saturated porous elastic solid was developed by Biot (1941). The equations governing the flow of a compressible fluid through a saturated isotropic elastic soil medium can be written as follows:

$$\frac{\partial \sigma_{ij}}{\partial x_j} + \rho_s b_i = 0 \quad (1)$$

$$\frac{\partial w_i}{\partial t} = \frac{k}{\eta} \left(\frac{\partial \pi}{\partial x_i} + \rho_f b_i\right) \quad (2)$$

where σ_{ij} is the total stress tensor; π is the porewater pressure; k is the coefficient of permeability; η is the fluid viscosity; w_i are the components of the vector of relative displacement between the soil skeleton and the pore fluid; ρ_s and ρ_f are respectively the solid and fluid densities and b_i are body forces. Also

$$w_i = f(u_s - u_f) \tag{3}$$

where f is the porosity and u_s and u_f are respectively displacements of the solid and fluid phases. The strain displacement relationship for the soil and the pore fluid take the forms

$$e_{ij} = \frac{1}{2}(u_{i,j} + u_{j,i}) \tag{4}$$

and

$$\xi = w_{i,i} \tag{5}$$

respectively, in which u_i and w_i are the displacements in the soil and the relative motion between the soil skeleton and the porewater respectively; e_{ij} and ξ are the components of the strain tensor of the soil skeleton and the volume change of fluid respectively. The linear elastic constitutive relationship for a porous elastic medium can be written as

$$\sigma_{ij} = 2Ge_{ij} + \lambda\delta_{ij}\,\delta_{kl}\,e_{kl} + \alpha\delta_{ij}\,\pi \tag{6a}$$

$$\pi = \alpha M\delta_{ij}\,e_{ij} + M\xi \tag{6b}$$

where G and λ are Lame's constants for the soil skeleton and α and M are the material properties defining the compressibility of the fluid and the solid particles of the soil skeleton. In the case of an incompressible fluid and solid particles we have α=1 and $M \rightarrow \infty$ and the equations (6) reduce to

$$\sigma_{ij} = 2Ge_{ij} + \lambda\delta_{ij}\,\delta_{kl}\,e_{kl} + \pi\delta_{ij} \tag{7a}$$

$$\xi = -e_{ij}\,\delta_{ij} \tag{7b}$$

The system of equations are solved subject to boundary and initial conditions prescribed in relation to the displacements, pore fluid pressures, soil skeletal tractions etc., i.e.

$$u_i(\underset{\sim}{x},t) = \tilde{u}_i(\underset{\sim}{x},t) \text{ on } S_1 \tag{8}$$

$$\sigma_{ij}(\underset{\sim}{x},t)\,n_j(\underset{\sim}{x}) = \tilde{T}_i(\underset{\sim}{x},t) \text{ on } S_2 \tag{9}$$

$$\pi(\underset{\sim}{x},t) = \tilde{\pi}(\underset{\sim}{x},t) \text{ on } L_1 \tag{10}$$

$$\frac{\partial w_i}{\partial t}(\underset{\sim}{x},t)\,n_j(\underset{\sim}{x}) = \tilde{Q}(\underset{\sim}{x},t) \text{ on } L_2 \tag{11}$$

where S_1, S_2, ... etc. are regions where the respective boundary conditions are prescribed. The equations (1) to (11) completely define the boundary value problem pertaining to a poroelastic medium.

3 FINITE ELEMENT MODELLING

The field equations defined in the previous section are converted into a form suitable for finite element analysis by employing the procedure developed by Sandhu and Pister (1970) which is based on the variational principles of Gurtin (1964). An excellent recent account of the application of finite element procedures to soil consolidation problems is also given by Lewis and Schrefler (1987). The procedure involves the transformation of the field equations into their integral (or weak Galerkin) form by taking Laplace transform of the variables involved. Gurtin's variational method is then used to construct the appropriate functional for the physical process. Following this procedure, a functional has been derived for the flow of compressible fluid through a porous elastic medium.

Based on this functional, a finite element analysis program has been developed for the consolidation analysis of soils. In summary, the finite element equations for the consolidation can be written in a matrix form as

$$\begin{bmatrix} \underset{\sim}{K} & \underset{\sim}{C} \\ \underset{\sim}{C}^T & -(\underset{\sim}{E}+\delta\Delta t\underset{\sim}{H}) \end{bmatrix} \begin{bmatrix} u_{t+\Delta t} \\ \pi_{t+\Delta t} \end{bmatrix} = \begin{bmatrix} F_{t+\Delta t} \\ G_{t+\Delta t} \end{bmatrix} \tag{12}$$

where
- $\underset{\sim}{K}$ = stiffness matrix of soil skeleton
- $\underset{\sim}{C}$ = stiffness matrix due to interaction between soil and pore fluid
- $\underset{\sim}{E}$ = compressibility matrix of fluid
- $\underset{\sim}{H}$ = permeability matrix
- Δt = increment of time
- δ = integration constant
- F,G = force vectors due to external traction, body forces and flows
- u,π = nodal displacement and pore pressures respectively

The program is applicable for obtaining the static response of linear-elastic and/or poro-elastic media under plane-stress, plane-strain, or axisymmetric loading conditions. It is capable of relating selected degrees-of-freedom (dof) through linear multipoint constraint conditions. Element library in the program is shown in Figure 1. Composite type elements with both displacements and pore pressures as nodal variables have been used in the program. They represent quadratic variation for displacements and linear variation for pore pressures. This is achieved by assigning pore pressures to the corner nodes only while displacements are defined at all nodes. This special technique, first introduced by Sandhu and Wilson (1969), ensures the same order of linear stress variation within the element for both the effective stress in the soil skeleton and pore water pressure. This special feature of composite element can also be used to obtain the instantaneous incompressible response of the systems for which the diagonal terms associated with the pore pressures are zero in equation (12). A simple technique to solve this case would be to assign the first few nodes to mid-side nodes which do not have pore pressure degrees of freedom. The underlying rationale is that by the time the Gaussian elimination process reaches the zero diagonal terms, they would have been changed to a non-zero number.

The development of interpolation functions for triangular and quadrilateral elements has been described in detail by Bathe (1982). (The Figures 5.5 and 5.25 of that reference present these functions for both the type of elements. In this connection, it may be noted that, the sign of the function h_q (Figure 5.5 of Bathe, (1982)) associated with the corner nodes should be "plus" instead of "minus" reported in the article).

The infinite elements implemented in the program are of the mapped infinite type discussed by Marques and Own (1984). In this approach, an infinite element in the physical plane is represented by a finite element of a regular shape in the natural plane by use of suitable coordinate transformation. The variation of the function itself within the element is represented by the conventional shape functions. Thus the visual Gauss-Legendre numerical integration scheme can be used for the formation of element matrices.

The uni-directional infinite elements represent infinity in the local ξ-direction. The mapping functions are constructed with a singularity at the positive ξ-face. Thus, $\xi=+1$ in the local coordinate system actually represents the global position at infinity. The poroelastic element is obtained by superposing the five-noded serendipity infinite element representing displacement field over two-noded superparametric element for the pore pressure field. This element has quadratic variation for displacements and linear variation for pore pressures ensuring the compatibility between the variations of effective stresses and pore pressures. The mapping and shape functions used model 1/r type of decay for the field variables.

Poroelastic bi-directional infinite element has been constructed by superposing 3-noded serendipity element representing the displacement field over 1-noded superparametric element for pore pressure field. Mapping functions for these elements are constructed with singularities at the positive faces of the local axes. Thus they model infinite domain in both the directions. Again, this element also ensures the compatibility between the variations of effective stresses and pore pressures. A complete treatment of the infinite element approach to the treatment of the consolidation problem will be given in a forthcoming article by Selvadurai and Gopal (1988).

The computer code has the capability to apply multipoint constraint conditions on the field variables. The constraint conditions can be applied in a general manner so that selected variables can be made linear functions of other variables. This feature of the problem can be used to simulate the behaviour of rigid footings in which the settlements under the footing are related to each other. The procedure following in implementing constraint conditions is that reported by Abel and Shephard (1979). The method involves the row and column manipulation of all the degrees of freedom affected by the constraint conditions. Major advantages of this method are that it does not increase the number of unknowns, does not involve any renumbering of equations and is very easy to implement in an already existing finite element code.

The accuracy of the finite element problem has been verified for various load cases against the corresponding analytical results available in the literature. It is seen that the finite element technique predicts both rate of consolidation and the amount of settlement to a high degree of accuracy. Infinite elements are found to significantly improve the accuracy of results for two-dimensional cases. Although, infinite soil media can be idealized with a large mesh of finite elements, the accuracy of solution does not increase in proportion to the additional computational effort. On the other hand, with the use of infinite elements, a relatively smaller mesh was found to predict the consolidation behaviour within acceptable limits of error.

4 INTERACTING RIGID FOOTINGS ON A CONSOLIDATING LAYER

We now focus on the plane-strain problem of two interacting rigid foundations which rest on a consolidating layer of finite thickness underlain by a smooth impervious base. Figure 2 shows the typical configuration of interacting foundations and its finite element model considered in the analysis. Soil was modelled by poro-elastic composite type finite and infinite elements. Isoparametric quadratic elements were used to model the soil up to six times the foundation width on both sides. Infinite elements were used to model the soil beyond this region. The foundations were modelled by inextensible beam elements. Their rigid behaviour was modelled by choosing a relative rigidity factor (see e.g. Selvadurai (1979)) between the soil and foundation as 1,000, which is defined as

$$K = \frac{E_{fdn}}{E_s}\,(1-\nu_s^2)\,(\frac{B}{t})^3 .$$

where E_{fdn} is the elastic modulus and t the thickness of the foundation.

The clay layer is supported on a smooth, rigid, and impermeable base. Foundations are assumed to be impermeable. The following boundary and drainage conditions were imposed.

(i) $u_z=0, \quad z=H, \quad -\infty<x<\infty, \quad t>0$

(ii) $\sigma_{zz}=q, \quad z=0, \quad \frac{\Gamma B}{2}<x<(\frac{\Gamma B}{2}+B), \quad t>0$

$\qquad z=0, \quad -(\frac{\Gamma B}{2}+B)<x<-\frac{\Gamma B}{2}, \quad t>0$

(Note that although tractions are prescribed within the footing region the presence of rigid beam elements ensure the rigid behaviour of the strip footing).

(iii) $\sigma_{rz}=0, \quad z=0, \quad -\infty<x<\infty, \quad t>0$

(iv) $\pi=0, \quad -<x<-(\frac{\Gamma B}{2}+B), \quad z=0, \quad t>0$

$\qquad -\frac{\Gamma B}{2}<x<\frac{\Gamma B}{2}, \quad z=0, \quad t>0$

$\qquad \frac{\Gamma B}{2}<x<\infty, \quad z=0, \quad t>0$

(v) $\frac{\partial\pi}{\partial n}=0, \quad \frac{\Gamma B}{2}<x<(\frac{\Gamma B}{2}+B), \quad z=0, \quad t>0$

$\qquad -(\frac{\Gamma B}{2}+B)<x<-\frac{\Gamma B}{2}, \quad z=0, \quad t>0$

$\qquad -\infty<x<\infty, \quad z=H, \quad t>0$

Various cases of spacing between the centres of foundations and the clay strata thicknesses were considered in the analysis. Both these distances were expressed in terms of the width of foundation. Table 1 presents the various cases of analysis. Independent behaviour of footings is represented by Γ=0 case.

H	Γ		
B	2	3	5
2B	2	3	5
4B	2	3	5

Table 1: Cases of Analysis

5 RESULTS AND DISCUSSION

The Figures 3, 4 and 5 present the normalized central displacement as a function of time factor for all the cases analysed. Time factor, T_v, has been defined as

$$T_v = \frac{ct}{B^2}$$

in which $c = \frac{2Gk}{\gamma_w}$

G = shear modulus of soil
k = coefficient of permeability of the soil
γ_w = unit weight of water
B = width of foundation, and
t = time

Normalized central displacement has been defined as Gw_o/Bq, in which w_o is the vertical displacement at the centre of the footing and q is the intensity of loading.

Differential settlement induced due to interaction has been expressed in terms of normalized rotation factors, defined as Gw_d/Bq, in which w_d is the difference in settlement at the two ends, (w_3-w_1). Time variation of differential settlement is shown in Figures 6, 7 and 8.

It can be seen that the interaction effects reduce as the distance between the footings increase or the thickness of clay stratum decreases. It can be observed from Figures 3 and 6 that the interaction effects are least for H=B case and maximum for H=4B case as shown in Figures 5 and 8. For $\Gamma>2$ in H=B case, the foundations can be considered to act independent of each other as shown by the negligible differential settlement in Figure 6. In fact the differential settlement for $\Gamma=5$ case was so small that it could not be plotted to any reasonable scale.

Figures 5 and 8 show the corresponding behaviour for H=4B case. Even in this case, the result for $\Gamma=5$ corresponds very closely with the independent behaviour of footings. It could again be observed that the differential settlement is maximum for $\Gamma=2$ case and gradually vanishes as the distance increases. Results for the intermediate depth case, H=2B, can be seen to lie in between these two extreme behaviours.

All the results presented point to the significant effect the depth of stratum has on the interaction behaviour. The effect is only felt if they are closer to each other than the depth of clay layer and the effect rapidly decays as they are situated further apart. It can be said as a general conclusion that the effects are very small if the footings are apart by more than two times the depth of stratum. In general the duration for total consolidation was found to be dependent only on the depth of the stratum and it is not significantly influenced by the spacing between the footings.

ACKNOWLEDGEMENTS

The work described in this paper was initiated as a part of the Geomechanics Research Programme at the Department of Civil Engineering, Carleton University and supported by a Research Grant A3866 awarded to the senior author by the Natural Sciences and Engineering Research Council of Canada. Dr. Gopal wishes to express his thanks to Drs. R.J. Bathurst and P.M. Jarret for their cooperation and encouragement to conduct additional work related to this paper at the Royal Military College.

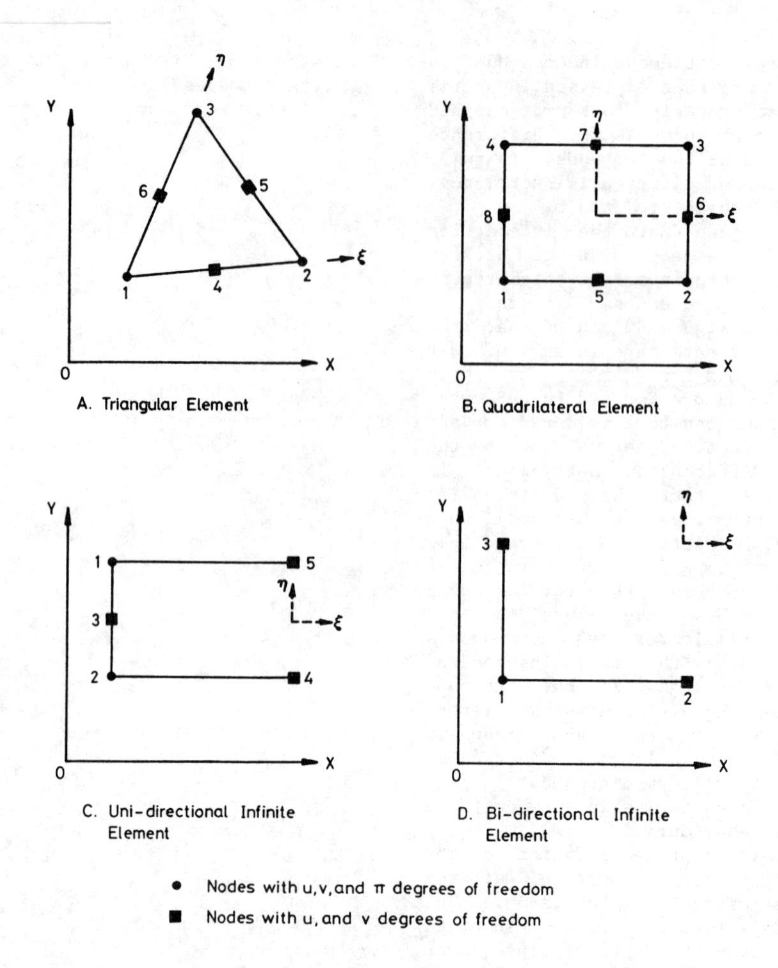

Figure 1: Element Library in the Computer Program

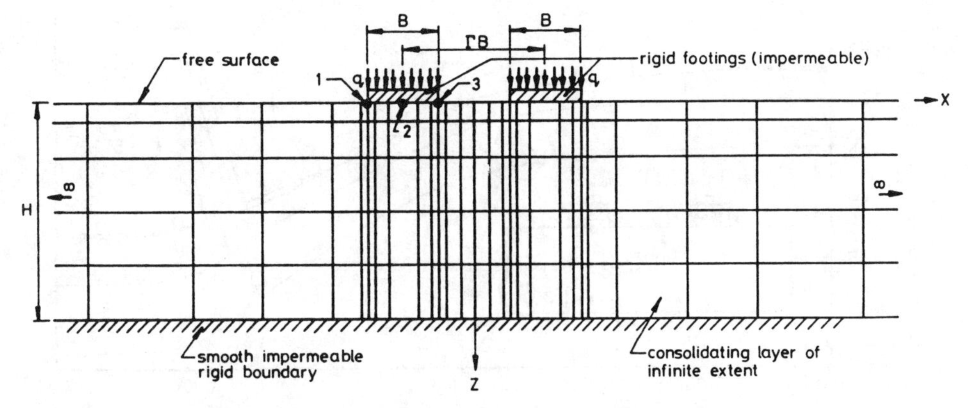

Figure 2: Typical Finite Element Discretization

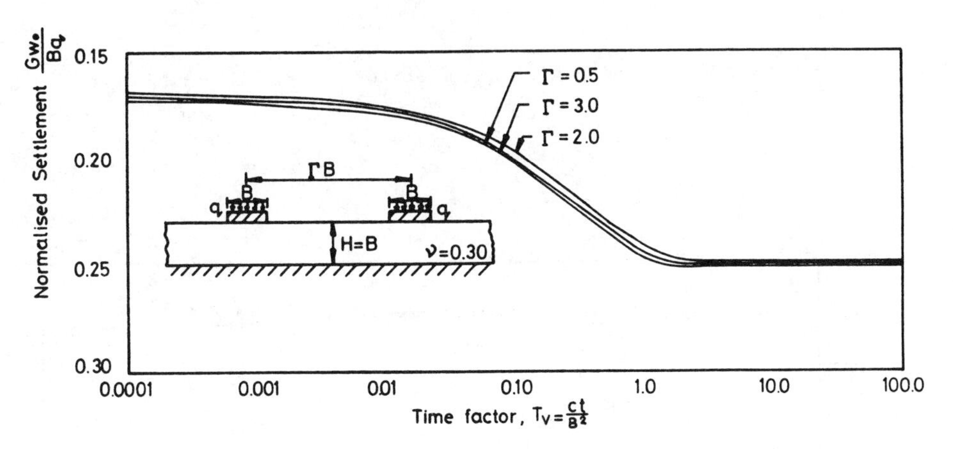

Figure 3: Normalized Settlement vs Time Factor for Rigid Footing (H/B=1)

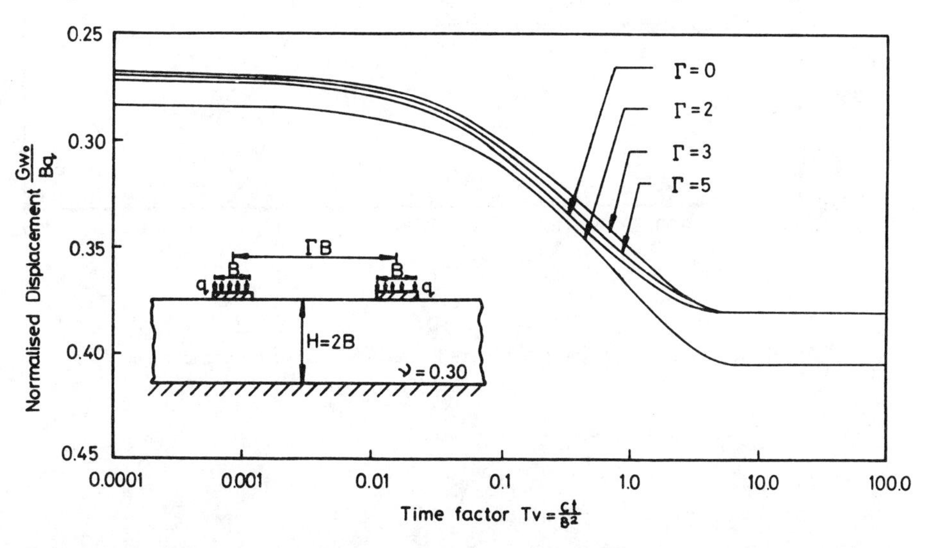

Figure 4: Normalized Settlement vs Time Factor for Rigid Footing (H/B=2)

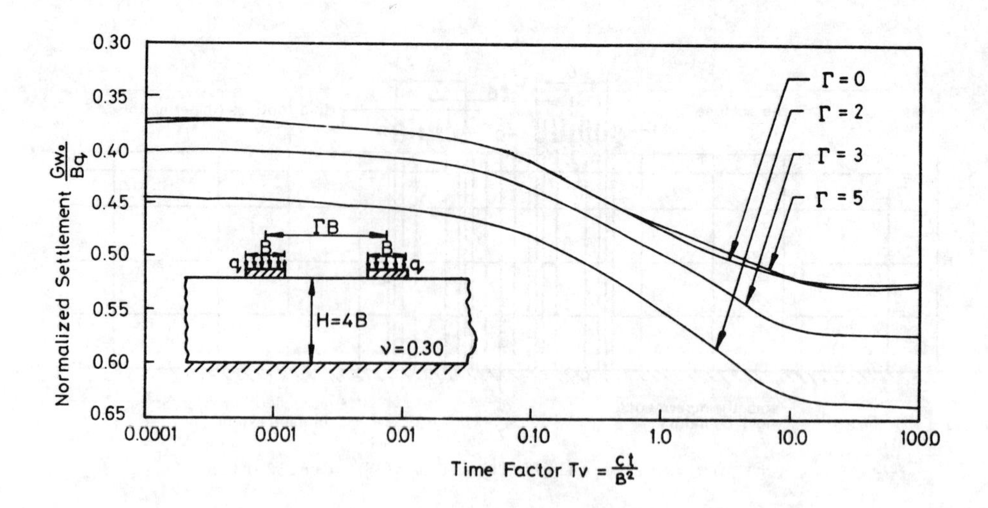

Figure 5: Normalized Settlement vs Time Factor for Rigid Footing (H/B=4)

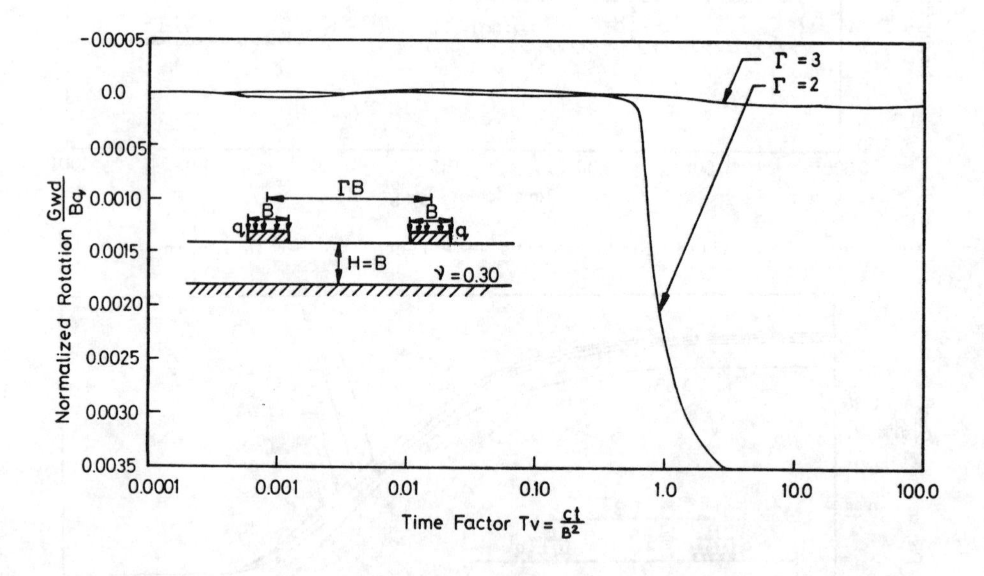

Figure 6: Normalized Differential Settlement for Rigid Footing (H/B=1)

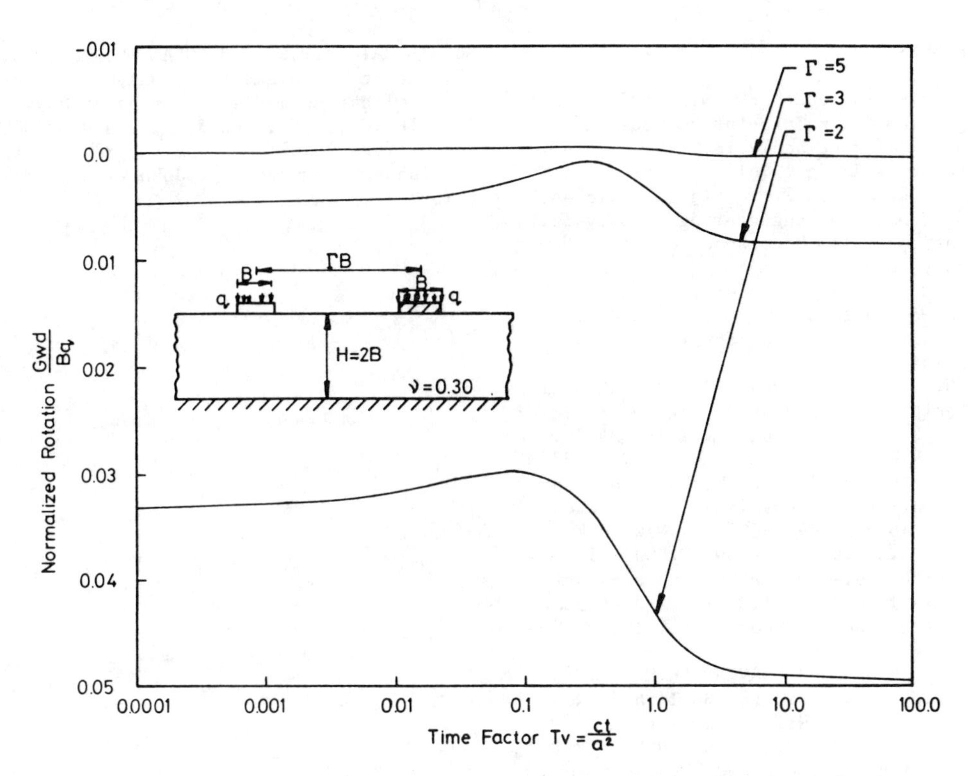

Figure 7: Normalized Differential Settlement for Rigid Footing (H/B=2)

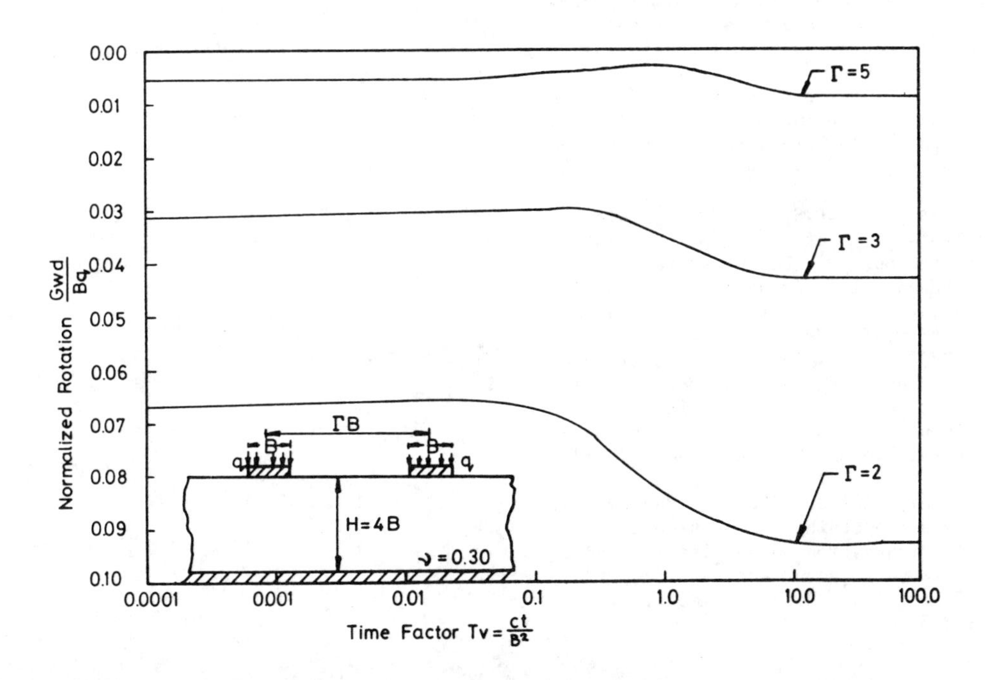

Figure 8: Normalized Differential Settlement for Rigid Footing (H/B=4)

REFERENCES

ABEL, J.F. and SHEPHARD, M.S. 1979. An algorithm for multipoint constraints in finite element analysis, Int. J. Num. Method. in Engng., Vol.3, 464-467.

BATHE, K.-J. 1982. Finite Element Procedures in Engineering Analysis, Prentice-Hall, Englewood Cliffs, New Jersey, U.S.A.

BIOT, M.A. 1941. General theory of three-dimensional consolidation, Journal of Applied Physics, Vol.12, 155-164.

GURTIN, M.E. 1964. Variational principles for linear elastodynamics Arch. Rat. Mech. Anal., Vol.16, 34-50.

LEWIS, R.W. and SCHREFLER, B.A. 1987. The Finite Element Method in the Deformation and Consolidation of Porous Media, John Wiley and Sons, New York.

MARQUES, J.M.M.C. and OWEN, D.R.J. 1984. Infinite elements in quasi-static materially non-linear problems. Computers and Structures, Vol.18, 739-751.

POULOS, H.G. and DAVIS, E.H. 1975. Elastic Solutions for Soil and Rock Mechanics, John Wiley, New York.

POULOS, H.G. and DAVIS, E.H. 1980. Pile Foundation Analysis and Design, John Wiley, New York.

SANDHU, R.S. and WILSON, E.L. 1969. Finite element analysis of seepage in elastic media, J. Eng. Mech. Div., Proc. ASCE, Vol.95, 641-652.

SANDHU, R.S. and PISTER, K.S. 1970. A variational principle for linear, coupled field problems in continuum mechanics, Int. J. Engng. Sci., Vol.8, 989-999.

SELVADURAI, A.P.S. 1979. Elastic Analysis of Soil-Foundation Interaction, Developments in Geomechanical Engineering, Vol.17, Elsevier Scientific Publishing, Amsterdam, The Netherlands.

SELVADURAI, A.P.S. 1982. The additional settlement of a rigid circular foundation on an isotropic elastic halfspace due to multiple distributed external loads, Geotechnique, Vol.32, 1-7.

SELVADURAI, A.P.S. 1983. Fundamental results concerning the settlement of a rigid foundation on an elastic medium due to an adjacent surface load, Int. J. Numer. Anal. Meth. Geomech., Vol.7, 209-223.

SELVADURAI, A.P.S. and GOPAL, K.R. 1988. Infinite element analysis of consolidating media (in preparation).

ZIENKIEWICZ, O.C., LEWIS, R.W. and STAGG, K.G. (eds.) 1978. Numerical Methods in Offshore Engineering, John Wiley, New York.

Numerical Methods in Geomechanics (Innsbruck 1988), Swoboda (ed.)
© 1988 Balkema, Rotterdam. ISBN 90 6191 809 X

Undrained bearing capacity of anisotropically consolidated clay

Akira Nishihara
Fukuyama University, Japan
Hideki Ohta
Kanazawa University, Japan

ABSTRACT: The undrained strength of Ko-consolidated clays sheared under plane strain condition is formulated on the basis of an elasto-plastic constututive equation. The equation for the anisotropic undrained strength is applied to the theory of slip lines for the bearing capacity of strip footings placed on anisotropically consolidated clays. The bearing capacity factors and the correction factors for effect of strength anisotropy are given for various shear test strengths. The validity of the present theory is examined by the comparison of calculated and the experimental results reported in the past.

1 INTRODUCTION

Experimental investigations indicate that an anisotropically consolidated clay changes its undrained shear strength depending on the experimental procedures and/or direction of shear. This anisotropy of undrained strength has been paid attention in connection with the analysis of undrained stability ([2],[3],[7],[10],[12]). Numerous equations for the undrained strength anisotropy have been proposed ([1],[4],[5],[6],[14],[15]) and applied to stability caluculations ([5],[14],[15],[17]). Most of these equations, however, were derived mathematically to fit experimental results and were not based on the behaviour of soils, and consequently they have some difficulty in application to practical problems, especially in the determination of parameters used.

In undrained loadings in two dimensinal problems, the shear failure occures under plane strain condition with no volume change (undrained condition). The plane strain condition and the undrained condition are restrictions on the deformations, and therefore, the relationship between stress and deformation i.e. a constitutive equation is needed to formulate the undrained behaviour of soils under plane strain condition. Ohta, Nishihara and Morita [21] formulated the plane strain condition and the undrained condition on the basis of an elasto-plastic constitutive equation proposed by Sekiguchi and Ohta [22], Ohta and Sekiguchi [19].

Table 1. Stress and Material parameters

MATERIAL PARAMETERS	
λ	$= 0.434Cc$ (Cc; compression index)
κ	$= 0.434Cs$ (Cs; swelling index)
D	= Dilatancy coefficient proposed by Shibata(1963)
Λ	$= 1-\kappa/\lambda$; Irreversibility ratio
M	$= \frac{\lambda-\kappa}{D(1+e_o)} = \frac{6\sin\phi'}{3-\sin\phi'}$; critical state parameter
Ko	= Coefficient of earth pressure at rest
β	$= \frac{\sqrt{3}\eta_o\Lambda}{2M}$
STRESS PARAMETERS	
η^*	$= \sqrt{\frac{3}{2}\left(\frac{s_{ij}}{p'} - \frac{s_{ijo}}{p'_o}\right)\left(\frac{s_{ij}}{p'} - \frac{s_{ijo}}{p'_o}\right)}$
p'	= effective mean principal stress
s_{ij}	$= \sigma'_{ij} - p'\delta_{ij}$ (δ_{ij}; Kronecker's delta) ; deviatric stress tensor
σ'_v	= effective overburden pressure
η_o	$= \frac{3(1-Ko)}{1+2Ko}$

Note; Subscript o specifies the value at the time of completion of Ko-consolidation. Subscript i specifies the value at the initial state prior to undrained loading.

From these conditions together with the failure condition, they derived a relation between the undrained shear strength and the direction of the major principal stress.

In this paper, this equation is applied to the analysis of the bearing capacity of strip footing on Ko-consolidated clays in the theory of slip lines. The parameters used in this paper are summarized in Table 1.

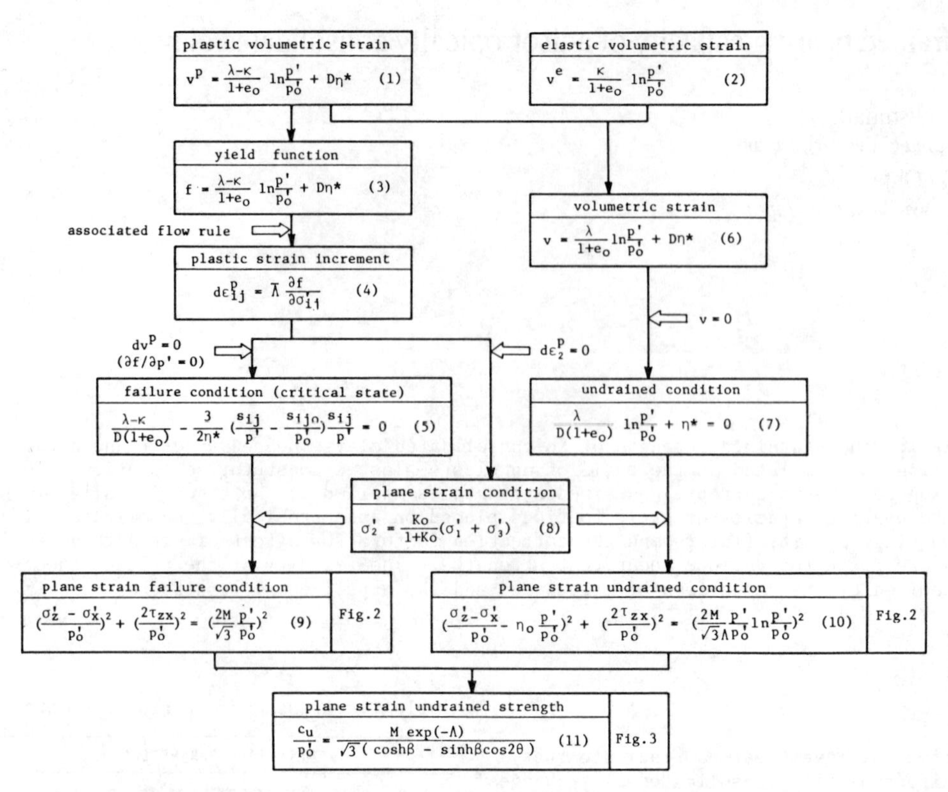

Fig.1 Process of deriving the equation for the undrained strength under plane strain condition (Ohta, Nishihara and Morita [21])

2 UNDRAINED SHEAR STRENGTH OF CLAY UNDER PLANE STRAIN CONDITION

Fig.1 shows the process of deriving the equation for the undrained strength under plane strain condition [21]. The undrained strength under plane strain condition is derived by simultaneously solving a set of equations, i.e. the failure condition, the undrained condition and plane strain condition. These conditions are derived based on the constitutive equation proposed by Sekiguchi and Ohta [22], Ohta and Sekiguchi [19]. Sekiguchi-Ohta Model is the extention of the constitutive equation proposed by Ohta[18] which is essentially identical to Cam Clay Model, taking the effect of anisotropic consolidation and rotation of the principal stresses into account by using a parameter η^* , see Table 1.

In Sekiguchi-Ohta Model, the plastic and elastic components of the volumetric strain of anisotropically consolidated clays are given by Eqs.(1) and (2) in Fig.1 respectively. Considering the plastic volumetric strain as the strain hardening parameter, the yield function is defined as Eq.(3). This yield function together with the associated flow rule gives the plastic strain increments as Eq.(4). The failure condition, Eq.(5), is obtained by substitution of definition of the critical state into plastic strain increment. The undrained condition is derived as Eq.(7) from the condition that the total volumetric strain given by Eq.(6) must be zero. The plane strain condition is approximated by zero increment of intermediate principal plastic strain. Eq.(8) is thus obtained plane strain condition.

Substitution of Eq.(8) into Eqs.(5) and (7) yields the failure condition and the undrained condition under plane strain condition as given by Eqs.(9) and (10) respectively. These equations are represented by two surfaces in the normalized stress space, as shown in Fig.2. The plane strain failure condition, Eq.(9), is a cone with an apex at the origin of the coordinate system. The plane strain undrained condition, Eq.(10), is in a skewed bullet shape, the cross section of which with p'/p'_0 = constant planes are circles with the centres on Ko-line. The cross section made by the failure surface and the

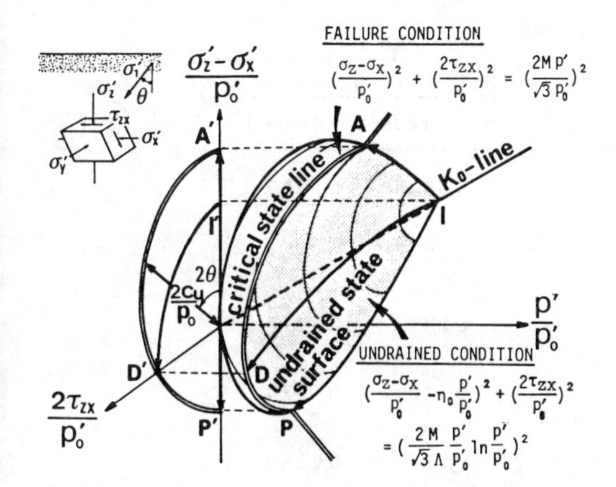

Fig.2 Plane strain failure and undrained condition

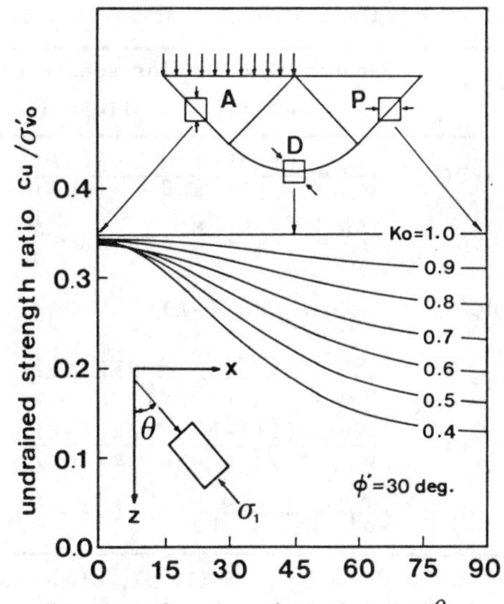

Fig.3 Variation of undrained strength with direction of major principal stress

undrained surface gives the critical state line (plane strain undrained failure condition) ADP. The initial stress state of Ko-consolidated clay foundation is represented by the point I in Fig.2. For undrained loading, the stress state at any point in the ground moves on the undrained surface with rotation of the principal stresses until it reachs the state of ultimate failure at critical state line ADP. The effective stress states at typical points A, D, P in Fig.3 may follow paths such as I→A, I→D, I→P in Fig.2.

The projection of the critical state line ADP onto the $(\sigma'_z - \sigma'_x)/p'_0$, $2\tau_{zx}/p'_0$ plane is a curve A'D'P' similar to an ellipse. The radius from the origin to the point on the curve A'D'P' represents the twice of the undrained strength while θ indicates the angle between the direction of the major principal stress and the vertical. The equation of the curve A'P'D' is derived from Eqs.(9) and (10) with some approximation, as given by Eq.(11). The soil parameters needed in calculating the undrained strength by Eq.(11) are Ko, M(ϕ') and Λ (Cc,Cs), see Table 1. If an empirical relation

$$\Lambda = M/1.75$$

reported by Karube [8] is used, the undrained strength can be estimated by using only two parameters Ko and ϕ'. Fig.3 shows the variation of undrained strength with direction of the major principal stress, calculated by Eq.(11) for values of Ko ranging from 0.4 to 1 with $\phi' = 30°$.

The undrained strengths at typical points A, D and P correspond to those measured by KoPUC(plane strain undrained compression test on Ko-consolidated clays), DSS(direct simple shear test) and KoPUE(plane strain undrained extension test on Ko-consolidated clays) respectively. These test strength are derived, from Eq.(11) by substituting 0, π/4 and π/2 for θ, as shown in Table 2. In table 2, the undrained strengths under axi-symmetric stress condition, KoUC (triaxial undrained compression test on Ko-consolidated clays), IUC(triaxial undrained compression test on isotropically consolidated clays) and KoUE(triaxial undrained extension test on Ko-consolidated clays) are listed as well. The equations for KoUC, IUC and KoUE are derived directly from Eqs.(5) and (7) with the axi-symmmetric stress condition [20]. On the right hand side of Table 2, the experimental data reported by Ladd [11] and Kinner and Ladd [10] are compared with theoretical values calculated by using the reported parameters ϕ' and Ko. The theoretical values are generally in good agreement with the experimental values.

The theory described here is easily extended to overconsolidated clays [20]. The undrained strength ratio of the overconsolidated clays is related to that of the normally consolidated clay by

$$\left(\frac{c_u}{\sigma'_{vi}}\right)_{OC} = \left(\frac{c_u}{\sigma'_{vo}}\right)_{NC} \cdot OCR^{\Lambda}$$

where OCR $(=\sigma'_{vo}/\sigma'_{vi})$ is the overconsolidation ratio.

Table 2 Undrained strength for various testing method

Test	Reduced equation for specified test on normally consolidated clay	Undrained strength	
		measured*	predicted
KoPUC	$\frac{c_u}{\sigma'_{vo}} = \frac{(1+2K_o)\ M \exp(-\Lambda)}{3\sqrt{3}(\cosh\beta - \sinh\beta)}$	0.34	0.347
KoUC	$\frac{c_u}{\sigma'_{vo}} = \frac{1+2K_o}{3}\frac{M}{2}\exp(-\Lambda + \frac{\Lambda}{M}\eta_o)$	0.33	0.318
IUC	$\frac{c_u}{\sigma'_{vo}} = \frac{M}{2}\exp(-\Lambda)$	0.31	0.311
DSS	$\frac{c_u}{\sigma'_{vo}} = \frac{(1+2K_o)\ M \exp(-\Lambda)}{3\sqrt{3}\cosh\beta}$	0.20	0.224
KoPUE	$\frac{c_u}{\sigma'_{vo}} = \frac{(1+2K_o)\ M \exp(-\Lambda)}{3\sqrt{3}(\cosh\beta + \sinh\beta)}$	0.19	0.165
KoUE	$\frac{c_u}{\sigma'_{vo}} = \frac{1+2K_o}{3}\frac{M}{2}\exp(-\Lambda - \frac{\Lambda}{M}\eta_o)$	0.155	0.135

* test data by Ladd(1973), Kinner and Ladd(1973): $\phi' = 33°$, $K_o = 0.5$

3 APPLICATION TO THE THEORY OF SLIP LINES

3.1 Stress field

The stresses at limiting equilibrium are governed by the equilibrium equations and failure condition. The equilibrium equations in a two dimensional problem are given by

$$\frac{\partial}{\partial x}(P - Q\cos 2\theta) + \frac{\partial}{\partial z}(Q\sin 2\theta) = 0$$

$$\frac{\partial}{\partial x}(Q\sin 2\theta) + \frac{\partial}{\partial z}(P + Q\cos 2\theta) = \gamma$$

where γ is the unit weight of the soil and

$$P = \frac{1}{2}(\sigma_1 + \sigma_3)\ ,\quad Q = \frac{1}{2}(\sigma_1 - \sigma_3)$$

In the plastic zone, Q is given by Eq.(11) as

$$Q = c_u = \frac{M \exp(-\Lambda)}{\sqrt{3}(\cosh\beta - \sinh\beta\cos 2\theta)} p'_o \quad (12)$$

The parameter β in denominator of Eq.(12) is a function of Ko, see Table 1, and expresses the effect of the stress induced anisotropy. Substituting Eq.(12) into equilibrium equations, we have

$$\begin{aligned}&\frac{\partial P}{\partial x} + 2Q(\sin 2\theta - Q_\theta\cos 2\theta)\frac{\partial\theta}{\partial x} + 2Q(\cos 2\theta + Q_\theta\sin 2\theta)\frac{\partial\theta}{\partial z} = 0\\&\frac{\partial P}{\partial z} + 2Q(\cos 2\theta + Q_\theta\sin 2\theta)\frac{\partial\theta}{\partial x} + 2Q(\sin 2\theta - Q_\theta\cos 2\theta)\frac{\partial\theta}{\partial z} = \gamma\end{aligned} \quad (13)$$

where,

$$Q_\theta = \frac{1}{2Q}\frac{\partial Q}{\partial\theta} = -\frac{\sinh\beta\sin 2\theta}{\cosh\beta - \sinh\beta\cos 2\theta}$$

The directions of stress characteristics are given by

$$\frac{dx}{dz} = \tan(\Psi \pm \frac{\pi}{4}) \quad (14)$$

as shown in Fig.4, where Ψ is an angle given by

$$\tan 2\Psi = \frac{\cosh\beta\sin 2\theta}{\cosh\beta\cos 2\theta - \sinh\beta} \quad (15)$$

For isotropically consolidated clays for which Ko=1 and hence $\beta = 0$, we have $\Psi = \theta$ and thus stress characteristics are angled at $\pm\pi/4$ to the direction of the major principal stress as is also derived from the conventional isotropic theory. The stress state along characteristics are obtained by substituting the slope of respective characteristics into Eq.(13) as

$$\begin{aligned}&dP + \frac{2\bar{c}\sqrt{\cosh 2\beta - \sinh 2\beta\cos 2\theta}}{(\cosh\beta - \sinh\beta\cos 2\theta)^2}d\theta = \gamma dz\ ;S_1\\&dP - \frac{2\bar{c}\sqrt{\cosh 2\beta - \sinh 2\beta\cos 2\theta}}{(\cosh\beta - \sinh\beta\cos 2\theta)^2}d\theta = \gamma dz\ ;S_2\end{aligned} \quad (16)$$

where

$$\bar{c} = \frac{p'_o}{\sqrt{3}} M \exp(-\Lambda) \quad (17)$$

3.2 Verocity field

The plastic strain imcrements at plane strain undrained failure are given, from Eq.(11) together with associated flow rule, by

$$d\varepsilon_x^p = \frac{\bar{\Lambda}}{2}\frac{\cosh\beta\cos2\theta - \sinh\beta}{\cosh\beta - \sinh\beta\cos2\theta}$$

$$d\varepsilon_z^p = - d\varepsilon_x^p$$

$$d\gamma_{zx}^p = \frac{\bar{\Lambda}}{2}\frac{\cosh\beta\sin2\theta}{\cosh\beta - \sinh\beta\cos2\theta}$$

These plastic strain increments together with the relations between strain increments and velocities yield the differential equations for the velocity field. The characteristics of the velocity field make angles at $\pm\pi/4$ to the direction of major principal strain and coincide with those of the stress field. However, the principal directions of strain and stress are not coincident except when $\theta = 0$ or $\theta = \pi/2$. Along velocity characteristics, we have Geiringer's equations,

$$dv_1 + v_2\, d\psi_1 = 0 \quad ; S_1$$

$$dv_2 + v_1\, d\psi_2 = 0 \quad ; S_2$$

as same as in the isotropic case, in which v_1 and v_2 are the velocity components along S_1 and S_2 lines and ψ is the angle of characteristics from the vertical.

3.3 Undrained shear strength on the slip line

Under undrained condition, the slip lines i.e. the velocity characteristics coincide the stress characteristics and are given by Eqs.(14) and (15). Denoting the angle between S_1-line and horizontal by ω as shown in Fig.4, $\omega = \pi/4 + \Psi$ and hence, from Eq.(15), we have

$$\cot2\omega = \frac{\cosh\beta\sin2\theta}{\cosh\beta\cos2\theta - \sinh\beta}$$

Solving this equation for θ, neglecting the small term of higher order, we obtain

$$\cos2\theta = \frac{\sinh\beta + \cosh\beta\sin2\omega}{\cosh\beta + \sinh\beta\sin2\omega} \qquad (18)$$

Substition of Eq.(18) into Eq.(11) yields

$$c_u = \bar{c}\,(\cosh\beta + \sinh\beta\sin2\omega) \qquad (19)$$

The shear strength on the S_1-line angled at $\omega + \theta$ to the major principal plane (see Fig.4) is given from the geometry of Fig.5 as

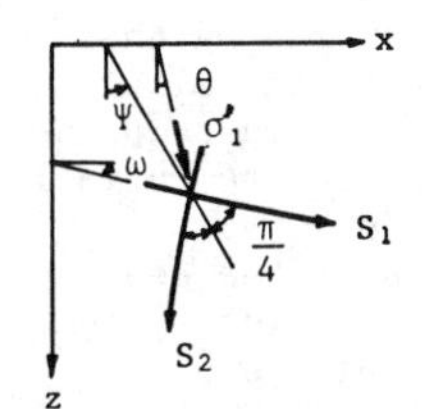

Fig.4 Direction of the slip lines

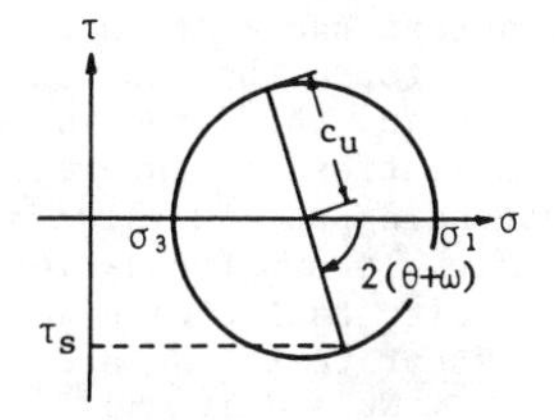

Fig.5 Shear strength on the slip line

$$\tau_s = c_u \sin2(\theta + \omega) \qquad (20)$$

Substituting Eqs.(18) and (19) into Eq.(20), we have

$$\tau_s = \bar{c}\,(\cos^2 2\omega + \sinh\beta\sin2\omega + \cosh\beta\sin^2 2\omega) \qquad (21)$$

Eq.(21) gives the relation between the undrained shear strength and direction of the slip line. This expression of the undrained shear strength is useful in stability analysis by the limit equilibrium method such as conventional circular slip line method. The undrained strength measured by shear box test(SBT) is that on the horizontal plane, and thus, putting $\omega=0$ into Eq.(21), we have

$$\tau(SBT) = \bar{c} = \frac{P\delta}{\sqrt{3}} M \exp(-\Lambda) \qquad (22)$$

4 BEARING CAPACITY OF ANISOTROPICALLY CONSOLIDATED CLAY

The ultimate bearing capacity of a strip footing on purely cohesive homogeneous clay is conventionally evaluated from

$$q_{ult} = Nc\, c_u$$

in which Nc is the bearing capacity factor. For homegeneous isotropic clay, Nc = 5.14 is used , however, for anisotropic clay Nc depends on the degree of anisotropy ([5], [14]). Nc also depends on the type of shear test since the undrained strength varies depending on the type of shear test. In this paper, the bearing capacity factor is given for various types of shear test.

The undrained bearing capacity is calculated by integlating the differential equations, Eq.(16), along the characteristics. The integlation of Eq.(16) can be easily performed by the finite-difference method. Since the parameter β involved in Eq.(16) is a function of Ko, the results of numerical solution depend on Ko. Fig.6 shows the slip lines and the direction of the major principal stress along the slip lines for the cases of Ko=0.5 and Ko=1 assuming constant undrained strength with depth. It is found that the shape of the slip lines and the direction of the major principal stress in the region acd are variable depending on the value of Ko.

Fig.7 shows the variation of the bearing capacity factor with Ko for various types of shear test. Nc are calculated by

$$N_c = q_{ult} / c_u$$

with undrained strengths from equations listed in Table 2 and Eq.(22). It should be noted that Nc for KoUC and IUC are not equal to 5.14 even if Ko = 1 since they are not test under plane strain condition.

Table 3 shows the comparison of the calculated results by present theory with the experimental results of the ultimate bearing capacity obtained by Kinner and Ladd [10] for model footings on clays with overconsolidation ratio of 1, 2 and 4. The theoretical values are calculated by using the bearing capacity factor in Fig.7 and the undrained strength data reported by Kinner and Ladd [10]. The ultimate bearing capacities calculated by present theory are in good agreement with the experimental results no matter what type of shear test is used. It should be noted that the results obtained in sections 2 and 3 are extendable to overconsolidated clays with a slight modification, see Ref.[21].

In table 3, the ultimate bearing capacities calculated with Nc = 5.14 given by conventional isotropic theory are listed as well. As shown in Fig.7, for a usual range of Ko from 0.4 to 0.8, the bearing capacity factors for SBT and DSS are somewhat larger than the conventional value of 5.14 and hence, the use of SBT or DSS strength assuming the isotropic strength

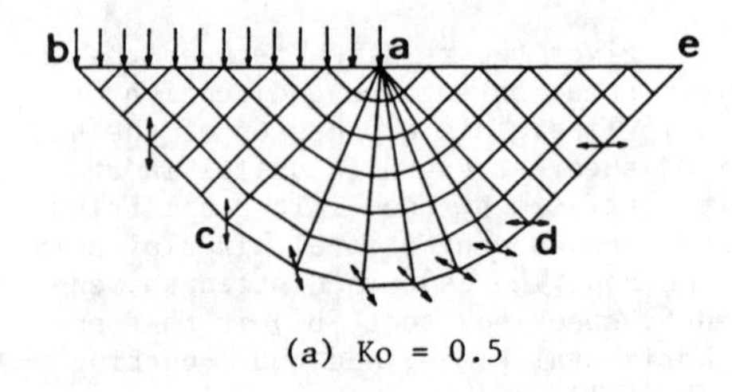

(a) Ko = 0.5

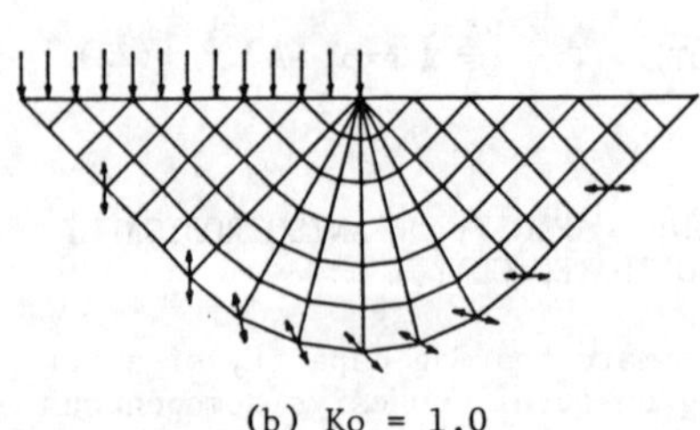
(b) Ko = 1.0

Fig.6 Slip lines and direction of major principal stress for (a) Ko = 0.5 and (b) Ko = 1.0

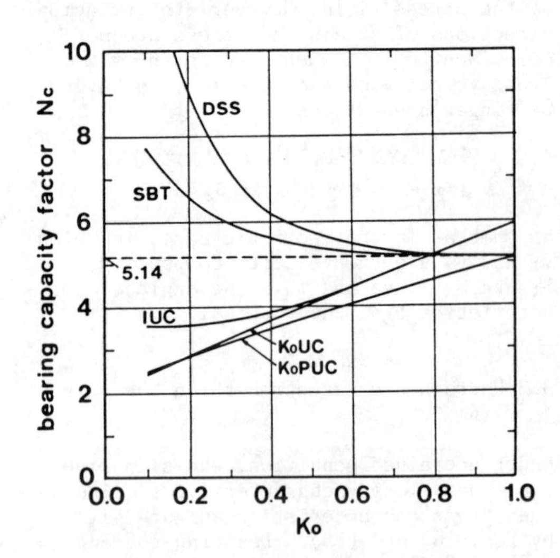

Fig.7 Bearing capacity factors for various types of test

Table3 Predicted and measured bearing capacity

OCR	measured q/σ'_{v1}*	method of prediction	average c_u/σ'_{v1}*	present theory Nc	predicted q/σ'_{v1}	ratio predicted/measured	isotropic theory Nc	predicted q/σ'_{v1}	ratio predicted/measured
1	1.34	KoPUC	0.34	3.72	1.26	0.94	5.14	1.75	1.31
		IUC	0.31	4.15	1.29	0.96		1.59	1.19
		DSS	0.20	5.77	1.15	0.86		1.03	0.77
2	2.43	KoPUC	0.57	3.72	2.12	0.87	5.14	2.93	1.21
		IUC	0.55	4.15	2.28	0.94		2.82	1.16
		DSS	0.37	5.77	2.13	0.88		1.90	0.78
4	4.20	KoPUC	0.95	3.72	3.53	0.84	5.14	4.88	1.16
		IUC	0.91	4.15	3.78	0.90		4.67	1.11
		DSS	0.61	5.77	3.52	0.84		3.14	0.75

* test data by Kinner and Ladd [10]; $\phi' = 33°$, Ko = 0.5

results in conservative estimation of undrained bearing capacity, while IUC, KoUC or KoPUC strength overestimates the bearing capacity of soils for Ko<0.8 if the strength is assumed to be isotropic, as demonsrated in Table 3. Therefore the undrained strength measured by the shear tests should be corrected in connection with the influence of the strength anisotropy when they are introduced in the conventional stability calculations. The influence of shear rate on the undrained strength should be also taken into account [2]. Thus the field strength to be introduced in stability analysis is derived from the undrained strength measured by the shear tests as follows.

$$(c_u)_{field} = (c_u)_{test} \times \mu \qquad (23)$$

$$\mu = \mu_a \times \mu_r$$

where μ_a is the correction factor for strength anisotropy and μ_r is the correction factor for the effect of shear rate. This idea was first proposed by Bjerrum [2] for the stability analysis of embankments based on the in-situ vane strength. From the back-analysis of embankment failures, Bjerrum [2],[3] derived empirical correction factors for the in-situ vane test strength based on plasticity index. Bjerrum's approach seems to be adequate for preliminary design purposes.

μ_a is considered to be a factor which reduces the undrained strength measured by the shear tests to the average strength mobilized along the slip line and hence, μ_a is given by the ratio of the average mobilized strength to the shear test strength. Fig.8 shows thus obtained correction factor μ_a for various shear test strengths. The average strength mobilized along the slip line is calculated from the ultimate bearing capacity q_{ult} through a relation

$$(c_u)_{ave.} = q_{ult}/5.14$$

while the undrained strengths of KoPUC, IUC, DSS and SBT are calculated by the equations listed in Table 2 and Eq.(22). The vane strength is estimated by euations as follows. For a rectangular vane with a hight twice of the diameter, the undrained strength is given by

$$\tau(\text{VANE}) = \frac{1}{7}(6\tau_v + \tau_h)$$

where τ_h is the undrained strength on the horizontal plane and is assumed to be equal to the SBT strength given by Eq.(22) since these are conjugate. The undrained strength on the vertical plane τ_v is cal-

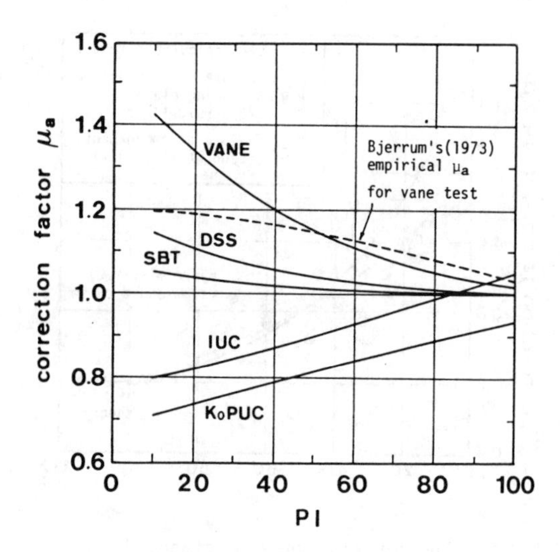

Fig.8 Correction factors μ_a for various types of shear test

culated by

$$\frac{\tau_v}{\sigma'_{vo}} = \frac{1+2K_o}{3\sqrt{3}}\sqrt{\left(\frac{M}{\Lambda}\frac{p'}{p'_o}\ln\frac{p'}{p'_o}\right)^2 - \eta_o^{\,2}\left(1-\frac{p'}{p'_o}\right)^2}$$

in which p'/p'_o is the solution of

$$\frac{p'}{p'_o}\left\{\left(\frac{M}{\Lambda}\ln\frac{p'}{p'_o}\right)^2 + \frac{M^2}{\Lambda}\ln\frac{p'}{p'_o}\right\} + \eta_o^{\,2}\left(1-\frac{p'}{p'_o}\right) = 0$$

These equations are derived from Eqs.(5) and (7) using the tortional stress system. In calculatiug the undrained strengths, the parameters Ko and ϕ'(and hence M) are determined based on plasticity index through empirical relations,

$$K_o = 0.44 + 0.42\,PI/100$$

proposed by Massarsch [16] and

$$\sin\phi' = 0.8 - 0.227\log PI$$

by Kenney [9]. Karube's empirical equation $\Lambda = M/1.75$ is also employed in deriving the parameter Λ.

We have here no theoretical basis for discussing the influence of the shear rate, however for laboratory shear tests, the influence of shear rate can be minimized if the tests are performed at sufficiently small strain rate, and hence only the influence of strength anisotropy may be taken into account. For in-situ vane tests, the influence of the shear rate cannot be disregarded since soils are sheared so rapidly as to hold the undrained condition, but fortunately, the empirical correction

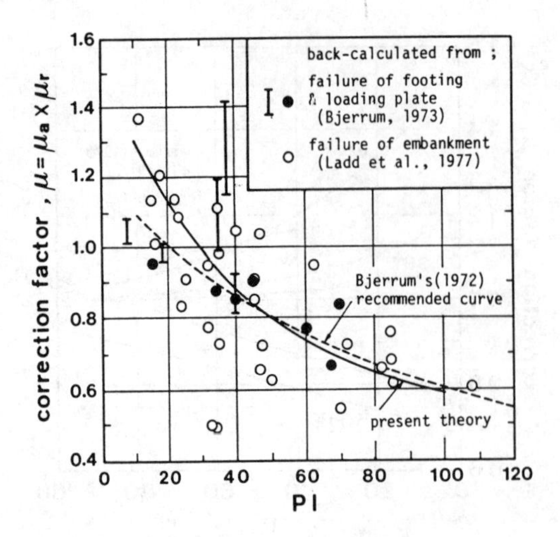

Fig.9 Theoretical and experimental correction factor for in-situ vane

factor μ_r given by Bjerrum [3] is applicable to the in-situ vane tests. Fig.9 shows the correction factor μ for the in-situ vane test calculated by Eq.(23) using the theoretical value of μ_a given in Fig.8 together with Bjerrum's correction factor μ_r. The plots ● are the correction factors back-analyzed from the failures of footings and loading plates reported by Bjerrum [3]. The plots ○ are the correction factors back-analyzed from the failures of embankments reported by Ladd et al.[13]. The curve recommended by Bjerrum [2] is also shown. The theoretical curve by present theory has a similar tendency to Bjerrum's curve and is in good agreement with back-analysis of field failures.

5 CONCLUSIONS

The bearing capacity of anisotropically consolidated clay is analyzed by the theory of slip lines taking the effect of anisotropy of the undrained strength into account. The variations of bearing capacity factors with Ko are given for various types of shear test. The bearing capacities calculated by the present theory show fairly good agreement with the experimental results obtained from the model footing tests.

The undrained strength measured by shear tests should be corrected in connection with the strength anisotropy and the effect of shear rate when they are introduced in conventional stability analysis. The correcion factors for strength anisotropy. are given by using present theory for various kinds of test strength. The validity of the present theory is examined by comparing the calculated correction factor with the experimental results reported in the literature.

REFERENCES

[1] Bishop, A.W.: The strength of soils as engineering material, Geotechnique 16, 91-128, 1966
[2] Bjerrum, L.: Embankments os soft ground, Proc. ASCE Spec. Conf. on Performance of Earth and Earth-Supported Structures, Vol.2, 1-54, 1972
[3] Bjerrum, L.: Problems of soil mechanics and construction on soft clays, State of the Art Report, Proc. 8th ICSMFE, Vol.3, 109-159, 1973
[4] Casagrande, A. and Carillo, N.: Shear failure of anisotropic soils, Proc. Boston Soc. Civ. Engineers, Vol.31, 74-87, 1944
[5] Davis, E.H. and Christian, J.T.: Bearing capacity of anisotropic cohesive soil. Proc. ASCE 97, SM5, 81-104, 1971
[6] Hansen, J.B. and Gibson, R.E.: Undrained shear strength of anisotropically consolidated clays. Geotechnique 1, No.3, 189-204, 1949
[7] Hanzawa,H., Matsuda,E., Suzuki,K. & Kishida, T.: Stability analysis and field behaviour of earth filles on alluvial marine clay, Soils and Foundations 20, No.4, 37-51, 1980
[8] Karube, D.: Nonstandarized triaxial testing method and its problems, 20th Symp. on Soil Eng. JSSMFE, 45-60, 1975
[9] Kenny ,T.C.: Discussion on Geotechnical properties of glacial lake clays, Proc. ASCE 85, SM3, 67-79, 1959
[10] Kinner, E.B. and Ladd, C.C.: Undrained bearing capacity of footing on clay, 8th ICSMFE, Vol.1, 209-215, 1973
[11] Ladd, C.C.: Panel discussion on Session 1, Shear strength of soft clay, Proc. Geotech. Conf on Shear Strength Properties of Natural Soils and Rocks, Vol.2, 112-115, 1967
[12] Ladd, C.C. and Foott, R.: New design procedure for stability of soft clays, Proc. ASCE 100, GT7, 763-786, 1974
[13] Ladd,C.C., Foott,R., Ishihara,K., Schlosser, F. and Poulos,H.G.: Stress-deformation and strength characteristics, 9th ICSMFE 2, 421-494, 1977
[14] Livneh, M. and Greenstein, J.: Plastic equilibrium in anisotropic cohesive clays, Soils and Foundations 14, No.4, 1-11, 1974
[15] Lo, K.Y.: Stability of slopes in anisotropic soils. Proc. ASCE 94 ,SM4, 85-106, 1965
[16] Massarsch, K.R.: Lateral earth pressure in normally consolidated clays. Proc. 7th Eur. Conf on SMFE, Vol.2, 245-249, 1975
[17] Menzies, B.K.: An approximate correction for the influence of stregth anisotropy on conventional shear vane measurements used to field bearing capcity. Geotechnigue 26, 631-634, 1976
[18] Ohta, H.: Analysis of deformations of soils based on theory of plasticity and its application to settlement of embankments, Dr. Eng. Thesis, Kyoto Univ., 1971
[19] Ohta,H. and Sekiguchi,H.: Constitutive equations considering anisotropy and stress reorientation in clay. Proc. 3rd ICONMIG, 475-484, 1979
[20] Ohta, H. and Nishihara, A.: Anisotropy of undrained shear strength of clays under axi-symmetric loading conditions. Soils and Foundations 25, No.3, 73-86, 1985
[21] Ohta,H.,Nishihara, A. and Morita,Y.: Undrained stability of Ko-consolidated clays, Proc. 11th ICSMFE, Vol.1, 613-616, 1985
[22] Sekiguchi, H. and Ohta, H.: Induced anisotropy and time dependency in clays, Proc. 9th ICSMFE, Spec. Session 9, 229-238, 1977

Numerical Methods in Geomechanics (Innsbruck 1988), Swoboda (ed.)
© 1988 Balkema, Rotterdam. ISBN 90 6191 809 X

Three-dimensional bearing capacity analysis of clays under partially drained condition

A.Asaoka & S.Ohtsuka
Department of Geotechnical Engineering, Nagoya University, Japan

ABSTRACT: An attempt is made to derive a general procedure of determining ultimate bearing capacity of normally and lightly overconsolidated clay foundations based on the critical state concept with the use of rigid plastic finite element method. The plastic flow at critical state is derived from Sekiguchi-Ohta's model which evaluates anisotropic deformation-strength characteristics of naturally deposited clays. Bearing capacity is calculated for two different drainage conditions: one is undrained condition in which instantaneous loading is assumed and the other, partially drained condition where consolidation during gradual load application is expected to increase the bearing capacity. The effect of footing property such as rigidity and roughness of footing on ultimate bearing capacity is also examined.

1 INTRODUCTION

In conventional stability analysis it is usual to assume instantaneous loading to obtain an ultimate bearing capacity of a clay foundation. However, soft clay engineering requires the prediction of ultimate load intensity as a function of both loading rate and drainage condition. On the other hand, naturally deposited clay exhibits anisotropic deformation-strength characteristics. The main object of the present study is to provide a numerical yet simple procedure of obtaining approximate values of ultimate load intensity taking into account both the anisotropic deformation-strength characteristics and the loading history as well.

Limit equilibrium analysis with the use of stress-strain rate relationship at critical state is made without the explicit concept of strength. Undrained bearing capacity of clay is determined from the following three effective stress states:

1. initial effective stress state which has regulated clay for long years during their sedimentation and consolidation process, and
2. current effective stress state just before instantaneous loading, and
3. critical stress state which undoubtedly depends loading condition.

2. METHOD OF ANALYSIS

Rigid plastic finite element analysis based on the upper bound theorem (Tamura et al.,1984) is applied to the estimation of the bearing capacity of a foundation. Main assumptions introduced in the analysis are as follows:

1. every soil element reaches a critical state in the whole region of a soil mass at failure, and
2. in the same region, continuous velocity field is also assumed.

2.1 Plastic flow at critical state

Sekiguchi-Ohta's model describes anisotropic deformation-strength characteristics of naturally deposited and anisotropically consolidated clays. Employing the critical state parameter M, the dilatancy parameter D and the strain hardening parameter ε_v^p (plastic volumetric strain), the yield function F (and/or f) is expressed as

$$F = f - \varepsilon_v^p = MD\ln\frac{p'}{p'_y} + D\eta^* = 0 \qquad (1)$$

in which

$$\eta^* = \sqrt{\frac{3}{2}(\eta_{ij}-\eta_{ij0})(\eta_{ij}-\eta_{ij0})} \qquad (2)$$

$$\eta_{ij} = s_{ij}/p', \quad \eta_{ij0} = s_{ij0}/p'_0$$

where s_{ij0}, p'_0 denotes deviatoric stress and mean effective stress at the end of anisotropic consolidation, respectively which are refered to initial stress state in the present study. As initial 'pressure-normalized' deviatoric stress tensor η_{ij0} determines the shape of the yield function, it is regarded as a material parameter which regulates anisotropic deformation-strength characteristics. When isotropically consolidated clay is considered, since η_{ij0} becomes zero and Eq.(1) gives the yield function of the original Cam clay model.

Applying the associate flow rule and letting the plastic strain rate $\dot{\varepsilon}^p_{ij}$ be indeterminate one obtains the equation of critical state condition in terms of effective stresses as follows:

$$M - \frac{3}{2\eta^*}\eta_{ij}(\eta_{ij}-\eta_{ij0}) = 0 \qquad (3)$$

The strain rate characteristics at the critical state is found that

$$\dot{\varepsilon}^p_v = 0 \qquad (4)$$

$$\dot{\varepsilon}^p_{ij} = \sqrt{\frac{3}{2}}\,\frac{\dot{e}}{\eta^*}(\eta_{ij}-\eta_{ij0}) \qquad (5)$$

in which

$$\dot{e} = \sqrt{\dot{\varepsilon}^p_{ij}\dot{\varepsilon}^p_{ij}}\,.$$

Therefore the rate of specific plastic energy dissipation defined in effective stresses becomes

$$D(\dot{\varepsilon}^p_{ij}) = \sigma'_{ij}\dot{\varepsilon}^p_{ij} = s_{ij}\dot{\varepsilon}^p_{ij} = \sigma_{ij}\dot{\varepsilon}^p_{ij} \qquad (6)$$

and there is no difference with that defined in total stress.

Eqs.(3)-(6) can be interpreted easily when the original Cam clay (i.e., $\eta_{ij0} = 0$) is considered. Although Eq.(3) has the same form of the extended Mises criterion, Eq.(5) still shows that the plastic flow at the critical state is absolutely similar to the plastic flow of Mises material with the yield function

$$s_{ij}s_{ij} = \sigma_0^2 \qquad (7)$$

where σ_0 is a material constant. In other words the critical state concept requires the limit analysis of inhomogeneous Mises continuum in which the Mises constant σ_0 is distributed according to the spatial distribution of p' in the soil mass, as indicated by Eq.(3).

The rate of specific plastic energy dissipation is calculated using Eqs.(3) and (5) as follows:

$$D(\dot{\varepsilon}^p_{ij}) = \sqrt{\frac{2}{3}}\,M\dot{e}p' \qquad (8)$$

which is a function of mean effective stress p' satisfying Eq.(3). As far as the plastic energy dissipation D is concerned there is no difference between Sekiguchi-Ohta's model and Cam clay model.

2.2 Rigid plastic finite element analysis

The fundamental equations for numerical analysis are as follows:

$$\Sigma\int_{V_k}(s_{ij}+\kappa_k\delta_{ij})\delta\dot{\varepsilon}^p_{ij}dV = \mu\int_S T_i\delta\dot{u}_i dS + \int_V F_i\delta\dot{u}_i dV \qquad (9)$$

$$\delta\kappa_k\int_{V_k}\dot{\varepsilon}^p_{ii}dV = 0 \qquad (10)$$

$$\delta\mu\left(\int_{S_\sigma}T_i\dot{u}_i dS - 1\right) = 0 \qquad (11)$$

in which $V = \Sigma V_k$ denotes the volume of soil mass concerned, V_k an arbitrary partition of V, S_σ the traction boundary, T_i the given shape of traction force having unit intensity applied on S_σ, F_i the body force and $\dot{u}_i$ the velocity compatible with the plastic strain rate $\dot{\varepsilon}^p_{ij}$, respectively. Eq.(9) represents the limit equilibrium equation resulting from the stationary condition of the rate of plastic energy dissipation. Lagrange multiplier κ_k corresponds to the mean total stress p at limit equilibrium in V while μT_i gives ultimate load intensity on S_σ. Eqs.(10) and (11) are constraint conditions so that $\dot{\varepsilon}^p_v = 0$ and the uniqueness of the solution may be satisfied. Substituting the stress-strain rate relationship of Sekiguchi-Ohta's model at the critical state into Eq.(9) one gets

$$\Sigma\int_{V_k}\left(\sqrt{\frac{2}{3}}Mp'\frac{\dot{\varepsilon}^p_{ij}}{\dot{e}}+\kappa_k\delta_{ij}\right)\delta\dot{\varepsilon}^p_{ij}dV$$
$$=\mu\int_{S_\sigma}T_i\delta\dot{u}_i dS+\int_V F_i\delta\dot{u}_i dV. \quad (12)$$

Employing finite element discretization technique Eqs.(10)-(12) yield nonlinear simultaneous equations with respect to $\dot{u}_i$, μ and κ_k, k= 1, 2, .., n. In this analysis procedure, the distribution of total stress satisfying limit equilibrium is consistent with the plastic flow which is derived from effective stress-strain rate relationship.

Eq.(12) suggests that one needs the distribution of mean effective stress p' at critical state in a whole region of a soil mass before proceeding rigid plastic finite element computation. The distribution of p' at failure should undoubtedly depends on loading history, boundary condition and so on. Here discussed first is an iterative method for determining p' distribution for undrained condition, which corresponds to an instantaneous load application.

2.3 Bearing capacity analysis under undrained condition

Letting λ and κ denote slopes of consolidation and swelling lines, respectively, undrained condition for normally consolidated clay can be generally expressed as follows:

$$\frac{M}{\Lambda}\ln\frac{p'}{p'_c}+\eta^*=0, \quad \Lambda=1-\frac{\kappa}{\lambda} \quad (13)$$

The parameter p'_c in Eq.(13) can be determined by solving Eq.(13) itself with respect to p'_c through substitution of the current effective stresses $\tilde{p}'$ and $\tilde{\eta}^*$.

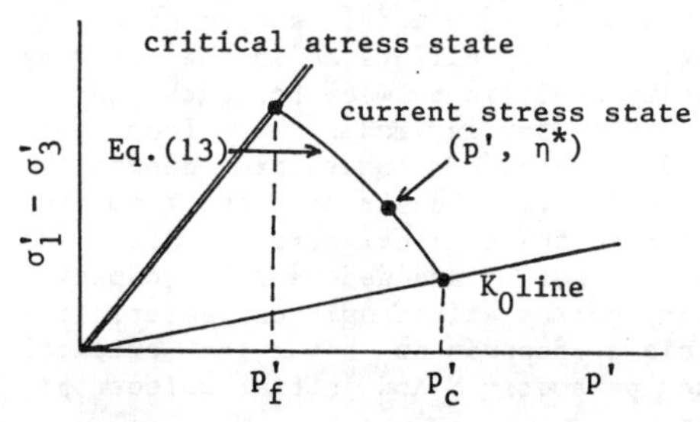

Fig.1 Determination of p'_c

Employing

1. Undrained condition, Eq.(13), and
2. Critical state condition, Eq.(3), and
3. Stress - strain rate relationship, Eq.(5),

the distribution of the mean effective stress p' can be obtained

$$p'=p'_c\exp\{\Lambda(-1+\sqrt{\frac{3}{2}}\frac{\dot{\varepsilon}^p_{kl}\eta_{kl0}}{\dot{e}^2})\} \quad (14)$$

according to plastic strain rate field that should have been determined by rigid plastic finite element analysis. Eq.(14) completes the iterative method for determining the spatial distribution of p' at failure. Remember that the Lagrange multiplier κ_k, k = 1, 2, .., n gives the spatial distribution of mean total stress p, one gets the distribution of excess pore pressure u as follows:

$$u=p-p'-\gamma_t z \quad (15)$$

in which $\gamma_t z$ denotes hydrostatic pressure at depth z.

When lightly overconsolidated clay foundation is considered, the behavior under undrained condition differs from that of normally consolidated clay. In this case the parameter p'_c is derived from overconsolidation ratio $n=p'_0/\tilde{p}'$, by

$$p'_c=\tilde{p}'\,n^{\Lambda}. \quad (16)$$

The analysis proceeds similarly to the case of normally consolidated clays.

As shown in Fig.2 the current effective stress state reflects strain hardening procedure corresponding to the loading history just before instantaneous loading to failure. This concept will be employed for determining bearing capacity under partially drained condition in section 4.

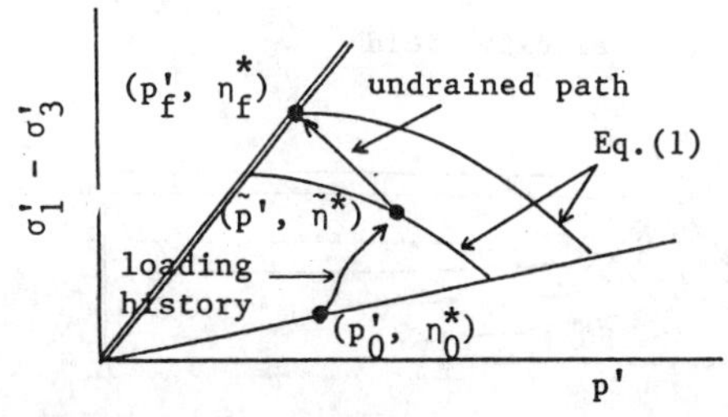

Fig.2 The hardening procedure of soils

Undrained bearing capacity is examined here for the case of axi-symmetric problem. Fig.3 shows the boundary condition employed while Fig.4, current stress distribution. Table 1 gives the soil constants. Firstly, undrained bearing capacity of a normally consolidated clay foundation is analyzed in which initial stress state is assumed to be given by the current stress state, for simplicity.

Table 1. Soil constants

λ	0.25	κ	0.1304
M	1.2	D	0.0326
K_0	0.7	γ_t	16.3kN/m^3

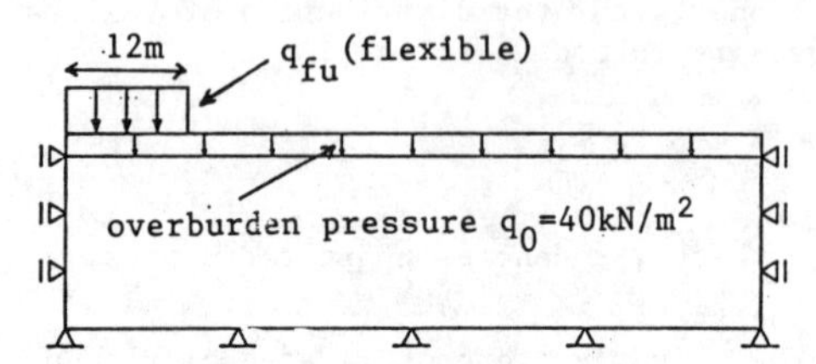

Fig.3 Boundary condition employed

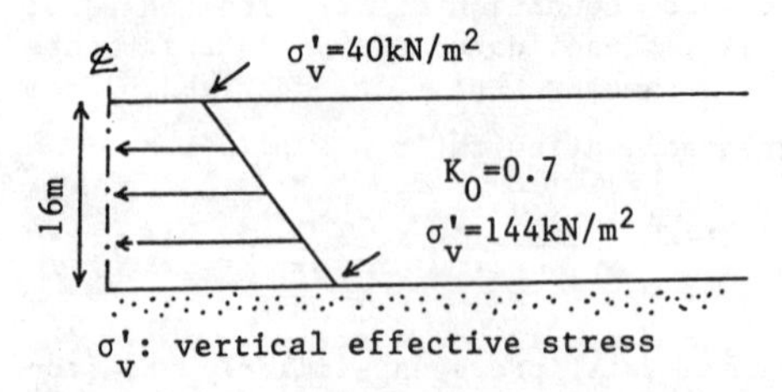

Fig.4 Current stress state

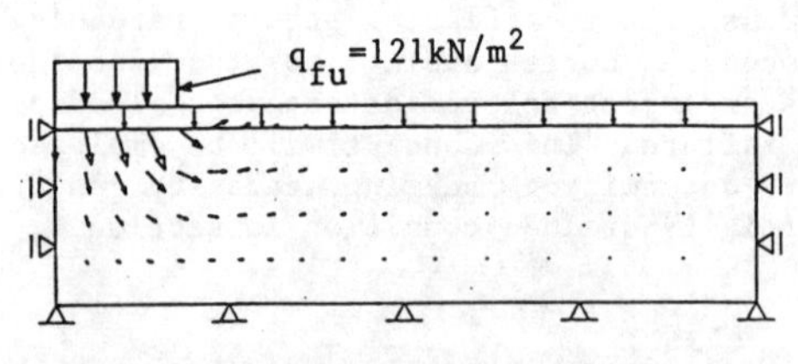

Fig.5(a) Velocity field

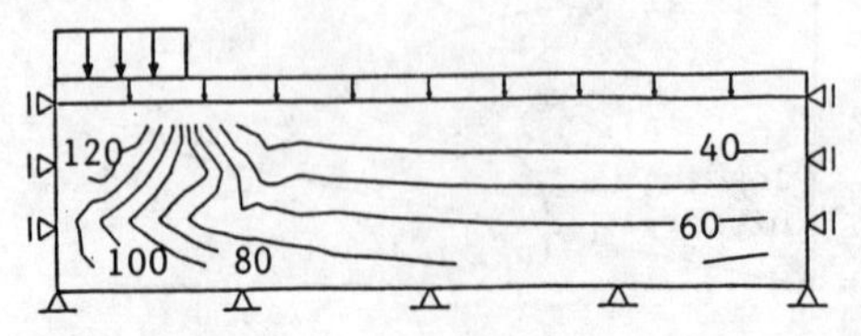

Fig.5(b) Excess pressure distribution

Computed reults of ultimate load intensity and velocity field are illustrated in Fig.5(a) while excess pore water pressure distribution is shown in Fig.5(b).

Next examined is the bearing capacity of lightly overconsolidated clay foundation, which requires the distribution of the initial stress state (Fig.6) other than the current stress state of Fig.4. The results are summarized in Fig.7(a) and (b). The plastic flow is quite similar to the case of Fig.5(a). However the ultimate load intensity is found to be about 10% higher than the case of a normally consolidated clay.

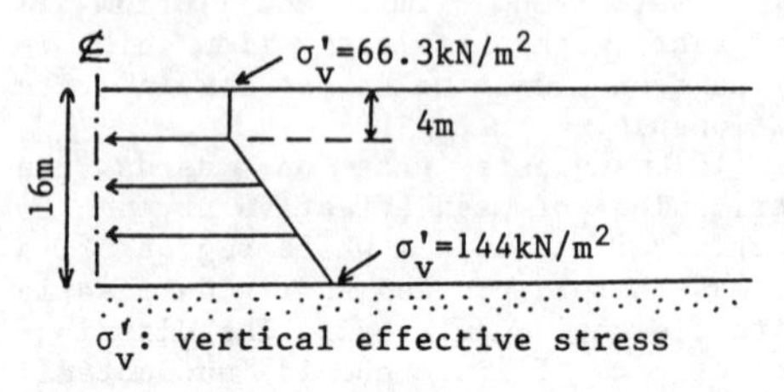

Fig.6 Initial stress state when over-consolidated clay is considered

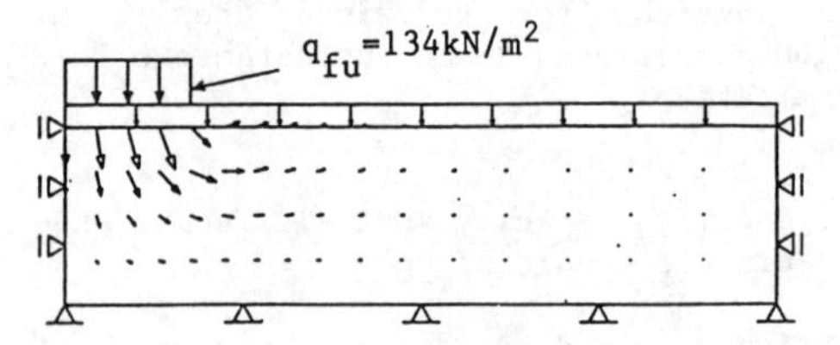

Fig.7(a) Velocity field of O. C. clay

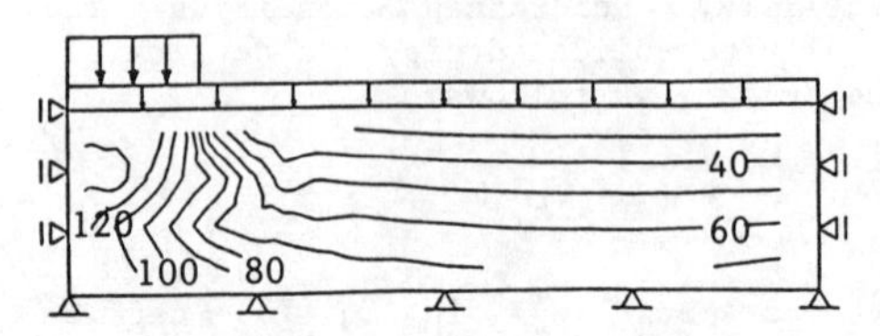

Fig.7(b) Excess pressure distribution of O.C. clay

The comparison between Sekiguchi-Ohta's model and Cam clay model is also given in Table 2. The difference in the bearing capacity analysis between Sekiguchi-Ohta's model and Cam clay model comes from only the difference of 'undrained condition' which adjusts the stress path from the current to the critical stress state. The reduction of undrained bearing capacity resulting from anisotropic characteristics of clays depends on both the critical state parameter M and initial anisotropic stress condition K_0.

Table 2. Comparative results

	N.C. clay	O.C. clay
Sekiguchi-Ohta	121	134
Cam clay	135	145

(kN/m^2)

3. RIGIDITY AND ROUGHNESS OF FOOTING

In the former section bearing capacity was obtained assuming that the external loading was fully flexible. In this section the effects of the rigidity and roughness of the footing on the increase of the bearing capacity will be examined.

Analysis are made under following two conditions:

1. rigid footing with completely rough surface, in which velocity distribution just below the footing should be absolutely uniform with vertical direction, and
2. rigid footing with completely smooth surface, where vertical component of the velocity just below footing should be distributed uniformly.

For the sake of comparative discussion with Fig.5(a), normally consolidated clay foundation is considered under axi-symmetric rigid circular footing. The same soil constants and the same current stress state as those of Table 1 and Fig.4 are used in the analysis. Boundary conditions are also the same as those of Fig.3 except the rigidity and roughness of the footing.

Fig.8(a) shows the computed results of ultimate velocity field when rigid footing with completely rough surface is

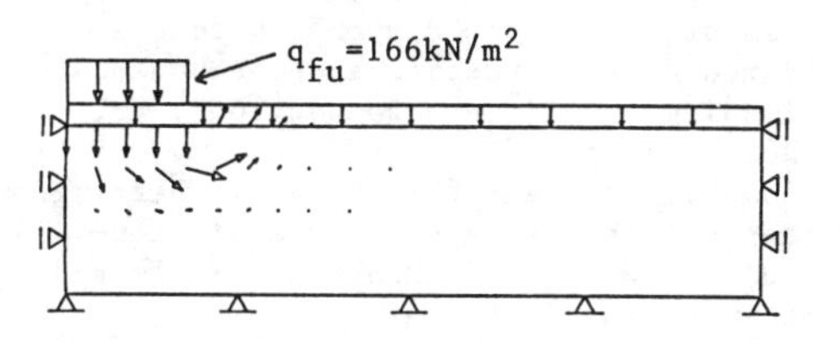

Fig.8(a) Velocity field in the case of completely rough surface

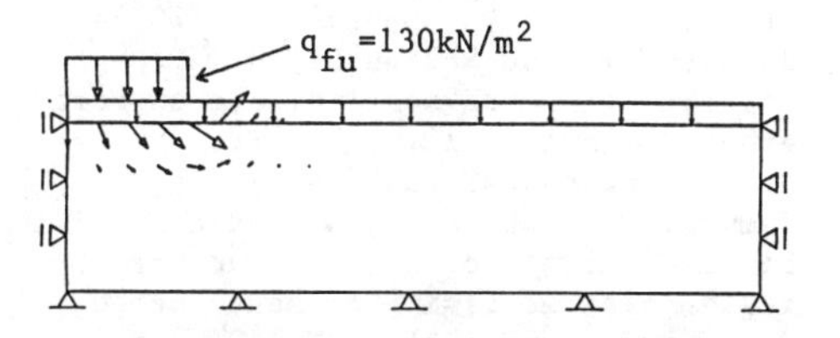

Fig.8(b) Velocity field in the case of completely smooth surface

considered. Fig.8(b) corresponds to the case of rigid footing with completely smooth surface. Since constraint condition on the admissible velocity field is most severe in the case of rigid footing with completely rough surface, then the biggest ultimate bearing capacity is obtained in Fig.8(a). The second biggest one is found in Fig.8(b) in which rigid footing with completely smooth surface.

4. BEARING CAPACITY UNDER PARTIALLY DRAINED CONDITION

When gradual load application with finite width such as an execution stage of embanking is considered, since the loading rate is finite consolidation proceeds partially during loading stage. Bearing capacity analysis, therefore, should be made taking into account partially drained condition. This can be done easily by combining elasto-plastic consolidation analysis with undrained bearing capacity analysis. The former analysis can provide current effective stress distribution within a clay foundation from time to time corresponding to a given loading history. As stated in the end of Section 2, undrained bearing capacity is obtained as a function of current effective stress state just before instantaneous loading.

To make the discussion simple suppose that external load intensity q is expressed as

$$q(t) = \dot{q}\cdot t \tag{17}$$

keeping the shape of load application constant. Applying the current effective stress state at time t, the undrained bearing capacity is also obtained as a function of time t:

$$q_{fu} = q_{fu}(t) \tag{18}$$

It is obvious that $q(t) < q_{fu}(t)$ during gradual load application before failure. Bearing capacity under partially drained condition is thus obtained by solving Eqs.(17) and (18) simultaneously with respect to time t.

A graphical method for obtaining this is shown in Fig.9. The ultimate load intensity q_f is obtained by the extrapolation of $q_{fu}(t) \sim t$ curve towards $q=\dot{q}\cdot t$ line, as shown in the figure. The difference between q_f and $q_{fu}(t=0)$ is refered to as partial drain effect on the increase of bearing capacity.

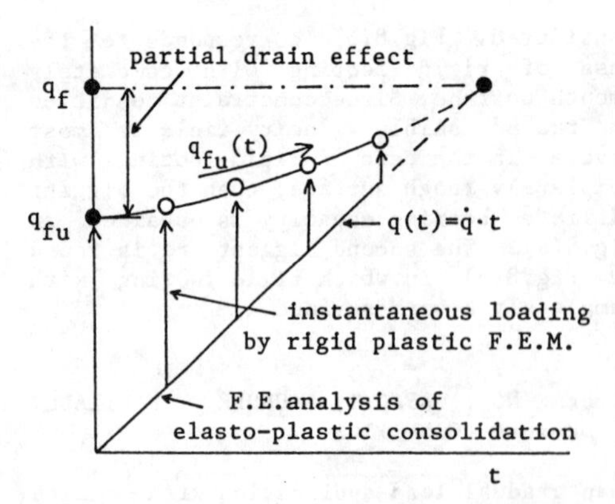

Fig.9 Graphical method for determining bearing capacity

The axi-symmetric bearing capacity problem given in Fig.3 is solved here under a given loading rate $\dot{q}$. Table 3 gives additional soil parameters required in the elasto-plastic consolidation analysis. Numerical results are shown in Fig.10. Partial drain effect on ultimate bearing capacity in terms of coefficient of consolidation c_v, loading rate $\dot{q}$ and maximum drain length H_d is discussed by Shibata and Sekiguchi (1981), and Asaoka and Ohtsuka (1986).

Fig.10 shows another example that different loading histories give different magnitudes of ultimate bearing capacity.

Table 3. Additional parameters

$\dot{q}$	$2kN/m^2day$	k	$5.88\times10^{-4}m/day$

* permeable top and bottom

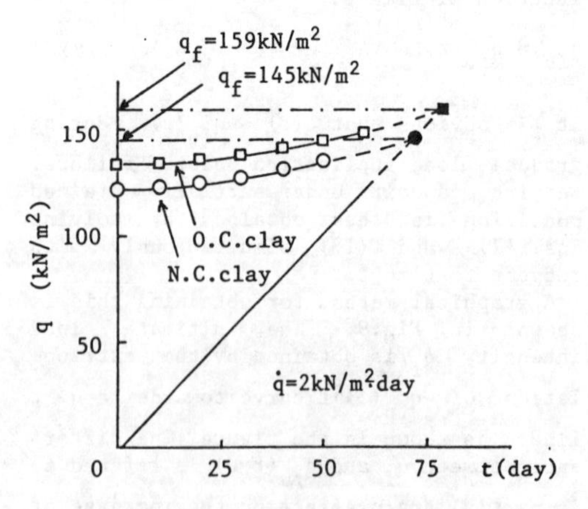

Fig.10 Computation of partial drain effect

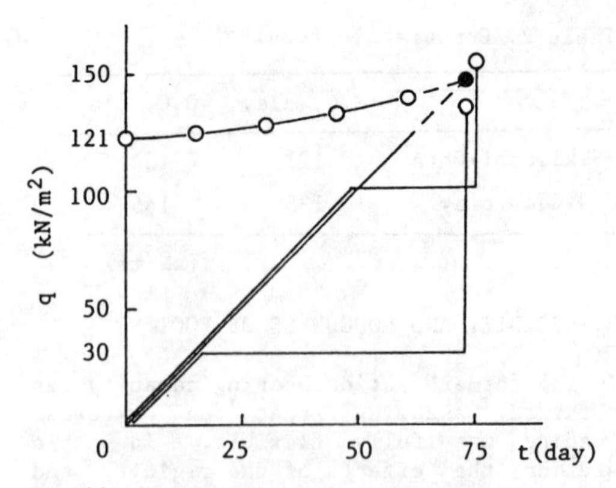

Fig.11 History dependent bearing capacity during loading

5. CONCLUSIONS

Limit equilibrium concept as the limit of the upper bound theorem in plasticity was newly applied based on the critical state concept of soils, to estimate the ultimate bearing capacities of normally consolidated and lightly overconsolidated clay foundations. The attempt was made to derive the general procedure for obtaining the ultimate load intensity as a function of loading history, along which innegligible amount of strain hardening effect of clay was expected. The graphical and then simple method for obtaining ultimate load intensity under partially drained condition was also provided.

REFERENCES

(1) Asaoka, A. and Ohtsuka, S.: The analysis of failure of a normally consolidated clay foundation under embankment loading., Soils and Foundations, Vol. 26, No.2, pp.47-59 (1986)
(2) Asaoka, A. and Ohtsuka, S.: Bearing capacity analysis of a normally consolidated clay foundation, Soils and foundations (1987)
(3) Sekiguchi, H. and Ohta, H.: Induced anisotropy and time dependency in clays, Proc. of Speciality Session 9, 9th ICSMFE, Tokyo, pp.229-238 (1977)
(4) Shibata, T. and Sekiguchi, H.: Prediction of embankment failure on soft ground, Proc. 10th ICSMFE, Vol.1, Stockholm, pp.247-250 (1981)
(5) Tamura, T., Kobayashi, S. and Sumi, T.: Limit analysis of soil structure by rigid plastic finite element method, Soils and Foundations, Vol.24, No.1, pp.34-42 (1984)

Numerical Methods in Geomechanics (Innsbruck 1988), Swoboda (ed.)
© 1988 Balkema, Rotterdam. ISBN 90 6191 809 X

The effect of anisotropy on consolidation in a soil layer

M.F.Randolph
University of Western Australia
J.R.Booker
University of Sydney, Australia

ABSTRACT: In routine design, it is customary to use simplified one-dimensional consolidation theory to estimate the rate of settlement of buildings or other structures. The actual drainage conditions will differ significantly from the one-dimensional model, and the pattern of excess pore pressures and effective stress changes will be three-dimensional. Anisotropic soil conditions will also affect the consolidation response. In this paper, an analytical solution for consolidation of a cross-anisotropic soil layer is outlined. The solution is based on combined Laplace and Fourier transforms of the governing differential equations. Results of the analysis are presented showing the progress of consolidation for different loading conditions, shape of loaded area, and degree of anisotropy of the soil. The results are compared with those obtained assuming one-dimensional consolidation and significant differences between the two approaches discussed.

1. INTRODUCTION

Analysis of the consolidation of soil under the action of surface loading is an important issue. Engineers need to be able to estimate the time dependent settlement of structures and the rate of strengthening of the soil due to increasing effective stress levels.

For onshore foundations, where the primary component of load is the dead weight of the structure, stability is generally most critical in the short term. Provided the initial strength of the soil is sufficient to carry the foundation loads, then the long term stability will be assured. However, the critical loading condition for offshore foundations is generally environmental loading. The rate of strengthening of the soil is critical in the design process, since, from a statistical viewpoint, the magnitude of the design storm increases with the length of time the structure is in service. Assuming that the structure is positioned during a season of quite weather, accurate estimates of the increase in effective stress level in the soil over the ensuing 6 months may be critical in the design process. If necessary, it may be required to accelerate the rate of effective stress increase, by applying underdrainage to the foundation.

In exploring the consolidation process, it has been customary to use either simple one-dimensional theory, as developed by Terzaghi (1923), or finite element analysis. The latter approach, although versatile and able to allow for plastic straining of the soil (Small et al (1976)) is computationally laborious. The introduction of Fourier transform techniques in the horizontal plane has enabled the number of spatial variables to be reduced to one, with a resultant saving in computer storage and time for each analysis (Booker and Small (1985)). More recently, combined Laplace and Fourier transforms have been used to provide a complete analytical solution in transformed variables, the final solution being obtained by numerical inversion techniques. These solutions have generally been confined to soil with isotropic elastic properties (Vardoulakis and Harnpattanapanich (1986), Booker and Small (1987)). However, a solution for crossanisotropic soil properties has been developed by Booker and Randolph (1984). This solution has been used to provide the results presented later in this paper. The basis of the solution will be outlined first.

2 ANALYTICAL SOLUTION

The solution for elastic consolidation of a crossanisotropic soil follows that developed by Biot (1956). The solution rests on the four basic equations of equilibrium, volume constraint, effective stress-strain relationships and Darcy's flow law. Assuming a Cartesian set of axes, x, y, z, with corresponding stress increments $\underline{\sigma} = (\sigma_{xx}, \sigma_{yy}, \sigma_{zz}, \sigma_{zy}, \sigma_{xz}, \sigma_{xy})^T$ and and a displacement vector $\underline{\delta} = (\delta_x, \delta_y, \delta_z)^T$, these equations are:

(1) Equilibrium

$$\partial^T \underline{\sigma} = 0 \qquad (1)$$

$$\text{where } \partial^T = \begin{bmatrix} \frac{\partial}{\partial x} & 0 & 0 & 0 & \frac{\partial}{\partial z} & \frac{\partial}{\partial y} \\ 0 & \frac{\partial}{\partial y} & 0 & \frac{\partial}{\partial z} & 0 & \frac{\partial}{\partial x} \\ 0 & 0 & \frac{\partial}{\partial z} & \frac{\partial}{\partial y} & \frac{\partial}{\partial x} & 0 \end{bmatrix}$$

(2) Volume constraint

$$\underline{\nabla}.\underline{v} = -\frac{\partial \epsilon_v}{\partial t} \qquad (2)$$

where $\underline{v}$ is the velocity vector and ϵ_v is the volumetric strain given by

$$\epsilon_v = \underline{a}^T \underline{\epsilon} \qquad (3)$$

with $\underline{a} = (1,1,1,0,0,0)^T$. The strain vector $\underline{\epsilon}$ is related to the displacements by

$$\underline{\epsilon} = \partial\, \underline{\delta} \qquad (4)$$

(3) Effective stress-strain relations

$$\underline{\sigma}' = \underline{\sigma} - \underline{a}u = M\underline{\epsilon} \qquad (5)$$

where u is the pore pressure and M is a matrix of elastic constants (e.g. Poulos and Davis (1974)).

(4) Darcy's Law

$$\underline{v} = -K\underline{\nabla}u \qquad (6)$$

where K is a matrix of permeability coefficients. Thus, for material with horizontal permeability k_h, vertical permeability k_v and unit weight of pore fluid γ_w,

$$K = \begin{bmatrix} k_h/\gamma_w & 0 & 0 \\ 0 & k_h/\gamma_w & 0 \\ 0 & 0 & k_v/\gamma_w \end{bmatrix} \qquad (7)$$

In order to arrive at a general solution, it is necessary to take repeated Fourier and Laplace transforms of the field quantities. Typically, we may write

$$(U, \underline{\Delta}, \underline{V}, \underline{S}, \underline{E}) = \frac{1}{4\pi^2}\int_{-\infty}^{\infty}\int_{-\infty}^{\infty}(u, \underline{\delta}, \underline{v}, \underline{\sigma}, \underline{\epsilon})\, e^{-i(\alpha x+\beta y)}\, dx\, dy \qquad (8)$$

and

$$(\bar{U}, \bar{\underline{\Delta}}, \bar{\underline{V}}, \bar{\underline{S}}, \bar{\underline{E}}) = \int_0^{\infty}(U, \underline{\Delta}, \underline{V}, \underline{S}, \underline{E})e^{-st}dt \qquad (9)$$

For convenience in the algebra that follows, the superscript bar is dropped.

The effect of these transformations is that the various partial derivatives may be replaced as shown below

$$\frac{\partial}{\partial x} \rightarrow i\alpha, \quad \frac{\partial}{\partial y} \rightarrow i\beta, \quad \frac{\partial}{\partial t} \rightarrow s$$

which reduces the relationships to a set of ordinary differential equations. As discussed by Booker and Randolph (1984), solution of the equations is greatly simplified by a coordinate transformation

$$H = \begin{bmatrix} \cos\epsilon & \sin\epsilon & 0 \\ -\sin\epsilon & \cos\epsilon & 0 \\ 0 & 0 & 1 \end{bmatrix} \qquad (10)$$

where $\cos\epsilon = \alpha/\rho$, $\sin\epsilon = \beta/\rho$ and $\rho = \sqrt{\alpha^2+\beta^2}$. Under the transformation, vector quantities become, typically

$$\underline{\Delta}^* = (\Delta_\xi, \Delta_\eta, \Delta_z) = H\underline{\Delta} \qquad (11)$$

while tensor quantities become

$$S^* = \begin{bmatrix} S_{\xi\xi} & S_{\xi\eta} & S_{\xi z} \\ S_{\eta\xi} & S_{\eta\eta} & S_{\eta z} \\ S_{z\xi} & S_{z\eta} & S_{zz} \end{bmatrix} = H \begin{bmatrix} S_{xx} & S_{xy} & S_{xz} \\ S_{yx} & S_{yy} & S_{yz} \\ S_{zx} & S_{zy} & S_{zz} \end{bmatrix} H^T \qquad (12)$$

The transformation leads to two separate sets of equations, one of which correspnds to a plane problem in the (ξ, z) plane, and the other of which corresponds to a shearing in the (η, z) plane (Booker and Randolph (1984)). The solution of the latter set of equations is independent of time (for constant surface loads). For the case of interest in this paper, where the applied loading is normal to the surface, the shearing solution is zero and so will not be considered further.

The equations for the plane problem lead to a set of three second order simultaneous differential equations in Δ_ξ, Δ_z and U :

$$\begin{bmatrix} fD^2-a\rho^2 & -\rho(c+f)D & -\rho \\ \rho(c+f)D & dD^2-f\rho^2 & D \\ s\rho & sD & (k_vD^2-k_h\rho^2)/\gamma_w \end{bmatrix} = \begin{bmatrix} i\Delta_\xi \\ \Delta_z \\ U \end{bmatrix} \quad (13)$$

where $D = d/dz$, $a = N_eE^*(1-\nu_{vh}\nu_{hv})/(1+\nu_{hh})$, $c = N_eE^*\nu_{vh}$, $d = E^*(1-\nu_{hh})$ and $f = G_{vh}$, with $N_e = E_h/E_v$ and $E^* = E_v/(1-\nu_{hh}-2\nu_{vh}\nu_{hv})$. The solution of these equations may be assumed in the form

$$i\Delta_\xi = Ae^{\phi z} \quad (14)$$

$$\Delta_z = Be^{\phi z} \quad (15)$$

$$U = Ce^{\phi z} \quad (16)$$

where ϕ^2 satisfies the cubic obtained from the determinant (replacing the operator D by ϕ). The functions A, B and C may then be obtained from any two of the simultaneous equations in (13); for example, the second and third equations lead to

$$A(\phi) = (d\phi^2-f\rho^2)(k_v\phi^2-k_h\rho^2)/\gamma_w - s\phi^2 \quad (17)$$

$$B(\phi) = -\rho\phi[(c+f)(k_v\phi^2-k_h\rho^2)/\gamma_w - s] \quad (18)$$

$$C(\phi) = \rho s[(c+f)\phi^2 - (d\phi^2-f\rho^2)] \quad (19)$$

It may be shown that, for the case of repeated roots $\phi_1 = \phi_2$, the solution for ϕ_2 may be replaced by, typically

$$A(\phi_2) \rightarrow zA(\phi_1) + A'(\phi_1) \quad (20)$$

where the prime denotes differentiation with respect to ϕ. For the case of isotropic elasticity, the elastic parameters $a = d$ and $f = (a-c)/2$; the roots of the cubic are then $\phi^2 = \rho^2$ (repeated) and $\phi^2 = \rho^2(k_h/k_v) + s/c_v$, with c_v being the coefficient of consolidation for vertical drainage, given by $c_v = dk_v/\gamma_w$.

Expressions for the stresses may be obtained from the displacements and pore pressure in the normal way. The full solution may then be expressed typically as

$$\Delta_z = \sum_{i=1}^{6} X_i(\Delta_z)_i \quad (21)$$

The constants X_i may be found from the boundary conditions at the top and bottom of the soil layer. The solution may form the basis of a transfer matrix for soil layers, in the way described by Booker and Small (1987) for isotropic soil. In the present paper, attention is restricted to a single soil layer, where the boundary conditions at the base of the layer are zero displacement and either zero pore pressure of zero flow.

For surface loading of the form shown in Fig. 1, the boundary conditions are

$$S_{\xi z} = 0 \quad (22)$$

$$S_{zz} = \frac{q}{4\pi^2}\int_{-\infty}^{\infty}\int_{-\infty}^{\infty} e^{-i(\alpha x+\beta y)}dxdy \quad (23)$$

$$U = \frac{u}{4\pi^2}\int_{-\infty}^{\infty}\int_{-\infty}^{\infty} e^{-i(\alpha x+\beta y)}dxdy \quad (24)$$

where q is the applied surcharge loading and u is the applied excess pore pressure (generally zero, or negative where underdrainage is applied). For a circular area, the integrals reduce to a Hankel transform, so that typically

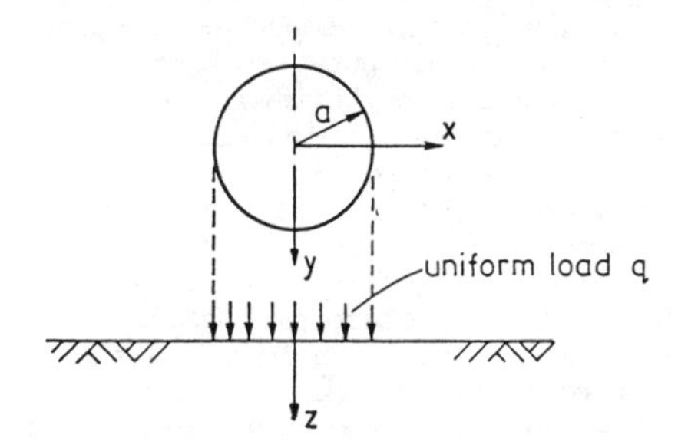

(a) Circular area

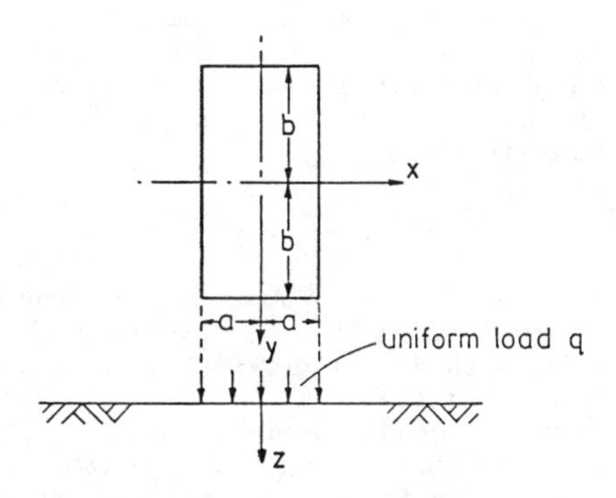

(b)Rectangular area

Figure 1 Uniform vertical loading of circular and rectangular regions

$$U = \frac{u}{2\pi}\int_0^a rJ_0(\rho r)dr = \frac{u}{2\pi}\frac{a}{\rho} J_1(\rho a) \quad (25)$$

while for a rectangular loaded area, the equivalent expression is

$$U = \frac{u}{\pi^2}\frac{\sin(\alpha a)}{\alpha}\frac{\sin(\beta b)}{\beta} \quad (26)$$

It is also possible to approximate the response of under a rigid circular loaded area by adopting a surcharge distribution of the form $0.5q/\sqrt{a^2-r^2}$ for $0 < r < a$, where q is the average applied pressure. This leads to an expression for the transformed vertical stress of

$$S_{zz} = \frac{q}{4\pi}\frac{\sin(\rho a)}{\rho} \quad (27)$$

2.1 Immediate and long term solutions

In implementing the solution (see later) it is helpful to evaluate separately the immediate and long term solutions. These solutions provide a check on the Laplace transform inversion. The long term solution also allows improved accuracy and efficiency of computation when considering times where consolidation is nearly complete. The immediate solution may conveniently be obtained from the solution above by allowing the Laplace transform variable to tend to infinity. Dividing through the last line of equation (13) by s, it may be seen that the determinant leads to a quadratic in ϕ^2, and considerable simplification of the functions A, B and C.

The long term solution may be obtained in a similar fashion by allowing s to tend to zero. The set of three simultaneous differential equations decouple into an equation for the excess pore pressure and a pair of equations for the displacements. The solution for the excess pore pressure takes the form

$$U = Ce^{\lambda\rho z} \quad (28)$$

where $\lambda^2 = k_h/k_v$. Again, a quadratic is obtained for ϕ^2, and equations (17) and (18) (with s = 0) provide the complementary functions for the displacements. Where the loaded area has non-zero excess pore pressure (for example, where underdrainage is to be analysed), the expressions for the displacements must include particular integrals. Some care is needed in evaluating these for cases of repeated roots (that is, where $\lambda\rho = \phi$).

3. IMPLEMENTATION OF SOLUTION

The solution outlined above has been implemented in the manner described by Booker and Randolph (1984), but with various improvements in accuracy and efficiency. For the case of anisotropic elasticity, the cubic in ϕ^2 is solved by Cardan's method, except for the case of small values of time. At small times, numerical inversion of the Laplace transform involves large values of s. For small values of ρ, and large s, the coefficients of the cubic may differ by 6 orders of magnitude or more. Even using (complex) double precision, round-off errors become excessive. It then becomes necessary to use an iterative (Newton-Raphson) approach to solve the cubic.

Inversion of the Laplace transform is achieved using the algorithm presented by Talbot (1979). The results appear remarkably stable and insensitive to the number of values of the parameter s that are used in the inversion. Generally, 10 values were found to be sufficient.

The Cartesian displacements and stress changes were obtained by applying the inverse co-ordinate transformation (see equation (12)), before inverting the Fourier transform. The latter was then evaluated using the polar co-ordinate form

$$(u, \underline{\delta}, \underline{v}, \underline{\sigma}, \underline{\epsilon}) = \quad (29)$$

$$\int_{-\infty}^{\infty}\int_{-\theta}^{2\pi-\theta} (U, \underline{\Delta}, \underline{V}, \underline{S}, \underline{E})\, e^{i\rho\cos(\theta-\epsilon)}\, d\epsilon\, d\rho$$

For axisymmetric loading, the integration in ϵ may be replaced by $2\pi J_0(\rho r)$, and the double Fourier inversion reduces to a single Hankel inversion. For rectangular loaded areas, the exponential term above was expanded as cosine and sine functions of αx and βy. Appropriate terms were then retained according to the behaviour of the rest of the integrand. Thus, since the loading term is even in α and β, it is necessary to retain remaining terms in the integrand that are also even in α and β. For example, in evaluating $\Delta_x = \cos\epsilon\, \Delta_\xi = (\alpha/\rho)\Delta_\xi$, with Δ_ξ being a function of ρ alone, it is necessary to retain the part of the exponential which is odd in α but even in β.

The numerical inversions were obtained by 20 point Gaussian quadrature, with a single interval for ϵ over the range 0 to $\pi/2$, and several intervals for ρ. The latter integration was truncated at a value $\rho = R$ that depended on the value of time, t. At small times (non-dimensionalised times $T = c_v t/a^2$ of about 0.01), R was set typically at 100a, while at large times, R was reduced proportionally with the square root of time. The overall interval was then divided into 1 to 10 subintervals for the 20 point Gaussian quadrature.

The use of a gradually decreasing integration domain, which proved very efficient, was introduced in parallel with a technique whereby the Laplace and Fourier inversions were effectively performed on the difference between the actual solution and the long term solution. Thus, for any given value of the Fourier parameter ρ, the long term solution (in transformed space) was evaluated and subtracted from the solution obtained in Laplace transform space. The numerical inversions were then performed before adding back in the long term solution. (The latter was calculated accurately, prior to evaluating the time dependent solution.) With increasing values of time, differences between actual and long term solution were confined to progressively smaller values of ρ, and it was therefore sufficient to truncate the integration sooner.

Computation times depended on the number of values of time and depth coordinate for which the solution was required. Typical solution times on an 80286 microcomputer varied from a minute or two up to about 20 minutes for a complete solution evaluated at a 10 x 10 grid of interior field points, at 10 different time values.

4. RESULTS OF ANALYSIS

As an illustration of the effect of different degrees of anisotropy, an example problem will be considered where uniform load is applied over a circular or rectangular area as shown in Figure 1. The depth of the soil layer is taken as equal to the width of the loaded area (that is, h = 2a), with the base boundary conditions being rough, rigid and impermeable. The aspect ratio of the rectangular area is taken as 5:1. Two different loading conditions have been considered, firstly a uniform surcharge applied over the loaded area, and secondly a uniform 'suction' (corresponding to underdrainage of a foundation). In both cases, the upper soil surface has been assumed fully permeable.

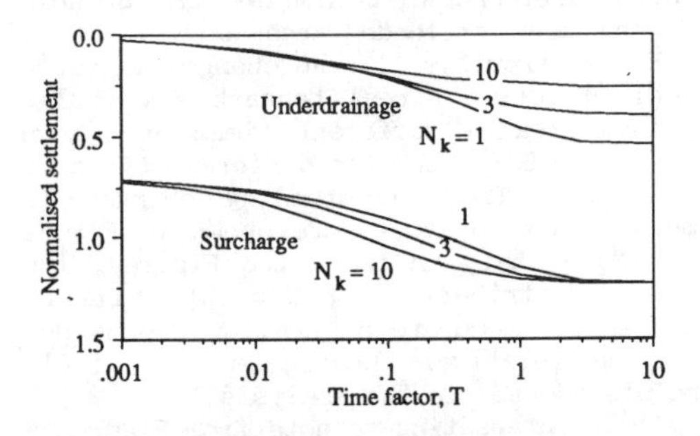

Figure 2 Settlement reponse for uniform circular region (isotropic elasticity)

4.1 Settlement Response

For the case of a circular loaded area on soil with isotropic elastic properties, the settlement response with time is as shown in Figure 2 (Poisson's ratio has been taken as 0.25). Three different ratios of horizontal to vertical permeability have been considered, with $N_k = k_h/k_v = 1$, 3 and 10. In each of the plots presented, the time factor T has been taken as

$$T = \frac{dk_v t}{\gamma_w a^2} = \frac{c_v t}{a^2} \qquad (30)$$

where d is the elastic parameter given previously (equivalent to the vertical one-dimensional modulus). In Figure 2, the normalised settlement is $\delta_z d/aq$ where a is the radius of the loaded area and q is the magnitude of the applied load (either surcharge or underdrainage).

As expected, there is an initial settlement under the surcharge loading of 2/3 the final settlement (since $\nu = 0.25$), while there is no immediate settlement due to the underdrainage. Increasing the permeability ratio leads to more rapid consolidation for both types of loading. However, in the case of underdrainage, the final settlement reduces significantly as the permeability ratio increases.

It may be shown that, for an isotropic elastic half space, the consolidation response of the soil (that is, changes in effective stress and settlement with time following any immediate change) is identical for either surcharge or underdrainage loading. A corrollary of this result is that no consolidation would occur if the loading took the form of equal total stress and pore pressure changes (for example, wave loading).

For a layer of finite depth, the equivalence is no longer strictly true. However, the consolidation response is still very similar under the two forms of loading, as may be seen from Figure 3, where the degree of consolidation, (given by $U = \delta_z(t)/\delta_z(\infty)$), is plotted against $\sqrt{T}$. The

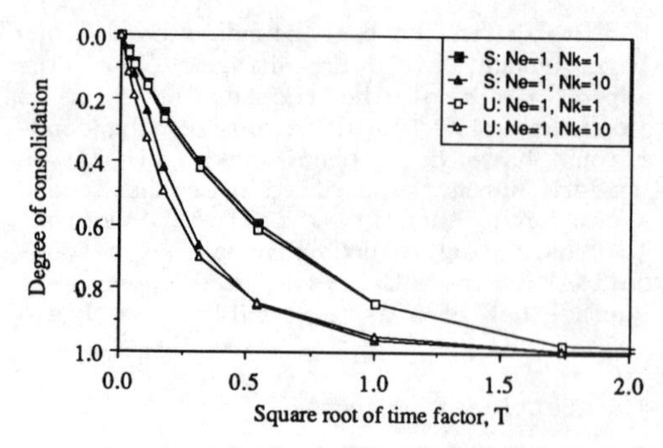

Figure 3 Comparison of consolidation for surcharge (S) and underdrainage (U)

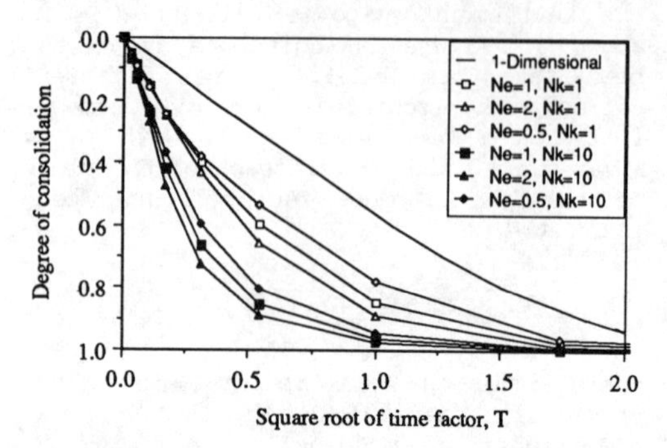

Figure 4 Effect of anisotropy on consolidation: circular load, surcharge loading

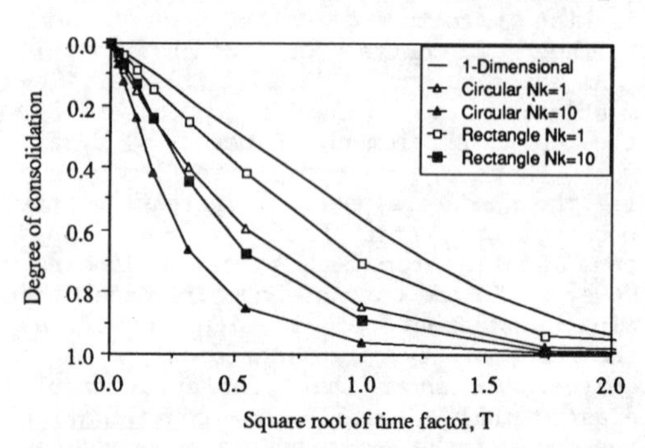

Figure 5 Effect of shape of loaded area on consolidation (isotropic elasticity)

curves for the different types of loading are almost identical, even where the permeability ratio is increased from 1 to 10.

The effect of anisotropic elastic properties is shown in Figure 4. The ratio of horizontal to vertical modulus is given by $N_e = E_h/E_v$. Values of Poisson's ratio have been adopted such that the one-dimensional moduli are in the same ratio as the Young's moduli. Thus ν_{hh} has been kept at 0.25, while the product $\nu_{vh}\nu_{hv}$ has been kept equal to $0.25^2 = 0.0625$. As might be expected, for a given permeability ratio, the effect of increasing the modulus ratio is to accelerate the consolidation and vice versa.

For comparison, the consolidation curve for one-dimensional consolidation (assuming an initial rectangular distribution of excess pore pressure) is also shown in Figure 4. Clearly, the one-dimensional assumption grossly underpredicts the rate of settlement. The underprediction arises partly due to ignoring horizontal drainage, and partly due to the assumption of an initial rectangular distribution of excess pore pressure.

The effect of the shape of the loaded area is shown in Figure 5, where the rate of consolidation for a cirular area is compared with that for a rectangular area of aspect ratio 5:1. Consolidation of the rectangular area is slower than for the circular area, due both to a reduced effect of horizontal drainage and to a more uniform initial distribution of excess pore pressure. These effects also lead to improved agreement between the one-dimensional solution and the exact solution for the rectangular loaded area (isotropic permeability).

4.2 Effective Stress Changes

The effective stress level in the soil beneath a foundation will increase as consolidation progresses, giving improved stability of the structure. This aspect may be of critical importance in the design of offshore gravity base structures. The effective stress regime beneath a foundation will vary in a three-dimensional (or axisymmetric) manner. However, as a guide to the effective stress levels, results are presented below for a position at a distance of a/2 off the centre line of the loaded area. Attention is restricted to circular loaded areas, although it has been found that similar trends hold for rectangular loaded areas as well.

For isotropic soil, the changes in vertical effective stress beneath a circular loaded area at different times are shown in Figure 6 for the two different types of loading. The corresponding changes in mean effective stress are shown in Figure 7. (Note that, in these two Figures, the stress distribution for T = 0.01 will be somewhat inaccurate at shallow depths due to the relatively coarse grid of field points adopted in the analysis.)

It is interesting to note from Figures 6 and 7 that the profiles of vertical effective stress changes are very different for

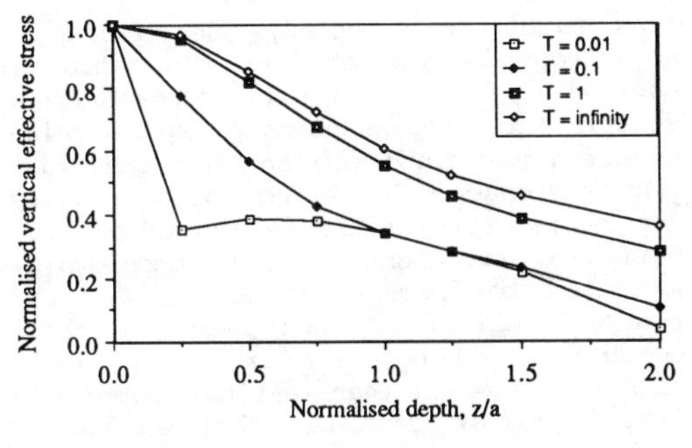

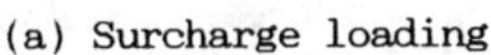
(a) Surcharge loading

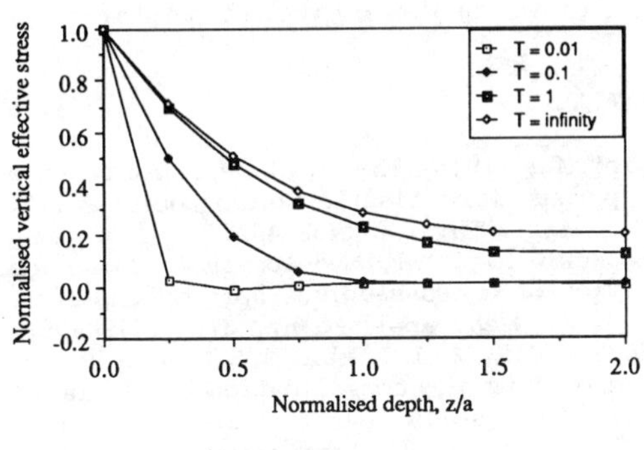

(b) Underdrainage

Figure 6 Development of vertical effective stress beneath circular loaded area

the two types of loading (surcharge or underdrainage), while the profiles of mean effective stress changes are similar.

The effect of anisotropy may be seen by comparing the distribution of effective stress change at a particular time. Figure 8 shows such a comparison for the case of changes in vertical effective stress at T = 0.1. It may be seen that the effective stress changes due to underdrainage are significantly less than those due to surcharge loading. This factor may be important when considering the improvement in short term stability of an offshore platform due to underdrainage.

It is clear from Figure 8 that the effect of anisotropic permeability is rather more pronounced than the effect of anisotropic elasticity. Indeed, for underdrainage, elastic anisotropy appears to have negligible effect.

The vertical effective stress change predicted from the one-dimensional solution is also shown on Figure 8. Note that the same stress change is predicted for both surcharge and underdrainage. Since the one-dimensional solution gives zero initial change in effective stress, the

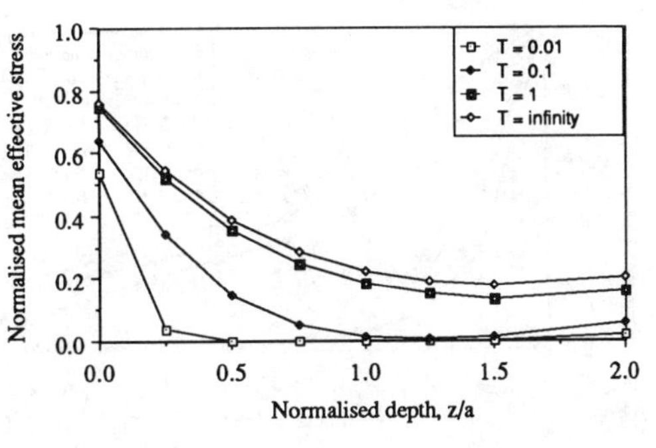

(a) Surcharge loading

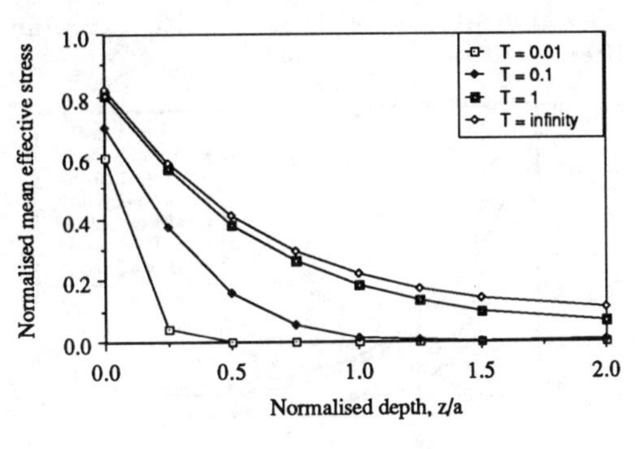

(b) Underdrainage

Figure 7 Devlopment of mean effective stress beneath circular loaded area

solution agrees rather better with the exact solution for underdrainage than with that for surcharge loading.

Corresponding profiles of mean effective stress change are shown in Figure 9 (again, for a time of T = 0.1). Here, the effect of the type of loading is much less pronounced, and the curves all lie reasonably close together. The largest increase in mean effective stress is obtained under surcharge loading, with permeability ratio N_k = 10, while the smallest increase is for underdrainage, again for N_k = 10.

5. CONCLUSIONS

This paper has outlined an analytical solution for the consolidation of a cross-anisotropic elastic soil layer. The solution uses Laplace and Fourier transforms in order to reduce the governing partial differential equations to a tractable differential equation involving the depth, z, alone. Computationally efficient and accurate numerical inversion schemes are then used to evaluate the solution at the required field points and times.

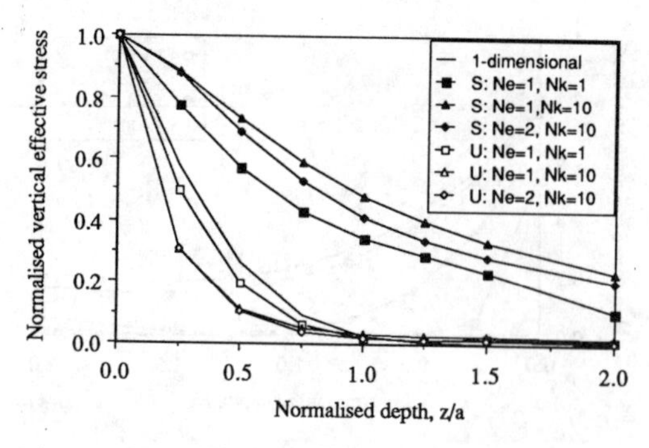

Figure 8 Effect of anisotropy on vertical effective stress profile beneath circular load (x/a = 0.5, T = 0.1)

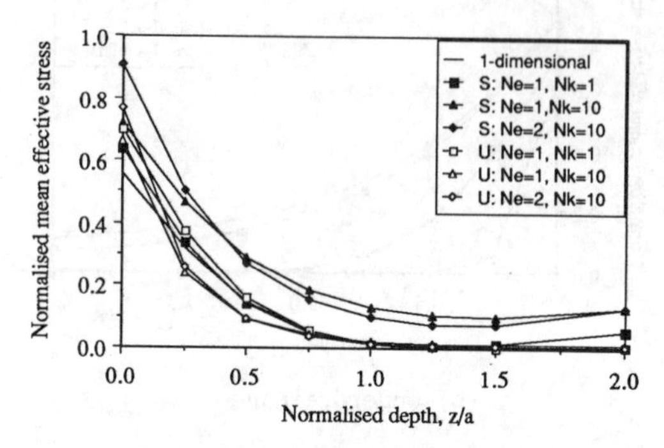

Figure 9 Effect of anisotropy on mean effective stress profile beneath circular load (x/a = 0.5, T = 0.1)

The solution has been used to investigate the consolidation response of circular and rectangular loaded areas resting on a soil layer of depth equal to the width of the loaded area. For such a geometry, it would not be uncommon to adopt a simple one-dimensional consolidation solution in engineering calculations. However, the results show that such an approach could lead to significant errors, overpredicting times for a given degree of consolidation settlement, and underpredicting changes in vertical effective stress at a given time after application of the load.

The relative effects of surcharge loading and underdrainage have been explored, and it has been shown that changes in effective stress and settlement <u>during consolidation</u> are similar for the two types of loading. However, the immediate settlement and changes in vertical effective stress that occur for surcharge loading give rise to different total values of settlement and effective stress change.

As might be expected, the effect of anisotropy is dominated by the ratio of horizontal to vertical permeability. Elastic anisotropy appears to have a relatively minor influence on the consolidation response. The effect of increasing the permeability ratio is to reduce effective stress changes due to underdrainage, but to increase those due to surcharge loading. Since most naturally occurring soil deposits exhibit markedly higher horizontal than vertical permeability, it may be concluded that the use of underdrainage to increase effective stress levels beneath a foundation is rather less effective than an equivalent surcharge.

REFERENCES

Biot M.A. 1956. Theory of deformation of a porous viscoelastic anisotropic solid. J. Appl. Phys. 27:459-467.

Booker J.R. & M.F.Randolph 1984. Consolida tion of a cross-anisotropic soil medium. Q. Jl Mech. appl. Math., 37(3):479-495.

Booker J.R. & J.C.Small 1987. A method of computing the consolidation behaviour of layered soils using direct numerical inversion of Laplace transforms. Int. J. Num. & Anal. Meth. in Geomechanics, 11(4):363-380.

Booker J.R. & J.C.Small 1985. Finite layer analysis of settlement, creep and consolidation using micro-computers. Proc. 5th Int. Conf. on Num. Meth. in Geomechanics, Nagoya, 1:3-18.

Poulos H.G. & E.H.Davis. 1974. Elastic solutions for soil and rock mechanics. Wiley.

Small J.C., J.P.Carter & E.H.Davis 1976. Elasto-plastic consolidation of soil. Int. J. Solids & Struct., 12:431-448.

Talbot A. 1979. The accurate numerical inversion of Laplace transforms. J. Inst. Math. Applic., 23:97-120.

Terzaghi 1923. Die Berechnung der Durch lassigkeitziffer des Tones aus dem Verlauf der hydrodynamischen Spannungsersheinungen. Original paper published in 1923, reprinted in From Theory to Practice in Soil Mechanics, Wiley, 133-146

Vardoulakis I. & T.Harnpattanapanich 1986. Numerical Laplace-Fourier transform inversion technique for layered soil consolidation problems: I. Fundamental solutions and validation. Int. J. Num. & Anal. Meth. in Geomech., 10(4):347365.

Numerical Methods in Geomechanics (Innsbruck 1988), Swoboda (ed.)
© 1988 Balkema, Rotterdam. ISBN 90 6191 809 X

Flow analysis of clay layer due to berth construction

T.Adachi
Kyoto University, Japan
F.Oka
Gifu University, Japan
M.Mimura
Disaster Prevention Research Institute of Kyoto University, Uji, Japan

ABSTRACT: Flow and consolidation of the very soft clay deposit due to berth construction is analyzed by elasto-viscoplastic finite element method. The monitored performance of deformation of the clay foundation is presented and compared with the calculated results and we also discuss the predictive ability of the overstress type elasto-viscoplastic constitutive model.

1 INTRODUCTION

With an increase in construction of offshore structures, issues dealing with the behavior of very soft clay ground due to local loading (i.e., berth, quay-wall, wharf, breakwater and so forth) have been raised.

A berth was constructed on a thick alluvial clay foundation here in Japan. According to the geological data, the bearing stratum (basement rock) appeared at the depth of -80m, and due to unavoidable circumstances, rapid construction was deemed necessary. The most serious problem encountered in this hasty berth construction was how to control the lateral movement of the ground. Sand drains were driven at this site, prior to filling, in order to suppress the lateral movement of the ground by promoting consolidation as soon as possible. Field measurements were started just after completion of the filling on the sea bed.

As is known, soft clay clearly shows rate-sensitive behavior, such as the strain rate dependency of strength, secondary compression and creep rupture. Therefore, in order to solve the boundary value problems and make accurate predictions of the settlement and lateral deformation of soft clay foundations, it is necessary to apply the finite element method coupled with elasto-viscoplastic constitutive model.

In this paper, deformation of the soft alluvial clay ground due to construction of the berth is analyzed by the elasto-viscoplastic finite element method under a plane-strain condition. We also discuss the predictive ability of the overstress type elasto-viscoplastic constitutive model by comparing the calculated results with the field measurement data.

2 ELASTO-VISCOPLASTIC CONSTITUTIVE MODEL

Adachi and Oka (1982) developed an elasto-viscoplastic constitutive model to describe the rate-dependent behavior of normally consolidated clay, by combining the Cam-clay model and Perzyna's overstress type viscoplasticity theory. The proposed theory has been experimentally verified. In this paper, an overstress elasto-viscoplastic constitutive model is used to analyze the deformation of the clay foundation during construction of the container berth. In this section, we will explain the theoretical structure of the overstress model proposed by Adachi and Oka (1982).

2.1 Overstress model

The viscoplastic flow rule for an overstress type elasto-viscoplastic constitutive model is generally expressed as:

$$\dot{\varepsilon}^p_{ij} = \gamma \langle \Phi(F) \rangle \frac{\partial f}{\partial \sigma_{ij}} \quad (1)$$

$$F = f(\sigma_{ij}, \varepsilon^p_{ij})/k_s - 1$$

where, γ is the coefficient of viscosity, k_s is the static hardening parameter, f is the static yield function,and Φ is the functional of overstress. The stress strain relation for Adachi and Oka's model, a typical overstress model, is expressed as follows:

$$\dot{\varepsilon}_{ij}=\dot{\varepsilon}^{e}_{ij}+\dot{\varepsilon}^{p}_{ij}=\frac{1}{2G}\dot{S}_{ij}+\frac{\kappa}{3(1+e_0)}\cdot\frac{\dot{\sigma}'_m}{\sigma'_m}\delta_{ij}+$$

$$\frac{1}{M^*\sigma'_m}\cdot\Phi(F)\frac{S_{ij}}{\sqrt{2J_2}}+\frac{1}{3'1^*\sigma'_m}\cdot\Phi(F)\ M^*-\frac{\sqrt{2J_2}}{\sigma'_m}\ \delta_{ij}\quad(2)$$

$$\Phi(F)=C\cdot M^*\cdot\sigma'_m\cdot\exp m'\{\frac{\sqrt{2J_2}}{M^*\sigma'_m}+\ln\frac{\sigma'_m}{\sigma'_{me}}-\frac{1+e_0}{\lambda-\kappa}v^p\}$$

where, $\sqrt{2J_2}$ is the second invariant of deviatoric stress tensor S_{ij}, δ_{ij} is Kronecker's delta, σ'_m is the mean effective stress, σ'_{me} is the mean effective stress at the end of consolidation, G is the elastic shear modulus, and M^* is the effective stress ratio in the critical state.

The superscripts, e and p, denote the elastic and viscoplastic components of the strain rate, respectively. There are six parameters for this model, namely, the compression and swelling(recompression) indices λ and κ, the critical stress ratio M^*, the elastic shear modulus G, the parameter that estimates the secondary compression m', and the viscoplastic parameter C. Parameters λ, κ, M^* and G can be determined by empirical methods, such as consolidation and swelling tests and strain rate-controlled undrained compression tests. The two viscoplastic parameters, m' and C, can be derived from triaxial compression tests under at least two different constant rates of strain.

2.2 Modified model

Several researchers have pointed out that the overstress elasto-viscoplastic constitutive model cannot describe the acceleration creep and creep rupture because of its theoretical structure. We have, however, already proved these phenomena mathematically (Adachi et al. 1987 a), and have modified the overstress constitutive model proposed by Adachi and Oka to describe acceleration creep and undrained creep rupture (Adachi et al. 1987 b). The modified model was experimentally verified for Osaka clay (Adachi et al. 1987, b). In this section, we briefly show the modified model, and the calculated results of performances for undrained creep using this modified model.

Let us consider the viscoplastic parameters m' and C for this model. Parameter m' can be expressed as the function of the secondary compression index, α, as follows:

$$m'=\frac{\lambda-\kappa}{1+e_0}\cdot\frac{1}{\alpha}\quad(3)$$

This parameter m' is a defined constant. The viscoplastic parameter, C, was also assumed to be constant for undrained shearing, under the condition of a constant rate of strain (Adachi and Oka, 1982). This is not necessarily true, however, under a constant sustained load. We will discuss the variation in this viscoplastic parameter C during the undrained creep process, by comparing its values with those of the experimental results.

The strain rate varies under constant creep stress during the undrained creep process. A decrease in the mean effective stress caused by excess pore water pressure generation contributes decisively to the undrained creep rupture of normally consolidated clay. Therefore we examined the relation between the strain rate, $\dot{\varepsilon}_{11}$, and the viscoplastic parameter, C, by estimating the effective stress state, (M-q/p').

A ln C - (M-q/p') relation for some creep stress levels is shown in Fig. 1. These undrained creep tests for Umeda clay were carried out by Murayama et al.

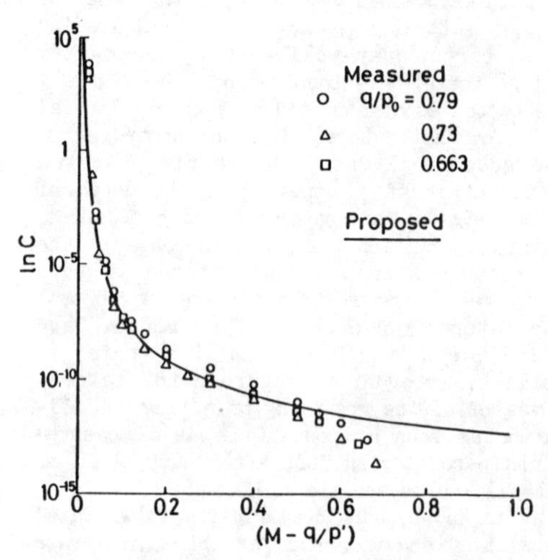

Fig. 1. Variation in viscoplastic parameter C, with the effective stress state (M-q/p').

The viscoplastic parameter C changes drastically with the stress state, (M-q/p'), near the critical state. The proposed curve shown in Fig.1 is determined from the experimental data using the least squares method. The relation between the viscoplastic parameter, C and the stress state, (M-q/p'), can be expressed as follows:

$$C = \exp\frac{\delta}{(M-q/p')} - \xi \qquad (4)$$

in which δ and ξ are material constants determined experimentally.

Calculations for triaxial undrained creep were made by the modified overstress elasto-viscoplastic constitutive model. The material constants used are those for Umeda clay reported by Sekiguchi (1984). Two more material constants in Eq.(3), namely, δ and ξ, which are required to describe the proposed relation between the viscoplastic parameter, C, and the stress state, (M-q/p'), were determined as 0.3 and 13.0 respectively, from the experimental results shown in Fig. 1 using the least squares method.

The creep strain - elapsed time curves are shown in Fig.2. The calculated results clearly show that undrained creep rupture follows acceleration creep. The calculated creep strain rate-elapsed time curves are shown in Fig.3. This rate decreases in the primary stage, and after the minimum rate, diverges to infinite, independent of creep stress.

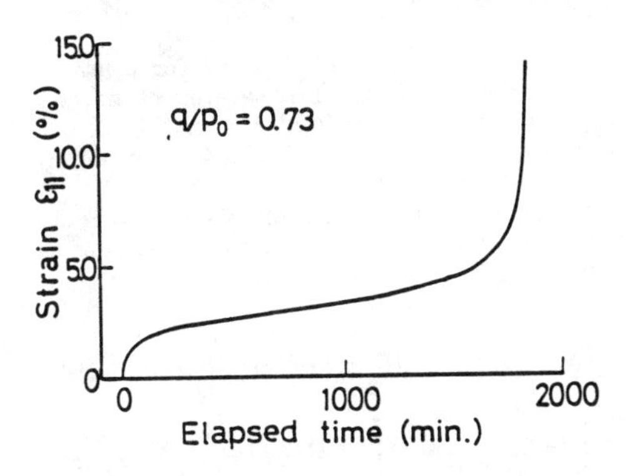

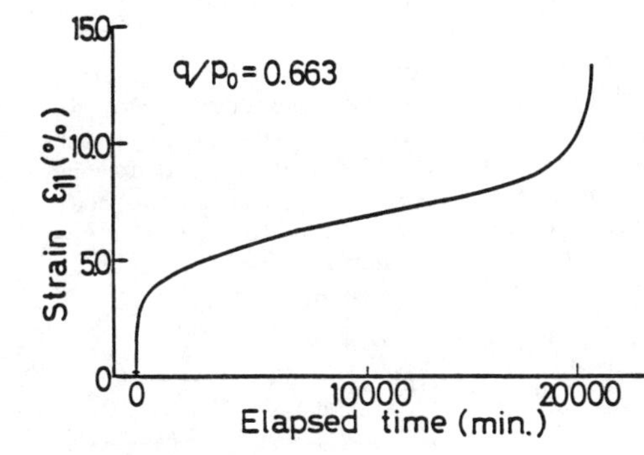

Fig. 2. Calculated results of creep strain-elapsed time relations for the modified model.

The modified constitutive model is based on the assumption that viscoplastic parameter C varies with the function of the stress ratio M-q/p', and thus, has been adapted in order to describe the undrained creep process, including acceleration creep and creep rupture.

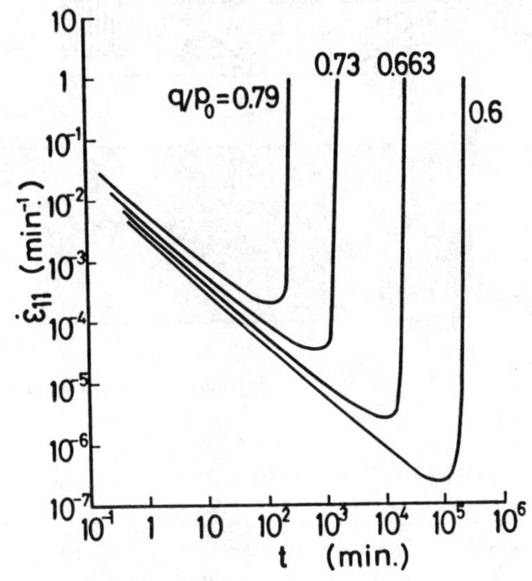

Fig. 3. Calculated results of log $\dot{\varepsilon}_{11}$ - log t relations for the modified model.

3 CONSTRUCTION OF CONTAINER BERTH

A berth for container ships was constructed in Kobe in 1978. The length of the quaywall is 1000 meters, and the berth has a capacity to moor three container ships of 35,000 dead weight tons at the same time.

3.1 Geotechnical condition of the site

A typical cross section of the berth and foundation is shown in Fig. 4. It is seen that the quaywall consists of a rubble mound and the caisson wall. The geotechnical profile is also shown in this figure. The alluvial clay layer is distributed with a thickness of 18 meters below the sea bed (at a depth of C.D.L. -12 m). The diluvial clay layer is distributed from the depth of C.D.L. -45 m to -80 m. According to the geological data, the diluvial clay is homogeneous and slightly overconsolidated (O.C.R.=1.5). Between the alluvial and diluvial layers, there is an alternation of strata, including the thin clay or silt layers with a thickness of 3 meters at most, and comparatively hard sand layers whose N-value for S.P.T. extends to 30. At a depth of C.D.L. -80 meters, the basement rock appears with an N-value of more than 50.

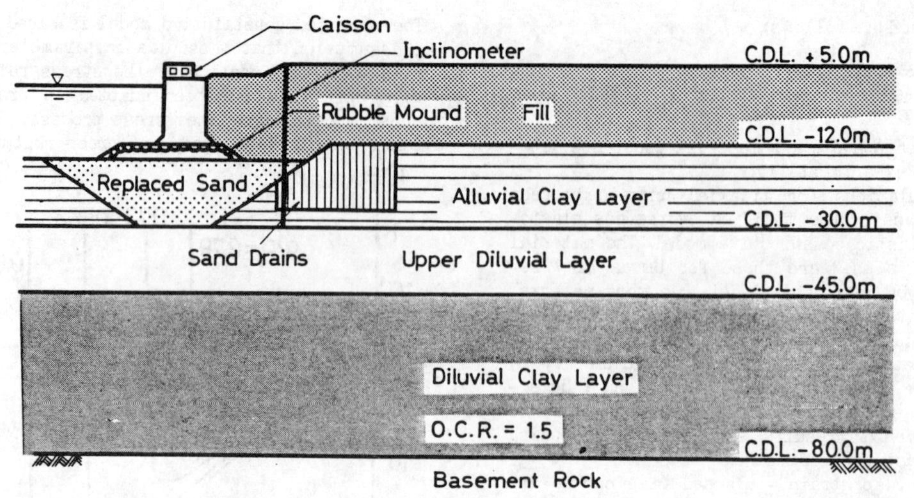

Fig. 4. Typical cross-section of the berth construction site.

3.2 Construction sequence

Removal of the on-site soft alluvial clay by dredging lasted from June to July of 1977. At the same time, the soft clay surrounding the site, where the container berth was proposed to be built, was replaced by sand, as shown in Fig.4. Construction of the replaced sand foundation for the quaywall caisson was also completed in July of 1977, as shown in Fig.4. After the sand replacing work, the formation of a rubble mound was finished in September of the same year, with a crest level of C.D.L. -12 meters. A stretch of the caisson wall was constructed by sinking prefabricated concrete caissons onto the rubble mound. Sand drains were driven to a depth of C.D.L. -25.5 meters with a rectangular arrangement. The diameter of each sand pile was 40 cm, and the pitch of them was 2.5 m. Fill construction at the back of the caisson was then started in December, 1977 and finished in September, 1978, with a crest level of C.D.L. +5.0 meters.

The berth for container ships was constructed in the above mentioned sequence and completed in September of 1978.

4. FINITE ELEMENT ANALYSIS AND MONITORED PERFORMANCE OF DEFORMATION

The behavior of clay deposits under plane strain conditions during the construction of the berth can be analyzed numerically using the finite element method coupled with the elasto-viscoplastic constitutive equations and Biot's consolidation theory. A more detailed numerical procedure will be explained in the following sub-section.

4.1 Numerical procedure

Many systems have been proposed for consolidation analysis by the finite element method. We have used a modification of Christian's method because it is simple, and the accuracy of calculation is relatively high.

Applying the principle of virtual work to the equilibrium equation gives the following relation for the load increment:

$$\int_v \{\Delta\varepsilon\}^T\{\Delta\sigma\}\,dv = \int_v \{\Delta\varepsilon\}^T\{\Delta\sigma'\}\,dv + \int_v \{\Delta\varepsilon\}^T\{\Delta u_w\}\,dv$$

$$= \int_v \{\Delta u\}^T\{\Delta F_b\}\,dv + \int_s \{\Delta u\}^T\{\Delta T\}\,ds \qquad (5)$$

in which $\{\Delta\varepsilon\}$ is the strain increment vector, $\{\Delta\sigma\}$ is the total stress increment, $\{\Delta\sigma'\}$ is the effective stress increment, $\{\Delta u_w\}$ is the pore water pressure increment, $\{\Delta u\}$ is the displacement increment, $\{\Delta F_b\}$ is the body force increment vector and $\{\Delta T\}$ is the applied surface load increment vector. $\{\Delta u\}$ in the element is expressed as

$$\{\Delta u\} = [N]\,\{\Delta u\} \qquad (6)$$

in which [N] is the matrix of the interpolation functions and $\{\Delta u\}$ denotes the vector of the displacement increment at the nodal point.

The strain increment vector, $\{\Delta\varepsilon\}$, and

volumetric strain increment in the element are given by

$$\{\Delta\varepsilon\}=[B]\{\Delta u\} \quad (7)$$

$$\Delta v=[C'^T]\{\Delta u\} \quad (8)$$

in which [B] is the strain nodal displacement matrix and [C'] is the volumetric strain nodal displacement matrix.

Equation (1) can be rewritten in a simple form as:

$$\dot{\varepsilon}_{ij}^{vp}=\beta_{ij}(\sigma_{kl},\varepsilon_{kl}^{vp}) \quad (9)$$

The viscoplastic strain rate is obtained by a fully implicit scheme (Owen and Damjanic, 1982) in the present paper. By applying a limited Taylor series expansion, the viscoplastic strain increment at the end of time interval ($\Delta t=t_{n+1}-t_n$) is obtained by

$$\{\Delta\varepsilon_n^{vp}\}=[S]\{\varepsilon_n^{vp}\}+[H_n]\{\Delta\sigma_n\}\Delta t \quad (10)$$

$$[S]=[[I]-\Delta t[J_n]]^{-1} \quad (11)$$

$$[J_n]=[\frac{\partial\beta}{\partial\varepsilon^{vp}}] \quad (12)$$

$$[H_n]=[\frac{\partial\beta}{\partial\sigma}] \quad (13)$$

([I] is a unit matrix)

From equation (10) and an elastic strain increment, the stress increment vector is obtained as:

$$\{\Delta\sigma_n\}=[D]\{\Delta\varepsilon_n\}-\{\Delta\sigma_n^r\} \quad (14)$$

$$\{\Delta\sigma_n^r\}=[D][S]\{\dot{\varepsilon}_n^{vp}\}\Delta t \quad (15)$$

$$[D]=([C^e]+[S][H_n]\Delta t) \quad (16)$$

where $[C^e]$ is an elastic coefficient matrix.

Substituting equations (6), (7), (8) and (14) in equation (5), the following relation is derived because the magnitude of $\{\Delta u\}$ is arbitrary:

$$[K]\{\Delta u\}+\{K_v\}\Delta U_w=\{\Delta F\} \quad (17)$$

$$K=\int_v[B]^T[C^e][B]\,dv \quad (18)$$

$$\{K_v\}=\int_v[C']\,dv \quad (19)$$

$$\{\Delta F\}=\int_v[N]^T\{\Delta F_b\}dv+\int_s[N]^T\{\Delta T\}ds+\int_v[B]^T\{\Delta\sigma^r\}dv \quad (20)$$

Next, one must formulate the equation of continuity. This is derived from the equation of the balance of mass and the equation of the motion of the fluid phase (Darcy's law).

$$\frac{d\varepsilon_{kk}}{dt}=-\frac{k}{\gamma_w}\frac{\partial^2u_w}{\partial x_i^2} \quad (21)$$

As the implicit backward finite difference scheme has been shown to be more stable numerically than the forward difference scheme, it is used in the approximation of equation (21). In equation (21), k is the coefficient of permeability, and γ_w is the density of the pore water.

The approximation of equation (21) is explained in detail in the reference(Oka et al., 1986). Combining equations (17) and (21), the generalized force-stiffness formulation for the element is

$$\begin{Bmatrix}[K]\{K_v\} & \{\Delta\bar{u}\} \\ \{K_v\}^T\beta & u_w(t+\Delta t)\end{Bmatrix}=\begin{Bmatrix}\{K_v\}u_w(t)+\{\Delta F\} \\ \sum_i\beta_iu_{wi}(t)\end{Bmatrix} \quad (22)$$

4.2 The geometry of the problem

The finite element mesh used in this problem is shown in Fig.5. The center line is located at the back of the caisson wall, and the distance from the center line to the lateral boundary, L, is taken as 188 meters. The thickness of the foundation is supposed to be 68 meters (from the sea bed C.D.L. -12m, and to the basement rock C.D.L. -80m).

The sand layer located in the upper diluvial layer at a depth of C.D.L. -30m, is considered to be the drainage layer. In addition, the basement rock and sea bed are assumed as the drainage boundary.

The loading due to back filling is conducted as shown in Fig.6. Although the actual construction sequence is rather complicated, as explained in the previous section, here we assume a constant loading rate for the filling processes. The total amount of load on the clay foundation will extends to 21 tonf/m^2.

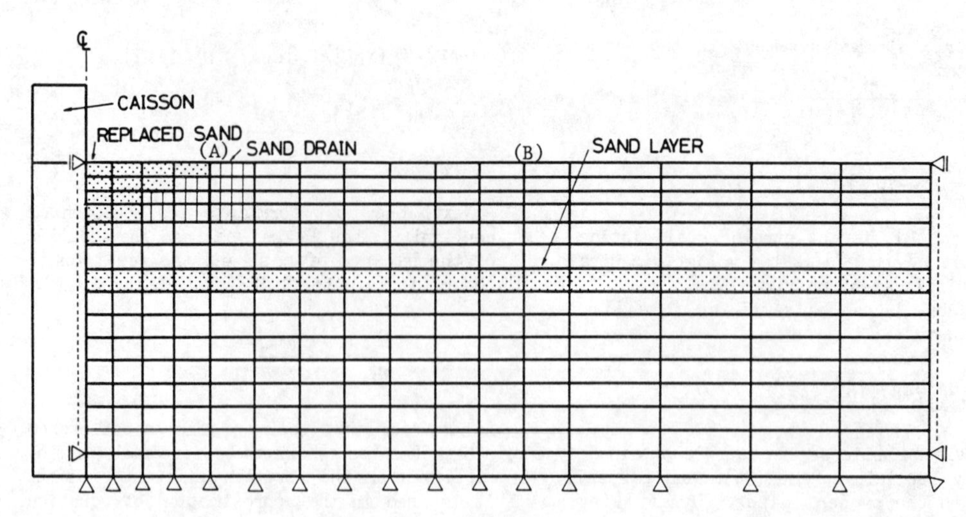

Fig. 5. Finite element mesh

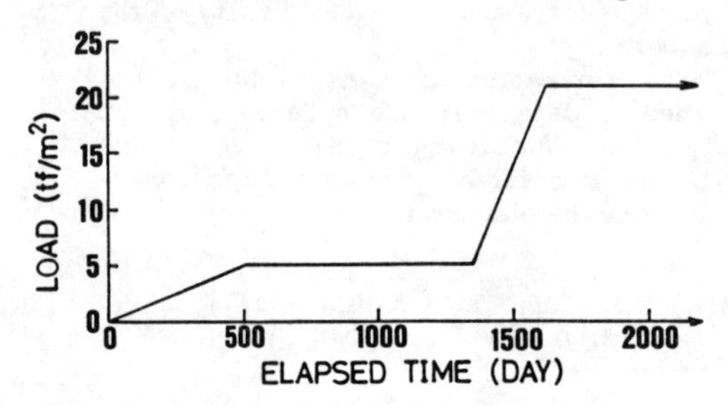

Fig. 6. Loading sequence

4.3 Material properties

The material parameters for alluvial clay and diluvial clay are listed in Table 1.
Material parameters such as compression index λ, swelling index κ, stress ratio at critical state M^*, viscoplastic parameters m' and C, coefficient of earth pressure at rest K_0, and λ_k, are assumed to be constant.

The initial overburden pressure σ'_{vo} is equal to $\gamma' z$. According to the data, while the alluvial clay layer can be regarded normally consolidated, the diluvial clay layer is lightly overconsolidated, with a value of O.C.R=1.5. Therefore, for the diluvial clay layer, the pre-consolidation stress σ'_{vc} is assumed to be $1.5\ \sigma'_{vo}$.

The initial void ratio e_0 is derived from the in-situ water content distribution.

The elastic shear modulus G_0 for clay can be calculated as a function of Poisson's ratio and the effective overburden pressure, σ'_{vo} or σ'_{vc} by the following relations.

$$G_0 = \frac{(1-2\nu')}{2(1-\nu')} \cdot \frac{(1+e_0)}{\kappa} \cdot \sigma'_{vo} \quad \text{(N.C)}$$

$$G_0 = \frac{(1-2\nu')}{2(1-\nu')} \cdot \frac{(1+e_0)}{\kappa} \cdot \sigma'_{vc} \quad \text{(O.C)}$$

On the other hand, sand was assumed to be an elastic material, with the value of Poisson's ratio, ν, equal to 0.33. The initial elastic shear modulus, G_0 was calculated from the N-value of S.P.T. in the following relation.

$$G_0 = 26\ N$$

Table 1. Material properties for alluvial and diluvial clay

	alluvial clay	diluvial clay
λ	0.450	0.600
κ	0.050	0.070
M^*	1.050	1.050
m'	20.20	19.90
C	1.11×10^{-10}	2.40×10^{-11}
λ_k	0.450	0.350
ν	0.360	0.360
K_0	0.560(N.C.)	0.636(O.C.)

4.4 Results and Discussion

Field measurements of the deformation caused by construction of the container berth were also started just after the completion of back filling at this site. This is because it is very difficult to observe the deformation of the sea bed, especially in the primary stage of the offshore construction.

In this paper, therefore, a comparison of the calculated results and the field measurement data will be limited to the period after the completion of back filling.

Settlement-time performance of sea bed

The measured settlement-time performances of the sea bed are shown in Fig.7. Selected points (A) and (B) are located 27 and 90 meters from the center line, respectively. The area including the point (A) is improved by sand drains, while the area including the point (B) is remained without improvement.

The solid lines shown in the figure represent the calculated results. As mentioned above, the field measurements were started just after the completion of the back filling, and comparisons of the calculated and measured performances are limited to the period between T=1620 day and T=2160 day. The calculated results underestimate the measured data for both points. But as a whole, we have good agreement between the calculated results and the measurement data qualitatively.

Then, let us examine the effect of the sand drains on the consolidation rate. In order to accurately estimate the drainage effect of sand drains, Sekiguchi introduced the method of macro-element. Here, however, we assume the equivalent value for the coefficient of permeability, k, for the element with sand drains.
The same overburden is loaded on the points (A) and (B). The difference in the conditions between the two points lies only in the existence or non-existence of sand drains. In Fig.7, the rate of settlement for (A) is much larger than that for (B). Moreover, the total settlement for (A) is also larger than that for (B) at this stage, namely, T=2160 day. These facts show that consolidation has been promoted due to the sand drains, and thus, the rate and the total amount of settlement for (A) become larger than that for (B).

Settlement profile of the sea bed

The settlement profiles of the sea bed are shown in Fig.8. The solid line represents the calculated settlement profile at the time of T=2160 day, and the hatched line represents the field measurement data observed by the settlement plates. It is seen that the calculated results for settlement profiles can predict the observed settlement profiles well, except for the replace sand. The calculated values for the settlement of the replaced sand overestimate the observed values. This is because some ambiguity remains in the determination of the material parameters for this replaced sand, and we ignore the interaction between the foundation and the caisson. However, on the whole, the numerical analysis can represent the deformation process of the clay foundation due to the berth construction.

Next, we examine the final settlement of the clay foundation due to the berth construction at this site.

The final settlement, ρ_f, for the alluvial clay layer is calculated invicidly by the one-dimensional consolidation theory, so that ρ_f is derived to be 6.0 meters for this problem. The largest settlement at T=2160 day is about 3.0 meters, which is more than twice what the settlement will be subsequently.

On the other hand, from completion of the final loading, it takes about 7 years to reach 90% of the degree of consolidation for the alluvial clay layer. The degree of con-

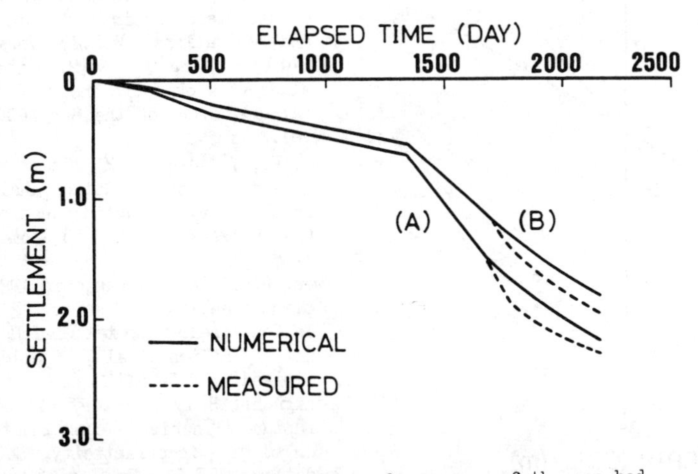

Fig. 7. Settlement - time performances of the sea bed

solidation at T=2160 day is calculated to be approximately 15-20%.

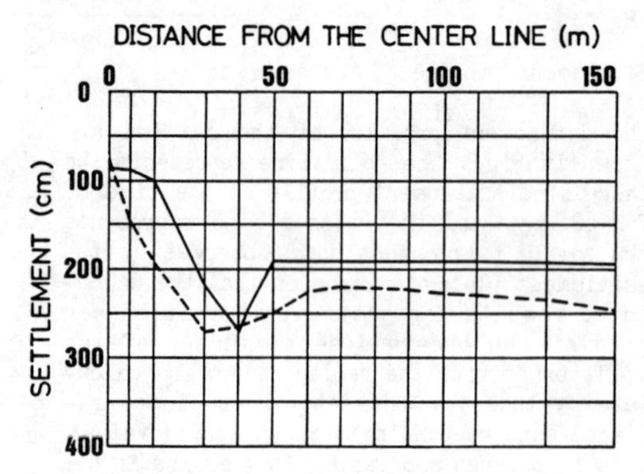

Fig. 8. Settlement profiles at the sea bed

Profile of lateral ground movement

The profiles of the foundation's lateral displacement are shown in Fig.9. The inclinometer was installed to a depth of C.D.L. -29m at completion of the back filling, and the observation of lateral displacement was started at T=1680 day. Therefore, both the calculated results (solid line) and the measured data (hatched line) for the lateral displacement shown in Fig.9 correspond to the increments from T=1680 day to T=2160 day. The inclinometer is located about 13.0 meters from the center line as shown in Fig.4.

While lateral displacement occurs towards the sea in the middle of the alluvial clay layer, soil moves toward the land near the sea bed. This is due to the effect of the replaced sand and the fill. The calculated results can trace the monitored data to some extent.

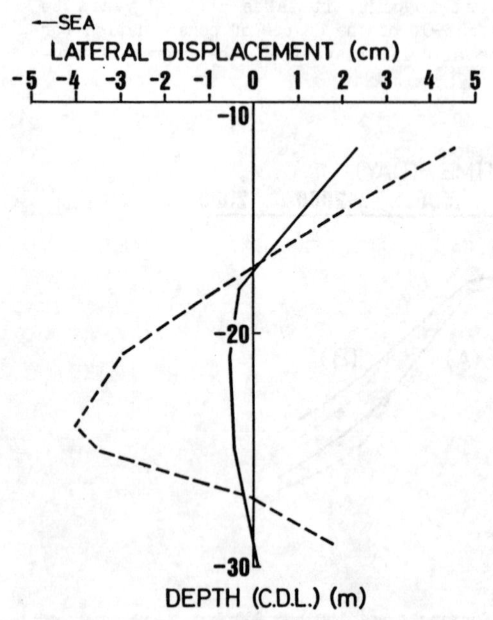

Fig. 9. Profiles of lateral displacement

Remaining problems

The deformation of diluvial clay has been ignored in the conventional analysis. Moreover, no field measurements have been taken for diluvial clay layers. With an increase in the construction of large structure, however, even diluvial clay layers will deform. In particular, delayed deformation (that which occurs a long time after the completion of structure) does exist.

5. CONCLUSION

Conclusions made from this study are summarized as follows.

1. Finite element equations coupled with the overstress elasto-viscoplastic constitutive model have been formulated and applied to boundary value problems.
2. Numerical analyses can predict the monitored performances of settlement and lateral ground movement of clay foundations due to the construction of berths.

REFERENCES

Adachi, T. & F. Oka 1982. Constitutive equations for normally consolidated clay based on elasto-viscoplasticity. Soils and Foundations, Vol.22. No.4. : 57-70.

Adachi, T., F. Oka & M. Mimura 1987. Mathematical structure of an overstress elasto-viscoplastic model for clay. Soils and Foundations. Vol.27. No.3 : 31-42.

Adachi, T., F. Oka & M. Mimura 1987. An elasto-viscoplastic theory for clay failure. Proc. of the 8th ARC on SMFE. Vol.1. : 5-8

Oka, F., T. Adachi & Y. Okano 1986. Two-dimensional consolidation analysis using an elasto-viscoplastic equation. Int. J. Nnm. and Anal. Method in Geomech. Vol.1.: 1-16

Owen, D.R.J. and F. Damjanic 1982. Visco-plastic analysis of soils. Recent Advances in Non-linear Computational Mechanics. Edt. by Hilton et al., Pineridge Press Limited. : 225-254.

Sekiguchi, H 1984. Theory of undrained creep rupture of normally consolidated clay based on elasto-viscoplasticity, Soils and Foundations. Vol.24. No.1. : 129-147.